T0221230

THE INTERNATIONAL DICTIONARY OF HEATING, VENTILATING AND AIR CONDITIONING

Air Conditioning
A practical introduction
D. V. Chadderton

Building Energy Management Systems
An application to heating and control
G. J. Levermore

Building Services Engineering
D. V. Chadderton

Combustion Engineering and Gas Utilisation
3rd edition
British Gas

Engineering Noise Control
Theory and practice
D. Bies and C. Hansen

Illustrated Encyclopedia of Building Services
D. Kut

Noise Control in Industry
3rd edition
Sound Research Laboratories Limited

Spon's Asia Pacific Construction Costs Handbook
Edited by Davis Langdon & Seah International

Spon's European Construction Costs Handbook
Edited by Davis Langdon & Everest

Spon's Mechanical and Electrical Services Price Book
Published annually
Edited by Davis Langdon & Everest

Ventilation of Buildings
H. B. Awbi

THE INTERNATIONAL DICTIONARY OF HEATING, VENTILATING AND AIR CONDITIONING

Second edition

Compiled by
The Publishing Committee of REHVA
(Federation of European Heating and
Airconditioning Associations)

Taylor & Francis
Taylor & Francis Group

LONDON AND NEW YORK

By Taylor & Francis,
2 Park Square, Milton Park, Abingdon, Oxon, OX14 4RN

First edition 1982
Second edition 1994

Transferred to Digital Printing 2006

© 1982, 1994 REHVA (Federation of European Heating and Airconditioning Associations)

ISBN 0 419 15390 x

A catalogue record for this book is available from the British Library

Library of Congress Cataloging-in-Publication Data
The International dictionary of heating, ventilating, and air
  conditioning/compiled by the Publishing Committee of REHVA
  [sic] (Federation of European Heating and Airconditioning
  Associations). – 2nd ed.
        p.   cm.
Rev. ed. of: The international dictionary of heating, ventilating,
  and air conditioning/Representatives of European Heating and
  Ventilating Associations. Documentation Committee. 1982.
    Includes indexes.
    ISBN 0-419-15390-X (alk. paper)
1. Heating – Dictionaries – Polyglot.   2. Ventilation – Dictionaries –
  Polyglot.   3. Air conditioning – Dictionaries – Polyglot.
  4. Dictionaries, Polyglot.   I. Federation of European Heating and
  Airconditioning Associations. Publishing Committee.
  II. Representatives of European Heating and Airconditioning
  Associations. Documentation Committee. International dictionary of
  heating, ventilating, and air conditioning.
  TH7007.R46   1993
  697′.0003 – dc20   92-46393
                                                                    CIP

Printed and bound by CPI Antony Rowe, Eastbourne

Contents

Preface

REHVA is an organization of professional associations of heating and air conditioning engineers from 18 European countries. It was inaugurated in 1963 with the aim of facilitating the development of heating and air conditioning services and improving their design and installation. Co-operation between members and the sharing of technical knowledge and expertise is a vital function of the organization.

The Documentation Committee, later renamed the Publishing Committee, plays a key role in enabling REHVA to realise its aims. One of the projects undertaken by the Committee was the first edition of this book – a multilingual dictionary of the terms used in the HEVAC industry. The first edition was published in 1982 and covered 10 languages. Close to 2000 copies of the book were sold.

Since 1982 the scope of HEVAC technology has broadened significantly, and the language related to the technology has evolved. Many new terms have been introduced and some others have fallen by the wayside. Moreover, with the onset of the single European market in 1992/3, the need to revise the dictionary became critical. This second edition has arisen from that need and now covers 11 languages as well as English. The Publishing Committee allotted responsibility for the second edition to a special Editorial Committee under the chairmanship of Mr Franco Stanzani, from the Italian Association, AICARR. This committee compiled the list of words which were to be included in the dictionary. In each of the participating countries a working party of professional engineers took care of the translation of the agreed word list into their own language, in order to ensure the accuracy and usability of the terms in each language.

The dictionary will be used primarily by those concerned with the industry in a practical way – manufacturers, contractors and consulting engineers – whose work involves them in projects or with organizations outside their own countries, or who need to read technical information in languages other than their own. It will also be of use to people involved in research and education.

Acknowledgements

The American Society of Heating, Ventilating, Refrigeration and Airconditioning Engineers (ASHRAE) gave permission to use the new edition of *ASHRAE's Terminology of Heating, Ventilation, Airconditioning and Refrigeration* in the compilation of this dictionary.

Thanks are given to all the members of the 11 working parties for their co-operation and hard work in the translation of the 4151 terms contained herein.

Finally acknowledgement is due to the efforts of the three senior members of the Editorial Committee – Walter Knoll, Wim J. F. Oudijn and Franco Stanzani – for their dedication and commitment to the difficult task of supervising and co-ordinating the compilation of the dictionary over a period of two years.

How to use this book

The dictionary is divided into two main parts. Part One is the main section and contains all the terms listed in alphabetical order in English. Each term has been given a number – the 'entry number' – which is used for reference in Part Two of the book. The corresponding terms in the other 11 languages are listed beneath the English word.

For convenience the name of each language has been abbreviated according to the ISO recommendations. Thus:

DA = Danish
DE = German
ES = Spanish
 F = French
 I = Italian
MA = Magyar (Hungarian)
NL = Dutch
PO = Polish
PY = Russian
SU = Finnish
SV = Swedish

The languages appear in the alphabetical order of the abbreviations used.

Where the English term has a more commonly used alternative then a cross-reference is given to the second term and no translations appear. However where there is more than one translation of the English term in another of the languages then all alternatives will appear together under the English term.

All terms are nouns except where an indication appears next to the English word: verbs are indicated by *vb* and adjectives by *adj*. Where different genders are used for nouns this is indicated by *m*, *f*, *n* or *c* appearing with the appropriate translation.

Part Two of the dictionary contains 11 separate indexes – one for each language except English. Each index lists all the terms which appear in the dictionary, in the alphabetical order of the language concerned. The index refers the reader to the **entry number** used in Part One. It is therefore very easy for someone who knows the term in one language to look it up in the appropriate index and then find the main entry for that term in the first part of the dictionary. The main entry then provides translations of that term into every other language including English.

The indexes appear in the same order as the language list above.

There are two appendices at the end of the dictionary. Appendix A lists the technical units of measurement used in the industry which require additional explanation. Appendix B lists in full some of the abbreviations which appear in the dictionary.

If further explanation and definition of the terms appearing here is required the reader is referred to *ASHRAE's Terminology of Heating, Ventilation, Airconditioning and Refrigeration* (ASHRAE, 1991; ISBN 0-910110-77-8).

Forord

REHVA er en sammenslutning af VVS-tekniske foreninger i 18 europæiske lande. REHVA blev grundlagt i 1963 for at fremme udviklingen af varme- og ventilationsanlæg. Samarbejde herom og udbredelse af teknologisk viden i øvrigt er vigtige brikker i REHVA's arbejde.

Dokumentations-komitéen (Udgivelses-komitéen) har en central placering i REHVA's bestræbelser på at nå sine mål. Et af komitéens realiserede projekter var den første udgave af nærværende bog – en flersproget ordbog for VVS-branchen. Førsteudgaven kom i 1982 og dækkede 10 sprog. Der blev solgt knap 2.000 eksemplarer af ordbogen.

I årene siden 1982 har VVS-fagene bredt sig over stadigt større områder, og det teknologiske sprogbrug har udviklet sig. Mange nye begreber er kommet til, og andre er udgået. Ved udsigten til etablering af EF's Indre Marked i 1992/93 blev en fornyelse af ordbogen påtrængende. Denne 2. udgave dækker i alt 12 sprog incl. engelsk. Udgivelseskomitéen overdrog ansvaret for det redaktionelle arbejde til et arbejdsudvalg under ledelse af Hr. Franco Stanzani fra den italienske forening AICARR. Udvalget sammenstillede ordlisten, og i de deltagende lande har kyndige ingeniører forestået oversættelsen med henblik på at sikre nøjagtighed og anvendelighed af de i ordbogen medtagne ord og begreber.

REHVA-ordbogen vil primært være nyttig for fabrikanter, entreprenører og rådgivende ingeniører, som beskæftiger sig med arbejde i andre lande, eller som har brug for at læse teknisk information på fremmede sprog. Også inden for forskning og uddannelse vil bogen finde anvendelse.

REHVA takker

Tak til ASHRAE (The American Society of Heating, Ventilating, Refrigeration and Airconditioning Engineers) for tilladelse til at anvende ASHRAE's terminologi-bog som grundlag for ordlisten.

Tak til oversættelsesgrupperne i de 11 deltagende lande for deres arbejde med at oversætte i alt 4.151 ord og begreber.

En særlig tak til udgivelseskomitéens 3 seniormedlemmer – Walter Knoll, Wim J. F. Oudijn og Franco Stanzani – for deres utrættelige arbejde i mere end 2 år med at overvåge og koordinere ordbogens tilblivelse.

Hvordan bruges denne ordbog

REHVA-ordbogen er delt i to afsnit. Første afsnit indeholder alle de oversatte ord/begreber i alfabetisk rækkefølge på engelsk. Hvert ord/begreb har et nummer ("indgangsnummer"), der anvendes som reference i bogens andet afsnit. De tilhørende ord/begreber på de andre 11 sprog er opført under den engelske betegnelse.

Navnene på de enkelte sprog er forkortet i henhold til ISO's anbefalinger:

DA = dansk
DE = tysk
ES = spansk
 F = fransk
 I = italiensk
MA = ungarsk
NL = hollandsk
PO = polsk
PY = russisk
SU = finsk
SV = svensk

Hvor den engelske betegnelse har et almindeligt anvendt alternativ, findes en krydsreference til betydning, uden oversættelse til andre sprog.

Hvor der gives mere end én oversættelse af engelske ord/begreber, optræder alle oversættelser samlet under den engelske betegnelse.

Alle ord/begreber er substantiver (navneord), undtagen hvor en anmærkning forekommer ved siden af det engelske ord: verber (udsagnsord) er markeret med *vb* og adjektiver (tillægsord) med *adj*. Hvor forskellige grammatiske køn for substantiver optræder, er dette markeret med *m, f, n* eller *c* ved den tillhørende oversættelse.

Ordbogens andet afsnit indeholder 11 separate indeks, et for hvert sprog, bortset fra engelsk. Hvert indeks oplister alle ord/begreber, som forekommer i ordbogen, i alfabetisk orden for det pågældende sprog. Disse indeks henviser læseren til indgangsnummeret i første afsnit. Det er derfor ganske let for en læser at finde et ord/begreb på ét sprog i indeks og derefter slå ordet op i første afsnit, hvor alle oversættelser findes efter den engelske betegnelse.

Indeks forekommer i samme rækkefølge som sprogforkortelserne ovenfor.

Bagerst i ordbogen findes to appendix. Appendix A oplister tekniske måleenheder, som kræver supplerende forklaring. Appendix B angiver den fulde betydning af nogle af de i ordbogen anvendte forkortelser.

Hvor der er behov for yderligere forklaring på forekommende ord/begreber, henvises læseren til *ASHRAE's Terminology of Heating, Ventilation, Airconditioning and Refrigeration* (ASHRAE, 1991; ISBN 0-910110-77-8).

Vorwort

Die REHVA ist ein Zusammenschluß von Interessenverbänden von Ingenieuren für Heizungs- und Klimatechnik aus 18 europäischen Ländern. Sie wurde 1963 mit dem Ziel gegründet, die Entwicklung von Heizungs- und Klimaanlagen zu unterstützen und deren Konstruktion und Installation zu verbessern. Die Kooperation zwischen den Mitgliedern und der Austausch von technischem Know-how und Erfahrungen sind wesentliche Aufgaben der Organisation.

Der Dokumentationsausschuß (Documentation Committee), später in Veröffentlichungsausschuß (Publishing Committee) umbenannt, spielt für die Realisierung der Ziele der REHVA eine Schlüsselrolle. Eins der von diesem Ausschuß durchgeführten Projekte war die Herausgabe der ersten Auflage dieses Buches, eines mehrsprachigen Wörterbuchs der in der HEVAC-Branche verwendeten Begriffe. Die erste Auflage war 10sprachig und erschien 1982; es wurden etwa 2000 Exemplare davon verkauft.

Seit 1982 hat sich das Arbeitsfeld der HEVAC-Technik erheblich erweitert, die zugehörige Fachsprache hat sich ebenfalls weiterentwickelt. Viele Begriffe sind neu hinzugekommen, andere sind veraltet. Außerdem hatte die Öffnung des EG-Markts in 1992/93 eine Überarbeitung des Wörterbuchs dringend erforderlich gemacht. Die vorliegende, zweite Auflage ist aus diesen Notwendigkeiten heraus entstanden, sie umfaßt jetzt neben Englisch 11 weitere Sprachen. Der Veröffentlichungsausschuß hatte die Verantwortung für die zweite Auflage an einen speziellen Herausgeberausschuß (Editorial Committee) unter dem Vorsitz von Herrn Franco Stanzani (vom italienischen Verband AICARR) übertragen. Dieser Ausschuß stellte die Liste der Begriffe zusammen, die in das Wörterbuch aufgenommen werden sollten. Danach sorgte in jedem Mitgliedsland eine Arbeitsgruppe von Fachingenieuren für die Übersetzung der festgelegten Begriffsliste in die jeweilige Muttersprache, wodurch die Genauigkeit und die Treffsicherheit der Begriffe in jeder Sprache sichergestellt wurde.

Das Wörterbuch ist hauptsächlich für solche Benutzer gedacht, die mitten in der Industriepraxis stehen (Hersteller, Zulieferer und beratende Ingenieure) und in Projekten oder mit Organisationen außerhalb ihres Heimatlandes mitarbeiten oder technische Dokumentationen in einer für sie fremden Sprache lesen müssen. Darüber hinaus ist das Buch auch für die in Forschung und Lehre Tätigen nützlich.

Danksagungen

Die American Society of Heating, Ventilating, Refrigeration and Airconditioning Engineers, ASHRAE (Amerikanische Vereinigung der Heizungs-, Lüftungs-, Kälte- und Klimatechnik-Ingenieure), gab die Erlaubnis, die neue Ausgabe ihres Buchs *ASHRAE's Terminology of Heating, Ventilation, Airconditioning and Refrigeration* (ASHRAEs Begriffslexikon für Heizungs-, Lüftungs-, Klima- und Kältetechnik) für die Zusammenstellung dieses Wörterbuchs zu benutzen.

Dank geht an alle Mitglieder der 11 Arbeitsgruppen für ihre gute Zusammenarbeit und für ihre harte Arbeit bei der Übersetzung der 4151 in diesem Buch enthaltenen Begriffe.

Dank geht schließlich auch an die drei leitenden Mitglieder des Herausgeberausschusses, Walter Knoll, Wim J. F. Oudijn und Franco Stanzani, für ihr Engagement und ihren Einsatz bei der schwierigen Aufgabe, die zwei Jahre dauernde Erstellung des Wörterbuchs zu überwachen und zu koordinieren.

Das Benutzen dieses Buchs

Das Wörterbuch besteht aus zwei Teilen. Teil eins ist der Hauptteil, der sämtliche Begriffe auf Englisch in alphabetischer Reihenfolge enthält. Jeder Begriff ist durch eine Nummer, die „Eintragsnummer" (entry number) gekennzeichnet, die den Bezug zu Teil zwei des Buchs herstellt.

Die entsprechenden Begriffe in den anderen 11 Sprachen stehen jeweils unmittelbar unterhalb des englischen Begriffs.

Aus praktischen Gründen werden die Namen aller Sprachen gemäß den ISO-Normen wie folgt abgekürzt:

DA = Dänisch
DE = Deutsch
ES = Spanisch
 F = Französisch
 I = Italienisch
MA = Ungarisch

NL = Holländisch
PO = Polnisch
PY = Russisch
SU = Finnisch
SV = Schwedisch

Die Begriffe erscheinen in einer Reihenfolge, die der alphabetischen Ordnung der o.a. Abkürzungen für die Sprachen entspricht.

Für Fälle, in denen es für einen englischen Begriff eine statt seiner häufiger benutzte Alternative gibt, wird ein Verweis auf den alternativen Begriff, aber keine Übersetzung angegeben. Wenn es jedoch für einen englischen Begriff in einer der anderen Sprachen mehr als eine Übersetzung gibt, so werden diese alle unterhalb des englischen Begriffs aufgeführt.

Alle englischen Begriffe sind Substantive, es sei dann, sie haben eine direkt nachfolgende Kennzeichnung: *vb* kennzeichnet Verben, *adj* Adjektive. Falls Substantive mit unterschiedlichem Genus auftauchen, wird dies durch *m, f, n* oder *c* bei der jeweiligen Übersetzung gekennzeichnet.

Teil zwei des Wörterbuchs besteht aus 11 getrennten Indexverzeichnissen, eins für jede Sprache außer Englisch. Jedes Indexverzeichnis enthält alle Begriffe, die das Wörterbuch enthält, und zwar in alphabetischer Reihenfolge in der betreffenden Sprache. Das Indexverzeichnis verweist den Leser mit Hilfe der Eintragsnummer auf den entsprechenden Eintrag in Teil eins. Wenn man einen Begriff in einer Sprache kennt, sucht man ihn zunächst im entsprechenden Indexverzeichnis und kommt so sehr einfach zum Haupteintrag dieses Begriffs in Teil eins des Wörterbuchs. Dieser Haupteintrag gibt einem dann die Bezeichnung des Begriffs in allen Sprachen einschließlich Englisch an.

Die verschiedenen Indexverzeichnisse sind entsprechend der obenstehenden Sprachentabelle geordnet.

Am Ende des Wörterbuchs finden sich zwei Anhänge. Anhang A führt solche in der Branche benutzten technischen Maßeinheiten auf, für die eine zusätzliche Erläuterung erforderlich ist. Anhang B enthält eine Aufstellung einiger im Wörterbuch benutzter Abkürzungen in ihrer ausgeschriebenen Form.

Für weitere Erläuterungen und Definitionen der in diesem Buch verwendeten Begriffe siehe das Buch *ASHRAE's Terminology of Heating, Ventilation, Airconditioning and Refrigeration* (ASHRAE, 1991; ISBN 0-910110-77-8).

Prólogo

REHVA es una organización de asociaciones profesionales de ingenieros en calefacción y aire acondicionado de 18 países europeos. Fue constituida en 1963 con el propósito de facilitar el desarrollo de los servicios de calefacción y aire acondicionado y mejorar su diseño e instalación. Es función vital de la organización la colaboración entre sus miembros y el intercambio de experiencias y conocimientos técnicos.

El Comité de Documentación, titulado últimamente Comité de Publicaciones, juega un papel clave de apoyo a REHVA para cumplir sus objetivos. Uno de los proyectos que asumió el Comité fue la primera edición de este libro – un diccionario multilingüe de los términos usados en la industria de ventilación, calefacción y aire acondicionado. La primera edición fue publicada en 1982 y comprendía 10 idiomas. Cerca de 2000 ejemplares del libro fueron vendidos.

Desde 1982, el panorama de la tecnología se ha ensanchado de forma significativa, y el lenguaje relativo a la tecnología ha evolucionado. Han sido introducidos muchos nuevos términos y muchos otros han caído en desuso. Por otra parte, con el inicio del mercado único europeo en 1992/93, la necesidad de revisar el diccionario se hizo crítica. Esta segunda edición nace de esta necesidad e incluye esta vez 11 idiomas además del inglés. El Comité de Publicaciones asignó la responsabilidad de la segunda edición a un Comité Editorial especial bajo la presidencia del Sr. Franco Stanzani, de la Asociación Italiana AICARR. Este Comité compiló la lista de palabras que se incluirían en el diccionario. En cada uno de los países participantes, un equipo de ingenieros expertos se hizo cargo de la traducción a su propio idioma de la lista de palabras acordada, al objeto de asegurar la exactitud y la utilidad de los términos en cada idioma.

El diccionario será de utilidad, en primera instancia para todos aquellos relacionados con la industria de forma directa – fabricantes, instaladores e ingenieros consultores – cuyo trabajo los implica en proyectos o con organizaciones fuera de sus propios países, o que precisan leer información técnica en idiomas distintos del suyo propio. Tambien será de utilidad para aquellas personas relacionadas con la investigación y la educación.

Agradecimientos

La Sociedad Americana de Ingenieros de Calefacción, Ventilación, Refrigeración y Aire Acondicionado (ASHRAE) concedió el permiso para la utilización de la nueva edición del *ASHRAE's Terminology of Heating, Ventilation, Airconditioning and Refrigeration* en la confección de este diccionario.

Expresamos nuestra gratitud a todos los miembros de los 11 grupos de trabajo por su colaboración y esfuerzo en la traducción de los 4151 términos aquí incluidos.

Finalmente, debemos agradecer los esfuerzos de los miembros permanentes del Comité Editorial – Walter Knoll, Wim J. F. Oudijn y Franco Stanzani – por su dedicación y compromiso con la difícil tarea de supervisar y coordinar la compilación del diccionario durante un período de dos años.

Cómo utilizar este libro

El diccionario se divide en dos partes principales. La primera sección contiene todos los términos en inglés, ordenados alfabéticamente. A cada término se le ha otorgado un número – "entry number" – que se utiliza como referencia en la segunda parte del libro. Los correspondientes términos en los otros 11 idiomas están listados debajo de cada palabra inglesa.

El nombre de cada idioma ha sido abreviado, de acuerdo con las recomendaciones ISO, como sigue:

DA = danés
DE = alemán
ES = español
F = francés
I = italiano
MA = húngaro
NL = holandés
PO = polaco
PY = ruso
SU = finlandés
SV = sueco

Los idiomas aparecen en el orden alfabético de las abreviaturas utilizadas.

Cuando el término inglés tiene una alternativa más comúnmente utilizada, se da una referencia al segundo término y no figuran traducciones. Sin embargo, cuando existe más de una traducción del término inglés en otro de los idiomas, todas las alternativas figuran juntas bajo el término inglés.

Todos los términos son sustantivos, excepto cuando figura una indicación junto a la palabra inglesa: los verbos se indican por *vb* y los adjetivos por *adj*. Cuando se utilizan diferentes géneros para los sustantivos, éstos se indican por *m*, *f*, *n* o *c* en su correspondiente traducción.

La segunda parte del diccionario contiene 11 índices separados – uno para cada idioma exceptuando el inglés. Cada índice relaciona todos los términos que aparecen en el diccionario, en el orden alfabético de su correspondiente idioma. El índice remite al lector al "entry number" utilizado en la primera parte. De esta forma, para cualquiera que conoce el término en un idioma, es muy fácil localizarlo en el índice correspondiente y encontrar la entrada del término en la primera parte del diccionario. Esta entrada proporciona la traducción del término a cualquier otro idioma incluido el inglés.

Los índices aparecen en el mismo orden de los idiomas antes listados.

Existen dos apéndices al final del diccionario. El apéndice A relaciona las unidades técnicas de medida, utilizadas en la industria, que requieren explicaciones adicionales. El apéndice B relaciona y explicita algunas de las abreviaturas que aparecen en el diccionario.

Si el lector necesita mayores explicaciones y definiciones de los términos que aquí aparecen, puede remitirse al *ASHRAE's Terminology of Heating, Ventilation, Airconditioning and Refrigeration* (ASHRAE, 1991; ISBN 0-910110-77-8).

Préface

REHVA est une organisation regroupant des ingénieurs qui appartiennent à des associations professionnelles de 18 pays européens dans le domaine du chauffage et de la climatisation. Elle fut créée en 1963, dans le but de faciliter le développement des services de ces deux domaines ainsi que pour améliorer les activités de conception et d'installation. La coopération entre ses membres et le partage des connaissances techniques sont d'une importance capitale pour le fonctionnement de l'organisation.

Le Comité de Documentation, baptisé plus tard Comité de Publication, joue un rôle primordial dans l'association en lui permettant d'atteindre les objectifs qu'elle s'est fixés. Un des projets entrepris par le Comité fut la première édition de ce livre – un dictionnaire multilingue des termes utilisés dans l'industrie du chauffage, de la ventilation et du conditionnement d'air – qui a été publiée en 1982 et qui regroupait 10 langues différentes. Environ 2000 exemplaires de ce livre ont été vendus.

Depuis 1982, le domaine de la technologie du chauffage, de la ventilation et du conditionnement d'air s'est élargi de manière significative et les termes la décrivant ont évolué. De nombreux termes nouveaux ont été introduits et quelques autres ne sont plus utilisés. De plus, avec l'importance prise par le marché unique européen en 1992/93, la nécessité de réviser le dictionnaire devenait urgente. Cette seconde édition en est le résultat et regroupe désormais 11 langues ainsi que l'anglais. Cette deuxième édition fut placée, par le Comité de Publication, sous la responsabilité d'un Comité Editorial spécial dirigé par M. Franco Stanzani de l'association italienne AICARR. Ce comité a établi la liste des mots qui devaient être inclus dans le dictionnaire. Dans chacun des pays participants, un groupe de travail d'ingénieurs spécialisés s'est chargé de la traduction de la liste des mots sélectionnés dans sa propre langue, afin d'assurer l'exactitude et l'emploi correct des dans chaque langue.

Le dictionnaire sera utilisé principalement par les personnes qui, dans l'industrie, font un travail technique – industriels, entrepreneurs, ingénieurs/consultants – et qui sont amenés à établir des projets ou à avoir des contacts avec des organismes étrangers, ou bien qui lisent de la documentation dans des langues étrangères. Il sera également utile pour les personnes qui font de la recherche ou qui travaillent dans l'enseignement.

Remerciements

L'entreprise américaine de chauffage, ventilation, réfrigération et de climatisation (ASHRAE) a donné l'autorisation d'utiliser la nouvelle édition de *ASHRAE's Terminology of Heating, Ventilation, Airconditioning and Refrigeration* dans la compilation de ce dictionnaire.

Nous remercions tous les membres des 11 groupes de travail pour leur coopération et leur travail assidu dans la traduction des 4151 termes contenus dans ce livre.

Enfin, nous remercions les efforts des trois personnes responsables du Comité Editorial – Walter Knoll, Wim J. F. Oudijn et Franco Stanzani – pour leur dédicace et leur participation à la difficile tâche de superviser et coordonner la compilation du dictionnaire pendant deux ans.

Comment utiliser ce livre

Le dictionnaire est divisé en deux parties principales.

La première partie est la plus importante et recense tous les termes classés par ordre alphabétique en anglais. Un numéro a été attribué à chaque terme – le " N d'entrée " – qui est utilisé comme référence dans la deuxième partie du livre. Les termes correspondants, dans toutes les autres langues, sont énumérés sous le mot anglais.

Par commodité, le nom de chaque langue a été abrégé selon les recommandations des normes ISO, ainsi :

DA = Danois
DE = Allemand
ES = Espagnol
 F = Français
 I = Italien
MA = Magyar (Hongrois)
NL = Hollandais
PO = Polonais
PY = Russe
SU = Finlandais
SV = Suédois

Les langues apparaissent dans l'ordre alphabétique des abréviations utilisées.

Lorsque le mot anglais a un synonyme utilisé plus communément, un renvoi est alors indiqué au second terme et il n'y a pas de traduction. Néanmoins, lorsqu'il y a plus d'une seule traduction du mot anglais dans une autre langue, toutes les possibilités sont mentionnées ensemble sous le mot anglais.

Tous les mots sont des substantifs sauf quand une indication apparaît à côté du mot anglais : les verbes sont indiqués par la mention *vb* et les adjectifs par la mention *adj*. Lorsque différents genres sont utilisés, cela est indiqué par les mentions *m, f, n* ou *c* apparaissant avec la traduction appropriée.

La deuxième partie du dictionnaire contient 11 index séparés, un pour chaque langue sauf l'anglais. Chaque index liste tous les termes qui apparaissent dans le dictionnaire dans l'ordre alphabétique de la langue concernée. L'index renvoie le lecteur au numéro d'entrée utilisé dans la première partie. Il est ainsi très facile pour quelqu'un qui connaît le terme dans une langue, de le rechercher dans l'index approprié et de trouver alors l'accès principal pour ce mot dans la première partie du dictionnaire. L'accès principal donne alors des traductions de ce mot dans chacune des autres langues, y compris l'anglais.

Les index sont classés dans le même ordre que la liste des langues mentionnée ci-dessus.

Il y a deux appendices à la fin du dictionnaire. L'appendice A liste les unités techniques de mesure, utilisées dans l'industrie, qui nécessitent une explication supplémentaire. L'appendice B liste intégralement quelques-unes des abréviations qui apparaissent dans le dictionnaire.

Si le lecteur désire une explication ou une définition plus complète des termes apparaissant dans ce dictionnaire, il peut se référer à *ASHRAE's Terminology of Heating, Ventilation, Airconditioning and Refrigeration* (ASHRAE, 1991 ; ISBN 0-910110-77-8).

Prefazione

REHVA è una federazione che unisce le associazioni dei technici del settore del riscaldamento e del condizionamento dell'aria di diciotto paesi Europei. Essa è stata creata nel 1963 con il fine di promuovere lo sviluppo del settore e migliorare la pratica della progettazione e costruzione degli impianti. La cooperazione fra gli associati e la diffusione di conoscenze ed esperienze è una funzione essenziale della federazione.

Il Comitato Documentazione (più recentemente denominato Comitato Editoriale) svolge un ruolo determinante nel consentire a REHVA di realizzare i propri fini. Uno dei progetti intrapresi dal Comitato è stata la prima edizione di questo libro – un dizionario multilingue della terminologia in uso nel settore della climatizzazione. La prima edizione risale al 1982 e comprendeva dieci lingue; di essa sono state vendute circa 2000 copie.

In quest'ultimo decennio, il campo di attività del settore si è notevolmente ampliato e il linguaggio degli impiantisti si è modificato: molti nuovi termini sono stati introdotti ed altri sono caduti in disuso. Inoltre, con la creazione del mercato unico europeo nel 1992/93, la necessità di rivedere il dizionario è diventata impellente. Questa seconda edizione risponde a tali esigenze e comprende ora undici lingue oltre all'inglese. Il Comitato Editoriale ha affidato la responsabilità della seconda edizione ad uno specifico Comitato di Redazione presieduto da Franco Stanzani dell'associazione italiana AICARR. Questo comitato ha compilato la lista di termini da includere nel dizionario e in ciascuno dei paesi partecipanti un gruppo di lavoro composto da tecnici del settore ha curato la traduzione nella propria lingua della lista di termini concordata, al fine di garantire la precisa corrispondenza e l'utilizzabilità di ciascun termine nelle varie lingue.

Il dizionario sarà utilizzato principalmente da coloro che operano professionalmente nel settore della climatizzazione (costruttori di componenti, imprese installatrici e consulenti) con attività contemplano la partecipazione a progetti internazionali, il contatto con organizzazioni straniere, o l'esame di documentazione tecnica scritta in lingue diverse dalla propria. Il dizionario potrà inoltre essere utile a coloro che operano nella ricerca e nell'insegnamento.

Riconoscimenti

La American Society of Heating, Refrigerating and Airconditioning Engineers (ASHRAE) ha consentito l'uso della nuova edizione del documento *ASHRAE's Terminology of Heating, Ventilation, Airconditioning and Refrigeration* per la compilazione di questo dizionario.

Ringraziamenti sono dovuti a tutti i membri degli undici gruppi di lavoro nazionali per la collaborazione e l'impegno nella traduzione dei 4151 termini qui contenuti.

Un particolare riconoscimento è dovuto all'impegno dei tre "membri anziani" del Comitato di Redazione – Walter Knoll, Wim J. F. Oudijn e Franco Stanzani – per la loro dedizione nel difficile compito di supervisione e coordinamento della compilazione del dizionario nell'arco di un periodo di due anni.

Come utilizzare questo libro

Il dizionario è suddiviso in due parti. La Parte Prima (Part One) è la sezione principale e contiene l'elenco, in ordine alfabetico, di tutti i termini in lingua inglese. A ciascun termine è stato assegnato un numero progressivo ("entry number") che è utilizzato quale riferimento nella Parte Seconda (Part Two) del libro. Le traduzioni nelle altre undici lingue sono riportate sotto al termine in inglese.

Per convenienza il nome di ciascuna lingua è stato abbreviato in accordo con le raccomandazioni ISO, per cui:

DA = Danese
DE = Tedesco
ES = Spagnolo
F = Francese
I = Italiano
MA = Magiaro (Ungherese)
NL = Olandese
PO = Polacco
PY = Russo
SU = Finlandese
SV = Svedese

Le traduzioni sono riportate secondo l'ordine alfabetico delle suddette abbreviazioni.

Qualora il termine inglese abbia un sinonimo di uso più frequente, viene riportato il riferimento a tale sinonimo e non compare alcuna traduzione. Tuttavia, quando in una delle altre lingue esista più di una traduzione del termine inglese, tutte le possibili traduzioni vengono riportate insieme sotto al termine inglese.

Tutti i termini vanno intesi come sostantivi, eccetto quando compare accanto al termine inglese una delle seguenti indicazioni: *vb* per i verbi e *adj* per gli aggettivi. Quando i sostantivi hanno genere diverso, si indica *m, f, n* oppure *c* accanto alla traduzione nella lingua appropriata.

La Parte Seconda del dizionario contiene undici indici separati – uno per ciascuna lingua eccetto l'inglese. Ciascun indice elenca tutti i termini che compaiono nel dizionario, in ordine alfabetico per la lingua di appartenenza. L'indice rimanda il lettore al numero progressivo ("entry number") utilizzato nella Parte Prima. E' pertanto molto facile per chi conosca il termine in una qualsiasi lingua individuarlo nell'indice appropriato e quindi risalire attraverso il numero progressivo alla sua collocazione nella Parte Prima, dalla quale è immediato ottenere la traduzione del termine in ciascun'altra lingua, compreso l'inglese.

Gli indici compaiono nello stesso ordine delle lingue sopra elencate.

Vi sono due appendici alla fine del dizionario: l'Appendice A elenca le unità di misura utilizzate nel settore che richiedono ulteriori spiegazioni; l'Appendice B riporta per esteso alcune delle abbreviazioni che compaiono nel dizionario.

Si rimanda il lettore desideroso di ulteriori definizioni o spiegazioni dei termini del dizionario alla consultazione del documento *ASHRAE's Terminology of Heating, Ventilation, Airconditioning and Refrigeration* (ASHRAE, 1991; ISBN 0-910110-77-8).

Előszó

A REHVA az épületgépészeti (fűtés és klímaberendezési) szak-mérnökök egyesületeinek szervezete 18 európai országból. 1963-ban alapították, azzal a céllal, hogy elősegítse a fűtés- és klímaberendezési szolgáltatások fejlesztését és tökéletesítse azok tervezését és kivitelezését. A tagok közötti együttműködés, a műszaki ismeretek és a szaktudás egymás közötti kicserélése a szervezet alapvető feladata.

A *"Dokumentációs Bizottság"* – melynek későbbi elnevezése "Kiadói Bizottság" lett – kulcsszerepet játszik abban, hogy a REHVA megvalósítsa a célkitűzéseit. A bizottság egyik fő tevékenysége ezen szótár első kiadása volt, mely a szakmában használatos kifejezéseket tartalmazta több nyelven. Az első kiadás 1982-ben jelent meg 10 nyelven és közel 2000 példányban fogyott el.

1982 óta a fűtés és a légtechnika területe jelentősen bővült és a szakmai nyelv is alaposan fejlődött. Sok új szakkifejezés került bevezetésre, néhány régebbi pedig elavult. Ezenkívül az Európai Közös Piac kialakulása 1992/3-ban is szükségessé tette a korábbi szótár javított és bővített kiadását. Ebből az igényből valósult meg ez a második kiadás és most már az angolon kívül 11 nyelvet tartalmaz. A Kiadói Bizottság a második kiadás felelőseként egy speciális Szerkesztő Bizottságot hozott létre Franco Stanzani vezetésével, aki az olasz AICARR egyesület képviselője. Ez a bizottság állította össze a szakkifejezések listáját, amely a szótárba bekerült. Az érintett országok-ban szakmérnökök munkacsoportja végezte el a megegyezés szerinti szó lista fordítását országuk nyelvére, hogy biztosítsák a kifejezések pontosságát és használhatóságát mindegyik nyelven.

A szótárt elsősorban azok a szakemberek használhatják, akik gyártó, tervező, kivitelező tevékenységük során más országokkal kerülnek kapcsolatba, illetve akik műszaki információikat szeretnék bővíteni idegen nyelvű forrásokból. Ezenkívül természetesen a kutató és oktató szakembereknek is hasznára lehet.

Köszönetnyilvánítások

Köszönet az "Amerikai Fütési, Szellőzési, Hűtési és Klimatizálási Mérnökök Szövetségének/ASHRAE/", akik engedélyezték, hogy ezen szótár összeállításánál felhasználjuk az *ASHRAE's Terminology of Heating, Ventilation, Airconditioning and Refrigeration* című könyvük legújabb kiadását.

Köszönet illeti a 11 munkacsoportot az itt tartalmazott 4151 kifejezés szakszerű fordításának kemény munkájáért.

Végül köszönet jár a szótár szerkesztésében és lektorálásában két éven át nyújtott teljesítményükért a szerkesztő bizottság három rangidős tagjának: Walter Knoll, Wim J. F. Oudijn és Franco Stanzani uraknak.

Hogyan használjuk ezt a könyvet

A szótár két fő fejezetre oszlik. Az első rész/Part One/a fontosabb fejezet, mely tartalmazza az összes kifejezést ABC sorrendben angolul. Minden kifejezésnek van egy "Hivatkozási száma", melyre a könyv második részében hivatkozunk. Az angol kifejezések alatt találhatók a további 11 nyelven a megfelelő szavak, az ISO ajánlásaiban rögzített rövidítések szerinti alábbi sorrendben:

DA = dán
DE = német
ES = spanyol
F = francia
I = olasz
MA = magyar
NL = holland
PO = lengyel
PY = orosz
SU = finn
SV = svéd

A nyelvek a használt rövidítések ABC sorrendjében fordulnak elő.

Ahol az angol kifejezésnek más, használatos alternatívája is van, ott feltüntettük ezt a második kifejezést is, de fordítások nélkül. Ha viszont az angol kifejezésnek az egyes nyelveken van több változata, ezeket mind feltüntetjük az angol címszó alatt.

Minden kifejezés általában főnév, kivéve ahol az angol szó mellett jelölve van, hogy *vb*/verb/az ige, vagy *adj.*/adjectives/ az jelző. Egyes nyelveknél a szó neme is jelölésre került, m = hímnemű, f = nőnemű, n = semlegesnemű.

A szótár második része/Part Two/különálló listát tartalmaz a 11 nyelv saját ABC sorrendje szerinti kifejezésekkel, mindegyiknél megadva a "Hivatkozási szám" – ot a könyv első részére vonatkozóan. Ezért nagyon könnyű bárkinek egy kifejezést megkeresni egy adott nyelv listájában és akkor az első fejezetnél megtalálja ezt a kifejezést a hivatkozási számnál lévő címszónál minden nyelven. Ezek a nyelvek szerinti listák az előbbiekben ismertetett sorrendben követik egymást a könyv második részében.

A szótár végén két függelék található. Az "A" függelék azokat az iparban használatos műszaki mértékegységeket sorolja fel, melyek további magyarázatra szorulnak. A "B" függelék néhány olyan rövidítés teljes kifejezését tartalmazza, melyek a szótárban előfordulnak.

Ha a szótárban lévő kifejezések további magyarázatát vagy meghatározását kívánja az olvasó, ajánljuk a már hivatkozott *ASHRAE's Terminology of Heating, Ventilation, Airconditioning and Refrigeration* kiadványt./ASHRAE, 1991. ISBN 0-910110-77-8/

Inleiding

REHVA is een federatie van technische verenigingen van verwarmings- en luchtbehandelingstechnici uit 18 Europese landen en werd opgericht in 1963 met het doel de bevordering van wetenschap en techniek op het gebied van verwarmings- en luchtbehandelingsinstallaties. Technische samenwerking en de uitwisseling van kennis en ervaring is een belangrijke activiteit van de organisatie.

De "Documentation Committee", later hernoemd in "Publishing Committee", neemt een sleutelpositie in bij het realiseren van deze doelstelling. Eén van de door deze commissie uitgevoerde taken was de publicatie van de eerste editie van dit boek, een veeltalig woordenboek van woorden, gebruikt in de verwarmings- en luchtbehandelings-industrie. De eerste uitgave verscheen in 1982 en omvatte 10 talen. Bijna 2000 exemplaren van dit woordenboek werden verkocht.

Sinds 1982 breidde het gebied van de HEVAC techniek zich in belangrijke mate uit en daarmede ook de gebruikte woordenschat. Veel nieuwe woorden werden ingevoerd en andere raakten in onbruik. Bovendien werd het noodzakelijk het woordenboek te herzien met het oog op de éénwording van de Europese markt in 1992/1993. Hieruit ontstond deze tweede editie, die nu behalve Engels elf talen omvat. De "Publishing Committee" droeg de verantwoordelijkheid voor de tweede editie op aan een hiertoe opgerichte "Editorial Committee" onder voorzitterschap van Mr Franco Stanzani, lid van de Italiaanse vereniging AICARR. De commissie stelde de lijst van op te nemen woorden samen. In elk van de deelnemende landen zorgde een werkgroep van beroepsdeskundigen voor de juiste en actuele vertaling van deze woorden in hun eigen taal.

Het woordenboek is bestemd voor allen die werkzaam zijn in het vakgebied als fabrikanten, installateurs en adviseurs, maar zeker niet in de laatste plaats voor hen betrokken bij het vaktechnisch onderwijs en onderzoek en voor beheerders en gebruikers van installaties in gebouwen.

Dankbetuiging

De Amerikaanse Vereniging van Ingenieurs op het gebied van Verwarming, Ventilatie, Koeling en Luchtbehandeling ASHRAE gaf toestemming tot gebruik van de jongste editie van *ASHRAE's Terminology of Heating, Ventilation, Airconditioning and Refrigeration* bij de samenstelling van dit woordenboek. Dank is eveneens verschuldigd aan de leden van de elf werkgroepen in de deelnemende landen voor hun medewerking aan de vertaling van de 4151 woorden, die in het woordenboek zijn opgenomen.

Tenslotte wordt ook dank betuigd aan drie seniorleden van de "Editorial Committee" – Walter Knoll, Wim J. F. Oudijn en Franco Stanzani – voor hun toewijding en inzet bij de moeilijke taak gedurende een periode van twee jaar leiding en coördinatie te geven aan de samenstelling van het woordenboek.

Hoe dit boek te gebruiken

Het woordenboek is verdeeld in twee delen. Deel I is het hoofdgedeelte en bevat alle woorden in alfabetische volgorde in het Engels. Ieder woord heeft een nummer – het "coderingsnummer" – dat gebruikt wordt voor verwijzing vanuit deel II van het boek. De overeenkomstige vertaling van ieder woord in de overige elf talen wordt gegeven onder het Engelse woord.

Uit practische overwegingen werd de naam van iedere taal afgekort overeenkomstig de ISO aanbevelingen, aldus:

DA = Deens
DE = Duits
ES = Spaans
 F = Frans
 I = Italiaans
MA = Hongaars
NL = Nederlands
PO = Pools
PY = Russisch
SU = Fins
SV = Zweeds

De talen worden vermeld in alfabetische volgorde van deze afkortingen.

Daar waar het Engelse woord een meer algemeen gebruikt alternatief heeft wordt geen vertaling gegeven, maar naar dit alternatief verwezen. Wanneer er echter meer vertalingen zijn van een Engels woord in één van de andere talen, dan worden deze alle gegeven onder het Engelse woord.

Alle woorden zijn zelfstandige naamwoorden behalve wanneer achter het Engelse woord een aanduiding wordt gegeven: werkwoorden zijn aangeduid met *vb* en bijvoeglijke naamwoorden met *adj*. Bij die talen, waar onderscheid wordt gemaakt in het geslacht van zelfstandige naamwoorden is dit vermeld met de toevoeging *m, f, n* of *c*.

Deel II van het woordenboek bevat elf lijsten, één voor iedere taal behalve Engels. Iedere lijst vermeldt de woorden in alfabetische volgorde van de betreffende taal. Deze lijsten verwijzen de gebruiker via het coderingsnummer naar het betreffende woord in Deel I. Het is aldus eenvoudig om een woord in een bepaalde taal in de betreffende lijst van die taal op te zoeken en in Deel I de vertaling te vinden in alle andere talen. De woordenlijsten van de diverse talen zijn overeenkomstig het hierboven vermelde in alfabetische volgorde opgenomen.

Er zijn twee aanhangsels aan het eind opgenomen. Aanhangsel A vermeldt enige eenheden die nadere toelichting behoeven. Aanhangsel B vermeldt de volledige term van enige afkortingen. Indien nadere verklaring of definitie van een woord wordt verlangd wordt de gebruiker verwezen naar de *ASHRAE's Terminology of Heating, Ventilation, Airconditioning and Refrigeration* (ASHRAE, 1991; ISBN 0-910110-77-8),

Przedmowa

REHVA jest organizacją zawodowych stowarzyszeń ogrzewnictwa i klimatyzacji 18 krajów Europy. Utworzona została w 1963 r. w celu wspierania rozwoju przemystu ogrzewnictwa i klimatyzacji oraz poprawy projektowania i instalacji. Zasadniczą funkcją organizacji jest współpraca pomiędzy jej członkami oraz wymiana wiedzy technicznej i doświadczeń.

Komitet Dokumentacji, przemianowany później na Komitet Publikacji, pełni kluczową rolę w umożliwianiu REHVA wypetniania tych celów. Jednym z przedsięwzięć podjętych przez Komitet było pierwsze wydanie tej książki – wielojęzycznego słownika terminów używanych w dziedzinie HEVAC. Pierwsze wydanie zostało opublikowane w 1982 r. i obejmowało 10 języków. Sprzedanych zostało prawie 2000 egzemplarzy.

Od roku 1982 zakres techniki HEVAC znacznie się poszerzył i nastąpił rozwój języka związanego z tą techniką. Wprowadzono szereg nowych terminów, zrezygnowano z używania innych. Ponadto po ustanowieniu w roku 1992/93 jednego rynku europejskiego potrzeba przeprowadzenia korekty słownika stała się niezwykle ważna. Niniejsze, drugie wydanie, powstało z takiej potrzeby, a obejmuje ono obecnie 11 języków oraz język angielski. Komitet Publikacji powierzył odpowiedzialność za drugie wydanie specjalnemu Komitetowi Wykonawczemu pod kierownictwem p. Franco Stanzani ze stowarzyszenia włoskiego AICARR. Komitet ten zestawił listę słów, które miały być zawarte w słowniku. W celu zapewnienia dokładności i użytkowości zawartych terminów, w każdym z krajów uczestniczących, przekładem uzgodnionych słów na swój własny język zajął się zespół roboczy zawodowych inżynierów.

Słownik będzie użytkowany głównie przez osoby związane z zawodem w sposób praktyczny – producentów, wykonawców, inżynierów konsultantów – których praca związana jest z projektami lub organizacjami poza ich własnym krajem lub osoby pragnące zdobyć informacje techniczne w innych językach. Słownik może być także używany przez osoby związane z badaniami i szkolnictwem.

Podziękowania

Amerykańskie Stowarzyszenie Inżynierów Ogrzewnictwa, Wentylacji, Chłodnictwa i Klimatyzacji (ASHRAE) wyraziło zgodę na wykorzystanie nowego wydania *ASHRAE Terminologii Ogrzewnictwa, Wentylacji, Klimatyzacji i Chłodnictwa* do zestawienia niniejszego słownika.

Składa się podziękowania wszystkim członkom 11 zespołów roboczych za ich współpracę i trud włożony w przetłumaczenie 4151 zawartych haseł.

Wreszcie podziękowanie należy się za wysiłek trzech czołowych członków Komitetu Wydawniczego – Waltera Knoll'a, Wima J. F. Oudijn'a i Franco Stanzani'ego – za ich 2-letnie poświęcenie i zaangażowanie w trudne zadanie nadzorowania i koordynacji zestawienia słownika.

Jak korzystać ze słownika

Słownik podzielono na dwie główne części. Część Pierwsza stanowi sekcję główną słownika i zawiera wszystkie hasła zestawione w porządku alfabetycznym w języku angielskim. Każde hasło posiada numer – "numer wpisu", który użyty jest jako odsyłacz w Części Drugiej słownika. Odpowiadające hasła w pozostałych 11 językach podano poniżej hasła w języku angielskim.

Dla ułatwienia nazwa każdego języka została użyta w skrócie, zgodnie z zaleceniem ISO:

DA = duński
DE = niemiecki
ES = hiszpański
F = francuski
I = włoski
MA = węgierski
NL = holenderski
PO = polski
PY = rosyjski
SU = fiński
SV = szwedzki

Hasła w odpowiednich językach pojawiają się w porządku alfabetycznym zastosowanych skrótów.

Gdy hasło w języku angielskim ma więcej powszechnie używanych odpowiedników, do takiego hasła zastosowano odnośnik, już bez przekładu. Jednak, gdy hasło podane w języku angielskim posiada więcej niż jeden przekład w innym języku, wówczas wszystkie odpowiedniki podane są pod hasłem w języku angielskim.

Wszystkie hasła są rzeczownikami z wyjątkiem, sytuacji gdy obok słowa w języku angielskim pojawia się wskaźnik: *vb* – czasowniki, *adj* – przymiotniki. Użycie rzeczowników różnych rodzajów oznaczono literami: *m* (męski), *f* (żeński), *n* (nijaki) lub *c*, umieszczonymi przy odpowiednim tłumaczeniu.

Część Druga słownika składa się z 11 oddzielnych indeksów – jeden dla każdego języka, z wyjątkiem języka angielskiego. Każdy indeks zawiera spis haseł zawartych w słowniku w porządku alfabetycznym danego języka. Indeks kieruje czytelnika do "numeru wpisu" użytego w Części Pierwszej. W ten sposób, dla kogoś znającego hasło w jednym języku jest bardzo łatwo spojrzeć do odpowiedniego indeksu i odnaleźć główny wpis dla tego hasła w pierwszej części słownika. Główny wpis podaje przekłady w pozostałych językach wraz z językiem angielskim.

Indeksy przedstawiono w tym samym porządku, co listę języków zamieszczona powyżej.

Na końcu słownika zamieszczono dwa dodatki. Dodatek A zestawia techniczne jednostki miar używanych w zawodzie, wymagające dodatkowych wyjaśnień. Dodatek B zawiera pełną listę skrótów użytych w słowniku.

Jeśli wymagane jest dalsze wyjaśnienie i określenie haseł pojawiających się w tym słowniku, czytelnik winien skorzystać z *ASHRAE's Terminology of Heating, Ventilation, Airconditioning and Refrigeration* (ASHRAE, 1991; ISBN 0-910110-77-8).

Введение

REHVA является международной ассоциацией обществ инженеров по отоплению и кондиционированию воздуха 18 европейских стран. Она основана в 1963 году с целью содействия развитию служб систем отопления и кондиционирования воздуха и улучшения качества проектирования и монтажных работ посредством сотрудничества и распространения технических ананий и зкспертизы.

Комиссия по документации, позже переименованная в Издательский комитет, сыграла важную роль в осуществлении целей REHVA. Один из проектов, выполненных этим комитетом состоял в издании многоязычного словаря терминов, употребляемых в этой отрасли промышленности. Впервые этот словарь был издан в 1982 году на 10 языках. Почти 2000 его экземпляров были распроданы.

С 1982 года масштаб технологии ОВК значительно расширился и ее язык сильно развился. Появилось много новых терминов, а некоторые старые определения стали использоваться реже. А с образованием единого европейского рынка в 1992/93 году, возникла острая необходимость в переиздании словаря. Появилось второе издание словаря, включающее термины на 11 языках, помимо английского. Издательский комитет поручил эту работу специальной Редакционной комиссии под руководством Г-на Franco Stanzani, члена итальянской ассоциации AICARR. Эта комиссия составила список терминов, которые позже вошли в словарь. Чтобы обеспечить точность перевода и практичность использования терминов, каждая из участвующих в этом проекте стран выделила группу специалистов для перевода этих терминов на родной язык.

Этот словарь составлен в первую очередь для тех, кто практически занят в отрасли отопления и вентиляции – производителей, предпринимателей, инженеров. Это издание будет полезно специалистам, работающим с компаниями и организациями других стран, нуждающимся в технической информации на других языках, а также специлистам, работающим в области исследований и образования.

Наша благодарность

Американское общество инженеров по отоплению, вентиляции, охлаждению и кондиционированию воздуха (ASHRAE) дало свое согласие на использование нового издания *Терминология ASHRAE по отоплению, вентиляции, охлаждению и кондиционированию воздуха* в качестве источника информации для данного словаря.

Следует выразить благодарность всем членам 11 рабочих групп за их сотрудничество и большую работу, проделанную по переводу 4151 терминов, включенных в словарь.

Особенную признательность нужно высказать членам Редакционной комиссии – Walter Knoll, Wim J. F. Oudijn и Franco Stanzani – за их усилия при выполнении трудной задачи по координации работ над словарем в течение 2-ух лет.

О пользовании словарем

Словарь состоит из 2-ух частей. Первая, основная, часть, содержит все термины, размещенные в алфавитном порядке на английском языке, причем, для каждого термина устанавливается тождество посредством справочного номера. Соответствующие термины на других 11 языках приводятся под английскнм термином. Термины на английском языке приводятся в алфавитном порядке согласно сокращеням Международной организации стандартов для названий языков, как показано ниже:

DA = датский
DE = немецкий
ES = испанский
 F = французский
 I = итальянский
MA = венгерский
NL = голландский
PO = польский
PY = русский
SU = финский
SV = шведский

В случае, когда английский термин имеет более широко используемый синоним, он приводится без перевода и дается ссылка на другой термин.

Однако, там, где английский термин имеет несколько значений на каком-либо другом языке, под английским словом даны все варианты перевода.

Глаголы обозначены через *vb*, прилагательные через *adj*. Под имен существительных приводится соответствующими сокращениями *m*, *f*, *n*, *c* (м, ж, с, к).

Вторая часть словаря состоит из 11 алфавитных указателей соответствующих языков. Каждый указатель включает в себя термины, представленные в словаре, в алфавитном порядке. Пользователь словарем посылается к основной части посредством справочного номера термина. Этот номер, являющийся общим для всех 12 языков, также значит, что этот словарь возможно легко расширить включением в него дополнительных языков.

Указатели даны в том же порядке, что и названия языков.

В приложениях A и B в конце словаря даны: в A – единицы измерения, используемые в индустрии, в B – пояснения к некоторым сокращениям, использованные в словаре.

При необходимости дополнительной информации о терминах, представценных в этом словаре, читателю следует обратиться к изданию *Терминология ASHRAE по отоплению, вентиляции, охлаждению и кондиционированию воздуха* (ASHRAE, 1991; ISBN 0-910110-77-8).

Esipuhe

REHVA on 18 ammatillisen LVI-alan yhdistyksen liitto. Siinä on jäseniä 18:sta Euroopan maasta. Se perustettiin vuonna 1963 edistämään lämmitys- ja ilmastointialan suunnittelu- ja asennuspalveluja. Jäsenten välinen yhteistyö sekä tiedon ja kokemusten vaihto on olennainen osa järjestön toimintaa.

Dokumentointikomitea, joka myöhemmin nimettiin julkaisukomiteaksi, on tärkeä REHVAN tavoitteiden toteuttamisen kannalta. Yksi komitean aikaansaannoksista oli tämän kirjan ensimmäinen painos – LVI-alan termien monikielinen sanakirja. Ensimmäinen painos, jossa oli kymmenen kieltä, julkaistiin vuonna 1982. Kirjaa myytiin lähes 2000 kpl.

Vuodesta 1982 LVI-tekniikka on laajentunut merkittävästi ja siihen liittyvä kieli on kehittynyt. Monia uusia termejä on otettu käyttöön ja joitakin vanhoja on poistettu. Lisäksi Euroopan yhtenäisten markkinoiden muodostuessa 1992–1993 uusitun sanakirjan tarve korostui. Sanakirjan toinen painos sisältää englannin lisäksi 11 muuta kieltä. Julkaisukomitea antoi toisen painoksen toimitusvastuun toimituskunnalle, jonka puheenjohtaja oli Franco Stanzani Italian LVI-yhdistyksestä, AICARR:sta. Komitea laati alustavan listan sanakirjaan sisällytettävistä sanoista. Jokaisen mukana olevan maan työryhmä hyväksyi sanalistan täydennysten jälkeen ja käänsi sen omalle kielelleen.

Sanakirja on tarkoitettu käytännön tehtävissä toimivien – valmistajien, urakoitsijoiden ja konsulttien – käyttöön, joiden työ edellyttää toimimista ulkomaisissa projekteissa tai organisaatioissa tai joiden tulee lukea teknistä kirjallisuutta muulla kuin omalla kielellään. Se soveltuu myös tutkimuksen ja opetuksen piirissä toimivien käyttöön.

Kiitokset

Amerikkalainen lämmitys-, ilmanvaihto-, jäähdytys- ja ilmastointi-insinöörien yhdistys (ASHRAE) antoi luvan käyttää uusinta painosta ASHRAE:n terminologiasanakirjasta (*ASHRAE's Terminology of Heating, Ventilation, Air Conditioning and Refrigeration*) tämän sanakirjan laadinnassa.

Kiitokset kaikille jäsenille 11 työryhmässä yhteistyöstä ja työpanoksesta käännettäessä sanakirjan 4151 termiä. Lopuksi kiitokset toimituskunnan jäsenille – Walter Knoll, Wim J. F. Oudijn ja Franco Stanzani – heidän panoksestaan valvonta- ja koordinointi-työn vaikeassa tehtävässä sanakirjan laadintatyössä yli kahden vuoden ajan.

Kuinka tätä kirjaa käytetään

Sanakirja on jaettu kahteen pääosaan. Osa 1 on päähakemisto. Se sisältää kaikki termit aakkosellisessa järjestyksessä englanniksi. Jokaiselle termille on annettu numero – "the entry number" – jota on käytetty referenssinä kirjan toisessa osassa. Vastaavat termit muilla yhdellätoista kielellä on lueteltu englantilaisen sanan alla. Kielten lyhenteinä on käytetty ISO:n mukaisia suosituksia. Siten:

DA = Tanska
DE = Saksa
ES = Espanja
 F = Ranska
 I = Italia
MA = Unkari
NL = Hollanti
PO = Puola
PY = Venäjä
SU = Suomi
SV = Ruotsi

Kielet esiintyvät lyhenteiden mukaisessa aakkosjärjestyksessä.

Silloin kun englanninkielisellä termillä on olemassa myös yleisemmin käytetty vaihtoehto on harvinaisemman termin kohdalla viitattu yleisempään eikä käännöstä esiinny. Kuitenkin kun englantilaisella termillä on useampia kuin yksi käännös, vaihtoehdot on lueteltu yhdessä englanninkielisen termin kanssa.

Kaikki termit ovat substantiiveja paitsi silloin kun englanninkielisen sanan yhteydessä on muuten osoitettu: verbit on ilmoitettu lyhenteellä *vb* ja adjektiivit lyhenteellä *adj*. Silloin kun substantiiveilla on eri sukuja, nämä on ilmoitettu lyhenteillä *m, f, n* tai *c*. Ne esiintyvät asianomaisen käännöksen yhteydessä.

Sanakirjan osassa 2 on 11 erillistä indeksiluetteloa – yksi kullekin kielelle paitsi englannille. Jokaisessa indeksiluettelossa sanakirjan kaikki termit esiintyvät aakkosjärjestyksessä. Luettelossa on esitetty sanakirjan osassa I käytetty viittausnumero (entry number). Tämän vuoksi jokaisen joka tuntee termin yhdessä kielessä on helppo etsiä asianomainen viitenumero ja sen jälkeen termi sanakirjan ensimmäisen osan päähakemistosta. Päähakemistosta löytyy tämän jälkeen termin käännös kaikille muille kielille, mukaanlukien englanti. Hakemistot ovat edellä esitetyn kieliluettelon mukaisessa järjestyksessä. Sanakirjan lopussa on kaksi liitettä: liitteessä A luetellaan teollisuuden käyttämät mittayksiköt, jotka kaipaavat lisäselvityksiä. Liitteessä B luetellaan sanakirjassa käytettyjä lyhenteitä. Jos tässä käytettyjen termien lisämäärittelyjä tarvitaan, niin tässä viitataan ASHRAE:n julkaisuun *ASHRAE's Terminology of Heating, Ventilation, Air Conditioning and Refrigeration* (ASHRAE, 1991; ISBN 0-910110-77-8).

Förord

REHVA är en organisation av professionella föreningar inom värme-ventilations och luftkonditioneringsområdet från 18 länder inom Europa. REHVA bildades 1963 med målet att medverka i utvecklingen av värme- ventilations- och luftkonditioneringstekniken och förbättring av konstruktions-utföranden och installationer. Samverkan mellan medlemsorganisationer och utbyte av tekniska kunskaper är en väsentlig del inom REHVA's arbete.

Dokumentationskommittén, senare nämnd Publikationskommittén, har stor betydelse för REHVA. Ett av projekten inom denna kommitté var att utge den första upplagan av det flerspråkiga lexikonet (multilingual dictionary). Denna första upplaga publicerades 1982. Den täckte tio språk. Ca 2000 exemplar har sålts.

Från 1982 och framåt har värme- och ventilationsteknikens arbetsområde breddats och de termer som använts har förändrats. Många nya termer har introducerats och andra har utgått. Utvecklingen av den europeiska marknaden 1992/1993 har understrukit behovet av att förnya ordboken med en andra upplaga som nu täcker in elva språk utöver engelska. Inom Publikationskommittén bildades en särskild kommitté som gavs ansvaret för att ta fram denna andra upplaga under ordförandeskap av Hr. Franco Stanzani från den italienska föreningen, AICARR. Inom denna kommitté sammanställdes förteckningar över de ord och termer som skulle ingå i ordboken. Inom vart och ett av de deltagande länderna bildades arbetsgrupper med uppgift att översätta dessa termer till det egna språket på ett korrekt sätt.

Ordboken avses i första hand bli använd av tillverkare, entreprenörer och konsulter för att underlätta arbetet inom flerspråkiga projekt och organisationer och för att lättare förstå teknisk information på olika språk. Även personer inom forskning och utveckling avses kunna använda den nya ordboken.

Tillkännagivanden

American Society of Heating, Ventilating, Refrigeration and Air-conditioning Engineers (ASHRAE) har gett tillåtelse att utnyttja

ASHRAE's Terminology of Heating, Ventilation, Airconditioning and Refrigeration vid utarbetandet av denna ordbok. Vidare ges ett tack till medlemmarna i de elva arbetsgrupper som genomfört ett omfattande arbete vid översättningen av de 4.150 olika termerna.

Tre personer ges särskild uppmärksamhet för deras deltagande som seniormedlemmar i kommittén nämligen Walter Knoll, Wim J. F. Oudijn och Franco Stanzani för deras intresse i den svåra uppgiften att övervaka och samordna arbetet över en 2-årsperiod.

Hur ordboken används

Ordboken är indelad i två huvuddelar. Del ett är huvuddelen som innehåller alla de olika termerna i alfabetisk ordning på engelska. Varje term har getts ett eget nummer – referensnumret – vilket används som referens i den andra delen. Motsvarande termer i de olika elva språken har listats under det engelska ordet.

För de olika språken har använts följande förkortningar:

DA = Danska
DE = Tyska
ES = Spanska
F = Franska
I = Italienska
MA = Ungerska
NL = Holländska
PO = Polska
PY = Ryska
SU = Finska
SV = Svenska

I de fall den engelska termen har ett allmänt använt alternativ har en korsreferens getts till den andra termen utan översättning. Om det finns mer än en översättning på den engelska termen till ett annat språk anges alla dessa under den engelska termen.

Termer som är verb anges med förkortningen *vb* och adjektiv med *adj.* Olika genus för substantiv indikeras med *m, f, n* eller *c*.

Ordbokens andra del innehåller elva separata index – ett för varje språk utom engelska. Varje index upptar alla de termer på det aktuella språket som finns i ordboken i alfabetisk ordning. Med referensnumret sker hänvisning till del ett. Om man känner till en term i ett visst språk är det därför lätt att via referensnumret finna motsvarande term i den första delen av ordboken med sina direkta översättningar.

Indexen anges i samma ordning som de olika språken enligt ovan.

Två bilagor finns. Bilaga A anger mätenheter. Bilaga B redovisar olika förkortningar och deras betydelse.

För ytterligare förklaring av olika termers innebörd hänvisas till den tidigare nämnda ASHRAE-publikationen *ASHRAE's Terminology of Heating, Ventilation, Airconditioning and Refrigeration* (ASHRAE, 1991; ISBN 0-910110-77-8).

American Society of Heating, Refrigerating and Air Conditioning Engineers

Publications for Tomorrow's HVAC&R Technology

ASHRAE Handbook Series
Known internationally as the definitive HVAC&R reference.

1993 Handbook— Fundamentals
- ☐ Inch-Pound Edition 81930 $119
- ☐ Metric Edition 81931 $119

1992 Handbook—HVAC Systems and Equipment
- ☐ Inch-Pound Edition 81920 $119
- ☐ Metric Edition 81921 $119

1991 Handbook—HVAC Applications
- ☐ Inch-Pound Edition 81910 $119
- ☐ Metric Edition 81911 $119

1990 Handbook—Refrigeration
- ☐ Inch-Pound Edition 81910 $119
- ☐ Metric Edition 91911 $119

Standard 90.1-1989
Energy Efficient Design of New Buildings Except Low-Rise Residential Buildings
Includes the printed standard, plus two computer programs.
- ☐ 5 1/4" disk 86237 Non-Member: $98.00 Member: $65.00
- ☐ 3 1/2" disk 86238 Non-Member: $98.00 Member: $65.00

Standard 62-1989
Ventilation for Acceptable Indoor Air Quality
The new ventilation standard of the 1990s!
- ☐ 86157 Non-Member: $42.00 Member: $28.00

To order mark the number of books you want beside each selection and note the total cost here: $_____.
There is no extra charge for shipping if books are sent via sea mail.
Allow 6-12 weeks for delivery. Charges for expedited shipping are available upon request.

International Associate members are extended the same discount rights and purchase privileges as ASHRAE members. Please list your society on the order form.

NAME	MEMBERSHIP NO.

STREET ADDRESS (No PO Boxes, please.)

CITY	STATE/PROVINCE

COUNTRY	ZIP

PHONE   INTERNATIONAL ASSOCIATE
- ☐ Payment Enclosed in US Funds.
- ☐ MasterCard ☐ VISA ☐ American Express ☐ Diners Club

CREDIT CARD NO.	EXP. DATE

SIGNATURE (required)

Mail this order form to: ASHRAE Publication Sales/1791 Tullie Circle, NE/ Atlanta, GA 30329-2305. For quick service, phone or fax credit card orders. PHONE: (404) 636-8400. FAX: (404) 321-5478. To receive a catalog listing all of ASHRAE's publications contact ASHRAE Publication Sales.

Part One

Alphabetical Listing, with all language equivalents

A

1 above ground
DA over jorden
DE
ES sobre el suelo
 superficial
F aérien
 non enterré
I fuori terra
MA füldfeletti
NL bovengronds
PO ponad terenem
PY надземный
SU maanpäällinen
SV ovan jord *c*

2 abrade *vb*
DA afskrabe
DE abnutzen
 abreiben
ES erosionar
F abraser
I erodere
 esercitare azione
 abrasiva
MA koptat
NL afschaven
 afschuren
PO ścierać
PY шлифовать
SU hioa
SV nöta

3 abrasion
DA slid
DE Abnutzung *f*
 Verschleiß *m*
ES abrasión *f*
F abrasion *f*
I abrasione *f*
MA kopás
 koptatás
NL schuren *n*
 schuurplek *f*
PO ścieranie *n*
PY износ *m*
SU kuluminen
SV nötning *c*

4 abrasive
DA slibemiddel
DE Abrieb *m*
 Schleifmittel *n*
ES abrasivo *m*
F abrasif *m*
I abrasivo *m*
MA koptató anyag
NL schuurmiddel *n*
 slijpmiddel *n*
PO materiał *m* ścierny
PY абразив *m*
 шлифующий
 материал *m*
SU hioma-aine
SV slipmedel *n*

5 abrasive *adj*
DA slibende
DE schleifend
 zerreibend
ES abrasivo
F abrasif
I abrasivo
MA koptató
NL schurend
 slijpend
PO ścierny
PY абразивный
 шлифующий
SU kuluttava
SV slitande

6 absolute *adj*
DA absolut
DE absolut
ES absoluto
F absolu
I assoluto
MA abszolút
NL absoluut
PO bezwzględny
PY абсолютный
SU absoluuttinen
SV absolut

7 absolute filter
DA svævestøv filter
DE Schwebstoffilter *m*
ES filtro absoluto *m*
 ultrafiltro *m*
F filtre *m* absolu
I filtro *m* assoluto
MA abszolút szűrő
NL absoluut filter *n*
PO filtr *m* absolutny
PY абсолютный
 фильтр *m*
SU absoluuttinen
 suodatin
SV absolutfilter *n*

8 absolute roughness
DA absolut ruhed
DE absolute Rauhig-
 keit *f*
ES rugosidad absoluta *f*
F rugosité *f* absolue
I rugosità *f* assoluta
MA abszolút érdesség
NL absolute ruwheid *f*
PO chropowatość *f*
 bezwzględna
PY абсолютная
 шероховатость *f*
SU absoluuttinen
 karheus
SV absolut råhet *c*

9 absorb *vb*
DA absorbere
DE absorbieren
ES absorber
F absorber
I absorbire
 assorbire
MA abszorbeál
 elnyel
NL absorberen
 opnemen
 opzuigen
PO pochłaniać
 wchłaniać
PY абсорбировать
 поглощать
SU absorpoitua
SV absorbera

10 absorbate
DA absorberende
   medium
DE Absorbat *n*
ES sustancia absorbida *f*
F substance *f* absorbée
I sostanza *f* assorbita
MA abszorbátum
NL geabsorbeerde stof *f*
PO substancja *f*
   absorbowana
PY абсорбируемая
   среда *f*
SU absorpoida
SV absorbat *n*

11 absorbent
DA absorptionsmiddel
DE Absorptionsmittel *n*
ES absorbente *m*
F absorbant *m*
I assorbente *m*
MA abszorpciós küzeg
NL absorbens *n*
   absorptiemiddel *n*
PO absorbent *m*
   pochłaniacz *m*
   substancja *f*
   absorbująca
PY абсорбент *m*
   поглотитель *m*
SU absorbentti
SV absorptionsmedel *n*

12 absorber
DA absorber
DE Absorber *m*
ES absorbedor *m*
   amortiguador *m*
F absorbeur *m*
I assorbitore *m*
MA abszorber
   elnyelő
NL absorptievat *n*
PO absorber *m*
   aparat *m*
   absorbcyjny
PY абсорбер *m*
SU absorboija
SV absorbator *c*

DA =	Danish
DE =	German
ES =	Spanish
F =	French
I =	Italian
MA =	Magyar (Hungarian)
NL =	Dutch
PO =	Polish
PY =	Russian
SU =	Finnish
SV =	Swedish

13 absorber area
DA absorptionsareal
DE Absorberfläche *f*
ES superficie de
   absorción *f*
F surface *f*
   d'absorption
I area *f*
   dell'assorbitore
MA elnyelő felület
NL absorbtieoppervlak
   n
PO powierzchnia *f*
   absorbująca
PY площадь *f*
   абсорбции
SU absorptioala
SV absorbatoryta *c*

14 absorptance (*see* absorption coefficient)

15 absorption
DA absorption
DE Absorption *f*
ES absorción *f*
F absorption *f*
I assorbimento *m*
MA abszorpció
NL absorptie *f*
   opneming *f*
   opzuiging *f*
PO absorpcja *f*
   pochłanianie *n*
PY абсорбция *f*
   поглощение *n*
SU absorptio
SV absorption *c*

16 absorption capacity
DA absorptionsevne
DE Absorptions-
   kapazität *f*
ES capacidad de
   absorción *f*
   poder absorbente *m*
F capacité *f*
   d'absorption
I capacità *f* di
   assorbimento
MA elnyelőképesség
NL absorptievermogen
   n
PO zdolność *f*
   absorpcyjna
PY поглотительная
   способность *f*
SU absorptiokapasiteetti
SV absorptions-
   förmåga *c*

17 absorption coefficient
DA absorptions-
   koefficient
DE Absorptions-
   koeffizient *m*
ES coeficiente de
   absorción *m*
F coefficient *m*
   d'absorption
I coefficiente *m* di
   assorbimento
MA abszorpciós tényező
NL absorptiecoëfficiënt *m*
PO współczynnik *m*
   absorpcji
PY коэффициент *m*
   поглощения
   (абсорбции)
SU absorptiokerroin
SV absorptions-
   koefficient *c*

18 absorption cooling
DA absorptionskøling
DE Absorptions-
   kühlung *f*
ES frigorífico de
   absorción *m*
   refrigeración por
   absorción *f*
F réfrigération *f*
   refroidissement *m*
   par absorption
I refrigerazione *f* ad
   assorbimento
MA abszorpciós hűtés
NL absorptiekoeling *f*
PO chłodzenie *n* absorp-
   cyjne
PY абсорбционное охл-
   аждение *n*
SU absorptiojäähdytys
SV absorptionskylning
   c

19 absorption medium
DA absorptionsmedium
DE Absorptionsmittel *n*
ES medio absorbente *m*
F matière *f* absorbante
I mezzo *m* assorbente
MA abszorpciós küzeg
NL absorptiemiddel *n*
PO czynnik *m* absorp-
   cyjny
PY абсорбирующая
   среда *f*
SU absorptioaine
SV absorptionsmedel *n*

20 absorptivity (*see* absorption capacity)

21 acceleration
DA accelleration
hastighedsforøgelse
DE Beschleunigung *f*
ES aceleración *f*
F accélération *f*
I accelerazione *f*
MA gyorsulás
NL acceleratie *f*
versnelling *f*
PO przyspieszenie *n*
PY ускорение *n*
SU kiihdytys
kiihtyvyys
SV acceleration *c*

22 accelerator (*see* circulation pump)

23 acceptance
DA godkendelse
DE Abnahme *f*
Übernahme *f*
ES recepción *f*
F réception *f*
I accettazione *f*
MA átvétel
NL keuring *f*
oplevering *f*
PO odbiór *n*
przyjęcie *n*
PY прием *m*
приемка *f*
SU hyväksyminen
SV godkännande *n*

24 acceptance angle
DA godkendelsesvinkel
DE Abnahme-
verfahren *n*
ES ángulo de admisión *m*
F angle *m* de réception
I angolo *m* di accettazione
MA átvételi szüg
NL acceptatiehoek *m*
PO kąt *m* dopuszczalny
PY допустимый угол *m*
SU kulmaraja
näkyvyyskulma
SV godkännandevinkel *c*

25 acceptance test
DA godkendelsesprøve
leveringsprøvning
DE Abnahmeversuch *m*
ES prueba de recepción *f*
F essai *m* de réception
I prova *f* di accettazione
MA átvételi vizsgálat
NL keuringsproef *f*
opleveringsproef *f*
PO sprawdzian *m*
PY приемное испытание *n*
SU vastaanottokoe
SV leveransprov *n*

26 access control
DA adgangskontrol
DE Zugangskontroll-
system *n*
ES control de acceso *m*
F contrôle d'accès
I controllo *m* di accesso
MA bemenő szabályozás
NL toegangscontrole *f*
PO kontrola *f* wejścia
PY контроль *m* на входе
SU kulunvalvonta
tietokoneyhteyden valvonta
SV inspektions-
kontroll *c*

27 access door
DA adgangslem
inspektionsluge
DE Bedienungstür *f*
Zugangstür *f*
ES portillo de servicio *m*
puerta de acceso *f*
F panneau *m* d'accès
porte *f* d'accès
I portello *m* di accesso
MA kezelőajtó
NL toegangsdeur *f*
PO drzwiczki *pl* kontrolne
PY входная дверь *f*
SU tarkastusluukku
SV åtkomstüppning *c*

28 access level
DA adgangsniveau
DE Bedienungsebene *f*
Zugangsebene *f*
ES nivel de acceso *m*
F niveau *m* d'accès
I livello *m* di accesso
MA kezelő szint
NL toegangsniveau *n*
PO poziom *m* wejścia
PY приблизительный уровень *m*
SU huoltotaso
SV inspektionsnivå *c*

29 accessory
DA tilbehør
DE Zubehör *n*
Zubehörteil *n*
ES accesorio *m*
F accessoire *m*
I accessorio *m*
MA tartozék
NL accessoire *n*
toebehoren *n*
PO osprzęt *m*
PY арматура *f*
принадлежность *f*
SU tarvike
SV tillbehür *n*

30 acclimatization
DA akklimatisering
tilvænning
DE Akklimatisation *f*
Anpassung *f*
ES aclimatación *f*
F acclimatation *f*
I acclimatazione *f*
MA aklimatizáció
NL acclimatiseren *n*
PO aklimatyzacja *f*
PY акклиматизация *f*
SU akklimatisaatio
SV acklimatisering *c*

31 accumulator
DA akkumulator
DE Akkumulator *m*
Sammler *m*
ES acumulador *m*
F accumulateur *m*
I accumulatore *m*
MA akkumulátor
gyüjtő
NL accumulator *m*
opslagvat *n*
PO akumulator *m*
zasobnik *m*
PY аккумулятор *m*
SU varaaja
SV ackumulator *c*

32 accuracy
DA præcision
DE Genauigkeit *f*
   Präzision *f*
ES exactitud *f*
   precisión *f*
F précision *f*
I precisione *f*
MA pontosság
NL nauwkeurigheid *f*
PO dokładność *f*
PY точность *f*
SU tarkkuus
SV noggrannhet *c*

33 acetylene
DA acetylen
DE Azetylen *n*
ES acetileno *m*
F acétylène *m*
I acetilene *m*
MA acetilén
NL acetyleen *n*
PO acetylen *m*
PY ацетилен *m*
SU asetyleeni
SV acetylen *c*

34 ACH (*see* air changes per hour)

35 acid
DA syre
DE Säure *f*
ES ácido *m*
F acide *m*
I acido *m*
MA sav
NL zuur *n*
PO kwas *m*
PY кислота *f*
SU happo
SV syra *c*

36 acid attack
DA syreangreb
DE Säureangriff *m*
   Säurewirkung *f*
ES ataque por un ácido *m*
F attaque *f* par un acide
I attacco *m* acido
MA savhatás
NL aantasting *f* door zuur
PO działanie *n* kwasu
PY агрессивность *f* кислоты
SU happohyükkäys
SV syraangrepp *n*

37 acid dew point
DA syredugpunkt
DE Säuretaupunkt *m*
ES punto de rocío del ácido *m*
F point *m* de rosée acide
I punto *m* di rugiada acida
MA sav harmatpont
NL zuurdauwpunt *n*
PO punkt *m* rosy kwasu
PY точка *f* росы кислоты
SU happokastepiste
SV syradaggpunkt *c*

38 acidity
DA surhedsgrad
DE Säuregrad *m*
ES acidez *f*
F acidité *f*
I acidità *f*
MA savtartalom
NL aciditeit *f*
   zuurgraad *m*
PO kwasowość *f*
   kwaśność *f*
PY кислотность *f*
SU happamuus
SV surhetsgrad *c*

39 acoustics
DA akustik
DE Akustik *f*
ES acústica *f*
F acoustique *f*
I acustica *f*
MA akusztika
NL akoestiek *f*
   geluidsleer *f*
   klankleer *f*
PO akustyka *f*
PY акустика *f*
SU akustiikka
SV akustik *c*

40 action
DA aktion
   virkning
DE Aktion *f*
   Wirkung *f*
ES acción *f*
F action *f*
I azione *f*
MA hatás
   müküdés
NL actie *f*
   werking *f*
PO działanie *n*
PY действие *n*
SU toiminta
   vaikutus
SV verkan *c*

41 activated alumina desiccant
DA aktivt aluminiumtørremiddel
DE Tonerde-Trocknungsmittel *n*
ES deshidratante de alúmina activada *m*
F déshydratant *m* à l'alumine activée
I essiccante *m* ad allumina attivata
MA aktív timfüld szárítóküzeg
NL actief aluminium-trioxyde-droogmiddel *n*
PO osuszający aktywowany tlenek *m* glinu
PY сорбент *m* из активированной окиси алюминия
SU aluminioksi-dikuivaaja
SV desickant aktiverad aluminiumoxid *c*

42 activated carbon
DA aktivt kul
DE Aktivkohle *f*
ES carbón activado *m*
F charbon *m* actif
I carbone *m* attivo
MA aktív szén
NL actieve kool *f*
   adsorptiekool *f*
PO węgiel *m* aktywowany
PY активированный уголь *m*
SU aktiivihiili
SV aktivt kol *n*

DA	=	Danish
DE	=	German
ES	=	Spanish
F	=	French
I	=	Italian
MA	=	Magyar (Hungarian)
NL	=	Dutch
PO	=	Polish
PY	=	Russian
SU	=	Finnish
SV	=	Swedish

43 actual *adj*
DA aktuel
faktisk
virkelig
DE wirklich
ES verdadero
efectivo
real
F actuel
effectif
réel
I attuale
MA tényleges
NL momenteel
werkelijk
PO rzeczywisty
PY действительный
SU
SV aktiv
verklig

44 actuator
DA aktuator
udløser
DE Schalter *m*
ES accionador *m*
actuador *m*
F contacteur *m*
I attuatore *m*
contattore
MA kapcsoló
NL aandrijver *m*
servomotor *m*
PO siłownik *m*
urządzenie *n*
uruchamiające
PY привод *m*
SU toimilaite
SV igångsättare *c*

45 actuator runtime
DA aktuator drifttid
DE Schalterlaufzeit *f*
ES tiempo de
funcionamiento de
un accionador *m*
F temps *m* de
fonctionnement
d'un actionneur
I tempo *m* di
funzionamento
dell'attuatore
MA kapcsolási idő
NL servomotor-
looptijd *m*
PO czas *m* działania
siłownika
PY время *n*
срабатывания
привода
SU toimielimen ajoaika
SV igångsättares *c*
gångtid *c*

46 adaptive control
DA adaptiv
(selvjusterende)
styring
DE adaptive Regelung *f*
ES control autoajustable
m
F régulation *f* auto-
adaptative
I controllo *m*
adattativo
MA illesztett szabályozás
NL adaptieve regeling *f*
PO adaptacyjny układ
m regulacji
PY наладка *f*
настройка *f*
SU adaptiivinen säätü
SV självjusterande
kontroll *c*

47 additive
DA additiv
tilsætningsmiddel
DE Zusatz *m*
ES aditivo *m*
F additif *m*
I additivo *m*
MA járulék
NL additief *n*
toevoegsel *n*
PO dodatek *m*
PY добавка *f*
добавочный
SU lisäaine
SV additiv *n*

48 addressable
DA tilgængelig
DE adressierbar
ES dirigible
F adressable
I indirizzabile
MA irányítható
NL adresseerbaar
PO adresowy
dający się
zaadresować
PY направленный
SU osoitteellinen
tietopaikka
osoitteistettavissa
SV tillgänglig

49 adhesive
DA klæbemiddel
lim
DE Klebstoff *m*
ES adhesivo *m*
F matière *f* adhésive
substance *f* adhésive
I adesivo *m*
MA ragasztó
NL kleefmiddel *n*
lijm *m*
plakmiddel *n*
PO klej *m*
spoiwo *n*
PY клей *m*
SU liima
SV adhesiv *n*

50 adhesive *adj*
DA klæbende
DE anhaftend
klebend
ES adhesivo
F adhésif
I adesivo
MA ragadó
NL klevend
plakkend
PO klejący
lepki
PY клейкий
SU tarttuva
SV klibbig

51 adiabatic *adj*
DA adiabatisk
DE adiabatisch
ES adiabático
F adiabatique
I adiabatico
MA adiabatikus
NL adiabatisch
isentropisch
PO adiabatyczny
PY адиабатный
SU adiabaattinen
SV adiabatisk

52 adjust *vb*
DA indstille
    justere
DE einregulieren
    einstellen
ES ajustar
    regular
F ajuster
    régler
I regolare
MA beállít
    szabályoz
NL aanpassen
    afstellen
    instellen
PO dostosować
    regulować
PY регулировать
SU asettaa
    säätää
SV reglera

53 adjustable *adj*
DA justerbar
DE einstellbar
    regulierbar
ES ajustable
    regulable
F réglable
I regolabile
MA szabályozható
NL instelbaar
    verstelbaar
PO dający sie
    dostosować
    dający się regulować
PY регулируемый
SU aseteltava
    säädettävä
SV reglerbar

54 adjustment
DA justering
DE Einstellung *f*
    Regulierung *f*
ES ajuste *m*
    regulación *f*
F réglage *m*
I regolazione *f*
MA beállítás
    szabályozás
NL aanpassing *f*
    afstelling *f*
    instelling *f*
PO dostosowanie *n*
    regulacja *f*
PY регулировка *f*
SU asetus
    säätü
SV reglering *c*

55 adjustment range
DA justeringsområde
DE Einstellbereich *m*
ES campo de
    regulación *m*
    escala de ajuste *f*
F plage *f* de réglage
I campo *m* di
    regolazione
MA beállítási tartomány
NL instelbereik *n*
PO zakres *m* regulacji
PY градуировочная
    шкала *f*
SU asetusalue
    säätüalue
SV reglerområde *n*

56 adsorbent
DA adsorbent
DE Adsorptionsmittel *n*
ES absorbente *m*
F absorbant
I adsorbente *m*
MA adszorpciós küzeg
NL adsorbtiemiddel *n*
PO adsorbent *m*
    substancja *f*
    adsorbująca
PY адсорбент *m*
SU absorbentti
SV adsorptionsmedel *n*

57 adsorption
DA adsorbtion
DE Adsorption *f*
ES adsorción *f*
F adsorption *f*
I adsorbimento *m*
MA adszorpció
NL adsorptie *f*
PO adsorpcja *f*
    pochłanianie *n*
    powierzchniowe
PY адсорбция *f*
SU adsorptio
SV adsorption *c*

58 aerate *vb*
DA aflufte
    lufte
DE belüften
ES airear
    ventilar
F aérer
I aerare
MA levegőztet
    szellőztet
NL beluchten
PO napowietrzać
PY проветривать
SU tuulettaa
SV lufta

59 aeration
DA luftning
DE Belüftung *f*
ES aireación *f*
    ventilación *f*
F aération *f*
I aerazione *f*
MA levegőztetés
NL beluchting *f*
PO aeracja *f*
    napowietrzanie *n*
PY аэрация *f*
SU tuuletus
SV luftning *c*

60 aerodynamic *adj*
DA aerodynamisk
DE aerodynamisch
ES aerodinámico
F aérodynamique
I aerodinamico
MA aerodinamikai
NL aerodynamisch
PO aerodynamiczny
PY аэродинамический
SU aerodynaaminen
SV aerodynamisk

61 aerogel
DA aerogel
DE Aerogel *n*
ES gel de sílice *m*
F silicagel *m* (gel de
    silice)
I essiccante *m* a
    biossido di silicio
MA aerogél
NL aerogel *n*
PO aerożel *m*
PY аэрогель *m*
SU aerogeeli
SV aerogel *c*

62 aerosol
DA aerosol
DE Aerosol *n*
ES aerosol *m*
F aérosol *m*
I aerosol *m*
MA aeroszol
NL aerosol *f*
PO aerozol *m*
PY аэрозоль *m*
SU aerosoli
SV aerosol *c*

63 after cooler
DA efterkøler
DE Nachkühler *m*
ES subenfriador *m*
F refroidisseur *m* aval
I postraffreddatore *m*
MA utóhűtő
NL nakoeler *m*
PO chłodnica *f* wtórna
PY доохладитель *m*
SU jälkijäähdytin
SV efterkylare *c*

64 after heater
DA eftervarmer
DE Nachwärmer *m*
ES recalentador *m*
F réchauffeur *m* aval
   post-réchauffeur
I postriscaldatore *m*
MA utófűtő
NL naverwarmer *m*
PO nagrzewnica *f*
   wtórna
PY догреватель *m*
SU jälkilämmitin
SV eftervärmare *c*

65 ageing
DA
DE Alterung *f*
ES envejecimiento *m*
F vieillissement *m*
I invecchiamento *m*
MA üregedés
NL veroudering *f*
PO starzenie
PY старение *n*
SU vanhentaminen
SV åldring *c*

66 aggregate
DA aggregat
DE Aggregat *n*
ES agregado *m*
   árido (hormigón) *m*
F agrégat *m*
I aggregato *m*
MA agregát
   gép
NL aggregaat *n*
PO agregat *m*
   skupienie *n*
   zespół *m*
PY агрегат *m*
SU karkea sora
   rakeinen aine
SV aggregat *n*

67 aggressive *adj*
DA aggressiv
DE aggressiv
ES agresivo
F agressif
I aggressivo
MA agresszív
NL aantastend
   agressief
PO agresywny
PY агрессивный
SU agressiivinen
SV aggressiv

68 aggressiveness
DA aggressivitet
DE Aggressivität *f*
ES agresividad *f*
F agressivité *f*
I aggressività *f*
MA agresszivitás
NL agressiviteit *f*
PO agresywność *f*
PY агрессивность *f*
SU agressiivisyys
SV aggressivitet *c*

69 aggressivity (*see* aggressiveness)

70 agitator
DA omrører
DE Rührer *m*
   Rührwerk *n*
ES agitador *m*
F agitateur *m*
I agitatore *m*
MA keverő
NL roerinrichting *f*
   roertoestel *n*
   roerwerk *n*
PO mieszadło *n*
   mieszalnik *m*
PY мешалка *f*
SU sekoituskone
SV blandare *c*

71 air
DA luft
DE Luft *f*
ES aire *m*
F air *m*
I aria *f*
MA levegő
NL lucht *f*
PO powietrze *n*
PY воздух *m*
SU ilma
SV luft *c*

72 air agitation system
DA luftblandesystem
DE Luftbewegungs-system *n*
ES sistema de agitación del aire *m*
F système *m* à insufflation d'air
I sistema *m* di agitazione dell'aria
MA légkeverési rendszer
NL roersysteem *n* met lucht
PO system *m* pobudzania powietrza
   układ *m* wprowadzający powietrze w ruch
PY воздухосмеситель-ьная система *f*
SU ilma-avusteinen jääntekojärje-stelmä
SV luftrürelsesystem *n*

73 air analysis
DA luftanalyse
DE Luftanalyse *f*
   Luftzusammen-setzung *f*
ES análisis del aire *m*
F analyse *f* d'air
I analisi *f* dell'aria
MA levegő elemzés
NL luchtanalyse *f*
PO analiza *f* składu powietrza
PY анализ *m* воздуха
SU ilma-analyysi
SV luftanalys *c*

DA =	Danish
DE =	German
ES =	Spanish
F =	French
I =	Italian
MA =	Magyar (Hungarian)
NL =	Dutch
PO =	Polish
PY =	Russian
SU =	Finnish
SV =	Swedish

74 air atomizing burner
DA luftforstøvnings-
     brænder
DE Luftzerstäubungs-
     brenner m
ES quemador de
     atomización de
     aire m
F brûleur m à
     atomisation
I bruciatore m con
     atomizzazione ad
     aria
MA légporlasztásos égő
NL drukluchtverstuivings-
     brander m
PO palnik m z
     rozpylaniem
     powietrza
PY пневматическая
     форсунка f
SU paineilmahajotteinen
     poltin
SV luftspridar-
     brännare c

75 air blast cooling
DA luftforceret køling
DE Gebläseluft-
     kühlung f
ES enfriamiento por aire
     forzado m
F refroidissement m à
     air pulsé
I raffreddamento m
     per espansione di
     aria
MA légbefúvásos hűtés
NL geforceerde
     luchtkoeling f
PO chłodzenie n
     podmuchowe
PY вентиляторное
     охлаждение n
SU ilmasuihkujäähdytys
SV tryckluftskylning c

76 airborne particles
DA luftbårne partikler
DE luftgetragene
     Teilchen f,pl
ES partículas
     aerodifusas f
F particules f
     aériennes
I particolato m
     aerodisperso
MA lebegő részecskék
NL zweefstof n
PO cząsteczki f
     przenoszone drogą
     powietrzną
PY частицы f
     (примеси) в
     воздухе
SU leijuvat hiukkaset
SV luftburna partiklar c

77 airborne sound
DA luftbåret lyd
DE Luftschall m
ES ruido llevado por el
     aire m
     sonido transmitido
     por el aire m
F bruit m aérien
I suono m trasmesso
     attraverso l'aria
MA léghang
NL luchtgeluid n
PO dźwięk m
     rozchodzący się w
     powietrzu
PY воздушный шум m
SU ilmaääni
SV luftburet ljud n

78 air brick
DA ventilationssten
DE Hohlziegelstein m
     Lochstein m
ES ladrillo (hueco) m
     ladrillo
     (perforado) m
F brique f alvéolée
I mattone m di
     ventilazione
MA üreges tégla
NL holle baksteen m
     ventilatiesteen m
PO pustak m
     wentylacyjny
PY пустотелый
     кирпич m
SU reikätiili
SV håltegel n

79 air bubble
DA luftboble
DE Luftblase f
ES bolsa de aire f
     burbuja de aire f
F bulle f d'air
I bolla f di aria
MA légbuborék
NL luchtbel f
PO pęcherzyk m
     powietrza
PY воздушный
     пузырек m
SU ilmakupla
SV luftblåsa c

80 air change
DA luftskifte
DE Luftwechsel m
ES renovación de aire f
F renouvellement m
     d'air
I ricambio m di aria
MA légcsere
NL luchtwisseling f
PO wymiana f
     powietrza
PY воздухообмен m
SU ilmanvaihto
SV luftväxling c

81 air change rate
DA luftskiftefrekvens
DE Luftwechselzahl f
ES tasa de renovación
     del aire f
F taux m de
     renouvellement
     d'air
I tasso m di ricambio
     dell'aria
MA légcsere mértéke
NL luchtwisselings-
     voud n
PO krotność f wymiany
     powietrza
PY кратность f
     воздухообмена
SU ilmanvaihtokerroin
     ilmanvaihtuvuus
SV luftväxlings-
     frekvens c

82 air changes per hour
DA luftskifte pr. time
DE Luftwechsel *m* pro Stunde *f*
ES renovaciones de aire por hora *f*
F taux *m* horaire de renouvellement de l'air
I ricambi *m,pl* di aria orari
MA óránkénti légcsere
NL luchtwisselingen *pl* per uur
PO krotność *f* wymiany powietrza na godzinę
PY воздухообмен *m* в час
SU ilmanvaihtoa tunnissa
SV luftväxlingar *c* per timme *c*

83 air circulation
DA luftcirkulation
DE Luftzirkulation *f*
ES circulación del aire *f*
F brassage *m* d'air
I circolazione *f* dell'aria
MA levegő keringés
NL luchtcirculatie *f*
PO cyrkulacja *f* powietrza
    ruch *m* powietrza
PY подвижность *f* воздуха
    циркуляция *f* воздуха
SU ilmankierrätys
SV luftcirkulation *c*

84 air circulation rate
DA luftcirkuleringsgrad
DE Luftwechselzahl *f*
ES tasa de circulación del aire *f*
F taux *m* de brassage d'air
I tasso *m* di circolazione dell'aria
MA levegő keringés mértéke
NL luchtcirculatie-voud *n*
PO wielkość *f* cyrkulacji powietrza
PY кратность *f* воздухообмена
SU kiertoilmavirta
SV luftcirkulations-frekvens *c*

85 air cleaner
DA luftrenser
DE Luftreiniger *m*
ES purificador de aire *m*
F dépoussiéreur *m*
I depuratore *m* di aria
MA légtisztító
NL luchtreiniger *m*
PO filtr *m* powietrza
    urządzenie *n* do oczyszczania powietrza
PY воздухоочиститель *m*
SU ilmanpuhdistin
SV luftrenare *c*

86 air cleaning
DA luftrensning
DE Luftreinigung *f*
ES depuración del aire *f*
    filtración del aire *f*
F épuration *f* de l'air
I depurazione *f* dell'aria
MA légtisztítás
NL luchtreiniging *f*
    luchtzuivering *f*
PO oczyszczanie *n* powietrza
PY очистка *f* воздуха
SU ilmanpuhdistus
SV luftrening *c*

87 air condition
DA lufttilstand
DE Luftbehandlung *f*
ES acondicionar climatizar
F climatiser
I condizione *f* dell'aria
MA légállapot
NL luchtconditie *f*
    luchttoestand *m*
PO stan *m* powietrza
PY состояние *n* воздуха
SU ilmastoida
SV luftbehandling *c*

88 air conditioner
DA klimaapparat
    luftbehandler
DE Klimagerät *n*
ES acondicionador de aire *m*
    aparato *m* de acondicion-amiento de aire *m*
    climatizador *m*
F climatiseur *m*
    conditionneur *m* d'air
I condizionatore *m* di aria
MA légkondicionáló készülék
NL luchtbehandelings-apparaat *n*
PO klimatyzator *m*
PY кондиционер *m* воздуха
SU ilmastointikone
SV klimatapparat *c*

89 air conditioning
DA luftkonditionering
DE Klimatisierung *f*
ES acondicionamiento de aire *m*
    climatización *f*
F conditionnement *m* d'air
I climatizzazione *f*
    condizionamento *m* dell'aria
MA légkondicionálás
NL airconditioning *f*
    luchtbehandeling *f*
PO klimatyzacja *f*
PY кондиционирование *n* воздуха
SU ilmastointi
SV luftkonditionering *c*

90 air conditioning process
DA klimaanlæg
   luftkonditionerings-
   anlæg
DE Klimaprozeß *m*
   Klimaverfahren *n*
ES proceso de
   acondiciona-
   miento *m*
F procédé *m* de
   conditionnement
   d'air
I processo *m* di condi-
   zionamento
MA légkondicionálási
   eljárás
NL luchtbehandelings
   proces *n*
PO proces *m*
   klimatyzacyjny
PY процесс *m*
   кондицион-
   ирования воздуха
SU ilmastointiprosessi
SV luftkonditionerings-
   process *c*

91 air conditioning system
DA luftbehandlings-
   system
DE Klimasystem *n*
ES sistema de
   climatización *m*
F système *m* de
   conditionnement
   d'air
I impianto *m* di
   condizionamento
MA légkondicionáló
   rendszer
NL luchtbehandelings-
   systeem *n*
PO system *m*
   klimatyzacji
PY система *f*
   кондицион-
   ирования воздуха
SU ilmastointijärje-
   stelmä
SV luftkonditionerings-
   system *n*

92 air consumption
DA luftforbrug
DE Luftverbrauch *m*
ES consumo de aire *m*
F consommation *f*
   d'air
I consumo *m* di aria
MA levegő fogyasztás
NL luchtverbruik *n*
PO zużycie *n* powietrza
PY расход *m* воздуха
SU ilmankulutus
SV luftförbrukning *c*

93 air contaminant
DA luftforurener
DE Luftverunreiniger *m*
ES contaminante del
   aire *m*
F polluant *m* de l'air
I contaminante *m*
   dell'aria
   inquinante *m*
   dell'aria
MA légszennyező
NL luchtvervuilende
   stof *m*
PO substancja *f*
   zanieczyszczająca
   powietrze
PY вредность *f*
   примесь *f* к
   воздуху
SU ilman epäpuhtaus
SV luftförorening *c*

94 air-cooled *adj*
DA luftkølet
DE luftgekühlt
ES enfriado por aire
F refroidi par air
I raffreddato ad aria
MA léghűtéses
NL luchtgekoeld
PO chłodzony
   powietrzem
PY охлажденный
   воздухом
SU ilmajäähdytetty
SV luftkyld

95 air-cooled condenser
DA luftkølet
   kondensator
DE luftgekühlter
   Verflüssiger *m*
ES condensador
   enfriado por
   aire *m*
F condenseur *m* à air
I condensatore *m*
   raffreddato ad aria
MA léghűtéses
   kondenzátor
NL luchtgekoelde
   condensor *m*
PO skraplacz *m*
   chłodzony
   powietrzem
PY конденсатор *m* с
   воздушным
   охлаждением
SU ilmalauhdutin
SV luftkyld
   kondensor *c*

96 air-cooled condensing unit
DA luftkølet
   kondensator-
   aggregat
DE luftgekühlte
   Verflüssiger-
   Einheit *f*
ES unidad
   condensadora
   enfriada por aire *f*
F unité *f* de
   condensation
   refroidie par air
I unità *f* condensante
   raffreddata ad aria
MA léghűtéses
   kondenzátor
   egység
NL luchtgekoelde
   condensor *m*
PO zespół *m* ze
   skraplaczem
   chłodzonym
   powietrzem
PY конденсаторная
   установка *f* с
   воздушным
   охлаждением
SU ilmalauhdutin
SV luftkyld
   kondensorenhet *c*

97 air cooler
DA luftkøler
DE Luftkühler *m*
ES enfriador de aire *m*
   refrigerador *m*
F refroidisseur *m* d'air
I refrigeratore *m* di
   aria
MA léghűtő
NL luchtkoeler *m*
PO chłodnica *f*
   powietrza
PY воздухоохладитель
   m
SU ilmajäähdytin
SV luftkylare *c*

98 air cooler battery
DA luftkølebatteri
DE Luftkühlungs-
   register *n*
ES batería frigorífica *f*
   batería refrigerante *f*
F batterie *f* de
   refroidissement à
   air
I batteria *f* di
   raffreddamento
   aria
MA léghűtőtest
NL luchtkoelbatterij *f*
   luchtkoelelement *n*
PO chłodnica *f*
   powietrza
PY воздухоохладитель-
   ьная установка *f*
SU jäähdytyspatteri
SV luftkylarbatteri *n*

99 air cooling
DA luftkøling
DE Luftkühlung *f*
ES enfriamiento del
   aire *m*
   refrigeración del
   aire *f*
F refroidissement *m*
   d'air
I raffreddamento *m*
   dell'aria
MA léghűtés
NL luchtkoeling *f*
PO chłodzenie *n*
   powietrza
PY воздушное
   охлаждение *n*
SU ilmajäähdytys
SV luftkylning *c*

100 air current
DA luftstrøm
DE Luftstrom *m*
ES corriente de aire *f*
F écoulement *m* d'air
I corrente *f* di aria
MA légáram
NL luchtstroom *m*
PO prąd *m* powietrzny
PY поток *m* воздуха
SU ilmavirta
   ilmavirtaus
SV luftström *c*

101 air curtain
DA lufttæppe
DE Luftvorhang *m*
ES cortina de aire *f*
F rideau *m* d'air
I cortina *f* di aria
   lama *f* di aria
MA légfüggüny
NL luchtgordijn *n*
PO kurtyna *f*
   powietrzna
   zasłona *f* powietrzna
PY воздушная завеса *f*
   воздушная
   заслонка *f*
SU ilmaverho
SV luftridå *c*

102 air cushion
DA luftpude
DE Luftkissen *n*
ES amortiguador de
   aire *m*
   colchón de aire *m*
F matelas *m* d'air
I cuscino *m* di aria
MA légpárna
NL luchtkussen *n*
PO poduszka *f*
   powietrzna
PY воздушная
   подушка *f*
SU ilmatyyny
SV luftkudde *c*

**103 air-cycle
   refrigeration**
DA luftkredsløbskøling
DE Luftkreislauf-
   kühlung *f*
ES enfriamiento por
   aire *m*
F réfrigération *f* à air
   (cycle à air)
I refrigerazione *f* a
   ciclo ad aria
MA légforgatásos hűtés
NL luchtkoeling *f* door
   expanderende
   lucht
   luchtkring-
   loopkoeling *f*
PO chłodzenie *n* w
   cyklu
   powietrznym
PY воздушный
   холодильный
   цикл *m*
SU ilmanpaisuntajääh
   dytys
SV luftkretskylning *c*

104 air damper
DA luftspjæld
DE Luftklappe *f*
ES compuerta de aire *f*
F amortisseur *m* à air
I serranda *f* per aria
MA légzsalu
NL luchtklep *f*
PO przepustnica *f*
   powietrza
PY воздушный (дроссе-
   льный) клапан *m*
SU sälepelti
   säätüpelti
SV luftspjäll *n*

**105 air
   dehumidification**
DA luftaffugtning
DE Luftentfeuchtung *f*
ES deshumectación del
   aire *f*
   deshumidificación
   del aire *f*
F déshumidification *f*
   d'air
I deumidificazione *f*
   dell'aria
MA levegő
   nedvességelvonás
   szárítás
NL luchtontvochtiging *f*
PO osuszanie *n*
   powietrza
PY осушение *n*
   воздуха
SU ilmankuivaus
SV luftavfuktning *c*

106 air dehumidifier
DA luftaffugter
DE Luftentfeuchter *m*
ES deshumectador de
aire *m*
deshumidificador de
aire *m*
F déshumidificateur *m*
d'air
I deumidificatore *m* di
aria
MA levegő
nedvességelvonó
szárító
NL luchtontvochtiger *m*
PO urządzenie *n* do
osuszania
powietrza
PY воздухоосушитель
m
SU ilmankuivaaja
SV luftavfuktare *c*

107 air diffusion
DA luftdiffusion
DE Luftdiffusion *f*
ES difusión del aire *f*
F diffusion *f* d'air
I diffusione *f* dell'aria
MA légelosztás
NL
PO rozdział *m*
powietrza w
pomieszczeniu
PY воздушная
диффузия *f*
распростанение *n*
воздуха
SU ilmanjako
SV luftspridning *c*

**108 air discharge
coefficient**
DA ventilations-
koefficient
DE Luftausströmungs-
koeffizient *m*
ES coeficiente de
descarga de aire *m*
F coefficient *m* de
contraction de
veine d'air soufflé
I coefficiente *m* di
efflusso dell'aria
MA légkibocsátási
tényező
NL luchtuitstromings-
coëfficiënt *m*
PO współczynnik *m*
wydatku
powietrza
PY коэффициент *m*
расхода
(отверстия)
SU purkautumiskerroin
SV ventilations-
koefficient *c*

109 air distribution
DA luftfordeling
DE Luftverteilung *f*
ES distribución de
aire *f*
F distribution *f* d'air
I distribuzione *f*
dell'aria
MA légeloszlás
légelosztás
NL luchtverdeling *f*
PO rozdział *m*
powietrza
rozprowadzenie *n*
powietrza
PY распределение *n*
воздуха
SU ilmanjako
SV luftdistribution *c*

**110 air distribution
envelope**
DA luftdistributions-
profil
DE Luftverteilungs-
hülle *f*
ES perfil de la
distribución del
aire *m*
F profils *m* des vitesses
d'air
I inviluppo *m* della
distribuzione
dell'aria
MA légelosztó burkolat
NL luchtstroom-
omhullend vlak *n*
PO kontur *m*
(przestrzenny)
rozdziału
powietrza
PY поверхность *f*
равных
скоростей при
распределении
воздуха
SU heittokuvio
SV luftdistributions-
profil *c*

111 air-dried *adj*
DA lufttørret
DE luftgetrocknet
ES secado al aire *m*
secado por aire *m*
F séché à l'air
I essiccato in aria
MA légszáraz
NL luchtgedroogd
PO osuszony
powietrzem
PY воздушно-сухой
SU ilmakuivattu
SV lufttorkad

DA = Danish
DE = German
ES = Spanish
F = French
I = Italian
MA = Magyar
(Hungarian)
NL = Dutch
PO = Polish
PY = Russian
SU = Finnish
SV = Swedish

112 air drop
DA luftnedfald
DE Luftabfall *m*
Luftabsinken *n*
ES caida del chorro de
aire *f*
F chute *f* d'air
I caduta *f* del getto di
aria
MA légesés
NL daling *f* van de
luchtstroom
PO odrywanie *n*
strumienia
powietrza
opadanie *n*
strumienia
powietrza
PY вертикальная
составляющая *f*
траектории
воздушной струи
падение *n* струи
SU ilmasuihkun taipuma
SV luftläcka *c*

113 air dryer
DA lufttørrer
DE Lufttrockner *m*
ES desecador del aire *m*
secador de aire *m*
F sécheur *m* d'air
I essiccatore *m* di aria
MA légszárító
NL luchtdroger *m*
PO osuszacz *m*
powietrza
PY воздухоосушитель
m
SU ilmankuivaaja
SV lufttorkare *c*

114 air drying
DA lufttørring
DE Lufttrocknung *f*
ES desecación al aire *f*
desecación del aire *f*
secado al aire *m*
F séchage *m* à l'air
I essiccazione *f*
dell'aria
MA légszárítás
NL luchtdroging *f*
PO suszenie *n* na
powietrzu
PY осушение *n*
воздуха
SU ilmankuivaus
SV lufttorkning *c*

115 air duct
DA luftkanal
DE Luftkanal *m*
Luftleitung *f*
ES conducto de aire *m*
F conduit *m* d'air
I condotto *m* per aria
MA légcsatorna
NL luchtkanaal *n*
luchtkoker *m*
PO kanał *m* powietrzny
przewód *m*
powietrzny
PY воздуховод *m*
SU ilmakanava
SV luftkanal *c*

116 air dumping
DA luftafkast
DE Luftabscheidung *f*
ES descarga de aire *f*
F chute *f* de l'air
I scarico *m* dell'aria
MA levegő lecsapódás
NL luchtval *m*
PO wyrzucanie *n*
powietrza
PY затапливание *n*
воздухом сверху
SU ilmasuihkun
tippuminen
SV luftdumpning *c*

117 air eliminator
DA luftudlader
DE Entlüfter *m*
Luftabscheider *m*
ES eliminador de aire *m*
purgador de aire *m*
F purgeur *m* d'air
I scaricatore *m* di aria
MA légtelenítő
NL luchtafscheider *m*
PO eliminator *m*
powietrza
odpowietrznik *m*
PY каплеотделитель *m*
SU ilmanpoistin
SV avluftare *c*

118 air exfiltration
DA luftlækage
(luft ud)
DE Luftabscheidung *f*
ES exfiltración de aire *f*
F exfiltration *f* d'air
I esfiltrazione *f*
dell'aria
MA légkiszűrődés
NL luchtexfiltratie *f*
PO eksfiltracja *f*
powietrza
PY эксфильтрация *f*
воздуха
SU ilman ulosvuoto
SV luftexfiltration *c*

119 air exhaust
DA luftaftræk
DE Luftabsaugung *f*
ES extracción de aire *f*
F air *m* extrait
I espulsione *f* dell'aria
MA légelszívás
NL luchtafvoer *m*
PO wyciąg *m* powietrza
wywiew *m*
powietrza
PY выпуск *m* воздуха
SU ilmanpoisto
SV luftevakuering *c*

120 air filter
DA luftfilter
DE Luftfilter *m*
ES filtro de aire *m*
F filtre *m* à air
I filtro *m* per aria
MA légszűrő
NL luchtfilter *m*
PO filtr *m* powietrza
PY воздушный фильтр
m
SU ilmasuodatin
SV luftfilter *n*

121 air flow
DA luftstrømning
DE Luftstrümung *f*
ES corriente de aire
(mejor) *f*
flujo de aire *m*
F écoulement *m* d'air
I flusso *m* di aria
MA légáram
NL luchtstroom *m*
PO przepływ *m*
powietrza
strumień *m*
powietrza
PY воздушный
поток *m*
SU ilmavirta
SV luftflöde *n*

122 air flow pattern
DA luftstrømsmønster
DE Luftströmungs-
modell *n*
ES configuración de la
corriente de
aire/diagrama del
flujo de aire *f*
configuración del
flujo de aire *f*
F schéma *m*
d'écoulement d'air
I andamento *m* del
flusso di aria
MA légáramlás módja
NL luchtstromings-
patroon *n*
PO charakter *m*
przepływu
powietrza
PY схема *f*
воздушного
потока
SU ilman virtauskuvio
SV luftflödesschema *n*

123 air flow rate
DA luftudstrømningsrate
DE Luftstrümungszahl *f*
ES caudal de aire *m*
F débit *m* d'air
I portata *f* del flusso
di aria
MA légáramlás mértéke
NL luchtstroomhoeveel-
heid *f*
PO wielkość *f*
strumienia
powietrza
PY скорость *f*
воздушного
потока
SU ilmamäärä
ilmavirran suuruus
SV luftflödeshastighet *c*

124 air flow resistance
DA luftstrømsmodstand
DE Luftströmungs-
widerstand *m*
ES resistencia al flujo de
aire *f*
F résistance *f*
aéraulique
I resistenza *f* al moto
dell'aria
MA légáramlási ellenállás
NL luchtweerstand *m*
PO opór *m* przepływają-
cego powietrza
PY аэродинамическое
сопротивление *n*
SU ilman virtausvastus
SV luftflödesmotstånd *n*

125 airfoil fan
DA aerodynamisk
ventilator
DE Flügelventilator *m*
ES ventilador de álabes
perfilados *m*
F ventilation *f* à aubes
profilées
I ventilatore *m* con
pale a profilo alare
MA profil szárnylapátos
ventilátor
NL ventilator *m* met
geprofileerde
schoepen
PO wentylator *m*
łopatkowy
PY крыльчатый
вентилятор *m*
SU aksiaalipuhallin
SV aerodynamisk
fläkt *c*

126 air handling unit
DA luftbehandlings-
aggregat
DE Luftbehandlungs-
gerät *n*
ES climatizador *m*
F unité *f* de traitement
d'air
I unità *f* di
trattamento
dell'aria
MA légkezelő egység
NL luchtbehandelings-
apparaat *n*
PO centrala *f* do
uzdatniania
powietrza
zespół *m* do
uzdatniania
powietrza
PY установка *f*
кондицион-
ирования воздуха
SU ilmastointikone
SV luftbehandlings-
apparat *c*

127 air heater
DA luftvarmer
DE Lufterhitzer *m*
Luftheizer *m*
ES calentador de aire *m*
F réchauffeur *m* d'air
I riscaldatore *m* di
aria
MA légfűtő
léghevítő
NL luchtverhitter *m*
PO nagrzewnica *f*
powietrza
PY воздухонагреватель
m
SU ilmalämmitin
SV luftvärmare *c*

128 air heater battery
DA luftvarmebatteri
DE Luftheizregister *n*
ES batería de
calefacción para
aire caliente *f*
F batterie *f* de
réchauffage d'air
I batteria *f* di
riscaldamento di
aria
MA léghevítő
NL luchtverwarmings-
batterij *f*
PO nagrzewnica *f*
powietrza
PY воздухонаревател-
ьная установка *f*
SU ilmalämmityspatteri
SV luftvärmeväxlare *c*

129 air heating
DA luftopvarmning
DE Luftheizung *f*
ES calefacción del aire *f*
calentamiento del
aire *m*
F chauffage *m* d'air
I riscaldamento *m* di
aria
MA légfűtés
NL luchtverwarming *f*
PO ogrzewanie *n*
powietrzne
PY воздушное
отопление *n*
SU ilmalämmitys
SV luftvärmning *c*

130 air humidification
DA luftbefugtning
DE Luftbefeuchtung *f*
ES humectación del
   aire *f*
   humidificación del
   aire *f*
F humidification *f*
   d'air
I umidificazione *f* di
   aria
MA légnedvesítés
NL luchtbevochtiging *f*
PO nawilżanie *n*
   powietrza
PY увлажнение *n*
   воздуха
SU ilmankostutus
SV luftfuktning *c*

131 air humidifier
DA luftbefugter
DE Luftbefeuchter *m*
ES humidificador de
   aire *m*
F humidificateur *m*
   d'air
I umidificatore *m* di
   aria
MA légnedvesítő
NL luchtbevochtiger *m*
PO nawilżacz *m*
   powietrza
PY увлажнитель *m*
   воздуха
SU ilmankostuttaja
SV luftfuktare *c*

132 air infiltration
DA luftinfiltration
   luftlækage (luft in)
DE Luftinfiltration *f*
ES infiltración de aire *f*
F infiltration *f* d'air
I infiltrazione *f* di aria
MA légbeszűrődés
NL luchtinfiltratie *f*
PO infiltracja *f*
   powietrza
PY инфильтрация *f*
   воздуха
SU vuotoilma
SV luftinfiltration *c*

133 air inlet
DA luftilførselssted
DE Lufteinlaß *m*
ES entrada de aire *f*
F entrée *f* d'air
I immissione *f* di aria
MA levegő bevezetés
NL luchtinlaat *m*
   luchttoevoer-
   opening *f*
PO wlot *m* powietrza
PY вход *m* воздуха
SU tuloilmalaite
SV luftintag *n*

134 air intake
DA luftindtag
DE Lufteinlaß *m*
   Lufteintritt *m*
ES toma de aire *f*
F prise *f* d'air
I presa *f* di aria
MA levegő belépés
NL luchtaanzuig-
   opening *f*
PO czerpnia *f* powietrza
PY впуск *m* воздуха
SU ulkoilma-aukko
SV luftintag *n*

135 air jet
DA luftstråle
DE Luftstrahl *m*
ES chorro de aire *m*
F jet *m* d'air
I getto *m* di aria
MA légfúvó
   légsugár
NL luchtstraal *m*
PO dysza *f* powietrzna
   strumień *m*
   powietrza
PY воздушная струя *f*
SU ilmasuihku
SV luftstråle *c*

136 air leak
DA luftlækage
   utæthed
DE Luftleckage *f*
   Luftundichtigkeit *f*
ES fuga de aire *f*
F fuite *f* d'air
I perdita *f* di aria
MA légrés
NL luchtlek *n*
PO wypływ *m* powietrza
   w wyniku
   nieszczelności
PY утечка *f* воздуха
SU ilmavuoto
SV luftläcka *c*

137 air leakage
DA luftlækage
   udsivning
DE Luftverlust *m*
ES escape de aire *m*
   pérdida de aire *f*
F fuite *f* d'air
I fuga *f* di aria
MA légveszteség
   tümítetlenség
NL luchtlekkage *f*
   luchtverlies *m*
PO wypływ *m* powietrza
   w wyniku
   nieszczelności
PY воздушная
   неплотность *f*
SU ilmavuoto
SV luftläckage *n*

138 air lock (*see* air
pocket)

139 air mixing unit
DA luftblandeunit
DE Luftmischer *m*
ES unidad de mezcla de
   aire *f*
F boîte *f* de mélange
   d'air
   mélangeur *m* d'air
I miscelatore *m* di
   aria
MA légkeverő egység
NL luchtmengkast *f*
PO komora *f* mieszania
   powietrza
   zespół *m* mieszający
   powietrza
PY устройство *n* для
   смешивания
   воздуха
SU ilmastointikoneen
   sekoitusosa
SV luftblandningsdon *n*

140 air motor
DA luftmotor
DE Luftmotor *m*
ES motor de aire *m*
F moteur *m* à air
I motore *m* ad aria
MA légmotor
NL luchtgedreven motor
   m
   pneumatische
   servomotor *m*
PO silnik *m* powietrzny
PY пневмопривод *m*
SU paineilmamoottori
SV luftmotor *c*

141 air movement
DA luftbevægelse
DE Luftbewegung *f*
ES movimiento del
     aire *m*
F mouvement *m* d'air
I moto *m* dell'aria
MA légmozgás
NL luchtbeweging *f*
PO ruch *m* powietrza
PY движение *n*
     воздуха
SU ilman liike
SV luftrörelse *c*

142 air outlet
DA luftudsugningssted
DE Luftauslaß *m*
     Luftaustritt *m*
ES salida de aire *f*
F sortie *f* d'air
I uscita *f* dell'aria
MA levegő kilépés
NL luchtuitlaat *m*
PO wylot *m* powietrza
PY выпуск *m* воздуха
SU poistoilma-aukko
SV luftutsläpp *n*

143 air passage
DA luftgennemgang
DE Luftdurchgang *m*
ES conducto de aire *m*
     conducto de
     aireación *m*
     paso de aire *m*
F passage *m* d'air
I passaggio *m* di aria
MA légút
NL luchtdoorlaat *m*
PO przelot *m* powietrza
PY проход *m* воздуха
SU ilmareitti
SV luftgenomgång *c*

144 air pocket
DA luftlomme
DE Luftsack *m*
ES bolsa de aire *f*
F bouchon *m* d'air
I sacca *f* di aria
MA légzsák
NL luchtbel *f*
     luchtslot *n*
PO korek *m* powietrzny
PY воздушный
     затвор *m*
SU ilmatasku
SV luftficka *c*

145 air pollution
DA luftforurening
DE Luftver-
     unreinigung *f*
ES contaminación del
     aire *f*
F pollution *f* de l'air
I inquinamento *m*
     dell'aria
MA légszennyeződés
NL luchtver-
     ontreiniging *f*
     luchtvervuiling *f*
PO zanieczyszczenie *n*
     powietrza
PY загрязнение *n*
     воздуха
SU ilman saastuminen
SV luftförorening *c*

**146 air pollution
        concentration**
DA luftforurenings-
     koncentration
DE Luftverunreinigung-
     konzentration *f*
ES tasa de
     contaminación *f*
F taux *m* de pollution
     de l'air
I concentrazione *f* di
     inquinanti
     nell'aria
MA légszennyeződési
     koncentráció
NL luchtverontreinigings-
     concentratie *f*
PO stężenie *n*
     zanieczyszczenia
     powietrza
PY концентрация *f*
     вредности в
     воздухе
SU ilman saastepitoisuus
SV luftförorenings-
     koncentration *c*

147 air pre-heater
DA luftforvarmer
DE Luftvorwärmer *m*
ES precalentador de
     aire *m*
F préchauffeur *m* d'air
I preriscaldatore *m* di
     aria
MA levegő előfűtő
NL luchtvoor-
     verwarmer *m*
PO nagrzewnica *f*
     wstępna
PY воздухонагреватель
     m
     предваритель-
     ьного подогрева
SU ilman esilämmitin
SV luftfürvärmare *c*

148 air pressure
DA lufttryk
DE Luftdruck *m*
ES presión del aire *f*
F pression *f* de l'air
I pressione *f* dell'aria
MA légnyomás
NL luchtdruk *m*
PO ciśnienie *n*
     powietrza
PY давление *n* воздуха
SU ilmanpaine
SV lufttryck *n*

149 air pump
DA luftpumpe
DE Luftpumpe *f*
ES bomba de aire *f*
F pompe *f* à air
I pompa *f* per aria
MA légszivattyú
NL luchtpomp *f*
PO pompa *f* powietrza
PY воздушный
     насос *m*
SU ilmapumppu
SV luftpump *c*

150 air purity
DA luftrenhed
DE Luftreinheit *f*
ES pureza del aire *f*
F pureté *f* de l'air
I purezza *f* dell'aria
MA levegő tisztaság
NL luchtzuiverheid *f*
PO czystość *f* powietrza
PY чистота *f* воздуха
SU ilman puhtaus
SV luftrenhet *c*

DA	=	Danish
DE	=	German
ES	=	Spanish
F	=	French
I	=	Italian
MA	=	Magyar (Hungarian)
NL	=	Dutch
PO	=	Polish
PY	=	Russian
SU	=	Finnish
SV	=	Swedish

151 air quality
DA luftkvalitet
DE Luftqualität *f*
ES calidad del aire *f*
F qualité *f* de l'air
I qualità *f* dell'aria
MA levegő minőség
NL luchtkwaliteit *f*
PO jakość *f* powietrza
PY количество *n*
  воздуха
SU ilmanlaatu
SV luftkvalitet *c*

152 air quality control
DA luftkvalitetskontrol
DE Luftqualitäts-
  kontrolle *f*
  Luftqualitäts-
  regelung *f*
ES control de la calidad
  del aire *m*
F contrôle *m* de la
  qualité de l'air
I controllo *m* della
  qualità dell'aria
MA levegő minőség
  szabályozás
NL luchtkwaliteits-
  regelaar *m*
  luchtkwaliteits-
  regeling *f*
PO kontrola *f* jakości
  powietrza
PY регулирование *n*
  количества
  воздуха
SU ilmanlaadunvalvonta
SV luftkvalitets-
  kontroll *c*

153 air quality sensor
DA luftkvalitetsføler
DE Luftqualitätssensor
  m
ES sonda de calidad del
  aire *f*
F capteur *m* de la
  qualité de l'air
I sensore *m* della
  qualità dell'aria
MA levegő minőség
  érzékelő
NL luchtkwaliteits-
  opnemer *m*
PO czujnik *m* jakości
  powietrza
PY датчик *m*
  количества
  воздуха
SU ilmanlaatuanturi
SV luftkvalitetsgivare *c*

**154 air regulating
  damper**
DA luftreguleringsspjæld
DE Luftregulierklappe *f*
ES compuerta de
  regulación de
  aire *f*
F volet *m* de réglage
  d'air
I serranda *f* di
  regolazione
  dell'aria
MA levegő szabályozó
  csappantyú
NL luchtregelklep *m*
PO przepustnica *f*
  regulująca
  przepływ
  powietrza
PY воздушный
  регулирующий
  клапан *m*
SU säätüpelti
SV regleringsspjäll *n*

155 air renewal
DA luftfornyelse
  luftskifte
DE Lufterneuerung *f*
ES renovación del aire *f*
F renouvellement *m*
  d'air
I ricambio *m* di aria
MA légfrissítés
NL luchtverversing *f*
PO wymiana *f*
  powietrza
PY обновление *n*
  воздуха
SU ilman uusiutuminen
SV luftväxling *c*

156 air requirement
DA luftbehov
DE Luftbedarf *m*
ES necesidades de
  aire *f,pl*
F besoins *m* d'air
I fabbisogno *m* di
  aria
MA levegő igény
NL luchtbehoefte *f*
PO zapotrzebowanie *n*
  na powietrze
PY потребность *f* в
  воздухе
SU ilmanvaihdon tarve
SV luftbehov *n*

157 air seal
DA lufttætning
DE Luftabdichtung *f*
ES junta estanca al
  aire *f*
F joint *m* étanche à
  l'air
I tenuta *f* all'aria
MA légszigetelés
NL luchtafdichting *f*
PO uszczelnienie *f*
  powietrzne
PY воздушное
  уплотнение *n*
SU ilmatiiviste
SV lufttätning *c*

158 air separation
DA luftudskillelse
DE Luftabscheidung *f*
ES separación del aire *f*
F séparation *f* d'air
I separazione *f*
  dell'aria
MA levegő leválasztás
NL
PO separacja *f*
  powietrza
PY отделение *n*
  воздуха
SU ilman erotin
SV luftavskiljning *c*

159 air space
DA luftrum
DE Lufthohlraum *m*
  Luftraum *m*
ES cámara de aire *f*
  cavidad de aire *f*
F espace *m* d'air
I camera *f* di aria
MA légtér
NL luchtruimte *f*
  luchtspleet *f*
  spouw *f*
PO kubatura *f*
  przestrzeń *f*
  powietrzna
PY воздушная
  прослойка *f*
  воздушное
  пространство *n*
SU ilmatila
SV luftutrymme *n*

160 air speed
DA lufthastighed
DE Luftgeschwindig-
keit *f*
ES velocidad del aire *f*
F vitesse *f* d'air
vitesse *f* de l'air
I velocità *f* dell'aria
MA légsebesség
NL luchtsnelheid *f*
PO prędkość *f*
powietrza
PY скорость *f* воздуха
SU ilman nopeus
SV lufthastighet *c*

161 air splitter
DA luftfordeler
DE Luftzerleger *m*
ES álabes direccionales
m
F aubes *f* directrices
I aletta *f* direttrice
MA légáram leválasztó
NL luchtverdeler *m*
PO rozdzielacz *m*
powietrza
PY воздушная
направляющая
лопатка *f*
дроссельное
устройство *n*
SU ilmanjakaja
SV luftdelare *c*

162 air spread
DA luftspredning
DE Luftausbreitung *f*
ES dispersión del aire *f*
F portée *f* latérale
d'une bouche de
soufflage
I dispersione *f* del
getto di aria
MA légnyaláb
NL luchtspreiding *f*
PO rozpraszanie *n*
powietrza
rozprzestrzenianie *n*
się powietrza
PY распространение *n*
потока воздуха
SU ilman leviäminen
SV luftspridning *c*

163 air stratification
DA luftlagdeling
DE Luftschichtung *f*
ES estratificación del
aire *f*
F stratification *f* de
l'air
I stratificazione *f*
dell'aria
MA levegő rétegződés
NL luchtlaagvorming *f*
luchtstratificatie *f*
PO uwarstwienie *n*
powietrza
PY расслоение *n*
воздуха
SU ilman kerrostuminen
SV luftskiktning *c*

164 air supply
DA luftforsyning
DE Luftzufuhr *f*
Luftzuleitung *f*
ES entrada de aire *f*
suministro de aire *m*
F admission *f* d'air
alimentation *f* en air
I erogazione *f* di aria
mandata *f* dell'aria
MA légellátás
NL luchttoevoer *m*
PO doprowadzenie *n*
powietrza
nawiew *m* powietrza
PY подача *f* воздуха
SU tuloilmalaitteisto
SV lufttillförsel *c*

165 air supply fixtures
DA luftforsynings-
armatur
DE Luftversorgungs-
installation *f*
ES dispositivos de
descarga de aire *m*
F équipements *m* de
diffusion d'air
I dispositivi *m,pl* di
immissione
dell'aria
MA légellátó
szerelvények
NL luchttoevoer-
ornamenten *pl*
PO elementy *m*
instalacji
nawiewnej
PY устройство *n* для
раздачи
приточного
воздуха
SU tuloilmalaite
SV lufttillförselfixtur *c*

166 air terminal device
DA luftindblæsnings-
organ
DE Luftauslaß *m*
Luftdurchlaß *m*
ES aparato terminal de
un circuito de
aire *m*
dispositivo terminal
de un circuito de
aire *m*
F appareil *m* terminal
d'un circuit d'air
I unità *f* terminale (di
un circuito di aria)
MA légkibocsátó
szerkezet
NL luchtuitblaas-
ornament *n*
uitblaasrooster *m*
PO element *m* końcowy
w instalacji
wentylacyjnej
PY воздухораспреде-
лительное
устройство *n*
SU päätelaite
SV luftdon *n*

167 air throughput
DA luftomsætning
DE Luftdurchlaß *m*
ES caudal de aire *m*
F débit *m* d'air
taux *m* de
renouvellement
d'air
I portata *f* di aria
MA légteljesítmény
NL luchtdoorvoer *m*
PO zdolność *f*
przepustowa
powietrza
PY воздухопроизводитель-
ность *f*
SU ilmanvaihto
SV luftomsättning *c*

168 air throw
DA kastelængde
DE Luftwurfweite *f*
ES alcance del chorro
de aire *m*
F portée *f* d'air (jet)
I gittata *f* dell'aria
lancio *m* dell'aria
MA levegő vetőtávolság
NL worp *m*
PO strumień *m*
powietrza
PY составляющая *f*
воздушной струи
SU heittopituus
SV luftstöt *c*

169 airtight *adj*
DA lufttæt
DE luftdicht
ES estanco al aire
   hermético al aire
F étanche à l'air
I a tenuta di aria
MA légtümür
NL luchtdicht
PO powietrznoszczelny
PY воздухопрони-
   цаемый
SU ilmatiivis
SV lufttät

170 air tightness
DA lufttæthed
DE Luftdichtheit *f*
ES estanquidad al aire *f*
F étanchéité *f* à l'air
I tenuta *f*
MA légtümürség
NL luchtdichtheid *f*
PO szczelność *f* dla
   powietrza
PY воздухопроницае-
   мость *f*
SU ilmatiiviys
SV lufttäthet *c*

171 air treatment
DA luftbehandling
DE Luftbehandlung *f*
ES tratamiento del aire
   m
F traitement *m* d'air
I trattamento *m*
   dell'aria
MA légkezelés
NL luchtbehandeling *f*
PO uzdatnianie *n*
   powietrza
PY обработка *f*
   воздуха
SU ilmankäsittely
SV luftbehandling *c*

172 air valve (*see* air
   damper)

173 air velocity (*see* air
   speed)

174 air vent
DA udluftning
DE Entlüftung *f*
ES orificio de
   ventilación *m*
   purgador de aire *m*
   ventanillo de
   aireación *m*
F purge *f* d'air
I sfiato *m* dell'aria
   sfogo *m* di aria
MA légtelenítő
NL beluchter *m*
   ontluchter *m*
PO odpowietrzenie *n*
   odpowietrznik *m*
PY удаление *n* воздуха
SU poistoilmalaite
   tuuletuslaite
SV avluftning *c*

175 air vessel
DA luftbeholder
DE Luftbehälter *m*
   Windkessel *m*
ES depósito de aire
   (comprimido) *m*
F réservoir *m* d'air
   (comprimé)
I serbatoio *m* di aria
MA légtartály
NL luchtketel *m*
   luchtvat *n*
   windketel *m*
PO zbiornik *m*
   powietrza
PY воздушный
   сосуд *m*
SU ilmasäiliü
SV luftbehållare *c*

176 air washer
DA luftvasker
DE Luftwäscher *m*
ES lavador de aire *m*
F laveur *m* d'air
I lavatore *m* di aria
MA légmosó
NL luchtbevochtiger *m*
   luchtwasser *m*
PO komora *f* zraszania
   płuczka *f*
   powietrzna
PY воздухопромыва-
   тель *m*
SU ilmapesuri
SV lufttvättare *c*

177 air well
DA ventilationsskakt
DE Luftschacht *m*
ES chimenea de
   ventilación *f*
   patinillo de
   ventilación *m*
   pozo de
   ventilación *m*
F courette *f*
I pozzo *m* di
   ventilazione
MA légakna
NL luchtkoker *m*
   luchtschacht *f*
PO szyb *m* wentylacyjny
PY источник *m*
   воздуха
SU ilmanvaihtokuilu
SV ventilationsschakt *n*

178 alarm
DA alarm
DE Alarm *m*
ES alarma *f*
F alarme *f*
   avertisseur *m*
I allarme *m*
MA riasztás
NL alarm *n*
   alarmsignaal *n*
PO alarm *m*
PY сигнал *m*
SU hälytys
SV larm *n*

179 alarm bell
DA alarmklokke
DE Alarmglocke *f*
ES timbre de alarma *m*
F sonnette *f* d'alarme
I campanello *m* di
   allarme
MA riasztó csengő
NL alarmbel *f*
PO dzwonek *m*
   alarmowy
PY сигнальный
   звонок *m*
SU hälytyskello
SV larmklocka *c*

DA	=	Danish
DE	=	German
ES	=	Spanish
F	=	French
I	=	Italian
MA	=	Magyar (Hungarian)
NL	=	Dutch
PO	=	Polish
PY	=	Russian
SU	=	Finnish
SV	=	Swedish

180 alcohol
DA alkohol
DE Alkohol *m*
ES alcohol *m*
F alcool *m*
I alcool *m*
MA alkohol
NL alcohol *m*
PO alkohol *m*
etanol *m*
PY алкоголь *m*
спирт *m*
SU alkoholi
SV alkohol *c*

181 algae
DA alger
DE Algen *f,pl*
ES algas *f,pl*
F algues *f*
I alga *f*
MA algák
NL algen *pl*
PO alga *f*
glon *m*
PY водоросли *f,pl*
SU levä
SV alger *c*

182 algaecide
DA algedræber
DE Algenverhütungs-
mittel *n*
ES algicida *m*
F algicide *m*
I algicida *m*
MA algaülő szer
NL algenbestrijdings-
middel *n*
PO środek *m* niszczący
glony
PY защита *f* от
водорослей
SU levämyrkky
SV algicid *c*

183 algae formation
DA algedannelse
DE Algenbildung *f*
ES formación de algas *f*
F formation *f* d'algues
I formazione *f* di
alghe
MA algaképződés
NL algenafzetting *f*
algengroei *m*
PO powstawanie *n*
glonów
PY отложение *n*
водорослей
SU levän muodostus
SV algformation *c*

184 algorithm
DA algoritme
DE Algorithmus *m*
Rechenvorschrift *f*
ES algoritmo *m*
F algorithme *m*
I algoritmo *m*
MA algoritmus
NL algoritme *n*
PO algorytm *m*
PY алгоритм *m*
SU algoritmi
SV algoritm *c*

185 alkaline *adj*
DA alkalisk
basisk
DE alkalisch
ES alcalino
F alcalin
I alcalino
MA lúgos
NL alkalisch
basisch
PO zasadowy
PY щелочный
SU emäksinen
SV alkalisk

186 alkalinity
DA alkalitet
DE Alkalinität *f*
ES alcalinidad *f*
F alcalinité *f*
I alcalinità *f*
MA lúgosság
NL alkaliteit *f*
PO zasadowość *f*
PY щелочность *f*
SU emäksisyys
SV alkalitet *c*

**187 all-air air
conditioning
system**
DA luftbehandlings-
system luft/luft
DE Nur-Luft-
Klimasystem *n*
ES sistema de
climatización
todo-aire *m*
F système *m* de
conditionnement
d'air tout air
I impianto *m* di
condizionamento
a tutt'aria
MA teljes
légkondicionáló
rendszer
NL all-air
luchtbehandelings-
systeem *n*
PO system *m*
klimatyzacji
jednoczynnikowy
(tylko powietrze)
PY система *f*
кондицион-
ирования воздуха
с централизо-
ванным
теплоснаб-
жением
SU ilmajärjestelmä
SV luftbehandlings-
system *n* (luft-luft)

188 alloy
DA legering
DE Legierung *f*
ES aleación *f*
F alliage *m*
I lega *f*
MA ütvüzet
NL legering *f*
PO stop *m* (metali)
PY сплав *m*
SU lejeerinki
SV legering *c*

189 alphanumeric *adj*
DA alfanumerisk
DE alphanumerisch
ES alfanumérico
F alphanumérique
I alfanumerico
MA alfanumerikus
NL alphanumeriek
PO alfanumeryczny
PY альфаметрический
SU alfanumeerinen
SV alfanumerisk

190 alternating current
DA vekselstrøm
DE Wechselstrom *m*
ES corriente alterna *f*
F courant *m* alternatif
I corrente *f* alternata
MA váltóáram
NL wisselstroom *m*
PO prąd *m* zmienny
PY переменный ток *m*
SU vaihtovirta
SV växelström *c*

191 alternative energy source
DA alternativ energikilde
DE alternative Energiequelle *f*
ES fuente de energía alternativa *f*
F source *f* alternative d'energie
I fonte *f* di energia alternativa
MA megújuló energiaforrások
NL alternatieve energiebron *f*
PO alternatywne źródło *n* energii
PY альтернативный источник *m* энергии
SU vaihtoehtoinen energialähde
SV alternativ energi-källa *c*

192 alumina
DA aluminiumoxid alun
DE Aluminiumoxid *n* Tonerde *f*
ES alúmina *f*
F alumine *f*
I allumina *f*
MA timfüld
NL aluinaarde *f* aluminium-trioxyde *n*
PO tlenek *m* glinu
PY окись *f* алюминия
SU kuivauksessa käytetty aluminioksidi
SV aluminiumoxid *c*

193 aluminium
DA aluminium
DE Aluminium *n*
ES aluminio *m*
F aluminium *m*
I alluminio *m*
MA alumínium
NL aluminium *n*
PO aluminium *n*
PY алюминий *m*
SU alumiini
SV aluminium *n*

194 ambient *adj*
DA omgivende
DE umgebend
ES ambiente
F ambiant
I ambiente
MA környezeti
NL omgevend
PO otaczający
PY окружающий
SU ympäröivä
SV omgivande

195 ambient air
DA omgivende luft
DE Umgebungsluft *f*
ES aire ambiente *m*
F air *m* ambiant
I aria *f* ambiente
MA külső levegő
NL omgevingslucht *f*
PO powietrze *n* otaczające powietrze *n* wewnętrzne
PY окружающий воздух *m*
SU ympäröivä ilma
SV omgivande luft *c*

196 ambient-compensated *adj*
DA udlignet med omgivelserne
DE umgebungsaus-gleichend
ES compensado ambientalmente
F ambiant compensé
I compensato in base alle condizioni ambiente
MA környezettel kiegyenlített
NL omgevings-temperatuur *f* gecompenseerd
PO kompensujący wpływ otoczenia
PY влияние *n* внешних условий
SU ympäristün vaikutuksilta kompensoitu
SV omgivnings-kompenserad

197 ambient pressure
DA omgivende tryk
DE Umgebungsdruck *m*
ES presión ambiental *f*
F pression *f* ambiante
I pressione *f* ambiente
MA külső nyomás
NL omgevingsluchtdruk *m*
PO ciśnienie *n* otoczenia
PY внешнее давление *n*
SU ympäristün paine
SV omgivande tryck *n*

198 ambient temperature
DA omgivende temperatur
DE Umgebungs-temperatur *f*
ES temperatura ambiental *f*
F température *f* ambiante
I temperatura *f* ambiente
MA külső hőmérséklet
NL omgevings-temperatuur *f*
PO temperatura *f* otoczenia
PY наружная температура *f*
SU ympäristün lämpütila
SV omgivande temp-eratur *c*

199 ammeter
DA ampéremeter
DE Ampermeter *n*
ES amperímetro *m*
F ampèremètre *m*
I amperometro *m*
MA ampermérő
NL ampèremeter *m*
PO amperomierz *m*
PY амперметр *m*
SU amperimittari
SV amperemeter *c*

200 ammonia
DA ammoniak
DE Ammoniak *n*
ES amoniaco *m*
F ammoniac *m*
I ammoniaca *f*
MA ammónia
NL ammoniak *m*
PO amoniak *m*
PY аммиак *m*
SU ammoniakki
SV ammoniak *c*

201 ammonia solution
DA ammoniakopløsning
salmiakspiritus
DE Ammoniaklüsung *f*
ES solución amoniacal
f
F ammoniaque *f*
I soluzione *f*
ammoniacale
MA ammónia oldat
NL ammoniakoplossing
f
PO roztwór *m*
amoniaku
PY аммиачный
раствор *m*
SU ammoniakkiliuos
SV ammoniaklösning *c*

202 amortization
DA amortisering
DE Amortisation *f*
ES amortización *f*
F amortissement *m*
I ammortamento *m*
ammortizzamento *m*
MA amortizáció
csillapítás
NL amortisatie *f*
PO amortyzacja *f*
PY амортизация *f*
SU kuoletus
SV amortering *c*

203 amortize *vb*
DA amortisere
DE amortisieren
ES amortizar
F amortir
I ammortare
ammortizzare
MA amortizál
csillapít
NL afschrijven
amortiseren
PO amortyzować
PY амортизировать
SU kuolettaa
SV amortera

204 amount
DA behov
mængde
DE Bedarf *m*
Menge *f*
ES cantidad *f*
importe *m*
porcentaje *m*
proporción *f*
F besoin *m*
quantité *f*
I quantità *f*
MA mennyiség
NL bedrag *n*
grootte *f*
hoeveelheid *f*
PO ilość *f*
kwota *f*
suma *f*
PY количество *n*
сумма *f*
SU määrä
SV mängd *c*

205 ampacity
DA ampérekapacitet for
strømførende
kabel
DE zulässige
Stromstärke *f*
ES intensidad admisible
f
F intensité *f*
admissible
I portata *f* di un
conduttore
MA vezetőképesség
NL toelaatbare
stroomsterkte *f*
PO obciążalność *f*
prądowa (w
amperach)
PY электропроводи-
мость *f*
SU virrankestävyys
SV amperekapacitet *c*
för strömförande
kabel *c*

206 amperage
DA strømstyrke
DE Ampere *n*
ES amperaje *m*
F ampérage *m*
I amperaggio *m*
MA áramerősség
NL amperage *f*
PO amperaż *m*
natężenie *n* prądu w
amperach
PY величина *f* тока
ток *m*
SU virranvoimakkuus
SV

207 amplification
DA forstærkning
DE Vergrößerung *f*
Verstärkung *f*
ES amplificación *f*
F amplification *f*
I amplificazione *f*
MA erősítés
NL versterking *f*
PO wzmocnienie *n*
zwiększenie *n*
PY усиление *n*
SU vahvistus
SV förstärkning *c*

208 amplifier
DA forstærker
DE Verstärker *m*
ES amplificador *m*
F amplificateur *m*
I amplificatore *m*
MA erősítő
NL versterker *m*
PO wzmacniacz *m*
PY усилитель *m*
SU vahvistin
SV förstärkare *c*

209 amplify *vb*
DA forstærke
DE vergrüßern
verstärken
ES amplificar
F amplifier
I amplificare
MA erősít
NL versterken
PO wzmacniać
zwiększać
PY увеличивать
усиливать
SU vahvistaa
SV förstärka

210 amplitude
DA amplitude
   udslag
DE Amplitude f
ES amplitud f
F amplitude f
I ampiezza f
MA amplitúdó
   kitérés
NL amplitude f
   uitslag m
PO amplituda f
PY амплитуда f
SU amplitudi
SV amplitud c

211 analog (USA) (*see* analogue)

212 analogue *adj*
DA analog
DE analog
ES análogo
F analogue
I analogico
MA analóg
   hasonló
NL analoog
PO analogowy
PY аналоговый
SU analoginen
SV analog

213 analogue data
DA analoge data
DE Analogwerte m,pl
ES datos analógicos m
F données f
   analogiques
I dato m analogico
MA analógérték
NL analoge gegevens pl
PO dane f analogowe
PY аналоговые данные
   n,pl
SU analogiatieto
SV analog data c

214 analogue display
DA analog visning
DE Analoganzeigefeld n
   Analoganzeigetafel f
ES pantalla analógica f
F affichage m
   analogique
I display m analogico
MA analóg display
NL analoge weergave f
PO monitor m
   obrazowy
   analogowy
   wyświetlacz m
   analogowy
PY аналоговое
   представление n
SU analoginen näyttü
SV analog indikering c

215 analogue indication
DA analog indikation
DE Analoganzeige f
ES indicación analógica
   f
F indication f
   analogique
I indicazione f
   analogica
MA analóg jel
NL analoge indicatie f
PO wskazanie n
   analogowe
PY аналоговая
   индикация f
SU analoginen osoitus
SV analog indikation c

216 analogue input
DA analog indgang
DE Analogeingabe f
ES entrada analógica f
F entrée f analogique
I ingresso m
   analogico
MA analóg bemenet
NL analoge invoer m
PO wejście n analogowe
PY входная величина f
SU analoginen
   sisäänmeno
SV analog inmatning c

217 analogue output
DA analog udgang
DE Analogausgabe f
ES salida analógica f
F sortie f analogique
I uscita f analogica
MA analóg kimenet
NL analoge uitvoer m
PO wyjście n analogowe
PY выходная
   величина f
SU analoginen ulostulo
SV analog utmatning c

218 analogue transmission
DA analog transmission
DE Analog-
   übertragung f
ES transmisión
   analógica f
F transmission f
   analogique
I trasmissione f
   analogica
MA analóg átvitel
NL analoge
   overdracht f
   analoge
   transmissie f
PO przekaz m
   analogowy
PY аналоговая
   передача f
SU analoginen
   viestinsiirto
SV analog
   transmission c

219 analyser
DA analyser
DE Analysator m
ES analizador m
F analyseur m
I analizzatore m
MA analizátor
   elemző
NL analysator m
PO analizator m
PY анализатор m
   генератор m
   абсорбционной
   холодильной
   машины
SU analysaattori
SV analysator c

220 analysis
DA analyse
DE Analyse *f*
ES análisis *m*
F analyse *f*
I analisi *f*
MA analízis
elemzés
NL analyse *f*
PO analiza *f*
PY анализ *m*
SU analyysi
SV analys *c*

221 anchor
DA anker (holdepunkt)
forankring
DE Festpunkt *m*
ES anclaje *m*
punto fijo *m*
riostra *f*
F ancrage *m*
point *m* fixe
I ancoraggio *m*
punto *m* fisso
MA fix megfogás
kihorgonyzás
NL anker *n*
PO kotew *m*
kotwica *f*
PY якорь *m*
SU ankkuri
kiintopiste
SV förankring *c*

222 anchor *vb*
DA forankre
DE befestigen
verankern
ES anclar
F ancrer
I ancorare
fissare
MA lehorgonyoz
NL ankeren
verankeren
PO zakotwić
PY поставить на якорь
SU ankkuroida
SV förankra

DA = Danish
DE = German
ES = Spanish
F = French
I = Italian
MA = Magyar
(Hungarian)
NL = Dutch
PO = Polish
PY = Russian
SU = Finnish
SV = Swedish

223 anemometer
DA anemometer
lufthastighedsmåler
DE Anemometer *n*
ES anemómetro *m*
F anémomètre *m*
I anemometro *m*
MA anemométer
sebességmérő
NL anemometer *m*
luchtsnelheids-
meter *m*
PO anemometr *m*
PY анемометр *m*
SU anemometri
SV anemometer *c*

224 aneroid barometer
DA aneroid barometer
DE Aneroidbarometer *n*
ES barómetro aneroide
m
F baromètre *m*
anéroide
I barometro *m*
aneroide
MA aneroid barométer
fém légnyomásmérő
NL aneroide
barometer *m*
membraan-
barometer *m*
PO barometr *m*
aneroidalny
PY анероидный
барометр *m*
SU rasiabarometri
SV aneroid barometer *c*

225 aneroid capsule
DA aneroid (lufttæt)
kapsel
DE Aneroidkapsel *f*
ES cápsula aneroide *f*
F capsule *f* anéroïde
I capsula *f* aneroide
MA aneroid tok
NL aneroide capsule *f*
membraandoos *f*
PO mieszek *m* sprężysty
PY анероидная
капсула *f*
SU painerasia
SV aneroid kapsel *c*

226 angle
DA vinkel
DE Winkel *m*
ES ángulo *m*
F angle *m*
I angolo *m*
MA szüg
NL hoek *m*
kant *m*
PO kąt *m*
PY угол *m*
SU kulma
SV vinkel *c*

227 angle factor
DA vinkelfaktor
DE Winkelverhältnis *n*
ES factor angular *m*
factor de forma *m*
F facteur *m* d'angle
facteur *m* de forme
I fattore *m* di forma
MA szügtényező
NL vormfactor *m*
zichtfactor *m*
PO współczynnik *m*
kątowy
PY угловой
коэффициент *m*
SU näkyvyyskerroin
SV vinkelförhållande *n*

228 angle iron
DA vinkeljern
DE Winkeleisen *n*
ES angular de acero *m*
hierro angular *m*
F cornière *f*
I angolare *m* di ferro
MA szügacél
NL hoekijzer *n*
hoekstaal *n*
PO kątownik *m* stalowy
PY уголковое железо *n*
SU kulmarauta
SV vinkeljärn *n*

229 angle of discharge
DA afkastvinkel
DE Ausblasewinkel *m*
ES ángulo de descarga
m
F angle *m* de diffusion
I angolo *m* di efflusso
MA kibocsátási szüg
NL uitblaashoek *m*
PO kąt *m* wypływu
PY угол *m* выпуска
SU hajoamiskulma
SV avlastningsvinkel *c*

230 angle of outlet
DA udløbsvinkel
DE Auslaßwinkel *m*
Austrittswinkel *m*
ES ángulo de salida *m*
F angle *m* de sortie
I angolo *m* di uscita
MA kifúvási szüg
NL uittredehoek *m*
PO kąt *m* wylotu
PY угол *m* выхода
SU lähtükulma
SV utloppsvinkel *c*

231 angle valve
DA vinkelventil
DE Eckventil *n*
ES válvula de escuadra *f*
F robinet *m* d'équerre
I valvola *f* a squadra
MA sarokszelep
NL haakse afsluiter *m*
PO zawór *m* kątowy
PY угловой вентиль *m*
SU kulmaventtiili
SV vinkelventil *c*

232 anion
DA anion
negativ ion
DE Anion *n*
ES anión *m*
F anion *m*
I anione *m*
MA anion
NL anion *n*
PO anion *m*
jon *m* ujemny
PY анион *m*
отрицательный
ион *m*
SU negatiivinen ioni
SV negativ jon *c*

233 anneal *vb*
DA udgløde
DE glühen
ES recocer
F adoucir (le métal, le
verre)
détremper (le métal)
recuire
I ricuocere
temprare
MA lágyít
temperál
NL ontharden
uitgloeien
PO wyżarzać
PY прокаливать
SU
SV glüdga

234 annular flow
DA ringformet
strømning
DE Ringströmung *f*
ES flujo anular *m*
F écoulement *m*
annulaire
I flusso *m* anulare
MA gyűrűs áramlás
NL ringvormige
tweefasen-
stroming *f*
PO strumień *m*
pierścieniowy
PY кольцевой поток *m*
SU rengasvirtaus
SV ringflöde *n*

235 anode
DA anode
DE Anode *f*
ES anodo *m*
F anode *f*
I anodo *m*
MA anód
NL anode *f*
PO anoda *f*
PY анод *m*
SU anodi
SV anod *c*

236 anthracite coal
DA antracit kul
DE Anthrazitkohle *f*
ES antracita *f*
F anthracite *m*
I antracite *f*
MA antracit szén
NL antraciet *f*
PO antracyt *m*
PY антрацит *m*
SU antrasiitti
SV antracit *c*

**237 anticipating
control**
DA selvstyring
DE Vorwegnahme-
Regelung *f*
ES control anticipativo
m
F contrôle *m*
anticipatif
I controllo *m*
previsionale
MA megelőző
szabályozás
NL anticiperende
regeling *f*
PO sterowanie *n*
wyprzedzające
PY предварительный
контроль *m*
SU ennakoiva säätü
SV självstyrande
kontroll *c*

238 anticipator
DA fremadretter
DE Beschleuniger *m*
ES anticipador *m*
F prédicteur *m*
I regolatore *m* di
anticipo
MA megelőző
NL voorspeller *m*
PO urządzenie *n*
przyspieszające
PY подогреватель *m*
термостата
SU ennakointivastus
SV anticipator *c*
(förutseende)

239 anti-corrosive *adj*
DA korrosion-
shæmmende
DE korrosionsverhütend
ES anticorrosivo
F anticorrosif
I anticorrosivo
MA korróziógátló
NL corrosiebestendig
corrosiewerend
PO antykorozyjny
przeciwkorozyjny
PY защищающий от
коррозии
SU korroosiolta
suojaava
SV korrosionshindrande

240 antifreeze
DA frostsikring
DE Frostschutz *m*
Gefrierschutz-
mittel *n*
ES anticongelante *m*
solución
incongelable *f*
F antigel *m*
I antigelo *m*
MA fagyálló
fagyást gátló
NL antivries *m*
vorstbestrijdings-
middel *n*
PO substancja *f* trudno
zamarzająca
substancja *f*
zapobiegająca
zamarzaniu
PY антифриз *m*
SU jäätymissuoja
SV frostskydd *n*

241 anti-friction lining
DA friktionsløs foring
DE Lagermetall *n*
ES revestimiento
antifricción *m*
F revêtement *m*
antifriction
I rivestimento *m*
antifrizione
MA csapágyfém
csúszócsapágy
NL antifrictie voering *f*
PO wykładzina *f*
przeciwcierna
wykładzina *f*
przeciwko tarciu
PY облицовка *f*,
защищающая от
трения
SU kitkan poistava
verhous
liukastusaine
SV friktionsminskande
beläggning *c*

242 antiscale
DA kedelstens-
hæmmende middel
DE Kesselsteinverhütungs-
mittel *n*
ES antiincrustante *m*
desincrustante *m*
F antitartre *m*
I antincrostante *m*
MA kazánkőoldó
NL anti-ketelsteen *n*
PO czynnik *m* przeciw
tworzeniu się
kamienia
kotłowego
PY средство *n* против
накипи
SU kattilakiven estoaine
SV pannstensskydds-
medel *n*

243 anti-siphon valve
DA antihævertventil
vakuumbryder
DE Rückschlagventil *n*
ES válvula antisifón *f*
F soupape *f* anti-
siphon (casse-vide)
I valvola *f* antisifone
MA visszaszívást gátló
szelep
NL antihevelafsluiter *m*
PO zawór *m*
przeciwsyfonowy
PY вакуумный вентиль
m
SU takaisinimusuoja
SV vakuumregler-
ventil *c*

**244 anti-vibration
mounting**
DA vibrationsdæmpende
montering
DE schwingungs-
dämpfende Aufstellung *f*
ES montaje
antivibratorio *m*
F dispositif *m*
antivibratoire
I supporto *m*
antivibrante
MA rezgésgátló szerelés
NL trillingdemper *m*
PO wibroizolacja *f*
zabezpieczenie *n*
antywibracyjne
PY амортизационное
устройство *n*
SU tärinänvaimennin
SV vibrationsdämpare *c*

245 apartment heating
DA lejligheds-
opvarmning
DE Etagenheizung *f*
ES calefacción
individual *f*
F chauffage *m*
d'appartement
I riscaldamento *m* di
appartamento
MA egyszinti fűtés
etázsfűtés
NL etageverwarming *f*
flatverwarming *f*
PO ogrzewanie *n*
mieszkaniowe
PY квартирное
отопление *n*
SU huoneistokohtainen
lämmitys
SV vånings-
uppvärmning *c*

246 aperture area
DA åbningsareal
DE Öffnungsfläche *f*
ES superficie de
apertura *f*
F aire *f* d'ouverture
I area *f* di apertura
MA nyílásfelület
NL toetredings-
oppervlak *n*
PO powierzchnia *f*
czynna otworu
PY площадь *f* проема
SU aukon ala
auringonkerääjän
tehollinen ala
SV hålarea *c*

247 apparatus
DA apparat
DE Apparat *m*
ES aparato *m*
F appareil *m*
I apparato *m*
apparecchio *m*
MA készülék
NL apparaat *n*
toestel *n*
PO aparat *m*
aparatura *f*
przyrząd *m*
PY аппарат *m*
SU koje
SV apparat *c*

248 apparatus dewpoint
DA apparats dugpunkt
DE Gerätetaupunkt *m*
ES temperatura de rocío del aparato *f*
F appareil *m* à point de rosée
I temperatura *f* di rugiada della batteria refrigerante
MA készülék harmatpont
NL dauwpunt *n* van uitgaande koelerlucht *m*
PO punkt *m* rosy przyrządu
PУ температура *f* точки росы аппарата
SU kojeen kastepiste
SV apparats daggpunkt *c*

249 apparent power
DA tilsyneladende effekt
DE Scheinleistung *f*
ES potencia aparente *f*
F puissance *f* apparente
I potenza *f* apparente
MA látszólagos teljesítmény
NL schijnbaar vermogen *n*
PO moc *f* pozorna
PУ кажущаяся мощность *f*
SU näennäinen teho
SV skenbar effekt *c*

250 apparent temperature
DA aflæsningstemperatur
DE scheinbare Temperatur *f*
ES temperatura aparente *f*
F température *f* apparente
I temperatura *f* apparente
MA látszólagos hőmérséklet
NL schijnbare temperatuur *f*
PO temperatura *f* pozorna
PУ радиационная температура *f*
SU näennäinen lämpütila
SV avläst temperatur *c*

251 appliance
DA anordning
DE Gerät *n* Vorrichtung *f*
ES aparato *m* dispositivo *m* instrumento *m* mecanismo *m*
F appareil *m*
I apparecchiatura *f*
MA készülék
NL toestel *n*
PO przyrząd *m* urządzenie *n*
PУ устройство *n*
SU laite
SV anordning *c*

252 application
DA anvendelse tilpasning
DE Anwendung *f* Verwendung *f*
ES aplicación *f*
F application *f*
I applicazione *f*
MA alkalmazás
NL toepassing *f*
PO zastosowanie *n* zgłoszenie *n*
PУ приложение *n* применение *n*
SU sovellus
SV tillämpning *c*

253 application rating
DA anvendelsesgrad
DE Anwendungsleistungs- beurteilung *f*
ES campo de aplicación *m*
F performance *f* pratique
I limiti *m,pl* di impiego
MA alkalmassági minősítés
NL vermogen *n* onder bedrijfsconditions
PO zastosowana wartość *f* znamionowa
PУ допустимые условия *n pl*
SU todellinen teho
SV tillämpningsdata *c*

254 apprentice
DA lærling
DE Lehrling *m*
ES aprendiz *m*
F apprenti *m*
I apprendista *m*
MA ipari tanuló
NL leerjongen *m* leerling *m*
PO praktykant *m* uczeń *m*
PУ практикант *m*
SU oppipoika
SV lärling *c*

255 approach
DA tilnærmelse
DE Anlauf *m* Näherung *f*
ES aproximación *f*
F approche *f*
I differenza *f* di temperatura tra fluido in uscita e in ingresso
MA megküzelítés
NL benadering *f* uitgaand temperatuur- verschil *n*
PO podejście *n* zbliżanie *n*
PУ подход *m* приближение *n* разность *f* (температуры)
SU lähestymistapa
SV tillförsel *c*

256 appurtenance
DA tilbehør
DE Zubehör *n*
ES accesorio *m*
F accessoire *m*
I accessorio *m*
MA tartozék
NL toebehoren *n*
PO aparat *m* mechanizm *m* przynależność *n*
PУ запасная часть *f*
SU lisälaite lisälaitteet
SV tillbehör *n*

257 aquastat
DA vandstandsmåler
DE Wassertemperatur-
regler *m*
ES termostato de
inmersión *m*
F aquastat *m*
I termostato *m* per
acqua (ad
immersione)
MA vízhőmérséklet
érzékelő
NL waterthermostaat *m*
PO regulator *m*
poziomu wody
PY термостат *m* для
воды
SU vedenkestävä
termostaatti
SV akvastat *c*

258 architect
DA arkitekt
DE Architekt *m*
ES arquitecto *m*
F architecte *m*
I architetto *m*
MA építész
NL architect *m*
PO architekt *m*
PY архитектор *m*
конструктор *m*
SU arkkitehti
SV arkitekt *c*

**259 architectural area
(of building)**
DA bygningsareal
DE architektonisch zu
gestaltende
Fläche *f*
ES superficie construida
f
F surface *f*
architecturale
(d'un bâtiment)
I superficie *f* costruita
MA beépített alapterület
NL gebruiksoppervlak *n*
van een gebouw
PO powierzchnia *f*
architektoniczna
budynku
PY архитектурная
площадь *f*
здания (общая
полезная
площадь)
SU bruttoala
SV byggnadsyta *c*

**260 architectural
volume**
DA bygningsvolumen
DE architektonisches
Volumen *n*
Bauvolumen *n*
ES volumen construido
m
F volume *m*
architectural
I volume *m* costruito
MA beépített térfogat
NL gebruiksinhoud *m*
van een gebouw
PO przestrzeń *f*
architektoniczna
PY архитектурный
объем *m*
строительный
объем *m*
SU bruttotilavuus
SV byggnadsvolym *c*

261 arc welding
DA buesvejsning
DE Lichtbogen-
schweißung *f*
ES soldadura por arco *f*
F soudure *f* à l'arc
I saldatura *f* ad arco
MA ívhegesztés
NL elektrisch lassen *n*
vlambooglassen *n*
PO spawanie *n* łukowe
PY дуговая сварка *f*
SU kaarihitsaus
SV bågsvetsning *c*

262 area
DA flade
zone
DE Fläche *f*
Zone *f*
ES área *f*
región *f*
superficie *f*
zona *f*
F aire *f*
superficie *f*
zone *f*
I area *f*
superficie *f*
MA terület
NL gebied *n*
oppervlak *n*
PO powierzchnia *f*
PY площадь *f*
поверхность *f*
SU ala
SV area *c*

263 ARI (*see* Appendix B)

264 arithmetic *adj*
DA beregnet
DE arithmetisch
ES aritmético
F arithmétique
I aritmetico
MA számtani
NL rekenkundig
PO arytmetyczny
PY арифметический
SU aritmeettinen
SV aritmetisk

265 asbestos
DA asbest
DE Asbest *m*
ES amianto *m*
asbesto *m*
F amiante *m*
I amianto *m*
MA azbeszt
NL asbest *n*
PO azbest *m*
PY асбест *m*
SU asbesti
SV asbest *c*

266 asbestos cord
DA asbestsnor
DE Asbestschnur *f*
ES cordón de
amianto *m*
F cordon *m* d'amiante
I cordone *m* di
amianto
MA azbesztzsinór
NL asbestkoord *n*
PO sznur *m* azbestowy
PY асбестовый
шнур *m*
SU asbestinaru
SV asbestväv *c*

267 as-built drawing
DA tegning
overensstemmende
med udførelse
DE Bestandszeichnung *f*
ES plano real definitivo
m
F dessin *m* de
recollement
I disegno *m*
dell'eseguito
MA megvalósulási tervek
NL revisietekening *f*
PO rysunek *m*
powykonawczy
PY исполнительный
чертеж *m*
SU luovutuspiirrustus
SV relationsritning *c*

268 ascending force
DA opdrift
DE Auftrieb *m*
   Auftriebskraft *f*
ES fuerza ascensional *f*
F force *f*
   ascensionnelle
   poussée *f*
   ascensionnelle
I forza *f* ascensionale
MA felhajtóerő
NL opwaartse kracht *f*
PO siła *f* wznoszenia
PY подъемная сила *f*
SU nostovoima
SV lyftkraft *c*

269 ASCII (*see* Appendix B)

270 ash
DA aske
DE Asche *f*
ES ceniza *f*
F cendre *f*
I cenere *f*
MA hamu
NL as *f*
PO popiół *m*
PY зола *f*
SU tuhka
SV aska *c*

271 ash box
DA askebeholder
   askeskuffe
DE Aschenkasten *m*
ES cenicero *m*
F cendrier *m*
I ceneraio *m*
MA hamuláda
NL asemmer *m*
   aslade *f*
PO zasobnik *m* popiołu
PY зольный ящик *m*
SU tuhkalaatikko
SV askbehållare *c*

272 ash content
DA askeindhold
DE Aschegehalt *m*
ES contenido de
   cenizas *m*
   porcentaje de
   cenizas *m*
   proporción de
   cenizas *f*
F teneur *f* en cendres
I contenuto *m* in
   cenere
MA hamutartalom
NL asgehalte *n*
PO zawartość *f* popiołu
PY содержание *n* золы
SU tuhkapitoisuus
SV askhalt *c*

273 ash door
DA askedør
DE Aschentür *f*
ES puerta del cenicero *f*
F porte *f* de cendrier
I portello *m* del
   ceneraio
MA hamuajtó
NL asdeur *f*
PO drzwiczki *pl*
   rewizyjne
   popielnika
PY дверь *f* для золы
SU tuhkaluukku
SV asklucka *c*

274 ash pit
DA askegrube
DE Aschengrube *f*
ES foso de cenizas *m*
F fosse *f* à cendres
I fossa *f* del ceneraio
MA hamugüdür
NL askuil *m*
PO popielnik *m*
PY зольник *m*
SU tuhkavarasto
SV askrum *n*

275 ASHRAE (*see* Appendix B)

276 ash removal
DA askefjernelse
DE Entaschung *f*
ES separación de
   cenizas *f*
F évacuation *f* des
   cendres
I rimozione *f* della
   cenere
MA kihamuzás
NL asverwijdering *f*
PO usuwanie *n* popiołu
PY золоудаление *n*
SU tuhkanpoisto
SV asktömning *c*

277 ash retention figure
DA asketal
DE Unverbranntes *n* in
   der Asche *f*
ES (proporción de)
   componentes
   inquemados de las
   cenizas *f*
F imbrûlés *m* dans les
   cendres
I numero *m* di
   incombusti nella
   cenere
MA hamu visszamaradási
   tényező
NL percentage *f*
   onverbrand in as
PO zawartość *f* części
   niepalnych w
   popiele
PY содержание *n* в
   золе частиц
   неполного
   сгорания
SU tuhkan erotusaste
SV asktal *n*

278 ash separator
DA askeudskiller
DE Aschenabscheider *m*
ES separador de
   cenizas *m*
F séparateur *m* de
   cendres
I separatore *m* di
   cenere
MA pernyeleválasztó
NL asafscheider *m*
PO oddzielacz *m*
   popiołu
PY сепаратор *m* золы
SU tuhkanerottaja
SV askavskiljare *c*

DA = Danish
DE = German
ES = Spanish
F = French
I = Italian
MA = Magyar
      (Hungarian)
NL = Dutch
PO = Polish
PY = Russian
SU = Finnish
SV = Swedish

279 aspect ratio
DA formatforhold
DE Längenverhältnis *n*
    Streckungs-
      verhältnis *n*
ES factor de forma *m*
F facteur *m* de forme
    (d'une bouche,
    d'un conduit)
I rapporto *m* di
    aspetto
MA oldalarány
NL lengte-breedte
    verhouding *f*
PO proporcja *f*
    wymiarów
PY соотношение *n*
    асбеста
SU sivusuhde
SV formatförhållande *n*

280 asphalt
DA asfalt
DE Asphalt *m*
ES asfalto *m*
F asphalte *m*
I asfalto *m*
MA aszfalt
NL asfalt *n*
PO asfalt *m*
PY асфальт *m*
SU asfaltti
SV asfalt *c*

281 aspirated
    hygrometer
DA aspirations-
    hygrometer
DE Aspirations-
    hygrometer *n*
ES higrómetro de
    aspiración *m*
F hygromètre *m* à
    aspiration
I igrometro *m* ad
    aspirazione
MA átszívásos
    nedvességmérő
NL hygrometer *m* met
    ventilator
PO higrometr *m*
    respiracyjny
PY аспирационный
    гигрометр *m*
SU tuuletettu
    psykrometri
SV aspirations-
    hygrometer *c*

282 aspiration
DA udsugning
DE Ansaugung *f*
    Aspiration *f*
ES aspiración *f*
F aspiration *f*
I aspirazione *f*
MA szívás
NL aanzuiging *f*
    inademing *f*
PO wsysanie *n*
    zasysanie *n*
PY аспирация *f*
SU imu
SV aspiration *c*

283 aspiration
    psychrometer
DA udsugningsfugtig-
    hedsmåler
DE Aspirations-
    psychrometer *n*
ES sicrómetro de
    aspiración *m*
F psychromètre *m* à
    aspiration
I psicrometro *m* ad
    aspirazione
MA átszívásos
    pszichrométer
NL psychrometer *m* met
    ventilator
PO psychrometr *m*
    respiracyjny
PY аспирационный
    психрометр *m*
SU tuuletettu
    psykrometri
SV aspirations-
    psykrometer *c*

284 aspirator
DA udsugningsapparat
DE Aspirator *m*
ES aspirador *m*
F aspirateur *m*
I aspiratore *m*
MA aspirátor
    szívókészülék
NL aspirator *m*
    zuigtoestel *n*
PO aspirator *m*
    urządzenie *n*
    zasysające
PY аспиратор *m*
SU tuuletin
SV aspirator *c*

285 assemble *vb*
DA montere
    samle
DE montieren
    zusammenbauen
ES armar
    montar
F assembler
I assemblare
    montare
MA szerel
NL assembleren
    monteren
    samenvoegen
PO montować
    składać
PY монтировать
SU koota
SV montera

286 assembly
DA montage
    samling
DE Montage *f*
    Zusammenbau *m*
ES conjunto (de piezas
    montadas o
    soldadas) *m*
    montaje *m*
F assemblage *m*
I assemblaggio *m*
    montaggio *m*
MA szerelés
NL assemblage *f*
    vergadering *f*
    verzameling *f*
PO montaż *n*
    montowanie *n*
    zespół *m*
    zestawienie *n*
PY монтаж *m*
SU kokoonpano
SV montering *c*

287 assembly instruction
DA monteringsinstruks
DE Montage-anweisung*f*
ES instrucciones de montaje *f,pl*
F notice *f* d'assemblage
notice *f* de montage
I istruzioni *f,pl* per il montaggio
MA szerelési utasítás
NL assemblagevoor-schrift *n*
montagevoor-schrift *n*
PO instrukcja *f* montażowa
PY инструкция *f* по монтажу
SU kokoonpano-ohje
SV monterings-anvisning *c*

288 ASTM (*see* Appendix B)

289 athermour barrier
DA Athermours afskærmning (af strålevarme)
DE Athermour-Sperre *f*
ES barrear térmica *f*
F écran *m* anti-rayonnement
I barriera *f* isolante
MA sugárzóhő gát
NL stralings-afscherming *f*
warmtestralings-schild *n*
PO zapora *f* Athermoura
PY защита *f* от лучистой теплоты
SU säteilysuoja
SV icketermisk barriär *c*

290 atmosphere
DA atmosfære
DE Atmosphäre *f*
ES atmósfera *f*
F atmosphère *f*
I atmosfera *f*
MA atmoszféra légkür
NL atmosfeer *f* dampkring *m*
PO atmosfera *f*
PY атмосфера *f*
SU ilmakehä
SV atmosfär *c*

291 atmosphere of reference
DA vejrdata
DE Referenz-atmosphäre *f*
ES atmósfera de referencia *f*
F atmosphère *f* de référence
I atmosfera *f* di riferimento
MA referencia légkür
NL referentie-atmosfeer *f*
PO atmosfera *f* odniesienia
PY рекомендуемые характеристики *f,pl* атмосферы
SU standardiolosuhteet
SV referensatmosfär *c*

292 atmospheric *adj*
DA atmosfærisk
DE atmosphärisch
ES atmosférico
F atmosphérique
I atmosferico
MA atmoszférikus
NL atmosferisch
PO atmosferyczny
PY атмосферный
SU atmosfäärinen
SV atmosfärisk

293 atmospheric burner
DA atmosfærisk brænder
DE atmosphärischer Brenner *m*
ES quemador atmosférico *m*
F brûleur *m* atmosphérique
I bruciatore *m* atmosferico
MA atmoszférikus égő
NL atmosferische brander *m*
PO palnik *m* atmosferyczny
palnik *m* o ciągu naturalnym
PY атмосферная горелка *f*
SU luonnonvetoinen kaasupoltin (ei palamisilmapu-hallinta)
SV atmosfärisk brännare *c*

294 atmospheric condenser
DA overrislings-kondensator
DE atmosphärischer Verflüssiger *m*
ES condensador atmosférico *m*
condensador de lluvia *m*
F condenseur *m* atmosphérique
I condensatore *m* atmosferico
MA atmoszférikus kondenzátor
NL atmosferische condensor *m*
open condensor *m*
verdampings-condensor *m*
PO skraplacz *m* powietrzny
PY конденсатор *m*, орошаемый в атмосферном воздухе
SU ilmalauhdutin
SV atmosfärisk kondensor *c*

33

295 atmospheric dust
DA atmosfærisk støv
DE atmosphärischer
Staub *m*
ES polvo atmosférico *m*
F poussière *f*
atmosphérique
I pulviscolo *m*
atmosferico
MA atmoszférikus por
NL atmosferisch stof *n*
PO pył *m* atmosferyczny
PY атмосферная
пыль *f*
SU ulkoilman leijuva
püly
SV atmosfäriskt stoff *n*

**296 atmospheric dust
spot efficiency
(ASHRAE)**
DA atmosfærisk
støvfaktor
DE Wirkungsgrad *m* des
Absetzvermögens
n atmosphärischen
Staubes *m*
ES eficiencia de filtrado
(método
opacimétrico) *f*
F rendement *m* moyen
à la tache
(ASHRAE)
I efficienza *f* di
filtrazione
(metodo
opacimetrico)
MA atmoszférikus
porfolt hatásfok
NL rendement *n* volgens
verkleuringstest-
methode
PO skuteczność *f* wg
metody pyłu
atmosferycznego
PY эффективность *f*
поглощения
пыли (ASHRAE)
SU pülytäpläerotusaste
SV

**297 atmospheric freeze
drying**
DA atmosfærisk fryse-
tørring
DE atmosphärische
Gefrier-
trocknung *f*
ES liofilización
atmosférica *f*
F cryodessiccation *f*
atmosphérique
I liofilizzazione *f* a
pressione
ambiente
MA atmoszférikus
fagyasztva porítás
NL vriesdrogen *n* onder
atmosferische
druk
PO atmosferyczne
suszenie *n*
sublimacyjne
PY вымораживание *n*
при
атмосферном
давлении
SU jäähdytyskuivaus
normaalipaineessa
SV atmosfärisk
frystorkning *c*

**298 atmospheric
pressure**
DA atmosfærisk tryk
DE atmosphärischer
Druck *m*
ES presión
atmosférica *f*
F pression *f*
atmosphérique
I pressione *f*
atmosferica
MA atmoszférikus
nyomás
NL atmosferische
druk *m*
PO ciśnienie *n*
atmosferyczne
PY атмосферное
давление *n*
SU ilmakehän paine
SV atmosfäriskt tryck *n*

299 atomize *vb*
DA forstøve
DE zerstäuben
ES atomizar
pulverizar
F atomiser
pulvériser
I nebulizzare
polverizzare
MA porlaszt
NL vernevelen
verstuiven
PO rozpylać
PY распыливать
SU sumuttaa
SV finfördela

**300 atomizing
humidifier**
DA forstøvningsbefugter
DE Zerstäubungs-
befeuchter *m*
ES humidificador por
atomización *m*
F pulvérisateur *m* à
air comprimé
I umidificatore *m* a
nebulizzazione
MA porlasztásos
nedvesítés
NL verstuivings-
bevochtiger *m*
PO nawilżacz *m*
drobnorozpylający
PY распылительный
увлажнитель *m*
SU sumuttava kostuttaja
SV trycklufts-
befuktare *c*

301 attendance
DA pasning
tilsyn
DE Bedienung *f*
Pflege *f*
Wartung *f*
ES mantenimiento *m*
servicio *m*
vigilancia *f*
F conduite *f*
service *m*
surveillance *f*
I conduzione *f*
sorveglianza *f*
MA kezelés
megjelenés
NL bediening *f*
toezicht *n*
PO dozór *m*
nadzór *m*
obsługa *f*
PY обслуживание *n*
SU läsnäolo
SV dämpning *c*
skötsel *c*

DA =	Danish
DE =	German
ES =	Spanish
F =	French
I =	Italian
MA =	Magyar (Hungarian)
NL =	Dutch
PO =	Polish
PY =	Russian
SU =	Finnish
SV =	Swedish

302 attenuation
DA fortynding
svækkelse
DE Dämpfung *f*
Verdünnung *f*
ES amortiguación *f*
atenuación *f*
disminución *f*
F atténuation *f*
I attenuazione *f*
MA tompítás
NL geluiddemping *f*
PO tłumienie *n*
PY затухание *n*
ослабление *n*
SU vaimennus

303 attic
DA loft
DE Dachgeschoß *n*
ES ático *m*
F comble *m*
grenier *m*
I attico *m*
solaio *m*
MA attika
oromzat
NL vliering *f*
zolder *m*
PO attyka *f*
poddasze *n*
strych *m*
PY аттик *m*
мансарда *f*
чердак *m*
SU ullakko
SV vind *c*

304 audibility
DA hørbarhed
DE Hörbarkeit *f*
ES audibilidad *f*
F audibilité *f*
I udibilità *f*
MA hallhatóság
NL hoorbaarheid *f*
PO słyszalność *f*
PY слышимость *f*
SU kuuluvuus
SV hörbarhet *c*

305 audible *adj*
DA hørbar
DE hörbar
ES audible
F audible
I udibile
MA hallható
NL hoorbaar
PO słyszalny
PY слышимый
SV hörbar

306 authority
DA autoritet
DE Autorität *f*
ES autoridad *f*
F autorité *f*
I autorità *f*
MA engedély
hatóság
NL autoriteit *f*
PO prawo *n*
uprawnienie *n*
PY степень *f*
регулирования
SU auktoriteetti
SV myndighet *c*

**307 autogenous
ignition
temperature**
DA gasantændelses-
temperatur
DE autogene
Zündtemperatur *f*
ES temperatura de
autoignición *f*
F température *f*
d'auto-
inflammation
I temperatura *f* di
autoaccensione
MA üngyulladási
hőmérséklet
NL zelfontbrandings-
temperatuur *f*
PO temperatura *f*
samozapłonu
PY температурный
режим *m*
газовой сварки
SU syttymislämpütila
SV gasantändnings-
temperatur *c*

308 automatic *adj*
DA automatisk
DE automatisch
ES automático
F automatique
I automatico
MA automatikus
NL automatisch
PO automatyczny
samoczynny
PY автоматический
SU automaattinen
SV automatisk

309 automation
DA automatisering
DE Automation *f*
ES automación *f*
F automation *f*
I automazione *f*
MA automatizálás
NL automatiseren *n*
automatisering *f*
PO automatyzacja *f*
PY автоматизация *f*
SU automaatio
SV automation *c*

310 auxiliary air
DA spædeluft
DE Zusatzluft *f*
ES aire auxiliar *m*
F air *m* auxiliaire
I aria *f* ausiliaria
MA pótlevegő
NL aanvullende
luchttoevoer *m*
PO powietrze *n*
dodatkowe
PY дополнительный
воздух *m*
SU apupuhallin
SV reservluft *c*

**311 auxiliary electrical
system**
DA hjælpe el-system
DE Zusatzelektro-
system *n*
ES sistema eléctrico
auxiliar *m*
F système *m* électrique
auxiliaire
I impianto *m* elettrico
ausiliario
MA tartalék elektromos
rendszer
NL elektrisch
hulpsysteem *n*
PO system *m*
elektryczny
pomocniczy
PY вспомогательная
электрическая
система *f*
SU varavoimajärje-
stelmä
SV reservelektriskt
system *n*

312 auxiliary machine
DA hjælpemaskine
DE Hilfsmaschine *f*
ES máquina auxiliar *f*
F machine *f* auxiliaire
I macchina *f* di
    riserva
MA tartalék gép
NL hulpmachine *f*
    reserve machine *f*
PO maszyna *f*
    pomocnicza
PY вспомогательный
    агрегат *m*
SU apukone
SV hjälpmaskin *c*

**313 auxiliary thermal
source**
DA hjælpevarmekilde
DE Zusatzwärme-
    quelle *f*
ES fuente térmica
    auxiliar *f*
F source *f* thermique
    auxiliaire
I sorgente *f* termica
    ausiliaria
MA kiegészítő
    energiaforrások
NL hulpwarmtebron *f*
PO źródło *n* ciepła
    dodatkowe
PY вспомогательный
    источник *m*
    теплоты
SU lisälämmünlähde
SV reservvärmekälla *c*

314 average *adj*
DA gennemsnitlig
DE durchschnittlich
ES media *f*
    promedio *m*
    término medio *m*
F moyenne *f*
I medio
MA átlagos
NL gemiddeld
PO przeciętny
    średni
PY средний
SU keskimääräinen
SV genomsnittligt

315 axial *adj*
DA aksial
DE axial
ES axial
F axial
I assiale
MA axiális
NL axiaal
PO osiowy
PY осевой
SU aksiaali-
    aksiaalinen
SV axialt

316 axial compensator
DA aksial kompensator
DE Axialkompen-
    sator *m*
ES compensador
    axial *m*
F compensateur *m*
    axial
I compensatore *m*
    assiale
MA tengelyirányú
    táguláskiegyenlítő
NL axiale compen-
    sator *m*
PO kompensator *m*
    osiowy
    wydłużka *f* osiowa
PY осевой компрессор
    m
SU aksiaalikompressori
SV axialkompensator *c*

317 axial fan
DA aksialventilator
DE Axialventilator *m*
ES ventilador axial *m*
    ventilador helicoidal
    m
F ventilateur *m* axial
I ventilatore *m* assiale
MA axiális ventilátor
NL axiaalventilator *m*
PO wentylator *m*
    osiowy
PY осевой
    вентилятор *m*
SU aksiaalipuhallin
SV axialfläkt *c*

**318 axial flow
compressor**
DA aksialkompressor
DE Axialverdichter *m*
ES compresor axial *m*
F compresseur *m* à
    flux axial
I compressore *m*
    assiale
MA axiális átümlésű
    kompresszor
NL axiaalcompressor *m*
PO sprężarka *f* o
    przepływie
    osiowym
PY турбокомпрессор *m*
SU aksiaalikompressori
SV axialflödes-
    kompressor *c*

319 axial velocity
DA aksial hastighed
DE Axialgeschwindig-
    keit *f*
ES velocidad axial *f*
F vitesse *f* axiale
I velocità *f* assiale
MA tengelyirányú
    sebesség
NL axiale snelheid *f*
PO prędkość *f* osiowa
PY осевая скорость *f*
SU aksiaalinopeus
SV axial hastighet *c*

320 axis
DA akse, midtlinie
DE Achse *f*
    Mittellinie *f*
ES eje (en
    general:geometría,
    geología,
    matemáticas,
    etc.) *m*
F axe *m*
I asse *m*
MA axis
    tengely
NL as *f*
    hartlijn *f*
PO oś *f* (w układzie
    współrzędnych)
PY ось *f*
SU akseli
    keskiviiva
SV axel *c*

321 axle
DA aksel
DE Achse *f*
ES eje (vehículos,
 máquinas) *m*
F essieu *m*
I assale *m*
MA tengely
NL as *f*
 spil *f*
PO oś *f*
 półoś *f*
 wał *m* osiowy
PY вал *m*
 ось *f*
SU akseli
SV axel *c*

**322 axonometric
 drawing**
DA aksiometrisk tegning
DE axiometrische
 Zeichnung *f*
ES plano axonométrico
 m
F dessin *m*
 axonométrique
I assonometria *f*
 disegno *m*
 assonometrico
MA axonometrikus rajz
NL perspectief
 tekening *f*
PO rysunek *m*
 aksonometryczny
PY аксонометрический
 чертеж
SU aksonometrinen
 piirros
SV isometrisk
 projektion *c*

323 azeotropic *adj*
DA med konstant
 kogepunkt
DE azeotropisch
ES azeotrópico
F azéotropique
I azeotropico
MA azeotropikus
NL azeotropisch
PO azeotropowy
PY азеотропный
SU azeotrooppinen
SV azeotropisk

DA	=	Danish
DE	=	German
ES	=	Spanish
F	=	French
I	=	Italian
MA	=	Magyar (Hungarian)
NL	=	Dutch
PO	=	Polish
PY	=	Russian
SU	=	Finnish
SV	=	Swedish

B

324 Bacharach number
DA Bacharachtal
sodtal
DE Bacharach-Zahl *f*
ES índice de
Bacharach *m*
número de
Bacharach *m*
F indice *m* de
Bacharach
I numero *m* di
Bacharach
MA Bacharach szám
NL Bacharach getal *n*
PO liczba *f* Bacharacha
PY число *n* Бахараха
SU Bacharach-luku
SV bacharachtal *n*

325 backdraught (*see* downdraught

326 backdraught damper
DA tilbageløbsspjæld
DE Rückstromklappe *f*
ES compuerta
antirretorno *f*
F registre *m* à volets à
contre-tirage
I serranda *f* di non
ritorno
MA visszaáramlásgátló
csappantyú
NL luchtkanaalterugslag-
klep *f*
PO przepustnica *f*
zwrotna
PY обратный клапан
m
SU alipainepelti
SV bakdragsspjäll *n*

327 background heating
DA grundvarme
DE Grundheizung *f*
ES calefacción de base *f*
F chauffage *m* de base
I riscaldamento *m* di
base
MA alapfűtés
NL basisverwarming *f*
PO ogrzewanie *n*
dyżurne
PY основное отопление
n
SU peruslämmitys
SV grundvärme *c*

328 background irradiance
DA baggrundsstråling
DE Grundstrahlung *f*
ES radiación de fondo *f*
F rayonnement *m* de
fond
I irradianza *f* di
fondo
MA háttérsugárzás
NL achtergrond-
straling *f*
PO promieniowanie *n*
tła
PY рассеянное
излучение *n*
SU taustasäteily
SV bakgrunds-
strålning *c*

329 back pressure
DA modtryk
DE Gegendruck *m*
ES contrapresión *f*
F contre-pression *f*
I contropressione *f*
MA ellennyomás
NL tegendruk *m*
zuigdruk *m*
PO ciśnienie *n* wsteczne
ciśnienie *n* zasysania
(przed sprężarką)
przeciwciśnienie *n*
PY противодавление *n*
SU vastapaine
SV mottryck *n*

330 back pressure regulator
DA modtryksregulator
DE Gegendruckregler *m*
ES presostato de
aspiración *m*
F régulateur *m* de
pression
d'aspiration
I regolatore *m* di
contropressione
(pressione
dell'evaporatore)
MA ellennyomás
szabályozó
NL tegendrukregelaar *m*
zuigdrukregelaar *m*
PO regulator *m*
ciśnienia
wstecznego
PY регулятор *m*
давления в
испарителе
SU vastapaineen säädin
SV mottrycks-
regulator *c*

331 back pressure valve
DA kontraventil
tilbageslagsventil
DE Gegendruckventil *n*
ES válvula de aspiración
f
F vanne *f* de pression
d'aspiration
I valvola *f* di contro-
pressione
MA expanziós szelep
NL zuigdrukregelklep *f*
PO zawór *m* zwrotny
PY защитный клапан
m испарителя
SU vastapaineventtiili
SV mottrycksventil *c*

332 back-up *adj*
DA sikkerhedskopiering
   (EDB)
DE rückwärts gerichtet
ES de reserva
F sauvegardé
I di riserva
MA támogató
NL reserve
PO rezerwowy
   wspomagający
   zapasowy
PY поддерживающий
SU vara-
SV hjälp-

333 back-up battery
DA reservebatteri
DE Hilfsbatterie *f*
ES batería de reserva *f*
F batterie *f* de secours
I batteria *f* di riserva
MA segítő gépcsoport
NL reserve batterij *f*
PO wymiennik *m*
   wspomagający
PY поддерживающая
   установка *f*
SU varavirtalähde
SV hjälpbatteri *n*

334 backward-curved
adj
DA bagudvendte
DE rückwärts gekrümmt
ES curvado hacia atrás
F incurvé vers l'arrière
I a pale rovesce
MA hátrahajló
NL achterover gebogen
PO zakrzywiony do tyłu
PY загнутая назад
SU taaksepäin kaartuva
   siipi
SV bakåtböjd

335 bacteria
DA bakterie
DE Bakterien *f,pl*
ES bacterias *f,pl*
F bactérie *f*
I batterio *m*
MA baktérium
NL bacteriën *pl*
PO bakteria *f*
PY бактерия *f*
SU bakteeri
SV bakterie *c*

336 bacterial *adj*
DA bakteriel
DE bakteriell
ES bacterial
F bactérien
I batterico
MA bakteriális
NL bacteriologisch
PO bakteryjny
PY бактериальный
SU bakteeripitoinen
SV bakteriell

337 bacterial action
DA bakterieangreb
DE Bakterientätigkeit *f*
ES acción bacteriana *f*
F action *f* bactérienne
I azione *f* batterica
MA bakteriális
   tevékenység
NL bacteriologische
   werking *f*
PO działanie *n*
   bakteryjne
PY бактериологическое
   действие *n*
SU bakteeritoiminta
SV bakterieaktivitet *c*

338 bacterial decay
DA bakteriel
   nedbrydning
DE Bakterienbefall *m*
   Bakterienfäulnis *f*
ES degradación
   bacteriana *f*
F altération *f*
   bactérienne
I decadimento *m*
   batterico
MA bakteriális rothadás
NL bacteriologische
   aantasting *f*
PO rozkład *m*
   bakteryjny
PY бактериальное
   разложение *n*
SU pilaantuminen
SV bakteriesönderfall *n*

339 bacterial growth
DA bakterievækst
DE Bakterien-
   wachstum *n*
ES desarrollo de
   bacterias *m*
F développement *m*
   des bactéries
I sviluppo *m* batterico
MA baktérium
   szaporodás
NL bacteriologische
   groei *m*
PO rozwój *m* bakterii
PY прирост *m*
   бактерий
SU bakteerikasvu
SV bakterietillväxt *c*

340 bactericidal *adj*
DA bakteriedræbende
DE keimtötend
ES bactericida
F bactéricida
I battericida
MA baktériumülő
NL bacteriedodend
   kiemdodend
PO bakteriobójczy
PY бактерицидный
SU bakteereita tappava
SV bakteriedödande

341 bactericide
DA bakteriedræbende
DE Desinfektions-
   mittel *n*
ES bactericida *m*
F bactéricide *m*
I battericida *m*
MA baktériumülő szer
NL bacteriedodende
   stof *m*
PO środek *m*
   bakteriobójczy
PY бактерицидный
   агент *m*
SU desinfioiva aine
SV baktericid *c*

342 baffle
DA baffel
  (lyd) skærm
DE Leitblech *n*
  Prallwand *f*
ES deflector *m*
F chicane *f*
  déflecteur *m*
I deflettore *m*
MA gát
  terelő
NL plaat *f*
  schot *n*
PO deflektor *m*
  przegroda *f*
PY отражательная
  поверхность *f*
SU lamelli
SV skärm *c*

343 baffle plate
DA ledeplade
  skærmplade
DE Ablenkplatte *f*
ES placa deflectora *f*
  tabique deflector *m*
F chicane *f*
I piastra *f* del
  deflettore
MA terelőlemez
NL geleidingsplaat *f*
  keerplaat *f*
PO płyta *f* deflektora
PY отбойный щиток *m*
SU lamellilevy
SV skärmplåt *c*

344 bag
DA pose
  sæk
DE Beutel *m*
  Sack *m*
  Tasche *f*
ES bolsa *f*
  saco *m*
F sac *m*
I sacco *m*
MA zsák
NL zak *m*
PO torba *f*
  worek *m*
PY мешок *m*
SU pussi
SV påse *c*

345 bag filter
DA posefilter
DE Schlauchfilter *m*
  Taschenfilter *m*
ES filtro de bolsa *m*
  manga para filtrar *f*
F filtre *m* à sac
I filtro *m* a sacco
MA zsákos szűrő
NL zakkenfilter *m*
PO filtr *m* kieszeniowy
  filtr *m* workowy
PY рукавный
  фильтр *m*
SU pussisuodatin
SV påsfilter *n*

346 balance *vb*
DA balancere
DE ausgleichen
ES equilibrar
F équilibrer
I equilibrare
MA kiegyenlít
NL balanceren
  inregelen
PO bilansować
  równoważyć
  utrzymywać w
  równowadze
PY выравнивать
SU tasapainottaa
SV balansera

347 balanced *adj*
DA afbalanceret
DE ausgeglichen
ES equilibrado
F équilibré
I equilibrato
MA kiegyenlített
NL evenwichtig
  ingeregeld
PO zrównoważony
PY выравненный
  сбалансированный
SU tasapainotettu
SV balanserad

**348 balanced draft
(USA)** (*see* balanced
draught)

349 balanced draught
DA afbalanceret træk
DE Zugausgleich *m*
ES tiro equilibrado *m*
F tirage *m* équilibré
I tiraggio *m*
  equilibrato
MA kiegyenlített huzat
NL gebalanceerde
  trek *m*
PO ciąg *m*
  zrównoważony
PY сбалансированная
  тяга *f*
SU tasapainotettu veto
SV balanserad
  ventilation *c*

350 balanced flow
DA afbalanceret
  strømning
DE ausgeglichene
  (abgestimmte)
  Strömung *f*
ES flujo equilibrado *m*
F débit *m* équilibré
I flusso *m* bilanciato
MA kiegyenlített áramlás
NL gebalanceerde
  stroming *f*
PO przepływ *m*
  zrównoważony
PY баланс *m* между
  притоком и
  вытяжкой
  сбалансированный
  поток *m*
SU tasapainotettu
  virtaus
SV balanserat flöde *n*

351 balanced flue
DA afbalanceret
 ventilation
DE ausgeglichene
 Abgasführung *f*
ES conducto de humos
 de corriente
 uniforme *m*
F circuit *m* de fumées
 équilibré
I circuito *m*
 equilibrato
MA kiegyenlített
 égéstermékelve-
 zetési rendszer
 (SEDUCT)
NL gebalanceerd
 rookgasafvoer-
 stuk *n*
 gebalanceerd
 rookgasafvoer-
 systeem *n*
PO czopuch *m* z
 naturalnym
 przepływem spalin
PY сбалансированный
 газоход *m*
 (дымоход *m*)
SU palamisilma- ja savu-
 kaasupuhaltimilla
 varustettu laitos
SV balanserad
 ventilation *c*

352 balance pipe
DA udligningsrør
DE Ausgleichsleitung *f*
ES conducto
 equilibrador *m*
 tubo compensa-
 dor *m*
 tubo equilibrador *m*
F tuyauterie *f*
 d'équilibre
I tubo *m* di
 bilanciamento
 tubo *m* di
 equalizzazione
MA kiegyenlítő cső
NL evenwichtsleiding *f*
 nivelleringsleiding *f*
PO rura *f* wyrównawcza
PY выравнивающая
 труба *f*
SU tasausputki
SV utjämningsrör *n*

353 balance point
 temperature
DA ligevægtstemperatur
DE Temperatur *f* im
 Einstellpunkt *m*
ES temperatura de
 equilibrio *f*
F température *f*
 d'équilibre
I temperatura *f* di
 equilibrio
MA egyensúlyi
 hőmérséklet
NL evenwichts-
 temperatuur *f*
PO temperatura *f*
 punktu
 równowagi
PY температура *f*
 обеспечивающая
 баланс расходов
 теплоты
SU tasapainolämpütila
SV balanstemperatur *c*

354 balance pressure
DA udligningstryk
DE Ausgleichsdruck *m*
ES presión de equilibrio
 f
F pression *f*
 d'équilibre
I pressione *f* di
 equilibrio
MA kiegyenlítő nyomás
NL evenwichtsdruk *m*
PO ciśnienie *n*
 równowagi
PY равенство *n*
 давлений
SU ulkoilman paine
 (systeemissä)
SV jämviktstryck *n*

355 balancer
DA udligningsanordning
 vægt
DE Ausgleichssystem *n*
 Schwinghebel *m*
ES estabilizador *m*
F organe *m*
 d'équilibrage
I bilanciere *m*
MA kiegyenlítő
NL inregelfirma *f*
 inregeltechnicus *m*
PO stabilizator *m*
 wyważarka *f*
PY наладчик *m*
SU tasapainotin
SV utjämningsdon *n*

356 balance tank
DA udligningsbeholder
DE Ausgleichs-
 behälter *m*
 Ausgleichstank *m*
ES depósito de
 equilibrado *m*
F réservoir *m*
 d'équilibre
I serbatoio *m* di
 bilanciamento
MA kiegyenlítő tartály
NL nivelleringstank *m*
PO zbiornik *m*
 wyrównawczy
PY выравнивающий
 резервуар *m*
 (бак)
SU tasaussäiliü
SV utjämningstank *c*

357 balancing
DA indregulering
 udligning
DE Ausgleich *m*
ES compensador *m*
 equilibrador *m*
F équilibrage *m*
I equilibratura *f*
MA kiegyenlítés
NL inregeling *f*
PO równoważenie *n*
 wyrównywanie *n*
PY балансировка *f*
 выравнивание *n*
SU tasapainottaminen
SV jämvikt *c*

358 balancing station
DA balancepunkt
DE Ausgleichsstation *f*
ES estación de
 equilibrado *f*
F poste *m*
 d'équilibrage
I stazione *f* di
 bilanciamento
MA kiegyenlítő állomás
NL inregeleenheid *f*
PO stacja *f*
 wyrównawcza
PY выравнивающий
 комплекс *m*
SU virtauksen
 säätüasema
SV utjämningsstation *c*

359 balancing valve
DA udligningsventil
DE Ausgleichsventil *n*
ES válvula de
    equilibrado *f*
F robinet *m*
    d'équilibrage
I valvola *f* di taratura
MA kiegyenlítő szelep
NL inregelafsluiter *m*
    inregelklep *f*
PO zawór *m*
    wyrównawczy
PY выравнивающий
    вентиль *m*
SU kertasäätüventtiili
SV utjämningsventil *c*

360 ballast
DA ballast
DE Ballast *m*
ES balasto *m*
F ballast *m*
I reattore *m* di
    lampade a scarica
    starter *m* di
    lampade a scarica
MA ballaszt
    nehezék
NL ballast *m*
PO obciążenie *n*
PY балласт *m*
SU kuristin
SV ballast *c*

361 ball bearing
DA kugleleje
DE Kugellager *n*
ES cojinete de bolas *m*
    rodamiento *m*
F roulement *m* à billes
I cuscinetto *m* a sfere
MA golyóscsapágy
NL kogellager *n*
PO łożysko *n* kulkowe
PY шариковый
    подшипник *m*
SU kuulalaakeri
SV kullager *n*

**362 ball bearing
    support**
DA kuglelejeunderlag
DE Kugellagerbock *m*
ES soporte de
    rodamiento *m*
    soporte del cojinete
    de bolas *m*
F palier *m* à billes
I supporto *m* per
    cuscinetto a sfere
MA golyóscsapágyas
    támasz
    gürdülő alátámasztás
NL kogellagerstoel *m*
PO podpora *f* z
    łożyskiem
    kulkowym
PY опора *f* для
    шарикового
    подшипника
SU laakeripesä
SV kullagerhus *n*

363 ball cock (*see* float
    valve)

364 ball joint
DA kugleled
DE Kugelgelenk *n*
ES junta de rótula *f*
    rótula *f*
F joint *m* à rotule
I giunto *m* a sfera
MA gümbcsukló
NL kogelgewricht *n*
PO przegub *m* kulowy
    złącze *n* kulowe
PY сферический
    шарнир *m*
SU pallonivel
SV kulled *c*

365 ball valve (*see* float
    valve)

366 bandwidth
DA båndbredde
DE Bandbreite *f*
ES amplitud de la
    banda *f*
F largeur *f* de bande
I ampiezza *f* di banda
MA sávszélesség
NL bandbreedte *f*
PO szerokość *f* pasma
    (częstotliwości)
PY ширина *f* полосы
    частот
SU kaistan leveys
SV bandbredd *c*

367 bar
DA bar
    stang
DE Bar(Druckeinheit) *n*
    Stab *m*
    Stange *f*
ES barra *f*
    rótula *f*
F barre *f*
I barra *f*
MA bar
    rúd
NL staaf *f*
    stang *f*
PO bar *m* (jednostka
    ciśnienia)
    pręt *m*
PY балка *f*
SU tanko
SV bom *c*

368 barometer
DA barometer
DE Barometer *n*
ES barómetro *m*
F baromètre *m*
I barometro *m*
MA barométer
    légnyomásmérő
NL barometer *m*
PO barometr *m*
PY барометр *m*
SU ilmapuntari
SV barometer *c*

**369 barometric
    condenser**
DA luftkondensator
DE barometrischer
    Verflüssiger *m*
ES condensador
    atmosférico *m*
F condenseur *m* à
    contact direct
I condensatore *m*
    barometrico
MA légküri kondenzátor
NL atmosferische
    stoom-
    condensor *m*
PO skraplacz *m*
    barometryczny
PY контактный
    конденсатор *m*,
    работающий при
    атмосферном
    давлении
SU vesilauhdutin
SV luftkondensator *c*

DA	=	Danish
DE	=	German
ES	=	Spanish
F	=	French
I	=	Italian
MA	=	Magyar
		(Hungarian)
NL	=	Dutch
PO	=	Polish
PY	=	Russian
SU	=	Finnish
SV	=	Swedish

370 barometric damper
DA lufttrykreguleret
   spjæld
DE barometrische
   Klappe *f*
   Luftklappe *f*
ES compuerta
   atmosférica *f*
   regulador de tiro *m*
F régulateur *m* de
   tirage
I valvola *f*
   barometrica
MA huzatszabályozó
   csappantyú
NL trekregelaar *m*
PO przepustnica *f*
   ciśnieniowa
PY атмосферная
   заслонка *f*
SU alipaineläppä
   vedonsäädin
SV lufttrycksreglerat
   spjäll *n*

371 barometric effect
DA barometervirkning
DE barometrischer
   Effekt *m*
ES efecto barométrico *m*
F effet *m* baro-
   métrique
I effetto *m*
   barometrico
MA légküri hatás
NL atmosferische
   drukverandering *f*
PO efekt *m*
   barometryczny
PY изменение *n*
   барометричес-
   кого давления
SU ilmanpaineen
   vaihtelun vaikutus
SV barometereffekt *c*

**372 barometric
   pressure**
DA barometrisk tryk
DE Luftdruck *m*
ES presión
   barométrica *f*
F pression *f*
   barométrique
I pressione *f*
   barometrica
MA légküri nyomás
NL barometrische
   druk *m*
PO ciśnienie *n*
   barometryczne
PY барометрическое
   давление *n*
SU ilmanpaine
SV barometertryck *n*

373 barrier
DA spærring
DE Schranke *f*
   Sperre *f*
ES barrera *f*
F barrière *f*
I barriera *f*
MA gát
   torló
NL afsluiting *f*
   barrière *f*
PO próg *m*
   przegroda *f*
PY изолятор *m*
   пароизоляционный
   слой *m*
SU sulku
SV spärr *c*

374 BAS (*see* building
   automation system)

375 base
DA base
   fundament
DE Basis *f*
   Fundament *n*
   Sockel *m*
ES base *f*
   cimiento *m*
   fundación *f*
   plancha de fondo *f*
F base *f*
   socle *m*
I base *f*
   zoccolo *m*
MA alap
   lábazat
NL basis *f*
   fundatie *f*
   grondvlak *n*
PO baza *f*
   podłoże *n*
   podstawa *f*
PY основание *n*
SU perus-
SV bas *c*

376 baseboard
DA fodpanel
DE Fußleiste *f*
   Fußleistenkonvektor
   m (USA)
ES tablero de base
   (montajes) *m*
F plinthe *f* chauffante
I zoccolo *m*
MA szegély
NL plint *n*
PO grzejnik *m* listwowy
   listwa *f*
   przypodłogowa
PY плинтусный
   нагревательный
   прибор *m*
SV golvpanel *c*

377 baseboard heater
DA fodpanelradiator
DE Fußleistenheizer *m*
ES calefactor de
   zócalo *m*
F plinthe *f* chauffante
I riscaldatore *m*
   perimetrale
MA szegélyfűtő
NL plintverwarmings-
   toestel *n*
PO grzejnik *m* listwowy
PY плинтусный
   нагреватель *m*
SV panelvärmare *c*

378 baseboard radiator
DA fodpanelradiator
DE Fußleisten-
    radiator *m*
ES radiador de zócalo *m*
F plinthe *f* chauffante
I radiatore *m* peri-
    metrale
MA szegélyfűtőtest
NL plint radiator *m*
PO grzejnik *m* listwowy
PY плинтусный
    радиатор *m*
SU listalämmitin
SV panelradiator *c*

**379 baseload
    generation**
DA grundlast generering
DE Grundlaststrom-
    erzeugung *f*
ES generación de base *f*
F production *f* de base
I generazione *f*
    elettrica di base
MA alapterheléses
    gerjesztés
NL grondlast-
    opwekking *f*
PO wytwarzanie *n*
    obciążenia
    podstawowego
PY расчетная
    электрическая
    нагрузка *f*
SU perustehontuotto
SV baslastalstring *c*

380 basement
DA kælderetage
DE Erdgeschoß *n*
    Kellergeschoß *n*
ES sótano *m*
F cave *f*
    sous-sol *m*
I scantinato *m*
MA alagsor
    alapozás
NL kelder *m*
    souterrain *n*
PO podziemie *n*
PY подвал *m*
SU pohjakerros
SV källarvåning *c*

381 base plate
DA sokkelplade
DE Grundplatte *f*
ES bancada *f*
    placa de base *f*
F plaque *f* d'assise
    socle *m*
I basamento *m*
MA alaplemez
NL fundatieplaat *f*
    voetplaat *f*
PO płyta *f* denna
    płyta *f*
    fundamentowa
PY фундаментная
    плита *f*
SU alusta
    perusta
SV bottenplatta *c*

382 bath
DA bad
DE Bad *n*
    Badewanne *f*
ES baño *m*
F bain *m*
I bagno *m*
MA fürdő
NL bad *n*
PO kąpiel *f*
    łaźnia *f*
    wanna *f*
PY ванна *f*
SU kylpy
SV bad *n*

383 bathroom
DA badeværelse
DE Badezimmer *n*
ES cuarto de baño *m*
F salle *f* de bains
I stanza *f* da bagno
MA fürdőszoba
NL badkamer *f*
PO łazienka *f*
PY ванная комната *f*
SU kylpyhuone
SV badrum *n*

384 batten
DA brødt
    planke
DE Verschalung *f*
ES eje *m*
    tabla *f*
F latte *f*
I asse *m*
    tavola *f*
MA szegélyléc
NL badding *m*
    tengel *m*
    vloerbint *n*
PO listwa *f*
    łata *f*
PY доска *f*
    рейка *f*
SU lista
SV fäste *n*

385 battery
DA batteri
DE Batterie *f*
ES batería *f*
F batterie *f*
I batteria *f*
MA akkumulátor
    gépcsoport
NL accu *m*
    batterij *f*
PO akumulator *m*
    bateria *f*
PY батарея *f*
SU patteri
SV batteri *n*

386 BCS (*see* building
    control system)

387 beam (building)
DA bjælke
DE Balken *m*
ES viga *f*
F encastrer
I trave *f*
MA gerenda
NL balk *m*
PO belka *f*
    (konstrukcyjna)
PY балка *f* (в
    строительстве)
SU palkki
SV bjälke *c*

388 beam (light)
DA stråle
DE Strahl *m*
ES rayo de luz *m*
F rayon *m* de lumière, de soleil
I raggio *m*
MA fénysugár
NL lichtstraal *m*
PO promień *m* (światła)
PY луч *m* (света)
SU säde
SV stråle *c*

389 beam irradiance
DA lysstråling
DE gerichtete Ausstrahlung *f*
ES radiación directa *f*
F rayonnement *m*
I irradianza *f* diretta
MA sugárnyaláb
NL straling *f* met loodrechte invalshoek
PO strumień *m* natężenia napromieniowania
PY рассеянная радиация *f*
SU suora säteily
SV utstrålning *c* (från ljusstråle)

390 beam valve (*see* reed valve)

391 bearing
DA leje
DE Lager *n*
ES cojinete *m* rodamiento *m*
F palier *m*
I cuscinetto *m*
MA csapágyazás
NL draagvlak *n* kussenblok *n* lager *n*
PO łożysko *n* podparcie *n* podpora *f*
PY опора *f* подшипник *m*
SU laakeri
SV lager *n*

392 bearing support
DA lejeblok
DE Lagerbock *m*
ES soporte de cojinete *m*
F palier *m*
I supporto *m* per cuscinetto
MA csapágyazott támasz
NL lagersteun *m* lagerstoel *m*
PO podpora *f*
PY опора *f* подшипника
SU laakeripesä
SV lagerstöd *n*

393 Beaufort scale
DA Beauforts skale vindstyrkeskala
DE Beaufort-Skala *f*
ES escala de Beaufort *f*
F échelle *f* de Beaufort
I scala *f* di Beaufort
MA Beaufort skála
NL Beaufort schaal *f* schaal *f* van Beaufort
PO skala *f* Beauforta
PY шкала *f* Бофорта
SU Beaufort asteikko
SV Beaufort-skala *c*

394 bed
DA seng
DE Bett *n*
ES cama *f* lecho *m* plancha de montaje *f*
F lit *m*
I letto *m*
MA ágy ágyazat
NL bedding *f* grondslag *m*
PO łoże *n* podłoże *n* warstwa *f* złoże *n*
PY основание *n* фундамент *m*
SU peti
SV bädd *c*

395 bed plate
DA fundamentsplade
DE Grundplatte *f*
ES placa de fondo *f* placa de fundación
F plaque *f* d'assise socle *m*
I piastra *f* a fondazione
MA alaplemez
NL fundatieplaat *f*
PO płyta *f* fundamentowa
PY фундаментная плита *f*
SU alusta perusta
SV bottenplatta *c*

396 bellows
DA blæsebælg bælg
DE Blasebalg *m* Gebläse *n*
ES fuelle *m*
F soufflet *m*
I soffietto *m*
MA hullámos fémtümlő
NL balg *m*
PO harmonijka *f* miech *m* mieszek *m*
PY сильфон *m*
SU palje
SV bälg *c*

397 bellows seal
DA bælgtætning
DE Balgendichtung *f*
ES empaquetadora de membrana *f*
F garniture *f* à soufflet
I tenuta *f* a soffietto
MA membrános tümítés
NL balgafdichting *f*
PO uszczelnienie *n* mieszkowe
PY сильфонное уплотнение *n*
SU tiivistyspalje
SV kompensatortätning *c*

398 bellows valve
DA bælgventil
 kompensator
DE Balgenventil *n*
ES válvula de
 membrana *f*
F vanne *f* à soufflet
I valvola *f* a soffietto
MA membrános szelep
NL balgafsluiter *m*
PO zawór *m* z dławicą
 mieszkową
PY сильфонный
 вентиль *m*
 (клапан)
SU paljeventtiili
SV kompensator *c*

399 belt
DA rem
DE Riemen *m*
ES correa *f*
 hilada saliente
 (arquitectura) *f*
F courroie *f*
I cinghia *f*
MA szíj
NL riem *m*
 transportband *m*
 V-snaar *f*
PO pas *m*
 taśma *f*
PY ремень *m*
SU hihna
SV rem *c*

400 belt drive
DA remtræk
DE Riementrieb *m*
ES accionamiento por
 correa *m*
F commande *f* par
 courroie
I trasmissione *f* a
 cinghia
MA szíjhajtás
NL riemaandrijving *f*
 snaaraandrijving *f*
PO napęd *m* pasowy
PY ременный
 привод *m*
SU hihnakäyttü
SV remdrift *c*

401 BEMCS (*see* building
 energy management
 control system)
ES

402 BEMS (*see* building
 energy management
 system)

403 bend
DA bøje
 forkrøbning
DE Bogen *m*
ES curva *f*
 inflexión *f*
F coude *m*
I curva *f*
 gomito *m*
MA ív
 künyük
NL bocht *f*
 bochtstuk *n*
PO kolano *n*
 łuk *m*
PY дуга *f*
SU mutka
SV böj *c*

404 bent pipe
DA knærør
 rørbøjning
DE Bogenrohr *n*
 Schenkelrohr *n*
ES tubo acodado *m*
 tubo curvado *m*
F tube *m* coudé
I tubo *m* a gomito
MA ívcső
 künyükcső
NL gebogen pijp *f*
PO kolano *n* rurowe
PY отвод *m*
 поворот *m* трубы
SU putkimutka
SV knärör *n*

405 bias
DA tendens til afvigelse
 fra korrekt værdi
DE (Gitter-)
 Vorspannung *f*
ES sesgo *m*
F déviation *f*
I bias *m*
MA torzítás
NL afwijking *f*
PO błąd *m*
 systematyczny
 napięcie *n* wstępne
 odchylenie *n*
 przesunięcie *n*
 punktu pracy
 skos *m*
PY отклонение *n*
 уклон *m*
SU systemaattinen virhe
SV avvikelse *c* från
 börvärde *n*

406 bifurcated fan
DA tvedelt ventilator
DE doppelseitiger
 Ventilator *m*
ES ventilador
 bifurcador *m*
F ventilateur *m*
 bifurqué
I ventilatore *m* a
 biforcazione
MA axiálventilátor a
 légáramból
 kiemelt motorral
NL axiaalventilator *m*
 met bifurcatie
PO wentylator *m*
 równolegle
 pracujący
 (bliźniaczy)
PY вентилятор *m* с
 двигателем на
 одной оси
 осевой вентилятор
 m с двигателем
 вне потока
SU kaksiaukkoinen
 puhallin
SV fläkt *c* med två
 utlopp *n*

407 bifurcation
DA afgrening
 tvedeling
DE Gabelung *f*
ES bifurcación *f*
F bifurcation *f*
I biforcazione *f*
MA elágazás
NL bifurcatie *f*
 dubbele vork *m*
PO rozgałęzienie *n*
 rozwidlenie *n*
PY разветвление *n*
SU haara
SV bifurkation *c*

408 bimetal
DA bimetal
DE Bimetall *n*
ES aleación bimetálica *f*
 bimetal *m*
F bimétal *m*
I bimetallo *m*
MA bimetál
 kettősfém
NL bi-metaal *n*
PO bimetal *m*
PY биметалл *m*
SU bi-metalli
SV bimetall *c*

409 bimetallic element
DA bimetalelement
DE Bimetall-Element n
ES bilámina f
  elemento
  bimetálico m
F bilame m
I elemento m
  bimetallico
MA bimetál elem
NL bimetaalelement n
  bimetaalstrip m
PO element m
  bimetaliczny
PY биметаллический
  элемент m
SU bi-metallielementti
SV bimetallelement n

410 bimetallic thermometer
DA bimetaltermometer
DE Bimetall-
  Thermometer n
ES termómetro
  bimetálico m
F thermomètre m
  bimétallique
I termometro m
  bimetallico
MA bimetálos hőmérő
NL bimetaal-
  thermometer m
PO termometr m
  bimetaliczny
PY биметаллический
  термометр m
SU bimetallil-
  äm pümittari
SV bimetall-
  termometer c

411 binary adj
DA binær
DE binär
  zweigliedrig
ES binario
F binaire
I binario
MA kettős
NL binair
  tweetallig
PO binarny
  dwójkowy
  dwuskładnikowy
  podwójny
PY бинарный
SU binääri
SV binär

412 binary vapor cycle
DA binær dampcyklus
DE Zweistoff-Dampf-
  Kreislauf m
ES ciclo de vapor
  binario m
F cycle m de vapeur
  binaire
I ciclo m a vapore
  binario
MA kettős gőz
  kürfolyamat
NL binaire
  koelkringloop m
PO obieg m parowy
  dwuczynnikowy
PY бинарный паровой
  цикл m
SU binääriseosprosessi
SV binär ångcykel c

413 bit
DA
DE Bit n
ES bit m
F fragment m
  morceau m
  pièce f
I bit m
MA véső
NL bit n
PO bit m
PY бит m
SU terä
SV bit c

414 bitumen paper
DA asfaltpap
  tagpap
DE Bitumenpapier n
ES papel bituminoso m
F papier m bitumé
I carta f bitumata
MA kátránypapír
NL asphaltpapier n
  bitumenpapier n
PO papa f
  papier m bitumiczny
PY битумная бумага f
  пергамин m
SU bitumipaperi
SV bitumenpapper n

415 bituminous coal
DA bituminøst kul
DE Pechkohle f
ES carbón
  bituminoso m
F charbon m
  bitumeux
I carbone m
  bituminoso
MA bitumenes szén
NL bitumineuze kool f
  vetkool f
PO węgiel m bitumiczny
PY уголь m с
  большим
  содержанием
  смолы
SU kaasuhiili
SV bituminüst kol n

416 bituminous paint
DA bituminøs maling
DE Bitumenanstrich m
ES pintura bituminosa f
F enduit m bitumeux
  peinture f bitumeuse
I vernice f bituminosa
MA bitumenes festés
NL bitumineuze verf f
PO farba f asfaltowa
  farba f bitumiczna
PY битумная
  покраска f
SU bitumimaali
SV bitumenfärg c

417 black body
DA sort legeme
DE schwarzer Körper m
ES cuerpo negro m
F corps m noir
I corpo m nero
MA fekete test
NL zwart lichaam n
PO ciało n doskonale
  czarne
PY абсолютно черное
  тело n
SU musta kappale
SV svart kropp c

DA	=	Danish
DE	=	German
ES	=	Spanish
F	=	French
I	=	Italian
MA	=	Magyar (Hungarian)
NL	=	Dutch
PO	=	Polish
PY	=	Russian
SU	=	Finnish
SV	=	Swedish

418 blackbody equivalent temperature
DA sort legemes ekvivalente temperatur
DE Schwarzkörper-äquivalent-temperatur *f*
ES temperatura equivalente del cuerpo negro *f*
F température *f* équivalente de corps noir
I temperatura *f* equivalente del corpo nero
MA fekete test egyenértékű hőmérséklet
NL zwartebol temperatuur *f*
PO temperatura *f* równoważna ciała doskonale czarnego
PY эквивалент *m* температуры абсолютно черного тела
SU mustaa kappaletta vastaava lämpütila
SV ekvivalent temperatur *c* med svart kropp *c*

419 blackness test
DA sværtningsprøve
DE Schwärzung *f* (nach Ringelmann-Skala)
ES prueba de opacidad de Ringelmann *f*
F essai *m* d'opacité
I prova *f* di Ringelmann
MA feketeségi próba
NL Ringelmanntest *m* zwartingsproef *f*
PO próba *f* czarności
PY испытание *n* на прозрачность (мутность)
SU nokilukukoe
SV svärtningsprov *n*

420 blade
DA propelvinge skovl
DE Schaufel *f*
ES álabe *m* pala *f*
F aube *f*
I pala *f*
MA lapát szárny
NL blad *n* schoep *f*
PO łopatka *f* wirnika ostrze *n* płaskownik *m*
PY лопатка *f*
SU siipi
SV skovel *c*

421 blanket type thermal insulation
DA måtteisolerings-materiale
DE deckenartige Wärme-dämmung *f*
ES aislamiento térmico de mantas *m*
F isolant *m* thermique en matelas
I isolamento *m* termico a cappotto
MA paplan hőszigetelés
NL dekenisolatie *f*
PO izolacja *f* cieplna płaszczowa
PY рулонная теплоизоляция *f* теплоизоляция *f* из гибких матов
SU lämpüeristysmatto
SV filtisolering *c*

422 blank flange
DA flange uden boltehuller
DE Blindflansch *m*
ES brida ciega *f*
F bride *f* non percée
I flangia *f* non forata
MA vakperem
NL ongeboorde flens *m*
PO zaślepka *f* kołnierzowa
PY глухой фланец *m* фланец *m* без отверстия
SU laippa ilman pultin reikiä
SV blindfläns *c*

423 blast coil
DA rørslynge for varmeoverføring
DE Gebläsewärme-tauscher *m*
ES serpentín de aire forzado *m*
F batterie *f* à air pulsé
I scambiatore *m* di calore ad aria a convezione forzata
MA léghevítő
NL luchtwarmte-wisselaar *m*
PO wężownica *f* z wymuszonym przepływem
PY омываемая поверхность *f* змеевика
SU rivoitettu pinta
SV utblåsningsrür *n*

424 blast cooler
DA lynkøler
DE Gebläsekühler *m*
ES enfriador por aire forzado *m*
F refroidisseur *m* à air pulsé
I raffreddatore *m* per espansione
MA léghűtő
NL luchtkoeler *m*
PO chłodnica *f* z wymuszonym przepływem
PY вентиляторный охладитель *m*
SU ilmasuihkujäähdytin
SV snabbkylare *c*

425 blast cooling (*see* air blast cooling)

426 blast freezer
DA lynfryser
DE Gebläsegefrier-
   anlage *f*
   Gebläsegefrier-
   gerät *n*
ES congelador por aire
   forzado *m*
F congélateur *m* à air
   pulsé
I congelatore *m* per
   espansione
MA befúvásos fagyasztó
NL snelvriesruimte *f*
PO chłodnia *f* z
   wymuszonym
   przepływem
PY вентиляторный
   замораживатель
   m
SU ilmasuihkupakastin
SV snabbfrys *c*

427 blast freezing
DA lynfrysning
DE Gebläsegefrieren *n*
ES congelación por aire
   forzado *f*
F congélation *f* par air
   pulsé
I congelamento *m* per
   espansione
MA befúvásos fagyasztás
NL snelvriezen
PO zamrażanie *n*
   wymuszone
PY вентиляторное
   замораживание
   n
SU ilmasuihkupakastus
SV snabbfrysning *c*

428 blast gate damper
DA afblæsningsspjæld
DE Gebläseauslaß-
   Absperrung *f*
ES compuerta de
   escape *f*
F registre *m* à glissire
I serranda *f* di
   regolazione aria
MA fojtócsappantyú
NL schuif *f* in
   luchtkanaal
PO przepustnica *f*
   odcinająca
   przeciwwy-
   buchowa
PY направляющий
   клапан *m*
SU liukupelti
SV utblåsningsspjäll *n*

429 blast heater
DA blæselampe
DE Gebläseerhitzer *m*
ES calentador soplante
   m
F batterie *f* à air
   chaud
I riscaldatore *m* di
   aria a tubo
   alettato
MA hőlégfúvó
NL luchtverhitter *m*
PO ogrzewacz *m* z
   wymuszonym
   przepływem
PY вентиляторный
   нагреватель *m*
SU lämminilmakone
SV värmebatteri *n*

430 bleed *vb*
DA udlufte
DE entnehmen
ES purgar
   sangrar
   soplar
F purger
   soutirer
I spurgare
MA leereszt
   leürít
NL aftappen
   spuien
PO opróżniać
   przeciekać
   spuszczać
PY отбирать
SU vuotaa
SV lufta

431 bleeder
DA afluftningsventil
DE Entnahmestelle *f*
ES sangrador *m*
   purgador *m*
F canalisation *f* de
   soutirage
I tubazione *f* di
   spillamento
MA leeresztő
   leürítő
NL aftapper *m*
PO spust *m*
   upust *m*
PY конденсатопровод
   m
   трубопровод *m*
   для отвода
   конденсата
SU ilmaus
   ohivirtaus
   ohivuoto
SV luftning *c*

**432 bleeder type
   condenser**
DA aftapnings-
   kondensator
DE Entnahme-
   kondensator *m*
ES condensador de
   extracción *m*
   grifo de purga de
   condensador *m*
F condenseur *m* à
   soutirage
I condensatore *m* a
   pioggia
MA megcsapolásos
   kondenzátor
NL aftapcondensor *m*
PO skraplacz *m*
   zraszany
PY оросительный
   конденсатор *m*
SU lauhdutin jossa pieni
   nestevirtaus
SV avtappnings-
   kondensor *c*

433 bleed off *vb*
DA udblæse
DE entnehmen
ES purgar
F soutirager
I prelevare
MA ürít
NL afblazen
   aftappen
PO spuszczać
   upuszczać
PY отбирать
SU
SV avtappa

434 bleed pipe
DA afblæsningsrør
   udtagsrør
DE Entnahmeleitung *f*
ES tubo de
   extracción *m*
   tubo de purga *m*
F purge *f* d'air
   tuyau *m* de
   soutirage
I tubo *m* di spurgo
MA ürítőcső
NL aftapleiding *f*
   spuileiding *f*
PO rura *f* spustowa
PY трубопровод *m*
   для отбора
SU juoksutusputki
SV luftningsledning *c*

435 bleed valve
DA aftapningsventil
DE Entnahmeventil *n*
ES válvula de purga *f*
F vanne *f* de soutirage
I valvola *f* di sfiato
MA ürítőszelep
NL aftapkraan *f*
PO zawór *m* spustowy
PY отводной клапан *m*
SU vuotoventtiili
SV avtappningsventil *c*

436 blender
DA blander
blanding
DE Mischer *m*
ES mezclador *m*
F mélangeur *m*
mitigeur *m*
I miscelatore *m*
MA keverő
NL mengtoestel *n*
PO mieszacz *m* z
automatycznym
dozowaniem
PY смесительный
насос *m*
SU sekoitin
SV blandare *c*

437 blending box
DA blandeboks
DE Mischbox *f*
Mischkammer *f*
ES cámara de mezcla *f*
F boîte *f* de mélange
I camera *f* di miscela
MA keverőedény
NL mengbox *m*
mengkast *f*
PO komora *f* mieszająca
PY смесительная
камера *f*
SU sekoituslaatikko
SV blandningsbox *c*

438 blind flange
DA blindflange
DE Blindflausen *m*
ES brida ciega *f*
obturador *m*
F bride *f* pleine
I flangia *f* cieca
MA vakperem
NL blindflens *m*
PO zaślepka *f*
kołnierzowa
PY заглушка *f*
SU umpilaippa
SV blindfläns *c*

439 block *vb*
DA blokere
DE blockieren
sperren
ES bloquear
F bloquer
I bloccare
MA eltorlaszol
lezár
NL afsluiten
blokkeren
PO blokować
zamknąć
PY блокировать
SU tukkia
SV blockera

440 blockage
DA blokering
DE Blockierung *f*
Sperrung *f*
ES bloqueo *m*
F blocage *m*
I bloccaggio *m*
MA lezárás
NL blokkering *f*
verstopping *f*
PO blokada *f*
zablokowanie *n*
PY блокировка *f*
SU tukos
SV blockering *c*

441 block heating
DA blokopvarmning
DE Blockheizung *f*
ES calefacción para
un grupo de
viviendas *f*
F chauffage *m* d'Mlot
I riscaldamento *m*
isolato
MA tümbfütés
NL blokverwarming *f*
wijkverwarming *f*
PO ogrzewanie *n*
systemu
zamkniętego
PY блочное
отопление *n*
квартирное
отопление *n*
SU aluelämmitys
keskuslämmitys
SV central-
uppvärmning *c*

**442 block-type thermal
insulation**
DA varmeisoleringsblok
DE blockartige
Wärme-
dämmung *f*
ES aislamiento térmico
de bloques *m*
F isolant *m* en
panneau rigide
I isolamento *m*
termico in blocchi
MA tümbüs hőszigetelés
NL blokvormige
isolatie *f*
PO izolacja *f* cieplna
obudowująca
PY теплоизоля-
ционный блок *m*
SU lämpüeristys-
elementti
SV värmeisolerings-
block *n*

443 block valve
DA blokeringsventil
stopventil
DE Sperrventil *n*
ES vávula de bloqueo *f*
F vanne *f* de blocage
I valvola *f* di blocco
MA zárószelep
NL afsluiter *m*
PO zawór *m* odcinający
PY сблокированный
вентиль *m*
(клапан)
SU liitäntäventtiili
SV blockventil *c*

444 blow *vb*
DA blæse
puste
DE blasen
ES insuflar
soplar
F souffler
I soffiare
MA fúj
NL blazen
spuien
waaien
PO dmuchać
wydmuchiwać
PY продувать
SU puhaltaa
SV blåsa

445 blowdown
DA bundblæse
DE ablassen
entwässern
ES descarga *f*
F purge *f* de
déconcentration
I spurgo *m*
MA lefúvatás
NL afblazen *n*
spuien *n*
PO wydmuch *m*
PY продувка *f*
SU tyhjennys
SV utsläpp *n*

446 blowdown valve
DA aftapningsventil
DE Ablaßventil *n*
ES válvula de descarga *f*
F robinet *m* de
décharge
robinet *m* de purge
I valvola *f* di scarico
MA lefúvató szelep
NL aftapkraan *f*
spuikraan *f*
PO zawór *m* spustowy
(kotłowy)
zawór *m*
wydmuchowy
PY продувочный
клапан *m*
SU tyhjennysventtiili
SV bottenblåsnings-
ventil *c*

447 blower
DA blæser
ventilator
DE Ventilator *m*
ES aventador *m*
fuelle *m*
ventilador
impelente *m*
F ventilateur *m*
I soffiante *f*
MA fúvó
NL ventilator *m*
PO dmuchawa *f*
wentylator *m*
PY воздуходувка *f*
SU puhallin
SV fläkt *c*

448 blower fan
DA blæseventilator
DE Gebläseventilator *m*
ES ventilador soplante
m
F ventilateur *m* de
soufflage
I soffiante *f*
MA fúvóventilátor
NL ventilator *m*
PO wentylator *m*
PY рабочее колесо *n*
вентилятора
SU potkuripuhallin
SV fläkt *c*

449 blowing
DA udblæsning
DE Aufblasen *n*
Blasen *n*
ES soplado *m*
F soufflage *m*
I soffiaggio *m*
MA fúvás
NL blazen *n*
PO dmuchanie *n*
wydmuchiwanie *n*
PY дутье *n*
SU puhallus
SV utblåsning *c*

450 blowoff valve
DA udblæsningsventil
DE Ablaßventil *n*
ES válvula de vaciado *f*
F vanne *f* de vidange
I valvola *f* di sfiato
MA kifúvó szelep
NL afblaaskraan *f*
PO zawór *m* spustowy
(kotłowy)
zawór *m*
wydmuchowy
PY предохранительный
клапан *m*
SU ylivirtausventtiili
SV utblåsningsventil *c*

451 BMS (*see* building
management system)

**452 board-type thermal
insulation**
DA pladevarmeisolering
DE tafelartige
Wärme-
dämmung *f*
ES aislamiento térmico
de paneles *m*
F isolation *f*
thermique plane
I isolamento *m*
termico in pannelli
rigidi
MA táblás hőszigetelés
NL plaatisolatie *f*
PO izolacja *f* cieplna
płytowa
PY теплоизоляционная
плита *f*
SU lämpüeristyslevy
SV värmeisolering *c* av
boardtyp *c*

453 body
DA karosseri
krop
DE Kürper *m*
ES cuerpo *m*
F corps *m*
I corpo *m*
MA test
NL lichaam *n*
motorblok *n*
PO bryła *f*
ciało *n*
kadłub *m*
korpus *m*
PY корпус *m*
SU runko
SV kropp *c*

454 boil *vb*
DA koge
DE kochen
sieden
ES hervir
F bouillir
I bollire
MA forral
NL koken
PO gotować
wrzeć
PY кипеть
SU kiehua
SV koka

455 boiler
DA kedel
DE Kessel *m*
   Wärmeerzeuger *m*
ES caldera *f*
F chaudière *f*
I caldaia *f*
MA kazán
NL ketel *m*
PO kocioł *m*
PY бойлер *m*
   котел *m*
SU kattila
SV panna *c*

456 boiler–burner unit
DA kedelbrænderunit
DE Kessel-Brenner-
   Einheit *f*
ES conjunto caldera-
   quemador *m*
F ensemble *m* brûleur-
   chaudière
I unità *f* monoblocco
   caldaia -
   bruciatore
MA kazánégő egység
NL ketel-brander
   eenheid *f*
PO zespół *m* palnikowy
   kotła
PY котельная
   установка *f* c
   устройством
   подачи топлива
SU kattilapoltinyksikkü
SV pannbrännare

457 boiler capacity
DA kedelydelse
DE Kesselleistung *f*
ES potencia de la
   caldera *f*
F puissance *f* de
   chaudière
I potenzialità *f* della
   caldaia
MA kazánteljesítmény
NL ketelcapaciteit *f*
PO wydajność *f* kotła
PY теплопроизводитель-
   ность *f* котла
SU kattilateho
SV pannkapacitet *c*

458 boiler casing
DA kedelbeklædning
DE Kesselmantel *m*
ES camisa de caldera *f*
   envolvente de
   caldera *f*
   revestimiento de
   caldera *m*
F casing *m*
   jaquette *f*
I mantello *m* della
   caldaia
MA kazánburkolat
NL ketelmantel *m*
PO obudowa *f* kotła
   płaszcz *m* kotła
PY обмуровка *f*
SU kattilan vaippa
SV pannmantel *c*

459 boiler cleaning
DA kedelrensning
DE Kesselreinigung *f*
ES limpieza de caldera *f*
F nettoyage *m* de
   chaudière
I pulitura *f* della
   caldaia
MA kazántisztítás
NL ketelreiniging *f*
PO czyszczenie *n* kotła
PY чистка *f* котла
SU kattilan puhdistus
SV pannrengöring *c*

460 boiler efficiency
DA kedelvirkningsgrad
DE Kesselwirkungs-
   grad *m*
ES rendimiento de
   caldera *m*
F rendement *m* de
   chaudière
I rendimento *m* della
   caldaia
MA kazánhatásfok
NL ketelrendement *n*
PO sprawność *f* kotła
PY производитель-
   ность *f* котла
SU kattilan hyütysuhde
SV pannverknings-
   grad *c*

461 boiler end plate
DA kedelendebund
DE Kesselrückwand *f*
ES placa de fondo *f*
   placa tubular
   (calderas) *f*
F plaque *f* arrière de
   chaudière
   plaque *f* de fond de
   chaudière
I piastra *f* terminale
   della caldaia
MA kazán véglemez
NL ketelfrontplaat *f*
PO dennica *f* kotła
   dno *n* kotła
PY передняя секция *f*
   котла
SU kattilan päätylevy
SV panngavelplåt *c*

462 boiler feed
DA kedelfødning
   vandpåfyldning
DE Kesselspeisung *f*
ES alimentación de la
   caldera *f*
F alimentation *f* de
   chaudière
I alimentazione *f*
   della caldaia
MA kazán táplálás
NL ketelvoeding *f*
PO zasilenie *n* kotła
PY питание *n* котла
SU kattilan syüttü
SV pannmatning *c*

463 boiler feed pump
DA kedelfødepumpe
DE Kesselspeisepumpe *f*
ES bomba de
   alimentación de la
   caldera *f*
F pompe *f*
   d'alimentation de
   chaudière
I pompa *f* di
   alimentazione
   della caldaia
MA kazán tápszivattyú
NL ketelvoedings-
   pomp *f*
PO pompa *f* zasilająca
   kocioł
PY питательный насос
   m котла
SU kattilan
   syüttüvesipumppu
SV pannmatarpump *c*

464 boiler feedwater
DA kedelfødevand
DE Kesselspeisewasser n
ES agua de alimentación de la caldera f
F eau f d'alimentation de chaudière
I acqua f di alimentazione della caldaia
MA kazán tápvíz
NL ketelvoedings-water n
PO woda f zasilająca kocioł
PY питающая котел вода f
SU kattilan syüttüvesi
SV pannmatarvatten n

465 boiler feedwater heater
DA kedelfødevands-forvarmer
DE Kesselspeisewasser-erhitzer m
ES recalentador del agua de alimentación de la caldera m
F réchauffeur m d'eau d'alimentation de chaudière
I riscaldatore m dell'acqua di alimentazione della caldaia
MA kazán tápvíz melegítő
NL ketelvoedingswater-verwarmer m
PO podgrzewacz m wody zasilającej kocioł
PY нагреватель m подпиточной воды
SU kattilan syüttüveden-lämmitin
SV värmare c dör pannmatar-vatten n

466 boiler foaming
DA kedeloverkogning
DE Schäumen n des Kessels m
ES espumaje en la caldera m
F primage m de chaudière
I formazione f di schiuma in un generatore di vapore
MA kazán habzás
NL schuimen van ketelwater n
PO burzenie n się wody w kotle
PY грязевик m
SU kattilakuonan kulkeutuminen menojohtoon kattilan vaahtoaminen
SV überkokning c

467 boiler generator
DA dampgenerator
DE Heißwasser-erzeuger m
ES generador de vapor m
F chaudière f
I generatore m di vapore
MA kazángenerátor
NL generator m van absorptiekoel-machine
PO wytwornica f pary
PY генератор m абсорбционной холодильной машины
SU keitin
SV panngenerator c

468 boiler heating surface
DA kedelhedeflade
DE Kesselheizfläche f
ES superficie de calefacción de la caldera f
F surface f de chauffe de chaudière
I superficie f riscaldante della caldaia
MA kazán fűtőfelület
NL verwarmend oppervlak n van een ketel
PO powierzchnia f ogrzewalna kotła
PY поверхность f нагрева котла
SU kattilan lämpüpinta
SV panneldyta c

469 boiler horsepower
DA kedelydelse
DE Kesselleistung f (in HP)
ES potencia de caldera f
F puissance f de chaudière
I unità f di misura pari a 9809.5 W
MA kazán teljesítmény
NL ketelvermogen n
PO moc f kotła (w koniach mechanicznych)
PY тепловая лошадиная сила f
SU kattilan teho hevosvoimissa
SV panneffekt c

470 boiler house
DA kedelhus
DE Kesselhaus n
ES edificio de calderas m sala de calderas f
F chaufferie f
I locale m caldaia
MA kazánház
NL ketelhuis n
PO kotłownia f
PY котельная f
SU kattilahuone
SV pannrum n

471 boiler jacket (*see* boiler casing)

DA	=	Danish
DE	=	German
ES	=	Spanish
F	=	French
I	=	Italian
MA	=	Magyar (Hungarian)
NL	=	Dutch
PO	=	Polish
PY	=	Russian
SU	=	Finnish
SV	=	Swedish

472 boiler mounting
DA kedelfundament
DE Kesselaufstellung *f*
ES accesorio de
    caldera *m*
    montaje de la
    caldera *m*
F socle *m* de chaudière
I basamento *m* della
    caldaia
MA kazán szerelés
NL ketelfundatie *f*
    ketelmontage *f*
PO mocowanie *n* kotła
    montaż *m* kotła
PY монтаж *m* котла
SU kattilan alusta
SV pannfundament *n*

473 boiler plant
DA kedelanlæg
DE Kesselanlage *f*
ES instalación de
    calderas *f*
    taller (fábrica) de
    calderas *m*
F installation *f* de
    chaufferie
I centrale *f* termica
MA kazántelep
NL ketelhuis *n*
    ketelinstallatie *f*
PO kotłownia *f*
PY котельная
    установка *f*
SU kattilalaitteisto
SV pannanläggning *c*

474 boiler plate
DA kedelplade
DE Kesselplatte *f*
ES chapa para
    calderas *f*
    plancha de caldera *f*
F plaque *f* de
    chaudière
I piastra *f* di caldaia
MA kazánlemez
NL ketelplaat *f*
PO blacha *f* kotłowa
    płyta *f* kotła
PY плита *f* котловая
SU kattilalevy
SV pannplåt *c*

475 boiler priming
DA kedeloverkogning
DE Spucken *n* des
    Kessels *m*
ES chisporroteo en la
    caldera *m*
F primage *m* de
    chaudière
I trasporto *m*
    eccessivo di gocce
    di acqua in un
    generatore
MA kazán táplálás
NL meesleuren van
    water *n* met
    stoom
PO porywanie *n* wody
    przez parę w kotle
PY гидравлический
    затвор *m* котла
SU kattilan täyttü
    vesipisaroiden pääsy
    hüyryyn
SV pannkokning *c*

476 boiler rating
DA kedelydelse
DE Kesselleistung *f*
ES potencia de calderas
    f
F puissance *f* de
    chaudière
I potenzialità *f* della
    caldaia
MA kazán méretezés
NL ketelcapaciteit *f*
PO obciążenie *n* kotła
    wydajność *f* znamio-
    nowa kotła
PY мощность *f* котла
SU kattilan nimellisteho
SV panneffekt *c*

477 boiler room
DA kedelrum
DE Heizraum *m*
ES sala de calderas *f*
F chaufferie *f*
I locale *m* caldaia
MA kazánhelyiség
NL ketelhuis *n*
PO kotłownia *f*
PY котельное
    помещение *n*
SU kattilahuone
SV pannrum *n*

478 boiler section
DA kedelelement
DE Kesselglied *n*
ES elemento de la
    caldera *m*
    haz tubular principal
    de la caldera *m*
F élément *m* de
    chaudière
    section *f* de
    chaudière
I elemento *m* della
    caldaia
MA kazántag
NL ketellid *n*
PO człon *m* kotła
    sekcja *f* kotła
PY секция *f* котла
SU kattilalohko
SV pannsektion *c*

479 boiler shell (*see*
    boiler casing)

480 boiler water leg
DA kedlens vandveje
DE Kesselwasser-
    abschnitt *m*
ES sección de agua de la
    caldera *f*
F lame *f* d'eau de
    chaudière
I volume *m* di acqua
    di una caldaia
MA kazán víztér
NL ketel-watermantel *m*
PO króciec *m* wodny
    kotła
PY пространство *n*
    для нагревания
    воды в котле
SU kattilan vesivaippa
SV pannvattenväg *c*

481 boiler waterline
DA kedelvandstand
DE Kesselwasserlinie *f*
ES nivel de agua de la
    caldera *m*
F niveau *m* d'eau de
    chaudière
I livello *m* dell'acqua
    in una caldaia
MA kazán vízszint
NL ketelwaterniveau *n*
PO linia *f* wodna kotła
PY уровень *m* воды в
    котле
SU kattilan
    syüttüvesijohto
SV pannvattenlinje *c*

482 boiler water temperature control
DA kedeltermostat
DE Kesselwasser-
temperaturmesser *m*
ES regulador de
temperatura de la
caldera *m*
F controle *m* de
température de
chaudière
I regolazione *f* della
temperatura del
bollitore
MA kazán
vízhőmérséklet
szabályozó
NL ketelwatertemperatuur-
regeling *f*
PO regulacja *f*
temperatury wody
w kotle
PY регулирование *n*
температуры
воды в котле
SU kattilaveden
lämpütilansäätü
SV pannvattentemperatur-
kontroll *c*

483 boiling *adj*
DA kogende
DE siedend
ES hervidero
F bouillant
I in ebollizione
MA forrásban lévő
NL kokend
PO wrzący
PY кипящий
SU
SV kokande

484 boiling point
DA kogepunkt
DE Siedepunkt *m*
ES punto de ebullición
m
F point *m* d'ébullition
I punto *m* di
ebollizione
MA forráspont
NL kookpunt *n*
PO punkt *m* wrzenia
PY точка *f* кипения
SU kiehumispiste
SV kokpunkt *c*

485 bolometer
DA bolometer
(strålevarmemåler)
DE Bolometer *n*
(Strahlungswärme-
meßgerät)
ES bolómetro (medidor
de radiación) *m*
F bolomètre *m*
I bolometro *m*
misuratore *m* di
energia raggiante
MA hősugárzásmérő
NL bolometer *m*
stralingsmeter *m*
PO bolometr *m*
miernik *m*
promieniowania
PY болометр *m*
устройство *n*
измерения
потока лучистой
теплоты
SU bolometri
säteilylämpümittari
SV bolometer *c*

486 bolt
DA bolt
DE Bolzen *m*
ES bulón *m*
pasador *m*
perno *m*
F boulon *m*
I bullone *m*
MA fejes csavar
NL bout *m*
PO sworzeń *m*
śruba *f*
PY болт *m*
SU pultti
SV bult *c*

487 bonding (*see* earthing)

488 bonnet
DA dæksel
kappe
DE Haube *f*
Kappe *f*
ES caperuza *f*
domo *m*
F jaquette *f*
I coperchio *m*
vitone *m*
MA fedél
kupak
NL deksel *n*
kap *f*
kraag *m*
PO nakładka *f*
osłona *f*
pokrywa *f*
PY колпак *m*
SU kansi
kupu
SV dom *c*

489 boost
DA forstærke
DE Auftrieb *m*
ES empuje *m*
F allure *f* poussée
I sovralimentazione *f*
spinta *f*
MA fokozás
NL aanjaging *f*
PO wzmaganie *n*
zwiększanie *n*
PY подкачка *f*
SU apu-
lisä-
SV ükning *c*

490 boost *vb*
DA forstærke
DE auftreiben
ES aumentar (la
presión)
sobrealimentar
F soulever par derrière
survolter
I sovralimentare
spingere
MA fokozni
NL
PO wzmagać
zwiększać
PY подкачивать
SU
SV üka

491 booster
DA forstærker
DE Verstärker *m*
Zusatz *m*
ES acelerador (bomba)
m
F surpresseur *m*
I dispositivo *m* per
aumentare le
prestazioni
MA fokozó
NL aanjager *m*
PO urządzenie *n*
wspomagające
PY подкачивающее
устройство *n*
SU apu-
lisä-
SV hjälpdon *n*

**492 booster
compressor**
DA lavtrykskompressor
DE Aufladeverdichter *m*
Zusatzverdichter *m*
ES compresor surpresor
m
F compresseur *m*
surpresseur
I compressore *m*
ausiliario
MA nyomásfokozó
kompresszor
NL aanjaagcompressor
PO sprężarka *f*
wspomagająca
PY вспомогательный
компрессор *m*
SU apukompressori
SV hjälpkompressor *c*

493 booster pump
DA trykforøgerpumpe
DE Druckerhöhungs-
pumpe *f*
ES bomba elevadora de
presión *f*
F pompe *f* auxiliaire
pompe *f* de
surpression
I pompa *f* ausiliare
pompa *f* di
alimentazione
MA nyomásfokozó
szivattyú
NL aanjaagpomp *f*
drukverhogings-
pomp *f*
PO pompa *f*
wspomagająca
PY подкачивающий
насос *m*
SU paineenlisäys-
pumppu
SV hjälppump *c*

494 boot
DA auto-bagagerum
DE Schutzkappe *f*
Unterlegung *f*
ES reducción *f*
F pièce *f* de
transformation
I raccordo *m* tra
canali di sezione
circolare e
rettangolare
MA átmeneti idom
NL kanaalverloopstuk *n*
PO bagażnik *m*
PY переход *m* с
круглого на
прямоугольное
сечение
SU pyüreän ja
suorakaide-
kanavan
liitoskappale
SV sko *c*

495 bore (of pipe)
DA lysning (af rør)
DE Innendurch-
messer *m*
Kaliber *n*
ES calibre *m*
diámetro interior de
un tubo *m*
F calibre *m* (de tuyau)
diamètre *m* (de
tuyau)
I diametro *m* interno
del tubo
MA cső belső átmérő
furat
NL binnendiameter *m*
PO średnica *f* otworu
rury
PY отверстие *n* (в
трубе)
SU sisähalkaisija
SV rördiameter *c*

496 bottled gas
DA flaskegas
DE Flaschengas *n*
ES gas embotellado *m*
F gaz *m* en bouteille
I gas *m* compresso (in
bombole)
MA palackozott gáz
NL flessengas *n*
PO gaz *m* płynny
PY газ *m* в емкости
SU nestekaasu
SV flaskgas *c*

497 boundary
DA grænse
DE Grenze *f*
ES límite *m*
F limite *f*
I limite *m*
MA határ
NL begrenzing *f*
grens *f*
PO granica *f*
obszar *m* graniczny
PY граница *f*
SU raja
SV gräns *c*

DA = Danish
DE = German
ES = Spanish
F = French
I = Italian
MA = Magyar
(Hungarian)
NL = Dutch
PO = Polish
PY = Russian
SU = Finnish
SV = Swedish

498 boundary conditions
DA grænsetilstand
DE Grenzbedingungen f,pl
ES condiciones límite f
F conditions f aux limites
I condizioni f,pl al contorno
MA határfeltételek
NL grenswaarden pl
PO warunki m graniczne
PY граничные условия n
SU reunaehdot
SV gränsfürhållanden n

499 boundary layer
DA grænselag
DE Grenzschicht f
ES capa límite f
F couche f limite
I strato m limite
MA határréteg
NL grenslaag f
PO warstwa f graniczna
PY пограничный слой m
SU rajakerros
SV gränsskikt n

500 Bourdon gauge
DA Bourdon manometer
 fjedermanometer
DE Bourdon-Manometer n
ES manómetro de Bourdon m
F tube m de Bourdon
I manometro m Bourdon
MA Bourdoncsüves manométer
NL veermanometer m
PO manometr m Bourdona
PY манометр m с пружиной Бурдона
SU bourdonkaari-painemittari
SV fjädermanometer c

501 Bourdon tube
DA bourdonrør
DE Bourdon-Rohr n
ES tubo de Bourdon m
F tube m de Bourdon
I tubo m Bourdon
MA Bourdon cső
NL Bourdon buis
PO rurka f Bourdona
PY трубка f Бурдона
SU Bourdon-putki
SV bourdonrör n

502 bow
DA bøjle
DE Bogen m
ES arco m
 codo m
 curvatura f
F coude m
I curva f
 gomito m
MA hurok
 ív
NL bocht f
 boog m
PO łuk m
 wygięcie n
PY отвод m
SU kaari
SV böj c

503 box
DA boks
 kasse
DE Büchse f
 Gehäuse n
 Kasten m
ES caja f
F boîte f
I camera f
 scatola f
MA doboz
 szekrény
NL box m
 doos f
 kist f
PO komora f
 skrzynka f
PY короб m
SU laatikko
SV låda c

504 bracket
DA bærejern
 konsol
DE Klammer f
 Konsole f
ES abrazadera f
 soporte m
F console f
 fixation f
 support m
I mensola f
 staffa f
 supporto m
MA bilincs
 konzol
NL beugel m
 console f
PO konsola f
 wspornik m
PY подпорка f
SU kannatin
 kiinnike
 konsoli
 pidin
SV hållare c

505 brackish water
DA brakvand
DE Brackwasser n
ES agua salobre f
F eau f saumâtre
I acqua f salmastra
MA brakkvíz
 édes-sós víz
NL brak water n
PO woda f słonawa
PY вода f с содержанием соли больше, чем в питьевой солончаковая вода f
SU murtovesi
SV bräckt vatten n

506 brake power
DA bremsekraft
DE Bremskraft f
ES potencia al freno f
F puissance f sur l'arbre
I potenza f all'albero
MA fékező erő
NL remvermogen n
PO moc f hamowania
PY действительная мощность f
SU akseliteho
SV bromskraft c

507 branch
DA afgrening
DE Abzweig *m*
ES derivación *f*
   ramal *m*
F dérivation *f*
   ramification *f*
I derivazione *f*
   ramo *m*
MA ág
   leágazás
NL aftakking *f*
   vertakking *f*
PO odgałęzienie *n*
   odnoga *f*
   rozgałężnik *m*
PY ответвление *n*
SU haara
SV förgrening *c*

508 branch duct
DA afgreningskanal
DE Abzweigluftkanal *m*
ES conducto
   secundario *m*
F branchement *m*
I canale *m* di
   derivazione
MA ágvezeték
NL aftakkanaal *n*
   zijkanaal *n*
PO kanał *m* odgałęźny
   przewód *m*
   odgałęźny
PY ответвление *n*
   воздуховода
SU haarakanava
SV grenkanal *c*

509 branching
DA afgrening
DE Abzweigung *f*
ES derivación *f*
   ramificación *f*
F embranchement *m*
I derivazione *f*
   diramazione *f*
MA elágazás
NL aftakking *f*
   splitsing *f*
   spruitstuk *n*
PO rozgałęzienie *n*
PY отвод *m*
SU haara
SV avgrening *c*

510 branch line
DA stikledning
DE Zweiglinie *f*
ES ramal *m*
F conduite *f*
   secondaire
I ramo *m* derivato
MA elágazó vonal
NL aftakking *f*
   aftakleiding *f*
PO linia *f* boczna
   przewód *m*
   odgałęźny
PY ответвительная
   линия *f*
SU haarajohto
SV sidobana *c*

511 branch take-off
DA rørafgrening
DE Rohrabzweigung *f*
ES entronque *m*
   injerto (tuberias) *m*
F piquage *m*
I raccordo *m* di
   derivazione
MA csőleágazás
NL aftakaansluiting *f*
PO miejsce *n*
   odgałęzienia
PY ответвление *n*
   провода
SU haarakohta
SV avgrening *c*

512 brass *adj*
DA messing
   rødgods
DE messingen (aus
   Messing)
ES latón *m*
F laiton *m*
I in ottone
MA sárgaréz
NL geelkoper
   messing
PO mosiężny
PY латунный
SU
SV mässings

513 braze *vb*
DA slaglodde
DE hartlöten
ES soldar (con latón)
F braser
I brasare
MA keményforraszt
NL hardsolderen
PO lutować lutem
   twardym
PY сваривать
   спаивать
SU kovajuottaa
SV hårdlöda

514 brazed *adj*
DA slagloddet
DE hartgelötet
ES soldado
F brasé
I brasato
MA keményforrasztott
NL hardgesoldeerd
PO lutowany lutem
   twardym
PY сваренный
   спаянный
SU kovajuotettu
SV hårdlödd

515 brazed joint
DA slaglodning
DE Hartlötverbindung *f*
ES junta soldada *f*
F joint *m* brasé
I giunto *m* brasato
MA keményforrasztású
   kütés
NL hardgesoldeerde
   verbinding *f*
PO połączenie *n*
   lutowane lutem
   twardym
PY соединение *n*
   сваркой
SU kovajuotettu liitos
SV hårdlödd skarv *c*

516 brazed tube
DA slagloddet rør
DE hartgelötetes Rohr *n*
ES tubo soldado *m*
F tube *m* brasé
I tubo *m* brasato
MA keményforrasztású
   cső
NL naadpijp *f*
PO rura *f*
   mosiądzowana
PY электросварная
   труба *f*
SU pitkittäin saumattu
   putki
SV hårdlött rör *n*

517 brazing
DA slaglodning
DE Hartlötung *f*
ES soldadura con metal
   de aportación *f*
   soldadura fuerte *f*
F brasure *f*
I brasatura *f* forte
MA keményforrasztás
NL hardsolderen
PO lutowanie *n* twarde
PY пайка *f* твердым
   припоем
SU kovajuotos
SV hårdlödning *c*

518 breakdown
DA driftforstyrrelse
   havari
DE Betriebsstörung *f*
   Panne *f*
ES avería *f*
   rotura *f*
F panne *f*
I guasto *m*
MA üzemzavar
NL defect *n*
   storing *f*
PO awaria *f*
   przebicie *n* izolacji
   uszkodzenie *n*
PY авария *f*
SU vika
SV driftstörning *c*

519 breaker
DA afbryder
DE Schalter *m*
   Unterbrecher *m*
ES disyuntor *m*
   interruptor *m*
F coupure *f* thermique
I barriera *f* isolante
   sezionatore *m*
MA kapcsoló
   megszakító
NL veiligheids-
   schakelaar *m*
PO przerywacz *m*
   wyłącznik *m*
PY аварийный
   прибор *m*
SU lämpükatko
   sulake
SV brytare *c*

520 breakout noise
DA støjudbredelse
DE Ausschaltgeräusch *n*
ES ruido de desconexión
   m
F bruit *m* émis
I scoppio *m*
MA kiszűrődő zaj
NL doorstralend
   geluid *n*
PO hałas *m*
   spowodowany
   wybuchem
PY аварийный
   звуковой
   сигнал *m*
SU kanavan siirtämä ja
   säteilemä melu
   sivutiesiirtymä
SV buller *n*

521 break-through
DA gennembrud
DE Durchbruch *m*
ES perforación *f*
F boyau *m* de mine
I innovazione *f*
   scoperta *f* inno-
   vativa
MA áttürés
NL stofdoorslag *m*
PO przebicie *n*
   przedarcie *n* się
PY разрыв *m*
SU läpäisy
SV genombrott *n*

522 breath
DA ånde
DE Atem *m*
ES empañamiento *m*
   respiración *f*
   vaho *m*
F respiration *f*
I appannatura *f*
   respiro *m*
MA lélegzés
NL adem *m*
   bries *f*
PO oddech *m*
PY дыхание *n*
SU hengitys
SV andning *c*

523 breather plug
DA udluftningsprop
DE Entlüftungs-
   stüpsel *m*
ES tapón de desaire *m*
F évent *m*
I apertura *f* di
   aerazione
MA kilégző dugó
NL beluchtingsplug *f*
   ontluchtingsplug *f*
PO korek *m*
   odpowietrznika
PY дыхательный
   клапан *m*
SU paineentasausaukko
   tulppa
SV ventilationspropp *c*

**524 breathing
apparatus**
DA snøfteventil
DE Atemschutzgerät *n*
ES aparato respira-
   torio *m*
F appareil *m*
   respiratoire
I apparecchio *m* per
   la respirazione
MA légzőkészülék
NL ademhalings-
   toestel *n*
PO aparat *m*
   oddechowy
PY дыхательный
   аппарат *m*
SU hengityskone
SV luftventil *c*

525 breeching
DA røgkanal
DE Schornsteinfuchs *m*
ES conducto de humos
   m
F carneaux *m* de
   fumée
I canale *m* da fumo
   collettore *m* dei
   fumi
MA rókatorok
NL rookgasaansluit-
   stuk *n*
PO czopuch *m* żelazny
PY боров *m*
SU hormi
SV underkanal *c*

526 brick
DA mursten
   teglsten
DE Ziegelstein *m*
ES ladrillo *m*
F brique *f*
I mattone *m*
MA tégla
NL baksteen *m*
   metselsteen *m*
PO cegła *f*
PY кирпич *m*
SU tiili
SV tegelsten *c*

527 brickwork
DA murværk
DE Backsteinmauer-
   werk *n*
ES albañilería *f*
   fábrica de ladrillo *f*
   mampostería *f*
F maçonnerie *f*
I muratura *f*
MA téglafal
NL metselwerk *n*
PO mur *m* z cegły
   obmurze *n*
   roboty *f,pl* murowe
PY кирпичная кладка *f*
SU muuraus
SV murverk *n*

528 brine
DA brine
   saltopløsning
DE Sole *f*
ES salmuera *f*
F saumure *f*
I salamoia *f*
MA sólé
NL pekel *m*
   zoutoplossing *f*
PO solanka *f*
PY рассол *m*
SU liuos
SV saltvatten *n*

529 brine circuit
DA brinekredsløb
DE Solekreislauf *m*
ES circuito de
   salmuera *m*
F circuit *m* de
   saumure
I circuito *m* di
   salamoia
MA sólé keringtetés
NL pekelcircuit *n*
PO obieg *m* solankowy
PY рассольный
   контур *m*
SU liuoskierto
SV saltvattens-
   cirkulation *c*

530 brine cooler
DA brinekøler
DE Solekühler *m*
ES enfriador de
   salmuera *m*
F refroidisseur *m* de
   saumure
I refrigeratore *m* di
   salamoia
MA sólé hűtő
NL pekelkoeler *m*
PO chłodnica *f*
   pośrednicząca
   chłodnica *f*
   solankowa
PY рассольный
   охладитель *m*
SU liuosjäähdytin
SV küldgenerator *c*

**531 brine spray
refrigerating
system**
DA brineoverrislings-
   køleanlæg
DE Solesprühgefrier-
   system *n*
ES sistema de
   refrigeración por
   pulverización de
   salmuera *m*
F système *m* de
   réfrigération à
   pulvérisation de
   saumure
I impianto *m*
   frigorifero a
   spruzzamento di
   salamoia
MA sólé porlasztásos
   hűtőrendszer
NL pekelsproeikoel-
   systeem *n*
PO system *m* chłodzenia
   rozpyloną solanką
PY рассольно-
   оросительная
   система *f*
   охлаждения
SU liuossumutus-
   jäähdytys
SV küldbärarspridnings-
   system *n*

532 brine system
DA brineanlæg
DE Solesystem *n*
ES sistema de salmuera
   m
F système *m* de
   saumure
I sistema *m* a
   salamoia
MA sólé rendszer
NL pekelsysteem *n*
PO system *m* solankowy
PY рассольная
   система *f*
SU liuosjärjestelmä
SV saltvattensystem *n*

DA	=	Danish
DE	=	German
ES	=	Spanish
F	=	French
I	=	Italian
MA	=	Magyar (Hungarian)
NL	=	Dutch
PO	=	Polish
PY	=	Russian
SU	=	Finnish
SV	=	Swedish

533 brine tank
DA brinebeholder
DE Soletank *m*
ES tanque de
   salmuera *m*
F bac *m* à saumure
I serbatoio *m* di
   salamoia
MA sólé tartály
NL pekelvat *n*
   tank *m*
PO zbiornik *m* solanki
PY резервуар *m* для
   возврата рассола
SU liuossäiliü
SV saltvattens-
   behållare *c*

534 briquette
DA briket
DE Brikett *n*
ES briqueta *f*
F briquette *f*
I mattonella *f* di
   carbone
MA brikett
NL briket *f*
PO brykiet *m*
   kostka *f* prasowana
PY брикет *m*
SU briketti
SV brikett *c*

535 British Standard
DA britisk standard
DE britischer
   Standard *m*
ES norma inglesa *f*
F norme *f* anglaise
I British Standard
MA angol szabvány
NL British Standard
PO norma *f* brytyjska
PY британский
   стандарт *m*
SU englantilainen
   standardi
SV brittisk standard *c*

536 broken coke
DA kokssmuld
DE Brechkoks *m*
ES coque troceado *m*
F coke *m* concassé
I coke *m* frantumato
MA darakoksz
NL gebroken cokes *f*
   parelcokes *f*
PO tłuczeń *m* koksowy
PY коксовый
   щебень *m*
SU koksimurska
SV krossad koks *c*

537 brown coal
DA brunkul
DE Braunkohle *f*
ES carbón
   bituminoso *m*
   lignito *m*
F lignite *m*
I lignite *f*
MA barnaszén
NL bruinkool *f*
PO lignit *m*
   węgiel *m* brunatny
PY бурый уголь *m*
SU ruskohiili
SV brunkol *n*

538 brush filter
DA børstefilter
DE Bürstenfilter *m*
ES filtro de cepillos *m*
F filtre *m* à brosses
I filtro *m* a spazzola
MA kefeszűrő
NL borstelfilter *n*
PO filtr *m* szczotkowy
PY сетчатый фильтр *m*
   ячейковый
   фильтр *m*
SU harjasuodatin
SV borstfilter *n*

539 BS (*see* British
   Standard)

540 BTU (*see* Appendix
   A)

541 bubble
DA blære
   boble
DE Blase *f*
ES burbuja *f*
F bulle *f*
I bolla *f*
MA buborék
NL bel *f*
PO pęcherzyk *m*
PY пузырек *m*
SU kupla
SV bubbla *c*

542 bubble flow
DA boblestrøm
DE Blasenströmung *f*
ES circulación de
   burbujas *f*
F écoulement *m* à
   bulles
I corrente *f* (bifase) a
   bolle
MA pezsgő áramlás
NL gas-vloeistof *f*
   emulsie stroom *m*
PO przepływ *m*
   pęcherzykowy
PY двухфазный поток
   m
   пузырьковый поток
   m
SU kuplavirtaus
SV bubbelflöde *n*

543 buffer
DA stødpude
DE Puffer *m*
ES amortiguador *m*
F tampon *m*
I buffer *m*
MA puffer
NL buffer *m*
PO bufor *m*
   pamięć *f* buforowa
   podkładka *f*
   sprężysta
   zderzak *m*
PY буфер *m*
SU tasain
SV buffert *c*

544 buffer tank
DA buffertank
   stødpudebeholder
DE Pufferbehälter *m*
   Speicherbehälter *m*
ES depósito pulmón *m*
F réservoir *m* tampon
I serbatoio *m*
   polmone
MA puffer tároló
NL buffertank *m*
PO zbiornik *m* wstępny
   zbiornik *m*
   wyrównawczy
PY регулирующая
   емкость *f*
SU tasaussäiliü
SV bufferttank *c*

545 building
DA bygning
DE Bau *m*
Gebäude *n*
ES edificio *m*
inmueble *m*
F bâtiment *m*
immeuble *m*
I costruzione *f*
edificio *m*
immobile *m*
stabile *m*
MA épület
NL bouwen *n*
gebouw *n*
PO budowla *f*
budownictwo *n*
budynek *m*
PY здание *n*
строительство *n*
SU rakennus
SV byggnad *c*

546 building area
DA bygningsareal
DE Baugelände *n*
ES superficie construida
f
F surface *f* du
bâtiment
I area *f* dell'edificio
MA beépített terület
NL bebouwd
oppervlak *n*
PO powierzchnia *f*
zabudowy
PY площадь *f*
застройки
пятно *n* застройки
SU rakennusala
SV byggnadsyta *c*

**547 building
automation system**
DA bygningsautomation
DE Gebäudeautomations-
system *n*
ES sistema de
automatización de
un edificio *m*
F automatisation *f* des
systèmes des
bâtiments
I sistema *m* di
automazione
dell'edificio
MA épületautomatizálási
rendszer
NL gebouwautomatiserings-
systeem *n*
PO budownictwo *n* z
prefabrykatów
system *m*
zautomatyzowania
bydynku
PY система *f*
автоматики
здания
SU rakennusautomaatio-
järjestelmä
SV byggnadsautomations-
system *n*

**548 building control
system**
DA bygningsregulerings-
system
DE zentrale
Leittechnik *f*
ES sistema de control de
un edificio *m*
F système *m* de
gestion technique
d'un bâtiment
I sistema *m* di
controllo
dell'edificio
MA épületfelügyeleti
rendszer
NL gebouwautomatiserings-
systeem *n*
PO system *m* sterowania
urządzeniami w
budynku
PY система *f*
регулирования
здания
SU valvontajärjestelmä
SV byggnadsreglerings-
system *n*

549 building element
DA byggeelement
DE Bauelement *n*
Bauteil *n*
ES elemento de
construcción *m*
F élément *m* de
construction
I elemento *m* da
costruzione
MA építőelem
NL bouwelement *n*
bouwsectie *f*
PO element *m* budynku
PY часть *f* здания
элемент *m* здания
SU rakennusosa
SV byggelement *n*

**550 building energy
management
control system**
DA energistyrings-
kontrolsystem for
bygninger
DE Energiekontroll-
system *n* eines
Gebäudes *n*
ES sistema de control y
gestión energética
de un edificio *m*
F système *m* de
contrôle et de
gestion de
l'energie des
bâtiments
I sistema *m* di
controllo della
gestione energetica
dell'edificio
MA épület
energiafelügyeleti
rendszer
NL gebouw-
automatiserings-
en energie-
beheersysteem *n*
PO system *m* kontroli
gospodarowania
energią w
budynku
system *m* regulacji
energooszczędnej
gospodarki w
budynku
PY система *f*
регулирования
потребления
энергии зданием
SU rakennuksen
energianhallinnan
säätöjärjestelmä
SV byggnadsenergistyr-
system *n*

551 building energy management system
DA energistyrings-kontrolsystem for bygninger
DE Energieverwaltungs-system *n* eines Gebäudes *n*
ES sistema de gestión energética de un edificio *m*
F système *m* de gestion de l'energie des bâtiments
I sistema *m* di supervisione dell'edificio
MA épület energiagazdál-kodási rendszer
NL gebouwenergiebeheer-systeem *n*
PO system *m* energooszczędnej gospodarki w budynku system *m* gospodarowania energią w budynku
PY система *f* управления энергией в здании
SU rakennuksen energianhallintajär-jestelmä
SV energistyrnings-system *n* för byggnad *c*

552 building envelope
DA klimaskærm
DE Gebäudehülle *f*
ES envolvente del edificio *f*
F enveloppe *f* du bâtiment
I involucro *m* dell'edificio
MA épület külső héjazat
NL gebouwschil *f*
PO obudowa *f* budynku
PY наружные ограждения *n,pl* здания
SU rakennuksen vaippa
SV klimatskärm *c*

553 building envelope void
DA klimaskærms termiske modstand
DE Gebäudehüllen-lücke *f*
ES abertura en la envolvente del edificio *f*
F défaut *m* dans l'enveloppe du bâtiment
I apertura *f* nell' involucro edilizio
MA épület héjazat hézag
NL koudebrug *f* in gebouwschil *f*
PO przestrzeń *f* powietrzna w obudowie budynku
PY полость *f* (пустота) в наружном ограждении
SU rakennuksen ulkovaipan tilavuus
SV klimatskärmens *c* termiskt motstånd *n*

554 building management system
DA bygningsstyresystem
DE Gebäudeverwaltungs-system *n*
ES sistema de gestión de un edificio *m*
F gestion *f* technique de bâtiment
I sistema *m* di gestione dell'edificio
MA épületirányítási rendszer
NL gebouwbeheer-systeem *n*
PO system *m* zarządzania budynkiem
PY система *f* управления зданием
SU rakennuksen hallintajärjestelmä
SV bygglednings-system *n*

555 building material
DA byggemateriale
DE Baumaterial *n*
ES material de construcción *m*
F matériel *m* de construction
I materiale *m* da costruzione
MA építőanyag
NL bouwmateriaal *n*
PO materiał *m* budowlany
PY строительный материал *m*
SU rakennusmateriaali
SV byggmaterial *n*

556 building project
DA byggeprojekt
DE Bauprojekt *n*
ES proyecto de edificio *m*
F projet *m* de bâtiment
I progetto *m* di un edificio
MA építési terv
NL bouwproject *n*
PO projekt *m* budowlany projekt *m* budynku
PY проект *m* здания
SU rakennusprojekti
SV byggprojekt *n*

557 building space
DA bygningsafstand
DE Bauvolumen *n*
ES espacio edificado *m*
F espace *m* de bâtiment
I ambiente *m* edificato
MA épület térfogat
NL bouwvolume *n*
PO kubatura *f* budynku
PY объем *m* здания
SU rakennuksen nettotilavuus
SV byggyta *c*

DA	=	Danish
DE	=	German
ES	=	Spanish
F	=	French
I	=	Italian
MA	=	Magyar (Hungarian)
NL	=	Dutch
PO	=	Polish
PY	=	Russian
SU	=	Finnish
SV	=	Swedish

558 building thermal load
DA bygnings
   varmebelastning
DE Gebäudewärme-
   bedarf *m*
ES carga térmica del
   edificio *f*
F déperditions *f*
   calorifiques du
   bâtiment
I carico *m* termico
   dell'edificio
MA épület hőterhelés
NL thermische
   gebouwbelasting *f*
PO obciążenie *n* cieplne
   budynku
PY тепловая нагрузка
   f здания
   тепловые потери
   f,pl здания
SU rakennuksen
   lämpükuorma
SV byggnads *c*
   värmelast *c*

559 building volume
(*see* building space)

560 built-in *adj*
DA indbygget
DE Einbau-
ES montado
   incorporado
F incorporé
I montato in loco
MA beépített
NL ingebouwd
PO utwierdzenie *n*
   wbudowanie *n*
PY встроенный
SU sisäänrakennettu
SV inbyggnad

561 bulb
DA kolbe
   kugle
   vulst
DE Fühler *m*
   Glaskolben *m*
ES bulbo *m*
F bulbe *m*
I bulbo *m*
MA hőmérő gümb
   kürte
NL gloeilamp *f*
   voeler *m*
PO bańka *f*
   zbiornik *m*
   termometru
PY колба *f*
SU anturi
   hehkulamppu
SV glob *c*
   kolv *c*

562 bulk
DA masse
   volumen
DE Größe *f*
   Masse *f*
ES masa *f*
   tamaño *m*
   volumen *m*
F masse *f*
I massa *f*
MA halom
   tümeg
NL grootte *f*
   massa *f*
   omvang *m*
   volume *n*
PO masa *f*
PY масса *f*
SU määrä
SV massa *c*

563 bulk density
DA massetæthed
   volumenvægt
DE Schüttdichte *f*
ES densidad aparente *f*
   masa por unidad de
   volumen *f*
F densité *f* en vrac
I densità *f* apparante
MA térfogatsűrűség
NL stortgewicht *n*
PO gęstość *f* nasypowa
PY объемная масса *f*
SU tiheys
SV skrymvolym *c*

564 bullseye
DA køøje
DE Sichtglas *n*
ES ojo de buey *m*
F hublot *m*
I occhio *m* di bue
MA betekintő ablak
NL kijkglas *n*
PO iluminator *m*
   kołnierz *m* elektrody
   okno *n* okrągłe
PY трубка *f* уровня
SU tarkastuslasi
SV oxöga *n*

565 bundle
DA bundt
DE Bündel *n*
ES haz *m*
   rollo de alambre *m*
F faisceau *m*
I fascio *m*
MA küteg
   nyaláb
NL bos *m*
   bundel *m*
   pak *n*
   rol *m*
PO wiązka *f*
PY связка *f*
SU nippu
SV bunt *c*

566 bunker
DA bunker
   kulrum
DE Bunker *m*
ES carbonera *f*
   tolva *f*
F soute *f*
I deposito *m* di
   combustibile
   ricovero *m* antiaereo
   o antiatomico
MA bunker
NL bunker *m*
PO bunkier *m*
   zasobnik *m*
PY бункер *m*
SU hiilivarasto
SV bunker *c*

567 burn *vb*
DA brænde
DE brennen
ES quemar
F brûler
I bruciare
MA ég
égett
NL branden
verbranden
PO spalać
PY гореть
SU polttaa
SV bränna

568 burner
DA brænder
DE Brenner *m*
ES quemador *m*
F brûleur *m*
I bruciatore *m*
MA égő
NL brander *m*
PO palnik *m*
PY горелка *f*
SU poltin
SV brännare *c*

569 burner flame-failure response time
DA flammekontrol
DE Wiedereinschaltzeit *f* nach Brennerausfall *m*
ES tiempo de respuesta al fallo de llama *m*
F temps *m* de réponse de sécurité de flamme
I tempo *m* di intervento a seguito di spegnimento della fiamma
MA égő lángkimaradási idő
NL reactietijd *m* bij vlamstoring
PO czas *m* reakcji na brak płomienia w palniku
PY время *n* затухания работы горелки
SU polttimen varolaitteiden vasteaika
SV flamkontroll *c*

570 burner flange
DA brænderflange
DE Brennerflansch *m*
ES brida del quemador *f*
F plaque *f* de brûleur
I flangia *f* del bruciatore
MA égő perem
NL branderflens *m*
PO kołnierz *m* palnika
PY фланец *m* горелки
SU polttimen laippa
SV brännarfläns *c*

571 burner ignition
DA brændertænding
DE Brennerzündung *f*
ES encendido del quemador *m*
F allumage *m* de brûleur
I accensione *f* del bruciatore
MA égő gyújtás
NL branderontsteking *f*
PO palnik *m* zapalający zapłonnik *m*
PY зажигание *n* горелки
SU polttimen sytytys
SV brännartändning *c*

572 burner nozzle
DA brænderdyse
DE Brennerdüse *f*
ES boquilla del quemador *f*
F tête *f* de brûleur
I ugello *m* del bruciatore
MA égő fúvóka
NL brandersproeier *m*
PO dysza *f* palnika
PY сопло *n* горелки
SU polttimon suutin
SV brännarmunstycke *n*

573 burner plate
DA brænderplade
DE Brennerplatte *f*
ES placa de la caldera (para el quemador) *f*
F plaque *f* de chaudière (de brûleur)
I piastra *f* del bruciatore
MA égő alaplemez
NL branderplaat *f*
PO płyta *f* palnika
PY плита *f* горелки
SU polttimen kiinnityslevy
SV brännarplåt *c*

574 burner pot
DA pottefyr
DE Brennerschale *f* Brennertopf *m*
ES cazuela del quemador *f*
F pot *m* de brûleur
I tarra *f* del bruciatore
MA égő serleg
NL branderbak *m*
PO obudowa *f* palnika
PY резервуар *m* горелки
SU hüyrystyspoltin
SV pottbrännare *c*

575 burner purge
DA brænderrensning
DE Brennerreinigung *f*
ES barrido de la cámara de combustión *m*
F postbalayage *m* de chaudière
I preventilazione *f* della camera di combustione
MA égő tisztítás
NL branderspoelperiode *f*
PO odpowietrzenie *n* palnika
PY прочистка *f* горелки
SU palamistilan huuhtelu
SV brännargenomspolning *c*

576 burner register
DA brænderregister
DE Brennerregister *n*
ES regulador de
   combustión *m*
F registre *m* de
   brûleur
I alette *f,pl* di
   regolazione
   dell'aria di
   combustione
MA égő terelőlemez
NL branderlucht-
   regelklep *f*
PO urządzenie *n*
   regulujące dopływ
   powietrza do
   palnika
   zasuwa *f* palnika
PY регулятор *m*
   процесса горения
SU palamisilman
   ohjaussäleikkü
SV brännarregister *n*

577 burner shroud (*see*
burner tip)

578 burner throat
DA brænderhoved
DE Brennermündung *f*
ES boca del quemador *f*
F nez *m* de brûleur
I gola *f* del bruciatore
MA égő torok
NL branderconus *m*
PO gardziel *f* palnika
   przewężenie *n*
   palnika
PY топливный канал
   m горелки
SU polttimen kurkku
SV brännarhals *c*

579 burner tip
DA brændermundstykke
DE Brennerkopf *m*
ES llave para la tuerca
   de la tobera *f*
   pantalla de la
   boquilla del
   quemador *f*
F tête *f* de brûleur
I testa *f* del
   bruciatore
   ugello *m* del
   bruciatore
MA égőfej
NL branderkop *m*
PO dysza *f* palnika
   końcówka *f* palnika
   wylot *m* palnika
PY сопло *n* горелки
SU polttimen suutin
SV brännarmunstycke *n*

**580 burner window
box**
DA brænder-skueglas
DE Brennersichtglas-
   gehäuse *n*
ES mirilla del quemador
   f
F voyant *m* de
   brûleur
I presa *f* di aria
   secondaria
MA égőház ellenőrző
   ablakkal
NL luchtkast *f* voor
   secundaire lucht
PO otwór *m* kontrolny
   palnika
PY камера *f* подачи
   воздуха на
   горение
SU sekundääri-ilman
   syüttükotelo
SV siktglas *n* för
   brännare *c*

581 burn out
DA brænde ud
DE Ausbrand *m*
   ausbrennen
ES quemador
   averiado *m*
   quemadura *f*
F brûler complètement
I bruciare
   fondere
MA kiégés
   leégés
NL defect *n* door
   verbranding
PO wypalenie
PY выжигать
SU polttaa loppuun
SV utbränning *c*

582 bursting disc
DA sikkerhedsskive
   sprængplade
DE Berstscheibe *f*
   Sicherheitsscheibe *f*
ES disco de
   seguridad *m*
F disque *m* de sécurité
   pastille *f*
I dischetto *m* di
   sicurezza
MA hasadótárcsa
NL breekplaat *f*
PO przepona *f*
   bezpieczeństwa
PY диск *m*
   безопасности
SU varokalvo
   varolevy
SV säkerhetsbleck *n*

583 bury *vb*
DA begrave
   tildække
DE eingraben
   vergraben
ES enterrar
   sumergir
F enfouir
   enterrer
I interrare
MA betemet
NL begraven
PO zakopać
PY закопать
SU haudata
SV gräva ned

584 bus (electrical)
DA bus, el-ledning
DE Sammelleitung *f*
   Sammelschiene *f*
ES colector (elec) *m*
F bus *m* (électrique)
I bus *m*
MA gyűjtővezeték
   sin
NL bus *f*
PO szyna *f* zbiorcza
PY щит *m*
   (электрический)
SU kytkentärima
SV buss *c*

585 bush
DA bøsning
DE Lagerbüchse *f*
ES casquillo de cojinete
m
virola *f*
F coussinet *m*
I boccola *f*
cuscinetto *m*
sbarra *f*
MA hüvely
persely
NL lagervoering *f*
manchet *f*
mof *f*
PO panewka *f*
tuleja *f*
PY гильза *f*
SU hela
hylsy
SV hylsa *c*

586 bush half
DA rørskål
DE Lagerschale *f*
ES medio cojinete *m*
semicojinete *m*
F demi-coussinet *m*
I semi-bussola *f*
MA félpersely
NL lagerschaal *f*
PO półpanewka *f*
PY полувкладыш *m*
полугильза *f*
SU laakerin puolikas
SV rörskål *c*

587 bushing
DA bøsning
DE Durchführung *f*
(elektr.)
Muffe *f*
ES manguito *m*
F pièce *f* de réduction
I bronzina *f*
MA átvezetés
NL verloopring *m*
PO złączka *f*
nakrętno-wkrętna
złączka *f* wkrętna
zwężkowa
PY прокладка *f* в
гильзе
SU laakerointi
SV foder *n*

588 butane
DA butan
DE Butan *n*
ES butano *m*
F butane *m*
I butano *m*
MA bután
NL butaan *n*
PO butan *m*
PY бутан *m*
SU butaani
SV butan *c*

589 butterfly damper
DA butterflyspjæld
drosselspjæld
DE Drosselklappe *f*
ES registro
de mariposa *m*
F registre *m* tournant
(papillon)
I serranda *f* a farfalla
MA pillangóscsappantyú
NL luchtsmoorklep *f*
luchtvlinderklep *f*
PO przepustnica *f*
motylkowa
PY дроссельный
клапан *m*
SU säätüläppä
SV trottelspjäll *n*

590 butterfly valve
DA butterflyventil
drosselventil
DE Drosselventil *n*
ES válvula de
mariposa *f*
F robinet *m* à papillon
vanne *f* à papillon
I serranda *f*
valvola *f* a farfalla
MA pillangószelep
NL smoorklep *f*
vlinderklep *f*
PO zawór *m* motylkowy
PY дроссельный
вентиль *m*
SU läppäventtiili
SV trottelventil *c*

591 butt-welded joint
DA stumpsvejst samling
DE stumpfgeschweißte
Verbindung *f*
ES unión soldada a
tope *f*
F raccordement *m* par
soudure bout à
bout
I giunto *m* di testa
MA tompahegesztésű
csatlakozás
NL stomplas *f*
PO spoina *f* w spawaniu
doczołowym
spoina *f* w spawaniu
na styk
PY сваренный встык
SU hitsattava liitos
puskusauma
SV stumfog *c*

592 butt welding
DA stumpsvejsning
DE Stumpfschweißung *f*
ES soldadura a tope *f*
F soudure *f* bout à
bout
I giunzione *f* di testa
MA tompahegesztés
NL stomplassen *n*
stuiklassen *n*
PO spawanie *n*
doczołowe
spawanie *n* na styk
zgrzewanie *n*
doczołowe
PY стыковая сварка *f*
SU (pusku)hitsaus
SV stumsvetsning *c*

593 by-pass
DA omløb
shunt
DE Beipaß *m*
ES derivación *f*
F by-pass *m*
dérivation *f*
I by-pass *m*
deviazione *f*
MA megkerülő ág
megkerülőjárat
NL omloop *m*
omloopleiding *f*
PO bocznik *m*
obejście *n*
PY байпас *m*
обвод *m*
SU ohitus
SV fürbigång *c*

594 by-pass *vb*
DA by-pass
   omgå
DE umfahren
   umgehen
ES desviación *f*
   desviar (del
      recorrido
      principal)
F amener en dérivation
I deviare
MA megkerül
NL omleiden
   overslaan
   passeren
PO bocznikować
   obchodzić
   omijać
PY обводить
SU ohittaa
SV leda förbi

595 by-pass damper
DA by-pass-spjæld
DE Beipaßklappe *f*
ES registro de
   derivación *m*
F volet *m* de
   dérivation
I serranda *f* di by-
   pass
MA megkerülőjárati
   csappantyú
NL omloopklep *f*
PO przepustnica *f*
   obejściowa
PY обводной клапан
   m
SU ohituspelti
SV förbigångsventil *c*

596 by-pass line
DA omløbsforbindelse
DE Beipaßleitung *f*
ES tubo de
   derivación *m*
F conduite *f* de
   dérivation
I tubazione *f* di by-
   pass
MA megkerülővezeték
NL omloopleiding *f*
PO przewód *m*
   obejściowy
PY обводная линия *f*
   обводная труба *f*
SU ohitusjohto
SV förbigångsledning *c*

597 by-pass valve
DA omløbsventil
   shuntventil
DE Beipaßventil *n*
ES válvula de
   derivación *f*
F vanne *f* de
   dérivation
I valvola *f* di by-pass
MA megkerülőjárati
   szelep
NL omloopafsluiter *m*
   omloopklep *f*
PO zawór *m* obejściowy
PY обводной
   вентиль *m*
SU ohitusventtiili
SV förbigångsventil *c*

598 byte
DA byte (EDB)
DE Byte *n*
ES byte *m*
F groupe *m* de
   positions
   binaires, de bits
   consécutifs
   multiplet *m*
I byte *m*
MA byte
NL byte *n*
PO bajt *m*
PY байт *m*
SU tavu
SV byte *c*

DA =	Danish
DE =	German
ES =	Spanish
F =	French
I =	Italian
MA =	Magyar (Hungarian)
NL =	Dutch
PO =	Polish
PY =	Russian
SU =	Finnish
SV =	Swedish

C

599 cable
DA kabel
DE Kabel *n*
ES cable *m*
  cadena de ancla *f*
F câble *m*
I cavo *m*
MA kábel
NL kabel *m*
PO kabel *m*
  lina *f* stalowa
PY кабель *m*
SU kaapeli
SV kabel *c*

600 cable connector
DA kabelforbindelse
DE Kabelverbinder *m*
ES conector de cable *m*
F connecteur *m* de
  câble
I connettore *m* del
  cavo
MA kábel csatlakozó
NL kabelplug *f*
PO łącznik *m* kablowy
PY кабельная муфта *f*
SU kaapeliliitin
SV kabelförbindning *c*

601 cable entry
DA kabelindføring
DE Kabeleingang *m*
ES entrada de cable *f*
F entrée *f* de câble
I ingresso *m* del cavo
MA kábel bevezetés
NL kabelinvoer *m*
PO wejście *n* kablowe
  wlot *m* kablowy
PY кабельный ввод *m*
SU kaapelin tuloaukko
SV kabelinfüring *c*

602 CAD (*see* computer-
  aided design)

603 calculation
DA beregning
DE Berechnung *f*
  Kalkulation *f*
ES cálculo *m*
F calcul *m*
I calcolo *m*
MA számítás
NL berekening *f*
  calculatie *f*
PO obliczanie *n*
PY расчет *m*
SU laskenta
SV beräkning *c*

604 calibration
DA justering
  kalibrering
DE Eichung *f*
ES calibración *f*
  tarado *m*
  verificación *f*
F calibrage *m*
I calibratura *f*
MA hitelesítés
NL ijking *f*
  kalibrering *f*
PO kalibrowanie *n*
  wzorcowanie *n*
PY калибровка *f*
SU kalibrointi
  tarkastus
SV kalibrering *c*

605 call for tenders
DA udbyde i licitation
DE Ausschreibung *f*
ES concurso *m*
  solicitud de ofertas *f*
F appel *m* d'offres
I gara *f* di appalto
MA versenytárgyalási
  felhívás
NL offerteaanvraag *f*
  prijsaanvraag *f*
PO ogłoszenie *n*
  przetargu
PY приглашение *n*
  подавать заявки
SU tarjouspyyntü
SV anbudsinfordran *c*

606 calorie
DA kalorie
DE Kalorie *f*
ES caloría *f*
F calorie *f*
I caloria *f*
MA kalória
NL calorie *f*
PO kaloria *f*
PY калория *f*
SU kalori
SV kalori *c*

607 calorific *adj*
DA varme-
DE wärmeerzeugend
ES calorífico
F calorifique
I calorifico
MA hő
  kalorikus
NL calorisch
PO cieplny
PY тепловой
SU lämpüä kehittävä
SV värmealstrande

**608 calorific value
  (gross)**
DA brændværdi (øvre)
DE Brennwert *m*
ES poder calorífico *m*
F pouvoir *m*
  calorifique
I potere *m* calorifico
  superiore
MA égéshő
  felső fűtőérték
  hőérték
NL calorische
  bovenwaarde *f*
PO ciepło *n* spalania
PY теплота *f* сгорания
  (верхняя)
SU ylempi lämpüarvo
SV värmevärde *n*
  (brutto)

609 calorific value (net)
DA brændværdi (nedre)
DE Heizwert *m*
ES poder calorífico *m*
F pouvoir *m* calorifique
I potere *m* calorifico inferiore
MA alsó fűtőérték fűtőérték
NL calorische onderwaarde *f*
PO wartość *f* opałowa
PY теплота *f* сгорания (нетто)
SU alempi lämpüarvo
SV värmevärde *n* (netto)

610 calorifier (heat exchanger)
DA varmeveksler
DE Wärme-austauscher *m*
ES cambiador de calor (calorífero) *m*
F échangeur *m* de chaleur
I scambiatore *m* di calore
MA kalorifer (hőcserélő)
NL tegenstroom-apparaat *n* warmtewisselaar *m*
PO wymiennik *m* ciepła
PY калорифер *m* (теплообменник)
SU lämmünsiirrin lämmünvaihdin
SV värmeväxlare *c*

611 calorimeter
DA kalorimeter varmemåler
DE Kalorimeter *n*
ES calorímetro *m*
F calorimètre *m*
I calorimetro *m*
MA hőmennyiségmérő
NL caloriemeter *m*
PO kalorymetr *m*
PY калориметр *m*
SU kalorimetri
SV kalorimeter *c*

612 cam
DA kam knast
DE Nocken *m*
ES leva *f*
F came *f*
I camma *f*
MA bütyük
NL schijf *f* tand *m*
PO krzywka *f*
PY кулачек *m*
SU nokka
SV kam *c* nock *c*

613 CAM (*see* computer-aided manufacturing)

614 cap
DA dæksel
DE Deckel *m* Kappe *f*
ES casquillo lámparas *m* tapón hembra *m*
F chapeau *m* couvercle *m*
I cappuccio *m* coperchio *m*
MA kupak
NL afsluitdop *m* afsluitkap *f* kap *f*
PO kołpak *m* pokrywa *f*
PY колпак *m*
SU tulppa
SV kåpa *c*

615 capacitance
DA kapacitans kapacitet
DE Kapazität *f* (elektr.)
ES capacitancia *f*
F capacitance *f*
I capacità *f*
MA kapacitás
NL capacitieve weerstand *m*
PO kapacytancja *f* reaktancja *f* pojemnościowa
PY емкость *f* (электрическая)
SU kapasitanssi
SV kapacitans *c*

616 capacitor (condenser)
DA kondensator
DE Kondensator *m* (elektr.)
ES condensador *m*
F capacité *f* (condensateur)
I condensatore *m*
MA kondenzátor
NL condensator *m*
PO kondensator *m*
PY конденсатор *m* (электрический)
SU kondensaattori
SV kondensator *c*

617 capacity
DA kapacitet ydeevne
DE Kapazität *f* Leistung *f*
ES capacidad *f* potencia *f*
F capacité *f* puissance *f* volume *m*
I capacità *f* potenza *f* potenzialità *f*
MA teljesítmény
NL capaciteit *f* inhoud *m*
PO pojemność *f* przepustowość *f* wydajność *f* zdolność *f*
PY мощность *f*
SU kapasiteetti tilavuus
SV kapacitet *c*

618 capacity control
DA kapacitetsregulering
DE Leistungsregelung *f*
ES control de capacidad *m*
F contrôle *m* de puissance
I controllo *m* di potenza
MA teljesítményszabá-lyozás
NL capaciteitsregeling *f* vermogensregeling *f*
PO regulacja *f* wydajności
PY регулирование *n* производитель-ности (мощности)
SU tehonsäätü
SV kapacitetskontroll *c*

619 capacity controller
DA kapacitetsregulator
DE Leistungsregler *m*
ES regulador de
capacidad *m*
F régulateur *m* de
puissance
I regolatore *m* di
potenzialità
MA teljesítményszabá-
lyozó
NL capaciteitsregelaar
m
vermogensregelaar
m
PO regulator *m*
wydajności
PY регулятор *m*
мощности
SU tehonsäädin
SV effektreglerdon *n*

620 capacity factor
DA kapacitetsfaktor
DE Leistungsfaktor *m*
ES factor de potencia *m*
F facteur *m* de
puissance
I coefficiente *m* di
utilizzazione
MA teljesítménytényező
NL arbeidsfactor *m*
PO współczynnik *m*
wydajności
PY фактор *m*
мощности
SU kapasiteettisuhde
SV kapacitansfaktor *c*

621 capacity reducer
DA effektregulator
DE Leistungsminderer
m
ES reductor de
capacidad *m*
F régulateur *m* de
puissance
I riduttore *m* di
capacità
riduttore *m* di
potenza
MA teljesítmény
csükkentő
NL vermogensterug-
regelaar *m*
PO reduktor *m*
wydajności
PY редуктор *m*
мощности
SU tehonsäädin
SV effektregulator *c*

622 capacity regulator
DA kapacitetsregulator
DE Leistungsregler *m*
ES regulador de
capacidad *m*
F dispositif *m* de
variation de
puissance
I regolatore *m* di
potenza
MA teljesítményszabá-
lyozó
NL capaciteitsregelaar
m
PO regulator *m*
wydajności
PY регулятор *m*
мощности
(производитель-
ности)
SU kylmäkompressorin
tehonsäätü-
laitteisto
SV effektregulator *c*

623 capillarity
DA hårrørsvirkning
kapillarvirkning
DE Kapillarität *f*
ES capilaridad *f*
F capillarité *f*
I capillarità *f*
MA hajszálcsüvesség
kapillaritás
NL capillariteit *f*
PO kapilarność *f*
włoskowatość *f*
PY капиллярный
эффект *m*
SU kapillaarisuus
SV kapillaritet *c*

624 capillary *adj*
DA kapillar
DE kapillar
ES capilar
F capillaire
I capillare
MA kapilláris
NL capillair
PO kapilarny
włoskowaty
PY капиллярный
SU kapillaarinen
SV kapillär

625 capillary action
DA hårrørsvirkning
kapillarvirkning
DE kapillare Aktion *f*
ES efecto capilar *m*
F capillarité *f*
I capillarità *f*
MA hajszálcső hatás
NL capillaire werking *f*
PO działanie *n*
kapilarne
działanie *n*
powierzchniowe
PY капиллярное
действие *n*
SU kapillaarinen imu
SV kapillaritet *c*

**626 capillary air
washer**
DA kapillar-luftvasker
DE Kapillarluftwäscher
m
ES humidificador por
capilaridad *m*
F laveur *m* d'air par
capillarité
I lavatore *m*
(umidificatore) di
aria a pacco
evaporante
MA hajszálcsüves
légmosó
NL capillaire
luchtbevochtiger
m
PO nawilżacz *m*
powietrza
kapilarny
PY аппарат *m* с
орошаемой
насадкой
SU kapillaarikostutin
SV kapillarlufttvättare *c*

627 capillary tube
DA kapillarrør
DE Kapillare *f*
Kapillarrohr *n*
ES capilar *m*
tubo capilar *m*
F tube *m* capillaire
I tubo *m* capillare
MA hajszálcső
kapilláris cső
NL capillaire buis *f*
PO kapilara *f*
rurka *f* włoskowata
PY обводная труба *f*
SU kapillaariputki
SV kapillärrür *n*

628 carbon (C)
DA kul
kulstof
DE Kohlenstoff *m*
ES carbono *m*
F carbone *m*
I carbonio *m*
MA szén
NL koolstof *f*
PO węgiel *m*
PY углерод *m*
SU hiili
SV kol *n*

629 carbon dioxide (CO$_2$)
DA kultveilte
DE Kohlendioxid *n*
ES anhídrido
carbónico *m*
bióxido de
carbono *m*
F anhydride *m*
carbonique
gaz *m* carbonique
I anidride *f* carbonica
MA széndioxid
NL koolstofdioxyde *n*
PO dwutlenek *m* węgla
PY углекислый газ *m*
SU hiilidioksidi
SV koldioxid *c*

630 carbon filter (USA)
(*see* charcoal filter)

631 carbonic acid (H$_2$CO$_3$)
DA kulsyre
DE Kohlensäure *f*
ES ácido carbónico *m*
F acide *m* carbonique
I acido *m* carbonico
MA szénsav
NL koolzuur *n*
PO kwas *m* węglowy
PY угольная кислота *f*
SU hiilihappo
SV kolsyra *c*

DA =	Danish
DE =	German
ES =	Spanish
F =	French
I =	Italian
MA =	Magyar
	(Hungarian)
NL =	Dutch
PO =	Polish
PY =	Russian
SU =	Finnish
SV =	Swedish

632 carbonization
DA forkulning
karbonisering
DE Verkohlung *f*
ES carbonización *f*
F carbonisation *f*
I carbonizzazione *f*
MA szenesedés
NL carbonisatie *f*
verkoling *f*
PO koksowanie *n*
zwęglanie *n*
PY карбонизация *f*
SU hiiletys
SV förkoksning *c*

633 carbon monoxide (CO)
DA kulilte
DE Kohlenmonoxid *n*
ES monóxido de
carbono *m*
F oxyde de carbone *m*
I monossido *m* di
carbonio
MA szénmonoxid
NL koolmonoxyde *n*
PO tlenek *m* węgla
PY окись *f* углерода
угарный газ *m*
SU hiilimonoksidi
SV koloxid *c*

634 cargo
DA fragt
skibslast
DE Fracht *f*
Ladung *f*
ES carga *f*
F cargo *m*
I mercantile *m*
MA szállítmány
NL lading *f*
PO ładunek *m*
towar *m*
PY груз *m*
SU lasti
rahti
SV last *c*

635 Carnot cycle
DA Carnot proces
DE Carnot-
Kreisprozeß *m*
ES ciclo de Carnot *m*
F cycle *m* de Carnot
I ciclo *m* di Carnot
MA Carnot kürfolyamat
NL Carnot-kringloop *m*
PO obieg *m* Carnota
PY цикл *m* карно
SU Carnot-prosessi
SV carnotprocess *c*

636 carrier frequency
DA bærefrekvens
DE Trägerfrequenz *f*
ES frecuencia portante *f*
F fréquence *f* porteuse
I frequenza *f* portante
MA hordozó frekvencia
NL draaggolf-
frequentie *f*
PO częstotliwość *f*
nośna
PY несущая частота *f*
SU kantoaallon taajuus
SV bärfrekvens *c*

637 carryover eliminator
DA overføringsudskiller
DE tragbarer
Netzanschluß *m*
(elektr.)
ES supresor de parásitos
(elec) *m*
F régulation
I separatore *m* di
gocce
MA ürvénylésgátló
NL druppelvanger *m*
PO eliminator *m*
przenoszenia
PY пластинчатый
каплеотделитель
m
SU siirtymisenestolaite
SV anordning *c* av
vätska för att
bindra
medryckning *c*

638 cartridge filter (*see*
cellular filter)

639 cartridge fuse
DA patronsikring
DE Sicherungspatrone *f*
ES fusible de cartucho
m
F fusible *m* à
cartouche
I fusibile *m* a
cartuccia
MA hüvelyes biztosíték
NL buissmeltveiligheid *f*
PO bezpiecznik *m*
wkładka *f* topikowa
zamknięta
PY плавкая вставка *f*
SU sulake
SV säkring *c*

640 cascade
DA kaskade
DE Kaskade *f*
ES cascada *f*
F cascade *f*
I cascata *f*
MA kaszkád
    lépcsős
NL achter elkaar
    cascade
PO kaskada *f*
    układ *m* kaskadowy
PY водопад *m*
    каскад *m*
SU kaskadi
SV kaskad *c*

641 cascade control
DA kaskadestyring
DE Kaskadenregelung *f*
ES regulación en
    cascada *f*
F régulation *f* en
    cascade
I regolazione *f* in
    cascata
MA kaszkád szabályozás
NL cascaderegeling *f*
PO regulacja *f*
    kaskadowa
    sterowanie *n*
    kaskadowe
PY каскадное
    регулирование *n*
SU kaskadisäätü
SV hastighetsreglering *c*

642 cascade controller
DA kaskaderegulator
DE Kaskadenregler *m*
ES regulador *m* en
    cascada *m*
F régulateur à cascade
I regolatore *m* in
    cascata
MA kaszkád szabályozó
NL cascaderegelaar *m*
PO regulator *m*
    kaskadowy
PY каскадный
    регулятор *m*
SU kaskadisäädin
SV hastighets-
    övervakare *c*

643 case
DA kasse
    omslag
    sag
DE Behälter *m*
    Kiste *f*
ES caja *f*
F boîte *f*
    caisse *f*
I cassa *f*
MA burkolat
    tok
NL doos *f*
    geval *n*
    mantel *m*
    omkasting *f*
PO kaseta *f*
    pudło *n*
    skrzynka *f*
PY ящик *m*
SU kotelo
    laatikko
    säiliü
SV huv *c*
    hölje *n*

644 case study
DA sagsundersøgelse
DE Fallstudie *f*
ES análisis del caso *m*
F étude *f* de cas
I analisi *f* del caso
MA esettanulmány
NL gevalsonderzoek *n*
PO badanie *n* dotyczące
    pewnego
    przypadku
PY исследование *n*
    корпуса
    (оболочки)
SU esimerkkitutkimus
SV fallstudie *c*

645 casing
DA hus
    kappe
DE Gehäuse *n*
    Umhüllung *f*
ES carcasa *f*
    envuelta *f*
    revestimiento
    (calderas) *m*
F cache-radiateur *m*
    enveloppe *f*
I custodia *f*
    mantello *m*
MA burkolás
NL mantel *m*
    omhulsel *n*
    omkasting *f*
PO obudowa *f*
    osłona *f*
PY обмуровка *f*
SU kotelo
SV kåpa *c*

646 cast iron
DA støbejern
DE Gußeisen *n*
ES arrabio *m*
    función *f*
    hierro colado *m*
    hierro fundido *m*
F fonte *f*
I ghisa *f*
MA üntüttvas
NL gietijzer *n*
PO żeliwo *n*
PY чугун *m*
SU valurauta
SV gjutjärn *n*

647 cast iron boiler
DA støbejernskedel
DE gußeiserner
    Kessel *m*
    Gußheizkessel *m*
ES caldera de
    fundición *f*
    caldera de hierro
    fundido *f*
F chaudière *f* en fonte
I caldaia *f* in ghisa
MA üntüttvas kazán
NL gietijzeren ketel *m*
PO kocioł *m* żeliwny
PY чугунный котел *m*
SU valurautakattila
SV gjutjärnspanna *c*

648 cast iron fitting
DA støbejernsfitting
DE gußeisernes
    Formstück *n*
ES accesorios de hierro
    colado *m,pl*
    accesorios de hierro
    fundido *m,pl*
F raccord *m* en fonte
I raccordo *m* di ghisa
MA üntüttvas idom
NL gietijzeren fitting *m*
PO armatura *f* żeliwna
PY чугунная фасонная
    часть *f*
SU valurautaiset putken
    osat
SV gjutjärnsrördel *c*

649 cast iron radiator
DA støbejernsradiator
DE Gußheizkörper *m*
ES radiador de
   fundición *m*
F radiateur *m* en fonte
I radiatore *m* di ghisa
MA üntüttvas radiátor
NL gietijzeren
   radiator *m*
PO grzejnik *m* żeliwny
PY чугунный
   радиатор *m*
SU valurautapatteri
SV gjutjärnsradiator *c*

650 casual heat gain
DA tilfældigt
   varmetilskud
DE zufälliger
   Wärmegewinn *m*
ES ganancia de calor
   imprevista *f*
F gain *m* de chaleur
   occasionnel
I apporto *m* di calore
   occasionale
MA esetenkénti
   hőnyereség
NL toevallige
   warmtewinst *f*
PO zysk *m* ciepła
   chwilowy
PY случайное
   поступление *n*
   теплоты
SU tilapäinen
   lämpükuorma
SV tillfällig
   värmeavgivning *c*

651 catalyst
DA katalysator
DE Katalysator *m*
ES catalizador *m*
F catalyseur *m*
I catalizzatore *m*
MA katalizátor
NL katalysator *m*
PO katalizator *m*
PY катализатор *m*
SU katalysaattori
SV katalysator *c*

652 cathode
DA katode
   negativ ion
DE Kathode *f*
ES cátodo *m*
F cathode *f*
I catodo *m*
MA katód
NL kathode *f*
PO katoda *f*
PY катод *m*
SU katodi
SV katod *c*

653 cathodic protection
DA katodisk beskyttelse
DE kathodischer
   Schutz *m*
ES protección
   catódica *f*
F protection *f*
   cathodique
I protezione *f*
   catodica
MA katódos védelem
NL kathodische
   bescherming *f*
PO zabezpieczenie *n*
   katodowe
PY катодная защита *f*
SU katodinen suojaus
SV katodiskt skydd *n*

654 cation
DA kation
DE Kation *n*
ES catión *m*
F cation *m*
I catione *m*
MA kation
NL kation *n*
PO jon *m* dodatni
   kation *m*
PY катион *m*
SU positiivinen ioni
SV katjon *c*

655 caulking (of boilersections)
DA kalfatring
DE Abdichten *n*
   Verstemmen *n*
ES calafateado *m*
F matage *m*
I sigillatura *f*
MA tümítés
   (kazántagoknál)
NL koken *n* van
   ketelleden
PO uszczelnianie *n*
   (członów kotła)
PY чеканка *f* (секций
   котла)
SU kattilaosien välinen
   tiivistys
SV tätning *c* (av
   pannsektioner)

656 cavitation
DA kavitation
DE Kavitation *f*
ES cavitación *f*
F cavitation *f*
I cavitazione *f*
MA kavitáció
NL cavitatie *f*
PO kawitacja *f*
PY кавитация *f*
SU kavitaatio
SV kavitation *c*

657 cavity
DA hulrum
DE Hohlraum *m*
ES cavidad *f*
F cavité *f*
I cavità *f*
MA üreg
   üreg
NL holte *f*
PO wgłębienie *n*
   wnęka *f*
PY впадина *f*
SU ontelo
SV kavitet *c*

658 ceiling
DA loft
DE Decke *f*
ES cielo raso *m*
   techo *m*
F plafond *m*
I soffitto *m*
MA mennyezet
NL plafond *n*
PO sufit *m*
PY потолок *m*
SU katto
SV tak *n*

659 ceiling coil
DA loftslange
DE Deckenrohr-
schlange *f*
ES batería horizontal de
techo *f*
F batterie *f* de plafond
(grille de plafond)
I pannello *m* radiante
a soffitto
MA mennyezeti csőkígyó
NL plafondregister *n*
PO wężownica *f*
zatopiona w
suficie
PY потолочный
змеевик *m*
SU kattopatteri
SV takbatteri *n*

660 ceiling diffuser
DA loftdiffuser
DE Deckenluftauslaß *m*
ES difusor de techo *m*
F diffuseur *m* de
plafond
diffuseur *m*
plafonnier
I diffusore *m* a
soffitto
MA mennyezeti
befúvórács
NL anemostaat *m*
plafonduitblaas-
element *n*
PO anemostat *m*
PY потолочный
диффузор *m*
потолочный
плафон *m*
SU kattohajottaja
SV takspridare *c*

661 ceiling grid (*see*
ceiling coil)

662 ceiling heating
DA loftsvarme
DE Deckenheizung *f*
ES calefacción por el
techo mediante
paneles
suspendidos *f*
F chauffage *m* par le
plafond
I riscaldamento *m* a
soffitto
MA mennyezetfűtés
NL plafond-
verwarming *f*
PO ogrzewanie *n*
sufitowe
PY потолочное
отопление *n*
SU kattolämmitys
SV takvärme *c*

663 ceiling outlet
DA loftaftræk
DE Deckenauslaß *m*
ES rejilla de techo *f*
F bouche *f* de reprise
en plafond
I presa *f* di corrente a
soffitto
MA mennyezeti
kifúvórács
NL plafonduitblaas-
element *n*
PO wylot *m* sufitowy
PY потолочное
отверстие *n*
SU poistoaukko katolla
SV takutlopp *n*

664 ceiling structure
DA loftkonstruktion
DE Deckenbauart *f*
ES estructura de techo *f*
F plancher *m* (haut)
I solaio *m*
struttura *f* del
soffitto
MA mennyezet szerkezet
NL plafondconstructie *f*
PO strop *m*
PY перекрытие *n*
SU kattorakenne
SV takkonstruktion *c*

665 ceiling ventilator
DA loftventilator
DE Deckenventilator *m*
ES ventilador de techo
m
F bouche *f* de plafond
I bocca *f* a soffitto
MA mennyezeti
szellőzőnyílás
NL plafondventilator *m*
PO wentylator *m*
sufitowy
PY потолочный
вентилятор *m*
SU huippuimuri
kattotuuletin
SV takfläkt *c*

666 ceiling void
DA lofthulrum
DE Deckenhohlraum *m*
ES espacio hueco sobre
el falso techo *m*
F espace *m* entre
plafond et faux
plafond
I spazio *m* tra
controsoffitto e
intradosso del
solaio
MA mennyezeti hézag
NL plafondplenum *n*
PO przestrzeń *f* stropu
podwieszonego
PY пространство *n* за
потолком
SU välikattotila
SV undertaks-
utrymme *n*

667 cell
DA celle
element
DE Element *n*
Zelle *f*
ES celdilla *f*
célula *f*
F cellule *f*
I cella *f*
cellula *f*
MA cella
sejt
NL cel *f*
PO komórka *f*
PY ячейка *f*
SU solu
SV cell *c*

668 cellar

668 cellar
DA kælder
DE Keller *m*
ES bodega *f*
    sótano *m*
F cave *f*
    sous-sol *m*
I cantina *f*
    scantinato *m*
MA pince
NL kelder *m*
PO piwnica *f*
    podziemie *n*
PY подвал *m*
SU kellari
SV källare *c*

669 cellular air filter
(*see* cellular filter)

670 cellular filter
DA cellefilter
    kassettefilter
DE Zellenfilter *m*
ES filtro de cartucho *m*
F filtre *m* cellulaire
    (filtre à cartouche)
I filtro *m* a cartuccia
MA szűrőcella
NL cassettefilter *m*
PO filtr *m* powietrza
    działkowy
PY ячейковый фильтр
    m
SU suodatinkasetti
SV kassettfilter *n*

671 cellular glass (*see*
foam glass)

**672 cellular rubber
thermal insulation**
DA celleopbygget
    isolering af gummi
DE Zellgummi-
    Wärme-
    dämmung *f*
ES aislamiento térmico
    de goma celular *m*
F isolation *f*
    thermique à
    caoutchouc
    cellulaire
I isolamento *m*
    termico in gomma
    cellulare
MA üreges gumi
    hőszigetelés
NL schuimrubber
    isolatie *f*
PO izolacja *f* cieplna z
    gumy piankowej
PY резино-ячеистая
    теплоизоляция *f*
SU solukumilämpüeriste
SV cellgummi
    värmeisolering *c*

673 cellular structure
DA cellestruktur
DE Zellenbauweise *f*
ES estructura celular *f*
F structure *f* cellulaire
I struttura *f* cellulare
MA sejtszerkezet
    üreges szerkezet
NL cellulaire structuur *f*
PO budowa *f*
    komórkowa
    konstrukcja *f*
    komórkowa
PY сотовая
    структура *f*
    ячеистая
    структура *f*
SU huokoinen rakenne
SV cellbyggnad *c*

**674 cellular thermal
insulation**
DA celletermoisolering
DE zellulare
    Wärme-
    dämmung *f*
ES aislamiento térmico
    celular *m*
F isolation *f*
    thermique
    cellulaire
I isolamento *m*
    termico cellulare
MA üreges hőszigetelés
NL cellulaire thermische
    isolatie *f*
PO izolacja *f* cieplna
    komórkowa
PY ячеистая
    теплоизоляция *f*
SU solumainen
    lämpüeriste
SV cellvärmeisolering *c*

675 cement
DA bindemiddel
    cement
DE Zement *m*
ES cemento *m*
    pegamento *m*
F ciment *m*
I cemento *m*
MA cement
NL cement *n*
PO cement *m*
PY цемент *m*
SU sementti
SV cement *c*

676 centigrade
DA celsiusgrader
DE Zentigrad *m*
ES centígrado
F centigrade *m*
I centigrado
MA Celsius fok
NL graad *m* Celcius
    honderddelig
PO stopień *m* Celsjusza
PY градус *m* Цельсия
SU celsiusaste (C)
SV celsiusgrad *c*

DA = Danish
DE = German
ES = Spanish
F = French
I = Italian
MA = Magyar
    (Hungarian)
NL = Dutch
PO = Polish
PY = Russian
SU = Finnish
SV = Swedish

677 central boiler plant
DA kedelcentral
DE Heizzentrale f
ES central de
   calefacción f
F chaufferie f centrale
I centrale f termica
MA küzponti kazántelep
NL centraal ketelhuis n
   centrale
   ketelinstallatie f
PO kotłownia f
   centralna
PY центральная
   котельная f
SU keskuslämmityslaitos
SV panncentral c

678 central control
DA centralstyring
DE Zentralregelung f
ES control centralizado
   m
F commande f
   centrale
   régulateur m central
I regolazione f
   centrale
MA küzponti szabályozás
NL centrale besturing f
   centrale regeling f
PO regulacja f centralna
   sterowanie n
   centralne
PY центральное
   регулирование n
SU keskitetty säätü
SV central kontroll c

**679 central control
panel**
DA centralstyrepanel
DE Zentralschaltwarte f
ES cuadro central de
   control m
F tableau m général de
   commande
I quadro m centrale
   di controllo
MA küzponti szabályozó
   tábla
NL centraal
   bedienings-
   paneel n
PO centralna tablica f
   sterownicza
PY центральная панель
   f управления
SU keskusvalvonta-
   paneeli
SV centralkontroll-
   panel c

680 central heating
DA centralvarme
DE Zentralheizung f
ES calefacción central f
F chauffage m central
I riscaldamento m
   centrale
MA küzponti fűtés
NL centrale
   verwarming f
PO centralne
   ogrzewanie n
PY центральное
   отопление n
SU keskuslämmitys
SV centralvärme c

**681 central heating
plant**
DA centralvarmeanlæg
DE Zentralheizungs-
   anlage f
ES central de
   calefacción f
F chaufferie f centrale
I impianto m di
   riscaldamento
   centrale
MA küzponti
   fűtőberendezés
NL centrale
   verwarmings-
   installatie f
PO instalacja f
   centralnego
   ogrzewania
   urządzenie n
   centralnego
   ogrzewania
PY установка f
   центрального
   отопления
SU lämpükeskus
SV centralvärme-
   anläggning c

**682 central heating
station**
DA varmecentral
DE Heizzentrale f
ES central de
   calefacción f
F chaufferie f centrale
   de chauffage
I centrale f termica
MA küzponti
   fűtőállomás
NL centraleverwarming-
   sonderstation n
PO kotłownia f
   centralnego
   ogrzewania
PY теплоцентраль f
SU lämpükeskus
SV värmecentral c

**683 central processor
unit**
DA centralbehandler-
   enhed
DE Zentralprozessor m
ES procesador central m
F processeur m central
I processore m
   centrale
MA küzponti
   processzoregység
NL centrale
   verwerkings-
   eenheid (CVE) f
PO centralny
   procesor m
PY центральный
   процессор m
SU keskusyksikkü
SV centralenhet c

**684 central station air
handling unit**
DA centralluftbehandlingsenhed
DE zentrales
   Luftbehandlungs-
   gerät n
ES climatizador central
   m
F centrale f de
   traitement d'air
I centrale f di
   trattamento
   dell'aria
MA küzponti légkezelő
   egység
NL centrale
   luchtbehandelings-
   kast f
PO zespół m
   centralnego
   uzdatniania
   powietrza
PY центральная
   установка f
   кондицион-
   ирования воздуха
SU keskusilmasto-
   intikone
SV central
   luftbehandlings-
   anläggning c

685 centre-line velocity
DA centerhastighed
DE Axialgeschwindig-
keit *f*
ES velocidad en el eje
longitudinal *f*
F vitesse *f* axiale
I velocità *f* assiale
MA küzépvonali sebesség
NL kernsnelheid *f*
snelheid *f* in
centrum van een
luchtstraal
PO prędkość *f* osiowa
PY осевая скорость *f*
SU nopeus keskiviivalla
SV axialhastighet *c*

686 centre of gravity
DA tyngdepunkt
DE Schwerpunkt *m*
ES centro de
gravedad *m*
F centre *m* de gravité
I centro *m* di gravità
MA súlypont
NL zwaartepunt *n*
PO środek *m* ciężkości
PY центр *m* тяжести
SU painopiste
SV gravitations-
centrum *n*

687 centrifugal *adj*
DA centrifugal
DE zentrifugal
ES centrífugo
F centrifuge
I centrifugo
MA centrifugális
NL centrifugaal
middelpuntvliedend
PO odśrodkowy
PY центробежный
SU
SV centrifugal

**688 centrifugal
compressor**
DA centrifugal-
kompressor
DE Kreiselverdichter *m*
ES compresor
centrífugo *m*
F compresseur *m*
centrifuge
I compressore *m*
centrifugo
MA centrifugál
kompresszor
NL centrifugaal-
compressor *m*
PO sprężarka *f*
odśrodkowa
PY турбо-
компрессор *m*
SU keskipako-
kompressori
SV centrifugal-
kompressor *c*

689 centrifugal fan
DA centrifugalventilator
DE Radialventilator *m*
ES ventilador centrífugo
m
F ventilateur *m*
centrifuge
I ventilatore *m*
centrifugo
MA centrifugál ventilátor
NL centrifugaal
ventilator *m*
PO wentylator *m*
odśrodkowy
PY радиальный
вентилятор *m*
центробежный
вентилятор *m*
SU keskipakopuhallin
SV centrifugalfläkt *c*

690 centrifugal force
DA centrifugalkraft
DE Zentrifugalkraft *f*
ES fuerza centrífuga *f*
F force *f* centrifuge
I forza *f* centrifuga
MA centrifugális erő
NL centrifugaal kracht *f*
middelpuntvliedende
kracht *f*
PO siła *f* odśrodkowa
PY центробежная
сила *f*
SU keskipakovoima
SV centrifugalkraft *c*

691 centrifugal pump
DA centrifugalpumpe
DE Kreiselpumpe *f*
ES bomba centrífuga *f*
F pompe *f* centrifuge
I pompa *f* centrifuga
MA centrifugál szivattyú
NL centrifugaalpomp *f*
PO pompa *f*
odśrodkowa
PY центробежный
насос *m*
SU keskipakopumppu
SV centrifugalpump *c*

**692 centrifugal
separator**
DA cyklonudskiller
DE Fliehkraft-
abscheider *m*
ES separador
centrífugo *m*
F séparateur *m*
centrifuge
I separatore *m*
centrifugo
MA centrifugális
leválasztó
NL centrifugaal
afscheider *m*
PO separator *m*
odśrodkowy
PY центробежный
сепаратор *m*
SU keskipakoerotin
sentrifugi
SV cyklonavskiljare *c*

693 centrifuge
DA centrifuge
DE Zentrifuge *f*
ES centrífugo
F centrifuge
I centrifuga *f*
MA centrifuga
NL centrifuge *f*
PO wirówka *f*
PY центрифуга *f*
SU sentrifugi
SV centrifug *c*

694 CER (*see* cooling
efficiency ratio)

695 certificate
DA certifikat
DE Zertifikat *n*
   Zeugnis *n*
ES certificado *m*
F attestation *f*
  certificat *m*
I attestato *m*
  certificato *m*
MA bizonylat
NL certificaat *n*
PO atest *m*
  certyfikat *m*
  świadectwo *n*
PY сертификат *m*
SU todistus
SV certifikat *n*

696 CFC (*see* chlorofluorocarbon)

697 cfm (*see* Appendix A)

698 cfs (*see* Appendix A)

699 chain grate stoker
DA kæderist
DE Wanderrost *m*
ES cargador automático
  de parrilla
  articulada
  (calderas) *m*
F foyer *m* à grille
  foyer *m* à grille
  mécanique
I focolare *m* a griglia
  meccanica
MA láncrostély
NL kettingroosterstook-
  inrichting *f*
PO ruszt *m* łańcuchowy
  mechaniczny
PY топка *f* с цепной
  колосниковой
  решеткой
SU ketjuarina
SV kedjerost *c*

700 chalk
DA kridt
DE Kreide *f*
ES chimenea *f*
F calcaire *m*
  craie *f*
I calcare *m*
  gesso *m*
MA kréta
  szívató
NL kalk *m*
PO kreda *f*
PY мел *m*
SU kalkki
SV krita *c*

701 chamber
DA kammer
  rum
DE Kammer *f*
  Raum *m*
ES cámara *f*
F chambre *f*
I camera *f*
MA kamra
NL kamer *f*
  vertrek *n*
PO komora *f*
PY камера *f*
SU kammio
SV rum *n*

702 change of state
DA tilstandsændring
DE Änderung *f* des
  Aggregat-
  zustandes *m*
ES cambio de estado *m*
  cambio de fase *m*
F changement *m*
  d'état
I cambiamento *m* di
  stato
MA állapotváltozás
NL toestands-
  verandering *f*
PO zmiana *f* stanu
  skupienia
PY изменение *n*
  состояния
SU muutostila
SV tillståndsändring *c*

703 changeover
DA omstilling
DE Umschaltung *f*
ES inversión *f*
F changement *m* de
  température été-
  hiver
  changement *m* été-
  hiver
I inversione *f*
MA átváltás
NL omschakeling *f*
PO przełączenie *n*
  przestawienie *n*
PY переход *m* с
  режима на режим
SU vaihtokytkentä
SV omkoppling *c*

704 changeover control
DA vekselstyring
DE Wechselregelung *f*
ES control de inversión
  m
F commande *f* de
  changement de
  température été-
  hiver
I controllo *m* di
  inversione
MA váltószabályozás
NL omschakelregeling *f*
PO regulacja *f*
  przełączania
PY изменение *n*
  режима работы
SU vaihtokytkentäsäätü
SV omkastarkontroll *c*

705 changeover switch
DA omskifter
DE Wechselschalter *m*
ES conmutador
  inversor *m*
F inverseur *m*
I commutatore *m*
MA váltókapcsoló
NL omschakelaar *m*
  wisselschakelaar *m*
PO przełącznik *m*
PY переключатель *m*
  режима
SU vaihtokytkin
  ylimenokytkin
SV omkastare *c*

706 channel
DA kanal
DE Kanal *m*
  Spur *f*
ES canal *m*
F canal *m*
I canale *m*
MA csatorna
NL kanaal *n*
PO kanał *m*
PY канал *m*
SU kanava
SV kanal *c*

707 characteristic
DA karakteristik
  kendetegn
DE Charakteristik *f*
ES característica *f*
F caractéristique *f*
I caratteristica *f*
MA jellemző
  karakterisztika
NL karakteristiek *f*
PO charakterystyka *f*
PY характеристика *f*
SU ominainen
SV karaktäristik *c*

708 characteristic curve
DA karakteristisk kurve
DE Kennlinie *f*
ES curva característica *f*
F courbe *f* caractéristique
I curva *f* caratteristica
MA jelleggürbe
NL karakteristieke kromme *f*
PO charakterystyka *f* liniowa
krzywa *f* charakterystyczna
krzywa *f* charakterystyki
PY графическая характеристика *f*
SU ominaiskäyrä
SV karaktäristisk kurva *c*

709 characteristic number
DA karakteristisk tal
DE Kennzahl *f*
ES número característico *m*
F nombre *m* caractéristique
I numero *m* caratteristico
MA jellemző szám
NL kengetal *n*
PO liczba *f* charakterystyczna
wartość *f* własna
PY характеристическое число *n*
SU ominaisluku
SV karaktäristiskt tal *n*

710 charcoal filter
DA kulfilter
DE Kohlefilter *m*
ES filtro de carbón vegetal *m*
F filtre *m* à charbon
I filtro *m* a carbone attivo
MA szénszűrő
NL koolfilter *m* noritfilter *m*
PO filtr *m* węglowy
PY угольный фильтр *m*
SU hiilisuodatin
SV kolfilter *n*

711 charge
DA last
opladning
DE Füllung *f*
Ladung *f*
Last *f*
ES carga *f*
F charge *f*
I carica *f*
MA tültés
NL belasting *f*
lading *f*
PO ładunek *m*
obciążenie *n*
PY загрузка *f*
заправка *f*
SU ladata
vastata
veloittaa
SV last *c*

712 charge *vb*
DA (op)lade
belaste
DE laden
ES carga *f*
F charger
I caricare
MA tülteni
NL belasten
laden
opladen
PO ładować
obciążać
PY загружать
заправлять
SU
SV belasta

713 charge controller
DA belastningsregulator
DE Laderegler *m*
ES regulador de carga *m*
F régulateur *m* de charge
I regolatore *m* di carica
MA tültésszabályozó
NL ladingsregelaar *m*
PO regulator *m* obciążenia
PY регулятор *m* заправки
SU kuormituksen säädin
SV belastnings- regulator *c*

714 chargehand
DA formand
sjakbajs
DE Vorarbeiter *m*
ES capataz *m*
encargado *m*
jefe de equipo *m*
F chef *m* d'équipe
I caposquadra *m*
MA előmunkás
NL chefmonteur *m*
leidend monteur *m*
PO kierownik *m* zespołu
PY руководитель *m* группы
SU esimies
SV förman *c*

715 charging (furnace or stove)
DA indfyring
DE Beschickung *f*
Füllen *n*
ES carga (de un horno) *f*
operación de carga *f*
F chargement *m*
I caricamento *m*
MA feltültés (kazánt)
NL vullen (oven of kachel)
PO ładowanie *n* (paleniska lub pieca)
PY нагрузка *f* (топки или печи)
SU täyttü (kattilan)
SV (panna *c* eller ugn *c*) påfyllning *c*

716 charging capacity (thermal storage)
DA ladekapacitet
DE Speicherkapazität *f*
ES capacidad de acumulación térmica *f*
F capacité *f* calorifique (stockage thermique)
I capacità *f* di accumulo
MA tültési kapacitás
NL oplaadcapaciteit *f* (thermische opslag)
PO zdolność *f* ładowania (akumulacja cieplna)
PY аккумуляционная емкость *f*
SU ladattu (lämpü)määrä
SV beskicknings-kapacitet *c*

717 charging connection
DA påfyldnings-forbindelse
DE Aufladeverbindung *f*
ES conexión para carga *f*
F système *m* de mise en charge
I raccordo *m* di carica
MA tültő csatlakozás
NL vulleiding *f*
PO połączenie *f* ładowania
PY заправочная линия *f*
SU täyttüyhde
SV påfyllningsrör *n*

718 charging door
DA fyrdør påfyldningslem
DE Fülltür *f*
ES puerta de carga *f*
F porte *f* de chargement
I portello *m* di caricamento
MA tültőajtó
NL vuldeur *f*
PO drzwi *pl* wsadowe drzwi *pl* zasypowe
PY загрузочный проем *m* монтажный проем *m*
SU täyttüluukku
SV påfyllningslucka *c*

719 charging rate
DA fyldningsgrad ladestrøm
DE Ladestrom *m* Stromaufnahme *f*
ES carga (cantidad de) *f*
F taux *m* de charge
I velocità *f* di accumulo
MA tültési fok
NL oplaadsnelheid *f*
PO wielkość *f* ładowania
PY расходная емкость *f*
SU latausteho
SV påfyllnings-hastighet *c*

720 charging set
DA opladningsaggregat
DE Ladeeinheit *f* Ladegerät *n* Ladesatz *m*
ES dispositivo de carga *m*
F chargement *m*
I dispositivo *m* di caricamento
MA tültő készlet
NL laadapparaat *n*
PO zespół *m* ładowania
PY аккумуляционная установка *f*
SU latauslaitteisto
SV laddningsaggregat *n*

721 check
DA kontrol prøvning
DE Kontrolle *f*
ES comprobación *f*
F contrôle *m*
I controllo *m*
MA ellenőrzés
NL controle *f*
PO sprawdzenie *n* test *m*
PY контроль *m*
SU tarkastaa
SV kontroll *c*

722 check *vb*
DA kontrollere prøve
DE kontrollieren prüfen
ES comprobar controlar
F contrôler
I controllare
MA ellenőrizni
NL controleren nameten
PO pękać sprawdzać zatrzymywać
PY контролировать
SU tarkastaa
SV kontrollera

723 check valve
DA kontraventil
DE Prüfventil *n* Rückschlagventil *n*
ES válvula de retención *f*
F clapet *m* de retenue
I valvola *f* di ritegno
MA ellenőrző szelep
NL terugslagklep *f*
PO zawór *m* jednokierunkowy zawór *m* zwrotny
PY обратный клапан *m*
SU takaiskuventtiili
SV kontrollventil *c*

DA	=	Danish
DE	=	German
ES	=	Spanish
F	=	French
I	=	Italian
MA	=	Magyar (Hungarian)
NL	=	Dutch
PO	=	Polish
PY	=	Russian
SU	=	Finnish
SV	=	Swedish

724 chemicals
DA kemikalier
DE Chemikalien *f,pl*
ES productos químicos
   m,pl
F produits *m*
   chimiques
I prodotti *m,pl*
   chimici
MA vegyszerek
NL chemicaliën *pl*
PO chemikalia *pl*
PY химические
   продукты *m,pl*
SU kemikaalit
SV kemikalier *c*

725 chest freezer
DA kassefryser
   kummefryser
DE Kastenkühler *m*
ES arcón congelador *m*
F coffre *m* congélateur
I cassone *m*
   congelatore
   domestico
MA hűtőláda
NL vrieskist *f*
PO zamrażarka *f*
   szafkowa
PY морозильная
   камера *f* со
   съемной крышей
SU pakastinarkku
SV frysbox *c*

726 chill *vb*
DA køle
DE abkühlen
   kühlen
ES enfriar
   refrigerar
F refroidir
I raffreddare
MA hűt
NL afkoelen
   koelen
PO chłodzić
   ochładzać
PY охлаждать
SU jäähdyttää
SV kyla

727 chilled cargo
DA kølelast
DE Kühlfracht *f*
ES carga refrigerada *f*
F cargo *m* froid
I mercantile *m*
   refrigerato
MA hűtütt szállítmány
NL gekoelde lading *f*
PO towar *m* schłodzony
PY охлажденный
   груз *m*
SU jäähdytetty lasti
SV kyllast *c*

728 chilled water
DA kølevand
DE gekühltes Wasser *n*
   Kaltwasser *n*
ES agua enfriada *f*
   agua refrigerada *f*
F eau *f* glacée
I acqua *f* refrigerata
MA hűtütt víz
NL gekoeld water *n*
PO woda *f* chłodząca
PY охлажденная
   вода *f*
SU jäähdytetty vesi
SV kylvatten *n*

729 chiller
DA køler
DE Kühler *m*
ES enfriador *m*
F réfrigérant *m*
   refroidisseur *m*
I refrigeratore *m*
MA hűtő
NL koeler *m*
PO agregat *m*
   chłodniczy
   chłodziarka *f*
PY водоохладитель *m*
SU vedenjäähdytin
SV kylare *c*

730 chilling (cooling)
DA afkøling
DE Abkühlung *f*
   Kühlung *f*
ES refrigeración
   (enfriamiento) *f*
F réfrigération *f*
   (refroidissement)
I refrigerazione *f*
MA hűtés
NL afkoeling *f*
PO chłodzenie *n*
PY охлаждение *n*
SU jäähdytys
SV kylning *c*

731 chilling effect
DA køleeffekt
DE Kühlwirkung *f*
ES efecto de
   refrigeración *m*
F effet *m* de
   refroidissement
I effetto *m*
   refrigerante
MA hűtőhatás
NL koeleffect *n*
PO efekt *m* chłodzenia
PY эффект *m*
   охлаждения
SU jäähdytysvaikutus
SV kyleffekt *c*

732 chilling room
DA kølerum
DE Abkühlraum *m*
   Kühlraum *m*
ES cámara frigorífica *f*
F chambre *f* froide
I camera *f* frigorifera
MA hűtőhelyiség
NL koelruimte *f*
PO komora *f* chłodząca
PY охлаждаемое
   помещение *n*
SU kylmähuone
SV kylrum *n*

733 chimney
DA skorsten
DE Kamin *m*
   Schornstein *m*
ES chimenea *f*
F cheminée *f*
I camino *m*
MA kémény
NL schoorsteen *m*
PO komin *m*
PY вытяжной канал *m*
   дымовая труба *f*
SU savupiippu
SV skorsten *c*

734 chimney base
DA skorstensfundament
DE Schornstein-
   fundament *n*
ES base de la
   chimenea *f*
F embase *f* de
   cheminée
I base *f* del camino
MA kéményalap
NL schoorsteenvoet *m*
PO cokół *m* komina
   podstawa *f* komina
PY основание *n*
   дымовой трубы
SU savupiipun perusta
SV skorstensfot *c*

735 chimney cleaning (sweeping)
- DA skorstensfejning
- DE Schornstein-reinigung *f*
- ES limpieza de chimeneas *f*
- F ramonage *m* de cheminée
- I pulizia *f* del camino
- MA kéménytisztítás
- NL schoorsteen-reiniging *f* schoorsteen-schoonmaak *m*
- PO czyszczenie *n* komina
- PY очистка *f* дымовой трубы
- SU nuohous
- SV sotning *c*

736 chimney draught
- DA skorstenstræk
- DE Schornsteinzug *m*
- ES tiro de chimenea *m*
- F tirage *m* de cheminée
- I tiraggio *m* del camino
- MA kéményhuzat
- NL schoorsteentrek *m*
- PO ciąg *m* kominowy
- PY тяга *f* дымовой трубы
- SU savupiipun veto
- SV skorstensdrag *n*

737 chimney effect
- DA skorstensvirkning
- DE Kaminwirkung *f*
- ES efecto de chimenea *m*
- F effet *m* de cheminée
- I effetto *m* camino
- MA kéményhatás
- NL schoorsteen-werking *f*
- PO efekt *m* kominowy
- PY эффект *m* вытяжного канала эффект *m* дымовой трубы
- SU savupiippuvaikutus
- SV skorstensverkan *c*

738 chimney intake at base
- DA skorstenstilslutning
- DE Schornsteinfuchs *m*
- ES admisión por la base de la chimenea *f* toma en la base de la chimenea *f*
- F carneau *m*
- I canale *m* da fumo
- MA rókatorok
- NL schoorsteen-aansluiting *f*
- PO wlot *m* do komina przy podstawie
- PY дымовой канал *m*
- SU savupiipun sisäänmeno
- SV skorstensintag *n*

739 chimney loss
- DA skorstenstab
- DE Schornstein-verlust *m*
- ES pérdidas de chimenea *f,pl*
- F perte *f* à la cheminée
- I perdita *f* al camino
- MA kéményveszteség
- NL schoorsteenverlies *n*
- PO strata *f* kominowa
- PY потери *f,pl* в дымовой трубе
- SU savukaasuhäviü
- SV skorstensfürlust *c*

740 chisel
- DA mejsel
- DE Meißel *m*
- ES cincel *m* escoplo *m* formón *m*
- F burin *m*
- I scalpello *m*
- MA véső
- NL beitel *m*
- PO dłuto *n* przecinak *m*
- PY зубило *n*
- SU meisseli
- SV mejsel *c*

741 chlorofluorocarbon
- DA CFC
- DE Fluorchlorkohlen-stoff *m*
- ES clorofluorcarbono *m*
- F chlorofluorocarbone *m* (CFC)
- I clorofluoro-carburo *m*
- MA CFC hűtőküzeg klórozott fluorkarbon
- NL chloorfluorkoolstof (CFK) *f*
- PO chlorofluoro-carbon *m*
- PY фреон *m* хлорфторуглерод *m*
- SU kloorifluorihiilivety
- SV klorfluorkarbon *n*

742 chute
- DA nedløb sliske
- DE Rutsche *f*
- ES canaleta *f* rampa *f* tobogán *m*
- F goulotte *f*
- I grembiale *m*
- MA ereszcsatorna vályú
- NL stortkoker *m* vultrechter *m*
- PO rynna *f* zsypowa
- PY ссыпание *n*
- SU kuilu siilo viettävä kouru
- SV glidskena *c*

743 cinder
- DA slagge
- DE Asche *f* ausgeglühte Kohle *f*
- ES cenizas *f,pl* escorias *f,pl*
- F imbrûlés *m* solides
- I cenere *f* tizzone *m*
- MA salak
- NL as *f* sintel *f* slak *f*
- PO żużel *m*
- PY зола *f* шлак *m*
- SU kuona murska tuhka
- SV slagg *c*

744 circuit
DA kredsløb
DE Kreislauf m
   Umdrehung f
ES circuito m
F circuit m
I circuito m
MA áramkür
   kür
NL circuit n
   kring m
   kringloop m
PO obwód m
PY контур m
   цепь f
SU kiertopiiri
SV strömkrets c

745 circuit interrupt
DA kredsløbsafbrydelse
DE Stromkreisunter-
   brechung f
ES cortocircuito m
F dispositif m de
   délestage
I interruzione f di un
   circuito
MA áramkür megszakítás
NL onderbrekings-
   schakelaar m
   stroomkringonder-
   breking f
PO przerwanie n
   obwodu
PY выключение n цепи
   разрыв m цепи
SU suojakytkin
SV kretsavbrott n

746 circulating fan
DA cirkulations-
   ventilator
DE Zirkulations-
   ventilator m
ES ventilador de
   circulación m
F ventilateur m
   brasseur d'air
I ventilatore m di
   circolazione
MA keringető ventilátor
NL plafondventilator m
   tropenventilator m
PO wentylator m
   obiegowy
PY рециркуляционный
   вентилятор m
SU huonetuuletin
SV cirkulationsfläkt c

747 circulation
DA cirkulation
DE Umlauf m
   Zirkulation f
ES circulación f
F circulation f
I circolazione f
MA keringés
NL circulatie f
PO cyrkulacja f
   krążenie n
PY циркуляция f
SU kierto
SV cirkulation c

**748 circulation
pressure**
DA cirkulationstryk
DE Umtriebsdruck m
ES presión de
   circulación f
F pression f de
   circulation
I pressione f di
   circolazione
MA keringető nyomás
NL circulatiedruk m
PO ciśnienie n obiegowe
PY давление n
   циркуляции
SU kiertopaine
SV cirkulationstryck n

749 circulation pump
DA cirkulationspumpe
DE Umwälzpumpe f
ES bomba de
   circulación f
F pompe f de
   circulation
I pompa f di
   circolazione
MA keringető szivattyú
NL circulatiepomp f
PO pompa f
   cyrkulacyjna
   pompa f obiegowa
PY циркуляционный
   насос m
SU kiertopumppu
SV cirkulationspump c

750 circulator
DA cirkulator
DE Umlaufpumpe f
   Zirkulator m
ES acelerador m
   bomba aceleradora f
F accélérateur m
   circulateur m
I circolatore m
MA keringtető
   (szivattyú)
NL circulatiepomp f
PO urządzenie n
   cyrkulacyjne
PY насос m
SU kiertopumppu
SV cirkulationspump c

751 circumference
DA omkreds
   periferi
DE Umfang m
   Umkreis m
ES circunferencia f
F circonférence f
I circonferenza f
MA kerület
NL omtrek m
PO obwód m
   okrąg m
PY периметр m
SU kehä
   piiri
SV omkrets c

752 cistern
DA cisterne
   reservoir
DE Wasserbehälter m
   Zisterne f
ES cisterna f
F citerne f
I cisterna f
MA tartály
NL reservoir n
   tank m
   vergaarbak m
PO cysterna f
   zbiornik m
PY цистерна f
SU säiliü
SV cistern c

753 city water
DA vandværksvand
DE Stadtwasser *n*
ES agua de red *f*
F eau *f* de ville
I acqua *f* di acque-
dotto
MA hálózati víz
NL drinkwater *n*
leidingwater *n*
PO woda *f* miejska
PY городская вода *f*
SU vesijohtoverkon vesi
SV kommunalt vatten *n*

754 cladding
DA beklædning
DE Verkleidung *f*
ES recubrimiento con
chapa metálica *m*
revestimiento
metálico *m*
F enveloppe *f*
jaquette *f*
I rivestimento *m*
MA küpeny
NL bekleding *f*
beschermende laag *f*
PO okładzina *f*
PY облицовка *f*
SU pinnoite
päällyste
SV inklädnad *c*

755 clamp
DA klampe
spændstykke
DE Klammer *f*
Schelle *f*
Schraubzwinge *f*
ES abrazadera *f*
F pièce *f* de serrage
I staffa *f*
MA bilincs
kapocs
NL beugel *m*
klamp *m*
kram *f*
PO docisk *m*
zacisk *m*
PY зажим *m*
SU pidin
tuki
SV klammer *c*
klämma *c*

756 clamp bracket
DA fastspændingsbæring
DE Rohrschelle *f*
ES brida de ajuste *f*
brida de apriete *f*
F collier *m* de serrage
I staffa *f* del collare
staffa *f* di serraggio
MA csőbilincs
NL klembeugel *m*
pijpbeugel *m*
PO wspornik *m*
zaciskowy
PY зажим *m* обоймы
SU pidin
SV rörsvep *n*

**757 clarification
drawing**
DA detailtegning
DE Erklärungs-
zeichnung *f*
ES dibujo explicativo *m*
F schéma *m* explicatif
I disegno *m*
esplicativo
MA felmérési rajz
NL verklarende
tekening *f*
PO rysunek *m*
wyjaśniający
PY графическая
интерпретация *f*
SU detaljipiirustus
yksityiskohtapi-
irustus
SV detaljritning *c*

758 classification
DA klassifikation
DE Einteilung *f*
Klassifizierung *f*
ES clasificación *f*
F classement *m*
classification *f*
I classificazione *f*
MA osztályozás
NL classificatie *f*
PO klasyfikacja *f*
PY классификация *f*
SU luokittelu
SV klassifikation *c*

759 clean air
DA ren luft
DE saubere Luft *f*
ES aire limpio *m*
F air *m* épuré
air *m* pur
I aria *f* pulita
MA tiszta levegő
NL schone lucht *f*
zuivere lucht *f*
PO powietrze *n* czyste
PY чистый воздух *m*
SU puhdas ilma
SV ren luft *c*

760 clean air act
DA luftforureningslov
DE Gesetz *n* zur
Reinhaltung *f* der
Luft *f*
ES leyes contra la
contaminación
atmosférica *f,pl*
reglamentación
contra la
contaminación
atmosférica *f*
F décret *m* sur la
pollution de l'air
I legge *f* antinquina-
mento
MA levegőtisztasági
türvény
NL wet *f* tegen
luchtver-
ontreiniging
PO ustawa *f* dotycząca
czystości
powietrza
PY Закон *m* о чистоте
воздуха
SU ilmansuojelulaki
SV luftvårdslag *c*

761 cleaning
DA rengøring
DE Reinigung *f*
ES limpieza *f*
purificación *f*
F épuration *f*
nettoyage *m*
ramonage *m*
I depurazione *f*
pulizia *f*
MA tisztítás
NL reiniging *f*
schoonmaak *m*
PO czyszczenie *n*
PY очистка *f*
SU puhdistus
SV rengöring *c*

762 cleaning door
DA renselem
DE Reinigungstür f
ES puerta de limpieza f
F porte f de nettoyage
  porte f de ramonage
I portello m di
  pulitura
MA tisztítóajtó
NL reinigingsdeur f
  schoonmaakdeur f
PO otwór m rewizyjny
  wyczystka f
PY лаз m для очистки
SU puhdistusluukku
SV renslucka c

763 clean room
DA rent rum
DE Reinraum m
ES extracción de
  escorias f
  sangría de la escoria
  (hornos) f
F chambre f propre
I camera f bianca
MA tiszta helyiség
NL schone ruimte f
  steriele ruimte f
PO pomieszczenie n
  czyste
PY чистая комната f
SU puhdas huone
SV rent rum n

764 clean space
DA rent område
DE reiner Raum m
ES sala blanca f
F salle f blanche
I camera f bianca
MA tiszta tér
NL schone ruimte f
  steriele ruimte f
PO przestrzeń f czysta
PY чистый объем m
SU puhdas alue
SV rent utrymme n

765 clean workstation
DA rent arbejdsområde
DE reiner Arbeits-
  platz m
ES zona de trabajo
  blanca f
F zone f de travail
  blanche
I camera f bianca
MA tisztaterű munkahely
NL schone werkplek f
  steriele werkplek f
PO stanowisko n pracy
  czyste
PY чистая рабочая
  зона f
SU puhdas työasema
SV rent-rumsstation c

766 clearance
DA spillerum
  tolerance
DE lichter Raum m
  Spiel n
  Zwischenraum m
ES tolerancia f
F jeu m
I gioco m
  luce f netta
MA belméret
  űrszelvény
NL speling f
  vrije ruimte f rond
  een apparaat
PO luz m
  prześwit m
PY зазор m
  просвет m
SU välys
SV spelrum n
  urtag n

767 clearance (in cylinder)
DA boring
  diameter
DE schädlicher Raum m
ES tolerancia de un
  cilindro f
F espace m nuisible
  (du cylindre)
I spazio m morto
MA káros tér
  (motorhengerben)
NL compressievolume n
  dode ruimte f
PO martwa przestrzeń f
  (w cylindrze)
PY зазор m (в
  цилиндре)
SU kuollut tila
SV spelrum n (i
  cylinder c)

768 clearance fraction (in cylinder)
DA afstandsforhold
DE Abstands-
  verhältnis n
ES espacio muerto m
  holgura f
F rapport m de
  l'espace nuisible à
  l'espace bal
I frazione f dello
  spazio morto
MA káros tér hányad
NL percentage f
  schadelijke ruimte
  (in cilinder)
PO udział m martwej
  przestrzeni (w
  cylindrze)
PY свободное
  пространство n
  (в цилиндре)
SU puristusuhteen
  käänteisarvo
  suhteellinen etäisyys
SV skadlig volym c

769 clearance volume
DA kompressions-
  volumen
DE schädlicher Raum m
ES volumen muerto m
F espace m nuisible
I volume m dello
  spazio morto
MA káros térfogat
NL schadelijke ruimte f
PO objętość f martwej
  przestrzeni
  objętość f szkodliwa
PY объем m габарита
SU haitallinen tilavuus
SV kompressionsrum n

770 clerk of works
DA byggeleder
  konduktør
DE Bauleiter m
ES encargado de la
  obra m
  vigilante de obras m
F conducteur m de
  travaux
I sovrintendente m
MA épitésvezető
NL project-
  administrateur m
PO kierownik m robót
PY руководитель m
  работ
SU rakennusmestari
SV byggledare c

DA = Danish
DE = German
ES = Spanish
F = French
I = Italian
MA = Magyar
  (Hungarian)
NL = Dutch
PO = Polish
PY = Russian
SU = Finnish
SV = Swedish

771 climate
DA klima
DE Klima *n*
ES clima *m*
F climat *m*
I clima *m*
MA éghajlat
   klíma
NL klimaat *n*
PO klimat *m*
PY климат *m*
SU ilmasto
SV klimat *n*

772 climatic test chamber
DA klimarum
DE Klimakammer *f*
   Klimaprüfraum *m*
ES cámara de ensayos
   climatológicos *f*
   cámara de pruebas
   de climatización *f*
F chambre *f* d'essais
   climatiques
I camera *f* climatica
MA trópusi
   vizsgálókamra
NL klimaatkamer *f*
PO komora *f* badań
   klimatycznych
PY климатическая
   испытательная
   камера *f*
SU klimakammio
SV klimatkammare *c*

773 clinker
DA slagge
DE Schlacke *f*
ES escorias (de hierro o
   de carbón) *f,pl*
   ladrillo holandés *m*
   ladrillo recocido *m*
F mâchefer *m*
   scorie *f*
I clinker *m*
   scoria *f*
MA klinker
   salak
NL sintel *f*
   slak *f*
PO klinkier *m*
   żużel *m* (zastygły)
PY котельный шлак *m*
SU klinkkeri
   kuona
   tulenkestävä tiili
SV klinker *c*
   slagg *c*

774 clinker bed
DA slaggegrube
DE Schlackenbett *n*
ES lecho de escorias *m*
F couche *f* de
   mâchefer
   lit *m* de mâchefer
I letto *m* di scoria
MA salakágy
NL slakkenbed *n*
PO podłoże *n* żużla
   zastygłego
PY слой *m* шлака
SU kuonan
   päästükammio
SV slaggbädd *c*

775 clinker removal
DA slaggefjerner
DE Schlackenabzug *m*
ES extracción de
   escorias *f*
   sangría de la escoria
   (hornos) *f*
F décrassage *m*
I asportazione *f* delle
   scorie
MA salak eltávolítás
NL ontslakking *f*
PO usuwanie *n* żużla
   zastygłego
PY удаление *n* шлака
SU kuonanpoisto
SV slaggborttagning *c*

776 clo (*see* Appendix A)

777 clogging
DA blokering
DE Verstopfung *f*
ES obstrucción *f*
F colmatage *m*
   encrassement *m*
   obstruction *f*
I inceppamento *m*
   intasamento *m*
MA dugulás
   torlasz
NL verstopping *f*
PO zatykanie *n* się
   przewodów
PY закупорка *f*
   пробка *f*
SU tukkeuma
SV ojämn

778 closed cycle
DA kredsproces
   lukket kredsløb
DE geschlossener
   Kreislauf *m*
ES ciclo cerrado *m*
F cycle *m* fermé
I ciclo *m* chiuso
MA zárt kürfolyamat
NL gesloten cyclus *m*
   kringloop *m*
PO cykl *m* zamknięty
PY закрытый цикл *m*
SU suljettu kierto
SV kretslopp *n*

779 closed loop control
DA lukket
   sløjferegulering
DE geschlossener
   Regelkreis *m*
ES control en circuito
   cerrado *m*
F contrôle *m* en
   boucle fermée
I controllo *m* con
   retroazione
MA zárt szabályozókür
NL regeling *f*
PO sterowanie *n* w
   obwodzie
   zamkniętym
   sterowanie *n* ze
   sprzężeniem
   zwrotnym
PY закрытый
   регулирующий
   контур *m*
SU takaisinkytketty
   säätü
SV styrning *c* med
   åtorkoppling *c*

780 closed system
DA lukket system
DE geschlossenes
   System *n*
ES sistema de ciclo
   cerrado *m*
   sistema de circuito
   cerrado *m*
F circuit *m* fermé
I sistema *m* chiuso
MA zárt rendszer
NL gesloten systeem *n*
PO obieg *m* zamknięty
   system *m* zamknięty
   układ *m* zamknięty
PY закрытая система *f*
SU suljettu järjestelmä
SV slutet system *n*

781 close nipple
DA nippelhætte
   rørprop
DE Verschlußstopfen *m*
ES tapón roscado *m*
F mamelon *m*
I manicotto *m*
   filettato
MA záró dugó
NL pijpnippel *m*
PO złączka *f* wkrętna z
   gwintem ciągłym
PY сгон *m*
SU pitkänippa
SV tät nippel *c*

782 coal
DA kul
DE Kohle *f*
ES carbón *m*
   hulla *f*
F charbon *m*
   houille *f*
I carbone *m*
MA szén
NL kool *f*
   steenkool *f*
PO węgiel *m*
PY уголь *m*
SU kivihiili
SV kol *n*

783 coal (hard coal)
DA kul
   stenkul
DE Steinkohle *f*
ES carbón
   antracitoso *m*
F anthracite *m*
I antracite *f*
MA kőszén
NL antraciet *f*
   steenkool *f*
PO antracyt *m*
   węgiel *m* kamienny
PY каменный уголь *m*
SU antrasiitti
SV antracit *c*

784 coal delivery
DA kulleverance
DE Kohlenlieferung *f*
ES suministro de carbón
   m
F fourniture *f* de
   charbon
I fornitura *f* di
   carbone
MA szénszállítás
NL aflevering *f* van
   kolen
   kolenleverantie *f*
PO dostawa *f* węgla
PY поставка *f* угля
SU hiilitoimitus
SV kolleverans *c*

785 coal dust
DA kulstøv
DE Kohlenstaub *m*
ES polvo de carbón *m*
F poussière *f* de
   charbon
I polvere *f* di carbone
MA szénpor
NL kolenstof *n*
PO pył *m* węglowy
PY пылеугольное
   топливо *n*
   угольная пыль *f*
SU hiilipüly
SV kolpulver *n*

786 coal-fired *adj*
DA kulfyret
DE kohlenbeheizt
ES combustión a carbón
F (réchauffeur *m* d'eau)
   au charbon
I alimentato a carbone
MA széntüzelésű
NL kolengestookt
PO opalany węglem
PY сжигающий уголь
SU
SV koleldad

787 coal-fired boiler
DA kulfyret kedel
DE Kohlenkessel *m*
ES caldera de carbón *f*
F chaudière *f* au
   charbon
I caldaia *f* a carbone
MA széntüzelésű kazán
NL kolengestookte
   ketel *m*
PO kocioł *m* opalany
   węglem
PY котел *m* с угольной
   топкой
SU kivihiilikattila
SV koleldad panna *c*

788 coal firing
DA kulfyring
DE Kohlenfeuerung *f*
ES combustión a carbón
   f
F chauffe *f* au
   charbon
I alimentazione *f* a
   carbone
   combustione *f* a
   carbone
MA széntüzelés
NL kolenstoken *n*
PO opalanie *n* węglem
PY угольная топка *f*
SU hiilen poltto
SV koleldning *c*

789 coal store
DA kullager
DE Kohlenbunker *m*
   Kohlenkeller *m*
ES carbonera *f*
F soute *f* à charbon
I carbonaia *f*
   carbonile *m*
MA széntároló
NL kolenopslag *m*
   kolenopslagplaats *f*
PO skład *m* węgla
PY угольный бункер *m*
SU hiilivarasto
SV kollager *n*

790 coarse-grained
   coal
DA grovkornet kul
DE grobkörnige Kohle *f*
ES carbón de grano
   basto *m*
   carbón de grano
   grueso *m*
F charbon *m* de gros
   calibre
I carbone *m* a grana
   grossa
MA daraszén
NL grofkorrelige
   steenkool *f*
PO węgiel *m*
   gruboziarnisty
PY крупнозернистый
   уголь *m*
SU karkearakeinen hiili
SV grovkornigt kol *n*

791 coating
DA belægning
DE Belag *m*
   Schicht *f*
   Überzug *m*
ES capa de pintura *f*
   recubrimiento *m*
   revestimiento
      protector *m*
F enduit *m*
   enrobage *m*
   revêtement *m*
I rivestimento *m*
MA bevonat
   burkolat
NL bekleding *f*
   coating *n*
   deklaag *f*
PO otulina *f*
   powłoka *f*
PY обшивка *f*
SU pinnoite
SV beläggning *c*

792 coating barrier
DA overtrækshindring
DE Beschichtungs-
      sperre *f*
ES barrera de vapor *f*
F parevapeur *f*
I barriera *f* al vapore
MA burkolat
   válaszfal
NL dampdichte
      bekleding *f*
   dampremmende
      laag *f*
PO osłona *f* otuliny
PY пароизоляция *f*
SU hüyrysulku
SV beläggningsspärr *c*

793 coaxial *adj*
DA koaksial
DE koaxial
ES coaxial
F coaxial (coax)
I coassiale
MA koaxiális
NL coaxiaal
PO współosiowy
PY коаксиальный
SU koaksiaali
SV koaxial

794 cock
DA
DE Absperrhahn *m*
   Hahn *m*
ES grifo *m*
   llave *f*
F robinet *m*
I rubinetto *m*
MA csap
NL kraan *f*
PO kurek *m*
PY кран *m*
SU hana
SV kran *c*

795 cock valve
DA toldehane
DE Absperrhahn *m*
ES grifo *m*
F robinet *m* à
      boisseau
I valvola *f* a sfera per
      liquido
MA csapszelep
NL plugkraan *f*
PO kurek *m*
   zawór *m* kurkowy
PY пробковый кран *m*
SU tulppahana
SV kranventil *c*

796 code
DA forskrift
   kode
DE Code *m*
   Kode *m*
ES código *m*
F code *m*
I codice *m*
   regolamento *m*
   regole *f,pl* dell'arte
MA kád
NL code *m*
   reglement *n*
   voorschrift *n*
PO kod *m*
   przepisy *m,pl*
PY код *m*
SU määräys
   säännüs
SV kod *c*

797 coefficient
DA koefficient
DE Beiwert *m*
   Koeffizient *m*
ES coeficiente *m*
F coefficient *m*
I coefficiente *m*
MA tényező
NL coëfficiënt *m*
PO współczynnik *m*
PY коэффициент *m*
SU kerroin
SV koefficient *c*

**798 coefficient of
   compressibility** (*see*
   compression factor)

**799 coefficient of
   conductivity**
DA ledningskoefficient
DE Leitfähigkeitszahl *f*
ES coeficiente de
      conductividad *m*
   conductividad *f*
F coefficient *m* de
      conductivité
I conducibilità *f*
      termica
   conduttività *f*
      termica
MA vezetőképességi
      tényező
NL geleidingscoëfficiënt *m*
PO współczynnik *m*
      przewodności
PY коэффициент *m*
      проводимости
SU lämmünjohtavuus
SV ledningskoefficient *c*

**800 coefficient of
   discharge**
DA afløbskoefficient
DE Ausström-
      koeffizient *m*
ES coeficiente de
      descarga *m*
F coefficient *m* de
      décharge
I coefficiente *m* di
      efflusso
MA kiáramlási tényező
NL uitstroomcoëfficiënt *m*
PO współczynnik *m*
      wydatku
PY коэффициент *m*
      расхода
SU kuroumakerroin
SV utloppskoefficient *c*

801 coefficient of expansion

DA udvidelseskoefficient
DE Ausdehnungs-
koeffizient *m*
ES coeficiente de
dilatación *m*
F coefficient *m* de
dilatation
I coefficiente *m* di
dilatazione
MA tágulási tényező
NL uitzettings-
coëfficiënt *m*
PO współczynnik *m*
rozszerzalności
PY коэффициент *m*
расширения
SU
SV expansions-
koefficient *c*

802 coefficient of performance

DA virkningsgrad
DE Leistungs-
koeffizient *m*
ES factor de
rendimiento *m*
F coefficient *m* de
performance (cop)
I coefficiente *m* di
prestazione
MA teljesítménytényező
NL prestatiecoëfficiënt *m*
PO współczynnik *m*
sprawności
PY коэффициент *m*
полезного
действия (кпд)
SU kylmäkerroin
lämpükerroin
SV verkningsgrad *c*

803 coefficient of surface conductance

DA varmeovergangs-
koefficient
DE Oberflächen-
leitzahl *f*
ES coeficiente de
película *m*
coeficiente de
transmisión
superficial *m*
F coefficient *m* de
conductance
surfacique
I coefficiente *m* di
conduttanza
superficiale
MA felületvezetési
tényező
NL oppervlaktegeleidings-
coëfficiënt *m*
PO współczynnik *m*
przewodzenia
powierzchniowego
PY коэффициент *m*
восприятия у
поверхности
SU lämmünsiirty-
miskerroin
SV ytkonduktans-
koefficient *c*

804 cogeneration

DA regenering
DE Kraft-Wärme-
Kopplung *f*
ES cogeneración *f*
F cogénération *f*
I cogenerazione *f*
MA kapcsolt
energiatermelés
NL co-generatie *f*
PO współwytwarzanie *n*
PY проебразование *n*
SU sähkün ja lämmün
samanaikainen
tuotto
SV samproduktion *c*

805 coil (pipe)

DA batteri
rørslange
DE Batterie *f*
Rohrschlange *f*
ES batería *f*
serpentín *m*
F serpentin *m*
I batteria *f*
rotolo *m*
serpentino *m*
MA bordáscsőküteg
csőkígyó
NL pijpenbundel *m*
pijpspiraal *f*
PO wężownica *f*
PY змеевик *m*
(трубный)
SU putkikierukka
SV rörslinga *c*

806 coil depth

DA batteridybde
DE Verdampfertiefe *f*
ES profundidad de
batería *f*
F profondeur *f* de
batterie
I numero *m* di ranghi
di un serpentino
profondità *f* di un
serpentino
MA csőkígyó vastagsága
NL batterijdiepte *f*
PO głębokość *f*
wężownicy
PY глубина *f* змеевика
число *n* рядов по
потоку
SU patterin syvyys
SV batteridjup *n*

807 coil face area

DA batterioverflade
DE Verdampferanström-
fläche *f*
ES superficie frontal de
batería *f*
F surface *f* frontale de
batterie
I area *f* frontale di un
serpentino
MA csőkígyó
homlokfelülete
NL aanstroomoppervlak
n van een batterij
PO powierzchnia *f*
czołowa
wężownicy
PY площадь *f*
змеевика (по
фасаду)
SU patterin
otsapintaala
SV batteriarea *c*

DA = Danish
DE = German
ES = Spanish
F = French
I = Italian
MA = Magyar
(Hungarian)
NL = Dutch
PO = Polish
PY = Russian
SU = Finnish
SV = Swedish

808 coil length
DA batterilængde
 spirallængde
DE Verdampferlänge *f*
ES longitud de batería *f*
F longueur *f* de
 batterie
I lunghezza *f* di un
 serpentino
MA csőkígyó hossza
NL batterijlengte *f*
PO długość *f* wężownicy
PY длина *f* змеевика
SU patterin pituus
SV flänsbatteriets
 längd *c*

809 coil width
DA batteribredde
DE Verdampferbreite *f*
ES ancho de batería *m*
F largeur *f* de batterie
I larghezza *f* di un
 serpentino
MA csőkígyó szélessége
NL batterijbreedte *f*
PO szerokość *f*
 wężownicy
PY ширина *f* змеевика
SU patterin leveys
SV flänsbatteriets
 bredd *c*

810 coke
DA koks
DE Koks *m*
ES coque *m*
F coke *m*
I coke *m*
MA koksz
NL cokes *f*
PO koks *m*
PY кокс *m*
SU koksi
SV koks *c*

811 coke-fired boiler
DA koksfyret kedel
DE Kessel *m* mit
 Koksfeuerung *f*
ES caldera caldeada con
 coque *f*
 caldera calentada
 por coque *f*
F chaudière *f* à coke
I caldaia *f* a coke
MA koksztüzelésű kazán
NL cokesgestookte
 ketel *m*
PO kocioł *m* opalany
 koksem
PY котел *m* с коксовой
 топкой
SU koksikattila
SV kokseldad panna *c*

812 coke firing
DA koksfyring
DE Koksfeuerung *f*
ES caldeo por coque *m*
F chauffe *f* au coke
I combustione *f* a
 coke
MA koksztüzelés
NL stoken van cokes
PO opalanie *n* koksem
PY коксовая топка *f*
SU koksinpoltto
SV kokseldning *c*

813 coke oven gas
DA gas fra koksovn
DE Koksofengas *n*
ES gas de hornos de
 coque *m*
F gaz *m* de four à
 coke
I gas *m* di cokeria
MA kokszkohógáz
NL cokesovengas *n*
PO gaz *m* koksowniczy
PY коксовый газ *m*
SU koksiuunikaasu
SV koksugnsgas *c*

814 coking
DA forkoksning
DE Verkokung *f*
ES coquización *f*
F cokéfaction *f*
I cokificazione *f*
MA kokszolás
NL vercokesing
PO koksowanie *n*
PY коксовый
SU koksaus
SV förkoksning *c*

815 coking coal
DA kokskul
DE Kokskohle *f*
ES carbón para
 coque *m*
 hulla para coque *f*
F charbon *m* cokéfiant
 charbon *m* à coke
I carbone *m* da coke
MA kokszolható szén
NL cokeskool *f*
PO węgiel *m*
 koksowniczy
PY коксующийся
 уголь *m*
SU koksikivihiili
SV kokskol *n*

816 cold
DA kold
DE Kälte *f*
ES frío *m*
F froid *m*
I freddo *m*
MA hideg
NL kou *f*
 koude *f*
PO chłód *m*
 zimno *n*
PY холод *m*
SU kylmä
SV kyla *c*

817 cold *adj*
DA kold
DE kalt
ES frío
F froid
I freddo
MA hideg
NL koud
PO chłodny
 zimny
PY холодный
SU kylmä
SV kall

818 cold air
DA kold luft
DE Kaltluft *f*
ES aire frío *m*
F air *m* froid
I aria *f* fredda
MA hideg levegő
NL koude lucht *f*
PO powietrze *n* zimne
PY холодный
 воздух *m*
SU kylmä ilma
SV kalluft *c*

819 cold chain
DA kølekræde
DE Kühlkette *f*
ES cadena del frío *f*
F chaîne *f* du froid
I catena *f* del freddo
MA hűtőlánc
NL koude keten *m*
PO system *m* kolejnego chłodzenia
PY хранение *n* продуктов в холоде
SU kylmäketju
SV kylkedja *c*

820 cold production
DA køleproduktion
DE Kälteerzeugung *f*
ES produccín de frío *f*
F production *f* de froid
I produzione *f* di freddo
MA hidegképződés
NL koudeopwekking *f* koudeproduktie *f*
PO produkcja *f* chłodu
PY производство *n* холода
SU kylmäntuotto
SV köldalstring *c*

821 cold-rated output
DA kølekapacitet
DE Leistung *f* bei kalter Umgebung
ES potencia frigorífica nominal *f* producción nominal de frío *f*
F puissance *f* frigorifique
I potenza *f* frigorifera
MA hidegjárati teljesítmény
NL koelcapaciteit *f*
PO wydajność *f* chłodnicza
PY производитель-ьность *f* по холоду
SU kylmäteho
SV kyleffekt *c*

822 cold resistance
DA kuldemodstand
DE Kältebeständigkeit *f* Kältewiderstand *m*
ES resistencia al frío *f*
F résistance *f* au froid
I resistenza *f* al freddo
MA fagyállóság
NL koudeweerstand *m*
PO odporność *f* na chłód
PY холодильное сопротивление *n*
SU kylmänkestävyys
SV köldmotstånd *n*

823 cold room
DA kølerum
DE Kälteraum *m* Kühlraum *m*
ES cámara frigorífica *f*
F chambre *f* froide
I cella *f* frigorifera
MA hűtőkamra
NL koelruimte *f* koude ruimte *f*
PO komora *f* chłodnicza
PY холодильная камера *f*
SU kylmähuone
SV kylrum *n*

824 cold source
DA kølekilde
DE Kältequelle *f*
ES foco frío *m* fuente de frío *f*
F source *f* froide
I fonte *f* fredda sorgente *f* fredda
MA hidegforrás
NL koudebron *f*
PO źródło *n* chłodu
PY источник *m* холода
SU kylmänlähde
SV köldgenerator *c*

825 cold storage
DA kølelagring
DE Kaltlagerung *f*
ES almacenamiento frigorífico *m*
F conservation *f* par le froid
I magazzino *m* frigorifero
MA hűtőtárolás
NL gekoelde opslag *m* koudeopslag *m*
PO magazynowanie *n* chłodu składowanie *n* w chłodni
PY холодильное хранение *n*
SU kylmävarasto
SV kyllagring *c*

826 cold store
DA kølehus kølelager
DE Kühlhaus *n* Kühlraum *m*
ES almacén frigorífico *m*
F entrepôt *m* frigorifique
I deposito *m* frigorifero
MA hűtőtároló
NL koelhuis *n*
PO chłodnia *f* składowa
PY холодный склад *m*
SU kylmänvarastointi
SV kylrum *n*

827 cold water
DA koldt vand
DE Kaltwasser *n*
ES agua fría *f*
F eau *f* froide
I acqua *f* fredda
MA hidegvíz
NL koud water *n*
PO woda *f* zimna
PY холодная вода *f*
SU kylmä vesi
SV kallvatten *n*

828 cold water supply
DA koldtvandsforsyning
DE Kaltwasser-
versorgung *f*
ES suministro de agua
fría *m*
F alimentation *f* en
eau froide
I alimentazione *f* di
acqua fredda
MA hidegvíz ellátás
NL koudwatervoor-
ziening *f*
PO zaopatrzenie *n* w
wodę zimną
PY подача *f* холодной
воды
SU kylmä vesijohtovesi
SV kallvatten-
försörjning *c*

829 collector
DA indgangselektrode
samlerør
DE Kollektor *m*
Sammler *m*
Verteiler *m*
ES colector *m*
F collecteur *m*
I collettore *m*
MA gyűjtő
kollektor
NL verdeelstuk *n*
verdeler *m*
verzamelaar *m*
PO kolektor *m*
rozdzielacz *m*
rura *f* zbiorcza
PY коллектор *m*
SU kerääjä
SV samlingsrör *n*

830 colorimetric *adj*
DA kolorimetrisk
DE kolorimetrisch
ES colorimétrico
F colorimétrique
I colorimetrico
MA kolorimetrikus
NL colorimetrisch
PO kolorymetryczny
PY
SU
SV colorimetrisk

831 colour
DA farve
DE Farbe *f*
ES color *m*
F couleur *f*
I colore *m*
MA szín
NL kleur *f*
PO barwa *f*
kolor *m*
PY окраска *f*
цвет *m*
SU väri
SV färg *c*

832 colour code
DA farvekode
DE Farbskala *f*
ES código de colores *m*
F code *m* des couleurs
normalisées
I codice *m* di colore
MA színkód
NL kleurcode *m*
PO kod *m* barw
PY цветовой код *m*
SU värikoodi
SV färgkod *c*

833 coloured smoke
DA farvet røg
DE Rauchfärbung *f*
ES humo coloreado *m*
F fumées *f* colorées
I fumo *m* colorato
MA színezett füst
NL gekleurde rook *m*
PO dym *m* zabarwiony
PY окрашенный
дым *m*
SU värillinen savu
SV färgad rök *c*

834 colour temperature
DA farvetemperatur
DE Farbtemperatur *f*
ES temperatura del
color *f*
F température *f* de
couleur
I temperatura *f* del
colore
MA színhőmérséklet
NL kleurtemperatuur *f*
PO temperatura *f* barwy
PY цветовая
температура *f*
SU värilämpütila
SV färgtemperatur *c*

835 column
DA søjle
DE Säule *f*
ES columna *f*
F colonne *f*
I colonna *f*
MA oszlop
NL kolom *f*
PO kolumna *f*
słup *m*
stojak *m*
PY столб *m*
SU pilari
SV pelare *c*

836 column of mercury
DA kviksølvsøjle
DE Quecksilbersäule *f*
ES columna de
mercurio *f*
F colonne *f* de
mercure
I colonna *f* di
mercurio
MA higanyoszlop
NL kwikkolom *f*
PO słup *m* rtęci
PY ртутный столб *m*
SU elohopeapatsas
SV kvicksilverpelare *c*

837 column of water
DA vandsøjle
DE Wassersäule *f*
ES columna de agua *f*
F colonne *f* d'eau
I colonna *f* di acqua
MA vízoszlop
NL waterkolom *f*
PO słup *m* wody
PY водяной столб *m*
SU vesipatsas
SV vattenpelare *c*

838 column radiator
DA søjleradiator
DE Säulenheizkörper *m*
ES radiador de
columnas *m*
F radiateur *m* à
colonnes
I radiatore *m* a
colonna
MA oszlopos radiátor
tagos hőleadó
NL kolomradiator *m*
ledenradiator *m*
PO grzejnik *m* słupowy
PY колонный
радиатор *m*
SU liitepatteri
SV sektionsradiator *c*

**839 combination
control**
DA kombinations-
regulering
DE Verbundregelung *f*
ES control combinado
m
F automatisme *m*
combinatoire
I controllo *m* di
combinazione
MA kombinatív
szabályozás
NL meerfunctie
regelaar *m*
PO regulacja *f*
kombinowana
PY комбинированное
регулирование *n*
SU yhdistelmäsäätü
SV kombinations-
kontroll *c*

**840 combined heat and
power station**
DA kraftvarmeværk
DE Heizkraftwerk *n*
ES central combinada
para calefacción y
energía *f*
F centrale *f* combiné
chaleur force
I centrale *f* di
produzione di
energia e calore
MA fűtőerőmű
NL warmtekracht-
centrale *f*
PO elektrociepłownia *f*
PY теплоэлектро-
централь *f*
(ТЭЦ)
SU vastapainevoi-
malaitos
SV kraftvärmeverk *n*

841 combustibility
DA brændbarhed
DE Brennbarkeit *f*
ES combustibilidad *f*
F combustibilité *f*
I combustibilità *f*
MA éghetőség
NL brandbaarheid *f*
PO palność *f*
zapalność *f*
PY воспламеняемость
f
горючесть *f*
SU syttyvyys
SV brännbarhet *c*

842 combustible *adj*
DA brændbar
DE brennbar
ES combustible
F combustible
I combustibile
MA éghető
NL brandbaar
PO palny
zapalny
PY горючий
SU palava
SV brännbar

**843 combustible gas
(or vapour)
detector**
DA røggasdetektor
DE Brenngasspürgerät *n*
ES detector de gas
combustible *m*
F détecteur *m* de gaz
combustible
I rivelatore *m* di gas
combustibile
MA éghető gáz (vagy
gőz) detektor
NL gasdetector *m*
PO detektor *m* palnego
gazu (lub pary)
PY детектор *m*
горючего газа
(или пара)
SU palamiskaasun
ilmaisin
SV gasdetektor *c*

844 combustion
DA forbrænding
DE Verbrennung *f*
ES combustión *f*
F combustion *f*
I combustione *f*
MA égés
NL verbranding *f*
PO spalanie *n*
PY горение *n*
сгорание *n*
SU palaminen
SV förbränning *c*

845 combustion air
DA forbrændingsluft
DE Verbrennungsluft *f*
ES aire de combustión
m
F combustion *f* d'air
I aria *f* di
combustione
MA égési levegő
NL verbrandings-
lucht *f*
PO powietrze *n* do
spalania
PY воздух *m* для
горения
SU palamisilma
SV förbränningsluft *c*

**846 combustion
arrangement**
DA forbrændings-
anordning
DE Verbrennungs-
einrichtung *f*
ES dispositivo de
combustión *m*
F dispositif *m* de
combustion
I dispositivo *m* di
combustione
MA égető berendezés
NL verbrandings-
inrichting *f*
PO układ *m* spalania
PY устройство *n* для
сжигания
SU polttojärjestely
SV förbrännings-
anordning *c*

**847 combustion
chamber**
DA forbrændings-
kammer
DE Brennkammer *f*
ES cámara de
combustión *f*
F chambre *f* de
combustion
I camera *f* di
combustione
MA égőkamra
NL verbrandings-
kamer *f*
PO komora *f*
paleniskowa
komora *f* spalania
PY камера *f* сгорания
SU palamiskammio
tulipesä
SV förbrännings-
kammare *c*

848 combustion chamber lining
- DA forbrændings-kammerforing
- DE Brennkammer-ausmauerung *f*
- ES revestimiento interior de la cámara de combustión *m*
- F garnissage *m* réfractaire
- I rivestimento *m* della camera di combustione
- MA égőkamra bélés
- NL vuurvaste bekleding *f* vuurvaste bemetseling *f*
- PO wykładzina *f* komory spalania
- PY обмуровка *f* камеры сгорания футеровка *f* топки
- SU tulenkestävä verhous
- SV eldfast inmurning *c*

849 combustion controller
- DA forbrændings-regulator
- DE Verbrennungs-regler *m*
- ES regulador de combustión *m*
- F régulateur *m* de combustion
- I regolatore *m* di combustione
- MA égésszabályozó
- NL verbrandings-regelaar *m*
- PO regulator *m* spalania
- PY регулятор *m* горения
- SU palamisen säädin
- SV förbrännings-styrdon *n*

850 combustion detector
- DA forbrændings-detektor
- DE Verbrennungs-fühler *m*
- ES sonda de combustión *f*
- F détecteur *m* de combustion
- I sensore *m* di fiamma
- MA égésjelző
- NL vlambeveiliging *f*
- PO czujka *f* spalania detektor *m* spalania
- PY индикатор *m* горения
- SU liekinvartija
- SV förbrännings-detektor *c*

851 combustion diagram
- DA forbrændings-diagram
- DE Verbrennungs-diagramm *n*
- ES diagrama de combustión *m*
- F diagramme *m* de combustion
- I diagramma *m* di combustione
- MA égési diagram
- NL verbrandings-diagram *n*
- PO wykres *m* spalania
- PY диаграмма *f* горения
- SU polttokaavio
- SV förbrännings-diagram *n*

852 combustion duration
- DA brændtid
- DE Brenndauer *f*
- ES duración de la combustión *f*
- F durée *f* de combustion
- I durata *f* della combustione
- MA égéstartam
- NL verbrandingsduur *m*
- PO czas *m* trwania spalania czasokres *m* spalania
- PY продолжительность *f* горения
- SU palamisen kesto
- SV brinntid *c*

853 combustion efficiency
- DA forbrændingsvirk-ningsgrad
- DE Verbrennungswirkungs-grad *m*
- ES rendimiento de la combustión *m*
- F rendement *m* de combustion
- I rendimento *m* della combustione
- MA égési hatásfok
- NL nuttig effect *n* van de verbranding verbrandings-rendement *n*
- PO sprawność *f* spalania
- PY эффективность *f* сгорания
- SU palamishyütysuhde
- SV förbrännings-verkningsgrad *c*

854 combustion heat
- DA forbrændingsvarme
- DE Verbrennungs-wärme *f*
- ES calor de combustión *m*
- F chaleur *f* de combustion
- I calore *m* di combustione
- MA égéshő
- NL verbrandings-warmte *f*
- PO ciepło *n* uzyskane ze spalania
- PY теплота *f* сгорания
- SU palamislämpü
- SV förbrännings-värme *n*

DA	=	Danish
DE	=	German
ES	=	Spanish
F	=	French
I	=	Italian
MA	=	Magyar (Hungarian)
NL	=	Dutch
PO	=	Polish
PY	=	Russian
SU	=	Finnish
SV	=	Swedish

855 combustion medium
DA brændsel
brændstof
DE Brennmaterial n
Brennstoff m
ES medio
comburente m
F comburant m
I comburente m
MA tüzelőanyag
NL verbrandings-
middel n
PO paliwo n
środek m opałowy
PY горючая среда f
горючее
вещество n
SU polttoaine
SV förbrännings-
medium n

856 combustion product
DA forbrændings-
produkt
DE Verbrennungs-
produkt n
ES productos de la
combustión m,pl
F produit m de
combustion
I prodotto m della
combustione
MA égéstermék
NL verbrandings-
produkt n
PO produkt m spalania
PY продукт m
сгорания
SU palamistuote
SV förbrännings-
produkt c

857 combustion regulator
DA forbrændings-
regulator
DE Verbrennungs-
regler m
ES regulador de
combustión m
F régulateur m de
combustion
I regolatore m di
combustione
MA égésszabályozó
NL verbrandings-
regelaar m
PO regulator m spalania
PY регулятор m
горения
SU palamisen säädin
SV förbrännings-
regulator c

858 combustion residue
DA forbrændingsrest
DE Verbrennungs-
rückstand m
ES residuos de
combustión m,pl
F imbrûlés m
résidus m de
combustion
I residuo m della
combustione
MA égési maradvány
NL verbrandingsrest f
PO pozostałość f po
spalaniu
PY остаток m горения
SU palamaton
polttoaine
SV förbränningsrest c

859 combustion space
DA forbrændingsrum
DE Brennkammer f
Verbrennungs-
raum m
ES cámara de
combustión f
F chambre f de
combustion
I camera f di
combustione
MA égőtér
NL verbrandings-
ruimte f
PO przestrzeń f spalania
PY камера f сгорания
SU tulipesä
SV förbränningsrum n

860 combustion temperature
DA forbrændings-
temperatur
DE Verbrennungs-
temperatur f
ES temperatura de
combustión f
F température f de
combustion
I temperatura f di
combustione
MA égési hőmérséklet
NL verbrandings-
temperatuur f
PO temperatura f
spalania
PY температура f
горения
SU palamislämpütila
SV förbrännings-
temperatur c

861 combustion (gas) test
DA forbrændingsanalyse
røgprøve
DE Abgastest m
ES análisis de los gases
de combustión m
F essais m de
combustion (gaz)
I analisi f dei gas di
combustione
MA égésvizsgálat
NL rookgasproef f
PO próba f (gazowa)
spalania
PY испытание n на
горение
SU savukaasuanalyysi
SV förbränningsprov n

862 combustion turbine
DA gasturbine
DE Gasturbine f
Verbrennungs-
turbine f
ES turbina de
combustión f
F turbine f de
combustion
I turbina f a
combustione
MA égési turbina
NL gasturbine f
PO turbina f spalinowa
PY газовая турбина f
SU kaasuturbiini
SV gasturbin c

863 combustion velocity

DA forbrændings-
   hastighed
DE Verbrennungs-
   geschwindigkeit *f*
ES velocidad de
   combustión *f*
F vitesse *f* de
   combustion
I velocità *f* di
   combustione
MA égési sebesség
NL verbrandings-
   snelheid *f*
PO prędkość *f* spalania
PY скорость *f* горения
SU palamisnopeus
SV förbrännings-
   hastighet *c*

864 combustion ventilation

DA forbrændingsluft-
   tilførsel
DE Verbrennungsluft-
   zufuhr *f*
ES alimentación de aire
   para la
   combustión *f*
F alimentation *f* en air
   comburant
I alimentazione *f* di
   aria comburente
MA égéslevegő
   hozzávezetés
NL verbrandingslucht-
   toevoer *m*
PO doprowadzenie *n*
   powietrza
   niezbędnego do
   spalania
PY подача *f* воздуха
   на гоение
SU palamisilmanvaihto
SV förbränningslufttill-
   försel *c*

865 combustion volume

DA forbrændings-
   volumen
DE Verbrennungs-
   volumen *n*
ES volumen de
   combustión *m*
F volume *m* de
   combustion
I volume *f* di
   combustione
MA égési tér
   égési térfogat
NL verbrandings-
   volume *n*
PO objętość *f* spalania
PY объем *m* для
   горения
SU palamistilavuus
SV förbrännings-
   volym *c*

866 combustive air

DA forbrændingsluft
DE Verbrennungsluft *f*
ES aire comburente *m*
F air *m* comburant
I aria *f* comburente
MA égési levegő
   égést tápláló levegő
NL verbrandingslucht *f*
PO powietrze *n* do
   spalania
PY воздух *m* для
   горения
SU palamisilma
SV förbränningsluft *c*

867 comfort

DA bekvemmelighed
   komfort
DE Behaglichkeit *f*
   Komfort *m*
ES bienestar *m*
   confort *m*
F confort *m*
I benessere *m*
   confort *m*
MA kényelem
   komfort
NL behaaglijkheid *f*
   comfort *n*
PO komfort *m*
PY комфорт *m*
SU viihtyisyys
   viihtyvyys
SV komfort *c*

868 comfort air conditioning system

DA komfortluftbehandlings-
   anlæg
DE Komfortklima-
   system *n*
ES sistema de
   climatización para
   confort *m*
F système *m* de
   conditionnement
   d'air de confort
I impianto *m* di
   condizionamento
   per il benessere
MA komfort
   légkondicionáló
   rendszer
NL behaaglijkheidslucht-
   behandelings-
   installatie *f*
PO klimatyzacja *f*
   komfortu
PY комфортная
   система *f*
   кондицион-
   ирования воздуха
SU comfort
   ilmastointijärje-
   stelmä
SV komfortklimat-
   anläggning *c*

869 comfort chart

DA komfortdiagram
DE Komfort-
   diagramm *n*
   Komfortkarte *f*
ES diagrama de confort
   m
F diagramme *m* de
   confort
I diagramma *m* del
   confort
MA komfort diagram
NL behaaglijkheids-
   diagram *n*
PO wykres *m* komfortu
PY карта *f*
   комфортных
   условий
SU viihtyisyyspiirros
SV komfortdiagram *n*

870 comfort cooling
DA komfortkøling
DE Komfortkühlung f
ES enfriamiento para
bienestar m
enfriamiento para
confort m
F rafraîchissement m
d'ambiance
rafraîchissement m
de confort
I raffrescamento m
per il benessere
MA komfort hűtés
NL behaaglijkheids-
koeling f
comfortkoeling f
PO chłodzenie n dla
komfortu
PY комфортное
охлаждение n
SU jäähdytys
SV komfortkylning c

871 comfort index
DA behagelighedsindeks
DE Behaglichkeits-
index m
ES índice de confort m
F indice m de confort
I indice m di
benessere
MA komfort mérőszám
NL behaaglijkheids-
index m
PO wskaźnik m
komfortu
PY индекс m
комфорта
SU viihtyvyysindeksi
SV komfortindex n

872 comfort zone
DA komfortzone
DE Behaglichkeits-
bereich m
ES zona de bienestar f
zona de confort f
F zone f de confort
I zona f di benessere
MA komfort zóna
NL behaaglijkheids-
zone f
PO strefa f komfortu
PY зона f комфорта
SU viihtyisyysalue
SV komfortzon c

873 commercial *adj*
DA kommerciel
DE gewerblich
kaufmännisch
ES comercial
F commercial
I commerciale
MA kereskedelmi
NL commercieel
zakelijk
PO handlowy
ogólnego
przeznaczenia
przemysłowy
techniczny
PY коммерческий
SU kaupallinen
SV kommersiell

**874 commercial
refrigerator**
DA kommercielt
kølesystem
DE gewerblicher
Kühlraum m
ES frigorífico comercial
m
F réfrigérateur m
commercial
I refrigeratore m
commerciale
MA kereskedelmi hűtő
NL openbare
koelruimte f
PO szafa f chłodnicza
przemysłowa
PY торговый
холодильник m
SU myymäläjääkaappi
SV kommersiell
kylare c

875 commercial system
DA erhvervsanlæg
DE gewerbliches
System n
ES sistema comercial m
F système m
commercial
I impianto m per il
terziario
MA kereskedelmi
rendszer
NL bedrijfssysteem n
PO system m
przemysłowy
PY инженерная
система f
общественного
здания
коммерческая
система f
SU liikerakennuksiin
tarkoitettu
järjestelmä
SV kommersiellt
system n

876 commissioning
DA idriftsættelse
DE Abnahme f
Inbetriebnahme f
ES puesta en servicio f
F essai m de réception
I collaudo m
messa f in funzione
MA üzembehelyezés
NL inbedrijfstelling f
oplevering f
PO odbiór m końcowy
PY пуско-наладочные
испытания n,pl
SU vastaanotto
SV igångkörning c

DA	=	Danish
DE	=	German
ES	=	Spanish
F	=	French
I	=	Italian
MA	=	Magyar (Hungarian)
NL	=	Dutch
PO	=	Polish
PY	=	Russian
SU	=	Finnish
SV	=	Swedish

877 commissioning authority
DA offentlig myndighed
DE Abnahmebehörde f
ES permiso de funcionamiento m
F commission f de réception
I commissione m di collaudo
MA üzembehelyező szakhatóság
NL opleverings-instantie f
PO urząd m uprawniający do odbioru np. urządzeń
PY пуско-наладочная служба f
SU vastaanottotark-astaja
SV igångsättnings-myndighet c

878 commissioning plan
DA driftsinstruks
DE Inbetriebnahme-plan m
ES plan de puesta en marcha m
F plan m de mise en route
I programma m di collaudo
MA üzembehelyezési terv
NL opleveringsplan n
PO plan m przekazania (odbioru)
PY проект m пуско-наладочных работ
SU vastaanotto-suunnitelma
SV igångsättningsplan c

879 common main
DA hovedledning
DE Hauptleitung f
ES conducto principal común m
F conduite f principale
I condotta f principale
MA fővezeték küzüs vezeték
NL hoofddienstleiding f
PO przewód m magistralny rurociąg m magistralny
PY главный трубопровод m магистраль f
SU pääjohto
SV huvudledning c

880 common neutral
DA fælles nulledning
DE gemeinsamer Nulleiter m
ES neutro común m
F neutre m commun
I neutro m comune
MA küzüs nullvezeték
NL nulleider m
PO zerowanie n
PY центральный нулевой провод m
SU o-vaihe
SV gemensam nolledare c nolledare c

881 communication network
DA kommunikations-netværk
DE Kommunikations-netz n
ES sistema de comunicación m
F réseau m de commu-nications
I rete f di telecomunicazione
MA távküzlési hálózat
NL communicatie netwerk n
PO sieć f łączności
PY сеть f коммуникаций сеть f трубопроводов
SU tietoliikenneverkko
SV kommunikations-nät n

882 communications-based system
DA overføringssystem
DE System n auf Kommunikations-basis f
ES sistema de gestión centralizada m
F système m de gestion technique centralisé
I sistema m telematico
MA számítógépes küzponti vezérlő rendszer
NL gecomputeriseerd (energiebeheer) systeem n
PO sieć f telekomunikacyjna system m łączności podstawowej
PY энергетическая система f, управляемая компьютером
SU rakennusautomaatiojärje-stelmä
SV kommunikations-baserat system n

883 companion flange
DA koblingsflange
DE Gegenflansch m
ES contrabrida f
F bride f fixe
I flangia f di boccaporto
MA perempár
NL tegenflens m
PO kołnierz m przyłączny tarcza f sprzęgająca
PY ответный фланец m
SU liitoslaippa
SV kopplingsfläns c

884 compatible adj
DA forenelig, kan passe Esammen
DE passend
ES compatible
F compatible
I compatibile
MA üsszeegyeztethető
NL compatibel verenigbaar
PO kompatybilny zgodny
PY совместимый согласование
SU yhteensopiva
SV kompatibilitet c

885 compensating coupling
DA kompensator-
    forbindelse
DE Ausdehnungsstück *n*
ES acoplamiento
    elástico *m*
    junta elástica *f*
F joint *m* de dilatation
I giunto *m* di
    dilatazione
MA kiegyenlítő kapcsolás
NL compensator *m*
    expansiestuk *n*
PO złącze *n*
    kompensacyjne
PY компенсационное
    соединение *n*
SU kompensaattori
SV kompensator-
    koppling *c*

886 compensating loop
DA ekspansionssløjfe
DE Ausdehnungs-
    bogen *m*
ES bucle de compen-
    sación *m*
F boucle *f* de
    dilatation
I curva *f* di
    dilatazione
MA kiegyenlítő hurok
NL expansiebocht *f*
PO pętla *f*
    kompensacyjna
    wydłużka *f* lirowa
PY компенсационная
    петля *f*
    компенсационное
    удлинение *n*
SU paisuntakaari
SV expansionslyra *c*

887 compensator
DA kompensator
DE Dehnungs-
    ausgleicher *m*
    Kompensator *m*
ES compensador *m*
F compensateur *m*
I compensatore *m*
MA kiegyenlítő
    kompenzátor
NL compensator *m*
PO kompensator *m*
    wydłużka *f*
PY компенсатор *m*
SU kompensaattori
SV kompensator *c*

**888 compensator
    (articulated)**
DA lyddæmper
DE Gelenk-
    kompensator *m*
ES compensador
    (articulado) *m*
F compensateur *m*
    articulé
I compensatore *m*
    articolato
MA csuklós kompenzátor
NL laterale
    compensator *m*
PO kompensator *m*
    segmentowy
    wydłużka *f*
    segmentowa
PY П-образный
    компенсатор *m*
SU niveltasaaja
SV länkkompensator *c*

**889 complete
    combustion**
DA fuldstændig
    forbrænding
DE vollkommene
    Verbrennung *f*
ES combustión
    completa *f*
F combustion *f*
    complète
I combustione *f*
    completa
MA tükéletes égés
NL volledige
    verbranding *f*
PO spalanie *n* całkowite
PY полное сгорание *n*
SU täydellinen
    palaminen
SV fullständig
    förbränning *c*

890 completion period
DA byggetid
DE Fertigstellungszeit *f*
ES plazo de ejecución
    m
    plazo de realización
    m
F délai *m* d'exécution
I tempo *m* di
    esecuzione
MA befejezési időszak
NL afwerkingsperiode *f*
    uitvoeringsperiode *f*
PO okres *m* realizacji
PY завершающий
    период *m*
SU rakennusaika
SV byggtid *c*

891 component
DA bestanddel
    komponent
DE Bestandteil *n*
    Komponente *f*
ES componente *m*
    elemento *m*
F élément *m*
I componente *m*
    parte *f*
MA alkotóelem
    komponens
NL bestanddeel *n*
    onderdeel *n*
PO składnik *m*
PY компонент *m*
    составная часть *f*
SU komponentti
    osa
SV beståndsdel *c*

**892 compound
    compression**
DA flertrins kompression
DE mehrstufige
    Verdichtung *f*
ES compresión
    compuesta *f*
    compresión
    escalonada *f*
F compression *f* à
    plusieurs étages
I compressione *f* a più
    stadi
MA lépcsős sűrítés
NL meertraps-
    compressie *f*
PO sprężanie *n*
    sprężżone
PY многоступенчатое
    сжатие *n*
SU moniportainen
    puristus
SV tvåstegs-
    kompression *c*

893 compound compressor
DA flertrinskompressor
DE Verbund-
   verdichter *m*
ES compresor de etapas
   m
F compresseur *m* bi-
   étagé (compound)
I compressore *m* a più
   stadi
MA lépcsős kompresszor
NL meertraps-
   compressor *m*
PO sprężarka *f*
   sprzężona
PY многоступенчатый
   компрессор *m*
SU moniportainen
   kompressori
SV tvåstegs-
   kompressor *c*

894 compound gage
DA trykmåler
DE Verbundmaß *n*
   Verbundmesser *m*
ES manómetro absoluto
   m
F manovacuomètre *m*
I misuratore *m* di
   pressione assoluta
MA lépcsős nyomásmérő
NL mano-
   vacuümmeter *m*
PO zespolony przyrząd
   m pomiarowy
PY мановакуумметр *m*
SU yli- ja alipainemittari
SV tryckmätare *c*

895 compressed air
DA trykluft
DE Druckluft *f*
ES aire comprimido *m*
F air *m* comprimé
   air *m* sous pression
I aria *f* compressa
MA sűrített levegő
NL druklucht *f*
   perslucht *f*
PO powietrze *n*
   sprężone
PY сжатый воздух *m*
SU paineilma
SV tryckluft *c*

896 compressed gas
DA komprimeret gas
DE Druckgas *n*
ES gas comprimido *m*
F gaz *m* comprimé
   gaz *m* sous pression
I gas *m* compresso
MA sűrített gáz
NL gecomprimeerd
   gas *n*
   samengeperst gas *n*
PO gaz *m* sprężony
PY сжатый газ *m*
SU painekaasu
SV komprimerad gas *c*

897 compressed liquid
DA komprimeret væske
DE Verdichter(Druck)-
   Flüssigkeit *f*
ES líquido comprimido
   m
F liquide *m* comprimé
I liquido *m* in
   pressione
MA sűrített folyadék
NL gecomprimeerde
   vloeistof *f*
   samengeperste
   vloeistof *f*
PO ciecz *f* sprężona
PY сжатая жидкость *f*
SU nesteytetty kaasu
   puristettu neste
SV komprimerad
   vätska *c*

898 compressibility
DA kompressionsevne
DE Verdichtbarkeit *f*
ES compresibilidad *f*
F compressibilité *f*
I comprimibilità *f*
MA üsszenyomhatóság
NL samendrukbaar-
   heid *f*
PO ściśliwość *f*
   zdolność *f* sprężania
PY сжимаемость *f*
SU kokoonpuristuvuus
SV kompressibilitet *c*

899 compressibility factor
DA kompressionsfaktor
DE Verdichtungs-
   verhältnis *n*
ES coeficiente de com-
   presibilidad *m*
F facteur *m* de
   compressibilité
I fattore *m* di
   comprimibilità
MA üsszenyomhatósági
   tényező
NL samendrukbaar-
   heidsfactor *m*
PO współczynnik *m*
   ściśliwości
   współczynnik *m*
   zdolności
   sprężania
PY коэффициент *m*
   сжимаемости
SU kompressibiliteetti
SV kompressibilitets-
   faktor *c*

900 compression
DA kompression
   sammentrykning
DE Kompression *f*
   Verdichtung *f*
ES compresión *f*
F compression *f*
I compressione *f*
MA kompresszió
   sűrítés
NL compressie *f*
PO sprężanie *n*
   ściskanie *n*
PY сжатие *n*
SU puristus
SV kompression *c*

901 compression cycle
DA kompressionsk-
   redscyklus
DE Verdichtungs-
   kreislauf *m*
ES ciclo de compresión
   m
F cycle *m* de
   compression
I ciclo *m* frigorifero a
   compressione
MA sűrítési kürfolyamat
NL compressiecyclus *m*
PO cykl *m* sprężania
PY холодильный цикл
   m
   цикл *m* сжатия
SU puristusvaihe
SV kompressionscykel *c*

902 compression efficiency
DA kompressor-
   virkningsgrad
DE Verdichtungs-
   wirkungsgrad *m*
ES rendimiento de
   compresión *m*
F rendement *m* de
   compression
I rendimento *m* di
   compressione
MA sűrítési hatásfok
NL compressie-
   rendement *n*
PO sprawność *f*
   sprężania
PY эффективность *f*
   сжатия
SU puristushyütysuhde
SV kompressions-
   verkningsgrad *c*

903 compression factor
DA kompressionsfaktor
DE Verdichtungs-
   faktor *m*
ES factor de compresión
   m
F facteur *m* de
   compression
I fattore *m* di
   compressione
MA sűrítési tényező
NL compressiefactor *m*
PO współczynnik *m*
   sprężania
PY коэффициент *m*
   сжатия
SU reaalisuuskerroin
   ideaalikaasuyh-
   tälüssä
SV kompressions-
   faktor *c*

DA = Danish
DE = German
ES = Spanish
F = French
I = Italian
MA = Magyar
     (Hungarian)
NL = Dutch
PO = Polish
PY = Russian
SU = Finnish
SV = Swedish

904 compression joint
DA kompressions-
   samling
DE Druckverbindung *f*
ES junta de compresión
   f
F joint *m* de
   compression
I giunto *m* a
   compressione
MA nyomásos
   csatlakozás
NL flareverbinding *f*
   klemkoppeling *f*
PO złącze *n*
   kompresyjne
PY уплотнение *n*
   стыка
SU puserrusliitos
SV kompressions-
   koppling *c*

905 compression ratio
DA kompressionsforhold
DE Verdichtungs-
   verhältnis *n*
ES relación de
   compresión *f*
F rapport *m* de
   compression
   taux *m* de
   compression
I rapporto *m* di
   compressione
MA sűrítési arány
NL compressie-
   verhouding *f*
PO stopień *m* sprężania
PY степень *f* сжатия
SU puristussuhde
SV kompressionsför-
   hållande *n*

906 compression stage
DA tryktrin
DE Verdichtungsphase *f*
ES etapa de compresión
   f
F étage *m* de
   compression
I stadio *m* di
   compressione
MA sűrítési lépcső
NL compressietrap *m*
PO faza *f* sprężania
PY каскадное сжатие *n*
   фаза *f* сжатия
SU puristusvaihe
SV kompressionssteg *n*

907 compression stroke
DA kompressionsslag
   (=takt)
DE Verdichtungshub *m*
ES embolada de
   compresión *f*
F course *f* de compres-
   sion
I fase *f* di
   compressione (di
   un ciclo)
MA sűrítési ütem
NL compressieslag *m*
PO skok *m* sprężarki
   suw *m* sprężania
PY рабочий ход *m*
   такт *m* сжатия
SU puristusisku
SV kompressionstakt *c*

908 compression tank
DA trykbeholder
DE Druckbehälter *m*
ES depósito a
   presión *m*
   recipiente a
   presión *m*
F réservoir *m* sous
   pression
I serbatoio *m* di
   compressione
MA nyomótartály
NL drukvat *n*
PO zbiornik *m*
   ciśnieniowy
PY напорный
   резервуар *m*
SU paineastia
SV kompressionstank *c*

909 compression test
DA trykprøve
DE Druckprobe *f*
ES prueba a la
   compresión *f*
   prueba de presión *f*
F essai *m* de
   compression
   essai *m* de pression
I prova *f* a pressione
   prova *f* di
   compressione
MA nyomáspróba
NL drukproef *f*
   persproef *f*
PO próba *f* sprężania
   próba *f* ściskania
PY испытание *n* на
   сжатие
SU painekoe
SV kompressionsprov *n*

910 compression type refrigerating system
DA kompressorkøleanlæg
DE Verdichter-Kälte-System *n*
ES sistema de refrigeración por compresión *m*
F système *m* de réfrigération à compression
I sistema *m* di refrigerazione a compressione
MA kompresszoros hűtőrendszer
NL compressiekoeling *f*
PO system *m* chłodzenia przy użyciu sprężarki
PY компрессионная холодильная система *f* компрессионная холодильная установка *f*
SU kompressorijäähdytys
SV kompressionskylsystem *n*

911 compression volume ratio
DA kompressionsvolumenforhold
DE Verdichtungsvolumenverhältnis *n*
ES relación de volumen de compresión *f*
F taux *m* volumique de compression
I rapporto *m* di compressione volumetrico
MA sűrítési térfogatarány
NL compressie volume verhouding *f*
PO stosunek *m* objętości sprężonej
PY объемная степень *f* сжатия
SU puristussuhde
SV kompressionsvolymfürhållande *n*

912 compressor
DA kompressor
DE Kompressor *m* Verdichter *m*
ES compresor *m*
F compresseur *m*
I compressore *m*
MA kompresszor
NL compressor *m*
PO kompresor *m* sprężarka *f*
PY компрессор *m*
SU kompressori
SV kompressor *c*

913 compressor capacity
DA kompressorkapacitet
DE Verdichterleistung *f*
ES capacidad del compresor *f*
F puissance *f* d'un compresseur
I potenzialità *f* del compressore
MA kompresszor teljesítmény
NL compressorcapaciteit *f*
PO wydajność *f* sprężarki
PY производительность *f* компрессора
SU kompressorin teho
SV kompressorkapacitet *c*

914 compressor discharge
DA kompressoraflastning
DE Verdichterauslaß *m*
ES descarga del compresor *f*
F refoulement *m* d'un compresseur
I scarico *m* del compressore
MA kompresszor nyomócsonk
NL compressorperszijde *f*
PO wydajność *f* sprężarki
PY выхлоп *m* компрессора
SU kompressorin painepuoli
SV kompressoravlastning *c*

915 compressor discharge stroke
DA kompressorrudstrømningsslag
DE Verdichterauslaßhub *m*
ES desplazamiento del compresor *m*
F course *f* de refoulement d'un compresseur
I fase *f* di scarico del compressore
MA kompresszor kinyomó ütem
NL compressieslag *m*
PO skok *m* wydajności sprężarki suw *m* rozprężania w sprężarce
PY рабочий ход *m* нагнетания компрессора такт *m* нагнетания компрессора
SU puristusiskun loppuvaihe
SV kompressoravlastningsslag *n*

916 compressor displacement
DA kompressorslagvolumen
DE Verdichterverdrängung *f*
ES desplazamiento del compresor *m*
F course *f* d'un compresseur
I cilindrata *f* del compressore
MA kompresszor lükettérfogat
NL verplaatst volume *n* per compressorslag
PO sprężarka *f* wyporowa
PY объем *m* газа, подаваемого компрессорм в единицу времени
SU kompressorin iskutilavuus
SV kompressordeplacement *n*

917 compressor economizing

DA kompressor-
  forvarmning
DE Verdichter-
  vorwärmung f
ES economizador del
  compresor m
F système m à
  compresseur
  économe
I economizzatore m
  del compressore
MA kompresszor
  hőhasznosítás
NL compressor
  rendements-
  verbetering f
PO sprężarka f ekono-
  miczna
PY повышение n
  экономичности
  компрессора
SU kompressorin
  ekonomaiserin
  käyttü
SV kompressor-
  förvärmning c

918 compressor heating effect (heat pump)

DA varmepumpeydelse
DE Wärmepumpen-
  wirkung f des
  Verdichters m
ES efecto calefactor del
  compresor (bomba
  de calor) m
F coefficient m de
  performance
  (pompe à chaleur)
I effetto m utile di
  una pompa di
  calore
MA kompresszor
  fűtőhatás
  (hőszivattyú)
NL warmtepompeffect n
PO efekt m cieplny
  sprężarki (pompa
  cieplna)
PY отопительный
  эффект m
  компрессора (в
  режиме
  теплового
  насоса)
SU kompressorin
  lämmitysteho
SV kompressionsvärme-
  effekt c

919 compressor surge

DA kompressorover-
  belastning
DE Verdichterdruck-
  schwankung f
ES sobretensión del
  compresor m
F surpression f du
  compresseur
I pressione f di picco
  del compressore
MA kompresszor
  lükéshullám
NL gasgebrek n van
  compressor
PO niestabilna praca f
  sprężarki
PY помпаж m
  компрессора
SU keskipako-
  kompressorin
  toimintaraja
SV kompressoröver-
  belastning c

920 compressor unit

DA kompressorunit
DE Verdichtereinheit f
ES unidad compresora
F compresseur m
I impianto m a
  compressione
MA kompresszor egység
NL compressoreenheid f
PO zespół m
  sprężarkowy
PY компрессорный
  агрегат m
SU kompressoriyksikkü
SV kompressorenhet c

921 compressor unloader

DA kompressoraflaster
DE Verdichter-
  entlaster m
ES descargador de
  compresor m
F dispositif m de
  délestage de
  compresseur
I valvola f di scarico
  del compressore
MA kompresszor
  tehermentesítő
NL decompressieklep f
PO reduktor m ciśnienia
  wylotowego
  sprężarki
PY регулятор m
  давления
  компрессора
SU kompressorin
  lepuutus
  kompressorin
  paineentasaaja
SV kompressor-
  avlastare c

922 computer

DA computer
  datamaskine
DE Computer m
  Rechner m
ES computador m
  ordenador m
F ordinateur m
I calcolatore m
  elaboratore m
MA komputer
  számítógép
NL computer m
PO komputer m
PY вычислительная
  машина f
  компьютер m
SU tietokone
SV dator c

923 computer-aided design
DA CAD, edb-tegnesystem
DE computerunter-stütztes Entwerfen n
ES diseño por ordenador m
F conception f assistée par ordinateur
I progettazione f assistita dal calcolatore
MA számítógéppel segített tervezés
NL computer ondersteund ontwerpen n
PO projektowanie n wspomagane komputerowo
PY проектирование n c использованием компьютеров
SU tietokoneavusteinen suunnittelu
SV computer-aided design c

924 computer-aided manufacturing
DA edb-styret fabrikation
DE computerunter-stützte Fabrikation f
ES fabricación robotizada f
F fabrication f assistée par ordinateur
I produzione f assistita dal calcolatore
MA számítógéppel segített gyártás
NL computer ondersteund fabriceren n
PO produkcja f wspomagana komputerowo
PY компьютерная обработка f
SU tietokoneavusteinen valmistus
SV computer-aided manufacturing c

925 computer-based system
DA computerstyret system
DE System n auf Computerbasis f
ES sistema computerizado m
F système m piloté par ordinateur
I sistema m basato su elaboratore
MA számítógépes rendszer
NL computer gestuurd systeem n
PO system m wspomagany komputerem
PY базовая система f компьютера
SU tietokoneohjattu järjestelmä
SV datorbaserat system n

926 computer code
DA computerkode
DE Computercode m
ES clave del ordenador f
F code m numérique
I codice m di calcolo
MA számítógép kód
NL computercode m
PO lista f rozkazów komputera
PY компьютерный код n
SU ohjelmakoodi
SV datorkod c

927 computer design
DA computer tegning
DE Computerentwurf m
ES proyecto de cálculo m
F étude f par ordinateur
I progettazione f del calcolatore
MA számítógépes tervezés
NL computerontwerp n
PO projektowanie n komputerowe
PY компьютерное проектирование n
SU tietokonesuunnittelu
SV datorkonstruktion c

928 computer hardware
DA computer hardware
DE Computergeräte n,pl
ES hardware m
F matériel m informatique
I hardware m di un elaboratore
MA számítógépes hardver
NL computer-apparatuur f
PO sprzęt m komputerowy
PY компьютерные технические средства n
SU tietokonelaite
SV datorhårdvara c

929 computer input
DA computer indgangsdata
DE Computereingabe f
ES entrada a ordenador f
F entrée f d'ordinateur
I dato m di ingresso di un elaboratore
MA számítógép bemeneti adat
NL computerinvoer m
PO wejscie n komputera (wprowadzanie danych)
PY вводное устройство n компьютера исходные данные n,pl компьютера
SU syüttütiedot
SV dators c ingångsdata c

930 computer memory
DA computer hukommelse
DE Computer-speicher m
ES memoria del ordenador f
F mémoire f d'ordinateur
I memoria f di un elaboratore
MA számítógép memória
NL computergeheugen n
PO pamięc f komputera
PY запоминающее устройство n компьютера компьютерная память f
SU tietokoneen muisti
SV datorminne n

931 computer output
DA computers
udgangsdata
DE Computerausgang
m
ES salida del ordenador
f
F sortie *f* d'ordinateur
I dato *m* in uscita da
un elaboratore
MA számítógép kimeneti
adat
NL computeruitvoer *m*
PO wyjście *n* komputera
(wynik pracy
komputera)
PY выводное
устройство *n*
компьютера
компьютерный
вывод *m* данных
SU tietokoneen tulostus
SV dators *c* utgångs-
data *c*

932 computer overflow
DA computeroverløb
DE Computer-
überlauf *m*
ES sobrecarga del
ordenador *f*
F débordement *m*
(surcharge)
d'ordinateur
I overflow *m* di un
elaboratore
MA számítógép
túlterhelés
NL computerover-
belasting *f*
PO przepełnienie *n*
pamięci
komputera
PY переполнение *n*
компьютера
SU ylivuoto
SV datoröverflöde *n*

933 computer print-out
DA computerudskrift
DE Computer-
ausdruck *m*
ES impresión del
ordenador *f*
F édition *f*
d'ordinateur
I uscita *f* a stampa
(tabulato) di un
elaboratore
MA számítógépes
nyomtatás
NL computerafdruk *m*
computerprint *n*
PO wydruk *m*
komputera
PY компьютерная
распечатка *f*
SU tietokonetuloste
SV datautskrift *c*

934 computer run
DA computerkørsel
DE Computerlauf *m*
ES funcionamiento del
ordenador *m*
F exécution *f* d'un
programme
informatique
I calcolo *m* con
elaboratore
MA számítógép futás
NL uitvoering *f* com-
puterprogramma
PO praca *f* komputera
PY компьютерный счет
m
SU tietokoneajo
SV datorkörning *c*

**935 computer
simulation**
DA computersimulering
DE Computer-
simulation *f*
ES simulación por
ordenador *f*
F simulation *f*
numérique
I simulazione *f*
all'elaboratore
MA számítógépes
szimuláció
NL computersimulatie *f*
PO symulacja *f*
komputera
PY компьютерное
моделирование *n*
SU tietokonesimulointi
SV datorsimulation *c*

936 computer software
DA computer
programmel
(software)
DE Computer-
programm *n*
ES software *m*
F logiciel *m*
I programma *m* per
elaboratore
MA számítógépes
szoftver
NL computer-
programma's *pl*
PO oprogramowanie *n*
komputera
PY программное
обеспечение *n*
компьютера
SU tietokoneohjelma
SV datormjukvara *c*

937 computer storage
DA computer lager
DE Computer-
speicherung *f*
ES almacenamiento del
ordenador *m*
F taille mémoire *f*
d'un ordinateur
I memoria *f* di un
elaboratore
MA számítógép memória
NL gegevensopslag *m*
PO pamięć *f* komputera
PY компьютерная
память *f*
SU tietokoneen muisti
SV datorinternminne *n*

938 computer switch
DA computerkontakt
(afbryder)
DE Computerschalter *m*
ES conmutador del
ordenador *m*
F interrupteur *m*
d'ordinateur
I interruttore *m* del
computer
switch *m* del
computer
MA számítógép kapcsoló
NL programma-
vertakking *f*
PO wyłącznik *m*
komputera
PY компьютерный
переключатель *m*
SU tietokoneohjelman
haarakohta
virtakytkin
SV datorkontakt *c*

DA	=	Danish
DE	=	German
ES	=	Spanish
F	=	French
I	=	Italian
MA	=	Magyar
		(Hungarian)
NL	=	Dutch
PO	=	Polish
PY	=	Russian
SU	=	Finnish
SV	=	Swedish

939 concentrating solar collector
DA koncentrerende solfanger
DE konzentrierender Sonnenkollektor *m*
ES colector solar concentrador *m*
F capteur *m* solaire à concentration
I collettore *m* solare a concentrazione
MA üsszpontosító napkollektor
NL stralingsconcentrerende zonnecollector *m*
PO kolektor *m* słoneczny zwierciadlany
PY концентрирующий солнечный коллектор *m*
SU keskittävä aurinkokerääjä
SV koncentrationssolkollektor *c*

940 concentration
DA koncentration
DE Konzentration *f*
ES concentración *f*
F concentration *f*
I concentrazione *f*
MA koncentráció tüménység
NL concentratie *f*
PO koncentracja *f* stężenie *n*
PY концентрация *f*
SU pitoisuus
SV koncentration *c*

941 concentration ratio
DA koncentrationsforhold
DE Konzentrationsverhältnis *n*
ES relación de concentración *f*
F concentration *f* (d'un produit)
I rapporto *m* di concentrazione
MA koncentráció arány
NL concentratieverhouding *f*
PO stopień *m* koncentracji
PY отношение *n* концентраций
SU pitoisuussuhde
SV koncentrationsförhållande *n*

942 concentric tubes
DA koncentrisk rør
DE konzentrische Rohre *n,pl*
ES tubos concéntricos *m,pl*
F tubes *m* concentriques
I tubi *m,pl* concentrici
MA koncentrikus csüvek
NL concentrische buizen *pl* concentrische pijpen *pl*
PO rury *f* współśrodkowe
PY концентрические трубы *f*
SU samakeskeiset putket
SV koncentriska rör *n*

943 concrete
DA beton
DE Beton *m*
ES hormigón *m*
F béton *m*
I calcestruzzo *m*
MA beton
NL beton *n*
PO beton *m*
PY бетон *m*
SU betoni
SV betong *c*

944 condensate
DA kondensat
DE Kondensat *n*
ES condensado *m*
F condensat *m* eau *f* condensée
I condensa *f*
MA kondenzátum lecsapódás
NL condensaat *n*
PO kondensat *m* skropliny *pl*
PY конденсат *m*
SU lauhde
SV kondensat *n*

945 condensate collecting vessel
DA kondensatbeholder
DE Kondensatsammelbehälter *m*
ES recipiente colector del condensado *m* tanque colector de condensados *m*
F bac *m* bâche *f* récepteur *m* d'eau condensée
I raccoglitore *m* di condensa serbatoio *m* di raccolta condensa
MA kondenzgyűjtő tartály
NL condensvat *n* condensverzamelaar *m*
PO zbiornik *m* kondensatu
PY сборный конденсационный бак *m*
SU lauhteen keräysastia
SV kondensatsamlare *c*

946 condensate line
DA kondensledning
DE Kondensatleitung *f*
ES línea de condensados *f*
F tuyauterie *f* d'eau condensée
I tubazione *f* della condensa
MA kondenzvezeték
NL condensaatleiding *f* condensleiding *f*
PO przewód *m* dla skroplin
PY конденсационная линия *f*
SU lauhdejohto
SV kondensatledning *c*

947 condensate meter
DA kondensatmåler
DE Kondensatmesser *m*
ES contador de condensados *m*
F compteur *m* d'eau condensée
I contatore *m* di condensa
MA kondenzvízmenn-yiségmérő
NL condensaatmeter *m*
PO miernik *m* ilości kondensatu
PY счетчик *m* конденсата
SU lauhdemittari
SV kondensatmätare *c*

948 condensate return pump
DA kondensatpumpe
DE Kondensatpumpe *f*
ES bomba de aspiración del condensado *f* bomba de retorno del condensador *f*
F pompe *f* de retour d'eau condensée
I pompa *f* di ricircola-zione condensa
MA kondenz szivattyú
NL condensaatpomp *f*
PO pompa *f* powrotna kondensatu
PY насос *m* для возврата конденсата
SU lauhdepumppu
SV kondensatpump *c*

949 condensation
DA fortætning kondensering
DE Kondensation *f* Verflüssigung *f*
ES condensación de agua (sobre una superficie fría) *f*
F condensation *f*
I condensazione *f*
MA kondenzáció
NL condensatie *f*
PO kondensacja *f* skraplanie *n*
PY конденсация *f*
SU kondenssi tiivistyminen
SV kondensation *c*

950 condensation point
DA kondensationspunkt
DE Kondensations-punkt *m*
ES punto de condensación *m*
F point *m* de condensation
I punto *m* di condensazione
MA kondenzációs pont
NL condensatiepunt *m* verzadigingspunt *m*
PO punkt *m* skraplania temperatura *f* skraplania
PY точка *f* конденсации
SU lauhtumispiste
SV kondensations-punkt *c*

951 condense *vb*
DA kondensere
DE kondensieren
ES condensar
F condenser
I condensare
MA kondenzál
NL concentreren condenseren verdichten
PO skraplać
PY конденсироваться
SU lauhtua
SV kondensera

952 condenser
DA kondensator
DE Kondensator *m* Verflüssiger *m*
ES condensador *m*
F condenseur *m*
I condensatore *m*
MA kondenzáló
NL condensator *m* condensor *m*
PO skraplacz *m*
PY конденсатор *m*
SU lauhdutin
SV kondensor *c*

953 condenser coil
DA kølespiral (slange)
DE Kondensator-schlange *f*
ES serpentín condensador *m*
F batterie *f* de condensation
I serpentino *m* del condensatore
MA kondenzátor csőkígyó
NL condensorbatterij *f*
PO wężownica *f* skraplacza
PY змеевик *m* конденсатора
SU lauhdutinpatteri
SV kondensorslinga *c*

954 condenser duty
(*see* condenser heat rejection effect)

955 condenser heat rejection effect
DA kondensator varmeomsætnings-effekt
DE Kondensatorwärmerückgabe-effekt *m*
ES efecto de disipación de calor del condensador *m*
F énergie *f* rejetée au condensateur
I dissipazione *f* termica al condensatore
MA kondenzátor hővisszatartó hatás
NL condensorwarmte-afgifte *f*
PO efekt *m* oddawania ciepła przez skraplacz
PY потеря *f* тепловой энергии в конденсаторе
SU lauhduttimen teho
SV värmeavgivande effekt (kondensor) *c*

956 condensing furnace
DA kondenserende kedel
DE Kondensations-turm *m*
ES horno de condensación *m*
F chaudière *f* à condensation
I caldaia *f* a condensazione
MA kondenzációs kazán
NL condenserende ketel *m*
PO palenisko *n* kondensacyjne piec *m* kondensacyjny
PY конденсационная топка *f*
SU kondenssikattila
SV kondenserings-panna *c*

957 condensing pressure
DA kondenseringstryk
DE Verflüssigungs-druck *m*
ES presión de condensación *f*
F pression *f* de condensation
I pressione *f* di con-densazione
MA kondenzációs nyomás
NL condensatiedruk *m*
PO ciśnienie *n* skraplania
PY давление *n* конденсации
SU lauhtumispaine
SV kondensortryck *n*

958 condensing pressure valve
DA vandudlader
DE Verflüssiger-Druckventil *n*
ES válvula de presión de condensación *f*
F régulateur *m* de pression de condensation
I valvola *f* di regolazione della pressione di condensazione
MA kondenzációs nyomószelep
NL condensdruk-klep *f*
PO zawór *m* ciśnieniowy kondensatu
PY клапан *m* высокого давления в конденсаторе
SU paineventtiili
SV kondensortryck-ventil *c*

959 condensing temperature
DA fortætnings-temperatur
DE Kondensations-temperatur *f*
ES temperatura de condensación *f*
F température *f* de condensation
I temperatura *f* di condensazione
MA kondenzációs hőmérséklet
NL condensatie-temperatuur *f*
PO temperatura *f* skraplania
PY температура *f* конденсации
SU lauhtumislämpütila
SV kondens-temperatur *c*

960 condensing unit
DA kondenseringsenhed
DE Verdichter-Kältesatz *m* Verflüssigersatz *m*
ES grupo compresor-condensador *m* unidad condensadora *f*
F groupe *m* condenseur
I unità *f* condensante
MA kondenzációs egység
NL condensorunit *n*
PO skraplacz *m*
PY конденсаторный агрегат *m*
SU lauhdutinyksikkü
SV kondensorenhet *c*

961 condensing unit capacity
DA kondenserings-kapacitet
DE Kälteerzeuger-leistung *f*
ES capacidad de la unidad condensadora *f*
F puissance *f* frigorifique
I potenzialità *f* del condensatore
MA kondenzációs egység teljesítménye
NL condensor-capaciteit *f*
PO pojemność *f* skraplacza
PY производитель-ьность *f* конденсационной установки
SU lauhduttimen teho
SV kondensor-kapacitet *c*

962 conditioned air
DA behandlet luft
DE behandelte Luft *f*
ES aire acondicionado *m*
F air *m* conditionné
I aria *f* condizionata
MA kezelt levegő kondicionált levegő
NL behandelde lucht *f* geconditioneerde lucht *f*
PO powietrze *n* uzdatnione
PY кондициониро-ванный воздух *m*
SU käsitelty ilma
SV behandlad luft *c*

DA = Danish
DE = German
ES = Spanish
F = French
I = Italian
MA = Magyar (Hungarian)
NL = Dutch
PO = Polish
PY = Russian
SU = Finnish
SV = Swedish

963 condition line
DA linie i
Ix-diagram
DE Zustandslinie *f*
ES conducto del
condensado *m*
F droite *f* de soufflage
I retta *f* di carico
MA kondicionálási gürbe
NL toestandslijn *f*
PO taśma *f*
(produkcyjna)
klimatyzowana
PY линия *f* состояний
приточного
воздуха (на
психрометри-
ческои карте или
диаграмме
молье)
SU kuormitussuora
SV

964 condition of the air
DA lufttilstand
DE Luftzustand *m*
ES estado del aire *m*
F état *m* d'air
I stato *m* dell'aria
MA levegő állapota
NL luchtconditie *f*
PO stan *m* powietrza
PY состояние *n*
воздуха
SU ilmankäsittely
SV luftkondition *c*

965 conductance
DA ledningsevne
DE Leitfähigkeit *f*
ES conductancia *f*
F conductance *f*
I conduttanza *f*
MA vezetőképesség
NL geleidbaarheid *f*
geleidings-
vermogen *n*
PO przewodnictwo *n*
przewodność *f*
PY проводимость *f*
SU konduktanssi
SV konduktans *c*

966 conduction
DA varmeledning
DE Leitung *f*
ES conducción *f*
F conduction *f*
I conduzione *f*
MA vezetés
NL geleiding *f*
PO przewodzenie *n*
PY проводимость *f*
теплопроводность
f
SU johtuminen
SV överföring *c*

967 conduction gain
DA varmelednings-
gevinst
DE Wärmeleitungs-
gewinn *m*
ES aporte (de calor) por
conducción *m*
ganancia (de calor)
por conducción *f*
F gain *m* par
conduction
I apporto *m* di calore
per conduzione
MA vezetési nyereség
NL warmtegeleidings-
winst *f*
PO zysk *m* od
przewodzenia
PY усиление *n*
проводимости
SU johtumislämpü-
kuorma
SV värmevinst *c*

968 conduction loss
DA varmeledningstab
DE Wärmeleitungs-
verlust *m*
ES pérdida(s) por
conducción *f*
F perte *f* par
conduction
I disperdimento *m* di
calore per
conduzione
MA vezetési veszteség
NL warmtegeleidings-
verlies *n*
PO strata *f* przez
przewodzenie
PY потеря *f* за счет
проводимости
SU johtumislämpühäviü
SV värmeförlust *c*

969 conduction of heat
DA ledning af varme
DE Wärmeleitung *f*
ES conducción
térmica *f*
F conduction *f*
thermique
I conduzione *f* del
calore
MA hővezetés
NL warmtegeleiding *f*
PO przewodzenie *n*
ciepła
PY теплопроводность
f
SU lämmün johtuminen
SV värmeledning *c*

970 conductivity
DA specifik ledeevne
DE Leitfähigkeit *f*
ES conductividad
(térmica) *f*
F conductivité *f*
I conduttività *f*
MA vezetőképesség
NL geleidbaarheid *f*
geleidings-
vermogen *n*
PO przewodność *f*
właściwa
PY удельная
проводимость *f*
SU johtavuus
SV konduktivitet *c*

971 conductor
DA leder
DE Ader *f*
Kabel *n*
Leiter *m*
ES conductor *m*
F conducteur *m*
I conduttore *m*
MA vezető
NL geleider *m*
PO przewodnik *m* (ciało
przewodzące)
rura *f* spustowa
PY проводник *m*
SU johdin
SV ledare *c*

972 conduit
DA ledning
    rørledning
DE Leitung *f*
    Rohrleitung *f*
ES canalización *f*
    conducto *m*
    tubería *f*
F canalisation *f*
    conduit *m*
    conduite *f*
I canale *m*
    condotto *m*
    tubatura *f*
MA csatorna
    vezeték
NL leiding *f*
PO kanał *m*
    przewód *m*
PY трубопровод *m*
SU johto
SV ledning *c*

973 configuration
DA konfiguration
DE Anordnung *f*
    Konfiguration *f*
ES configuración *f*
F configuration *f*
I configurazione *f*
MA alak
    alakzat
NL configuratie *f*
PO kształt *m*
    układ *m*
    ukształtowanie *n*
PY конфигурация *f*
    форма *f*
SU kokoonpano
    rakenne
SV konfiguration *c*

974 configuration factor
DA konfigurationsfaktor
DE Anordnungs-
    faktor *m*
ES factor de forma *m*
F facteur *m* de forme
I fattore *m* di forma
MA alakténуező
NL configuratiefactor *m*
PO współczynnik *m*
    kształtu
PY формфактор *m*
SU näkyvyyskerroin
SV formfaktor *c*

975 connect *vb*
DA forbinde
DE verbinden
ES conectar
    empalmar
F connecter
I collegare
    connettere
MA csatlakoztat
    üsszeküt
NL aansluiten
    koppelen
    verbinden
PO łączyć
    połączyć
PY соединять
SU yhdistää
SV ansluta

976 connected load
DA tilslutningslast
DE angeschlossene
    Last *f*
ES carga conectada *f*
F charge *f* connectée
I carico *m* collegato
MA csatolt terhelés
NL totale belasting *f*
PO obciążenie *n*
    połączone
PY суммарная
    нагрузка *f*
SU kytketty kuorma
SV ansluten last *c*

977 connecting clamp
DA forbindelsesklemme
DE Verbindungs-
    klammer *f*
ES mordaza de unión *f*
F connecteur *m*
I staffa *f* di
    collegamento
MA csatlakozóidom
NL aansluitklem *f*
    koppelingsbeugel *m*
PO zacisk *m*
    połączeniowy
PY зажим *m*
    соединения
SU yhdistäjä
SV anslutnings-
    klämma *c*

978 connecting flange
DA forbindelsesflange
DE Verbindungs-
    flansch *m*
ES brida de conexión *f*
F bride *f* de connexion
I flangia *f* di
    collegamento
MA csatlakozóelem
NL aansluitflens *m*
    koppelingsflens *m*
    verbindingsflens *m*
PO kołnierz *m*
    połączeniowy
PY соединительный
    фланец *m*
SU liitoslaippa
SV anslutningsfläns *c*

979 connecting piece
DA forbindelsesstykke
DE Verbindungsstück *n*
ES pieza de conexión *f*
    pieza de unión *f*
F connecteur *m*
    (appareil de
    connexion)
I giunto *m* di
    collegamento
MA üsszekütő idom
NL aansluitstuk *n*
    koppelingsstuk *n*
    verbindingsstuk *n*
PO łącznik *m*
    złączka *f*
PY соединительная
    деталь *f*
    соединительный
    участок *m*
SU liitoskappale
SV anslutningsdon *n*

980 connecting pipe
DA forbindelsesrør
DE Verbindungsrohr *n*
ES tubo de
    comunicación *m*
    tubo de conexión *m*
F tube *m* de
    connexion
    tube *m* de
    raccordement
I tubo *m* di
    collegamento
MA üsszekütő cső
NL aansluitleiding *f*
    verbindingspijp *f*
PO rura *f* łącząca
PY соединительная
    труба *f*
SU liitosjohto
SV anslutningsrör *n*

981 connecting rod
DA pejlestang
DE Verbindungskabel n
ES biela f
F bielle f
I biella f
MA hajtókar
NL verbindingsstaaf f
verbindingsstang f
PO korbowód m
łącznik m
PY соединительный
шток m
шатун m
SU liitäntätanko
SV dragstång c

982 connecting terminal
DA tilslutningsterminal
DE Anschlußklemme f
ES borna de conexión f
terminal de conexión m
F borne f
I morsettiera f
MA kapocsdeszka
NL aansluitklem f
PO zacisk m liniowy
PY клемма f
соединительный
зажим m
SU kytkentäpiste
liitin
SV anslutnings-
klämma c

983 connection
DA tilslutning
DE Verbindung f
ES conexión f
enlace m
unión f
F connexion f
I collegamento m
MA csatlakozás
NL aansluiting f
koppeling f
verbinding f
PO łącznik m
połączenie n
złącze n
PY соединение n
SU liitos
SV anslutning c

984 connection in parallel
DA parallelforbindelse
DE Parallelverbindung f
ES conexión en paralelo f
F connexion f en parallèle
I collegamento m in parallelo
MA párhuzamos csatlakozás
NL parallelschakeling f
PO połączenie n równoległe
PY параллельное соединение n
SU rinnankytkentä
SV parallellkoppling c

985 connection in series
DA serieforbindelse
DE Verbindung f in Serie f
ES conexión en serie f
F connexion f en série
I collegamento m in serie
MA soros csatlakozás
NL serieschakeling f
PO połączenie n szeregowe
PY последовательное соединение n
SU sarjaankytkentä
SV seriekoppling c

986 connector
DA forbindelse
stik
DE Verbindungsstück n
ES conector m
empalme m
enlace m
F connecteur m
raccord m
I connettore m
morsetto m
raccordo m
serrafilo m
MA csatlakozó
NL aansluitstuk n
koppelstuk n
verbindingsstuk n
PO łącznik m
złączka f
PY соединитель m
SU liitin
SV anslutningsdon n

987 conservation of energy law
DA energisparelov
DE Energieerhaltungs-
satz n
ES ley de conservación de la energía f
F loi f de la conservation de l'énergie
I legge f di conservazione dell'energia
MA energiamegmaradási türvény
NL wet f van behoud van energie
PO prawo n zachowania energii
PY закон m сохранения энергии
SU energian häviämättü-
myyden laki
SV lagen c oförstörbarhet om energins c

988 console air conditioner
DA ophængt luftbehandler
DE Konsolklimagerät n
ES climatizador tipo cónsola m
F conditionneur m d'air en console
I condizionatore m per banco di comando
MA parapet légkondicionáló
NL wandkamerkoeler m
PO klimatyzator m indywidualny
PY консольный кондиционер m воздуха
SU ilmastointikone-
paketti
SV upphängd luftkonditionerings-
apparat c

989 constant
DA konstant
DE Konstante *f*
ES constante *f*
F constante *f*
I costante *f*
MA állandó
   konstans
NL constante *f*
PO stała *f*
PY константа *f*
   постоянная *f*
SU vakio
SV konstant *c*

990 constant *adj*
DA konstant
DE konstant
ES constante
F constant
I costante
MA konstans
NL constant
PO stały
   trwały
PY постоянный
SU vakio
SV konstant

**991 constant level
   valve**
DA niveauventil
DE Konstantniveau-
   ventil *n*
ES válvula de nivel
   constante *f*
F vanne *f* à niveau
   constant
I valvola *f* di livello
   costante
MA szinttartó szelep
NL niveauregelklep *f*
PO zawór *m* stałego
   poziomu
PY клапан *m*
   постоянного
   уровня
SU uimuriventtiili
SV konstantnivåventil *c*

**992 constant pressure
   valve**
DA trykreguleringsventil
DE Konstantdruckregel-
   ventil *n*
ES válvula de presión
   constante *f*
F régulateur *m* de
   pression
I valvola *f* a pressione
   costante
MA nyomásszabályozó
   szelep
NL drukregelventiel *n*
   reduceerventiel *n*
PO zawór *m* stałego
   ciśnienia
PY клапан *m*
   постоянного
   давления
SU vakiopaineventtiili
SV tryckregulator *c*

**993 constant value
   control**
DA konstant
   værdikontrol
DE Konstantwert-
   regelung *f*
ES control de valor
   constante *m*
F régulation *f* à valeur
   de consigne fixe
I regolazione *f* a
   punto fisso
   regolazione *f* a
   valore costante
MA állandó érték
   szabályozás
NL constante-waarde
   regeling *f*
PO regulator *m*
   stałowartościowy
   stała *f* regulacja
   wartości
PY регулирование *n*
   по постоянной
   величине
SU vakioarvosäätü
SV konstantvärdes-
   kontroll *c*

994 construction
DA konstruktion
DE Bau *m*
   Konstruktion *f*
ES construcción *f*
F construction *f*
I costruzione *f*
MA szerkezet
NL bouwwerk *n*
   constructie *f*
PO budowa *f*
   konstrukcja *f*
PY конструкция *f*
SU konstruktio
   rakennus
SV konstruktion *c*

**995 constructional
   component**
DA konstruktionsdel
DE Bauelement *n*
ES elemento de
   construcción *m*
F élément *m* de
   construction
I elemento *m* di
   costruzione
MA szerkezeti elem
NL bouwelement *n*
   constructiedeel *n*
PO składnik *m*
   konstrukcyjny
PY конструктивный
   элемент *m*
SU rakenneosa
SV byggnadselement *n*

**996 consulting
   engineer**
DA rådgivende ingeniør
DE beratender
   Ingenieur *m*
ES ingeniero consultor
   m
F ingénieur *m* conseil
I ingegnere *m* libero
   professionista
MA tanácsadó mérnök
NL raadgevend
   ingenieur *m*
PO doradca *m*
   techniczny
   inżynier *m*
   projektant
PY консультирующий
   инженер *m*
SU konsultti
   neuvotteleva
   insinüüri
SV konsulterande
   ingenjör *c*

DA = Danish
DE = German
ES = Spanish
F = French
I = Italian
MA = Magyar
   (Hungarian)
NL = Dutch
PO = Polish
PY = Russian
SU = Finnish
SV = Swedish

997 consumer connection
DA forbrugerforbindelse
DE Verbraucher-
anschluß *m*
ES acometida *f*
F branchement *m*
particulier
I derivazione *f* utente
MA fogyasztói
csatlakozás
NL verbruikers-
aansluiting *f*
PO przyłączenie *n*
odbiorcy
PY присоединение *n*
абонента
SU kuluttajaliitäntä
SV enskild anslutning *c*

998 consumption
DA forbrug
DE Verbrauch *m*
ES consumo *m*
F consommation *f*
I consumo *m*
MA fogyasztás
NL consumptie *f*
verbruik *n*
PO zużycie *n*
PY расход *m*
SU kulutus
käyttü
SV förbrukning *c*

999 contact
DA kontakt
DE Kontakt *m*
ES contacto *m*
F contact *m*
I contatto *m*
MA érintkezés
NL contact *n*
PO styk *m*
zetknięcie *n*
PY контакт *m*
SU kosketus
SV kontakt *c*

1000 contactor
DA kontaktor
DE Schütz *n*
ES contactor *m*
F contacteur *m*
I contattore *m*
MA mágneskapcsoló
NL relais *n*
PO stycznik *m*
PY переключатель *m*
SU kontaktori
SV kontaktor *c*

1001 contact point
DA kontaktpunkt
DE Kontaktpunkt *m*
ES punto de contacto *m*
F point *m* de contact
I punto *m* di contatto
MA érintkező csúcs
NL contactpunt *n*
PO punkt *m* kontaktu
punkt *m* zazębienia
punkt *m* zetknięcia
PY контакт *m*
точка *f* контакта
SU kosketuspiste
SV kontaktpunkt *c*

1002 contact rating
DA kontaktgrad
DE Kontaktleistung *f*
ES capacidad de
contacto *f*
F pouvoir *m* de
coupure
I grado *m* di contatto
MA érintkezési minősítés
NL contactbelasting *f*
schakelvermogen *n*
PO stykowa wartość *f*
znamionowa
PY контактный
параметр *m*
SU
SV kraftstyrning *c*

1003 contact thermometer
DA kontakttermometer
DE Kontakt-
thermometer *n*
ES termómetro de
contacto *m*
F thermomètre *m* à
contact
I termometro *m* a
contatto
MA érintő hőmérő
tapintó hőmérő
NL contact-
thermometer *m*
PO termometr *m*
kontaktowy
termometr *m*
stykowy
PY контактный
термометр *m*
SU pintalämpümittari
SV kontakt-
termometer *c*

1004 container
DA beholder
container
DE Behälter *m*
ES contenedor *m*
recipiente *m*
F conteneur *m*
récipient *m*
I contenitore *m*
MA konténer
tároló
NL bak *m*
container *m*
laadkist *f*
PO kontener *m*
pojemnik *m*
zasobnik *m*
PY коллектор *m*
контейнер *m*
резервуар *m*
SU säiliü
SV behållare *c*

1005 contaminant
DA forurene
DE Verunreiniger *m*
ES contaminante *m*
impurezas *f,pl*
F polluant *m*
I contaminante *m*
impurezza *f*
MA szennyezőanyag
NL besmettende stof *n*
verontreiniger *m*
PO substancja *f*
zanieczyszczająca
PY загрязнение *n*
SU epäpuhtaus
SV förorening *c*

1006 contaminate *vb*
DA forurener
DE verunreinigen
ES contaminar
F contaminer
polluer
I contaminare
inquinare
MA szennyez
NL besmetten
verontreinigen
PO skażać
zanieczyszczać
PY загрязнять
SU liata
saastuttaa
SV förorena

1007 contents
DA indhold
DE Gehalt *m*
Inhalt *m*
ES contenido *m*
F contenu *m*
teneur *f*
I contenuto *m*
MA tartalom
NL inhoud *m*
PO zawartość *f*
PY содержание *n*
SU sisällys
sisällysluettelo
sisältü
SV innehåll *n*

1008 continuous control
DA kontinuerlig
regulering
DE kontinuierliche
Regelung *f*
ES control continuo *m*
F contrôle *m* continu
I controllo *m*
continuo
MA folyamatos
szabályozás
NL continuregeling *f*
PO regulacja *f* ciągła
sterowanie *n* ciągłe
PY постоянный
контроль *m*
SU jatkuva säätü
SV kontinuerlig
kontroll *c*

**1009 continuous
operation**
DA kontinuerlig drift
DE Dauerbetrieb *m*
ES funcionamiento
continuo *m*
marcha continua *f*
servicio
permanente *m*
F marche *f* continue
I funzionamento *m*
continuo
MA folyamatos műküdés
NL continubedrijf *n*
PO działanie *n* ciągłe
praca *f* ciągła
PY постоянная
работа *f*
SU jatkuva käyttü
SV kontinuerlig drift *c*

1010 contract
DA kontrakt
DE Vertrag *m*
ES contrato *m*
F contrat *m*
I contratto *m*
MA szerződés
NL contract *n*
overeenkomst *f*
verbintenis *f*
PO kontrakt *m*
umowa *f*
PY договор *m*
контракт *m*
SU sopimus
SV kontrakt *n*

1011 contract heating
DA fjernvarme
DE Fernwärme-
versorgung *f*
ES calefacción por
contrato (servicio
contratado) *f*
F exploitation *f*
(contractuelle) de
chauffage
I riscaldamento *m*
contrattuale
MA fűtésszolgáltatás
NL verwarming *f* op
contractbasis *f*
PO ogrzewanie *n*
dyżurne
PY заказанное
отопление *n*
SU lämmünmyynti
SV fjärrvärme-
försörjning *c*

1012 contractor
DA entreprenør
DE Lieferant *m*
Unternehmer *m*
ES contratista *m*
instalador *m*
F entrepreneur *m*
I appaltatore *m*
imprenditore *m*
impresario *m*
MA kivitelező
NL aannemer *m*
installateur *m*
PO kontraktor *m*
przedsiębiorca *m*
wykonawca *m* robót
PY предприниматель
m
SU urakoitsija
SV entreprenür *c*

1013 contra-rotation
DA modrotation
DE Gegendrehung *f*
ES contrarrotación *f*
F contra-rotation *f*
I controrotazione *f*
MA ellentétes forgás
NL tegengestelde
draairichting *f*
PO przeciwbieżność *f*
PY противооборот *m*
SU pyürimissuuntaa
vastaan
SV motrotation *c*

1014 control
DA kontrol
regulering
styring
DE Regelung *f*
ES control *m*
regulación *f*
F commande *f*
régulation *f*
I regolazione *f*
MA szabályozás
vezérlés
NL besturing *f*
controle *f*
regeling *f*
PO kontrola *f*
regulacja *f*
sterowanie *n*
PY контроль *m*
регулирование *n*
управление *n*
SU ohjaus
säätü
SV kontroll *c*

1015 control *vb*
DA regulere
styre
DE regeln
steuern
ES controlar
regular
F commander
régler
I controllare
regolare
MA ellenőriz
szabályoz
vezérel
NL besturen
controleren
regelen
PO regulować
sterować
PY контролировать
регулировать
управлять
SU ohjata
säätää
SV kontrollera

1016 control action
DA styrefunktion
DE Regelungswirkung *f*
ES acción de control *f*
F action *f* de
   régulation
I azione *f* di
   regolazione
MA ellenőrző
   tevékenység
NL regelactie *f*
PO działanie *n*
   sterowane
   mechanizm *m*
   sterowany
PY регулирующее
   (управляющее)
   воздействие *n*
SU säätötoiminta
SV kontrollåtgärd *c*

1017 control algorithm
DA regneoperation
DE Regelalgorithmus *m*
ES algoritmo de control
   m
F algorithme *m* de
   contrôle
I algoritmo *m* di
   controllo
MA szabályozó
   algoritmus
NL regelalgoritme
PO algorytm *m* regulacji
   algorytm *m*
   sterowania
PY алгоритм *m*
   управления
SU säätöalgoritmi
SV kontrollalgoritm *c*

1018 control cabin
DA kontrolrum
DE Steuerkabine *f*
ES cabina de control *f*
   cabina de mandos *f*
   mueble de control *m*
F cabine *f* de
   commande
I cabina *f* di comando
   cabina *f* di controllo
MA vezérlőfülke
NL controlekamer *f*
   regelkamer *f*
PO kabina *f* sterownicza
PY кабина *f*
   управления
SU valvomo
SV manöverrum *n*

1019 control desk
DA kontrol (styre)-pult
DE Schaltpult *n*
   Steuerpult *n*
ES pupitre de
   mandos *m*
F pupitre *m* de
   commande
I quadro *m* di
   comando
MA vezérlőpult
NL regellessenaar *m*
   schakellessenaar *m*
PO tablica *f* sterownicza
PY пульт *m*
   управления
SU säätöpulpetti
SV manöverbord *n*

1020 control deviation
DA afvigelse fra
   indstillingsværdi
DE Regelabweichung *f*
ES desviación de
   regulación *f*
F écart *m* de réglage
I scarto *m* di
   regolazione
MA szabályozási eltérés
NL regelafwijking *f*
PO odchyłka *f*
   sterowania
PY регулируемое
   отклонение *n*
SU säätöpoikkeama
SV avvikelse *c*

1021 control device
DA kontrolapparat
   reguleringsanordning
DE geregelte
   Einrichtung *f*
ES dispositivo de
   control *m*
F équipement *m* de
   commande
I strumento *m* di
   regolazione
MA szabályozó készülék
NL regelapparaat *n*
PO urządzenie *n*
   sterujące
PY регулирующее
   устройство *n*
SU säätölaite
SV reglerdon *n*

1022 control element
DA styreelement
DE Regelelement *n*
ES elemento de control
   m
F élément *m* terminal
   de commande
I attuatore *m* di
   controllo
MA szabályozó elem
NL reglement *n*
PO element *m*
   sterowniczy
PY регулирующий
   (управляющий)
   элемент *m*
SU säätölaite
SV styrdon *n*

1023 control function
DA kontrolfunktion
   reguleringsfunktion
DE Regelfunktion *f*
ES función de control *f*
F fonction *f* de
   contrôle-
   commande
I funzione *f* di
   controllo
MA szabályozási feladat
NL regelfunctie *f*
PO funkcja *f* kontroli
   funkcja *f* sterowania
PY функция *f*
   управления
SU säätöfunktio
SV kontrollfunktion *c*

1024 controlled atmosphere
DA reguleret atmosfære
DE geregelte
   Atmosphäre *f*
ES atmósfera
   controlada *f*
F atmosphère *f*
   contrôlée
I atmosfera *f*
   controllata
MA szabályozott
   légállapot
NL geregelde
   omgevingslucht *f*
PO atmosfera *f*
   kontrolowana
PY регулируемая
   атмосфера *f*
SU säädetyt
   ympäristüolot
SV kontrollerad
   atmosfär *c*

1025 controlled medium
DA reguleret medium
DE geregeltes Medium n
ES medio controlado m
F milieu m contrôlé
I mezzo m controllato
MA szabályozott küzeg
NL geregeld medium n
PO czynnik m regulowany
czynnik m sterowany
PY регулируемая среда f
SU säädettävä aine
SV styrt flöde n

1026 controlled variable
DA reguleringsvariabel
DE geregelte Variable f
ES variable controlada f
F variable f régulée
I variabile f controllata
MA szabályozott változó
NL geregelde grootheid f
PO zmienna f regulowana
zmienna f sterowana
PY регулируемая переменная f
SU säätümuuttuja
SV kontrollvariabel c

1027 controller
DA regulator
DE Regler m
ES aparato de regulación m
regulador m
F régulateur m
I regolatore m
MA szabályozó
NL controleur m
regelapparaat n
PO kontroler m
regulator m
urządzenie n sterujące
PY регулятор m
SU säätäjä
SV regulator c

1028 control loop
DA reguleringssløjfe
DE Regelschleife f
ES circuito de control m
F boucle f de contrôle
I catena f di controllo
MA szabályozó hurok
NL regelkring m
PO pętla f regulacji
pętla f sterowania
PY контур m регулирования (управления)
SU säätüpiiri
SV kontrollslinga c

1029 control panel
DA betjeningstavle
DE Schaltpult n
Schalttafel f
ES cuadro de mandos m
F panneau m de régulation
tableau m de régulation
I quadro m di comando
MA vezérlőtábla
NL regelpaneel n
schakelkast f
PO tablica f sterownicza
PY панель f управления
SU säätüpaneeli
SV manöverpanel c

1030 control point
DA kontrolpunkt
DE Regelpunkt m
ES punto de control m
F point m de contrôle
I punto m di controllo
MA szabályozási pont
NL regelpunt n
PO punkt m kontrolny
punkt m sterowania
PY контрольная точка f
SU säätüpiste
SV kontrollpunkt c

1031 control power element
DA effektregulator
DE Regelgeräte-antrieb m
ES fuerza motriz del control f
F élément m de régulation de puissance
I elemento m di controllo della potenza
MA szabályozás müküdtető elem
NL bedienend element n van het regelorgaan
PO element m sterowania mocą
PY исполнительный механизм m (в системе регулирования)
SU toimilaite toimimoottori
SV kraftelement n

1032 control range
DA reguleringsområde
DE Regelbereich m
ES campo de control m
F plage f de régulation
I campo m di regolazione
MA szabályozási tartomány
NL regelbereik n
PO zakres m regulacji
PY диапазон m регулирования
SU säätüalue
SV kontrollomfång n

1033 control strategy
DA styringsstrategi
DE Regelstrategie f
ES estrategia de control f
F stratégie f de contrôle
I strategia f di controllo
MA szabályozási stratégia
NL regelprogramma n
regelstrategie f
PO kierunki m,pl regulacji
strategia f sterowania
PY стратегия f управления
SU säätüstrategia
SV kontrollmetod c

1034 convection
DA konvektion
varmeledning
DE Konvektion f
ES convección f
F convection f
I convezione f
MA áramlás
konvekció
NL convectie f
PO konwekcja f
unoszenie n
PY конвекция f
SU konvektio
kuljettuminen
SV konvektion c

1035 convection heater
DA konvektionsvarme-
flade
DE Konvektions-
heizgerät n
ES calentador por
convección m
F appareil m de
chauffage par
convection
I apparecchio m di
riscaldamento a
convezione
MA konvekciós fűtés
NL convectieverwarmings-
toestel n
PO grzejnik m
konwekcyjny
PY конвективный
нагреватель m
SU konvektori
SV konvektions-
värmare c

1036 convective adj
DA konvektions-
DE konvektiv
ES convectivo
F convective
I convettivo
MA konvektív
NL convectief
PO konwekcyjny
PY конвективный
SU konvektiivinen
SV konvektiv

1037 convective current
DA konvektionsstrøm
DE Konvektivstrom m
ES corriente de
convección f
F courant m de
convection
I corrente f di
convezione
MA konvektív áramlás
NL convectiestroom m
PO prąd m
konwekcyjny
PY конвективный
поток m
SU konvektiovirtaus
SV konvektionsström c

1038 convector
DA konvektor
DE Konvektor m
ES convector m
F convecteur m
I convettore m
MA konvektor
NL convector m
PO konwektor m
PY конвектор m
SU konvektori
SV konvektor c

1039 conversion
DA omdannelse
DE Umwandlung f
ES conversión f
F conversion f
I conversione f
MA átalakítás
NL conversie f
omzetting f
PO konwersja f
przekształcanie n się
przemiana f
PY преобразование n
SU muunnos
SV omformning c

1040 conversion burner
DA omstilbar brænder
DE Umstellbrenner m
ES quemador inversor
m
F brûleur m
polycombustible
I bruciatore m per
combustibile
diverso da quello
originario
MA átállítható égő
NL brander m voor
diverse
brandstoffen
PO palnik m na różne
paliwa
PY комбинированная
горелка f
SU kaasutuspoltin
monipolttoainepoltin
muutospoltin
SV konverterings-
brännare c

1041 converter
DA omformer
DE Umformer m
ES convertidor m
transformador m
F convertisseur m
I convertitore m
MA áramátalakító
NL omzetter m
PO konwerter m
przetwornica f
przetwornik m
PY преобразователь m
SU muunnin
SV omformare c

1042 convertible boiler
DA omstillelig kedel
DE Umstellbrand-
kessel m
ES caldera convertible f
F chaudière f
convertible
I caldaia f
convertibile
MA átállítható kazán
NL ketel m voor diverse
brandstoffen
PO kocioł m na różne
paliwa
PY котел m со
сменной топкой
SU muunneltava kattila
SV omställbar panna c

DA	=	Danish
DE	=	German
ES	=	Spanish
F	=	French
I	=	Italian
MA	=	Magyar (Hungarian)
NL	=	Dutch
PO	=	Polish
PY	=	Russian
SU	=	Finnish
SV	=	Swedish

1043 conveyor
DA transportør
DE Aufzug *m*
    Förderband *n*
ES transportador *m*
F transporteur *m*
I trasportatore *m*
MA konvejor
    szállítóberendezés
NL transportband *m*
    transporteur *m*
PO przenośnik *m*
PY транспортер *m*
SU kuljetin
SV transportör *c*

1044 conveyor belt
DA transportbånd
DE Förderband *n*
ES cinta
    transportadora *f*
F bande *f*
    transporteuse
I nastro *m*
    trasportatore
MA szállító szalag
NL transportband *m*
PO pas *m* transmisyjny
    taśma *f* przenośnika
PY ленточный
    транспортер *m*
SU kuljetinhihna
SV transportband *n*

1045 cool *vb*
DA køle
DE abkühlen
    kühlen
ES enfriar
    refrigerar
F refroidir
I raffreddare
    raffrescare
MA hűt
NL afkoelen
    koelen
PO chłodzić
PY охлаждать
SU jäähdyttää
SV kyla

1046 coolant
DA kølemiddel
DE Kühlmittel *n*
ES refrigerante *m*
F fluide *m*
    frigoporteur
I refrigerante *m*
MA hűtőküzeg
NL koelmiddel *n*
PO chłodziwo *n*
    czynnik *m*
    chłodzący
PY хладагент *m*
SU jäähdyke
SV kylmedel *n*

1047 cooled air
DA kølet luft
DE gekühlte Luft *f*
ES aire enfriado *m*
    aire refrigerado *m*
F air *m* refroidi
I aria *f* raffreddata
MA hűtütt levegő
NL gekoelde lucht *f*
PO powietrze *n*
    ochłodzone
PY охлажденный
    воздух *m*
SU jäähdytetty ilma
SV kylluft *c*

1048 cooler
DA køler
DE Kühler *m*
ES cámara de
    refrigeración *f*
    enfriador *m*
F réfrigérant *m*
    refroidisseur *m*
I raffreddatore *m*
    refrigeratore *m*
MA hűtő
NL koeler *m*
PO agregat *m*
    chłodniczy
PY охладитель *m*
SU jäähdyttäjä
SV kylare *c*

1049 cooler battery
DA kølebatteri
DE Kühlaggregat *n*
ES batería de
    refrigeración *f*
F batterie *f* de
    refroidissement
I batteria *f* di
    raffreddamento
MA hűtőtest
NL koelbatterij *f*
PO chłodnica *f*
    zespół *m* chłodzący
PY холодильная
    батарея *f*
SU jäähdytyspatteri
SV kylbatteri *n*

1050 cooling
DA køling
DE Kühlung *f*
ES enfriamiento *m*
F refroidissement *m*
I raffreddamento *m*
MA hűtés
NL afkoeling *f*
    koeling *f*
PO chłodzenie *n*
    ochładzanie *n*
PY охлаждение *n*
SU jäähdytys
SV kylning *c*

1051 cooling agent
DA kølemiddel
DE Kühlmittel *n*
ES fluido frigorífico *m*
    medio
    refrigerante *m*
    refrigerante *m*
F réfrigérant *m*
I mezzo *m* di
    raffreddamento
MA hűtőküzeg
NL koelmiddel *n*
    koelvloeistof *f*
PO środek *m* chłodniczy
PY холодильный
    агент *m*
SU jäähdytysaine
SV köldmedium *n*

1052 cooling air
DA køleluft
DE Kühlluft *f*
ES aire de
  enfriamiento *m*
  aire refrigerado *m*
F air *m* de
  refroidissement
  air *m*
  rafraichissant
I aria *f* di
  raffreddamento
MA hűtő levegő
NL koellucht *f*
PO powietrze *n*
  chłodzące
PY охлаждающий
  воздух *m*
SU jäähdytysilma
SV kylluft *c*

1053 cooling capacity
DA kølekapacitet
DE Kühlleistung *f*
ES potencia frigorífica *f*
F puissance *f* de
  refroidissement
I potenzialità *f* di
  raffreddamento
MA hűtőteljesítmény
NL koelcapaciteit *f*
PO wydajność *f*
  chłodzenia
  zdolność *f*
  chłodzenia
PY холодильная
  производитель-
  ность *f*
SU jäähdytysteho
SV kyleffekt *c*

1054 cooling coil
DA kølebatteri
  køleflade
DE Kühlschlange *f*
ES serpentín de
  enfriamiento *m*
F serpentin *m* de
  refroidissement
I batteria *f* di
  raffreddamento
MA hűtő hőcserélő
NL koelbatterij *f*
  koelspiraal *f*
PO chłodnica *f*
  wężownica *f*
  chłodząca
PY охлаждающий
  змеевик *m*
SU jäähdytyspatteri
SV kylslinga *c*

1055 cooling curve
DA kølekurve
DE Kühlkurve *f*
ES curva de
  enfriamiento *f*
F courbe *f* de
  refroidissement
I curva *f* di
  raffreddamento
MA hűtési gürbe
NL koelcurve *f*
PO krzywa *f* chłodzenia
PY кривая *f*
  охлаждения
SU jäähdytyskäyrä
SV kylningskurva *c*

1056 cooling degree day
DA kølegraddag
DE Kältegradtag *m*
ES grado-día de
  enfriamiento *m*
F degrés jours *m* de
  refroidissement
I grado-giorno *m* di
  raffreddamento
MA hűtési napfok
NL koelgraaddag *m*
PO dobostopień *m*
  chłodzenia
PY суточный уровень
  m температурной
  разности
SU jäähdytyksen
  astepäiväluku
SV kylgraddag *c*

1057 cooling down
DA nedkøle
DE Abkühlung *f*
ES enfriamiento *m*
F refroidissement *m*
  naturel
I raffreddamento *m*
  naturale
  riduzione *f* di
  temperatura
MA lehűtés
NL afkoeling *f*
PO chłodzenie *n*
  naturalne
PY остывание *n*
SU jäähdytys
SV nedkylning *c*

1058 cooling effect
DA køleeffekt
DE Kühleffekt *m*
  Kühlwirkung *f*
ES potencia frigorífica
  (específica) *f*
F effet *m* de
  refroidissement
I effetto *m* di
  raffreddamento
MA hűtőhatás
NL koeleffect *n*
PO efekt *m* chłodzenia
PY холодильный
  эффект *m*
SU jäähdytysteho
SV kyleffekt *c*

1059 cooling efficiency ratio
DA kølevirkningsgrad
DE Kühlwirkungsgrad-
  verhältnis *n*
ES rendimiento de
  enfriamiento *m*
F coefficient *m* d'effet
  frigorifique
I effetto *m* frigorifero
MA hűtési hatásfok
NL koelrendement *n*
PO stosunek *m*
  wydajności
  chłodzenia
  współczynnik *m*
  wydajności
  chłodzenia
PY коэффициент *m*
  эффективности
  охлаждения
SU kylmäkerroin
SV kylverkningsgrads-
  frhållande *n*

1060 cooling element
DA køleelement
DE Kühlelement *n*
ES elemento enfriador
  m
F élément *m* de
  refroidissement
I elemento *m*
  raffreddante
MA hűtő elem
NL koelelement *n*
PO element *m*
  chłodzący
PY охлаждающий
  элемент *m*
SU jäähdytyselementti
SV kylelement *n*

1061 cooling jacket
DA kappekøler
DE Kühlmantel *m*
ES camisa enfriadora *f*
F chemise *f* de
   refroidissement
I camicia *f* di
   raffreddamento
MA hűtő küpeny
NL koelmantel *m*
PO płaszcz *m* chłodzący
PY охлаждающая
   рубашка *f*
SU jäähdytysvaippa
SV kylmantel *c*

1062 cooling load
DA kølelast
DE Kühllast *f*
ES carga frigorífica *f*
F charge *f* frigorifique
I carico *m* frigorifero
MA hűtési terhelés
NL koellast *m*
PO obciążenie *n*
   chłodnicze
PY расход *m* холода
SU jäähdytyskuorma
SV kylbehov *n*

1063 cooling loss
DA køletab
DE Abkühlungs-
   verlust *m*
ES pérdida de frío *f*
F perte *f* de froid
I perdita *f* di freddo
MA hűtési veszteség
NL koelverlies *n*
PO strata *f* chłodzenia
PY потери *f,pl* холода
SU jäähdytyshäviü
SV köldförlust *c*

1064 cooling medium
DA kølemiddel
DE Kältemittel *n*
ES medio
   refrigerante *m*
F réfrigérant *m*
I mezzo *m* di
   raffreddamento
MA hűtőküzeg
NL koelmedium *n*
PO czynnik *m*
   chłodzący
PY холодоноситель *m*
SU jäähdytysaine
SV köldmedium *n*

1065 cooling plant
DA køleanlæg
DE Kälteanlage *f*
ES instalación
   frigorífica *f*
F installation *f*
   frigorifique
I impianto *m*
   frigorifero
MA hűtőberendezés
NL koelinrichting *f*
   koelmachineruimte *f*
PO maszynownia *f*
   chłodnicza
PY холодильная
   установка *f*
SU jäähdytyslaitos
SV kylanläggning *c*

1066 cooling range
DA køleområde
DE Kältebereich *m*
ES campo de
   enfriamiento *m*
F chute *f* de
   température
   entrée-sortie
I campo *m* di
   raffreddamento
MA hűtési tartomány
NL koelbereik *n*
PO zakres *m* chłodzenia
PY перепад *m*
   температуры в
   охладителе
SU jäähdytysnesteen
   lämpütilaero
SV kylomräde *n*

1067 cooling rate
DA afkølingsgrad
DE Abkühlgeschwindig-
   keit *f*
ES velocidad de
   enfriamiento *f*
F vitesse *f* de
   refroidissement
I velocità *f* di
   raffreddamento
MA hűtési fok
NL koelsnelheid *f*
   koelvermogen *n*
PO prędkość *f*
   chłodzenia
   stopień *m*
   chłodzenia
PY темп *m*
   охлаждения
SU jäähtymisnopeus
SV kylningshastighet *c*

1068 cooling surface
DA køleflade
DE Kühlfläche *f*
ES superficie de
   enfriamiento *f*
F surface *f* de
   refroidissement
I superficie *f* di
   raffreddamento
MA hűtőfelület
NL koelend oppervlak *n*
PO powierzchnia *f*
   chłodząca
PY поверхность *f*
   охлаждения
SU jäähdytyspinta
SV kylyta *c*

1069 cooling system
DA kølesystem
DE Kühlsystem *n*
ES sistema de
   enfriamiento *m*
   sistema frigorífico *m*
F système *m* de
   refroidissement
I sistema *m* di
   raffreddamento
MA hűtőrendszer
NL koelsysteem *n*
PO system *m* chłodzenia
PY система *f*
   охлаждения
SU jäähdytysjärjestelmä
SV kylsystem *n*

1070 cooling tower
DA køletårn
DE Kühlturm *m*
ES torre de enfriamiento
   f
   torre de refrigeración
   f
F tour *f* de
   refroidissement
I torre *f* di
   raffreddamento
MA hűtőtorony
NL koeltoren *m*
PO chłodnia *f*
   kominowa
   wieża *f* chłodnicza
PY башенная градирня
   f
SU jäähdytystorni
SV kyltorn *n*

1071 cooling tower filling

DA køletårnspak-materiale
DE Kühlturmfüll-material *n*
ES relleno de la torre de refrigeración *m*
F garnissage *m* de tour de ruissellement
I riempimento *m* della torre evaporativa
MA hűtőtorony tültet
NL koeltorenvulling *f*
PO wypełnienie *n* chłodni kominowej wypełnienie *n* wieży chłodniczej
PY орошаемая насадка *f* градирни
SU jäähdytystornin täyte
SV kyltornsfyllning *c*

1072 cooling tower fogging

DA køletårnståge
DE Kühlturmnebel *m*
ES nebulosidad de la torre de refrigeración *f*
F brouillard *m* de tour de refroidissement
I nebbia *f* di torre di raffreddamento
MA hűtőtorony párolgása
NL mistvorming *f* rond koeltoren
PO zamglenie *n* w wyniku pracy chłodni kominowej zamglenie *n* w wyniku pracy wieży chłodniczej
PY образование *n* тумана над градирней
SU jäähdytystornin sum-unmuodostus
SV kyltorns-dimbildning *c*

1073 cooling tower packing (*see* cooling tower filling)

1074 cooling tower plume

DA dampfane fra køletårn
DE Kühlturmschwaden-emission *f*
ES penacho de la torre de refrigeración *m*
F panache *m* de tour de refroidissement
I pennacchio *m* della torre di raffreddamento
MA hűtőtorony párolgása
NL koeltorenmist *m*
PO opary *pl* unoszące się nad chłodnią kominową opary *pl* unoszące się nad wieżą chłodniczą
PY наблюдаемый выпуск *m* из градирни
SU jäähdytystornin poistovirtaus
SV kyltornsdimma *c*

1075 cooling tunnel

DA køletunnel
DE Kühltunnel *m*
ES tunel de refrigeración *m*
F tunnel *m* de réfrigération
I tunnel *m* frigorifero
MA hűtő alagút
NL koeltunnel *m*
PO tunel *m* chłodniczy
PY тоннельный охладитель *m*
SU jäähdytystunneli
SV kyltunnel *c*

1076 cooling unit

DA køleunit
DE Kältesatz *m*
ES equipo frigorífico *m*
F bloc *m* refroidisseur
I blocco *m* frigorifero
MA hűtőegység
NL koelapparaat *n* koelunit *n*
PO agregat *m* chłodniczy zespół *m* chłodzący
PY охлаждающий агрегат *m*
SU jäähdytysyksikkü
SV kylapparat *c*

1077 cooling water

DA kølevand
DE Kühlwasser *n*
ES agua de enfriamiento *f* agua de refrigeración *f*
F eau *f* de refroidissement
I acqua *f* di raffreddamento
MA hűtővíz
NL koelwater *n*
PO woda *f* chłodząca
PY охлаждающая вода *f*
SU jäähdytysvesi
SV kylvatten *n*

1078 cool storage

DA kølelager
DE Kältelager *n* Kühllager *n*
ES almacenamiento de frío *m*
F stockage *m* froid
I accumulo *m* di freddo
MA hűtütt tárolás
NL gekoelde opslag *m* koudeopslag *m*
PO akumulacja *f* energii chłodniczej
PY аккумулятор *m* холода
SU viileä varasto
SV kylförråd *n*

1079 co-ordinated design

DA samordnet projektering
DE koordinierter Entwurf *m*
ES proyecto coordinado *m*
F plan *m* d'ensemble
I progetto *m* coordinato
MA üsszhangba hozott tervezés
NL gecoördineerd ontwerp *n*
PO projektowanie *n* skoordynowane
PY скоординированное проектирование *n*
SU koordinoitu suunnittelu
SV samordnad projektering *c*

1080 COP (*see* coefficient of performance)

1081 copper
DA kobber
DE Kupfer *n*
ES cobre *m*
F cuivre *m*
I rame *m*
MA réz
NL roodkoper *n*
PO miedź *f*
PY медь *f*
SU kupari
SV koppar *c*

1082 copper pipe (tube)
DA kobberrør
DE Kupferrohr *n*
ES tubo de cobre *m*
F tube *m* de cuivre
I tubo *m* di rame
MA rézcső
NL roodkoperen pijp *f*
PO rura *f* miedziana
PY медная труба *f*
SU kupariputki
SV kopparrör *n*

1083 copper plating
DA kobberbeklædning
DE Kupferauflage *f*
   Kupferplattierung *f*
ES chapado de cobre *m*
F revêtement *m* de
   cuivre
I ramatura *f*
MA rézborítás
NL koperbekleding *f*
PO miedziowanie *n*
   powlekanie *n*
   miedzią
PY медное
   гальваническое
   покрытие *n*
SU kuparipinnoite
SV kopparplätering *c*

1084 core
DA kærne
DE Kern *m*
ES núcleo *m*
F coeur *m*
   noyau *m*
I anima *f*
   nucleo *m*
MA kábelér
   mag
NL kern *f*
PO rdzeń *m*
PY кабельная жила *f*
   sterjenç *m*
   ядро *n*
SU sydän
   ydin
SV kärna *c*

1085 core area
DA kærneareal
DE Kernfläche *f*
ES superficie del núcleo
   f
F zone *f* centrale
I area *f* interna (di un
   edificio
   climatizzato)
MA belső mag területe
NL kerngebied *n*
   kernoppervlak *n*
PO powierzchnia *f*
   przekroju
   poprzecznego
   (strumienia)
PY площадь *f* сечения
   (ядра, стержня,
   жилы)
SU rakennuksen
   ikkunaton
   keskiosa
SV kärnyta *c*

1086 cork
DA kork
DE Kork *m*
ES corcho *m*
F liège *m*
I sughero *m*
MA parafa
NL kurk *n*
PO korek *m*
PY пробка *f*
SU korkki
SV kork *c*

**1087 corkboard thermal
   insulation**
DA korkpladevarme-
   isolering
DE Korkplattenwärme-
   dämmung *f*
ES aislamiento térmico
   con coquillas de
   corcho *m*
F isolation *f* par
   plaque de liège
I isolante *m* termico
   in pannelli di
   sughero
MA parafa hőszigetelés
NL kurkplaat isolatie *f*
PO izolacja *f* cieplna z
   płyt korkowych
PY пробковая
   теплоизоляция *f*
SU korkkieriste
SV värmeisolering *c* av
   kork *c*

1088 corner burner
DA vinkelbrænder
DE Eckbrenner *m*
ES quemador
   angular *m*
F brûleur *m* d'angle
I bruciatore *m* ad
   angolo
MA sarokégő
NL hoekbrander *m*
PO palnik *m* kątowy
PY угловая горелка *f*
SU nurkkapoltin
SV hörnbrännare *c*

1089 corner valve
DA vinkelventil
DE Eckventil *n*
ES llave angular *f*
F robinet *m* d'équerre
I valvola *f* di angolo
MA sarokszelep
NL haakse afsluiter *m*
   hoekafsluiter *m*
PO zawór *m* kątowy
PY угловой вентиль *m*
   (клапан)
SU kulmaventtiili
SV vinkelventil *c*

**1090 corrected effective
   temperature**
DA korrigeret effektiv
   temperatur
DE korrigierte
   Effektiv-
   temperatur *f*
ES temperatura efectiva
   corregida *f*
F température *f*
   effective corrigée
I temperatura *f*
   effettiva corretta
MA helyesbített tényleges
   hőmérséklet
NL gecorrigeerde
   effectieve
   temperatuur *f*
PO temperatura *f*
   efektywna
   skorygowana
PY скорректированная
   эффективная
   температура *f*
SU korjattu tehollinen
   lämpütila
SV korrigerad effektiv
   temperatur *c*

1091 correcting variable
DA reguleret variabel
DE Korrekturvariable f
ES variable corrección f
F variable f de
   correction
I variabile f di
   correzione
MA helyesbített változó
NL corrigerende
   variabele f
PO poprawka f
   zmienna f
   korekcyjna
PY корректирующий
   сигнал m
SU takaisinkytketty
   suure
SV kontrollvariabel c

1092 corrective maintenance
DA forberende
   vedligehold
DE vorbeugende
   Wartung f
ES mantenimiento
   correctivo m
F maintenance f
   corrective
I manutenzione f
   correttiva
MA javító karbantartás
NL correctief
   onderhoud m
PO obsługa f naprawcza
PY учет m отказов при
   эксплуатации
SU ehkäisevä huolto
SV reparation c

1093 corrode vb
DA korrodere
DE korrodieren
ES corroer
F corroder
I corrodere
MA korrodál
NL aantasten
   corroderen
PO korodować
PY корродировать
SU syüpyä
SV korrodera

1094 corrosion
DA korrosion
   tæring
DE Korrosion f
ES corrosión f
F corrosion f
I corrosione f
MA korrózió
NL aantasting f
   corrosie f
PO korozja f
PY коррозия f
SU korroosio
SV korrosion c

1095 corrosion inhibitor
DA korrosions-
   hæmmende stof
DE Korrosions-
   schutzmittel n
ES inhibidor de
   corrosión m
   medio
   anticorrosivo m
F inhibiteur m de
   corrosion
I inibitore m di
   corrosione
MA korróziógátló szer
   korróziós inhibítor
NL corrosiebestrijder m
   corrosie-inhibitor m
PO inhibitor m korozji
PY средство n
   антикоррозийной
   защиты
SU korroosionestoaine
SV korrosions-
   inhibitor c

1096 corrosion protection
DA korrosions-
   beskyttelse
DE Korrosionsschutz m
ES protección contra la
   corrosión f
F protection f contre
   la corrosion
I protezione f
   anticorrosione
MA korrózióvédelem
NL corrosie-
   bescherming f
PO zabezpieczenie n
   przeciwkorozyjne
PY антикоррозийная
   защита f
SU korroosiosuojaus
SV korrosionsskydd n

1097 corrosion-resistant adj
DA korrosionsmod-
   standsdygtig
DE
ES anticorrosivo
   (resistente a la
   corrosión)
F résistant à la
   corrosion
I resistente alla
   corrosione
MA korrózióálló
NL corrosievast
PO odporny na korozję
PY коррозион-
   ностойкий
SU korroosiota kestävä
SV korrosionshärdig

1098 corrosive adj
DA korroderende
   tærende
DE korrosiv
ES corrosivo
F corrosif
I corrosivo
MA korróziós
NL aantastend
   corrosief
PO korozyjny
PY корродирующий
SU korrodoiva
SV korrosiv

1099 corrugated bend
DA rynkebøjning
DE Faltenrohrbogen m
ES curva corrugada f
F coude m en tube
   plissé
I curva f corrugata
   gomito m corrugato
MA redős ívcső
NL plooibocht f
PO kolano n faliste
PY гофрированное
   колено n
SU taipuisa mutka
SV korrugerad böj c

DA	=	Danish
DE	=	German
ES	=	Spanish
F	=	French
I	=	Italian
MA	=	Magyar (Hungarian)
NL	=	Dutch
PO	=	Polish
PY	=	Russian
SU	=	Finnish
SV	=	Swedish

1100 corrugated elbow
DA rynket knærør
DE Wellrohrbogen *m*
ES codo corrugado *m*
F coude *m* en tube plissé
I curva *f* corrugata gomito *m* corrugato
MA redős künyük
NL plooi-elleboog *m*
PO łuk *m* falisty
PY гофрированный отвод *m*
SU taipuisa kulma
SV korrugerad böj *c*

1101 cost(s)
DA omkostning udgift
DE Ausgabe *f* Kosten *f*
ES coste(s) *m* costo(s) *m* gasto(s) *m*
F coût *m* dépense *f* prix *m*
I costo *m* spesa *f*
MA kültség
NL kosten *pl*
PO koszt *m* (koszty)
PY стоимость *f*
SU kustannus
SV kostnad(er) *c*

1102 cost in use
DA driftomkostning
DE Betriebskosten *f,pl*
ES coste de explotación *m*
F coût *m* d'exploitation
I costo *m* di esercizio
MA üzemeltetési kültség
NL bedrijfskosten *pl*
PO koszt *m* eksploatacji koszt *m* użytkowania
PY стоимость *f* в эксплуатации
SU käyttükustannus
SV driftkostnad *c*

1103 counter check
DA efterkontrol
DE Nachprüfung *f*
ES contraprueba *f*
F contre-épreuve *f*
I controprova *f*
MA ellenhatás
NL contra-expertise *f* tegenproef *f*
PO badanie *n* końcowe kontrola *f* dodatkowa sprawdzenie *n* kontrolne
PY противоточное течение *n*
SU jälkitarkastus
SV efterkontroll *c*

1104 counter flange
DA modflange
DE Gegenflansch *m*
ES contrabrida *f*
F contre-bride *f*
I controflangia *f*
MA ellenperem
NL tegenflens *m*
PO przeciwkołnierz *m*
PY контрфланец *m*
SU vastalaippa
SV motfläns *c*

1105 counter flow
DA modstrøm
DE Gegenstrom *m*
ES contracorriente *f*
F contre-courant *m*
I flusso *m* contrapposto
MA ellenáram
NL tegenstroom *m*
PO przepływ *m* przeciwprądowy
PY противоток *m*
SU vastavirta
SV motström *c*

1106 counter flow circulation
DA modstrøms-cirkulation
DE Gegenstrom-zirkulation *f*
ES circulación en contracorriente *f*
F circulation *f* à contre-courant
I circolazione *f* controcorrente
MA ellenáramú cirkuláció
NL tegenstroom-circulatie *f*
PO cyrkulacja *f* przeciwprądowa
PY противоточное течение *n*
SU vastavirtakierto
SV motströms-cirkulation *c*

1107 counter flow heat exchanger
DA modstrømsvarme-veksler
DE Gegenstromwärme-austauscher *m*
ES cambiador de calor a contracorriente *m*
F échangeur *m* de chaleur à contre-courant
I scambiatore *m* di calore in controcorrente
MA ellenáramú hőcserélő
NL tegenstroomwarmte-wisselaar *m*
PO wymiennik *m* ciepła przeciwprądowy
PY противоточный теплообменник *m*
SU vastavirtalämmün-siirrin
SV motströmsvärme-växlare *c*

1108 counter slope
DA modfald
DE Gegengefälle *n*
ES contrapendiente *f*
F contre-pente *f*
I contropendenza *f*
MA ellenirányú lejtés
NL tegenschot *n*
PO przeciwspadek *m*
PY контруклон *m*
SU nousu
SV motfall *n*

1109 coupling
DA kobling
DE Kupplung *f*
    Verbindung *f*
ES acoplamiento *m*
F accouplement *m*
    couplage *m*
    raccordement *m*
I accoppiamento *m*
    giunto *m*
MA csatolás
    kapcsolás
NL koppeling *f*
PO połączenie *n*
    sprzężenie *n*
PY соединение *n*
SU kytkentä
    liitos
SV koppling *c*

1110 coupling in parallel
DA parallelkobling
DE Parallelschaltung *f*
ES acoplamiento en
    paralelo *m*
F couplage *m* en
    parallèle
I accoppiamento *m* in
    parallelo
MA párhuzamos
    kapcsolás
NL parallelkoppeling *f*
    parallelschakeling *f*
PO połączenie *n*
    równoległe
PY параллельное
    соединение *n*
SU rinnankytkentä
SV parallellkoppling *c*

1111 coupling in series
DA seriekobling
DE Hintereinander-
    schaltung *f*
    Reihenschaltung *f*
    Serienschaltung *f*
ES acoplamiento en
    serie *m*
F couplage *m* en série
I accoppiamento *m* in
    serie
MA soros kapcsolás
NL seriekoppeling *f*
    serieschakeling *f*
PO połączenie *n*
    szeregowe
PY последовательное
    соединение *n*
SU sarjaankytkentä
SV seriekoppling *c*

1112 coupling sleeve
DA koblingsmuffe
DE Muffe *f*
ES manguito de
    acoplamiento *m*
F manchon *m* de
    raccordement
I manicotto *m*
MA karmantyú
NL verbindings-
    manchet *n*
    verbindingsmof *f*
PO tuleja *f* łącząca
PY соединительная
    муфта *f*
SU liitosmuhvi
SV kopplingshylsa *c*

1113 cover
DA dæksel
DE Belag *m*
    Deckel *m*
ES cubierta *f*
    tapa *f*
F capot *m*
    couvercle *m*
I coperchio *m*
MA burkolat
    fedél
NL afdekking *f*
    bedekking *f*
    deksel *n*
PO pokrywa *f*
PY крышка *f*
SU kansi
    luukku
SV kåpa *c*

1114 covering
DA beklædning
DE Überzug *m*
    Verkleidung *f*
ES cubierta *f*
    funda *f*
    recubrimiento *m*
    revestimiento *m*
F couverture *f*
    revêtement *m*
I copertura *f*
MA burkolás
NL bekleding *f*
PO przykrycie *n*
    przykrywanie *n*
PY перекрытие *n*
SU peitto
SV beklädnad *c*

1115 cover plate
DA afdækningsplade
DE Abdeckplatte *f*
ES chapa de
    recubrimiento *f*
    placa de cierre *f*
F tôle *f* de couverture
I piastra *f* di
    copertura
MA burkoló lemez
NL dekplaat *f*
PO płyta *f*
    przykrywająca
PY перекрывающая
    плита *f*
SU kansilevy
SV täckplåt *c*

1116 cowl
DA røghætte
DE Schornstein-
    aufsatz *m*
ES sombrerete de
    chimenea *m*
F antirefouleur *m*
    aspirateur *m*
    statique
    chapeau *m* de
    cheminée
I aspiratore *m* statico
MA kéménysisak
NL gek *m*
    schoorsteenkap *f*
PO deflektor *m*
    nasada *f* kominowa
    osłona *f*
PY дефлектор *m*
SU savupiipun hattu
SV rökhuv *c*

1117 CPU (*see* central
processor unit)

1118 crack
DA revne
DE Fuge *f*
    Riß *m*
    Spalt *m*
ES fisura *f*
F fente *f*
    fissure *f*
I fessura *f*
    incrinatura *f*
MA hézag
    repedés
NL barst *m*
    scheur *f*
    spleet *f*
PO rysa *f*
    szczelina *f*
PY трещина *f*
SU rako
SV spricka *c*

1119 crackage
DA revnedannelse
DE Spaltung *f*
ES fisuración *f*
F fissuration *f*
I fessurazione *f*
MA repesztés
NL scheuring *f*
PO pęknięcie *n*
     spękanie *n*
PY разрыв *m*
SU rakoilu
SV sprickbildning *c*

1120 crack length
DA revnelængde
DE Fugenlänge *f*
ES longitud de fisura *f*
F longueur *f* de fentes
I lunghezza *f* di
     incrinatura
MA hézaghossz
NL scheurlengte *f*
     spleetlengte *f*
PO długość *f* pęknięcia
     długość *f* szczeliny
PY длина *f* трещины
SU rakopituus
SV sprickslängd *c*

1121 crank
DA forkrøbning
DE Kurbel *f*
ES manivela *f*
     manubrio *m*
     placa de apoyo
     (hornos) *f*
F manivelle *f*
I manovella *f*
MA forgatás
     forgattyú
NL krukas *f*
     slinger *m*
PO korba *f*
     ramię *n*
     wahacz *m*
PY кривошип *m*
SU veivi
SV vev *c*

1122 crank *vb*
DA forkrøppe
DE ankurbeln
     anwerfen
ES girar
     virar
     voltear
F mettre en marche à
     la manivelle
I avviare
     far girare
MA forgat
NL aanslingeren
PO obracać korbą
PY приводить в
     движение
     рукояткой
SU veivata
SV starta med vev *c*

1123 crankcase
DA krumtaphus
DE Kurbelgehäuse *n*
ES cárter *m*
F carter *m*
I carter *m*
MA forgattyúház
NL carter *n*
PO karter *m*
     skrzynia *f* korbowa
PY картер *m*
SU kampikammio
SV vevhus *n*

1124 crankpin
DA krumtappind
DE Kurbelzapfen *m*
ES bulón *m*
F maneton *m*
I perno *m* di biella
MA forgattyús csap
NL krukpen *f*
PO sworzeń *m* korbowy
PY палец *m*
     кривошипа
     шейка *f*
     коленчатого вала
SU varmistussokka
SV vevtapp *c*

1125 crankshaft
DA krumtapaksel
DE Kurbelwelle *f*
ES cigüeñal *m*
F vilebrequin *m*
I albero *m* a gomiti
MA forgattyús tengely
     főtengely
NL krukas *f*
PO wał *m* korbowy
PY коленчатый вал *m*
SU kampiakseli
SV vevaxel *c*

1126 critical *adj*
DA kritisk
DE kritisch
ES crítico
F critiqué
I critico
MA kritikus
NL kritisch
PO krytyczny
PY критический
SU kriittinen
SV kritisk

1127 cross (*see* cross-piece)

1128 cross-current
DA krydsstrøm
DE Kreuzstrom *m*
ES corriente cruzada *f*
F courants *m* croisés
I corrente *f* incrociata
MA keresztáram
NL dwarsstroom *m*
     kruisstroom *m*
PO prąd *m* krzyżowy
PY перекрестный
     ток *m*
SU ristivirta
SV korsströmm *c*

1129 cross-flow
DA krydsstrømning
DE Kreuzstrom *m*
ES corriente cruzada *f*
     flujo cruzado *m*
F courants *m* croisés
I flusso *m* incrociato
MA keresztáramlás
NL kruisstroom *m*
PO przepływ *m*
     krzyżowy
     przepływ *m*
     poprzeczny
PY перекрестное
     течение *n*
     перекрестный
     поток *m*
SU ristivirta
SV tvärström *c*

DA	=	Danish
DE	=	German
ES	=	Spanish
F	=	French
I	=	Italian
MA	=	Magyar
		(Hungarian)
NL	=	Dutch
PO	=	Polish
PY	=	Russian
SU	=	Finnish
SV	=	Swedish

1130 cross-flow circulation
DA krydsstrøms-cirkulation
DE Kreuzstrom-zirkulation f
ES circulación en corrientes cruzadas f
F circulation f à courants croisés
I circolazione f a flusso incrociato
MA keresztáramú cirkuláció
NL kruisstroom-circulatie f
PO obieg m o przepływie krzyżowym
PY перкрестное течение n
SU ristivirtakierto
SV tvärströms-cirkulation c

1131 cross-flow fan
DA krydsstrøms-ventilator
DE Tangential-ventilator m
ES ventilador tangencial m
F ventilateur m tangentiel
I ventilatore m tangenziale
MA keresztáramú ventilátor
NL kruisstroom-ventilator m
PO wentylator m o przepływie poprzecznym
PY тангенциальный вентилятор m
SU poikittaisvirtaus-puhallin
SV tvärströmsfläkt c

1132 cross-flow heat exchanger
DA krydsstrømsvarme-veksler
DE Kreuzstromwärme-austauscher m
ES intercambiador de calor a contracorriente m
F échangeur m à courants croisés
I scambiatore m di calore a flussi incrociati
MA keresztáramú hőcserélő
NL kruisstroomwarmte-wisselaar m
PO wymiennik m ciepła krzyżowy
   wymiennik m ciepła o przepływie poprzecznym
PY теплообменник m с перекрестным током
SU ristivirtalämmün-siirrin
SV korsströmsvärme-växlare c

1133 cross-flow tower
DA krydsstrømstårn
DE Kreuzstromkühl-turm m
ES torre a contracorriente f
F tour f à courants croisés
I torre f di raffreddamento a flussi incrociati
MA keresztáramú hűtőtorony
NL kruisstroomkoel-toren m
PO wieża f o przepływie krzyżowym
PY градирня f с перекрестным током
SU ristivirtajäähdy-tystorni
SV tvärströmstorn n

1134 cross-piece
DA skråbånd
DE Kreuzstück n
ES cruceta f
   pieza transversal f
   travesaño m
F croix f
I raccordo m a croce
MA keresztirányú merevítés
NL kruisstuk n
PO czwórnik m
   krzyżak m
PY крестовина f
SU ristikappale
SV tvärstycke n

1135 cross-section
DA tværsnit
DE Querschnitt m
ES sección transversal f
F section f transversale
I sezione f trasversale
MA keresztmetszet
NL doorsnede f
PO przekrój m poprzeczny
PY поперечное сечение n
SU leikkaus
SV tvärsnitt n

1136 cross-sectional area
DA tværsnitsareal
DE Querschnittfläche f
ES área de sección transversal f
F aire f transversale section f transversale
I area f trasversale
MA keresztmetszeti felület
NL doorsnede-oppervlak n
PO pole n przekroju poprzecznego
PY площадь f поперечного сечения
SU poikkipinta-ala
SV tvärsnittsarea c

1137 cross-talk
DA overhøring
DE Übersprechen *n*
ES comprobación *f*
F pont *m* phonique
I diafonia *f*
MA áthallás
NL overspraak *f*
PO przenik *m*
przesłuch *m*
PY перекрестный
сигнал *m*
SU häiriüsignaali
SV överhörning *c*

1138 cross-tube boiler
DA tværrørskedel
DE Quersiedelkessel *m*
ES caldera vertical con
el hogar
atravesado por
tubos de gas *f*
F chaudière *f* à
bouilleurs croisés
I caldaia *f* a tubi
sovrapposti
MA keresztcsüves kazán
NL dwarspijpketel *m*
PO kocioł *m* z
poprzecznymi
rurkami
PY котел *m* с
поперечными
трубами
SU poikittaisputkikattila
SV tvärrörspanna *c*

1139 cross-ventilation
DA krydsventilation
DE Querlüftung *f*
ES ventilación
transversal *f*
F ventilation *f*
transversale
I ventilazione *f*
trasversale
MA keresztszellőzés
NL dwarsventilatie *f*
PO wentylacja *f*
poprzeczna
PY вентиляция *f* со
встречным
расположением
приточных и
вытяжных
устройств
SU ristiveto
SV tvärventilation *c*

1140 cryogenic *adj*
DA kryogen
DE kryogen
ES criogénico
F cryogénique
I criogenico
MA hűtütt
kriogén
NL cryogeen
PO niskotemperaturowy
PY криогенный
SU kryo-
SV kryogenisk

1141 cube ice
DA isterning
DE Würfeleis *n*
ES cubito de hielo *m*
F cube *m* de glace
I ghiaccio *m* in
cubetti
MA jégkocka
NL blokjesijs *n*
PO lód *m* w kostkach
PY куб *m* льда
кубик *m* льда
SU jääkuutio
SV tärningsis *c*

1142 cup anemometer
DA klokkeanemometer
DE Schalenkreuzwind-
messer *m*
ES anemómetro de copa
m
F anémomètre *m* à
coupelles
I anemometro *m* a
coppa
MA kanalas
szélsebességmérő
NL halvebol-
anemometer *m*
PO anemometr *m*
czaszowy
PY чашечный
анемометр *m*
SU kuppianemometri
SV skovelanemometer *c*

1143 cup burner
DA pottefyr
DE Schalenbrenner *m*
ES quemador de
cazoleta
(cubeta) *m*
F coupelle *f*
I bruciatore *m* a
coppa
bruciatore *m* a tazza
MA csészés égő
serleges égő
NL roterende brander *m*
PO dysza *f* palnika
palnik *m* kielichowy
PY сопло *n* горелки
SU hüyrystyspoltin
SV frångnings-
brännare *c*

1144 current
DA strøm (el, vand)
DE Strömung *f*
Strom *m*
ES corriente *f*
F courant *m*
I corrente *m*
MA áram
NL stroom *m*
stroomsterkte *f*
PO prąd *m*
PY ток *m*
SU virta
SV ström *c*

1145 current capacity
(*see* ampacity)

1146 current intensity
(*see* amperage)

1147 curve
DA bue
kurve
DE Kurve *f*
ES curva *f*
F courbe *f*
I curva *f*
MA gürbe
NL curve *f*
kromme *f*
PO krzywa *f*
PY кривая *f*
характеристика *f*
SU kaari
kaarre
käyristys
SV kurva *c*

1148 cut-in
DA indkoble
DE einschalten
ES conexión f
F déclenchement m
I intervento m
MA beavatkozás
NL inschakeling f
PO ścinanie n
PY врезание n
врезка f
SU aloitus
SV inkoppling c

1149 cut-in point
DA indkoblingspunkt
DE Einschaltpunkt m
ES punto de conexión m
F point m
d'enclenchement
I punto m di
intervento
MA beavatkozási pont
NL inschakelpunt n
inschakelwaarde f
PO punkt m ścięcia
PY начальная
величина f
начальная точка f
SU aloitusarvo
SV inkopplingspunkt c

1150 cut-off valve
DA afspærringsventil
DE Absperrventil n
ES válvula de cierre f
válvula de corte f
F vanne f d'isolement
I valvola f di arresto
MA lezáró szelep
NL stopkraan f
PO zawór m odcinający
PY отключающий
клапан m
SU sulkuventtiili
SV avstängningsventil c

1151 cut-out
DA afbryder
DE Ausschaltung f
ES corte m
desconexión f
interruptor m
F coupure f
I arresto m
disinserimento m
distacco m
interruzione f
MA kiiktatás
kikapcsolás
NL uitschakeling f
PO wycięcie n
wyłącznik m
zawór m
PY выключатель m
SU lopetus
SV frånslag n

1152 cut out vb
DA afbryde
frakoble
DE ausschalten
unterbrechen
ES desconectar
F couper
interrompre
I disinserire
MA kiiktat
kikapcsol
NL uitschakelen
PO wycinać
wyłączać
PY выключать
SU sulkea
SV bryta

1153 cut-out point
DA frakoblingspunkt
DE Ausschaltpunkt m
ES punto de
desconexión m
F point m de
déclenchement
I punto m di arresto
MA kikapcsolási pont
NL uitschakelpunt n
uitschakelwaarde f
PO punkt m wyłączenia
PY точка f
отключения
SU lopetusarvo
SV brytpunkt c

1154 cut-out setting
DA frakoblingsindstilling
DE Ausschaltstellung f
ES punto de corte m
F position f de
coupure
I punto m di arresto
MA kikapcsolt állás
NL instelwaarde f
uitschakelpunt
PO nastawa f
wyłącznika
nastawa f zaworu
PY выключающее
устройство n
SU lopetustila
SV brytläge n

1155 cycle
DA cyklus
periode
DE Kreislauf m
Periode f
Zyklus m
ES ciclo m
periodo m
F cycle m
période f
I ciclo m
MA ciklus
kürfolyamat
ütem
NL cyclus m
kringloop m
periode f
PO cykl m
okres m
PY цикл m
SU jakso
kiertoprosessi
SV cykel c

1156 cycling
DA kredsløb
DE periodischer
Betrieb m
ES ciclaje m
funcionamiento
cíclico m
F opération f cyclique
opération f
périodique
I oscillazione f
MA szakaszos üzemelés
NL periodiek bedrijf n
PO działanie n
okresowe
praca f cykliczna
praca f okresowa
PY периодический
процесс m
SU jaksottainen
SV kretslopp n

1157 cycling life
DA periodisk livslængde
DE Dauer *f* des
   periodischen
   Betriebs *m*
ES vida periódica *f*
F durée *f* de vie
I durata *f* del ciclo
MA ciklus időtartam
NL levensduur *m* (in
   aantal cycli)
PO czas *m* trwania
   pracy okresowej
PY период *m* цикла
SU käyttüikä
SV periodisk livslängd *c*

1158 cyclone
DA cyklon
DE Zyklon *m*
ES ciclón *m*
F cyclone *m*
I ciclone *m*
MA ciklon
   porleválasztó
NL cycloon *m*
PO cyklon *m*
PY циклон *m*
SU sykloni
SV cyklon *c*

1159 cylinder
DA cylinder
   valse
DE Zylinder *m*
ES cilindro *m*
F cylindre *m*
I cilindro *m*
MA henger
NL cilinder *m*
PO cylinder *m*
   walec *m*
PY цилиндр *m*
SU sylinteri
SV cylinder *c*

1160 cylinder head
DA cylinder topstykke
DE Zylinderkopf *m*
ES culata *f*
F tête *f* de cylindre
I testa *f* del cilindro
MA hengerfej
NL cilinderkop *m*
PO głowica *f* cylindra
PY головка *f*
   цилиндра
SU sylinterin kansi
SV cylinderhuvud *n*

DA	=	Danish
DE	=	German
ES	=	Spanish
F	=	French
I	=	Italian
MA	=	Magyar (Hungarian)
NL	=	Dutch
PO	=	Polish
PY	=	Russian
SU	=	Finnish
SV	=	Swedish

D

1161 daily output
DA daglig ydelse
   døgnproduktion
DE Tagesleistung *f*
ES producción diaria *f*
   rendimiento diario
   m
F débit *m* journalier
I portata *f* giornaliera
MA napi teljesítmény
NL dagproduktie *f*
PO wydajność *f* dobowa
PY суточный расход *m*
SU päivittäinen tuotos
SV dygnsproduktion *c*

1162 daily range
DA dagligt interval
   (mængde)
DE Tagesgang *m*
   Tagesrate *f*
ES tasa diaria *f*
F variation *f*
   quotidienne
I escursione *f* termica
   giornaliera
MA napi
   hőmérséklet-
   külünbség
NL dagtemperatuur-
   bereik *n*
PO zakres *m* dobowy
PY перепад *m*
   температуры за
   день
SU päivittäinen vaihtelu-
   alue
SV daglig mängd *c*

1163 daily service tank
DA dagtank
DE Tagesbehälter *m*
ES tanque de servicio
   (diario) *m*
F réservoir *m*
   journalier
I serbatoio *m*
   giornaliero
MA napitartály
NL dagtank *m*
PO zbiornik *m* o
   pojemności
   dobowej
PY расходный
   резервуар *m*
SU päiväsäiliü
SV dagtank *c*

1164 damage
DA skade
DE Schaden *m*
ES avería *f*
   daño *m*
F défaut *m*
   dommage *m*
I danno *m*
MA kár
   sérülés
NL beschadiging *f*
   schade *f*
PO uszkodzenie *n*
PY повреждение *n*
SU vahinko
SV skada *c*

1165 damp
DA fugt
DE Feuchte *f*
ES humedad *f*
F humidité *f*
I umidità *f*
MA fojtás
   nedvesség
NL vocht *n*
   vochtigheid *f*
PO wilgoć *f*
   wilgotność *f*
PY влажность *f*
SU kosteus
SV fuktighet *c*

1166 damp *vb*
DA fugte
   dømpe
DE feuchten
ES húmedo
F assourdir
   étouffer
I ammortizzare
   bagnare
   inumidire
   smorzare
MA fojt
   nedvesít
NL bevochtigen
   dempen
   smoren
PO dławić
   tłumić drgania
PY демпфировать
   смачивать
   увлажнять
SU kostuttaa
   vaimentaa
SV fukta

1167 dampen *vb*
DA fugte, blive fugtig
DE befeuchten
   dämpfen
ES amortiguar
   humectar
   humedecer
   insonorizar
F amortir (oscillations)
I ammortizzare
   bagnare
   inumidire
   smorzare
MA fojt
   nedvesít
NL dempen
   smoren
PO dławić
   nawilżać
   tłumić
   zwilżać
PY демпфировать
   смачивать
   увлажнять
SU kostuttaa
SV fukta

1168 damper
DA spjæld
DE Drosselklappe *f*
ES compuerta *f*
   registro *m*
F registre *m* à volets
   volet *m* de réglage
I serranda *f*
MA fojtócsappantyú
NL luchtklep *f*
   smoorklep *f*
PO przepustnica *f*
   zasuwa *f*
PY дроссельный
   клапан *m*
SU säätüpelti
SV spjäll *n*

1169 damper actuator
DA spjældstyring
DE Klappenschalter *m*
ES actuador de una
   compuerta *m*
F moteur *m* de registre
I attuatore *m* della
   serranda
MA csappantyúmozgató
NL luchtklepservo-
   motor *m*
PO siłownik *m*
   przepustnicy
PY привод *m*
   дроссельного
   клапана
SU pellin toimilaite
SV manöverorgan *n* för
   spjäll *n*

1170 dampermotor
DA spjældmotor
DE Klappenstell-
   motor *m*
ES motor de compuerta
   m
F moteur *m* de registre
I serranda *f*
   motorizzata
MA csappantyúmozgató
   motor
NL luchtklepmotor *m*
PO siłownik *m*
   przepustnicy
PY привод *m*
   дроссельного
   клапана
SU peltimoottori
SV spjällmotor *c*

**1171 damping
(oscillation)**
DA svingningsdæmpning
DE Dämpfung *f*
ES amortiguación *f*
F amortissement *m*
   (oscillation)
I smorzamento *m*
MA csillapítás
   tompítás
NL demping (trilling) *f*
PO gaśnięcie *n* (fal)
   tłumienie *n* (drgań)
PY амортизация *f*
SU vaimennus
   (värähtely)
SV befuktning *c*

1172 dampness
DA fugtighed
DE Feuchtigkeit *f*
ES lugar de elevada
   humedad *m*
F humidité *f*
I umidità *f*
MA nedvesség
NL vochtigheid *f*
PO wilgoć *f*
   wilgotność *f*
PY влажность *f*
SU kosteus
SV fuktighet *c*

1173 damp proofing
DA damptæt
DE Feuchte-
   beständigkeit *f*
ES impermeabilización *f*
F imperméabilisation *f*
I protezione *f*
   dall'umidità
MA nedvességszigetelés
NL vochtwerend
   maken *n*
PO odporny na wilgoć
   przeciwwilgot-
   nościowy
PY защита *f* от
   влажности
   повышение *n*
   влагонепроницае-
   мости
SU hüyrysulku
   kosteuseriste
SV fukttätning *c*

1174 data
DA data
DE Daten *f,pl*
ES datos *m*
F donnée *f*
I dato *m*
MA adat
NL gegevens *pl*
PO dane *f*
PY данные *n*
   показатели *m*
SU tieto
SV data *c*

1175 databank
DA databank
DE Datenbank *f*
ES banco de datos *m*
F banque *f* de données
I banca *f* dati
MA adatbank
NL gegevensbestand *m*
PO bank *m* danych
PY банк *m* данных
SU tietopankki
SV databank *c*

1176 database
DA database
DE Datenbasis *f*
ES base de datos *f*
F base *f* de données
I archivio *m*
MA adatbázis
NL gegevensbestand *m*
PO baza *f* danych
PY база *f* данных
SU tietokanta
SV databas *c*

1177 data converter
DA data transformer
DE Datenumwandler *m*
ES convertidor de datos
   m
F convertisseur *m* de
   données
I convertitore *m* di
   dati
MA adatátalakító
NL gegevensomzetter *m*
PO przetwornik *m*
   danych
PY преобразователь *m*
   данных
SU datamuunnin
SV dataomvandlare *c*

1178 data display module
DA datavisningsmodul
DE Datensichtmodul *n*
ES pantalla de datos *f*
F module *m* d'affichage de données
I modulo *m* di visualizzazione dati
MA adat kijelző modul
NL gegevensweergavemoduul *m*
PO moduł *m* wyświetlania danych
PY данные *n* на дисплее
SU näyttüyksikkü
SV datadisplaymodul *c*

1179 data logging
DA dataregistrering
DE Datenspeicherung *f*
ES transcripción de datos *f*
F concentrateur *m* de données
I acquisizione *f* dati
MA adatrügzítés
NL gegevensvastlegging *f*
PO centralna rejestracja *f* danych
PY регистрируемые данные *n*
SU tiedonkeruu
SV dataregistrering *c*

1180 data processing device
DA databehandlingsudstyr
DE Datenverarbeitungsmaschine *f*
ES procesador de datos *m*
F traitement *m* de données (équipement)
I dispositivo *m* per l'elaborazione dei dati
MA adatfeldolgozó készülék
NL gegevensverwerkingsapparaat *n*
PO urządzenie *n* (maszyna) do przetwarzania danych
PY устройство *n* обработки данных
SU tiedonkäsittelyjärjestelmä
SV dator *c*

1181 data table
DA
DE Datentafel *f*
ES tabla de datos *f*
F table *f* de données
I tabella *f* di dati
MA adattábla
NL gegevenstabel *f*
PO zestawienie *n* danych
PY таблица *f* данных
SU taulukkomuotoinen tieto
SV datatabell *c*

1182 day extension
DA dagtelefonnummer
DE Tagesverlängerung *f*
ES extensión diurna *f*
F
I durata *f* del dì
MA nap hosszabbodás
NL dagverlenging *f*
PO przedłużony dzień *m* pracy
PY продолжительность *f* дня
SU päiväkäytünpidennys
SV

1183 DDC (*see* direct digital control)

1184 dead-air pocket
DA luftlomme
DE Windschatten *m*
ES aire viciado *m* zona muerta *f*
F bouchon *m* d'air poche *f* d'air
I
MA holttér légzsák
NL gebied *n* met stilstaande lucht
PO przestrzeń *f* nieruchomego powietrza strefa *f* martwa (powietrza)
PY воздушная тень *f*
SU ilmatasku
SV luftficka *c*

1185 dead band (*see* dead zone)

1186 dead leg
DA frakoblet ledning
DE Blindstrang *m*
ES derivación ciega *f*
F branchement *m* en attente branchement *m* fermé
I braccio *m* morto
MA holtág
NL dode leiding *f*
PO odgałęzienie *n* nieczynne (zaślepione) odgałęzienie *n* rezerwowe
PY глухое ответвление *n*
SU tulpattu haara
SV frånkopplad ledning *c*

1187 dead time
DA dødtid pause spildtid
DE Totzeit *f*
ES tiempo muerto *m*
F temps *m* mort
I tempo *m* morto
MA holtidő
NL dode tijd *m*
PO czas *m* jałowy czas *m* opóźnienia czas *m* zwłoki
PY время *f* запаздывания
SU kuollut aika
SV spilltid *c*

DA = Danish
DE = German
ES = Spanish
F = French
I = Italian
MA = Magyar (Hungarian)
NL = Dutch
PO = Polish
PY = Russian
SU = Finnish
SV = Swedish

1188 dead zone
DA dødzone
DE Totzone *f*
ES zona muerta *f*
F zone *f* morte
I banda *f* morta
MA holttér
NL dode zone *f*
PO martwa przestrzeń *f*
strefa *f* nieczułości
PY застойная зона *f*
неподвижный
слой *m*
SU kuollut alue
SV düd zon *c*

1189 de-aeration
DA afluftning
DE Luftabscheidung *f*
ES desaireación *f*
eliminación del
aire *f*
evacuación del aire *f*
purga del aire *f*
F désaération *f*
purge *f* d'air
I spurgo *m* di aria
MA légtelenítés
NL ontluchting *f*
PO odpowietrzanie *n*
PY деаэрация *f*
SU ilmaus
SV luftning *c*

1190 debug *vb*
DA finregulere
korrigere
rette
DE Fehler *m* beseitigen
ES descontaminar
F déverminer
I eliminare gli errori di
un codice di
calcolo
MA próbaüzemel
NL corrigeren van
fouten
PO uruchomić program
wykrywać i usuwać
usterki
PY отлаживать
SU poistaa virheitä
SV korrigera

1191 decay
DA nedbrydning
DE Verfall *m*
Zerlegung *f*
ES descomposición *f*
F décomposition *f*
I decadimento *m*
decomposizione *f*
MA bomlás
NL verval *n*
PO gnicie *n*
rozkład *m*
rozpad *m*
zanik *m*
PY затухание *n*
разрушение *n*
SU aleneminen
hajoaminen
vaimeneminen
SV nedbrytning *c*

1192 decay rate
DA nedbrydningstid
DE Verfallsrate *f*
Zerfallsrate *f*
ES grado de deterioro *m*
F taux *m* de
décroissance
I velocità *f* di
decadimento
MA bomlási arány
NL vervalsnelheid *f*
PO szybkość *f*
przemiany
promieniotwórczej
szybkość *f* rozpadu
PY темп *m* затухания
SU alenemisnopeus
SV nedbrytningstid *c*

1193 deceleration
DA hastighedsformindskelse
DE Geschwindigkeitsverminderung *f*
ES deceleración *f*
F décélération *f*
I decelerazione *f*
MA lassulás
NL vertraging *f*
PO opóźnienie *n*
przyspieszenie *n*
ujemne
PY замедление *n*
SU hidastuminen
SV hastighetsminskning *c*

1194 decibel (dB)
DA decibel
DE Dezibel *n*
ES decibelio *m*
F décibel *m*
I decibel *m*
MA decibel
NL decibel *m*
PO decybel *m*
PY децибел *m* (дБ)
SU desibeli
SV decibel *c*

1195 decipol
DA decipol
DE Dezipol *n*
ES decipol *m*
F décipol *m*
I decipol *m*
MA decipol
NL decipol
PO decypol *m*
PY деципол *m*
SU desipol
SV decipol *c*

1196 declination
DA indfaldsvinkel
DE Deklination *f*
ES declinación *f*
F déclinaison *f*
I declinazione *f*
MA elhajlás
eltérés
NL afwijking *f*
declinatie *f*
PO deklinacja *f*
odchylenie *n*
magnetyczne
PY склонение *n*
SU deklinaatio
kallistuma
SV deklination *c*

1197 decomposition
DA nedbrydning
DE Abbau *m*
Zersetzung *f*
ES descomposición *f*
F décomposition *f*
I decomposizione *f*
MA bomlás
NL ontbinding *f*
ontleding *f*
PO gnicie *n*
rozkład *m*
rozpad *m*
PY разложение *n*
распад *m*
SU hajoaminen
SV nedbrytning *c*

1198 decompressor
DA dekompressionstank
DE Dekompressor *m*
ES descompresor *m*
F décompresseur *m*
I decompressore *m*
MA nyomáscsükkentő
NL decompressie-
    toestel *n*
    ontspannings-
    toestel *n*
PO dekompresor *m*
PY декомпрессор *m*
SU paineentasaus
SV dekompressions-
    anordning *c*

1199 decrement factor
DA reduktionsfaktor
DE Verminderungs-
    faktor *m*
ES coeficiente de
    amortiguación *m*
    factor de
    amortiguación *m*
F facteur *m* de
    dépréciation
I fattore *m* di
    ammortamento
MA csillapítási tényező
NL verminderings-
    factor *m*
PO współczynnik *m*
    ubytku
    współczynnik *m*
    zmniejszenia
PY коэффициент *m*
    уменьшения
SU vaimennuskerroin
SV dekrementfaktor *c*

1200 de-dusting
DA støvfjernelse
DE Entstaubung *f*
ES desempolvado *m*
F dépoussiérage *m*
I depolverazione *f*
MA portalanítás
NL ontstoffing *f*
PO odpylanie *n*
PY пылеочистка *f*
SU pülynpoisto
SV avdamning *c*

1201 deep freezing
DA dybfrysning
DE Tiefkühlung *f*
ES ultracongelación *f*
F basse température *f*
    froid *m*
I congelamento *m* a
    bassa temperatura
MA mélyhűtés
NL diepvriezen
PO mrożenie *n* głębokie
PY глубокое
    замораживание
    n
SU pakastus
SV djupfrysning *c*

1202 deep vacuum
DA kraftig vakuum
DE Hochvakuum *n*
    niedriges Vakuum *n*
ES ultravacío *m*
F vide *m* poussé
I vuoto *m* spinto
MA mély vákuum
NL hoog vacuüm *n*
PO głęboka próżnia *f*
PY глубокий вакуум *m*
SU tyhjü
SV djupt vakuum *n*

1203 default value
DA normalværdi
DE Fehlergrüße *f*
ES tasa de deficiencia *f*
F valeur *f* de défaut
I valore *m* di default
MA hibaérték
NL standaardwaarde *f*
PO wielkość *f* błędu
PY недостаточная
    величина *f*
SU oletusarvo
SV felvärde *n*

1204 deflecting vane anemometer
DA anemometer med
    vindretningsviser
DE Flügelrad-
    anemometer *n*
ES anemómetro de
    corriente lateral *m*
F anémomètre *m* à
    palette
I anemometro *m* a
    pale orientabili
MA szárnykerekes
    anemométer
NL vaan-anemometer *m*
PO anemometr *m*
    skrzydełkowy
PY крыльчатый
    анемометр *m*
SU siipianemometri
SV anemometer *c* med
    riktningsvisare *c*

1205 deflector
DA ledeplade
    luftretter
DE Ablenkblech *n*
    Deflektor *m*
ES deflector *m*
F déflecteur *m* (aube
    de guidage)
I deflettore *m*
MA deflektor
    visszaverő ernyő
NL keerschot *n*
PO deflektor *m*
PY направляющая
    лопатка *f*
SU ohjaussiipi
SV deflektor *c*

1206 defrost *vb*
DA afrime
DE abtauen
    entfrosten
ES desescarchar
F dégivrer
I sbrinare
MA fagytalanít
    leolvaszt
NL ontdooien
PO odszraniać
    rozmrażać
    zapobiegać
    oszronieniu
PY оттаивать
SU sulattaa
SV avfrosta

1207 defrost control
DA afrimningsstyring
DE Abtauregelung *f*
ES control de
desescarche *m*
F contrôle *m* de
dégivrage
I controllo *m* dello
sbrinamento
MA leolvasztás
szabályozó
NL ontdooiregelaar *m*
ontdooiregeling *f*
PO sterowanie *n*
procesem
rozmrażania
PY регулирование *n*
оттаивания
SU sulatuksen säätü
SV avfrostnings-
kontroll *c*

1208 defrosting
DA afrimning
DE Abtauen *n*
ES desescarche *m*
F dégivrage *m*
I sbrinamento *m*
MA leolvasztás
NL ontdooiing *f*
PO odmrażanie *n*
odszranianie *n*
PY оттаивание *n*
SU sulattaminen
SV avfrostning *c*

1209 defrosting cycle
DA afrimningscyklus
DE Abtauperiode *f*
ES ciclo de desescarche
m
F cycle *m* de dégivrage
I ciclo *m* di
sbrinamento
MA leolvasztási ciklus
NL ontdooi-cyclus *m*
ontdooiperiode *f*
PO cykl *m* rozmrażania
PY цикл *m* оттаивания
SU sulatusjakso
SV avfrostningscykel *c*

1210 defrosting process
DA afrimningsproces
DE Abtauprozeß *m*
ES proceso de
desescarche *m*
F procédé *m* de
dégivrage
I processo *m* di
sbrinamento
MA leolvasztási eljárás
NL ontdooiingsproces *n*
PO proces *m*
odmrażania
PY процесс *m*
оттаивания
SU sulatusprosessi
SV avfrostnings-
process *c*

1211 defrosting system
DA afrimningssystem
DE Abtausystem *n*
ES sistema de
desescarche *m*
F système *m* de
dégivrage
I sistema *m* di
sbrinamento
MA leolvasztó rendszer
NL ontdooiinrichting *f*
PO system *m*
odmrażania
PY система *f*
оттаивания
SU sulatusjärjestelmä
SV avfrostnings-
system *n*

**1212 defrost period
(time)**
DA afrimningsperiode
DE Abtauzeit *f*
ES tiempo de
desescarche *m*
F durée *f* de dégivrage
I periodo *m* di
sbrinamento
MA leolvasztási időszak
NL ontdooiperiode *f*
PO okres *m*
rozmrażania
PY период *m*
оттаивания
SU sulatusjakso
SV avfrostningstid *c*

1213 degassing
DA afgasning
DE Entgasung *f*
ES eliminación de
gases *f*
purga de gas *f*
F dégazage *m*
I degasaggio *m*
MA gáztalanítás
NL ontgassen *n*
ontgassing *n*
PO odgazowywanie *n*
PY дегазация *f*
SU kaasunpoisto
SV avgasning *c*

1214 degree
DA grad
DE Grad *m*
ES grado *m*
F degré *m*
I grado *m*
MA fok
NL graad *m*
PO stopień *m*
PY градус *m*
степень *f*
SU aste
SV grad *c*

1215 degree-day
DA graddag
DE Gradtag *m*
ES grado-día *m*
F degré-jour *m*
I grado-giorno *m*
MA napfok
NL graaddag *m*
PO stopniodzień *m*
PY градусодень *m*
SU astepäivä
SV graddag *c*

1216 degree-day consumption
- DA graddagsforbrug
- DE Gradtags-
  verbrauch *m*
- ES consumo por
  grados-día *m*
- F consommation *f* par
  degré-jour
- I consumo *m* per
  grado giorno
- MA napfok fogyasztás
- NL graaddagverbruik *f*
- PO zużycie *n*
  (zapotrzebowanie)
  przypadające na 1
  dobostopień
- PY расход *m* на
  градусодень
- SU kulutus astepäivää
  kohden
- SV graddags-
  fürbrukning *c*

1217 degree of purity
- DA renhedsgrad
- DE Reinheitsgrad *m*
- ES grado de pureza *m*
- F degré *m* di pureté
- I grado *m* di purezza
- MA tisztaságfok
- NL zuiverheidsgraad *m*
- PO stopień *m* czystości
- PY степень *f* чистоты
- SU puhtausaste
- SV renhetsgrad *c*

1218 degree of superheat
- DA overophedningsgrad
- DE Überhitzungsgrad *m*
- ES grado de
  recalenta-
  miento *m*
- F degré *m* de
  surchauffe
- I grado *m* di
  surriscaldamento
- MA túlhevítési fok
- NL oververhittingsgraad
  m
- PO stopień *m*
  przegrzania
- PY степень *f* перегрева
- SU tulistusaste
- SV överhettningsgrad *c*

1219 degrees of freedom
- DA frihedsgrad
- DE Freiheitsgrad *m*
- ES grados de libertad *m*
- F degrés *m* de liberté
- I gradi *m,pl* di libertà
- MA szabadságfok
- NL vrijheidsgraden *pl*
- PO stopnie *m* swobody
- PY степени *f* свободы
- SU vapausaste
- SV frihetsgrad *c*

1220 dehumidification
- DA affugtning
- DE Entfeuchtung *f*
- ES deshumectación *f*
  deshumidificación *f*
- F déshumidification *f*
- I deumidificazione *f*
- MA nedvességelvonás
- NL ontvochtiging *f*
- PO osuszanie *n*
- PY осушение *n*
- SU kuivaus
- SV avfuktning *c*

1221 dehumidifier
- DA affugter
  fugtfjerner
- DE Entfeuchter *m*
- ES deshumidificador *m*
- F déshumidificateur *m*
- I deumidificatore *m*
- MA nedvességelvonó
- NL ontvochtiger *m*
- PO osuszacz *m*
- PY осушитель *m*
- SU kuivaaja
- SV avfuktare *c*

1222 dehumidify *vb*
- DA affugte
- DE entfeuchten
- ES deshumidificar
- F déshumidifier
- I deumidificare
- MA nedvességet von el
- NL ontvochtigen
- PO osuszać
- PY осушать
- SU kuivattaa
- SV avfukta

1223 dehumidifying effect
- DA affugtningseffekt
- DE Entfeuchtungs-
  effekt *m*
- ES efecto
  deshumidificador
  m
- F effect *m* de
  déshumidification
- I effetto *m*
  deumidificante
- MA nedvességelvonó
  hatás
- NL ontvochtigings-
  effect *n*
- PO efekt *m* osuszania
  zjawisko *m*
  osuszania
- PY осушительный
  эффект *m*
- SU kuivausteho
- SV avfuktningseffekt *c*

1224 dehydrate *vb*
- DA dehydrere
- DE entwässern
- ES deshidratar
- F déshydrater
- I disidratare
- MA szárít
  víztelenít
- NL ontwateren
  water onttrekken
- PO odwadniać
- PY дренировать
- SU kuivattaa
- SV avfukta

1225 dehydration
- DA dehydrering
- DE Entwässerung *f*
  Trocknung *f*
- ES deshidratador *m*
- F déshydratation *f*
- I disidratazione *f*
- MA víztelenítés
- NL dehydratie *f*
  ontwatering *f*
- PO dehydratacja *f*
  odwadnianie *n*
  (suszenie)
- PY дегидротация *f*
  обезвоживание *n*
- SU kuivaus
- SV avfuktning *c*

1226 dehydrator
DA dehydrator
DE Trockner *m*
ES deshidratación *m*
F déshydrateur *m*
(sécheur)
I essiccatore *m*
MA dehidrátor
víztelenítő
NL droger *m*
PO odwadniacz *m*
urządzenie *n*
odwadniające
PY осушитель *m*
SU kuivaaja
SV avfuktare *c*

1227 de-icing
DA afisning
DE Enteisung *f*
ES deshielo *m*
F fusion *f* de glace
I rimozione *f* del
ghiaccio
MA jégtelenít
NL ijsvrij maken
PO odladzanie *n*
(usuwanie
oblodzenia)
PY оттаивание *n*
SU jäätymisen esto
SV avisning *c*

1228 delivery
DA leverance
DE Lieferung *f*
ES entrega *f*
suministro *m*
F fourniture *f*
livraison *f*
I consegna *f*
fornitura *f*
MA szállítás
NL aflevering *f*
levering *f*
oplevering *f*
PO dostarczanie *n*
dostawa *f*
podawanie *n*
PY доставка *f*
питание *n*
подача *f*
SU toimittaa
SV leverans *c*

1229 delivery temperature
DA udløbstemperatur
DE Liefertemperatur *f*
ES temperatura de
alimentación *f*
F température *f*
délivrée
I temperatura *f* di
distribuzione
MA szállítási hőmérséklet
NL aanvoer-
temperatuur *f*
PO temperatura *f*
zasilenia
PY температура *f*
подачи
SU lähtülämpütila
SV tillfürsel-
temperatur *c*

1230 delivery time
DA leveringstidspunkt
DE Lieferzeit *f*
ES plazo de entrega *m*
F délai *m* de livraison
I termine *m* di
consegna
MA szállítási határidő
NL levertijd *m*
PO czas *m* dostawy
PY время *n* запитки
время *n* поставки
SU toimitusaika
SV leveranstid *c*

1231 delivery valve (*see*
discharge valve)

1232 demand
DA behov
DE Bedarf *m*
Belastung *f*
ES demanda *f*
F puissance *f*
électrique appelée
I domanda *f*
fabbisogno *m*
MA igény
küvetelmény
NL vraag *f*
PO zapotrzebowanie *n*
PY потребление *n*
SU teho
SV behov *n*

1233 demineralizing
DA afsaltning
demineralisering
DE Entsalzung *f*
ES desmineralización *f*
F déminéralisant
I demineralizzazione *f*
MA sótalanítás
NL demineralisatie
PO demineralizacja *f*
PY деминерализация *f*
обессоливание *n*
SU pehmentää
SV avsaltning *c*

1234 demodulation
DA demodulation
DE Demodulation *f*
ES desmodulación *f*
F démodulation *f*
I demodulazione *f*
MA demoduláció
NL demodulatie *f*
PO demodulacja *f*
PY выделение *n*
низкочастотного
сигнала
демодуляция *f*
SU modulointi
SV demodulering *c*

1235 density
DA massefylde
tæthed
DE Dichte *f*
ES densidad *f*
F densité *f*
I densità *f*
MA sűrűség
NL dichtheid *f*
PO gęstość *f*
PY плотность *f*
SU tiheys
SV täthet *c*

DA	=	Danish
DE	=	German
ES	=	Spanish
F	=	French
I	=	Italian
MA	=	Magyar (Hungarian)
NL	=	Dutch
PO	=	Polish
PY	=	Russian
SU	=	Finnish
SV	=	Swedish

1236 deodorant
DA deodorant
    lugtbekæmpelses-
    middel
DE Geruchsbeseitigungs-
    mittel *n*
ES desodorante *m*
F déodorant *m*
    désodorisant *m*
I deodorante *m*
MA szagtalanító
NL reukverdrijvend
    middel *n*
    stankbestrijdings-
    middel *n*
PO dezodorant *m*
    środek *m*
    odwaniający
PY дезодорант *m*
SU deodorantti
    hajustin
SV deodorant *c*

1237 deposit
DA aflejring
    bundfald
DE Ablagerung *f*
    Niederschlag *m*
ES depósito *m*
    sedimento *m*
F dépôt *m*
    précipitation *f*
    précipité *m*
I deposito *m*
MA lerakódás
NL bezinksel *n*
PO depozyt *m*
    osad *m*
PY осадок *m*
SU kerrostuma
    käsiraha
    sakka
    talletus
SV avsättning *c*

1238 deposit *vb*
DA aflejre
    bundfælde
DE ablagern
ES depositar
F déposer
I depositare
MA lerakódik
NL aanslibben
    neerslaan
PO osadzać
PY осаждать
SU kerrostua
    sakkautua
    tallettaa
SV avsätta

1239 deposition
DA aflejring
DE Ablagerung *f*
    Ausfällung *f*
    Ausscheidung *f*
ES depósito (de
    sedimento) *m*
    precipitación *f*
    sedimentación *f*
F précipitation *f*
I deposizione *f*
MA lerakódás
    üledék
NL afzetting *f*
PO osadzanie *n*
PY осаждение *n*
    отложение *n*
SU laskeuma
    laskeutuminen
SV avsättning *c*

1240 depreciation
DA afskrivning
    værdiforringelse
DE Abschreibung *f*
    Verringerung *f*
ES amortización *f*
    depreciación *f*
F dépréciation *f*
I ammortamento *m*
    deprezzamento *m*
MA elavulás
    értékcsükkenés
NL afschrijving *f*
PO deprecjacja *f*
    dewaluacja *f*
    obniżenie *n* wartości
PY обесценивание *n*
SU arvon lasku
SV avskrivning *c*

1241 de-scale *vb*
DA fjerne kedelsten
DE Kesselstein *m*
    entfernen
ES desincrustar
F désincruster
I disincrostare
MA kazánkövet eltávolít
NL ketelsteen
    verwijderen
    ontkalken
PO usuwać kamień
    kotłowy
PY удалять отложения
SU puhdistaa
    kattilakivestä
SV rena från pannsten *c*

1242 description
DA beskrivelse
DE Beschreibung *f*
ES características *f,pl*
    descripción *f*
    diseño *m*
F description *f*
I descrizione *f*
MA leírás
NL omschrijving *f*
PO opis *m*
PY описание *n*
SU kuvaus
SV beskrivning *c*

1243 desiccant
DA tørremiddel
DE Trocknungsmittel *n*
ES desecante *m*
    deshidratante *m*
F déshydratant
    désséchant *m*
    sécheur *m*
I essiccante *m*
MA szárítóküzeg
NL droogmiddel *n*
PO środek *m* osuszający
PY осушитель *m*
SU kuivausaine
SV torkmedel *n*

1244 desiccation
DA indtørring
DE Trocknung *f*
ES desecación *f*
F dessication *f*
I essiccazione *f*
MA szárítás
NL droging *f*
PO osuszanie *n*
    wysuszanie *n*
PY обезвоживание *n*
    осушение *n*
SU kuivaus
SV uttorkning *c*

1245 design
DA konstruktion
   udformning
DE Absicht *f*
   Entwurf *m*
   Plan *m*
   Projekt *n*
ES proyecto *m*
F conception *f*
   dessin *m*
   étude *f*
   plan *m*
   projet *m*
I progetto *m*
MA terv
   tervezés
NL ontwerp *n*
PO projektowanie *n*
PY проект *m*
SU suunnittelu
SV konstruktion *c*

1246 design *vb*
DA konstruere
   tegne
   udforme
DE entwerfen
   planen
ES proyectar
F projeter
I progettare
MA tervez
NL ontwerpen
PO projektować
PY конструировать
   проектировать
SU suunnitella
SV konstruera

1247 design air flow
DA projekteret luftstrøm
DE geplante
   Luftströmung *f*
ES flujo de aire de
   proyecto *m*
F débit *m* nominal
I portata *f* di aria di
   progetto
MA tervezett légáram
NL ontwerpluchthoeveel-
   heid *f*
PO obliczeniowy
   strumień *m*
   (przepływ)
   powietrza
PY расчетный расход
   m воздуха
SU suunniteltu ilmavirta
SV beräknat luftflüde *n*

1248 design conditions
DA projekterings-
   forudsætninger
DE Entwurfs-
   bedingungen *f*
ES condiciones de
   proyecto *f,pl*
F données *f* du projet
I condizioni *f,pl* di
   progetto
MA tervezett állapotok
NL ontwerpcondities *pl*
PO warunki *m*
   projektowania
PY проектные
   условия *n*
SU suunnitteluolo-
   suhteet
SV projekterings-
   villkor *n*

1249 design criteria
DA projekterings-
   kriterium
DE Entwurfskriterien *f*
ES criterios del proyecto
   m,pl
F caractéristiques *f* du
   projet
I criteri *m,pl* di
   progetto
MA tervezési feltétel
NL ontwerpcriteria *pl*
PO założenia *n*
   projektowe
PY проектные
   требования *n*
SU suunnitteluperusteet
SV projekterings-
   kriteria *n*

1250 design heat loss
DA dimensionerende
   varmetab
DE Wärmebedarfs-
   berechnung *f*
ES pérdida(s) de calor
   calculadas *f,pl*
   pérdida(s) de calor
   previstas (en el
   proyecto) *f,pl*
F déperditions *f*
   thermiques
   calculées
I disperdimento *m*
   secondo progetto
MA tervezett hőveszteség
NL warmteverlies *n*
   (berekend)
PO obliczeniowa strata
   f ciepła
   strata *f* ciepła w
   warunkach
   obliczeniowych
PY расчетные
   теплопотери *f,pl*
SU mitoituslämpühäviüt
SV dimensionerande
   värmefürlust *c*

1251 design load
DA dimensionerende
   belastning
DE geplante Belastung *f*
   geplante
   Entwurfslast *f*
ES carga de proyecto *f*
F charge *f*
   prévisionnelle
I carico *m* di progetto
MA tervezett terhelés
NL ontwerpbelasting *f*
PO obciążenie *n*
   obliczeniowe
PY расчетная
   нагрузка *f*
SU suunniteltu kuorma
SV dimensionerande
   belastning *c*

1252 design study
DA udkast
DE Entwurfsstudie *f*
ES proyecto *m*
F étude *f* du projet
I studio *m* di progetto
MA tanulmányterv
NL ontwerpstudie *f*
PO studium *n*
   projektowe
PY расчетные
   исследования
   n,pl
SU ehdotus
   luonnos
SV utkast *n*

1253 design temperature
DA dimensionerende temperatur
DE Temperaturannahme *f*
ES temperatura de proyecto *f*
F température *f* imposée
I temperatura *f* di progetto
MA tervezett hőmérséklet
NL ontwerptemperatuur *f*
PO temperatura *f* obliczeniowa
PY расчетная температура *f*
SU mitoituslämpütila
SV dimensionerande temperatur *c*

1254 design working pressure
DA dimensionerende arbejdstryk
DE Entwurfsarbeitsdruck *m*
ES presión de trabajo de proyecto *f*
F pression *f* nominale de travail
I pressione *f* di lavoro di progetto
MA tervezett üzemnyomás
NL ontwerpwerkdruk *m*
PO obliczeniowe ciśnienie *n* robocze
PY расчетное рабочее давление *n*
SU suunniteltu työpaine
SV dimensionerande arbetstryck *n*

1255 desorption
DA desorption
DE Desorption *f*
ES desorción *f*
F désorption *f*
I desorbimento *m*
MA deszorpció (elnyelt gáz felszabadulás)
NL desorptie *f*
PO desorpcja *f*
PY десорбция *f*
SU desorptio
SV desorption *c*

1256 desuperheater
DA køler for overhedet damp
DE Heißdampfkühler *m*
ES desrecalentador *m* preenfriador *m*
F désurchauffeur *m*
I desurriscaldatore *m*
MA túlhevítés megszüntető
NL oververhittestoomkoeler *m*
PO schładzacz *m* (pary przegrzanej)
PY пароохладитель *m*
SU hüyrynjäähdytin
SV kylare *c* för överhettad ånga *c*

1257 desuperheating coil
DA kølespiral for overhedet damp
DE Überhitzungskühlerschlange *f*
ES batería de desrecalentamiento *f*
F batterie *f* de désurchauffe
I serpentino *m* di desurriscaldamento
MA gőzhűtő túlhevítés megszüntető hőcserélő
NL oververhitgaskoeler *m*
PO wężownica *f* schładzacza
PY пароохладителььный змеевик *m*
SU tulistuslämmün jäähdytyspatteri
SV ångkylarslinga *c*

1258 detail
DA detalje
DE Einzelheit *f*
ES detalle *m*
F détail *m*
I dettaglio *m*
MA részlet
NL detail *n*
PO detal *m* szczegół *m*
PY деталь *f*
SU yksityiskohta
SV detalj *c*

1259 detail *vb*
DA detaljere
DE ausführlich beschreiben
ES detallar
F détailler
I dettagliare
MA részletez
NL detailleren
PO wyszczególniać
PY детализировать
SU viimeistellä
SV detaljera

1260 detail drawing
DA detailtegning
DE Einzelteilzeichnung *f*
ES plano de detalle *m*
F dessin *m* de détail
I disegno *m* particolareggiato
MA részletrajz
NL detailtekening *f*
PO rysunek *m* części rysunek *m* szczegółowy
PY детальный чертеж *m*
SU detaljipiirustus yksityiskohtapiirustus
SV detaljritning *c*

1261 detailed estimate
DA detaljeret kalkule
DE detaillierter Kostenvoranschlag *m*
ES cálculo detallado *m* presupuesto detallado *m*
F devis *m* détaillé
I offerta *f* dettagliata
MA részletes kültségvetés
NL gespecificeerde begroting *f*
PO kosztorys *m* szczegółowy
PY подробная смета *f*
SU yksityiskohtainen arvio
SV detaljerad kalkyl *c*

1262 detecting element
DA detektorelement
DE Geber *m*
   Meßfühler *m*
ES elemento detector *m*
F élément *m* détecteur
I elemento *m*
   rivelatore
MA érzékelő elem
NL opneemelement *n*
   voeler *m*
PO człon *m*
   wykrywający
   czujnik *m*
PY подробно
   разработанный
   элемент *m*
SU tunnistin
SV detektorelement *n*

1263 detector
DA detektor
   føler
DE Fühler *m*
ES detector *m*
F détecteur *m*
I sonda *f*
MA detektor
   érzékelő
NL detector *m*
   verklikker *m*
PO detektor *m*
   wykrywacz *m*
PY детектор *m*
SU tuntoelin
SV detektor *c*

1264 detergent
DA rensemiddel
DE Reinigungsmittel *n*
ES detergente *m*
F détergent *m*
I detergente *m*
MA tisztítószer
NL reinigingsmiddel *n*
PO detergent *m*
   środek *m* piorący
PY детергент *m*
SU puhdistusaine
SV tvättmedel *n*

1265 deterioration
DA forringelse
DE Verschlechterung *f*
   Zerfall *m*
ES deterioro *m*
F détérioration *f*
I degrado *m*
MA romlás
NL achteruitgang *m*
   verslechtering *f*
PO pogorszenie *n* się
   jakości
   psucie *n* się
PY брак *m*
   ущерб *m*
SU pilaantuminen
SV fürslitning *c*

1266 deviation
DA afvigelse
DE Abweichung *f*
ES desviación *f*
F déviation *f*
I deviazione *f*
MA eltérés
NL afwijking *f*
PO dewiacja *f*
   odchylenie *n*
PY отклонение *n*
SU poikkeama
SV avdrift *c*

1267 device
DA anordning
   apparat
DE Gerät *n*
   Vorrichtung *f*
ES dispositivo *m*
F dispositif *m*
I dispositivo *m*
MA készülék
   szerkezet
NL apparaat *n*
   instrument *n*
PO przyrząd *m*
   urządzenie *n*
PY устройство *n*
SU laite
SV anordning *c*

1268 dew
DA dug
DE Tau *m*
ES rocío *m*
F rosée
I rugiada *f*
MA harmat
NL dauw *m*
   vocht *n*
PO rosa *f*
PY отпотевание *n*
   роса *f*
SU tiivistynyt vesi
SV dagg *c*

1269 dew-point
DA dugpunkt
DE Taupunkt *m*
ES punto de rocío *m*
F point *m* de rosée
I punto *m* di rugiada
MA harmatpont
NL dauwpunt *n*
PO punkt *m* rosy
PY точка *f* росы
SU kastepiste
SV daggpunkt *c*

1270 dew-point depression
DA dugpunkts-
   nedsættelse
DE Taupunkt-
   erniedrigung *f*
ES descenso del punto
   de rocío *m*
F écart *m* du point de
   rosée
I abbassamento *m* del
   punto di rugiada
MA harmatponti hőmérs-
   ékletkülünbség
NL dauwpunts-
   verlaging *f*
PO depresja *f* punktu
   rosy
   obniżenie *n* punktu
   rosy
PY гигрометрическая
   разность *f*
   разность *f* между
   температурой по
   сухому
   термометру и
   тем (етч)
SU ilman kuiva- ja
   märkälämpütilan
   ero
SV daggpunktssänkning

DA	=	Danish
DE	=	German
ES	=	Spanish
F	=	French
I	=	Italian
MA	=	Magyar (Hungarian)
NL	=	Dutch
PO	=	Polish
PY	=	Russian
SU	=	Finnish
SV	=	Swedish

1271 dew-point hygrometer
DA dugpunkts-
hygrometer
DE Taupunkt-
hygrometer *n*
ES higrómetro de punto
de rocío *m*
F hygromètre *m* à
point de rosée
I igrometro *m* a
punto di rugiada
MA harmatponti
légnedvességmérő
NL dauwpunts-
hygrometer *m*
PO higrometr *m* punktu
rosy
PY конденсационный
гигрометр *m*
SU kastepistemittari
SV daggpunkts-
hygrometer *c*

1272 dew-point rise
DA dugpunktshævning
DE Taupunkt-
erhühung *f*
ES elevación del punto
de rocío *f*
F élévation *f* du point
de rosée
I innalzamento *m* del
punto di rugiada
MA harmatponti
nedvességtartalom
emelkedés
NL dauwpunts-
verhoging *f*
PO podwyższenie *n*
punktu rosy
PY повышение *n*
температуры
точки росы
увеличение *n*
влагосодержания
SU kosteuden nousu
kastepisteenä
ilmaistuna
SV daggpunkts-
stegring *c*

1273 diagram
DA diagram
DE Diagramm *n*
Schaubild *n*
ES diagrama *m*
F diagramme *m*
I diagramma *m*
MA diagram
NL diagram *n*
schema *n*
PO schemat *m*
wykres *m*
PY диаграмма *f*
SU diagrammi
kaavio
SV diagram *n*

1274 dial
DA skala
skive
DE Wähler *m*
Wählscheibe *f*
ES esfera (de un
medidor) *f*
F cadran *m*
I quadrante *m*
MA számlap
tárcsa
NL wijzerplaat *f*
PO skala *f* tarczowa
tarcza *f* numerowa
tarcza *f* z podziałką
PY круговая шкала *f*
SU osoitin
SV skala *c*
tavla *c*

1275 dial thermometer
DA skivetermometer
DE Zeiger-
thermometer *n*
ES termómetro de
cuadrante *m*
F thermomètre *m* à
cadran
I termometro *m* a
quadrante
MA számlapos hőmérő
NL wijzer-
thermometer *m*
PO termometr *m*
tarczowy
PY дисковый
термометр *m*
SU osoittava
lämpümittari
SV visartermometer *c*

1276 diameter
DA diameter
tværmål
DE Durchmesser *m*
ES diámetro *m*
F diamètre *m*
I diametro *m*
MA átmérő
NL diameter *m*
middellijn *f*
PO średnica *f*
PY диаметр *m*
SU halkaisija
SV diameter *c*

1277 diaphragm
DA membran
DE Membrane *f*
ES diafragma *m*
membrana *f*
F membrane *f*
I diaframma *m*
membrana *f*
MA diafragma
membrán
NL diafragma *n*
membraan *n*
PO kryza *f*
membrana *f*
przepona *f*
PY диафрагма *f*
мембрана *f*
SU kalvo
SV membran *n*

1278 diaphragm compressor
DA membran
kompressor
DE Membranverdichter
m
ES compresor de
diafragma *m*
F compresseur *m* à
membrane
I compressore *m* a
membrana
MA membrános
kompresszor
NL membraan-
compressor *m*
PO sprężarka *f* przepon-
owa
PY мембранный
компрессор *m*
SU kalvokompressori
kalvopumppu
SV diafragmakom-
pressor *c*

1279 diaphragm valve
DA membranventil
DE Membranventil *n*
ES válvula de diafragma
f
F robinet *m* à
membrane
I valvola *f* a
membrana
MA membránszelep
NL membraan-
afsluiter *m*
PO zawór *m*
membranowy
PY диафрагмовый
клапан *m*
мембранный
клапан *m*
SU kalvoventtiili
SV membranventil *c*

1280 diathermanous *adj*
DA strålevarmeover-
førende
DE infrarotdurchlässig
wärmedurchlässig
ES diatérmico
F diathermane
I diatermico
MA hősugáráteresztő
NL warmtedoorlatend
PO diatermiczny
PY теплопроводный
SU lämmünsiirrossa
ympäristün kanssa
lämpüä läpäisevä
SV värmeledande

1281 die
DA matrice
stempel
DE Gußform *f*
Matritze *f*
Würfel *m*
ES matriz *f*
troquel *m*
F matrice *f*
I filiera *f*
matrice *f*
stampo *m*
MA matrica
süllyeszték
NL matrijs *f*
snijkussen *n*
PO forma *f* odlewnicza
matryca *f*
PY матрица *f*
пуансон *m*
SU taltta
terä
SV matris *c*

1282 dielectric *adj*
DA dielektrisk
isolerende
DE dielektrisch
ES dieléctrico
F diélectrique
I dielettrico
MA dielektromos
NL dilectrisch
PO dielektryczny
PY диэлектрический
SU dielektrinen
SV dielektrisk

1283 dielectric constant
DA isoleringskonstant
DE dielektrische
Konstante *f*
ES constante dieléctrica
f
F constante *f*
diélectrique
I costante *f* dielettrica
MA dielektromos állandó
NL dilektrische
constante *f*
PO stała *f* dielektryczna
PY диэлектрическая
постоянная *f*
SU dielektrisyysvakio
SV isoleringskonstant *c*

1284 dielectric strength
DA gennemslagsstyrke
DE dielektrische Kraft *f*
dielektrische
Stärke *f*
ES intensidad dieléctrica
f
F tenue *f* diélectrique
I resistenza *f*
dielettrica
MA átütési feszültség
NL doorslagwaarde *f*
PO wytrzymałość *f*
dielektryczna
PY диэлектрическое
сопротивление *n*
SU jännitekestoisuus
läpilyüntilujuus
SV maximal ellast *c*

1285 dielectric thawing
DA elektrisk optøning
DE dielektrisches
Auftauen *n*
ES calentamiento
dieléctrico *m*
F décongélation *f*
diélectrique
I scongelamento *m*
dielettrico
MA dielektromos
olvasztás
NL dilektrische
ontdooiing *f*
PO rozmrażanie *n*
dielektryczne
PY диэлектрическое
оттаивание *n*
SU dielektrinen sulatus
induktiolämmitys
SV dielektrisk
värmekabel *c*

1286 diesel engine
DA dieselmotor
DE Dieselmaschine *f*
ES motor diesel *m*
F moteur *m* diesel
I motore *m* Diesel
MA dízel gép
dízelmozdony
NL dieselmotor *m*
PO silnik *m* Diesla
silnik *m*
wysokoprężny
PY дизель *m*
дизельный
двигатель *m*
SU dieselmoottori
SV dieselmotor *c*

1287 diesel motor
DA dieselmotor
DE Dieselmotor *m*
ES motor diesel *m*
F moteur *m* diesel
I motore *m* Diesel
MA dízelmotor
NL dieselmotor *m*
PO silnik *m*
wysokoprężny
PY дизельный
двигатель *m*
SU dieselmoottori
SV dieselmotor *c*

1288 difference
DA differens
forskel
DE Differenz *f*
Unterschied *m*
ES diferencia *f*
F différence *f*
I differenza *f*
MA külünbség
NL verschil *m*
PO różnica *f*
PY перепад *m*
разница *f*
разность *f*
SU ero
SV skillnad *c*

1289 differential
DA differential
DE Differential *n*
ES diferencial *m*
F différentiel *m*
I differenziale *m*
MA differenciál
kiegyenlítő
NL schakelverschil *n*
PO przekładnia *f*
różnicowa
różniczka *f*
PY дифференциал *m*
SU differentiaali-
ero
SV differential *c*

**1290 differential
controller**
DA differentialregulator
DE Differentialregler *m*
ES controlador
diferencial *m*
F controleur *m*
différentiel
I regolatore *m*
differenziale
MA kiegyenlítő
szabályozó
NL verschilregelaar *m*
PO urządzenie *n*
sterujące różnicowe
PY дифференциальный
регулятор *m*
SU erosäätü
SV differential-
regulator *c*

1291 differential grille
DA jalousispjæld
DE Luftgitter *n* mit
veränderlichem
Widerstand *m*
ES rejilla diferencial *f*
F bouche *f* à section
variable
I griglia *f* a sezione
variabile
MA kiegyenlítő rács
NL rooster *m* met instel-
bare weerstand
PO kratka *f* różnicowa
PY дифференциальная
решетка *f*
SU säleikkü
SV jalusispjäll *n*

**1292 differential
pressure**
DA differentialtryk
DE Differenzdruck *m*
ES diferencia de
presión *f*
presión diferencial *f*
F pression *f*
différentielle
I pressione *f*
differenziale
MA kiegyenlítő nyomás
NL verschildruk *m*
PO ciśnienie *n*
różnicowe
PY перепад *m*
давления
SU paine-ero
SV differentialtryck *n*

**1293 differential
pressure control**
DA differenstryks
regulator
DE Differenzdruck-
regler *m*
ES control de presión
diferencial *m*
F contrôle *m* de
pression
différentielle
I regolazione *f* di
pressione
differenziale
MA kiegyenlítő
nyomásszabály-
ozás
NL verschildruk-
regeling *f*
PO regulacja *f* ciśnienia
różnicowego
PY регулирование *n*
по разности
давления
SU paine-erosäädin
SV differentialtryck-
kontroll *c*

**1294 differential
pressure gauge**
DA differentialtrykmåler
DE Differenzdruck-
messer *m*
ES manómetro
diferencial *m*
F manomètre *m*
différentiel
I manometro *m*
differenziale
MA kiegyenlítő
nyomásmérő
NL verschildrukmeter *m*
PO manometr *m*
różnicowy
PY дифференциальный
манометр *m*
SU paine-eromittari
SV differenstryck-
mätare *c*

1295 differential temperature
DA differential-temperatur
DE Differenz-temperatur *f*
ES temperatura diferencial *f*
F différentiel *m* de température
I temperatura *f* differenziale
MA kiegyenlítő hőmérséklet
NL temperatuur-differentiaal *n*
PO temperatura *f* różnicowa
PY перепад *m* температуры
SU lämpütilaero
SV differential-temperatur *c*

1296 diffuse *adj*
DA diffus spredt
DE diffundierend diffus durchdringend
ES difuso
F diffus
I diffuso
MA szétszórt szórt
NL diffuus verstrooid
PO rozproszony
PY веерный диффузионный
SU diffuusi
SV diffus

1297 diffuser
DA diffuser spreder
DE Diffusor *m*
ES difusor *m*
F diffuseur *m*
I diffusore *m*
MA átmeneti idom légbefúvó
NL anemostaat *m* luchtuitblaas-ornament *n*
PO anemostat *m* dyfuzor *m*
PY диффузор *m*
SU diffuusori hajoitin
SV diffusor *c*

1298 diffuse reflectance
DA diffus reflektion
DE diffuse Reflektion *f*
ES reflexión difusa *f*
F reflection *f* diffuse
I riflettanza *f* diffusa
MA szórt visszaverődés
NL diffuse reflectie *f*
PO dyfuzyjny współczynnik *m* odbicia
PY отражательная способность *f* рефлектора
SU diffuusiheijastus
SV diffus reflektion *c*

1299 diffuser radius of diffusion
DA diffuser-spredningsvinkel
DE Diffusorradius *m* der Diffusion *f*
ES radío de difusión diseminada *m*
F rayon *m* de diffusion d'un diffuseur
I raggio *m* di diffusione del diffusore
MA szóródási sugár
NL worp van plafonduitlaat *m*
PO anemostat *m* kołowy rozpraszający
PY радиус *m* плафона
SU heittopituus
SV diffusionsradie *c*

1300 diffusion
DA diffusion
DE Diffusion *f*
ES difusión *f*
F diffusion *f*
I diffusione *f*
MA diffúzió szétszóródás
NL diffusie *f* verspreiding *f*
PO dyfuzja *f* rozproszenie *m*
PY диффузия *f* рассеивание *n*
SU diffuusio
SV diffusion *c*

1301 diffusion area
DA diffusionsareal
DE Diffusionsfläche *f*
ES superficie de difusión *f*
F aire *f* de diffusion
I area *f* di diffusione
MA szétszóródási terület
NL verspreidings-gebied *n*
PO powierzchnia *f* dyfuzji
PY площадь *f* рассеивания эффективная площадь *f* вентиляционных струй
SU heittokuvion ala
SV diffusionsarea *c*

1302 diffusion coefficient
DA diffusionskoefficient
DE Diffusionszahl *f*
ES coeficiente de difusión *m*
F coefficient *m* de diffusion
I coefficiente *m* di diffusione
MA diffúziós együtthato
NL diffusiecőfficint *m*
PO współczynnik *m* dyfuzji
PY коэффициент *m* диффузии
SU diffuusiokerroin
SV diffusions-koefficient *c*

1303 diffusion–absorption system

1303 diffusion–absorption system
DA diffusions absorptionssystem
DE Diffusionsabsorptions- system *n*
ES sistema de difusión *m* absorción
F système *m* à absorption- diffusion
I sistema *m* a diffusione- assorbimento
MA diffúziós-abszorpciós (hűtő) rendszer
NL diffusie- absorptie- systeem *n*
PO system *m* dyfuzyjno- absorpcyjny
PY диффузионно- абсорбционная система *f*
SU diffuusiokoneisto
SV diffusions- absorptions- system *n*

1304 digital *adj*
DA digital
DE digital
ES digital
F digital numérique
I digitale
MA digitális számjegyes
NL digitaal
PO cyfrowy
PY цифровой
SU digitaalinen
SV digital

1305 digital transmission
DA taloverføring
DE digitale Übertragung *f*
ES transmisión digital *f*
F transmission *f* numérique
I trasmissione *f* digitale
MA digitális adatátvitel
NL digitale overdracht *f*
PO cyfrowy przekaz *m* (danych)
PY цифровая передача *f* (данных)
SU digitaalinen tiedonsiirto
SV digital transmission *c*

1306 diluent
DA fortynder
DE Verdünner *m* Verdünnungs- mittel *n*
ES diluente *m*
F diluant *m*
I diluente *m*
MA hígító
NL verdunner *m*
PO czynnik *m* rozcieńczający
PY растворитель *m*
SU laimennin
SV fürtunningsmedel *n*

1307 diluting medium
DA fortyndingsmiddel
DE Verdünnungs- mittel *n*
ES diluente *m*
F substance *f* diluante
I diluente *m*
MA hígítószer
NL verdunnings- middel *n*
PO rozcieńczalnik *m* środek *m* rozcieńczający
PY подмешиваемая среда *f* разбавитель *m*
SU laimennusaine ohenne
SV fürtunningsmedel *n*

1308 dilution
DA fortynding
DE Verdünnung *f*
ES dilución *f*
F dilution *f*
I diluizione *f*
MA hígítás
NL verdunning *f*
PO rozcieńczenie *n*
PY разжижение *n* растворение *n*
SU laimennus
SV fürtunning *c*

1309 dimension
DA dimension
DE Abmessung *f* Dimension *f*
ES dimensión *f*
F dimension *f*
I dimensione *f*
MA kiterjedés méret
NL afmeting *f*
PO wymiar *m*
PY размер *m*
SU mitta
SV dimension *c*

1310 dimensioned drawing
DA måltegning
DE Maßbild *n* Maßzeichnung *f*
ES dibujo acotado *m* plano acotado *m*
F dessin *m* côté plan *m* côté
I disegno *m* dimensionale
MA léptékhelyes rajz
NL maatschets *f* maattekening *f*
PO rysunek *m* zwymiarowany
PY чертеж *m* с размерами
SU mittapiirros
SV ritning *c*

DA = Danish
DE = German
ES = Spanish
F = French
I = Italian
MA = Magyar (Hungarian)
NL = Dutch
PO = Polish
PY = Russian
SU = Finnish
SV = Swedish

1311 dipstick
DA pejlestok
DE Meßstab *m*
ES varilla indicadora de
  nivel *f*
F jauge *f*
I calibro *m* di
  profondità
MA mérőpálca
NL peilstok *m*
PO głębokościomierz *m*
  zwilżany prętowy
  prętowy wskaźnik *m*
  poziomu (oleju)
PY погружная шкала *f*
SU mittatikku
SV mätsticka *c*

1312 dip tank
DA pejletank
DE Tauchtank *m*
ES baño de inmersión *m*
F bac *m* de démoulage
I vasca *f* di disgelo
MA mártókád
NL dompeltank *m*
PO zbiornik *m*
  zamokowy
  zbiornik *m*
  zanurzeniowy
PY заглубленная
  емкость *f*
SU upotusallas
SV doppningskar *n*

**1313 direct-contact heat
  exchanger**
DA direkte
  kontaktvarme-
  veksler
DE Direktkontaktwärme-
  tauscher *m*
ES intercambiador de
  contacto directo *m*
F échangeur *m*
  thermique à
  contact direct
I scambiatore *m* di
  calore a contatto
  diretto
MA küzvetlen
  érintkezésű
  hőcserélő
NL warmtewisselaar *m*
  met rechtstreekse
  overdracht
PO wymiennik *m* ciepła
  bezprzeponowy
PY контактный
  теплообменник
  m
SU kontaktilämmün-
  siirrin
SV direktkontakt-
  värmeväxlare *c*

1314 direct cooling
DA direkte køling
DE direkte Kühlung *f*
ES enfriamiento directo
  m
F système *m* de
  refroidissement à
  contact direct
I raffrescamento *m*
  diretto
MA küzvetlen hűtés
NL directe koeling
PO chłodzenie *n*
  bezpośrednie
PY непосредственное
  охлаждение *n*
SU suora jäähdytys
SV direkt kylning *c*

**1315 direct cooling
  system**
DA direkte kølesystem
DE Direktkühlungs-
  system *n*
ES sistema de
  enfriamiento
  directo *m*
F système *m* de
  refroidissement
  direct
I sistema *m* a
  raffreddamento
  diretto
MA küzvetlen
  hűtőrendszer
NL direct-expansie
  koeling *f*
PO system *m*
  bezpośredniego
  chłodzenia
PY система *f*
  непосред-
  ственного
  (контактного)
  охлаждения
SU suora
  jäähdytysjärje-
  stelmä
SV direktkylnings-
  system *n*

1316 direct current
DA jævnstrøm
DE Gleichstrom *m*
ES corriente continua *f*
F courant *m* continu
I corrente *f* continua
MA egyenáram
NL gelijkstroom *m*
PO prąd *m* stały
PY постоянное
  течение *n*
SU tasavirta
SV likstrüm *c*

1317 direct cylinder
DA varmtvandstilbereder
DE Warmwasser-
  speicher *m*
  Warmwasservorrats-
  behälter *m*
ES cuerpo de bomba de
  acción directa *m*
F réservoir *m* d'eau à
  chauffe direct
I cilindro *m* diretto
MA fürdőhenger
NL direct-verwarmd
  warmwater-
  voorraadvat *n*
PO zbiornik *m*
  bezpośredni
PY бойлер *m*
SU lämpimän
  käyttüveden
  varaaja
SV varmvatten-
  beredare *c*

1318 direct digital control
DA direkte
  digitalregulering
DE DDC-Regelung *f*
ES control digital
  directo *m*
F régulation *f*
  numérique directe
I regolazione *f*
  digitale diretta
MA DDC
  küzvetlen digitális
  szabályozás
NL digitale regeling en
  sturing *f* (DDC)
PO sterowanie *n*
  cyfrowe
  bezpośrednie
PY прямое цифровое
  управление *n*
SU suora tietokonesäätü
SV direkt digital
  kontroll *c*

1319 direct drive
DA direkte
  kraftoverføring
DE Direktantrieb *m*
ES accionamiento por
  acoplamiento
  directo *m*
  acoplamiento
  directo *m*
F commande *f* directe
  entraînement *m*
  direct
I comando *m* diretto
  trazione *f* diretta
MA küzvetlen meghajtás
NL directe aandrijving *f*
PO napęd *m*
  bezpośredni
PY привод *m* на одной
  оси
  прямая передача *f*
SU suora käyttü
SV direktdrift *c*

1320 direct exhaust system
DA direkte
  udsugningssystem
DE Direktabsaugungs-
  system *n*
ES sistema de descarga
  directa *m*
F système *m*
  d'extraction direct
I sistema *m* a scarico
  diretto
MA beépített elszívó
  rendszer
NL directe
  rookgasafvoer *m*
PO system *m*
  bezpośredniego
  odciągu
PY система *f* вытяжки,
  совмещенная с
  выбросом
  дымовых газов
SU suora savukaasujen
  poisto
SV direktutsugnings-
  system *n*

1321 direct expansion coil
DA køleflade for direkte
  ekspansion
DE Direktverdampferrohr-
  schlange *f*
ES batería de expansión
  directa *f*
F batterie *f* à détente
  directe
  évaporation *f* à
  détente directe
  serpentin *m* à
  détente directe
I batteria *f* a
  espansione diretta
MA küzvetlen expanziós
  hőcserélő
NL directe expansie
  batterij *f*
  directe expansie
  spiraal *f*
PO wężownica *f*
  bezpośredniego
  odparowania
PY змеевик *m*
  непосред-ст-
  венного
  расширения
SU suoran paisunnan
  patteri
SV direktfürångare *c*

1322 direct-expansion refrigerating system
DA kølesystem med direkte ekspansion
DE direktverdampfendes Kühlsystem *n*
ES sistema de refrigeración por expansión directa *m*
F système *m* de réfrigération à détente directe
I sistema *m* di refrigerazione ad espansione diretta
MA küzvetlen expanziós hűtőrendszer
NL expansiekoelsysteem *n* (open)
PO urządzenie *n* chłodnicze bezpośredniego odparowania
PY холодильная система *f* непосредственного расширения
SU suoran paisunnan jäähdytysjärjestelmä
SV kylsystem *n* med direktexpansion *c*

1323 direct expansion system
DA system med direkte ekspansion
DE direktes Ausdehnungssystem *n*
ES sistema de expansión directa *m*
F système *m* à détente directe
I sistema *m* ad espansione diretta
MA küzvetlen expanziós rendszer
NL direct expansie systeem *n*
PO system *m* bezpośredniego rozprężania
PY система *f* непосредственного расширения
SU suoran paisunnan järjestelmä
SV direktexpansionssystem *n*

1324 direct-fired air heater
DA direkte fyret luftvarmer
DE direktbefeuerter Lufterhitzer *m*
ES aerotermo de combustión directa *m* calentador de aire de caldeo directo *m*
F aérotherme *m* à chauffe directe
I aerotermo *m* a combustione diretta
MA küzvetlen tüzelésű léghevítő
NL directgestookte luchtverwarmer *m*
PO grzejnik *m* powietrzny bezpośrednio ogniowy
PY огневой воздухонагреватель *m*
SU kuumailmapuhallin
SV varmluftspanna *c*

1325 direct-fired heater
DA direkte fyret varmeovn
DE direktbefeuertes Heizgerät *n*
ES calentador de llama directa *m*`
F
I riscaldatore *m* a combustione diretta
MA küzvetlen tüzelésű fűtőkályha
NL ruimteverwarmer *m* met open verbranding
PO grzejnik *m* bezpośrednio ogniowy
PY огневой нагреватель *m*
SU suorapolttolämmitin
SV direkteldad panna *c*

1326 direct flow valve
DA ligeløbsventil
DE Durchflußventil *n*
ES válvula de paso directo *f*
F robinet *m* à passage direct
I valvola *f* a flusso diretto
MA egyenes szelep
NL doorstroomafsluiter *m*
PO zawór *m* przepływowy prosty
PY проходной клапан *m* (без изменения направления течения)
SU läpivirtausventtiili
SV rak ventil *c*

1327 direct heating system
DA direkte varmesystem
DE Direktheizungssystem *n*
ES sistema de calentamiento directo *m*
F système *m* de chauffe direct
I sistema *m* a riscaldamento diretto
MA küzvetlen fűtési rendszer
NL direct warmteuitwisselend systeem *n*
PO system *m* bezpośredniego ogrzewania system *m* ogrzewania bezprzeponowego
PY местная отопительная система *f*
SU suora lämmitysjärjestelmä
SV direktuppvärmningssystem *n*

1328 direct hot water supply
DA direkte varmtvandsforsyning
DE direkte Warmwasserversorgung *f*
ES suministro directo de agua caliente *m*
suministro instantáneo de agua caliente *m*
F chauffe-eau *m* instantané
I scaldacqua *m* istantaneo
MA küzvetlen melegvízellátás
NL doorstroomwarmwatervoorziening *f*
PO bezpośrednie zasilanie *n* w wodę ciepłą
PY непосредственная подача *f* горчей воды
SU suora lämpimän käyttüveden jakelu
SV beredare *c* med direkt varmvatten *n*

1329 direction
DA retning
DE Richtung *f*
ES dirección *f*
sentido *m*
F direction *f*
I direzione *f*
senso *m*
MA irány
NL richting *f*
PO kierunek *m*
PY направление *n*
SU suunta
SV riktning *c*

1330 directional supply air terminal device
DA apparat til indblæsning af retningsbestemt luft
DE gerichteter Luftauslaß *m*
ES boca de salida de aire orientable *f*
terminal de aire orientable *m*
F bouche *f* de soufflage orientable
I bocchetta *f* di mandata orientabile
MA irányított légbefúvó
NL instelbaar luchtuitblaasmondstuk *n*
PO nawiewnik *m* powietrza z łopatkami kierującymi
urządzenie *n* do kierunkowego nawiewu powietrza
PY регулируемое приточное устройство *n*
SU suunnattava tuloilmalaite
SV tilluftsdon *n*

1331 directional thermal emittance
DA direkte varmestråling
DE gerichtete Wärmestrahlung *f*
ES emisión térmica direccional *f*
F émittance *f* thermique directionnelle
I emettenza *f* termica direzionale
MA irányított hősugárzás
NL richtingsafhankelijke emissiefactor *m*
PO ukierunkowana emitancja *f* ciepła
PY направленное тепловое излучение *n*
SU suunnattu emissiivisyys
SV riktad värmeavgivning *c*

1332 direction switch
DA retningsomskifter
DE Richtungsschalter *m*
ES conmutador *m*
F commutateur *m*
I commutatore *m*
MA irányváltó kapcsoló
NL wisselschakelaar *m*
PO wyłącznik *m* kierunkowy
PY прямой переключатель *m*
SU suuntakytkin
SV riktningsstrümbrytare *c*

1333 direct refrigerating system
DA direkte kølesystem
DE Direktkältesystem *n*
ES sistema de refrigeración directa *m*
F système *m* de refroidissement à détente directe
I sistema *m* a raffreddamento diretto
MA küzvetlen hűtőrendszer
NL expansiekoelsysteem *n* (gesloten)
PO system *m* bezpośredniego chłodzenia
PY холодильная система *f* с расположением испарителя (конденсатора) испарителя в обслуживаемом лространстве
SU suora jäähdytysjärjestelmä
SV direktkylsystem *n*

1334 direct vent system
DA direkte
   lufttilførselssystem
DE Direktventilations-
   system *n*
ES sistema de ventilado
   directo *m*
F système *m* de mise à
   l'atmosphère
I sistema *m* a sbocco
   diretto
   sistema *m* a sfogo
   diretto
MA beépített szellőzésű
   rendszer
NL gesloten
   verbrandings-
   systeem *n*
PO odpowietrzanie *n* w
   systemie
   bezpośrednim
PY дутьевая система *f*
SU palamisilman
   tuonnilla
   varustettu
   järjestelmä
SV direktventilsystem *n*

1335 dirt
DA smuds
   snavs
DE Schmutz *m*
ES mugre *f*
   suciedad *f*
F saleté *f*
I sporcizia *f*
MA szennyeződés
NL vuil
PO brud *m*
   zanieczyszczenie *n*
PY грязь *f*
   шлам *m*
SU lika
SV smuts *c*

1336 disassemble *vb*
DA adskille
   demontere
DE auseinandernehmen
   demonticren
ES desmontar
F démonter
I smantellare
   smontare
MA leszerel
   szétszerel
NL demonteren
PO demontować
   rozbierać na części
PY демонтировать
SU purkaa
SV demontera

1337 disc
DA skive
DE Scheibe *f*
ES disco *m*
F disque *m*
I disco *m*
MA tárcsa
NL schijf *n*
PO dysk *m*
   krążek *m*
   tarcza *f*
PY диск *m*
SU levy
SV skiva *c*

1338 discharge
DA aflastning
   udstrømning
DE Abfluß *m*
   Ablauf *m*
   Entladung *f*
ES descarga *f*
F décharge *f*
   évacuation *f*
   refoulement *m*
I evacuazione *f*
   scarico *m*
MA kifolyás
   ürítés
NL afvoer *m*
   ontlading *f*
   uitstroming *f*
PO odpływ *m*
   wydajność *f*
   wyładowanie *n*
   wypływ *m*
PY выброс *m*
   выхлоп *m*
SU purku
   tyhjennys
SV avlastning *c*

1339 discharge *vb*
DA aflaste
   udlede
DE ablassen
   ausfließen
   entladen
ES descargar
   evacuar
F évacuer
I evacuare
   scaricare
MA kiünt
   ürít
NL afvoeren
   lossen
   ontladen
PO wyładowywać
   wypływać
PY выбрасывать
   выгружать
SU purkaa
   tyhjentää
SV avlasta

**1340 discharge
   coefficient**
DA afløbskoefficient
DE Auslaßbeiwert *m*
ES coeficiente de
   descarga *m*
F coefficient *m* de
   décharge
I coefficiente *m* di
   efflusso
MA kiümlési tényező
NL uitstroomcoëfficiënt *m*
PO współczynnik *m*
   wydatku
   współczynnik *m*
   wypływu
PY коэффициент *m*
   выброса
SU kuroumakerroin
SV utloppskoefficient *c*

1341 discharge device
DA aftapningsanordning
DE Auslaßvorrichtung *f*
ES dispositivo de
   descarga *m*
F dispositif *m* de
   décharge
I dispositivo *m* di
   scarico
MA kibocsátó szerkezet
NL afvoerapparaat *n*
PO urządzenie *n*
   odpływowe
PY разгрузочное
   устройство *n*
SU tyhjennyslaite
SV avtappnings-
   anordning *c*

1342 discharge gauge
DA trykmanometer
DE Auslaßmanometer *n*
ES calibre de descarga
   m
F manomètre *m* de
   refoulement
I manometro *m* di
   scarico
MA magasnyomásmérő
   (hűtőgépnél)
NL persdrukmeter *m*
PO przyrząd *m* do
   pomiaru wielkości
   wypływu
   (wydatku)
PY измерительное
   устройство *n* на
   выбросе
SU painepuolen
   painemittari
SV utloppsmätning *c*

1343 discharge head
- DA faldhøjde
- DE Fallhühe *f*
- ES altura de descarga *f*
  altura de
    impulsión *f*
  presión de
    descarga *f*
  presión de
    impulsión *f*
- F hauteur *f* de
    refoulement
- I altezza *f* di mandata
  prevalenza *f*
- MA kifúvófej
- NL uittrededruk *m*
- PO wysokość *f* odpływu
  wysokość *f* wypływu
- PY зонт *m* над
    выбросом
  флюгарка *f*
- SU painepuolen
    nostokorkeus
- SV fallhüjd *c*

1344 discharge line
- DA udløbsledning
- DE Druckleitung *f*
  Heißgasleitung *f*
- ES línea de descarga *f*
- F conduite *f* de
    refoulement
    (conduite de gaz
    chaud)
- I linea *f* di scarico
- MA magasnyomású
    vezeték
    (hűtőgépnél)
- NL hete gasleiding *f*
- PO przewód *m*
    odpływowy
- PY разгрузочная
    линия *f*
- SU kuumakaasuputki
- SV utloppsledning *c*

1345 discharge opening
- DA udløbsåbning
- DE Auslaßüffnung *f*
- ES boca de descarga *f*
- F bouche *f*
    d'évacuation
- I bocca *f* di scarico
- MA kifolyónyílás
- NL uitlaatopening *f*
  uitlaatpoort *f*
- PO otwór *m* wylotowy
- PY выбросное
    отверстие *n*
- SU purkuaukko
- SV utloppsüppning *c*

1346 discharge pressure
- DA afgangstryk
- DE Austrittsdruck *m*
- ES presión de
    descarga *f*
  presión de
    impulsión *f*
- F pression *f* de
    décharge
- I pressione *f* di
    scarico
- MA nyomóoldali nyomás
- NL persdruk *m*
- PO ciśnienie *n* wylotowe
  ciśnienie *n* wypływu
- PY давление *n* на
    выбросе
- SU purkupaine
- SV utloppstryck *n*

1347 discharge rate
- DA tømningsgrad
- DE Abflußleistung *f*
- ES caudal de
    descarga *m*
  intensidad de
    descarga *f*
  velocidad de
    descarga *f*
- F débit *m*
    d'évacuation
- I portata *f* di scarico
- MA kiürítés mértéke
- NL afvoercapaciteit *f*
  afvoersnelheid *f*
- PO wielkość *f* odpływu
  wielkość *f* wypływu
- PY расход *m* на
    выбросе
  скорость *f*
    истечения
- SU purkunopeus
- SV utloppsflüde *n*

1348 discharge shaft
- DA udløbsbrønd
  udløbsskakt
- DE Abluftschacht *m*
- ES conducto de
    descarga *m*
  conducto de salida
    m
- F gaine *f* d'évacuation
- I canale *m* di scarico
- MA kifúvókürtő
- NL afvoerkoker *m*
  schacht *f*
- PO kanał *m* odpływowy
  szyb *m* odpływowy
- PY вытяжная шахта *f*
- SU purkukuilu
- SV utloppsschakt *n*

1349 discharge stop valve
- DA udløbsstopventil
- DE Druckabsperr-
    ventil *n*
- ES válvula de cierre de
    descarga *f*
- F robinet *m* de
    vidange
- I rubinetto *m* di
    scarico
- MA nyomás lezáró szelep
- NL afvoerleiding-
    afsluiter *m*
  persleiding-
    afsluiter *m*
- PO zawór *m* odcinający
    na odpływie
- PY клапан *m* для
    перекрытия
    выброса
- SU tyhjennysventtiili
- SV stoppventil *c* för
    avlopp *n*

1350 discharge stroke
- DA afløbsslag
- DE Auslaßhub *m*
- ES embolada de
    descarga *f*
- F course *f* de
    refoulement
- I fase *f* di scarico
- MA nyomóütem
    (hűtő-
    kompresszornál)
- NL persslag *m*
  uitlaatslag *m*
- PO suw *m* wypływu
- PY длина *f* хода
    поршня
  ход *m* нагнетания
- SU huuhteluisku
- SV utloppstakt *c*

DA	=	Danish
DE	=	German
ES	=	Spanish
F	=	French
I	=	Italian
MA	=	Magyar
		(Hungarian)
NL	=	Dutch
PO	=	Polish
PY	=	Russian
SU	=	Finnish
SV	=	Swedish

1351 discharge temperature
DA udløbstemperatur
DE Auslaßtemperatur *f*
ES temperatura de descarga *f*
F température *f* de refoulement
I temperatura *f* di mandata
temperatura *f* di scarico
MA nyomóoldali hőmérséklet
NL perstemperatuur *f* uittrede-temperatuur *f*
PO temperatura *f* odpływu
temperatura *f* powrotu
PY температура *f* выброса
SU purkulämpütila
SV utloppstemperatur *c*

1352 discharge valve
DA tømmeventil
DE Ablaßventil *n*
ES válvula de descarga *f*
válvula de impulsión *f*
F soupage *f* de décharge
I valvola *f* di scarico
MA ürítőszelep
NL afvoerafsluiter *m*
persklep *f*
PO zawór *m* odpływowy
zawór *m* upustowy
PY выпускной клапан *m*
SU poistoventtiili
tyhjennysventtiili
SV utloppsventil *c*

1353 discharging
DA losning
tømning
DE Entladung *f*
Entleerung *f*
ES de descarga
F décharge *f* (stockage thermique)
I evacuazione *f*
scarico *m*
MA kisülés
NL ontlading *f*
onttrekking *f*
PO rozładowanie *n*
PY разгрузка *f*
SU purkaa varastoa
SV avlastning *c*

1354 discharging capacity
DA afledningskapacitet
DE Entladungs-kapazität *f*
ES capacidad de descarga *f*
F capacité *f* de stockage (stockage thermique)
I capacità *f* di scarico
MA kisülési teljesítmény
NL ontladings-capaciteit *f*
onttrekkings-capaciteit *f*
PO zdolność *f* rozładowania
PY потери *f,pl* (энергии) при разгрузке (заполнении)
SU purkausteho
SV avledningsfürmåga *c*

1355 discoloration
DA affarvning
misfarvning
DE Entfärbung *f*
ES cambio de color *m*
decoloración *f*
F décoloration *f*
I decolorazione *f*
scoloramento *m*
MA elszíneződés
NL verkleuring *f*
PO odbarwienie *n*
PY обесцвечивание *n*
SU värivirhe
SV missfärgning *c*

1356 disc valve
DA skiveventil
DE Tellerventil *n*
ES válvula de disco *f*
F vanne *f* à disque
I rubinetto *m* a disco
valvola *f* a disco
MA membránszelep
tányérszelep
NL schuifafsluiter *m*
PO zawór *m* talerzowy
zawór *m* tarczowy
PY дисковый клапан *m*
SU luistiventtiili
SV slidventil *c*

1357 dished end-plate
DA endebund
DE schalenfürmige Endplatte *f*
ES fondo bombeado *m* placa terminal abombada *f*
F plaque *f* de fond
I piastra *f* terminale concava
MA domborított tartályfenék
NL bolle bodem *m*
PO dennica *f*
PY трубная доска *f* (в теплообменнике)
SU lämmünsiirtimen päätylevy
SV bottenplåt *c*

1358 disinfection
DA desinfektion
DE Desinfektion *f*
ES desinfección *f*
F désinfection *f*
I disinfezione *f*
MA fertőtlenítés
NL ontsmetting *f*
PO dezynfekcja *f*
PY дезинфекция *f*
SU desinfiointi
SV desinfektion *c*

1359 dispatching cold store
DA ekspeditions-
  kølelager
DE Auslieferungs-
  kühlhaus n
  Verteilungs-
  kühlhaus n
ES cámara frigorífica
  comercial f
F chambre f froide
  pour distribution
  locale
I magazzino m
  frigorifero di
  distribuzione
MA elosztó hűtött raktár
NL distributiekoelhuis n
PO rozbiór m
  zakumulowanego
  zimna
PY диспетчерская
  служба f
  холодохрани-
  лища
SU tarjoilukylmätila
SV mellanlager n

1360 dispersed flow
DA spredt strømning
DE aufgelüste
  Strümung f
ES flujo disperso m
F écoulement m
  dispersé
I flusso m disperso
MA diszperziós áramlás
NL dispersiestroming f
PO przepływ m
  rozproszony
  strumień m
  rozproszony
PY двухфазный
  (газожид-
  костный)
  поток m
SU kaksifaasivirtaus
SV spridningsflüde n

1361 display
DA display
  visning
DE Anzeige f
ES pantalla f
F affichage m
I display m
  visore m
MA display
  kijelző
NL signaaltableau n
  uitleesvenster n
PO monitor m
  obrazowy
  wyświetlacz m
PY дисплей m
SU näyttü
SV bildskärm c

1362 display vb
DA vise
DE anzeigen
ES exhibir
F afficher
I visualizzare
MA kijelez
NL afbeelden
  uitstallen
PO obrazować
  przedstawić
PY показывать
SU näyttää
SV visa

1363 disposable filter
DA engangsfilter
DE Wegwerffilter m
ES filtro de repuesto m
F filtre m à éléments
  non régénérables
I filtro m a perdere
MA eldobható szűrő
NL wegwerpfilter m
PO filtr m jednorazowy
  filtr m
  nieodnawialny
PY фильтр m без
  фильтрующего
  материала
SU kertakäyttüsuodatin
SV engångsfilter n

1364 dissolved gas
DA opløst gas
DE gelüstes Gas n
ES gas disuelto m
F gaz m dissous
I gas m disciolto
MA oldott gáz
NL opgelost gas n
PO gaz m rozpuszczony
PY растворенный
  газ m
SU liuennut kaasu
SV upplüst gas c

1365 distance
DA afstand
DE Abstand m
ES distancia f
  intervalo m
F distance f
  écartement m
  intervalle m
I distanza f
MA távolság
NL afstand m
PO dystans m
  odległość f
PY расстояние n
SU etäisyys
SV avstånd n

1366 distance reading thermometer
DA fjerntermometer
DE Fernthermometer n
ES teletermómetro m
  termómetro a
  distancia m
  termómetro de
  lectura a
  distancia m
F téléthermomètre m
  thermomètre m à
  lecture à distance
I termometro m a
  distanza
MA távleolvasós hőmérő
NL afstands-
  thermometer m
PO termometr m o
  zdalnym odczycie
PY дистанционный
  термометр m
SU etäislämpümittari
SV fjärrtermometer c

1367 distillation
DA destillation
DE Destillation f
ES destilación f
F distillation f
I distillazione m
MA desztilláció
NL distillatie f
PO destylacja f
PY дистилляция f
SU tislaus
SV destillation c

1368 distilled water
DA destilleret vand
DE destilliertes
Wasser *n*
ES agua destilada *f*
F eau *f* distillée
I acqua *f* distillata
MA desztillált víz
NL gedistilleerd water *n*
PO woda *f* destylowana
PY дистиллированная
вода *f*
SU tislattu vesi
SV destillerat vatten *n*

1369 distribution
DA fordeling
DE Verteilung *f*
ES distribución *f*
F distribution *f*
I distribuzione *f*
MA elosztás
NL verdeling *f*
verspreiding *f*
PO rozdział *m*
rozprowadzenie *n*
PY распределение *n*
SU jako
SV distribution *c*

1370 distribution box
DA fordelerdåse
DE Verteilerkasten *m*
ES caja de
distribución *f*
F boîte *f* de
distribution
I scatola *f* di
distribuzione
MA elosztódoboz
NL verdeeldoos *f*
verdeelkast *f*
PO komora *f*
rozdzielcza
skrzynka *f*
rozdzielcza
skrzynka *f*
rozgałęźna
PY распределительная
камера *f*
SU jakolaatikko
SV fürdelningslåda *c*

1371 distribution duct
DA fordelingskanal
DE Verteilerkanal *m*
Verteilleitung *f*
ES conducto de
distribución *m*
F gaine *f* de
distribution
I condotta *f* di
distribuzione
MA elosztóvezeték
NL verdeelkanaal *n*
PO kanał *m* rozdzielczy
przewód *m*
rozdzielczy
PY распределительный
канал *m*
SU jakokanava
SV fürdelningskanal *c*

1372 distribution from below
DA nedre fordeling
DE untere Verteilung *f*
ES distribución de abajo
arriba *f*
distribución desde
abajo *f*
F distribution *f*
inférieure
I distribuzione *f* dal
basso
MA alsóelosztás
NL onderverdeling *f*
PO rozdział *m* dolny
PY нижняя раздача *f*
распределение *n*
снизу
SU alajako
SV undre fürdelning *c*

1373 distribution network
DA fordelingsledningsnet
DE Verteilungsnetz *n*
ES red de distribución *f*
F réseau *m* de
distribution
I rete *f* di
distribuzione
MA elosztó hálózat
NL verdeelnet *n*
PO sieć *f* rozdzielcza
sieć *f*
rozprowadzająca
PY распределительная
сеть *f*
SU jakoverkko
SV distributionsnät *n*

1374 distribution piping
DA fordelingsledning
DE Versorgungs-
leitung *f*
ES tubería de
distribución *f*
F conduite *f* de
distribution
I circuito *m* di
distribuzione
rete *f* di
distribuzione
MA elosztó csővezeték
NL verdeelleiding *f*
PO przewody *m*
rozdzielcze
przewody *m*
rozprowadzające
PY распределительный
трубопровод *m*
SU jakoputkisto
SV distributions-
ledning *c*

1375 distribution system
DA fordelingssystem
DE Vertriebssystem *n*
ES sistema de
distribución *m*
F système *m* de
distribution
I sistema *m* di
distribuzione
MA elosztó rendszer
NL verdeelsysteem *n*
PO system *m*
rozprowadzający
PY распределительная
система *f*
SU jakelujärjestelmä
SV distributions-
system *n*

1376 distributor
DA fordeler
DE Verteiler *m*
ES distribuidor *m*
F distributeur *m*
I collettore *m*
distributore *m*
MA elosztó
NL verdeler *m*
PO rozdzielacz *m*
PY распределитель *m*
SU jakaja
SV fürdelare *c*

1377 district cooling
DA fjernkøling
DE Fernkühlung *f*
ES distribución urbana de frío *f*
   enfriamiento urbano *m*
F distribution *f* de froid à distance refroidie
   distribution *f* urbaine de froid
I raffreddamento *m* centralizzato (di quartiere)
MA távhűtés
NL afstandskoeling *f*
PO chłodzenie *n* zdalaczynne
PY районное охлаждение *n*
SU kaukojäädytys
SV fjärrkylning *c*

1378 district heating
DA fjernvarme
DE Fernheizung *f*
ES calefacción urbana por agua caliente *f*
F chauffage *m* à distance par eau chaude sous pression
I riscaldamento *m* centralizzato (di quartiere)
MA távfűtés
NL afstands-verwarming *f*
   stadsverwarming *f*
PO ciepłownictwo *n*
   ogrzewanie *n* zdalaczynne
PY районное теплоснабжение *n*
SU kaukolämmitys
SV fjärrvärme *c*

1379 district heating main
DA fjernvarmehoved-ledning
DE Fernwärmehaupt-leitung *f*
ES canalización principal de calefacción urbana *f*
   conducto (principal) de calefacción urbana *m*
F conduite *f* principale de chauffage à distance
I condotta *f* principale di riscaldamento centralizzato
MA távfűtő vezeték
NL hoofdleiding *f* afstands-verwarming
PO magistrala *f* cieplna
PY магистральный трубопровод *m* районного теплоснабжения
SU kaukolämmün pääjohto
SV fjärrvärmehuvud-ledning *c*

1380 district heating station
DA fjernvarmeværk
DE Fernheizzentrale *f*
ES central de calefacción urbana *f*
F centrale *f* de chauffage à distance centrale *f*
I centrale *f* termica di riscaldamento centralizzato
MA távfűtő
NL onderstation *n* afstands-verwarming
PO ciepłownia *f* węzeł *m* ciepłowniczy
PY районная котельная *f*
SU kaukolämpükeskus
SV fjärrvärmeverk *n*

1381 district heating supply
DA fjernvarmeforsyning
DE Fernwärme-versorgung *f*
ES suministro de calefacción urbana *m*
F distribution *f* de chaleur à distance
I alimentazione *f* di riscaldamento centralizzato
MA távhőellátás
NL afstandsverwarmings-aanvoer *m*
PO zasilanie *n* z sieci cieplnej
PY питание *n* от районного теплоснабжения
SU kaukolämmmün-jakelu
SV fjärrvärme-fürsürjning *c*

1382 district heating system
DA fjernvarmeanlæg
DE Fernheizsystem *n*
ES sistema de calefacción urbana *m*
F installation *f* de chauffage urbain
I impianto *m* di riscaldamento centralizzato (di quartiere)
MA távfűtő rendszer
NL afstandsverwarmings-systeem *n*
PO system *m* ciepłowniczy
   system *m* ogrzewania zdalaczynnego
PY система *f* районного теплоснабжения
SU kaukolämpüjärje-stelmä
SV fjärrvärmesystem *n*

DA = Danish
DE = German
ES = Spanish
F = French
I = Italian
MA = Magyar (Hungarian)
NL = Dutch
PO = Polish
PY = Russian
SU = Finnish
SV = Swedish

1383 diversion fitting
DA afgreningsfitting
DE Ablenkungsfitting *n*
Verteilungsfitting *n*
ES accesorio de
desviación *m*
F té *m* de distribution
I raccordo *m*
divergente
MA elágazó idom
NL injectie *f* teestuk
PO armatura *f* do
zmiany kierunku
(przepływu)
PY тройник *m* (на
трубопроводе)
SU venturilla varustettu
T-kappale
SV avledningsarmatur *c*

1384 diversity factor
DA samtidighedsfaktor
DE Ungleichfürmig-
keitsgrad *m*
ES factor de
utilización *m*
F facteur *m* de
simultanéité
I fattore *m* di
simultaneità
MA időkülünbüzeti
tényező
NL gelijktijdigheids-
factor *m*
PO współczynnik *m*
niejednoczesności
(obciążenia)
współczynnik *m*
nierównomier-
ności (rozbioru)
PY коэффициент *m*
одновременности
SU samanaikaisuus-
kerroin
SV matarvatten *n*
spridningsfaktor *c*

1385 diverter valve
DA tovejsventil
DE Umschaltventil *n*
ES válvula diversora *f*
F robinet *m* de
dérivation
vanne *f* de
dérivation
I valvola *f* di
deviazione
MA kétutú szelep
NL verdeelklep *f*
wisselklep *f*
PO zawór *m* rozdzielczy
PY распределительный
клапан *m*
SU jako(säätü)venttiili
SV tvåvägsventil *c*

1386 diverting circuit
DA afledningskredsløb
DE Nebenkreislauf *m*
ES circuito de
desviación *m*
F circuit *m* de
dérivation
I circuito *m*
divergente
MA elterelő áramkür
NL verdelend circuit *n*
PO obieg *m*
podmieszania
obieg *m* upustowy
PY отводной контур *m*
SU jakopiiri
SV

1387 dome
DA damphat
overdel
DE Dom *m*
Kuppel *f*
ES domo *m*
F dôme *m*
I cupola *f*
MA dóm
kupola
NL gewelf *n*
koepel *m*
PO kołpak *m* parowy
kopuła *f*
PY купол *m*
SU kupoli
kupu
SV dom *c*

1388 domestic *adj*
DA hus-
DE häuslich
ES doméstico
F domestique
I domestico
MA háztartási
NL huishoudelijk
PO domowy
PY внутренний
внутридомовой
SU asuinrakentamiseen
liittyvä
asumiseen liittyvä
SV hushålls-

1389 domestic heating
DA boligopvarmning
DE Wohnungsheizung *f*
ES calefacción
doméstica *f*
F chauffage *m*
domestique
I riscaldamento *m*
domestico
MA lakásfűtés
NL woningverwarming *f*
PO ogrzewanie *n*
mieszkaniowe
PY домовое
отопление *n*
SU asuntolämmitys
SV bostads-
uppvärmning *c*

1390 door fastening
DA dørbeslag
DE Türverschluß *m*
ES cerrojo *m*
dispositivo de
bloqueo de
puerta *m*
F verrouillage *m* de
porte
I serratura *f* del
portello
MA ajtó retesz
NL deurvergrendeling *f*
PO mocowanie *n* drzwi
PY закрепление *n*
двери
SU ovensulkija
SV dürrstängare *c*

1391 DOP test
DA DOP-prøve
DE DOP-Prüfung *f*
ES prueba de ftalato de
    dioctilo *f*
    prueba del DOP *f*
F essai *m* DOP
I prova *f* DOP
MA DOP teszt
    (dioctylphthalate)
NL DOP test *m*
PO test *m* DOP
PY диоктилфталиевый
    тест *m*
SU DOP-testi
    (mikrosuodatti-
    melle)
SV DOP-prov *n*

1392 dosage
DA dosering
DE Dosierung *f*
ES dosificación *f*
F dosage *m*
I dosaggio *m*
MA adagolás
NL dosering *f*
PO dawkowanie *n*
    dozowanie *n*
PY дозирование *n*
SU annos
SV dosering *c*

1393 dose *vb*
DA dosere
DE dosieren
ES dosificar
F doser
I dosare
MA adagol
NL doseren
PO dawkować
    dozować
PY дозировать
SU annostaa
SV dosera

1394 dosing water
DA fødevand
DE Dosierwasser *n*
ES agua de
    aportación *f*
    agua de relleno *f*
    agua de reposición *f*
    agua suplementaria
    f
F eau *f* d'appoint
I acqua *f* di raffredda-
    mento
MA adagolt víz
NL suppletiewater *n*
PO woda *f* dawkowana
PY подаваемая вода *f*
SU syüttüvesi
SV

**1395 double-acting
compressor**
DA dobbeltvirkende
    kompressor
DE doppeltwirkender
    Verdichter *m*
ES compresor de doble
    acción *m*
F compresseur *m* à
    double effet
I compressore *m* a
    doppio effetto
MA kettős műküdésű
    kompresszor
NL dubbelwerkende
    compressor *m*
PO sprężarka *f*
    dwusuwowa
    sprężarka *f*
    obustronnego
    działania
PY компрессор *m*
    двойного
    действия
SU kaksitoiminen
    kompressori
SV dubbelverkande
    kompressor *c*

1396 double break
DA dobbeltafbryder
DE Doppelunter-
    brechung *f*
ES freno doble *m*
F double coupure *f*
I doppia
    interruzione *f*
MA kettős megszakító
NL tweepolige
    uitschakeling *f*
PO przerwa *f* podwójna
PY двойной
    выключатель *m*
SU kaksoiskytkin
SV dubbel brytare *c*

**1397 double bundle
condenser**
DA dobbelt kondensator
DE Doppelrohrbündel-
    verflüssiger *m*
ES condensador de
    devanado doble *m*
F condensateur *m* à
    double faisceau
I condensatore *m* a
    doppio fascio
MA kettős csőnyalábú
    kondenzátor
NL condensor *m* met
    twee
    pijpenbundels
PO skraplacz *m* z
    dwoma wiązkami
    rur
PY конденсатор *m* с
    двумя рядами
    труб
SU kaksiosainen
    lauhdutin
SV kondensor *c* med två
    rörknippen *n*

**1398 double glazed
window**
DA vindue med to lag
    glas
DE Doppelscheiben-
    fenster *n*
ES ventana con doble
    acristalamiento *f*
    ventana de cristal
    doble *f*
F double fenêtre *f*
I finestra *f* a doppi
    vetri
MA kettős üvegezésű
    ablak
NL dubbelglas raam *n*
PO okno *n* podwójnie
    szklone
PY окно *n* с двойным
    остеклением
SU kaksilasinen ikkuna
SV tvåglasfünster *n*

1399 double inlet fan
DA ventilator med
dobbelt indløb
DE Ventilator *m* mit
doppeltem
Einlaß *m*
ES ventilador de doble
oído *m*
F ventilateur *m* à
double ouïe
I ventilatore *m* a
doppia aspirazione
MA kétoldalt szívó
ventilátor
NL tweezijdig
aanzuigende
ventilator *m*
PO wentylator *m*
dwustronnie ssący
PY вентилятор *m* с
двухсторонним
всасыванием
SU kaksipuolinen
puhallin
SV dubbelsidigt sugande
fläkt *c*

1400 double line break
DA dobbeltlednings-
afbryder
DE Zweileiter-
Unterbrechung *f*
ES freno de doble línea
m
F coupure *f* bipolaire
I doppia interruzione
f di linea
MA kétkürüs megszakító
NL tweepolige
onderbreking *f*
PO wyłącznik *m*
podwójny
PY двухлинейный
выключатель *m*
SU kaksoiskytkin
SV dubbelbrytning *c*

1401 double pipe condenser
DA to-rørskondensator
DE Doppelrohr-
verflüssiger *m*
ES condensador de
doble tubo *m*
condensador de tubo
en tubo *m*
F condenseur *m* à
tubes
concentriques
I condensatore *m* a
doppio tubo
MA kétcsüves
kondenzátor
NL condensor met
concentrische
pijpen *pl*
pijp in pijp
condensor *m*
PO skraplacz *m*
przeciwprądowy
(dwururowy)
skraplacz *m* typu
rura w rurze
PY конденсатор *m*
типа 'труба в
трубе'
SU kaksoisvaippalau-
hdutin
SV dubbelrürs-
kondensor *c*

1402 double pipe heat exchanger
DA dobbeltrørs-
varmeveksler
DE Doppelrohr-
Wärmetauscher *m*
ES intercambiador de
calor bitubular *m*
F échangeur *m* à tubes
concentriques
I scambiatore *m* di
calore a tubi
concentrici
MA kétcsüves hőcserélő
NL pijp in pijp
warmte-
wisselaar *m*
warmtewisselaar *m*
met concentrische
pijpen
PO wymiennik *m* ciepła
typu rura w rurze
PY теплообменник *m*
типа "труба в
трубе"
SU kaksoisputkilämmün-
siirrin
koaksiaalilämmün-
siirrin
SV dubbelrürsvärme-
växlare *c*

1403 double pole switch
DA to-pols-kontakt
DE Zweikontakt-
schalter *m*
ES conmutador bipolar
m
F contacteur *m*
bipolaire
I interruttore *m*
bipolare
MA kétsarkú kapcsoló
NL tweepolige
schakelaar *m*
PO wyłącznik *m*
dwubiegunowy
PY двухполюсный
выключатель *m*
SU kaksoiskytkin
SV tvåpolig
strümställare *c*

1404 double-seated valve
DA dobbeltsædet ventil
DE Doppelsitzventil *m*
ES válvula de asiento
doble *f*
F robinet *m* à double
siège
I valvola *f* a doppia
sede
MA kétülésű szelep
NL dubbelzitting-
afsluiter *m*
PO zawór *m*
dwugniazdowy
PY клапан *m* с двумя
седлами
SU kaksi-istukkainen
venttiili
SV dubbelsätesventil *c*

DA	=	Danish
DE	=	German
ES	=	Spanish
F	=	French
I	=	Italian
MA	=	Magyar (Hungarian)
NL	=	Dutch
PO	=	Polish
PY	=	Russian
SU	=	Finnish
SV	=	Swedish

1405 double suction compressor

1405 double suction compressor
- DA dobbeltsugende kompressor
- DE zweifach ansaugender Verdichter *m*
- ES compresor de doble aspiración *m*
- F compresseur *m* à double aspiration
- I compressore *m* a doppia aspirazione
- MA kettős szívású kompresszor
- NL compressor *m* met twee zuigaansluitingen
- PO sprężarka *f* dwustrumieniowa
- PY компрессор *m* двойного всасывания
- SU kaksi-imuinen kompressori
- SV dubbelsug-kompressor *c*

1406 double suction fan
- DA ventilator med dobbelt indsugning
- DE zweiseitig saugender Ventilator *m*
- ES ventilador de doble aspiración *m*
- F ventilateur *m* à double aspiration
- I ventilatore *m* a doppia aspirazione
- MA kétoldalt szívó ventilátor
- NL dubbelaanzuigende ventilator *m* tweezijdig-aanzuigende ventilator *m*
- PO wentylator *m* dwustronnie ssący
- PY вентилятор *m* с двухсторонним всасыванием
- SU kaksipuolinen puhallin
- SV dubbelsidigt sugande fläkt *c*

1407 double suction riser
- DA dobbeltansugende stigrør
- DE doppelte Saugleitung *f*
- ES presurizador de doble aspiración *m*
- F conduite *f* ascendante double
- I montante *m* a doppia aspirazione
- MA kétoldalt szívó felszálló
- NL dubbele aanzuigstijg-leiding *f*
- PO przewód *m* pionowy dwustrumieniowy
- PY дублированный масляный трубопровод *m* (в компрессоре)
- SU kaksipuolinen imuputki
- SV dubbelsidigt sugrür *n*

1408 double throw switch
- DA dobbelt kontakt
- DE Zweifachschalter *m*
- ES conmutador de doble vuelta *m*
- F inverseur *m* électrique
- I interruttore *m* a doppia posizione
- MA váltókapcsoló
- NL wisselschakelaar *m*
- PO przełącznik *m* wyłącznik *m* dwupołożeniowy
- PY переключатель *m*
- SU vaihtokytkin
- SV tvåvägs strümställare *c*

1409 downdraught
- DA kuldenedfald nedslag (røg)
- DE Zugerscheinung *f*
- ES corriente de aire descendiente *f* tiro de sentido descendente *m*
- F courant *m* d'air froid descendant contre-tirage
- I corrente *f* discendente di aria
- MA lefelé irányuló légáramlás visszaáramlás
- NL koudeval *m*
- PO ciąg *m* odwrotny (w kominie)
- PY обратная тяга *f*
- SU takaveto
- SV bakdrag *n*

1410 down-feed heating system
- DA varmeanlæg med øvre fordeling
- DE Heizungssystem *n* mit unterer Verteilung *f*
- ES sistema de calefacción descendente *m*
- F distribution *f* en parapluie
- I sistema *m* di riscaldamento a distribuzione dal basso
- MA felsőelosztású fűtési rendszer
- NL bovenverdelings-systeem *n*
- PO system *m* ogrzewania z rozdziałem dolnym
- PY система *f* отопления с нижней разводкой
- SU yläjakoinen lämmitysjärje-stelmä
- SV djupmatningsvärme-system *n*

1411 downstream
DA nedstrøm
DE stromabwärts
ES aguas abajo *f,pl*
  corriente abajo *f*
F aval *m*
I a valle
  secondo corrente
MA folyásirányban
  lefelé áramlás
NL stroomafwaarts
PO przepływ *m*
  współprądowy
  z prądem
PY ниже по потоку
SU alavirtaan
SV nedstrüm *c*

1412 downward combustion
DA underforbrænding
DE unterer Abbrand *m*
ES combustión
  descendente *f*
  combustión en
  sentido
  descendente *f*
F combustion *f*
  inversée (de haut
  en bas)
I combustione *f*
  dall'alto in basso
  combustione *f*
  discendente
MA lefelé égés
NL onderafbrand
PO spalanie *n* ze strugą
  skierowaną w dół
PY нижнее сжигание *n*
SU käännetty poltto
SV omvänd
  fürbränning *c*

1413 draft (USA) (*see* draught)

1414 drag
DA modstand
DE Widerstand *m*
ES arrastre *m*
  resistencia *f*
F résistance *f* à
  l'écoulement
I attrito *m*
  aerodinamico
  resistenza *f* allo
  scorrimento
  trascinamento *m*
MA küzegellenállás
NL sleeprem *f*
PO opór *m*
  (aerodynamiczny)
PY сопротивление *n*
SU vastus
SV motstånd *n*

1415 drag coefficient
DA modstandskoefficient
DE Widerstands-
  beiwert *m*
ES coeficiente de
  demora *m*
F coefficient *m* de
  traînée
I coefficient *m* di
  attrito
  aerodinamico
  coefficiente *m* di
  trascinamento
MA ellenállási tényező
NL vormweerstands-
  cöfficint *m*
PO współczynnik *m*
  oporu
PY коэффициент *m*
  сопротивления
SU vastuskerroin
SV motstånds-
  koefficient *c*

1416 drain
DA afløb
  dræn
DE Ablaß *m*
ES drenaje *m*
  vaciado *m*
F décharge *f*
  écoulement *m*
  vidange *f*
I spurgo *m*
MA csatorna
  lefolyócső
NL aftap *m*
  afvoer *m*
PO spust *m*
  ściek *m*
PY спуск *m*
SU tyhjennys
SV avlopp *n*

1417 drain *vb*
DA dræne
  tømme
DE abfließen
  ablassen
  entleeren
ES drenar
  vaciar
F drainer
I spurgare
MA csatornáz
NL aftappen
PO drenować
  odprowadzać ciecz
  odwadniać
PY дренировать
SU tyhjentää
SV tümma

1418 drainage
DA dræning
  kloakering
DE Entwässerung *f*
ES sistema de
  drenaje *m*
F drainage *m*
I drenaggio *m*
MA csatornázás
NL afwatering *f*
PO drenaż *m*
  odprowadzenie *n*
  cieczy
  odwodnienie *n*
PY дренаж *m*
SU tyhjennys
  viemäruinti
SV dränering *c*

1419 drain back
DA tilbageløb (i
  solfanger)
DE Rückfluß *m*
ES contrapendiente de
  un desagüe *f*
F vidange *f* de retour
I sistema *m* di
  drenaggio del
  collettore solare
  (drain back)
MA visszafolyás
NL terugloop *m*
  tcrugvloeien *n*
PO odwodnienie *n*
PY дренажная
  засыпка *f*
SU itsestään tyhjentyvä
  (esim.
  auringonkeräysjärje-
  stelmästä)
SV avrinning *c* bakåt

1420 drain cock
DA tømmehane
DE Ablaßhahn *m*
ES grifo de purga *m*
    grifo de vaciado *m*
F robinet *m* de purge
    robinet *m* de
      vidange
I rubinetto *m* di
    spurgo
MA leeresztőcsap
NL aftapkraan *f*
    spuikraan *f*
PO kurek *m* spustowy
PY спускной кран *m*
SU tyhjennyshana
SV avtappningskran *c*

1421 drain down
DA væsketilbageløb (i
    solfanger)
DE Ablauf *m*
    Entwässerung *f*
ES bajante de desagüe
    m
F
I sistema *m* di
    drenaggio del
    collettore solare
    (drain down)
MA lefolyás
NL aftappen *n*
    laten leeglopen *n*
PO odwadnianie *n*
    spuszczanie *n*
PY нижний дренаж *m*
    (в гелиосистемах
    - аккумулятор
    ниже коллектора)
SU tyhjentyvä (esim.
    auringonkeräysjärje-
    stelmästä)
SV avrinning *c* nedåt

1422 drain pan
DA afløbsbeholder
DE Ablaufblech *n*
    Ablaufpfanne *f*
ES bandeja de desagüe *f*
F bac *m* de
    récupération
I vaschetta *f* di
    raccolta
MA csepptálca
    kondenztálca
NL lekbak *m*
PO miska *f* spustowa
PY дренажный
    лоток *m*
SU tippuvesiastia
SV avtappningskärl *n*

1423 drain pipe
DA afløbsrør
DE Abflußrohr *n*
ES tubo de vaciado *m*
F drain *m*
    tuyauterie *f* de
      vidange
I tubo *m* di scarico
MA lefolyócső
NL aftapleiding *f*
    spuileiding *f*
PO rura *f* spustowa
    rura *f* ściekowa
PY дренажная труба *f*
SU tyhjennysputki
SV dräneringsrür *n*

1424 drain valve
DA aftapningsventil
DE Entleerungsventil *n*
ES válvula de vaciado *f*
F robinet *m* de
    puissage
I valvola *f* di spurgo
MA leeresztőszelep
NL aftapafsluiter *m*
    spuiafsluiter *m*
PO zawór *m* spustowy
PY дренажный
    клапан *m*
SU tyhjennysventtiili
SV avtappningsventil *c*

1425 draught
DA træk
DE Zug *m*
ES tiro *m*
F courant *m* d'air
    tirage *m*
I tiraggio *m*
MA huzat
NL tocht *m*
    trek *m*
PO ciąg *m*
PY тяга *f*
SU veto
SV drag *n*

1426 draught gauge
DA trækmåler
DE Zugmesser *n*
ES deprimómetro *m*
    medidor de tiro *m*
F déprimomètre *m*
I deprimometro *m*
    manometro *m*
    misuratore *m* del
      tiraggio
MA huzatmérő
NL trekmeter *m*
PO ciągomierz *m*
    przyrząd *m* do
      pomiaru wielkości
      ciągu
PY тягомер *m*
SU vetomittari
SV dragmätare *c*

1427 draught hood
DA emhætte
    røgfang
DE Abzugshaube *f*
ES sombrerete de
    chimenea *m*
F
I cappa *f* di tiraggio
MA áramlásbiztosító
    deflektor
    huzatmegszakító
NL trekonderbreker *m*
PO okap *m*
    okap *m* wyciągowy
PY шибер *m* на
    газоходе котла
SU huuva
SV dragkåpa *c*

1428 draught indicator
DA trækindikator
DE Zuganzeiger *m*
ES indicador de tiro *m*
F indicateur *m* de
    tirage
I indicatore *m* di
    tiraggio
MA huzatmérő
NL trekmeter *m*
PO wskaźnik *m* ciągu
PY тягомер *m*
SU vedon osoittaja
SV dragindikator *c*

DA	=	Danish
DE	=	German
ES	=	Spanish
F	=	French
I	=	Italian
MA	=	Magyar
		(Hungarian)
NL	=	Dutch
PO	=	Polish
PY	=	Russian
SU	=	Finnish
SV	=	Swedish

1429 **draught inducer**
DA trækforøger
DE Zugerzeuger *m*
ES inductor de tiro *m*
F accélérateur *m* de
      tirage
   ventilateur *m* de
      tirage
I ventilatore *m* di
      tiraggio ad
      induzione
MA huzatfokozó
NL rookgasventilator *m*
PO pobudzacz *m* ciągu
PY побудитель *m* тяги
SU vedon lisääjä
SV dragfürstärknings-
      anordning *c*

1430 **draught limiting
     device**
DA trækbegrænser
DE Zugbegrenzer *m*
ES limitador de tiro *m*
F accélérateur *m* de
      tirage
   limiteur *m* de tirage
I limitatore *m* del
      tiraggio
MA huzatszabályozó
      szerkezet
NL trekonderbreker *m*
   trekregelaar *m*
PO ogranicznik *m* ciągu
   urządzenie *n*
      ograniczające
      wielkość ciągu
PY ограничитель *m*
      тяги
SU vedon rajoittaja
SV dragregulator *c*

1431 **draught
     requirement**
DA trækbehov
DE Zugbedarf *m*
ES exigencia de tiro *f*
F besoins *m* de tirage
I tiraggio *m* richiesto
MA huzatigény
NL vereiste trek *m*
PO zapotrzebowanie *n*
      na ciąg
PY требуемая тяга *f*
SU vetovaatimus
SV dragbehov *n*

1432 **draught stabilizer**
DA trækstabilisator
DE Zugregler *m*
ES estabilizador de
      tiro *m*
F régulateur *m* de
      tirage
I regolatore *m* del
      tiraggio
MA huzatstabilizátor
NL trekregelaar *m*
PO stabilizator *m* ciągu
PY регулятор *m* тяги
SU vedon tasaaja
SV dragregulator *c*

1433 **drawing**
DA tegning
DE Skizze *f*
   Zeichnung *f*
ES croquis *m*
   dibujo *m*
F schéma *m*
I disegno *m*
MA rajz
NL tekening *f*
PO rysunek *m*
PY набросок *m*
   рисунок *m*
SU piirustus
SV ritning *c*

1434 **draw-off point**
DA aftapningspunkt
DE Zapfstelle *f*
ES punto de goma *m*
F point *m* de puisage
I punto *m* di prelievo
MA ürítési pont
   ürítő
NL aftapplaats *f*
   aftappunt *n*
PO punkt *m* czerpalny
PY точка *f* отбора
SU tyhjennyspiste
SV tappställe *n*

1435 **draw-off valve**
DA tapventil
DE Zapfventil *n*
ES válvula de toma *f*
F robinet *m* de
      puisage
I rubinetto *m* a collo
MA ürítőszelep
NL aftapkraan *f*
   spuikraan *f*
PO zawór *m* czerpalny
PY питательный кран
      m (клапан)
SU tyhjennysventtiili
SV tappventil *c*

1436 **drier**
DA affugter
DE Entfeuchter *m*
   Trockner *m*
ES deshidratador *m*
   secador *m*
F sécheur *m*
I essiccatore *m*
MA szárító
NL droger *m*
   ontvochtiger *m*
PO osuszacz *m*
   suszarka *f*
PY осушитель *m*
SU kuivaaja
SV avfuktare *c*

1437 **drier coil**
DA tørrespiral
DE Trocknerschlange *f*
ES batería de secado *f*
F batterie *f* froide
      sèche
I serpentino *m*
      dell'essiccatore
MA szárító csőkígyó
NL naverwarmings-
      batterij *f*
PO suszarka *f*
   wężownica *f*
      osuszacza
PY осушительный
      змеевик *m*
SU kuivauspatteri
SV torkslinga *c*

1438 **drift**
DA strømning
DE Strümung *f*
   Strom *m*
ES sedimentos *m*
F eau *f* entrainée
I deriva *f*
   spinta *f*
MA áramlás
   ürvénylés
NL meegevoerde
      waterdruppels *pl*
   verloop *n*
PO powolna zmiana *f*
   przesuwanie *n*
   unoszenie *n*
PY изменение *n*
      характеристик
SU kulkeuma
   liukuma
SV drift *c*

1439 drift eliminator (cooling tower)

1439 drift eliminator (cooling tower)
DA udkoblingselement
DE Tropfen-
abscheider *m*
ES eliminador de
sedimentos (torre
de refrigeración) *m*
F séparateur *m* de
gouttelettes
(tour de
refroidissement)
I separatore *m* di
gocce
MA áramlás
egyenirányító
(hűtőtoronynál)
NL druppelvanger *m*
PO eliminator *m*
unoszenia (w
wieży chłodniczej)
PY отбойный слой *m*
(в градирнях)
SU pisaran erotin
SV droppeliminator *c*

1440 drinking water
DA drikkevand
DE Trinkwasser *n*
ES agua potable *f*
F eau *f* potable
I acqua *f* potabile
MA ivóvíz
NL drinkwater *n*
PO woda *f* pitna
PY питьевая вода *f*
SU juomavesi
SV dricksvatten *n*

1441 drinking water cooler
DA drikkevandskøler
DE Trinkwasser-
kühler *m*
ES enfriador de agua
potable *m*
F refroidisseur *m*
d'eau potable
I refrigeratore *m* per
acqua potabile
MA ivóvíz hűtő
NL drinkwaterkoeler *m*
PO chłodziarka *f* wody
pitnej
PY охладитель *m*
питьевой воды
SU juomaveden
jäähdytin
SV dricksvattenkylare *c*

1442 drip
DA dryp
DE Tropfen *m*
ES goteo *m*
F exsudat *m*
I goccia *f*
MA csepegés
NL druipend water *n*
lekwaterafvoer *m*
PO kapanie *n*
ściekanie *n*
PY капля *f*
SU kondenssivesi
lauhteen
palautusjohto
SV dropp *n*

1443 drip tray
DA drypbakke
DE Tropfschale *f*
ES bandeja de goteo *f*
F égouttoir *m* (bac de
condensation)
I gocciolatoio *m*
MA csepptálca
NL lekbak *m*
PO taca *f* ociekowa
PY поддон *m* для
капель
SU tippuvesiastia
SV droppskål *c*

1444 drive
DA transmission
DE Antrieb *m*
ES accionamiento *m*
transmisión *f*
F commande *f*
entraînement *m*
transmission *f*
I comando *m*
MA hajtás
NL aandrijving *f*
PO napęd *m*
PY привод *m*
SU käyttü
SV drift *c*

1445 drive *vb*
DA overføre
DE antreiben
ES conducir
F actionner
entrainer
I condurre
guidare
MA hajt
NL aandrijven
besturen
PO napędzać
PY приводить в
движение
SU käyttää
SV driva

1446 drive shaft
DA drivaksel
DE Antriebswelle *f*
ES árbol *m*
eje motor *m*
transmisión *f*
F arbre *m*
I albero *m* motore
MA hajtótengely
NL aandrijfas *f*
PO wał *m* napędowy
PY карданный вал *m*
SU käyttüakseli
SV drivaxel *c*

1447 droop
DA spændingsfald
DE Senke *f*
ES caída *f*
inclinación *f*
F chute *f* de tension
I posizione *f*
abbassata
MA esés
lejtés
NL belastings-
afhankelijke
afwijking *f*
PO opadnięcie *n*
zwis *m*
PY понижение *n*
SU säätüsuureen
siirtymä
suhteellisessa
säädüssä
SV sänka *c*

1448 drop
DA drop
dråbe
fald
DE Tropfen *m*
ES caída *f*
gota *f*
F chute *f*
goutte *f*
I caduta *f*
goccia *f*
MA csepp
NL druppel *m*
PO kropla *f*
spadek *m*
zmniejszenie *n*
PY капля *f*
расстояние *n*
(сверху вниз)
SU pisara
tippuminen (esim.
ilmasuihkusta)
SV droppe *c*

1449 droplet
DA lille dråbe
DE Trüpfchen *n*
ES gota (fina) *f*
   gotita *f*
F gouttelette *f*
I gocciolina *f*
MA cseppecske
NL druppeltje *n*
PO kropelka *f*
PY капелька *f*
SU pisara
SV droppe *c*

1450 droplet condensation
DA dråbekondensering
DE Trüpfchen-kondensation *f*
ES condensación en gotas *f*
F condensation *f* de gouttelettes
I condensazione *f* a gocce
MA cseppkondenzáció
NL druppel-condensatie *f*
PO kondensacja *f* kropelkowa
PY капельная конденсация *f*
SU pisaralauhtuminen
SV dropp-kondensation *c*

1451 droplet separator
DA dråbeudskiller
DE Tropfen-abscheider *m*
ES separador de gotas *m*
F séparateur *m* de goutellettes
I separatore *m* di gocce
MA cseppelválasztó
NL druppelvanger *m*
PO odkraplacz *m*
PY каплеотделитель *m* сепаратор *m*
SU pisaran erotin
SV droppavskiljare *c*

1452 droplet size
DA dråbestørrelse
DE Tropfengrüße *f*
ES tamaño de la gota *m*
F dimension *f* de gouttelettes
I dimensione *f* della goccia
MA cseppméret
NL druppelgrootte *f*
PO wielkość *f* kropelki wymiar *m* kropelki
PY размер *m* капли
SU pisarakoko
SV droppstorlek *c*

1453 drop off *vb*
DA dryppe
DE abtropfen
ES gotear
F égoutter
I gocciolare
MA lehull
NL afdruipen
PO wykraplać
PY конденсировать
SU tiputtaa
SV droppa

1454 dropped ceiling
DA nedhængt loft
DE abgehängte Decke *f*
ES hielo raso *m*
F faux plafond *m*
I soffitto *m* ribassato
MA függesztett mennyezet
NL verlaagd plafond *n*
PO strop *m* perforowany
PY подвесной потолок *m*
SU alaslaskettu katto
SV sänkt tak *n*

1455 drum
DA tromle valse
DE Trommel *f*
ES depósito cilíndrico *m* tambor *m*
F ballon *m* réservoir *m*
I serbatoio *m* tamburo *m*
MA dob
NL reservoir *m* vat *n*
PO bęben *m* walczak *m*
PY барабан *m*
SU rumpu
SV trumma *c*

1456 drum cooler
DA tromlekøler
DE Trommelkühler *m*
ES enfriador de tambor *m*
F refroidisseur *m* à tambour
I refrigeratore *m* a tamburo
MA dobhűtő
NL trommelkoeler *m*
PO chłodnica *f* bębnowa
PY цилиндрический охладитель *m*
SU rumpujäähdytin
SV trumkylare *c*

1457 dry *adj*
DA tør
DE trocken
ES seco
F sec
I asciutto secco
MA száraz
NL droog
PO suchy
PY сухой
SU kuiva
SV torr

1458 dry *vb*
DA tørre
DE trocknen
ES secar
F sécher
I essiccare
MA szárít
NL drogen
PO suszyć
PY сушить
SU kuivata
SV torka

1459 dry air
DA tør luft
DE trockene Luft *f*
ES aire seco *m*
F air *m* sec
I aria *f* secca
MA száraz levegő
NL droge lucht *f*
PO powietrze *n* suche
PY сухой воздух *m*
SU kuiva ilma
SV torr luft *c*

1460 dry air cooler
DA tør luftkøler
DE Trockenluft-
   kühler *m*
ES enfriador de aire
   seco *m*
F refroidisseur *m* d'air
   du type sec
I refrigeratore *m* di
   aria secca
MA szárazlevegős hűtő
NL droge luchtkoeler *m*
PO chłodnica *f*
   powietrza
   pracująca na
   sucho
PY сухой
   воздухоох-
   ладитель *m*
   (работающий в
   режиме сухого
   охлаждения)
SU kuiva jäähdytin
SV torrluftskylare *c*

**1461 dry bulb
   temperature**
DA tør temperatur
DE Trockenkugel-
   temperatur *f*
ES temperatura del
   bulbo seco *f*
   temperatura del
   termómetro seco *f*
   temperatura seca *f*
F température *f* au
   thermomètre
   température *f* sèche
I temperatura *f* a
   bulbo asciutto
   (secco)
MA szárazhőmérséklet
NL drogebol
   temperatuur *f*
PO temperatura *f*
   termometru
   suchego
PY температура *f* по
   сухому
   термометру
SU ilman
   kuivalämpütila
SV torr temperatur *c*

**1462 dry bulb
   thermometer**
DA tørtermometer
DE trockenes
   Thermometer *n*
ES termómetro de bulbo
   seco *m*
F thermomètre *m* à
   bulbe sec
I termometro *m* a
   bulbo secco
MA száraz hőmérő
NL drogebol
   thermometer *m*
PO termometr *m* suchy
PY сухой
   термометр *m*
SU kuivalämpümittari
SV torr termometer *c*

1463 dry compression
DA tør kompression
DE trockene (ülfreie)
   Verdichtung *f*
ES compresión seca *f*
F compression *f* en
   régime de
   surchauffe
I compressione *f*
   secca
MA száraz kompresszió
NL compressie *m* van
   droge damp
PO sprężanie *n* suche
PY сухое сжатие *n*
SU kuiva puristus
SV torr kompression *c*

**1464 dry expansion
   (direct)**
DA tør ekspansion
DE trockene
   Verdampfung *f*
ES expansión seca
   (directa) *f*
F détente *f* directe
I espansione *f* secca
   (diretta)
MA száraz expanzió
NL directe expansie *f*
PO rozprężanie *n* suche
   (bezpośrednie)
PY сухое (прямое)
   испарение *n*
SU kuiva suorapaisunta
SV torr expansion *c*

**1465 dry expansion
   evaporator**
DA tør
   ekspansions-
   fordamper
DE Verdampfer *m*
   mit Direkt-
   verdampfung *f*
ES evaporador de
   expansión seca *m*
F évaporateur *m* à
   détente directe
I evaporatore *m* ad
   espansione secca
MA száraz expanziós
   elpárologtató
NL droge verdamper *m*
PO parownik *m* suchy
   rozprężny
PY испаритель *m*
   полного
   испарения
SU kuiva hüyrystin
SV torrexpansions-
   evaporator *c*

1466 dry filter
DA tørfilter
DE Trockenfilter *m*
ES filtro seco *m*
F filtre *m* sec
I filtro *m* di aria a
   secco
MA száraz hűtő
NL droge filter *m*
PO filtr *m* suchy
PY сухой фильтр *m*
SU kuiva suodatin
SV torrt filter *n*

1467 dry ice (solid CO$_2$)
DA tøris
DE Trockeneis *n*
ES anhídrido carbónico,
   sólido *m*
   hielo seco *m*
F neige *m* carbonique
I ghiaccio *m* secco
MA szárazjég
NL droog ijs *n* (vast
   koolzuur)
PO lód *m* suchy
PY сухой лед *m*
   (твердый CO$_2$)
SU hiilihappojää
SV kolsyre is *c*

1468 drying
DA tørring
DE Trocknung f
ES deshidratación f
secado m
F séchage m
I essiccazione f
MA szárítás
NL droging f
PO suszenie n
PY сушка f
SU kuivaus
SV torkning c

1469 drying cupboard
DA tørreskab
DE Trockenschrank m
ES armario secador m
F étuve f de séchage
I armadio m di essiccazione
MA szárítószekrény
NL droogkast f
PO szafa f suszarnicza
PY осушительный аппарат m
SU kuivauskaappi
SV torkskåp n

1470 drying out
DA udtørring
DE Austrocknung f
ES desecación f
F séchage m d'un bâtiment
I essiccamento m
MA kiszárítás
NL uitdroging f
PO wysuszenie n
PY осушение n
SU kuivuminen
SV uttorkning c

1471 drying plant
DA tørreanlæg
DE Trocknungsanlage f
ES instalación de deshidratación f
instalación de secado f
F installation f de séchage
I impianto m di essiccamento
MA szárítóberendezés
NL drooginstallatie f
PO suszarnia f
PY сушильня f
SU kuivaamo
SV torkanläggning c

1472 dry layer filter
DA tørt rullefilter
DE Filter m mit trockener Filterschicht f
ES filtro de capa seca m
F filtre m à couche sèche
I filtro m a strato secco
MA száraz réteges hűtő
NL droge filter m
PO filtr m suchy warstwowy
PY сухой (тканевый) фильтр m
SU kuiva suodatin
SV torrskiktsfilter n

1473 dryness
DA tørhed
DE Trockenheit f
ES sequedad f
F sécheresse f
siccité f
I secchezza f
MA szárazság
NL droogte f
PO stan m suchy suchość f
PY сухость f
SU kuivuus
SV torrhet c

1474 dryness fraction (of steam)
DA relativ dampfugtighed
DE Trockenheitsgrad m
ES grado de sequedad m
porcentaje de vapor seco m
relación de sequedad f
título del vapor m
F taux m de siccité
I titolo m del vapore
MA szárazgőztartalom
NL droogheidsgraad m (stoom)
PO udział m suchości (pary)
PY коэффициент m сухости (перегрева)
SU hüyrypitoisuus
SV relativ ångfuktighet c

1475 dry piston compressor
DA tørstempelkompressor
DE ülfreier Verdichter m
ES compresor de pistón seco m
F compresseur m à piston sec
I compressore m alternativo a secco
MA szárazdugattyús kompresszor
NL olievrije compressor m
PO sprężarka f suchatłokowa
PY поршневой безсмазочный компрессор m
SU üljytün kompressori
SV torrkolv-kompressor c

1476 dry resultant temperature
DA tør resulterende temperatur
DE trockene resultierende Temperatur f
ES temperatura seca resultante f
F température f résultante sèche
I temperatura f secca risultante
MA száraz eredő hőmérséklet
NL droge resulterende temperatuur f
PO temperatura f wynikowa sucha
PY сухая результирующая температура f
SU resultoiva lämpütila
SV torr temperatur c

DA	=	Danish
DE	=	German
ES	=	Spanish
F	=	French
I	=	Italian
MA	=	Magyar (Hungarian)
NL	=	Dutch
PO	=	Polish
PY	=	Russian
SU	=	Finnish
SV	=	Swedish

1477 dry return
DA tør returledning
DE trockener
  Rücklauf *m*
ES tubería del retorno
  sobre el nivel del
  agua (caliente) *f*
F retour *m* sec
I ritorno *m* a secco
MA száraz
  kondenzvezeték
NL droge
  condensaat-
  leiding *f*
PO przewód *m*
  powrotny
  (kondensacyjny)
  niezalany
PY сухой возврат *m*
SU kuiva paluujohto
SV torr returledning *c*

1478 dry-return heating system
DA varmesystem med
  tør retur
DE Heizungssystem *n*
  mit trockenem
  Rücklauf *m*
ES sistema de
  calefacción de
  retorno seco *m*
F système *m* de
  chauffage à vapeur
  avec retour
I sistema *m* di
  riscaldamento a
  ritorno secco
MA száraz
  kondenzvezetékes
  fűtési rendszer
NL verwarmingssysteem
  n met droge
  condensaatleiding
PO ogrzewanie *n* z
  niezalanym
  przewodem
  powrotnym
  (kondensacyjnym)
PY (замкнутая)
  система *f* парового
  отопления с
  сухим
  конденсато-
  проводом
  (замкнчмая)
SU hüyrylämmitys
SV torreturvärme-
  system *n*

1479 dry saturated steam
DA tør mættet damp
DE trockengesättigter
  Dampf *m*
ES vapor saturado
  seco *m*
F vapeur *f* saturée
  sèche
I vapore *m* saturo
  secco
MA száraz telített gőz
NL droge verzadigde
  stoom *m*
PO para *f* sucha
  nasycona
PY сухой насыщенный
  пар *m*
SU kuiva kylläinen
  hüyry
SV (torr) mättad ånga *c*

1480 dry steam
DA tør damp
DE trockener Dampf *m*
ES vapor seco *m*
F vapeur *f* sèche
I vapore *m* secco
MA száraz gőz
NL droge stoom *m*
PO para *f* sucha
PY сухой пар *m*
SU kuiva hüyry
SV torr ånga *c*

1481 dry-type air cooler
DA tør luftkøler
DE trockener
  Luftkühler *m*
ES enfriador de aire
  tipo seco *m*
F refroidisseur *m* d'air
  du type sec
I refrigeratore *m* di
  aria a secco
MA száraz típusú
  léghűtés
NL droge luchtkoeler *m*
  (niet bevochtigd)
PO chłodnica *f*
  powietrza
  pracująca na
  sucho
PY сухой
  (поверхностный)
  воздухоохла-
  дитель *m*
SU kuiva ilmanjäähdytin
SV torrluftskylare *c*

1482 dry-type evaporator
DA tør fordamper
DE trockener
  Verdampfer *m*
ES evaporador tipo seco
  m
F évaporateur *m* sec
I evaporatore *m* a
  secco
MA száraz típusú
  elpárologtató
NL droge verdamper *m*
PO parownik *m* typu
  suchego
PY сухой
  испаритель *m*
SU kuiva hüyrystin
SV torrtypsevaporator *c*

1483 dual compression
DA dobbeltkompression
DE doppeltwirkende
  Verdichtung *f*
ES compresión doble *f*
F compression *f* à
  double aspiration
I compressione *f*
  bistadio
MA kettős kompresszió
NL compressie *f* van
  twee zuigdrukken
PO sprężanie *n* z
  doładowaniem
  czynnika
PY двойное сжатие *n*
SU yhdistelmäpuristus
SV dubbel-
  kompression *c*

1484 dual duct air conditioning system
DA tokanal luftbehandlingssystem
DE Zweikanalklimasystem *n*
ES sistema de climatización de doble conducto *m*
F système *m* de climatisation à double conduit
I impianto *m* di condizionamento a doppio canale
MA kétcsatornás légkondicionáló rendszer
NL tweekanalen luchtbehandelingssysteem *n*
PO system *m* klimatyzacji dwuprzewodowej
PY двухканальная система *f* кондиционирования воздуха
SU kaksikanavailmastointijärjestelmä
SV tvåkanals luftkonditioneringssystem *n*

1485 dual duct system
DA to-kanal-system
DE Zweikanalsystem *n*
ES sistema (de acondicionamiento) de doble conducto *m*
F système *m* à deux conduits
I impianto *m* a doppio canale
MA kétvezetékes rendszer
NL tweekanalensysteem *n*
PO system *m* dwuprzewodowy
PY двухканальная система *f*
SU kaksikanavajärjestelmä
SV tvåkanalssystem *n*

1486 dual effect compressor
DA to-cylinder kompressor
DE doppeltwirkender Verdichter *m*
ES compresor de doble efecto *m*
F compresseur *m* à double aspiration
I compressore *m* a doppio effetto
MA kettős hatású kompresszor
NL compressor *m* met dubbele aanzuiging
PO sprężarka *f* z doładowaniem czynnika
PY компрессор *m* с двумя уровнями давления
SU kaksitehoinen kompressori
SV dubbeleffektkompressor *c*

1487 dual effect control
DA dobbeltstyring
DE doppeltwirkende Regelung *f*
ES control de doble efecto *m*
F contrôle *m* haute et basse pression
I regolazione *m* a doppio effetto
MA kettős hatású szabályozás
NL regelaar *m* met twee ingangen
PO regulacja *f* dwóch parametrów
PY контроль *m* в двух точках
SU kaksitehosäätü
SV dubbeleffektkontroll *c*

1488 dual fuel boiler
DA dobbeltkedel
DE Wechselbrandkessel *m*
ES caldera para dos combustibles *f*
F chaudière *f* à deux combustibles
I caldaia *f* a doppia combustione
MA alternatív kazán két tüzelőanyaggal működő kazán
NL ketel *m* voor twee brandstoffen
PO kocioł *m* na dwa rodzaje paliwa
PY котел *m* на два вида топлива
SU kaksoiskattila
SV dubbelpanna *c*

1489 dual fuel burner
DA dobbeltbrænder
DE Zweistoffbrenner *m*
ES quemador policombustible *m*
F brûleur *m* à deux combustibles
I bruciatore *m* dual-fuel
MA alternatív égő két tüzelőanyaggal működő égő
NL brander *m* voor twee brandstoffen
PO palnik *m* na dwa rodzaje paliwa
PY нефтегазовая горелка *f*
SU kahden polttoaineen poltin
SV dubbelbrännare *c*

1490 dual pressure control
DA dobbelt trykstyring
DE Zweidruckregelung f
ES control de presión doble m
F contrôle m haute et basse pression
I controllo m di minima e di massima pressione
MA kettős nyomásszabá- lyozás
NL tweedruksregelaar m
PO regulacja f dwuciśnieniowa
PY контроль m на двух уровнях давления
SU kaksitasoinen paineen säätü
SV dubbeltryck- kontroll c

1491 dual thermostat
DA dobbelt termostat
DE Doppelthermostat n
ES termostato doble m
F thermostat m à deux contacts
I termostato m a doppio effetto
MA kettős termosztát
NL dubbel- thermostaat m
PO termostat m dwupozycyjny
PY термостат m с двумя уровнями температуры
SU kaksoistermostaatti
SV dubbeltermostat c

1492 duct
DA kanal
DE Kanal m
   Leitung f
ES canalización f
   conducto m
F canal m
   caniveau m
   conduit m
   gaine f
I canale m
   condotto m
MA csatorna
   légcsatorna
   vezeték
NL luchtkanaal n
PO kanał m
   przewód m
PY канал m
   трубопровод m
SU kanava
SV kanal c

1493 duct breakout noise
DA kanallyd
DE Kanalgeräusch n
ES ruido de aire en un conducto m
F bruit m émis par les conduits d'air
I rumore m di sbocco del canale
MA csatornából kiszűrődő zaj
NL geluiduitstraling f van luchtkanaal
PO hałas m od wibracji przewodów
PY шум m от трубопровода
SU kanavamelun siirtyminen
SV kanalljud n

1494 duct distribution
DA kanalfordeling
DE Kanalverteilung f
ES distribución por conductos f
F distribution f d'air par réseau aéraulique
I distribuzione f in canali
MA vezetékes elosztás
NL luchtkanalennet n
PO rozprowadzenie n kanałów rozprowadzenie n przewodów
PY распределительный воздуховод m (трубопровод)
SU ilmanjakelu
SV kanaldistribution c

1495 duct elbow
DA kanalbøjning
DE Kanalbogen m
   Kanalkrümmer m
ES codo de un conducto m
F coude m
I curva f
   gomito m di canale
MA légcsatorna künyük
NL luchtkanaal- bochtstuk n
PO kolano n kanału łuk m na kanale
PY колено n трубопровода отвод m
SU kanavamutka
SV kanalbüj c

1496 ductility
DA sejhed
DE Dehnbarkeit f
ES ductibilidad f
F ductibilité f
I duttilità f
MA nyujthatóság
NL bewerkbaarheid f buigzaamheid f
PO ciągliwość f kowalność f plastyczność f
PY вязкость f пластичность f
SU taottavuus
SV tänjbarhet c

1497 ducting (see ductwork)

1498 duct noise
DA kanalstøj
DE Kanalgeräusch *n*
ES ruido en
   conductos *m*
F dimension *f* de
   conduit
I rumore *m* nei canali
MA csatornazaj
NL luchtkanaalgeluid *n*
PO hałas *m* pochodzący
   z kanałów w
   czasie przepływu
   powietrza
PY шум *m*
   трубопровода
SU kanavamelu
SV kanalbuller *n*

1499 duct radiation
DA kanalstråling
DE Kanalabstrahlung *f*
ES radiación en
   conductos *f*
F rayonnement *m* des
   conduits d'air
I radiazione *f* del
   canale
MA csatornasugárzás
NL geluiduitstraling *f*
   van luchtkanaal
PO promieniowanie *n*
   kanału
   promieniowanie *n*
   przewodu
PY распространение *n*
   шума от
   трубопровода
SU kanavamelun
   siirtyminen
SV kanalstrålning *c*

1500 duct section
DA kanalsektion
DE Kanalabschnitt *m*
ES sección del
   conducto *f*
F tronçon *m* de
   conduit
I tronco *m* di canale
MA csatornaszakasz
NL luchtkanaalsectie *f*
PO część *f* kanału
   (przewodu)
   przekrój *m* kanału
   (przewodu)
PY отрезок *m*
   трубопровода
   секция *f*
   воздуховода
SU kanavaosa
SV kanaldel *c*

1501 duct size
DA kanaldimension
DE Kanalabmessung *f*
ES dimensiones del
   conducto *f,pl*
   tamaño del
   conducto *m*
F dimension *f* de
   conduit
I dimensione *f* del
   canale
MA csatorna méret
NL luchtkanaal-
   afmeting *f*
PO wymiar *m* kanału
   wymiar *m* przewodu
PY размер *m* канала
   размер *m*
   трубопровода
SU kanavakoko
SV kanaldimension *c*

1502 duct sizing
DA kanaldimensionering
DE Kanalbemessung *f*
ES dimensiones de
   conductos *f,pl*
F dimensionnement *m*
   de conduits
I dimensionamento *m*
   del canale
MA csatorna méretezés
NL luchtkanaal-
   dimensionering *f*
PO wymiarowanie *n*
   kanału
   wymiarowanie *n*
   przewodu
PY прокладка *f*
   трубопровода
SU kanavamitoitus
SV kanal-
   dimensionering *c*

1503 duct system
DA kanalsystem
DE Kanalsystem *n*
ES sistema de
   conductos *m*
F réseau *m* de
   conduits
I canalizzazione *f*
   rete *f* di canali
MA csatornarendszer
NL luchtkanalen-
   systeem *n*
PO sieć *f* kanałów
   sieć *f* przewodów
PY система *f*
   трубопроводов
   (воздуховодов)
SU kanavajärjestelmä
SV kanalsystem *n*

**1504 duct transition
   section**
DA kanalovergangs-
   stykke
DE Kanalübergangs-
   stück *n*
ES reducción de un
   conducto *f*
F pièce *f* de
   raccordement
   aéraulique
I sezione *f* di
   transizione del
   canale
MA átmeneti idom
NL luchtkanaalover-
   gangsstuk *n*
PO przekrój *m* przez
   kształtkę
   przejściową na
   kanale
PY гильза *f* для
   прохода
   коммуникаций
SU kanavan
   muutoskappale
SV kanals *c*
   övergångsdel *c*

1505 ductwork
DA kanalarbejde
DE Kanalsystem *n*
ES red de conductos *f*
F canalisation *f*
I canalizzazione *f*
MA csatornahálózat
NL kanaalwerk *n*
   luchtkanaalwerk *n*
PO sieć *f* kanałów
   sieć *f* przewodów
PY раздающий
   (удаляющий)
   трубопровод *m*
   (воздуховод)
SU kanavisto
SV kanalsystem *n*

1506 dunnage
DA afstivning
DE Stauholz *n*
ES desecante *m*
F lattes *f* d'arrimage
I fardaggio *m*
MA alátét
NL stophout *n*
PO przekładka *f*
   drewniana np.
   pomiędzy
   skrzyniami
PY напольный стеллаж
   m
SU välirima
SV mellanlägg *n*

1507 duplex

1507 duplex
DA dobbelt
DE Doppel *n*
ES doble *m*
F duplex *m*
I duplex *m*
MA kettős
NL dubbel
  tweevoudig
PO układ *m* podwójny
PY дублер *m*
SU kaksisuuntainen
  kaksivaiheinen
SV duplex *n*

1508 durability
DA holdbarhed
DE Dauerhaftigkeit *f*
  Standzeit *f*
ES durabilidad *f*
F durabilité *f*
I durabilità *f*
MA tartósság
NL duurzaamheid *f*
PO trwałość *f*
PY долговечность *f*
SU kestävyys
SV hållbarhet *c*

1509 duration (time)
DA varighed
DE Dauer *f*
ES duración *f*
F durée *f*
I durata *f*
  tempo *m*
MA időtartam
NL duur *m*
PO czas *m* trwania
PY продолжительность
  f
SU kesto
  pysyvyys
SV varaktighet *c*

1510 dust
DA støv
DE Müll *m*
  Staub *m*
ES polvo *m*
F poussière *f*
I polvere *f*
MA por
NL stof *n*
PO kurz *m*
  pył *m*
PY пыль *f*
SU pöly
SV stoft *n*

1511 dust content
DA støvindhold
DE Staubgehalt *m*
ES contenido de
  polvo *m*
F teneur *f* en
  poussières
I contenuto *m* in
  polvere
MA portartalom
NL stofgehalte *n*
PO skład *m* pyłu
  zawartość *f* pyłu
PY содержание *n* пыли
SU pölypitoisuus
SV stoftinnehåll *n*

1512 dust deposit
DA støvansamling
DE Staubablagerung *f*
ES depósito de polvo *m*
  precipitación de
  polvo *f*
F dépôt *m* de
  poussières
  précipitation *f* de
  poussières
I accumulo *m* di
  polvere
MA porlerakódás
NL stofafzetting *f*
PO osad *m* pyłu
PY осадок *m* пыли
SU pölynlaskeuma
SV stoftavlagring *c*

1513 dust eliminator
DA støvfjerner
DE Staubabscheider *m*
ES eliminador de polvo
  m
F dépoussiéreur *m*
I separatore *m* di
  polvere
MA porleválasztó
  portalanító
NL stofafscheider *m*
PO eliminator *m* pyłu
  odpylacz *m*
PY пылеотделитель *m*
SU pölynpoistin
SV stoftavskiljare *c*

**1514 dust extracting
plant** (*see* dust
removal)

1515 dust extraction
DA støvudskillelse
DE Staubabscheidung *f*
ES extracción de
  polvo *f*
  filtrado para la
  eliminación del
  polvo *m*
F dépoussiérage *m*
I estrazione *f* delle
  polveri
MA porleválasztó
NL stofafscheiding *f*
PO oczyszczanie *n*
  spalin
  odpylanie *n*
  usuwanie *n* kurzu
PY удаление *n* пыли
SU pölynpoisto
SV stoftavskiljning *c*

1516 dust filter
DA støvfilter
DE Staubfilter *m*
ES filtro de polvo *m*
F filtre *m* à poussières
I filtro *m* da polvere
MA porszűrő
NL stoffilter *m*
PO filtr *m*
  przeciwpyłowy
PY пылевой фильтр *m*
SU pölysuodatin
SV stoftfilter *n*

**1517 dust holding
capacity**
DA støvakkumulerings-
  evne
DE Staubspeicher-
  fähigkeit *f*
ES capacidad de
  retención de
  polvo *f*
F capacité *f* de
  rétention de
  poussières
I capacità *f* di
  ritenzione della
  polvere
MA portárolási kapacitás
NL stofopname-
  vermogen *n*
PO zdolność *f*
  zatrzymywania
  pyłu
PY пылеемкость *f*
SU pölynsitomiskyky
SV stofthållnings-
  fürmåga *c*

1518 dust measurement
DA støvmåling
DE Staubmessung *f*
ES medición del
    contenido de
    polvo *f*
F mesure *f* de la
    teneur en
    poussières
I misurazione *f* del
    contenuto in
    polvere
MA por mérés
NL stofmeting *f*
PO pomiar *m* zapylenia
PY измерение *n*
    содержания пыли
SU pülynmittaus
SV stoftmätning *c*

1519 dust particles
DA støvpartikler
DE Staubteilchen *f,pl*
ES partículas de
    polvo *f,pl*
F particules *f* de
    poussières
I particelle *f,pl* di
    polvere
MA porszemcsék
NL stofdeeltjes *pl*
PO cząsteczki *f* pyłu
PY частицы *f* пыли
SU pülyhiukkanen
SV stoftpartiklar *c*

1520 dust proof *adj*
DA støvtæt
DE staubdicht
ES estanco al polvo
    hermético al polvo
F étanche à la
    poussière
I a tenuta di polvere
MA pormentes
NL stofdicht
PO pyłoszczelny
PY пыленепрони-
    цаемый
SU pülytiivis
SV dammtät

1521 dust removal
DA støvfjernelse
DE Entstaubung *f*
ES desempolvado *m*
    eliminación del
    polvo *f*
    extracción del
    polvo *f*
F dépoussiérage *m*
I depolverazione *f*
MA por eltávolítás
NL stofverwijdering *f*
PO usuwanie *n* pyłu
PY удаление *n* пыли
SU pülynpoisto
SV stoftrening *c*

1522 dust removal efficiency
DA støvudskilningsgrad
DE Entstaubungsgrad *m*
ES eficiencia de
    eliminación de
    polvo *f*
F rendement *m* de
    dépoussiérage
I efficienza *f* della
    depolverazione
MA portalanítási
    hatásfok
NL ontstoffingsgraad *m*
PO skuteczność *f*
    odpylania
PY эффективность *f*
    пылеудаления
SU pülynerotusaste
SV stoftreningsgrad *c*

1523 dust removal plant
DA støvfjerningsanlæg
DE Entstaubungs-
    anlage *f*
ES instalación de
    desempolvado *f*
    instalación de
    eliminación de
    polvo *f*
F installation *f* de
    dépoussiérage
I impianto *m* di
    depolverazione
MA portalanító
    berendezés
NL stofverwijderings-
    installatie *f*
PO instalacja *f* do
    usuwania pyłu
    instalacja *f*
    odpylająca
PY установка *f* для
    удаления пыли
SU pülynpoistolaitos
SV stoftrenings-
    anläggning *c*

1524 dust separation equipment
DA støvudskilnings-
    udstyr
DE Entstaubungs-
    ausrüstung *f*
ES equipo separador de
    polvo *m*
    equipo(s) de
    separación de
    polvo *m*
F installation *f* de
    dépoussiérage
I apparecchio *m* per
    la separazione
    delle polveri
MA porleválasztó
    készülék
NL ontstoffings-
    installatie *f*
    stofafscheidings-
    installatie *f*
PO odpylacz *m*
    urządzenie *n*
    odpylające
PY пылеочистное
    оборудование *n*
SU pülynerotuslaite
SV stoftavskiljnings-
    utrustning *c*

1525 dust separator
DA støvudskiller
DE Staubabscheider *m*
ES separador de
    polvo *m*
F dépoussiéreur *m*
    séparateur *m* de
    poussières
I separatore *m* di
    polvere
MA porleválasztó
NL stofafscheider *m*
PO odpylacz *m*
PY пылеотделитель *m*
SU pülynerotin
SV stoftavskiljare *c*

1526 dust spot opacity
DA støvplet
   uigennemsigtighed
DE Staubfleck-(Licht)-
   Undurchlässig-
   keit *f*
ES opacidad de la
   mancha de polvo *f*
F opacité *f* à la tache
I opacità *f* (metodo
   Ashrae dust spot)
MA porfolt homályosság
NL verkleuringstest-
   methode *f*
PO nieprzezroczystość *f*
   plamki pyłu
PY прозрачность *f*
   пылевого пятна
SU pülytäplän
   läpäisevyys
SV fläktgenoms-
   ynlighet *c*

1527 duty cycling
(electrical)
DA pausestyring
DE Arbeitszyklus *m*
ES ciclo de trabajo *m*
F intermittence *f* de
   fonctionnement
   électrique
I ciclo *m* di servizio
MA időszakos
   kikapcsolás
NL periodieke
   afschakeling *f*
   (elektrisch)
PO praca *f* cykliczna
PY дежурный режим
   m работы
SU sähkütehon ohjaus
SV pulskvot *c*

1528 dye
DA farve
DE Farbe *f*
ES color *m*
   tinte *m*
F teinture *f*
I colorante *m*
   tintura *f*
MA festék
NL kleurstof *f*
PO barwnik *m*
   farba *f*
PY краска *f*
SU väri
SV färg *c*

1529 dyestuff
DA farvestof
DE Farbstoff *m*
ES colorante *m*
F matière *f* colorante
   teinture *f*
I sostanza *f* colorante
MA festékanyag
NL organische
   kleurstof *f*
PO barwnik *m*
PY краситель *m*
SU väriaine
SV färgämne *n*

1530 dynamic *adj*
DA dynamisk
DE dynamisch
ES dinámico
F dynamique
I dinamico
MA dinamikus
NL dynamisch
PO dynamiczny
PY динамический
SU dynaaminen
SV dynamisk

1531 dynamic behaviour
DA dynamisk adfærd
DE dynamisches
   Verhalten *n*
ES comportamiento
   dinámico *m*
F comportement *m*
   dynamique
I comportamento *m*
   dinamico
MA dinamikus viselkedés
NL dynamisch gedrag *n*
PO zachowanie *n* się
   dynamiczne
PY динамическое
   состояние *n*
SU dynaaminen
   käyttäytyminen
SV

1532 dynamic graphics
DA dynamisk
   fremstilling
   (tegning)
DE dynamische
   Grafiken *f*
ES gráficos dinámicos *m*
F représentation *f*
   graphique
   dynamique
I grafica *f* animata
   grafica *f* dinamica
MA dinamikus grafika
NL dynamische grafische
   afbeeldingen *pl*
PO grafika *f*
   dynamiczna
PY динамическая
   характеристика *f*
SU dynaaminen
   grafiikka
   visualisointi
SV

1533 dynamic head loss
(*see* dynamic loss)

1534 dynamic loss
DA dynamisk tryktab
DE dynamischer
   Verlust *f*
ES pérdida de presión
   dinámica *f*
   pérdida dinámica *f*
F perte *f* de pression
   dynamique
I perdita *f* di
   pressione
   dinamica
MA dinamikus
   nyomásveszteség
NL dynamisch
   drukverlies *n*
PO strata *f* ciśnienia
   dynamicznego
PY динамические
   потери *f,pl*
SU dynaamisen paineen
   häviü
SV dynamisk fürlust *c*

1535 dynamic pressure
DA dynamisk tryk
DE dynamischer
  Druck *m*
ES presión dinámica *f*
F pression *f*
  dynamique
I pressione *f* dinamica
MA dinamikus nyomás
NL dynamische druk *m*
PO ciśnienie *n*
  dynamiczne
PY динамическое
  давление *n*
SU dynaaminen paine
SV dynamiskt tryck *n*

1536 dynamic viscosity
DA dynamisk viskositet
DE dynamische Zähig-
  keit *f*
ES viscosidad
  dinámica *f*
F viscosité *f*
  dynamique
I viscosità *f* dinamica
MA dinamikai
  viszkozitás
NL dynamische
  viscositeit *f*
PO lepkość *f*
  dynamiczna
PY динамическая
  вязкость *f*
SU dynaaminen
  viskositeetti
SV dynamisk
  viskositet *c*

1537 dynamometer
DA dynamometer
  kraftmåler
DE Kraftmesser *m*
ES dinamómetro *m*
F dynamomètre *m*
I dinamometro *m*
MA erőmérő műszer
NL dynamometer *m*
PO dynamometr *m*
PY динамометр *m*
SU dynamometri
SV dynamometer *c*

DA	=	Danish
DE	=	German
ES	=	Spanish
F	=	French
I	=	Italian
MA	=	Magyar (Hungarian)
NL	=	Dutch
PO	=	Polish
PY	=	Russian
SU	=	Finnish
SV	=	Swedish

E

1538 earth (electrical)
DA jordforbindelse
DE Erdleitung *f*
  Erdung *f*
ES conexión a tierra *f*
  puesta a tierra *f*
F mise *f* à la terre
I terra *f* (elettrico)
MA füld
NL aarde *f*
  aardverbinding *f*
PO uziom *m*
  zwarcie *n* doziemne
PY заземление *n*
SU maa
SV jord (el) *c*

1539 earthing
DA jording
DE Erdung *f*
ES toma de tierra *f*
F mise *f* à la terre
I messa *f* a terra
MA füldelés
NL aarding *f*
PO uziemienie *n*
PY заземление *n*
SU maadoitus
SV jordning *c*

1540 ebullition
DA overkogning
DE Aufschäumen *n*
  Überschäumen *n*
ES ebullición *f*
F ébullition *f*
I ebollizione *f*
MA forrás (folyadéké)
NL dampbelvorming *f*
PO wrzenie *n*
PY пузырьковое
  кипение *n*
SU kiehuminen
SV kokning *c*
  sjudning *c*

1541 eccentric *adj*
DA ekcentrisk
DE exzentrisch
ES excéntrico
F excentrique
I eccentrico
MA excentrikus
  külpontos
NL excentrisch
PO mimośrodowy
PY эксцентрический
SU epäkeskeinen
SV excentrisk

1542 economizer
DA economizer
DE Speisewasser-
  vorwärmer *m*
ES economizador *m*
F économiseur *m*
I economizzatore *m*
MA előmelegítő
  hőhasznosító
NL economiser
PO ekonomizer *m*
  podgrzewacz *m*
  wody
PY экономайзер *m*
SU lämmün
  talteenottolaite
SV ekonomiser *c*

1543 economy
DA besparelse
  økonomi
DE Wirtschaft *f*
  Wirtschaftlichkeit *f*
ES economía *f*
F économie *f*
I economia *f*
MA gazdaság
  megtakarítás
NL economie *f*
PO ekonomia *f*
  gospodarka *f*
  oszczędność *f*
PY экономия *f*
SU talous
SV ekonomi *c*

1544 eddy current
DA Foucault strøm
  hvirvelstrøm
DE Wirbelstrom *m*
ES corriente de
  Foncanet *f*
F écoulement *m*
  turbulent
I corrente *f*
  turbolenta
MA ürvényáram
NL wervelstroom *m*
PO prąd *m* wirowy
PY вихревое течение *n*
  электромагнитная
  индукция *f*
SU pyürrevirta
SV virvelstrüm *c*

1545 eddy current test
DA hvirvelstrømstest
DE Wirbelstrom-
  prüfung *f*
  Wirbelstrom-
  versuch *m*
ES prueba de la
  corriente de
  Foncanet *f*
F essai *m* de
  turbulence
I prova *f* di flusso
  turbolento
MA ürvényáramos
  vizsgálat
NL wervelstroomtest *f*
PO test *m* prądu
  wirowego
PY испытание *n*
  индуктивным
  методом
SU pyürrevirtatestaus
SV virvelstrümsprov *n*

1546 eddy flow (*see*
  turbulent flow)

1547 edge
DA kant
  æg
DE Ecke *f*
  Kante *f*
  Schärfe *f*
ES borde *m*
  canto *m*
F bord *m*
I bordo *m*
  spigolo *m*
MA él
  szegély
NL hoek *m*
  kant *m*
  rand *m*
PO brzeg *m*
  krawędź *f*
  ostrze *n*
PY лезвие *n*
SU kulma
  reuna
SV kant

1548 EER (*see* energy
  efficiency ratio)

1549 effect
DA effekt
  nyttevirkning
  ydelse
DE Effekt *m*
  Wirkung *f*
ES efecto *m*
F effet *m*
I effetto *m*
MA hatás
  jelenség
NL effect *n*
PO skutek *m*
  wynik *m*
PY эффект *m*
SU teho
SV effekt *c*

1550 effective *adj*
DA effektivt
DE effektiv
ES efectivo
F effectif
  efficace
I effettivo
MA hatásos
  tényleges
NL effectief
  werkzaam
PO efektywny
  skuteczny
  użyteczny
PY эффективный
SU efektiivinen
  tehollinen
SV effektiv

1551 effective area
DA arbejdsflade
  nytteareal
DE effektive Fläche *f*
ES area efectiva *f*
F surface *f* effective
I superficie *f* effettiva
MA hasznos
  keresztmetszet
NL effectief oppervlak *n*
  netto doorlaat *m*
PO powierzchnia *f*
  czynna
  powierzchnia *f*
  efektywna
PY эффективная
  площадь *f*
SU tehollinen ala
SV

1552 effective capacity
  (*see* net capacity)

1553 effective efficiency
  (*see* overall efficiency)

1554 effective life
DA levetid
DE effektive
  Lebensdauer *f*
ES duración efectiva *f*
  duración real *f*
  vida efectiva *f*
F durée *f* de vie réelle
I durata *f*
  vita *f* effettiva
MA tényleges élettartam
NL nuttige
  levensduur *m*
PO trwałość *f*
  żywotność *f*
  użyteczna
PY срок *m* службы
SU taloudellinen
  käyttüikä
SV livslängd *c*

**1555 effective
  temperature** (*see*
  operative temperature)

1556 efficiency
DA virkningsgrad
DE Leistungsfähigkeit *f*
  Wirkungsgrad *m*
ES rendimiento *m*
F rendement *m*
I efficienza *f*
  rendimiento *m*
  resa *f*
MA hatásfok
NL nuttig effect *n*
  rendement *n*
PO skuteczność *f*
  sprawność *f*
  wydajność *f*
PY коэффициент *m*
  полезного
  действия
  производитель-
  ьность *f*
  эффективность *f*
SU hyütysuhde
SV verkningsgrad *c*

1557 effluent water
DA spildevand
DE Abwasser *n*
ES aguas residuales *f*
F eau *f* usée
I acqua *f* effluente
MA kifolyó víz
  szennyvíz
NL afvalwater *n*
  afvoerwater *n*
PO ścieki *m, pl*
  woda *f* odpływowa
PY сточная вода *f*
SU jätevesi
SV avloppsvatten *n*

1558 efflux
DA afløb
  udstrømning
DE Abwasser *n*
  Ausfluß *m*
  Ausströmung *f*
ES evacuación *f*
F évacuation *f*
I efflusso *m*
MA kifolyás
NL uitstromende stof *n*
  uitstroming *f*
PO ścieki *m, pl*
  wyciek *m*
  wypływ *m*
PY вытекание *n*
  истечение *n*
SU ulosvirtaus
SV utflüde *n*

1559 ejector
DA ejektor
DE Ejektor *m*
ES eyector *m*
F éjecteur *m*
I eiettore *m*
MA ejektor
 sugárcső
NL ejecteur *m*
PO eżektor *m*
 strumienica *f* ssąca
PY эжектор *m*
SU ejektori
SV ejektor *c*

**1560 ejector cycle
 refrigerating
 system**
DA ejektor køleanlæg
DE Dampfstrahlkälte-
 system *n*
ES sistema de
 refrigeración de
 ciclo de eyección
 m
F système *m*
 frigorifique à
 éjection de vapeur
I sistema *m* di
 refrigerazione ad
 eiettore
MA gőzsugaras hűtési
 rendszer
NL stoomstraal
 koelsysteem *n*
PO system *m* chłodzenia
 w cyklu
 eżektorowym
PY пароструйная
 холодильная
 установка *f*
SU ejektorijäähdytys
SV ejektorcykelkyl-
 system *n*

1561 elasticity
DA elasticitet
DE Elastizität *f*
ES elasticidad *f*
F élasticité *f*
I elasticità *f*
MA rugalmasság
NL elasticiteit *f*
PO elastyczność *f*
 sprężystość *f*
PY упругость *f*
SU elastisuus
SV elasticitet *c*

1562 elastomer
DA elastomer
DE Elastomer *n*
ES elastómetro *m*
F élastomère *m*
I elastomero *m*
MA rugalmas anyag
NL elastomeer *f*
PO elastomer *m*
PY эластомер *m*
SU elastinen materiaali
SV elastomer *c*

1563 elbow
DA bøjning
 knærør
DE Bogen *m*
 Krümmer *m*
ES codo *m*
F coude *m*
I curva *f*
 gomito *m*
MA ívcső
 künyük
NL bocht *f*
 kniestuk *n*
PO kolano *n*
 łuk *m*
PY колено *n*
SU käyrä
SV büj *c*

1564 electric *adj*
DA elektrisk
DE elektrisch
ES eléctrico
F électrique
I elettrico
MA elektromos
 villamos
NL elektrisch
PO elektryczny
PY электрический
SU sähküinen
SV elektrisk

1565 electrical *adj*
DA elektrisk
DE elektrisch
ES eléctrico
F électrique
I elettrico
MA elektromos
 villamos
NL elektrisch
PO elektryczny
PY электрический
SU sähküinen
SV elektrisk

1566 electrical demand
DA el-behov
DE Strombedarf *m*
ES demanda eléctrica *f*
F énergie *f* électrique
 requise (ou
 puissance etc...)
I carico *m* elettrico
MA elektromos terhelés
 teljesítményszük-
 séglet
NL elektriciteits-
 behoefte *f*
PO zapotrzebowanie *n*
 na energię
 elektryczną
PY потребление *n*
 электроэнергии
SU sähkütehon tarve
SV elbehov *n*

1567 electrical diagram
DA el-diagram
DE Stromdiagramm *n*
ES esquema eléctrico *m*
F diagramme *m*
 électrique
I schema *m* elettrico
MA elektromos
 kapcsolási vázlat
NL elektrisch schema *n*
PO schemat *m*
 elektryczny
PY электрическая
 диаграмма *f*
SU kytkentäkaavio
SV eldiagram *n*

1568 electrical fault
DA el-fejl
DE Stromausfall *m*
ES fallo eléctrico *m*
F défaut *m* électrique
I guasto *m* elettrico
MA rüvidzárlat
NL elektrische storing *f*
 kortsluiting *f*
PO zakłócenie *n*
 elektryczne
PY короткое
 замыкание *n*
 (электрическое)
SU oikosulku
SV elfel *n*

1569 electrical ground
DA el-jordforbindelse
DE Erdung *f*
ES toma de tierra *f*
F terre *f* électrique
I messa *f* a terra
MA villamos fülddelés
NL aarding *f*
PO przewód *m*
   uziemiający
PY электрическое
   заземление *n*
SU o-potentiaali
   maa
SV jordning *c*

1570 electrical resistor
DA el-modstand
DE elektrischer
   Widerstand *m*
ES resistencia eléctrica *f*
F résistance *f*
   électrique
I resistenza *f*
MA villamos ellenállás
   villamos fűtő
NL elektrische
   weerstand *m*
PO opornik *m*
   elektryczny
PY электрорезистор *m*
SU sähküvastus
SV elmotstånd *n*

**1571 electric contact
   thermometer**
DA el-kontakt-
   termometer
DE elektrisches
   Kontakt-
   thermometer *n*
ES termómetro de
   contactos
   eléctricos *m*
F thermomètre *m* à
   contacts
   électriques
I termometro *m*
   elettrico a
   contatto
MA villamos kontakt
   hőmérő
NL elektrische
   contact-
   thermometer *m*
PO termometr *m*
   elektryczny
   stykowy
PY электроконтактный
   термометр *m*
SU kontaktiläm-
   pümittari
SV elkontakt-
   termometer *c*

1572 electric convector
DA el-konvektor
DE Elektrokonvektor *m*
ES convector
   eléctrico *m*
F convecteur *m*
   électrique
I convettore *m*
   elettrico
MA villamos fűtőtest
   villamos konvektor
NL elektrische
   convector *m*
PO konwektor *m*
   elektryczny
   ogrzewacz *m*
   przewiewowy
PY электрический
   конвектор *m*
SU sähkükonvektori
SV elkonvektor *c*

1573 electric fire
DA el-opvarmning
DE Elektrostrahler *m*
   Heizsonne *f*
ES estufa eléctrica *f*
F feu *m* électrique
I incendio *m* di
   origine elettrica
MA villamos hősugárzó
NL elektrische kachel *f*
PO piec *m* elektryczny
   promiennik *m*
   elektryczny
PY электрический
   излучатель *m*
SU sähkülämmitin
SV elkamin *c*

1574 electric heating
DA el-varme
DE Elektroheizung *f*
ES calefacción
   eléctrica *f*
F chauffage *m*
   électrique
I riscaldamento *m*
   elettrico
MA villamos fűtés
NL elektrische
   verwarming *f*
PO ogrzewanie *n*
   elektryczne
PY электрическое
   отопление *n*
SU sähkülämmitys
SV elvärme *c*

1575 electricity
DA el
   elektricitet
DE Elektrizität *f*
ES electricidad *f*
F électricité *f*
I elettricità *f*
MA villamosság
NL elektriciteit *f*
PO elektryczność *f*
PY электричество *n*
SU sähkü
SV elektricitet *c*

1576 electric motor
DA el-motor
DE Elektromotor *m*
ES motor eléctrico *m*
F moteur *m* électrique
I motore *m* elettrico
MA elektromotor
   villamos motor
NL elektromotor *m*
PO silnik *m* elektryczny
PY электрический
   двигатель *m*
SU sähkümoottori
SV elmotor *c*

1577 electric potential
DA el-potentiale
DE elektrisches
   Potential *n*
ES potencial eléctrico *m*
F potentiel *m*
   électrique
I potenziale *m*
   elettrico
MA feszültség
   villamos potenciál
NL elektrisch
   potentiaal *n*
PO potencjał *m*
   elektryczny
PY электрический
   потенциал *m*
SU sähküpotentiaali
SV elpotential *c*

DA =	Danish
DE =	German
ES =	Spanish
F =	French
I =	Italian
MA =	Magyar (Hungarian)
NL =	Dutch
PO =	Polish
PY =	Russian
SU =	Finnish
SV =	Swedish

1578 electric power absorbed

1578 electric power absorbed
DA indgående el-effekt
DE aufgenommene elektrische Leistung *f*
ES energía eléctrica absorbida *f* potencia eléctrica abosorbida *f*
F puissance *f* électrique absorbée
I energia *m* elettrica assorbita
MA felvett villamos teljesítmény
NL opgenomen elektrisch vermogen *n*
PO moc *f* elektryczna pobrana
PY потребляемая электрическая мощность *f*
SU sähkün ottoteho
SV ingående eleffekt *c*

1579 electric power demand interval
DA el-kraft intervalkrav
DE elektrisches Strombedarfs-intervall *n*
ES intérvalo de demanda de electricidad *m*
F intervalle *m* des besoins de puissance électrique
I intervallo *m* di richiesta di potenza elettrica
MA villamos terhelési időszak
NL meetperiode *f* opgenomen electrisch vermogen
PO przedział *m* zapotrzebowania na moc elektryczną
PY интервал *m* проведения обслуживания электроустановки
SU sähkün huipputehon mittausaika
SV elbehovsuppehåll *n*

1580 electric power demand load
DA el-energibehov
DE elektrische Strombedarfs-last *f*
ES carga de demanda eléctrica *f*
F charge *f* de la demande de puissance électrique
I carico *m* elettrico
MA tényleges villamos terhelés
NL momentane elektrische belasting *f*
PO obciążenie *n* zapotrzebowania na moc elektryczną
PY суммарная нагрузка *f* электрической сети
SU sähkütehon tarve
SV elbehovsbelastning *c*

1581 electric power demand period
DA el-energi behovperiode
DE elektrische Strombedarfs-periode *f*
ES periodo de demanda de electricidad *m*
F durée *f* de la demande de puissance électrique
I periodo *m* di richiesta di potenza elettrica
MA villamos terhelési időszak
NL meetperiode *f* opgenomen elektrisch vermogen
PO okres *m* zapotrzebowania na moc elektryczną
PY период *m* потребления электроэнергии
SU sähkütehon käyttüaika
SV elbehovsperiod *c*

1582 electric power factor
DA el-energifaktor
DE elektrischer Leistungsfaktor *m*
ES factor de potencia (elec) *m*
F facteur *m* électrique de puissance
I fattore *m* di potenza
MA villamos teljesítménytén-yező
NL arbeidsfactor *m* cos phi *n*
PO współczynnik *m* mocy elektrycznej
PY фактор *m* электрической мощности
SU sähkün tehokerroin
SV elkraftsfaktor *c*

1583 electric power generation
DA el-fremstilling
DE elektrische Stromerzeugung *f*
ES generación de electricidad *f*
F production *f* de puissance électrique
I generazione *f* di potenza elettrica
MA áramfejlesztés
NL stroomopwekking *f*
PO wytwarzanie *n* mocy elektrycznej
PY производство *n* электроэнергии
SU sähküntuotto
SV elgenerering *c*

1584 electric power load factor
DA el-lastfaktor
DE elektrischer Stromlastfaktor *m*
ES factor de carga (elec) *m*
F facteur *m* de charge de puissance électrique
I fattore *m* di carico
MA villamos terhelési tényező
NL gemiddelde arbeidsfactor *m*
PO wielkość *f* zysków ciepła od urządzeń elektrycznych
PY фактор *m* электрической мощности
SU sähkötehon kuormituskerroin
SV elbelastningsfaktor *c*

1585 electric power load shedding
DA el-lastsænkning
DE elektrischer Stromlast-abwurf *m*
ES difusión de la carga eléctrica *f*
F
I livellamento *m* del carico elettrico
MA villamos terhelésvédelem
NL vermogens-afschakeling *f*
PO rozpraszanie *n* zysków od zastosowanych urządzeń elektrycznych
PY защита *f* от перегрузки в электрической сети
SU huipputehon rajoitus
SV ellastreduktion *c*

1586 electric precipitator (*see* electrostatic precipitator)

1587 electric refrigerator
DA el-køleskab
DE Elektrokühl-schrank *m*
ES frigorífico eléctrico *m*
F réfrigérateur *m* électrique
I refrigeratore *m* elettrico
MA hűtőszekrény villamos hűtőgép
NL elekrische koelkast *f*
PO lodówka *f* elektryczna
PY электрический холодильник *m*
SU jääkaappi
SV elkylskåp *n*

1588 electric terminal
DA el-tilslutning pol
DE Klemme *f* Pol *m*
ES borna *f* borne *m* terminal *m*
F borne *f* électrique
I morsetto *m* elettrico terminale *m* elettrico
MA kapocsdeszka
NL aansluitklem *f*
PO zacisk *m* elektryczny
PY электрический ввод *m*
SU sähkürasia
SV elanslutning *c*

1589 electrode
DA elektrode
DE Elektrode *f*
ES electrodo *m*
F électrode *f*
I elettrodo *m*
MA elektróda
NL elektrode *f*
PO elektroda *f*
PY электрод *m*
SU clektrodi
SV elektrod *c*

1590 electrode boiler
DA elektrodekedel
DE Elektrodenkessel *m*
ES caldera de electrodos *f*
F chaudière *f* à électrodes
I caldaia *f* ad elettrodi
MA elektróda kazán
NL elektrode ketel *m*
PO kocioł *m* elektrodowy
PY электрический котел *m*
SU elektrodikattila
SV elpanna *c*

1591 electrolytic *adj*
DA elektrolytisk
DE elektrolytisch
ES electrolítico
F électrolytique
I elettrolitico
MA elektrolitikus
NL elektrolytisch
PO elektrolityczny
PY электролитический
SU elektrolyyttinen
SV elektrolytisk

1592 electrolytic couple
DA elektrolytisk
DE elektrolytisches Paar *n*
ES par electrolítico *m*
F couple *m* électrolytique
I coppia *f* elettrolitica
MA elektrolitikus pár
NL elektrolytisch koppel *n*
PO ogniwo *n* elektrolityczne
PY электролитический элемент *m*
SU sähkükemiallinen pari
SV elektrolytiskt par *n*

1593 electromagnet
DA elektromagnet
DE Elektromagnet *m*
ES electroimán *m*
F électro-aimant *m*
I elettromagnete *m*
MA elektrómágnes
NL elektromagneet *m*
PO elektromagnes *m*
PY электромагнит *m*
SU sähkümagneetti
SV elektromagnet *c*

1594 electromotive force
DA elektromotorisk kraft
DE elektromotorische Kraft *f*
ES fuerza electromotriz *f*
F force *f* électromotrice
I forza *f* elettromotrice
MA elektromotoros erő
NL elektromotorische kracht (emk) *f*
PO siła *f* elektromotoryczna
PY электродвижущая сила *f* (эдс)
SU sähkümotoorinen voima
SV

1595 electronic *adj*
DA elektronisk
DE elektronisch
ES electrónico
F électronique
I elettronico
MA elektronikus
NL elektronisch
PO elektroniczny
PY электронный
SU elektroninen
SV elektronisk

1596 electronic amplifier
DA elektronisk forstærker
DE elektronischer Verstärker *m*
ES amplificador electrónico *m*
F amplificateur *m* électronique
I amplificatore *m* elettronico
MA elektronikus erősítő
NL elektronische versterker *m*
PO wzmacniacz *m* elektroniczny
PY электронный усилитель *m*
SU elektroninen vahvistin
SV elektronisk fürstärkare *c*

1597 electronic cleaner
(*see* electrostatic filter)

1598 electronics
DA elektronik
DE Elektronik *f*
ES electrónica *f*
F électronique *f*
I elettronica *f*
MA elektronika
NL elektronica *f*
PO elektronika *f*
PY электроника *f*
SU elektroniikka
SV elektronik *c*

1599 electro-pneumatic *adj*
DA el-pneumatisk
DE elektronisch-pneumatisch
ES electro-neumático
F électropneumatique
I elettro-pneumatico
MA elektropneumatikus
NL elektropneumatisch
PO elektro-pneumatyczny
PY электро-пневматический
SU
SV elpneumatisk

1600 electrostatic filter
DA elektrostatisk filter
DE elektrostatischer Filter *m*
ES filtro electrostático *m*
F filtre *m* électrostatique
I filtro *m* elettrostatico
MA elektrosztatikus szűrő
NL elektrostatisch filter *n*
PO filtr *m* elektrostatyczny
PY электростатический фильтр *m*
SU sähküsuodatin
SV elektrostatiskt filter *n*

1601 electrostatic precipitator
DA elektrostatisk udskiller
DE Elektrofilter *m* elektrostatischer Abscheider *m*
ES filtro electrostático *m*
F dépoussiéreur *m* électrostatique filtre *m* électrostatique
I precipitatore *m* elettrostatico
MA elektrosztatikus leválasztó
NL elektrostatisch filter *n*
PO elektrofiltr *m* filtr *m* elektrostatyczny odpylacz *m* elektrostatyczny
PY электростатический осадитель *m* пыли
SU sähküsuodatin
SV elektrostatiskt filter *n*

1602 element
DA element grundstof
DE Element *n*
ES elemento *m*
F élément *m*
I elemento *m*
MA elem
NL element *n*
PO element *m* ogniwo *n* składnik *m*
PY элемент *m*
SU elementti osa
SV element *n*

1603 elevation drawing
DA lodret tegning
 opstalt
DE Aufrißzeichnung *f*
ES plano de alzado *m*
 F dessin *m* en
 élévation
 I prospetto *m*
MA függőleges csőterv
 oldalnézet
NL verticale
 projectie-
 tekening *f*
PO rysunek *m* elewacji
 rzut *m* pionowy
PY вертикальный
 разрез *m*
 профиль *m*
SU leikkauspiirustus
SV elevationsritning *c*

1604 eliminator
DA udskiller
DE Abscheider *m*
ES eliminador de
 gotas *m*
 F séparateur *m*
 I separatore *m*
MA cseppleválasztó
 rezgéscsillapító
NL afscheider *m*
PO eliminator *m*
 oddzielacz *m*
PY отделитель *m*
 элиминатор *m*
SU erotin
SV avskiljare *c*

1605 embedded *adj*
DA indstøbt
DE eingebettet
 verdeckt
ES empotrado
 F encastré
 enfoncé
 I annegato
MA beágyazott
NL ingestort
 verankerd
PO osadzony
 wbudowany
PY забетонированный
SU upotettu
SV inbyggd

1606 embedded panel
DA indstøbt panel
DE unter Putz liegende
 Tafel *f*
ES panel embebido *m*
 panel empotrado *m*
 panel encastrado *m*
 F panneau *m*
 incorporé
 I pannello *m*
 annegato
MA betonfűtőtest
NL verwarmingsspiraal
 f in beton *n*
PO płyta *f* wbudowana
 tablica *f*
 wbudowana
PY бетонная панель *f*
SU upotettu paneeli
SV inbyggnadspanel *c*

1607 embossed plate
DA relief-plade
DE bossierte Platte *f*
ES chapa repujada *f*
 F circuit *m* membouti
 I piastra *f* goffrata
MA domborított lemez
NL geprofileerde plaat *f*
PO płyta *f* wytłaczana
PY бетонная плита *f*
SU meistetty kilpi
 pakotettu kilpi
SV pressad plåt *c*

1608 EMCS (*see* energy
 management control
 system)

**1609 emergency relief
 valve**
DA nødsikkerhedsventil
DE Sicherheitsentlastungsventil *n*
ES válvula de escape de
 emergencia *f*
 F vanne *f* de sécurité
 et de décharge
 I valvola *f* di scarico
 di emergenza
MA vész ürítő szelep
NL veiligheidsklep *f*
PO zawór *m*
 nadmiarowy
 szybkiego
 działania
PY аварийный
 предохранитель-
 ьный клапан *m*
SU käsikäyttüinen
 varoventtiili
SV nüdavlastnings-
 ventil *c*

1610 EMF (*see*
 electromotive force)

1611 emission
DA emission
 stråling
DE Emission *f*
ES emisión *f*
 F émission *f*
 I emissione *f*
MA emisszió
 kibocsátás
NL afgifte *f*
 emissie *f*
 uitstoot *m*
PO emisja *f*
PY излучение *n*
 эмиссия *f*
SU emissio
SV emission *c*

1612 emissivity
DA strålingsevne
DE Emissions-
 vermügen *n*
ES emisividad (total) *f*
 F émissivité *f*
 I emissività *f*
MA sugárzóképesség
NL afgifte vermogen *n*
PO emisyjność *f*
 zdolność *f* emisyjna
PY коэффициент *m*
 излучения
 эмиссионная
 способность *f*
SU emissiivisyys
SV emissionsfürmåga *c*

1613 emittance
DA stråling
DE spezifische
 Ausstrahlung *f*
ES emitancia *f*
 F émittance *f*
 I emettenza *f*
MA fajlagos kisugárzás
NL stralingsfactor *m*
PO emitancja *f*
 promienność *f*
PY излучение *n*
SU emissiivisyys
SV avgivning *c*

1614 emptying
DA tømning
DE Entleerung *f*
ES agotamiento *m*
  vaciado *m*
F vidange *f*
I svuotamento *m*
MA ürítés
NL aftappen
PO opróżnianie *n*
PY опорожнение *n*
SU tyhjennys
SV tümning *c*

1615 emulsion
DA emulsion
DE Emulsion *f*
ES emulsión *f*
  envolvente *f*
F émulsion *f*
I emulsione *f*
MA emulzió
NL emulsie *f*
PO emulsja *f*
PY эмульсия *f*
SU emulsio
SV emulsion *c*

1616 emulsion flow (*see* bubble flow)

1617 enclosure
DA indbygning
  indkapsling
DE Einfassung *f*
  Umhüllung *f*
ES caja (de un aparato) *f*
  carcasa *f*
  envuelta *f*
  espacio cerrado *m*
  recinto cerrado *m*
F enceinte *f*
  enveloppe *f*
I cavità *f*
  involucro *m*
MA burkolat
NL bijlage *f*
  omsluiting *f*
PO obudowa *f*
  osłona *f*
PY оболочка *f*
SU kotelo
  tila
SV inkapsling *c*

1618 endothermic *adj*
DA varmeabsorberende
DE endothermisch
ES endotérmico
F endothermique
I endotermico
MA endoterm
  hőelnyelő
NL endothermisch
PO endotermiczny
PY эндометрический
SU endoterminen
SV endotermisk

1619 energy
DA energi
DE Energie *f*
ES energía *f*
F énergie *f*
I energia *f*
MA energia
NL energie *f*
PO energia *f*
PY энергия *f*
SU energia
SV energi *c*

1620 energy audit
DA energiregnskab
DE Energiebilanz *f*
ES auditoria energética *f*
F audit *m* énergétique
I diagnosi *f* energetica
  energy *f* audit
MA energiamérleg
NL energiegebruik-sonderzoek *n*
PO kontrola *f* wielkości energii
PY ревизия *f* расходования энергии
SU energiakatselmus
SV energirevision *c*

1621 energy availability
DA disponibel energi
DE Energieverfügbar-keit *f*
ES disponibilidad de energía *f*
F disponibilité *f* en énergie
I disponibilità *f* energetica
MA energia rendelkezésreállás
NL energiebeschik-baarheid *f*
PO dostępność *f* energii
PY энергетическая возможность *f* (доступность)
SU energian saatavuus
SV energitillgång *c*

1622 energy conservation
DA energibesparelse
DE Energieeinsparung *f*
ES conservación de la energía *f*
F conservation *f* de l'énergie
I conservazione *f* dell'energia
MA energiamegtakarítás
NL energiebesparing *f*
PO oszczędność *f* energii
  zachowanie *n* energii
PY консервация *f* энергии
SU energian säästü
SV energibesparing *c*

1623 energy consumption
DA energiforbrug
DE Energieverbrauch *m*
ES consumo de energía *m*
F consommation *f* d'énergie
I consumo *m* di energia
MA energiafogyasztás
NL energiegebruik *n*
PO zużycie *n* energii
PY расход *m* энергии
SU energian kulutus
SV energifürbrukning *c*

DA	=	Danish
DE	=	German
ES	=	Spanish
F	=	French
I	=	Italian
MA	=	Magyar (Hungarian)
NL	=	Dutch
PO	=	Polish
PY	=	Russian
SU	=	Finnish
SV	=	Swedish

1624 energy cost
DA energipris
DE Energiekosten *f*
ES coste de energía *m*
F prix *m* de l'énergie
I costo *m* dell'energia
MA energiakültség
NL energiekosten *pl*
PO koszt *m* energii
РУ стоимость *f* энергии
SU energiakustannus
SV energikostnad *c*

1625 energy economy
DA energiøkonomi
DE Energiewirtschaft *f*
ES economía energética *f* economización de energía *f*
F économie *f* d'énergie
I economia *f* energetica risparmio *m* energetico
MA energiagazdálkodás
NL energie-economie *f*
PO gospodarka *f* energią oszczędność *f* energii
РУ экономия *f* энергии энергетическое хозяйство *n*
SU energiatalous
SV energihushållning *c*

1626 energy efficiency ratio
DA energivirkningsgrad
DE energetischer Wirkungsgrad *m* Energienutzungszahl *f*
ES coeficiente de eficiencia energética *m*
F rendement *m* énergétique
I effetto *m* frigorigeno specifico
MA energiahasznosítási hányad
NL energie-rendements-factor *m*
PO skuteczność *f* (sprawność) wykorzystania energii
РУ степень *f* энергетической эффективности
SU energiatehokkuus
SV energiutbytes-fürhållande *n*

1627 energy factor
DA energifaktor
DE Energiefaktor *m* Leistungszahl *f*
ES factor energético *m*
F facteur *m* énergétique
I fattore *m* di energia
MA teljesítménytényező
NL energie-factor *m*
PO czynnik *m* energetyczny
РУ энергетический фактор *m*
SU energiatehokkuus
SV energifaktor *c*

1628 energy level
DA energiniveau
DE Energieniveau *n*
ES nivel energético *m*
F état *m* d'énergie
I livello *m* energetico
MA energiaszint
NL energie-niveau *n*
PO poziom *m* energetyczny
РУ энергетический уровень *m*
SU energiataso
SV energinivå *c*

1629 energy loss
DA energitab
DE Energieverlust *m*
ES pérdida de energía *f*
F perte *f* d'énergie
I perdita *f* di energia
MA energiaveszteség
NL energieverlies *n*
PO strata *f* energii
РУ потеря *f* энергии
SU energiahäviü
SV energifürlust *c*

1630 energy management
DA energistyring
DE Energie-management *n*
ES gestión energética *f*
F gestion *f* de l'énergie
I gestione *f* dell'energia
MA energia felügyelet
NL energiebeheer *n*
PO gospodarowanie *n* energią zarządzanie *n* energią
РУ управление *n* энергией
SU energian hallinta
SV energidrift *c*

1631 energy management control system
DA energistyringssystem
DE Energiemanagement-Kontrollsystem *n*
ES sistema de control de la gestión energética *m*
F système *m* de gestion énergétique
I sistema *m* di supervisione
MA energiafelügyeleti szabályozórend-szer
NL energie beheer- en regelsysteem *n*
PO system *m* kontroli gospodarowania energią
РУ система *f* контроля и управления энергией
SU energian hallintajärjestelmä
SV kontrollsystem *n* för energidrift *c*

1632 energy management function
DA energistyrings-
   funktion
DE Energiemanagement-
   aufgabe *f*
ES función de gestión
   energética *f*
F gestion *f* de l'énergie
I funzione *f* di
   gestione
   dell'energia
MA energiafelügyeleti
   tevékenység
NL energiebeheer-
   functie *f*
PO funkcja *f*
   zarządzania
   (gospodarowania)
   energią
PY эффективное
   использование *n*
   энергии (в
   зданиях)
SU energian tehokas
   käyttü
SV energidrifts-
   funktion *c*

1633 energy potential
DA energipotentiale
DE Energiepotential *n*
ES energía potencial *f*
F énergie *f* potentielle
I potenziale *m* di
   energia
MA energiapotenciál
NL energiepotentieel *n*
PO potencjał *m*
   energetyczny
PY потенциал *m*
   энергии
SU energiapotentiaali
SV energipotential *c*

1634 energy recovered
DA energigenvinding
DE rückgewonnene
   Energie *f*
ES recuperación
   energética *f*
F énergie *f* récupérée
I energia *f* recuperata
MA visszanyert energia
NL teruggewonnen
   energie *f*
PO energia *f* odzyskana
   odzyskiwanie *n*
   energii
PY утилизация *f*
   энергии
SU talteenotettu energia
SV energiåtervinning *c*

1635 energy source
DA energikilde
DE Energiequelle *f*
ES fuente de energía *f*
F source *f* d'énergie
I risorsa *f* energetica
MA energiaforrás
NL energiebron *f*
PO źródło *n* energii
PY источник *m*
   энергии
SU energialähde
SV energikälla *c*

1636 energy state
DA energiniveau
DE Energiezustand *m*
ES situación energética *f*
F état *m* d'énergie
I stato *m* energetico
MA energia állapot
NL energie-niveau *n*
PO stan *m* energetyczny
PY энергетический
   уровень *m*
SU energiatila
SV kraftsituation *c*

1637 energy transmittance ratio
DA energitransmissions-
   evne
DE Energiedurchlaß-
   grad *m*
ES coeficiente de
   transmisión
   energética *m*
F facteur *m* de
   transmission par
   rayonnement
I coefficiente *m* di
   trasmissione della
   radiazione
MA energia átbocsátási
   arány
NL energieoverdrachts-
   verhouding *f*
PO zdolność *f*
   przesyłania energii
PY эффективность *f*
   передачи энергии
SU energian
   läpäisysuhde
SV energiüverfürings-
   grad *c*

1638 energy value target
DA energiværdimål-
   sætning
DE Energiewertzahl *n*
ES objetivo de utilidad
   energética *m*
F performance *f*
   énergétique prévue
I valore *m* obiettivo
   di prestazione
   energetica
MA tervezett
   energiaigény
NL streefwaarde
   energiegebruik *n*
PO wymagana wielkość
   f energii
PY требуемая
   величина *f*
   энергии
SU energian
   käyttütavoite
SV energimål *n*

1639 engine
DA maskine
   motor
DE Maschine *f*
ES máquina *f*
F machine *f*
I macchina *f*
   motore *m*
MA gép
   motor
NL machine *f*
   motor *m*
PO maszyna *f*
   silnik *m*
PY машина *f*
SU kone
SV motor *c*

1640 engineer
DA ingeniør
   maskinmester
DE Ingenieur *m*
ES ingeniero *m*
F ingénieur *m*
I ingegnere *m*
MA mérnük
NL ingenieur *m*
   machinist *m*
   technicus *m*
PO inżynier *m*
PY инженер *m*
   механик *m*
SU insinüüri
SV ingenjür *c*

1641 engineering
DA ingeniørarbejde
  maskinkonstruktion
DE Ingenieurwesen *n*
  Maschinenbau *m*
ES ingeniería *f*
F ingéniérie *f*
I ingegneria *f*
MA gépészet
  mérnüki munka
NL kunde *f*
  techniek *f*
  technisch uitwerken
PO inżynieria *f*
PY машиностроение *n*
  строительство *n*
  техника *f*
SU insinüüritiede
  insinüürityü
SV ingenjürskonst *c*

1642 engine (gas) exhaust
DA motorudstødning
DE Auspuffgas *n*
ES escape de gas *m*
  orificio de escape de gases *m*
F gaz *m*
  d'échappement
I gas *m* di scarico del motore
MA kipufogógáz
NL motoruitlaat *m*
PO spaliny *pl* z silnika
PY выхлоп *m* двигателя
SU pakokaasu
SV motoravgaser *c*

1643 enrichment
DA berigelse
DE Anreicherung *f*
ES enriquecimiento *m*
F enrichissement *m*
I arricchimento *m*
MA dúsulás
NL verrijking *f*
PO wzbogacanie *n*
PY обогащение *n* (газа)
SU rikastus
SV anrikning *c*

1644 enterprise
DA foretagende
  projekt
DE Unternehmen *n*
ES empresa *f*
F entreprise *f*
I impresa *f*
MA vállalat
NL onderneming *f*
PO przedsiębiorstwo *n*
  przedsięwzięcie *n*
PY предприятие *n*
SU yritys
SV füretag *n*

1645 enthalpy
DA entalpi
DE Enthalpie *f*
ES entalpía *f*
F enthalpie *f*
I entalpia *f*
MA entalpia
NL enthalpie *f*
PO entalpia *f*
PY энтальпия *f*
SU entalpia
SV entalpi *c*

1646 enthalpy chart (*see* enthalpy diagram)

1647 enthalpy diagram
DA entalpidiagram
DE Enthalpie-Diagramm *n*
ES diagrama de entalpia *m*
F diagramme *m* enthalpique
I diagramma *m* entalpico
MA entalpia diagram
NL enthalpie diagram *n*
PO wykres *m* entalpowy
PY энтальпийная диаграмма *f*
SU entalpiapiirros
SV entalpidiagram *n*

1648 entrain *vb*
DA medrive
DE beimischen
  verladen
ES arrastrar
F entraîner
I trascinare
MA magával ragad
NL meesleuren
  meevoeren
PO pociągać
  porywać
PY увлекать
SU indusoitua
  lastata
SV medejektera

1649 entrainment
DA medrivning
DE Beimischung (von Luft) *f*
  Sekundärluft *f*
ES arrastre *m*
  embarque *m*
  inducción *f*
F entraînement *m*
  d'air (induction)
I trascinamento *m*
MA indukció
  magával ragadás
NL meenemen
  meeslepen
PO porywanie *n* (kropelek cieczy przez strumień)
PY поток *m*, эжектируемый струей
  унос *m*
SU induktio
  mukana vieminen
SV inblandning *c*

1650 entrainment ratio
DA medrivningsgrad
DE Beimischungs-verhältnis *n*
ES coeficiente de arrastre *m*
F taux *m* d'entraînement
I rapporto *m* di trascinamento
MA indukció arány
NL inductie-verhouding *f*
PO stopień *m* porywania
PY степень *f* эжекции
SU induktiosuhde
SV inblandnings-fürhållande *n*

1651 entropy
DA entropi
DE Entropie *f*
ES entropía *f*
F entropie *f*
I entropia *f*
MA entrópia
NL entropie *f*
PO entropia *f*
PY энтропия *f*
SU entropia
SV entropi *c*

1652 entropy chart (*see* entropy diagram)

1653 entropy diagram
DA entropidiagram
DE Entropie-
   diagramm *n*
ES diagrama entrópico
   m
F diagramme *m*
   entropique
I diagramma *m*
   entropico
MA entrópia diagram
NL entropie diagram *n*
PO wykres *m*
   entropowy (np.
   ciepła)
PY энтропийная
   диаграмма *f*
SU entropiapiirros
SV entropidiagram *n*

1654 entry
DA indløb
DE Eintritt *m*
ES entrada *f*
F accès *m*
   entrée *f*
I ingresso *m*
MA belépés
   beümlés
NL ingang *m*
   inschrijving *f*
   invoer *m*
PO wejście *n*
   wlot *m*
PY вход *m*
SU menoaukko
SV ingång *c*

1655 entry loss
DA indløbstab
DE Eintrittsverlust *m*
ES pérdida en la
   entrada *f*
F perte *f* de charge à
   l'entrée
I perdita *f* all'ingresso
MA belépési veszteség
NL instromingsverlies *n*
   intredeverlies *n*
PO strata *f* na wlocie
PY потери *f,pl* на
   входе
SU sisäänvirtaushäviü
SV ingångsfürlust *c*

1656 envelope
DA hylster
   konvolut
   skærm
DE Umhüllung *f*
ES envolvente *f*
   envuelta *f*
F enveloppe *f*
I busta *f*
   involucro *m*
MA burkolat
NL omhullende
   omhulling *f*
PO koperta *f*
   otoczka *f*
   powłoka *f*
PY обмуровка *f*
   покрытие *n*
SU vaippa
SV mantel *c*

1657 environment
DA omgivelse
DE Umgebung *f*
ES ambiente *m*
F environnement *m*
I ambiente *m*
MA környezet
NL milieu *n*
   omgeving *f*
PO otoczenie *n*
   środowisko *n*
PY окружающая
   среда *f*
SU ympäristü
SV miljü *c*

**1658 environmental
     chamber**
DA klimarum
DE Raumsimulator *m*
   Umweltsimulations-
   kammer *f*
ES cámara ambiental *f*
F chambre *f*
   climatique
I camera *f* climatica
MA klímakamra
NL klimaatkamer *f*
PO komora *f*
   klimatyczna z
   parametrami
   środowiska
PY климатическая
   камера *f*
SU jäähuone
   klimakammio
SV miljükammare *c*

**1659 environmental
     conditions**
DA klimabetingelser
DE Umwelt-
   bedingungen *f*
ES condiciones
   ambientales *f*
F conditions *f*
   d'ambiance
I condizioni *f,pl*
   ambientali
MA környezeti feltételek
NL omgevings-
   condities *pl*
PO warunki *m*
   otoczenia
   warunki *m*
   środowiska
PY окружающие
   условия *n*
SU ympäristüolot
SV miljüfürhållanden *n*

**1660 environmental
     design**
DA klimaprojektering
DE Umweltplan *m*
ES proyecto ambiental
   m
F conception *f*
   environnementale
I progettazione *f*
   ambientale
MA környezettervezés
NL milieubewust
   ontwerp *n*
PO projekt *m* ochrony
   środowiska
PY прогнозирование *n*
   окружающих
   условий
   проектирование *n*
   окружающих
   условий
SU ympäristüsuunnittelu
SV miljükonstruktion *c*

DA =	Danish
DE =	German
ES =	Spanish
F =	French
I =	Italian
MA =	Magyar
	(Hungarian)
NL =	Dutch
PO =	Polish
PY =	Russian
SU =	Finnish
SV =	Swedish

1661 environmental engineering
DA miljøteknik
DE Umwelttechnik *f*
ES ingeniería ambiental *f*
F génie *m* climatique
I ingegneria *f* ambientale
MA kürnyezettervező mérnüki tevékenység
NL milieutechniek *f*
PO inżynieria *f* środowiska
PY техника *f* кондиционирования воздуха
SU ilmastointitekniikka ympäristütekniikka
SV miljüteknik *c*

1662 environmental temperature
DA omgivende temperatur
DE Umgebungstemperatur *f*
ES temperatura ambiente *f*
F température *f* ambiante
I temperatura *f* ambiente
MA kürnyezeti hőmérséklet
NL omgevingstemperatuur *f*
PO temperatura *f* otoczenia temperatura *f* środowiska
PY температура *f* окружающего воздуха
SU ympäristün lämpütila
SV omgivningstemperatur *c*

1663 environmental test chamber
DA klimakammer
DE Klimakammer *f*
ES cámara de ensayos climáticos *f*
F chambre *f* d'essais climatiques
I camera *f* climatica
MA kísérleti klímakamra
NL klimaatkamer *f*
PO komora *f* pomiarowa warunków środowiska
PY климатическая исследовательская камера *f*
SU klimakammio säähuone
SV miljüprovkammare *c*

1664 equal friction method
DA metode til trykfaldsudligning
DE Methode *f* der gleichen Reibungswiderstände *mpl*
ES método de igual fricción *m* método de igualación de pérdidas de carga *m*
F méthode *f* de calcul à pertes de charges réparties constantes
I metodo *m* ad attrito costante
MA egyenlő súrlódás módszere
NL gelijke weerstandsmethode *f*
PO metoda *f* równoważnego tarcia
PY метод *m* эквивалентных потерь напора
SU vakiokitkamenetelmä
SV metod *c* för dimensionering *c* med lika friktionstrykfall *n*

1665 equalize *vb*
DA kompensator udligne
DE ausgleichen gleichstellen
ES equilibrar igualar
F égaliser
I equalizzare equilibrare
MA kiegyenlít
NL vereffenen
PO wyrównywać zrównywać
PY уравнивать
SU tasoittaa
SV utjämna

1666 equalizer
DA udligningsspjæld
DE Ausgleicher *m* Stabilisator *m*
ES equilibrador *m*
F égaliseur *m*
I equalizzatore *m*
MA kiegyenlítő
NL vereffeningsinrichting *f*
PO stabilizator *m* urządzenie *n* wyrównawcze
PY уравниватель *m*
SU stabilisaattori tasoitin
SV utjämningsdon *n*

1667 equalizer tank
DA udligningsbeholder
DE Ausgleichsbehälter *m*
ES depósito equilíbrador *m*
F réservoir *m* égalisateur
I serbatoio *m* equalizzatore
MA kiegyenlítő tartály
NL vereffeningstank *m*
PO zbiornik *m* wyrównawczy
PY уравнивающая емкость *f*
SU tasaajasäiliü
SV utjämningstank *c*

1668 equalizing damper
DA udligningsspjæld
DE Ausgleichsklappe *f*
ES registro
  equilibrador *m*
F volet *m*
  d'équilibrage
I serranda *f* di
  taratura
MA kiegyenlítő
  csappantyú
NL vereffeningsklep *f*
PO przepustnica *f*
  wyrównawcza
PY уравнивающий
  клапан *m*
SU ylivirtauspelti
SV utjämningsspjäll *n*

1669 equalizing pressure
DA udligningstryk
DE Druckausgleich *m*
ES presión de
  equilibrio *f*
F équilibrage *m* de
  pression
I pressione *f* di
  equilibrio
MA kiegyenlítő nyomás
NL vereffeningsdruk *m*
PO ciśnienie *n*
  wyrównawcze
PY уравнивание *n*
  давления
SU paineentasaus
SV tryckutjämning *c*

1670 equation
DA ligning
DE Gleichung *f*
ES ecuación *f*
F égalisation *f*
  équation *f*
I equazione *f*
MA egyenlet
NL vergelijking *f*
PO równanie *n*
PY выравнивание *n*
  уравнение *n*
SU yhtälü
SV utjämning *c*

1671 equation of state
DA tilstandsligning
DE Zustandsgleichung *f*
ES ecuación de estado *f*
F équation *f* d'état
I equazione *f* di stato
MA állapotegyenlet
NL toestands-
  vergelijking *f*
PO równanie *n* stanu
PY уравнение *n*
  состояния
SU tilasuureet
SV tillståndslikhet *c*

1672 equilibrium
DA ligevægt
DE Beharrungs-
  zustand *m*
  Gleichgewicht *n*
ES equilibrio *m*
F équilibre *m*
I equilibrio *m*
MA egyensúly
NL evenwicht *n*
PO równowaga *f*
PY равновесие *n*
SU tasapainotila
SV jämvikt *c*

**1673 equilibrium
  temperature**
DA ligevægtstemperatur
DE Beharrungs-
  temperatur *f*
ES temperatura de
  equilibrio *f*
F température *f*
  d'équilibre
I temperatura *f* di
  equilibrio
MA egyensúlyi
  hőmérséklet
NL evenwichts-
  temperatuur *f*
PO temperatura *f*
  równowagi
PY равновесная
  температура *f*
SU tasapainolämpütila
SV jämvikts-
  temperatur *c*

1674 equipment
DA udstyr
DE Ausrüstung *f*
  Ausstattung *f*
  Einrichtung *f*
ES equipo(s) *m*
  material *m*
F appareillage *m*
  équipement *m*
I apparecchiatura *f*
  attrezzatura *f*
  macchinario *m*
MA berendezés
  készülék
NL apparatuur *f*
  uitrusting *f*
PO sprzęt *m*
  wyposażenie *n*
PY оборудование *n*
SU laite
SV utrustning *c*

1675 equipment room
DA apparatrum
DE Maschinenraum *m*
  Zentrale *f*
ES sala de aparatos *f*
  sala de máquinas *f*
F local *m* technique
I macchine
MA gépház
NL technische ruimte *f*
PO pomieszczenie *n*
  techniczne
PY машинный зал *m*
  помещение *n* для
  размещения
  оборудования
SU konehuone
SV apparatrum *n*

1676 equivalent *adj*
DA ekvivalent
DE äquivalent
ES equivalente
F équivalente
I equivalente
MA egyenértékű
NL gelijkwaardig
PO równoważny
PY эквивалентный
SU samanarvoinen
SV ekvivalent

1677 erection
DA montering
   opførelse
DE Aufbau *m*
   Aufstellung *f*
ES levantamiento (de un
   edificio) *m*
   montaje en obra *m*
F mise *f* en place
   montage *m*
I assemblaggio *m*
   costruzione *f*
   montaggio *m*
MA felépítés
   szerelés
NL montage *f*
   opstelling *f*
PO montaż *m*
   wznoszenie *n*
PY монтаж *m*
   сборка *f*
SU kokoonpano
   pystytys
SV montering *c*

**1678 erection
instructions** (*see*
mounting instructions)

1679 erosion
DA erosion
DE Erosion *f*
ES erosión *f*
   sobrepresión *f*
F érosion *f*
I erosione *f*
MA erózió
   lepusztulás
NL erosie *f*
   uitslijping *f*
PO erozja *f*
PY эрозия *f*
SU eroosio
SV erosion *c*

1680 error
DA fejl
DE Fehler *m*
ES equivocación *f*
   error *m*
F erreur *f*
I errore *m*
MA hiba
NL fout *f*
   miswijzing *f*
PO błąd *m*
PY ошибка *f*
SU virhe
SV fel *n*

1681 error band
DA fejlområde
DE Fehlerbreite *f*
ES margen de error *m*
F plage *f* d'erreur
I banda *f* di
   tolleranza
MA hibasáv
NL afwijkingsgebied *n*
PO przedział *m* błędu
PY доверительный
   интервал *m*
SU virhealue
SV felband *n*

1682 estimate
DA budget
   vurdering
DE Kosten-
   voranschlag *m*
   Voranschlag *m*
ES cálculo *m*
   estimación *f*
   presupuesto *m*
F devis *m*
I preventivo *m*
   stima *f*
MA előirányzat
   kültségvetés
NL begroting *f*
   raming *f*
   schatting *f*
PO ocena *f*
   oszacowanie *n*
PY смета *f*
SU arvio
SV beräkning *c*

1683 estimate *vb*
DA vurdere
DE schätzen
   voranschlagen
ES estimar
   presupuestar
F estimer
I stimare
   valutare
MA előirányoz
NL begroten
   ramen
PO oceniać
   szacować
PY осмечивать
   оценивать
SU arvioida
SV beräkna

1684 estimation
DA vurdering
DE Schätzung *f*
ES estimación *f*
   valoración *f*
F estimation *f*
   évaluation *f*
I stima *f*
   valutazione *m*
MA becslés
   előirányzat
NL schatting *f*
   taxatie *f*
PO ocenianie *n*
   szacowanie *n*
PY оценка *f*
SU arviointi
SV beräkning *c*

1685 eupatheoscope
DA komfortmåler
DE Behaglichkeits-
   messer *m*
ES eupateoscopio *m*
   eupateóscopo *m*
F eupathéoscope *m*
I eupateoscopio *m*
MA komfortmérő műszer
NL behaaglijkheids-
   meter *m*
   eupatheoscoop *m*
PO eupateoskop *m*
PY эвпатоскоп *m*
SU viihtyvyysmittari
SV komfortmätare *c*

1686 eutectic *adj*
DA let smeltelig
DE eutektisch
ES eutéctico
F eutectique *m*
I eutettico
MA eutektikus
NL eutectisch
PO eutektyczny
PY эвтектический
SU eutektinen
SV eutektisk

1687 evacuated *adj*
DA evakueret
   udsuget
DE evakuiert
ES evacuado
 F évacué
   explusé
 I sotto vuoto
MA kiürített
   légritkított
NL gevacumeerd
   luchtledig gemaakt
PO opróżniony
   usunięty
PY вакуумный
   откаченный
SU tyhjä-
   tyhjü-
SV bortfürd

1688 evacuation
DA evakuering
   udsugning
DE Evakuierung *f*
ES evacuación *f*
   realización del
   vacío *f*
 F évacuation *f*
 I evacuazione *f*
   scarico *m*
MA kiszívás
   kiürítés
NL lediging *f*
   luchtledig maken *n*
   vacumeren *n*
PO opróżnianie *n*
   usuwanie *n*
PY опорожнение *n*
   откачивание *n*
SU tyhjentää
SV utsugning *c*

1689 evaporate *vb*
DA fordampe
DE verdampfen
ES evaporar
 F évaporer
 I evaporare
MA elpárologtat
NL verdampen
PO odparowywać
   parować
PY испарять
SU haihduttaa
SV förånga

1690 evaporation
DA fordampning
DE Verdampfung *f*
ES evaporación *f*
 F évaporation *f*
 I evaporazione *f*
MA elpárologtatás
NL verdamping *f*
PO odparowanie *n*
   parowanie *n*
PY испарение *n*
SU haihtuminen
SV förångning *c*

1691 evaporation coil
DA dampspiral
DE Verdampfers-
   chlange *f*
ES serpentín evaporador
   m
 F batterie *f*
   d'évaporation
 I evaporatore *m*
   (batteria)
MA elpárologtató
   csőkígyó
NL verdampings-
   spiraal *f*
PO wężownica *f*
   parownika
   wężownica *f*
   wyparki
PY змеевик *m*
   испарителя
SU hüyrystyspatteri
SV förångarslinga *c*

1692 evaporation loss
DA fordampningstab
DE Verdunstungs-
   verlust *m*
ES
 F perte *f* par
   évaporation
   (tour de
   refroidissement)
 I perdita *f* per
   evaporazione
MA elpárologtatási
   veszteség
NL verdampings-
   verlies *n*
PO straty *f*, *pl* wskutek
   odparowania
PY потери *f,pl* на
   испарение
SU haihtumishäviü
SV fürångningsfürlust *c*

**1693 evaporation
   pressure**
DA fordampningstryk
DE Verdampfungs-
   druck *m*
ES presión de
   evaporación *f*
   presión de
   vaporización *f*
 F pression *f*
   d'évaporation
 I pressione *f* di
   evaporazione
MA elpárologtatási
   nyomás
NL verdampingsdruk *m*
PO ciśnienie *n*
   odparowania
PY давление *n*
   испарения
SU hüyrystyspaine
SV fürångningstryck *n*

1694 evaporation rate
DA fordampningsgrad
DE Verdampfungsrate *f*
ES grado de
   evaporación *m*
 F taux *m*
   d'évaporation
 I tasso *m* di
   evaporazione
MA elpárologtatási arány
NL verdampings-
   snelheid *f*
PO stopień *m*
   odparowania
PY интенсивность *f*
   испарения
SU haihtumisnopeus
SV fürångnings-
   hastighet *c*

1695 evaporation temperature
DA fordampnings-
   temperatur
DE Verdampfungs-
   temperatur *f*
ES temperatura de
   evaporación *f*
   temperatura de
   evaporización *f*
F température *f*
   d'évaporation
I temperatura *f* di
   evaporazione
MA elpárologtatási
   hőmérséklet
NL verdampings-
   temperatuur *f*
PO temperatura *f*
   parowania
PY температура *f*
   испарения
SU hüyrystyslämpütila
SV furångnings-
   temperatur *c*

1696 evaporative air conditioner
DA luftbehandler med
   dampindblæsning
DE Verdunstungs-
   klimaanlage *f*
ES acondicionador
   evaporativo *m*
F conditionneur *m*
   d'air évaporatif
I condizionatore *m* di
   aria evaporativo
MA elpárologtatós
   légkondicionáló
NL verdampingskoel-
   apparaat *n*
PO klimatyzator *m* z
   czynnikiem
   odparowującym
PY испарительный
   кондиционер *m*
SU haihdutusjäähdytin
SV evaporativ
   klimatapparat *c*

1697 evaporative condensor
DA fordampnings-
   kondensator
DE Verdunstungs-
   kondensator *m*
ES condensador
   evaporativo *m*
F condenseur *m*
   évaporatif
I condensatore *m*
   evaporativo
MA elpárologtatós
   kondenzátor
NL verdampings-
   condensor *m*
PO skraplacz *m*
   wyparny
PY испарительный
   конденсатор *m*
SU haihdutuslauhdutin
SV furångnings-
   kondensor *c*

1698 evaporative cooler
DA fordampningskøler
DE Verdunstungs-
   kühler *m*
ES enfriador
   evaporativo *m*
F refroidisseur *m* par
   évaporation
I refrigeratore *m*
   evaporativo
MA elpárologtatós hűtő
NL verdampings-
   koeler *m*
PO urządzenie *n*
   chłodnicze
   wyparne
PY испарительный
   охладитель *m*
SU haihdutusjäähdytin
SV avdunstnings-
   kylare *c*

1699 evaporative cooling
DA fordampningskøling
DE Verdunstungs-
   kühlung *f*
ES enfriamiento
   evaporativo *m*
   enfriamiento por
   evaporación de
   agua *m*
F refroidissement *m*
   par évaporation
I raffreddamento *m*
   per evaporazione
MA elpárologtatós hűtés
NL verdampings-
   koeling *f*
PO chłodzenie *n*
   wyparne
PY испарительное
   охлаждение *n*
SU haihdutusjäähdytys
SV avdunstnings-
   kylning *c*

1700 evaporative equilibrium
DA fordampnings-
   ligevægt
DE Verdunstungs-
   gleichgewicht *n*
ES equilibrio de
   evaporación *m*
F température *f*
   humide d'équilibre
   (mesurée)
I condizione *f* di
   equilibrio tra
   liquido e vapore
   (saturazione)
MA elpárolgási egyensúly
NL verdampings-
   evenwicht *n*
PO stan *m* równowagi
   przy odparowaniu
PY испарительное
   равновесие *n*
SU märkälämpümittarin
   tasapainotila
SV evaporativ jämvikt *c*

1701 evaporative humidifier
DA dampbefugter
DE Dampfbefeuchter *m*
ES humidificador por evaporación *m*
F humidificateur *m* à évaporation
I umidificatore *m* ad evaporazione
MA párologtatós nedvesítő
NL verdampings-bevochtiger *m*
PO nawilżacz *m* wyparny
PY испарительный увлажнитель *m*
SU haihduttava kostuttaja
SV evaporativ fuktare *c*

1702 evaporative meter
DA fordampningsmåler
DE Verdunstungs-messer *m*
ES evaporímetro *m* vaporímetro *m*
F compteur *m* de chaleur à évaporation
I misuratore *m* di evaporazione
MA párolgásmérő
NL verdampings-meter *m*
PO miernik *m* odparowania
PY измеритель *m* испарения
SU haihduttava mittari
SV evaporativ mätare *c*

1703 evaporator
DA evaporator fordamper
DE Verdampfer *m*
ES evaporador *m*
F évaporateur *m*
I evaporatore *m*
MA elpárologtató
NL verdamper *m*
PO parownik *m* wyparka *f*
PY испаритель *m*
SU haihdutin
SV förångare *c*

1704 evaporator coil (*see* evaporation coil)

1705 evaporator pressure regulator
DA damptrykregulator
DE Verdampferdruck-regler *m*
ES regulador de presión de evaporación *m*
F régulateur *m* de pression d'évaporation
I regolatore *m* di pressione dell'evaporatore
MA elpárologtató nyomásszabályozó
NL verdamperdruk-regelaar *m*
PO regulator *m* ciśnienia parowania
PY регулятор *m* давления в испарителе
SU paisuntaventtiili
SV förångares *c* tryckregulator *c*

1706 evaporator turbulator
DA fordampnings-turbulator
DE Verdampfer-Turbulator *m*
ES turbulador de evaporación *m*
F turbulateur *m* d'évaporateur
I turbolatore *m* dell'evaporatore
MA elpárologtató keverő
NL verdamper-turbulator *m*
PO turbulizator *m* odparowania urządzenie *n* intensyfikujące odparowanie
PY турбулизатор *m* испарителя
SU hüyrystimen turbulaattori
SV förångarturbulator *c*

1707 evaporator unit
DA evaporatorunit
DE Verdampfereinheit *f*
ES unidad evaporadora *f*
F unité *f* d'évaporation
I unità *f* evaporante
MA elpárologtató egység
NL verdamper-compressor-eenheid *f*
PO parownik *m* wyparka *f*
PY испарительный агрегат *m*
SU hüyrystinyksikkü
SV förångare *c*

1708 excess
DA overskridelse overskud
DE Überschuß *m*
ES excedente *m* exceso *m*
F excès *m*
I eccesso *m*
MA felesleg
NL overmaat *f*
PO nadmiar *m* nadwyżka *f*
PY избыток *m*
SU ylijäämä ylimäärä
SV överskott *n*

1709 excess air
DA overskudsluft
DE Luftüberschuß *m*
ES aire sobrante *m*
F excès *m* d'air
I eccesso *m* di aria
MA légfelesleg
NL luchtovermaat *f*
PO nadmiar *m* powietrza
PY избыток *m* воздуха
SU ilmaylimäärä
SV överskottsluft *c*

1710 excess pressure
DA overtryk
DE Überdruck *m*
ES sobrepresión *f*
F surpression *f*
I sovrapressione *f*
MA túlnyomás
NL te hoge druk *m*
PO nadmiar *m* ciśnienia
PY избыточное давление *n* подпор *m*
SU ylipaine
SV övertryck *n*

1711 excess temperature
DA overtemperatur
DE Übertemperatur *f*
ES exceso de temperatura *m*
sobrecalentamiento *m*
F excès *m* de température
surchauffe *f*
I sovratemperatura *f*
MA túlhőmérséklet
NL te hoge temperatuur *f*
PO nadmierna temperatura *f*
nadwyżka *f* temperatury
PY избыточная температура *f*
SU ylilämpütila
SV üvertemperatur *c*

1712 exchange
DA udveksling
DE Austausch *m*
ES cambio *m*
intercambio *m*
F échange *m*
I scambio *m*
MA csere
NL uitwisseling *f*
PO wymiana *f*
PY замена *f*
изменение *n*
SU vaihto
SV utbyte *n*

1713 exchange *vb*
DA udveksle
DE austauschen
ES cambiar
F échanger
I scambiare
MA cserél
NL uitwisselen
wisselen
PO wymieniać
PY заменять
изменять
SU siirtää
vaihtaa
SV utbyta

1714 exchanger
DA veksler
DE Austauscher *m*
ES cambiador (de calor) *m*
intercambiador *m*
F échangeur *m*
I scambiatore *m*
MA cserélő
NL wisselaar *m*
PO wymiennik *m*
PY обменник *m*
SU siirtäjä
vaihtaja
SV växlare *c*

1715 excitation
DA magnetisering
DE Ausgang *m*
Erregung *f*
ES excitación *f*
F excitation *f*
I eccitazione *f*
MA gerjesztés
NL bekrachtiging *f*
opwekking *f*
PO pobudzanie *n*
wzbudzanie *n*
PY возбуждение *n*
электризация *f*
SU heräte
SV magnetisering *c*

1716 execution
DA udførelse
DE Ausführung *f*
ES ejecución *f*
F exécution *f*
I esecuzione *f*
MA kivitel
NL uitvoering *f*
PO wykonanie *n*
PY исполнение *n*
SU toimeenpano
SV utfürande *n*

1717 exergy
DA exergi
DE Exergie *f*
ES exergía *f*
F éxergie *f*
I exergia *f*
MA energiaveszteség
exergia
NL exergie *f*
PO egzergia *f*
PY эксергия *f*
SU exergia
SV exergi *c*

1718 exfiltration
DA luftexfiltration
DE Ausfilterung *f*
Austritt *m*
ES exfiltración *f*
F exfiltration *f*
I exfiltrazione *f*
MA kiszűrődés
NL uitstroming *f*
PO eksfiltracja *f*
PY эксфильтрация *f*
SU ulosvuoto
SV luftexfiltration *c*

1719 exhaust
DA udsugning
DE Abdampf *m*
Abgas *n*
Auspuff *m*
ES escape *m*
F extraction *f*
I espulsione *f*
estrazione *f*
scarico *m*
smaltimento *m*
MA elszívás
NL afvoer *m*
uitlaat *m*
PO wyciąg *m*
wyrzut *m*
wywiew *m*
PY вытяжка *f*
SU poisto
SV utblåsning *c*
utsugning *c*

1720 exhaust *vb*
DA udsuge
DE absaugen *vb*
ES evacuar
extraer
F évacuer
I scaricare
MA elszív
NL afvoeren
PO usuwać
wyrzucać
wywiewać
PY удалять (воздух)
SU poistaa
SV utsuga

1721 exhaust air
DA udsugningsluft
DE Abluft *f*
   Fortluft *f*
   oberirdisch
ES aire evacuado *m*
   aire extraído *m*
F air *m* extrait
   air *m* rejeté
I aria *f* di estrazione
MA elszívott levegő
NL afvoerlucht *f*
PO powietrze *n*
   usuwane
   powietrze *n*
   wywiewane
PY уходящий
   (удаляемый)
   воздух *m*
SU poistoilma
SV frånluft *c*

1722 exhaust air equipment
DA luftudsugningsudstyr
DE Ablufteinrichtung *f*
   Abluftgerät *n*
ES equipo de
   evacuación de
   aire *m*
   equipo de extracción
   de aire *m*
F dispositif *m* de rejet
   d'air
I dispositivo *m* di
   estrazione di aria
MA elszívó berendezés
NL luchtafzuig-
   inrichting *f*
PO urządzenie *n* do
   wywiewu
   powietrza
PY воздуховыбросное
   устройство *n*
SU poistoilmalaite
SV frånluftsutrustning *c*

1723 exhaust air filter
DA udsugningsluftfilter
DE Abluftfilter *m*
ES filtro de aire de
   extracción *m*
F filtre *m* sur air
   extrait
   filtre *m* sur air rejeté
I filtro *m* di aria di
   estrazione
MA elszívott levegő szűrő
NL afvoerluchtfilter *m*
PO filtr *m* powietrza
   wywiewanego
PY фильтр *m* для
   очистки
   вытяжного
   воздуха
SU poistoilmasuodatin
SV frånluftsfilter *n*

1724 exhaust air system
DA luftudsugnings-
   system
DE Abluftanlage *f*
   Abluftsystem *n*
ES sistema de extracción
   de aire *m*
F installation *f*
   d'extraction d'air
I impianto *m* di
   estrazione di aria
MA elszívásos rendszer
NL afvoerlucht-
   systeem *n*
   afzuigsysteem *n*
PO instalacja *f*
   wywiewna
PY система *f* вытяжки
   воздуха
SU poistoilmajärjestelmä
SV frånluftssystem *n*

1725 exhaust air terminal device
DA udsugnings-
   anordning
DE Abluftauslaß *m*
   Abluftgitter *n*
ES boca de extracción
   de aire *f*
F bouche *f*
   d'extraction d'air
I bocchetta *f* di
   estrazione di aria
MA elszívó szerkezet
NL luchtafvoerrooster *n*
PO wyrzutnia *f*
   powietrza
   wywiewnik *m*
PY вытяжное
   устройство *n*
   (для воздуха)
SU poistoilmalaite
SV frånluftsdon *n*

1726 exhaust duct
DA aftrækskanal
DE Abluftkanal *m*
   Abluftleitung *f*
ES conducto de
   extracción *m*
F conduit *m*
   d'évacuation
   conduit *m*
   d'extraction
I canale *m* di
   estrazione
MA elszívó légcsatorna
NL afvoerkanaal *n*
PO kanał *m* wywiewny
   przewód *m*
   wywiewny
PY вытяжной канал *m*
SU poistoilmakanava
SV frånluftskanal *c*

1727 exhaust fan
DA udsugningsventilator
DE Abluftventilator *m*
ES aspirador *m*
   extractor *m*
   ventilador de
   extracción *m*
F ventilateur *m*
   d'extraction d'air
I ventilatore *m* di
   estrazione di aria
MA elszívó ventilátor
NL afvoerventilator *m*
   afzuigventilator *m*
PO wentylator *m*
   wyciągowy
   wentylator *m*
   wywiewny
PY вытяжной
   вентилятор *m*
SU poistoilmapuhallin
SV frånluftsfläkt *c*

1728 exhaust gas
DA udstødsgas
DE Abgas *n*
ES gas de escape *m*
F gaz *m*
   d'échappement
   gaz *m* extrait
   gaz *m* rejeté
I gas *m* di scarico
MA égéstermék
   füstgáz
NL uitlaatgas *n*
PO gazy *m*, *pl*
   spalinowe
   gazy *m*, *pl*
   wydechowe
   spaliny *pl*
PY уходящий газ *m*
SU pakokaasu
SV avgas *c*

1729 exhaust gas connection
DA aftræksforbindelse
DE Abgasanschluß *m*
ES tubo de escape de gases *m*
F raccordement *m* d'évacuation de gaz
I raccordo *m* per il gas di scarico
MA égéstermék csatlakozás
NL uitlaatgas-aansluiting *f*
PO złączka *f* przewodu spalinowego
PY присоединение *n* для вытяжки газа
SU pakokaasuliitäntä
SV avgasanslutning *c*

1730 exhaust gas damper
DA aftræksspjæld
DE Abgasklappe *f*
ES registro de evacuación de gases *m*
   registro de gases de escape *m*
   registro de gases eva-cuados *m*
F clapet *m* d'évacuation de gaz
I serranda *f* per il gas di scarico
MA égéstermék csappantyú
NL uitlaatgasklep *f*
PO przepustnica *f* spalin
PY клапан *m* на вытяжке газа
SU pakokaasupelti
SV avgasspjäll *n*

1731 exhaust gas recirculation
DA røgcirkulation
DE Abgasrückführung *f*
ES recirculación de los gases de escape *f*
F recirculation *f* d'air recyclage *m* d'air
I ricircolazione *f* del gas di scarico
MA égéstermék visszakeringetés
NL uitlaatgas-recirculatie *f*
PO recyrkulacja *f* gazu wylotowego
PY рециркуляция *f* (газа)
SU pakokaasukierto
SV avgasrecirkulation *c*

1732 exhaust hood
DA udsugningshætte
DE Absaughaube *f*
ES campana de extracción *f*
   caperuza de escape de gases *f*
   caperuza de evacuación *f*
F hotte *f* d'évacuation hotte *f* d'extraction
I cappa *f* di estrazione
MA elszívó ernyő
NL afzuigkap *f*
PO okap *m* wyciągowy
PY вытяжной зонт *m*
SU poistohuuva
SV frånluftshuv *c*

1733 exhaust opening
DA udblæsningsåbning
DE Absaugüffnung *f*
ES abertura de extracción *f*
F bouche *f* d'air extrait
I apertura *f* di espulsione
MA elszívó nyílás
NL afzuigopening *f*
PO otwór *m* wywiewny
PY вытяжное отверстие *n*
SU poistoilma-aukko
SV avgasüppning *c*

1734 exhaust steam
DA udblæsningsdamp
DE Abdampf *m*
ES vapor de escape *m*
F évacuation *f* de vapeur
   vapeur *f* d'échappement
I vapore *m* di scarico
MA fáradt gőz
NL afgewerkte stoom *m*
PO para *f* odlotowa
PY отработавший пар *m*
SU ulospuhallushüyry
SV utloppsånga *c*

1735 exhaust system
DA udsugningssystem
DE Entlüftungssystem *n*
ES sistema de evacuación *m*
   sistema de extracción *m*
F système *m* d'extraction d'air
I sistema *m* di scarico
MA elszívó rendszer
NL afvoersysteem *n*
PO system *m* wywiewny
PY система *f* вытяжки (воздуха)
SU poistojärjestelmä
SV frånluftssystem *n*

1736 exhaust valve
DA udsugningsventil
DE Abluftventil *n*
ES válvula de escape *f*
F soupape *f* d'échappement
   soupape *f* d'évacuation
I valvola *f* di scarico
MA elszívó szelep
NL afvoerklep *f*
PO zawór *m* wywiewny
PY вытяжной клапан *m*
SU poistoventtiili ulospuhallusventtiili
SV avgasventil *c*

DA	=	Danish
DE	=	German
ES	=	Spanish
F	=	French
I	=	Italian
MA	=	Magyar (Hungarian)
NL	=	Dutch
PO	=	Polish
PY	=	Russian
SU	=	Finnish
SV	=	Swedish

1737 exhaust ventilation
DA udsugnings-
 ventilation
DE Entlüftung *f*
ES extracción de aire *f*
 ventilación por
 depresión *f*
F ventilation *f* par
 extraction
I ventilazione *m* per
 depressione
MA elszívásos
 szellőztetés
NL afzuigventilatie *f*
PO wentylacja *f*
 wyciągowa
 wentylacja *f*
 wywiewna
PY вытяжная
 вентиляция *f*
SU poistoilmanvaihto
 ulospuhallus
SV frånlufts-
 ventilation *c*

1738 exit air
DA afgangsluft
DE Fortluft *f*
ES aire de salida *m*
F air *m* rejeté
I aria *f* espulsa
MA kilépő levegő
NL afvoerlucht *f*
 uittredende lucht *f*
PO powietrze *n*
 wylotowe
PY уходящий
 воздух *m*
SU poistoilma
SV frånluft *c*

1739 exit velocity
DA afgangshastighed
DE Austrittsgesch-
 windigkeit *f*
ES velocidad de salida *f*
F vitesse *f* de sortie
I velocità *f* di uscita
MA kilépési sebesség
NL uittreesnelheid *f*
PO prędkość *f*
 wylotowa
PY скорость *f* на
 выходе
SU ulospuhallusnopeus
SV utloppshastighet *c*

1740 expanded plastics
DA skumplast
DE Schaumstoffe *m,pl*
ES plásticos expandidos
 m,pl
F plastiques *m*
 expansés
I materie *f,pl*
 plastiche espanse
MA habosított műanyag
NL schuimplastic *n*
PO tworzywo *n* sztuczne
 porowate
PY расширяющиеся
 пластмассы *f*
SU solumuovi
SV skumplast *c*

1741 expansion
DA ekspansion
 udvidelse
DE Ausdehnung *f*
ES dilatación *f*
 expansión *f*
F dilatation *f*
 expansion *f*
I dilatazione *f*
 espansione *f*
MA tágulás
NL expansie *f*
 uitzending *f*
PO rozprężanie *n*
 rozszerzanie *n* się
PY расширение *n*
SU paisunta
SV expansion *c*

1742 expansion bellows
DA ekspansionsbælg
DE Dehnungsbälge *m,pl*
ES fuelle de
 dilatación *m*
 fuelle de
 expansión *m*
F compensateur *m* de
 dilatation
 soufflet *m* de
 dilatation
I soffietto *m* di
 dilatazione
MA harmonikás
 táguláskiegyenlítő
NL balgcompensator *m*
PO mieszek *m*
 kompensacyjny
 wydłużka *f*
 mieszkowa
PY сильфонный
 компенсатор *m*
SU paljetasain
SV expansionsbälg *c*

1743 expansion bend
DA ekspansionsbøjning
DE Dehnungsbogen *m*
ES codo de
 dilatación *m*
F coude *m* de
 dilatation
 lyre *f* de dilatation
I curva *f* di
 dilatazione
MA táguláskiegyenlítő ív
NL expansiebocht *f*
PO kolano *n*
 kompensacyjne
PY компенсатор *m*
SU paisuntamutka
SV expansionskrük *c*

1744 expansion coefficient
DA udvidelseskoefficient
DE Ausdehnungs-
 koeffizient *m*
ES coeficiente de
 expansión *m*
F coefficient *m* de
 dilatation
I rapporto *m* di
 espansione
MA tágulási együttható
NL uitzettings-
 coëfficiënt *m*
PO współczynnik *m*
 rozszerzalności
PY коэффициент *m*
 расширения
SU paisuntakerroin
SV expansions-
 koefficient *c*

1745 expansion coil
DA ekspansionsrørspiral
DE Ausdehnungs-
 bogen *m*
 Ausdehnungs-
 schlange *f*
ES bateria de expansión
 f
F batterie *f* de détente
 directe
I serpentino *m* di
 espansione
MA táguláskiegyenlítő
 spirálcső
NL expansiespiraal *f*
PO parownik *m*
 wężownicowy
PY змеевик *m*
 расширения
 кожухотрубный
 испаритель *m*
SU kierukkahüyrystin
SV expansionsslinga *c*

1746 expansion float type valve
DA flydeventil
DE Expansions-Schwimmer-ventil *n*
ES válvula de expansión de flotador *f*
F régleur *m* à flotteur
I valvola *f* di espansione a galleggiante
MA úszógolyós szelep
NL expansievlotter-ventiel *n*
PO zawór *m* rozprężny pływakowy
PY вентиль *m* с расширительной уравнительной трубкой
SU uimuriventtiili
SV expansionsflottür-ventil *c*

1747 expansion gland
DA ekspansionsbøsning
DE Stopfbüchsendehnungs-ausgleicher *m*
ES prensaestopas de dilatación *m*
F presse-étoupe *m* de dilatation
I compensatore *m* di dilatazione a cannocchiale
MA tümszelencés táguláskiegyenlítő
NL schuif-compensator *m*
PO dławik *m* kompensacyjny dławnica *f*
PY сальниковый компенсатор *m*
SU liukutasain
SV expansions-packning *c*

1748 expansion joint
DA ekspansionsfuge
DE Dehnungsfuge *f* Kompensator *m*
ES junta de dilatación *f*
F joint *m* de dilatation
I giunto *m* di dilatazione
MA dilatációs hézag
NL dilatatievoeg *f*
PO złącze *n* kompensacyjne
PY компенсационная муфта *f*
SU paisuntaliitos
SV expansionsfog *c*

1749 expansion loop
DA ekspansionssløjfe
DE Dehnungsbogen *m*
ES lira de dilatación *f*
F boucle *f* de dilatation lyre *f* de dilatation
I curva *f* di dilatazione
MA táguláskiegyenlítő csőlíra
NL expansiebocht *m*
PO wydłużalnik *m* rurowy pętlicowy zamknięty
PY петлевой компенсатор *m* П-образный компенсатор *m*
SU paisuntakaari
SV expansionsslinga *c*

1750 expansion pipe
DA ekspansionsrør
DE Sicherheits-ausdehnungsleitung *f*
ES tubería de expansión *f* tubo compen-sador *m*
F tube *m* de sécurité
I tubo *m* di sicurezza
MA tágulási vezeték
NL expansieleiding *f*
PO kompensator *m* rurowy
PY компесационный трубопровод *m*
SU paisuntaputki
SV expansionsrür *n*

1751 expansion stroke
DA ekspansionsslag
DE Ausdehnungshub *m* Ausdehnungstakt *m*
ES golpe de expansión *m*
F course *f* de détente
I fase *m* di espansione
MA dugattyú kifúvó üteme
NL arbeidsslag *m*
PO suw *m* roboczy suw *m* rozprężania
PY ход *m* расширения
SU paisuntavaihe
SV expansionsslag *n*

1752 expansion tank
DA ekspansionsbeholder
DE Ausdehnungs-gefäß *n*
ES recipiente de expansión *m* tanque de expansión *m*
F vase *m* d'expansion
I vaso *m* di espansione
MA tágulási tartály
NL expansietank *m*
PO naczynie *n* wzbiorcze zbiornik *m* wyrównawczy
PY расширительный бак *m* (сосуд)
SU paisunta-astia
SV expansionskärl *n*

1753 expansion valve
DA ekspansionsventil
DE Expansionsventil *n*
ES válvula de expansión *f* válvula de laminación *f* válvula de regulación *f*
F soupape *f* de sécurité
I valvola *f* di espansione
MA expanziós szelep hűtésnél
NL expansieventiel *n*
PO zawór *m* rozprężny
PY расширительный вентиль *m* терморегули-рующий вентиль *m* (ТРВ)
SU paisuntaventtiili
SV expansionsventil *c*

1754 expansion valve capacity
DA ekspansionsventil-kapacitet
DE Expansionsventil-grüße *f*
ES capacidad de la válvula de expansión *f*
F débit *m* frigorifique d'un détenteur
I potenzialità *f* della valvola di espansione
MA expanziós szelep teljesítménye
NL expansieventiel-capaciteit *f*
PO przepustowość *f* zaworu rozprężnego
PY производитель-ьность *f* терморегули-рующего вентиля
SU paisuntaventtiilin koko
SV expansionsventil-kapacitet *c*

1755 expendable refrigerant system
DA ekspansionskøle-system
DE verbrauchbares Kältemittelsystem *n*
ES sistema de refrigerante expandible *m*
F système *m* frigorifique à compression
I sistema *m* di refrigerazione a perdita totale
MA fogyó hűtőküzeges rendszer
NL koelsysteem *n* met verbruik van koelmedium
PO system *m* wykorzystujący środek chłodniczy jednorazowego użytku
PY система *f* с безвозвратно расходуемым хладагентом
SU avoin jäähdytysjärje-stelmä
SV expansionskyl-system *n*

1756 expense
DA udgift
DE Ausgabe *f* Kosten *f*
ES gasto *m* pérdida *f*
F coût *m* dépense *f* prix *m*
I spesa *f*
MA kültség
NL kosten *pl* uitgaven *pl*
PO koszt *m* wydatek *m*
PY затрата *f* расход *m*
SU
SV kostnad *c*

1757 experiment
DA forsøg
DE Experiment *n* Versuch *m*
ES ensayo *m* experimento *m*
F essai *m* expérience *f*
I esperimento *m*
MA kísérlet
NL experiment *n* proefneming *f*
PO doświadczenie *n* eksperyment *m*
PY эксперимент *m*
SU koe
SV experiment *n*

1758 experiment *vb*
DA forsøge
DE erproben versuchen
ES experimentar
F expérimenter
I sperimentare
MA kísérletez
NL experimenteren
PO eksperymentować
PY эксперимент-ировать
SU kokeilla
SV experimentera

1759 expert system
DA ekspertsystem (EDB)
DE Expertensystem *n*
ES sistema experimentado *m*
F systémé *m* expert
I sistema *m* esperto
MA szakértői rendszer
NL expert systeem *n* kennissysteem *n*
PO system *m* ekspertowy system *m* specjalistyczny
PY экспертная система *f*
SU asiantuntijajärje-stelmä
SV expertsystem *n*

1760 explosion
DA eksplosion
DE Explosion *f* Verpuffung *f*
ES explosión *f*
F explosion *f*
I esplosione *f*
MA robbanás
NL explosie *f*
PO eksplozja *f* wybuch *m*
PY взрыв *m*
SU räjähdys
SV explosion *c*

1761 exponent
DA eksponent karakteristik
DE Exponent *m*
ES exponente *m*
F explicateur *m* interprète *m*
I esponente *m*
MA kitevő
NL exponent *m*
PO wykładnik *m* potęgi
PY показатель *m* степени экспонента *f*
SU exponentti
SV exponent *c*

1762 exposed *adj*
DA blotlagt
DE freistehend
  ungeschützt
ES al descubierto
  descubierto
  expuesto
F exposé
I esposto
  scoperto
MA kitett
NL blootgesteld aan
  onbeschut
PO wystawiony
PY открытый
  поверхностный
SU altistettu
SV exponerad

1763 extended surface
DA forstørret overflade
DE vergrößerte
  Oberfläche *f*
ES superficie adicional *f*
F paroi *m* d'échange à
  grande surface de
  contact
I superficie *f* estesa
MA kiterített felület
NL vergroot
  oppervlak *n*
PO powierzchnia *f*
  rozwinięta
PY наружная
  поверхность *f*
  расширенная
  поверхность *f*
SU ripa
SV ytfürstoring *c*

1764 exterior
DA udvendig
DE äußerlich
  Außen
ES exterior
F extérieur *m*
I ambiente *m* esterno
MA külső
NL buitenkant *m*
PO strona *f* zewnętrzna
PY вид *m* снаружи
SU ulkopuoli
SV yttre *n*

1765 external *adj*
DA ekstern
DE äußere
ES externo
F externe
I esterno
MA külső
NL extern
  uitwendig
PO zewnętrzny
PY внешний
  наружный
SU ulkoinen
SV yttre

1766 external diameter
DA udvendig diameter
DE Außendurch-
  messer *m*
ES diámetro exterior *m*
F diamètre *m* extérieur
I diametro *m* esterno
MA külső átmérő
NL buitendiameter *m*
  uitwendige
  diameter *m*
PO średnica *f*
  zewnętrzna
PY наружный
  диаметр *m*
SU ulkohalkaisija
SV ytterdiameter *c*

1767 external sensor
DA udeføler
DE Außenfühler *m*
ES sonda exterior *f*
F capteur *m* extérieur
I sensore *m* esterno
MA külső érzékelő
NL buitenvoeler *m*
PO czujnik *m*
  zewnętrzny
PY наружный
  датчик *m*
SU ulkoinen tuntoelin
SV utegivare *c*

**1768 external
  temperature**
DA udetemperatur
DE Außentemperatur *f*
ES temperatura
  exterior *f*
F température *f*
  extérieure
I temperatura *f*
  esterna
MA külső hőmérséklet
NL buitentemperatuur *f*
PO temperatura *f*
  zewnętrzna
PY наружная
  температура *f*
SU ulkolämpütila
SV utetemperatur *c*

1769 external wall
DA ydermur
DE Außenwand *f*
ES pared exterior *f*
F mur *m* extérieur
  paroi *f* extérieure
I muro *m* esterno
  parete *f* esterna
MA külső fal
NL buitenmuur *m*
PO ściana *f* zewnętrzna
PY наружная стена *f*
SU ulkoseinä
SV yttervägg *c*

1770 extinguish *vb*
DA slukke
DE auslüschen
  lüschen
ES apagar
  extinguir
F arrêter
  éteindre
I estinguere
  spegnere
MA kiolt
NL blussen
  doven
PO gasić
  wygaszać
PY тушить
SU sammuttaa
SV släcka

1771 extinguisher
DA ildslukker
DE Lüscher m
ES extintor m
F éteignoir m
(appareil)
extincteur m
I estintore m
MA tűzoltó készülék
NL blusapparaat n
brandblus-
apparaat n
PO gaśnica f
przeciwpożarowa
PY огнетушитель m
SU sammutin
SV släckare c

1772 extract vb
DA uddrage
DE ausziehen
ES extraer
F extraire
I estrarre
MA elszív
kivon
NL afzuigen
onttrekken
PO wyciągać
wywiewać
PY вытягивать
SU
SV suga ut

1773 extractor
DA centrifuge
DE Abluftventilator m
Absauger m
ES aspirador m
extractor m
F extracteur m
I estrattore m
MA elszívó
NL afzuiger m
PO ekstraktor m
wyciąg m
PY вытяжное
устройство n
экстрактор m
SU poistolaite
SV utsugnings-
anordning c

F

1774 fabric filter
DA stoffilter
DE Tuchfilter *m*
ES filtro de tela *m*
F filtre *m* à média
I filtro *m* a manica
MA szűrőszövet
NL doekfilter *m*
PO filtr *m* tkaninowy
PY матерчатый
  (тканевый)
  фильтр *m*
SU kuitusuodatin
SV textilfilter *n*

1775 facade
DA facade
DE Fassade *f*
ES fachada *f*
F façade *f*
I facciata *f*
MA homlokzat
NL gevel *m*
PO elewacja *f*
  fasada *f*
PY фасад *m*
SU julkisivu
SV fasad *c*

1776 face area
DA frontareal
DE Ausstrümfläche *f*
  Frontfläche *f*
ES superficie frontal *f*
F surface *f* frontale
I area *f* frontale
MA homlokfelület
NL frontoppervlak *n*
PO powierzchnia *f*
  czołowa
PY площадь *f* живого
  сечения
SU otsapinta-ala
SV frontyta *c*

1777 face velocity
DA fronthastighed
DE Geschwindigkeit *f*
  an der Vorderseite
ES velocidad frontal *f*
F vitesse *f* frontale
I velocità *f* frontale
MA homlokfelületi
  sebesség
NL aanstroomsnelheid *f*
  frontale snelheid *f*
PO prędkość *f* czołowa
PY скорость *f* в
  сечении
SU otsapintanopeus
SV fronthastighet *c*

1778 facility cost
DA anlægsudgift
DE Anlagekosten *f,pl*
ES coste de la ayuda *m*
F facilité *f* de
  financement
I costo *m* di impianto
MA berendezés kültsége
NL totale kosten *pl*
  voor een
  voorziening
PO koszt *m* urządzenia
PY стоимость *f*
  оборудования
SU kokonaiskäyttüku-
  stannus
SV anläggnings-
  kostnad *c*

1779 factor
DA faktor
DE Faktor *m*
ES factor *m*
F facteur *m*
I fattore *m*
MA faktor
  tényező
NL factor *m*
PO czynnik *m*
  współczynnik *m*
PY коэффициент *m*
  показатель *m*
  фактор *m*
SU kerroin
SV faktor *c*

1780 factor of safety
DA sikkerhedsfaktor
DE Sicherheitsfaktor *m*
ES factor de seguridad
  m
F facteur *m* de sécurité
I fattore *m* di
  sicurezza
MA biztonsági tényező
NL veiligheidsfactor *m*
PO współczynnik *m*
  bezpieczeństwa
PY коэффициент *m*
  надежности
  фактор *m*
  надежности
SU varmuuskerroin
SV säkerhetsfaktor *c*

1781 factory
DA fabrik
DE Fabrik *f*
ES fábrica *f*
  factoría *f*
F usine *f*
I fabbrica *f*
  stabilimento *m*
MA gyár
NL fabriek *f*
PO fabryka *f*
  wytwórnia *f*
PY завод *m*
SU tehdas
SV fabrik *c*

DA	=	Danish
DE	=	German
ES	=	Spanish
F	=	French
I	=	Italian
MA	=	Magyar
		(Hungarian)
NL	=	Dutch
PO	=	Polish
PY	=	Russian
SU	=	Finnish
SV	=	Swedish

1782 factory-assembled system
DA præfabrikeret system
DE System *n* mit
    werkseitiger
    Montage *f*
ES sistema montado en
    fábrica
    (autónomo) *m*
F matériel *m* assemblé
    en usine
I impianto *m*
    premontato
MA előregyártott
    rendszer
NL geprefabriceerd
    systeem *n*
PO instalacja *f*
    prefabrykowana
    urządzenie *n*
    montowane w
    fabryce
PY система *f*
    заводской сборки
SU tehtaaalla koottu
    järjestelmä
SV fürtillverka system *n*

1783 factory-mounted *adj*
DA præfabrikeret
DE werkseitig montiert
ES instalado en fábrica
F monté en usine
I assemblato in
    fabbrica
MA gyárilag üsszeszerelt
NL fabrieksgemonteerd
PO fabrycznie
    zmontowane
PY собранный на
    заводе
SU tehdasvalmisteinen
    tehtaalla asennettu
SV fabriksmonterad

1784 factory-set *adj*
DA fabriksindstillet
DE werkseitig eingestellt
ES montada en fábrica
F monté en usine
I regolato in fabbrica
MA sorozatgyártású
NL fabrieksingesteld
PO prefabrykowane
PY укомплектованный
    на заводе
SU tehdasasennettu
    tehtaan asetus
SV fabriksutrustad

1785 failsafe
DA fejlsikret
DE ausfallsicher
    pannensicher
ES seguridad de fallo *f*
F rail *m*
I a prova di guasto
MA hibamentes
    müküdésbiztos
NL veilig bij storing *f*
PO odporność *f* na
    uszkodzenia
PY надежный *adj*
    прочный *adj*
SU toimintavarma
SV felsäkerhet *c*

1786 failure response
DA fejlmeldesystem
DE Fehlerantwort *f*
ES respuesta a fallo *f*
F réponse *f* signifiant
    un non
    fonctionnement
I risposta *f* a guasto
MA hiánypótlási
    kütelezettség
NL storingsmaat-
    regelen *pl*
    voorzieningen *pl*
    voor storing
PO odpowiedź *f* na
    uszkodzenie
PY ремонтопригод-
    ность *f*
SU huoltotoimi
SV felsignal *c*

1787 falling film cooler
DA køler med
    filmstrømning
DE
ES enfriador pelicular *m*
F refroidisseur *m* par
    ruissellement
I raffreddatore *m* a
    film liquido
MA permetezéses hűtő
NL Baudelotkoeler *m*
    vloeistoffilm-
    koeler *m*
PO chłodnia *f* z
    powierzchnią
    zraszaną
PY гравитационный
    пленочный
    охладитель *m*
SU valuvakalvojäädytin
SV filmkylare *c*

1788 false ceiling (*see* dropped ceiling)

1789 false floor
DA falsk gulv
    løftet gulv
DE Zwischenboden *m*
ES suelo falso *m*
F faux plancher *m*
I pavimento *m*
    sopraelevato
MA kettős padló
NL verhoogde vloer *m*
PO podłoga *f* pływająca
PY фальшпол *m*
SU korotettu lattia
SV undergolv *n*

1790 fan
DA blæser
    ventilator
DE Ventilator *m*
ES soplador *m*
    ventilador *m*
F ventilateur *m*
I ventilatore *m*
MA ventilátor
NL ventilator *m*
PO wentylator *m*
PY вентилятор *m*
    воздуходувка *f*
SU puhallin
SV fläkt *c*

1791 fan air density
DA ventilatorlufttæthed
DE Ventilator-
    luftdichte *f*
ES densidad de aire en
    ventilador *f*
F densité *f* d'air du
    ventilateur à
    l'entrée
I densità *f* dell'aria
    del ventilatore
MA ventilátor által
    beszívott levegő
    sűrűsége
NL dichtheid
    ventilator-
    inlaatlucht *f*
PO gęstość *f* powietrza
    w wentylatorze
PY давление *n* на
    выхлопе
    вентилятора
    плотность *f*
    воздуха после
    вентилятора
SU ilman tiheys
    puhaltimen
    imuaukossa
SV fläktlufttäthet *c*

1792 fan appurtenances
DA ventilatortilbehør
DE Ventilatorzubehür *n*
ES accesorios del
ventilador *m*
F accessoires *m* du
ventilateur
I accessori *m,pl* del
ventilatore
MA ventilátor tartozékok
NL ventilator-
toebehoren *n*
PO osprzęt *m*
wentylatora
PY запасные части *f*
вентилятора
принадлежности *f*
вентилятора
SU puhaltimen
lisälaitteet
SV fläkttillbehür *n*

1793 fan baffle
DA ventilatorledeskovl
DE Ventilatorleit-
schaufel *f*
ES álabe director (de un
ventilador) *m*
F aube *f* directrice
I deflettore *m* del
ventilatore
MA ventilátor terelőlapát
NL ventilatorgeleide-
schot *n*
PO deflektor *m*
wentylatora
kierownica *f*
wentylatora
PY глушитель *m*
вентилятора
SU puhaltimen
johtosiipi
SV fläktledskena *c*

1794 fan blade
DA ventilatorvinge
DE Ventilatorlauf-
schaufel *f*
ES álabe del
ventilador *m*
paleta del
ventilador *f*
F aube *f* de ventilateur
I pala *f* del
ventilatore
MA ventilátor lapát
NL ventilatorschoep *f*
PO łopatka *f*
wentylatora
PY лопатка *f*
вентилятора
SU puhaltimen siipi
SV fläktskovel *c*

1795 fan casing
DA ventilatorhus
DE Ventilatorgehäuse *n*
ES caja del ventilador *f*
caracola del
ventilador *f*
envolvente del
ventilador *f*
F enveloppe *f* de
ventilateur
I carcassa *f* del
ventilatore
MA ventilátor ház
NL slakkenhuis *n*
ventilatorhuis *n*
PO obudowa *f*
wentylatora
PY кожух *m*
вентилятора
SU puhallinkaapu
SV fläktkåpa *c*

1796 fan coil unit
DA luftkøler med
ventilator
DE Ventilator-
konvektor *m*
ES fancoil *m*
ventiloconvector *m*
F ventilo-
convecteur *m*
I ventilconvettore *m*
MA fan coil egység
ventilátoros
konvektor
NL ventilator-
convector *m*
PO konwektor *m*
wentylatorowy
PY вентиляторный
конвектор *m*
SU puhallinkonvektori
SV fläktkonvektor *c*

1797 fan curve
DA ventilator-
karakteristik
DE Ventilator-
kennlinie *f*
ES curva característica
del ventilador *f*
F courbe *f*
caractéristique de
ventilateur
I curva *f* caratteristica
del ventilatore
MA ventilátor jelleggürbe
NL ventilator-
karakteristiek *f*
PO krzywa *f*
charakterystyki
wentylatora
PY характеристика *f*
вентилятора
SU puhaltimen
ominaiskäyrä
SV fläktkurva *c*

1798 fan inlet
DA ventilatorindløb
DE Ventilatoreinlaß *m*
ES boca de admisión del
ventilador *f*
oído *m*
F ouïe *f* d'aspiration
de ventilateur
I ingresso *m* del
ventilatore
MA ventilátor szívócsonk
NL ventilatoraanzuig-
opening *f*
PO przewód *m* ssawny
wentylatora
wlot *m* do
wentylatora
PY всасывающее
отверстие *n*
вентилятора
вход *m*
вентилятора
SU puhaltimen
imuaukko
SV fläktinlopp *n*

1799 fan input power boundary
DA grænse for tilført ventilatoreffekt
DE Ventilatorstromaufnahme-begrenzer *m*
ES límite de potencia del ventilador *m*
F puissance *f* limite du ventilateur à l'entrée
I alimentazione *f* di potenza del ventilatore
MA ventilátor belépő csatlakozás
NL vermogensverlies *n* in ventilator
PO zakres *m* mocy zainstalowanej wentylatora
PY потери *f,pl* мощности в вентиляторе
SU puhaltimen ottaman tehon raja
SV ineffektgräns *c* för fläkt *c*

1800 fan laws
DA ventilatorbestem-melser
DE Ventilator-Gesetze *n,pl*
ES leyes físicas de los ventiladores *f,pl*
F lois *f* fondamentales des ventilateurs
I leggi *f,pl* dei ventilatori
MA ventilátor türvényszerűségek
NL ventilatortheorie *f* ventilatorwetten *pl*
PO prawa *n* wentylatora
PY теория *f* вентилятора
SU puhallinlait
SV fläktnormer *c*

1801 fan performance curve
DA ventilatorydelses-kurve
DE Ventilatorleistungs-kurve *f*
ES curva característica del ventilador *f*
F courbe *f* caractéristique de ventilateur
I curva *f* caratteristica del ventilatore
MA ventilátor teljesítménygürbe
NL ventilatorprestatie-curve *f*
PO krzywa *f* sprawności wentylatora
PY характеристика *f* вентилятора в координатах 'давление-расход'
SU puhaltimen suoritusarvokäyrä
SV fläkteffektkurva *c*

1802 fan power
DA ventilatoreffekt
DE Ventilatorleistung *f*
ES potencia del ventilador *f*
F puissance *f* de ventilateur
I portata *f* del ventilatore
MA ventilátor teljesítmény
NL ventilator-vermogen *n*
PO moc *f* wentylatora
PY производитель-ьность *f* вентилятора
SU puhaltimen teho
SV fläkteffekt *c*

1803 fan pressurization test
DA ventilatortrykprøve
DE Ventilator-Druckprüfung *f*
ES prueba depresión del ventilador *f*
F essais *m* de ventilateur
I test *m* di pressurizzazione del ventilatore
MA ventilátoros nyomásvizsgálat
NL overdruklektest *f*
PO próba *f* ciśnieniowa wentylatora
PY испытание *n* здания (помещения) на герметичность
SU rakennuksen vuotokoe puhaltimella paineistaen
SV üvertrycksprov *n* för fläkt *c*

1804 fan propeller
DA propelventilator
DE Ventilator-propeller *m*
ES rodete del ventilador *m*
F ventilateur *m* hélicoïde
I girante *f* del ventilatore
MA ventilátor szárnykerék
NL propeller *m* ventilatorwaaier *m*
PO wirnik *m* wentylatora
PY рабочее колесо *n* вентилятора
SU puhaltimen siipipyürä
SV fläktpropeller *c*

1805 fan shroud
DA ventilatorbeklædning
DE Ventilatorhülle *f*
ES envolvente del
  ventilador *f*
F protection *f* du
  ventilateur
I coclea *f* del
  ventilatore
  voluta *f* del
  ventilatore
MA ventilátor ház
NL ventilatorhuis *n*
PO tarcza *f*
  wzmacniająca
  wentylatora
PY обод *m* рабочего
  колеса
  вентилятора
SU puhaltimen
  suojakotelo
SV fläktskydd *n*

1806 fan sound power
DA ventilatorlydtryk
DE Ventilator-
  lautstärke *f*
ES potencia sonora del
  ventilador *f*
F puissance *f*
  acoustique d'un
  ventilateur
I potenza *f* sonora del
  ventilatore
MA ventilátor
  hangnyomásszintje
NL ventilatorgeluids-
  vermogen *n*
PO moc *f* akustyczna
  wentylatora
PY звуковая мощность
  f вентилятора
SU puhaltimen äänen
  tehotaso
SV fläktljudeffekt *c*

1807 fan wheel cone
DA ventilatorhjulskonus
DE Ventilatorlaufrad-
  konus *m*
ES oído del ventilador
  m
F
I bocca *f* di
  aspirazione conica
  del ventilatore
MA ventilátor járókerék
  szívókúp
NL ventilatorwaaier-
  konus *m*
PO stożkowa
  przekładnia *f*
  wentylatora
PY входной конфузор
  m вентилятора
SU puhaltimen
  siipipyörän napa
SV fläkthjulkon *c*

1808 fastening
DA binding
  sikring
DE Befestigung *f*
ES atadura *f*
  fijación *f*
  sujeción *f*
F fixation *f*
I fissaggio *m*
MA rügzítés
NL bevestiging *f*
  sluiting *f*
PO mocowanie *n*
  utwierdzanie *n*
PY закрепление *n*
SU kiinnitys
SV fastsättning *c*

1809 faucet (USA) (*see*
  tap)

1810 fault
DA fejl
  mangel
DE Fehler *m*
ES espita *f*
  grifo *m*
F défaut *m*
  défectuosité *f*
I difetto *m*
MA hiba
NL defect *n*
  fout *f*
  storing *f*
PO błąd *m*
  skaza *f*
  uszkodzenie *n*
PY ошибка *f*
SU virhe
SV fel *n*

1811 fee
DA honorar
DE Gebühr *f*
ES cuota *f*
  honorarios *m*
F honoraire *m*
I tariffa *f*
MA díj
NL honorarium *n*
  tarief *n*
PO wynagrodzenie *n*
  zapłata *f*
PY оплата *f*
SU maksu
  tariffi
SV tariff *c*

1812 feed
DA tilførsel, føde-
DE Beschickung *f*
  Einspeisung *f*
ES alimentación *f*
F alimentation *f*
I alimentazione *f*
MA táplálás
NL aanvoer *m*
  toevoer *m*
  voeding *f*
PO podawanie *n*
  zasilanie *n*
PY загрузка *f*
  питание *n*
SU syüttü
SV matning *c*

1813 feed *vb*
DA tilføre
DE speisen
  zuführen
ES alimentar
F alimenter
I alimentare
MA táplál
NL aanvoeren
PO doprowadzać
  dostarczać
  zasilać
PY загружать
SU syüttää
SV mata

DA	=	Danish
DE	=	German
ES	=	Spanish
F	=	French
I	=	Italian
MA	=	Magyar (Hungarian)
NL	=	Dutch
PO	=	Polish
PY	=	Russian
SU	=	Finnish
SV	=	Swedish

1814 feedback
DA tilbagekobling
   tilbagemelding
DE Rückführung *f*
   Rückkopplung *f*
ES realimentación *f*
F contre réaction *f*
I feedback *m*
   retroazione *f*
MA visszacsatolás
   visszatáplálás
NL terugkoppeling *f*
PO sprzężenie *n* zwrotne
PY возврат *m* части
   рециркуляция *f*
SU takaisinkytkentä
SV återfüring *c*

1815 feedback system
DA tilbageføringssystem
DE Rückführungs-
   system *n*
ES sistema de
   realimentación *m*
F système *m* de contre
   réaction
I sistema *m* a
   retroazione
MA visszacsatolós
   rendszer
NL terugkoppel-
   systeem *n*
PO system *m* sprzężenia
   zwrotnego
PY рециркуляционная
   система *f*
SU takaisinkytketty
   järjestelmä
SV återfüringssystem *n*

1816 feed check valve
DA fødeventil
DE Regelventil *n* für
   Einspeisung *f*
ES válvula de
   alimentación de
   cierre automático *f*
   válvula de retorno
   de alimentación *f*
F robinet *m*
   d'alimentation
   robinet *m* de réglage
I valvola *f* di
   alimentazione e
   ritegno
MA tápszelep
   tültőszelep
NL voedingskeerklep *f*
PO zawór *m* zwrotny na
   zasileniu
PY питательный
   регулирующий
   клапан *m*
SU syüttüventtiili
SV matarvattenventil *c*

1817 feed control
DA tilbageførings-
   regulering
DE Beschickungs-
   regelung *f*
ES regulación de la
   alimentación *f*
F commande *f*
   d'alimentation
I regolazione *f* di
   portata
MA tültésszabályozás
NL voedingsregeling *f*
PO regulacja *f* zasilania
PY регулирование *n*
   подачи
SU syütün säätü
SV matarkontroll *c*

1818 feeder (electrical)
DA fødeledning
DE Speiseleitung *f*
   Versorgungs-
   leitung *f*
ES conductor principal
   m
   alimentación (elec) *f*
F conduit *m* principal
   d'alimentation
   électrique
I alimentatore *m*
   (elettrico)
MA tültő
NL voedingskabel *m*
PO linia *f* zasilająca
   przewód *m*
   zasilający
PY фидер *m*
SU syüttü
SV matarledning *c*

1819 feed water
DA fødevand
DE Speisewasser *n*
ES agua de
   alimentación *f*
F eau *f* d'alimentation
I acqua *f* di
   alimentazione
MA tápvíz
NL voedingswater *n*
PO woda *f* zasilająca
PY питательная вода *f*
SU syüttüvesi
SV matarvatten *n*

**1820 feed water
economizer**
DA fødevandsforvarmer
DE Speisewasser-
   vorwärmer *m*
ES economizador de
   alimentación de
   agua *m*
F économiseur *m*
   d'eau alimentaire
I economizzatore *m*
   dell'acqua di
   alimentazione
MA tápvíz előmelegítő
NL voedingswater-
   economiser
PO podgrzewacz *m*
   wody zasilającej
PY экономайзер *m* на
   линии питающей
   воды
SU syüttüveden
   esilämmitys
SV matarvatten-
   ekonomiser *c*

1821 feed water tank
DA fødevandstank
DE Speisewasser-
      behälter *m*
ES tanque del agua de
      alimentación *m*
F   bâche *f* alimentaire
I   vaso *m* di
      alimentazione
MA tápvíz tartály
NL voedingswater-
      tank *m*
PO zbiornik *m* wody
      zasilającej
PY бак *m*
      подпитывающей
      воды
SU syüttüvesiastia
SV matarvattencistern *c*

1822 felt
DA filt
DE Filz *m*
ES fieltro *m*
F   feutre *m*
I   feltro *m*
MA filc
NL vilt *n*
PO filc *m*
      wojłok *m*
PY войлок *m*
SU huopa
SV filt *c*

1823 female connection
DA hunkobling
DE Verbindungsstück *n*
ES conexión hembra *f*
      rácor hembra *m*
F   connexion *m* femelle
I   attacco *m* femmina
MA belsőmenetes
      csatlakozás
NL binnendraad-
      verbinding *f*
PO połączenie *n* z
      gwintem
      wewnętrznym
PY внутреннее
      сопротивление *n*
SU naarasliitin
SV anslutningsstycke *n*

1824 female thread
DA indvendigt gevind
DE Innengewinde *n*
ES rosca hembra *f*
      rosca interior *f*
F   filetage *m* femelle
      taraudage *m*
I   filettatura *f* femmina
MA belsőmenet
NL binnendraad *m*
PO gwint *m* wewnętrzny
PY внутренняя
      резьба *f*
SU sisäkierre
SV innergänga *c*

1825 fenestration
DA vinduesplacering
DE Fensterwerk *n*
ES ventanaje *m*
F   surface *f* vitrée
I   finestratura *f*
MA ablakkiosztás
      ablakozás
NL beglaasd
      oppervlak *n*
      vensterindeling *f*
PO rozmieszczenie *n*
      okien
      rozmieszczenie *n*
      otworów
PY степень *f* (доля)
      остекления
SU ikkunat
SV fünster *n*

**1826 fenestration
components**
DA vindueskomponenter
DE Fensterbauteile *n,pl*
ES componentes de
      ventanaje *m*
F   composants *m* de
      surface vitrée
I   componenti *m,pl*
      vetrati
MA ablakelemek
NL raamsysteem-
      elementen *pl*
PO składowe *f*
      rozmieszczania
      otworów
PY остекление *n*
SU ikkunan osat
SV fünster-
      komponenter *c*

1827 fiber optic
DA fiberoptik
DE Faseroptik *f*
ES fibra óptica *f*
F   fibre *f* optique
I   fibra *f* ottica
MA száloptika
NL glasvezeltechniek *f*
PO technika *f*
      swiatłowodowa
PY волоконная
      оптика *f*
SU kuituopitiikka
SV fiberoptik *c*

1828 fibre
DA fiber
      tave
DE Faser *f*
ES fibra *f*
F   fibre *f*
I   fibra *f*
MA rost
      szál
NL vezel *f*
PO włókno *n*
PY волокно *n*
      фибра *f*
SU kuitu
SV fiber *c*

1829 fibrous dust
DA fibrøst støv
DE Faserstaub *m*
ES polvo de fibras *m*
F   poussière *f* fibreuse
I   polvere *f* fibrosa
MA szálas por
NL vezelachtige stof *n*
PO pył *m* włóknisty
PY волокнистая
      пыль *f*
SU kuitumainen püly
SV fibrüst stoff *n*

1830 fibrous filter
DA fiberfilter
DE Faserfilter *m*
ES filtro de fibras *m*
F   filtre *m* à média
      fibreux
I   filtro *m* in materiale
      fibroso
MA szálanyagos szürő
NL vezelfilter *n*
PO filtr *m* włóknisty
PY волокнистый
      фильтр *m*
SU kuitusuodatin
SV fiberfilter *n*

1831 field device
DA feltudstyr
DE Feldgerät n
ES aparato de campo m
F instrument m
portable de terrain
I apparecchiatura f di campo
MA helyszíni szerkezet
NL apparaat n voor gebruik op het werk
PO urządzenie n polowe
PY область f использования устройства переносное устройство n
SU kenttälaite
SV fältanordning c

1832 field engineer
DA arbejdspladsingeniør
DE Außendienst-ingenieur m
ES ingeniero de campo m
F ingénieur m de chantier
I tecnico m di cantiere
MA helyszíni művezető mérnük
NL technicus m op het werk
PO inżynier m budowy
PY инженер-инспектор m
SU työmaainsinüüri
SV fältingenjür c

1833 field-installed device
DA feltinstalleret udstyr
DE am Bau m installiertes Gerät n
ES aparato instalado en obra m
F équipement m installé sur site
I apparecchiatura f di campo
MA helyszínen szerelt készülék
NL geinstalleerd apparaat f
PO urządzenie n zainstalowane na budowie
PY место n установки прибора
SU kentälle asennettu laite
SV platsinstallerad anordning c

1834 field of view
DA synsfelt
DE Sichtfeld n
ES campo de visión m
F champ m visuel
I campo m visivo
MA látómező
NL gezichtsveld n
PO pole n widzenia
PY поле n зрения
SU näkyvyysalue näkükenttä
SV synfält n

1835 figure of merit
DA godhedstal
DE Verdienstspanne f
ES imagen de calidad f
F facteur m de mérite
I figura f di merito
MA minőségi mutatószám
NL waarderingscijfer n
PO kryterium n jakości współczynnik m jakości
PY показатель m качества
SU edullisuus
SV godhetstal n

1836 filler
DA fyldmasse spartelmasse
DE Trichter m
ES relleno m
F appareil m de remplissage
I materiale m di riempimento
MA tülcsér
NL vulmiddel n vulstof f
PO wypełniacz m
PY наполнитель m
SU täytemassa
SV fyllnadsmassa c

1837 filling (see initial charge)

1838 fill thermal insulation
DA varmeisolerende granulat
DE einzufüllende Wärme-dämmung f
ES aislamiento térmico de relleno m
F isolation f thermique en vrac
I materiale m isolante sciolto
MA tültelékanyagos hőszigetelés
NL thermische isolatie f met losse vulstof
PO izolacja f cieplna wypełniająca
PY теплоизоляционное заполнение n
SU irtonainen lämpüeriste
SV fyllnadsvärme-isolering c

1839 film
DA film hinde
DE Film m Grenzschicht f
ES película f
F pellicule f de film
I pellicola f
MA film hártya
NL film m laag f
PO błonka f warstewka f
PY пленка f
SU kalvo
SV film c

1840 film boiling
DA kogning i grænselag
DE Filmsieden n
ES combustión pelicular f
F ébullition m par film
I ebollizione m a film
MA filmforrás
   hártyás forrás
NL koken n in
   grenslaag
PO wrzenie n błonkowe
PY пленочное
   кипение n
SU kalvokiehuminen
SV filmkokning c

1841 film coefficient of heat transfer
DA varmetransmissions-
   koefficient i grænselag
DE Wärmeübertragungs-
   koeffizient m der
   Grenzschicht f
ES coeficiente de
   película m
   coeficiente de
   transmisión
   superficial (de
   calor) m
F coefficient m de
   convection
   superficiel
I coefficiente m di
   trasmissione del
   calore liminare
MA hőátadás határréteg
   tényezője
NL warmteoverdrachts-
   coëfficiënt m
PO współczynnik m
   wymiany ciepła
   przez błonkę
PY коэффициент m
   теплоотдачи
SU lämmünsiirtymis-
   kerroin
SV filmvärmetransmissions-
   koefficient c

1842 film conductance
DA varmeledning i
   grænselag
DE Wärmeleitfähigkeit f
   der Grenz-
   schicht f
ES conductancia en
   película f
   conductancia
   (térmica)
   peculiar f
F transmittance f
   thermique
   superficielle
I conduttanza f
   liminare
MA film vezetőképessége
NL grenslaagwarmte-
   geleiding f
PO przewodność f
   błonkowa
PY поверхностная
   проводимость f
SU pinnan konduktanssi
SV gränsskiktsvärme-
   ledning c

1843 film cooling tower
DA filmkøletårn
DE Film-Kühlturm m
ES torre de refrigeración
   pelicular f
F tour f de
   refroidissement
   par film
I torre f di
   raffreddamento a
   pacco evaporante
MA vízfilmes hűtőtorony
NL koeltoren met
   filmverdamping f
PO chłodnia f
   kominowa
   ociekowa
PY пленочная
   градирня f
SU jäähdytystorni
SV filmkyltorn n

1844 film forming
DA grænselagsdannende
DE Filmbildung f
ES formación de
   película f
F
I formazione f di film
MA filmképződés
NL filmvorming f
PO tworzenie n się
   warstewki
PY формирование n
   пленки
SU kalvonmuodostus
   esim.
   lauhtumisessa
SV skiktbildning c

1845 film resistance
DA grænselagsmodstand
DE Wärmeleit-
   widerstand m der
   Grenzschicht f
ES resistencia peculiar f
F résistance f
   thermique
   superficielle
I resistenza f
   pellicolare
MA hártyaellenállás
NL oppervlakte-
   weerstand m
PO opór m błonki
   (warstwy)
PY поверхностное
   сопротивление n
SU pinnan
   lämmünsiirtymis-
   vastus
SV gränsskikts n
   värmemotstånd n

1846 filter
DA filter
DE Filter m
ES filtro m
F filtre m
I filtro m
MA szűrő
NL filter n
PO filtr m
PY фильтр m
SU suodatin
SV filter n

1847 filter *vb*
DA filtrere
DE filtern
   filtrieren
ES filtrar
F filtrer
I filtrare
MA szűr
NL filteren
   filtreren
PO filtrować
PY фильтровать
SU suodattaa
SV filtrera

1848 filter area
DA filterareal
DE Filterfläche *f*
ES superficie filtrante *f*
F surface *f* filtrante
I superficie *f* filtrante
MA szűrőfelület
NL filteroppervlak *n*
PO powierzchnia *f*
   filtracji
   powierzchnia *f* filtru
PY площадь *f*
   фильтрации
   фильтрующая
   поверхность *f*
SU suodattimen
   pintaala
SV filterarea *c*

1849 filter bag
DA posefilter
DE Filtersack *m*
   Filtertasche *f*
ES bolsa filtrante *f*
   manga filtradora *f*
   manga para filtrar *f*
F sac *m* filtrant
I sacco *m* filtrante
MA szűrőzsák
NL filterzak *m*
PO worek *m* filtracyjny
PY мешочный
   (рукавный)
   фильтр *m*
SU suodatinpussi
SV påsfilter *n*

1850 filter cartridge (*see* filter element)

1851 filter cell (*see* filter element)

1852 filter chamber
DA filterkammer
DE Filterkammer *f*
ES cámara de filtrado *f*
F chambre *f* de
   filtration
I camera *f* di
   filtrazione
MA szűrőkamra
NL filterkamer *f*
PO komora *f* filtracyjna
PY фильтрационная
   камера *f*
SU suodatinkammio
SV filterkammare *c*

1853 filter cleaning
DA filterrensning
DE Filterreinigung *f*
   Filtersäuberung *f*
ES limpieza de filtros *f*
F nettoyage *m* de filtre
I pulizia *f* del filtro
MA szűrő tisztítás
NL filterreiniging *f*
PO czyszczenie *f* filtru
PY очистка *f* фильтра
SU suodattimen
   puhdistus
SV filterrengüring *c*

1854 filter cloth
DA filterdug
DE Filtertuch *n*
ES tela filtrante *f*
F toile *f* filtrante
I tessuto *m* filtrante
MA szűrőszüvet
NL filterdoek *m*
PO tkanina *f* filtracyjna
PY фильтрующая
   ткань *f*
SU suodatinkangas
SV filterduk *c*

1855 filter dehydrator
DA filtertørrer
DE Filtertrockner *m*
ES filtro deshidratador
   m
F filtre *m*
   déshydrateur
I essiccatore *m* del
   filtro
MA szűrőszárító
NL drogingsfilter *m*
PO odwadniacz *m* filtru
PY фильтр-
   осушитель *m*
SU kuivaaja
SV filteravfuktare *c*

1856 filter drier (*see* filter dehydrator)

1857 filter element
DA filterelement
DE Filterelement *n*
ES elemento filtrante *m*
F élément *m* filtrant
I elemento *m* filtrante
MA szűrőelem
NL filterelement *n*
PO działka *f* filtracyjna
   wkład *m* filtracyjny
PY фильтрующая
   установка *f*
   фильтрующий
   элемент *m*
SU suodatinkasetti
SV filterelement *n*

1858 filter frame
DA filterramme
DE Filterrahmen *m*
ES bastidor del filtro *m*
F cadre *m* de filtre
I telaio *m* del filtro
MA szűrőkeret
NL filterframe *n*
   filterraam *n*
PO rama *f* filtru
PY каркас *m* фильтра
   рама *f* фильтра
SU suodatinkehys
SV filterram *c*

1859 filter medium
DA filtermateriale
DE Filtermaterial *n*
ES materia filtrante *f*
   medio filtrante *m*
F matière *f* filtrante
I materiale *m* filtrante
MA szűrőküzeg
NL filtermateriaal *n*
   filtermedium *n*
PO materiał *m*
   filtracyjny
PY фильтрующая
   среда *f*
   фильтрующий
   материал *m*
SU suodatinmateriaali
SV filtermaterial *n*

1860 filter paper
DA filterpapir
DE Filterpapier *n*
ES papel de filtrar *m*
  papel filtro *m*
F papier *m* filtrant
I carta *f* filtro
MA szűrőpapír
NL filterpapier *m*
PO bibuła *f* filtracyjna
  papier *m* filtracyjny
PY фильтрующая
  бумага *f*
SU suodatinpaperi
SV filterpapper *n*

**1861 filter separation
efficiency**
DA filterudskilningsgrad
DE Filterabscheide-
  grad *m*
ES eficiencia de filtrado
  m
F rendement *m* d'un
  filtre
I efficienza *f* di
  filtrazione
MA szűrő leválasztási
  hatásfok
NL filterrendement *n*
PO sprawność *f* filtracji
PY эффективность *f*
  очистки
SU suodattimen
  erotusaste
SV filterseparations-
  verkningsgrad *c*

1862 filter tank
DA filterbeholder
DE Filterbecken *n*
ES depósito de
  filtración *m*
F réservoir *m* à
  filtration
  réservoir *m*
  d'épuration
I serbatoio *m* del
  filtro
MA szűrőtartály
NL filtreertank *m*
PO zbiornik *m*
  filtracyjny
PY емкость *f* сбора
  пыли
  пылеосадочная
  камера *f*
SU suodatinsäiliü
SV filterbehållare *c*

1863 filter unit (*see* filter
element)

1864 filtration
DA filtrering
DE Filtrierung *f*
ES filtración *f*
  filtrado *m*
F filtrage *m*
I filtrazione *f*
MA szűrés
NL filtratie *f*
PO filtracja *f*
PY фильтрация *f*
  фильтрование *n*
SU suodattaminen
SV filtrering *c*

1865 fin
DA finne
  lamel
  ribbe
DE Lamelle *f*
  Rippe *f*
ES aleta *f*
F ailette *f*
  lamelle *f*
I aletta *f*
MA borda
  lamella
NL lamel *f*
  ribbe *f*
PO żeberko *n*
  żebro *n*
PY ребро *n*
SU lamelli
  ripa
SV fläns *c*

1866 final pressure
DA sluttryk
DE Enddruck *m*
ES presión final *f*
F pression *f* finale
I pressione *f* finale
MA végnyomás
NL einddruk *m*
PO ciśnienie *n* końcowe
PY давление *n* в конце
  процесса
  остаточное
  давление *n*
SU loppupaine
SV sluttryck *n*

1867 fin efficiency
DA lamelvirkningsgrad
DE Rippenwirkungs-
  grad *m*
ES efectividad de la
  aleta *f*
  rendimiento de la
  aleta *m*
F rendement *m*
  d'ailette
I rendimento *m*
  dell'aletta
MA borda hatás
NL lamelrendement *n*
PO sprawność *f* żebra
PY эффективность *f*
  ребра
SU ripahyütysuhde
SV flänsverkningsgrad *c*

1868 finned *adj*
DA med lameller
DE berippt
ES aleteado
F à ailettes
I alettato
MA bordás
NL van lamellen
  voorzien
PO ożebrowany
  użebrowany
PY оребренный
SU ripa-
  rivallinen
SV flänsad

1869 finned tube
DA ribberør
DE Lamellenrohr *n*
  Rippenrohr *n*
ES tubo de aletas *m*
F tube *m* aileté
  tuyau *m* à ailettes
I tubo *m* alettato
MA bordáscső
NL ribbenbuis *m*
PO rura *f* ożebrowana
PY оребренная труба *f*
  ребристая труба *f*
SU ripaputki
SV kamflänsrür *n*

DA	=	Danish
DE	=	German
ES	=	Spanish
F	=	French
I	=	Italian
MA	=	Magyar (Hungarian)
NL	=	Dutch
PO	=	Polish
PY	=	Russian
SU	=	Finnish
SV	=	Swedish

1870 finned-tube radiator
DA ribberørsradiator
DE Rippenrohr-
 heizkürper *m*
ES radiador de tubos
 aleteados *m*
F radiateur *m* à tubes
 à ailettes
I radiatore *m* a tubi
 alettati
MA bordáscsüves
 fűtőtest
NL ribbenbuis-
 radiator *m*
PO grzejnik *m* z rur
 ożebrowanych
PY радиатор *m* из
 ребристых труб
SU ripaputkipatteri
SV kamflänsrür-
 radiator *c*

1871 fin pitch
DA ribbehøjde
DE Rippenhühe *f*
ES paso de las aletas *m*
 separación entre
 aletas *f*
F pas *m* d'ailettes
I passo *m*
 dell'alettatura
MA bordaosztás
NL lamelafstand *m*
 lamelsteek *m*
PO podziałka *f* żebra
 rozstaw *m* żeber
PY шаг *m* ребер
SU ripajako
SV flänshüjd *c*

1872 fin spacing
DA ribbeafstand
DE Rippenabstand *m*
ES separación de las
 aletas *f*
F nombre *m* d'ailettes
 par unité de
 longueur
I distanziatura *f* delle
 alette
MA bordatávolság
NL lameltussenruimte *f*
PO odstęp *m* między
 żebrami
PY расстояние *n*
 между ребрами
SU ripaväli
SV flänsavstånd *n*

1873 fire
DA brand
 ild
DE Feuer *n*
 Glut *f*
ES fuego *m*
F feu *m*
I fuoco *m*
MA tűz
NL brand *m*
 vuur *n*
PO ogień *m*
 pożar *m*
PY огонь *m*
SU tuli
SV brand *c*

1874 fire alarm system
DA brandalarmsystem
DE Feuerwarnsystem *n*
ES sistema de alarma
 contra incendio *m*
F avertisseur *m*
 d'incendie
I sistema *m* di allarme
 antincendio
MA tűzriasztó rendszer
NL brandalarm-
 systeem *n*
PO instalacja *f* alarmu
 pożarowego
PY система *f*
 предупреждения
 о пожаре
SU palohälytysjärje-
 stelmä
SV brandalarmsystem *n*

1875 fire bar
DA fyrrist
DE Feuerrost *m*
ES parrilla hogar *f*
F grille *f* de cheminée
I griglia *f* a barrotti
 del focolare
MA rostély
NL roosterstaaf *f*
PO ruszt *m* paleniskowy
PY ребро *n*
 колосниковой
 решетки
SU arina
SV eldgaller *n*

1876 fire bar element
DA ristelement
DE Roststab *m*
ES barrote de
 parrilla *m*
F barreau *m* de grille
I barrotto *m* di griglia
MA rostély elem
NL roosterelement *n*
 roosterstaaf *f*
PO element *m* rusztu
 paleniskowego
PY элемент *m*
 колосниковой
 решетки
SU arinarauta
SV roststav *c*

1877 firebox
DA fyrboks
DE Brennkammer *f*
ES cámara de
 combustión *f*
F chambre *f* de
 combustion
I camera *f* di
 combustione
 focolare *m*
MA égőtér
 tűztér
NL vuurhaard *m*
 vuurkist *f*
PO komora *f*
 paleniskowa
 komora *f* spalania
PY топливник *m*
 топочная камера *f*
SU tulipesä
SV brännkammare *c*

1878 firebrick
DA chamottesten
 ildfast sten
DE Schamottestein *m*
ES ladrillo refractario
 m
F brique *f* réfractaire
I mattone *m*
 refrattario
MA tűzálló tégla
NL vuurvaste steen *m*
PO cegła *f* ogniotrwała
PY огнеупорный
 кирпич *m*
SU tulenkestävä tiili
SV eldfast tegel *n*

1879 fire bridge
DA fyrbro
DE Feuerbrücke *f*
ES altar del horno *m*
   tranco *m*
F autel *m*
I altare *m* del focolare
MA tűzhíd
NL vuurbrug *f*
PO mostek *m* ogniowy
   próg *m* ogniowy
PY топочный порог *m*
SU paloporras
   palosilta
SV eldbrygga *c*

1880 fire clay
DA chamotte
   ildfast ler
DE Schamotte *f*
ES arcilla refractaria *f*
F argile *f* réfractaire
   chamotte *f*
I argilla *f* refrattaria
MA samott
   tűzálló agyag
NL vuurvaste klei *f*
PO glina *f* ogniotrwała
   szamota *f*
PY огнеупорная
   глина *f*
SU tulenkestävä tiili
SV eldfast lera *c*

1881 fire damper
DA brandspjæld
DE Feuerschutzklappe *f*
ES compuerta
   cortafuegos *f*
F clapet *m* coupe-feu
I serranda *f*
   tagliafuoco
MA tűzvédelmi
   csappantyú
NL brandklep *f*
PO przepustnica *f*
   ogniowa
   przepustnica *f*
   przeciwpożarowa
PY топочная
   заслонка *f*
SU palopelti
SV brandspjäll *n*

1882 fire door
DA branddør
DE Feuerschutztür *f*
   Feuertür *f*
ES puerta del hogar *f*
F porte *f* de coupe-feu
   porte *f* de foyer
I porta *f* tagliafuoco
MA tűzvédelmi ajtó
NL branddeur *f*
   vuurhaarddeur *f*
PO drzwi *pl*
   przeciwpożarowe
PY топочная дверь *f*
SU palo-ovi
SV branddürr *c*

1883 fire extinguisher
DA ildslukker
DE Feuerlüscher *m*
ES apagafuegos *m*
   extintor de
   incendios *m*
F extincteur *m*
I estintore *m*
MA tűzoltó készülék
NL brandblusser *m*
PO gaśnica *f*
   przeciwpożarowa
PY огнетушитель *m*
SU palonsammutin
SV brandsläckare *c*

1884 fireplace
DA ildsted
DE Feuerstätte *f*
   Kamin *m*
ES chimenea francesa *f*
   hogar *m*
F cheminée *f*
   d'appartement
I caminetto *m*
MA kandalló
   tűzhely
NL open haard *m*
   stookplaats *f*
PO kominek *m*
PY камин *m*
   топка *f*
SU takka
SV eldstad *c*

1885 fire point
DA flammepunkt
DE Flammpunkt *m*
ES punto de
   inflamación *m*
F point *m*
   d'inflammabilité
I temperatura *f* di
   infiammabilità
MA gyulladáspont
NL ontbrandingspunt *n*
PO punkt *m* zapłonu
PY температура *f*
   загорания
SU syttymispiste
SV brinnpunkt *c*

1886 fireproof *adj*
DA brandsikker
   ildfast
DE feuerfest
ES resistente al fuego
F incombustible
I incombustibile
MA tűzálló
NL vuurbestendig
   vuurvast
PO ognioodporny
   ogniotrwały
PY огнеупорный
SU palonkestävä
SV brandsäker

1887 fireproof *vb*
DA brandsikre
DE feuerfest machen
   flammsicher machen
ES ignifugar
F ignifuger
I ignifugare
   proteggere
   antifiamma
MA tűzállóvá tesz
NL brandwerend maken
PO czynić
   ognioodpornym
PY придавать
   огнеупорность
SU tehdä
   palonkestäväksi
SV güra brandsäker

1888 fireproofing
DA brandsikret
DE Feuerfestigkeit *f*
   feuersicher
ES ignifugación *f*
F ignifugeage *m*
I ignifugazione *f*
   protezione *f*
   antifiamma
MA tűzmentesítés
NL brandwerend
   maken *n*
PO ognioodporność *f*
PY покрытие *n*
   огнеупорным
   составом
SU palontorjunta
SV brandskydd *c*

1889 fire resistance
DA brandmodstand
DE Feuerbeständigkeit *f*
ES resistencia al fuego *f*
F résistance *f* au feu
I resistenza *f* al fuoco
MA tűzálló
   tűzbiztos
NL brandwerendheid *f*
PO ognioodporność *f*
   ogniotrwałość *f*
PY огнеупорность *f*
SU palonkestävyys
SV brandmotstånd *n*

1890 fire retarding *adj*
DA branddrøj
DE feuerdämmend
   feuerhemmend
ES retardador de
   incendios *m*
   retardador del
   fuego *m*
F pare-feu
I resistente al fuoco
MA tűzgátló
NL brandvertragend
PO opóźniający palenie
   opóźniający zapłon
   zmniejszający
   palność
PY огнезадер-
   живающий
SU paloahidastava
SV eldbegränsande

1891 fire tube
DA ildkanal
   røgrør
DE Rauchrohr *n*
ES tubo de humos *m*
F tube *m* de fumée
I tubo *m* di fumo
MA tűzcső
NL rookbuis *f*
   vlampijp *f*
PO płomieniówka *f*
PY дымогарная
   труба *f*
SU tuliputki
SV eldstadsrür *n*

1892 fire tube boiler
DA røgrørskedel
DE Flammrohrkessel *m*
   Rauchrohrkessel *m*
ES caldera de tubos de
   agua *f*
F chaudière *f* à tubes
   de fumée
I caldaia *f* a tubi di
   fumo
MA tűzcsüves kazán
NL vlampijpketel *m*
PO kocioł *m*
   płomieniówkowy
PY котел *m* с
   дымогарными
   трубами
SU tuliputkikattila
SV flamrürspanna *c*

1893 fire valve
DA brandventil
DE Brandventil *n*
ES válvula
   contrafuegos *f*
   válvula interruptora
   para caso de
   incendio *f*
F robinet *m* d'incendie
I valvola *f* a fuoco
MA tűzcsap
NL brandkraan *f*
PO zawór *m* pożarowy
PY пожарный
   клапан *m*
SU paloposti
SV brandventil *c*

1894 firing
DA tænding
DE Feuerung *f*
ES caldeo *m*
   disparo *m*
   encendido *m*
F chauffe *f*
I accensione *f*
   combustione *f*
MA tüzelés
NL aansteken *n*
   stoken *n*
PO opalanie *n*
   spalanie *n*
PY топка *f*
SU poltto
SV eldning *c*

1895 firing schedule
DA fyringsprogram
DE Feuerführung *f*
ES programa de
   encendido *m*
F commande *f* en
   cascade des
   chaudières
I programma *m* di
   accensione caldaia
MA tüzelési ütemterv
NL stookprogramma *n*
PO program *m* opalania
   kotła
PY график *m* топки
   расписание *n* топки
SU polttoaikataulu
SV eldningsprogram *n*

1896 first cost
DA anskaffelsespris
DE Einkaufspreis *m*
ES coste de
   instalación *m*
F dépenses *f*
   d'investissement
I costo *m* iniziale
MA beruházási kültség
NL initile kosten *pl*
   investerings-
   kosten *pl*
PO koszt *m*
   inwestycyjny
PY начальная
   стоимость *f*
SU hankintakustannus
SV anskaffnings-
   kostnad *c*

1897 fitter
DA montør
DE Installateur *m*
   Monteur *m*
ES montador *m*
F monteur *m*
I installatore *m*
   montatore *m*
MA szerelő
NL gasfitter *m*
   monteur *m*
PO monter *m*
PY слесарь-монтажник
   m
SU asentaja
SV montür *c*

1898 fitting
DA fitting
   tilpasning
DE Fitting *n*
   Formstück *n*
ES conexión *f*
   junta *f*
   montaje *m*
   rácor *m*
F raccord *m*
I raccordo *m*
MA szerelvény
NL fitting *m*
PO łącznik *m*
   złączka *f*
PY фасонная часть *f*
SU putken osa
SV armatur *c*

1899 fixed-directional
   grille
DA fast vandret rist
DE feststehendes
   Gitter *n*
ES rejilla con deflectores
   fijos *f*
F bouche *f* à lames
   fixes
I griglia *f* ad alette
   fisse
MA fixzsalu
NL rooster *n* met vaste
   lamellen
PO kratka *f* z
   kierownicami
   nieruchomymi
PY решетка *f* с
   фиксирующими
   направляющими
SU suunnattu
   kiinteäsiipinen
   säleikkü
SV fast rost *c*

1900 fixed point
DA fikspunkt
DE Festpunkt *m*
ES punto fijo *m*
F point *m* de consigne
   fixe
I punto *m* fisso
MA fix pont
NL vast punt *m*
PO punkt *m* stały
   punkt *m*
   zamocowania
PY неподвижная
   опора *f*
   фиксированная
   точка *f*
SU kiinteä arvo
   kiinteä piste
SV fast punkt *c*

1901 fixed setting
   control
DA fastindstillet styring
DE Regelung *f* mit
   fester
   Einstellung *f*
ES control de posición
   fija *m*
F régulation *f* à point
   de consigne fixe
I controllo *m* a
   configurazione
   fissa
MA értéktartó
   szabályozás
NL regelaar *m* met vaste
   waarde
PO regulacja *f*
   stałowartościowa
   regulacja *f* w
   oparciu o wartości
   nastawione
PY регулирование *n*
   по
   фиксированной
   точке
SU säätü kiinteillä
   asetusarvoilla
SV fast inställd
   kontroll *c*

1902 fixing bolt
DA fastspændingsbolt
DE Führungsstift *m*
ES pasador de
   fijación *m*
   perno de sujeción *m*
F boulon *m* de
   fixation
I bullone *m* di
   fissaggio
MA rügzítócsavar
NL bevestigingsbout *m*
PO śruba *f* mocująca
   śruba *f* ustalająca
PY закрепляющий
   болт *m*
SU kiinnityspultti
SV styrpinne *c*

1903 flake ice
DA flis-is
   kunstig sne
DE Flockeneis *n*
   Schuppeneis *n*
ES hielo en escamas *m*
F glace *f* en éclats
I ghiaccio *m* in scaglie
MA darabos jég
NL scherfijs *n*
   schilferijs *n*
PO lód *m* w postaci
   płatków
PY чешуйчатый лед *m*
SU jäähile
SV konstgjord snü *c*

1904 flame
DA flamme
DE Flamme *f*
ES llama *f*
F flamme *f*
I fiamma *f*
MA láng
NL vlam *f*
PO płomień *m*
PY пламя *n*
SU liekki
SV flamma *c*

DA	=	Danish
DE	=	German
ES	=	Spanish
F	=	French
I	=	Italian
MA	=	Magyar
		(Hungarian)
NL	=	Dutch
PO	=	Polish
PY	=	Russian
SU	=	Finnish
SV	=	Swedish

1905 flame failure device
DA flammesvigtsikring
DE Flammenüber-
wachung f
ES dispositivo (de
seguridad) para el
caso de apagarse
la llama m
F contrôle m de
flamme
I dispositivo m di
controllo dello
spegnimento della
fiamma
MA lángőr
NL vlambeveiligings-
apparaat n
PO urządzenie n
zabezpieczające
przed zgaśnięciem
płomienia
wziernik m do
kontroli płomienia
PY устройство n для
контроля
пламени
SU liekinvartija
SV flamvakt c

1906 flame impingement
DA flammestråling
DE Flammstoß m
ES control de llama m
F impact m de flamme
I contatto m di
fiamma
MA lángütküzés
NL vlamcontact n (met
vuurhaardwand)
PO uderzenie n
płomienia
PY действие n
пламени
SU liekin kosketus
SV direktinverkan c av
flamma c

1907 flame safeguard control
DA flammesikkerheds-
kontrol
DE Zündsicherheits-
kontrolle f
ES control de seguridad
de llama m
F contrôle m de
flamme
I controllo m di
sicurezza di
fiamma
MA lángbiztosító
szabályozás
NL vlambeveiliging f
vlambeveiligings-
systeem n
PO kontrola f
zabezpieczenia
płomienia przed
zgaśnięciem
PY пламязащитный
контроль m
SU liekinvartija
SV flamskydds-
kontroll c

1908 flame separation
DA flammedeling
DE Abheben n der
Flamme
ES desprendimiento (de
la llama) m
F décollement m de
flamme
I separazione f di
fiamma
MA láng leválás
NL weglopen n van de
vlam
PO rozdział m
płomienia
PY отделение n
пламени
SU liekin levitys
SV flamspridning c

1909 flammability
DA antændelighed
brændbarhed
DE Entzündbarkeit f
Zündfähigkeit f
ES inflamabilidad f
F inflammabilité f
I infiammabilità f
MA éghetőség
NL brandbaarheid f
ontvlambaarheid f
PO palność f
PY горючесть f
SU syttyvyys
SV antändbarhet c

1910 flammable adj
DA antændelig
DE brennbar
entflammbar
ES inflamable
F inflammable
I infiammabile
MA éghető
gyúlékony
NL brandbaar
ontvlambaar
PO palny
PY сгораемый
SU syttyvä
SV antändbar

1911 flange
DA flange
DE Flansch m
ES brida f
F bride f
I flangia f
MA karima
perem
NL flens m
PO kołnierz m
PY фланец m
SU laippa
SV fläns c

1912 flange connection
DA flangesamling
DE Flanschverbindung f
ES acoplamiento por
bridas m
F raccordement m par
brides
I attacco m a flangia
MA karimás csatlakozás
NL flensverbinding f
PO połączenie n
kołnierzowe
PY фланцевое
соединение n
SU laippaliitos
SV flänsfürband n

1913 flange joint (see flange connection)

1914 flanking transmission
DA flanketransmission
DE Flanken-
 übertragung *f*
ES transmisión lateral *f*
F transmission *f*
 latérale
I trasmissione *f*
 laterale
MA oldalirányú
 átbocsátás (zaj)
NL flankerende
 geluids-
 overdracht *f*
PO przekładnia *f* zębata
PY отражение *n*
 (звука)
SU sivutiesiirtymä
SV flanktransmission *c*

1915 flapper valve
DA klapventil
DE Flossenventil *n*
 Klappenventil *n*
ES válvula de disco *f*
F clapet *m* fléchissant
 soupape *f* à
 languettes
I valvola *f* a clapet
MA rugós szelep
NL afsluiter met
 tongklep *f*
PO zawór *m* klapowy
PY откидной клапан *m*
SU läppäventtiili
SV klaffventil *c*

1916 flare
DA blus
 krave
DE Aufflackern *n*
 Lichtsignal *n*
ES ensanchamiento
 cónico *m*
 llamarada *f*
F déflagration
I svasatura *f*
MA jelzőfény
NL fakkel *f*
 gloed *m*
PO błysk *m*
 rozbłysk *m*
 wybuch *m*
PY раструб *m*
SU loimu
 roihu
SV flamma *c*

1917 flared joint
DA kravesamling
DE konische
 Verbindung *f*
ES junta abocardada *f*
F joint *m* mandriné
I giunto *m* svasato
MA hollandi csatlakozás
NL conische
 klemfitting *m*
 flareverbinding *f*
PO połączenie *n*
 błyskawiczne
PY раструбное
 соединение *n*
SU puserrusliitos
SV flänsfürband *n*

1918 flare fitting
DA kravefitting
DE konisches Fitting *n*
ES acoplamiento
 abocardado *m*
F collet *m*
 joint *m* conique
I accoppiamento *m*
 svasato
MA hollandi szerelvény
NL conische
 klemfitting *m*
 flarefitting *f*
PO złącze *n*
 błyskawiczne
PY раструбная
 фасонная часть *f*
SU puserrusliitos
SV flänsad rürdel *c*

1919 flare nut
DA kravemøtrik
DE konische Schraube *f*
ES tuerca cónica *f*
F
I dado *m* svasato
MA hollandi anya
NL conische
 wartelmoer *f*
 flaremoer *f*
PO nakrętka *f*
 błyskawiczna
PY раструбная
 муфта *f*
SU kiristysmutteri
SV flänsmutter *c*

1920 flaring block
DA kraveværktøj
DE Erweiterungs-
 werkzeug *n*
ES tope cónico *m*
F dudgeonnière *f*
I utensile *m* per
 svasare
MA peremező szerszám
NL flareblok *n*
 flare-ijzer *n*
PO roztłaczarka *f*
 rozwiertarka *f*
PY оправка *f*
 устройство *n* для
 формования
 раструба
SU putken
 levitystyükalu
SV uppflänsnings-
 verktyg *n*

1921 flaring tool (*see* flaring block)

1922 flash chamber
DA vædskeudskiller
DE Entspannungs-
 kammer *f*
ES cámara de
 separación (detrás
 de la válvula de
 expansión) *f*
F chambre *f* de
 détente
I camera *f* di
 evaporazione
MA sarjúgáz kamra
NL afdamper *m*
 ontspanningsvat *n*
PO oddzielacz *m*
 nieskroplonych
 gazów
 odpowietrzacz *m*
PY ресивер *m*
 холодильной
 машины
SU hüyrysäiliü
 kaasusäiliü (kts.
 flash gas)
SV fürångare *c*

1923 flash gas
DA flashgas
DE Gas *n* durch
plützlichen
Druckabfall *m*
entstanden
ES gas instantáneo
(desprendido por
el refrig.al
expansionarse) *m*
F évaporation *f* causée
par chute brusque
de pression
I gas *m* di accensione
MA sarjúgáz
NL flashgas *n*
PO odparowanie *n*
spowodowane
gwałtownym
spadkiem ciśnienia
PY вспышка *f* газа
SU äkillisen paineen
alenemisen vuoksi
hüyrystynyt kaasu
SV strypgas *c*

1924 flash intercooler
DA flashkøler
DE
ES enfriador
instantáneo *m*
F refroidisseur *m*
intermédiaire à
détente
I raffreddatore *m*
intermedio per
espansione
MA küzbenső
elpárologtató
NL verdampingstussen-
koeler *m*
PO dochładzacz *m*
odparowujący
PY регенеративный
теплообменник
m (холодильной
машины)
SU säiliü jossa neste
hüyrystyy äkillisen
paineen laskun
vuoksi
SV momentanmellan-
kylare *c*

1925 flash point
DA flammepunkt
DE Flammpunkt *m*
ES punto de
evaporación *m*
punto de
evaporización *m*
punto de
inflamación *m*
F point *m* d'éclair
I punto *m* di
accensione
MA lobbanáspont
NL ontvlammingspunt *n*
PO temperatura *f*
zapłonu
PY точка *f*
воспламенения
(вспышки)
SU leimahduspiste
SV flampunkt *c*

1926 flash steam
DA flashdamp
genfordampnings-
damp
DE entspannter
Dampf *m*
ES vapor instantáneo *m*
F autovaporisation *f*
I vapore *m* nascente
MA sarjúgőz
NL flashstoom *m*
PO samoodparowanie *n*
PY утечка *f* пара
SU korkeapaineisen
lauhteen paineen
laskusta syntyvä
hüyry
SV strypånga *c*

1927 flash tank
DA genfordampnings-
beholder
DE Kondensat-
sammler *m*
ES cámara de
separación (detrás
de la válvula de
expansión) *f*
F réservoir *m* pour
autovaporisation
I serbatoio *m* a
evaporazione
rapida
MA elpárologtató tartály
NL afdamper *m*
ontspanningsvat *n*
PO zbiornik *m*
szybkiego
odparowania
PY расширительный
бак *m*
SU hüyrysäiliü (kts.
flash steam)
SV stryptank *c*

1928 flash vaporization
DA genfordampning
DE Spritzverdampfung *f*
ES vaporización
instantánea *f*
F vaporisation *f*
instantánée
I generazione *f* di
vapore di flash
MA hirtelen
elpárologtatás
NL verdamping *f* door
drukverlaging
PO odparowanie *n*
szybkie
PY расширительное
испарение *n*
SU paineen äkillisestä
laskusta aiheutuva
hüyrystyminen
SV momentanfürång-
ning *c*

1929 flash vessel (*see* flash
chamber)

1930 flat bar frame
DA pladejernsramme
DE Flachrahmen *m*
ES bastidor de hierro plano *m*
F cadre *m* en fers plats
I telaio *m* in ferro piatto
MA laposacél perem
NL frame uit staalstrip *m*
PO rama *f* z płaskownika
PY рама *f* из полосового металла
SU lattarautakehys
SV plattjärnsram *c*

1931 flat plate collector
DA plan solfanger
DE Flachplatten-kollektor *m*
ES colector de placa plana *m*
F collecteur *m* plan
I collettore *m* a piastra piana
MA síklemezes kollektor
NL vlakke plaatcollector *m*
PO kolektor *m* płaski
PY плоский коллектор *m* (в гелиосистемах)
SU tasokeräin
SV plan solfångare *c*

1932 flat rate charge
DA fast pris
DE Einfachtarif *m* Pauschalgebühr *f*
ES carga en porcentaje fijo *f* carga uniforme *f*
F tarif *m* forfaitaire tarif *m* uniforme
I tariffa *f* forfettaria
MA lakbér
NL uniform tarief *n*
PO opłata *f* ryczałtowa
PY предварительная оплата *f*
SU perusmaksu
SV grundkostnad *c*

1933 flexibility
DA elastictet fleksibilitet
DE Biegsamkeit *f* Flexibilität *f*
ES flexibilidad *f*
F flexibilité *f*
I flessibilità *f*
MA hajlékonyság
NL buigzaamheid *f* flexibiliteit *f*
PO elastyczność *f* giętkość *f*
PY гибкость *f*
SU joustavuus
SV flexibilitet *c*

1934 flexible *adj*
DA fleksibel
DE biegsam
ES flexible
F flexible
I flessibile
MA flexibilis hajlékony
NL flexibel
PO elastyczny giętki
PY гибкий
SU joustava
SV flexibel

1935 flexible coupling
DA elastisk kobling
DE flexible Kupplung *f*
ES acoplamiento flexible *m*
F accouplement *m* élastique
I accoppiamento *m* elastico accoppiamento *m* flessibile
MA rugalmas csatlakozás
NL flexibele verbinding *f*
PO połączenie *n* elastyczne
PY гибкое соединение *n*
SU joustava liitos
SV elastisk koppling *c*

1936 flexible duct
DA bøjelig kanal
DE flexible Rohrleitung *f*
ES conducto flexible *m*
F conduit *m* d'air flexible
I tubo *m* flessibile
MA flexibilis csővezeték
NL flexibel kanaal *n*
PO kanał *m* elastyczny przewód *m* elastyczny
PY гибкая вставка *f* гибкий участок *m* трубопровода
SU joustava kanava taivutettava kanava
SV flexibel kanal *c*

1937 flexible ducting (*see* flexible duct)

1938 flexible joint
DA fleksibel samling
DE flexible Verbindung *f* Schlauch-verbindung *f*
ES junta flexible *f*
F raccord *m* flexible
I raccordo *m* flessibile
MA rugalmas csatlakozás
NL flexibele verbinding *f*
PO złącze *n* elastyczne
PY гибкий стык *m* шарнир *m*
SU joustava liitos
SV flexibel fürbindning *c*

1939 flexible pipe
DA bøjeligt rør
DE Schlauch *m*
ES tubo flexible *m*
F tube *m* flexible tuyau *m* flexible
I tubo *m* flessibile
MA flexibilis cső
NL flexibele pijp *f*
PO rura *f* clastyczna
PY гибкая труба *f*
SU joustava putki
SV flexibelt rür *n*

1940 float
DA flyder
svømmer
DE Schwimmer *m*
ES flotador *m*
F flotteur *m*
I galleggiante *m*
MA lebegő
úszó
NL vlotter *m*
PO pływak *m*
PY поплавок *m*
SU uimuri
SV flottür *c*

1941 floating action
DA flydebevægelse
DE Pendelung *f*
ES acción flotante *f*
F action *f* flottante
I azione *f* flottante
MA változó működés
NL zwevende
regelactie *f*
PO ruch *m* płynny
PY астатическое
действие *n*
SU ohjaus
SV flottürverkan *c*

1942 floating control
DA svømmeregulering
DE Schwimmer-
regelung *f*
ES control flotante *m*
F régulation *f*
flottante
I controllo *m* flottante
MA változó szabályozás
NL zwevende regeling *f*
PO sterowanie *n*
astatyczne
PY астатическое
регулирование *n*
поплавковое
регулирование *n*
SU liukuva säätü
SV flüdesreglering *c*

1943 float switch
DA svømmereafbryder
DE Schwimmer-
schalter *m*
ES interruptor de
flotador *m*
F interrupteur *m* à
flotteur
I interruttore *m* a
galleggiante
MA úszókapcsoló
NL vlotterschakelaar *m*
PO wyłącznik *m*
pływakowy
PY поплавковый
выключатель *m*
SU uimurikytkin
SV flottürbrytare *c*

1944 float trap
DA flydedampudlader
DE Kondenstopf *m*
ES purgador de
flotador *m*
F purgeur *m* à flotteur
I scaricatore *m* di
condensa a
galleggiante
MA kondenzedény
NL vlottercondenspot *m*
PO odwadniacz *m*
pływakowy
PY поплавковый трап
m (затвор)
SU automaattinen
lauhteen erotin
SV flottürångfälla *c*

1945 float valve
DA svømmerventil
DE Schwimmerventil *n*
ES válvula de flotador *f*
F robinet *m* à flotteur
I valvola *f* a
galleggiante
MA úszószelep
NL vlotterkraan *f*
PO zawór *m* pływakowy
PY поплавковый
клапан *m*
SU uimuriventtiili
SV flottürventil *c*

1946 flock point
DA flokningspunkt
DE Flockungspunkt *m*
ES punto de misión *m*
F point *m* de floc
I punto *m* di
intorbidamento
MA pelyhesedési pont
NL uitvlokkingspunt *n*
PO punkt *m* flokulacji
PY температура *f*
помутнения
(выпадения
масла из
хладагента)
SU lämpütila jossa
kylmäkoneüljyn ja
R22:n (20%) seos
samenee
samepiste
SV flockpunkt *c*

1947 flooded evaporator
DA vandfyldt evaporator
DE überfluteter
Verdampfer *m*
ES evaporador
inundado *m*
F évaporateur *m* noyé
I evaporatore *m*
allagato
MA elárasztott
elpárologtató
NL verzopen
verdamper *m*
PO parownik *m* zalany
PY испаритель *m*
затопленного
типа
SU säiliühüyrystin
SV vattenfylld
evaporator *c*

1948 floor (structure)
DA etage
gulv
DE Fußboden *m*
Geschoß *n*
ES forjado *m*
F plancher *m* (bas)
I pavimento *m*
MA emelet
padló
NL verdieping *f*
vloer *m*
PO podłoga *f*
PY пол *m*
(конструкция)
SU lattia
SV golv *n*

1949 **floor area**
DA gulvareal
DE Geschoßfläche *f*
ES superficie *f*
F surface *f* de plancher
I area *f* del pavimento
MA padlófelület
NL vloeroppervlak *n*
PO powierzchnia *f* podłogi
PY площадь *f* пола
SU rakennusala
SV golvyta *c*

1950 **floor heating**
DA gulvvarme
DE Fußbodenheizung *f*
ES calefacción bajo el pavimento *f* calefacción por calentamiento del piso *f*
F chauffage *m* par le sol
I riscaldamento *m* a pannelli a pavimento
MA padlófűtés
NL vloerverwarming *f*
PO ogrzewanie *n* podłogowe
PY напольное отопление *n*
SU lattialämmitys
SV golvvärme *c*

1951 **floor rack**
DA gulvtræk
DE Doppelboden *m*
ES bancada *f*
F caillebotis *m*
I raffreddamento *m* ad aria sottopavimento
MA szellőztetett padló
NL roostervloer *m* voor gekoelde opslag
PO podłoga *f* podparta (podłoga z przepływem powietrza)
PY стеллажный пол *m* (для вентиляции)
SU korotettu lattia
SV golvdrag *n*

1952 **flotation**
DA flotation
DE Schweben *n* Schwimmen *n*
ES flotación *f*
F traitement *m* par flotation
I galleggiamento *m*
MA úsztatás
NL flotatie *f*
PO flotacja *f* pływanie *n*
PY флотация *f*
SU kelluminen
SV flotation *c*

1953 **flow**
DA strømning
DE Ausfluß *m* Strümung *f* Vorlauf *m*
ES corriente *f* flujo *m*
F débit *m* écoulement *m*
I corrente *f* flusso *m*
MA áramlás
NL stroom *m*
PO przepływ *m* ruch *m* cieczy strumień *m*
PY поток *m* течение *n*
SU virta virtaus
SV flüde *n*

1954 **flow area**
DA gennemstrømnings-areal
DE Strümungs-querschnitt *m*
ES superficie de flujo *f*
F section *f* de passage
I area *f* di flusso
MA átáramlási keresztmetszet
NL stromingsgebied *n*
PO powierzchnia *f* przepływu
PY площадь *f* потока
SU virtausala
SV genomloppsarea *c*

1955 **flow capacity**
DA gennemstrømnings-kapacitet
DE Durchflußmenge *f*
ES capacidad de flujo *f*
F débit *m* nominal (d'un filtre)
I capacità *f* di flusso
MA átáramló teljesítmény
NL debiet *n*
PO zdolność *f* przepustowa
PY объемный расход *m*
SU nimellisvirtaus
SV flüdeskapacitet *c*

1956 **flow connection (on boiler)**
DA rørforbindelse (på kedel)
DE Vorlaufanschluß *m*
ES conexión de salida (de una caldera) *f*
F raccordement *m* de départ (des chaudières)
I attacco *m* di mandata
MA csatlakozó csonk
NL aanvoeraansluiting *f*
PO przyłączenie *n* zasilania (w kotle)
PY гидравлический затвор *m* (на котле)
SU putkiliitäntä (kattila)
SV rüranslutning *c*

1957 **flow control**
DA strømningsregulering
DE Durchflußregelung *f*
ES control de flujo *m*
F régulation *f* de débit
I controllo *m* di flusso
MA áramlás szabályozás
NL debietregeling *f* hoeveelheids-regeling *f*
PO regulacja *f* przepływu
PY регулирование *n* течения
SU virtauksen säätü
SV flüdeskontroll *c*

1958 flow control valve
DA indreguleringsventil
DE Durchflußregel-
ventil *n*
ES válvula de regulación
de flujo *f*
F vanne *f* de
régulation de débit
I valvola *f* di
controllo di flusso
MA áramlás szabályozó
szelep
NL debietregel-
afsluiter *m*
PO zawór *m* regulacyjny
przepływu
PY клапан *m*
регулирования
течения
SU virtauksen
säätüventtiili
SV flüdeskontroll-
ventil *c*

1959 flow diagram
DA flow diagram
DE Durchfluß-
diagramm *n*
ES diagrama *f*
F ordinogramme *m*
organigramme *m*
I diagramma *m* di
flusso
MA folyamatábra
NL stromingsdiagram *n*
PO schemat *m* blokowy
schemat *m* działania
programu
schemat *m* technolo-
giczny
sieć *f* działań
PY диаграмма *f*
течения
SU virtauskaavio
SV flüdesdiagram *n*

1960 flow direction
DA strømningsretning
DE Strümungsrichtung *f*
ES dirección de flujo *f*
sentido de flujo *m*
F direction *f* de
l'écoulement
I direzione *f* del flusso
MA áramlási irány
NL stromingsrichting *f*
PO kierunek *m*
przepływu
PY направление *n*
течения
SU virtaussuunta
SV strümnings-
riktning *c*

1961 flow equalizer
DA strømningsudligner
DE Strümungsgleich-
richter *m*
ES equilibrador de
flujo *m*
F égalisateur *m*
répartiteur *m*
d'écoulement
I equalizzatore *m* di
flusso
MA áramlás
egyenirányító
NL stromingsgelijk-
richter *m*
PO stabilizator *m*
przepływu
PY стабилизатор *m*
течения
SU virtauksentasaaja
SV flüdesutjämnare *c*

1962 flow factor (kv/cv)
DA strømningsfaktor
DE Strümungsfaktor *m*
ES factor de flujo *m*
F facteur *m*
d'écoulement
I fattore *m* di flusso
MA áramlási tényező
NL kv *m*
stromingsfactor *m*
PO współczynnik *m*
przepływu
PY коэффициент *m*
потока
(отношение
действительной
скорости к
критической)
SU kapasiteettikerroin
kv-arvo
virtauskerroin
SV flüdesfaktor *c*

1963 flow meter
DA flowmeter
strømningsmåler
DE Durchflußmesser *m*
Strümungsmesser *m*
ES fluxómetro *m*
medidor de flujo *m*
F débitmètre *m*
I misuratore *m* di
portata
MA áramlásmérő
NL debietmeter *m*
PO licznik *m* przepływu
przepływomierz *m*
PY расходомер *m*
SU virtausmittari
SV flüdesmätare *c*

1964 flow nozzle
DA måledyse
DE Meßdüse *f*
ES tobera *f*
F ajutage *m*
tuyère *f*
I ugello *m* di flusso
MA fúvóka
NL uitstroom-
mondstuk *n*
uitstroomtuit *f*
PO dysza *f* pomiarowa
dysza *f*
przepływowa
PY расходомерное
сопло *n*
струйная
форсунка *f*
SU suutin
SV flüdesmunstycke *n*

1965 flow pattern
DA strømningsbillede
DE Strümungsbild *n*
ES configuración de
flujo *f*
disposición de
corriente *f*
F schéma *m*
d'écoulement
I andamento *m* dei
filetti fluidi
MA áramkép
NL stromingsbeeld *n*
stromingspatroon *n*
PO charakter *m*
przepływu
PY схема *f* течения
SU virtauskuvio
SV flüdesbild *c*

1966 flow rate
DA strømningsgrad
DE Durchflußrate *f*
Menge *f*
ES caudal *m*
medida del gasto *f*
F débit *m*
d'écoulement
I portata *f*
MA áramlási arány
NL debiet *n*
doorstromende
hoeveelheid *f*
PO natężenie *n*
przepływu
wielkość *f*
przepływu
PY расход *m* (потока)
SU virtausmäärä
SV flüdeshastighet *c*

1967 flow rate
controller
DA strømningsmængde-
regulator
DE Durchflußregler *m*
ES regulador de
caudal *m*
F régulateur *m* de
débit
I regolatore *m* di
portata
MA áramlási arány
szabályozó
NL debietregelaar *m*
PO regulator *m*
wielkości
przepływu
PY регулятор *m*
расхода
SU virtauksen säädin
SV flüdesregulator *c*

1968 flow regulating
valve
DA reguleringsventil
DE Durchflußregel-
ventil *n*
ES válvula de regulación
de caudal *f*
F robinet *m* de réglage
vanne *f* de réglage
I valvola *f* di
regolazione di
portata
MA áramlásszabályozó
szelep
NL debietregelafsluiter
m
PO zawór *m* regulujący
przepływ
PY клапан *m*,
регулирующий
течение
SU virtauksen
säätüventtiili
SV flüdesregulator-
ventil *c*

1969 flow resistance
DA strømningsmodstand
DE Strümungs-
widerstand *m*
ES resistencia de flujo *f*
F résistance *f* à
l'écoulement
I resistenza *f* al flusso
MA áramlási ellenállás
NL stromings-
weerstand *m*
PO opór *m* przepływu
PY сопротивление *n*
течению
SU virtausvastus
SV flüdesmotstånd *n*

1970 flow temperature
DA fremløbstemperatur
DE Vorlauftemperatur *f*
ES temperatura de
salida (flujo) *f*
temperatura inicial
(flujo) *f*
F température *f* de
départ
I temperatura *f* di
mandata
MA áramlási hőmérséklet
NL aanvoer-
temperatuur *f*
PO temperatura *f*
przepływu
temperatura *f*
strumienia
PY температура *f*
потока
SU virtauksen lämpütila
SV flüdestemperatur *c*

1971 flow velocity
DA strømningshastighed
DE Strümungs-
geschwindigkeit *f*
ES velocidad de flujo *f*
velocidad de la
corriente *f*
F vitesse *f*
d'écoulement
I velocità *f* di flusso
MA áramlási sebesség
NL stroomsnelheid *f*
PO prędkość *f*
przepływu
prędkość *f*
strumienia
PY скорость *f* течения
SU virtauksen nopeus
SV flüdeshastighet *c*

1972 flow work
DA fortrængningsarbejde
DE Strümungsarbeit *f*
ES conducción *f*
F énergie *f*
d'écoulement
I lavoro *m* di
spostamento
MA áramlási munka
NL stromingsenergie *f*
PO praca *f* strumienia
PY энергия *f* течения
SU virtauksen
synnyttämä tyü
SV

1973 flue
DA aftræk
røgkanal
DE Feuerzug *m*
Rauchfang *m*
ES conducto de
humos *m*
tubo de humos *m*
F carneau *m*
cheminée *f*
conduit *m* de fumée
I canna *f* fumaria
MA füstcső
kémény
NL rookkanaal *n*
vlampijp *f*
PO czopuch *m*
kanał *m* dymowy
płomienica *f*
PY дымовая труба *f*
SU savukaasu
SV rökgaskanal *c*

1974 flue brush
DA kedelbørste
DE Rußreinigungs-
bürste *f*
ES escobillón *m*
F écouvillon *m*
I scovolo *m* per canna
fumaria
MA kéménytisztító kefe
NL borstel *m* voor
rookkanaal
schoorsteen-
borstel *m*
PO szczotka *f*
czyszcząca
czopuch
wycior *m* czopucha
PY приспособление *n*
для чистки
дымовой трубы
SU nokiharja
SV viska *c*

DA	=	Danish
DE	=	German
ES	=	Spanish
F	=	French
I	=	Italian
MA	=	Magyar (Hungarian)
NL	=	Dutch
PO	=	Polish
PY	=	Russian
SU	=	Finnish
SV	=	Swedish

1975 flue collar
DA røgkanalmuffe
DE Abgasstutzen *m*
ES collar de chimenea *m*
F
I collare *m* di sbocco dei fumi
MA kémény sisak
NL afvoergas-aansluitkraag *m*
   afvoergas-aansluitstuk *n*
PO kołnierz *m* przyłączny czopucha
PY устье *n* дымовой трубы
SU savuhormin liitoskaulus
SV skorstenskrage *c*

1976 flue connection
DA røgkanal
DE Abgasanschluß *m*
ES conexión de humos *f*
F carneau *m*
   raccordement *m*
I condotto *m* ascendente
   condotto *m* dei fumi
MA rókatorok
NL rookgaskanaal-aansluiting *f*
PO połączenie *n* czopucha
PY соединение *n* с дымовой трубой
SU kattilan liitäntä savupiippuun
SV rökgasanslutning *c*

1977 flue draught
DA røgtræk
DE Schornsteinzug *m*
ES tiro de la chimenea *m*
F tirage *m* de cheminée
I tiraggio *m* del camino
MA kéményhuzat
NL schoorsteentrek *m*
PO ciąg *m* kominowy
PY тяга *f* дымовой трубы
SU savupiipun veto
SV skorstensdrag *n*

1978 flue gas
DA røggas
DE Abgas *n*
   Rauchgas *n*
ES humo *m*
   gas de combustión *m*
F fumée *f*
I gas *m* di combustione
MA égéstermék füstgáz
NL rookgas *n*
PO gaz *m* spalinowy spaliny *pl*
PY дымовые газы *m,pl*
SU savukaasu
SV rökgas *c*

1979 flue gas loss
DA røggastab
DE Abgasverlust *m*
ES pérdida(s) por gases de la combustión *f,pl*
F pertes *f* à la cheminée
   pertes *f* par les fumées
I perdita *f* di fumi al camino
MA füstgázveszteség
NL schoorsteenverlies *n*
PO strata *f* przy przepływie spalin
PY потери *f,pl* теплоты с уходящими дымовыми газами
SU savukaasuhäviü
SV rökgasförlust *c*

1980 flue gas removal
DA røgfjerner
DE Abgasbeseitigung *f*
ES eliminación de gases de la combustión *f*
   eliminación de humos *f*
F évacuation *f* des fumées
   évacuation *f* des gaz
I scarico *m* dei gas di combustione (fumi)
MA égéstermék elvezetés
NL rookgasafvoer *m*
PO usuwanie *n* spalin
PY удаление *n* дымовых газов
SU savukaasun poisto
SV bortförande *n* av rökgas *c*

1981 flue gas vent
DA røgaftræk
DE Rauchgasabzug *m*
ES barrido de gases *m*
F conduit *m* de fumée
I scarico *m* dei prodotti di combustione
MA égéstermékelvezető kürtő
NL rookgasuitlaat *m* naar buitenlucht
PO wylot *m* spalin
PY канал *m* для выхода дымовых газов в атмосферу
SU pakoputki savuhormi
SV rökgasventil *c*

1982 flueless
DA røgfri
DE schornsteinlos
ES sin tubos de gases *m*
F sans gaine d'évacuation
I apparecchio *m* a scarico libero
MA kémény nélküli
NL schoorsteenloos
PO bezczopuchowy
PY нагреватель *m* без дымовой трубы
SU savupiiputon lämmitin
SV ej skorstensansluten

1983 flue liner
DA skorstensindsats
DE Schornsteinfutter *n*
ES aislamiento de la chimenea *m*
   revestimiento (tubular) de la chimenea *m*
F tubage *m* de cheminée
I rivestimento *m* della canna fumaria
MA füstcsőbélés
NL schoorsteenvoering *f*
PO wykładzina *f* czopucha
PY футеровка *f* дымовой трубы
SU savupiipun verhous
SV rökkanal *c*

1984 flue outlet
DA røgafgang
DE Rauchgasauslaß *m*
ES salida de gases *f*
F débouché *m* de
conduit de fumée
I sbocco *m* della
canna fumaria
MA kémény kilépő nyílás
NL rookgasuitlaat *m*
PO wylot *m* czopucha
wylot *m* kanału
spalinowego
PY выходное отверстие
n дымовой трубы
SU savukaasun
poistokohta
SV rökgasutlopp *n*

1985 flue temperature
DA røgtemperatur
DE Abgastemperatur *f*
ES temperatura de
humos *f*
F température *f* de
fumée
I temperatura *f* dei
prodotti di
combustione
MA kémény hőmérséklet
NL rookgas-
temperatuur *f*
schoorsteen-
temperatuur *f*
PO temperatura *f* spalin
w czopuchu
temperatura *f* spalin
w kanale
spalinowym
PY температура *f*
дымовых газов
SU savukaasun
lämpütila
SV rökgastemperatur *c*

1986 fluid
DA væske
DE Flüssigkeit *f*
Fluid *n*
Gas *n*
ES fluido *m*
F fluide *m*
I fluido *m*
MA folyadék
folyékony
NL vloeistof *f*
PO ciecz *f*
płyn *m*
PY жидкость *f*
SU neste
SV vätska *c*

1987 fluidics
DA strømningslære
DE Fluide *n,pl*
fluide Medien *n,pl*
ES fluídos *m*
F fluidique *f*
I fluidica *f*
MA folyadéktan
NL fluidica *pl*
PO automatyka *f*
strumieniowa
technika *f*
strumieniowa
PY изучение *n* и
использование *n*
свойств жидких
и газовых-
потоков
SU fluidistoritekniikka
SV fluidik *c*

**1988 fluidized bed
combustion**
DA forbrænding
svævende (fluidized
bed)
DE Wirbelschicht-
verbrennung *f*
ES combustión en lecho
fluidificado *f*
F combustion *f* par lit
fluidisé
I combustione *f* a
letto fluido
MA fluidágyas égés
NL wervelbed-
verbranding *f*
PO spalanie *n* na złożu
fluidalnym
PY горение *n* в
псевдос-
жиженном слое
SU leijukerrospoltto
SV fluidiserad bädd *c*
förbränning *c*

**1989 fluidized bed
freezer**
DA dispersionsfryser
DE Wirbelschicht-
gefrieren *n*
ES congelador de lecho
fluído *m*
F congélateur *m* à lit
fluidisé
I congelatore *m* a
letto fluido
MA fluidágyas fagyasztó
NL wervelbedvriezer *m*
PO zamrażarka *f*
fluidyzacyjna
PY холодильное
устройство *n* на
основе
псевдо-
сжиженного слоя
SU leijupakastin
SV fluidiserad bädd *c*
frysapparat *c*

1990 fluid mechanics
DA strømningslære
DE Strümungs-
mechanik *f*
ES mecánica de fluídos *f*
F mécanique *f* des
fluides
I meccanica *f* dei
fluidi
MA folyadékok
mechanikája
NL stromings-
mechanica *f*
PO mechanika *f* płynów
PY гидроаэро-
динамика *f*
гидромеханика *f*
SU virtausmekaniikka
SV hydromekanik *c*

1991 fluorocarbon
DA fluor
DE Fluorkohlenstoff *m*
ES fluorcarbono *m*
F fluorocarbure *m*
I fluorocarburo *m*
MA fluorkarbon
(hűtőküzeg)
NL fluorkoolstof *f*
PO fluoropochodna *f*
węglowodorów
PY фторуглерод *m*
SU fluorattu hiilivety
SV fluorkol *n*

1992 flush *vb*
DA udskylle
DE ausspülen
durchspülen
spülen
ES limpiar a presión
(mediante un
líquido o un gas)
limpiar con un fluido
a presión
F rincer
I lavare
pulire
MA eláraszt
üblít
NL doorspoelen
spoelen
PO spłukać
spłukiwać
PY промывать
SU huuhtoa
SV spola

1993 flush-fitting door
DA vandtæt lem
DE dichtschließende
Tür *f*
ES puerta estanca *f*
I porta *f* a scomparsa
MA üblítőnyílás
NL vlakliggende deur *f*
PO drzwi *pl* wpuszczane
klamka *f*
PY плотно пригнанная
дверь *f*
SU seinäpinnan tasoon
sovitettu ovi
SV

1994 flushing cistern
DA skyllecisterne
DE Spülkasten *m*
ES cisterna de
inodoro *f*
F réservoir *m* de
chasse
I cassetta *f* di cacciata
MA üblítőtartály
NL stortbak *m*
PO płuczka *f* ustępowa
zbiornik *m*
spłukujący
(klozetowy)
PY бак *m* для
промывки
SU huuhtelusäiliü
SV spolcistern *c*

1995 flush valve
DA spuleventil
DE Spülventil *n*
ES válvula de descarga
automática *f*
F robinet *m* de chasse
I valvola *f* di scarico
MA üblítőszelep
NL spoelklep *f*
PO zawór *m* spłukujący
PY промывной
клапан *m*
SU huuhteluventtiili
SV spolventil *c*

1996 flux
DA flus
loddepulver
strøm
DE Flußmittel *n*
Strümung *f*
Strom *m*
ES flujo (de calor,
magnético, etc.) *m*
fundente *m*
F flux *m*
métal *m* d'apport
pour soudure
I disossidante *m*
flusso *m*
metallo *m* di
apporto per
saldatura
MA fluxus
NL stroom *m*
vloeimiddel *n*
PO strumień *m*
topnik *m*
PY поток *m*
SU virta
virtaus
SV fluss *n*

1997 fly ash
DA flyveaske
DE Flugasche *f*
ES cenizas volantes *f,pl*
F cendres *f* volantes
I cenere *f* volatile
MA szállópernye
NL vliegas *f*
PO popiół *m* lotny
PY копоть *f*
летучая зола *f*
SU lentotuhka
SV flygaska *c*

1998 flywheel
DA svinghjul
DE Schwunggrad *m*
ES volante *m*
F volant *m* mécanique
I volano *m*
MA lendkerék
NL vliegwiel *n*
PO koło *n* zamachowe
PY маховик *m*
SU vauhtipyürä
SV svänghjul *n*

1999 foam
DA skum
DE Schaum *m*
ES espuma *f*
F écume *f*
mousse *f*
I schiuma *f*
MA hab
NL schuim *n*
PO piana *f*
PY пена *f*
SU vaahto
SV skum *n*

2000 foam *vb*
DA skumme
DE schäumen
ES espumar
F écumer
mousser
I schiumare
MA habzik
NL schuimen
PO pienić się
PY вспенивать
SU vaahdota
SV skumma

2001 foamed-in-place thermal insulation
DA skumisolering fremstillet på stedet
DE Ortschaumwärme-dämmung *f*
ES aislamiento térmico expandido *m*
F isolation *f* expansée in situ
I isolamento *m* termico di schiuma formata in loco
MA helyszínen habosított hőszigetelés
NL in situ geschuimde isolatie *f*
PO izolacja *f* cieplna piankowa natryskiwana na budowie
PY теплоизоляция *f* из вспененного материала
SU lämpüeristevaahto
SV skumvärme-isolering *c*

2002 foam glass
DA opskummet isoleringsmateriale (foam glass)
DE Schaumglas *n*
ES espuma de vidrio *f*
F
I vetro *m* cellulare
MA üveghab
NL schuimglas *n*
PO szkło *n* piankowe
PY вспененное стекло *n*
SU vaahdotettu lasi
SV glasfiber *c*

2003 fog
DA tåge
DE Nebel *m*
ES niebla *f*
F brouillard *m*
I nebbia *f*
MA küd
NL mist *m* nevel *m*
PO mgła *f*
PY туман *m*
SU sumu
SV dimma *c*

2004 fogging
DA tågedannelse
DE Vernebelung *f*
ES niebla *f*
F nébulisation *f*
I formazione *f* di nebbia
MA küdüsítés
NL verneveling *f*
PO zadymienie *n* zamglenie *n* zaparowanie *n*
PY образование *n* тумана
SU sumunmuodostus
SV dimbildning *c*

2005 foil thermal insulation (*see* reflective thermal insulation)

2006 footstep bearing
DA fodleje
DE Spitzenlager *n* unteres Zapfenlager *n*
ES tejuelo *m*
F crapaudine *f*
I cuscinetto *m* reggispinta
MA talpcsapágy
NL onderste druklager *n* taatslager *n*
PO łożysko *n* stopowe łożysko *n* wzdłużne dolne
PY опорный подшипник *m* подпятник *m*
SU aksiaalilaakeri
SV fotsteg *n*

2007 force
DA kraft
DE Kraft *f*
ES fuerza *f*
F force *f*
I forza *f*
MA erő
NL kracht *f*
PO siła *f*
PY сила *f*
SU voima
SV kraft *c*

2008 forced air
DA overtryksluft
DE Druckluft *f*
ES aire impulsado *m*
F air *m* pulsé
I aria *f* forzata
MA kényszerített áramlás
NL mechanische ventilatie *f*
PO powietrze *n* o wymuszonym przepływie powietrze *n* tłoczone
PY нагнетаемый воздух *m*
SU pakotettu ilmavirtaus
SV tryckluft *c*

2009 forced circulation
DA tvangscirkulation
DE Zwangsumlauf *m*
ES circulación forzada *f*
F circulation *f* accélérée circulation *f* forcée
I circolazione *f* forzata
MA kényszeráramlás
NL mechanische circulatie *f*
PO cyrkulacja *f* wymuszona obieg *m* wymuszony
PY принудительная циркуляция *f*
SU pakotettu kierto
SV tvångscirkulation *c*

2010 forced-circulation air cooler
DA luftkøler med tvungen cirkulation
DE Ventilatorzug-Luftkühler *m*
ES enfriador de aire forzado *m*
F refroidisseur *m* d'air à convection forcée
I refrigeratore *m* di aria a circolazione forzata
MA kényszeráramlású léghűtő
NL luchtkoeler *m* met gedwongen luchtstroom
PO chłodnica *f* z wymuszonym obiegiem powietrza
PY воздухоохладитель *m* с принудительной циркуляцией
SU pakkokiertoinen jäähdytin
SV tvångscirkulations-luftkylare *c*

2011 forced convection
DA tvungen konvektion
DE erzwungene Konvektion *f*
ES convección forzada *f*
F convection *f* forcée
I convezione *f* forzata
MA kényszerkonvekció
NL gedwongen convectie *f*
PO konwekcja *f* wymuszona
PY вынужденная конвекция *f*
SU pakotettu konvektio
SV forcerad konvektion *c*

DA = Danish
DE = German
ES = Spanish
F = French
I = Italian
MA = Magyar (Hungarian)
NL = Dutch
PO = Polish
PY = Russian
SU = Finnish
SV = Swedish

2012 forced draught
DA kunstigt træk
DE künstlicher Zug *m*
ES tiro forzado por impulsión *m*
F tirage *m* mécanique
I tiraggio *m* meccanico forzato
MA mesterséges légáram túlnyomás
NL mechanische trek *m*
PO ciąg *m* sztuczny ciąg *m* wymuszony
PY вынужденный поток *m*
SU puhallinveto (ylipaine)
SV forcerat drag *n*

2013 forced draught burner
DA overtryksbrænder
DE Gebläsebrenner *m*
ES quemador de tiro forzado *m*
F brûleur *m* à air soufflé
I bruciatore *m* ad aria soffiata bruciatore *m* pressurizzato
MA túlnyomásos égő
NL overdrukbrander *m*
PO palnik *m* wentylatorowy palnik *m* z wymuszonym ciągiem
PY горелка *f* с принудительным дутьем
SU ylipainepoltin
SV brännare *c* med fläkt *c*

2014 forced-draught combustion chamber
DA overtryksforbrændings-kammer
DE Überdruckbrenn-kammer *f*
ES cámara de combustión de tiro forzado *f*
F foyer *m* à tirage mécanique (foyer soufflé)
I camera *f* di combustione a tiraggio forzato
MA túlnyomásos égőtér
NL verbrandingskamer *f* met mechanische trek
PO komora *f* spalania z ciągiem wymuszonym
PY топка *f* с принудительным дутьем
SU ylipainetulipesä
SV panna *c* med forcerat drag *n*

2015 forced draught condenser
DA kondensator med tvungen konvektion
DE Ventilatorzug-Kondensator *m*
ES condensador de tiro forzado *m*
F condenseur *m* à air à convection forcée
I condensatore *m* ad aria forzata
MA túlnyomásos kondenzátor
NL condensor *m* met gedwongen luchtstroom
PO skraplacz *m* z wymuszonym ciągiem
PY конденсатор *m* с охлаждением вынужденным потоком воздуха
SU pakkokiertoinen lauhdutin
SV trycklufts-kondensor *c*

2016 forced draught cooling
DA køling med tvungen konvektion
DE Ventilatorzug-Kühlung *f*
ES enfriamiento de tipo forzado *m*
F refroidissement *m* à convection forcée
I raffreddamento *m* ad aria forzata
MA túlnyomásos hűtés
NL koeling *f* met gedwongen luchtcirculatie
PO chłodzenie *n* z wymuszonym ciągiem
PY охлаждение *n* в вынужденном потоке
SU pakkokiertoinen jäähdytys
SV tryckluftskylning *c*

2017 forced-draught water cooling tower
DA køletårn med tvungen vandcirkulation
DE Ventilatorzug-Kühlsystem *m*
ES torre de refrigeración de agua de tiro forzado *f*
F tour *f* de refroidissement à air pulsé à convection forcée
I torre *f* di raffreddamento ad aria forzata
MA túlnyomásos vízhűtő torony
NL koeltoren *m* met gedwongen luchtcirculatie
PO wieża *f* chłodnicza z wymuszonym ciągiem
PY градирня *f* с охлаждением вынужденным потоком воздуха
SU pakkokiertoinen jäähdytystorni
SV vattenkyltorn *n* med tryckluft *c*

2018 forced feed oiling
(*see* forced lubrication)

2019 forced lubrication
DA tryksmøring
DE Druckschmierung *f*
ES lubricación forzada *f*
F lubrification *f* forcée
I lubrificazione *f* forzata
MA kényszerkenés
NL druksmering *f*
PO smarowanie *n* ciśnieniowe
PY смазка *f* под давлением
SU painevoitelu
SV trycksmürjning *c*

2020 forced thermal convection (*see* forced convection)

2021 forced warm air furnace
DA varmluftovn med blæser
DE Ventilator-Warmluft-erzeuger *m*
ES horno de convección *m*
F générateur *m* d'air chaud à air pulsé
I generatore *m* di aria calda a tiraggio forzato
MA hőlégfúvó
NL directgestookte luchtverhitter *m* met ventilator
PO palenisko *n* z wymuszonym przepływem ciepłego powietrza
PY топка *f* с наддувом
SU puhallinkiertoinen ilmanlämmitin
SV forcerad varmluftspanna *c*

2022 forecooler
DA forkøler
DE Vorkühler *m*
ES pre-enfriador *m*
F prérefroidisseur *m*
I prerefrigeratore *m*
MA előhűtő
NL voorkoeler *m*
PO chłodnica *f* wstępna
PY предохладитель *m*
SU esijäähdytin
SV förkylare *c*

2023 forms of energy
DA energiformer
DE Energieformen *f*
ES formas de energía *f,pl*
F forme *f* d'énergie
I forme *f,pl* di energia
MA energiafajta
NL energievormen *pl*
PO postacie *f* energii
PY формы *f* энергии
SU energiamuodot
SV energiformer *c*

2024 forward curved *adj*
DA fremadrettet
DE vorwärtsgekrümmt
ES curvado hacia delante
F incurvé vers l'avant
I a pale in avanti
MA előrehajló
NL voorwaarts gebogen
PO wygięty do przodu
PY загнутый вперед
SU eteenpäin kaartuva siipi
SV framåtbüjd

2025 fossil fuel
DA fossilt brændsel
DE fossiler Brennstoff *m*
ES combustible fósil *m*
F combustible *m* fossile
I combustibile *m* fossile
MA fosszilis tüzelőanyag
NL fossiele brandstof *f*
PO paliwo *n* kopalne
PY ископаемое топливо *n*
SU fossiilinen polttoaine
SV fossilt bränsle *n*

2026 fouling
DA forurening
DE Verschmutzung *f*
ES ensuciamiento *m* incrustación *f* suciedad *f*
F encrassement *m*
I incrostazione *f* sporcamento *m*
MA elszennyeződés
NL aangroeiing *f*
PO osad *m* śluzowaty zanieczyszczenie *n*
PY загрязнение *n*
SU likaantuminen
SV fürsmutsning *c*

2027 fouling factor
DA forureningsfaktor
DE Verschmutzungs-
faktor *m*
ES coeficiente de
ensuciamiento *m*
F facteur *m*
d'encrassement
I fattore *m* di
incrostazione
MA elszennyeződési
tényező
NL aangroeiings-
factor *m*
vervuilingsfactor *m*
PO współczynnik *m*
zanieczyszczenia
PY степень *f*
загрязнения
SU likaantumiskerroin
SV nedsmutsnings-
faktor *c*

2028 foundation
DA fundament
DE Fundament *n*
ES bancada *f*
cimentación *f*
cimientos *m,pl*
F fondation *f*
I fondamenta *f,pl*
fondazione *f*
MA alap
alapozás
NL fundament *n*
grondplaat *f*
PO fundament *m*
podłoże *n*
PY основание *n*
фундамент *m*
SU perustus
SV grundkonstruktion *c*

2029 four pipe system
DA firrørs
ventilationssystem
DE Vierleitersystem *n*
ES sistema a cuatro
tubos *m*
F système *m* à quatre
tuyaux
I sistema *m* a quattro
tubi
MA négyvezetékes
rendszer
NL vierpijpssysteem *n*
PO system *m*
czterorurowy
PY четырехтрубная
система *f*
SU neliputkijärjestelmä
SV fyrrörssystem *n*

2030 fragmented ice
DA krystal is
DE Scherbeneis *n*
ES hielo fragmentado *m*
F glace *f* en paillettes
I ghiaccio *m*
frammentato
MA jégszilánk
NL scherfijs *n*
schilferijs *n*
PO lód *m* rozdrobniony
PY колотый лед *m*
SU levyjää
SV kristallis *c*

2031 frame
DA ramme
stel
DE Rahmen *m*
ES bastidor *m*
F bâti *m*
cadre *m*
I cornice *f*
telaio *m*
MA keret
váz
NL chassis *n*
frame *n*
PO rama *f*
ramownica *f*
PY рама *f*
SU kehys
SV ram *c*

2032 frame construction
DA rammekonstruktion
DE Rahmen-
konstruktion *f*
ES estructura *f*
F ossature *f*
I incastellazione *f*
ossatura *f*
MA keretszerkezet
NL geraamte *n*
PO konstrukcja *f*
ramowa
PY рамная
конструкция *f*
SU kehysrakenne
SV ramkonstruktion *c*

2033 framework
DA skelet
struktur
understel
DE Fachwerk *n*
Gerüst *n*
ES armazón *f*
bastidor *m*
estructura *f*
F charpente *f*
échafaudage *m*
I intelaiatura *f*
traliccio *m*
MA vázszerkezet
NL raamwerk *n*
stelling *f* (bouw)
PO kratownica *f*
szkielet *m*
konstrukcji
PY каркас *m*
конструкции
SU kehys
runko
SV ramverk *n*

**2034 free blow air
conditioner**
DA fritblæsende
luftbehandler
DE freiblasendes
Klimagerät *n*
ES climatizador de
condal variable *m*
F conditionneur *m*
d'air à soufflage
direct
I condizionatore *m* di
ambiente senza
presa di aria
esterna
MA szabadkifúvású
légkondicionáló
NL directuitblazend
luchtbehandelings-
apparaat *n*
PO klimatyzator *m* ze
swobodnym
wylotem powietrza
PY рециркуляционный
кондиционер *m*
SU vapaasti puhaltava
ilmastointilaite
SV friblåsande
luftbehandlare *c*

2035 free convection number
DA tal for naturlig konvektion
DE Wärmeübergangszahl *f* bei freier Konvektion *f*
ES número de convección libre *m*
F coefficient *m* de convection naturel
I numero *m* di Grashof
MA Grashof szám
NL Grashofgetal *n* vrije convectiegetal *n*
PO liczba *f* swobodnej konwekcji
PY число *n* свободной конвекции (Грасгофа)
SU Grashofin luku vapaan konvektion luku
SV tal för naturlig konvektion *c*

2036 free-delivery-type air cooler
DA luftkøler med direkte leverance
DE freiausblasender Luftkühler *m*
ES enfriador de aire de capacidad variable *m*
F refroidisseur *m* d'air à soufflage direct
I refrigeratore *m* di aria a getto libero
MA szabadkifúvású léghűtő
NL directuitblazende luchtkoeler *m*
PO chłodnica *f* powietrza bezkanałowa
PY воздухоохладитель *m* рециркуляционного типа
SU vapaasti puhaltava ilmastointilaite
SV luftkylare *c* med direktleverans *c*

2037 free thermal convection (*see* natural convection)

2038 freezant
DA frysemedium
DE Gefriermittel *n*
ES congelante *m*
F gelé
I congelante *m*
MA hűtőküzeg
NL vriesmiddel *n*
PO czynnik *m* mrożący
PY рассол *m* холодоноситель *m*
SU jäähdyke kylmäliuos
SV frysmedium *n*

2039 freeze *vb*
DA fryse
DE frieren tiefkühlen
ES congelar
F congeler
I congelare
MA fagyaszt
NL vriezen
PO mrozić zamarzać zamrażać
PY замораживать
SU pakastaa
SV frysa

2040 freeze dryer
DA fryse tørrer
DE Gefriertrockner *m*
ES secador congelador *m*
F lyophilisateur *m*
I impianto *m* di liofilizzazione
MA hűtőszárító
NL vriesdroger *m*
PO suszarka *f* sublimacyjna
PY осушитель-вымораживатель *m*
SU pakastuskuivain
SV frystorknings-apparat *c*

2041 freeze drying
DA fryse tørring
DE Gefriertrocknung *f*
ES liofilización *f*
F lyophilisation *f*
I essiccazione *f* per congelamento liofilizzazione *f*
MA hűtőszárítás liofilizálás
NL droging vriesdrogen
PO suszenie *n* przez wymrażanie suszenie *n* sublimacyjne
PY сублимационная сушка *f* сушка *f* вымораживанием
SU pakastuskuivaus
SV frystorkning *c*

2042 freeze out *vb*
DA nedfryse
DE ausfrieren
ES congelar
F séparer par congélation
I congelare
MA kifagyaszt
NL scheiding door bevriezing uitvriezen
PO wymrażać
PY разделять вымораживанием
SU erottaa pakastamalla
SV nedfrysa

2043 freezer
DA fryser
DE Gefriergerät *n*
ES congelador *m*
F congélateur *m*
I congelatore *m*
MA fagyasztó
NL vriesapparaat *n* vriesruimte *f*
PO zamrażalnik *m* zamrażarka *f*
PY вымораживатель *m*
SU pakastin
SV frysapparat *c*

2044 freeze-resistant *adj*
DA frostsikker
DE frostbeständig
   frostsicher
ES incongelable
   (resistente a la
   congelación)
F incongelable
I anticongelante
   resistente al gelo
MA fagyálló
NL vorstbestendig
PO niezamrażalny
   odporny na
   mrożenie
PY морозостойкий
SU jäätymistä kestävä
SV frostbeständig

2045 freeze up
DA nedfryse
DE zufrieren
ES congelado *m*
F obturation *f* par
   congélation
I congelamento *m* in
   superficie
MA ráfagyás
NL aanvriezen
   dichtvriezen
   opvriezen
PO zamarznięcie *n*
PY отказ *m* терморегу-
   лирующего
   вентиля
SU umpeenjäätyminen
SV infrysning *c*

2046 freeze-up control
DA nedfrysningsstyring
DE Frostschutz *m*
ES control de
   congelación *m*
F contrôle *m* antigel
I controllo *m* del
   congelamento in
   superficie
MA fagyvédő
   szabályozás
NL aanvriesbeveiliging *f*
PO kotrola *f*
   zamrażania
PY регулирование *n*
   замораживания
SU jäätymissuojasäätü
SV infrysnings-
   kontroll *c*

2047 freezing
DA frysning
DE Gefrieren *n*
ES congelación *f*
F congélation *f*
I congelamento *m*
MA fagyás
NL bevriezen
   invriezen
PO mrożenie *n*
   zamrażanie *n*
PY замораживание *n*
SU pakastaminen
SV frysning *c*

**2048 freezing
compartment**
DA fryserum
DE Gefrierabteil *n*
ES compartimiento
   congelador *m*
F compartiment *m* de
   congélation
I cella *f* di
   congelamento
MA fagyasztókamra
NL vriesruimte *f*
   vriesvak *n*
PO komora *f* mrożenia
PY отделение *n* для
   замораживания
SU pakastelokero
SV frysfack *n*

2049 freezing equipment
DA fryseudstyr
DE Gefrierapparat *m*
ES equipo
   congelador *m*
   equipo de
   congelación *m*
F installation *f* de
   congélation
I macchinario *m* per il
   congelamento
MA fagyasztó berendezés
NL vriesinstallatie *f*
PO urządzenie *n*
   zamrażające
PY оборудование *n*
   для
   замораживания
SU pakastuslaitteisto
SV frysapparat *c*

2050 freezing mixture
DA fryseblanding
DE Kältemischung *f*
ES mezcla
   congeladora *f*
F mélange *m*
   frigorifique
I miscela *f* congelante
MA hűtőkeverék
NL koudmakend
   mengsel *n*
PO mieszanina *f*
   mrożąca
PY морозильная
   смесь *f*
   рассол *m*
SU alhaisen lämpütilan
   tuottava jään ja
   suolan seos
SV küldblandning *c*

2051 freezing plant
DA fryseanlæg
DE Gefrieranlage *f*
ES instalación de
   congelación *f*
F installation *f* de
   congélation
I impianto *m* di
   congelamento
MA fagyasztó üzem
NL diepvriesfabriek *f*
   vriesinstallatie *f*
PO instalacja *f*
   mrożeniowa
PY морозильная
   установка *f*
SU pakastuslaitos
SV frysanläggning *c*

2052 freezing point
DA frysepunkt
DE Gefrierpunkt *m*
ES puente de
   congelación *m*
F point *m* de
   congélation
I punto *m* di
   congelamento
MA fagypont
NL vriespunt *n*
PO punkt *m* zamarzania
   temperatura *f*
   krzepnięcia
PY точка *f* замерзания
SU jäätymispiste
SV fryspunkt *c*

2053 freezing process
DA frysemetode
DE Gefrierverfahren *n*
ES método de
　congelación *m*
　proceso de
　congelación *m*
F procédé *m* de
　congélation
I processo *m* di
　congelamento
MA fagyasztó eljárás
NL vriesproces *n*
PO proces *m* mrożenia
　proces *m*
　zamarzania
PY процесс *m*
　замораживания
SU pakastusmenetelmä
SV frysmetod *c*

2054 freezing speed
DA frysehastighed
DE Gefrier-
　geschwindigkeit *f*
ES velocidad de
　congelación *f*
F vitesse *f* de
　congélation
I velocità *f* di
　congelamento
MA hűtési sebesség
NL invriessnelheid *f*
　vriessnelheid *f*
PO szybkość *f* mrożenia
　szybkość *f*
　zamarzania
PY скорость *f*
　замораживания
SU pakastusnopeus
SV fryshastighet *c*

2055 freezing time
DA frysetid
DE Gefrierzeit *f*
ES tiempo de
　congelación *m*
F temps *m* de
　congélation
I tempo *m* di
　congelamento
MA hűtési idő
NL invriestijd *m*
PO czas *m* mrożenia
　czas *m* zamarzania
PY время *n*
　замораживания
SU pakastusaika
SV frystid *c*

2056 freon
DA freon
DE Freon *n*
ES difluordicloro-
　metano *m*
　freón
　(inadecuada) *m*
　R-12
　refrigerante-12 *m*
F fréon *m*
I freon *m*
MA freon
NL freon *n*
PO freon *m*
PY фреон *m*
SU Freon
SV freon *n*

2057 frequency
DA frekvens
　svingningshastighed
DE Frequenz *f*
ES frecuencia *f*
F fréquence *f*
I frequenza *f*
MA frekvencia
NL frequentie *f*
PO częstość *f*
　częstotliwość *f*
PY частота *f*
SU taajuus
SV frekvens *c*

**2058 frequency
response**
DA frekvensgengivelse
DE Frequenzantwort *f*
ES respuesta de
　frecuencia *f*
F réponse *f* en
　fréquence
I risposta *f* in
　frequenza
MA frekvenciafüggő
NL frequentie
　responsie *f*
PO odpowiedź *f*
　częstotliwościowa
PY частотная
　характеристика *f*
SU taajuusvaste
SV frekvenskurva *c*

2059 fresh *adj*
DA fersk
　frisk
DE frisch
ES fresco
F frais
I fresco
MA friss
NL fris
　vers
PO świeży
PY свежий
SU raikas
　tuore
SV frisk

2060 fresh air
DA frisk luft
DE Außenluft *f*
　Frischluft *f*
ES aire exterior *m*
　aire fresco *m*
F air *m* frais
　air *m* neuf
I aria *f* di rinnovo
　aria *f* esterna
MA frisslevegő
NL buitenlucht *f*
　verse lucht *f*
PO powietrze *n* świeże
PY свежий воздух *m*
SU raikas ilma
　ulkoilma
SV friskluft *c*

2061 fresh air make up
DA friskluftsupplering
DE Frischluftanteil *m*
ES renovación de aire
　fresco *f*
F apport *m* d'air frais
I ricambio *m* di aria
MA frisslevegő pótlás
NL verse lucht
　aandeel *n*
PO uzdatnianie *n*
　świeżego
　powietrza
PY приготовление *n*
　свежего воздуха
SU ulkoilmavirta
SV friskluftstillsats *c*

2062 freshness
DA friskhed
DE Frische *f*
ES frescura *m*
  pureza *f*
F fraîcheur *f*
  pureté *f*
I freschezza *f*
MA frissesség
NL frisheid *f*
PO świeżość *f*
PY свежесть *f*
SU raikkaus
SV friskhet *c*

2063 freshness index
DA friskluftandel
DE Außenluftrate *f*
ES índice de frescor *m*
F indice *m* d'air frais
  proportion *f* d'air
  frais
I indice *m* di
  freschezza
MA frisslevegő arány
NL frisheidsindex *m*
PO wskaźnik *m*
  świeżości
PY индекс *m* свежести
SU raikkausindeksi
SV friskluftsandel *c*

2064 friction
DA friktion
  gnidning
DE Reibung *f*
ES rozamiento *m*
F frottement *m*
I attrito *m*
MA súrlódás
NL brandstof-
  transport *m*
  wrijving *f*
PO tarcie *n*
PY трение *n*
SU kitka
SV friktion *c*

2065 frictional resistance
DA friktionsmodstand
DE Reibungs-
  widerstand *m*
ES resistencia al
  rozamiento *f*
F résistance *f* de
  frottement
I resistenza *f* per
  attrito
MA súrlódási ellenállás
NL wrijvings-
  weerstand *m*
PO opór *m* tarcia
PY сопротивление *n*
  трения
SU kitkavastus
SV friktionsmotstånd *n*

2066 friction factor
DA friktionsfaktor
DE Reibungs-
  koeffizient *m*
ES coeficiente de
  rozamiento *m*
F coefficient *m* de
  frottement
I coefficiente *m* di
  attrito
MA súrlódási tényező
NL wrijvingscoëfficint *m*
PO współczynnik *m*
  tarcia
PY коэффициент *m*
  трения
SU kitkakerroin
SV friktionsfaktor *c*

2067 friction loss
DA friktionstab
DE Reibungsverlust *m*
ES pérdida por
  rozamiento *f*
F perte *f* par
  frottement
I perdita *f* per attrito
MA súrlódási veszteség
NL wrijvingsverlies *n*
PO strata *f* na tarcie
PY потеря *f* на трение
SU kitkahäviü
SV friktionsfürlust *c*

2068 friction ring
DA friktionsring
DE Reibring *m*
ES anillo de fricción *m*
F bague *f* d'étanchéité
I anello *m* di tenuta
MA tengelytümítő gyűrű
NL frictiering *m*
PO pierścień *m* cierny
  pierścień *m* trący
PY шлифовальный
  круг *m*
SU akselin tiiviste
SV friktionsring *c*

2069 front plate
DA forplade
  frontplade
DE Frontplatte *f*
ES placa anterior (de
  hogar) *f*
  placa frontal (de
  hogar) *f*
F plaque *f* avant de
  foyer
I griglia *f* di focolare
MA homloklap
NL frontplaat *f*
PO płyta *f* czołowa
PY передняя
  (фронтальная)
  плита *f*
SU etulevy
SV frontplåt *c*

2070 front section
DA fronlelement
DE Vorderglied *n*
ES sección frontal *f*
F élément *m* de façade
I sezione *f* frontale
MA homlokmetszet
  homlokrész
NL voorlid *n*
PO człon *m* przedni
  sekcja *f* czołowa
PY фронтальная
  секция *f*
SU etuosa
SV frontparti *n*

DA = Danish
DE = German
ES = Spanish
F = French
I = Italian
MA = Magyar (Hungarian)
NL = Dutch
PO = Polish
PY = Russian
SU = Finnish
SV = Swedish

2071 frost
DA frost
  rim
DE Frost *m*
ES escarcha *f*
  hielo *m*
F gel *m*
I brina *f*
  gelo *m*
MA fagy
  zúzmara
NL rijp *m*
  vorst *m*
PO mróz *m*
PY мороз *m*
SU huurre
  routa
SV frost *c*

2072 frost damage
DA frostskade
DE Frostschaden *m*
ES accidente causado
  por
  congelación *m*
  deterioro por
  congelación *m*
F dégât *m* dQ au gel
I danno *m* da brina
  danno *m* da gelo
MA fagykár
NL vorstschade *f*
PO uszkodzenie *n* w
  wyniku działania
  mrozu
PY повреждение *n* от
  замораживания
SU jäätymisvaurio
  routavaurio
SV frostskada *c*

2073 frost deposit
DA rimlag
DE Gefrierablagerung *f*
ES depósito de escarcha
  m
F dépôt *m* de givre
I deposito *m* da
  disgelo
MA jéglerakódás
NL ijsafzetting *f*
PO szron *m*
PY отложение *n* инея
SU huurteen muodostus
SV frostbeläggning *c*

2074 frost formation
DA rimdannelse
DE Eisbildung *f*
  Reifbildung *f*
ES formación de
  escarcha *f*
F givrage *m*
I formazione *f* di
  brina
  formazione *f* di
  ghiaccio
MA jégképződés
NL ijsvorming *f*
PO tworzenie *n* szronu
PY покрытие *n* инеем
SU jäänmuodostus
SV frostbildning *c*

2075 frost protection
DA frostbeskyttelse
DE Gefrierschutz *m*
ES protección contra el
  hielo *f*
  protección contra la
  escarcha *f*
  protección contra las
  heladas *f*
F protection *f*
I protezione *f* antigelo
MA fagy elleni védelem
NL vorstbescherming *f*
PO zabezpieczenie *n*
  przed
  zamarzaniem
PY защита *f* от
  мороза
SU jäätymissuojaus
SV frostskydd *n*

2076 fuel
DA brændsel
  brændstof
DE Brennstoff *m*
ES combustible
  líquido *m*
F combustible *m*
I combustibile *m*
MA fűtőanyag
  tüzelőanyag
  üzemanyag
NL brandstof *f*
PO paliwo *n*
PY топливо *n*
SU polttoaine
SV bränsle *n*

2077 fuel *vb*
DA tanke
DE tanken
ES proveer combustible
  líquido
F alimenter en
  combustible
I rifornire (di
  combustibile)
MA tüzelőanyaggal ellát
NL brandstof innemen
  tanken
PO tankować
  zaopatrywać w
  paliwo
PY отапливать
SU tankata
  täyttää
SV tanka

2078 fuel analysis
DA brændselsanalyse
DE Brennstoffanalyse *f*
ES análisis del
  combustible *m*
F analyse *f* de
  combustible
I analisi *f* del
  combustibile
MA tüzelőanyag elemzés
NL brandstofanalyse *f*
PO analiza *f* paliwa
PY анализ *m* топлива
SU polttoaineanalyysi
SV bränsleanalys *c*

2079 fuel choice
DA brændstofvalg
DE Brennstoffwahl *f*
ES selección del
  combustible *f*
F choix *m* de
  combustible
I scelta *f* del
  combustibile
MA tüzelőanyag
  választék
NL brandstofkeuze *f*
PO dobór *m* paliwa
  wybór *m* paliwa
PY подбор *m* топлива
SU polttoainevalinta
SV bränsleval *n*

2080 fuel consumption

2080 fuel consumption
DA brændselsforbrug
DE Brennstoff-
    verbrauch *m*
ES consumo de
    combustible *m*
F consommation *f* de
    combustible
I consumo *m* di
    combustibile
MA tüzelőanyag
    fogyasztás
NL brandstofverbruik *n*
PO zużycie *n* paliwa
PY расход *m* топлива
SU polttoaineenkulutus
SV bränsle-
    fürbrukning *c*

2081 fuel feed
DA brændselstilførsel
DE Brennstoffzufuhr *f*
ES carga de
    combustible *f*
    suministro de
    combustible *m*
F alimentation *f* en
    combustible
I alimentazione *f* del
    combustibile
MA tüzelőanyag táplálás
NL brandstoftoevoer *m*
PO zasilenie *n* paliwem
PY подвод *m* топлива
SU polttoaineensyüttü
SV bränsletillfürsel *c*

2082 fuel filter
DA brænstoffilter
DE Brennstoffilter *m*
ES filtro de
    combustible *m*
F filtre *m* à
    combustible
I filtro *m* del
    combustibile
MA tüzelőanyag szűrő
NL brandstoffilter *m*
PO filtr *m* paliwa
PY топливный
    фильтр *m*
SU polttoainesuodatin
SV bränslefilter *n*

2083 fuel gas
DA forbrændingsgas
    gas
DE Brenngas *n*
ES gas combustible *m*
F gaz *m* combustible
I gas *m* combustibile
MA fűtőgáz
NL stookgas *n*
PO gaz *m* opałowy
PY гоючий газ *m*
SU kaasumainen
    polttoaine
SV (värme) gas *c*

2084 fuel handling
DA brændselshåndtering
DE Brennstoff-
    behandlung *f*
ES manipulación del
    combustible *f*
F manutention *f* de
    combustible
I trasporto *m* di
    combustibile
MA tüzelőanyag kezelés
NL brandstof-
    transport *n*
PO dostawa *f* paliwa
    zaopatrywanie *n* w
    paliwo
PY транспортировка *f*
    и уход *m* за
    топливом
SU polttoaineenkäsittely
SV bränslehantering *c*

2085 fuel oil
DA brændselsolie
    fuelolie
DE Heizül *n*
ES aceite
    combustible *m*
F fuel *m*
I olio *m* combustibile
MA fűtőolaj
NL stookolie *f*
PO olej *m* opałowy
PY жидкое топливо *n*
SU polttoüljy
SV eldningsolja *c*

2086 fuel preparation
DA brændselsbehandling
DE Brennstoff-
    aufbereitung *f*
ES tratamiento previo
    del combustible *m*
F préparation *f* de
    combustible
I preparazione *f* del
    combustibile
MA tüzelőanyag
    előkészítés
NL brandstofbereiding *f*
PO przygotowanie *n*
    paliwa
PY подготовка *f*
    топлива
SU polttoaineenkäsittely
SV bränslepreparering *c*

2087 fuel store
DA brændselslager
DE Brennstofflager *n*
ES depósito de
    combustible *m*
F soute *f* à
    combustible
I deposito *m* del
    combustibile
MA tüzelőanyag tároló
NL brandstofopslag *m*
PO skład *m* paliwa
PY склад *m* топлива
SU polttoainevarasto
SV bränslelager *n*

2088 fuel supply
DA brændselsforsyning
DE Brennstoffbedarf *m*
    Brennstofflieferung *f*
    Brennstoff-
    versorgung *f*
ES suministro de
    combustible *m*
F approvisionnement
    m en combustible
I rifornimento *m* di
    combustibile
MA tüzelőanyag ellátás
NL brandstofaanvoer *m*
    brandstofvoor-
    ziening *f*
PO zaopatrzenie *n* w
    paliwo
PY поставка *f* топлива
SU polttoainetoimitus
SV bränsletillfürsel *c*

2089 fuel transport
DA brændselstransport
DE Brennstoff-
transport *m*
ES transporte de
combustible *m*
F transport *m* du (de)
combustible
I trasporto *m* di
combustibile
MA tüzelőanyag szállítás
NL brandstof-
transport *m*
PO transport *m* paliwa
PY транспортировка *f*
топлива
SU polttoainekuljetus
SV bränsletransport *c*

2090 fuel type
DA brændselsart
DE Brennstoffsorte *f*
ES tipo de
combustible *m*
F type *m* de
combustible
I tipo *m* di
combustibile
MA tüzelőanyag típus
NL brandstofsoort *v*
PO rodzaj *m* paliwa
PY вид *m* топлива
SU polttoainetyyppi
SV bränsletyp *c*

**2091 full air
conditioning**
DA fuld luftbehandling
DE Klimatisierung *f* für
alle Räume *m,pl*
ES climatización total *f*
F climatisation *f* totale
I condizionamento *m*
integrale
MA teljes
légkondicionálás
NL volledige
lucht-
behandeling *f*
PO klimatyzacja *f* pełna
PY полное
кондицион-
ирование *n*
воздуха
SU täysilmastointi
SV fullständig
luft-
konditionering *c*

2092 full central heating
DA centralvarme
DE Zentralheizung *f*
ES calefacción central *f*
F chauffage *m* central
général
I riscaldamento *m*
centralizzato
MA teljes küzponti fűtés
NL volledige centrale
verwarming *f*
PO centralne ogrzewanie
n pełne
PY центральное
отопление *n*
SU keskuslämmitys
SV centralvärme *c*

2093 full load
DA fuld last
DE Vollast *f*
ES plena carga *f*
F pleine charge *f*
I pieno carico *m*
MA teljes terhelés
NL vollast *m*
PO obciążenie *n* pełne
PY полная нагрузка *f*
SU täyskuorma
SV fullast *c*

2094 full modulation
DA fuld-modulerende
DE volle Modulation *f*
ES modulación total *f*
F modulation *f*
complète
I modulazione *f*
completa
MA teljes szabályozás
NL continuregeling *f*
PO modulacja *f*
całkowita
pełna modulacja *f*
PY полная
модуляция *f*
SU täysmodulointi
SV fullmodulering *c*

2095 full way cock (*see*
through way cock)

2096 fume
DA dunst
røg
DE Dunst *m*
ES humo *m*
vapor *m*
F fumée *f*
vapeur *f*
I fumo *m*
MA füst
pára
NL damp *m*
rook *m*
PO dym *m*
opary *pl*
PY дым *m*
SU huuru
hüyry
SV rükgas *c*

2097 fume cupboard
DA stinkskab
DE Abzugsschrank *m*
Digestorium *n*
ES campana de
humos *f*
F hotte *f* de
laboratoire
I cappa *f*
MA vegyifülke
NL zuurkast *f*
PO digestorium *n*
wyciąg *m*
laboratoryjny
PY вытяжной шкаф *m*
SU vetokaappi
SV dragskåp *n*

2098 funnel
DA røgfang
DE Trichter *m*
ES embudo *m*
tolva *f*
F entonnoir *m*
trémie *f*
I imbuto *m*
tramoggia *f*
MA kémény
NL koker *m*
trechter *m*
tunnel *m*
PO komin *m* metalowy
lej *m*
PY воронка *f*
дымоход *m*
канал *m*
SU suppilo
tratti
SV rükfång *n*

2099 furnace
DA fyr
DE Brennofen *m*
   Ofen *m*
   Schmelzofen *m*
ES cámara de
      combustión *f*
   hogar *m*
   horno *m*
F chambre *f* de
      combustion
   foyer *m*
I caldaia *f*
   forno *m*
MA kályha
   kemence
   tűzhely
NL oven *m*
   vuurhaard *m*
PO palenisko *n*
   piec *m*
PY печь *f*
SU lämminilmakehitin
SV eldstad *c*

2100 furnace firebox
DA forbrændingsrum
   fyrboks
DE Heizkessel-
      Feuerungsraum *m*
ES cámara de
      combustión *f*
F chambre *f* de
      combustion d'un
      générateur de
      chaleur
I camera *f* di
      combustione
   focolare *m*
MA kazán égőkamra
NL verbrandings-
      kamer *f*
   vuurhaard *m*
PO komora *f*
      paleniskowa kotła
PY топливник *m* печи
SU tulipesä
SV brännkammare *c*

2101 fuse
DA smeltesikring
DE Sicherung *f*
ES fusible *m*
F fusible *m*
I fusibile *m*
MA olvadó biztosíték
NL smeltveiligheid *f*
   zekering *f*
PO bezpiecznik *m*
      topikowy
PY плавкая вставка *f*
   плавкий
      предохранитель
      m
SU sulake
SV smältsäkring *c*

2102 fusible *adj*
DA smeltelig
DE schmelzbar
ES fusible *m*
F fusible
I fusibile
MA olvadó
NL smeltbaar
PO łatwotopliwy
   niskotopliwy
PY плавкий
SU sulava
SV smältbar

2103 fusion
DA smeltning
DE Fusion *f*
   Schmelzen *n*
ES fusión *f*
F fusion *f*
I fusione *f*
MA fúzió
   olvadás
NL smelting *f*
   versmelting *f*
PO masa *f* stopiona
   stopienie *n*
   topienie *n*
PY плавление *n*
SU sulaminen
SV smältning *c*

2104 fusion heat
DA smeltevarme
DE Schmelzwärme *f*
ES calor de fusión *m*
F chaleur *f* de fusion
I calore *m* di fusione
MA olvadáshő
NL smeltwarmte *f*
PO ciepło *n* topnienia
PY теплота *f*
      плавления
SU latenttilämpü
SV smältvärme *c*

DA	=	Danish
DE	=	German
ES	=	Spanish
F	=	French
I	=	Italian
MA	=	Magyar
		(Hungarian)
NL	=	Dutch
PO	=	Polish
PY	=	Russian
SU	=	Finnish
SV	=	Swedish

G

2105 gage (USA) (*see gauge*)

2106 gage glass (USA) (*see gauge glass*)

2107 gain
DA forstærkning
gevinst
DE Gewinn *m*
ES aporte *m*
entrada *f*
ganancia *f*
F gain *m*
I apporto *m*
guadagno *m*
MA nyereség
NL toename *f*
winst *f*
PO zysk *m*
PY прибыль *f*
SU hyöty
kuorma
vahvistus
SV fürstärkning *c*

2108 galvanize *vb*
DA galvanisere
DE galvanisieren
verzinken
ES galvanizar
F galvaniser
I zincare
MA galvanizál
horganyoz
NL galvaniseren
verzinken
PO cynkować
galwanizować
ocynkować
PY оцинковывать
SU galvanoida
SV galvanisera

2109 galvanized *adj*
DA galvaniseret
DE galvanisiert
ES galvanizado
F galvanisé
I zincato
MA horganyzott
NL gegalvaniseerd
verzinkt
PO galwanizowany
ocynkowany
PY оцинкованный
SU galvanoitu
SV galvaniserad

2110 galvanizing
DA galvanisering
DE Galvanisation *f*
Verzinkung *f*
ES galvanización *f*
F galvanisation *f*
I zincatura *f*
MA horganyzás
NL galvanisering
verzinken
PO cynkowanie *n*
galwanizowanie *n*
PY оцинкование *n*
SU galvanointi
SV galvanisering *c*

2111 gas
DA gas
luftart
DE Gas *n*
ES gas *m*
F gaz *m*
I gas *m*
MA gáz
NL gas *n*
PO gaz *m*
PY газ *m*
SU bensiini
kaasu
SV gas *c*

2112 gas analysis
DA gasanalyse
DE Gasanalyse *f*
ES análisis de gases *m*
F analyse *f* de gaz
I analisi *f* del gas
MA gázelemzés
NL gasanalyse *f*
PO analiza *f* gazu
PY анализ *m* газа
SU kaasuanalyysi
SV gasanalys *c*

2113 gas boiler
DA gaskedel
DE Gaskessel *m*
ES caldera de gas *f*
F chaudière *f* à gaz
I caldaia *f* a gas
MA gázkazán
NL gasketel *m*
PO kocioł *m* gazowy
PY газовый котел *m*
SU kaasukattila
SV gaspanna *c*

2114 gas burner
DA gasbrænder
DE Gasbrenner *m*
ES quemador de gas *m*
F brûleur *m* à gaz
I bruciatore *m* a gas
MA gázégő
NL gasbrander *m*
PO palnik *m* gazowy
PY газовая горелка *f*
SU kaasupoltin
SV gasbrännare *c*

2115 gas burner universal
DA universel gasbrænder
DE Allgasbrenner *m*
ES quemador de gas universal *m*
F brûleur *m* tous gaz
I bruciatore *m* a gas universale
MA általános gázégő univerzális gázégő
NL universeel gasbrander *m*
PO palnik *m* gazowy uniwersalny
PY универсальная газовая горелка *f*
SU yleiskaasupoltin
SV allgasbrännare *c*

2116 gas cleaning
DA gasrensning
DE Gasreinigung *f*
ES depuración de gases *f*
F épuration *f* de gaz
I depurazione *f* del gas
MA gáztisztítás
NL gasreiniging *f*
PO oczyszczanie *n* gazu
PY очистка *f* газа
SU kaasunpuhdistus
SV gasrening *c*

2117 gas coal
DA gaskul
DE Gaskohle *f*
ES carbón de gas *m*
F charbon *m* à gaz
I carbone *m* da gas
MA gázszén
NL gaskool *f*
PO węgiel *m* gazowany
PY коксующийся уголь *m*
SU kaasuhiili
SV gaskol *n*

2118 gas concentration
DA gaskoncentration
DE Gaskonzentration *f*
ES concentración de gas *f*
F concentration *f* de gaz
I concentrazione *f* di gas
MA gázkoncentráció
NL gasconcentratie *f*
PO stężenie *n* gazu
PY концентрация *f* газа
SU kaasupitoisuus
SV gaskoncentration *c*

2119 gas consumption
DA gasforbrug
DE Gasverbrauch *m*
ES consumo de gas *m*
F consommation *f* de gaz
I consumo *m* di gas
MA gázfogyasztás
NL gasverbruik *n*
PO zyżycie *n* gazu
PY расход *m* газа
SU kaasunkulutus
SV gasfürbrukning *c*

2120 gas cooler
DA gaskøler
DE Gaskühler *m*
ES enfriador de gases *m*
F refroidisseur *m* de gaz
I raffreddatore *m* di gas
MA gázhűtő
NL gaskoeler *m*
PO ochładzacz *m* gazu
PY газовый холодильник *m*
SU kaasun jäähdytin
SV gaskylare *c*

2121 gas distribution
DA gasdistribution
DE Gasverteilung *f*
ES distribución de gas *f*
F distribution *f* de gaz
I distribuzione *f* del gas
MA gázelosztás
NL gasdistributie *f* gasvoorziening *f*
PO rozdział *m* gazu rozprowadzenie *n* gazu
PY распределение *n* газа
SU kaasunjakelu
SV gasdistribution *c*

2122 gaseous *adj*
DA gasformigt
DE gasfürmig
ES gaseoso
F gazeux
I gassoso
MA gáznemű
NL gasachtig gasvorming
PO gazowy
PY газовый газообразный
SU kaasumainen
SV i gasform

2123 gas explosion
DA gaseksplosion
DE Gasexplosion *f*
ES explosión de gas *f*
F déflagration *f* de gaz
I esplosione *f* di gas scoppio *m* di gas
MA gázrobbanás
NL gasexplosie *f*
PO wybuch *m* gazu
PY взрыв *m* газа
SU kaasuräjähdys
SV gasexplosion *c*

2124 gas failure safety device
DA gasmangelsikring
DE Gasmangelsicherung *f*
ES dispositivo de seguridad contra la falta de gas *m*
F dispositif *m* de sécurité en cas de manque de gaz
I dispositivo *m* di sicurezza per mancata erogazione del gas
MA gázhiánybiztosító szerelvény
NL gasgebrekbeveiliging *f*
PO urządzenie *n* zabezpieczające w przypadku braku dopływu gazu
PY защита *f* от утечки газа
SU kaasusyütün varolaite
SV gasbristsäkring *c*

2125 gas filter
DA gasfilter
DE Gasfilter *m*
ES filtro para gases *m*
F filtre *m* à gaz
I filtro *m* per gas
MA gázszűrő
NL gasfilter *m*
PO filtr *m* gazu
PY газовый фильтр *m*
SU kaasusuodatin
SV gasfilter *n*

2126 gas fire
DA gasfyr
DE Gasfeuer *n*
Gasradiator *m*
ES estufa *f*
F radiateur *m* à gaz
I radiatore *m* a gas
MA gázkályha
NL gashaard *m*
gaskachel *f*
PO piecyk *m* gazowy
PY газовый
нагреватель *m*
SU kaasuliekki
SV gaseldning *c*

2127 gas-fired air heater
DA gasfyret luftvarmer
DE Gaslufterhitzer *m*
Gasluftheizer *m*
ES aerotermo de gas *m*
F aérotherme *m* à gaz
réchauffeur *m* d'air
à gaz
I aerotermo *m* a gas
MA gáztüzelésű léghevítő
NL gasgestookte
luchtverhitter *m*
PO podgrzewacz *m*
powietrza zasilany
gazem
PY газовый
воздухонагре-
ватель *m*
SU kaasukäyttüinen
ilmalämmitin
SV gaseldad
luftvärmare *c*

2128 gas-fired air heating
DA gasfyret
luftopvarmning
DE Gasluftheizung *f*
ES calefacción por aire
caliente a gas *f*
F chauffage *m* à air
chaud au gaz
I riscaldamento *m* di
aria a gas
MA gáztüzelésű légfűtés
NL gasgestookte
luchtverwarming *f*
PO ogrzewanie *n*
powietrzne
zasilane gazem
PY газовоздушное
отопление *n*
SU kaasukäyttüinen
ilmalämmitys
SV luftuppvärmning *c*
med gas *c*

2129 gas-fired central heating
DA gasfyret
centralvarme
DE Gaszentralheizung *f*
ES calefacción central a
gas *f*
F chauffage *m* central
au gaz
I riscaldamento *m*
centrale a gas
MA gáztüzelésű központi
fűtés
NL gasgestookte centrale
verwarming *f*
PO centralne ogrzewanie
n zasilane gazem
PY центральное
отопление *n* с
использованием
газа (в виде
топлива)
SU kaasukäyttüinen
keskuslämmitys
SV centralvärme *c* med
gas *c*

2130 gas-fired hot water heating
DA gasfyret
brugsvands-
opvarmning
DE Gaswarmwasser-
heizung *f*
ES calefacción por agua
caliente a gas *f*
F chauffage *m* à eau
chaude au gaz
I riscaldamento *m* ad
acqua calda a gas
MA gáztüzelésű
melegvízfűtés
NL gasgestookte
warmwater-
voorziening *f*
PO ogrzewanie *n* gorącą
wodą zasilane
gazem
PY водяное
отопление *n* с
использованием
газа в виде
топлива
SU kaasukäyttüinen
lämpimän
käyttüveden
lämmitys
SV vattenvärmning *c*
med gas *c*

2131 gas firing
DA gasfyring
DE Gasfeuerung *f*
ES caldeo por
combustión de
gas *m*
calentamiento por
gas *m*
F chauffe *f* au gaz
I combustione *f* a gas
MA gáztüzelés
NL gasstoken *n*
PO spalanie *n* gazu
PY работа *f* на
газовом топливе
SU kaasunpoltto
SV gaseldning *c*

DA	=	Danish
DE	=	German
ES	=	Spanish
F	=	French
I	=	Italian
MA	=	Magyar (Hungarian)
NL	=	Dutch
PO	=	Polish
PY	=	Russian
SU	=	Finnish
SV	=	Swedish

2132 gas flue
DA gasaftræk
DE Abgasschornstein *m*
ES chimenea de
evacuación de
gases quemados *f*
F conduit *m*
d'évacuation des
gaz brûlés
I camino *m*
ciminiera *f*
MA gáz égéstermék
NL rookgaskanaal *n*
PO czopuch *m*
kanał *m* spalinowy
PY газоход *m*
дымоход *m*
SU savukaasun
poistoputki
SV gaskanal *c*

2133 gas furnace
DA gasovn
DE Gasofen *m*
ES generador de aire
caliente por gas *m*
horno de gas *m*
F foyer *m* à gaz
I focolare *m*
forno *m* a gas
MA gázkemence
NL gasoven *m*
(industrie)
PO palenisko *n* gazowe
piec *m* gazowy
PY газовая печь *f*
SU kaasu-uuni
SV gaseldstad *c*

2134 gas heating
DA gasopvarmning
DE Gasheizung *f*
ES calefacción de gas *f*
F chauffage *m* au gaz
I riscaldamento *m* a
gas
MA gázfűtés
NL gasverwarming *f*
PO ogrzewanie *n*
opalane gazem
PY газовое
отопление *n*
SU kaasulämmitys
SV gasuppvärmning *c*

2135 gasification
DA forgasning
DE Vergasung *f*
ES gasificación *f*
F gazéification *f*
I gasificazione *m*
MA elgázosítás
gázosítás
NL vergassing *f*
PO gazyfikacja *f*
PY газификация *f*
SU kaasutus
SV fürgasning *c*

2136 gasket
DA pakning
DE Dichtung *f*
ES empaquetadura *f*
junta *f*
material para
junta *m*
F joint *m*
I guarnizione *f*
MA tümítés
NL pakking *f*
PO uszczelka *f*
uszczelnienie *n*
PY прокладка *f*
SU tiiviste
SV packning *c*

2137 gas leak detector
DA gasdetektor
DE Gasspurenmesser *m*
ES detector de fugas de
gas *m*
F détecteur *m* de gaz
I cercafughe *m* per
gas
MA gázszivárgás kereső
NL gaslekdetector *m*
PO czujka *f* wycieku
gazu
wykrywacz *m*
nieszczelności
gazu
PY детектор *m* утечки
газа
SU kaasuvuodon
ilmaisin
SV gasdetektor *c*

2138 gas main
DA gashovedledning
DE Hauptgasleitung *f*
ES canalización de
gas *f*
conducto principal
de gas *m*
F conduite *f* principale
de distribution de
gaz
I condotta *f* di gas a
grande distanza
MA gáz fővezeték
NL hoofdgasleiding *f*
PO magistrala *f* gazowa
przewód *m* główny
gazu
PY магистральный
газопровод *m*
SU kaasun pääjohto
SV gashuvudledning *c*

**2139 gas making
equipment** (*see* gas
producer)

2140 gas meter
DA gasmåler
DE Gaszähler *m*
ES gasómetro *m*
F compteur *m* à gaz
I contatore *m* di gas
MA gázmérő
NL gasmeter *m*
PO gazomierz *m*
PY газовый счетчик *m*
SU kaasumittari
SV gasmätare *c*

2141 gas mixture
DA gasblanding
DE Gasgemisch *n*
ES mezcla gaseosa *f*
F mélange *m* de gaz
I miscela *f* gassosa
MA gázkeverék
NL gasmengsel *n*
PO mieszanina *f* gazowa
PY смесь *f* газов
SU kaasusekoitus
SV gasblandning *c*

2142 gas pipe
DA gasrør
DE Gasleitung *f*
  Gasrohr *n*
ES tubería de gas *f*
 F conduite *f* de gaz
 I tubazione *f* di gas
MA gázcső
NL gasleiding *f*
PO przewód *m* gazowy
  rura *f* gazowa
PY газовая труба *f*
SU kaasuputki
SV gasledning *c*

2143 gas producer
DA gasgenerator
DE Gaserzeuger *m*
ES productor de gas *m*
 F gazière *f*
  gazogène *m*
  (appareil)
 I generatore *m* di gas
MA gáztermelő
NL gasproducent *m*
PO wytwornica *f* gazu
  wytwórca *m* gazu
PY производитель *m*
  газа
SU kaasun tuottaja
SV gastillverkare *c*

2144 gas production
DA gasproduktion
DE Gaserzeugung *f*
ES producción de gas *f*
 F production *f* de gaz
 I produzione *f* di gas
MA gáztermelés
NL gasproductie *f*
PO wytwarzanie *n* gazu
PY производство *n*
  газа
SU kaasun valmistus
SV gastillverkning *c*

2145 gas quality
DA gaskvalitet
DE Gasqualität *f*
ES calidad de gas *f*
 F qualité *f* du gaz
 I qualità *f* del gas
MA gázminőség
NL gaskwaliteit *f*
PO jakość *f* gazu
PY качество *n* газа
SU kaasun laatu
SV gaskvalitet *c*

2146 gas radiant heater
DA gasstrålevarmer
DE Gasdecken-
  strahler *m*
ES calentador por
  infrarrojos a
  gas *m*
 F panneaux *m*
  radiants au gaz
 I radiatore *m* a gas
MA gázüzemű hősugárzó
NL gasgestookte
  plafondstraler *m*
PO grzejnik *m* gazowy
  promiennikowy
  promiennik *m*
  gazowy
PY газовый
  излучатель *m*
SU kaasukäyttüinen
  säteilylämmitin
SV gasstrålvärmare *c*

2147 gas room heater
DA gaskamin
DE Gasraumheizer *m*
ES radiador de gas *m*
 F poêle *m* à gaz
 I stufa *f* a gas
MA gáz fűtőkészülék
NL gaskachel *f*
PO grzejnik *m* gazowy
  pokojowy
PY газовый
  комнатный
  нагреватель *m*
SU kaasukäyttüinen
  huonelämmitin
SV gaskamin *c*

2148 gas storage
DA gasbeholder
DE Gasspeicher *m*
ES almacenamiento de
  gas *m*
 F stockage *m* de gaz
 I accumulo *m* di gas
MA gáztárolás
NL gasopslag *m*
PO magazynowanie *n*
  gazu
PY газовое
  хранилище *n*
SU varastointi
  kontrolloidussa
  kaasuympäristüssä
SV gaslagring *c*

2149 gas stove
DA gasovn
DE Gasofen *m*
ES estufa de gas *f*
 F four *m* à gaz
 I forno *m* a gas
  stufa *f* a gas
MA gázkályha
  gáztűzhely
NL gasfornuis *n*
PO piec *m* gazowy
PY газовая печь *f*
SU kaasuliesi
SV gasspis *c*

2150 gas supply
DA gasforsyning
DE Gasversorgung *f*
ES distribución de gas *f*
  suministro de gas *m*
 F distribution *f* de gaz
 I fornitura *f* di gas
MA gázellátás
NL gaslevering *f*
  gasvoorziening *f*
PO zaopatrzenie *n* w
  gaz
PY газоснабжение *n*
  распределение *n*
  газа
SU kaasunjakelu
SV gasfürsürjning *c*

2151 gas-tight *adj*
DA gastæt
DE gasdicht
ES estanco a los gases
  hermético a los gases
 F étanche au gaz
 I a tenuta di gas
MA gáztümür
NL gasdicht
PO gazoszczelny
PY газоплотный
SU kaasutiivis
SV gastät

2152 gas turbine
DA gasturbine
DE Gasturbine *f*
ES turbina de
  combustión
  interna *f*
  turbina de gases *f*
 F turbine *f* à gaz
 I turbina *f* a gas
MA gázturbina
NL gasturbine *f*
PO turbina *f* gazowa
PY газовая турбина *f*
SU kaasuturbiini
SV gasturbin *c*

2153 gas washing plant
DA gasvaskeanlæg
   skrubberanlæg
DE Gaswaschanlage *f*
ES instalación de lavado
   de gases *f*
F installation *f* de
   lavage de gaz
I impianto *m* di
   lavaggio del gas
MA gázmosó berendezés
NL gaswasinstallatie *f*
PO płuczka *f* gazowa
   urządzenie *n*
   płukania gazu
PY устройство *n* для
   мокрой очистки
   газа
SU kaasunpesulaitteisto
SV gastvättare *c*

2154 gas welding
DA gassvejsning
DE Autogen-
   schweißung *f*
   Gasschweißung *f*
ES soldadura por llama
   de gas *f*
F soudage *m* au
   chalumeau
I saldatura *f* a
   cannello
MA gázhegesztés
NL autogeen lassen *n*
PO spawanie *n* gazowe
PY газовая сварка *f*
SU kaasuhitsaus
SV gassvetsning *c*

2155 gas yield
DA gasudbytte
DE Gasausbeute *f*
ES prendimiento en la
   producción de
   gas *m*
F rendement *m* de
   production en gaz
I resa *f* del gas
MA gázhozam
NL gasopbrengst *f*
PO wydajność *f* gazu
PY добыча *f* газа
SU kaasunsaanto
SV gasutbyte *n*

2156 gate valve
DA skydeventil
DE Schieber *m*
ES válvula de
   compuerta *f*
F robinet *m* vanne
   vanne *f* à passage
   direct
I valvola *f* a
   saracinesca
MA tolózár
NL schuifafsluiter *m*
PO zasuwa *f*
PY проходной
   клапан *m*
SU luistiventtiili
SV skjutspjäll *n*

2157 gateway
DA portåbning
DE Tor *n*
ES puerta *f*
F canal *m* de décharge
   canal *m* de fuite
I entrata *f*
   passaggio *m*
   portone *m*
MA folyosó
   kapubejárat
NL gateway *m*
   protocolomzetter *m*
PO brama *f*
   chodnik *m*
   (górniczy)
PY проход *m*
SU portti
SV gång *c*

2158 gauge
DA måleapparat
DE Lehre *f*
   Maß *n*
   Messer *m*
   Meßgerät *n*
ES aparato de
   medición *m*
   calibre *m*
   galga *f*
   manómetro *m*
F appareil *m* de
   mesure (calibre)
I calibro *m*
   misura *f*
MA mérő
NL meetinstrument *n*
   meter *m*
PO przyrząd *m*
   pomiarowy
   wskaźnik *m*
PY измерительный
   прибор *m*
SU mittari
SV mätinstrument *n*

2159 gauge glass
DA måleglas
DE Meßglas *n*
ES medidor de vidrio *m*
F indicateur *m* de
   niveau
   tube *m* de niveau
I indicatore *m* di
   livello
MA mérőüveg
NL peilglas *n*
PO rurka *f*
   wodowskazowa
   szkło *n*
   wodowskazowe
PY мерное стекло *n*
SU nestelasi
   vedenkorkeuslasi
SV vätskeståndsglas *n*

2160 gear pump
DA tandhjulspumpe
DE Zahnradpumpe *f*
ES bomba de
   engranajes *f*
F pompe *f* à
   engrenage
I pompa *f* a
   ingranaggio
MA fogaskerékszivattyú
NL tandradpomp *f*
PO pompa *f* zębata
PY шестеренчатый
   насос *m*
SU hammasratas-
   pumppu
SV kugghjulspump *c*

2161 generation
DA frembringelse
DE Erzeugung *f*
ES producción *f*
F production *f*
I produzione *f*
MA fejlesztés
NL opwekking *f*
PO wytwarzanie *n*
PY изготовление *n*
   образование *n*
   (пара)
SU tuotto
SV generering *c*

2162 generator
DA generator
DE Generator *m*
ES generador *m*
F générateur *m*
I generatore *m*
MA generátor
NL dynamo *m*
    generator *m*
PO generator *m*
    wytwornica *f*
PY генератор *m*
SU generaattori
SV generator *c*

2163 germ
DA bakterie
    kim
    mikrobe
DE Keim *m*
ES germen *m*
F germe *m*
I germe *m*
MA baktérium
    csíra
NL bacterie *f*
    kiem *f*
PO drobnoustrój *m*
    chorobotwórczy
    zarazek *m*
    zarodek *m*
PY микроб *m*
SU bakteeri
SV bakterie *c*

2164 germ free *adj*
DA bakteriefri
    steril
DE keimfrei
ES aséptico
    estéril
    esterilizado
F stérile
I sterile
MA csíramentes
NL kiemvrij
    steriel
PO wolny od
    drobnoustrojów
PY стерильный
SU bakteeriton
SV steril

DA	=	Danish
DE	=	German
ES	=	Spanish
F	=	French
I	=	Italian
MA	=	Magyar (Hungarian)
NL	=	Dutch
PO	=	Polish
PY	=	Russian
SU	=	Finnish
SV	=	Swedish

2165 germicidal *adj*
DA kimdræbende
DE keimtütend
ES germicida
F germicide
I germicida
MA csíraülő
NL bacteriedodend
    kiemdodend
PO bakteriobójczy
PY бактерицидный
SU bakteereita tappava
SV bakteriedüdande

2166 geyser
DA badeovn
    gennemstrømnings-
    ovn
DE Durchlauferhitzer *m*
ES gelser *m*
F chauffe-eau *m*
    instantané
I scaldabagno *m*
    istantaneo
MA átfolyós vízmelegítő
NL geiser *m*
PO gejzer *m*
PY прямоточный
    подогреватель *m*
SU läpivirtauslämmitin
SV genomstrümnings-
    beredare *c*

2167 gland
DA stopbøsning
DE Stoffbüchsbrille *f*
    Stoffbüchse *f*
ES prensaestopas *m*
F bouchon *m*
    chapeau *m* de
    presse-étoupe
I pressotreccia *f*
MA tümszelence
NL pakkingbus *f*
PO dławik *m*
    dławnica *f*
    uszczelka *f*
    dławikowa
PY дроссель *m*
SU tiivistysholkki
SV bussning *c*

2168 glass
DA glas
DE Glas *n*
ES vidrio *m*
F verre *m*
I vetro *m*
MA üveg
NL barometer *m*
    glas *n*
PO szkło *n*
PY стекло *n*
SU lasi
SV glass

2169 glass fibre
DA glasfibre
DE Glasfaser *f*
ES fibra de vidrio *f*
F fibre *f* de verre
I fibra *f* di vetro
MA üvegszál
NL glasvezel *f*
PO włokno *n* szklane
PY стекловолокно *n*
SU lasikuitu
SV glasfiber *c*

2170 glass wool
DA glasuld
DE Glaswolle *f*
ES lana de vidrio *f*
F laine *m* de verre
I lana *f* di vetro
MA üveggyapot
NL glaswol *f*
PO wata *f* szklana
    wełna *f* szklana
PY стекловата *f*
SU lasivilla
SV glasull *c*

2171 globe temperature
DA globetemperatur
DE Globustemperatur *f*
ES temperatura de ter-
    mómetro de globo *f*
F température *f* au
    thermomètre globe
I temperatura *f* di
    globotermometro
MA gümbhőmérséklet
NL globetemperatuur *f*
PO temperatura *f*
    termometru
    kulistego
PY температура *f*
    шарового
    термометра
SU pallolämpütila
SV globtemperatur *c*

2172 globe thermometer
DA globetermometer
DE Globus-
    thermometer *n*
ES termómetro de
    globo *m*
F thermomètre *m*
    globe
I globotermometro *m*
MA gümbhőmérő
NL globe-
    thermometer *m*
PO termometr *m* kulisty
PY шаровый
    термометр *m*
SU pallolämpümittari
SV globtermometer *c*

2173 glow
DA glød
DE Glut *f*
ES brillo *m*
    incandescencia *f*
F incandescence *f*
I incandescenza *f*
MA izzás
    parázs
NL gloed *m*
PO jarzenie *n*
    żar *m*
PY накаливание *n*
    свечение *n*
SU hehku
SV glüd *c*

2174 glow *vb*
DA gløde
DE glühen
ES brillar
    inflamarse
F porter à
    l'incandescence
    porter au rouge
I brillare
    essere incandescente
MA hevít
    izzít
NL gloeien
PO żarzyć się
PY накаляться
SU hehkua
SV glüda

2175 glowing *adj*
DA glødende
DE glühend
ES incandescente
F incandescent
I incandescente
MA izzítás
NL gloeiend
PO żarzący się
PY накаливающийся
SU hehkuva
SV glüdande

2176 gradient
DA gradient
    hældning
DE Anstieg *m*
    Gefälle *n*
    Gradient *m*
ES gradiente *m*
F gradient *m*
I gradiente *m*
MA gradiens
NL gradient *m*
PO gradient *m*
PY градиент *m*
SU gradientti
SV lutning *c*

2177 graph
DA diagram
    graf
DE Diagramm *n*
    Kurvenblatt *n*
ES ábaco *m*
    diagrama *m*
    gráfico *m*
F abaque *m*
    diagramme *m*
I diagramma *m*
    grafico *m*
MA grafikon
NL grafiek *f*
    grafische
    voorstelling *f*
PO wykres *m*
PY граф *m*
SU diagrammi
    kaavio
SV diagram *n*

2178 graph *vb*
DA tegne
DE zeichnen
ES representar
    gráficamente
F mettre sous forme
    graphique
I tracciare
MA ábrázol
NL grafisch voorstellen
    grafisch weergeven
PO kreślić
    rysować
PY чертить график
SU piirtää
SV rita

2179 graphics
DA kurver
DE Graphik *f*
ES gráficos *m*
F graphique *f*
I grafica *f*
MA grafika
NL grafische
    voorstellingen *pl*
PO grafika *f*
    przesłanie *n*
    informacji w
    postaci obrazów
PY график *m*
SU grafiikka
SV grafik *c*

2180 grate
DA gitter
    rist
DE Rost *m*
ES parrilla *f*
F grille *f*
I griglia *f*
MA rács
    rostély
NL staafrooster *n*
PO krata *f*
    ruszt *m*
PY колосник *m*
    решетка *f*
SU arina
    ritilä
SV galler *n*

2181 grate area
DA ristareal
DE Rostfläche *f*
ES superficie de
    parrilla *f*
F surface *f* de grille
I superficie *f* di griglia
MA rostélyfelület
NL roosteroppervlak *n*
PO powierzchnia *f*
    rusztu
PY поверхность *f*
    решетки
SU arinan pinta-ala
SV rostyta *c*

2182 grate bar
DA ristestang
DE Roststab *m*
ES barrote de
    parrilla *m*
F barreau *m* de grille
I barrotto *m* di griglia
MA rostélypálca
NL roosterstaaf *f*
PO pręt *m* rusztu
PY стержень *m*
    решетки
SU arinarauta
SV roststav *c*

2183 grate loading
DA ristebelastning
DE Rostbelastung *f*
ES carga de la parrilla *f*
F charge *f* de la grille
I carica *f* della griglia
MA rostélyterhelés
NL roosterbelasting *f*
PO obciążenie *n* rusztu
PY нагрузка *f* решетки
SU arinakuormitus
SV rostbelastning *c*

2184 grating (*see* grate)

2185 gravimetric efficiency
DA gravimetrisk effekt
DE gravimetrischer Wirkungsgrad *m*
ES rendimiento gravimétrico *m*
F rendement *m* pondéral
I rendimento *m* ponderale
MA gravimetriás hatásfok
NL gravimetrisch rendement *n*
PO sprawność *f* grawimetryczna
PY эффективность *f* гравитации
SU massahyütysuhde
SV gravimetrisk verkningsgrad *c*

2186 gravity
DA tyngdekraft
DE Gewicht *n*
ES gravedad *f*
F gravité *f*
I gravità *f*
MA gravitáció
NL zwaartekracht *f*
PO ciążenie *n* ciężkość *f*
PY гравитация *f*
SU painovoima
SV tyngdkraft *c*

2187 gravity circulation
DA selvcirkulation tyngdekraft (selv-)
DE Schwerkraft-zirkulation *f*
ES circulación por gravedad *f*
F circulation *f* par thermosiphon
I circolazione *f* a gravità
MA gravitációs keringés
NL natuurlijke circulatie *f* zwaartekracht-circulatie *f*
PO obieg *m* grawitacyjny obieg *m* naturalny
PY гравитационная циркуляция *f* естественная циркуляция *f*
SU painovoimainen kierto
SV självcirkulation *c*

2188 gravity circulation heating
DA opvarmning ved selvcirkulation
DE Schwerkraft-heizung *f*
ES calefacción por aire caliente *f* calefacción por termosifón *f*
F chauffage *m* par thermosiphon
I riscaldamento *m* a circolazione naturale
MA gravitációs fűtés
NL verwarming *f* door middel van natuurlijke circulatie
PO ogrzewanie *n* grawitacyjne
PY гравитационное отопление *n* отопление *n* с естественной циркуляцией
SU painovoimakier-toinen lämmitys
SV uppvärmning *c* med självcirkulation *c*

2189 grease filter
DA fedtfilter
DE Fettfilter *m*
ES filtro de grasas *m*
F filtre *m* à graisses
I filtro *m* antigrasso
MA zsírszűrő
NL vetvangfilter *m*
PO filtr *m* tłuszczowy
PY масляный фильтр *m*
SU rasvasuodatin
SV fettfilter *n*

2190 grille
DA gitter rist
DE Gitter *n*
ES parrilla *f* rejilla *f* rejilla de aire *f*
F bouche *f* d'air grillage *m* grille *f*
I griglia *f*
MA rács
NL rooster *n*
PO kratka *f*
PY решетка *f*
SU säleikkü
SV galler *n*

2191 grille differential (pressure)
DA trykfald over rist
DE Gitterwiderstand *m*
ES pérdida de presión a través de la rejilla *f*
F perte *f* de charge de bouche d'air
I differenziale *m* di pressione al turbolatore
MA kiegyenlítő rács
NL drukverschil *n* over rooster
PO spadek *m* ciśnienia w kratce
PY перепад *m* давления на решетке
SU säleikün painehäviü
SV tryckfall *n* över galler *n*

2192 grind *vb*
DA slibe
DE schleifen
ES triturar
F affûter meuler roder
I macinare
MA csiszol őrül
NL slijpen vermalen
PO rozdrabniać szlifować
PY размалывать шлифовать
SU hioa
SV mala

2193 grit
DA grus sand
DE Flugasche *f* Sand *m*
ES cenizas volantes *f,pl* partículas arrastradas por los humos de combustión *f,pl*
F cendres *f* volantes
I fuliggine *f*
MA pernye szemcse
NL scherp zand *n* staalkorrel *m*
PO grys *m* popiół *m* lotny żwirek *m*
PY летучая зола *f*
SU sora
SV grus *n*

2194 grit arrester
DA sandfang
DE Sandfänger *m*
ES captador de
    hollín *m*
F capte-suie *m*
I captatore *m* di
    fuliggine
MA pernyefogó
    szemcseleválasztó
NL gritvanger *m*
PO urządzenie *n* do
    zatrzymywania
    popiołu lotnego
PY золоуловитель *m*
SU sorapesä
    soratasku
SV grusficka *c*

2195 gross capacity
DA bruttokapacitet
DE Bruttoleistung *f*
ES capacidad *f*
F volume *m* brut de
    chambre froide
I capacità *f* lorda
MA brutto teljesítmény
NL bruto capaciteit *f*
PO pojemność *f* brutto
    pojemność *f*
    całkowita
PY объемная
    производитель-
    ьность *f*
SU bruttoteho
    kokonaisteho
SV bruttokapacitet *c*

2196 ground
DA grund
    terræn
DE Boden *m*
    Erde *f*
    Grund *m*
ES suelo *m*
    tierra *f*
F sol *m*
    terre *f*
I terra *f*
MA füld
NL aarde *f*
    grond *m*
PO grunt *m*
    podłoże *n*
    ziemia *f*
PY грунт *m*
SU maanpinta
SV mark *c*

2197 ground fault
DA jordforskastning
DE Erdungsfehler *m*
ES fuga a tierra *f*
F défaut *m* de mise à
    la terre
I messa *f* a terra
MA füldelési hiba
NL aardsluiting *f*
PO ziemnozwarcie *n*
PY электрическое
    заземление *n*
SU maadoitusvika
SV markfel *n*

2198 grounding (USA)
(*see* earthing)

2199 ground water
DA grundvand
DE Grundwasser *n*
ES aguas freáticas *f,pl*
    aguas subterráneas
    f,pl
F eau *f* souterraine
    nappe *f* aquifère
    nappe *f* phréatique
I acqua *f* di
    sottosuolo
    falda *f* acquifera
MA talajvíz
NL grondwater *n*
PO woda *f* gruntowa
PY грунтовая вода *f*
SU pohjavesi
SV grundvatten *n*

2200 ground water level
DA grundvandstand
DE Grundwasser-
    spiegel *m*
ES nivel freático *m*
F niveau *m* phréatique
I livello *m* di falda
    sotterranea
    (freatica)
MA talajvízszint
NL grondwaterpeil *n*
PO poziom *m* wody
    gruntowej
PY уровень *m*
    грунтовых вод
SU pohjaveden pinta
SV grundvattennivå *c*

2201 group heating (*see*
block heating)

2202 guarantee
DA garanti
DE Garantie *f*
ES garantía *f*
F garantie *f*
I garanzia *f*
MA garancia
NL garantie *f*
PO gwarancja *f*
PY гарантия *f*
SU takuu
SV garanti *c*

2203 guide plate
DA ledeplade
    ledeskinne
DE Leitblech *n*
ES álabe director *m*
    deflector *m*
    placa de guía *f*
    placa directriz *f*
F chicane *f*
    déflecteur *m*
    tôle *f* de guidage
I deflettore *m*
MA terelőlemez
NL deflector *m*
    leidschot *n*
PO łopatka *f*
    kierunkowa
    prowadnica *f*
PY направляющая
    лопасть *f*
SU ohjaussiipi
SV ledskena *c*

2204 guide rail
DA styreskinne
DE Führungsschiene *f*
ES carril de guía *m*
F rail *m* de guidage
I binario *m* di guida
MA vezetősín
NL geleidingsrail *f*
PO prowadnica *f*
    szyna *f* kierunkowa
    szyna *f* prowadnicza
PY направляющий
    рельс *m*
SU ohjauskisko
SV styrskena *c*

2205 guide vane (*see* guide
plate)

2206 guide vane assembly

DA ledskolvsenhed
DE Leitblech-
Anordnung *f*
ES conjunto deflector *m*
F ensemble *m* d'aubes
directrices
I alette *f,pl* direzionali
MA terelőlemezes
egyenirányító
NL schoepenkrans-
regelaar *m*
PO zestaw *m* łopatek
kierunkowych
PY направляющий
аппарат *m*
(вентилятора)
SU johtosiipisäätü-
laitteisto
SV ledarskovel-
montering *c*

2207 guillotine damper

DA hejsespjæld
DE Jalousietyp-
Stellklappe *f*
ES compuerta de
guillotina *f*
F registre *m* à
guillotine
I valvola *f* a
ghigliottina
MA tolattyú
NL luchtschuif *f*
PO zasuwa *f* jednopłasz-
czyznowa
PY гильотинный
клапан *m*
диафрагмовый
клапан *m*
SU liukupelti
SV fallspjäll *n*

DA	=	Danish
DE	=	German
ES	=	Spanish
F	=	French
I	=	Italian
MA	=	Magyar (Hungarian)
NL	=	Dutch
PO	=	Polish
PY	=	Russian
SU	=	Finnish
SV	=	Swedish

H

2208 hair crack
- DA hårdfin revne
- DE Haarriß *m*
- ES fisura capilar *f*
  grieta capilar *f*
  grieta fina *f*
  pelo (piezas
  fundidas) *m*
- F fissure *f* capillaire
- I fessura *f* capillare
- MA hajszálrepedés
- NL haarscheurtje *n*
- PO pęknięcie *n*
  włoskowate
  ryska *f*
- PY волосяная
  трещина *f*
- SU hiushalkeama
- SV spricka *c*

2209 hair hygrometer
- DA hår-hygrometer
- DE Haarhygrometer *n*
- ES higrómetro de
  cabello *m*
- F hygromètre *m* à
  cheveux
- I igrometro *m* a
  capello
- MA hajszálas
  légnedvességmérő
- NL haarhygrometer *m*
- PO higrometr *m*
  włosowy
- PY волосяной
  гигрометр *m*
- SU hiushygrometri
- SV hårhygrometer *c*

2210 halogenated *adj*
- DA halogen
- DE halogeniert
- ES halogenado
- F halogène
- I alogenato
- MA halogenizált
- NL gehalogeneerd
- PO chlorowcowany
  halogenowy
- PY галогенный
- SU halogenoitu
- SV halogenerad

2211 hand-hole
- DA lille mandehul
  luge
- DE Handloch *n*
- ES agujero de lavado *m*
  registro *m*
- F trou *m* de poing
- I foro *m* di passaggio
- MA ellenőrző nyílás
- NL handgat *n*
- PO otwór *m*
  wyczystkowy
  wyczystka *f*
- PY лючок *m*
  смотровое
  отверстие *n*
- SU tarkastusluukku
- SV inspektionshål *n*

2212 handle
- DA håndtag
- DE Griff *m*
  Halter *m*
- ES empuñadura *f*
- F manche *m*
  manette *f*
  poignée *f*
- I maniglia *f*
  volantino *m*
- MA fogantyú
  kilincs
- NL handgreep *m*
  hendel *n*
  zwengel *m*
- PO rękojeść *f*
  trzonek *m*
  uchwyt *m*
- PY рукоятка *f*
- SU kahva
- SV handtag *n*

2213 hand-wheel
- DA håndhjul
  rat
- DE Handrad *n*
- ES volante de mano *m*
- F volant *m* à main
- I volantino *m*
- MA kézikerék
- NL handwiel *n*
- PO kółko *n* ręczne
  pokrętło *n*
- PY маховик *m*
  штурвал *m*
- SU käsipyürä
- SV ratt *c*

2214 hanger
- DA bøjle
  strop
- DE Aufhängung *f*
  Halterung *f*
- ES barra de
  suspensión *f*
  cuelgue *m*
- F support *m*
- I sostegno *m*
  staffa *f*
  supporto *m*
- MA függesztő
- NL ophanging *f*
- PO wieszak *m*
- PY подвеска *f*
- SU kannatin
  koukku
- SV upphängnings-
  anordning *c*

2215 hardness of water
- DA vands hårhed
- DE Wasserhärte *f*
- ES dureza del agua *f*
- F dureté *f* de l'eau
- I durezza *f* dell'acqua
- MA vízkeménység
- NL waterhardheid *f*
- PO twardość *f* wody
- PY жесткость *f* воды
- SU veden kovuus
- SV vattenhårdhet *c*

2216 hard soldering
DA slaglodning
DE Hartlötung *f*
ES soldadura fuerte *f*
F brasure *f* forte
I brasatura *f*
MA keményforrasztás
NL hardsolderen *n*
PO lutowanie *n* twarde
PY твердая пайка *f*
SU kovajuotos
SV hårdlödning *c*

2217 head
DA hoved
  tryk
DE Druck *m*
  Hauptabschnitt *m*
  Hühe *f*
ES altura estática *f*
  cabezal *m*
  presión *f*
F hauteur *f* de charge
I carico *m*
  pressione *f*
MA dinamikus nyomás
  nyomómagasság
NL opvoerhoogte *f*
PO głowica *f*
  łeb *m*
  wysokość *f* ciśnienia
PY высота *f*
  давление *n*
  напор *m*
SU nostokorkeus
  paine
SV hüjd *c*
  tryck *n*

2218 header
DA binder
  samleledning
DE Hauptleitung *f*
  Sammelrohr *n*
  Sammler *m*
ES cabezal *m*
  colector *m*
  distribuidor *m*
F collecteur *m*
  distributeur *m*
I collettore *m*
  distributore *m*
MA elosztófej
  gyüjtő
NL verdeler *m*
  verzamelaar *m*
PO kolektor *m*
  rozdzielacz *m*
PY коллектор *m*
  магистраль *f*
SU jakotukki
SV samlingsledning *c*

2219 header tank
DA samletank
  tryktank
DE Druckbehälter *m*
ES tanque de
  alimentación por
  gravedad *m*
F réservoir *m* en
  charge
I serbatoio *m* di
  carico
MA gyüjtőtartály
NL expansievat *n* (open)
PO zbiornik *m* opadowy
  zbiornik *m*
  wyrównawczy
PY напорный бак *m*
SU jakoastia
SV vattentank *c*

2220 head loss
DA trykfald
DE Druckverlust *m*
ES pérdida de presión *f*
F chute *f* de pression
  dynamique
I perdita *f* di carico
MA dinamikus
  nyomásveszteség
NL drukverlies *n*
PO spadek *m* ciśnienia
  strata *f* ciśnienia
PY потери *f,pl*
  динамического
  напора
SU painehäviü
SV tryckfall *n*

**2221 head pressure
  control** (*see* high
  pressure control)

2222 hearth
DA esse
  herd
  ildsted
DE Herd *m*
ES hogar *m*
  placa de asiento *f*
  solera de horno *f*
F plaque *f* d'assise
  plaque *f* de base
I focolare *m*
  suola *f*
MA kemence
  tűzhely
NL haard *m*
  stookplaats *f*
PO płyta *f* dolna
  paleniska
PY основание *n*
SU ahjo
SV härd *c*

2223 hearth fire
DA åbent ildsted
DE Herdfeuer *n*
ES chimenea abierta *f*
  chimenea francesa *f*
F feu *m* ouvert
I focolare *m*
MA kemencetüzelés
NL haardvuur *n*
PO płomień *m* paleniska
PY камин *m*
SU hehkuva liekki
SV üppen härd *c*

2224 heat
DA varme
DE Hitze *f*
  Wärme *f*
ES calor *m*
F chaleur *f*
I calore *m*
MA hő
NL hitte *f*
  warmte *f*
PO ciepło *n*
PY теплота *f*
SU lämpü
SV värme *c*

2225 heat *vb*
DA opvarme
DE erhitzen
  heizen
ES calentar
F chauffer
  échauffer
I riscaldare
MA fűt
NL verhitten
  verwarmen
PO ogrzewać
PY нагревать
SU lämmittää
SV värma

2226 heat anticipation
DA forvented varme
DE Wärmeempfindung *f*
ES anticipación térmica
  f
F chaleur *f* présumée
I anticipo *m* del
  calore
MA előfűtés
NL warmteversnelling *f*
PO wyprzedzenie *n*
  pracy ogrzewania
PY точность *f*
  термостатиро-
  вания
SU ennakointivastus
  termostaatissa
SV värmemotstånd *n*

2227 heat balance
DA varmebalance
DE Wärmebilanz *f*
ES balance térmico *m*
F bilan *m* thermique
I bilancio *m* termico
MA hőmérleg
NL warmtebalans *f*
PO bilans *m* ciepła
PY тепловой баланс *m*
SU lämpütasapaino
SV värmebalans *c*

2228 heat bridge
DA kuldebro
DE Wärmebrücke *f*
ES puente térmico *m*
F pont *m* thermique
I ponte *m* termico
MA hőhíd
NL koudebrug *f*
PO mostek *m* cieplny
PY тепловой мостик *m*
SU kylmäsilta
SV värmebrygga *c*

2229 heat capacity
DA varmekapacitet
DE Wärmeleistung *f*
ES capacidad térmica *f*
F capacité *f* thermique
I capacità *f* termica
MA hőteljesítmény
NL warmtecapaciteit *f*
PO pojemność *f* cieplna
PY теплоемкость *f*
SU lämpükapasiteetti
SV värmekapacitet *c*

2230 heat conduction
DA varmeledning
DE Wärmeleitung *f*
ES conducción térmica *f*
F conduction *f* de la chaleur
I conduzione *f* del calore
MA hővezetés
NL warmtegeleiding *f*
PO przewodzenie *n* ciepła
PY теплопроводность *f*
SU lämmün johtuminen
SV värmeledning *c*

2231 heat conductor
DA varmeleder
DE Wärmeleiter *m*
ES conductor térmico *m*
F conducteur *m* de chaleur
I termoconduttore *m*
MA hővezető
NL warmtegeleider *m*
PO przewodnik *m* ciepła
PY проводник *m* теплоты
SU lämmünjohdin
SV värmeledare *c*

2232 heat content
DA varmeindhold
DE Wärmeinhalt *m*
ES capacidad calorífica *f*
F capacité *f* calorifique
I capacità *f* termica contenuto *m* termico
MA hőtartalom
NL warmte-inhoud *m*
PO entalpia *f* pojemność *f* cieplna
PY теплосодержание *n*
SU lämpüsisältü
SV värmeinnehåll *n*

2233 heat demand (*see* heat requirement)

2234 heated ceiling
DA loftvarme
DE Deckenheizung *f*
ES techo calefactado *m* techo calefactor *m*
F plafond *m* chauffant
I soffitto *m* riscaldato
MA fűtőmennyezet
NL stralingsplafond *n*
PO sufit *m* grzejny
PY потолочное отопление *n*
SU lämmitetty katto
SV värmetak *n*

2235 heater
DA varmeapparat varmeovn
DE Erhitzer *m* Heizkürper *m* Heizvorrichtung *f*
ES calentador *m*
F réchauffeur *m*
I riscaldatore *m*
MA fűtőtest
NL verwarmings-apparaat *n* verwarmings-lichaam *n*
PO grzejnik *m* nagrzewnica *f* podgrzewacz *m*
PY нагреватель *m*
SU lämmitin
SV värmare *c*

2236 heater battery
DA varmeakkumulator
DE Heizregister *n*
ES batería de calefacción *f*
F batterie *f* de chauffage
I batteria *f* riscaldante
MA fűtőbattéria fűtőkészülék
NL verwarmings-batterij *f*
PO nagrzewnica *f*
PY нагревательная батарея *f*
SU lämmityspatteri
SV värmebatteri *n*

2237 heat exchange
DA varmeveksling
DE Wärmeaustausch *m*
ES intercambio de calor *m* intercambio térmico *m*
F échange *m* de chaleur
I scambio *m* termico
MA hőcsere
NL warmte-uitwisseling *f*
PO wymiana *f* ciepła
PY теплообмен *m*
SU lämmünsiirto
SV värmeväxling *c*

2238 heat exchanger
DA varmeveksler
DE Wärme-
   austauscher *m*
ES intercambiador de
   calor *m*
F échangeur *m* de
   chaleur
I scambiatore *m* di
   calore
MA hőcserélő
NL warmtewisselaar *m*
PO wymiennik *m* ciepła
PY теплообменник *m*
SU lämmünsiirrin
   lämmünvaihdin
SV värmeväxlare *c*

2239 heat flow
DA varmestrøm
DE Wärmestrom *m*
ES flujo de calor *m*
   paso de calor *m*
F flux *m* de chaleur
I flusso *m* di calore
MA hőáram
NL warmtestroming *f*
   warmtestroom *m*
PO przepływ *m* ciepła
   strumień *m* ciepła
PY тепловой поток *m*
SU lämpüvirta
SV värmeflüde *n*

2240 heat gain
DA varmeindvinding
   (gratis varme)
DE Wärmegewinn *m*
ES aporte de calor *m*
   entradas de calor
   f,pl
   ganancia de calor *f*
F gain *m* de chaleur
I apporto *m* di calore
MA hőnyereség
NL warmtewinst *f*
PO zysk *m* ciepła
PY поступление *n*
   теплоты
SU lämpükuorma
SV värmevinst *c*

```
DA = Danish
DE = German
ES = Spanish
F  = French
I  = Italian
MA = Magyar
     (Hungarian)
NL = Dutch
PO = Polish
PY = Russian
SU = Finnish
SV = Swedish
```

2241 heating
DA opvarmning
DE Beheizung *f*
   Heizung *f*
ES calefacción *f*
F chauffage *m*
I riscaldamento *m*
MA fütés
NL verwarming *f*
PO ogrzewanie *n*
PY отопление *n*
SU lämmitys
SV uppvärmning *c*

2242 heating appliance
DA varmetilbehør
DE Heizgerät *n*
   Heizkürper *m*
ES aparato de
   calefacción *m*
   calorífero *m*
F appareil *m* de
   chauffage
   corps *m* de chauffe
I corpo *m* scaldante
MA fütőberendezés
NL verwarmings-
   toestel *n*
PO urządzenie *n*
   ogrzewcze
PY отопительный
   прибор *m*
SU lämmityslaite
SV värmeapparat *c*

2243 heating coil
DA varmeflade
DE Heizschlange *f*
ES serpentín de
   calefacción *m*
F batterie *f* de
   chauffage
I batteria *f* di
   riscaldamento
   serpentina *f* di
   riscaldamento
MA fütőcsőkígyó
NL verwarmings-
   spiraal *f*
PO nagrzewnica *f*
   wężownica *f* grzejna
PY отопительный
   змеевик *m*
SU lämpüpatteri
SV värmeslinga *c*

2244 heating costs
DA opvarmningsudgift
DE Heizkosten *f*
ES coste de
   calefacción *m*
F frais *m* de chauffage
I costi *m,pl* di
   riscaldamento
MA fütési költségek
NL verwarmings-
   kosten *pl*
PO koszty *m*
   ogrzewania
PY стоимость *f*
   отопления
SU lämmitysku-
   stannukset
SV värmekostnader *c*

2245 heating curve
DA varmekurve
DE Heizkurve *f*
ES curva de calefacción
   f
F courbe *f* de
   chauffage
I curva *f* di
   riscaldamento
MA fütési gürbe
NL stooklijn *f*
PO krzywa *f*
   ogrzewania
   krzywa *f*
   regulacyjna
   ogrzewania
PY отопительный
   график *m*
SU lämmityksen
   ohjauskäyrä
   lämmityskäyrä
SV värmekurva *c*

2246 heating degree day
DA varmegraddag
DE Heizgradtag *m*
ES grados-día de
   calefacción *m*
F degré-jour *m* de
   chauffage
I grado-giorno *m* di
   riscaldamento
MA fütési napfok
NL verwarmings-
   graaddag *m*
PO stopniodzień *m*
   ogrzewania
PY отопительный
   градусодень *m*
SU lämmityksen
   astepäiväluku
SV värmegraddag *c*

2247 heating economics
DA varmeøkonomi
DE Wärmewirtschaft f
ES economía térmica f
termoeconomía f
F économie f du
chauffage
I studio m economico
del riscaldamento
MA fűtés gazdaságosság
NL warmte-economie f
PO gospodarka f
cieplna
oszczędzanie n
ciepła
PY тепловое
хозяйство n
SU lämpütalous
SV värmeekonomi c

2248 heating element
DA elektrisk kolbe
DE Heizelement n
ES elemento
calefactor m
F élément m chauffant
I elemento m
riscaldante
MA fűtőelem
NL verwarmings-
element n
PO element m grzejny
PY нагревательный
элемент m
SU lämmityselementti
SV värmeelement n

2249 heating fluid
DA varmebærer
DE Heizflüssigkeit f
Heizmittel n
ES fluido calefactor m
F fluide m caloporteur
I fluido m scaldante
MA fűtőfolyadék
NL verwarmings-
vloeistof f
PO czynnik m grzejny
PY жидкий
теплоноситель m
SU lämmünsiirtoneste
SV värmebärare c

2250 heating installation
DA varmeinstallation
DE Heizungsanlage f
ES instalación de
calefacción f
F installation f de
chauffage
I impianto m di
riscaldamento
MA fűtésszerelés
NL verwarmings-
installatie f
PO instalacja f
ogrzewania
PY отопительная
установка f
SU lämmityslaitteisto
SV värmeinstallation c

2251 heating load
DA varmelast
DE Heizlast f
ES carga calorífica f
carga por
calefacción f
F charge f calorifique
I carico m termico
MA fűtésterhelés
NL warmtebehoefte f
warmtelast m
PO obciążenie n
ogrzewania
PY отопительная
нагрузка f
SU lämmüntarve
SV värmebelastning c

2252 heating medium
DA varmemedium
DE Heizmittel n
Wärmeträger m
ES medio calefactor m
F fluide m de
chauffage
I mezzo m scaldante
MA fűtőküzeg
NL verwarmings-
medium n
PO czynnik m grzejny
PY теплоноситель m
SU lämmünsiirtoaine
SV värmemedium n

2253 heating method
DA opvarmningsform
DE Heizungsart f
ES método de
calefacción m
F mode m de
chauffage
I sistema m di
riscaldamento
MA fűtési mód
NL verwarmings-
methode f
PO metoda f
ogrzewania
sposób m
ogrzewania
PY способ m
отопления
SU lämmitystapa
SV uppvärmnings-
metod c

2254 heating oil
DA fyringsolie
DE Heizül n
ES aceite de
calefacción m
petroleo para
calefacción m
F fuel m de chauffage
I olio m di
riscaldamento
MA fűtőolaj
NL huisbrandolie f
PO olej m grzejny
PY отопительное
масло n
SU lämmitysüljy
SV eldningsolja c

2255 heating panel
DA varmepanel
DE Flachheizkürper m
ES panel radiante m
F panneau m
chauffant
I pannello m di
riscaldamento
MA fűtőpanel
NL verwarmings-
paneel n
PO grzejnik m płytowy
płyta f grzejna
PY отопительная
панель f
SU lämmityspaneeli
SV värmepanel c

2256 heating plant
DA varmeanlæg
DE Heizungsanlage *f*
ES instalación de
calefacción *f*
F installation *f* de
chauffage
I impianto *m* di
riscaldamento
sistema *m* di
riscaldamento
MA fűtőberendezés
NL ketelinstallatie *f*
verwarmings-
installatie *f*
PO instalacja *f*
ogrzewcza
PY отопительная
установка *f*
отопительный
агрегат *m*
SU lämpükeskus
SV värmeanläggning *c*

2257 heating season
DA fyringssæson
DE Heizperiode *f*
Heizzeit *f*
ES temporada de
calefacción *f*
F saison *f* de
chauffage
I stagione *f* di
riscaldamento
MA fűtési idény
NL stookseizoen *n*
PO sezon *m* ogrzewczy
PY отопительный
сезон *m*
SU lämmityskausi
SV eldningssäsong *c*

2258 heating stack loss
DA røggastab
skorstenstab
DE Schornsteinwärme-
verlust *m*
ES pérdida de calor en
chimenea *f*
F pertes *f* d'énergie
aux fumées
I perdita *f* al camino
MA fűtési
kéményveszteség
NL schoorsteenverlies *n*
PO strata *f* ciepła
kominowa
PY потеря *f* теплоты
трубопроводами
SU savukaasuhäviü
SV skorstensfürlust *c*

2259 heating stove
DA varmeovn
DE Heizofen *m*
ES estufa *f*
F poêle *m* de chauffage
I stufa *f*
MA fűtőkályha
NL kachel *f*
PO piec *m* ogrzewczy
PY отопительная
печь *f*
SU lämmitysuuni
SV värmeugn *c*

2260 heating surface
DA hedeflade
DE Heizfläche *f*
ES superficie
calefactora *f*
F surface *f* de
chauffage
surface *f* de chauffe
I superficie *f*
scaldante
MA fűtőfelület
NL verwarmend
oppervlak *n*
PO powierzchnia *f*
ogrzewalna
PY поверхность *f*
нагрева
SU lämpüpinta
SV värmeyta *c*

2261 heating system
DA varmesystem
DE Heizungssystem *n*
ES sistema de
calefacción *m*
F système *m* de
chauffage
I sistema *m* di
riscaldamento
MA fűtőrendszer
NL verwarmings-
systeem *n*
PO system *m*
ogrzewania
PY система *f*
отопления
SU lämmitysjärjestelmä
SV värmesystem *n*

2262 heating up
DA opvarmning
DE Aufheizung *f*
ES puesta en régimen de
la calefacción *f*
F mise *f* en
température
I messa *f* a regime
MA felfűtés
NL opstoken *n*
opwarmen *n*
PO nagrzewanie *n*
rozgrzewanie *n*
PY подогрев *m*
SU palautuslämmitys
SV uppvärmning *c*

2263 heating-up time
DA opvarmningstid
DE Aufheizzeit *f*
ES tiempo de puesta en
régimen de la
calefacción *m*
F durée *f* de mise en
température
I tempo *m* di messa a
regime
MA felfűtési idő
NL aanwarmtijd *m*
opwarmtijd *m*
PO czas *m* rozgrzewania
PY время *n* подогрева
SU palautuslämmity-
saika
SV uppvärmningstid *c*

2264 heating value
DA varmeevne
DE Heizwert *m*
ES tasa de calefacción *f*
F pouvoir *m*
calorifique
I potere *m* calorifico
MA fűtőérték
NL verbrandings-
waarde *f*
PO ciepło *n* spalania
wartość *f* grzejna
wartość *f* opałowa
PY теплотворная
способность *f*
(топлива)
SU lämpüarvo
SV värmevärde *n*

2265 heating wire
DA varmekabel
DE Heizdraht *m*
ES hilo calefactor *m*
hilo térmico *m*
F fil *m* chauffant
I cavo *m* riscaldante
MA fűtőhuzal
NL verwarmingskabel *m*
PO drut *m* grzejny
PY нагревательная
спираль *f*
нить *f* накаливания
SU lämmityskaapeli
vastuslanka
SV glüdtråd *c*

2266 heat input
DA varmetillførsel
DE Wärmezufuhr *f*
ES calor absorbido *m*
F apport *m* de chaleur
I fornitura *m* di
calore
MA hőráfordítás
NL warmtetoevoer *m*
PO ciepło *n*
doprowadzone
PY теплопотребление
n
SU lämmünsyüttü
SV värmeintag *n*

**2267 heat insulating
material**
DA varmeisolerings-
materiale
DE Wärmedäm-
material *n*
ES aislante térmico *m*
material de
aislamiento *m*
F isolant *m* thermique
I materiale *m*
coibente
MA hőszigetelő anyag
NL warmte-isolerend
materiaal *n*
PO materiał *m* izolacji
cieplnej
PY теплоизоля-
ционный материал *m*
SU lämpüeriste
SV värmeisoler-
material *n*

2268 heat insulation
DA varmeisolering
DE Wärmedämmung *f*
ES aislamiento
térmico *m*
instalación de
calefacción *f*
F isolation *f*
thermique
I coibentazione *f*
termica
isolamento *m*
termico
MA hőszigetelés
NL warmte-isolatie *f*
PO izolacja *f* cieplna
PY тепловая
изоляция *f*
SU lämpüeristys
SV värmeisolering *c*

2269 heat lag
DA varmebeklædning
(på kedel)
DE Wärme-
verzügerung *f*
ES refractario *m*
F retard *m* thermique
I ritardo *m* di un
fenomeno termico
sfasamento *m* di un
fenomeno termico
MA hőkésleltetés
NL warmtevertraging *f*
PO opóźnienie *n* fali
cieplnej
PY тепловая инерция *f*
SU lämpükapasiteetin
aiheuttama viive
SV värmefürdrüjning *c*

2270 heat load
DA varmebelastning
DE Heizlast *f*
Wärmelast *f*
ES carga calorífica *f*
F charge *f* thermique
I carico *m* termico
MA hőterhelés
NL warmtebelasting *f*
PO obciążenie *n* cieplne
PY тепловая
нагрузка *f*
SU lämpükuorma
SV värmebelastning *c*

2271 heat loss
DA varmetab
DE Wärmeverlust *m*
ES pérdida de calor *f*
F déperdition *f*
calorifique
perte *f* de chaleur
I dispersione *f* di
calore
MA hőveszteség
NL warmteverlies *n*
PO strata *f* ciepła
PY потери *f,pl*
теплоты
теплопотери *f,pl*
SU lämpühäviü
SV värmefürlust *c*

**2272 heat loss
calculation**
DA varmetabsberegning
DE Wärmebedarfsberech-
nung *f*
ES cálculo de las
pérdidas de calor *f*
F calcul *m* des
déperditions
calorifiques
I calcolo *m* delle
dispersioni
MA hőveszteségszámítás
NL warmteverlies-
berekening *f*
PO obliczenie *n* strat
ciepła
PY расчет *m*
теплопотерь
SU lämpühäviülaskelma
SV värmefürlust-
beräkning *c*

2273 heat loss rate
DA varmetab
DE Wärmeverlustrate *f*
ES proporción de
pérdida de calor *f*
F perte *f* de chaleur
par unité de temps
I potenza *f* termica
dispersa
MA hőveszteség mértéke
NL warmteverlies-
coëfficiënt *m*
PO wielkość *f* straty
ciepła
PY норма *f* тепловых
потерь
SU häviülämpüvirta
SV värmefürlustgrad *c*

2274 heat meter
DA varmemåler
DE Wärmemesser *m*
ES calorímetro *m*
F compteur *m* de calories
  compteur *m* de chaleur
I contatore *m* di calore
MA hőfogyasztásmérő
  hőmennyiségmérő
NL warmtemeter *m*
PO licznik *m* ciepła
PY счетчик *m* теплоты
SU lämpümäärämittari
SV värmemätare *c*

2275 heat output (rating)
DA varmeafgivelse
DE Heizleistung *f*
  Wärmeabgabe *f*
ES potencia calorífica *f*
F débit *m* calorifique
  puissance *f* calorifique
I potenzialità *f*
  resa *f* termica
MA fűtőteljesítmény
NL warmteproduktie *f*
PO wydajność *f* cieplna
PY теплопроизводитель-ность *f*
SU nimellislämmitysteho
SV värmeavgivning *c*

2276 heat pipe
DA varmerør
DE Wärmeübertragungs-rohr *n*
ES tubo de calor *m*
F caloduc *m*
I tubo *m* di calore
MA fűtőcső
NL heatpipe *m*
  warmtepijp *f*
PO rura *f* grzejna
PY нагревательная труба *f*
SU lämpüputki
SV värmerür *n*

2277 heat power station
DA kraft-varme-værk
DE Heizkraftwerk *n*
ES central termoeléctrica *f*
F centrale *f* de chaleur/force
I centrale *f* termoelettrica
MA fűtőerőmű
NL warmtecentrale *f*
  warmtekracht-centrale *f*
PO elektrociepłownia *f*
PY теплоэлектро-централь *f*
SU lämpüvoimalaitos
SV värmekraftverk *n*

2278 heat production
DA varmeproduktion
DE Wärmeerzeugung *f*
ES producción de calor *f*
F production *f* de chaleur
I produzione *f* di calore
MA hőtermelés
NL warmteproductie *f*
PO wytwarzanie *n* ciepła
PY теплопрдукция *f*
SU lämmün tuotanto
SV värmealstring *c*

2279 heat pump
DA varmepumpe
DE Wärmepumpe *f*
ES bomba de calor *f*
F pompe *f* à chaleur
I pompa *f* di calore
MA hőszivattyú
NL warmtepomp *f*
PO pompa *f* ciepła
PY тепловой насос *m*
SU lämpüpumppu
SV värmepump *c*

2280 heat rate
DA varmemængde
DE Wärmepreis *m*
  Wärmezahl *f*
ES proporción de calor *f*
F puissance *f* thermique
I potenza *f* termica
MA fűtési díj
NL warmtegetal *n*
PO jednostkowe zużycie *n* ciepła
PY тепловая эффективность *f*
SU voimalaitoksen lämmünkäytün ja sähküntuoton suhde
SV värmevärde *n*

2281 heat recovery
DA varmegenvinding
DE Wärmerück-gewinnung *f*
ES recuperación de calor *f*
F récupération *f* de chaleur
I recupero *m* del calore
MA hővisszanyerés
NL warmteterug-winning *f*
PO odzyskiwanie *n* ciepła
PY теплоутилизация *f*
SU lämmüntalteenotto
SV värmeåtervinning *c*

DA = Danish
DE = German
ES = Spanish
F = French
I = Italian
MA = Magyar (Hungarian)
NL = Dutch
PO = Polish
PY = Russian
SU = Finnish
SV = Swedish

2282 heat recovery system
DA varmegenvindings-anlæg
DE Wärmerückgewinnungs-system *n*
ES sistema de recuperación de calor *m*
F système *m* de récupération de chaleur
I sistema *m* di recupero del calore
MA hővisszanyerő rendszer
NL warmteterugwinnings-systeem *n*
PO system *m* odzyskiwania ciepła
PY система *f* утилизации теплоты устройство *n* для утилизации теплоты
SU lämmün talteenottojärje-stelmä
SV värmeåtervinnings-system *n*

2283 heat requirement
DA varmebehov
DE Wärme-anforderung *f* Wärmebedarf *m*
ES necesidades de calor *f*
F demande *f* de chaleur
I fabbisogno *m* di calore
MA hőigény
NL warmtebehoefte *f*
PO zapotrzebowanie *n* na moc cieplną
PY потребность *f* в теплоте
SU lämmüntarve
SV värmebehov *n*

2284 heat resistant *adj*
DA varmebestandig
DE wärmebeständig
ES estable al calor refractaria termorresistente
F réfractaire résistant au feu
I refrattario
MA hőálló
NL hittebestendig warmtebestendig
PO ciepłoodporny
PY теплостойкий
SU lämpüä kestävä
SV värmebeständig

2285 heat science
DA varmelære
DE Wärmelehre *f*
ES termotecnia *f*
F thermique *f*
I termotecnica *f*
MA hőtan
NL warmtekunde *f* warmtetechniek *f*
PO nauka *f* o cieple
PY наука *f* о теплоте
SU lämpüoppi
SV värmelära *c*

2286 heat source
DA varmekilde
DE Wärmequelle *f*
ES foco de calor *m* fuente de calor *f*
F source *f* de chaleur
I sorgente *f* di calore
MA hőforrás
NL warmtebron *f*
PO źródło *n* ciepła
PY источник *m* теплоты
SU lämmünlähde
SV värmekälla *c*

2287 heat station
DA varmecentral
DE Heizzentrale *f*
ES central de calefacción *f* central térmica *f*
F centrale *f* de chauffage
I centrale *f* termica
MA fűtőmű
NL onderstation *n* voor verwarming
PO ciepłownia *f*
PY тепловая станция *f*
SU lämpükeskus
SV värmecentral *c*

2288 heat storage
DA varmelager
DE Wärmespeicherung *f*
ES acumulación de calor *f* almacenamiento de calor *m*
F accumulateur *m* de chaleur
I accumulo *m* di calore
MA hőtárolás
NL warmteopslag *m*
PO akumulacja *f* ciepła
PY аккумуляция *f* теплоты
SU lämmünvarastointi
SV värmelagring *c*

2289 heat transfer
DA varmeoverføring
DE Wärme-übertragung *f*
ES transmisión de calor *f* transmisión térmica *f*
F transfert *m* de chaleur
I trasmissione *f* del calore
MA hőátadás
NL warmteoverdracht *f*
PO przenikanie *n* ciepła wymiana *f* ciepła
PY теплопередача *f*
SU lämmünsiirtyminen
SV värmeüverfüring *c*

2290 heat transfer coefficient
DA varmeoverførings-koefficient
DE Wärmeübertragungs-zahl *f*
ES coeficiente de transmisión de calor *m*
F coefficient *m* de transfert thermique
I coefficiente *m* di trasmissione del calore
MA hőátadási tényező
NL warmteoverdrachts-coëfficiënt *m*
PO współczynnik *m* przenikania ciepła
PY коэффициент *m* теплопередачи
SU lämmünläpäisy-kerroin
SV värmeüverfürings-koefficient *c*

2291 heat transfer oil
DA varmetransmissions-
    olie
DE Wärmeträgerül *n*
ES aceite transmisor de
    calor *m*
    termofluido
    (aceite) *m*
 F fluide *m* caloporteur
 I olio *m* diatermico
MA olaj hőhordozó
NL thermische olie *f*
PO nośnik *m* ciepła
    olejowy
PY масло *n* для
    переноса теплоты
SU lämmünsiirtoüljy
SV värmeüverfürings-
    olja *c*

2292 heat transmission
DA varmeoverføring
DE Wärmedurchgang *m*
ES transmisión de calor
    f
 F transmission *f* de
    chaleur
 I trasferimento *m* di
    calore
MA hőátadás
NL warmtetransmissie *f*
PO przenikanie *n* ciepła
    wymiana *f* ciepła
PY теплопереход *m*
SU lämmünläpäisy
SV värmeüverfüring *c*

2293 heat trap
DA varmefælde
DE Wärmeverschluß *m*
ES trampa de calor *f*
 F
 I trappola *f* di calore
MA kondenzedény
NL thermosyphon-
    blokkering *f*
PO syfon *m* cieplny
PY тепловой затвор *m*
    термосифон *m*
SU luonnonkierronesto
    veden
    lämmittimessä
SV värmefälla *c*

2294 height
DA højde
DE Hühe *f*
ES altura *f*
 F hauteur *f*
 I altezza *f*
MA magasság
NL hoogte *f*
PO wysokość *f*
PY высота *f*
SU korkeus
SV hüjd *c*

2295 helical *adj*
DA spiralformet
DE
ES helicoidal
 F hélicoïde
 I elicoidale
MA csigavonalú
    spirális
NL schroefvormig
PO spiralny
    śrubowy
PY пропеллерный
SU spiraalimainen
SV spiralgående

2296 hemp
DA hamp
DE Hanf *m*
ES cáñamo *m*
 F chanvre *m*
 I canapa *f*
MA kender
NL hennep *m*
PO konopie *pl*
PY конопля *f*
    пенька *f*
SU hamppu
SV hampa *c*

2297 HEPA filter (*see*
    absolute filter)

2298 hermetic *adj*
DA hermetisk
DE hermetisch
ES hermético
 F hermétique
 I ermetico
MA hermetikus
    légtümür
NL hermetisch
PO hermetyczny
    szczelny
PY герметичный
SU hermeettinen
SV hermetisk

2299 hermetical (*see*
    hermetic *adj*)

2300 hermetically-
    sealed *adj*
DA hermetisk forsejlet
DE hermetisch
    abgedichtet
ES herméticamente
    sellado
 F scellé
    hermétiquement
 I sigillato
    ermeticamente
MA hermetikusan
    tümített
NL hermetisch gesloten
PO hermetycznie
    szczelny
PY герметически
    заваренный
SU hermeettisesti
    tiivistetty
SV hermetiskt sluten

2301 hermetic
    compressor
DA hermetisk
    kompressor
DE geschlossener
    Verdichter *m*
ES compresor
    hermético *m*
 F compresseur *m*
    hermétique
 I compressore *m*
    ermetico
MA hermetikus
    kompresszor
NL gesloten
    compressor *m*
    hermetische
    compressor *m*
PO sprężarka *f*
    hermetyczna
PY герметичный
    компрессор *m*
SU hermeettinen
    kompressori
SV hermetisk
    kompressor *c*

2302 high discharge temperature cut out
DA overkogningssikring
DE Ausschalter *m* für hohe Austritts- temperatur *f* Übertemperatur- schalter *m*
ES desconector por alta temperatura de descarga *m*
F thermostat *m* de sécurité au refoulement
I limitazione *f* di sicurezza della temperatura di scarico
MA nagy távozó hőmérsékletnél kikapcsoló
NL hogeperstemperatuur- beveiliger *m* hogeperstemperatuur- beveiliging *f*
PO ogranicznik *m* temperatury
PY термореле *n* автоматической защиты компрессора
SU ylikuumenemissuoja
SV überkoknings- säkring *c*

2303 high efficiency filter
DA mikrofilter
DE Hochleistungs- filter *m*
ES filtro de alta eficacia *m* filtro de alto rendimiento *m*
F filtre *m* à haute efficacité
I filtro *m* ad alta efficienza
MA nagyteljesítményű szűrő
NL hoogrendements- filter *m*
PO filtr *m* wysokosprawny
PY высокоэффек- тивный фильтр *m*
SU mikrosuodatin
SV mikrofilter *n*

2304 high–low control
DA høj-lav-styring
DE Zweipunkt- regelung *f*
ES control de alta/baja *m*
F contrôle *m* haute et basse pression
I regolazione *f* alto- basso
MA kétszintű szabályozás
NL hoog-laag regeling *f*
PO kontrola *f* między wartością górną i dolną
PY двухпозиционное регулирование *n*
SU high-low säätü kaksitehosäätü
SV hüg-lågkontroll *c*

2305 high point
DA højdepunkt
DE Hühepunkt *m*
ES vértice *m*
F point *m* haut
I punto *m* alto
MA csúcspont
NL hoogste punt *n*
PO punkt *m* górny punkt *m* szczytowy
PY высшая точка *f*
SU lakipiste
SV hüjdpunkt *c*

2306 high pressure
DA højtryk
DE Hochdruck *m*
ES alta presión *f*
F haute pression *f*
I alta pressione *f*
MA nagynyomás
NL hoge druk *m*
PO ciśnienie *n* wysokie
PY высокое давление *n*
SU korkeapaine
SV hügtryck *n*

2307 high pressure boiler
DA højtrykskedel
DE Hochdruckkessel *m*
ES caldera de alta presión *f*
F chaudière *f* haute pression
I caldaia *f* ad alta pressione
MA nagynyomású kazán
NL hogedrukketel *m*
PO kocioł *m* wysokociśnie- niowy
PY котел *m* высокого давления
SU korkeapainekattila
SV hügtryckspanna *c*

2308 high pressure control
DA højtryksregulering
DE Hochdruck- regelung *f*
ES control de alta presión *m*
F contrôle *m* haute pression
I regolazione *f* di alta pressione
MA nagynyomású szabályozás
NL hogedrukregeling *f*
PO kontrola *f* wysokiego ciśnienia
PY контроль *m* высокого давления
SU korkeapaineen säätü yläpaineen säätü
SV hügtryckskontroll *c*

2309 high pressure hot water
DA hedt vand
DE Heißwasser *n*
ES agua caliente a alta presión *f*
agua sobre-calentada *f*
F eau *f* chaude haute pression
eau *f* surchauffée
I acqua *f* surriscaldata
MA nagynyomású forróvíz
NL heetwater *n*
PO woda *f* gorąca wysokiego ciśnienia
PY горячая вода *f* под высоким давлением
перегретая вода *f*
SU korkeapaineinen kuumavesi
SV hetvatten-uppvärmning *c*

2310 high-pressure hot water heating
DA hedtvands-opvarmning
DE Heißwasserheizung *f*
ES calefacción por agua sobrecalentada *f*
F chauffage *m* à eau chaude sous pression
I riscaldamento *m* ad acqua surriscaldata
MA nagynyomású forróvízfűtés
NL heetwater-verwarming *f*
PO ogrzewanie *n* wodą gorącą o wysokim ciśnieniu
PY водяное отопление *n* высокого давления
SU kuumavesilämmitys
SV hetvatten-uppvärmning *c*

2311 high pressure safety cut-out
DA højtrykssikring
DE Hochdruck-Sicherheits-ausschalter *m*
Überdruck-schalter *m*
ES desconector de seguridad de alta presión *m*
F pressostat *m* de sécurité haute pression
I limitazione *f* di sicurezza della pressione
MA biztonsági lezáró nagynyomás esetén
NL hogedrukbeveiligings-schakelaar *m*
PO ogranicznik *m* wysokiego ciśnienia
wyłącznik *m* bezpieczeństwa wysokiego ciśnienia
PY предохранительное устройство *n* от превышения давления
SU ylipainesuoja
SV hügtryckssmält-skydd *n*

2312 high pressure steam
DA højtryksdamp
DE Hochdruckdampf *m*
ES vapor (de agua) a alta presión *m*
F vapeur *f* haute pression
I vapore *m* ad alta pressione
MA nagynyomású gőz
NL hogedrukstoom *m*
PO para *f* wysokiego ciśnienia
para *f* wysokoprężna
PY пар *m* высокого давления
SU korkeapainehöyry
SV hügtrycksånga *c*

2313 high rise cold store
DA fleretages kølehus
DE Kühlhochhaus *n*
ES silo frigorífico *m*
F chambre *f* frigorifique de grandeur hauteur
I magazzino *m* frigorifero
MA hűtött magasraktár
NL hoge koelruimte *f*
PO chłodnia *f* składowa wielokondygna-cyjna
PY грузоподъемное холодохран-илище *n*
SU korkea kylmävarasto
SV flervåningskylhus *n*

2314 high temperature water
DA hedt vand
DE Heißwasser *n*
ES agua sobre-calentada *f*
F eau *f* haute température
I acqua *f* ad alta temperatura
MA forróvíz
NL warm tapwater *n*
warmwater *n*
PO woda *f* o wysokiej temperaturze
PY высокотем-пературная вода *f*
SU kuuma vesi
SV hetvatten *n*

2315 high vacuum
DA kraftig vakuum
DE Hochvakuum *n*
ES alto vacío *m*
F vide *m* poussé
I vuoto *m* spinto
MA nagy vákum
NL diepvacuüm *n*
hoogvacuüm *n*
PO głęboka próżnia *f*
wysoka próżnia *f*
PY глубокий вакуум *m*
SU tyhjü
SV hügvakuum *n*

2316 high-velocity air conditioning
DA højhastigheds-
   ventilation
DE Hochgeschwindigkeits-
   klimaanlage f
ES acondicionamiento
   de aire a alta
   velocidad m
F conditionnement m
   d'air à haute
   vitesse
I condizionamento m
   dell'aria ad alta
   velocità
MA nagysebességű
   légkondicionálás
NL hogedruk
   lucht-
   behandeling f
PO klimatyzacja f
   wysokopręd-
   kościowa
PY высокоскоростная
   система f
   кондицион-
   ирования воздуха
SU korkeapaineil-
   mastointi
SV hüghastighetsklimat-
   anläggning c

2317 high velocity fan
DA højhastigheds-
   ventilator
DE Hochdruck-
   ventilator m
ES ventilador de alta
   presión m
F ventilateur m à
   haute vitesse
I ventilatore m per
   impianto ad alta
   velocità
MA nagysebességű
   ventilátor
NL hogedruk-
   ventilator m
PO wentylator m o
   dużej prędkości
PY высокоскоростной
   вентилятор m
SU korkeapainepuhallin
SV hüghastighetsfläkt c

2318 high volatile coal
DA fede kul
DE Fettkohle f
   Gasflammkohle f
ES carbón de alto
   contenido de
   volátiles m
F charbon m à haute
   teneur en matières
   volatiles
I carbone m a lunga
   fiamma
   carbone m ricco
MA illékony szén
NL vetkool f
   vlamkool f
PO węgiel m o dużej
   zawartości części
   lotnych
PY уголь m с
   большим
   содержанием
   летучих
SU lihava hiili
SV feta kol n

2319 hinge
DA hængsel
   led
DE Scharnier n
ES bisagra f
F charnière f
   gond m
I cardine m
   cerniera f
MA forgópánt
   zsanér
NL scharnier n
PO przegub m
   zawiasa f
PY шарнир m
SU sarana
SV gångjärn n

2320 hinged lid
DA hængslet låg
DE Verschlußklappe f
ES tapa abisagrada f
F clapet m battant
   couvercle m
I sportello m a
   cerniera
MA nyitható fedél
NL scharnierend
   deksel n
PO pokrywa f na
   zawiasach
PY шарнирная
   крышка f
SU saranoitu kansi
SV svänglucka c

2321 hole
DA hul
DE Loch n
   Öffnung f
ES agujero m
F trou m
I buco m
   foro m
MA lyuk
   nyílás
NL gat n
   opening f
PO otwór m
PY отверстие n
SU reikä
SV hål n

2322 holes (in structures for services)
DA udsparing
DE Mauer-
   aussparungen f,pl
ES pasamuros m
F percements m
   réservations f
I fori m,pl passanti
MA kezelőnyílás
NL sparingen pl
PO otwory m
   montażowe
PY монтажное
   отверстие n
SU läpivienti
SV genomfürings-
   üppningar c

2323 hood
DA kappe
   kætte
DE Abzug m
   Deckel m
   Haube f
ES sombrerete m
F hotte f
I coperchio m
MA ernyő
   fedél
NL kap f
PO okap m
PY колпак m
SU huuva
SV huv c

DA	=	Danish
DE	=	German
ES	=	Spanish
F	=	French
I	=	Italian
MA	=	Magyar (Hungarian)
NL	=	Dutch
PO	=	Polish
PY	=	Russian
SU	=	Finnish
SV	=	Swedish

2324 hook
DA krog
DE Haken *m*
ES gancho *m*
F crochet *m*
I gancio *m*
   uncino *m*
MA horog
NL haak *m*
PO hak *m*
PY крюк *m*
SU koukku
SV krok *c*

2325 hopper
DA silo
   tragtbeholder
DE Trichter *m*
ES silo *m*
   tolva *f*
F silo *m*
   trémie *f*
I silo *m*
   tramoggia *f*
MA garat
NL silo *m*
   stortkoker *m*
PO lej *m*
   samowyładowczy
   lej *m* zasypowy
   silos *m*
   zbiornik *m*
PY бункер *m*
   воронка *f*
SU siilo
SV magasin *n*

2326 hopper-fed boiler
DA magasinfyret kedel
DE Füllschachtkessel *m*
ES caldera con carga
   por tolva *f*
F chaudière *f* à trémie
   de combustible
I caldaia *f* alimentata
   da tramoggia
MA tültőgaratos kazán
NL ketel *m* met
   brandstofvul-
   trechter
PO kocioł *m* z górnym
   zasilaniem
PY котел *m* с
   бункером для
   угля
SU polttoainesäiliüllä
   varustettu kattila
SV magasinseldad
   panna *c*

2327 hopper firing
DA magasinfyring
   stokerfyring
DE Schachtfeuerung *f*
ES caldeo con
   alimentación por
   tolva *m*
F chauffe *f* avec
   alimentation par
   trémie
I combustione *f* a
   tramoggia
MA tültőgaratos tüzelés
NL stookmethode *f* met
   brandstofvul-
   trechter
PO palenisko *n* z
   górnym zasilaniem
PY топка *f* с бункером
   (для угля)
SU polttaminen
   polttoainesäiliün
   avulla
SV magasinseldning *c*

2328 horizontal flue
DA liggende røgkanal
   vandret aftræk
DE waagerecht
   Feuerzug *m*
ES conducto horizontal
   de humos *m*
F carneau *m*
I canale *m* da fumo
   condotto *m*
   orizzontale
MA rókatorok
   vízszintes
   füstcsatorna
NL horizontaal
   rookkanaal *n*
PO czopuch *m*
   poziomy kanał *m*
   spalinowy
PY боров *m*
   горизонтальный
   газоход *m*
SU vaakasuora
   savukanava
SV liggande rükkanal *c*

2329 horsepower
DA hestekraft
DE Pferdestärke (PS) *f*
ES caballo vapor *m*
F cheval *m* vapeur
   puissance *f*
   mécanique
I cavallo *m* vapore
MA lóerő
NL paardekracht *f*
PO koń *m* mechaniczny
PY лошадиная сила *f*
SU hevosvoima
SV hästkraft *c*

2330 hose
DA slange
DE Schlauch *m*
ES manguera *f*
F tuyau *m* flexible
I manichetta *f*
   tubo *m* flessibile
MA tümlő
NL brandslang *f*
   buigzame buis *f*
   slang *f*
PO przewód *m*
   elastyczny
   wąż *m*
PY шланг *m*
SU letku
SV slang *c*

2331 hose connection
DA slangeforbindelse
DE Schlauch-
   verbindung *f*
ES conexión de
   manguera *f*
F raccordement *m* par
   tuyau flexible
I raccordo *m*
   portagomma
MA tümlővéges
   csatlakozás
NL slangaansluiting *f*
   slangverbinding *f*
PO złączka *f* do węża
PY гибкое
   соединение *n*
SU letkuliitin
SV slangkoppling *c*

2332 hot *adj*
DA varm
DE heiß
ES caliente
F chaud
I bollente
   caldo
MA forró
   meleg
NL heet
   warm
PO gorący
PY горячий
SU kuuma
SV het

2333 hot air
DA varm luft
DE Heißluft *f*
Warmluft *f*
ES aire caliente *m*
F air *m* chaud
I aria *f* calda
MA meleglevegő
NL hete lucht *f*
PO gorące powietrze *n*
PY горячий воздух *m*
SU kuuma ilma
SV hetluft *c*

2334 hot dipped galvanized (*adj*)
DA varmt galvaniseret
DE tauchbadgalvanisiert
ES galvanizado en caliente
F galvanisé au bain
I zincato a caldo
MA forró horganyfürdőbe mártott
NL in volbad verzinkt
PO ocynkowany na gorąco
PY оцинкованный горячим способом
SU kuumagalvanoitu
SV varmfürzinkad

2335 hot plate
DA kogeplade varmeplade
DE Heizplatte *f*
ES placa calefactora *f* placa caliente *f*
F plaque *f* chauffante
I piastra *f* scaldante
MA főzőlap
NL kookplaat *f* warmhoudplaat *f*
PO płyta *f* grzejna
PY нагревательная плита *f*
SU lämpülevy
SV värmeplatta *c*

2336 hot room
DA varmluftrum
DE Heißluftraum *m*
ES cuarto estufa *m* habitación caliente *f*
F pièce chaude
I camera *f* calda
MA meleg helyiség
NL hete ruimte *f*
PO komora *f* gorąca pomieszczenie *n* gorące
PY горячий цех *m*
SU kuumahuone
SV varmluftsrum *n*

2337 hot water
DA varmt vand
DE Heißwasser *n* Warmwasser *n*
ES agua caliente *f*
F eau *f* chaude
I acqua *f* calda
MA forróvíz
NL warm tapwater *n*
PO woda *f* gorąca
PY горячая вода *f*
SU kuuma vesi lämmin vesi
SV hetvatten *n*

2338 hot water boiler
DA varmtvandskedel
DE Warmwasser- kessel *m*
ES caldera de agua caliente *f*
F chaudière *f* à eau chaude
I generatore *m* ad acqua calda
MA melegvízkazán
NL warmwaterketel *m*
PO kocioł *m* wody gorącej
PY котел *m* горячего водоснабжения
SU lämminvesikattila
SV hetvattenpanna *c*

2339 hot water demand
DA varmtvandsbehov
DE Warmwasser- bedarf *m*
ES demanda de agua caliente *f* necesidades de agua caliente *f,pl*
F besoins *m* d'eau chaude
I fabbisogno *m* di acqua calda
MA melegvízigény
NL warmwater- behoefte *f* warmwatervraag *f*
PO zapotrzebowanie *n* na wodę gorącą
PY потребность *f* в горячей воде
SU lämpimän veden tarve
SV hetvattenbehov *n*

2340 hot water distribution
DA varmtvands- distribution
DE Warmwasser- verteilung *f*
ES distribución de agua caliente *f*
F distribution *f* d'eau chaude
I distribuzione *f* di acqua calda
MA melegvíz elosztás
NL warmwater- verdeling *f*
PO rozprowadzenie *n* wody gorącej
PY распределение *n* горячей воды
SU lämpimän veden jakelu
SV hetvatten- distribution *c*

2341 hot water storage
DA varmtvandslager
DE Warmwasser-
speicher *m*
ES acumulación de agua
caliente *f*
F stockage *m* d'eau
chaude
I accumulo *m* di
acqua calda
MA melegvíz tárolás
NL warmwateropslag *m*
warmwater-
voorraad *m*
PO akumulacja *f* wody
gorącej
PУ аккумуляция *f*
горячей воды
SU lämpimän veden
varaaja
SV hetvattenlagring *c*

**2342 hot water storage
cylinder**
DA varmtvandsbeholder
DE Boiler *m*
Vorratswarmwasser-
bereiter *m*
ES acumulador de agua
caliente *m*
F ballon *m* d'eau
chaude
I serbatoio *m*
dell'acqua calda
MA melegvíztároló
henger
NL warmwatervoor-
raadtank *m*
PO zasobnik *m* wody
gorącej
PУ аккумулятор *m*
горячей воды
SU lämminvesisäiliü
SV hetvatten-
ackumulator *c*

2343 hot water supply
DA varmtvandsforsyning
DE Warmwasserver-
sorgung *f*
ES alimentación de agua
caliente *f*
F alimentation *f* en
eau chaude
I alimentazione *f* di
acqua calda
MA melegvízellátás
NL warmwater-
vooziening *f*
PO zasilenie *n* w wodę
gorącą
PУ горячее
водоснабжение *n*
SU lämpimän veden
tuotto
SV hetvattenfürsürjning
c

**2344 hot water supply
boiler**
DA kedel for varmt
brugsvand
DE Warmwasserbereiter
m
ES calentador de agua
m
F chaudière *f* de
préparation d'eau
chaude sanitaire
I caldaia *f* per acqua
calda
MA melegvízellátó kazán
NL warmwatervoorzienings-
ketel *m*
PO kocioł *m* przygotow-
ujący wodę gorącą
PУ котел *m* горячего
водоснабжения
SU lämminvesikattila
SV hetvattenpanna *c*

2345 hot water system
DA varmtvandssystem
DE Warmwassersystem
n
ES sistema de agua
caliente *m*
F service *m* d'eau
chaude
I impianto *m* ad
acqua calda
MA melegvíz rendszer
NL warmwatersysteem
n
PO instalacja *f* wody
gorącej
PУ система *f* горячего
водоснабжения
SU lämminvesijärje-
stelmä
SV hetvattensystem *n*

2346 hot well
DA varmekilde
DE Wärmequelle *f*
ES depósito de agua cal-
iente *m*
F source *f* de chaleur
souterraine
I pozzo *m* caldo
MA hőforrás
NL heetwaterbron *f*
PO zbiornik *m* kondens-
atu
zbiornik *m* skroplin
PУ источник *m*
теплоты
SU lämmünlähde
SV värmekälla *c*

**2347 hot wire
anemometer**
DA varmtvandsanemo-
meter
DE Hitzdrahtanemo-
meter *n*
ES anemómetro de hilo
caliente *m*
F anémomètre *m* à fil
chaud
I anemometro *m* a
filo caldo
MA hődrótos
anemométer
NL hittedraad
anemometer *m*
PO anemometr *m* z
gorącym włóknem
PУ анемометр *m* с
нитью
накаливания
SU kuumalanka-
anemometri
SV varmvattenanemo-
meter *c*

2348 house connection
DA stikledning
DE Hausanschluß *m*
ES acometida *f*
F branchement *m*
d'immeuble
I collegamento *m* di
edificio
MA házi csatlakozás
NL huisaansluiting *f*
PO przyłącze *n* domowe
PУ домовой ввод *m*
SU rakennuksen
liittyminen
SV servisledning *c*

2349 housing (*see* casing)

2350 hub
DA nav
DE Nabe *f*
ES cubo (rueda) *m*
F moyeu *m*
I mozzo *m*
MA agy
NL naaf *f*
PO kielich *m* rury
piasta *f*
PY раструб *m*
SU napa
SV nav *n*

2351 humid *adj*
DA fugtig
DE feucht
ES húmedo
F humide
I umido
MA nedves
NL vochtig
PO wilgotny
PY влажный
SU kostea
SV fuktig

2352 humidification
DA befugtning
DE Befeuchtung *f*
ES humidificación *f*
F humidification *f*
I umidificazione *f*
MA nedvesítés
NL bevochtiging *f*
PO nawilżanie *n*
PY увлажнение *n*
SU kostutus
SV befuktning *c*

2353 humidifier
DA befugter
DE Befeuchter *m*
ES humidificador *m*
F humidificateur *m*
I umidificatore *m*
MA nedvesítő
NL bevochtiger *m*
PO nawilżacz *m*
PY увлажнитель *m*
SU kostutin
SV befuktare *c*

2354 humidify *vb*
DA befugte
DE befeuchten
ES humidificar
F humidifier
I umidificare
MA nedvesít
NL bevochtigen
PO nawilżać
PY увлажнять
SU kostuttaa
SV befukta

2355 humidifying equipment
DA befugtningsudstyr
DE Befeuchtungsgerät *n*
ES equipo de
humidificación *m*
F humidificateur *m*
I apparecchiatura *f* di
umidificazione
MA nedvesítő készülék
NL bevochtigings-
apparatuur *f*
PO urządzenie *n*
nawilżające
PY оборудование *n*
для увлажнения
SU kostutuslaite
SV befuktnings-
utrustning *c*

2356 humidistat (*see* hygrostat)

2357 humidity
DA fugtighed
DE Feuchte *f*
Feuchtigkeit *f*
ES humedad *f*
F humidité *f*
I umidità *f*
MA nedvesség
NL vochtigheid *f*
PO wilgotność *f*
PY влажность *f*
SU kosteus
SV fuktighet *c*

2358 humidity control
DA fugtighedsregulering
DE Feuchteregelung *f*
ES regulación de la
humedad *f*
F régulation *f*
d'humidité
I controllo *m*
dell'umidità
MA nedvességszabá-
lyozás
NL vochtigheids-
regeling *f*
PO regulacja *f*
wilgotności
PY регулирование *n*
влажности
SU kosteuden säätü
SV fuktighetskontroll *c*

2359 humidity controller
DA befugtningsauto-
matik
DE Feuchteregler *m*
ES control de humedad
m
F régulateur *m*
d'humidité
I regolatore *m* di
umidità
MA nedvesség
szabályozó
NL vochtigheids-
regelaar *m*
PO regulator *m*
wilgotności
PY регулятор *m*
влажности
SU kosteuden säädin
SV fuktighets-
regulator *c*

2360 humidity sensor
DA fugtighedsføler
DE Feuchtefühler *m*
ES sonda de humedad *f*
F capteur *m*
d'humidité
I sensore *m* di
umidità
MA nedvesség érzékelő
NL vochtigheids-
opnemer *m*
vochtigheids-
voeler *m*
PO czujnik *m*
wilgotności
PY датчик *m*
влажности
SU kosteusanturi
SV fuktighets-
avkännare *c*

2361 hunting
DA pendling
DE Oszillieren n
   Pendeln n
ES fluctuación f
   hostilación periódica
     anormal f
   inestabilidad f
F pompage m en
   régulation
I oscillazione f (di un
   regolatore)
MA lengés
NL oscilleren n
   pendelen n
PO niestateczność f
   regulatora
PY колебание n
SU värähtelevä
SV svängning c

2362 hydraulic adj
DA hydraulisk
DE hydraulisch
ES hidráulico
F hydraulique
I idraulico
MA hidraulikus
NL hydraulisch
PO hydrauliczny
PY гидравлический
SU hydraulinen
SV hydraulisk

2363 hydrocarbon
DA kulbrinte
DE Kohlenwasser-
   stoff m
ES hidrocarburo f
F hydrocarbure m
I idrocarburo m
MA szénhidrogén
NL koolwaterstof f
PO węglowodór m
PY углеводород m
SU hiilivety
SV kolväte n

2364 hydrostatic adj
DA hydrostatisk
DE hydrostatisch
ES hidroestático
F hydrostatique
I idrostatico
MA hidrosztatikus
NL hydrostatisch
PO hydrostatyczny
PY гидростатический
SU hydrostaattinen
SV hydrostatisk

2365 hygienic adj
DA hygiejnisk
DE hygienisch
ES higiénico
F hygiénique
I igienico
MA higiénikus
NL hyginisch
PO higieniczny
PY гигиенический
SU hygieeninen
SV hygienisk

2366 hygrometer
DA hygrometer
DE Hygrometer n
ES higrómetro m
F hygromètre m
I igrometro m
MA higrométer
   nedvességmérő
NL hygrometer m
   vochtigheidsmeter m
PO higrometr m
PY гигрометр m
SU hygrometri
SV hygrometer c

2367 hygroscopic adj
DA hygroskopisk
DE hygroskopisch
ES higroscópico
F hygroscopique
I igroscopico
MA higroszkópikus
   nedvszívó
NL hygroscopisch
PO higroskopijny
PY гигроскопический
SU hygroskooppinen
SV hygroskopisk

2368 hygrostat
DA hygrostat
DE Feuchtefühler m
   Hygrostat m
ES hidrostato m
F hygrostat m
I igrostato m
MA higrosztát
   nedvességérzékelő
NL hygrostaat m
   vochtigheids-
   regelaar m
PO higrostat m
   humidostat m
PY гигростат m
SU hygrostaatti
SV hygrostat c

2369 hyperbolic adj
DA hyperbolisk
DE hyperbolisch
ES hiperbólico
F hyperbolique
I iperbolico
MA hiperbolikus
NL hyperbolisch
PO hyperboliczny
PY гиперболический
SU hyperbolinen
SV hyperbolisk

2370 hypothermia
DA underafkøling
DE Hypothermie f
   Unterkühlung f
ES hipotermia f
F hypothermie f
I ipotermia f
MA túlhűtés
NL hypothermie f
PO hipotermia f
PY гипотермия f
SU alijäähtyminen
   hypotermia
SV hypotermi c

DA	=	Danish
DE	=	German
ES	=	Spanish
F	=	French
I	=	Italian
MA	=	Magyar (Hungarian)
NL	=	Dutch
PO	=	Polish
PY	=	Russian
SU	=	Finnish
SV	=	Swedish

I

2371 IAQ (*see* indoor air quality)

2372 ice
DA is
DE Eis *n*
ES hielo *m*
F glace *f*
I ghiaccio *m*
MA jég
NL ijs *n*
PO lód *m*
PY лед *m*
SU jää
SV is *c*

2373 ice bank
DA islager
DE Eisspeicher *m*
ES montón de hielo *m*
F stockage *m* de glace
I accumulo *m* di ghiaccio
MA jégtárolós rendszer
NL ijsbank *f*
   koude-accumulatie *f* via ijs
PO wolnolodowy zasobnik *m* zimna
   zasobnik *m* lodowy
PY образование *n* льда
SU veden jäätymistä hyväksi käyttävä kylmävarasto
SV islagring *c*

2374 ice block
DA isblok
DE Eisblock *m*
ES barra de hielo *f*
   bloque de hielo *m*
F bloc *m* de glace
I blocco *m* di ghiaccio
MA jégtümb
NL ijsblok *n*
PO blok *m* lodu
PY блок *m* льда
SU jäälohkare
SV isblock *n*

2375 icebox
DA isboks
DE Eisbehälter *m*
ES nevera (de hielo) *f*
   refrigerador por hielo *m*
F boîte *f* à glace
I ghiacciaia *f*
MA jégszekrény
NL ijskist *f*
PO lodówka *f*
PY контейнер *m* для льда
SU jääastia
SV isbehållare *c*

2376 ice crystal
DA iskrystal
DE Eiskristall *m*
ES cristal de hielo *m*
F cristal *m* de glace
I cristallo *m* di ghiaccio
MA jégkristály
NL ijskristal *n*
PO kryształ *m* lodu
PY кристалл *m* льда
SU jääkide
SV iskristall *c*

2377 ice cube
DA isterning
DE Eiswürfel *m*
ES cubito de hielo *m*
F cube *m* de glace
I cubo *m* di ghiaccio
MA jégkocka
NL ijsblokje *n*
PO kostka *f* lodu
PY брикет *m* льда
SU jääkuutio
SV iskub *c*

2378 iced water
DA isvand
DE Eiswasser *n*
ES agua helada *f*
F eau *f* glacée
I acqua *f* gelata
MA jeges víz
NL gekoeld drinkwater *n*
   ijswater *n*
PO woda *f* lodowa
PY ледяная вода *f*
SU jäävesi
SV isvatten *n*

2379 iced water cooling
DA isvandskøling
DE Eiswasserkühlung *f*
ES enfriamiento por agua helada *m*
F refroidissement *m* par eau glacée
I raffreddamento *m* con acqua gelata
MA jegesvizes hűtés
NL ijswaterkoeling *f*
PO chłodzenie *n* przy użyciu wody lodowej
PY охлаждение *n* ледяной водой
SU jäähdytys jäävedellä
SV isvattenkylning *c*

2380 ice formation
DA isdannelse
DE Eisbildung *f*
ES escarchado *m*
   formación de hielo *f*
F formation *f* de glace
I formazione *f* di ghiaccio
MA jégképződés
NL ijsvorming *f*
PO tworzenie *n* się lodu
PY образование *n* льда
SU jäänmuodostus
SV isbildning *c*

2381 ice layer
DA islag
DE Eisschicht *f*
ES capa de hielo *f*
F couche *f* de glace
I strato *m* di ghiaccio
MA jégréteg
NL ijslaag *f*
PO warstwa *f* lodu
PY слой *m* льда
SU jääkerros
SV isskikt *n*

2382 ice-making plant
DA isproduktionsanlæg
DE Eiserzeugungs-
    anlage *f*
ES instalación de
    fabricación de
    hielo *f*
F installation *f* de
    production de
    glace
I impianto *m* per la
    fabbricazione di
    ghiaccio
MA jégkészítő üzem
NL ijsinstallatie *f*
PO instalacja *f* do
    wytwarzania lodu
PY генератор *m* льда
SU keinojäälaitos
SV istillverknings-
    anläggning *c*

**2383 identification
    colour**
DA kendingsfarve
DE Kennfarbe *f*
ES color de
    identificación *m*
F teinte *f*
    conventionnelle
I colore *m* di
    identificazione
MA azonosító színjelzés
NL kenkleur *f*
PO barwa *f*
    identyfikacyjna
PY отличительная
    окраска *f*
SU tunnisteväri
SV igenkänningsfärg *c*

2384 idle time
DA tomgangstid
DE Stillstandszeit *f*
ES tiempo de paro *m*
F temps *m* d'arrêt
I tempo *m* di arresto
MA holtidő
NL leeglooptijd *m*
PO czas *m* przerwy
    czas *m* przestoju
PY холодный
    период *m*
SU joutoaika
SV stillestándstid *c*

2385 idling
DA tomgang
DE Leerlauf *m*
ES marcha lenta *f*
    ralantí *m*
F marche *f* à vide
I marcia *f* a vuoto
MA üresjárat
NL leegloop *m*
PO bieg *m* jałowy
PY холостой ход *m*
SU tyhjäkäynti
SV tomgång *c*

2386 ignition
DA antændelse
DE Zündung *f*
ES encendido *m*
    ignición *f*
F allumage *m*
I accensione *f*
MA gyújtás
NL ontsteking *f*
PO zapłon *m*
PY вспышка *f*
    зажигание *n*
SU sytytys
SV antändning *c*

2387 ignition electrode
DA tændelektrode
DE Zündelektrode *f*
ES electrodo de
    cebado *m*
    electrodo de
    encendido *m*
F électrode *m*
    d'allumage
I elettrodo *m* di
    accensione
MA gyújtó elektróda
NL ontstekings-
    elektrode *f*
PO elektroda *f* zapłonu
PY электрод *m*
    зажигания
SU sytytyselektrodi
SV tändelektrod *c*

2388 ignition point
DA antændelsespunkt
DE Zündpunkt *m*
ES punto de ignición *m*
F point *m*
    d'inflammation
I punto *m* di
    accensione
MA gyulladáspont
NL ontstekingspunt *n*
PO punkt *m* zapłonu
PY точка *f* зажигания
SU syttymispiste
SV antändningspunkt *c*

**2389 ignition safety
    device**
DA tændsikringsudstyr
DE Zündsicherung *f*
ES dispositivo de
    seguridad en la
    ignición *m*
F dispositif *m* de
    sécurité
    d'allumage
I dispositivo *m* di
    sicurezza per
    l'accensione
MA gyújtásbiztosító
    szerkezet
NL ontstekings-
    beveiliging *f*
PO urządzenie *n*
    zabezpieczające
    zapłon
PY устройство *n* для
    защиты от
    вспышки
SU syttymisen
    varmistaja
SV tändsäkring *c*

DA	=	Danish
DE	=	German
ES	=	Spanish
F	=	French
I	=	Italian
MA	=	Magyar
		(Hungarian)
NL	=	Dutch
PO	=	Polish
PY	=	Russian
SU	=	Finnish
SV	=	Swedish

2390 ignition temperature
DA antændelsestemperatur
DE Zündtemperatur *f*
ES temperatura de ignición *f*
F température *f* d'ignition
I temperatura *f* di ignizione
MA gyulladási hőmérséklet
NL ontstekingstemperatuur *f*
PO temperatura *f* zapłonu
PY температура *f* вспышки
температура *f* зажигания
SU syttymislämpütila
SV antändningstemperatur *c*

2391 illuminance
DA belysning
DE Beleuchtungsstärke *f*
ES iluminancia *f*
F luminance *f*
I illuminamento *m*
MA világító kisugárzás
NL lichtgevende straling *f*
luminantie *f*
PO natężenie *n* oświetlenia
PY световое излучение *n*
SU valaistusvoimakkuus
SV belysning *c*

2392 immerse *vb*
DA dyppe
DE eintauchen
ES sumergir
F immerger
plonger
I immergere
MA márt
merít
NL dompelen
PO zanurzać
PY погружать
SU upottaa
SV doppa

2393 immersion heater
DA dypkoger
DE Tauchsieder *m*
ES calentador de inmersión *m*
F réchauffeur *m* à immersion
réchauffeur *m* plongeur
I resistenza *f* ad immersione
MA merülőforraló
NL verwarmingsdompelaar *m*
PO grzałka *f* nurnikowa
grzejnik *m* nurnikowy
PY погружной нагреватель *m*
SU uppokuumennin
SV doppvärmare *c*

2394 immersion thermostat
DA dyptermostat
DE Tauchthermostat *m*
ES termostato de inmersión *m*
F thermostat *m* plongeur
I termostato *m* ad immersione
MA merülőhőmérő
NL dompelthermostaat *m*
PO termostat *m* zanurzeniowy
PY погружной термостат *m*
SU uppotermostaatti
SV dopptermostat *c*

2395 impact filter
DA trykfilter
DE Stoßfilter *m*
ES filtro de impacto *m*
F filtre *m* à choc
I filtro *m* inerziale
MA ütküzéses szűrő
NL traagheidsfilter *m*
PO filtr *m* uderzeniowy
filtr *m* wstrząsowy
PY инерционный пылеотделитель *m*
SU dynaaminen suodatin
türmäyssuodatin
SV stütfilter *n*

2396 impeller
DA skovlhjul
DE Laufrad *n*
ES rodete *m*
turbina *f*
F propulseur *m*
I ventola *f*
MA forgórész
járókerék
NL schoepenrad *n*
waaier *m*
PO wirnik *m*
PY рабочее колесо *n*
SU siipipyürä
SV fläkthjul *n*

2397 impeller tip
DA skovltop
DE Laufradspitze *f*
ES borde de los álabes (del rodete) *m*
extremo de los álabes (del rodete) *m*
F extrémité *f* d'aube
I estremità *f* della girante
MA járókerék csúcs
NL waaieruiteinde *n*
PO nasada *f* wirnika
PY верх *m* лопатки (рабочего колеса)
SU juoksusiiven kärki
SV skoveltopp *c*

2398 impermeable *adj*
DA tæt
uigennemtrængelig
DE undurchlässig
ES impermeable
F imperméable
I impermeabile
MA át nem eresztő
NL ondoordringbaar
ondoorlatend
PO nieprzemakalny
nieprzepuszczalny
PY непроницаемый
SU läpäisemätün
SV vattentät

2399 incandescence
DA glødning
DE Weißglut *f*
ES incandescencia *f*
F incandescence *f*
I incandescenza *f*
MA izzás
NL gloed *m*
gloeiing *f*
PO żarzenie *n*
PY накаливание *n*
SU hehkutus
SV glüdande

2400 incident *adj*
DA tilfældig
DE vorkommend
ES fortuito
F incident
qui arrive
I incidente
MA váratlan
véletlen
NL invallend
PO chwilowy
związany (z czymś)
PY случайный
SU tulo-
SV tillfällig

2401 incidental heat gain
DA tilfældig
varmegenvinding
DE zufälliger
Wärmegewinn *m*
ES aporte fortuito del
calor *m*
entradas fortuitas
del calor *f,pl*
ganancia fortuita del
calor *f*
F gains *m* de chaleur
variable
I apporto *m* di calore
accidentale
MA alkalmi hőnyereség
NL toevallige
warmtewinst *f*
PO zysk *m* ciepła
chwilowy
PY случайное
теплопосту-
пление *n*
SU satunnainen
lämpükuorma
SV variabel
värmetillfürsel *c*

2402 incinerator
DA affaldsforbrænd-
ingsovn
DE Verbrennungs-
ofen *m*
ES incinerador *m*
F incinérateur *m*
I inceneritore *m*
MA szemétégető
NL verbrandingsoven *m*
PO piec *m* do
spopielania
PY мусоросжигател-
ьная печь *f*
SU polttolaitos
SV incinerator *c*

2403 inclined-tube boiler
DA skrårørskedel
DE Schrägrohrkessel *m*
ES caldera de tubos
inclinados *f*
F chaudière *f* à tubes
inclinés
I caldaia *f* a tubi
inclinati
MA ferdecsüves kazán
NL ketel *m* met
schuinliggende
waterpijpen
PO kocioł *m* z rurą
pochyłą
PY котел *m* с
наклонными
трубами
SU kaltevaputkinen
kattila
SV panna *c* med lutande
tuber *c*

2404 inclined-tube manometer
DA skrårørsmanometer
DE Schrägrohrmano-
meter *n*
ES manómetro de tubo
inclinado *m*
F manomètre *m* à
tube incliné
I manometro *m* a tubi
inclinati
MA ferdecsüves
manométer
NL schuinebuismano-
meter *m*
PO manometr *m* z
rurką pochyłą
PY манометр *m* с
наклонной
трубкой
SU mikromanometri
SV lutande
manometer *c*

2405 incombustible *adj*
DA ubrændbar
DE unverbrennbar
ES incombustible
F incombustible
I incombustibile
ininfiammabile
MA éghetetlen
NL onbrandbaar
PO niepalny
PY несгораемый
SU palamaton
SV obrännbar

2406 incomplete combustion
DA ufuldstændig
forbrændning
DE unvollständige
Verbrennung *f*
ES combustión
incompleta *f*
F combustion *f*
incomplète
I combustione *f*
incompleta
MA tükéletlen égés
NL onvolledige
verbranding *f*
PO spalanie *n*
niecałkowite
PY неполное
сгорание *n*
SU epätäydellinen
palaminen
SV ofullständig
fürbränning *c*

2407 incondensible *adj*
DA ej kondenserbar
DE nicht kondensierbar
ES incondensable
F incondensable
I incondensabile
MA nem kondenzálható
NL niet condenseerbaar
PO nieskraplający się
PY неконденси-
рующийся
SU tiivistymätün kaasu
SV okondenserbar

2408 incrustation
DA kedelstensdannelse
DE Inkrustation *f*
Kesselstein *m*
ES incrustación *f*
F incrustation *f*
I incrostazione *f*
MA kazánkő
lerakódás
NL incrustratie *f*
ketelsteenvorming *f*
PO kamień *m* kotłowy
PY образование *n*
накипи
SU kattilakivi
SV slaggbildning *c*

2409 index
DA indeks
   tal
DE Index *m*
ES índice *m*
F indice *m*
I indice *m*
MA index
   mutató
NL index *m*
   lijst *f*
PO indeks *m*
   wskaźnik *m*
PY индекс *m*
   стрелка *f*
   указатель *m*
SU indeksi
SV index *n*

2410 indicating light
DA signallys
DE Signallampe *f*
ES lámpara
   indicadora *f*
   luz indicadora *f*
F lampe *f* de
   signalisation
   lampe *f* témoin
I lampada *f* di
   segnalazione
MA jelzőlámpa
NL signaallamp *f*
PO lampa *f*
   sygnalizacyjna
PY индикаторная
   лампа *f*
SU merkkivalo
SV signallampa *c*

2411 indication
DA indikering
   tegn
DE Anzeige *f*
ES indicación *f*
F indication *f*
   signalisation *f*
I indicazione *f*
   segnale *m*
MA jelzés
NL aanwijzing *f*
   signalering *f*
PO wskazanie *n*
   wskazywanie *n*
PY индикация *f*
   сигнализация *f*
SU merkki
SV indikering *c*

2412 indication range
DA måleområde
DE Anzeigebereich *m*
ES escala de indicación *f*
F plage *f*
I intervallo *m* di
   indicazione
MA jelzési tartomány
NL aanwijsbereik *n*
PO zakres *m* wskazań
PY диапазон *m*
   измерения
SU merkkialue
SV indikeringsområde *n*

2413 indicator
DA viser
DE Anzeiger *m*
   Indikator *m*
ES indicador *m*
F indicateur *m*
I indicatore *m*
MA jelző
   mutató
NL aanwijzer *m*
   signaalapparaat *n*
   wijzer *m*
PO czujnik *m*
   wskaźnik *m*
PY индикатор *m*
SU osoittaja
SV indikator *c*

2414 individual heating
DA individuel
   opvarmning
DE Einzelheizung *f*
ES calefacción
   individual *f*
F chauffage *m*
   individuel
I riscaldamento *m*
   individuale
MA egyedi fűtés
NL individuele
   verwarming *f*
   verwarming *f* per
   vertrek
PO ogrzewanie *n*
   indywidualne
   ogrzewanie *n*
   miejscowe
PY индивидуальное
   отопление *n*
SU talokohtainen
   lämmitys
SV individuell
   uppvärmning *c*

2415 indoor air
DA indeluft
   rumluft
DE Raumluft *f*
ES aire ambiente
   interior *m*
F air *m* intérieur
I aria *f* interna
MA helyiséglevegő
NL binnenlucht *f*
PO powietrze *n*
   wewnętrzne
PY внутренний
   воздух *m*
SU sisäilma
SV inneluft *c*

2416 indoor air quality
DA rumluftkvalitet
DE Raumluftqualität *f*
ES calidad del aire
   interior *m*
F qualité *f* d'air
   intérieur
I qualità *f* dell'aria
   interna
MA belső terek
   légminősége
NL binnenlucht-
   kwaliteit *f*
PO jakość *f* powietrza
   wewnętrznego
PY качество *n*
   внутреннего
   воздуха
SU sisäilman laatu
SV inneluftkvalitet *c*

2417 induced *adj*
DA kunstig
DE induziert
ES inducido
F induit
I indotto
MA indukált
   kényszerített
NL geënduceerd
PO indukowany
   wymuszony
   wzbudzony
PY индуцированный
SU indusoitunut
SV inducerad

2418 induction
DA induktion
DE Induktion *f*
ES inducción *f*
F induction *f*
I induzione *f*
MA indukció
NL inductie *f*
PO indukcja *f*
wzbudzenie *n*
PY индукция *f*
SU induktio
SV induktion *c*

2419 induction heating
DA induktiv opvarmning
DE Induktionsheizung *f*
ES calefacción por
inducción *f*
F chauffage *m* par
éjecto-convecteur
I riscaldamento *m* a
induzione
MA indukciós fűtés
NL inductie
luchtverwarming *f*
PO ogrzewanie *n*
indukcyjne
PY индукционный
нагрев *m*
SU induktiolämmitin
SV induktiv
uppvärmning *c*

2420 induction ratio
DA induktionsrate
DE Induktionsverhältnis
n
ES relación de
inducción *f*
F taux *m* d'induction
I rapporto *m* di
induzione
MA indukciós hányad
NL inductie-
verhouding *f*
PO stopień *m* indukcji
stosunek *m* indukcji
PY коэффициент *m*
индукции
SU induktiosuhde
SV induktions-
fürhållande *n*

2421 induction unit
DA induktionsapparat
DE Induktionsgerät *n*
ES inductor *m*
F éjecto-convecteur *m*
I apparecchio *m* ad
induzione
induttore *m*
MA indukciós készülék
NL inductie-apparaat *n*
PO urządzenie *n*
indukcyjne
PY индукционное
устройство *n*
местный
эжекционный
кондиционер *m*
SU induktioyksikkü
SV induktionsapparat *c*

**2422 industrial air
conditioning**
DA teknisk
luftbehandling
DE Industrie-
Klimatisierung *f*
ES aire acondicionado
industrial *m*
F conditionnement *m*
d'air industriel
I condizionamento *m*
dell'aria
industriale
MA ipari
légkondicionálás
NL industriele
lucht-
behandeling *f*
PO klimatyzacja *f*
przemysłowa
PY производственное
кондицион-
ирование *n*
воздуха
SU teollisuusilmastointi
SV industriluftkondition-
ering *c*

2423 industrial water
DA industrivand
DE Industriewasser *n*
ES agua para usos
industriales *f*
F eau *f* industrielle
I acqua *f* industriale
MA ipari víz
NL industriewater *n*
PO woda *f* przemysłowa
PY промышленная
вода *f*
SU teollisuusvesi
SV industrivatten *n*

2424 inert *adj*
DA inaktiv
DE träge
ES inerte
F inerte
I inerte
MA inert
küzümbüs
NL inert
traag
PO bezwładny
obojętny
PY инертный
SU inertti
SV inert

2425 inertia
DA inerti
træghed
DE Beharrungs-
vermögen *n*
Trägheit *f*
ES inercia *f*
F inertie *f*
I inerzia *f*
MA tehetetlenség
NL inertie *f*
traagheid *f*
PO bezwładność *f*
PY инерция *f*
SU hitaus
inertia
SV trüghet *c*

2426 infection
DA infektion
DE Infektion *f*
ES infección *f*
F infection *f*
I infezione *f*
MA fertőzés
NL besmetting *f*
infectie *f*
PO infekcja *f*
zakażenie *n*
PY инфекция *f*
SU infektio
SV infektion *c*

2427 infiltration
DA gennemsivning
infiltration
DE Infiltrierung *f*
ES infiltración *f*
F infiltration *f*
I infiltrazione *f*
MA beszivárgás
beszűrődés
infiltráció
NL infiltratie *f*
PO infiltracja *f*
PY инфильтрация *f*
SU vuoto
SV infiltration *c*

2428 infiltration through cracks
DA gennemsivning gennem revner
DE Fugenlüftung *f*
ES infiltración por rendijas *f*
F ventilation *f* par infiltration
I infiltrazione *f* da fessura
MA beszivárgás réseken át
NL luchtlekkage *f* door spleten of kieren
PO infiltracja *f* przez nieszczelności
PY инфильтрация *f* через неплотности
SU rakovuoto
SV genom springor *c* infiltration *c*

2429 infitting door (*see* flush-fitting door)

2430 influx
DA indstrømning
DE Zufluß *m* Zufuhr *f*
ES admisión *f* afluencia *f* aflujo *m*
F admission *f* amenée *f*
I influsso *m*
MA beáramlás hozzáfolyás
NL instroom *m* toevloed *m*
PO napływ *m*
PY приток *m*
SU sisäänvirtaus
SV inströmning *c*

DA = Danish
DE = German
ES = Spanish
F = French
I = Italian
MA = Magyar (Hungarian)
NL = Dutch
PO = Polish
PY = Russian
SU = Finnish
SV = Swedish

2431 information exchange
DA informationsudveksling
DE Informationsaustausch *m*
ES intercambio de información *m*
F échange *m* d'information
I scambio *m* di informazioni
MA információ csere
NL informatie-uitwisseling *f*
PO wymiana *f* informacji
PY информационный обмен *m*
SU tiedonvaihto
SV informationsutbyte *n*

2432 infrared heating
DA infrarød opvarmning
DE Infrarotheizung *f*
ES calefacción por infrarrojos *f*
F chauffage *m* infrarouge
I riscaldamento *m* ad infrarossi
MA infravörös fűtés
NL infrarood-verwarming *f*
PO ogrzewanie *n* promieniami podczerwonymi
PY инфракрасное отопление *n*
SU infrapunalämmitys
SV infravärmning *c*

2433 infrared survey
DA undersøgelse ved infrarød belysning
DE Infrarotmessung *f*
ES inspección por infrarrojos *f*
F examen *m* infrarouge
I rilevamento *m* all'infrarosso
MA infravörös vizsgálat
NL infrarood inspectie *f* onderzoek *n* met infraroodstraling
PO pomiar *m* promieniami podczerwonymi
PY инфракрасная съемка *f*
SU infrapunakuvaus
SV infrarüd bild *c*

2434 ingress (of air)
DA (luft)-indløb
DE Lufteintritt *m* Luftzutritt *m*
ES entrada de aire *f*
F infiltration *f* d'air
I ingresso *m* di aria
MA levegő beáramlás
NL luchttoetreding *f*
PO dopływ *m* powietrza wlot *m* powietrza
PY вход *m* воздуха
SU ilman sisääntulo
SV luftinlopp *n*

2435 inhibitor
DA hæmningsstof inhibitor
DE Inhibitor *m*
ES inhibidor *m*
F inhibiteur *m*
I inibitore *m*
MA gátló anyag inhibitor
NL inhibitor *m* vertrager *m*
PO czynnik *m* hamujący inhibitor *m*
PY ингибитор *m*
SU inhibiittori
SV inhibitor *c*

2436 initial charge
DA første påfyldning
DE Erstbefüllung *f*
ES carga inicial *f*
F charge *f* initiale
I carica *f* iniziale
MA első feltültés
NL eerste vulling *f*
PO obciążenie *n* początkowe
PY начальная нагрузка *f*
SU ensitäyttü
SV initialfyllning *c*

2437 initial temperature
DA begyndelsestemperatur
DE Anfangstemperatur *f*
ES temperatura inicial *f*
F température *f* initiale
I temperatura *f* iniziale
MA kezdeti hőmérséklet
NL begintemperatuur *f*
PO temperatura *f* początkowa
PY начальная температура *f*
SU alkulämpütila
SV initialtemperatur *c*

2438 injection atomizer
DA injektionsforstøver
DE Injektions-
zerstäuber *m*
ES atomizador por
inyección *m*
F gicleur *m*
I ugello *m*
MA porlasztásos befúvó
NL injectieverstuiver *m*
PO rozpylacz *m*
wtryskowy
PY инжекторный
распылитель *m*
SU sumutin
suutin
SV spridarmunstycke *n*

2439 injection nozzle
DA injektionsdyse
DE Einspritzdüse *f*
ES boquilla de
inyección *f*
tobera de
inyección *f*
F gicleur *m*
I iniettore *m*
ugello *m* di iniezione
MA fúvóka
NL injectieverstuivings-
kop *m*
PO dysza *f* wtryskowa
PY инжекторное
сопло *n*
SU hajoitussuutin
SV insprutnings-
munstycke *n*

2440 injection pipe
DA injektionsrør
DE Einspritzrohr *n*
ES tubo de inyección *m*
F conduite *f*
d'injection
I tubo *m* di iniezione
MA befecskendező cső
NL injectiepijp *f*
PO rura *f* wtryskowa
PY инжекторная
труба *f*
SU sumutusputki
SV insprutningsrür *n*

2441 injector
DA injektor
DE Injektor *m*
Strahlpumpe *f*
ES inyector *m*
F injecteur *m*
I iniettore *m*
MA injektor
NL injecteur *m*
PO inżektor *m*
wtryskiwacz *m*
PY инжектор *m*
SU injektori
SV injektor *c*

2442 injector burner
DA injektionsbrænder
DE Injektorbrenner *m*
ES quemador de
inyección *m*
F brûleur *m* à
injection
I bruciatore *m* ad
iniezione
MA injektoros égő
NL injectiebrander *m*
PO palnik *m*
inżektorowy
palnik *m* wtryskowy
PY инжекционная
горелка *f*
SU injektoripoltin
SV injektorbrännare *c*

2443 inlet
DA indløb
tilløb
DE Einlaß *m*
Eintritt *m*
ES entrada *f*
F entrée *f*
I entrata *f*
ingresso *m*
MA belépés
beümlés
NL inlaat *m*
intree *f*
toevoer *m*
PO wlot *m*
PY впуск *m*
SU syüttü
SV inlopp *n*

2444 inlet opening
DA indløbsåbning
DE Eintrittsüffnung *f*
ES orificio de
admisión *m*
orificio de
entrada *m*
F orifice *m* d'entrée
ouverture *f* d'entrée
I apertura *f* di
ingresso
MA beümlőnyílás
szívónyílás
NL inlaatopening *f*
toevoeropening *f*
PO otwór *m* wlotowy
PY впускное
отверстие *n*
SU sisäänvirtausaukko
SV inloppsüppning *c*

2445 inlet valve
DA indløbsventil
DE Einlaßventil *n*
ES válvula de
admisión *f*
válvula de
aspiración *f*
F registre *m* d'entrée
(d'air)
soupape *f*
d'admission
I valvola *f* di
ammissione
MA szívószelep
NL inlaatafsluiter *m*
PO zawór *m* wlotowy
PY впускной клапан *m*
SU syüttüventtiili
SV inloppsventil *c*

2446 inner fin
DA indre lamel
DE Innenrippe *f*
ES aleta interior *f*
F ailette *f* intérieure
I aletta *f* interna
MA belső borda
NL inwendige lamel *f*
PO żeberko *n*
wewnętrzne
PY внутренне ребро *n*
SU sisäpuolinen ripa
SV inre fläns *c*

2447 inside air
DA indeluft
DE Innenluft *f*
Raumluft *f*
ES aire interior *m*
F air *m* intérieur
I aria *f* interna
(ambiente)
MA belső levegő
NL binnenlucht *f*
PO powietrze *n*
wewnętrzne
PY внутренний
воздух *m*
SU sisäilma
SV inneluft *c*

2448 insolation
DA solindfald
DE Sonnen-
einstrahlung *f*
ES insolación *f*
F ensoleillement *m*
I insolazione *f*
MA benapozás
NL zoninstraling *f*
PO nasłonecznienie *n*
PY инсоляция *f*
SU auringon säteily
SV solstrålning *c*

2449 inspection hole
DA inspektionshul
DE Schauloch *n*
ES mirilla de
inspección *f*
ventanillo de
inspección *m*
F regard *m*
trou *m* de visite
I foro *m* di ispezione
MA ellenőrző nyílás
NL inspectie-opening *f*
kijkgat *n*
PO otwór *m* rewizyjny
PY контрольное
отверстие *n*
SU tarkastusaukko
SV inspektionslucka *c*

2450 install *vb*
DA installere
DE einbauen
installieren
montieren
ES instalar
F installer
I installare
MA beépít
felszerel
NL installeren
PO instalować
montować
PY монтировать
SU asentaa
SV installera

2451 installation
DA anlæg
installation
DE Anlage *f*
Einrichtung *f*
Installation *f*
ES instalación *f*
F installation *f*
I impianto *m*
installazione *f*
MA felszerelés
installáció
NL installatie *f*
installeren *n*
PO instalacja *f*
PY установка *f*
SU asennus
SV installation *c*

**2452 instantaneous
water heater**
DA gennemstrømningsvand-
varmer
DE Durchlauferhitzer *m*
ES calentador de agua
instantáneo *m*
F chauffe-eau *m*
instantané
I scaldacqua *m*
istantaneo
MA átfolyós vízmelegítő
NL doorstroom-
warmwater-
bereider *m*
PO podgrzewacz *m*
wody
natychmiastowego
działania
PY скоростной
водоподогре-
ватель *m*
SU läpivirtauslämmitin
SV genomströmnings-
varmvattenberedare *c*

2453 instructions for use
DA (brugs-) vejledning
DE Gebrauchs-
anweisung *f*
ES instrucciones de
servicio *f,pl*
F instructions *f*
mode *m* d'emploi
prescriptions *f*
I istruzioni *f,pl* per
l'uso
MA használati utasítás
NL bedieningsvoor-
schrift *n*
gebruiksaanwijzing *f*
PO instrukcja *f* obsługi
PY инструкция *f* по
обслуживанию
SU käyttöohjeet
SV bruksanvisningar *c*

2454 insulant
DA isolerende materiale
DE Isoliermaterial *n*
Wärmedäm-
material *n*
ES aislante *m*
F isolant *m*
I coibente *m*
isolante *m*
MA szigetelőanyag
NL isolatie *f* (materiaal)
PO materiał *m*
izolacyjny
PY изолятор *m*
SU eriste
SV isolermaterial *n*

**2455 insulate
(electric/acoustic)**
vb
DA isolere (el, lyd)
DE isolieren
ES aislar
(acústicamente)
aislar
(eléctricamente)
F isoler
I coibentare
isolare
MA szigetel
NL isoleren
(elektrisch/
akoustisch)
PO izolować
(elektrycznie-
akustycznie)
PY изолировать
(электрически/
акустически)
SU eristää
SV isolera

2456 insulate (thermal)
vb
DA isolere (varme)
DE isolieren
wärmedämmen
ES aislar (térmicamente)
F calorifuger
isoler
I isolare
MA hőszigetel
NL isoleren (thermisch)
PO izolować (cieplnie)
PY изолировать
(термически)
SU eristää
SV isolera

2457 insulating jacket
DA isolationskappe
DE Isoliermantel *m*
ES camisa aislante *f*
envolvente aislante *f*
F jaquette *f* calorifuge
I mantello *m* isolante
MA szigetelő burkolat
NL isolatiemantel *m*
PO płaszcz *m* izolacyjny
PY изоляционная
оболочка *f*
SU eristävä peite
SV isolermaterial *n*

2458 insulating material
DA isolationsmateriale
DE Dämmstoff *m*
Isoliermaterial *n*
ES material aislante *m*
F matériau *m* isolant
I materiale *m* isolante
MA szigetelőanyag
NL isolatiemateriaal *n*
PO materiał *m*
izolacyjny
PY изоляционный
материал *m*
SU eristysaine
SV isolermaterial *n*

2459 insulation
DA isolering
DE Isolierung *f*
Wärmedämmung *f*
ES aislamiento *m*
F calorifuge *m*
isolation *f*
I coibentazione *f*
isolamento *m*
MA szigetelés
NL isolatie *f*
PO izolacja *f*
PY изоляция *f*
SU eristäminen
SV isolering *c*

2460 intake air
DA tilluft
udeluft
DE Zuluft *f*
ES aire de admisión *m*
aire de entrada *m*
F air *m* aspiré
air *m* introduit
I aria *f* introdotta
MA beszívott levegő
NL aanzuiglucht *f*
toevoerlucht *f*
PO powietrze *n* wlotowe
PY впуск *m* воздуха
SU tuloilma
ulkoilma
SV tilluft *c*

2461 integral control
DA integralregulering
DE Integralregelung *f*
ES regulación integral *f*
F régulation *f*
intégrale
I regolazione *f*
integrale
MA integrál szabályozás
NL integrerende
regeling *f*
PO sterowanie *n*
całkowe
PY астатическое
регулирование *n*
интегральное
управление *n*
SU integroitu säätü
SV integralstyrning *c*

2462 integrated *adj*
DA integreret
DE integriert
ES integrado
F intégré
I integrato
MA integrált
üsszegzett
NL geëntegreerd
PO scalony
zintegrowany
PY интегральный
SU integroitu
SV samordnad

2463 integrator
DA integrerende tæller
DE Integrator *m*
ES integrador *m*
F intégrateur *m*
I integratore *m*
MA integrátor
üsszegző
NL integrator *m*
PO integrator *m*
przyrząd *m*
całkujący
PY интегратор *m*
SU integraattori
SV integrator *c*

2464 intensity
DA intensitet
styrke
DE Intensität *f*
Stärke *f*
ES intensidad *f*
F intensité *f*
I intensità *f*
MA erősség
intenzitás
NL intensiteit *f*
PO intensywność *f*
nasilenie *n*
natężenie *n*
PY интенсивность *f*
SU intensiteetti
SV intensitet *c*

**2465 intensity of
illumination**
DA belysningsstyrke
DE Beleuchtungs-
stärke *f*
ES intensidad de
iluminación *f*
F intensité *f* lumineuse
I intensità *f* luminosa
MA megvilágítási erősség
NL verlichtingssterkte *f*
PO natężenie *n*
oświetlenia
PY освещенность *f*
SU valaistusvoimakkuus
SV belysningsstyrka *c*

2466 interchange ability
DA udskiftelighed
DE Austauschbarkeit *f*
ES intercambiabilidad *f*
F interchangeabilité *f*
I intercambiabilità *f*
MA egymással
felcserélhetőség
NL uitwisselbaarheid *f*
PO możliwość *f*
wymiany
PY заменяемость *f*
SU vaihdettavuus
SV utbytbarhet *c*

2467 intercooler
DA mellemkøler
DE Zwischenkühler *m*
ES enfriador
  intermedio *m*
F refroidisseur *m*
I raffreddatore *m*
  intermedio
MA küzbenső hűtő
NL tussenkoeler *m*
PO chłodnica *f*
  międzystopniowa
PY промежуточный
  охладитель *m*
SU välijäähdyttäjä
SV mellankylare *c*

2468 interference
DA indblanding
DE Interferenz *f*
ES interferencia *f*
F interférence *f*
I interferenza *f*
MA beavatkozás
  interferencia
NL interferentie *f*
  tussenkomst *f*
PO interferencja *f* (fal)
  luz *m* ujemny
  zakłócenie *n*
PY помеха *f*
SU interferenssi
SV inblandning *c*

2469 interior
DA indre
DE innen
  innere
ES interior *m*
F intérieur *m*
I ambiente *m* interno
MA belső
NL interieur *n*
  inwendige *n*
PO wnętrze *n*
PY интерьер *m*
SU sisustus
  sisä-
SV inre *n*

2470 interlock
DA forrigle, gribe ind i
  hinanden
DE Verblockung *f*
ES sincronizador *m*
F intermittant *m*
I dispositivo *m* di
  blocco
MA üsszekütés
NL vergrendeling *f*
PO blokada *f*
  sprężanie *n*
PY блокировка *f*
  смыкание *n*
SU sisäinen lukitus
SV fürregling *c*

2471 intermediate layer
DA mellemlåg
DE Zwischenschicht *f*
ES capa intermedia *f*
F couche *f*
  intermédiaire
I strato *m* intermedio
MA küzbenső réteg
NL tussenlaag *f*
PO warstwa *f* pośrednia
PY промежуточный
  слой *m*
SU välikerros
SV mellanskikt *n*

**2472 intermediate
  superheater**
DA mellemoverheder
DE Zwischen-
  überhitzer *m*
ES cambiador de calor
  intermedio *m*
  recalentador de
  intermedio *m*
F resurchauffeur *m*
I surriscaldatore *m*
  intermedio
MA küzbenső túlhevítő
NL tussen-
  oververhitter *m*
PO przegrzewacz *m*
  pośredni
PY промежуточный
  перегреватель *m*
SU välitulistin
SV mellanüverhettare *c*

2473 intermittent *adj*
DA afbrudt
DE intermittierend
ES intermitente
F intermittent
I intermittente
MA szakaszos
NL intermitterend
PO nieciągły
  przerywany
PY периодический
SU jaksottainen
  lämmitys
SV intermittent

2474 internal *adj*
DA intern
DE inner(lich)
ES interno
F interne
I interno
MA belső
NL inwendig
PO wewnętrzny
PY внутренний
SU sisä-
  sisäinen
SV inre

2475 internal dimension
DA indvendig dimension
DE Innenmaß *n*
ES medida interior *f*
F dimension *f*
  intérieure
I dimensione *f* interna
MA belméret
NL binnenafmeting *f*
  inwendige
  afmeting *f*
PO wymiar *m*
  wewnętrzny
PY внутренний
  размер *m*
SU sisämitta
SV innermått *n*

2476 internal wall
DA indermur
DE Innenwand *f*
ES pared interior *f*
F paroi *f* intérieure
I parete *f* interna
MA belső fal
NL binnenmuur *m*
  binnenwand *m*
PO ściana *f* wewnętrzna
PY внутренняя стена *f*
SU sisäseinä
SV innervägg *c*

2477 interrupt
DA afbryde
DE Unterbrechung *f*
ES
F interrompre
I interrupt *m*
MA megszakítás
NL onderbreking *f*
PO przerywanie *n*
PY прерывать
SU keskeyttää
SV avbrott *n*

2478 interval
DA interval
DE Intervall *n*
ES interruptor *m*
F intervalle *m*
I intervallo *m*
MA időküz
NL interval *n*
tussenruimte *f*
PO odstęp *m*
przedział *m*
PY интервал *m*
SU väliaika
välimatka
SV intervall *n*

2479 invalid *adj*
DA ugyldig
DE ungültig
ES inválido
F invalide
I non accettabile
non valido
MA rokkant
NL ongeldig
PO nieważny
PY недействительный
ошибочный
SU kelvoton
SV invalidiserad
ogiltig

2480 ion
DA ion
DE Ion *n*
ES ión *m*
F ion *m*
I ione *m*
MA ion
NL ion *n*
PO jon *m*
PY ион *m*
SU ioni
SV jon *c*

2481 ion exchange
DA ionbytning
DE Ionenaustausch *m*
ES intercambio de
iones *m*
F échange *m* d'ions
I scambio *m* di ioni
MA ioncsere
NL ionenuitwisseling *f*
PO wymiana *f* jonowa
PY ионный обмен *m*
SU ionivaihtaja
SV jonbyte *n*

2482 ionization
DA ionisering
DE Ionisierung *f*
ES ionización *f*
F ionisation *f*
I ionizzazione *f*
MA ionizáció
NL ionisatie *f*
PO jonizacja *f*
PY ионизация *f*
SU ionisaatio
SV jonisering *c*

2483 iron
DA jern
DE Eisen *n*
ES hierro *m*
F fer *m*
I ferro *m*
MA vas
NL ijzer *n*
PO żelazo *n*
PY железо *n*
SU rauta
SV järn *n*

2484 irradiation
DA bestråling
stråling
DE Bestrahlung *f*
ES irradiación *f*
F rayonnement *m*
I irraggiamento *m*
radiazione *f*
MA besugárzás
NL bestraling *f*
PO napromieniowanie *n*
PY облучение *m*
SU säteilytys
SV strålning *c*

2485 isenthalpic *adj*
DA isentalpisk
DE adiabatisch
ES isoentálpico
F isenthalpique
I isentalpico
MA izentalpikus
NL isenthalpisch
met gelijke
enthalpie *f*
PO izentalpowy
PY изоэнтальпийный
SU vakioentalpia-
SV isentalpisk

2486 isobar
DA isobar
DE Isobare *f*
ES isóbara *f*
F isobare *f*
I isobara *f*
MA egyenlő nyomású
izobar
NL isobaar
PO izobara *f*
PY изобара *f*
SU isobaari
SV isobar *c*

2487 isolating valve
DA afspærringsventil
DE Absperrventil *n*
ES válvula de
aislamiento *f*
válvula de
seccionamiento *f*
F robinet *m*
d'isolement
I rubinetto *m* di
arresto
MA leválasztó szelep
NL scheidings-
afsluiter *m*
PO zawór *m* odcinający
PY изолирующий
клапан *m*
SU sulkuventtiili
SV avstängningsventil *c*

DA	=	Danish
DE	=	German
ES	=	Spanish
F	=	French
I	=	Italian
MA	=	Magyar (Hungarian)
NL	=	Dutch
PO	=	Polish
PY	=	Russian
SU	=	Finnish
SV	=	Swedish

2488 isometric drawing
DA isometrisk tegning
DE isometrische
   Zeichnung *f*
ES plano isométrico *m*
F dessin *m*
   isométrique
I assonometria *f*
   isometrica
MA izometrikus rajz
NL isometrische
   tekening *f*
PO schemat *m*
   izometryczny
PY изометрический
   чертеж *m*
SU isometrinen piirustus
SV isometrisk ritning *c*

2489 isotherm
DA isoterm
DE Isotherme *f*
ES isotermo
F isotherme *f*
I isoterma *f*
MA izoterma
NL isotherm *m*
PO izoterma *f*
PY изотерма *f*
SU isotermi
SV isoterm *c*

2490 isothermal *adj*
DA isotermisk
DE isothermisch
ES isotérmico
F isotherme
I isotermico
   isotermo
MA azonos hőmérsékletű
   izotermikus
NL isothermisch
PO izotermiczny
PY изотермный
SU isoterminen suihku
SV isotermisk

2491 isotope
DA isotop
DE Isotop *n*
ES isótopo *m*
F isotope *m*
I isotopo *m*
MA izotóp
NL isotoop *m*
PO izotop *m*
PY изотоп *m*
SU isotooppi
SV isotop *c*

DA	=	Danish
DE	=	German
ES	=	Spanish
F	=	French
I	=	Italian
MA	=	Magyar (Hungarian)
NL	=	Dutch
PO	=	Polish
PY	=	Russian
SU	=	Finnish
SV	=	Swedish

J

2492 jacket
DA kappe
  kapsel
DE Mantel *m*
  Umhüllung *f*
ES camisa *f*
  doble pared *f*
  envuelta *f*
F enveloppe *f*
  jaquette *f*
I mantello *m*
MA burkolat
  küpeny
NL bus *f*
  mantel *m*
  omhulsel *n*
PO osłona *f*
  otulina *f*
  płaszcz *m*
PY кожух *m*
SU peite
  peitto
SV mantel *c*

2493 jacketed *adj*
DA indkapslet
DE ummantelt
ES revestido
F
I rivestito con
  materiale isolante
MA burkolt
NL omkast
  ommanteld
PO izolowany
  osłonięty
  otulony
PY обшитый
SU peitetty
  ympäriverhottu
SV mantlad

2494 jet
DA stråle
DE Strahl *m*
ES chorro *m*
F jet *m*
I getto *m*
MA fúvóka
  sugár
NL jet *m*
  straal *m*
PO dysza *f*
  rozpylacz *m*
  strumień *m*
PY струя *f*
SU suihku
SV stråle *c*

2495 jet cooling
DA strålekøling
DE Strahlkühlung *f*
ES enfriamiento a
  chorro *m*
F refroidissement *m*
  par jet d'air froid
I raffreddamento *m* a
  getto
MA légsugaras hűtés
NL geforceerde
  luchtkoeling *f*
PO chłodzenie *n*
  strumieniowe
PY струйное
  охлаждение *n*
SU ilmasuihkujäähdytys
SV strålkylning *c*

2496 jet momentum
DA stråleimpuls
DE Strahlimpuls *m*
ES impulso del
  chorro *m*
F quantité *f* de
  mouvement d'un
  jet
I forza *f* del getto
MA sugárimpulzus
NL luchtstraalimpuls *m*
PO pęd *m* strumienia
PY импульс *m* струи
  количество *n*
  движения в струе
SU suihkun liikemäärä
SV strålimpuls *c*

2497 jet range (*see* jet
  throw)

2498 jet throw
DA kastelængde
DE Wurfweite *f*
ES alcance del chorro *m*
F portée *f* d'un jet
I distanza *f* del getto
MA vetőtávolság
NL luchtstraalworp *m*
  worp van de
  straal *m*
PO zasięg *m* strumienia
PY дальнобойность *f*
  струи
SU suihkun
  heittopitoisuus
SV kastlängd *c*

2499 jet ventilation
DA stråleventilation
DE Strahllüftung *f*
ES ventilación forzada
  por chorro *f*
F aération *f* par jet
  ventilation *f* par
  entrainement
I ventilazione *f* a
  getto
MA légsugár szellőzés
NL ventilatie *f* door
  middel van injectie
PO wentylacja *f*
  strumieniowa
PY струйная
  вентиляция *f*
SU ilmanvaihto suihkuja
  käyttäen
SV jetventilation *c*

2500 joint
DA kobling
   sammenføjning
DE Anschluß *m*
   Dichtung *f*
   Verbindung *f*
ES empalme *m*
   junta *f*
   unión *f*
F joint *m*
   raccord *m*
I giunto *m*
   guarnizione *f*
   raccordo *m*
MA csatlakozás
   kütés
NL aansluiting *f*
   verbinding *f*
PO połączenie *n*
   spółka *f*
   złącze *n*
PY стык *m*
SU liitos
SV koppling *c*

2501 jointing compound
DA tætningsmasse
DE Dichtungsmasse *f*
ES mástico para
   juntas *m*
F mastic *m* pour joint
I mastice *m* per
   giunzione
MA kiüntő massza
NL afdichtings-
   materiaal *n*
   pakkingsmateriaal *n*
PO masa *f*
   uszczelniająca
   złącze
PY уплотнитель *m*
   стыка
SU tiivistysaine
SV kopplingstätning *c*

2502 joint protection
DA tætningsbeskyttelse
DE Dichtungsschutz *m*
ES tapajuntas *m*
F couvre-joint *m*
I coprigiunto *m*
MA egyesített védelem
NL afdichtings-
   bescherming *f*
PO osłona *f* złącza
   zabezpieczenie *n*
   złącza
PY защита *f* стыка
SU tiivistyssuoja
SV tätningsskydd *n*

2503 jumper
DA krydsforbindelse
DE Kurzschlußbrücke *f*
ES puente (elec) *m*
F saut *m*
I collegamento *m*
   volante
MA áthidalóvezeték
   kiváltógerenda
NL doorverbinding *f*
PO łącznik *m*
   przewód *m*
   połączeniowy
   (elektr.)
PY перемычка *f*
SU iskupora
   yhdysjohto
SV kryssfürbindelse *c*

2504 junction box
DA fordelerdåse
DE Abzweigdose *f*
ES caja de conexiones *f*
F boîte *f* de jonction
I scatola *f* di
   derivazione
MA elágazódoboz
NL aansluitdoos *f*
   verdeeldoos *f*
PO puszka *f*
   połączeniowa
   skrzynka *f*
   przyłączna
PY присоединительная
   коробка *f*
SU kytkentälaatikko
SV kopplingsdosa *c*

K

2505 katathermometer
DA katatermometer
DE Katathermometer *n*
ES catatermómetro *m*
F catathermomètre *m*
I katatermometro *m*
MA katahőmérő
NL katathermometer *m*
PO katatermometr *m*
PY кататермометр *m*
SU katatermometri
SV katatermo-
    meter *c*

2506 keyboard
DA tastatur
DE Tastatur *f*
ES teclado *m*
F clavier *m*
I tastiera *f*
MA billentyűzet
NL toetsenbord *n*
PO klawiatura *f*
PY коммутатор *m*
    щит *m* управления
SU näppäimistü
SV tangentbord *n*

2507 kinetic *adj*
DA kinetisk
DE kinetisch
ES cinético
F cinétique
I cinetico
MA kinetikus
    mozgási
NL kinetisch
PO kinetyczny
PY кинетический
SU kineettinen
SV kinetisk

2508 knob
DA greb
    håndtag
    knop
DE Griff *m*
    Knopf *m*
ES empuñadura *f*
    mango *m*
    pomo *m*
F poignée *f*
I bottone *m*
    imbottitura *f*
    pomo *m*
MA fogantyú
    gomb
NL handgreep *m*
    knop *m*
PO pokrętło *n*
    uchwyt *m*
PY вороток *m*
SU nuppi
SV knopp *c*

2509 knowledge-based
system (*see* expert
system)

DA	=	Danish
DE	=	German
ES	=	Spanish
F	=	French
I	=	Italian
MA	=	Magyar (Hungarian)
NL	=	Dutch
PO	=	Polish
PY	=	Russian
SU	=	Finnish
SV	=	Swedish

L

2510 laboratory
DA laboratorium
DE Labor *n*
   Laboratorium *n*
ES laboratorio *m*
F laboratoire *m*
I laboratorio *m*
MA laboratórium
NL laboratorium *n*
PO laboratorium *n*
PY лаборатория *f*
SU laboratorio
SV laboratorium *n*

2511 labyrinth seal
DA labyrint-forsegling
DE Labyrinthdichtung *f*
ES sello laceríntico *m*
F joint *m* à labyrinthe
I tenuta *f* a labirinto
MA labirinttümítés
NL labyrintafdichting *f*
PO uszczelnienie *n*
   labiryntowe
PY лабиринтный
   затвор *m*
SU labyrinttitiiviste
SV labyrinttätning *c*

2512 lacquer
DA lak
DE Lack *m*
ES barniz *m*
   laca *f*
F laque *f*
   vernis *m*
I vernice *f*
MA lakk
NL lak *n*
   vernis *n*
PO lakier *m*
PY лак *m*
SU lakka
SV lack *n*

2513 laminar *adj*
DA laminar
DE laminar
ES laminar
F laminaire
I laminare
MA lamináris
   réteges
NL laminair
PO laminarny
PY ламинарный
SU laminaarinen
SV laminär

2514 lamp
DA lampe
   lygte
DE Lampe *f*
   Licht *n*
ES lámpara *f*
F lampe *f*
I lampada *f*
MA lámpa
NL lamp *f*
PO lampa *f*
PY лампа *f*
SU lamppu
SV lampa *c*

2515 LAN (*see* local area
network)

2516 latent *adj*
DA latent
DE latent
ES latente
F latent
I latente
MA rejtett
NL latent
PO utajony
PY скрытый
SU latentti-
SV latent

2517 latent heat
DA bunden varme
   latent varme
DE nichtfühlbare
   Wärme *f*
ES calor latente *m*
F chaleur *f* latente
I calore *m* latente
MA rejtett hő
NL latente warmte *f*
PO ciepło *n* utajone
PY скрытая теплота *f*
SU latenttilämpü
SV latent värme *c*

**2518 latent heat of
melting**
DA bunden smeltevarme
DE Schmelzwärme *f*
ES calor latente de
   fusión *m*
F chaleur *f* latente de
   fusion
I calore *m* latente di
   fusione
MA olvadáshő
NL smeltwarmte *f*
PO ciepło *n* utajone
   topnienia
PY скрытая теплота *f*
   плавления
SU sulamislämpü
SV latent smältvärme *c*

2519 latent heat of vaporization
DA fordampningsvarme
DE Verdampfungs-wärme f
ES calor latente de vaporización m
F chaleur f latente d'évaporation
I calore m latente di evaporazione
MA párolgáshő
rejtett párolgási hő
NL verdampings-warmte f
PO ciepło n utajone parowania
PY скрытая теплота f испарения
SU hüyrystyslämpü
SV ångbildningsvärme c

2520 layer
DA lag
DE Schicht f
ES capa f
F couche f
lame f
I strato m
MA réteg
NL laag f
PO warstwa f
PY слой m
SU kerros
SV lager n

2521 layer thickness
DA lagtykkelse
DE Schichtstärke f
ES espesor de la capa m
F épaisseur f de couche
I spessore m dello strato
MA rétegvastagság
NL laagdikte f
PO grubość f warstwy
PY толщина f слоя
SU kerrospaksuus
SV skikttjocklek c

2522 laying underground
DA forlægning i jord
DE Erdverlegung f
ES colocación bajo tierra f
instalación bajo tierra f
tendido subterráneo m
F pose m souterraine
I posa f sottoterra
MA füldbefektetés
NL ondergrondse ligging f
PO układanie n podziemne
PY подземная прокладка f
SU maanalainen
SV fürläggning c
i mark c

2523 lead
DA bly
DE Blei n
ES plomo m
F plomb m
I piombo m
MA ólom
NL lood
PO ołów m
PY свинец m
SU lyijy
SV bly n

2524 lead packing
DA blypakning
DE Bleidichtung f
ES junta de plomo f
F joint m de plomb
I guarnizione f di piombo
MA ólomtümítés
NL loodafdichting f
loodpakking f
PO uszczelnienie n na ołów
PY свинцовое уплотнение n
SU lyijytiiviste
SV blydiktning c

2525 leak (leakage)
DA læk (lækage)
DE Leck n
Leckstelle f
ES fuga f
F fuite f
I perdita f
MA hézag
lék
NL lek n
PO nieszczelność f
przeciek m
PY утечка f
SU vuoto
SV läcka c

2526 leak vb
DA lække
DE lecken
undicht sein
ES fugar
F fuir
I perdere
spandere
MA szivárog
NL lekken
PO przeciekać
przepuszczać
PY течь
утекать
SU vuotaa
SV läcka c

2527 leakage test
DA tæthedsprøve
DE Dichtheitsprüfung f
ES prueba de estanquidad f
F épreuve f d'étanchéité
essai m d'étanchéité
I prova f di tenuta
MA szivárgásvizsgálat
NL dichtheids-onderzoek n
lektest m
PO próba f szczelności
PY испытание n на плотность
SU tiiviyskoe
SV läckageprov n

2528 leak detector
DA læksporingsapparat
DE Lecksucher *m*
ES detector de fugas *m*
F détecteur *m* de fuite
I rivelatore *m* di
   perdite
MA szivárgáskereső
NL lekdetector *m*
PO wykrywacz *m*
   nieszczelności
   wykrywacz
   *m*przecieków
PY индикатор *m*
   утечки
SU vuodonilmaisin
SV läcksükare *c*

2529 lean gas
DA generatorgas
DE Generatorgas *n*
ES gas de gasógeno *m*
   gas pobre *m*
F gaz *m* d'épreuve
I gas *m* povero
MA sovány gáz
NL arm gas *n*
   gas *n* met lage
   verbrandings-
   waarde
PO gaz *m*
   niskokaloryczny
   gaz *m* ubogi
PY низкокалорийный
   газ *m*
SU generaattorikaasu
SV generatorgas *c*

2530 LED (*see* light emitting diode)

2531 length
DA længde
DE Länge *f*
ES longitud *f*
F longueur *f*
I lunghezza *f*
MA hossz
NL afstand *m*
   duur *m*
   lengte *f*
PO długość *f*
PY длина *f*
SU pituus
SV längd *c*

2532 level
DA niveau
DE Niveau *n*
   Stand *m*
ES nivel *m*
F levier *m*
I livella *f*
   livello *m*
MA szint
NL niveau *n*
   peil *n*
PO libella *f*
   poziom *m*
PY уровень *m*
SU pinta
   taso
SV nivå *c*

2533 level controller
DA niveauregulator
DE Niveauregler *m*
ES regulador de nivel *m*
F régulateur *m* de
   niveau
I regolatore *m* di
   livello
MA szintszabályozó
NL niveauregelaar *m*
PO regulator *m*
   poziomu
PY регулятор *m*
   уровня
SU pinnansäätäjä
SV nivåregulator *c*

2534 level gauge
DA niveaumåler
DE Niveaumesser *m*
ES indicador de nivel *m*
F indicateur *m* de
   niveau
I indicatore *m* di
   livello
MA szintmérő
NL peilglas *n*
PO wskaźnik *m*
   poziomu
PY датчик *m* уровня
SU pinnankorkeuden-
   ilmaisin
SV nivåmätare *c*

2535 lever
DA arm
DE Hebel *m*
ES palanca *f*
F levier *m*
I leva *f*
MA emelő
   fogantyú
NL hefboom *m*
   kruk *f*
   zwengel *m*
PO dźwignia *f*
PY рычаг *m*
SU vipu
SV spak *c*

2536 licensed *adj*
DA medbevilling
DE zugelassen
ES liberado
   autorizado
F licencié
I abilitato
MA engedélyezett
NL bevoegd
   toegelaten
PO licencjonowany
   uprawniony
   uznany
PY имеющий
   разрешение
   лицензированный
   (дипломиро-
   ванный)
SU laillistettu
   valtuutettu
SV licensierad

2537 lid
DA dæksel
   låg
DE Deckel *m*
ES tapa *f*
F couvercle *m*
I coperchio *m*
MA fedél
NL deksel *n*
PO pokrywa *f*
   wieko *n*
PY крышка *f*
SU kansi
SV lock *n*

DA	=	Danish
DE	=	German
ES	=	Spanish
F	=	French
I	=	Italian
MA	=	Magyar (Hungarian)
NL	=	Dutch
PO	=	Polish
PY	=	Russian
SU	=	Finnish
SV	=	Swedish

2538 life
DA levetid
　　liv
DE Lebensdauer f
ES duración f
　　vida f
F durée f
　vie f
I durata f
MA élet
　　élettartam
NL levensduur m
PO okres m trwania
　　trwałość f
　　żywotność f
PY живучесть f
　　прочность f
　　срок m службы
SU elinikä
SV livslängd c

2539 life cycle
DA levetid
DE Lebenszyklus m
　　während der
　　Lebensdauer f
ES periodo de duración
　　m
F cycle m de vie
I ciclo m di vita
MA életciklus
NL levenscyclus m
　　levensduur m
PO okres m
　　użytkowania
PY длительность f
　　цикла
SU elinkaari
SV livscykel c

2540 life cycle cost
DA levetidsudgift
DE Gesamtbetriebs-
　　kosten f,pl
ES coste del periodo de
　　duración m
F coût m du cycle de
　　vie
I costo m del ciclo di
　　vita
MA élettartam kültség
NL levensduurkosten pl
PO koszt m okresu
　　użytkowania
PY остаточная
　　стоимость f
SU elinikäinen
　　kustannus
SV livscykel kostnad c

2541 lift
DA hejseværk
DE Aufzug m
　　Hub m
ES ascensor m
　　carrera f
　　elevación f
F ascenseur m
I ascensore m
MA felhajtóerő
　　felvonó
NL hefvermogen n
　　lift m
　　opvoerhoogte f
PO dźwig m
　　winda f
PY высота f водяного
　　столба
　　лифт m
SU hissi
SV hiss c

2542 light
DA lys
DE Licht n
ES luz f
F lumière f
I luce f
MA fény
　　világítás
NL lamp f
　　licht n
PO światło n
PY освещение n
　　свет m
SU valo
SV ljus n

**2543 light emitting
diode**
DA lysdiode
DE lichtemittierende
　　Diode f
ES diodo encendido m
F diode f lumineuse
I diodo m luminoso
　　LED m
MA fénykibocsátó dióda
NL led f
　　licht emitterende
　　diode f
PO dioda f
　　elektro-
　　luminescencyjna
　　dioda f emisyjna
　　dioda f świecąca
PY индикаторная
　　лампа f
　　электрической
　　нагрузки
SU valodiodi
SV ljusemitterande
　　diod c

**2544 lighting
engineering**
DA belysningsteknik
DE Lichttechnik f
ES ingeniería de
　　iluminación f
F éclairagisme m
I illuminotecnica f
MA világítástechnika
NL lichttechniek f
PO technika f
　　oświetleniowa
PY светотехника f
SU valaistustekniikka
SV ljusteknik c

2545 lighting tube
DA lysrør
DE Leuchtrühre f
ES tubo fluorescente m
F tube m fluorescent
I tubo m fluorescente
MA fénycső
NL t.l.buis f
PO świetlówka f
PY люминесцентная
　　лампа f
SU loisteputki
SV lysrür n

2546 light intensity
DA lysstyrke
DE Lichtstärke f
ES intensidad
　　luminosa f
　　luminosidad f
F intensité f lumineuse
　　luminosité f
I intensità f luminosa
MA fényerősség
NL lichtsterkte f
PO natężenie n
　　oświetlenia
PY интенсивность f
　　освещения
SU valovoima
SV ljusstyrka c

2547 light sensor
DA lysføler
DE Lichtfühler m
ES sonda lumínica f
F détecteur m de
　　lumière
I sensore m di luce
MA fényérzékelő
NL lichtdetector m
　　lichtgevoelige cel f
PO czujnik m światła
PY светочувствитель-
　　ьный датчик m
SU valokenno
SV ljusavkännare c

2548 lignite
DA brunkul
DE Lignit *m*
ES lignito *m*
F lignite *m*
I lignite *f*
MA lignit
NL bruinkool *f*
PO lignit *m*
PY бурый уголь *m*
лигнит *m*
SU ruskohiili
SV brunkol *n*

2549 lignite coal
DA brunkul
DE Braunkohle *f*
Lignitkohle *f*
ES
F lignite *m*
I lignite *f*
MA barnaszén
NL bruinkool *f*
PO węgiel *m* lignitowy
PY бурый уголь *m*
SU ligniitti
SV brunkol *n*

2550 limit control
DA begrænsnings-
regulering
DE Begrenzer *m*
ES control límite *m*
F contrôle *m* de
dépassement
I controllo *m* di limite
MA határoló szabályozás
NL grenswaarde
regelaar *m*
PO sterowanie *n*
graniczne
PY предел *m*
регулирования
SU raja-arvosäätü
SV gränskontroll *c*

2551 limiting device
DA begrænsningsudstyr
DE Begrenzungsvor-
richtung *f*
ES dispositivo de
limitación *m*
limitador *m*
F limiteur *m*
I dispositivo *m*
limitatore
MA határoló szerkezet
NL begrenzer *m*
PO ogranicznik *m*
urządzenie *n*
ograniczające
PY ограничитель *m*
SU rajoitin
SV begränsningsdon *n*

2552 linear diffuser
DA lineær spredning
DE linearer Diffusor *m*
ES difusor lineal *m*
F diffuseur *m* linéaire
I diffusore *m* lineare
MA átmeneti idom
befúvónyílás
NL lineair
uitblaas-
ornament *n*
PO dyfuzor *m* liniowy
PY линейный
диффузор *m*
SU rakohajoittaja
SV spaltdiffusor *c*

2553 linear expansion
DA længdeudvidelse
DE lineare
Ausdehnung *f*
ES dilatación lineal *f*
F dilatation *f* linéaire
I dilatazione *f* lineare
MA lineáris nyúlás
NL lineaire uitzetting *f*
PO rozszerzalność *f*
liniowa
PY линейное
расширение *n*
SU pituuden
laajeneminen
SV längdutvidgning *c*

2554 line printer
DA linieskriver
DE Zeilendrucker *m*
ES impresora de líneas *f*
F imprimante *f* en
ligne
I stampante *f*
MA sornyomtató
NL regeldrukker *m*
PO drukarka *f*
wierszowa
PY построчно-
печатающее
устройство *n*
SU rivikirjoitin
SV radskrivare *c*

2555 lining
DA foring
DE Ausmauerung *f*
ES forro *m*
revestimiento *m*
F garnissage *m*
I rivestimento *m*
MA bélelés
NL bekleding *f*
voering *f*
PO okładzina *f*
PY обшивка *f*
SU verhous
vuoraus
SV beklädnad *c*

2556 liquefaction
DA kondensering
smeltning
DE Verflüssigung *f*
ES licuación *f*
licuefacción *f*
F liquéfaction *f*
I liquefazione *f*
MA cseppfolyósítás
NL vloeibaar maken
PO skraplanie *n* gazu
PY сжижение *n*
SU nesteytys
SV kondensering *c*

2557 liquefier
DA kondensator
DE Verflüssiger *m*
ES licuador *m*
F liquéfacteur *m*
I liquefattore *m*
MA cseppfolyósító
NL condensor *m*
verdichter *m*
PO skraplacz *m*
PY конденсатор *m*
SU nesteytin
SV kondensor *c*

2558 liquid
DA væske
DE Flüssigkeit *f*
ES líquido *m*
F liquide *m*
I liquido *m*
MA folyadék
NL vloeistof *f*
PO ciecz *f*
PY жидкость *f*
SU neste
SV vätska *c*

2559 liquid *adj*
DA flydende
DE flüssig
ES líquido
F liquide
I liquido
MA folyékony
NL vloeibaar
PO ciekły
    płynny
PY жидкий
SU nestemäinen
SV flytande

2560 liquid injection
DA væskeinjektion
DE Flüssigkeits-
    einspritzung *f*
ES inyección de
    líquido *f*
F injection *f* de liquide
I iniezione *f* di liquido
MA folyadék
    befecskendezés
NL vloeistofinspuiting *f*
PO wtrysk *m* cieczy
PY впрыскивание *n*
    жидкости
SU nesteen ruiskutus
SV vätskeinsprutning *c*

2561 liquid level
DA væskeniveau
DE Flüssigkeits-
    spiegel *m*
ES nivel de líquido *m*
F niveau *m* de liquide
I livello *m* del liquido
MA folyadékszint
NL vloeistofniveau *n*
    vloeistofpeil *n*
PO poziom *m* cieczy
PY уровень *m*
    жидкости
SU nestepinta
SV vätskenivå *c*

**2562 liquid level
    indicator**
DA væskeniveauviser
DE Flüssigkeitsspiegel-
    anzeiger *m*
ES indicador de nivel de
    líquido *m*
F indicateur *m* de
    niveau
I indicatore *m* di
    livello
MA folyadékszintmutató
NL vloeistofniveau-
    aanwijzer *m*
PO wskaźnik *m*
    poziomu cieczy
PY уровнемер *m*
SU nestepinnanosoitin
SV nivåindikator *c*

2563 liquid mixture
DA væskeblanding
DE Flüssigkeits-
    gemisch *n*
ES mezcla de líquidos *f*
    mezcla líquida *f*
F mélange *m* de
    liquides
I miscela *f* liquida
MA folyadék keverék
NL vloeibaar mengsel *n*
    vloeistofmengsel *n*
PO mieszanina *f* cieczy
PY смесь *f* жидкости
SU nestesekoitus
SV vätskeblandning *c*

2564 liquid separator
DA væskeseparator
DE Flüssigkeits-
    abscheider *m*
ES separador de
    líquido *m*
F séparateur *m* de
    liquide
I separatore *m* di
    liquidi
MA folyadék leválasztó
NL vloeistof-
    afscheider *m*
PO separator *m* cieczy
PY сепаратор *m*
    жидкости
SU nesteenerotin
SV vätskeavskiljare *c*

2565 lithium bromide
DA litium bromid
DE Lithiumbromid *n*
ES bromuro de litio *m*
F bromure *m* de
    lithium
I bromuro *m* di litio
MA lítiumbromid
NL lithium bromide *n*
PO bromek *m* litu
PY бромистый
    литий *m*
SU litiumbromidi
SV litiumbromid *c*

2566 litre
DA liter
DE Liter *m*
ES litro *m*
F litre *m*
I litro *m*
MA liter
NL liter *m*
PO litr *m*
PY литр *m*
SU litra
SV liter *c*

2567 live steam
DA frisk damp
DE Frischdampf *m*
ES vapor vivo *m*
F vapeur *f* vive
I vapore *m* vivo
MA élesgőz
NL verse stoom *m*
PO para *f* żywa
PY острый пар *m*
SU tuorehüyry
SV direktånga *c*

2568 load
DA belastning
    last
DE Belastung *f*
    Last *f*
ES carga *f*
F allure *f*
    charge *f*
I carico *m*
MA terhelés
NL belasting *f*
    vulling *f*
PO ładunek *m*
    obciążenie *n*
PY нагрузка *f*
SU kuorma
SV belastning *c*

2569 load cycling (*see*
    load shedding)

2570 load factor
DA belastningsfaktor
DE Belastungsfaktor *m*
ES factor de carga *m*
F facteur *m* de charge
I fattore *m* di carico
MA terhelési tényező
NL belastingsfactor *m*
PO współczynnik *m* obciążenia
PY коэффициент *m* загрузки
SU kuormitusaste
SV belastningsfaktor *c*

2571 load-free start
DA aflastet start
DE entlasteter Anlauf *m*
ES arranque en vacío *m*
F démarrage *m* à vide
I avviamento *m* a vuoto
MA terhelésmentes indítás
NL onbelaste start *m*
PO rozruch *m* bez obciążenia
PY неопределенность *f* нагрузки во времени
SU kuormittamaton käynnistys
SV avlastad start *c*

2572 loading chamber
DA påfyldningsrum
DE Füllschacht *m*
ES cámara de carga *f*
F chambre *f* de chargement
I camera *f* di carico
MA tültőkamra
NL vulschacht *f*
PO komora *f* obciążania
PY загрузочная камера *f*
SU lastaustila
SV laddnings-kammare *c*

2573 load levelling
DA belastnings-nivellering
DE Lastausgleich *m*
ES nivelación de la carga *f*
F nivelage *m* de charge
I livellamento *m* del carico
MA terheléskiegyenlítés
NL belastingsverdeling *f*
PO wyrównywanie *n* obciążenia
PY равномерная нагрузка *f*
SU tehontasaus
SV belastnings-nivellering *c*

2574 load limit
DA belastningsgrænse
DE Lastbereich *m*
ES carga límite *f* límite de carga *m*
F limite *f* de charge limite *f* de puissance
I limite *m* di carico
MA terhelési határ
NL grensbelasting *f* toelaatbare belastingsgrens *f*
PO granica *f* obciążenia
PY предел *m* нагрузки
SU kuormitusraja
SV belastningsgräns *c*

2575 load shedding
DA aflastning
DE Entlastung *f*
ES reducción de carga *f* restricción de la carga *f*
F réduction *f* de charge réduction *f* de débit
I riduzione *f* di carico
MA terheléscsükkentés
NL belastingvermin-dering *f*
PO redukcja *f* obciążenia
PY ограничение *n* нагрузки
SU kuorman tasaus kuormituksen ohjaus
SV avlastning *c*

2576 load variation
DA belastningsvariation
DE Lastschwankung *f*
ES variación de la carga *f*
F variation *f* de charge/débit
I variazione *f* di carico
MA terhelésingadozás
NL belastingvariatie *f*
PO zmienność *f* obciążenia
PY изменение *n* нагрузки
SU kuorman vaihtelu
SV belastnings-variation *c*

2577 local area network
DA lokalt netværk
DE lokales Flächen-netzwerk *m*
ES red local *f*
F réseau *m* local
I rete *f* locale
MA helyi hálózat
NL lan *n* lokaal netwerk *n*
PO
PY местная сеть *f*
SU paikallisverkko
SV lokalt nätverk *n*

2578 local resistance
DA engangsmodstand enkeltmodstand
DE Einzelwiderstand *m*
ES resistencia local *f*
F résistance *f* locale
I resistenza *f* localizzata
MA helyi ellenállás
NL plaatselijke weerstand *m*
PO opór *m* miejscowy
PY местное сопротивление *n*
SU paikallisvastus
SV engångsmotstånd *n*

DA	=	Danish
DE	=	German
ES	=	Spanish
F	=	French
I	=	Italian
MA	=	Magyar (Hungarian)
NL	=	Dutch
PO	=	Polish
PY	=	Russian
SU	=	Finnish
SV	=	Swedish

2579 lock
DA lås
DE Verschluß *m*
ES cerradura *f*
cierre *m*
F fermeture *f*
serrure *f*
I serratura *f*
MA zár
zsilip
NL slot *n*
sluis *f*
PO zamek *m*
zamknięcie *n*
PY затвор *m*
SU lukko
SV lås *n*

2580 logging printer
DA dataudskriver
DE Berichtsdrucker *m*
ES impresora de
registros *f*
F imprimante *f*
d'acquisition de
données
I stampante *f* di
resoconto
MA naplónyomtató
NL gegevensprinter *m*
PO drukarka *f* rejestru
danych
PY регистрирующий
принтер *m*
SU kirjoitin
SV skrivare *c*

**2581 log mean
temperature
difference**
DA logaritmisk
middeltemperatur-
forskel
DE logarithmische
Temperatur-
differenz *f*
ES diferencia media
logarítmica de
temperatura *f*
F écart *m* moyen
logarithmique de
température
I differenza *f* di
temperatura media
logaritmica
MA logaritmikus
küzéphőmérséklet
külünbség
NL logarithmisch
gemiddeld
temperatuur-
verschil *n*
PO średnia *f*
logarytmiczna
różnica
temperatury
PY логарифмическая
разность *f*
температуры
SU logaritminen
keskilämpütila
SV logaritmisk
medeltemperatur-
skillnad *c*

2582 long-flame coal
DA gaskul
langtflammende kul
DE Gasflammkohle *f*
ES carbón de llama
larga *m*
hulla de llama
larga *f*
F flénu *adj*
I carbone *m* a fiamma
lunga
MA hosszúlángú szén
NL vlamkool *f*
PO węgiel *m*
długopłomienny
PY длиннопламенный
уголь *m*
SU kaasuhiili
SV gasflamkol *n*

2583 longitudinal fin
DA langsgående ribbe
DE Längsrippe *f*
ES aleta longitudinal *f*
F ailette *f*
longitudinale
I aletta *f*
longitudinale
MA hosszanti borda
NL lamel *f* in lengte-
richting
lengteribbe *f*
PO żeberko *n* wzdłużne
PY продольное
ребро *n*
SU pitkittäinen ripa
SV längslamell *c*

2584 long side
DA langside
DE Längsseite *f*
ES fachada
longitudinal *f*
F façade *f*
longitudinale
I lato *m* longitudinale
MA hosszoldal
NL lange zijde *f*
PO elewacja *f* podłużna
PY продольный
фасад *m*
SU pitkä sivu
SV långsida *c*

2585 loop
DA sløjfe
DE Schleife *f*
ES bucle *m*
lira *f*
F boucle *f*
I anello *m*
circuito *m*
MA hurok
NL kring *m*
lus *f*
PO pętla *f*
PY петля *f*
SU lenkki
silmukka
SV slinga *c*
ügla *c*

2586 loss
DA tab
DE Verlust *m*
ES pérdida *f*
F déperdition *f*
　perte *f*
I perdita *f*
MA veszteség
NL verlies *n*
PO strata *f*
　ubytek *m*
PY потеря *f*
SU häviü
SV fürlust *c*

2587 loss of head
DA tryktab
DE Druckverlust *m*
ES pérdida de carga *f*
　pérdida de presión *f*
F perte *f* de charge
　perte *f* de pression
I perdita *f* di carico
MA nyomómagasság
　veszteség
NL drukverlies *n*
PO strata *f* ciśnienia
　hydrostatycznego
PY потеря *f* давления
SU painehäviü
SV tryckfürlust *c*

2588 loudness
DA lydstyrke
DE Lautstärke *f*
ES intensidad sonora *f*
　sonoridad *f*
F intensité *f* sonore
I intensità *f* sonora
MA hangosság
NL geluidssterkte *f*
PO głośność *f*
PY громкость *f*
SU äänenvoimakkuus
SV ljudstyrka *c*

2589 loudness level
DA lydniveau
DE Schallpegel *m*
ES nivel de intensidad
　sonora *m*
　nivel de
　sonoridad *m*
F niveau *m* sonore
I livello *m* di intensità
　sonora
MA hangosságszint
NL geluidsniveau *n*
PO poziom *m* głośności
PY уровень *m*
　громкости
SU äänentaso
SV ljudnivå *c*

2590 louvre
DA jalousi
DE Luftschlitz *m*
ES rejilla (de paso de
　aire) *f*
F déflecteur *m*
I persiana *f*
MA zsalu
NL jalouzie *f*
　leidschoep *f*
PO żaluzja *f*
PY дефлектор *m*
　жалюзи *n*
SU säleikkü
SV galler *n*

2591 louvred grille
DA jalousirist
DE Jalousiegitter *n*
ES rejilla apersianada *f*
　rejilla de persiana *f*
F bouche *f* d'aération
　à lames
I griglia *f* di aerazione
MA zsalus rács
NL jalouzierooster *m*
PO krata *f* z żaluzjami
PY жалюзийная
　решетка *f*
SU säätüsäleikkü
SV reglerbart spjäll *n*

2592 louvred shutter
DA lamelspjæld
DE Jalousieverschluß *m*
ES obturador de
　persiana *m*
F fermetures *f* à
　persiennes
　jalousie *f*
I chiusura *f* a
　persiana
MA zsalus lezáró
NL jalouziesluiter *m*
PO zamknięcie *n*
　żaluzjowe
PY жалюзийный
　затвор *m*
SU sälepelti
SV jalusispjäll *n*

2593 low grade fuel
DA lavkvalitetsbrændsel
DE Brennstoff *m* mit
　niedrigem
　Heizwert *m*
ES combustible de baja
　calidad *m*
　combustible de bajo
　poder calorífico *m*
　combustible pobre
　m
F combustible *m*
　pauvre
I combustibile *m*
　povero
MA alacsony fűtőértékű
　tüzelőanyag
NL brandstof *f* met lage
　verbrandings-
　waarde
PO paliwo *n*
　niskokaloryczne
PY низкокалорийное
　топливо *n*
SU lämpüarvoltaan
　alhainen
　polttoaine
SV lågvärdigt bränsle *n*

2594 low level cut-out
DA lavniveauafbryder
DE Wassermangel-
　sicherung *f*
ES interruptor de
　seguridad para
　nivel mínimo *m*
F dispositif *m* de
　sécurité de
　manque d'eau
I arresto *m* per basso
　livello
MA vízhiánykapcsoló
NL laagwaterstand-
　beveiliging *f*
　laagwaterstand-
　schakelaar *m*
PO wyłącznik *m*
　dolnego poziomu
PY выключатель *m*
　низкого
　напряжения
SU kuiviinkiehumissuoja
SV vattenbristsäkring *c*

2595 low pressure
DA lavtryk
DE Niederdruck *m*
ES baja presión *f*
presión de baja *f*
F basse pression *f*
I bassa *f* pressione
MA kisnyomás
NL lage druk *m*
PO ciśnienie *n* niskie
PY низкое давление *n*
SU matalapaine
SV lågtryck *n*

2596 low-pressure air conditioning system
DA lavtryksklimaanlæg
DE Niederdruckklima-anlage *f*
ES sistema de acondiciona-miento de aire a baja presión *m*
F installation *f* de conditionnement d'air à basse pression
I impianto *m* di condizionamento dell'aria a bassa pressione
MA kisnyomású légkondicionáló rendszer
NL lagedruk luchtbehan-delingssysteem *n*
PO system *m* klimatyzacji niskiego ciśnienia
PY система *f* кондицион-ирования воздуха низкого давления
SU matalapaineilmastointijärje-stelmä
SV lågtrycksklimat-anläggning *c*

2597 low pressure boiler
DA lavtrykskedel
DE Niederdruckdampf-kessel *m*
ES caldera de baja presión *f*
F chaudière *f* basse pression
I caldaia *f* a bassa pressione
MA kisnyomású kazán
NL lagedrukketel *m*
PO kocioł *m* niskiego ciśnienia
PY котел *m* низкого давления
SU matalapainekattila
SV lågtryckspanna *c*

2598 low pressure control
DA lavtryksregulering
DE Niederdruck-regelung *f*
ES control de baja presión *m*
F contrôle *m* basse pression
I regolazione *f* a bassa pressione
MA kisnyomású szabályozás
NL lagedrukregeling *f*
PO sterowanie *n* niskociśnieniowe
PY регулирование *n* на стороне низкого давления
SU alipainesäätü
SV lågtryckskontroll *c*

2599 low pressure fan
DA lavtryksventilator
DE Niederdruck-ventilator *m*
ES ventilador de baja presión *m*
F ventilateur *m* basse pression
I ventilatore *m* a bassa pressione
MA kisnyomású ventilátor
NL lagedruk ventilator *m*
PO wentylator *m* niskociśnieniowy
PY вентилятор *m* низкого давления
SU matalapainepuhallin
SV lågtrycksfläkt *c*

2600 low pressure heating
DA lavtryksopvarmning
DE Niederdruckdampf-heizung *f*
ES calefacción por vapor a baja presión *f*
F chauffage *m* à vapeur basse pression
I riscaldamento *m* a vapore a bassa pressione
MA kisnyomású fűtés
NL lagedruk verwarming *f*
PO ogrzewanie *n* niskociśnieniowe
PY отопление *n* низкого давления
SU matalapainehüyry-lämmitys
SV lågtrycks-uppvärmning *c*

2601 low-pressure hot water heating
DA lavtryksvarmtvands-opvarmning
DE Warmwasser-tiefdruckheizung *f*
ES calefacción por agua caliente a baja presión *f*
F chauffage *m* à eau chaude basse pression
I acqua *f* calda a bassa pressione
MA kisnyomású melegvízfűtés
NL warmwater-verwarming *f*
PO ogrzewanie *n* wodne niskociśnieniowe
PY водяное отопление *n* низкого давления
SU lämpimän käyttüveden lämmitys
SV lågtrycksvarmvatten-uppvärmning *c*

2602 low pressure safety cut-out

DA lavtrykssikkerheds-
   afbryder
DE Niederdrucksicherheits-
   ausschalter *m*
ES desconector de
   seguridad por baja
   presión *m*
F pressostat *m* de
   sécurité basse
   pression
I limitazione *f* di
   sicurezza della
   pressione minima
MA biztonsági
   alsónyomáskapcsoló
NL lagedruk-
   beveiliging *f*
PO wyłącznik *m*
   bezpieczeństwa
   niskiego ciśnienia
PY предохранитель *m*
   падения давления
   (в компрессоре)
SU alipainesuoja
SV lågtryckssäkerhets-
   brytare *c*

2603 low pressure steam

DA lavtryksdamp
DE Niederdruck-
   dampf *m*
ES vapor a baja
   presión *m*
F vapeur *f* basse
   pression
I vapore *m* a bassa
   pressione
MA kisnyomású gőz
NL lagedrukstoom *m*
PO para *f* niskiego
   ciśnienia
   para *f* niskoprężna
PY пар *m* низкого
   давления
SU matalapainehüyry
SV lågtrycksånga *c*

2604 lubricant

DA smøremiddel
DE Schmiermittel *n*
ES lubricante *m*
F lubrifiant *m*
I lubrificante *m*
MA kenőanyag
NL smeermiddel *n*
PO smar *m*
PY смазка *f*
SU voiteluaine
SV smürjmedel *n*

M

2605 machine
DA bearbejde på
    maskine
    maskine
DE Maschine *f*
ES máquina *f*
F machine *f*
I macchina *f*
MA gép
NL machine *f*
PO maszyna *f*
PY машина *f*
    механизм *m*
SU kone
SV maskin *c*

2606 mac value (*see*
    maximum allowable
    concentration)

2607 magnetic valve
DA magnetventil
DE Magnetventil *n*
ES válvula de
    solenoide *f*
    válvula
    electromagnética *f*
F vanne *f* magnétique
I valvola *f* a solenoide
MA mágnesszelep
NL magneetklep *f*
PO zawór *m*
    magnetyczny
PY магнитный
    клапан *m*
SU magneettiventtiili
SV magnetventil *c*

2608 main
DA hoved-
DE Haupt-
    Hauptrohr *m*
ES conducto
    principal *m*
    tubería principal *f*
F canalisation *f*
    principale
    conduite *f*
I colonna *f*
    tubazione *f*
    principale
MA fővezeték
NL hoofdleiding *f*
PO przewód *m* główny
PY главный
    трубопровод *m*
    магистраль *f*
SU pää-
SV huvud-

2609 main duct
DA hovedkanal
DE Hauptkanal *m*
    Hauptleitung *f*
ES conducto
    principal *m*
F conduite *f* principale
I canale *m* principale
MA főcsatorna
NL hoofdkanaal *n*
PO kanał *m* główny
    przewód *m* główny
PY магистральный
    канал *m*
SU pääkanava
SV huvudledning *c*

2610 main pipe
DA hovedledning
DE Hauptrohr *n*
ES tubería principal *f*
F tuyauterie *f*
    principale
I tubazione *f*
    principale
MA főcsővezeték
NL hoofdleiding *f*
PO rurociąg *m* główny
PY магистральная
    труба *f*
SU pääjohto
SV huvudledning *c*

2611 maintenance
DA vedligeholdelse
DE Instandhaltung *f*
ES conservación *f*
    entretenimiento *m*
    mantenimiento *m*
F entretien *m*
I manutenzione *f*
MA karbantartás
NL onderhoud *n*
PO konserwacja *f*
    utrzymanie *n*
    (instalacji - urządzeń)
PY текущий ремонт *m*
    уход *m*
    эксплуатация *f*
SU huolto
SV underhåll *n*

DA	=	Danish
DE	=	German
ES	=	Spanish
F	=	French
I	=	Italian
MA	=	Magyar (Hungarian)
NL	=	Dutch
PO	=	Polish
PY	=	Russian
SU	=	Finnish
SV	=	Swedish

2612 maintenance alarm
DA driftalarm
DE Wartungsalarm *m*
ES alarma de
mantenimiento *f*
F alarme *f* de
maintenance
I allarme *m* di
manutenzione
MA javítási riasztás
NL onderhouds-
melding *f*
onderhouds-
signalering *f*
PO alarm *m*
konserwacyjny
PY сигнал *m*
неисправности
SU huoltohälytys
SV underhållslarm *n*

2613 make up air
DA sekundær luft
DE Zusatzluft *f*
ES aire adicional *m*
aire de aporte *m*
aire suplementario
m
F air *m* d'appoint
I aria *f* di ricambio
MA pótlevegő
NL suppletielucht *f*
PO powietrze *n*
uzdatnione
PY дополнительно
подаваемый
воздух *m*
SU korvausilma
SV tillskottsluft *c*

2614 make up air unit
DA sekundær luftunit
DE Außenluftaufbereitungs-
gerät *n*
ES unidad de renovación
de aire *f*
F unité *f* de traitement
d'air
I unità *f* di reintegro
aria
MA pótlevegő készülék
NL toevoerlucht-
apparaat *n*
PO urządzenie *n* do
uzdatniania
powietrza
PY установка *f* для
подачи снаружи
нагретого
воздуха
SU tuloilmakone
SV tillskottsluftdon *n*

2615 make up water
DA spædevand
DE Zusatzwasser *n*
ES agua de aporte *f*
agua de reposición *f*
agua suplemen-
taria *f*
F eau *f* d'appoint
I acqua *f* di rabbocco
acqua *f* di ricambio
MA pótvíz
NL suppletiewater *n*
PO woda *f* uzdatniona
PY подпиточная
вода *f*
SU lisävesi
SV tillsatsvatten *n*

2616 male connection
DA hanstik
DE Verbindungsstück *n*
(außen)
ES acoplamiento
macho *m*
conexión macho *f*
F raccord *m* mâle
I attacco *m* maschio
MA külsőmenetes
csatlakozás
NL verbinding met
buitendraad *m*
PO połączenie *n*
zewnętrzne
PY соединение *n* типа
'труба в трубе'
SU urosliitin
SV koppling *c* (handel)

2617 male thread
DA udvendigt gevind
DE Außengewinde *n*
ES rosca macho *f*
F filetage *m* mâle
I filetto *m* maschio
MA külsőmenet
NL buitendraad *m*
PO gwint *m* zewnętrzny
PY наружная резьба *f*
SU ulkopuolinen kierre
SV yttergänga *c*

2618 malleable cast iron
DA aducerjern
tempergods
DE Temperguß *m*
ES fundición maleable *f*
F fonte *f* malléable
I ghisa *f* malleabile
MA temperüntvény
NL smeedbaar
gietijzer *n*
PO żeliwo *n* ciągliwe
PY ковкий чугун *m*
SU taottava valurauta
SV aducerjärn *n*

**2619 malleable cast iron
fittings**
DA aducerfittings
blødstøbte fittings
DE Temperguß-
fittings *n,pl*
ES accesorios de
fundición
maleable *m,pl*
F raccord *m* en fonte
malléable
I raccordi *m,pl* in
ghisa malleabile
MA temperüntésű
idomok
NL smeedbaar
gietijzeren
fittingen *pl*
PO osprzęt *m* z żeliwa
ciągliwego
PY фасонные части *f*
из ковкого
чугуна
SU valurautaiset putken
osat
SV aducergods-
armatur *c*

**2620 management
programme**
DA driftprogram
DE Management-
programm *n*
ES programa de gestión
m
F programme *m* de
gestion
I programma *m* di
gestione
MA irányítási program
NL beheerprogramma *n*
management-
programma *n*
PO program *m*
zarządzania
PY руководящая
программа *f*
SU hallintaohjelma
SV driftprogram *n*

2621 manhole
DA mandehul
DE Mannloch n
ES boca de hombre f
F regard m
  trappe f
I foro m
  passo m d'uomo
MA búvónyílás
NL mangat n
PO otwór m włazu
  studzienka f
  rewizyjna
  właz m
PY лаз m
  смотровой колодец m
SU tarkastusluukku
SV manlucka c

2622 manifold
DA forgrening
  grenrør
DE Verteiler m
ES colector m
  colector múltiple
  (tubos) m
  distribuidor m
F collecteur m de
  départ
I collettore m
MA gyűjtő
  osztó
NL verdeelstuk n
  verzamelleiding f
PO rozdzielacz m
PY коллектор m
  разделитель m
  тройник m
SU kammio
  tukki
SV fürgrening c

2623 manometer
DA manometer
DE Manometer n
ES manómetro m
F manomètre m
I manometro m
MA feszmérő
  manométer
NL manometer m
PO manometr m
PY манометр m
SU manometri
SV manometer c

2624 manometric *adj*
DA manometrisk
DE manometrisch
ES manométrico
F manométrique
I manometrico
MA manometrikus
NL manometrisch
PO manometryczny
PY манометрический
SU paine-
  pienipaineinen
SV manometer-

2625 manual *adj*
DA manuel
DE manuell
ES manual
F manuel
I manuale
MA kézi
NL handbediend
PO ręczny
PY ручной
SU käsikäyttüinen
SV manuell

2626 manual control
DA manuel regulering
DE Handregelung f
ES control manual m
  regulación manual f
F réglage m manuel
I regolazione f
  manuale
MA kézi vezérlés
NL handbediening f
PO regulacja f ręczna
  sterowanie n ręczne
PY ручное
  регулирование n
SU käsisäätü
SV manuell kontroll c

2627 manual operation
DA håndbetjent
DE Handbedienung f
ES accionamiento a
  mano m
  maniobra manual f
F commande f
  manuelle
I comando m
  manuale
MA kézi müküdtetés
NL handbediening f
  handbedrijf n
PO uruchamianie n
  ręczne
PY ручное
  управление n
SU käsikäyttü
SV manuell drift c

2628 manual override
DA manuel overstyring
DE manuell überlagern
ES sobrerrecorrido de
  maniobra manual
  m
F déverrouillage m
  manuel
I impostazione f
  manuale
MA kézi hatálytalanítás
NL overname door
  handbediening f
PO sterowanie n ręczne
  kasujące
  nastawienie
  urządzenia
PY ручное
  отключение n
SU käsiohitus
SV

2629 manufacture
DA fabrikat
DE Fabrikation f
  Herstellung f
ES fabricación f
F fabrication f
I fabbricazione f
MA gyártás
NL fabrikaat n
  vervaardiging f
PO produkcja f
  wyrób m
  wytwarzanie n
PY производство n
SU valmistaminen
SV tillverkning c

2630 manufacture *vb*
DA fabrikat
DE herstellen
ES fabricar
F fabriquer
I fabbricare
  manufatturare
MA gyárt
NL fabriceren
  vervaardigen
PO produkować
  wyrabiać
  wytwarzać
PY производить
SU valmistaa
SV tillverka

2631 manway (see
  manhole)

2632 marine air conditioning
DA skibsventilation
DE Schiffs-
 klimatisierung f
ES acondicionamiento
 del aire para
 buques m
 climatización de
 buques f
F conditionneur m
 d'air des navires
I condizionamento m
 di aria navale
MA hajó
 légkondicionálás
NL scheepslucht-
 behandeling f
PO klimatyzacja f dla
 okrętów
PY судовое
 кондицион-
 ирование n
 воздуха
SU laivailmastointi
SV fartygsventilation c

2633 mass
DA masse
 mængde
DE Masse f
ES masa f
F masse f
I massa f
MA tümeg
NL massa f
PO masa f
PY масса f
SU massa
SV massa c
 mängd c

2634 mass flow rate
DA massestrøm
DE Massenstrom m
ES caudal másico m
F débit m
 masse f
I portata f di massa
MA tümegáram
NL massastroom m
PO masowe natężenie n
 przepływu
PY удельный поток m
 массы
SU massavirta
SV massflüde n

2635 mass transfer
DA masseoverføring
DE Stoffübergang m
ES transmisión de
 masa f
 transporte de
 masa m
F transfert m de masse
I trasmissione f di
 massa
MA anyagátadás
NL stofoverdracht f
PO wymiana f masy
PY массоперенос m
SU aineensiirto
SV massüverfüring c

2636 master controller
DA hovedregulator
DE Hauptregler m
ES regulador master m
F régulateur m de
 masse
I regolatore m
 principale
MA fő szabályozó
NL hoofdregelaar m
PO główny regulator m
 główny układ m
 sterowniczy
PY образцовый
 (эталонный)
 регулятор m
SU pääsäädin
SV huvudstyrnings-
 don n

2637 material of construction
DA byggemateriale
DE Baumaterial n
 Baustoff m
ES material de
 construcción m
F matériel m de
 construction
I materiale m da
 costruzione
MA szerkezeti anyag
NL bouwmateriaal n
PO materiał m
 budowlany
PY строительный
 материал m
SU rakennusaine
SV byggnadsmaterial n

2638 matter
DA emne
 stof
DE Materie f
 Substanz f
ES materia f
F matière f
I materia f
MA anyag
 dolog
NL materie f
 stof f
PO materia f
 substancja f
PY материя f
SU asia
SV ämne n

2639 maximum *adj*
DA maksimum
DE maximal
ES máximo
F maximum
I massimo
MA maximális
NL maximaal
PO maksymalny
PY максимальный
SU maksimaalinen
SV maximum

2640 maximum allowable concentration
DA størtst tilladelige koncentration
DE hüchstzulässige Konzentration f
ES concentración máxima permisible f
F concentration f maximum admise
I concentrazione f massima ammissibile
MA maximális megengedett koncentráció
NL MAC-waarde f maximaal toelaatbare concentratie f
PO najwyższe dopuszczalne stężenie n stężenie n dopuszczalne maksymalne
PY предельно- допустимая концентрация f (ПДК)
SU enimmäispitoisuus haitalliseksi tunnettu pitoisuus
SV maxtillåten koncentration c

2641 maximum load
DA max. last
DE Maximallast f
ES carga límite f carga máxima f límite máximo de carga m
F charge f maximale
I carico m massimo
MA maximális terhelés
NL maximale belasting f
PO obciążenie n maksymalne
PY максимальная нагрузка f
SU maksimikuorma
SV maxbelastning c

2642 maximum permissible
DA tilladeligt maksimum
DE hüchstzulässig
ES máximo admisible m
F maximun m admissible
I massimo m permissibile
MA maximálisan megengedett
NL maximaal toelaatbaar
PO wielkość f maksymalna dopuszczalna
PY предельно- допустимый
SU korkein sallittu
SV hügsta tillåtna värde n

2643 mean *adj*
DA gennemsnitlig middel
DE durchschnittlich mittler
ES medio significativo
F moyen
I medio
MA küzepes
NL gemiddeld
PO średni
PY средний
SU keskimääräinen
SV medel-

2644 mean pressure
DA middeltryk
DE Durchschnitts- druck m
ES presión media f
F pression f admise
I pressione f media
MA küzépnyomás
NL gemiddelde druk m
PO ciśnienie n średnie
PY среднее давление n
SU keskimääräinen paine
SV medeltryck n

2645 mean radiant temperature
DA middelstrålings- temperatur
DE mittlere Strahlungs- temperatur f
ES temperatura radiante media f
F température f maximale de servie
I temperatura f media radiante
MA küzepes sugárzási hőmérséklet
NL gemiddelde stralings- temperatuur f
PO średnia temperatura f promieniowania
PY среднерадиа- ционная температура f
SU keskimääräinen säteilylämpütila
SV medelstrålnings- temperatur c

2646 mean temperature difference
DA middeltemperatur- differense
DE mittlere Temperatur- differenz f
ES diferencia media de temperatura f
F écart m moyen de température
I scarto m medio di temperatura
MA küzepes hőmérsékletkülün- bség
NL gemiddeld temperatuur- verschil n
PO średnia różnica f temperatury
PY средняя разность f температуры
SU keskimääräinen lämpütilaero
SV medeltemperatur- differens c

2647 mean time between failure
- DA gennemsnitlig tid mellem svigt
- DE mittlere stürungsfreie Zeit *f*
- ES tiempo medio entre fallos *m*
- F temps *m* moyen entre pannes
- I tempo *m* medio tra due guasti
- MA hibák küzütti küzepes idő
- NL gemiddelde storingsinterval *n*
- PO średni czas *m* międzyawaryjny
- PY среднее время *n* между отказами
- SU keskimääräinen vikaantumisväli
- SV medeltid *c* mellan fel *n*

2648 mean time to repair
- DA gennemsnitlig reparationstid
- DE mittlere Reparaturdauer *f*
- ES tiempo medio de reparación *m*
- F durée *f* moyenne de réparation
- I tempo *m* medio di riparazione
- MA javítások küzütti küzepes idő
- NL gemiddelde reparatietijd *m*
- PO średni czas *m* między naprawami
- PY средняя продолжитель-ьность *f* ремонта
- SU keskimääräinen korjausaika
- SV medelreparations-tid *c*

DA	=	Danish
DE	=	German
ES	=	Spanish
F	=	French
I	=	Italian
MA	=	Magyar (Hungarian)
NL	=	Dutch
PO	=	Polish
PY	=	Russian
SU	=	Finnish
SV	=	Swedish

2649 measurement
- DA måling
- DE Messung *f*
- ES medición *f*
- F mesure *f*
- I misura *f* misurazione *f*
- MA mérés
- NL afmeting *f* maat *f* meting *f*
- PO pomiar *m*
- PY измерение *n*
- SU mittaus
- SV mätning *c*

2650 measuring equipment
- DA måleudstyr
- DE Meßeinrichtung *f* Meßgerät *n*
- ES aparatos de medición *m,pl* equipo de medida *m*
- F équipement *m* de mesure
- I apparecchio *m* di misura
- MA mérőeszküz
- NL meetapparatuur *f* meetinrichting *f*
- PO sprzęt *m* pomiarowy
- PY измерительная установка *f*
- SU mittalaite
- SV mätutrustning *c*

2651 measuring instrument
- DA måleinstrument
- DE Meßinstrument *n*
- ES instrumento de medición *m* instrumento de medida *m*
- F instrument *m* de mesure
- I strumento *m* di misura
- MA mérőkészülék
- NL meetinstrument *n*
- PO przyrząd *m* pomiarowy
- PY измерительный инструмент *m*
- SU mittari
- SV mätinstrument *n*

2652 measuring socket
- DA målestuds
- DE Meßmuffe *f* Meßstutzen *m*
- ES cápsula para medición *f*
- F
- I presa *f* di misura
- MA mérőcsonk
- NL meetaansluiting *f* meetplug *f*
- PO gniazdo *n* pomiarowe
- PY измерительная трубка *f* измерительный патрубок *m*
- SU mittauspistoke mittayhde
- SV mätsockel *c*

2653 measuring technique
- DA måleteknik
- DE Meßtechnik *f*
- ES técnica de medición *f*
- F technique *f* de mesure
- I tecnica *f* di misurazione
- MA méréstechnika
- NL meettechniek *f*
- PO technika *f* pomiarowa
- PY измерительная техника *f*
- SU mittaustekniikka
- SV mätteknik *c*

2654 mechanical *adj*
- DA mekanisk
- DE mechanisch
- ES mecánico
- F mécanique
- I meccanico
- MA mechanikus
- NL mechanisch
- PO mechaniczny
- PY механический
- SU mekaaninen
- SV mekanisk

2655 mechanical atomizing burner
DA mekanisk forstøvningsbrænder
DE Gebläsebrenner *m*
ES quemador de pulverización mecánica *m*
F brûleur *m* à atomisation mécanique
I bruciatore *m* ad atomizzazione meccanica
MA mechanikus porlasztású égő
NL drukverstuivingsbrander *m*
PO palnik *m* rozpylania mechanicznego
PY горелка *f* с механическим распыливанием
SU mekaanisesti hajoittava poltin
SV mekanisk tryckluftsbrännare *c*

2656 mechanical draft water cooling tower
DA tvangsventilert koletårn
DE Ventilatorkühlturm *m*
ES torre de refrigeración de agua de tiro mecánico *f*
F tour *f* de refroidissement à tirage forcé
I torre *f* di raffreddamento a ventilazione meccanica
MA mechanikus szellőzésű vízhűtőtorony
NL waterkoeltoren *m* met geforceerde trek
PO wieża *f* chłodnicza wodna ze sztucznym ciągiem
PY градирня *f* с механическим побуждением (движения воздуха)
SU puhallinkäyttüinen jäähdytystorni
SV kyltorn *n* med fläckt *c*

2657 mechanical equivalent of heat
DA varmens mekaniske ekvivalent
DE mechanisches Wärmeäquivalent *n*
ES equivalente mecánico del calor *m*
F équivalent *m* mécanique de la chaleur
I equivalente *m* meccanico del calore
MA mechanikus hőegyenérték
NL mechanisch warmte-equivalent *n*
PO mechaniczny równoważnik *m* ciepła
PY механический эквивалент *m* теплоты
SU lämmün mekaaninen ekvivalentti
SV mekanisk värmeekvivalent *c*

2658 mechanical seal
DA mekanisk tætning
DE mechanische Dichtung *f*
ES sello mecánico *m*
F joint *m* mécanique
I tenuta *f* meccanica
MA mechanikus tümítés
NL mechanische asafdichting *f*
PO uszczelnienie *n* mechaniczne
PY механический затвор *m*
SU mekaaninen tiiviste
SV mekanisk fürsegling *c*

2659 medium pressure
DA mellemtryk middeltryk
DE Durchschnittsdruck *m* Mitteldruck *m*
ES presión intermedia *f* presión media *f*
F pression *f* moyenne
I pressione *f* media
MA küzepes nyomás
NL middeldruk *m*
PO ciśnienie *n* średnie
PY среднее давление *n*
SU keskipaine
SV medeltryck *n*

2660 medium temperature
DA middeltemperatur
DE Mitteltemperatur *f*
ES temperatura del medio *f*
F température *f* moyenne
I temperatura *f* media
MA küzepes hőmérséklet
NL middeltemperatuur *f*
PO temperatura *f* średnia
PY средняя температура *f*
SU keskilämpütila
SV medeltemperatur *c*

2661 medium volatile coal
DA mediumflammekul
DE Flammkohle *f*
ES carbón con proporción media de volátiles *m*
F charbon *m* flambant
I carbone *m* bituminoso
MA küzepesen illó szén
NL cokeskool *f*
PO węgiel *m* średniolotny
PY уголь *m* со средним содержанием летучих
SU lieskahiili
SV flamkol *n*

2662 melt *vb*
DA smelte
DE schmelzen
ES fundir
F mélanger
I fondere
MA olvadt
NL smelten
PO topić topnieć
PY плавить(ся) таять
SU sulaa sulattaa
SV smälta

2663 melting point
DA smeltepunkt
DE Schmelzpunkt *m*
ES punto de fusión *m*
F point *m* de fusion
I punto *m* di fusione
MA olvadáspont
NL smeltpunt *n*
PO punkt *m* topnienia
PY точка *f* плавления
SU sulamispiste
SV smältpunkt *c*

2664 membrane
DA membran
DE Membran *f*
ES membrana *f*
F membrane *f*
I membrana *f*
MA membrán
NL membraan *n*
PO membrana *f* przepona *f*
PY мембрана *f*
SU kalvo
SV membran *n*

2665 membrane expansion vessel
DA membranekspansions-beholder
DE Membranausdehnungs-gefäß *n*
ES vaso de expansión cerrado *m*
F vase *m* de dilatation à membrane
I vaso *m* di espansione a membrana
MA membrános tágulási tartály
NL membraan-expansievat *n*
PO naczynie *n* rozszerzalne membranowe naczynie *n* wzbiorcze membranowe
PY мембранный расширительный сосуд *m*
SU kalvopaisunta-astia
SV expansionskärl *n* med membrane *n*

2666 membrane valve
DA membranventil
DE Membranventil *n*
ES válvula de membrana *f*
F soupape *f* à membrane
I valvola *f* a membrana
MA membránszelep
NL membraan-afsluiter *m*
PO zawór *m* membranowy
PY мембранный клапан *m*
SU kalvoventtiili
SV membranventil *c*

2667 mercury
DA kviksølv
DE Quecksilber *n*
ES mercurio *m*
F mercure *m*
I mercurio *m*
MA higany
NL kwik *n*
PO rtęć *f*
PY ртуть *f*
SU elohopea
SV kvicksilver *n*

2668 mercury switch
DA kviksølvafbryder
DE Quecksilber-schalter *m*
ES interruptor de mercurio *m*
F interrupteur *m* à mercure
I interruttore *m* a mercurio
MA higanykapcsoló
NL kwikschakelaar *m*
PO przełącznik *m* o styku rtęciowym wyłącznik *m* rtęciowy
PY ртутный выключатель *m*
SU elohopeakytkin
SV kvicksilverbrytare *c*

2669 mercury thermometer
DA kviksølvtermometer
DE Quecksilberthermo-meter *n*
ES termómetro de mercurio *m*
F thermomètre *m* à mercure
I termometro *m* a mercurio
MA higanyos hőmérő
NL kwikthermometer *m*
PO termometr *m* rtęciowy
PY ртутный термометр *m*
SU elohopealäm-pümittari
SV kvicksilvertermo-meter *c*

2670 mesh
DA maske
DE Masche *f*
ES malla *f*
F maille *f*
I maglia *f*
MA dróthál szita
NL maas *f* plaatgaas *n*
PO oczko *n* (sita - siatki)
PY звено *n* очко *n*
SU verkko
SV maska *c*

2671 mesh width
DA maskevidde
DE Maschenweite *f*
ES ancho de malla *m*
F dimension *f* de maille
I lunghezza *f* della maglia
MA lyukbőség
NL maaswijdte *f*
PO szerokość *f* oczka (siatki)
PY размер *m* звена
SU verkon silmäkoko
SV maskvidd *c*

2672 metabolic heat
DA stofskiftevarme
DE Stoffwechselwärme *f*
ES calor metabólico *m*
F métabolisme *m*
I calore *m* metabolico
MA metabolikus hő
NL metabolische warmte *f* stofwisselings- warmte *f*
PO ciepło *n* metaboliczne
PY метаболическая теплота *f*
SU aineenvaihdunnan lämmüntuotto aineenvaihdunnan lämpüteho
SV metabolism *c*

2673 metal filter
DA metalfilter
DE Metallfilter *m*
ES filtro metálico *m*
F filtre *m* métallique
I filtro *m* metallico
MA fémszűrő
NL metalen filter *n*
PO filtr *m* metalowy
PY металлический фильтр *m*
SU metallisuodatin
SV metallfilter *n*

DA	= Danish
DE	= German
ES	= Spanish
F	= French
I	= Italian
MA	= Magyar (Hungarian)
NL	= Dutch
PO	= Polish
PY	= Russian
SU	= Finnish
SV	= Swedish

2674 metallic hose
DA metalslange
DE Metallschlauch *m*
ES manguera metálica *f* tubo metálico flexible *m*
F tuyau *m* flexible métallique
I tubo *m* flessibile metallico
MA fémtümlő
NL metalen slang *f*
PO przewód *m* elastyczny metaliczny
PY металлический шланг *m*
SU metalliletku
SV metallslang *c*

2675 meteorology
DA metereologi
DE Meteorologie *f*
ES meteorología *f*
F météorologie *f*
I meteorologia *f*
MA meteorológia
NL meteorologie *f* weerkunde *f*
PO meteorologia *f*
PY метеорология *f*
SU meteorologia
SV meteorologi *c*

2676 meter
DA måler
DE Meter *m* Zähler *m*
ES contador *m* medidor *m* metro *m*
F compteur *m*
I contatore *m* metro *m*
MA mérő méter
NL meetinstrument *n* meter *m*
PO metr *m* miernik *m* przyrząd *m* pomiarowy
PY счетчик *m*
SU mittari
SV meter *c*

2677 method
DA metode
DE Methode *f*
ES método *m*
F méthode *f*
I metodo *m*
MA módszer
NL methode *f* werkwijze *f*
PO metoda *f*
PY метод *m* способ *m*
SU menetelmä
SV metod *c*

2678 methylene blue test
DA metylblåtprøve
DE Methylenblau- Prüfung *f*
ES prueba con azul de metileno *f*
F essai *m* au bleu de méthylène
I prova *f* al blu di metilene
MA metilénkék vizsgálat
NL mythyleen- blauwproef *f*
PO test *m* błękitem metylowym
PY проба *f* с помощью метиловй синьки
SU metyylisinikoe
SV metylenblått test *n*

2679 microbore system
DA mikrokalibersystem
DE Kleinrohrsystem *n*
ES sistema de tubos de diámetro muy fino *m*
F chauffage *m* par tuyauteries de très faibles diamètres
I sistema *m* di riscaldamento a tubo di piccolo diametro
MA kisátmérőjű csüves rendszer
NL systeem *n* met buizen van kleine diameter
PO instalacja *f* z rurociągami o małej średnicy
PY система *f* с трубками малого диаметра
SU pienputkijärjestelmä
SV klenrürssystem *n*

2680 microclimate
DA mikroklima
DE Mikroklima *n*
ES microclima *m*
F microclimat *m*
I microclima *m*
MA mikroklíma
NL microklimaat *n*
PO mikroklimat *m*
PY микроклимат *m*
SU mikroilmasto
SV mikroklimat *n*

2681 micron
DA mikron
DE Mikron *n*
ES micra *f*
F micron *m*
I micron *m*
MA mikron
NL micrometer *m*
PO mikron *m*
PY микрон *m*
SU mikrometri
SV mikron *n*

2682 microprocessor
DA mikroprocessor
DE Mikroprozessor *m*
ES microprocesador *m*
F microprocesseur *m*
I microprocessore *m*
MA mikroprocesszor
NL computer *m*
   verwerkings-
   eenheid *f*
PO mikroprocesor *m*
PY микропроцессор *m*
SU mikroprosessori
SV mikroprocessor *c*

2683 mid season
DA højsæson
DE Übergangszeit *f*
ES entretiempo *m*
   época entre dos
   estaciones *f*
   época entre dos
   temporadas *f*
F demi-saison *f*
I mezza stagione *f*
MA átmeneti évszak
NL tussenseizoen *n*
PO okres *m* przejściowy
PY среднее *n* за сезон
SU ylimenokausi
SV mellansäsong *c*

2684 mine
DA mine
DE Bergwerk *n*
   Grube *f*
ES mina *f*
F mine *f*
I miniera *f*
MA bánya
NL mijn *f*
PO kopalnia *f*
PY шахта *f*
SU kaivos
SV gruva *c*

2685 mineral oil
DA mineralolie
DE Mineralül *n*
ES aceite mineral *m*
F huile *f* minérale
I olio *m* minerale
MA ásványolaj
NL minerale olie *f*
PO olej *m* mineralny
PY минеральное
   масло *n*
SU mineraaliüljy
SV mineralolja *c*

2686 mineral wool
DA mineraluld
DE Mineralwolle *f*
ES lana de escorias *f*
   lana mineral *f*
F laine *f* de roche
   laine *f* minérale
I lana *f* di roccia
MA ásványgyapot
NL minerale wol *f*
PO wełna *f* mineralna
PY минеральная вата *f*
SU mineraalivilla
SV mineralull *c*

2687 minimum air quantity
DA minimum
   luftmængde
DE Mindestluftmenge *f*
ES cantidad mínima de
   aire *f*
F débit *m* d'air
   minéral
I portata *f* minima
   dell'aria
MA minimális
   légmennyiség
NL minimale
   luchthoeveelheid *f*
PO minimalna ilość *f*
   powietrza
PY минимальное
   количество *n*
   воздуха
SU vähimmäisilmavirta
SV minimiluftmängd *c*

2688 mist
DA dis
   em
DE Nebel *m*
   Schwaden *m*
ES niebla *f*
F brouillard *m*
I brina *f*
   foschia *f*
   nebbia *f*
MA küd
NL mist *m*
   nevel *m*
PO mgła *f*
PY туман *m*
SU sumu
SV dimma *c*

2689 mist separator
DA emseparator
DE Schwaden-
   abscheider *m*
ES separador de
   niebla *m*
F séparateur *m* de
   goutellettes
I separatore *m* di
   gocce
   separatore *m* di
   prodotti
   nebulizzati
MA küdleválasztó
NL nevelafscheider *m*
PO odkraplacz *m*
PY сепаратор *m*
   капель
SU pisaran erotin
SV dimavskiljare *c*

2690 mixed air
DA blandet luft
DE Mischluft *f*
ES aire de mezcla *m*
   aire mixto *m*
   mezcla de aire *f*
F mélange *m* d'air
I aria *f* di miscela
MA kevert levegő
NL luchtmengsel *n*
   menglucht *f*
PO powietrze *f*
   zmieszane
PY воздушная смесь *f*
SU sekoitettu ilma
SV blandad luft *c*

2691 mixed flow fan
DA blandingsventilator
DE Mischströmungs-
ventilator *m*
ES ventilador centrífugo
helicoidal *m*
F ventilateur *m* hélico-
centrifuge
I ventilatore *m* a
flusso misto
MA keresztáramú
ventilátor
NL ventilator *m* met
tangentiale en
radiale stroming
PO wentylator *m* o
przepływie
mieszanym
PY смешивающий
вентилятор *m*
SU sekavirtauspuhallin
SV blandningsfläkt *c*

2692 mixed gas
DA blandet gas
DE Mischgas *n*
ES gas mixto *m*
mezcla de gases *f*
mezcla gaseosa *f*
F mélange *m* de gaz
I miscela *f* di gas
MA gázkeverék
kevert gáz
NL gasmengsel *n*
PO gaz *m* zmieszany
PY газовая смесь *f*
SU kaasuseos
SV blandgas *c*

2693 mixing box
DA blandeboks
DE Mischbox *f*
Mischkasten *m*
ES caja de mezcla *f*
caja mezcladora *f*
F boîte *f* de mélange
I sezione *f* di miscela
MA keverőszekrény
NL mengkast *f*
PO skrzynka *f*
mieszania
PY смесительное
устройство *n*
SU sekoituslaatikko
SV blandningslåda *c*

2694 mixing chamber
DA blandekammer
DE Luftmischkammer *f*
Mischkammer *f*
ES cámara de mezcla *f*
F chambre *f* de
mélange
plénum *m* de
mélange
I camera *f* di miscela
MA keverőkamra
NL mengkamer *f*
PO komora *f* mieszania
PY смесительная
камера *f*
SU sekoituskammio
SV blandnings-
kammare *c*

2695 mixing circuit
DA blandekredsløb
DE Mischkreislauf *m*
ES circuito de mezcla *m*
F circuit *m* malaxeur
circuit *m* mélangeur
I circuito *m* di miscela
MA keverő áramkür
NL mengcircuit *n*
PO obieg *m* mieszający
obieg *m* zmieszania
PY смесительный
контур *m*
SU sekoituspiiri
SV

2696 mixing control
DA blandeautomatik
DE Mischregelung *f*
ES regulación de la
mezcla *f*
F régulation *f* de
contrôle
I controllo *m* di
oscillazione
MA keverő szabályozás
NL mengregeling *f*
PO regulacja *f*
mieszania
PY регулирование *n*
смешивания
SU sekoitussäätü
SV blandnings-
regulator *c*

2697 mixing duct
DA blandekanal
DE Mischkanal *m*
ES conducto de aire de
mezcla *m*
F conduit *m* d'air
mélangé
I condotto *m* di
miscela
MA keverő csatorna
NL mengkanaal *n*
PO kanał *m* mieszający
przewód *m*
mieszający
PY смесительный
трубопровод *m*
(канал)
SU sekoitusilmakanava
SV blandningskanal *c*

2698 mixing pump
DA blandepumpe
DE Mischpumpe *f*
ES bomba mezcladora *f*
F pompe *f* de mélange
I pompa *f* di miscela
MA keverő szivattyú
NL mengpomp *f*
PO pompa *f* mieszająca
PY смесительный
насос *m*
SU sekoituspumppu
SV blandarpump *c*

2699 mixing ratio
DA blandingsforhold
DE Mischungs-
verhältnis *n*
ES relación de mezcla *f*
F rapport *m* de
mélange
I rapporto *m* di
miscela
MA keverési arány
NL mengverhouding *f*
PO stopień *m*
zmieszania
stosunek *m*
mieszania
PY степень *f* смешения
SU sekoitussuhde
SV blandnings-
fürhållande *n*

2700 mixing valve
DA blandeventil
DE Mischventil *n*
ES válvula
    mezcladora *f*
F vanne *f* de mélange
I valvola *f*
    miscelatrice
MA keverőszelep
NL mengklep *f*
PO zawór *m* mieszajacy
PY смесительный
    клапан *m*
SU sekoitusventtiili
SV blandningsventil *c*

2701 mixture
DA blanding
DE Mischung *f*
    Zusammensetzung *f*
ES mezcla *f*
F mélange *m*
I miscela *f*
MA keverék
NL mengsel *n*
PO mieszanina *f*
PY смесь *f*
SU sekoitus
SV blandning *c*

2702 mnemonic
DA huskesymbol
DE Gedächtnis-
ES mnemotécnica *f*
F aide-mémoire *m*
    mnémonique *f*
I mnemonica *f*
MA emlékezeterősítő
NL alpha-numerieke
    computer-
    instructie *f*
    geheugenleer *f*
PO mnemotechniczny
PY мнемоника
SU muistamista
    helpottava

2703 model
DA model
DE Modell *n*
    Muster *n*
ES modelo *m*
F modèle *m*
I campione *m*
    modello *m*
MA minta
    modell
NL model *n*
    voorbeeld *n*
PO model *m*
PY модель *f*
SU malli
SV modell *c*

2704 modelling
DA modellering
DE Modellieren *n*
    Nachbilden *n*
ES modelado *m*
F modélisation *f*
I modellizzazione *f*
MA mintázás
    modellezés
NL mathematische
    processimulatie *f*
    modelleren *n*
PO modelowanie *n*
PY моделирование *n*
SU mallintaminen
SV modellering *c*

2705 modem
DA modem
DE Signalumsetzer *m*
ES modem *m*
F Modem
    modulateur-
    démodulateur
I modem *m*
MA modem
    modulátor-
    demodulátor
    készülék
NL modem *n*
PO modem *m*
    modulator-
    demodulator *m*
PY модем *m*
    переходное
    устройство *n*
SU modemi
SV modem *n*

2706 modular *adj*
DA modulær
DE modular
ES modular
F modulaire
I modulare
MA modulos
NL modulair
PO modularny
    modułowy
PY модульный
SU modulaarinen
SV modul-

**2707 modular air
conditioning
system**
DA modulluftbehandlings-
    anlæg
DE Baukasten-
    Klimasystem *n*
ES sistema de
    climatización
    modular *m*
F centrale *f* modulaire
    de climatisation
I impianto *m* di
    condizionamento
    modulare
MA építőelemes
    légkondicionáló
    rendszer
NL modulair
    opgebouwd
    luchtbehandelings-
    systeem *n*
PO system *m*
    klimatyzacji
    modułowy
PY система *f*
    кондицион-
    ирования
    воздуха,
    собираемая из
    отдельных
    секций
SU moduloiva
    ilmastointijärje-
    stelmä
SV moduluppbyggt
    luftkonditionerings-
    system *n*

2708 modulating *adj*
DA modulerende
DE modulierend
ES modulante
F modulé
I modulante
MA modulált
NL modulerend
PO modulujacy
PY модулирующий
    пропорциональный
SU moduloiva
SV modulerande

2709 modulating gain
DA modulerende
volumen
DE Modulations-
gewinn *m*
ES aportación
modulante *f*
F gain *m* modulateur
I guadagno *m*
modulante
MA modulált erősítés
NL modulatie
versterking *f*
PO zysk *m* modularny
PY коэффициент *m*
моделирования
коэффициент *m*
усиления
SU moduloinnin
vahvistus
SV moduleringsük-
ning *c*

2710 module
DA modul
DE Modul *m*
ES módulo *m*
F module *m*
I modulo *m*
MA modul
NL moduul *m*
PO moduł *m*
zespół *m*
znormalizowany
wymienny
PY модуль *m*
SU moduuli
SV modul *c*

2711 moist *adj*
DA fugtig
DE feucht
ES húmedo
F humide
moite
I umido
MA nedves
NL vochtig
PO wilgotny
PY влажный
SU kostea
SV fuktig

DA = Danish
DE = German
ES = Spanish
F = French
I = Italian
MA = Magyar
(Hungarian)
NL = Dutch
PO = Polish
PY = Russian
SU = Finnish
SV = Swedish

2712 moisture
DA fugt
DE Feuchte *f*
ES humedad *f*
F humidité *f*
I umidità *f*
MA nedvesség
NL vocht *n*
vochtigheid *f*
PO wilgoć *f*
wilgotność *f*
PY влажность *f*
SU kosteus
SV fukt *c*

2713 moisture balance
DA fugtbalance
DE Feuchtegleich-
gewicht *n*
ES balance de humedad
m
equilibrio de
humedad *m*
F équilibre *m*
d'humidité
I bilancio *m* di
umidità
MA nedvességegyensúly
NL vochtbalans *f*
PO bilans *m* wilgoci
stan *m* równowagi
wilgoci
PY влажностный
баланс *m*
SU kosteustasapaino
SV fuktighetsjämvikt *c*

2714 moisture content
DA fugtindhold
DE Feuchtegehalt *m*
ES contenido de
humedad *m*
fracción máxima de
humedad *f*
humedad
específica *f*
F teneur *f* en eau
teneur *f* en humidité
I contenuto *m* di
umidità
MA nedvességtartalom
NL vochtgehalte *n*
PO zawartość *f* wilgoci
PY влагосодержание *n*
SU vesipitoisuus
SV fukthalt *c*

**2715 moisture
measurement**
DA fugtmåling
DE Feuchtemessung *f*
ES medición de la
humedad *f*
F mesure *f* d'humidité
I misura *f*
dell'umidità
MA nedvességmérés
NL vochtmeting *f*
PO pomiar *m*
wilgotności
PY измерение *n*
влажности
SU kosteusmittaus
SV fuktmätning *c*

**2716 moisture
protection**
DA fugtbeskyttelse
DE Feuchteschutz *m*
ES protección contra la
humedad *f*
F protection *f* contre
l'humidité
I protezione *f* contro
l'umidità
MA nedvesség elleni
védelem
NL vochtbescherming *f*
PO ochrona *f*
przeciwwilgociowa
PY защита *f* от
сырости
SU kosteussuojaus
SV fuktskydd *n*

2717 moisture retarder
DA fugtkatalysator
DE Feuchtigkeits-
sperre *f*
ES retardador de
humedad *m*
F retardeur *m*
d'humidité
I barriera *f*
all'umidità
MA nedvességgátló
NL vochtremmende
laag *f*
PO osuszacz *m*
PY влагопоглотитель
m
замедлитель *m*
увлажнения
SU kosteussulku
SV fuktspärr *c*

2718 moisture separator
DA fugtudskiller
DE Feuchte-
abscheider *m*
ES separador de
gotas *m*
F séparateur *m* de
goutellettes
I separatore *m* di
gocce
MA nedvességleválasztó
NL vochtafscheider *m*
PO oddzielacz *m* wilgoci
odkraplacz *m*
PY сепаратор *m* влаги
SU vedenerotin
SV fuktavskiljare *c*

2719 moisture transfer
DA fugtoverføring
DE Feuchtetransport *m*
ES transporte de
humedad *m*
F transfert *m*
d'humidité
I trasporto *m* di
umidità
MA nedvesség átadás
NL vochtoverdracht *f*
PO wymiana *f* wilgoci
PY перенос *m* влаги
SU kosteuden
siirtyminen
SV fuktvandring *c*

**2720 moisture
transmission**
DA fugttransmission
DE Feuchte-
übertragung *f*
ES transmisión de la
humedad *f*
F transfert *m*
d'humidité
I trasmissione *f* di
umidità
MA nedvesség áteresztés
NL vochtoverdracht *f*
PO przenikanie *n*
wilgoci
PY влагопередача *f*
SU kosteudenläpäisy
SV fuktgenomgång *c*

2721 Mollier diagram
DA Molliers diagram
DE Mollier-
Diagramm *n*
ES diagrama de Mollier
m
F diagramme *m* de
Mollier
I diagramma *m* di
Mollier
MA Mollier diagram
NL Mollierdiagram *n*
PO wykres *m* Molliera
PY диаграмма *f*
Молье
SU Mollier-piirros
SV Mollier-diagram *n*

2722 momentum
DA bevægelsesmængde
DE Impuls *m*
Moment *n*
ES cantidad de
movimiento *f*
F quantité *f* de
mouvement
I quantità *f* di moto
MA impulzus
nyomaték
NL arbeidsvermogen *n*
van beweging
PO pęd *m*
PY импульс *m*
количество *n*
движения
SU liikemäärä
momentti
SV moment *n*

2723 monitoring
DA overvågning
DE Überwachung *f*
ES instrucción *f*
F mesures *f*
automatiques
I monitoraggio *m*
MA felügyelet
megfigyelés
NL bewaken
opvragen
weergeven
PO kontrola *f*
ostrzeganie *n*
sygnalizacja *f*
ostrzegawcza
PY контроль *m*
управление *n*
SU monitorointi
SV üvervakning *c*

2724 motion
DA bevægelse
DE Bewegung *f*
ES movimiento *m*
F marche *f*
mouvement *m*
I marcia *f*
movimento *m*
MA mozgás
NL beweging *f*
PO ruch *m*
PY движение *n*
SU liike
SV rürelse *c*

2725 motive force
DA drivkraft
DE Antriebskraft *f*
ES fuerza motriz *f*
F force *f* motrice
I forza *f* motrice
MA mozgatóerő
NL beweegkracht *f*
drijvende kracht *f*
PO siła *f* napędowa
PY движущая сила *f*
SU liikkeelle paneva
voima
SV drivkraft *c*

2726 motor
DA motor
DE Motor *m*
ES motor *m*
F moteur *m*
I motore *m*
MA motor
NL motor *m*
PO silnik *m*
PY двигатель *m*
мотор *m*
SU moottori
SV motor *c*

2727 motor-bed plate
DA motorfundaments-
plade
DE Motorgrundplatte *f*
ES bancada del motor *f*
F socle *m* de moteur
I basamento *m*
MA motor alaplemez
NL motorfundatie-
plaat *f*
PO podstawa *f* silnika
PY станина *f*
двигателя
SU mottorialusta
SV motorfundament *n*

2728 motorized valve
DA motorventil
  servoventil
DE Motormischventil n
ES válvula
  motorizada f
F vanne f motorisée
I valvola f
  motorizzata
MA motoros szelep
NL gemotoriseerde
  afsluiter m
PO zawór m z
  siłownikiem
PY клапан m c
  двигателем
SU moottoriventtiili
SV servoventil c

2729 motor noise
DA maskinstøj
DE Motorgeräusch n
ES ruido del motor m
F bruit m de moteur
I rumore m di motore
MA motorzaj
NL motorgeluid n
PO hałas m silnika
PY шум m двигателя
SU moottorin melu
SV motorbuller n

2730 motor rating
DA motoreffekt
DE Motorleistung f
ES potencia del régimen
  del motor f
  potencia nominal del
  motor f
F puissance f de
  moteur
I potenza f del
  motore
MA motor méretezés
NL motorvermogen n
PO moc m silnika
PY мощность f
  двигателя
SU moottorin
  nimellisteho
SV motoreffekt c

2731 motor steptime
DA motorgangtid
DE Motorschrittzeit f
ES duración de la etapa
  de un motor f
F moteur m pas à pas
I tempo m di passo di
  un motore
MA motor léptetési idő
NL servomotor-
  pulstijd m
PO wyłącznik m
  czasowy silnika
PY число n оборотов
  двигателя
SU askelaika
SV motorgångtid c

2732 motor winding
DA motorbevikling
DE Motorwicklung f
ES bobinado del
  motor m
  devanado m
F bobinage m de
  moteur
I avvolgimenti m,pl
  del motore
MA motor tekercselés
NL motorwikkeling f
PO uzwojenie n silnika
PY перемотка f
  двигателя
SU moottorin käämitys
SV motorlindning c

2733 moulding
DA formstykke
  støbning
DE Formstück n
ES moldeado m
F pièce f moulée
I fusione f
  getto m
MA üntőminta
  üntvény
NL vormen n
  vormstuk n
PO formowanie n
  odlew m
PY профиль m
SU muotokappale
SV formstycke n

2734 mounting
DA montering
DE Aufstellung f
  Montage f
ES montaje m
F attache f
  montage m
I montaggio m
  sostegno m
MA felszerelés
NL montage f
  onderstel n
  opstelling f
PO instalowanie n
  montaż m
  zamocowanie n
PY монтаж m
SU asennus
SV montering c

2735 mounting hole
DA monteringsåbning
DE Montageöffnung f
ES abertura para el
  montaje f
F ouverture f pour
  montage
I foro m di
  montaggio
MA szerelőnyilás
NL montage-
  opening f
PO otwór m
  montażowy
PY монтажное
  отверстие n
SU asennusaukko
SV fästhål n

**2736 mounting
  instructions**
DA monteringsinstruks
DE Montage-
  anweisung f
ES instrucciones de
  montaje f,pl
F instruction f de
  montage
I istruzioni f,pl per il
  montaggio
MA szerelési utasítások
NL montagevoor-
  schriften pl
PO instrukcja f
  montażu
PY инструкция f по
  монтажу
SU asennusohjeet
SV monterings-
  füreskrifter c

2737 movement
DA bevægelse
DE Bewegung *f*
ES movimiento *m*
F mouvement *m*
I movimento *m*
MA mozgás
    működés
NL beweging *f*
PO ruch *m*
PY движение *n*
SU liike
SV rürelse *c*

2738 MTBF (*see* mean time
    between failure)

2739 MTTR (*see* mean time
    to repair)

2740 muffler
DA lyddæmper
DE Schalldämpfer *m*
ES silenciador *m*
F silencieux *m*
I silenziatore *m*
MA hangtompító
NL geluiddemper *m*
    knaldemper *m*
    knalpot *m*
PO tłumik *m*
PY глушитель *m*
SU äänenvaimennin
SV ljuddämpare *c*

**2741 muffler noise
    damper**
DA lyddæmperspjæld
DE Auspufftopf-
    Schalldämpfer *m*
ES compuesta de
    silenciador *f*
F silencieux *m*
I camera *f* di
    espansione
    marmitta *f*
MA kipufogó
    hangtompító
NL geluiddemper *m*
    knalpot *m*
PO tłumik *m* hałasu
PY шумоглушитель *m*
SU äänenvaimennin
SV ljuddämparspjäll *n*

2742 multiblade damper
    (*see* multileaf damper)

2743 multifuel boiler
DA kedel til flere slags
    brændsel
DE Umstellbrand-
    kessel *m*
ES caldera
    policombustibles *f*
F chaudière *f* multi-
    combustible
I caldaia *f* poli-
    combustibile
MA kazán tübbféle
    tüzelőanyaghoz
NL ketel *m* voor diverse
    brandstoffen
PO kocioł *m*
    wielopaliwowy
PY котел *m* на разные
    виды топлива
SU monipolttoaine-
    kattila
SV panna *c* för
    alternativa
    bränslen *n*

2744 multifuel firing
DA fyring med
    forskellige
    brændsler
DE Mehrbrennstoff-
    Feuerung *f*
ES hogar
    multi-
    combustibles *m*
F foyer *m* multi
    combustible
I focolare *m* a
    combustibili
    multipli
MA alternatív tüzelés
    tübbféle
    tüzelőanyagú
    tüzelés
NL stoken *n* van
    diverse
    brandstoffen
PO palenisko *f*
    wielopaliwowe
PY топка *f* на разные
    виды топлива
SU usean polttoaineen
    samanaikainen
    polttaminen
SV eldning *c* med
    alternativ bränslen
    c

2745 multijet burner
DA brænder med flere
    dyser
DE Mehrstrahl-
    brenner *m*
ES quemador de varias
    toberas *m*
F brûleur *m* à
    plusieurs becs
I bruciatore *m* a getti
    multipli
MA tübbfuvókás égő
NL brander *m* met meer
    dan één vlam
PO palnik *m*
    wielostrumieniowy
PY многоструйная
    горелка *f*
SU monisuuttiminen
    poltin
SV brännare *c* med flera
    munstycken *n*

2746 multileaf damper
DA jalousispjæld
DE Jalousieklappe *f*
ES compuerta de álabes
    múltiples *f*
F registre *m* à lames
    multiples
    registre *m* à
    persiennes
I serranda *f* ad alette
    multiple
MA zsalus csappantyú
NL jalouzieklep *f*
PO przepustnica *f*
    wielopłaszczy-
    znowa
PY многоствочатый
    клапан *m*
SU sälepelti
SV flerbladsspjäll *n*

**2747 multilouvre
    damper** (*see* multileaf
    damper)

DA	=	Danish
DE	=	German
ES	=	Spanish
F	=	French
I	=	Italian
MA	=	Magyar
		(Hungarian)
NL	=	Dutch
PO	=	Polish
PY	=	Russian
SU	=	Finnish
SV	=	Swedish

2748 multipass boiler
DA flertrækskedel
DE Mehrzugkessel *m*
ES caldera de pasos
   múltiples *f*
F chaudière *f* à
   plusieurs parcours
I caldaia *f* a passaggi
   multipli
MA tübbhuzamú kazán
NL meertreksketel *m*
PO kocioł *m*
   wielokanałowy
   kocioł *m*
   wieloprzepływowy
PY многоходовой
   котел *m*
SU monivetoinen kattila
SV tubpanna *c*

**2749 multipurpose cold
store**
DA kølelager til flere
   formål
DE Mehrzweckkühl-
   raum *m*
ES cámara frigorífica
   polivalente *f*
F entrepôt *m*
   frigorifique
   polyvalent
I magazzino *m*
   frigorifero con
   diverse
   temperature di
   stoccaggio
MA tübbcélú hűtőraktár
NL koelhuis *n* voor
   diverse producten
PO chłodnia *f* ogólnego
   przeznaczenia
PY многоцелевое
   холодное
   хранилище *n*
SU monikäyttüinen
   kylmävarasto
SV universalkyllager *n*

**2750 multishell
condenser**
DA kondensator med
   flere kapper
DE Mehrmantel-
   kondensator *m*
ES condensador de
   envolvente
   múltiple *m*
F condenseur *m* à
   calendres
   multiples
I condensatore *m* a
   mantello multiplo
MA tübbküpenyes
   kondenzátor
NL meervoudige
   condensor *m*
PO skraplacz *m*
   wielopowłokowy
PY пластинчатый
   конденсатор *m*
   секционный
   конденсатор *m*
SU moniputkilauhdutin
SV multipelkondensor *c*

2751 multistage *adj*
DA flertrins
DE mehrstufig
ES de varios escalones
   de varios saltos,
   escalonado
   múltiple
F à plusieurs étages
I stadi multipli
MA tübbfokozatú
NL meertraps
PO kaskadowy
   wielostopniowy
PY многоступенчатый
SU moniportainen
SV flerstegs-

**2752 multistage
compressor**
DA flertrinskompressor
DE mehrstufiger
   Verdichter *m*
ES compresor de varias
   etapas *m*
   compresor
   múltiple *m*
F compresseur *m*
   multi-étage
I compressore *m*
   multistadio
MA tübbfokozatú
   kompresszor
NL meertraps-
   compressor *m*
PO sprężarka *f*
   wielostopniowa
PY многоступенчатый
   компрессор *m*
SU moniportainen
   kompressori
SV flerstegs-
   kompressor *c*

**2753 multivane rotary
compressor**
DA mangebladet
   rotations-
   kompressor
DE Vielzellenrotations-
   verdichter *m*
ES compresor rotativo
   de álabes múltiples
   m
F compresseur *m*
   rotatif
   multicellulaire
I compressore *m*
   rotativo con più
   palette
MA tübblapátos forgó
   kompresszor
   turbokompresszor
NL roterende
   schotten-
   compressor *m*
PO sprężarka *f*
   rotacyjna
   wielołopatkowa
PY ротационный
   многолопато-
   чный компрессор
   m
SU lamellikompressori
SV rotorkompressor *c*
   med flera blad *n*

2754 multizone *adj*
DA flerzone
DE mehrere Zonen
betreffend
ES multizona
F multi-zones
I multizone
zone multiple
MA tübbzónás
NL met meer zones
PO wielostrefowy
PY многозональный
SU monivyühykkeinen
SV flerzons-

DA	=	Danish
DE	=	German
ES	=	Spanish
F	=	French
I	=	Italian
MA	=	Magyar (Hungarian)
NL	=	Dutch
PO	=	Polish
PY	=	Russian
SU	=	Finnish
SV	=	Swedish

N

2755 natural convection
DA naturlig konvektion
DE natürliche
Konvektion *f*
ES convección natural *f*
F convection *f*
naturelle
I convezione *f*
naturale
MA természetes
konvekció
NL natuurlijke
convectie *f*
PO konwekcja *f*
swobodna
PY естественная
конвекция *f*
SU luonnollinen
konvektio
SV egenkonvek-
tion *c*

**2756 natural convection
air cooler**
DA luftkøler med
naturlig
konvention
DE Naturzug-
luftkühler *m*
ES enfriador de aire de
convección natural
m
F refroidissement *m*
d'air à tirage
naturel
I refrigeratore *m* di
aria a convezione
naturale
MA természetes
konvekciós
léghűtő
NL luchtkoeler *m* met
natuurlijke
convectie
PO chłodnica *f*
powietrza
konwekcyjna
PY воздухоохладитель
m, работающий в
условиях
естественной
конвекции
SU painovoimakier-
toinen ilmanjäähd-
ytin
SV egenkonvektionsluft-
kylare *c*

**2757 natural draught
burner** (*see*
atmospheric burner)

**2758 natural-draught
water cooling
tower**
DA vandkøletårn med
naturligt træk
DE Naturzug-
kühlturm *m*
ES torre de refrigeración
de agua de tiro
natural *f*
F tour *f* de
refroidissement à
tirage naturel
I torre *f* di
raffreddamento a
ventilazione
naturale
MA természetes huzatú
vízhűtőtorony
NL koeltoren *m* met
natuurlijke
luchtcirculatie
PO wieża *f* chłodnicza o
ciągu naturalnym
PY градирня *f* с
естественным
движением
воздуха
SU luonnonkiertoinen
jäähdytystorni
SV vattenkyltorn *n*

2759 natural frequency
DA egenfrekvens
DE Eigenfrequenz *f*
ES frecuencia propia *f*
F fréquence *f* propre
I frequenza *f* naturale
MA rezonancia
frekvencia
NL eigen frequentie *f*
eigen trillingsgetal *n*
PO częstotliwość *f*
własna
PY собственная
частота *f*
SU ominaistaajuus
SV egenfrekvens *c*

2760 natural gas
DA naturgas
DE Erdgas *n*
ES gas natural *m*
F gaz *m* naturel
I gas *m* naturale
MA füldgáz
NL aardgas *n*
PO gaz *m* ziemny
PY природный газ *m*
SU maakaasu
SV naturgas *c*

2761 nebulize *vb*
DA finfordele
forstøvet vand
DE vernebeln
versprühen
ES nebulizar
F nébuliser
I nebulizzare
MA küdüsít
NL vernevelen
verstuiven
PO rozpylić na mgłę
PY распыливать
SU muodostaa
pisaranytimiä
SV finfürdela

2762 needle bearing
DA nåleleje
DE Nadellager *n*
ES cojinete de agujas *m*
F roulement *m* à
aiguilles
I cuscinetto *m* ad aghi
MA tűgürgős csapágy
NL naaldlager *n*
PO łożysko *n* igiełkowe
PY роликовый
подшипник *m*
малого диаметра
SU neulalaakeri
SV nållager *n*

2763 needle valve
DA nåleventil
DE Nadelventil *n*
ES válvula de aguja *f*
F robinet *m* à
pointeau
I valvola *f* a spillo
MA tűszelep
NL naaldventiel *n*
PO zawór *m* iglicowy
PY игольчатый
клапан *m*
SU neulaventtiili
SV nålventil *c*

2764 negative pressure
DA undertryk
DE Unterdruck *m*
ES depresión *f*
F dépression *f*
I depressione *f*
MA depresszió
negatív nyomás
NL onderdruk *m*
PO ciśnienie *n* ujemne
podciśnienie *n*
PY вакуумметрическое
давление *n*
SU alipaine
SV undertryck *n*

2765 net capacity
DA nettoydelse
DE Nettoleistung *f*
ES capacidad neta *f*
F volume *m* utile
d'une chambre
froide
I capacità *f* netta
MA nettó teljesítmény
NL netto capaciteit *f*
PO pojemność *f*
użyteczna
PY эффективная
производитель-
ьность *f* (нетто)
SU nettoteho
SV nettokapacitet *c*

2766 network
DA ledningsnet
DE Netz *n*
ES red *f*
F réseau *m*
I rete *f*
MA hálózat
NL netwerk *n*
PO sieć *f*
PY сеть *f*
SU verkosto
SV nät *n*

2767 neutral zone
DA neutral zone
DE neutrale Zone *f*
ES zona neutra *f*
F zone *f* neutre
I zona *f* neutra
MA semleges zóna
NL neutrale zone *f*
PO strefa *f* neutralna
strefa *f* nieczułości
PY нейтральная зона *f*
SU neutraalitaso
neutraalivyühyke
SV neutral zon *c*

2768 niche
DA niche
DE Nische *f*
ES hornacina *f*
nicho *m*
F niche *f*
I nicchia *f*
MA fülke
NL nis *f*
PO nisza *f*
wnęka *f*
PY ниша *f*
SU syvennys
SV nisch *c*

2769 night setback
DA natsænkning
DE Nachtabsenkung *f*
ES reducción nocturna *f*
F abaissement *m* du
point de consigne
pour la nuit
I abbassamento *m*
notturno del set-
point del
termostato
MA éjszakai csükkentett
üzem
NL nachtverlaging *f*
PO osłabienie *n* nocne
PY ночное снижение *n*
(температуры)
SU yüaikainen
lämpütilan lasku
SV nattsänkning *c*

2770 nipple
DA nippel
DE Nippel *m*
ES manguito roscado *m*
niple *m*
rácor *m*
F bague *f*
d'assemblage
I raccordo *m* filettato
MA karmantyú
NL nippel *m*
PO złączka *f* wkrętna
PY ниппель *m*
SU nippa
nippeli
SV nippel *c*

2771 nitrogen
DA kvælstof
DE Stickstoff *m*
ES nitrógeno *m*
F azote *m*
I azoto *m*
MA nitrogén
NL stikstof *f*
PO azot *m*
PY азот *m*
SU typpi
SV kväve *n*

2772 noise
DA støj
DE Geräusch *n*
   Lärm *m*
ES ruido *m*
F bruit *m*
I rumore *m*
MA zaj
NL geluid *n*
   geruis *n*
   lawaai *n*
PO hałas *m*
PY шум *m*
SU melu
SV buller *n*

2773 noise attenuation
(*see* noise reduction)

2774 noise criteria curves
DA støjkriteriekurver
DE Geräuschbewertungs-
   kurven *f,pl*
ES curvas (NC) *f,pl*
F courbes *f* de
   criètres
   acoustiques
I curve *f,pl*
   caratteristiche del
   rumore
MA zaj jelleggürbék
NL geluidsnorm-
   grafieken *pl*
PO krzywe *f*
   charakterystyczne
   hałasu
PY характеристические
   кривые *f* шума
SU melun rajakäyrä
   NR-käyrä
SV bullerkriterie-
   kurvor *c*

2775 noiseless *adj*
DA lydløs
   støjfri
DE geräuschlos
ES insonorizado
   silencio *m*
F silencieux *m*
I silenzioso
MA zajtalan
NL geruisloos
PO bezgłośny
   bezszumowy
   cichy
PY бесшумный
SU meluton
SV ljudlüs

2776 noise level
DA støjniveau
DE Geräuschpegel *m*
   Schallpegel *m*
ES nivel sonoro *m*
F niveau *m* sonore
I livello *m* sonoro
MA zajszint
NL geluidniveau *n*
PO poziom *m* hałasu
PY уровень *m* шума
SU melutaso
SV bullernivå *c*

2777 noise reduction
DA støjreduktion
DE Lärmminderung *f*
   Schalldämpfung *f*
ES reducción de sonido
   f
F réduction *f* des
   bruits
I attenuazione *f*
   sonora
MA zajcsükkentés
NL geluiddemping *f*
PO tłumienie *n* hałasu
PY уменьшение *n*
   шума
SU
SV bullerminskning *c*

2778 no-load condition
DA tomgangstilstand
DE lastfreier Zustand *m*
ES situación de carga
   nula *f*
F
I condizione *f* di
   assenza di carico
MA terheletlen állapot
NL nullasttoestand *m*
PO stan *m* jałowy
PY отсутствие *n*
   нагрузки
SU kuormaton tila
SV obelastat tillstånd *n*

2779 no-load start
DA tomgangsstart
DE lastfreier Anlauf *m*
ES arranque sin carga *m*
F démarrage *m* à vide
I avviamento *m* a
   vuoto
MA terheletlen indítás
NL onbelaste start *m*
PO rozruch *m* bez
   obciążenia
PY пуск *m* без
   нагрузки
   пуск *m* на
   холостом ходу
SU käynnistys
   kuormattomana
SV tomgångsstart *c*

2780 nominal diameter
DA nominel diameter
DE Nenndurchmesser *m*
ES diámetro nominal *m*
F diamètre *m* nominal
I diametro *m*
   nominale
MA névleges átmérő
NL nominale
   diameter *m*
PO średnica *f*
   nominalna
PY номинальный
   диаметр *m*
SU nimellishalkaisija
SV nominell diameter *c*

2781 nominal pressure
DA nominelt tryk
DE Nenndruck *m*
ES presión de régimen *f*
   presión nominal *f*
F pression *f* nominale
I pressione *f* nominale
MA névleges nyomás
NL nominale druk *m*
PO ciśnienie *n*
   nominalne
PY номинальное
   давление *n*
SU nimellispaine
SV nominellt tryck *n*

DA	=	Danish
DE	=	German
ES	=	Spanish
F	=	French
I	=	Italian
MA	=	Magyar (Hungarian)
NL	=	Dutch
PO	=	Polish
PY	=	Russian
SU	=	Finnish
SV	=	Swedish

2782 non-bitumenous coal
DA magre kul
DE Magerkohle *f*
ES carbón magro *m*
carbón no bituminoso *m*
hulla seca *f*
F charbon *m* maigre
I carbone *m* non bituminoso
MA sovány szén
NL magere kool *f*
PO węgiel *m* chudy
węgiel *m* niebitumiczny
PY тощий уголь *m*
SU laiha hiili
SV magra kol *n*

2783 non-changeover
DA ikke omstillelig
DE Nichtumstellung *f*
ohne Umstellung *f*
ES no conmutable
F non permutable
I non inversione *f*
MA felcserélhetetlen
NL niet omschakelbaar
PO nieprzełączalny
PY работа *f* в постоянном режиме
SU ilman vaihtokytkentää
SV

2784 non-coking coal
DA ikke-koksende kul
DE nichtverkokende Kohle *f*
ES carbón no coquizable *m*
F charbon *m* non cokéfiant
I carbone *m* non da cokefazione
MA nem kokszolható szén
NL niet-vercokesbare kool *f*
PO węgiel *m* niekoksujący
PY некоксующийся уголь *m*
SU koksaantumaton hiili
SV ickekoksandkol *n*

2785 non-depletable *adj*
DA vedvarende
DE unerschüpflich
ES no vaciable
F
I rinnovabile
MA kimeríthetetlen
NL duurzaam onuitputtelijk
PO niewyczerpalny niezubożony
PY возобновляемый неистощимый
SU ehtymätün uusiutuva
SV outtümbar

2786 non-flammable *adj*
DA uantændelig
DE nicht entzündbar
ES infamable
F ininflammable
I ininfiammabile
MA nem éghető
NL niet ontvlambaar onbrandbaar
PO niepalny
PY невоспламеняющийся
SU syttymätün
SV svårantändlig

2787 non-return damper
DA kontraklap
DE Rückschlagklappe *f*
ES registro de antirretorno *m*
registro de retención *m*
F volet *m* anti-retour
I serranda *f* di ritegno
MA visszacsapócsappantyú
NL terugslagklep *f* in luchtkanaal
PO przepustnica *f* zwrotna
PY обратный клапан *m*
SU takaiskuläppä
SV backspjäll *n*

2788 non-return valve
DA kontraventil
DE Rückschlagventil *n*
ES válvula antirretorno *f*
F clapet *m* de non retour
soupape *f* de retenue
I valvola *f* di non ritorno
MA visszacsapószelep
NL terugslagklep *f*
PO zawór *m* zwrotny
PY обратный клапан *m* (вентильного типа)
SU yksisuuntaventtiili
SV backventil *c*

2789 non steady-state
DA overgangstilstand
DE Übergangszustand *m*
ES régimen variable *m*
F régime *m* variable
I regime *m* instabile
stato *m* transitorio
MA nem szilárd állapot
NL niet-stationaire toestand *m*
PO stan *m* niestały
stan *m* nieustalony
stan *m* przejściowy
PY неустойчивое состояние *n*
SU epäjatkuvuustila
SV üvergångstillstånd *n*

2790 noxious *adj*
DA skadelig
DE schädlich
ES nocivo
F délétère
I nocivo
MA ártalmas káros
NL ongezond schadelijk
PO szkodliwy
PY вредный
SU haitallinen
SV skadlig

2791 nozzle
DA dyse
DE Düse *f*
ES tobera *f*
F bec *m*
　 gicleur *m*
I ugello *m*
MA fúvóka
NL sproeier *m*
　 uitstroomtuit *f*
　 verstuiver *m*
PO dysza *f*
PY сопло *n*
SU suutin
SV munstycke *n*

2792 nozzle outlet
DA dyseudløb
DE Düsenauslaß *m*
ES orificio de boquilla
　 m
F ajustage *m* de
　 soufflage
I scarico *m* dell'ugello
MA fúvóka nyílás
NL sproeieruitlaat *m*
PO otwór *m* wylotowy
　 dyszy
PY сопловое
　 отверстие *n*
SU ulosvirtaussuutin
SV munstycksutlopp *n*

2793 nuclear *adj*
DA kærne
DE nuklear
ES nuclear
F nucléaire
I nucleare
MA nukleáris
NL nucleair
PO nuklearny
PY ядерный
SU ydin-
SV kärn-

2794 number
DA antal
DE Anzahl *f*
　 Zahl *f*
ES número *m*
F nombre *m*
　 numéro *m*
I numero *m*
MA szám
NL aantal *n*
　 getal *n*
　 nummer *n*
PO liczba *f*
　 numer *m*
PY число *n*
SU numero
SV antal *n*

2795 numerical *adj*
DA numerisk
DE numerisch
ES numérico
F numérique
I numerico
MA numerikus
NL numeriek
PO liczbowy
　 numeryczny
PY цифровой
SU numeerinen
SV numerisk

2796 nut
DA møtrik
DE Schraubenmutter *f*
ES tuerca *f*
F écrou *m*
I dado *m*
MA csavaranya
NL moer *f*
　 noot *f*
PO nakrętka *f*
PY гайка *f*
SU mutteri
SV mutter *c*

O

2797 obstruction
DA forstoppelse
    spærring
DE Verstopfung *f*
ES obstrucción *f*
F obstruction *f*
I ostruzione *f*
MA dugulás
NL hindernis *f*
    versperring *f*
    verstopping *f*
PO przeszkoda *f*
PY засорение *n*
SU este
SV hinder *n*

2798 occupancy detector
DA belægningsføler
DE Belegungssensor *m*
ES detector de
    ocupación *m*
F détecteur *m* de
    présence
I sensore *m* di
    occupazione
MA foglaltság mutató
NL aanwezigheids-
    detector *m*
PO czujnik *m*
    wypełnienia
    wskaźnik *m*
    zapełnienia
PY задействованный
    датчик *m*
SU läsnäoloanturi
SV närvarosensor *c*

2799 occupancy sensor
(*see* occupancy
detector)

2800 occupant
DA beboer
DE Besitzer *m*
    Bewohner *m*
    Inhaber *m*
ES ocupante *m*
F occupant *m*
I occupante *m*
MA benntartózkodó
    személy
NL bewoner *m*
    bezitter *m*
    gebruiker *m*
PO użytkownik *m*
PY пользователь *m*
SU asukas
    henkilü
    ihminen
SV hyresgäst *c*

2801 occupied space
DA opholdsrum
DE Aufenthaltsraum *m*
ES espacio ocupado *m*
F local *m* occupé
I spazio *m* occupato
MA tartózkodási tér
NL bezette ruimte *f*
PO przestrzeń *f* zajęta
PY занятое
    пространство *n*
SU käytüssä oleva tila
SV utnyttjat utrymme *n*

2802 occupied zone
DA opholdszone
DE Aufenthaltszone *f*
ES zona ocupada *f*
F zone *f* d'occupation
I zona *f* occupata
MA tartózkodási zóna
NL verblijfsgebied *n*
    verblijfszone *f*
PO strefa *f* zajęta
PY занятая зона *f*
    рабочая зона *f*
SU käytüssä oleva
    vyühyke
SV vistelsezon *c*

2803 odour
DA lugt
DE Geruch *m*
ES olor *m*
F odeur *f*
I odore *m*
MA szag
NL geur *m*
    reuk *m*
PO zapach *m*
PY запах *m*
SU haju
SV lukt *c*

2804 odour filter
DA lugtfilter
DE Geruchsfilter *m*
ES filtro
    desodorizante *m*
F filtre *m* désodorisant
I filtro *m* deodorante
MA szagszűrő
NL reukfilter *m*
PO filtr *m*
    przeciwzapachowy
PY фильтр *m*,
    поглощающий
    запахи
SU hajusuodatin
SV luktfilter *n*

DA	=	Danish
DE	=	German
ES	=	Spanish
F	=	French
I	=	Italian
MA	=	Magyar (Hungarian)
NL	=	Dutch
PO	=	Polish
PY	=	Russian
SU	=	Finnish
SV	=	Swedish

2805 off-cycle defrosting
DA uregelmæssig afrimning
DE Abtauen *n* in Stillstandszeit
ES desescarche fuera de puntas *m*
F dégivrage *m* non-cyclique
I sbrinamento *m* con interruzione del ciclo frigorifero
MA üzemszüneti leolvasztás
NL ontdooien *n* buiten bedrijfsperiode
PO rozmrażanie *n* poza cyklem pracy
PY размораживание *n* вне цикла
SU seisokkiajan sulatus
SV avfrostning *c* under stillestånd *n*

2806 off-peak period
DA periode uden spidslast
DE Schwachlast-periode *f*
ES boquilla de quemador *f*
F période *f* hors pointe
I periodo *m* fuori picco
MA csúcsüzemen kívüli időszak
NL periode *f* buiten piekuren
PO okres *m* pozaszczytowy
PY внепиковый период *m*
SU huipputehon ulkopuolinen aika
SV period *c* utan toppbelastning *c*

2807 off-peak storage heating
DA opvarmning med akkumuleret varme
DE Speicherheizung *f* außerhalb der Spitzenbelastung *f*
ES calefacción por acumulación de calor en horas valle *f*
F chauffage *m* par accumulation d'énergie hors pointe
I riscaldamento *m* ad accumulo al di fuori delle ore di picco
MA csúcsüzemen kívüli tárolós fűtés
NL nachtstroom *m* bufferverwarming
PO ogrzewanie *n* akumulacyjne poz-aszczytowe
PY аккумуляционное отопление *n* за счет использования провала нагрузки
SU varaava lämmitys
SV uppvärmning *c* med ackumulering *c*

2808 offtake
DA afgang
DE Abzweig *m*
ES acometida *f* derivación *f* toma *f*
F branchement *m*
I diramazione *f*
MA elágazás
NL aftakking *f*
PO odgałęzienie *n* odprowadzenie *n*
PY ответвление *n*
SU haara
SV gasuttag *n*

2809 oil
DA olie
DE Öl *n*
ES aceite *m*
F huile *f*
I nafta *f* olio *m*
MA olaj
NL olie *f*
PO olej *m*
PY мазут *m* масло *n*
SU öljy
SV olja *c*

2810 oil bath
DA oliebad
DE Ölbad *n*
ES baño de aceite *m*
F bain *m* d'huile
I bagno *m* di olio
MA olajfürdő
NL oliebad *n*
PO kąpiel *f* olejowa
PY масляная ванна *f*
SU öljykylpy
SV oljebad *n*

2811 oil bath air filter
DA oliebadsfilter
DE ülbenetzter Filter *m*
ES filtro de aire en un baño de aceite *m*
F filtre *m* à air à bain d'huile
I filtro *m* di aria a bagno di olio
MA olajba mártott légszűrő
NL oliebad luchtfilter *m*
PO filtr *m* powietrza olejowy
PY воздушный фильтр *m* с поддоном для масла
SU öljytetty ilmasuodatin
SV oljefilter *n* (för luft *c*)

2812 oil burner
DA oliebrænder
DE Ölbrenner *m*
ES quemador de aceite *m*
   quemador de petroleo *m*
F brûleur *m* à fuel
I bruciatore *m* di nafta
MA olajégő
NL oliebrander *m*
PO palnik *m* olejowy
PY мазутная форсунка *f*
SU öljypoltin
SV oljebrännare *c*

2813 oil burner nozzle
DA oliebrænderdyse
DE Ölbrennerdüse *f*
ES boquilla de quemador *f*
F gicleur *m* de brûleur à fuel
I ugello *m* del bruciatore di nafta
MA olajégő fúvóka
NL oliebranderverstuiver *m*
PO dysza *f* palnika olejowego
PY сопло *n* мазутной форсунки
SU öljypolttimen suutin
SV oljebrännarmunstycke *n*

2814 oil consumption
DA olieforbrug
DE Ölverbrauch *m*
ES consumo de aceite *m*
   consumo de petróleo *m*
F consommation *f* de fuel
I consumo *m* di nafta
MA olajfogyasztás
NL olieverbruik *n*
PO zyżycie *n* oleju
PY расход *m* масла (мазута)
SU öljynkulutus
SV oljefürbrukning *c*

2815 oil cooler
DA oliekøler
DE Ölkühler *m*
ES enfriador de aceite *m*
F refroidisseur *m* d'huile
I refrigeratore *m* di olio
MA olajhűtő
NL oliekoeler *m*
PO chłodnica *f* olejowa
PY маслоохладитель *m*
SU öljynjäähdytin
SV oljekylare *c*

2816 oil cooling
DA oliekøling
DE Ölkühlung *f*
ES enfriamiento de aceite *m*
F refroidissement *m* d'huile
I raffreddamento *m* dell'olio
MA olajhűtés
NL oliekoeling *f*
PO chłodzenie *n* olejowe
PY охлаждение *n* масла
SU öljynjäähdytys
SV oljekylning *c*

2817 oil drain
DA olieaftapning
DE Ölablaß *m*
ES orificio de descarga del aceite *m*
   purgador de aceite *m*
F vidange *m* d'huile
I scarico *m* di nafta
   scarico *m* di olio
MA olajelfolyó
NL olie-aftap *m*
PO spust *m* oleju
PY спуск *m* масла
SU öljyn tyhjennys
SV oljeavtappning *c*

2818 oil feed control
DA olietilførselsregulering
DE Ölzufuhrregelung *f*
ES regulación de alimentación de aceite *f*
   regulación de alimentación de fueloil *f*
   regulación de alimentación de petroleo *f*
F commande *f* d'alimentation en huile
I regolazione *f* della portata di nafta
MA olajtápláló szabályozó
NL olietoevoerregeling *f*
PO regulacja *f* zasilania olejem
PY регулирование *n* расхода масла
SU öljynsyütün säätü
SV oljeregulator *c*

2819 oil-fired boiler
DA oliefyret kedel
DE Kessel *m* mit Ölfeuerung *f*
   Ölkessel *m*
ES caldera caldeada por aceite combustible *f*
   caldera caldeada por petróleo *f*
F chaudière *f* à fuel
I caldaia *f* a nafta
MA olajtüzelésű kazán
NL oliegestookte ketel *m*
PO kocioł *m* opalany olejem
PY котел *m* с мазутной топкой
SU öljykattila
SV oljeeldad panna *c*

2820 oil firing
DA oliefyring
DE Ölfeuerung *f*
ES caldeo con aceite combustible *m*
caldeo con petroleo *m*
F chauffe *f* au fuel
I combustione *f* a nafta
MA olajtüzelés
NL oliestook *f*
PO opalanie *n* olejem
PY мазутная топка *f*
SU üljynpoltto
SV oljeeldning *c*

2821 oil-free *adj*
DA uden olie
DE ülfrei
ES exento de aceite
F éxempt d'huile
sans huile
I senza lubrificazione
MA olajmentes
NL olievrij
PO bezolejowy
PY обезжиренный
SU üljytün
üljyvapaa
SV oljefri

2822 oil heating
DA olieopvarmning
DE Ölheizung *f*
ES calefacción por aceite *f*
F chauffage *m* au fuel
I riscaldamento *m* a nafta
MA olajfűtés
NL olieverwarming *f*
verwarming *f* met oliestook
PO ogrzewanie *n* olejowe
PY мазутное отопление *n*
SU üljylämmitys
SV oljeeldning *c*

2823 oil level
DA olieniveau
DE Ölstand *m*
ES nivel de aceite *m*
F niveau *m* d'huile
I livello *m* dell'olio
MA olajszint
NL oliepeil *n*
PO poziom *m* oleju
PY уровень *m* масла
SU üljynpinnan korkeus
SV oljenivå *c*

2824 oil mist
DA olietåge
DE Ölnebel *m*
ES separador de aceite *m*
F brouillard *m* d'huile
I nafta *f* nebulizzata
MA olajküd
NL olienevel *m*
PO mgła *f* olejowa
PY масляный туман *m*
SU üljysumu
SV oljedimma *c*

2825 oil pressure cut-out control
DA olietryksregulering
DE Öldruckabschalt-reglung *f*
ES control por presión de aceite *m*
F contrôle *m* par pressostat de sécurité d'huile
I limitazione *f* della pressione dell'olio
MA olajnyomás kikapcsoló
NL oliedrukbeveiliging *f*
PO ogranicznik *m* ciśnienia oleju regulacja *f* odcięcia w zależności od ciśnienia oleju
PY контроль *m* давления масла (в компрессоре)
SU üljynpaineen säätü
SV säkringskontroll *c* för oljetryk *n*

2826 oil separator
DA olieudskiller
DE Ölabscheider *m*
ES separador de aceite *m*
F déshuileur *m* séparateur *m* d'huile
I separatore *m* di olio
MA olajleválasztó
NL olie-afscheider *m*
PO oddzielacz *m* oleju odolejacz *m*
PY маслоуловитель *m*
SU üljynerotin
SV oljeavskiljare *c*

2827 oil storage tank
DA olielagertank
DE Ölbehälter *m*
Öltank *m*
ES tanque de almacenamiento de aceite *m*
tanque de almacenamiento de petróleo *m*
F réservoir *m* de fuel
I serbatoio *m* di nafta
MA olajtároló tartály
NL olie-opslagtank *m*
olietank *m*
PO zbiornik *m* oleju
PY бак *m* для хранения масла
SU üljysäiliü
SV oljecistern *c*

2828 oil stove
DA olieovn
DE Ölofen *m*
ES estufa de petróleo *f*
horno de petróleo *m*
F poêle *m* à mazout
I stufa *f* a gasolio
MA olajkályha
NL oliekachel *f*
PO piec *m* olejowy
PY мазутная печь *f*
SU üljykamiina
SV oljekamin *c*

2829 oil supply
DA olieforsyning
DE Ölversorgung *f*
ES suministro de aceite *m*
suministro de petróleo *m*
F alimentation *f* de fuel
I alimentazione *f* di nafta
MA olajellátás
NL olievoorziening *f*
PO zasilanie *n* olejem
PY подпитка *f* масла
SV oljefürsürjning *c*

2830 oil tanker (road)
DA tankbil
DE Öltankwagen *m*
ES camión cisterna
(para el transporte
de petróleo) *m*
F camion-citerne *m* de
fuel
I autocisterna *f*
MA olajszállító
tartálykocsi
NL tankauto *m*
PO cysterna *f* na olej
PY цистерна *f* для
масла
SU säiliüauto
SV tankbil *c*

**2831 oil temperature
cut-out control**
DA olietemperatur-
regulering
DE Öltemperaturabschalt-
regelung *f*
ES control de corte por
temperatura de
aceite *m*
F contrôle *m* par
thermostat de
sécurité d'huile
I limitazione *f* della
temperatura
dell'olio
MA olajhőmérsékletről
szabályozás
NL olietemperatuur-
beveiliging *f*
PO ogranicznik *m*
temperatury oleju
regulacja *f* odcięcia
w zależności od
temperatury oleju
PY контроль *m*
температуры
масла (в
компрессоре)
SU üljyn lämpütilan
säätü
SV säkringskontroll *c*
för oljetemperatur
c

DA = Danish
DE = German
ES = Spanish
F = French
I = Italian
MA = Magyar
(Hungarian)
NL = Dutch
PO = Polish
PY = Russian
SU = Finnish
SV = Swedish

2832 olf
DA olf
DE Olf *m*
ES olf *m*
F olf *m*
I olf *m*
MA olf
NL olf *m*
PO olf *m*
PY олф *m*
(антропогенное
загрязнение
воздуха одним
человеком)
SU hajuemission
yksikkü
olf
SV olf *c*

2833 one pipe heating
DA et-rørs-varmeanlæg
DE Einrohrheizung *f*
ES calefacción
monotubo *f*
F chauffage *m* à un
tuyau
chauffage *m*
monotube
I impianto *m* di
riscaldamento
monotubo
MA egycsüves fűtés
NL eenpijps-
verwarming *f*
PO ogrzewanie *n*
jednorurowe
PY однотрубная
система *f*
отопления
SU yksiputkilämmitys
SV ettrörssystem *n*

2834 on–off control
DA on-off-regulering
DE Auf-Zu-Regelung *f*
ES regulación *f*
F régulation *f* par tout
ou rien
I regolazione *f* tutto-
niente
MA ki-bekapcsolásos
szabályozás
NL aan-uit regeling *f*
PO regulacja *f*
dwupołożeniowa
regulacja *f*
dwustawna
PY двухпозиционное
регулирование *n*
SU kaksiasentosäätü
on-off -säätü
SV till- och från-
reglering *c*

2835 on-peak current
DA spidslaststrøm
DE Hochtarifstrom *m*
ES corriente en/a las
horas punta *f*
F courant *m* de pointe
I corrente *f* di punta
MA csúcsáram
NL stroom *m* tijdens
piekuren
PO prąd *m* szczytowy
PY пиковый поток *m*
SU huippuvirta
SV maxstrüm *c*

2836 on-peak period
DA spidslastperiode
DE Spitzenlastzeit *f*
ES período punta *m*
F période *f* de pointe
I periodo *m* di picco
MA csúcsidőszak
NL periode *f* tijdens
piekuren
PO okres *m* szczytowy
PY период *m* пиковой
нагрузки
SU huipputehon aika
SV topperiod *c*

2837 on-site generation
DA generering på stedet
DE Stromerzeugung *f*
auf dem
Grundstück *n*
ES generación in situ *f*
F production *f* sur site
I generazione *f* locale
MA helyszíni
áramfejlesztés
NL energie-opwekking *f*
op eigen terrein
PO wytwarzanie *n* na
budowie
PY выработка *f*
электроэнергии
на группу
потребителей
SU rakennuskohtainen
sähküntuotto
SV generering *c*
på plats *c*

2838 opacity
DA gennemskinnelighed
klarhed
DE Trübung *f*
ES opacidad *f*
F opacité *f*
I opacità *f*
MA átlátszatlanság
zavarosság
NL ondoorzichtigheid *f*
PO nieprzezroczystość *f*
PY затененность *f*
непрозрачность *f*
SU valonläpäisevyys
SV opacitet *c*

2839 opacity factor
DA gennemskinneligheds-
faktor
DE Trübungsfaktor *f*
ES coeficiente de
opacidad *m*
F facteur *m* d'opacité
I fattore *m* di opacità
MA átlátszatlansági
tényező
NL ondoorzichtigheids-
factor *m*
PO współczynnik *m*
nieprzezroczy-
stości
PY коэффициент *m*
непрозрачности
SU läpäisevyystekijä
SV opacitetsfaktor *c*

2840 opaque *adj*
DA uigennemsigtig
DE undurchsichtig
ES opaco
F opaque
I opaco
MA átlátszatlan
NL ondoorzichtig
PO nieprzezroczysty
PY светонепрони-
цаемый
SU säteilyä
läpäisemätün
SV ogenomskinlighet *c*

2841 opening
DA åbning
DE Öffnung *f*
ES abertura *f*
F ouverture *f*
I apertura *f*
MA nyílás
NL openen *n*
opening *f*
PO otwór *m*
PY отверстие *n*
SU aukko
SV üppning *c*

**2842 open-type
compressor**
DA åben kompressor
DE offener Verdichter *m*
ES compresor abierto *m*
F compresseur *m* de
type ouvert
I compressore *m*
aperto
MA nyitott típusú
kompresszor
NL open compressor *m*
PO sprężarka *f* typu
otwartego
PY сальниковый
компрессор *m*
SU avoin kompressori
SV üppen kompressor *c*

**2843 operating
conditions**
DA drifttilstand
DE Betriebszustand *m*
ES régimen de
funcionamiento *m*
F conditions *f* de
fonctionnement
I condizioni *f,pl*
operative
MA üzemelési feltételek
NL bedrijfsomstandig-
heden *pl*
PO warunki *m* działania
warunki *m*
eksploatacji
PY рабочие условия *n*
SU käyttüolot
SV driftfürhållanden *n*

2844 operating costs
DA driftudgift
DE Betriebskosten *f,pl*
ES coste de explotación
m
gastos de
explotación *m,pl*
F frais *m*
d'exploitation
I costo *m* di esercizio
MA üzemelési kültségek
NL bedrijfskosten *pl*
PO koszt *m* eksploatacji
PY эксплуатационные
расходы *m*
SU käyttükustannukset
SV driftkostnad *c*

2845 operating data
DA driftdata
DE Betriebsdaten *f,pl*
ES datos de explotación
m,pl
F bases *f*
d'exploitation
I dato *m* di esercizio
MA üzemelési kültségek
NL bedrijfsgegevens *pl*
PO dane *f*
eksploatacyjne
PY эксплуатационные
показатели *m*
SU käyttütiedot
SV driftdata *c*

**2846 operating
differential**
DA driftudsving
DE Betriebsabstufung *f*
ES diferencial de
operación *m*
F différentiel *m* de
fonctionnement
I differenziale *m* di
lavoro
MA müküdési külünbség
NL bedrijfs-
differentiaal *n*
PO różniczka *f*
sterowania pracą
PY рабочая разность *f*
SU eroalue
SV arbetsdifferential *c*

**2847 operating
instructions**
DA driftinstruktion
DE Betriebsanleitung *f*
ES instrucciones de
manejo *f,pl*
instrucciones sobre
funcionamiento
f,pl
F instructions *f*
d'exploitation
I istruzioni *f,pl* per
l'uso
MA müküdési utasítások
NL bedieningsvoor-
schriften *pl*
bedrijfsvoor-
schriften *pl*
PO instrukcje *f*
eksploatacyjne
instrukcje *f* obsługi
PY инструкция *f* по
обслуживанию
SU käyttüohjeet
SV skütsel-
instruktioner *c*

2848 operating load
DA driftbelastning
DE Betriebslast *f*
ES carga de régimen *f*
carga de servicio *f*
carga normal de
trabajo *f*
F allure *f* de marche
I carico *m* di esercizio
MA üzemi terhelés
NL bedrijfsbelasting *f*
PO obciążenie *n*
eksploatacyjne
obciążenie *n*
robocze
PY рабочая нагрузка *f*
SU käyttükuorma
SV belastning *c* under
drift *c*

2849 operating pressure
DA drifttryk
DE Betriebsdruck *m*
ES presión de
funcionamiento *f*
presión de servicio *f*
F pression *f* de marche
I pressione *f* di
esercizio
MA üzemnyomás
NL bedrijfsdruk *m*
PO ciśnienie *n* pracy
ciśnienie *n* robocze
PY рабочее давление *n*
SU käyttüpaine
SV drifttryck *n*

**2850 operating
temperature**
DA drifttemperatur
DE Betriebstemperatur *f*
ES temperatura de
régimen *f*
temperatura de
servicio *f*
F température *f* de
fonctionnement
I temperatura *f*
operante
MA üzemi hőmérséklet
NL bedrijfs-
temperatuur *f*
PO temperatura *f*
operacyjna
temperatura *f* pracy
temperatura *f*
robocza
PY рабочая
температура *f*
SU käyttülämpütila
SV drifttemperatur *c*

2851 operating time
DA drifttid
DE Betriebszeit *f*
ES tiempo de
funcionamiento *m*
tiempo de servicio *m*
F durée *f* de marche
durée *f* de vie
I durata *f* di servizio
MA üzemidő
NL bedrijfstijd *m*
PO czas *m* eksploatacji
czas *m* pracy
PY рабочее время *n*
SU käyttüaika
SV drifttid *c*

2852 operation
DA betjening
drift
funktion
DE Betrieb *m*
ES explotación *f*
funcionamiento *m*
servicio *m*
F marche *f*
opération *f*
I esercizio *m*
funzionamento *m*
servizio *m*
MA működés
üzemelés
NL bediening *f*
bedrijf *n*
werking *f*
PO działanie *n*
eksploatacja *f*
praca *f*
PY действие *n*
SU käyttü
SV drift *c*

**2853 operative
temperature**
DA operativ temperatur
DE Betriebstemperatur *f*
ES temperatura de
operación *f*
F température *f* de
fonctionnement
I temperatura *f* di
lavoro
MA operatív hőmérséklet
NL operatieve
temperatuur *f*
PO temperatura *f*
działania
temperatura *f*
efektywna
PY действительная
температура *f*
SU operatiivinen
lämpütila
SV drifttemperatur *c*

2854 operator interface
DA operatør grænseflade
DE Bedienungsschnitt-
stelle *f*
ES interfase operadora *f*
F interface *f*
d'opérateur
I interfaccia *f*
operatore
MA kezelő
kapcsolóállomás
NL bedienings-
terminal *m*
PO część *f*
oprogramowania
operatora
(użytkownika)
PY поверхность *f*
раздела
SU käyttäjäliityntä
SV operatürs-
gränssnitt *n*

**2855 opposed blade
damper**
DA spjæld med
modgående blade
DE gegenläufige
Klappe *f*
ES registro de lamas
opuestas *m*
F registre *m* à volets
opposés
I serranda *f* a lame
contrapposte
MA ellentétesen működő
zsalu
NL luchtklep *f* met
contra-roterende
schoepen
PO przepustnica *f* z
łopatkami
przeciwbieżnymi
PY многостворчатый
клапан *m* с
обратными
лопатками
SU sälepelti vastakkain
kääntyvin sälein
SV spjäll *n* med
motgående blad *n*

2856 order of magnitude
DA størrelsesorden
DE Grüßenanordnung *f*
ES orden de magnitud
 m
F ordre *m* de grandeur
I ordine *m* di
 grandezza
MA nagyságrend
NL grootte-orde *f*
PO rząd *m* wielkości
PY порядок *m*
 величины
SU suuruusluokka
SV storlek *c*

2857 orientation
DA orientering
DE Orientierung *f*
 Richtung *f*
ES orientación *f*
F orientation *f*
I orientamento *m*
MA tájolás
NL orintatie *f*
 plaatsbepaling *f*
PO orientacja *f*
PY ориентация *f*
SU suuntaus
SV riktning *c*

2858 orifice
DA blende
 åbning
DE Blende *f*
 Meßblende *f*
ES orificio *m*
F diaphragme *m*
 orifice *m*
 ouverture *f*
I orifizio *m*
MA nyílás
NL opening *f*
PO kryza *f*
 otwór *m*
PY диафрагма *f*
SU suutin
SV mynning *c*

2859 orifice plate
DA blendplade
DE Blendenplatte *f*
ES orificio calibrado *m*
F diaphragme *m*
 plaque *f*
 d'obturation
I piastra *f* di
 otturazione
MA diafragma
 fojtótárcsa
NL meetschijf *f*
PO płytka *f* przesłony
PY пластинка *f*
 диафрагмы
SU kuristuslaippa
SV strypfläns *c*

2860 O-ring
DA O-ring
DE O-Ring *m*
ES junta tórica *f*
F joint *m* torique
I anello *m* di tenuta
 O-ring *m*
MA O-gyűrű
NL O-ring *m*
PO pierścień *m* o
 przekroju
 okrągłym
 pierścień *m*
 samouszczel-
 niający
PY пластичная
 деталь *f*
SU O-rengas
SV O-ring *c*

2861 Orsat apparatus
DA orsatsapparat
DE Orsat-Apparat *m*
ES aparato de Orsat
 (humos) *m*
F appareil *m* d'Orsat
I apparecchiatura *f* di
 Orsat
MA Orsat készülék
NL Orsatapparaat *n*
PO aparat *m* Orsata
PY прибор *m* Орса
SU Orsat-laite
SV Orsat-apparat *c*

2862 oscillation
DA oscillation
 svingning
DE Oszillation *f*
 Schwingung *f*
ES oscilación *f*
F oscillation *f*
I oscillazione *f*
MA lengés
NL slingering *f*
 trilling *f*
PO oscylacja *f*
PY колебание *n*
 осциллирование *n*
SU huojunta
SV oscillation *c*

2863 Otto cycle
DA Otto-kredsproces
DE Otto-Kreislauf *m*
ES ciclo Otto *m*
F cycle *m* d'Otto
I ciclo *m* Otto
MA Otto kürfolyamat
NL Ottokringloop *m*
PO obieg *m* Otto
PY цикл *m* Отто
SU Otto-prosessi
SV Otto-cykel *c*

2864 outdoor air
DA udeluft
DE Außenluft *f*
ES aire de
 renovación *m*
 aire exterior *m*
 aire fresco *m*
F air *m* extérieur
I aria *f* esterna
MA külső levegő
NL buitenlucht *f*
PO powietrze *n*
 zewnętrzne
PY наружный воздух
 m
SU ulkoilma
SV uteluft *c*

2865 outdoor appliance
DA udeudstyr
DE Außengerät *n*
ES aparato para el
   exterior *m*
F équipement *m*
   extérieur
I apparecchio *m* per
   installazione
   esterna
MA külső berendezés
NL apparaat *n* voor
   buitenopstelling
PO urządzenie *n*
   zewnętrzne
PY прибор *m* для
   установки
   снаружи
SU ulkokäyttüün
   tarkoitettu laite
SV utomhus-
   anordning *c*

2866 outdoor temperature
DA udetemperatur
DE Außentemperatur *f*
ES temperatura
   exterior *f*
F température *f*
   extérieure
I temperatura *f*
   esterna
MA külső hőmérséklet
NL buitentemperatuur *f*
PO temperatura *f*
   zewnętrzna
PY наружная
   температура *f*
SU ulkolämpütila
SV utomhus-
   temperatur *c*

2867 outlet
DA aftræk
   udløb
DE Ablauf *m*
   Auslaß *m*
ES boca de descarga *f*
   boca de salida *f*
   salida *f*
F bouche *f* de
   soufflage
   sortie *f*
I bocca *f* di uscita
   uscita *f*
MA kifúvónyílás
   kilépés
NL afvoer *m* (opening)
   uitlaat *m* (opening)
PO wylot *m*
PY выходное
   отверстие *n*
   сток *m*
SU ulosvirtaus
SV utlopp *n*

2868 outlet air
DA aftræksluft
DE Abluft *f*
ES aire de salida *m*
F air *m* de soufflage
I aria *f* di immissione
MA távozó levegő
NL afvoerlucht *f*
PO powietrze *n*
   wywiewne
PY вытяжной
   воздух *m*
SU poistoilma
SV frånluft *c*

2869 outlet grille
DA udsugningsrist
DE Ausblasgitter *n*
ES rejilla de salida *f*
F bouche *f* de
   soufflage
   grille *f* de soufflage
I griglia *f* di uscita
MA kifúvórács
NL afvoerrooster *m*
PO kratka *f* wywiewna
PY вытяжная
   решетка *f*
SU poistoilmasäleikkü
SV utloppsgaller *n*

2870 outlet nozzle
DA afgangsstuds
DE Ausblasdüse *f*
ES boquilla de salida *f*
F buse *f* de soufflage
I ugello *m* di uscita
MA kilépő fúvóka
NL afvoertuit *f*
PO dysza *f* wylotowa
PY вытяжное сопло *n*
SU poistosuutin
SV utloppsmunstycke *n*

2871 outlet opening
DA udløbsåbning
DE Auslaßüffnung *f*
ES abertura de salida *f*
F grille *f* de soufflage
   orifice *m* de soufflage
I apertura *f* di uscita
MA kilépőnyílás
NL afvoeropening *f*
   uitlaatopening *f*
PO otwór *m* wylotowy
PY вытяжное
   отверстие *n*
SU poistoaukko
SV utloppsüppning *c*

2872 outlet piece
DA udløbsstykke
DE Auslaßstück *n*
ES tubería de
   descarga *f*
   tubo de salida *m*
F tubulure *f* de
   soufflage
I punto *m* di uscita
MA kifúvócsonk
   kilépőidom
NL afvoerstuk *n*
   uitlaatstuk *n*
PO króciec *m* wylotowy
PY вытяжной
   штуцер *m*
SU poistolaite
SV utloppsdon *n*

DA	=	Danish
DE	=	German
ES	=	Spanish
F	=	French
I	=	Italian
MA	=	Magyar (Hungarian)
NL	=	Dutch
PO	=	Polish
PY	=	Russian
SU	=	Finnish
SV	=	Swedish

2873 outlet velocity
DA udløbshastighed
DE Austrittsgeschwindig-
   keit *f*
ES velocidad de salida *f*
F vitesse *f*
   d'échappement de
   sortie
I velocità *f* di uscita
MA kilépési sebesség
NL afvoersnelheid *f*
   uitlaatsnelheid *f*
PO prędkość *f*
   wylotowa
PY скорость *f* на
   выходе
SU poistonopeus
SV utloppshastighet *c*

2874 output load
DA udløbsladning
DE Ausgangslast *f*
ES carga útil *f*
F production *f*
   frigorifique
I carico *m* di uscita
MA teljesítményterhelés
NL beschikbaar
   vermogen *n*
PO obciążenie *n*
   wyjściowe
PY результирующая
   нагрузка *f*
SU lähtüteho
   ulostuloteho
SV uteffekt *c*

2875 outside air
DA udeluft
DE Außenluft *f*
ES aire exterior *m*
F air *m* extérieur
I aria *f* esterna
MA külső levegő
NL buitenlucht *f*
PO powietrze *n*
   zewnętrzne
PY наружный
   воздух *m*
SU ulkoilma
SV uteluft *c*

2876 outside dimension
DA udvendig dimension
DE Außenmaß *n*
ES dimensiones
   exteriores *f,pl*
   medida exterior *f*
F cote *f* extérieure
I dimensione *f* esterna
MA külső méret
NL buitenmaat *f*
   buitenwerkse
   afmeting *f*
PO wymiar *m*
   zewnętrzny
PY наружный
   размер *m*
SU ulkomitta
SV yttermått *n*

2877 outside sensor
DA udeføler
DE Außenfühler *m*
ES sonda exterior *f*
F capteur *m* extérieur
I sensore *m* esterno
MA külső érzékelő
NL buitenvoeler *m*
PO czujnik *m*
   zewnętrzny
PY наружный
   датчик *m*
SU ulkoilma-anturi
SV utomhuskännare *c*

2878 outside storage
DA udendørslager
DE Außenlager *n*
ES almacenamiento al
   aire libre *m*
F stockage *m* extérieur
I deposito *m* esterno
MA külső tároló
NL opslag *m* in de
   buitenlucht
PO składowanie *n*
   zewnętrzne
PY наружный склад *m*
SU ulkovarasto
SV utomhuslager *n*

2879 outward flow
DA udadgående
   strømning
DE Ausströmung *f*
ES corriente de salida *f*
   corriente hacia
   afuera *f*
F écoulement *m* vers
   l'extérieur
I flusso *m* verso
   l'esterno
MA kifolyás
NL stroming *f* naar
   buiten
PO wypływ *m*
PY истечение *n*
SU ulkopuolinen virtaus
SV utströmning *c*

2880 overall efficiency
DA totalvirkningsgrad
DE
ES rendimiento
   efectivo *m*
   rendimiento total *m*
F rendement *m* global
I rendimento *m*
   globale
MA üsszhatásfok
NL totaal rendement *n*
PO sprawność *f*
   całkowita
   sprawność *f* ogólna
PY общая
   производитель-
   ьность *f*
SU kokonaishyütysuhde
SV total
   verkningsgrad *c*

2881 overall length
DA total længde
DE Gesamtlänge *f*
ES longitud total *f*
F longueur *f* totale
I lunghezza *f* totale
MA üsszes hossz
NL totale lengte *f*
PO długość *f* całkowita
PY общая длина *f*
SU kokonaispituus
SV totallängd *c*

2882 overfeed stoker
DA fødestoker
DE obere Beschickung *f*
ES hogar con
    alimentación
    superior *m*
F foyer *m* à
    chargement par
    dessus
I alimentatore *m* del
    fuoco (caricatore
    superiore)
MA felsőadagolású
    tolórostély
NL stookinrichting *f*
    met bovenaanvoer
PO ruszt *m*
    mechaniczny
    nasypowy
    ruszt *m*
    mechaniczny
    zasilany z góry
PY топка *f* с верхней
    подачей
SU yläsyüttüinen
SV üvermatnings-
    stoker *c*

2883 overflow
DA overløb
DE Überlauf *m*
ES derrame *m*
    rebosadero *m*
F trop-plein *m*
I troppo-pieno *m*
MA túlcsordulás
    túlfolyás
NL overloop *m*
PO przelew *m*
PY перелив *m*
SU ylivirtaus
SV üverströmning *c*

2884 overflow *vb*
DA flyde over
DE überfließen
    überlaufen
ES rebosar
F déborder
I inondare
    traboccare
MA túlcsordul
    túlfolyik
NL overlopen
PO przelewać
PY переливать
SU virrata yli
SV strümma üver

2885 overflow valve
DA overløbsventil
DE Überlaufventil *n*
ES válvula de rebosa *f*
    válvula de
    sobrecarga *f*
F vanne *f* de trop
    plein
I valvola *f* di troppo
    pieno
MA túlfolyószelep
NL overloopafsluiter *m*
    overstortafsluiter *m*
PO zawór *m*
    przelewowy
PY переливной клапан
    m
SU ylivirtausventtiili
SV üverströmnings-
    ventil *c*

2886 overhead coil (*see*
ceiling coil)

**2887 overhead
distribution**
DA øvrefordeling
DE obere Verteilung *f*
ES distribución por la
    parte superior *f*
F distribution *f* en
    parapluie
    distribution *f* par le
    haut
I distribuzione *f* a
    pioggia
MA felső elosztás
NL bovenverdeling *f*
PO rozdział *m* górny
PY верхняя разводка *f*
SU yläjako
SV övre fördelning *c*

2888 overheating
DA overopvarmning
DE Überheizen *n*
ES calentamiento
    anormal,
    sobrecalenta-
    miento *m*
F surchauffe *m*
I surriscaldamento *m*
MA túlfűtés
NL ververhitting *f*
PO przegrzanie *n*
PY перегрев *m*
SU ylilämmitys
SV üverhettning *c*

2889 overload
DA overbelastning
DE Überlast *f*
    Überlastung *f*
ES sobrecarga *f*
F surcharge *f*
I sovraccarico *m*
MA túlterhelés
NL overbelasting *f*
PO przeciążenie *n*
PY перегрузка *f*
SU ylikuorma
SV üverbelastning *c*

2890 overshoot
DA overskride
DE Hinausschießen *n*
ES exceso *m*
F amplitude *f* de
    suroscillation
I overshoot *m*
MA túllengés
NL overregeling *f*
PO przeregulowanie *n*
    (przejście ponad
    wartość
    nastawioną)
PY отклонение *n* от
    установленного
    значения
SU yliampuminen
    ylikorjaus
SV üverskridande *n*

2891 owning cost
DA anskaffelsesudgift
DE Selbstkosten *f*
ES costo de
    adquisición *m*
F prix *m* coûtant
I costo *m* globale
    prezzo *m* di costo
MA ünkültség
NL eigendomskosten *pl*
PO koszt *m* własny
PY себестоимость *f*
SU kokonaiskustannus
SV totalkostnad *c*

2892 oxidation
DA iltning
    oxydering
DE Oxidation *f*
ES oxidación *f*
F oxydation *f*
I ossidazione *f*
MA oxidáció
NL oxidatie *f*
PO utlenianie *n*
PY окисление *n*
SU hapettuminen
SV oxidation *c*

2893 oxyacetylene welding
DA autogensvejsning
DE Autogen-
schweißung *f*
ES soldadura
oxiacetilénica *f*
F soudure *f* oxy-
acétylénique
I saldatura *f* autogena
MA autogénhegesztés
NL autogeen lassen
PO spawanie *n*
acetyleno-tlenowe
PY ацетилено-
кислородная
сварка *f*
SU happi-
asetyleenihitsaus
SV acetylensvetsning *c*

2894 oxygen
DA ilt
oxygen
DE Sauerstoff *m*
ES oxígeno *m*
F oxygène *m*
I ossigeno *m*
MA oxigén
NL zuurstof *f*
PO tlen *m*
PY кислород *m*
SU happi
SV syre *n*

2895 oxygen content
DA iltindhold
DE Sauerstoffgehalt *m*
ES contenido de
oxígeno *m*
F teneur *f* en oxygène
I tenore *m* di ossigeno
MA oxigéntartalom
NL zuurstofgehalte *n*
PO zawartość *f* tlenu
PY содержание *n*
кислорода
SU happipitoisuus
SV syrehalt *c*

2896 ozone
DA ozon
DE Ozon *n*
ES ozono *m*
F ozone *m*
I ozono *m*
MA ózon
NL ozon *n*
PO ozon *m*
PY озон *m*
SU otsoni
SV ozon *n*

P

2897 packaged *adj*
DA færdigmonteret
DE einbaufertig
ES autónomo de tipo
　　monobloque *m*
F équipé
I monoblocco
MA üsszeépített
NL bedrijfsklaar
PO kompletny
　　zblokowany
　　zwarty
PY компактный
　　сборный
SU käyttüvalmis
　　tehtaalla koottu
SV monteringsfärdig

**2898 packaged air
conditioner**
DA færdigmonteret
　　luftbehandler
DE Raumklimagerät *n*
ES acondicionador de
　　aire autónomo *m*
F conditionneur *m*
　　d'air monobloc
I condizionatore *m*
　　monoblocco
　　autonomo
MA üsszeépített
　　légkondicionáló
NL bedrijfsklaar
　　luchtbehandelings-
　　apparaat *n*
PO klimatyzator *m*
　　zwarty
PY компактный
　　кондиционер *m*
SU käyttüvalmis
　　ilmastointikone
SV driftklar
　　klimat-
　　anläggning *c*

2899 packaged boiler
DA færdigmonteret kedel
DE Kessel *m* als betriebs-
　　fertige Einheit *f*
ES caldera monoblor *f*
F chaudière *f* équipée
I caldaia *f*
　　monoblocco
MA kompakt kazán
NL brandercombinatie *f*
　　ketel *m*
PO kocioł *m*
　　zblokowany
PY компактный
　　котел *m*
SU pakettikattila
SV driftklar panna *c*

**2900 packaged terminal
air conditioning
system**
DA færdigmonteret
　　luftbehandlings-
　　anlæg
DE Klimaanlage *f* mit
　　betriebsfertiger
　　Zentrale *f*
ES sistema de
　　climatización con
　　unidades
　　autónomas *m*
F conditionneur *m*
　　d'air autonome
I impianto *m* di
　　condizionamento
　　a terminali
　　monoblocco
MA kompakt
　　klimarendszer
NL bedrijfsklare
　　ruimtekoeler *m*
PO system *m*
　　klimatyzacji ze
　　zwartym
　　urządzeniem
　　końcowym
PY агрегатированная
　　установка *f*
　　кондицион-
　　ирования воздуха
SU seinän läpäisevä
　　käyttüvalmis
　　ilmastointikone
SV monteringsfärdigt
　　luftkonditionerings-
　　system *n*

2901 packaged unit
DA færdigmonteret
enhed
DE Einbaueinheit *f*
einbaufertige
Einheit *f*
ES grupo autónomo *m*
sistema (frigorífico)
autónomo *m*
F appareil *m* totale-
ment équipé
I apparecchio *m*
monoblocco
MA üsszeépített készülék
NL bedrijfsklaar
apparaat *n*
PO urządzenie *n*
zblokowane
PY компактный
агрегат *m*
SU käyttüvalmis
yksikkü
SV monteringsfärdig
enhet *c*

2902 packed chamber
DA pakbøsning
DE Stopfbüchse *f*
ES caja de
empaquetadura *f*
F boîte *f* à garniture
chambre *f* de presse-
étoupe
I camera *f* di tenuta
MA tümszelence
NL stopbus *f*
PO dławnica *m*
komora *f*
dławikowa
PY герметизированная
камера *f*
SU läpivientiputki
tiivistysholkki
SV packbussning *c*

2903 packing
DA pakning
DE Abdichtung *f*
Dichtung *f*
Packung *f*
ES empaquetadura *f*
guarnición *f*
relleno *m*
F garniture *f*
I dispositivo *m* di
tenuta
guarnizione *f*
MA tümítés
NL pakking *f*
PO opakowanie *n*
szczeliwo *n*
uszczelka *f*
PY набивка *f*
прокладка *f*
упаковка *f*
уплотнение *n*
SU tiiviste
SV packning *c*

2904 packing density
DA pakningstæthed
DE Packungsdichte *f*
ES densidad de
empaquetadura *f*
densidad de
relleno *f*
F densité *f* de
garniture
I densità *f* di
guarnizione
MA tümési sűrűség
NL pakkingsdichtheid *f*
PO gęstość *f* zapisu
spoistość *f*
uszczelnienia
PY плотность *f*
набивки
SU pakkaustiiviys
SV packningstäthet *c*

2905 packing material
DA pakmateriale
DE Dichtungsmaterial *n*
ES material de
empaquetadura *m*
F matériel *m* de
garniture
I materiale *m* di
guarnizione
MA tümítöanyag
NL pakkingmateriaal *n*
PO materiał *m*
uszczelniający
PY набивочный
материал *m*
упаковочный
материал *m*
SU tiivistysaine
SV tätningsmaterial *n*

2906 packing ring
DA foringsring
DE Dichtungsring *m*
ES anillo de
estanqueidad *m*
F anneau *m* de
garniture
I anello *m* di
guarnizione
MA tümítőgyűrű
NL pakkingring *m*
PO pierścień *m*
uszczelniający
PY уплотнительное
кольцо *n*
SU tiivisterengas
SV tätningsring *c*

2907 packless
DA uden pakning
DE ohne Packung *f*
ES sin empaquetadura *f*
F non équipé
I senza riempimento
MA tümítetlen
NL pakking(bus)loos
PO bez uszczelnienia
PY неплотность *f*
SU tiivisteetün
SV packboxlüs

2908 paint mist
DA farvetåge
DE Farbnebel *m*
ES neblina de pintura *f*
niebla de pintar al
duco *f*
F brouillard *m* de
peinture
I vernice *f* nebulizzata
MA festékküd
színes füst
NL verfnevel *m*
PO mgła *f* lakiernicza
PY распыленная
краска *f*
SU maalisumu
SV färgdimma *c*

335

2909 paint spraying booth
DA sprøjtemalekabine
DE Farbspritzkabine f
ES cabina para pintar a pistola f
F cabine m de peinture
I cabina f di verniciatura
MA festékszóró fülke
NL spuitcabine f (verf)
PO komora f malowania natryskowego
PY окрасочная камера f
SU ruiskumaalauskoppi
SV sprutbox c

2910 pane
DA plade rude
DE Scheibe f
ES cristal (de ventana) m entrepaño m
F vitre f
I vetrata f
MA tábla
NL ruit f
PO szyba f
PY оконное стекло n
SU ikkunalasi
SV glasruta c

2911 panel
DA panel
DE Schalttafel f Tafel f
ES panel m
F panneau m
I pannello m
MA panel tábla
NL paneel n schakelbord n
PO płyta f tablica f
PY панель f плита f
SU paneeli
SV panel c

2912 panel cooling
DA panelkøling
DE Flächenkühlung f
ES enfriamiento por paneles m
F refroidissement m par panneau
I raffreddamento m a pannelli
MA panelhűtés
NL koeling f door middel van panelen
PO chłodzenie n płaszczyznowe
PY панельное охлаждение n
SU paneelijäähdytys
SV panelkylning c

2913 panel heating
DA panelopvarmning
DE Flächenheizung f
ES calefacción por paneles f
F chauffage m par panneau
I riscaldamento m a pannelli
MA panelfűtés
NL paneelverwarming f stralings- verwarming f
PO ogrzewanie n płaszczyznowe
PY панельное отопление n
SU paneelilämmitys
SV panelvärmning c

2914 panel radiator
DA panelradiator
DE Plattenradiator m
ES panel radiante m
F radiateur m panneau
I radiatore m a piastra
MA lapradiátor
NL paneelradiator m plaatradiator m
PO grzejnik m płaszczyznowy grzejnik m płytowy
PY панельный радиатор m
SU paneelipatteri
SV panelradiator c

2915 parallel blade damper
DA jalousispjæld
DE Jalousieklappe f
ES compuerta de lamas paralelas f
F registre m à volets parallèles
I serranda f a lame parallele
MA párhuzamos műküdésű zsalu
NL luchtklep f met parallelschoepen
PO przepustnica f z łopatkami współbieżnymi
PY клапан m с параллельными лопатками
SU säätüpelti yhdensuuntaisin sälein
SV jalusispjäll n

2916 parallel connection
DA parallelforbindelse
DE Parallelschaltung f
ES conexión en paralelo f
F couplage m en parallèle raccordement m parallèle
I collegamento m parallelo
MA párhuzamos kapcsolás
NL parallelschakeling f
PO połączenie n równoległe
PY параллельное присоединение n параллельное соединение n
SU rinnankytkentä
SV parallellkoppling c

2917 parallel flow
DA parallelstrømning
DE Parallelstrümung *f*
ES flujo
    equicorriente *m*
F circulation *f* à
    courants parallèles
    courant *m* parallèle
I flusso *m* parallelo
MA párhuzamos áramlás
NL gelijkstroom *m*
    parallelstroming *f*
PO przepływ *m*
    równoległy
PY параллельное
    течение *n*
SU rinnakkainen virtaus
SV parallellstrüm *c*

2918 parallel flow burner
DA parallelstrøms-
    brænder
DE Parallelstrom-
    brenner *m*
ES horno de flujo
    equicorriente *m*
    quemador de flujo
    equicorriente *m*
F brûleur *m* à
    courants parallèles
I bruciatore *m* a
    correnti parallele
MA párhuzamos
    áramlású égő
NL parallelstromings-
    brander *m*
PO palnik *m* z
    przepływem
    równoległym
PY форсунка *f* с
    параллельными
    потоками
SU kaksoispoltin
    rinnakkaisvirtaus-
    poltin
SV parallellflüdes-
    brännare *c*

2919 parameter
DA parameter
DE Kenngrüße *f*
    Parameter *m*
ES parámetro *m*
F paramètre *m*
I parametro *m*
MA jellemző
    paraméter
NL parameter *m*
PO parametr *m*
PY параметр *m*
SU parametri
SV parameter *c*

2920 partial pressure
DA partialtryk
DE Partialdruck *m*
ES presión parcial *f*
F pression *f* partielle
I pressione *f* parziale
MA parciális nyomás
    résznyomás
NL partíle druk *m*
PO ciśnienie *n*
    cząstkowe
PY парциальное
    давление *n*
SU osapaine
SV partiellt tryck *n*

2921 particle
DA fnug
    partikel
DE Partikel *n*
    Teilchen *n*
ES partícula *f*
F particule *f*
I particella *f*
MA részecske
NL deeltje *n*
PO cząsteczka *f*
    cząstka *f*
PY частица *f*
SU hiukkanen
SV partikel *c*

2922 particle size
DA kornstørrelse
DE Teilchengrüße *f*
ES tamaño de las
    partículas *m*
F taille *f* de particule
I grandezza *f* della
    particella
MA részecskeméret
NL deeltjesgrootte *f*
PO wymiar *m* cząstki
PY размер *m* частицы
SU hiukkaskoko
SV partikelstorlek *c*

2923 part load
DA dellast
DE Teillast *f*
ES carga fraccionada *f*
    carga incompleta *f*
    carga reducida *f*
F charge *f* partielle
I carico *m* parziale
MA részterhelés
NL deellast *m*
PO obciążenie *n*
    częściowe
PY неполная
    нагрузка *f*
SU osakuorma
SV dellast *c*

2924 pattern
DA model
    mønster
DE Modell *n*
    Muster *n*
ES esquema *m*
    modelo *m*
    patrón *m*
F modèle *m*
I campione *m*
    modello *m*
MA minta
NL indeling *f*
    model *n*
    patroon *n*
PO model *m*
    próbka *f*
    wzór *m*
PY модель *f*
    образец *m*
    схема *f*
    шаблон *m*
SU kuvio
SV münster *n*

2925 peak
DA spids
DE Spitze *f*
ES punta *f*
F pointe *f*
I picco *m*
    punta *f*
MA csúcs
NL piek *f*
    spits *f*
PO szczyt *m*
    wartość *f* szczytowa
PY пик *m*
SU huippu
SV topp *c*

2926 peak load
DA spidslast
DE Spitzenlast *f*
ES carga punta *f*
F charge *f* de pointe
I carico *m* di punta
MA csúcsterhelés
NL piekbelasting *f*
    spitsbelasting *f*
PO obciążenie *n*
    szczytowe
PY пиковая нагрузка *f*
SU huippukuorma
    kulutushuippu
SV toppbelastning *c*

2927 peat
DA tørv
DE Torf *m*
ES turba *f*
F tourbe *f*
I torba *f*
MA tőzeg
NL turf *m*
PO torf *m*
PY торф *m*
SU turve
SV torv *c*

2928 pedestal bearing
DA sokkelbæring
DE Bocklager *n*
Stehlager *n*
ES soporte de cojinete *m*
F palier *m* support auxiliaire
I cuscinetto *m* di basamento
MA csúszócsapágy
NL steunlager *n*
PO łożysko *n* stojące (oczkowe)
PY опорный подшипник *m*
SU laakeri
SV konsollager *n*

2929 peep hole
DA inspektionshul skuehul
DE Schauloch *n*
ES mirilla (de inspección) *f*
F regard *m* de visite
I foro *m* di ispezione
MA kémlelőnyílás
NL kijkgat *n*
PO wziernik *m*
PY смотровой люк *m*
SU tarkastuslasi
SV tittglugg *c*

2930 peer-to-peer communication system
DA protokol (EDB)
DE Gleich-zu-Gleich-Kommunikations-system *n*
ES sistema de comunicación de igual a igual *m*
F communication *f* sur systèmes de même nature
I sistema *m* di comunicazione da pari a pari
MA egyenrangú kapcsolatrendszer
NL communicatie *f* tussen gelijkwaardige deelnemers
PO system *m* komunikacji obrazowej
PY система *f* обратной связи
SU yhteensopiva tietoliikennejärje-stelmä
SV

2931 Peltier effect
DA Peltier effekt
DE Peltier-Effekt *m*
ES efecto Peltier *m*
F effet *m* Peltier
I effetto *m* Peltier
MA Peltier effektus
NL Peltiereffect *n*
PO efekt *m* Peltiera
PY эффект *m* Пельтье
SU Peltier-ilmiü
SV Peltier-effekt *c*

2932 perfect combustion
DA fuldstændig forbrænding
DE vollkommene Verbrennung *f*
ES combustión perfecta *f*
F combustion *f* parfaite
I combustione *f* completa
MA tükéletes égés
NL stoichoimetrische verbranding *f* verbranding *f* met theoretische luchthoeveelheid
PO spalanie *n* całkowite spalanie *n* zupełne
PY полное сгорание *n*
SU täydellinen palaminen
SV total fürbränning *c*

2933 perforated *adj*
DA perforeret
DE durchlüchert
ES perforado
F perforé
I perforato
MA lyukacsos perforált
NL geperforeerd
PO perforowany
PY перфорированный
SU rei'itetty
SV perforerad

2934 perforated ceiling
DA perforeret loft
DE Lochdecke *f*
ES cielo raso perforado *m* techo perforado *m*
F plafond *m* perforé
I soffitto *m* forato
MA perforált mennyezet
NL geperforeerd plafond *n*
PO strop *m* perforowany
PY перфорированный потолок *m*
SU rei'itetty katto
SV perforerat tak *n*

DA =	Danish
DE =	German
ES =	Spanish
F =	French
I =	Italian
MA =	Magyar (Hungarian)
NL =	Dutch
PO =	Polish
PY =	Russian
SU =	Finnish
SV =	Swedish

2935 perforated grille
DA perforeret rist
DE Lochgitter *n*
ES rejilla perforada *f*
F grille *f* perforée
I griglia *f* a lamiera
forata
MA perforáltlemezes rács
NL geperforeerd
rooster *m*
PO kratka *f*
perforowana
PY перфорированная
решетка *f*
SU rei'itetty hajottaja
SV perforerat galler *n*

2936 perforated sheet
DA perforeret tyndplade
DE Lochblech *n*
ES chapa perforada *f*
placa perforada *f*
F plaque *f* perforée
I lamiera *f* forata
MA perforáltlemez
NL geperforeerde
plaat *f*
PO płyta *f* perforowana
PY перфорированная
плита *f*
SU rei'itetty levy
SV perforerad plåt *c*

2937 perforation
DA gennemhulning
perforering
DE Durchbohrung *f*
Perforation *f*
ES perforación *f*
F perforation *f*
I perforazione *f*
MA perforálás
NL perforatie *f*
PO perforacja *f*
PY перфорация *f*
SU rei'itys
SV perforering *c*

2938 performance
DA ydelse
DE Leistung *f*
ES rendimiento *m*
F performance *f*
I prestazione *f*
MA teljesítmény
NL prestatie *f*
werking *f*
PO charakterystyka *f*
pracy
działanie *n*
wydajność *f* pracy
PY действие *n*
исполнение *n*
SU suorituskyky
toiminta
SV utfürande *n*

**2939 performance
coefficient**
DA ydelseskoefficient
DE Leistungs-
koeffizient *m*
ES coeficiente de
eficacia
energética *m*
coeficiente de
rendimiento *m*
F coefficient *m* de
performance (cop)
I coefficiente *m* di
resa
MA teljesítménytényező
NL rendement *n*
PO współczynnik *m*
prawidłowości
działania
PY коэффициент *m*
полезного
действия (кпд)
SU tehokerroin
SV effektivitetsfaktor *c*

2940 periodic *adj*
DA periodisk
DE periodisch
ES periódico
F périodique
I periodico
MA időszakos
periodikus
NL periodiek
PO okresowy
PY периодический
SU jaksottainen
SV periodisk

2941 peripheral *adj*
DA periferisk
DE peripher
ES periférico
F circonférentiel
périphérique
tangentiel
I periferico
MA kerületi
periférikus
NL aan de omtrek *m*
gelegen
perifeer
PO obwodowy
peryferyjny
PY периферийный
SU lisä-
oheis-
SV periferisk

2942 permeability
DA gennemtrængelighed
permeabilitet
DE Durchlässigkeit *f*
ES permeabilidad *f*
F perméabilité *f*
I permeabilità *f*
MA átbocsátóképesség
NL doorlaatbaarheid *f*
PO przenikalność *f*
przepuszczalność *f*
PY проницаемость *f*
SU läpäisevyys
permeabiliteetti
SV permeabilitet *c*

2943 permitted duty *(see*
permitted load)

2944 permitted load
DA tilladt last
DE Genehmigungs-
leistung *f*
ES carga admisible *f*
F charge *f* admissible
I carico *m*
ammissibile
MA megengedett terhelés
NL toelaatbare
belasting *f*
PO obciążenie *n*
dopuszczalne
PY допустимая
нагрузка *f*
SU sallittu kuormitus
SV tillåten belastning *c*

2945 perspective drawing
- DA perspektivtegning
- DE perspektivische Zeichnung *f*
- ES plano de perspectiva *m*
- F dessin *m* en perspective
- I disegno *m* in prospettiva
- MA perspektivikus rajz
- NL perspectief tekening *f*
- PO rysunek *m* perspektywiczny
- PY изображение *n* в перспективе
- SU perspektiivinen piirustus
- SV perspektivritning *c*

2946 phase
- DA fase
- DE Phase *f*
- ES fase *f*
- F phase *f*
- I fase *f*
- MA fázis
- NL fase *f*
- PO faza *f*
- PY фаза *f*
- SU vaihe
- SV fas *c*

2947 PI control
- DA PI-regulering
- DE PI-Regler *m* Proportional-Integral-Regler *m*
- ES control proporcional-integral *m*
- F régulation *f* proportionnelle-intégrale (PI)
- I regolazione *f* proporzionale-integrale
- MA PI szabályozás
- NL PI-regelaar *m* PI-regeling *f*
- PO regulacja *f* proporcjonalno-całkująca
- PY регулирование *n* по точке
- SU PI-säätü
- SV PI-reglering *c*

2948 pier (building)
- DA mole stræbepille
- DE Pfeiler *m*
- ES pilar (edit) *m*
- F pilier *m*
- I banchina *f* pilastro *m* pilone *m*
- MA pillér
- NL penant *n*
- PO filar *m*
- PY свая *f* (в строительстве)
- SU pilari
- SV brygga *c*

2949 pilot light
- DA kontrollampe
- DE Kontrollampe *f*
- ES lámpara piloto *f*
- F lumière *f* de contrôle
- I spia *f* luminosa
- MA őrláng
- NL signaallampje *n* waakvlam *f*
- PO płomyk *m* stały zapalający światło *n* kontrolne
- PY запальный факел *m*
- SU merkkivalo
- SV signallampa *c*

2950 pipe
- DA rør
- DE Leitung *f* Rohr *n*
- ES cañería *f* tubería *f* tubo *m*
- F tube *m* tuyau *m*
- I condotta *f* tubazione *f* tubo *m*
- MA cső
- NL buis *f* leiding *f* pijp *f*
- PO rura *f*
- PY труба *f*
- SU putki
- SV rür *n*

2951 pipe bend
- DA rørbøjning
- DE Rohrbogen *m*
- ES codo de tubo *m* curva de tubo *f*
- F coude *m* de tuyauterie
- I curva *f* di tubo
- MA csőív
- NL pijpbocht *f*
- PO łuk *m* rury
- PY колено *n* трубы отвод *m*
- SU putkikaari
- SV rürbüj *c*

2952 pipe bracket
- DA rørkonsol
- DE Rohrbefestigung *f*
- ES abrazadera (tubos) *f* collar de fijación (tubo) *m*
- F collier *m*
- I collare *m* di tubazione
- MA csőbilincs
- NL pijpsteun *m*
- PO wspornik *m* rury
- PY фланец *m* на трубе
- SU putkipidin
- SV rürklammer *c*

2953 pipe coil
- DA rørslange rørspiral
- DE Rohrschlange *f*
- ES serpentín *m*
- F serpentin *m*
- I serpentino *m*
- MA csőkígyó
- NL pijpregister *n* pijpspiraal *f*
- PO wężownica *f*
- PY змеевик *m*
- SU putkikierukka
- SV rürslinga *c*

2954 pipe diameter
- DA rørdiameter
- DE Rohrdurchmesser *m*
- ES diámetro del tubo *m*
- F diamètre *m* de tube
- I diametro *m* del tubo
- MA csőátmérő
- NL leidingdiameter *m* pijpdiameter *m*
- PO średnica *f* rury
- PY диаметр *m* трубы
- SU putken halkaisija
- SV rürdiameter *c*

2955 pipe elbow
DA knærør
DE Kniestück *n*
ES codo *m*
F coude *m*
I raccordo *m* a
   gomito
MA csökünyük
NL kniestuk *n* (van
   pijp)
PO kolano *n* rury
PY колено *n* трубы
SU putkikulma
SV rürbüj *c*

2956 pipe friction
DA rørfriktion
DE Rohrreibung *f*
ES rozamiento dentro
   del tubo *m*
F perte *f* de charge
   dans un tube
I attrito *m* del tubo
MA csősurlódás
NL leidingweerstand *m*
PO tarcie *n* w rurze
PY трение *n* в трубе
SU putkikitka
SV rürfriktion *c*

2957 pipe hanger
DA rørbærer
DE Rohraufhängung *f*
ES cuelgatubos *m*
   percha para tubos *f*
F collier *m* de
   tuyauterie
   support *m* de
   tuyauterie
I mensola *f* per tubo
   supporto *m* per
   tubo
MA csőfüggesztő
NL pijpbeugel *m*
   pijpophanging *f*
PO wieszak *m* rury
PY крюк *m* для трубы
SU putkikannake
SV rürupphängning *c*

2958 pipeline
DA rørledning
DE Rohrleitung *f*
ES tubería de
   conducción (para
   líquidos y gases) *f*
F pipe-line *f*
   tuyauterie *f*
I rete *f*
   tubazione *f*
MA csővezeték
NL pijpleiding *f*
PO przewód *m* rurowy
PY трубопровод *m*
SU putkijohto
SV rürledning *c*

2959 pipe network
DA rørnet
DE Rohrnetz *n*
ES red de tuberías *f*
F tuyauterie *f*
I rete *f* di tubazioni
MA csőhálózat
NL leidingnet *n*
PO sieć *f* przewodów
   rurowych
PY сеть *f*
   трубопроводов
SU putkisto
SV rürnät *n*

2960 pipe section
DA rørstykke
DE Rohrabschnitt *m*
ES sección de tubos *f*
   tramo de tubería *m*
F tronçon *m* de tube
I sezione *f* di tubo
   spezzone *m* di tubo
MA csőszakasz
NL leidinggedeelte *n*
PO odcinek *m* rury
   przekrój *m* rury
PY секция *f* трубы
   участок *m* трубы
SU putken
   poikkileikkaus
SV rürsektion *c*

2961 pipe sizing
DA rørdimensionering
DE Rohrdimensionier-
   ung *f*
ES dimensionado de
   tubos *m*
F dimensionnement *m*
   de tuyauterie
I dimensionamento *m*
   del tubo
MA csőméretezés
NL leidingberekening *f*
PO wymiarowanie *n*
   rury
PY измерение *n* трубы
SU putkiston mitoitus
SV rürdimensionering *c*

2962 pipe support
DA rørunderstøtning
DE Rohraufhängung *f*
ES soportatubos *m*
   soporte para
   tubos *m*
F support *m* de tube
I supporto *m* per
   tubo
MA csőtámasz
NL pijpondersteuning *f*
PO podpora *f* rury
PY опора *f* для трубы
SU putken tuki
SV rürstüd *n*

2963 pipe system
DA rørsystem
DE Rohrsystem *n*
ES sistema de
   tuberías *m*
F système *m* de
   tuyauteries
I rete *f* di tubazioni
MA csőrendszer
NL leidingsysteem *n*
PO system *m* rurowy
PY система *f*
   трубопроводов
SU putkijärjestelmä
SV rürsystem *n*

2964 pipe thread
DA rørgevind
DE Rohrgewinde *n*
ES rosca para tubos *f*
F filetage *m* de tubes
I filetto *m* del tubo
MA csőmenet
NL pijpdraad *n*
PO gwint *m* rurowy
PY трубная резьба *f*
SU putkikierre
SV rürgänga *c*

2965 pipe tracer
DA rør-fejlsøger
DE Rohrsuchgerät *n*
ES buscador de
    tubos *m*
    buscatubos *m*
    trazador (para
    buscar tubos) *m*
F traceur *m* de tubes
I tracciante *m* per
    tubi
MA csőkereső
NL curvimeter *m*
    leidingzoek-
    apparaat *n*
PO urządzenie *n* do
    wykrywania rur
PY трубный
    разметчик *m*
SU putkenetsintälaite
SV felsökare *c* för rör *n*

2966 piston
DA stempel
DE Kolben *m*
ES émbolo *m*
    pistón *m*
F piston *m*
I pistone *m*
MA dugattyú
NL plunjer
    zuiger *m*
PO tłok *m*
PY поршень *m*
SU mäntä
SV kolv *c*

**2967 piston
    displacement**
DA cylindervolumen
DE Kolben-
    verdrängung *f*
ES desplazamiento del
    pistón *m*
F cylindrée *f*
I cilindrata *f*
MA dugattyúelmozdulás
NL zuigerslagvolume *n*
    zuigerverplaatsing *f*
PO objętość *f* skokowa
    cylindra
PY замена *f* поршня
SU iskutilavuus
SV slagvolym *c*

2968 piston flow
DA stempelstrømning
DE Kolbenstrümung *f*
ES flujo *m*
F
I flusso *m* a pistone
MA dugattyúszerű
    áramlás
NL verdringings-
    stroming *f*
PO przepływ *m* tłokowy
PY двухфазный
    поток *m*
SU mäntävirtaus
SV pluggflüde *n*

2969 pit
DA grube
DE Grube *f*
ES foso *m*
    pozo *m*
F fosse *f*
I fossa *f*
MA akna
    güdür
NL mijn *f*
    put *m*
    werkkuil *m*
PO dół *m*
    wykop *m*
PY шахта *f*
SU kuoppa
SV grop *c*

2970 pitch
DA afstand
    beg
DE Abstand *m*
ES paso *m*
    separación *f*
F distance *f*
    d'espacement
    pas *m*
I passo *m*
MA dőlés
    osztás
NL pek *n*
    spoed *m*
    steek *m*
    verdiepingshoogte *f*
PO skok *m* (gwintu)
    wysokość *f* tonu
PY шаг *m* (резьбы)
SU jako
    kaltevuus
SV delning *c*

2971 Pitot tube
DA pitotrør
DE Pitot-Rohr *n*
    Staurohr *n*
ES tubo de Pitot *m*
F tube *m* de pitot
I tubo *m* Pitot
MA Pitot cső
NL Pitotbuis *f*
PO rurka *f* Pitota
PY трубка *f* Пито
SU Pitot-putki
SV pitotrür *n*

2972 pitting
DA grubetæring
DE Lochfraß *m*
    punktfürmige
    Korrosion *f*
ES corrosión alveolar
    superficial *f*
    corrosión puntual *f*
    picadura *f*
F corrosion *f*
    ponctuelle
    pitting *m*
I corrosione *f*
    puntiforme
MA pontkorrózió
NL putcorrosie *f*
PO korozja *f* punktowa
    korozja *f* wżerowa
PY точечная
    коррозия *f*
SU pistekorroosio
SV punktfrätning *c*

2973 plan
DA plan
    tegning
DE Entwurf *m*
    Plan *m*
ES plan *m*
    plano *m*
    proyecto *m*
F plan *m*
I disegno *m*
    piano *m*
    pianta *f*
MA rajz
    terv
NL plan *n*
    plattegrond *m*
    schets *f*
PO plan *m*
    rzut *m* poziomy
PY план *m*
SU pohjapiirros
SV plan *n*

2974 plant
DA anlæg
   værk
DE Anlage *f*
   Betriebsanlage *f*
   Fabrikanlage *f*
ES instalación *f*
   planta *f*
F centrale *f*
   usine *f*
I impianto *m*
MA berendezés
   telep
NL bedrijf *n*
   installatie *f*
PO instalacja *f*
   urządzenie *n*
PY завод *m*
   силовая
      установка *f*
SU laitos
SV anläggning *c*

2975 plant cost
DA anlægsudgift
DE Anlagekosten *f,pl*
ES coste de la
      instalación *m*
F coût *m* de centrale
I costo *m* di impianto
MA berendezés kültsége
NL bedrijfskosten *pl*
   installatiekosten *pl*
PO koszt *m* instalacji
   koszt *m* urządzeń
PY стоимость *f*
   установки
SU asennuskustannus
SV anläggnings-
      kostnad *c*

2976 plant room
DA maskinrum
DE Betriebsraum *m*
   Maschinenraum *m*
ES sala de herrametal *f*
   sala de maquinaria *f*
   sala de material *f*
F local *m* de centrale
I sala *f* impianti
MA gépház
NL machinekamer *f*
   technische ruimte *f*
PO maszynownia *f*
   pomieszczenie *n*
      instalacyjne
PY помещение *n* для
      установки
SU konehuone
SV maskinrum *n*

2977 plastic
DA plast
DE Kunststoff *m*
ES material plástico *m*
   plástico *m*
F plastique *m*
I plastica *f*
MA műanyag
NL kunststof *f*
   plastic *n*
PO tworzywo *n* sztuczne
PY пластмасса *f*
SU muovi
SV plast *c*

2978 plastic foam
DA plastskum
DE Kunststoffschaum *m*
ES espuma plástica *f*
   plástico celular *m*
F plastique *m* expansé
      (isolation)
I resina *f* plastica
      schiumosa
MA műanyaghab
NL schuimplastic *n*
PO pianka *f* z tworzywa
      sztucznego
PY вспененная
      пластмасса *f*
SU vaahtomuovi
SV skumplast *c*

2979 plastic pipe (*see*
   plastic tube)

2980 plastic tube
DA plastrør
DE Kunststoffleitung *f*
   Kunststoffrohr *n*
ES tubería de plástico *f*
F tube *m* en matière
      plastique
I tubo *m* di plastica
MA műanyag cső
NL kunststofpijp *f*
   plastic pijp *f*
PO rura *f* z tworzywa
      sztucznego
PY пластмассовая
      труба *f*
SU muoviputki
SV plastrür *n*

2981 plate
DA lamel
   plade
DE Lamelle *f*
   Platte *f*
ES chapa *f*
   placa (metálica) *f*
F plaque *f* (métal)
I piastra *f*
MA lap
   lemez
NL plaat *f*
PO płyta *f*
   płytka *f*
PY пластина *f*
   плита *f*
SU levy
SV platta *c*
   plåt *c*

**2982 plate cooling
   tower packing**
DA pakning til
      pladekøletårn
DE Kühlturm-
      Plattenberiese-
      lungseinbau *m*
ES relleno de placas de
      torre de
      refrigeración *m*
F plaques *f* de
      remplissage de
      tour de
      refroidissement
I torre *f* di
      raffreddamento a
      piatti
MA lemezes hűtőtorony
      tültet
NL plaatvormige
      koeltorenvulling *f*
PO wieża *f* chłodnicza
      płytowa
PY градирня *f* с
      пластинчатой
      насадкой
SU jäähdytystornin
      täytelevyt
SV packning *c* till
      plattkyltorn *n*

DA	=	Danish
DE	=	German
ES	=	Spanish
F	=	French
I	=	Italian
MA	=	Magyar (Hungarian)
NL	=	Dutch
PO	=	Polish
PY	=	Russian
SU	=	Finnish
SV	=	Swedish

2983 plate fin
DA pladefinne
DE flache Kühlrippe *f*
ES aleta plana *f*
F plaque *f* ailette
I alettatura *f* a piastra
MA lemezborda
NL plaatlamel *f*
PO lamelka *f*
żeberko *n* płytowe
PY пластинчатое
ребро *n*
SU ripalevy
SV plåtfläns *c*

2984 plate freezer
DA pladefryser
DE Plattengefrierer *m*
ES congelador de placas *m*
F congélateur *m* à plaques
I congelatore *m* a piatto
MA laphűtő
NL platenvrieselement *n*
PO zamrażarka *f* płytowa
PY пластинчатый морозильный аппарат *m*
SU levypakastin
SV plattkylare *c*

2985 plate liquid cooler
DA plade-væske køler
DE Plattenflüssigkeits-kühler *m*
ES enfriador de placas *m*
F refroidisseur *m* à plaques
I refrigeratore *m* di liquido a piastra
MA lemezes folyadékhűtő
NL platenvloeistof-koeler *m*
PO płytowy ochładzacz *m* płynu
PY водо-водяной пластинчатый охладитель *m*
SU levymäinen jäähdytin
SV platt-vätskekylare *c*

2986 plate-type condenser
DA pladekondensator
DE Platten-kondensator *m*
ES condensador de placas *m*
F condensateur *m* à plaques
I condensatore *m* a piastra
MA lemezes típusú kondenzátor
NL platencondensor *m*
PO skraplacz *m* płytowy
PY пластинчатый конденсатор *m*
SU levylauhdutin
SV platt-typs kondensor *c*

2987 plate-type evaporator
DA pladeevaporator
DE Platten-verdampfer *m*
ES evaporador de placas *m*
F évaporateur *m* à plaques
I evaporatore *m* a piastra
MA lemezes típusú elpárologtató
NL platenverdamper *m*
PO parownik *m* płytowy
PY пластинчатый испаритель *m*
SU levyhüyrystin
SV platt-typs evaporator *c*

2988 PLC (*see* programmable logic control)

2989 plenum
DA fælles
DE Plenum *n* vollkommen ausgefüllter Raum *m*
ES cámara de distribución de aire *f* cámara de mezcla de aire *f* cámara de sobre-presión *f*
F plénum *f* plenum *m*
I camera *f* di pressione plenum *m*
MA légtér
NL holle ruimte *f* plenum *n*
PO komora *f* wyrównawcza plenum *n*
PY нагнетательный *adj* область *f* повышенного давления
SU kammio
SV samlings- *c*

2990 plug
DA stikprop
DE Stecker *m* Verschluß *m* Zapfen *m*
ES bujía *f* enchufe macho *m* obturador *m* tapón macho *m*
F tampon *m* prise *f* de courant bouchon *m*
I otturatore *m* spina *f* tappo *m*
MA dugó
NL plug *f* stekker *m* stop *m*
PO korek *m* wtyczka *f* (elektr.) zaślepka *f*
PY заглушка *f* пробка *f*
SU tulppa
SV propp *c*

2991 plug *vb*
DA tilproppe
DE verstopfen
zustopfen
ES taponar
F tamponner
I otturare
sigillare
MA dugaszol
NL afstoppen
dichtstoppen
PO korkować
zatykać
PY закупоривать
SU tulpata
SV proppa

2992 plug in *vb*
DA indkoble
indstikke
DE einschalten
einstecken
ES enchufar
(electricidad)
F bouchonner
tamponner
I innestare
inserire
tamponare
tappare (elettricità)
MA bedugaszol
(elektromosan)
NL insteken
PO wetknąć
wtykać
PY разъемный *adj*
штепсельный *adj*
SU kytkeä
SV koppla in

2993 plume
DA fjer
røgfane
DE Rauchfahne *f*
ES penacho de
chimenea *m*
penacho de
humos *m*
F panache *m* de fumée
I pennacchio *m* di
fumo
MA füstcsóva
NL rookpluim *f*
PO pióropusz *m* dymu
PY факел *m*
SU savuvana
(konvektio)
SV rükfana *c*

2994 plume dispersal
DA spredning af røgfane
DE Rauchverteilung *f*
ES dispersión de
humos *f*
F dispersion *f* des
fumées
I dispersione *m* dei
fumi
MA szétterjedő füstcsóva
NL rookverspreiding *f*
PO rozrzut *m*
pióropusza dymu
PY распыливание *n*
факелом
SU savuvanan
leviäminen
SV rükspridning *c*

2995 plume rise
DA stighøjde for røgfane
DE Rauchhühe *f*
ES altura de penachos
de humos *f*
F hauteur *f* du
panache de fumée
I sopraelevazione *m*
del pennacchio
MA felszálló füstcsóva
NL rookpluimhoogte *f*
PO uniesienie *n*
pióropusza dymu
PY подъем *m* факела
SU savuvanan
nousukorkeus
SV stighöjd *c* för
rökfana *c*

2996 pneumatic *adj*
DA pneumatisk
DE pneumatisch
ES neumático
F pneumatique
I pneumatico
MA pneumatikus
NL luchtgedreven
pneumatisch
PO pneumatyczny
PY пневматический
SU pneumaattinen
SV pneumatisk

**2997 pneumatic electric
control**
DA pneumatisk el-
styring
DE elektro-
pneumatische
Regelung *f*
ES control neumático-
eléctrico *m*
F régulation *f* électro-
magnétique
I regolazione *f*
elettro-pneumatica
MA elektropneumatikus
szabályozás
NL pneumatisch-
elektrische
regelaar *m*
pneumatisch-
elektrische
regeling *f*
PO sterowanie *n*
elektro-
pneumatyczne
PY пневмо-
электрическое
регулирование *n*
SU sähküpneumaattinen
säätü
SV pneumatisk elektrisk
kontroll *c*

2998 point
DA punkt
DE Punkt *m*
ES punto *m*
F point *m*
I punto *m*
MA pont
NL punt *m*
uiteinde *n*
PO ostrze *n*
punkt *m*
PY точка *f*
SU piste
SV punkt *c*

2999 poker
DA ildtang
DE Feuerhaken *m*
Schüreisen *n*
ES atizador *m*
F ringard *m*
I attizzatoio *m*
MA tűzpiszkáló
NL pook *m*
PO pogrzebacz *m*
PY кочерга *f*
шуровочный
лом *m*
SU kohennustanko
SV eldgaffel *c*

3000 pollutant
DA forurener
DE Verschmutzer *m*
ES contaminante *m*
F polluant *m*
I contaminante *m*
   inquinante *m*
MA szennyezőanyag
NL verontreiniger *m*
   vervuiler *m*
PO substancja *f*
   zanieczyszczająca
   środowisko
PY загрязняющий
   фактор *m*
SU saaste
SV fürorening *c*

3001 pollutant *adj*
DA forurenet
DE verschmutzend
ES contaminante
F polluant
I contaminante
   inquinante
MA szennyező
NL verontreinigend
   vervuilend
PO zanieczyszczający
   (środowisko)
PY загрязняющий
SU saasteinen
SV fürorenande

3002 pollution
DA forurening
DE Verunreinigung *f*
ES contaminación *f*
F pollution *f*
I inquinamento *m*
MA szennyezés
NL verontreiniging *f*
   vervuiling *f*
PO zanieczyszczenie *n*
   (środowiska)
PY загрязнение *n*
SU saastuminen
SV fürorening *c*

3003 polytropic *adj*
DA polytropisk
DE polytropisch
ES politrópico
F polytropique
I politropico
MA politropikus
NL polytropisch
PO politropowy
PY политропный
SU polytrooppinen
SV polytropisk

3004 port
DA havn
   åbning
DE Öffnung *f*
ES abertura (de válvula)
   f
F port *m*
I porta *f*
MA torkolat
NL doorlaat *m*
   opening *f*
PO otwór *m* przelotowy
PY проем *m*
   проход *m*
SU portti
   sisääntulo
SV üppning *c*

3005 portable *adj*
DA bærbar
   transportabel
DE tragbar
ES portátil
F portatif
   transportable
I portatile
MA hordozható
NL draagbaar
PO przenośny
PY переносной
   портативный
SU siirrettävä
SV portabel

3006 position
DA indstilling
   stilling
DE Lage *f*
   Stellung *f*
ES posición *f*
   situación *f*
F position *f*
   situation *f*
I posizione *m*
MA helyzet
NL houding *f*
   ligging *f*
   positie *f*
   stand *m*
PO położenie *n*
   pozycja *f*
PY положение *n*
   расположение *n*
SU asema
SV läge *n*

3007 positioning relay
DA indstillingsrelæ
DE Positionierungs-
   relais *n*
ES relé posicionador *m*
F relai *m* de
   positionnement
I rel *m* di
   posizionamento
MA helyzetjelző relé
NL stelrelais *n*
PO przekaźnik *m*
   pozycyjny
PY реле *n* положения
SU rajakytkin
SV inställningsrelä *n*

3008 positive-displacement compressor
DA fortrængnings-
   kompressor
DE Positivverdrängungs-
   kompressor *m*
ES compresor de
   desplazamiento
   positivo *m*
F compresseur *m*
   volumétrique
I compressore *m*
   volumetrico
MA pozitív lükető
   kompresszor
NL compressor *m* met
   mechanische
   volumeverkleining
PO sprężarka *f*
   wyporowa
PY поршневой
   прямоточный
   компрессор *m*
SU syrjäytyskompressori
SV positiv
   deplacement-
   kompressor *c*

3009 potable water
DA drikkevand
DE Trinkwasser *n*
ES agua potable *f*
F eau *f* potable
I acqua *f* potabile
MA ivóvíz
NL drinkwater *n*
PO woda *f* pitna
PY питьевая вода *f*
SU juotava vesi
SV dricksvatten *n*

3010 pot burner
DA fordampnings-
    brænder
    pottefyr
DE Topfbrenner *m*
    Verdampfungs-
    brenner *m*
ES quemador de
    crisol *m*
    quemador de
    horno *m*
F brûleur *m* à coupelle
    brûleur *m* à pot
I bruciatore *m* a
    pentola
MA serleges égő
NL bakbrander *m*
PO palnik *m* kuchenny
PY кухонная горелка *f*
SU hüyrystävä poltin
SV fürångnings-
    brännare *c*

3011 potential energy
DA potentiel energi
DE Lageenergie *f*
    Potentialenergie *f*
ES energía potencial *f*
F énergie *f* potentielle
I energia *f* potenziale
MA helyzeti energia
    potenciális energia
NL potentile energie *f*
PO energia *f*
    potencjalna
PY потенциальная
    энергия *f*
SU potentiaalienergia
SV potentiell energi *c*

3012 power
DA effekt
    kraft
DE Energie *f*
    Kraft *f*
    Leistung *f*
ES energía *f*
    fuerza *f*
    potencia *f*
F puissance *f*
I energia *f*
    potenza *f*
MA erő
    teljesítmény
NL kracht *f*
    vermogen *n*
PO moc *f*
PY мощность *f*
    энергия *f* для трубы
SU teho
SV effekt *c*
    kraft *c*

3013 power consumption
DA energiforbrug
    kraftforbrug
DE Leistungs-
    aufnahme *f*
ES consumo de
    energía *m*
    energía consumida *f*
F énergie *f*
    consommée
I consumo *m* di
    energia
MA energiafogyasztás
    teljesítményfelvétel
NL vermogensafname *f*
PO pobór *m* mocy
PY расход *m*
    мощности
    расход *m* энергии
SU tehontarve
SV kraftfürbrukning *c*

3014 power curve
DA effektkurve
DE Leistungskurve *f*
ES curva de potencia *f*
F courbe *f* de
    puissance
I curva *f* di potenza
MA teljesítménygürbe
NL vermogenskromme *f*
PO charakterystyka *f*
    mocy
    krzywa *f* mocy
PY кривая *f* мощности
SU tehokäyrä
SV effektkurva *c*

3015 power economy
DA energibesparelse
DE Energiewirtschaft *f*
ES ahorro de energía *m*
    economía
    energética *f*
F économie *f* d'énergie
I economia *f* di
    energia
MA energiagazdálkodás
NL vermogenshuis-
    houding *f*
PO oszczędność *f* mocy
PY экономия *f*
    мощности
    экономия *f* энергии
SU sähkütalous
SV energibesparing *c*

3016 powered *adj*
DA el-drevet
DE motorisch betrieben
ES potenciado
F actionné
    mécaniquement
I alimentato
MA táplált
NL aangedreven
PO napędzany
    (elektrycznie)
PY энерго-
    обеспеченный
SU
SV spänningsatt

3017 power failure
DA strømsvigt
DE Leistungsabfall *m*
ES fallo de potencia *m*
F panne *f* de courant
I interruzione *f*
    dell'erogazione di
    energia elettrica
MA energiahiány
NL spanningswegval *m*
PO przerwa *f* w
    dopływie energii
    elektrycznej
PY недостаток *m*
    мощности
SU tehonsyüttühäiriü
SV kraftbrist *c*

3018 power line carrier
DA transmission af
    styresignaler via
    effektledning
DE Starkstromleitung-
    sträger *m*
ES acometida de
    corriente *f*
F support *m* de ligne
    électrique
I linea *f* di
    alimentazione di
    potenza
MA távvezeték
    tartóoszlop
NL krachtstroom-
    kabel *m*
PO linia *f*
    elektro-
    energetyczna
PY энергопередающая
    линия *f*
SU verkkokäsky
SV eltransport *c*

3019 power plant
DA kraftværk
DE Kraftwerk *n*
ES central eléctrica *f*
F centrale *f* de production d'énergie
I impianto *m* di potenza
MA erőmű
NL elektrische centrale *f* krachtbron *f*
PO siłownia *f* energetyczna zespół *m* silnikowy
PY силовая станция *f*
SU voimalaitos
SV kraftstation *c*

3020 power point
DA kraftudtag
DE Leistungspunkt *m*
ES toma de fuerza *f*
F prise *f* de force
I punto *m* di presa energia
MA betáplálási pont
NL krachtstroom-aansluiting *f*
PO punkt *m* poboru mocy
PY точка *f* отбора мощности
SU sähkülittäntä
SV kraftuttag *n*

3021 power requirement
DA effektbehov
DE Kraftbedarf *m*
ES potencia requerida *f*
F besoin *m* d'énergie
I fabbisogno *m* di energia
MA energiaigény
NL vermogensbehoefte *f*
PO zapotrzebowanie *n* energii zapotrzebowanie *n* na moc
PY потребность *f* в мощности
SU sähkütehon tarve
SV effektbehov *n*

3022 power station
DA kraftværk
DE Kraftwerk *n*
ES central eléctrica *f*
F centrale *f* électrique
I centrale *f* elettrica
MA erőmű
NL elektrische centrale *f*
PO elektrownia *f* siłownia *f* energetyczna
PY энергетическая (силовая) станция *f*
SU voimalaitos
SV kraftstation *c*

3023 power-up (*see* start)

3024 precipitation
DA bundfældning
DE Ablagerung *f* Abscheidung *f*
ES depósito *m* precipitación *f*
F dépôt *m* précipitation *f*
I precipitazione *f*
MA ülepítés
NL neerslaan *n* neerslag *m*
PO opad *m* atmosferyczny unoszenie *n* wytrącanie *n*
PY осаждение *n*
SU sade sademäärä
SV fällning *c*

3025 precipitator
DA udfælde
DE Abscheider *m*
ES filtro *m* precipitador *m* separador *m*
F séparateur *m*
I precipitatore *m*
MA ülepítő
NL afscheider *m* bezinkbak *m*
PO oddzielacz *m* separator *m*
PY коагулянт *m*
SU erotin
SV avskiljare *c*

3026 precision work
DA præcisionsarbejde
DE Feinmechanik *f*
ES trabajo de precisión *m*
F mécanique *f* de précision
I lavoro *m* di precisione
MA precíziós munka
NL precisiewerk *n*
PO mechanika *f* precyzyjna praca *f* precyzyjna
PY точная (прецизионная) работа *f*
SU hienomekaaninen tyü
SV precisionsarbete *n*

3027 pre-cleaner
DA forfilter forrensning
DE Vorabscheider *m*
ES predepuración *m*
F pré-filtre *m*
I prefiltro *m*
MA előszűrő előtisztító
NL voorreiniger *m*
PO filtr *m* wstępny
PY устройство *n* предварительной очистки
SU esipuhdistin
SV för-renare *c*

3028 pre-cooler
DA forkøler
DE Vorkühler *m*
ES prefiltro *m*
F prérefroidisseur *m*
I preraffreddatore *m*
MA előhűtő
NL voorkoeler *m*
PO chłodnica *f* wstępna
PY предварительный охладитель *m*
SU esijäähdytin
SV förkylare *c*

DA	=	Danish
DE	=	German
ES	=	Spanish
F	=	French
I	=	Italian
MA	=	Magyar (Hungarian)
NL	=	Dutch
PO	=	Polish
PY	=	Russian
SU	=	Finnish
SV	=	Swedish

3029 pre-cooling
DA forkøling
DE Vorkühlung *f*
ES pre-enfriamiento *m*
F prérefroidissement *m*
I preraffreddamento *m*
MA előhűtés
NL voorkoeling *f*
PO chłodzenie *n* wstępne
PY предварительное охлаждение *n*
SU esijäähdytys
SV förkylning *c*

3030 predictive control
DA EDB-program der baserer sig på erfaring og aktuelle målinger
DE vorausschauende Regelung *f*
ES control predictivo *m*
F contrôle *m* prédictif
I controllo *m* predittivo
MA előzetes ellenőrzés
NL voorspellende regelaar *m*
PO regulacja *f* z wyprzedzeniem
PY программное регулирование *n*
SU ennakoiva säätü
SV adaptiv styrning *c*

3031 pre-filter
DA forfilter
DE Vorfilter *m*
ES pre-filtro *m*
F pré-filtre *m*
I prefiltro *m*
MA előszűrő
NL voorfilter *m*
PO filtr *m* wstępny
PY фильтр *m* предварительной очистки
SU csisuodatin
SV für-filter *n*

3032 pre-firing
DA forfyring
DE Vorfeuerung *f*
ES preencendido *m*
F avant-foyer *m*
I pre-accensione *f*
MA előtüzelés
NL voorontsteking *f*
PO przedpalenisko *n*
PY предварительное растапливание *n* предварительный подогрев *m*
SU esipoltto
SV für-värmning *c*

3033 pre-heat coil
DA forvarmeslange
DE Vorerhitzer-schlange *f*
ES batería de precalentamiento *f*
F batterie *f* de préchauffage
I serpentino preriscaldatore
MA előmelegítő csőkígyó
NL voorverwarmer *m*
PO nagrzewnica *f* wstępna
PY змеевик *m* предварительного подогрева
SU esilämmityspatteri
SV für-värmeslinga *c*

3034 pre-heater
DA forvarmer
DE Vorwärmer *m*
ES precalentador *m*
F préchauffeur *m* réchauffeur *m*
I preriscaldatore *m*
MA előfűtő
NL voorverwarmer *m*
PO podgrzewacz *m* wstępny
PY теплообменник *m* предварительного подогрева
SU esilämmitin
SV für-värmare *c*

3035 pre-heating
DA foropvarmning
DE Vorwärmung *f*
ES precalentamiento *m*
F préchauffage *m*
I pre-riscaldamento *m*
MA előfűtés
NL voorverwarming *f*
PO ogrzewanie *n* wstępne
PY предварительный подогрев *m*
SU esilämmitys
SV für-värmning *c*

3036 pre-heating time
DA foropvarmningstid
DE Anheizdauer *f* Vorwärmzeit *f*
ES tiempo de precalenta-miento *m*
F durée *f* de mise en température
I tempo *m* di pre-riscaldamento
MA felfűtési idő
NL aanwarmtijd *m* voorverwarmings-tijd *m*
PO czas *m* ogrzewania wstępnego
PY время *n* натопа время *n* предварител-ьного подогрева
SU esilämmitysaika
SV für-värmningstid *c*

3037 preliminary drawing
DA foreløbig tegning
DE Vorentwurfs-zeichnung *f*
ES plano preliminar *m*
F dessin *m* de projet
I disegno *m* preliminare
MA előzetes vázlat
NL schetstekening *f* voorlopige tekening *f*
PO rysunek *m* wstępny
PY предварительная осушка *f*
SU alustava piirustus
SV skiss *c*

349

3038 preparation
DA forberedelse
DE Vorbereitung *f*
ES preparación *f*
F préparation *f*
I preparazione *f*
MA előkészület
NL preparaat *n*
voorbereiding *f*
PO przygotowanie *n*
PY подготовка *f*
SU valmistelu
SV beredning *c*

3039 pre-set *adj*
DA forindstille
DE voreingestellt
ES previamente
ajustado
F préréglé
I prefissato
MA előbeállított
NL vast ingesteld
vooraf ingesteld
PO wstępnie nastawiony
wstępnie regulowany
PY заданный
SU esisäädetty
SV för-inställning *c*

3040 pre-setting
DA forindstilling
DE Voreinstellung *f*
ES preajuste *m*
F préréglage *m*
I pre-taratura *f*
MA előzetes beállítás
NL voorinstelling *f*
PO nastawienie *n*
wartości wstępnej
PY предварительное
осаждение *n*
SU esisäätü
SV für-inställning *c*

3041 pressure
DA spænding
tryk
DE Druck *m*
ES presión *f*
F pression *f*
I pressione *f*
MA nyomás
NL druk *m*
PO ciśnienie *n*
PY давление *n*
SU paine
SV tryck *n*

3042 pressure atomizing burner
DA trykforstøvnings-
brænder
DE Druckzerstäubungs-
brenner *m*
ES quemador
atomizador a
presión *m*
F brûleur *m* sous
pression à
atomisation
I bruciatore *m* a
iniezione pneumatica
MA nyomásporlasztásos
égő
NL drukverstuivings-
brander *m*
PO palnik *m*
ciśnieniowy
rozpylający
PY форсунка *f*
механического
распыливания
SU paineüljypoltin
SV tryckatomiserings-
brännare *c*

3043 pressure controller
DA trykregulator
DE Druckregler *m*
ES regulador de presión
m
F régulateur *m* de
pression
I regolatore *m* di
pressione
MA nyomásszabályozó
NL drukregelaar *m*
PO regulator *m*
ciśnienia
PY регулятор *m*
давления
SU paineensäädin
SV tryckregulator *c*

3044 pressure cooling
DA trykkøling
DE Druckkühlung *f*
ES enfriamiento a
presión *m*
F refroidissement *m*
par pression d'air
I raffreddamento *m*
sotto pressione
MA nyomásos hűtés
NL koeling *f* met lucht
onder druk
PO chłodzenie *n*
ciśnieniowe
PY давление *n*
охлаждения
SU jäähdytys
paineellisella
ilmalla
SV tryckkylning *c*

3045 pressure cut-out
DA overtryckssikring
DE Druckausfall *m*
ES presostato *m*
F pression *f* d'arrêt
I interruzione *f* di
pressione
limitazione *f* di pres-
sione
pressione *f* di
arresto
MA nyomáskioldó
NL drukbeveiligings-
schakelaar *m*
PO wyłącznik *m*
ciśnieniowy
PY отключатель *m*
давления
SU ylipainesuoja
SV üvertryckssäkring *c*

3046 pressure difference
DA trykforskel
DE Druckdifferenz *f*
ES diferencia de
presión *f*
F différence *f* de
pression
I differenza *f* di
pressione
MA nyomáskülünbség
NL drukverschil *n*
PO różnica *f* ciśnienia
PY разность *f*
давления
SU paine-ero
SV tryckdifferens *c*

3047 pressure differential cut-out
DA trykdifferenssikring
DE Differenzdruck-schalter *m*
ES desconector de presión diferencial *m*
F pressostat *m* différentiel
I arresto *m* per pressione differenziale
MA nyomáskülünbség kapcsoló
NL drukverschilbeveiligings-schakelaar *m*
PO wyłącznik *m* różnicowy ciśnienia
PY отключатель *m* по разности давления
SU paine-erokytkin
SV tryckdifferens-säkring *c*

3048 pressure drop
DA trykfald
DE Druckabfall *m*
ES caída de presión *f*
F chute *f* de pression
I caduta *f* di pressione
MA nyomásesés
NL drukval *m*
PO spadek *m* ciśnienia
PY падение *n* давления
SU painehäviü
SV tryckfall *n*

3049 pressure equalization
DA trykudligning
DE Druckausgleich *m*
ES equilibrado de presiones *m*
F équilibrage *m* de pression
I equilibratura *f* della pressione
MA nyomáskiegyenlítés
NL drukegalisatie *f* drukvereffening *f*
PO wyrównanie *n* ciśnienia
PY выравнивание *n* давления
SU paineentasaaja
SV tryckutjämning *c*

3050 pressure feed
DA tryktilførsel
DE Druckeinspritzung *f*
ES alimentación a presión *f*
F alimentation *f* sous pression
I alimentazione *f* di pressione
MA nyomásos táplálás
NL voeding *f* onder druk
PO zasilanie *n* pod ciśnieniem
PY подвод *m* давления
SU paineen syüttü
SV tryckreglering *c*

3051 pressure-fired boiler
DA overtrykskedel
DE Kessel *m* mit Überdruck-feuerung *f*
ES caldera a sobrepresión *f*
F chaudière *f* à foyer surpressé
I caldaia *f* pressurizzata
MA nyomástüzeléses kazán
NL overdrukketel *m*
PO kocioł *m* opalany pod ciśnieniem
PY котел *m* с наддувом
SU ylipainekattila
SV üvertryckseldad panna *c*

3052 pressure gage (USA) (*see* pressure gauge)

3053 pressure gauge
DA manometer trykmåler
DE Druckmesser *m* Manometer *n*
ES manómetro *m*
F manomètre *m*
I indicatore *m* di pressione manometro *m*
MA nyomásmérő
NL drukmeter *m* manometer *m*
PO ciśnieniomierz *m* manometr *m* wskaźnik *m* ciśnienia
PY индикатор *m* давления
SU painemittari
SV tryckmätare *c*

3054 pressure gradient
DA trykgradient
DE Druckgefälle *n*
ES gradiente de presión *m*
F gradient *m* de pression
I gradiente *m* di pressione
MA nyomásesés
NL drukgradînt *m* drukverval *n*
PO gradient *m* ciśnienia
PY градиент *m* давления
SU painegradientti
SV tryckgradient *c*

3055 pressure head
DA trykhøjde
DE Druckhühe *f*
ES altura manométrica *f* altura piezométrica *f*
F hauteur *f* de charge hauteur *f* manométrique pression *f* au refoulement
I altezza *f* manometrica pressione *f*
MA nyomómagasság
NL drukhoogte *f*
PO wysokość *f* ciśnienia hydrostatycznego
PY высота *f* столба давление *n* в любой точке трубопровода
SU painekorkeus
SV tryckhüjd *c*

3056 pressure jet burner
DA trykforstøvnings-
  brænder
DE Gebläsebrenner *m*
ES quemador de chorro
  a presión *m*
F brûleur *m* à injection
I bruciatore *m* a
  pressione
MA túlnyomásos égő
NL drukverstuivings-
  brander *m*
PO palnik *m*
  inżektorowy
PY инжекционная
  форсунка *f*
SU paineüljypoltin
SV tryckoljebrännare *c*

**3057 pressure limiting
  device**
DA trykbegrænsnings-
  regulator
DE Druckbegrenzer *m*
ES limitador de
  presión *m*
F limiteur *m* de
  pression
I limitatore *m* di
  pressione
MA nyomáshatároló
  készülék
NL drukbegrenzings-
  apparaat *n*
PO ogranicznik *m*
  ciśnienia
PY ограничитель *m*
  давления
SU paineen rajoittaja
SV tryckregulator *c*

3058 pressure loss
DA tryktab
DE Druckverlust *m*
ES pérdida de presión *f*
F perte *f* de pression
I perdita *f* di
  pressione
MA nyomásveszteség
NL drukverlies *n*
PO strata *f* ciśnienia
PY потеря *f* давления
SU painehäviü
SV tryckfall *n*

**3059 pressure
  measurement**
DA trykmåling
DE Druckmessung *f*
ES medida de la
  presión *f*
F mesure *f* de pression
I misurazione *f* della
  pressione
MA nyomásmérés
NL drukmeting *f*
PO pomiar *m* ciśnienia
PY измерение *n*
  давления
SU painemittaus
SV tryckmätning *c*

3060 pressure rating
DA trykklasse
DE Druckleistung *f*
ES grado de presión *m*
F
I pressione *f* di targa
MA nyomás számítás
NL drukklasse *f*
PO ciśnienie *m*
  znamionowe
PY оценка *f* давления
SU
SV tryck-
  dimensionering *c*

**3061 pressure reducing
  point**
DA trykreduktionspunkt
DE Druckreduzier-
  punkt *m*
ES punto de reducción
  de la presión *m*
F point *m* de détente
I punto *m* di
  riduzione di
  pressione
MA nyomáscsükkentési
  pont
NL drukverminderings
  punt *n*
PO punkt *m* redukcji
  ciśnienia
PY точка *f* снижения
  давления
SU paineenalennuspiste
SV tryckreducerings-
  punkt *c*

**3062 pressure reducing
  station**
DA trykreduktions-
  station
DE Druckreduzier-
  station *f*
ES central de reducción
  de presión *f*
F station *f* de détente
I stazione *f* di
  riduzione di
  pressione
MA nyomáscsükkentő
  állomás
NL drukreduceer-
  toestel *n*
PO stacja *f* redukcyjna
  ciśnienia
PY станция *f* снижения
  давления
SU paineenvähenny-
  sasema
SV tryckreduktions-
  station *c*

**3063 pressure reducing
  valve**
DA trykreduktionsventil
DE Druckminderer *m*
  Druckreduzier-
  ventil *n*
ES válvula reductora de
  presión *f*
F réducteur *m* de
  pression
I valvola *f* di
  riduzione di
  pressione
MA nyomáscsükkentő
  szelep
NL reduceerafsluiter *m*
  reduceerventiel *n*
PO reduktor *m* ciśnienia
  zawór *m* redukcyjny
PY редукционный
  клапан *m*
SU paineenalennus-
  venttiili
SV tryckreducerings-
  ventil *c*

3064 pressure reduction
DA trykreduktion
DE Druckreduzierung *f*
ES reducción de presión
 f
F détente *f*
 réduction *f* de
 pression
I riduzione *f* di
 pressione
MA nyomáscsükkentés
NL drukverlaging *f*
PO redukcja *f* ciśnienia
PY снижение *n*
 давления
SU
SV tryckreducering *c*

3065 pressure regulator
DA trykregulator
DE Druckregler *m*
ES regulador de
 presión *m*
F régulateur *m* de
 pression
I regolatore *m* di
 pressione
MA nyomásszabályozó
NL drukregelaar *m*
PO regulator *m*
 ciśnienia
PY регулятор *m*
 давления
SU paineensäädin
SV tryckregulator *c*

**3066 pressure relief
 valve**
DA tryksikkerhedsventil
DE Sicherheitsventil *n*
ES válvula de
 seguridad *f*
 válvula limitadora de
 presión *f*
F soupape *f* de
 sécurité
I valvola *f* di
 sicurezza
MA biztonsági szelep
NL drukveiligheids-
 klep *f*
 veiligheidsklep *f*
PO zawór *m*
 bezpieczeństwa
 zawór *m*
 nadmiarowy
 ciśnieniowy
PY предохранительный
 клапан *m*
SU ylipaineventtiili
SV tryckreducerings-
 ventil *c*

3067 pressure stage
DA tryktrin
DE Druckstufe *f*
ES escalón de
 presión *m*
F étage *m* de pression
I stadio *m* di
 pressione
MA nyomáslépcső
NL druktrap *m*
PO stopień *m* ciśnienia
PY ступень *f* давления
SU paineporras
SV trycksteg *n*

3068 pressure switch
DA pressostat
DE Druckschalter *m*
ES presostato *m*
F manostat *m*
 pressostat *m*
I pressostato *m*
MA nyomáskapcsoló
NL drukschakelaar *m*
PO wyłącznik *m*
 ciśnieniowy
PY выключатель *m*
 давления
SU painekytkin
SV tryckströmställare *c*

3069 pressure test
DA trykprøve
DE Druckprobe *f*
ES prueba de presión *f*
F épreuve *f* de
 pression
 essai *m* de pression
I prova *f* di pressione
MA nyomáspróba
NL drukproef *f*
 persproef *f*
PO próba *f* ciśnieniowa
PY испытание *n* под
 давлением
SU painekoe
SV tryckprovning *c*

3070 pressure test pump
DA trykprøvepumpe
DE Prüfdruckpumpe *f*
ES bomba para pruebas
 de presión *f*
F pompe *f* d'épreuve
I pompa *f* di prova di
 pressione
MA próbaszivattyú
NL perspomp *f*
PO pompa *f* próbna
 ciśnieniowa
PY определение *n*
 давления,
 развиваемого
 насосом
SU koepainepumppu
SV tryckprovnings-
 pump *c*

3071 pressure-tight *adj*
DA tryktæt
DE druckfest
ES a prueba de presión
 estanco (a la
 presión)
F étanche
I a tenuta stagna
MA nyomásálló
NL drukbestendig
PO szczelny
PY герметичный
SU paineenkestävä
SV tryckhållfast

3072 pressure vessel
DA trykbeholder
DE Druckbehälter *m*
 Druckkessel *m*
ES recipiente a
 presión *m*
F réservoir *m* sous
 pression
I recipiente *m* a
 pressione
MA nyomótartály
NL drukketel *m*
 drukvat *n*
PO naczynie *n*
 ciśnieniowe
PY сосуд *m* под
 давлением
SU paineastia
SV tryckkärl *n*

3073 pressurization
DA sætte under tryk
DE Druckhaltung f
ES presurización f
F pressurisation f
I pressurizzazione f
MA nyomás alatt tartás
NL onder druk brengen
PO utrzymywanie n pod ciśnieniem
PY герметизация f наддув m
SU paineistaminen
SV tryckhållning c

3074 pressurized combustion chamber
DA overtryksforbrændings-kammer
DE Überdruckbrenn-kammer f
ES cámara de combustión a sobrepresión f
F foyer m sous pression
I camera f di combustione a pressione
MA túlnyomásos égőtér
NL overdruk-vuurhaard m
PO komora f spalania ciśnieniowa
PY топка f с наддувом
SU paineistettu polttokammio
SV rum n för övertryck-seldning c

3075 pressurizer
DA overtrykspumpe
DE Überdruck-erzeuger m
ES presurizador m
F surpresseur m
I pressurizzatore m
MA túlnyomás előállító
NL drukverhoger m
PO urządzenie n utrzymujące ciśnienie
PY компенсатор m объема устройство n для поддержания давления
SU paineistaja
SV trycksättnings-anordning c

3076 preventive maintenance
DA forebyggende vedligeholdelse
DE vorbeugende Instandhaltung f
ES mantenimiento preventivo m
F maintenance f préventive
I manutenzione f preventiva
MA megelőző karbantartás
NL preventief onderhoud n
PO konserwacja f profilaktyczna przegląd m konserwacyjny
PY профилактическое обслуживание n
SU
SV fürebyggande underhåll n

3077 primary air
DA primærluft
DE Primärluft f
ES aire primario m
F air m primaire
I aria f primaria
MA primer levegő
NL primaire lucht f
PO powietrze n pierwotne
PY приточный воздух m
SU primääri ilma
SV primärluft c

3078 primary circuit
DA primærkreds
DE Primärkreislauf m
ES circuito primario m
F circuit m primaire
I circuito m primario
MA primer áramkür
NL primair circuit n primaire stroomkring m
PO obwód m pierwotny
PY первичная цепь f
SU primääripiiri
SV primärkrets c

3079 primary coolant
DA primærkølemiddel
DE Primärkühlmittel n
ES refrigerante primario m
F fluide m frigoporteur primaire
I refrigerante m principale
MA primer hűtőküzeg
NL primair koelmiddel n
PO czynnik m chłodzący pierwszego stopnia
PY первичный хладагент m
SU primääri jäähdyke
SV primärkylmedel n

3080 primary heating surface
DA primær hedeflade
DE direkte Heizfläche f
ES superficie de calefacción directa f
F surface f de chauffe directe
I superficie f riscaldante primaria
MA primer fűtőfelület
NL direct verwarmings-oppervlak n primair verwarmings-oppervlak n
PO powierzchnia f ogrzewania wstępnego
PY поверхность f нагрева
SU primäärilämpüpinta
SV primär uppvärmning-syta c

3081 prime *vb*
DA koge over
DE mit Wasser auffüllen
(bei nicht
selbstansaugenden
Pumpen)
in Tätigkeit setzen
ES cebar
imprimar
F amorcer (une
pompe)
I adescare (una
pompa)
MA feltült
NL gronden
opkoken
vullen ten behoeve
van aanzuigen
PO uruchamiać
zalać pompę (przed
uruchomieniem)
PY приготовлять
SU johtaa
keskipako-
pumppuun vettä
sen
käyttüünotossa
tulistaa
SV flüda

3082 prime contractor
DA entreprenør som
udfører udkogning
DE Hauptkontraktor *m*
ES contratista principal
m
F entreprise *m*
principale
I impresa *f*
capocommessa
MA fővállakozó
NL hoofdaannemer *m*
PO główny
wykonawca *m*
PY генеральный
подрядчик *m*
SU pääurakoitsija
SV huvudentreprenür *c*

DA =	Danish
DE =	German
ES =	Spanish
F =	French
I =	Italian
MA =	Magyar (Hungarian)
NL =	Dutch
PO =	Polish
PY =	Russian
SU =	Finnish
SV =	Swedish

3083 prime professional
DA faglig udkogning
DE Haupthand-
werker *m*
ES capataz *m*
F
I professionista *m*
capocommessa
MA vezető szakértő
NL adviseur *m* van
opdrachtgever
PO specjalista *m*
PY генеральный
эксплуата-
ционник *m*
SU rakennuttaja-
konsultti
SV yrkeskunnig *c*

3084 priming
DA overkogning
DE Spucken *n* eines
Kessels *m*
Zündung *f*
ES cebado *m*
imprimación *f*
F primage *m*
I adescamento *m* (di
una pompa)
MA feltültés
NL grondverven
opkoken
vullen ten behoeve
van aanzuigen
PO zalewanie *n* (pompy)
PY грунтовка *f*
заправка *f*
SU tulistus
(hüyrykattilasta)
SV flüde *n*

3085 printer
DA printer
DE Drucker *m*
ES impreso *m*
F imprimante *f*
imprimeur *m*
I stampante *f*
MA nyomtató
NL drukker *m*
printer *m*
PO drukarka *f*
PY принтер *m*
SU kirjoitin
SV printer *c*

3086 probe
DA sonde
DE Sonde *f*
ES probeta *f*
sonda *f*
F sonde *f*
I sonda *f*
MA szonda
NL sonde *f*
taster *m*
PO elektroda *f* czujnika
sonda *f*
PY зонд *m*
SU anturi
SV sond *c*

3087 process
DA metode
proces
DE Prozeß *m*
ES proceso *m*
F procédé *m*
process *m*
I processo *m*
MA eljárás
folyamat
NL proces *n*
PO proces *m*
PY процесс *m*
SU prosessi
SV process *c*

3088 process air conditioning
DA procesluftbehandling
DE Prozeßlufttechnik *f*
ES acondicionamiento
de aire
industrial *m*
F conditionnement *m*
d'air industriel
I condizionamento *m*
dell'aria per proce-
ssi industriali
MA technológiai
légkondicionálás
NL industriële
lucht-
behandeling *f*
PO proces *m*
klimatyzacji
PY процесс *m*
кондицион-
ирования воздуха
SU prosessi-ilmastointi
SV processluft-
behandling *c*

3089 process control
DA processtyring
DE Prozeßregelung *f*
ES control de proceso *m*
F régulation *f* de
 process
I controllo *m* di
 processo
MA folyamatszabályozás
NL industrïle regelaar *m*
 procesregelaar *m*
PO regulacja *f* procesu
 sterowanie *n*
 procesem
PY процесс *m*
 управления
 (регулирования)
SU prosessin säätü
SV processkontroll *c*

3090 producer gas
DA generatorgas
DE Generatorgas *n*
ES gas pobre *m*
F gaz *m* de gazogène
I gas *m* di città
MA generátorgáz
NL generatorgas *n*
PO gaz *m* czadnicowy
 gaz *m* generatorowy
 gaz *m* miejski
PY генераторный
 газ *m*
SU generaattorikaasu
SV generatorgas *c*

**3091 production
 schedule**
DA produktionsplan
DE Produktions-
 programm *n*
ES programa de
 fabricación *m*
 programa de
 producción *m*
F programme *m* de
 fabrication
I programma *m* di
 fabbricazione
MA gyártási program
NL produktie-
 programma *n*
PO program *m*
 produkcji
PY программа *f*
 производства
SU tuotantoaikataulu
SV produktions-
 schema *n*

**3092 professional
 engineer**
DA civilingeniør
DE beratender
 Ingenieur *m*
 freiberuflicher
 Ingenieur *m*
ES ingeniero
 especializado *m*
F ingénieur *m*
I ingegnere *m* libero
 professionista
MA szakmérnük
NL ingenieur *m* met
 erkenning
PO inżynier *m*
 dyplomowany
PY профессиональный
 (дипломиро-
 ванный)
 инженер *m*
SU hyväksytty insinüüri
 laillistettu insinüüri
SV yrkesingenjür *c*

3093 programmable
DA programmerbar
DE programmierbar
ES programable
F programmable
I programmabile
MA programozható
NL programmeerbaar
PO programowalny
 programowy
PY программиро-
 ванный
SU ohjelmoitava
SV programeringsbar

**3094 programmable
 logic control**
DA programmerbar
 logisk kontrol
DE programmierbare
 logische
 Regelung *f*
ES control lógico
 programable *m*
F automate *m*
 programmable
I controllo *m* a logica
 programmabile
MA programozható
 logikai
 szabályozás
NL programmeerbare
 logische
 schakeling *f*
PO programowalny
 logiczny układ *m*
 sterujący
 regulacja *f*
 programowa
 sterowanie *n*
 programowe
PY программиро-
 ванное
 логическое
 регулирование *n*
SU ohjelmoitava
 looginen säätü
SV programeringsbar
 kontroll *c*

**3095 programme
 controller**
DA programregulator
DE Programmregler *m*
ES programador *m*
F régulateur *m* à
 programme
I regolatore *m* di
 programma
MA programszabályozás
NL programma-
 regelaar *m*
PO regulator *m*
 programu
PY программный
 регулятор *m*
SU ohjaussäätü
SV programkontrollür *c*

3096 project architect
DA projekterende
  arkitekt
DE Projektarchitekt *m*
ES arquitecto
  proyectista *m*
F architecte *m* de
  projet
I architetto *m*
  capoprogetto
MA építésvezető
NL projectarchitect *m*
PO architekt-
  projektant *m*
PY архитектор-
  проектировщик
  m
SU projektiarkkitehti
SV projektarkitekt *c*

3097 proofed *adj*
DA afprøvet
  tæt
DE dicht
ES impermeabilizado
  probado
F étanche
I stagno
MA ellenálló
  tümür
NL beproefd
  bestand gemaakt
  tegen
PO szczelny
PY плотный
SU kyllästetty
  tiivistetty
SV tät

3098 propane
DA propan
DE Propan *n*
ES propano *m*
F propane *m*
I propano *m*
MA propán
NL propaan *n*
PO propan *m*
PY пропан *m*
SU propaani
SV propan *n*

3099 propeller
DA propel
  skibsskrue
DE Propeller *m*
ES hélice *f*
F hélice *f*
I elica *f*
MA szárnykerék
NL propeller *m*
  schroef *f*
PO śmigło *n*
  śruba *f*
PY винт *m*
SU potkuri
SV propeller *c*

3100 propeller fan
DA aksialventilator
DE Axialventilator *m*
  Schraubenlüfter *m*
ES ventilador axial *m*
  ventilador
  helicoidal *m*
F ventilateur *m*
  hélicoïde
I ventilatore *m*
  elicoidale
MA csavarventilátor
NL schroefventilator *m*
PO wentylator *m*
  osiowy
  wentylator *m*
  śmigłowy
PY осевой
  вентилятор *m*
SU potkuripuhallin
SV propellerfläkt *c*

3101 proportional *adj*
DA proportional
DE proportional
ES proporcional
F proportionnel
I proporzionale
MA arányos
  proporcionális
NL proportioneel
PO proporcjonalny
PY пропорциональный
SU suhteellinen
SV proportionell

3102 proportional band
DA proportionalbånd
DE Proportionalband *n*
ES banda proporcional
  f
F bande *f*
  proportionnelle
I banda *f*
  proporzionale
MA arányos tartomány
NL proportionele
  band *m*
PO zakres *m*
  proporcjonalności
PY зона *f*
  пропорциональ-
  ьности
SU suhdealue
SV proportionell
  frekvens *c*

**3103 proportional band
  control**
DA proportionalbånd-
  styring
DE Proportionalband-
  regelung *f*
ES control de banda
  proporcional *m*
F régulation *f*
  proportionnelle
I regolazione *f* a
  banda
  proporzionale
MA arányos
  tartományszabá-
  lyozás
NL proportionele-
  bandregelaar *m*
PO regulacja *f* zakresu
  proporcjonalności
PY полоса *f*
  пропорциональ-
  ьного регулирования
SU suhteellinen säätü
SV proportionell
  frekvenskontroll *c*

3104 proportional control

DA proportionalstyring
DE Proportional-
   regelung f
ES regulación
   proporcional f
F régulateur m
   proportionnel
I regolazione f
   proporzionale
MA arányos szabályozás
NL proportionele
   regeling f
PO regulacja f
   proporcjonalna
PY пропорциональное
   регулирование n
SU suhteellinen säätü
SV proportionell
   kontroll c

3105 protection

DA beskyttelse
DE Schutz m
ES protección f
F protection f
I protezione f
MA védelem
NL bescherming f
   beveiliging f
PO ochrona f
   zabezpieczenie n
PY защита f
SU suojaus
SV skydd n

3106 protection from cold

DA beskyttelse mod
   kulde
DE Kälteschutz m
ES protección térmica f
F protection f
   thermique
I protezione f dal
   freddo
MA hideg elleni védelem
NL bescherming f tegen
   koude
PO izolacja f cieplna
   ochrona f przed
   zimnem
PY защита f от холода
SU suojaus kylmältä
SV küldskydd n

3107 protection from heat

DA beskyttelse mod
   varme
DE Wärmeschutz m
ES protección térmica f
F protection f
   thermique
I protezione f dal
   caldo
MA hővédelem
NL bescherming f tegen
   warmte
PO ochrona f cieplna
PY защита f от
   теплоты
   (перегрева)
SU lämpüsuojaus
SV värmeskydd n

3108 protective adj

DA beskyttende
DE schützend
   Schutz m
ES protector
F protecteur
I protettivo
MA védő
NL beschermend
PO ochronny
   zabezpieczający
PY защитный
   предохранительный
SU suojaava
SV skyddande

3109 protective coating

DA beskyttelsesovertræk
DE Schutzüberzug m
   Schutzverkleidung f
ES recubrimiento
   protector m
F enduit m de
   protection
I rivestimento m di
   protezione
MA védőburkolat
NL beschermende
   kleding f
   beschermingslaag f
PO płaszcz m ochronny
   pokrycie n ochronne
PY защитная окраска f
SU suojaava pinnoite
SV skyddsbeläggning c

3110 protective device

DA beskyttelses-
   anordning
DE Schutzvorrichtung f
ES dispositivo de
   protección m
F dispositif m de
   protection
I dispositivo m di
   protezione
MA védőkészülék
NL beveiligings-
   inrichting f
PO urządzenie n
   zabezpieczające
PY защитное
   устройство n
SU suojalaite
SV skyddsanordning c

3111 protective layer

DA beskyttelseslag
DE Schutzschicht f
ES capa protectora f
F couche f de
   protection
I strato m protettivo
MA védőréteg
NL beschermingslaag f
PO warstwa f ochronna
PY защитный слой m
SU suojaava kerros
SV skyddande
   beklädnad c

3112 protective screen

DA beskyttelsesskærm
DE Schutzgitter n
ES pantalla
   protectora f
F écran m de
   protection
I schermo m di
   protezione
MA védőrács
NL beveiligingsscherm n
PO ekran m ochronny
PY защитный экран m
SU suojaverkko
SV skyddsskärm c

3113 protective tube
DA beskyttelsesrør
DE Schutzrohr *n*
ES tubo de
   protección *m*
F tube *m* de
   protection
I tubo *m* di
   protezione
MA védőcső
NL beschermingspijp *f*
   mantelbuis *f*
PO rura *f* ochronna
   tuleja *f*
PY предохранительная
   труба *f*
SU suojaputki
SV skyddsrür *n*

3114 psychrometer
DA fugtighedsmåler
DE Psychrometer *n*
ES sicrómetro *m*
F psychromètre *m*
I psicrometro *m*
MA légnedvességmérő
   pszichrométer
NL luchtvochtigheids-
   meter *m*
   psychrometer *m*
PO psychrometr *m*
PY психрометр *m*
SU psykrometri
SV psykrometer *c*

**3115 psychrometric
   chart**
DA fugtighedsdiagram
DE psychrometrisches
   Diagramm *n*
ES diagrama
   sicrométrico *m*
F diagramme *m*
   psychrométrique
I diagramma *m*
   psicrometrico
MA pszichrometrikus
   diagram
NL Mollierdiagram *n*
   psychrometrische
   kaart *f*
PO wykrcs *m* psychro-
   metryczny
PY психрометрическая
   карта *f*
SU psykrometrinen
   piirros
SV psykrometer-
   diagram *n*

3116 pulsating *adj*
DA pulserende
DE pulsierend
ES pulsante
F pulsatoire
I pulsante
MA lüktető
NL pulserend
PO pulsujący
   tętniący
PY пульсирующий
SU pulssimaisesti
   vaihteleva
SV pulserande

3117 pulse counting
DA pulstælling
DE Pulszählung *f*
ES contaje de impulsos
   m
F comptage *m*
   d'impulsions
I conteggio *m* di
   impulsi
MA lüktetésszámlálás
NL pulstelling *f*
PO liczenie *n* impulsów
   zliczanie *n* impulsów
PY частота *f*
   пульсации
SU pulssilaskenta
SV pulsräkning *c*

3118 pulverized coal
DA kulstøv
DE Kohlenstaub *m*
   Staubkohle *f*
ES carbón
   pulverizado *m*
F charbon *m* pulvérisé
I polvere *f* di carbone
MA porszén
NL poederkool *f*
PO pył *m* węglowy
PY порошкообразный
   уголь *m*
SU hiilipüly
SV kolpulver *n*

3119 pump
DA pumpe
DE Pumpe *f*
ES bomba *f*
F pompe *f*
I pompa *f*
MA szivattyú
NL pomp *f*
PO pompa *f*
PY насос *m*
SU pumppu
SV pump *c*

3120 pump body (*see*
   pump housing)

3121 pump casing (*see*
   pump housing)

3122 pump housing
DA pumpehus
DE Pumpengehäuse *n*
   Pumpenkürper *m*
ES carcaja de bomba *f*
F corps *m* de pompe
I corpo *m* di pompa
MA szivattyúház
NL pomphuis *n*
PO korpus *m* pompy
PY корпус *m* насоса
SU pumpun runko
SV pumphus *n*

3123 purge *vb*
DA rense
   udblæse
DE entgasen
   entlüften
   reinigen
ES purgar
F purger
I purgare
   scaricare
MA tisztít
NL doorspoelen
   leegblazen
   spuien
PO odpowietrzać
   przedmuchiwać
PY продувать
SU ilmata
   puhdistaa
SV rena

3124 purification
DA rensning
DE Reinigung *f*
ES depuración *f*
F épuration *f*
I depurazione *f*
MA tisztítás
NL zuivering *f*
PO oczyszczanie *n*
PY очистка *f*
SU puhdistus
SV rening *c*

3125 push-button control
DA trykknapstyring
DE Druckknopf-
   regelung f
ES mando por
   pulsador m
   maniobra por
   pulsadores f
F commande f par
   bouton poussoir
I regolazione f a
   pulsante
MA nyomógombos
   vezérlés
NL drukknop-
   bediening f
PO sterowanie n
   przyciskowe
PY кнопочное
   управление n
SU painonappiohjaus
SV tryckknapps-
   kontroll c

3126 putty
DA kit
DE Kitt m
ES masilla f
F mastic m
I mastice m
MA kitt
NL plamuur m
   stopverf f
PO kit m
   szpachlówka f
PY замазка f
SU kitti
SV kitt n

3127 pyrometer
DA pyrometer
DE Pyrometer n
ES pirómetro
F pyromètre m
I pirometro m
MA nagyhőmérséklet-
   mérő
   pirométer
NL pyrometer m
PO pirometr m
PY пирометр m
SU pyrometri
SV pyrometer c

Q

3128 quality
DA kvalitet
DE Qualität *f*
ES calidad *f*
F qualité *f*
I qualità *f*
MA minőség
NL kwaliteit *f*
PO jakość *f*
PY качество *n*
SU laatu
SV kvalitet *c*

3129 quality assurance
DA kvalitetssikring
DE Qualitätssicherung *f*
ES seguro de calidad *m*
F assurance *f* qualité
I assicurazione *f* di qualità
MA minőséggarancia
NL kwaliteitsgarantie *f*
PO gwarancja *f* jakości
PY гарантия *f* качества
SU laadunvarmistus
SV kvalitetsfürsäkring *c*

3130 quality control
DA kvalitetsstyring
DE Qualitätskontrolle *f*
ES control de calidad *m*
F contrôle *m* qualité
I controllo *m* di qualità
MA minőségi szabályozás
NL kwaliteitscontrole *f*
PO kontrola *f* jakości
PY контроль *m* качества
SU laadunvalvonta
SV kvalitetskontroll *c*

3131 quantity
DA kvantitet
DE Menge *f*
Quantität *f*
ES cantidad *f*
F quantité *f*
I quantità *f*
MA mennyiség
NL hoeveelheid *f*
kwantiteit *f*
PO ilość *f*
PY количество *n*
SU määrä
SV kvantitet *c*

3132 quick-action valve
DA hurtigvirkende ventil
DE Schnellschluß-
ventil *n*
ES válvula de acción
rápida *f*
F soupape *f* à action
rapide
I valvola *f* ad azione
rapida
MA gyorslezáró szelep
NL snelsluitklep *f*
PO zawór *m* szybkiego
działania
PY быстродейс-
твующий
клапан *m*
SU pikaventtiili
SV snabbverkande
ventil *c*

**3133 quick-release
coupling**
DA lynkobling
DE Schnellschluß-
kupplung *f*
ES acoplamiento de
desembrague
rápido *m*
enlace rápido *m*
rácor rápido *m*
F raccord *m* pompier
raccord *m* rapide
I raccordo *m* rapido
MA gyorskioldású
csatlakozás
NL snelkoppeling *f*
snelslipkoppeling *f*
PO połączenie *n*
błyskawiczne
PY быстроразъемное
соединение *n*
SU pikaliitäntä
SV snabbkoppling *c*

DA	=	Danish
DE	=	German
ES	=	Spanish
F	=	French
I	=	Italian
MA	=	Magyar (Hungarian)
NL	=	Dutch
PO	=	Polish
PY	=	Russian
SU	=	Finnish
SV	=	Swedish

R

3134 rack mounting
DA stativ
DE Gerüstmontage *f*
ES montaje de la
  bancada *m*
F montage *m* en rack
I montaggio *m* a
  cremagliera
  montaggio *m* in
  scaffale
MA állványra szerelés
NL op rek gemonteerd
  rekmontage *f*
PO podparcie *n*
  połączenie *n*
  zawieszenie *n*
PY монтаж *m* в
  каркасе
  монтаж *m* на
  стенде
SU hyllyasennus
SV

3135 radiant *adj*
DA udstrålende
DE strahlend
ES radiante
F radiant
  rayonnant
I radiante
MA sugárzó
NL stralend
PO promienisty
  promieniujący
PY лучистый
SU säteilevä
  säteily-
SV strålande

**3136 radiant comfort
heating**
DA komfort-strålevarme
DE Strahlungswohnraum-
  heizung *f*
ES calefacción de
  confort por
  radiación *f*
F chauffage *m* de
  confort par
  rayonnement
I riscaldamento *m* per
  irraggiamento
MA sugárzó
  komfortfűtés
NL comfortstralings-
  verwarming *f*
PO ogrzewanie *n*
  komfortowe przez
  promieniowanie
PY лучистое
  комфортное
  отопление *n*
SU säteilylämmitys
SV strålnings-
  uppvärmning *c*

3137 radiant cooling
DA strålekøling
DE Strahlungskühlung *f*
ES enfriamiento por
  radiación *m*
F refroidissement *m*
  par rayonnement
I refrigerazione *f* a
  pannelli radianti
MA sugárzó hűtés
NL stralingskoeling *f*
PO chłodzenie *n* przez
  promieniowanie
PY радиационное
  охлаждение *n*
SU säteilyjäähdytys
SV strålnings-
  kylsystem *n*

**3138 radiant cooling
system**
DA strålekøleanlæg
DE Strahlungskühl-
  system *n*
ES sistema de
  enfriamiento por
  radiación *m*
F système *m* de
  rafraîchissement
  par rayonnement
I impianto *m* a
  refrigerazione per
  irraggiamento
MA sugárzó
  hűtőrendszer
NL stralingskoel-
  systeem *n*
PO system *m* chłodzenia
  przez
  promieniowanie
PY система *f*
  радиационного
  охлаждения
SU säteilyjäähdytysjärje-
  stelmä
SV strålningskyl-
  system *n*

3139 radiant gas heater
DA gasstrålevarmeovn
DE Gasstrahler *m*
ES radiador de gas *m*
F appareil *m* radiant à
  gaz
I riscaldatore *m*
  radiante a gas
MA gázüzemű hősugárzó
NL gasstraalkachel *f*
  gaswarmtestraler *m*
PO promiennik *m*
  gazowy
PY газовый
  излучающий
  нагреватель *m*
SU kaasusäteilylämmitin
SV gaseldad
  strålnings-
  värmare *c*

3140 radiant heat
DA strålevarme
DE Strahlungswärme *f*
ES calor radiante *m*
F chaleur *f* rayonnée
I calore *m* radiante
MA sugárzó hő
NL stralingswarmte *f*
PO ciepło *n*
  promieniowania
PY лучистая теплота *f*
SU lämpüsäteily
SV strålningsvärme *c*

3141 radiant heater
DA strålevarmeflade
DE Strahlungsheiz-
  kürper *m*
ES radiador *m*
F appareil *m* de
  chauffage par
  rayonnement
I riscaldatore *m* a
  energia radiante
MA hősugárzó
NL straalkachel *f*
  warmtestraler *m*
PO grzejnik *m*
  promieniujący
  promiennik *m*
PY излучающий
  нагреватель *m*
SU säteilylämmitin
SV strålningsvärmare *c*

3142 radiant heating
DA stråleopvarmning
DE Strahlungsheizung *f*
ES calefacción
  radiante *f*
F chauffage *m* par
  rayonnement
I riscaldamento *m* ad
  irraggiamento
  riscaldamento *m* a
  pannelli radianti
MA sugárzó fűtés
NL stralings-
  verwarming *f*
PO ogrzewanie *n* przez
  promieniowanie
PY лучистое
  отопление *n*
SU säteilylämmitys
SV strålnings-
  uppvärmning *c*

3143 radiant panel
DA strålingspanel
DE Plattenheizkürper *m*
  Strahlplatte *f*
ES panel radiante *m*
F panneau *m*
  rayonnant
I pannello *m* radiante
MA lapradiátor
NL stralingspaneel *n*
PO płyta *f*
  promieniująca
  promiennik *m*
  płytowy
PY излучающая
  панель *f*
SU säteilypaneeli
SV strålningspanel *c*

3144 radiant strip heater
DA strålevarmestrip
DE Bandstrahler *m*
ES cinta calefactora *f*
F cordon *m* chauffant
I riscaldatore *m*
  radiante a strisce
MA sávsugárzó
NL bandwarmte-
  straler *m*
PO taśma *f*
  promieniująca
PY ленточный
  излучающий
  нагреватель *m*
SU nauhamainen
  säteilylämmitin
SV strålningspanel *c*

3145 radiant temperature
DA strålingstemperatur
DE Strahlungs-
  temperatur *f*
ES temperatura radiante
  f
F température *f*
  radiante
I temperatura *f*
  radiante
MA sugárzási
  hőmérséklet
NL stralings-
  temperatuur *f*
PO temperatura *f*
  promieniowania
PY радиационная
  (лучистая)
  температура *f*
SU säteilylämpütila
SV strålnings-
  temperatur *c*

3146 radiation
DA stråling
DE Strahlung *f*
ES radiación *f*
F rayonnement *m*
I irraggiamento *m*
  radiazione *f*
MA sugárzás
NL straling *f*
PO promieniowanie *n*
PY излучение *n*
  радиация *f*
SU säteily
SV strålning *c*

3147 radiator
DA radiator
DE Gliederheizkürper *m*
  Heizkürper *m*
  Radiator *m*
ES radiador *m*
F corps *m* de chauffe
  radiateur *m*
I corpo *m* scaldante
  radiatore *m*
MA radiátor
  sugárzó
NL radiator *m*
PO grzejnik *m*
PY радиатор *m*
SU lämpüpatteri
  radiaattori
SV radiator *c*

3148 radiator casing
DA radiatorskærm
DE Heizkürper-
  verkleidung *f*
ES cubrerradiador *m*
F cache-radiateur *m*
I copri-calorifero *m*
MA radiátorburkolat
NL radiatoromkasting *f*
PO obudowa *f* grzejnika
PY корпус *m*
  радиатора
SU jäähdyttimen vaippa
  patterin verhous
SV radiatormantel *c*

3149 radiator valve
DA radiatorventil
DE Heizkürperventil *n*
ES válvula de
 radiador *f*
F robinet *m* de
 radiateur
I valvola *f* per
 radiatore
MA radiátorszelep
NL radiatorkraan *f*
PO zawór *m*
 grzejnikowy
PY радиаторный
 кран *m*
SU patteriventtiili
SV radiatorventil *c*

3150 rake
DA hakke
 ildrager
DE Feuerhaken *m*
 Schürhaken *m*
ES hurgón *m*
 rastrillo *m*
F raclette *f*
I scovolo *m*
MA kaparó
NL hark *f*
 pook *m*
PO grabie *pl*
 kąt *m* natarcia
 pochylenie *n*
 pogrzebacz *m*
PY кочерга *f*
 скребок *m*
 уклон *m*
SU harava
SV raka *c*

3151 rake *vb*
DA rage
 rive
DE schüren
ES hurgar
 rastrillar
F décrasser
I disincrostare
MA kapar
NL poken
PO oczyszczać
 zgarniać
PY шуровать
SU haravoida
 kaapia
 kallistaa
SV raka

3152 rake out *vb*
DA rode frem
DE abschlacken
ES limar
F dégager
 enlever
 retirer
I disincrostare
MA kikapar
NL ontslakken
PO wygarniać popiół
PY выгребать
SU poistaa tuhka
SV raka ut

3153 raking
DA ransagning
 rivning
DE Schürung *f*
ES rastrillado *m*
F décrassage *m*
I disincrostazione *f*
MA kaparás
NL ontslakken *n*
 poken *n*
PO odpopielanie *n*
 odżużlanie *n*
PY шуровка *f*
SU haravointi
 tuhkanpoisto
SV slaggning *c*

3154 RAM (*see* random
 access memory)

**3155 random-access
 memory**
DA EDB-lagring
DE direkter
 Zugriffs-
 speicher *m*
ES memoria de acceso
 aleatoria *f*
F mémoire *f* vive
I memoria *f* ad
 accesso casuale
 memoria *f* ram
MA véletlen hozzáférésű
 memória
NL RAM *n*
 toegankelijk
 geheugen *n*
PO pamięć *f* o dostępie
 swobodnym
PY оперативное
 запоминающее
 устройство *n*
SU RAM-muisti
SV random-access
 memory *n*

3156 random error
DA tilfældig (hændelig)
 fejl
DE Zufallsfehler *m*
ES error aleatorio *m*
F erreur *f* aléatoire
I errore *m* casuale
MA véletlen tévedés
NL toevallige fout *f*
PO błąd *m*
 przypadkowy
PY ошибка *f*
 случайная
 погрешность *f*
SU satunnaisvirhe
SV slumpmässigt fel *n*

3157 random sampling
DA stikprøve
DE Probeentnahme *f*
ES muestreo al azar *m*
 muestreo
 aleatorio *m*
F prélèvement *m*
 d'échantillon
I prelievo *m* da
 campioni
MA szúrópróbaszerű
 mintavétel
NL steekproef *f*
PO pobieranie *n* losowe
 próbek
PY выборочный отбор
 m проб
SU satunnaisotanta
SV slumpmässigt
 urval *n*

3158 range
DA felt
 område
DE Bereich *m*
 Entfernung *f*
ES banda *f*
 gama *m*
 límites *m,pl*
 zona *f*
F domaine *m*
 gamme *f*
 plage *f*
 portée *f*
I gamma *f*
MA tartomány
NL bereik *n*
 gebied *n*
PO obszar *m*
 przedział *m*
 zakres *m*
PY амплитуда *f*
 диапазон *m*
 расстояние *n*
SU alue
SV omfång *n*

3159 Rankine cycle
DA Rankines kredsløb
DE Rankine-
Kreisprozeß *m*
ES ciclo de Rankine *m*
F cycle *m* de Rankine
I ciclo *m* Rankine
MA Rankine
kürfolyamat
NL Rankine-
kringloop *m*
Rankineproces *n*
PO cykl *m* Rankina
PY цикл *m* Рэнкина
SU Rankineprosessi
SV Rankine-cykel *c*

3160 rated capacity
DA nominel ydelse
DE Nennleistung *f*
ES potencia nominal *f*
F puissance *f*
nominale
I potenza *f* nominale
MA méretezett
teljesítmény
NL nominaal
vermogen *n*
PO pojemność *f*
znamionowa
wydajność *f*
znamionowa
PY номинальная
мощность *f*
SU nimellisteho
SV märkeffekt *c*

3161 rated heat output
DA nominel varmeydelse
DE Nennwärme-
leistung *f*
ES potencia calorífica
nominal *f*
F puissance *f* calori-
fique nominale
I potenza *f* calorifica
nominale
MA méretezett
hőteljesítmény
NL nominale
warmteafgifte *f*
PO wydajność *f* ciepła
nominalna
PY номинальная
теплопроизводитель-
ьность *f*
SU nimellislämpüteho
SV nominell
värmeeffekt *c*

3162 rate of heat flow
DA varmestrømnings-
mængde
DE Nennwärmestrom *m*
ES flujo de calor *m*
flujo térmico *m*
F flux *m* de chaleur
I flusso *m* di calore
MA hőáram mértéke
NL warmtestroom *m*
PO wielkość *f*
strumienia ciepła
PY величина *f*
теплового потока
SU lämpüvirran suuruus
SV värmeflüde *n*

3163 rating
DA klasse
normalydelse
DE Leistung *f*
ES potencia de
régimen *f*
potencia de
servicio *f*
F puissance *f*
I dato *m* di targa
potenza *f*
MA méretezés
NL toelaatbare
belasting *f*
PO wartość *f*
znamionowa
PY номинальная
производитель-
ьность *f*
SU luokittelu
nimellisarvo
SV dimensionering *c*

3164 ratio
DA forhold
proportion
DE Verhältnis *n*
ES proporción *f*
razón *f*
relación *f*
F rapport *m*
I rapporto *m*
MA arány
NL verhouding *f*
PO proporcja *f*
stosunek *m*
PY коэффициент *m*
отношение *n*
степень *f*
SU suhde
SV fürhållande *n*

3165 reaction
DA reaktion
DE Reaktion *f*
ES reacción *f*
F réaction *f*
I reazione *f*
MA reakció
NL reactie *f*
PO oddziaływanie *n*
reakcja *f*
PY реакция *f*
SU reaktio
SV reaktion *c*

3166 reactive power
DA blindeffekt
reaktiv kraft
DE Reaktionskraft *f*
ES potencia reactiva *f*
F puissance *f* réactive
I potenza *f* reattiva
MA visszaható erő
NL blind vermogen *n*
PO moc *f* bierna
PY реактивная
мощность *f*
SU loisteho
SV reaktiv effekt *c*

3167 reactor
DA reaktor
DE Reaktor *m*
ES reactor *m*
F réacteur *m*
I reattore *m*
MA reaktor
NL reactievat *n*
PO reaktor *m*
PY реактор *m*
стабилизатор *m*
SU reaktori
SV reaktor *c*

3168 reading
DA aflæsning
visning
DE Ablesung *f*
Anzeige *f*
ES lectura indicación *f*
F indication *f*
lecture *f*
mesure *f*
I indicazione *f*
lettura *f*
segnale *m*
MA leolvasás
NL aflezing *f*
PO odczyt *m*
wskazanie *n* danych
PY отсчет *m*
показание *n*
SU lukema
SV avläsning *c*

3169 read-only memory
DA hukommelse
  læselager
DE Lesespeicher *m*
ES memoria solo legible
  f
F mémoire *f* morte
I memoria *f* di sola
  lettura
MA csak (ki)olvasható
  memória
NL leesgeheugen *n*
  ROM *n*
PO pamięć *f* stała
PY постоянное
  запоминающее
  устройство *n*
SU ROM-muisti
SV fast minne *n*

3170 ready for use *adj*
DA brugsklar
DE betriebsfertig
ES listo para funcionar
  listo para su
  utilización
F en état *m* de marche
I pronto all'uso
MA üzemkészség
NL gebruiksklaar
PO gotowy do pracy
PY работоспособный
SU käyttüvalmis
SV driftklar

3171 real power
DA sand kraft
DE Wirkleistung *f*
ES potencia real *f*
F puissance *f* réelle
I potenza *f* attiva
MA tényleges
  teljesítmény
NL werkelijk
  vermogen *n*
PO moc *f* czynna
  moc *f* rzeczywista
PY активная
  мощность *f*
  располагаемая
  мощность *f*
SU pätüteho
SV aktiv effekt *c*

3172 real time
DA sand tid
DE Echtzeit *f*
ES tiempo real *m*
F temps *m* réel
I tempo *m* reale
MA tényleges idő
NL real time *n*
PO czas *m* bieżący
  czas *m* rzeczywisty
PY действительное
  время *n*
SU reaaliaika
SV realtid *c*

3173 receiver
DA beholder
  modtager
  samler
DE Sammelbehälter *m*
  Sammler *m*
ES colector *m*
  recipiente *m*
  recipiente de líquido
  m
F bâche *f*
  collecteur *m*
  réservoir *m*
I recipiente *m*
  ricevitore *m*
MA gyűjtő
NL reservoir *n*
  verzamelaar *m*
PO odbiornik *m*
  zbiornik *m*
PY ресивер *m*
SU säiliü
SV mottagare *c*

3174 receiver condenser
DA kondensbeholder
DE Kondensator-
  Sammler *m*
ES recipiente de líquido
  m
F condenseur *m*
  réservoir
  (condenseur
  bouteille)
I condensatore *m* con
  accumulo in fase
  solida
MA kondenzátorgyűjtő
NL opslagcondensor *m*
PO skraplacz *m*
  odbiorczy
PY емкость *f*
  конденсатора
SU lauhdutinvaraaja
  säiliülauhdutin
SV mottagar-
  kondensor *c*

**3175 reciprocating
compressor**
DA stempelkompressor
DE Kolben-
  kompressor *m*
  Kolbenverdichter *m*
ES compresor
  alternativo *m*
  compresor de
  émbolo *m*
  compresor de pistón
  m
F compresseur *m* à
  piston
I compressore *m*
  alternativo
MA dugattyús
  kompresszor
NL zuigercompressor *m*
PO sprężarka *f* tłokowa
PY поршневой
  компрессор *m*
SU mäntäkompressori
SV kolvkompressor *c*

**3176 reciprocating
pump**
DA stempelpumpe
DE Kolbenpumpe *f*
ES bomba
  aspirante-
  impelente *f*
  bomba de émbolo *f*
  rejilla fija *f*
F pompe *f* à piston
I pompa *f* a pistone
MA dugattyús szivattyú
NL plunjerpomp *f*
  zuigerpomp *f*
PO pompa *f* tłokowa
PY поршневой
  насос *m*
SU mäntäpumppu
SV kolvpump *c*

3177 recirculated air
DA recirkuleret luft
DE Umluft *f*
ES aire recirculado *m*
F air *m* recyclé
I aria *f* ricircolata
MA visszakeringtetett
  levegő
NL recirculatielucht *f*
PO powietrze *n*
  obiegowe
  powietrze *n*
  recyrkulacyjne
PY рециркуляционный
  воздух *m*
SU palautusilma
SV återluft *c*

3178 recirculation
DA recirkulation
DE Umwälzung *f*
Wiederumwälzung *f*
ES recirculación *f*
F recirculation *f*
recyclage *m*
I ricircolazione *f*
ricircolo *m*
MA visszakeringtetés
NL recirculatie *f*
PO obieg *m*
recyrkulacja *f*
PY рециркуляция *f*
SU ilman kierrättäminen
SV recirkulation *c*

3179 recooling
DA genkøling
DE Rückkühlung *f*
ES post-enfriamiento *m*
F post-refroidissement *m*
I postraffreddamento *m*
MA visszahűtés
NL nakoeling *f*
PO chłodzenie *n* wtórne
PY доохлаждение *n*
SU jälkijäähdytys
SV återkylning *c*

3180 record drawing
DA registreret tegning
DE Bestandszeichnung *f*
ES plano del expediente *m*
F
I disegno *m* di archivio
MA rajzoló rügzítő
NL revisietekening *f*
PO rysunek *m* wykresowy (przyrządu rejestrującego)
PY зарегистрированная схема *f*
принятая схема *f*
SU luovutuspiirustus
SV registrerad ritning *c*

3181 recorder
DA registreringsapparat skriver
DE Registrierapparat *m*
Schreiber *m*
ES aparato registrador *m*
contador *m*
registrador *m*
F enregistreur *m*
I registratore *m*
MA írószerkezet
NL registrerende meter *m*
PO przyrząd *m* rejestrujący
rejestrator *m*
PY регистрирующий прибор *m*
самопишущее устройство *n*
SU piirturi
SV skrivare *c*

3182 recoverable *adj*
DA genbrugelig
DE wiederherstellen
ES recuperable
F récupérable
I recuperabile
MA visszanyerhető
NL recupereerbaar terugwinbaar
PO odzyskiwalny regenerowalny
PY пригодный для утилизации утилизируемый
SU talteenotettava
SV återvinnbar

3183 recovery
DA genvinding udbytte
DE Rückgewinnung *f*
ES recuperación *f*
regeneración *f*
F récupération *f*
I recupero *m*
MA visszanyerő
NL terugwinning *f*
PO odzyskiwanie *n*
rekuperacja *f*
PY регенерация *f*
утилизация *f*
SU talteenotto
SV återvinning *c*

3184 rectifier
DA ensretter
DE Gleichrichter *m*
ES rectificador *m*
F redresseur *m*
I raddrizzatore *m*
MA egyenirányító
NL gelijkrichter *m*
PO prostownik *m*
PY выпрямитель *m*
ректификатор *m*
SU tasasuuntaaja
SV likriktare *c*

3185 reduced pressure
DA reduceret tryk
DE reduzierter Druck *m*
ES presión reducida *f*
F pression *f* réduite
I pressione *f* ridotta
MA csükkentett nyomás
NL gereduceerde druk *m*
PO ciśnienie *n* zredukowane
PY редуцированное давление *n*
сниженное давление *n*
SU alennettu paine
SV reducerat tryck *n*

3186 reducing valve
DA reduktionsventil
DE Reduzierventil *n*
ES válvula reductora *f*
F détendeur *m*
I detentore *m*
riduttore *m* di pressione
MA csükkentőszelep
NL reduceerafsluiter *m*
PO zawór *m* redukcyjny
PY редукционный клапан *m*
SU paineenalennusventtiili
SV reducerventil *c*

3187 reed valve
DA rørventil
DE Rohrventil *n*
ES válvula de lengüeta *f*
F
I valvola *f* a lamella
MA sípszelep
NL tongklep *f*
vingerklep *f*
PO zawór *m* dźwigniowy
PY лепестковый клапан *m*
SU liuskaventtiili
SV

3188 reference cycle
DA referenceperiode
DE Bezugsperiode *f*
ES ciclo de referencia *m*
F cycle *m* de référence
I ciclo *m* di
   riferimento
MA referencia
   kürfolyamat
NL referentiecyclus *m*
   vergelijkings-
   periode *f*
PO cykl *m* odniesienia
   obieg *m*
   porównawczy
PY исходный цикл *m*
   эталонный цикл *m*
SU vertailujakso
SV referensperiod *c*

3189 reference point
DA fikspunkt
   referencepunkt
DE Bezugspunkt *m*
   Festpunkt *m*
ES punto de referencia
   m
F point *m* de référence
I punto *m* fisso di
   riferimento
MA referencia pont
NL referentiepunt *n*
PO punkt *m* odniesienia
PY исходная
   (фиксированная)
   точка *f*
SU vertailupiste
SV fixpunkt *c*

3190 reference pointer
DA indstillingsviser
DE Bezugszeiger *m*
ES índice de potencia *m*
F index *m*
I indice *m* di
   riferimento
MA referencia mutató
NL referentiewijzer *m*
PO wskaźnik *m*
   odniesienia
PY исходный
   указатель *m*
SU pitoarvon osoitin
SV inställningsvisare *c*

3191 reflectance
DA reflektionskoefficient
DE Reflexions-
   vermügen *n*
ES reflectancia *f*
F facteur *m* de
   réflection
   pouvoir *m* réflecteur
I riflettanza *f*
MA visszaverődés
NL reflectiecöfficint *m*
PO współczynnik *m*
   odbicia
PY отражение *n*
SU heijastussuhde
SV reflektans *c*

**3192 reflective thermal
   insulation**
DA reflekterende
   varmeisolering
DE Reflexionswärme-
   dämmung *f*
ES aislamiento térmico
   reflectivo *m*
F calorifugeage *m*
   réflectif
I isolamento *m*
   termico riflettente
MA visszaverő
   hőszigetelés
NL stralingsisolatie *f*
PO izolacja *f* cieplna
   odblaskowa
PY отражающая
   теплоизоляция *f*
SU heijastava
   lämpüeriste
SV reflekterande
   värmeisolering *c*

3193 reflectivity
DA reflektionsevne
DE Reflexionsgrad *m*
ES coeficiente de
   reflexión *m*
   reflectancia *f*
F pouvoir *m*
   réfléchissant
   réflectivité *f*
I riflettività *f*
MA visszaverőképesség
NL reflectiecöfficint *m*
   reflectievermogen *n*
PO współczynnik *m*
   odbicia
   zdolność *f* odbijania
PY отражательная
   способность *f*
SU heijastussuhde
SV reflexionsfürmåga *c*

3194 reflector
DA refleksglas
   reflektor
DE Reflektor *m*
ES reflector *m*
F réflecteur *m*
I riflettore *m*
MA reflektor
   visszaverő
NL reflector *m*
PO reflektor *m*
PY отражатель *m*
   рефлектор *m*
SU heijastin
SV reflektor *c*

3195 refractoriness
DA ildfasthed
DE Feuerbeständigkeit *f*
   Widerstandskraft *f*
ES poder refractario *m*
   resistencia al fuego *f*
F résistance *f* au feu
   tenue *f* au feu
I resistenza *f* al fuoco
MA tűzállóság
NL vuurvastheid *f*
PO ognioodporność *f*
   ogniotrwałość *f*
PY огнеупорность *f*
   тугоплавкость *f*
SU tulenkestävyys
SV eldfasthet *c*

3196 refractory *adj*
DA ildfast
DE feuerfest
ES refractario
F réfractaire
I refrattario
MA tűzálló
NL vuurvast
PO ognioodporny
   ogniotrwały
PY огнеупорный
   тугоплавкий
SU tulenkestävä
SV eldfast

DA	=	Danish
DE	=	German
ES	=	Spanish
F	=	French
I	=	Italian
MA	=	Magyar (Hungarian)
NL	=	Dutch
PO	=	Polish
PY	=	Russian
SU	=	Finnish
SV	=	Swedish

3197 refractory brickwork
DA ildfast murværk
DE feuerfeste Ausmauerung f
ES ladrillo refractario m
F maçonnerie f réfractaire
I muratura f refrattaria
MA tűzálló tégla
NL vuurvaste bemetseling f
PO mur m ognioodporny mur m z cegły szamotowej
PY огнестойкая керамика f
SU tulenkestävä muuraus
SV eldfast murverk n

3198 refractory lining
DA ildfast foring
DE Brennkammer-ausmauerung f
ES revestimiento de la cámara de combustión m revestimiento refractario m
F garnissage m de chaudière garnissage m réfractaire
I camicia f refrattaria rivestimento m refrattario
MA tűzálló bélés
NL vuurvaste bekleding f
PO wykładzina f ognioodporna
PY огнеупорная облицовка f
SU tulenkestävä verhous
SV eldfast beläggning c

3199 refrigerant
DA kølemiddel
DE Kältemittel n
ES fluido frigorífico m refrigerante m
F réfrigérant m
I refrigerante m
MA hűtőküzeg
NL koelmiddel n
PO czynnik m chłodniczy
PY холодильный агент m
SU kylmäaine
SV küldmedium n

3200 refrigerant circuit
DA kølekredsløb
DE Kältemittel-kreislauf m
ES circuito del fluido frigorífico m circuito del refrigerante m circuito frigorífico m
F circuit m frigorifique
I circuito m del refrigerante
MA hűtőküzeg kür
NL koelcircuit n
PO obieg m chłodniczy
PY схема f циркуляции холодильного агента
SU kylmäpiiri
SV kylkrets c

3201 refrigerant water
DA kølevand
DE Kühlwasser n
ES agua de enfriamiento f agua refrigerante f
F eau f de réfrigération
I acqua f di refrigerazione
MA hűtővíz víz hűtőküzeg
NL koelwater n
PO woda f chłodnicza
PY вода f в качестве холодильного агента
SU jäähdytysvesi
SV kylvatten n

3202 refrigerate *vb*
DA køle
DE kühlen
ES enfriar (artificialmente) refrigerar
F réfrigérer
I raffreddare refrigerare
MA hűt
NL koelen
PO chłodzić
PY охлаждать
SU jäähdyttää
SV kyla

3203 refrigerated cargo
DA kølet last
DE gekühlte Fracht f
ES carga refrigerada f
F cargo m réfrigéré
I mercantile m refrigerato
MA hűtütt rakomány
NL gekoelde lading f
PO ładunek m chłodzony
PY охлаждаемый груз m
SU jäähdytetty lasti
SV kyllast c

3204 refrigerating compressor
DA kølekompressor
DE Kältekompressor m
ES compresor frigorífico m
F compresseur m frigorifique
I compressore m frigorifico
MA hűtőkompresszor
NL koelcompressor m
PO sprężarka f chłodnicza
PY холодильный компрессор m
SU kylmäkompressori
SV kylkompressor c

3205 refrigerating effect
DA køleeffekt
DE Kühlwirkung f
ES potencia frigorífica f
F effet m de refroidissement
I effetto m frigorifero
MA hűtőhatás
NL koeleffect m
PO efekt m chłodzenia
PY холодильный эффект m
SU kylmäteho
SV kyleffekt c

3206 refrigerating machine
- DA kølemaskine
- DE Kältemaschine *f*
- ES máquina frigorífica *f*
- F machine *f* frigorifique
- I macchina *f* frigorifera
- MA hűtőgép
- NL koelmachine *f*
- PO maszyna *f* chłodnicza
- PY холодильная машина *f*
- SU kylmäkone
- SV kylmaskin *c*

3207 refrigeration
- DA køling
- DE Kälteerzeugung *f*
- ES enfriamiento (artificial) *m* frío artificial *m*
- F réfrigération *f*
- I refrigerazione *f*
- MA hűtés
- NL koeling *f*
- PO chłodnictwo *n* chłodzenie *n*
- PY охлаждение *n*
- SU jäähdytys
- SV kylning *c*

3208 refrigeration controller
- DA kølestyring
- DE Kälteregler *m* Kühlregler *m*
- ES regulador de refrigeración *m*
- F régulateur *m* de réfrigération
- I regolatore *m* di refrigerazione
- MA hűtésszabályozó
- NL kouderegelaar *m*
- PO regulator *m* chłodzenia
- PY реле *n* разности давления холодильной машины
- SU jäähdytyksen säätü
- SV kylüvervakare *c*

3209 refrigeration cycle
- DA kølekredsløb
- DE Kältekreislauf *m*
- ES ciclo frigorífico *m*
- F cycle *m* frigorifique
- I ciclo *m* frigorifero
- MA hűtési folyamat
- NL koelcyclus *m*
- PO cykl *m* chłodniczy
- PY холодильный цикл *m*
- SU kompressori
- SV kylperiod *c*

3210 refrigeration plant
- DA køleanlæg
- DE Kälteanlage *f*
- ES instalación frigorífica *f*
- F installation *f* de réfrigération
- I impianto *m* frigorifero
- MA hűtőház
- NL koelinstallatie *f*
- PO instalacja *f* chłodnicza urządzenie *n* chłodnicze
- PY холодильная установка *f*
- SU kylmälaitos
- SV kylanläggning *c*

3211 refrigeration technology
- DA køleteknologi
- DE Kältetechnik *f*
- ES ingeniería del frío *f* técnica del frío *f* técnica frigorífica *f*
- F technologie *f* ou technique *f* du froid
- I tecnologia *f* del freddo tecnologia *f* frigorifera
- MA hűtőtechnológia
- NL koeltechniek *f*
- PO technika *f* chłodzenia technologia *f* chłodzenia
- PY холодильная технология *f*
- SU kylmätekniikka
- SV kylteknologi *c*

3212 refrigerator
- DA køleskab
- DE Kühlschrank *m*
- ES refrigerador *m*
- F réfrigérateur *m*
- I refrigeratore *m*
- MA hűtő
- NL koelelement *n* koelkast *f* koelruimte *f*
- PO chłodziarka *f* szafa *f* chłodnicza
- PY холодильник *m*
- SU jääkaappi
- SV kylskåp *n*

3213 refrigerator baffle
- DA køleskabsbaffel
- DE Kühlerleitblech *n*
- ES separador de frigorífico *m*
- F chicane *f* de réfrigérateur
- I deflettore *m* del refrigeratore
- MA hűtő terelőlap
- NL beschermschot *n* voor koelprodukt leidschot *n* voor gekoelde lucht
- PO owiewka *f* chłodząca (obudowę)
- PY устройство *n*, защищающее содержимое холодильника
- SU jäähdytyslamelli jäähdytyspaneeli
- SV kylskåpsbaffel *c*

3214 regain
- DA genvinding
- DE Rückgewinnung *f*
- ES recuperación *f*
- F récupération *f*
- I recupero *m*
- MA visszanyerés
- NL terugwinnen *n* terugwinning *f*
- PO odzyskiwanie *n*
- PY возврат *m*
- SU takaisinsaanti
- SV återvinning *c*

3215 regeneration
DA regenering
DE Regenerierung *f*
ES regeneración *f*
F régénération *f*
I rigenerazione *f*
MA regeneráció
NL regeneratie *f*
PO odzyskiwanie *n*
regeneracja *f*
PY регенерация *f*
SU regenerointi
SV regenerering *c*

3216 regenerative cooling
DA regenerativ køling
DE regenerative Kühlung *f*
ES enfriamiento regenerativo *m*
F refroidissement *m* par récupération
I raffreddamento *m* rigenerativo
MA regeneratív hűtés
NL koeling *f* met warmteterug-winning
PO chłodzenie *n* regeneracyjne (z odzyskiwaniem)
PY регенеративное охлаждение *n*
SU regeneratiivinen jäähdytys
SV regenerativ kylning *c*

3217 register
DA register spjæld
DE Register *n*
ES parrilla de registro *f* registro *m*
F registre *m*
I bocchetta *f* registro *m*
MA jegyzék regiszter
NL register *n* ventilatierooster *m*
PO rejestr *m* urządzenie *n* rejestrujące zasuwa *f* kominowa
PY регистр *m*
SU säleikkü
SV register *n* spjäll *n*

3218 register *vb*
DA registrere
DE anzeigen registrieren
ES indicar inscribir
F indiquer
I registrare
MA feljegyez regisztrál
NL inschrijven registreren
PO rejestrować
PY показывать регистрировать
SU rekisterüidä
SV registrera

3219 register burner
DA register- (flertrins) brænder
DE Registerbrenner *m*
ES registro de quemador *m*
F registre *m* de brûleur
I bruciatore *m* a registro
MA regiszteres égő
NL brander *m* met verstelbare luchtgeleider
PO palnik *m* fluidyzacyjny
PY регулятор *m* работы горелки
SU palamisilman ohjaussäleikkü
SV registerbrännare *c*

3220 registered engineer
(*see* professional engineer)

3221 regulate *vb*
DA regulere
DE regeln regulieren stellen
ES regular
F régler
I regolare
MA szabályoz
NL regelen
PO regulować
PY регулировать
SU säätää
SV reglera

3222 regulating tee
DA regulerings-tee
DE Regulier-T-Stück *n*
ES T de regulación *f*
F té *m* de réglage
I valvola *f* a tre vie di regolazione
MA szabályozó T idom
NL regeltee *n*
PO trójnik *m* regulacyjny
PY регулирующий тройник *m*
SU säädettävä kolmitieventtiili
SV reglerbar trevägsventil *c*

3223 regulating valve
DA reguleringsventil
DE Regulierungsventil *n* Stellventil *n*
ES válvula de regulación *f*
F robinet *m* de réglage
I valvola *f* di regolazione
MA szabályozószelep
NL regelafsluiter *m*
PO zawór *m* regulacyjny
PY регулирующий клапан *m*
SU säätüventtiili
SV reglerventil *c*

3224 regulation
DA regulering
DE Regulierung *f*
ES regulación *f*
F régulation *f*
I regolazione *f*
MA szabályozás
NL regelen *n* regeling *f* voorschrift *n*
PO regulacja *f*
PY регулирование *n*
SU säännüs
SV reglering *c*

3225 reheat
DA genopvarmning
DE Nachwärme *f*
ES recalentamiento *m*
F réchauffage *m*
I postriscalda-mento *m*
MA újra fűtés
NL naverwarming *f*
PO ogrzewanie *n* wtórne
PY отогрев *m* подогрев *m*
SU jälkilämmitys
SV tillsatsvärme *c*

3226 reheat vb
DA genopvarme
DE nachheizen
   wiedererwärmen
ES recalentar
F réchauffer
I postriscaldare
MA újra fűt
NL naverwarmen
PO dogrzewać
PY отогревать
   перегревать
SU jälkilämmittää
SV återupphetta

3227 reheat coil
DA eftervarme
DE Nacherhitzer-
   schlange f
ES batería de
   recalentamiento f
F batterie f de
   réchauffage
I batteria f di post-
   riscaldamento
MA újrafűtő csőkígyó
NL naverwarmings-
   spiraal f
PO nagrzewnica f
   wtórna
PY змеевик-
   подогреватель m
SU jälkilämmityspatteri
SV återupphettnings-
   slinga c

3228 reheater
DA eftervarmer
DE Nacherhitzer m
   Nachwärmer m
ES recalentador m
F post-réchauffeur m
   réchauffeur m
   secondaire
I postriscaldatore m
MA léghevítő
   újrafűtő
NL naverhitter m
   naverwarmer m
PO podgrzewacz m
   wtórny
   przegrzewacz m
   międzystopniowy
PY отогреватель m
   подогреватель m
SU jälkilämmitin
SV eftervärmnings-
   apparat c

3229 REHVA (see
   Appendix B)

3230 reinforce vb
DA forstærke
DE verstärken
ES reforzar
F renforcer
I rinforzare
MA erősített
NL versterken
   wapenen
PO wzmacniać
   (konstrukcję)
PY усиливать
SU vahvistaa
SV fürstärka

3231 reinforcing frame
DA forstærkningsramme
DE Verstärkungs-
   rahmen m
ES bastidor de
   refuerzo m
   estructura de
   refuerzo f
F cadre m de renfort
I struttura f di
   rinforzo
MA erősített keret
NL versterkingsraam-
   werk n
PO rama f
   wzmacniająca
PY усиленная рама f
SU vahvistava kehys
SV fürstärkningsram c

3232 relative adj
DA relativ
DE relativ
ES relativo
F relatif
I relativo
MA relatív
   viszonylagos
NL betrekkelijk
   relatief
PO względny
PY относительный
SU suhteellinen karheus
SV relativ

3233 relative humidity
DA relativ fugtighed
DE Regelung f der
   relativen
   Feuchte f
ES humedad relativa f
F humidité f relative
I umidità f relativa
MA relatív nedvesség
NL relatieve
   vochtigheid f
PO wilgotność f
   względna
PY относительная
   влажность f
SU suhteellinen kosteus
SV relativ fuktighet c

3234 relay
DA relæ
DE Relais n
ES relé m
   relevador m
F relais m
I rel m
MA relé
NL overdracht f
   relais n
PO przekaźnik m
PY реле n
SU rele
SV relä n

3235 reliability
DA pålidelighed
DE Zuverlässigkeit f
ES fiabilidad f
F sécurité f
I affidabilità f
   sicurezza f
MA megbízhatóság
NL betrouwbaarheid f
PO niezawodność f
PY безотказность f
SU luotettavuus
SV tillfürlitlighet c

DA	=	Danish
DE	=	German
ES	=	Spanish
F	=	French
I	=	Italian
MA	=	Magyar
		(Hungarian)
NL	=	Dutch
PO	=	Polish
PY	=	Russian
SU	=	Finnish
SV	=	Swedish

3236 relief valve
DA aflastningsventil
DE Entlastungsventil *n*
    Überströmventil *n*
ES válvula de escape *f*
    válvula de
    seguridad *f*
F soupape *f* de
    décharge
I valvola *f* di scarico
MA biztonsági szelep
NL ontlastklep *f*
PO zawór *m*
    nadmiarowy
    zawór *m*
    odciążający
PY предохранительный
    клапан *m*
SU ylipaineventtiili
SV avlastningsventil *c*

3237 remainder
DA rest
DE Rückstand *m*
ES remanente *m*
    residuo *m*
    resto *m*
F résidu *m*
I residuo *m*
    resto *m*
MA maradék
NL overblijfsel *n*
    residu *n*
PO pozostałość *f*
    reszta *f*
PY остаток *m*
SU jäännüs
SV rest *c*

3238 remote control
DA fjernstyring
DE Fernbedienung *f*
ES control a
    distancia *m*
    mando a
    distancia *m*
    telemando *m*
F commande *f* à
    distance
    régulation *f* à
    distance
I comando *m* a
    distanza
MA távvezérlés
NL afstandsbediening *f*
PO sterowanie *n* zdalne
PY дистанционное
    управление *n*
SU kaukosäätü
SV fjärrkontroll *c*

**3239 remote indicating
    thermometer**
DA fjerntermometer
DE Fernanzeigethermo-
    meter *n*
ES teletermómetro *m*
    termómetro a
    distancia *m*
    termómetro de
    lectura a
    distancia *m*
F thermomètre *m*
    indicateur à
    distance
I termometro *m* a
    distanza
MA távhőmérő
NL afstands-
    thermometer *m*
    thermometer *m* met
    aanwijzing op
    afstand
PO termometr *m* ze
    zdalnym odczytem
PY показывающий
    дистанционный
    термометр *m*
SU kaukolämpümittari
SV fjärrtermometer *c*

3240 remote sensor
DA fjernføler
DE Fernfühler *m*
ES sonda remota *f*
F capteur *m* à distance
    (décentralisé)
I sensore *m* remoto
MA távérzékelő
NL voeler *m* op afstand
PO czujnik *m* zdalnego
    sterowania
PY дистанционный
    датчик *m*
SU etäistuntoelin
SV fjärravkännare *c*

3241 removal
DA bortfjernelse
DE Beseitigung *f*
ES eliminación *f*
    evacuación *f*
F élimination *f*
I evacuazione *f*
MA eltávolítás
NL afvoer *m*
    verhuizing *f*
    verwijdering *f*
PO usuwanie *n*
PY удаление *n*
    устранение *n*
SU poisto
SV borttagning *c*

**3242 rendering
    (building)**
DA puds
    pudsning
DE Übergabe *f*
ES enlucido (edif) *m*
F enduit *m* (bâtiment)
I restituzione *f* grafica
MA vakolat
NL artist impression
    tekening *f*
    beraping *f*
PO pierwsza warstwa *f*
    tynku
PY штукатурка *f*
    (здания)
SU rappaus
SV rappning *c*

3243 replacement cost
DA genanskaffelsespris
DE Austauschkosten *f*
ES corte de sustitución
    m
F coût *m* de
    remplacement
I costo *m* di
    sostituzione
MA pótlási kültség
NL vervangingskosten
    pl (eigentijdse
    uitvoering)
PO koszt *m* wymiany
PY стоимость *f*
    реконструкции
SU uusintakustannus
SV utbyteskostnad *c*

3244 reproduction cost
DA genfremstillingspris
DE Wiederherstellungs-
    kosten *f,pl*
ES corte de
    reproducción *m*
F coût *m* de
    reproduction
I costo *m* di
    riproduzione
MA reprodukciós kültség
NL vervangingskosten
    pl (zelfde
    uitvoering)
PO koszt *m*
    odtworzenia
PY стоимость *f*
    воспроизведения
SU uusintakustannus
SV reproduktions-
    kostnad *c*

3245 research
DA forskning
DE Forschung *f*
ES investigación *f*
F recherche *f*
I ricerca *f*
MA kutatás
NL onderzoek *n*
speurwerk *n*
PO badanie *n* naukowe
PY исследование *n*
SU tutkimus
SV forskning *c*

3246 reservoir
DA beholder
reservoir
DE Speicher *m*
ES acumulador *m*
F réservoir *m*
I serbatoio *m*
MA tároló
NL reservoir *n*
verzamelbak *m*
PO zasobnik *m*
zbiornik *m*
PY резервуар *m*
SU säiliü
SV reservoar *c*

3247 reset *vb*
DA tilbagestille
DE wieder einstellen
ES rearmar
F mettre en place
replacer
I azzerare
ritarare
MA visszaállít
NL herstellen
initialiseren
PO ponownie nastawiać
(przyrząd)
przestawiać
PY устанавливать
заново
SU palauttaa
SV nollställa

3248 reset control
DA tilbagestillings-
regulering
DE wiedereinzustellender
Regler *m*
ES control de rearme *m*
F changement *m* de
point de consigne
I riposizionamento *m*
ritaratura *f*
MA visszaállító
szabályozás
NL cascaderegeling *f*
PO regulacja *f* w
układzie
zamkniętym
regulacja *f* z
przestawieniem
regulacja *f* ze
sprzężeniem
zwrotnym
PY регулирование *n*
пропусками
SU asetusarvosäätü
SV återställnings-
kontroll *c*

**3249 residential air
conditioning**
DA boligventilation
DE Wohnungsklimati-
sierung *f*
ES climatización
residencial *f*
F climatisation *f*
d'habitation
I condizionamento *m*
residenziale
MA lakóépületek
lékondicionálása
NL woonhuislucht-
behandeling *f*
PO klimatyzacja *f*
pomieszczeń
mieszkalnych
PY кондиционирование
n воздуха в
жилых зданиях
SU asuntoilmastointi
SV luftkonditionering
c i bostadshus *n*

3250 residential heating
DA boligopvarmning
DE Wohnungsheizung *f*
ES calefacción
residencial *f*
F chauffage *m*
résidentiel
I riscaldamento *m*
residenziale
MA lakásfűtés
NL woonhuis-
verwarming *f*
PO ogrzewanie *n*
pomieszczeń
mieszkalnych
PY отопление *n* жилых
зданий
SU asuntolämmitys
SV uppvärmning *c* i
bostadshus *n*

3251 resistance
DA modstand
modstandsevne
DE Widerstand *m*
ES resistencia *f*
F résistance *f*
I resistenza *f*
MA ellenállás
NL weerstand *m*
PO opór *m*
PY сопротивление *n*
SU vastus
SV motstånd *n*

3252 resistance electric heating
DA elektrisk modstands-opvarmning
DE elektrische Widerstands-heizung *f*
ES calefacción por resistencia eléctrica *f*
F chauffage *m* électrique par résistance
I riscaldamento *m* a resistenza
MA elektromos ellenállásfűtés
NL elektrische weerstands-verwarming *f*
PO ogrzewanie *n* elektryczne oporowe
PY электрическое отопление *n* с использованием электрического сопротивления
SU sähkülämmitys
SV direktelvärme *c*

3253 resistance welding
DA modstandssvejsning
DE Widerstands-schweißung *f*
ES soldadura eléctrica por resistencias *f*
F soudure *f* électrique par résistance
I saldatura *f* a resistenza
MA ellenálláshegesztés
NL weerstandslassen *n*
PO spawanie *n* oporowe
PY контактная сварка *f*
SU vastushitsaus
SV motståndssvetsning *c*

3254 resonance
DA genklang resonans
DE Resonanz *f*
ES resonancia *f*
F résonance *f*
I risonanza *f*
MA rezonancia
NL resonantie *f*
PO rezonans *m*
PY резонанс *m*
SU resonanssi
SV resonans *c*

3255 restriction
DA begrænsning
DE Beschränkung *f* Einschränkung *f*
ES limitación *f* reducción *f* restricción *f*
F ajutage *m*
I restrizione *f* strozzatura *f*
MA korlátozás
NL beperking *f* restrictie *f* voorbehoud *n*
PO ograniczenie *n*
PY ограничение *n*
SU rajoitus
SV restriktion *c*

3256 resultant *adj*
DA resulterende
DE resultierend
ES resultante
F résultant
I risultante
MA eredő
NL resulterend
PO wynikowy wypadkowy
PY результирующий
SU resultoiva
SV resulterande

3257 return
DA retur
DE Rücklauf *m*
ES retroceso *m*
F retour *m*
I ritorno *m*
MA visszaáramlás
NL retourleiding *f*
PO powrót *m*
PY возврат *m*
SU paluu
SV återledning *c*

3258 return air
DA returluft
DE Abluft *f* Rückluft *f*
ES aire de retorno *m*
F air *m* repris
I aria *f* di ricircolo aria *f* di ritorno
MA visszaáramló levegő
NL retourlucht *f*
PO powietrze *n* powrotne
PY обратный воздушный поток *m* рециркуляционный воздух *m*
SU kiertoilma palautusilma
SV återluft *c*

3259 return air inlet
DA returlufttilførsel
DE Abluftöffnung *f* Rückluftöffnung *f*
ES rejilla de retorno del aire *f*
F bouche *f* de reprise d'air
I presa *f* di ritorno
MA visszatérő levegő belépése
NL retourluchtopening *f*
PO wywiewnik *m*
PY вход *m* рециркуля-ционного воздуха
SU palautusilma-aukko
SV återluftsintag *n*

3260 return bend
DA U-bøjning
DE Rücklaufbogen *m*
ES codo de 180° *m* codo en U *m*
F coude *m* en U
I curva *f* a U
MA 180 fokos ívcső
NL dubbele bocht *f*
PO łuk *m* w kształcie litery U
PY калач *m*
SU 180:n kaari
SV returbüj *c*

3261 return connection
DA returforbindelse
DE Rücklaufanschluß *m*
ES conexión de
retorno *f*
F branchement *m* de
retour
I collegamento *m* di
ritorno
MA visszatérő
csatlakozás
NL retouraansluiting *f*
PO króciec *m*
przyłączny
powrotu
przyłącze *n* powrotu
PY обратная связь *f*
ответный
патрубок *m*
SU paluujohdon liitos
SV återledning *c*

**3262 return-flow
compressor**
DA vekselstrøms-
kompressor
DE Rückströmungs-
verdichter *m*
ES compresor de reflujo
m
F compresseur *m* à
contre-courant (à
flux inversé)
I compressore *m* a
flusso inverso
MA visszafolyásos
kompresszor
NL tegenstroom-
compressor *m*
PO sprężarka *f* o obiegu
zamkniętym
sprężarka *f* o
przepływie
nawrotnym
PY непрямоточный
компрессор *m*
SU palautusvirtaus-
kompressori
SV återströmnings-
kompressor *c*

3263 return line
DA returledning
DE Rücklaufleitung *f*
Rücklaufstrang *m*
ES conducto de
retorno *m*
F tuyauterie *f* de
retour
I condotta *f* di
ritorno
MA visszatérő vezeték
NL retourleiding *f*
PO przewód *m*
powrotny
PY обратная линия *f*
SU paluujohto
SV returledning *c*

3264 return manifold
DA retur-samlerør
DE Rücklaufsammler *m*
ES colector de
retorno *m*
F collecteur *m* de
retour
I collettore *m* di
ritorno
MA visszatérő gyűjtő
NL retour-
verzamelaar *m*
PO rozdzielacz *m*
powrotny
PY обратная
магистраль *f*
SU paluukammio
paluutukki
SV retursamlings-
ledning *c*

3265 return pipe
DA returrør
DE Rücklaufrohr-
leitung *f*
ES tubería de retorno *f*
F tuyauterie *f* de
retour
I tubo *m* di ritorno
MA visszatérő cső
NL retourleiding *f*
PO rura *f* powrotna
PY обратная труба *f*
SU paluuputki
SV återledning *c*

3266 return riser
DA returstigledning
DE Fallstrang *m*
ES columna de
retorno *f*
F colonne *f* de retour
I colonna *f* di ritorno
MA visszatérő felszálló
NL retourstrang *m*
PO pion *m* powrotny
PY обратный стояк *m*
SU nousujohto (paluu)
SV stigare *c* (retur)

3267 return temperature
DA returtemperatur
DE Rücklauf-
temperatur *f*
ES temperatura de
retorno *f*
F température *f* de
retour
I temperatura *f* di
ritorno
MA visszatérő
hőmérséklet
NL retourtemperatuur *f*
PO temperatura *f*
powrotu
PY температура *f*
возврата
(например,
обратной воды)
SU paluulämpütila
SV återgångs-
temperatur *c*

3268 return water
DA returvand
DE Rücklaufwasser *n*
ES agua de retorno *f*
F eau *f* de retour
I acqua *f* di ricircolo
MA visszatérő víz
NL retourwater *n*
PO woda *f* powrotna
PY обратная вода *f*
SU paluuvesi
SV returvatten *n*

3269 reverberation chamber
DA efterklangsrum
DE Hallraum *m*
ES cámara de reverberación *f*
F chambre *f* réverbérante
I camera *f* riverberante
MA utózengő szoba
NL nagalmkamer *f*
PO komora *f* pogłosowa
PY реверберационная камера *f*
SU kaiuntahuone
SV efterklangsrum *n*

3270 reverberation time
DA efterklangstid
DE Nachhallzeit *f*
ES tiempo de reverberación *m*
F temps *m* de réverbération
I tempo *m* di riverberazione
MA utózengési idő visszaverődési idő
NL nagalmtijd *m*
PO czas *m* pogłosu
PY время *n* затухания время *n* реверберации
SU jälkikaiunta-aika
SV efterklangstid *c*

3271 reverse cycle
DA omvendt kredsløb
DE Rückwärtslauf *m* Umkehrkreislauf *m*
ES ciclo inverso *m*
F cycle *m* inversé
I ciclo *m* inverso
MA megfordítható kürfolyamat
NL omgekeerde kringloop *m*
PO obieg *m* odwrotny
PY обратный цикл *m*
SU käänteinen prosessi
SV omkastningscykel *c*

3272 reverse cycle defrosting
DA reversibel afrimning
DE Umkehrkreislauf-Abtauung *f*
ES desecarche por inversión de ciclo *m*
F dégivrage *m* par inversion de cycle
I sbrinamento *m* a inversione di ciclo
MA megfordítható folyamatú leolvasztás
NL heetgasontdooiing *f*
PO odmrażanie *n* w obiegu odwrotnym
PY размораживание *n* обратным циклом
SU sulatus käänteisellä prosessilla
SV omkastningscykel-avfrostning *c*

3273 reverse cycle refrigeration
DA reversibel køling
DE umkehrbarer Kältekreislauf *m*
ES refrigeración por inversión del ciclo *f*
F cycle *m* de réfrigération inversé
I refrigerazione *f* a ciclo inverso
MA megfordítható folyamatú fagyasztás
NL omgekeerde koelkringloop *m*
PO chłodzenie *n* w obiegu odwrotnym
PY обратный цикл *m* охлаждения
SU käännetty kylmäprosessi
SV omkastningscykel-kylning *c*

3274 reversed return system
DA omvendt retursystem
DE Tichelmann-System *n* umgekehrtes Rücklaufsystem *n*
ES sistema Tichelmann (de retorno invertido) *m*
F boucle *f* de Tichelmann retour *m* à boucle inversée
I ritorno *m* invertito
MA Tichelmann rendszer
NL Tichelmann-systeem *n*
PO system *m* Tichelmanna
PY реверсивная система *f*
SU käännetyn paluun järjestelmä
SV reverserat retursystem *n*

3275 reversible *adj*
DA reversibel
DE umkehrbar
ES reversible
F réversible
I reversibile
MA megfordítható
NL omkeerbaar
PO odwracalny
PY обратный
SU palautuva
SV reversibel

3276 revolution
DA omdrejninger
DE Umdrehung *f*
ES número de revoluciones *m*
F tour *m*
I giro *m*
MA fordulatszám
NL omwenteling *f*
PO obrót *m*
PY вращение *n*
SU kierros kierrosaika
SV varv *n*

3277 revolving vane anemometer
DA løbehjulsanemometer
DE Drehschalenanemometer *n*
ES anemómetro de vena giratoria *m*
F anémomètre *m* à moulinet
I anemometro *m* a ventola
MA forgólapátos anemométer
NL vleugelradanemometer *m*
PO anemometr *m* skrzydełkowy obrotowy
PY крыльчатый анемометр *m*
SU siipipyüräanemometri
SV anemometer *c* med roterande propellerblad *n*

3278 rigid plastic foam
DA stiv pladeisolering
DE Hartschaumstoff *m*
ES espuma plástica rígida *f* plástico celular rígido (aislamientos) *m*
F plastique *m* expansé rigide
I schiuma *f* plastica rigida
MA szilárd műanyag hab
NL hard schuimplastic *n*
PO pianka *f* ze sztywnego tworzywa sztucznego
PY твердая пластмассовая пена *f*
SU jäykkä vaahtomuovieriste
SV skumplast *c*

3279 rise *vb*
DA stige
DE ansteigen
ES subir
F monter
I aumentare salire
MA felszáll
NL stijgen
PO wznosić
PY нарастать подниматься
SU nousta
SV stiga

3280 riser
DA stigledning
DE Steigleitung *f*
ES columna vertical *f* tubería ascendente *f*
F colonne *f* montante
I colonna *f* montante griglia *f* montante
MA felszálló
NL stijgleiding *f*
PO pion *m* pion *m* wznośny
PY стояк *m*
SU nousujohto
SV stigare *c*

3281 rising main
DA hovedstigrør
DE Hauptsteigleitung *f*
ES conducto (principal) ascendente *m*
F colonne *f* montante
I colonna *f* montante
MA főfelszálló
NL stijgende hoofdleiding *f*
PO przewód *m* wznośny główny
PY главный стояк *m*
SU päänousujohto
SV stigarledning *c*

3282 rising pipe
DA stigledning
DE Steigrohr *n*
ES tubo ascendente *m*
F tuyauterie *f* montante
I tubo *m* montante
MA felszálló vezeték
NL stijgleiding *f*
PO rura *f* wznośna
PY подающая труба *f*
SU nousujohto
SV stigrür *n*

3283 rivet
DA nagle nitte
DE Niete *f*
ES remache *m*
F rivet *m*
I rivetto *m*
MA szegecs
NL klinknagel *m* popnagel *m*
PO nit *m*
PY заклепка *f*
SU niitti
SV nit *c*

3284 rivet *vb*
DA nitte
DE nieten vernieten
ES remachar
F river
I rivettare
MA szegecsel
NL klinken
PO nitować
PY заклепывать
SU niitata
SV nita

3285 rock bed regeneration
DA regenerativ stenvarmeveksler
DE Gesteinsschicht-Regenerierung *f*
ES regeneración de fondos *f*
F régénération *f* sur lit de graviers
I rigenerazione *f* a letto di ghiaia
MA kőágyas regenerálás
NL grintbed *n* thermische terugwinning en opslag
PO regeneracja *f* ze złożem kamiennym
PY грунтово-гравийная аккумуляция *f*
SU kivivaraajaa hyväksikäyttävä lämmüntalte-enotto
SV geotermisk regenerering *c*

3286 rocking grate
DA rysterist
DE Rüttelrost *m*
  Schüttelrost *m*
ES parrilla oscilante *f*
F grille *f* oscillante
I griglia *f* oscillante
MA mozgórostély
NL schudrooster *m*
PO ruszt *m* wahliwy
PY качающаяся
  колосниковая
  решетка *f*
SU täryarina
SV skakrost *c*

3287 roller bearing
DA rulleleje
DE Rollenlager *n*
ES cojinete de bolas *m*
F roulement *m* à
  rouleaux
I cuscinetto *m* a rulli
MA gürgős csapágy
NL rollager *n*
PO łożysko *n* rolkowe
  łożysko *n*
  wałeczkowe
PY роликовый
  подшипник *m*
SU rullalaakeri
SV rullager *n*

3288 roll filter
DA rullefilter
DE Rollbandfilter *m*
ES filtro enrrolable *m*
F filtre *m* à
  déroulement
I filtro *m* a rullo
MA tekercsszűrő
NL rolfilter *m*
PO filtr *m* obrotowy
PY рулонный
  фильтр *m*
SU rullasuodatin
SV rullfilter *n*

**3289 rolling piston
  compressor**
DA skrue-stempel-
  kompressor
DE Drehkolben-
  verdichter *m*
ES compresor de pistón
  oscilante *m*
F compresseur *m* à
  piston rotatif
I compressore *m* a
  pistone rotativo
MA forgódugattyús
  kompresszor
NL roterende
  zuiger-
  compressor *m*
PO sprężarka *f* tłokowa
  obrotowa
PY компрессор *m* с
  катящимся
  поршнем
SU pyürivämäntäinen
  kompressori
SV rolling piston
  kompressor *c*

3290 ROM (*see* see read-
  only memory)

3291 roof
DA tag
DE Dach *n*
ES techo *m*
  tejado *m*
F toit *m*
I tetto *m*
MA tető
NL dak *n*
PO dach *m*
PY крыша *f*
SU katto
SV tak *n*

3292 roof spray cooling
DA køling ved
  besprøjtning af tag
DE Dachsprüh-
  kühlung *f*
ES enfriamiento por
  rociado de techo
  m
F refroidissement *m*
  en toiture par
  aspersion
I raffreddamento *m*
  con tetto bagnato
MA tetőpermetező hűtés
NL dakberegenings-
  koeling *f*
PO chłodzenie *n* dachu
  poprzez zraszanie
PY охлаждение *n*
  крыши
  распыливанием
  (воды)
SU katonkastelu-
  jäähdytys
SV taksprejkylning *c*

**3293 roof-top air
  conditioner**
DA tagmonteret
  klimaaggregat
DE Dachklimagerät *n*
  Dachlüftungsgerät *n*
ES acondicionador de
  aire de tejado *m*
F conditionneur *m*
  d'air en toiture
I condizionatore *m*
  roof-top
MA tetőtéri
  légkondicionáló
NL dakluchtbehandelings-
  centrale *f*
PO klimatyzator *m*
  dachowy
PY крышный
  кондиционер *m*
SU katolle asennettava
  ilmastointikone
SV takluftbehandlare *c*

3294 roof ventilator
DA tagventilator
DE Dachlüfter m
   Dachventilator m
ES ventilador de
   techo m
F ventilateur m de
   toiture
I torrino m
   ventilatore m a tetto
MA tetőventilátor
NL dakventilator m
PO wentylator m
   dachowy
PY крышный
   вентилятор m
SU kattotuuletin
SV takventilator c

3295 room
DA rum
DE Raum m
ES cámara f
   habitación f
   local m
   vivienda f
F chambre f
   espace m
   local m
   pièce f
   salle f
I camera f
   stanza f
MA helyiség
NL kamer f
   ruimte f
   vertrek n
PO pokój m
   pomieszczenie n
PY зал m
   комната f
   цех m
SU huone
SV rum n

3296 room air
DA rumluft
DE Raumluft f
ES aire ambiente
   interior m
F air m intérieur
I aria f ambiente
MA helyiséglevegő
NL vertreklucht f
PO powietrze n
   wewnętrzne
PY воздух m
   помещения
SU huoneilma
SV rumsluft c

**3297 room air
conditioner**
DA rumluftbehandler
DE Raumklimagerät n
ES acondicionador de
   aire doméstico m
F conditionnement m
   d'air de pièce
I condizionatore m di
   aria ambiente
MA helyiség
   légkondicionáló
   szobaklíma
NL kamerkoeler m
PO klimatyzator m
   pokojowy
PY комнатный
   кондиционер m
SU huonekohtainen
   ilmastointikone
SV rumsklimat-
   aggregat n

3298 room air humidifier
DA rumluftbefugter
DE Raumluft-
   befeuchter m
ES humidificador de
   aire m
F humidificateur m de
   pièce
I umidificatore m di
   aria ambiente
MA helyiség légnedvesítő
NL kamerlucht-
   bevochtiger m
PO nawilżacz m
   powietrza
   pokojowy
PY комнатный
   увлажнитель m
   воздуха
SU huonekohtainen
   ilmankostutin
SV rumsluftfuktare c

3299 room calorimeter
DA rumvarmemåler
DE Raumwärme-
   messer m
ES calorímetro local m
F chambre f
   calorimétrique
I calorimetro m
   ambiente
MA helyiség
   hőmennyiségmérő
NL calorimetriekamer f
PO kalorymetr m
   pokojowy
PY комнатный
   калориметр m
SU tehonmittaushuone
SV rumskalorimeter c

**3300 room
characteristic**
DA rumkarakteristika
DE Raumkenngrüße f
ES características del
   local f,pl
F caractéristique f du
   local
I caratteristica f del
   locale
MA helyiség jellemző
NL ruimte-
   karakteristiek f
PO charakterystyka f
   pomieszczenia
PY характеристика f
   помещения
SU huoneen ominaisuus
SV rums-
   karakteristika n

3301 room constant
DA rumkonstant
   (efterklangstid)
DE Raumkonstante f
ES constante del local f
F constante f de
   réverbération
I costante f del locale
MA helyiségállandó
NL ruimteconstante f
   (nagalmtijd)
PO stała f
   pomieszczenia
PY константа f
   помещения
SU huonevakio
SV rumskonstant c

3302 room control
DA rumregulering
DE Raumregelung f
ES control local m
F régulation f de local
I regolazione f
   ambiente
MA helyiség szabályozás
NL regeling f
   ruimtelucht
   ruimteregelaar m
PO regulacja f
   parametrów
   pomieszczenia
PY регулирование n
   параметров в
   помещении
SU huonesäätü
SV rumskontroll c

3303 room heater
DA rumvarmeapparat
DE Raumheizer *m*
ES estufa *f*
F poêle *m*
I riscaldatore *m* ambiente
MA helyiség fűtő
NL kamerverwarmings-apparaat *n*
PO podgrzewacz *m* pokojowy
PY комнатный нагреватель *m*
SU huonelämmitin
SV rumsvärmare *c*

3304 room height
DA rumhøjde
DE Raumhühe *f*
ES altura del local *f*
F hauteur *f* de local
I altezza *f* del locale
MA helyiség magasság
NL vertrekhoogte *f*
PO wysokość *f* pomieszczenia
PY высота *f* помещения
SU huonekorkeus
SV rumshüjd *c*

3305 room temperature
DA rumtemperatur
DE Raumtemperatur *f*
ES temperatura del local *f*
F température *f* intérieure du local
I temperatura *f* ambiente
MA helyiséghőmérséklet
NL vertrek-temperatuur *f*
PO temperatura *f* wewnętrzna
PY температура *f* помещения
SU huonelämpütila
SV rumstemperatur *c*

3306 rotary *adj*
DA roterende
DE drehend rotierend
ES rotativo
F rotatif
I rotativo
MA forgó
NL draaiend roterend
PO obrotowy rotacyjny
PY вращающийся
SU pyürivä
SV roterande

3307 rotary atomizing burner
DA rotationsforstøvnings-brænder
DE Rotationszerstäuber-brenner *m*
ES quemador atomizador rotativo *m*
F brûleur *m* rotatif à atomisation
I bruciatore *m* con atomizzazione rotativa
MA forgóporlasztásos égő
NL cupbrander *m* roterende brander *m*
PO palnik *m* obrotowy rozpylający
PY форсунка *f* с дисковым распыливанием
SU pyürivä poltin
SV rotationsbrännare *c*

3308 rotary burner
DA rotationsbrænder
DE Rotationsbrenner *m*
ES quemador rotativo *m*
F brûleur *m* à coupelle rotative
I bruciatore *m* rotativo
MA forgóégő
NL cupbrander *m* roterende brander *m*
PO palnik *m* obrotowy
PY поворотная горелка *f*
SU pyürivä poltin
SV rotationsbrännare *c*

3309 rotary compressor
DA rotationskompressor
DE Rotations-verdichter *m* Turboverdichter *m*
ES compresor rotativo *m*
F compresseur *m* rotatif compresseur *m* rotatif volumétrique
I compressore *m* rotativo
MA forgókompresszor
NL roterende compressor *m*
PO sprężarka *f* rotacyjna
PY ротационный компрессор *m*
SU keskipako-kompressori kiertomäntä-kompressori
SV rotations-kompressor *c*

3310 rotary filter
DA rullefilter tromlefilter
DE Umlauffilter *m*
ES filtro rotativo *m*
F filtre *m* rotatif
I filtro *m* rotativo
MA forgó szűrő
NL roterend filter *m*
PO filtr *m* obrotowy
PY вращающийся фильтр *m*
SU rullasuodatin
SV rullfilter *n*

DA	=	Danish
DE	=	German
ES	=	Spanish
F	=	French
I	=	Italian
MA	=	Magyar (Hungarian)
NL	=	Dutch
PO	=	Polish
PY	=	Russian
SU	=	Finnish
SV	=	Swedish

3311 rotary heat exchanger

DA rotationsvarme-veksler
DE Rotationswärme-austauscher *m*
ES cambiador de calor rotativo *m*
F échangeur *m* rotatif
I scambiatore *m* di calore rotativo
MA forgó hőcserélő
NL roterende warmte-wisselaar *m*
PO wymiennik *m* ciepła obrotowy
PY вращающийся теплообменник *m*
SU pyürivä lämmünsiirrin
SV regenerativ värmeväxlare *c*

3312 rotary pump

DA kapselpumpe
DE Rotationspumpe *f*
ES bomba rotativa *f*
F pompe *f* rotative
I pompa *f* rotativa
MA kürforgószivattyú
NL centrifugaal pomp *f*
PO pompa *f* obrotowa
   pompa *f* rotacyjna
PY циркуляционный насос *m*
SU rotaatiopumppu
SV rotationspump *c*

3313 rotation

DA rotation
DE Rotation *f*
   Umdrehung *f*
ES rotación *f*
F rotation *f*
I rotazione *f*
MA forgás
NL draaiing *f*
   omwenteling *f*
PO obrót *m*
   ruch *m* obrotowy
PY вращение *n*
SU pyüriminen
SV rotation *c*

3314 rotational flow

DA hvirvelstrøm
DE Drehstrümung *f*
ES flujo rotativo *m*
F écoulement *m* en spirales
I flusso *m* rotazionale
MA forgóáramlás
NL wervelstroom *m*
PO przepływ *m* wirowy
PY переменное течение *n*
SU pyürrevirtaus
SV rotationsflüde *n*

3315 rotor

DA rotor
DE Läufer *m*
   Rotor *m*
ES rotor *m*
F rotor *m*
I rotore *m*
MA forgórész rotor
NL rotor *n*
PO rotor *m*
   wirnik *m*
PY ротор *m*
SU roottori
SV rotor *c*

3316 roughness

DA ruhed
DE Rauhigkeit *f*
ES rugosidad *f*
F rugosité *f*
I rugosità *f*
MA érdesség
NL ruwheid *f*
PO chropowatość *f*
   szorstkość *f*
PY шероховатость *f*
SU karheus
SV råhet *c*

3317 roughness factor

DA ruhedsfaktor
DE Rauhigkeits-koeffizient *m*
ES coeficiente de rugosidad *m*
F coefficient *m* de rugosité
I fattore *m* di rugosità
MA érdességi tényező
NL ruwheidsfactor *m*
PO współczynnik *m* chropowatości
   współczynnik *m* szorstkości
PY коэффициент *m* шероховатости
SU karheuskerroin
SV råhetskoefficient *c*

3318 rubber gasket

DA gummipakning
DE Gummidichtung *f*
ES empaquetadura de caucho *f*
   junta de caucho *f*
F garniture *f* en caoutchouc
   joint *m* en caoutchouc
I guarnizione *f* di gomma
MA gumitümítés
NL rubberpakking *f*
PO uszczelka *f* gumowa
PY резиновая оболочка *f*
   резиновое уплотение *n*
SU kumitiiviste
SV gummitätning *c*

3319 rubber hose

DA gummislange
DE Gummischlauch *m*
ES manguera de caucho *f*
   tubería de caucho *f*
F tube *m* en caoutchouc
I tubo *m* in gomma
MA gumitümlő
NL rubberslang *f*
PO przewód *m* gumowy elastyczny
PY резиновый шланг *m*
SU kumiletku
SV gummislang *c*

3320 rubber tube (*see* rubber hose)

3321 run-around coil

DA omløbsspiral
DE Umlauf-Rohrschlange *f*
ES recuperador por baterías *m*
F batterie *f* de bipass
I batterie *f,pl* gemelle
MA hővisszanyerő csőkígyó
NL twee elementen warmteuitwisselings-systeem *n*
PO układ *m* z dwoma wężownicami
PY циркуляционный змеевик *m*
SU epäsuora lämmünsiirrin
SV omloppsslinga *c*

3322 running cost
DA driftudgift
DE Betriebskosten *f,pl*
ES coste de
   explotación *m*
F coût *m*
   d'exploitation
I costo *m* di esercizio
MA üzemelési kültség
NL bedrijfskosten *pl*
PO koszt *m* eksploatacji
PY стоимость *f*
   эксплуатации
SU käyttükustannus
SV driftkostnad *c*

3323 running the installation
DA drift af anlæg
DE Betreiben *n* der
   Anlage *m*
ES explotación de la
   instalación *f*
F conduite *f* de
   l'installation
I conduzione *f*
   dell'impianto
MA berendezés
   müküdése
   üzemeltetés
NL inbedrijfhouding *f*
   van de installatie
PO eksploatacja *f*
   instalacji
PY эксплуатация *f*
   установки
SU laitoksen käyttü
SV drift *c* av
   anläggning *c*

3324 running-up time
DA starttid
DE Anlaufzeit *f*
ES tiempo de puesta en
   marcha *m*
F temps *m* de mise en
   marche
I tempo *m* di messa in
   marcia
MA felfutási idő
NL aanlooptijd *m*
PO czas *m* rozruchu
PY время *n*
   приведения в
   действие
SU käynnistysaika
SV starttid *c*

3325 rust
DA rust
DE Rost *m*
ES corrosión *f*
   herrumbre *f*
   roya *f*
F rouille *f*
I ruggine *f*
   (corrosione)
MA rozsda
NL roest *m*
PO rdza *f*
PY ржавчина *f*
SU ruoste
SV rost *c*

S

3326 sacrificial anode
DA offeranode
DE Opferanode *f*
ES ánodo protector
fungible *m*
F anode *f* sacrificielle
I anodo *m* solubile
MA fogyó anód
NL opofferingsanode *f*
PO anoda *f*
protektorowa
anoda *f*
rozpuszczalna
PY растворимый
анод *m*
SU uhrattava anodi
SV galvanisk
anod *c*

3327
safety circuit
DA sikkerhedskreds
DE Nulleiter *m*
Sicherheitsleitung *f*
ES circuito de seguridad
m
F circuit *m* de sécurité
I circuito *m* di
sicurezza
MA biztonsági áramkür
NL veiligheidscircuit *n*
PO obwód *m*
bezpieczeństwa
PY защитный
контур *m*
SU turvakytkin
SV skyddskrets *c*

DA	=	Danish
DE	=	German
ES	=	Spanish
F	=	French
I	=	Italian
MA	=	Magyar
		(Hungarian)
NL	=	Dutch
PO	=	Polish
PY	=	Russian
SU	=	Finnish
SV	=	Swedish

3328 safety connection
DA sikkerhedskobling
DE Sicherheitsleitung *f*
Sicherheits-
verbindung *f*
ES conexión de
seguridad *f*
F raccordement *m* de
sécurité
I raccordo *m* di
sicurezza
MA biztonsági kapcsolás
NL veiligheids-
koppeling *f*
PO połączenie *n*
zabezpieczające
PY надежное
соединение *n*
SU turvakytkentä
SV säkerhetskoppling *c*

3329 safety control
DA sikkerhedsstyring
DE Sicherheits-
regelung *f*
ES control de seguridad
m
F dispositif *m* de
sécurité par
coupure
I regolazione *f* di
sicurezza
MA biztonsági
szabályozás
NL beveiliging *f*
veiligheids-
schakelaar *f*
PO ogranicznik *m*
regulacja *f*
bezpieczeństwa
wyłącznik *m*
bezpieczeństwa
PY контроль *m*
безопасности
SU suojakytkin
SV säkerhetskontroll *c*

3330 safety cut-out (*see*
safety control)

3331 safety device
DA sikkerhedsanordning
DE Sicherheits-
einrichtung *f*
ES dispositivo de
seguridad *m*
órgano de
seguridad *m*
F dispositif *m* de
sécurité
I dispositivo *m* di
sicurezza
MA biztonsági szerelvény
NL beveiligings-
apparaat *n*
PO urządzenie *n*
zabezpieczające
PY защитное
приспособление
n
SU turvalaite
SV säkerhets-
anordning *c*

3332 safety factor
DA sikkerhedsfaktor
DE Sicherheitsfaktor *m*
ES coeficiente de
seguridad *m*
F coefficient *m* de
sécurité
I fattore *m* di
sicurezza
MA biztonsági tényező
NL veiligheidsfactor *m*
PO współczynnik *m*
bezpieczeństwa
PY коэффициент *m*
надежности
SU varmuuskerroin
SV säkerhetsfaktor *c*

3342 saturation line

3333 safety head

- DA sikkerhedstryk
- DE Sicherheitsdruck-höhe *f*
- ES cabezal de seguridad *m*
  presión límite de seguridad *f*
- F pression *f* limite de sécurité
- I carico *m* di sicurezza
- MA biztonsági nyomómagasság
- NL toelaatbare druk *m*
  veilige overdruk *m*
- PO dopuszczalna wielkość *f* ciśnienia hydrostatycznego
  wielkość *f* ciśnienia bezpiecznego
- PY допустимое давление *n*
- SU avautumispaine
- SV säkerhetsgräns-tryck *c*

3334 safety in operation

- DA driftssikkerhed
- DE Betriebssicherheit *f*
- ES seguridad de funcionamiento *f*
- F sécurité *f* d'exploitation
- I sicurezza *f* di impiego
- MA üzembiztonság
- NL bedrijfszekerheid *f*
- PO bezpieczeństwo *n* eksploatacji
- PY эксплуатационная надежность *f*
- SU käyttüturvallisuus
- SV driftsäkerhet *c*

3335 safety valve

- DA sikkerhedsventil
- DE Sicherheitsventil *n*
- ES válvula de seguridad *f*
  válvula limitadora de presión *f*
- F soupape *f* de sûreté
- I valvola *f* di sicurezza
- MA biztonsági szelep
- NL veiligheidsklep *f*
- PO zawór *m* bezpieczeństwa
- PY предохранительный клапан *m*
- SU varoventtiili
- SV säkerhetsventil *c*

3336 safety working stress

- DA tilladeligt arbejdstryk
- DE zulässige Beanspruchung *f*
- ES carga de trabajo admisible *f*
- F charge *f* admissible
- I tensione *f* di lavoro ammissibile
- MA biztonságos megengedett igénybevétel
- NL toelaatbare spanning *f*
- PO naprężenie *n* robocze dopuszczalne
- PY допустимая нагрузка *f*
- SU sallittu kuormitus
- SV tillåten belastning *c*

3337 sample

- DA prøve
- DE Muster *n*
  Probe *f*
- ES muestra *f*
- F échantillon *m*
- I campione *m*
- MA minta
- NL monster *n*
  steekproef *f*
- PO próbka *f*
- PY проба *f*
- SU näyte
- SV prov *n*

3338 saturated air

- DA mættet luft
- DE gesättigte Luft *f*
- ES aire saturado *m*
- F air *m* saturé
- I aria *f* satura
- MA telített levegő
- NL verzadigde lucht *f*
- PO powietrze *n* nasycone
- PY насыщенный воздух *m*
- SU kylläinen ilma
- SV mättad luft *c*

3339 saturated steam

- DA mættet damp
- DE Sattdampf *m*
- ES vapor saturado *m*
- F vapeur *f* saturée
- I vapore *m* saturo
- MA telített gőz
- NL verzadigde stoom *m*
- PO para *f* nasycona
- PY насыщенный пар *m*
- SU kylläinen hüyry
- SV mättad ånga *c*

3340 saturation

- DA mætning
- DE Sättigung *f*
- ES saturación *f*
- F saturation *f*
- I saturazione *f*
- MA telítés
- NL verzadiging *f*
- PO nasycanie *n*
  nasycenie *n*
- PY насыщение *n*
- SU kyllästyminen
- SV mättning *c*

3341 saturation curve

- DA mætningskurve
- DE Sättigungskurve *f*
- ES curva de saturación *f*
- F courbe *f* de saturation
- I curva *f* di saturazione
- MA telítési gürbe
- NL verzadigings-kromme *f*
- PO krzywa *f* nasycenia
- PY кривая *f* насыщения
- SU kyllästyskäyrä
- SV mättningskurva *c*

3342 saturation line

- DA mætningslinie
- DE Sättigungslinie *f*
- ES línea de saturación *f*
- F ligne *f* de saturation
- I linea *f* di saturazione
- MA telítési hatás
- NL verzadigingslijn *f*
- PO linia *f* nasycenia
- PY граница *f* насыщения
- SU kyllästyssuora
- SV mättningslinje *c*

3343 saturation percentage
DA mætningsgrad
DE Sättigungsgrad m
ES grado de saturación m
F degré m de saturation
I grado m di saturazione
MA telítési százalék
NL verzadigingsgraad m
PO procent m nasycenia stopień m nasycenia
PY процент m насыщения
SU kyllästysaste
SV mättningsgrad c

3344 saturation point
DA mætningspunkt
DE Sättigungspunkt m
ES punto de saturación m
F point m de saturation
I punto m di saturazione
MA telítési pont
NL verzadigingspunt n
PO punkt m nasycenia punkt m rosy stężenie n graniczne
PY точка f насыщения
SU kyllästyspiste
SV mättnadspunkt c

3345 saturation pressure
DA mætningstryk
DE Sättigungsdruck m
ES presión de saturación f
F pression f de saturation
I pressione f di saturazione
MA telítési nyomás
NL verzadigingsdruk m
PO ciśnienie n nasycenia
PY давление n насыщения
SU kyllästyspaine
SV mättnadstryck n

3346 SBS (see sick building syndrome)

3347 scaffolding
DA stillads
DE Gerüst n Rüstzeug n
ES andamiaje m
F échafaudage m
I impalcatura f
MA állványozás
NL steiger m stelling f
PO rusztowanie n
PY подмости n
SU telineet
SV (byggnads c) ställning c

3348 scale
DA kedelsten skala vægt
DE Kesselstein m Maßstab m Skala f
ES incrustación f suciedad f
F dépôt m échelle f (dessin) tartre m
I bilancia f incrostazione f scala f (disegno)
MA skála vízkő
NL ketelsteen n schaal f
PO kamień m kotłowy podziałka f skala f waga f zgorzelina f
PY котельная накипь f
SU kattilakivi
SV pannsten c

3349 scale formation
DA kedelstensdannelse
DE Kesselsteinbildung f Steinbildung f
ES incrustación f
F entartrage m
I formazione f di incrostazione
MA vízkőképződés
NL aanslagvorming f ketelsteenvorming f
PO tworzenie n się kamienia kotłowego
PY образование n накипи в котле
SU kattilakiven muodostus
SV pannstensbildning c

3350 scale removal
DA kedelstensfjernelse
DE Kesselstein-entfernung f
ES desincrustación f
F desincrustation f détartrage m
I disincrostazione f
MA vízkőeltávolítás
NL aanslag-verwijdering f ketelsteen-verwijdering f
PO usuwanie n kamienia kotłowego
PY удаление n котельной накипи
SU kattilakiven poisto
SV avlägsnande n av pannsten c

3351 scotch boiler
DA ildkanalkedel skotsk kedel
DE Schotten-Kessel m Scotch-Kessel m
ES estufa escocesa f
F chaudière f à tube de fumée
I caldaia f scozzese
MA tűzcsüves (skót) kazán
NL schotse ketel m vlampijpketel m
PO kocioł m płomienicowo-płomieniówkowy
PY жаротрубный котел m
SU sylinterimäinen tuliputkikattila
SV skotsk ångpanna c (sjü-)

3352 screen
DA skærm
DE Blende f
    Schirm m
    Sieb n
ES tamiz m
F écran m
    tamis m
I schermo m
    setaccio m
MA árnyékoló
    ernyő
NL scherm n
    zeef f
PO ekran m
    osłona f
PY сито n
    экран m
SU verkko
SV filter n

3353 screw
DA propel
    skrue
DE Schraube f
ES tornillo m
F vis f
I vite f
MA csavar
NL schroef f
PO śruba f
    wkręt m
PY болт m
    винт m
    шуруп m
SU ruuvi
SV skruv c

3354 screw vb
DA skrue
DE schrauben
    schraubenfürmig
ES atornillar
    roscar
F visser
I avvitare
MA csavar
NL schroeven
    vastschroeven
PO łączyć śrubami
    wkręcać śrubę
PY завинчивать
SU ruuvata
SV skruva

3355 screw cap
DA kapsel
    møtrik
DE Schraubkapsel f
    Überwurfmutter f
ES tapón roscado m
F bouchon m taraudé
    écrou m borgne
I spina f a vite
MA kupak
NL schroefdeksel n
    schroefdop m
PO nakrętka f
    gwintowana
PY головка f болта
SU kierrekansi
SV skruvlock n

3356 screw compressor
DA skruekompressor
DE Schrauben-
    verdichter m
ES compresor de
    tornillo m
F compresseur m
    hélicoïdal
I compressore m a
    vite
MA csavarkompresszor
NL schroef-
    compressor m
PO sprężarka f śrubowa
PY винтовой
    компрессор m
SU ruuvikompressori
SV skruvkompressor c

3357 screwed joint
DA skruekobling
DE Schraub-
    verbindung f
ES junta roscada f
F raccord m vissé
I raccordo m filettato
MA csavaros csatlakozás
NL schroefdraad-
    verbinding f
PO połączenie n
    gwintowe
    połączenie n
    śrubowe
PY винтовое
    соединение n
SU ruuviliitos
SV skruvfürband n

3358 screw plug
DA skrueprop
DE Gewindestopfen m
ES tapón roscado m
F bouchon m fileté
I tappo m filettato
MA becsavarható dugó
NL plug f met
    schroefdraad
PO korek m gwintowy
    wkrętka f
PY патрубок m для
    очистки
SU kierteellinen tulppa
SV gängad propp c

3359 scroll compressor
DA sneglekompressor
DE Rollen-
    kompressor m
ES compresor de espiral
    m
F rouleau m
    compresseur
I compressore m a
    spirale
MA scroll kompresszor
    spirálkompresszor
NL scrollcompressor m
PO sprężarka f
    bębnowa
    sprężarka f spiralna
PY улитка f
    компрессора
SU kierukkakompressori
SV scroll-kompressor c

3360 scrubber
DA gasvasker
    skrubber
    vådudskiller
DE Scheuerbürste f
    Schrubber m
ES descarbonizador m
    lavador (de aire, de
    gases) m
F laveur m
I lavatore m
MA gázmosó
    súrolókefe
NL gaswasser m
    schrobber m
PO płuczka f wieżowa
    skruber m
PY скруббер m
SU märkäerottaja
    pesuri
SV skrubber c

3361 seal
DA forsegling
  tætning
DE Abdichtung *f*
  Dichtung *f*
ES cierre de cigüeñal *m*
  junta de
    estanqueidad *f*
  junta hermética *f*
  retén *m*
  sello *m*
F bourrage *m*
  joint *m*
  scellement *m*
I giunto *m*
  guarnizione *f*
  tenuta *f*
MA tümítés
NL afdichting *f*
  stempel *n*
  zegel *n*
PO uszczelka *f*
  uszczelnienie *n*
PY затвор *m*
SU tiiviste
SV tätning *c*

3362 seal *vb*
DA forsegle
  tætne
DE abdichten
ES cerrar
    herméticamente
  sellar
F sceller
I sigillare
MA tümít
NL afdichten
  stempelen
PO uszczelniać
PY закрывать
SU tiivistää
SV täta

**3363 sealed condensing
  unit**
DA forseglet
  kondenseringsunit
DE hermetische
  Kondensator-
    einheit *f*
ES unidad
  condensadora
  hermética *f*
F groupe *m*
  compresseur-
  condenseur
  hermétique
I condensatore *m*
  sigillato
MA tümített
  kondenzegység
NL gesloten
  compressor-
    eenheid *f*
PO skraplacz *m*
  hermetyczny
  skraplacz *m*
  zamknięty
PY герметичный
  компрессорно-
  конденсаторный
  агрегат *m*
SU hermeettinen
  lauhdutinyksikkü
SV sluten
  kondensorenhet *c*

3364 sealed system
DA lukket system
DE geschlossenes
  System *n*
ES sistema estanco *m*
  sistema hermético *m*
F circuit *m* étanche à
  vase fermé
I sistema *m* a tenuta
  sistema *m* chiuso
MA tümített rendszer
NL gesloten systeem *n*
PO system *m*
  hermetyczny
  system *m* zamknięty
PY герметичная
  система *f*
  закрытая система *f*
SU tiivistetty järjestelmä
SV slutet system *n*

**3365 sealed-unit
  compressor**
DA hermetisk
  kompressor
DE hermetischer
  Verdichter *m*
ES compresor hermético
  m
F compresseur *m*
  hermétique
I compressore *m*
  ermetico
MA hermetikus
  kompresszor
NL gesloten compressor-
  motoreenheid *f*
PO sprężarka *f*
  hermetyczna
PY герметичный
  компрессор *m*
SU hermeettinen
  kompressori
SV hermetiskt sluten
  kompressor *c*

3366 sealing compound
DA tætningsmasse
DE Dichtungsmasse *f*
ES masilla de
  estanqueidad *f*
F garniture *f* de joint
I guarnizione *f* di
  tenuta
MA tümítőanyag
NL afdichtingskit *f*
  afdichtingspasta *m*
PO masa *f*
  uszczelniająca
PY набивка *f*
SU tiivistemassa
SV tätningsmassa *c*

3367 seam
DA søm
DE Fuge *f*
  Naht *f*
ES costura *f*
  soldadura *f*
F soudure *f*
I cordone *m* (di
  saldatura)
  giunto *m* saldato
MA varrat
NL naad *m*
PO połączenie *n* blach
  na zakładkę
  spaw *m*
  szew *m*
PY шов *m*
SU sauma
SV fog *c*

DA = Danish
DE = German
ES = Spanish
F = French
I = Italian
MA = Magyar
     (Hungarian)
NL = Dutch
PO = Polish
PY = Russian
SU = Finnish
SV = Swedish

3368 seamless *adj*
DA sømløs
DE nahtlos
ES sin soldadura
F sans soudure *f*
I senza saldatura
MA varrat nélküli
NL naadloos
PO bez szwu
PY бесшовный
SU saumaton
SV sümlüs

3369 secondary air
DA sekundær luft
DE Sekundärluft *f*
ES aire secundario *m*
F air *m* secondaire
I aria *f* secondaria
MA másodlagos levegő
 szekunder levegő
NL secundaire lucht *f*
PO powietrze *n* wtórne
PY вторичный
 воздух *m*
SU toisioilma
SV sekundärluft *c*

3370 secondary condenser
DA sekundær
 kondensator
DE Sekundär-
 kondensator *m*
ES condensador
 secundario *m*
F condenseur *m*
 secondaire
I condensatore *m*
 secondario
MA szekunder
 kondenzátor
NL secundaire
 condensor *m*
PO skraplacz *m* wtórny
PY вторичный
 конденсатор *m*
SU jälkilauhdutin
 toisiolauhdutin
SV sekundär-
 kondensor *c*

3371 secondary coolant
DA sekundært
 kølemiddel
DE Sekundär-
 kühlmittel *n*
ES refrigerante
 secundario *m*
F fluide *m*
 frigoporteur
 secondaire
I fluido *m* refrigerante
 secondario
MA szekunder
 hűtőfolyadék
NL secundair
 koelmedium *n*
PO czynnik *m*
 chłodniczy
 drugiego stopnia
PY вторичный
 холодоноситель
 m
SU jälkijäähdytin
 toisiojäähdytin
SV sekundärkylmedel *n*

3372 secondary drying
DA sekundær tørring
DE Sekundär-
 trocknung *f*
ES secado secundario *m*
F dessication *f*
 secondaire
I essiccamento *m*
 secondario
MA szekunder szárítás
NL secundaire droging *f*
PO dosuszanie *n*
 suszenie *n* powtórne
PY вторичное
 осушение *n*
SU jälkikuivaus
 toisiokuivaus
SV sekundärtorkning *c*

3373 secondary heating surface
DA indirekte varmeflade
DE indirekte
 Heizfläche *f*
ES superficie de
 calefacción
 indirecta *f*
F surface *f* de chauffe
 secondaire
I superficie *f*
 scaldante
 secondaria
MA szekunder fűtőfelület
NL secundair
 verwarmend
 oppervlak *n*
PO powierzchnia *f*
 ogrzewalna
 dodatkowa
PY вторичная
 отопительная
 поверхность *f*
SU sekundäärinen
 lämpüpinta
 toisiolämpüpinta
SV indirekt värmeyta *c*

3374 section
DA afsnit
 sektion
DE Abschnitt *m*
 Glied *n*
 Zone *f*
ES elemento *m*
 sección *f*
 tramo *m*
F aire *f*
 élément *m*
 section *f*
 surface *f*
I elemento *m*
 sezione *f*
MA szakasz
NL doorsnede *f*
 lid *n*
 sectie *f*
PO człon *m*
 odcinek *m*
 przekrój *m*
 sekcja *f*
PY отрезок *m*
 разрез *m*
 секция *f*
SU leikkaus
SV sektion *c*

3375 sectional *adj*
DA sektionsdelt
DE Abschnitts-
Durchschnitts-
ES seccional
parcial
F à petits éléments
sectionnel
I compartimentato
MA szakaszos
NL geleed
uit delen opgebouwd
PO dzielony
sekcyjny
składany
PY секционный
SU osasto-
SV sektions-

3376 section drawing
DA deltegning
snittegning
DE Schnittzeichnung *f*
ES plano de sección *m*
F dessin *m* en coupe
I sezione *f*
MA részletrajz
NL doorsnedetekening *f*
PO przekrój *m*
(rysunek)
PY разрез *m* на
чертеже
SU leikkauspiirustus
SV sektionsritning *c*

3377 sediment
DA bundfald
slam
DE Sediment *n*
ES sedimento *m*
F dépôt *m*
sédiment *m*
I sedimento *m*
MA üledék
NL afzetsel *n*
bezinksel *n*
PO osad *m*
PY осадка *f*
осадок *m*
SU sakka
SV fällning *c*

3378 selector switch
DA vælgerkontakt
DE Wählschalter *m*
ES conmutador selector
m
F commutateur *m*
I interruttore *m*
selettore
MA választókapcsoló
NL keuzeschakelaar *m*
PO przełącznik *m*
wybierakowy
wybierak *m*
PY групповой
выключатель *m*
SU valintakytkin
SV väljaromkopplare *c*

**3379 self-contained air
conditioning
system**
DA kompakt
klimaaggregat
DE automatisches
Klimasystem *n*
ES sistema de
climatización con
unidades
autónomas *m*
F système *m*
monobloc de
conditionnement
d'air
I impianto *m* di
condizionamento
autonomo
MA egyedi
légkondicionáló
rendszer
NL apparaat *n* voor
volledige
luchtbehandeling
PO system *m*
klimatyzacji
autonomiczny
PY автономная
система *f*
кондицион-
ирования воздуха
автономный
кондиционер *m*
SU käyttüvalmis
ilmastointikone
paketti-
ilmastointikone
SV sluten
luftbehandlings-
anläggning *c*

**3380 self-operated
control**
DA selvvirkende
regulering
DE selbsttätige
Regelung *f*
ES control automático
m
F régulation *f* directe
I controllo *m*
autoattuato
MA ünműködő
szabályozás
NL regelaar *m* zonder
externe
hulpenergie
PO regulacja *f*
bezpośredniego
działania
PY автономное
управление *n*
SU omavoimainen säätü
SV självverkande
kontroll *c*

3381 self priming pump
DA selvansugende
pumpe
DE selbstansaugende
Pumpe *f*
ES bomba
autoaspirante *f*
F pompe *f* à amorçage
automatique
I pompa *f*
autoadescante
MA ünfelszívó szivattyú
NL zelfaanzuigende
pomp *f*
PO pompa *f*
samozasysająca
PY самовсасывающий
насос *m*
SU itseimevä pumppu
SV självsugande pump
c

3382 semi-automatic *adj*
DA halvautomatisk
DE halbautomatisch
ES semiautomático
F semi-automatique
I semiautomatico
MA félautomatikus
NL halfautomatisch
PO półautomatyczny
PY полуавтомати-
ческий
SU puoliautomaattinen
SV halvautomatisk

3383 semiconductor
DA halvleder
DE Halbleiter *m*
ES semiconductor *m*
F semiconducteur *m*
I semiconduttore *m*
MA félvezető
NL halfgeleider *m*
PO półprzewodnik *m*
PY полупроводник *m*
SU puolijohde
SV halvledare *c*

3384 semi-hermetic *adj*
DA semi-hermetisk
DE halbhermetisch
ES semihermético
F semi-hermétique
I semiermetico
MA félhermetikus
NL semi-hermetisch
PO półhermetyczny
PY полугерметичный
SU puolihermeettinen
SV halvhermetisk

3385 sensible *adj*
DA følelig
DE fühlbar
ES sensible
F sensible
I sensibile
MA érezhető
NL voelbaar
PO jawny
odczuwalny
PY ощутимый
SU tuntuva jäähdytys
SV sensibel

3386 sensing element
DA følerelement
DE Fühler *m*
ES sensor *m*
sonda *f*
F élément *m* sensible
sonde *f*
I elemento *m* sensibile
sonda *f*
MA érzékelő elem
NL detector *m*
voeler *m*
PO czujnik *m*
PY датчик *m*
чувствительный
элемент *m*
SU tuntoelin
SV känselkropp *c*

3387 sensor
DA føler
DE Fühler *m*
Sensor *m*
ES sonda *f*
F capteur *m*
I sensore *m*
MA érzékelő
NL detector *m*
opnemer *m*
voeler *m*
PO czujnik *m*
PY датчик *m*
SU tuntoelin
SV avkännare *c*

3388 separation
DA udskilning
DE Abscheidung *f*
ES separación *f*
F séparation *f*
I separazione *f*
MA leválasztás
NL afscheiding *f*
scheiding *f*
PO oddzielanie *n*
rozdzielanie *n*
separacja *f*
PY отделение *n*
сепарация *f*
SU erotus
SV avskiljning *c*

**3389 separation
efficiency**
DA udskilnings-
effektivitet
DE Abscheideleistung *f*
ES rendimiento de un
filtro *m*
F rendement *m* de
séparation
I efficienza *f* di
separazione
rendimento *m* di
separazione
resa *f* di separazione
MA leválasztási hatásfok
NL scheidings-
rendement *n*
PO skuteczność *f*
oddzielania
PY эффективность *f*
сепарации
SU erotusaste
SV avskiljnings-
effektivitet *c*

3390 separator
DA udskiller
DE Abscheider *m*
ES separador *m*
F séparateur *m*
I separatore *m*
MA leválasztó
NL afscheider *m*
PO oddzielacz *m*
separator *m*
PY сепаратор *m*
SU erotin
SV separator *c*

3391 sequence
DA rækkefølge
sekvens
DE Folge *f*
Reihenfolge *f*
Sequenz *f*
ES secuencia *f*
F séquence *f*
I sequenza *f*
MA folyamat
NL volgorde *f*
PO kolejność *f*
następstwo *n*
sekwencja *f*
PY порядок *m*
последовател-
ьность *f*
SU järjestys
SV sekvens *c*

3392 sequence control
DA sekvensregulering
DE Sequenzregelung *f*
ES control por
secuencia *f*
F régulation *f*
séquentielle
I controllo *m*
sequenziale
MA folyamatszabályozás
NL volgorderegeling *f*
PO regulacja *f*
sekwencyjna
PY порядок *m*
регулирования
последовател-
ьность *f* регулирования
SU sarjasäätü
SV sekvenskontroll *c*

3393 serial port
DA serieåbning
DE serienmäßiger
Auslaßstutzen *m*
serienmäßiger
Einlaßstutzen *m*
ES aberturas en serie *f*
F port série
I porta *f* seriale
MA nyílássor
NL seríle poort *f*
PO otwór *m* szeregowo
występujący
port *m* szeregowy
(komp.)
PY типовое (серийное)
отверстие *n*
SU sarjaportti
SV serieüppning *c*

3394 series connection
DA serieforbindelse
DE Hintereinander-
schaltung *f*
Reihenschaltung *f*
ES acoplamiento en
serie *m*
F couplage *m* en série
raccordement *m* en
série
I accoppiamento *m* in
serie
collegamento *m* in
serie
MA soros kapcsolás
NL serieschakeling *f*
PO połączenie *n*
szeregowe
PY последовательное
соединение *n*
SU sarjaankytkentä
SV seriekoppling *c*

3395 series operation
DA kaskadekobling
seriedrift
DE Kaskaden-
schaltung *f*
ES funcionamiento en
serie *m*
F opérations *f* en
cascade
I operazione *f* in
sequenza
MA soros műküdés
NL seriebedrijf *n*
werking *f* in cascade
PO działanie *n*
szeregowe
PY последовательная
(порядковая)
операция *f*
SU sarjakäyttü
SV seriedrift *c*

3396 service
DA betjening
DE Bedienung *f*
Betrieb *m*
ES servicio *m*
F service *m*
I servizio *m*
MA karbantartás
szerviz
NL dienst *m*
dienstverlening *f*
PO obsługa *f*
serwis *m*
PY обслуживание *n*
SU huolto
SV service *c*

3397 serviceability
DA brugbarhed
holdbarhed
DE Betriebsfähigkeit *f*
ES servicialidad *f*
F disponibilité *f*
technique
état *m* satisfaisant
(du point de vue
du fonctionne-
ment)
facilité *f* d'entretien
I durabilità *f*
MA szervizelhetőség
NL onderhoudbaarheid
f
PO dostępność *f* obsługi
możliwość *f* obsługi
przydatność *f* do
użytku
PY пригодность *f*
SU huollettavuus
toimivuus
SV tjänsteduglighet *c*

3398 service connection
DA hustilslutning
DE Hausanschluß *m*
Hauszuleitung *f*
ES acometida *f*
F raccordement *m*
I attacco *m* di servizio
MA szervizkapcsolás
NL dienstaansluiting *f*
PO połączenie *n* dla
obsługi
połączenie *n*
obsługowe
PY домовой ввод *m*
SU huoltoliitäntä
SV servisintag *n*

3399 service instructions
DA betjeningsinstruks
DE Bedienungs-
anweisung *f*
ES instrucciones de
servicio *f,pl*
F instructions *f* de
service
I istruzioni *f,pl* di
servizio
MA kezelési utasítások
NL bedieningsvoor-
schrift *n*
PO instrukcja *f* obsługi
PY инструкция *f* по
обслуживанию
SU huolto-ohjeet
SV skütsel-
instruktioner *c*

3400 service manual
DA betjeningshåndbog
DE Bedienungshand-
buch *n*
ES manual de
servicio *m*
F manuel *m* de service
I manuale *m* di
servizio
MA kezelési kézikünyv
NL bedienings-
handboek *n*
PO instrukcja *f* obsługi
PY ручное
управление *n*
SU huoltokäsikirja
SV servicehandbok *c*

3401 servicing
DA vedligeholdelse
DE Instandhaltung *f*
ES entretenimiento *m*
F entretien *m*
I manutenzione *f*
MA szervizelés
NL onderhoud
verrichten *n*
PO obsługa *f* techniczna
PY обслуживание *n*
SU huolto
SV service *c*

3402 servo control
DA servostyring
DE Hilfskraft-
  Regelung *f*
ES servocontrol *m*
F servocommande *f*
I servocontrollo *m*
MA szervoszabályozás
NL servobesturing *f*
PO sterowanie *n* przy
  użyciu siłownika
PY сервоконтроль *m*
  следящая
  автоматическая
  система *f*
  управления
SU servosäätü
SV servokontroll *c*

3403 servo motor
DA servomotor
DE Servomotor *m*
ES servomotor *m*
F servo-moteur *m*
I servo-motore *m*
MA szervomotor
NL servomotor *m*
PO siłownik *m*
PY серводвигатель *m*
  сервомотор *m*
SU servomoottori
SV servomotor *c*

3404 set *vb*
DA indstille
DE einstellen
ES ajustar
  fijar
F afficher
I fissare
  regolare
MA beállít
NL instellen
  plaatsen
PO nastawiać
  ustawiać
PY подгонять
  устанавливать
SU asettaa
SV ställa in

3405 set point
DA indstillingspunkt
DE Einstellpunkt *m*
ES punto de ajuste *m*
F point *m* de consigne
I punto *m* di taratura
  set point *m*
MA beállítási pont
NL instelpunt *n*
PO nastawa *f*
  sygnał *m* zadany
PY место *n* установки
SU asetusarvo
  ohjearvo
SV inställningspunkt *c*

3406 setting
DA indstilling
DE Einrichtung *f*
  Einstellung *f*
ES ajuste *m*
F mise *f* au point
  mise *f* en place
I messa *f* a punto
  taratura *f*
MA beállítás
NL instelling *f*
PO nastawianie *n*
  ustawianie *n*
PY наладка *f*
  подготовка *f*
SU asetus
SV inställning *c*

3407 setting accuracy
DA indstillings-
  nøjagtighed
DE Einstellgenauigkeit *f*
ES precisión de ajuste *f*
F précision *f*
  d'affichage
I precisione *f* di
  taratura
MA beállítási pontosság
NL instelnauwkeurig-
  heid *f*
PO dokładność *f*
  nastawienia
PY точность *f*
  настройки
  точность *f*
  установки
SU asettelutarkkuus
SV inställnings-
  noggrannhet *c*

3408 setting range (*see*
  adjustment range)

3409 setting up
DA montage
DE Montage *f*
ES montaje *m*
F montage *m*
I montaggio *m*
  preparazione *f*
MA felszedés
NL montage *f*
  opstellen *n*
PO montaż *m*
  zestawienie *n*
PY монтаж *m*
  наладка *f*
  сборка *f*
SU asettaa
  säätää
SV montering *c*

3410 set value
DA indstillingsværdi
DE Einstellwert *m*
ES valor de ajuste *m*
F valeur *f* de consigne
I settaggio *m*
  valore *m* di taratura
MA beállítási érték
NL instelwaarde *f*
PO wartość *f*
  nastawiona
  wartość *f* zadana
  wielkość *f* nastawy
PY заданная
  величина *f*
SU asetusarvo
  ohjearvo
SV inställningsvärde *n*

3411 sewer
DA afløbsledning
  kloak
DE Abwasserkanal *m*
  Abzugskanal *m*
ES alcantarilla *f*
  alcantarillado *m*
  canal de desagüe *m*
F égout *m*
I canale *m* di fogna
  fogna *f*
MA szennyvízcsatorna
NL riool *n*
  rioolbuis *f*
PO kanał *m* ściekowy
  ściek *m*
PY коллектор *m*
  канализации
SU viemäri
SV avloppsledning *c*

3412 shade *vb*
DA skygge
　　skærm
DE beschatten
　　schützen gegen
　　Sonnenstrahlen
　　m,pl
ES proteger contra el
　　sol
　　sombrear
F protéger du soleil
I riparare dal sole
MA árnyékol
NL afschermen tegen de
　　zon
　　beschaduwen
PO ocieniać
　　zacieniać
　　zasłaniać
PY затенять
SU varjostaa
SV skugga

3413 shading coefficient
DA afskærmningsfaktor
DE Beschattungs-
　　koeffizient *m*
ES coeficiente de
　　sombra *m*
F coefficient *m* de
　　protection du
　　soleil
I coefficiente *m* di
　　ombreggiamento
MA árnyékolási tényező
NL beschaduwings-
　　cöfficint *m*
PO współczynnik *m*
　　ocienienia
PY коэффициент *m*
　　затенения
SU suojauskerroin
SV skuggnings-
　　koefficient *c*

3414 shading device
DA solafskærmning
DE Sonnenschutz-
　　vorrichtung *f*
ES parasol *m*
F dispositif *m* pare-
　　soleil
I schermo *m*
MA árnyékoló szerkezet
NL zonnescherm *n*
PO urządzenie *n*
　　ocieniające
PY затеняющее
　　устройство *n*
SU auringon suoja
SV skuggnings-
　　anordning *c*

3415 shaft
DA aksel
　　skaft
DE Achse *f*
ES árbol *m*
　　eje *m*
F arbre *m*
　　cheminée *f*
　　puits *m*
　　trémie *f*
I albero *m*
　　camino *m*
　　condotta *f*
　　vano *m* tecnico
MA akna
　　kürtő
　　tengely
NL as *f*
　　schacht *f*
PO szyb *m*
　　wał *m*
　　wałek *m*
PY шахта *f*
SU akseli
　　kuilu
SV schakt *n*

3416 shaft seal
DA akseltætning
DE Achsendichtung *f*
ES empaquetadura de
　　eje *f*
F
I guarnizione *f*
　　dell'albero
MA tengelytümítés
NL asafdichting *f*
PO uszczelnienie *n*
　　szybu
　　uszczelnienie *n* wału
PY затвор *m* шахты
SU akselitiiviste
SV axeltätning *c*

3417 shaking grate
DA rysterist
DE Rüttelrost *m*
ES parrilla oscilante *f*
F grille *f* oscillante
I griglia *f* a
　　scuotimento
MA rázórostély
NL schudrooster *m*
PO ruszt *m* wstrząsowy
PY встряхиваемая
　　колосниковая
　　решетка *f*
SU täryarina
SV skakrost *c*

3418 sheet
DA ark
　　plade
DE Blatt *n*
　　Platte *f*
　　Scheibe *f*
ES hoja *f*
F feuille *f*
　　lame *f*
I foglio *m*
MA lemez
NL plaat *f*
　　vel *n*
PO arkusz *m*
　　blacha *f* cienka
PY пластина *f*
SU levy
SV skiva *c*

3419 sheet metal
DA blik
　　plade
DE Blech *n*
ES chapa *f*
　　plancha *f*
F feuille *f* de métal
　　tôle *f*
I lamiera *f* di metallo
　　lastra *f* di metallo
MA acéllemez
NL plaatstaal *n*
PO arkusz *m* blachy
PY лист *m* металла
SU pelti
SV metallskiva *c*

3420 sheet metal duct
DA pladekanal
DE Blechkanal *m*
ES conducto de
　　chapa *m*
F conduit *m* en tôle
I canale *m* di lamiera
MA acéllemez
　　légcsatorna
NL plaatstalen kanaal *n*
PO kanał *m* blaszany
PY воздуховод *m* из
　　листового
　　металла
　　трубопровод *m* из
　　листового
　　металла
SU peltikanava
SV plåtkanal *c*

3421 sheet thickness
DA pladetykkelse
DE Blechstärke *f*
ES espesor de chapa *m*
F épaisseur *f* de tôle
I spessore *m* della lamiera
MA lemezvastagság
NL plaatdikte *f*
PO grubość *f* blachy
PY толщина *f* листа
SU pellin paksuus
SV plåttjocklek *c*

3422 shell and coil condenser
DA rørslange-kondensator
DE Rohrschlangen-kondensator *m*
ES condensador multitubular *m*
F condenseur *m* à calandre et serpentin
I condensatore *m* a fascio tubiero
MA csőkígyós kondenzátor
NL condensor *m* met spiraal
PO skraplacz *m* płaszczowo-wężownicowy
PY кожухотрубный змеевиковый конденсатор *m*
SU kierukkalauhdutin
SV tubpanne-kondensor *c*

3423 shell and coil evaporator
DA rørslange-evaporator
DE Rohrschlangen-verdampfer *m*
ES evaporador multitubular *m*
F évaporateur *m* à calandre et serpentin
I evaporatore *m* a fascio tubiero
MA csőkígyós elpárologtató
NL verdamper *m* met spiraal
PO parownik *m* płaszczowo-wężownicowy
PY кожухотрубный змеевиковый испаритель *m*
SU kierukkahüyrystin
SV tubpanne-evaporator *c*

3424 shell and tube condenser
DA rørkedel-kondensator
DE Rohrbündel-verflüssiger *m*
ES condensador *m* condensador multitubular de envolvente *m*
F condenseur *m* à faisceau tubulaire
I condensatore *m* a tubi e mantello
MA csőküteges kondenzátor
NL condensor *m* met pijpenbundel
PO skraplacz *m* płaszczowo-rurowy
PY кожухотрубный конденсатор *m*
SU moniputkilauhdutin
SV tubpanne-kondensor *c*

3425 shell and tube evaporator
DA rørkedel-fordamper
DE Rohrbündel-verdampfer *m*
ES evaporador multitubular de envolvente *m*
F évaporateur *m* à faisceau tubulaire
I evaporatore *m* a tubi e mantello
MA csőküteges elpárologtató
NL verdamper *m* met pijpenbundel
PO parownik *m* płaszczowo-rurowy
PY кожухотрубный испаритель *m*
SU moniputkihüyrystin
SV tubpanne-evaporator *c*

3426 shell and tube exchanger
DA rørkedel-veksler
DE Rohrbündelwärme-austauscher *m*
ES cambiador multitubular de envolvente *m*
F échangeur *m* à faisceau tubulaire
I scambiatore *m* a tubi e mantello
MA csőküteges hőcserélő
NL warmtewisselaar *m* met pijpenbundel
PO wymiennik *m* płaszczowo-rurowy
PY кожухотрубный теплообменник *m*
SU moniputkilämmün-siirrin
SV tubpannevärme-växlare *c*

3427 shelter
DA læ
DE Schutzraum *m*
ES envolvente *f*
F abri *m*
I rifugio *m* riparo *m*
MA óvóhely
NL beschutting *f* schuilplaats *f*
PO schron *m*
PY укрытие *n*
SU suoja
SV skyddsrum *n*

3428 shelter *vb*
DA beskytte
DE beschützen
    schützen
ES abrigar
    proteger (de la
    intemperie)
F abriter
I proteggere
    riparare
MA megvéd
NL beschermen
    beschutten
PO chronić
PY укрывать
    хранить
SU suojata
SV skydda

3429 shock absorber
DA støddæmper
DE Stoßdämpfer *m*
ES amortiguador *m*
F amortisseur *m*
I ammortizzatore *m*
MA lengéscsillapító
NL schokdemper *m*
PO amortyzator *m*
    wstrząsów
PY амортизатор *m*
SU iskunvaimentaja
SV stütdämpare *c*

3430 shop drawing
DA arbejdstegning
DE Werkstatt-
    zeichnung *f*
ES plano de taller *m*
F dessin *m* d'atelier
    plan *m* d'atelier
I disegno *m* di officina
MA műhelyrajz
NL werktekening *f*
PO rysunek *m*
    warsztatowy
PY проектная
    документация *f*
SU asennuspiirustus
SV verkstadsritning *c*

3431 short circuit
DA kortslutning
DE Kurzschluß *m*
ES cortocircuito *m*
F court-circuit *m*
I corto circuito *m*
MA rüvidzárlat
NL kortsluiting *f*
PO krótkie spięcie *n*
    zwarcie *n*
PY короткое
    замыкание *n*
SU oikosulku
SV kortslutning *c*

3432 short cycling
DA pendling
DE Kurzzyklus *m*
ES ciclo interrumpido *m*
F cycle *m* court
I ciclo *m* abbreviato
MA rüvid megszakításos
    üzem
NL pendelen *n*
PO obieg *m* krótki
    obieg *m*
    przyspieszony
PY величины
    кратковременное
    периодическое
    изменение *n*
    регулируемой
SU usein muuttuva
SV kort cykel *c*

**3433 short-term
    operation**
DA korttidsdrift
DE Kurzbetrieb *m*
ES funcionamiento de
    corta duración *m*
    trabajo a corto
    plazo *m*
F service *m* de courte
    durée
I funzionamento *m* di
    breve durata
MA rüvidideju műküdés
NL kortetermijn-
    bedrijf *n*
PO działanie *n*
    krótkotrwałe
PY кратковременный
    режим *m*
SU lyhytaikainen käyttü
SV korttidsdrift *c*

3434 shunt
DA shunt
DE Nebenanschluß *m*
    Parallelschaltung *f*
    Shunt *m*
ES derivación *f*
F shunt *m*
I derivazione *f*
    shunt *m*
MA párhuzamos
    kapcsolás
    sünt
NL aftakking *f*
    bypass *n*
    omloop *m*
PO bocznik *m*
PY ответвление *n*
    шунт *m*
SU ohitus
SV shunt *c*

3435 shunt valve
DA shuntventil
DE Beipaßventil *n*
    Umgehungsventil *n*
ES válvula de
    derivación *f*
F robinet *m* shunt
I valvola *f* a shunt
MA megkerülőszelep
NL omloopafsluiter *m*
PO zawór *m*
    bocznikowy
PY переключающий
    клапан *m*
SU ohitusventtiili
SV shuntventil *c*

3436 shutdown
DA afbrud
    standsning
DE Betriebseinstellung *f*
    Stillegung *f*
ES interrupción (de la
    corriente, etc.) *f*
    paro *m*
F arrêt *m*
I arresto *m*
    fermata *f*
MA leállás
    üzemszünet
NL buiten werking
    stellen *n*
    stilleggen *n*
PO przestój *m*
    zamknięcie *n*
    zatrzymanie *n*
PY промежуток *m*
SU käyttühäiriü
    sulkeminen
SV avbrott *n*

3437 shut-off valve
DA afspærringsventil
DE Absperrventil *n*
ES válvula de cierre *f*
F robinet *m* d'arrêt
    vanne *f* d'arrêt
I rubinetto *m* di
    arresto
    saracinesca *f*
MA elzáró szelep
NL afsluitklep *f*
PO zawór *m* odcinający
PY выключающий
    клапан *m*
SU sulkuventtiili
SV avstängningsventil *c*

3438 shutter
DA klap
  skodde
DE Fensterladen *m*
  Klappe *f*
ES cierre *m*
  compuerta *f*
  registro *m*
F obturateur *m*
  volet *m*
I chiusura *f*
  saracinesca *f*
  serranda *f*
MA redőny
  zsalu
NL rolluik *n*
  schuifdeksel *n*
PO okiennica *f*
  zasłona *f*
  żaluzja *f*
PY задвижка *f*
  заслонка *f*
SU suljin
SV lucka *c*
  propp *c*

3439 sick building syndrome
DA syg bygning
DE Sick-Building-Syndrom *n*
ES síndrome de edificio enfermo *m*
F bâtiment *m* malsain (syndrome des bâtiments malsains)
I sindrome *f* dell'edificio malato
MA beteg épület jelenség
NL SBS
  ziekmakendgebouw-syndroom *n*
PO syndrom *m* budynku chorego
PY синдром *m* закрытых помещений
SU sairasrakennu-songelma
SV sjuk byggnad-syndromet *n*

3440 sieve
DA sigte
  sold
DE Sieb *n*
ES criba *f*
  tamiz *m*
F tamis *m*
I setaccio *m*
MA szita
NL zeef *f*
PO sito *n*
PY сито *n*
SU seula
SV såll *n*

3441 sight glass
DA skueglas
DE Schauglas *n*
ES mirilla *f*
  visor *m*
F viseur *m*
I vetro *m* spia
MA nézőüveg
NL kijkglas *n*
PO okienko *n* kontrolne wziernik *m*
PY смотровое стекло *n*
SU tarkastuslasi
SV inspektionsglas *n*

3442 signal
DA signal
DE Signal *n*
ES señal *f*
F signal *m*
I segnale *m*
MA jel
NL signaal *n*
  teken *n*
PO sygnał *m*
PY сигнал *m*
SU signaali
SV signal *c*

3443 signal bus
DA signalledning
DE Signalsammel-leitung *f*
ES colector de señales *m*
F bus *m* de données
I bus *m* di segnale
MA jelgyűjtő jeltovábbító
NL signaalbus *f*
PO szyna *f* zbiorcza sygnałowa
PY сигнальная шина *f*
SU väylä
SV signalledning *c*

3444 silencer
DA lyddæmper
  lydpotte
DE Schalldämpfer *m*
ES silenciador *m*
F silencieux *m*
I silenziatore *m*
MA hangtompító
NL geluiddemper *m*
PO tłumik *m* dźwięków
  tłumik *m* hałasu
PY глушитель *m*
SU äänenvaimentaja
SV ljuddämpare *c*

3445 silica gel
DA silikagel
DE Kieselgel *n*
  Silikagel *n*
ES gel de sílice *f*
F silicagel *m* (gel de silice)
I gel *m* di silice
MA szilikagél
NL silicagel *m*
PO silikażel *m*
  żel *m* krzemionkowy
PY силикагель *m*
SU piihappogeeli
SV silicagel *n*

3446 sill-mounted grille
DA brystningsmonteret rist
DE Brüstungsgitter *n*
ES rejilla de antepecho *f*
F grille *f* en tablette de fenêtre
I griglia *f* a barrotti
MA parapetrács
NL vensterbank-rooster *m*
PO kratka *f* podokienna
PY подоконная решетка *f*
SU korvausilmaventtiili
SV fünsterventil *c*

DA	=	Danish
DE	=	German
ES	=	Spanish
F	=	French
I	=	Italian
MA	=	Magyar (Hungarian)
NL	=	Dutch
PO	=	Polish
PY	=	Russian
SU	=	Finnish
SV	=	Swedish

3447 simplex circuit
DA enkelt strømkreds
DE Simplex-Schaltung *f*
ES circuito simple *m*
F circuit *m* électrique simple (2 fils dont I à la terre)
I circuito *m* a singolo conduttore
MA egyszerű kapcsolás
NL enkelvoudig circuit *n*
PO obwód *m* sympleksowy
PY одинарный (простой) контур *m*
SU yksinkertainen piiri
SV enkelkrets *c*

3448 simultaneous *adj*
DA samtidig
DE gleichzeitig simultan
ES simultáneo
F simultané
I simultaneo
MA egyidejű szimultán
NL gelijktijdig
PO jednoczesny
PY одновременный
SU samanaikainen
SV samtidig

3449 single acting compressor
DA enkeltvirkende kompressor
DE einfachwirkender Verdichter *m*
ES compresor de acción simple *m*
F compresseur *m* à simple effet
I compressore *m* a semplice azione
MA egyszeres működésű kompresszor
NL enkelvoudig-werkende compressor *m*
PO sprężarka *f* jednostronnego działania
PY однотактный компрессор *m*
SU yksitoiminen kompressori
SV enkelverkande kompressor *c*

3450 single-duct air conditioning plant
DA enkeltkanals ventilationsanlæg
DE Einkanalklima-anlage *f*
ES instalación de acondicionamiento de aire mono-conducto *f*
F installation *f* de climatisation à un conduit
I impianto *m* di condizionamento dell'aria monocondotta
MA egycsöves légkondicionáló rendszer
NL éénkanaalslucht-behandelingsinstallatie *f*
PO instalacja *f* klimatyzacji jednoprzewodowej
PY одноканальная установка *f* кондицион-ирования воздуха
SU yksikanavailmasto-intilaitos
SV luftkonditionerings-anläggning *c*

3451 single duct system
DA enkeltkanalsystem
DE Einkanalsystem *n*
ES sistema de un solo conducto *m*
F système *m* à un conduit
I sistema *m* mono-condotto
MA egyvezetékes rendszer
NL éénkanaalssysteem *n*
PO system *m* jednoprzewodowy
PY одноканальная система *f*
SU yksikanavailmastointijärje-stelmä
SV enkanalsystem *n*

3452 single glazed window
DA vindue med et lag glas
DE Einfachfenster *n*
ES ventana de cristal simple *f*
F fenêtre *f* simple
I finestra *f* a vetro semplice
MA egyszeres üvegezésű ablak
NL enkelbeglaasd raam *n* enkelglas raam *n*
PO okno *n* pojedyńczo szklone
PY окно *n* с одинарным остеклением
SU yksilasinen ikkuna
SV ettglasfünster *n*

3453 single leaf damper
DA enkeltbladet spjæld
DE Einflügelklappe *f*
ES compuerta de una sola hoja *f*
F registre *m* à papillon (à volet unique mobile)
I serranda *f* a lama singola
MA egylapos csappantyú pillangószelep
NL vlinderklep *f*
PO przepustnica *f* jednopłaszczy-znowa
PY одностворчатый клапан *m*
SU yksiosainen säätüpelti
SV enkelspjäll *n*

3454 single (one) pipe system
DA et-rørs-system
DE Einrohrsystem *n*
ES sistema de tubería única *m* sistema mono-tubular *m*
F système *m* monotube
I sistema *m* monotubo
MA egycsöves rendszer
NL éénpijpssysteem *n*
PO system *m* jednorurowy
PY однотрубная система *f*
SU yksiputkijärjestelmä
SV ettrörssystem *n*

3455 single-stage *adj*
DA et-trins
DE
ES de un etapa
F mono-étagé
I monostadio
MA egyfokozatú
NL ééntraps
enkeltraps
PO jednostopniowy
PY одноступенчатый
SU yksiportainen
SV enstegs-

3456 single-vane *adj*
DA encellet
DE
ES de un álabe
F mono-pale
I a singola aletta
MA egylapátos
NL enkelbladig
enkelschots
PO jednołopatkowy
PY однолопаточный
SU yksisiipinen
SV enskovlig

3457 sink
DA køkkenvask
DE Ablaufkanal *m*
Spülstein *m*
ES fregadero *m*
sumidero *m*
vertedero *m*
F évier *m*
puisard *m*
I scarico *m*
MA mosogató
süllyedés
NL gootsteen *m*
zinkput *m*
PO zlew *m*
zlewozmywak *m*
PY слив *m*
SU pesupüytä
viemärikaivo
SV avlopp *n*

3458 sintering
DA sintring
DE Sinterung *f*
ES sinterización *f*
vitrificación *f*
F frittage *m*
I sinterizzazione *f*
MA zsugorodás
NL sinteren *n*
PO spiekanie *n*
PY спекание *n*
SU sintraus
SV sintring *c*

3459 site
DA grund
DE Bauplatz *m*
Baustelle *f*
ES emplazamiento *m*
F lieu *m* de mise en place
site *m*
I luogo *m*
posto *m*
MA helyszín
NL bouwplaats *f*
lokatie *f*
PO plac *m* budowy
teren *m*
PY место *n*
расположения
SU rakennuspaikka
SV tomt *c*

3460 site agent
DA bygherre-repræsentant
DE Bauaufsicht-führender *m*
Beauftragter *m* des Bauherrn *m*
ES agente a pie de obra *m*
encargado de las obras *m*
F contrôleur *m* de chantier
I ispettore *m* dei lavori
rappresentante *m* del proprietario in cantiere
MA helyszíni műszaki ellenőr
NL vertegenwoordiger *m* op de bouwplaats
PO inspektor *m* nadzoru na budowie
PY инспектор *m* рабочего места (охраны труда)
SU asiamies
SV byggherre *c*

3461 site assembly
DA byggepladsmontage
DE Baustellenmontage *f*
ES montaje en obra *m*
F montage *m* sur chantier
I montaggio *m* in cantiere
MA helyszíni szerelés
NL montage *f* op de bouwplaats
PO montaż *m* na budowie
PY монтаж *m* на рабочем месте
SU paikallakoottu
SV platsmontering *c*

3462 size
DA størrelse
DE Grüße *f*
Maß *n*
ES dimensiones *f,pl*
tamaño *m*
F dimension *f*
grandeur *f*
mesure *f*
I dimensione *f*
grandezza *f*
misura *f*
MA méret
NL grootte *f*
maat *f*
PO rozmiar *m*
wielkość *f*
wymiar *m*
PY размер *m*
SU koko
suuruus
SV storlek *c*

3463 skirting heater (*see* baseboard heater)

3464 slack
DA kulstøv
DE Feinkohle *f*
Kohlengruß *m*
ES menudo de carbón *m*
F fines *f* de charbon
I fini *m,pl* di carbone
MA aprószén
NL fijnkool *f*
kolengruis *n*
PO węgiel *m* drobny
PY угольная мелочь *f*
SU murskahiili
SV kolstybb *c*

3465 slag (*see* clinker)

3466 slag discharge
DA slaggefjernelse
DE Schlackenabzug *m*
ES evacuación de
    escorias *f*
F évacuation *f* de
    scories
I scarico *m* di scorie
MA salak eltávolítás
NL slakkenafvoer *m*
PO odprowadzenie *n*
    żużla
PY отделение *n* шлака
SU kuonanpoisto
SV avlägsnande *n* av
    slagg *c*

3467 slag formation
DA slaggedannelse
DE Schlackenbildung *f*
ES escorificación *f*
    formación de
    escorias *f*
F formation *f* de
    scories
I formazione *f* di
    scorie
MA salakképződés
NL slakvorming *f*
PO tworzenie *n* się żużla
PY образование *n*
    шлака
SU kuonanmuodostus
SV slaggbildning *c*

3468 slag layer
DA slaggegrube
DE Schlackenschicht *f*
ES lecho de escorias *m*
F lit *m* de scories
I strato *m* di scorie
MA salakréteg
NL slakkenlaag *f*
PO warstwa *f* żużla
PY слой *m* шлака
SU kuonakerros
SV slaggbädd *c*

3469 slag tongs
DA slaggetang
DE Schlackenzange *f*
ES tenazas para escorias
    f,pl
F pince *f* à machefer
I pinza *f* per scorie
MA salakfogó
NL slakkentang *f*
PO szczypce *pl* do żużla
PY грейфер *m* для
    шлака
SU kuonapihdit
SV slaggtáng *c*

3470 slag wool
DA mineraluld
DE Schlackenwolle *f*
ES lana de escorias *f*
F laine *f* de scories
I lana *f* di scorie
MA salakgyapot
NL slakkenwol *f*
PO wełna *f* żużlowa
PY минеральная вата *f*
    шлаковата *f*
SU kuonavilla
SV mineralull *c*

3471 slave controller
DA sekundær styring
    slavestyring
DE nachgeschalteter
    Regler *m*
ES regulador esclavo *m*
F régulateur *m* esclave
I regolatore *m*
    asservito
MA segédszabályozás
NL volgregelaar *m*
PO regulator *m*
    nadążny
    regulator *m*
    podporządkowany
    regulator *m*
    podrzędny
PY регулятор *m* с
    приводом
SU orjasäätü
SV servoregulator *c*

3472 slave unit
DA sekundær enhed
DE nachgeschaltetes
    Gerät *n*
ES unidad esclava *f*
F unité *f* esclave
I unità *f* asservita
MA mellékegység
NL volgregeleenheid *f*
PO jednostka *f*
    podporządkowana
    urządzenie *n*
    nadążne
PY агрегат *m* с
    приводом
SU orjayksikkü
SV undercentral *c*

3473 sleeve
DA bøsning
    foring
DE Hülse *f*
    Manschette *f*
    Muffe *f*
ES casquillo *m*
    forro *m*
    manguito *m*
F douille *f*
    manchette *f*
    manchon *m*
I manicotto *m*
MA hüvely
    karmantyú
NL huls *f*
    manchet *f*
    mof *f*
PO tuleja *f*
PY муфта *f*
SU hylsy
    kaulus
    muhvi
SV hylsa *c*
    muff *c*

3474 sleeve bearing
DA glideleje
DE Lagerbüchse *f*
ES cojinete de
    fricción *m*
    cojinete liso *m*
F coussinet *m*
I cuscinetto *m* a
    bronzine
MA csapágypersely
NL glijdlager *n*
PO łożysko *n* tulejowe
PY муфтовый
    подшипник *m*
SU liukulaakeri
SV glidlager *n*

**3475 sleeve-bearing
    support**
DA glidelejebæring
DE Lagerstuhl *m*
ES soporte de cojinete
    liso *m*
F palier *m* lisse
I supporto *m* per
    cuscinetto a
    bronzine
MA csúszó alátámasztás
NL glijdlageronder-
    steuning *f*
PO podpora *f* łożyska
    tulejowego
PY опора *f* муфтового
    подшипника
SU laakeripesä
SV glidlagerstüd *n*

3476 sleeve filter
DA slangefilter
DE Schlauchfilter *m*
ES filtro de manga *m*
   filtro de tubo
   flexible *m*
   filtro tubular
   flexible *m*
F filtre *m* à manches
I filtro *m* a manica
MA tümlős szűrő
NL zakkenfilter *n*
PO filtr *m* rękawowy
PY муфтовый
   фильтр *m*
SU letkusuodatin
SV slangfilter *n*

3477 slide damper
DA skydespjæld
DE Jalousie-
   Stellklappe *f*
ES registro de
   guillotina *m*
F registre *m* à coulisse
I serranda *f* a
   ghigliottina
MA tolattyú
NL luchtschuif *f*
   schuivende
   luchtklep *f*
PO zasuwa *f*
PY заслонка-движок *f*
SU liukupelti
SV skjutspjäll *n*

3478 sliding damper
DA skydespjæld
DE Gleitschieber *m*
ES compuerta corredera
   f
F registre *m* coulissant
I serranda *f* a coulisse
   (scorrevole)
MA tolattyú
NL luchtschuif *f*
PO zasuwa *f* ślizgowa
PY скользящий
   (сдвигаемый)
   клапан *m*
SU tyüntüpelti
SV skjutspjäll *n*

3479 sliding expansion joint
DA glideekspansions-
   forbindelse
DE Gleitkompensator *m*
ES compensador de
   dilatación
   deslizante *m*
   junta de dilatación
   deslizante *f*
F compensateur *m* à
   joints glissants
I compensatore *m* di
   dilatazione a
   scorrimento
MA teleszkópos táguló
   csatlakozás
NL axiale
   compensator *m*
PO wydłużka *f*
   przesuwna
   wydłużka *f* ślizgowa
PY скользящий
   компенсатор *m*
SU liukutasain
SV expansions-
   koppling *c*

3480 sliding support
DA glidefod
DE Gleitstütze *f*
ES soporte deslizante *m*
F support *m*
   coulissant
I supporto *m*
   scorrevole
MA csúszó alátámasztás
NL glijdende
   ondersteuning *f*
   glijdende
   oplegging *f*
PO podpora *f* ślizgowa
PY скользящая
   опора *f*
SU ohjauspiste
SV glidstüd *n*

3481 sliding vane compressor
DA lamelkompressor
DE Gleitschaufel-
   verdichter *m*
ES compresor de álabe
   deslizante *m*
F compresseur *m* à
   palettes
I compressore *m* a
   palette
MA csúszólapátos
   kompresszor
NL schotten-
   compressor *m*
PO sprężarka *f*
   łopatkowa
PY ротационный
   компрессор *m* с
   эксцентрическим
   ротором
SU lamellikompressori
SV lamellkompressor *c*

3482 sling psychrometer
DA slyngpsykrometer
DE Schleuderpsychro-
   meter *n*
ES higrómetro de
   honda *m*
   psicómetro de
   honda *m*
F psychromètre *m*
   fronde
I psicrometro *m* a
   fionda
MA parittyás
   nedvességmérő
NL slingerpsychro-
   meter *m*
PO psychrometr *m*
   pętlowy
PY переносной
   психрометр *m*
SU linkopsykrometri
SV slingpsykrometer *c*

3483 slip flange
DA glideflange
DE loser Flansch *m*
ES brida deslizante *f*
   brida móvil *f*
F bride *f* folle
I flangia *f* folle
MA tolókarima
NL overschuifflens *m*
PO kołnierz *m* wolny
PY скользящий
   фланец *m*
SU irtolaippa
SV glidfläns *c*

3484 slot
DA kærv
 slids
DE Nut *f*
 Schlitz *m*
 Spalt *m*
ES muesca *f*
 ranura *f*
F fente *f*
I fessura *f*
MA rés
NL sleuf *f*
 sponning *f*
PO rowek *m*
 szczelina *f*
PY паз *m*
 прорезь *f*
SU rako
SV ränna *c*

3485 slot diffuser
DA spaltediffuser
DE Schlitzauslaß *m*
ES difusor lineal *m*
F diffuseur *m* à fente
I diffusore *m* lineare
MA kifúvórés
NL lijnrooster *m*
 lineair
 luchtuitblaas-
 ornament *n*
PO anemostat *m*
 szczelinowy
PY щелевой
 диффузор *m*
SU rakohajoittaja
SV spaltdiffusor *c*

3486 slotted outlet
DA spaltet udløb
DE geschlitzter
 Auslaß *m*
 Schlitzauslaß *m*
ES ranura de salida *f*
F bouche *f* linéaire
I diffusore *m* a
 feritoia
MA hézagos kilépőnyílás
NL sleufrooster *m*
PO wylot *m* szczelinowy
PY выходное щелевое
 отверстие *n*
SU säleikkü
SV slitsutlopp *n*

3487 sludge
DA slam
DE Schlamm *m*
ES fango *m*
 sedimento *m*
F boue *f*
I fango *m*
MA iszap
NL slib *n*
PO muł *m*
 osad *m* ściekowy
 szlam *m*
PY шлам *m*
SU viemäriliete
SV slam *n*

3488 sludge filter
DA slamfilter
DE Schlammfang *m*
ES filtro de fangos *m*
F filtre *m* à boues
I filtro *m* dei fanghi
MA iszapszűrő
NL slibfilter *m*
PO filtr *m* osadu
 odmulnik *m*
PY грязевик *m*
 шламоотделитель
 m
SU lietesuodatin
SV slamfilter *n*

3489 sludge removal
DA slamfjernelse
DE Entschlammen *n*
ES extracción de
 fangos *f*
F évacuation *f* des
 boues
I rimozione *f* dei
 fanghi
MA iszapeltávolítás
NL slibafvoer *m*
PO usuwanie *n* szlamu
PY удаление *n* шлама
SU lietteenpoisto
SV slamborttagning *c*

3490 slug flow
DA slamstrøm
DE Ventilstrümung *f*
ES flujo laminar *m*
F écoulement *m*
 diphasique ondulé
 et à bouchons
I corrente *f* a
 stantuffo
 corrente *f* a tappi
MA lassú áramlás
NL verdringings-
 stroming *f*
PO wypływ *m* powolny
PY поток *m* пульпы
SU tulppavirtaus
SV pluggflüde *n*

3491 small bore system
DA system med små rør
DE Kleinrohrsystem *n*
ES sistema de tubos
 de pequeño
 diámetro *m*
F chauffage *m* par
 tuyauterie de petit
 diamètre
I sistema *m* di
 riscaldamento
 minitubo
MA kisátmérőjű
 csőrendszer
NL leidingsysteem *n* met
 geringe diameter
PO system *m* z
 przewodami o
 małych średnicach
PY система *f* с
 тонкими трубами
SU pienputkijärjestelmä
SV klenrürssystem *n*

3492 smoke box
DA røgkammer
DE Rauchgaskasten *m*
ES caja de humos *f*
F boîte *f* à fumée
I camera *f* dei fumi
MA füstszekrény
NL rookkast *f*
PO dymnica *f*
PY дымовая коробка *f*
SU savukaasukammio
SV rükbox *c*

3493 smoke density
DA røgtæthed
DE Rauchgasdichte *f*
ES densidad del humo *f*
F densité *f* de fumée
 opacité *f* de fumée
I densità *f* del fumo
 opacità *f* del fumo
MA füst sűrűség
NL rookdichtheid *f*
PO gęstość *f* dymu
PY плотность *f* дыма
SU savun tiheys
SV rüktäthet *c*

DA	=	Danish
DE	=	German
ES	=	Spanish
F	=	French
I	=	Italian
MA	=	Magyar
		(Hungarian)
NL	=	Dutch
PO	=	Polish
PY	=	Russian
SU	=	Finnish
SV	=	Swedish

3494 smoke detector
DA røgdetektor
DE Rauchgasmelder *m*
ES detector de humos *m*
F détecteur *m* de fumée
I rivelatore *m* di fumo
MA füstérzékelő
NL rookdetector *m*
PO czujka *f* dymowa
wykrywacz *m* dymu
PY детектор *m* дыма
SU savuhälytin
SV rükdetektor *c*

3495 smoke flue
DA røgkanal
DE Rauchgasabzugs-kanal *m*
ES chimenea *f*
salida de humos *f*
F carneau *m* de fumée
cheminée *f*
I tubo *m* di fumo
MA füstgáz
NL rookgaskanaal *n*
PO czopuch *m*
kanał *m* dymowy
PY боров *m*
SU savukaasukanava
SV rükkanal *c*

3496 smoke number (*see* soot number)

3497 smokes
DA røgarter
DE Rauchgase *n,pl*
ES humos *m*
F fumées *f*
I fumi *m,pl*
MA füstük
NL dampen *pl*
PO dymy *m*
PY частицы *f* дыма
SU savu
SV rük *c*

3498 smoke stack (*see* chimney)

3499 smoke test
DA røgprøve
DE Rauchgasprobe *f*
Rauchtest *m*
ES ensayo de humos *m*
F essai *m* d'opacité de fumée
I controllo *m* dei fumi
MA füstpróba
NL roettest *m*
PO próba *f* szczelności przy użyciu dymu
PY проба *f* дыма
SU savukaasuanalyysi
SV rükgasprov *n*

3500 smoke tube boiler
(*see* fire tube boiler)

3501 smooth bend
DA blød bøjning
DE Glattrohrbogen *m*
ES codo suave *m*
curva *f*
F coude *m* en tube lisse
I curva *f* liscia
MA sima ív
NL gladde bocht *f*
PO łuk *m* gładki
PY плавный отвод *m*
SU virtaviivainen mutka
SV jämn büj *c*

3502 smudge
DA smuds
DE Schmutz *m*
Schmutzfleck *m*
ES suciedad adherente *f*
tiznadura de hollín o de carbón *f*
F dépôt *m* de suie ou de poussières
I deposito *m* di polvere e fuliggine
sbaffo *m*
traccia *f* di sporco
MA piszok
NL vlek *f*
vuile plek *f*
PO brud *m*
plama *f*
powłoka *f* zabezpieczająca (np. przy lutowaniu)
PY пятно *n*
SU lika
SV smuts *c*

3503 smuts
DA sodflager
DE Ruß *m*
Rußflocken *f*
ES partículas de hollín *f,pl*
F flocons *m* de suies
I fiocchi *m,pl* di fuliggine
MA korompehely
NL roetvlokken *pl*
PO sadze *f*
PY сажа *f*
SU noki
SV smuts *c*

3504 snap action
DA hurtigtvirkende
DE Schnapp-mechanismus *m*
ES actuación rápida *f*
F action *f* à déclic
I azione *f* rapida
MA csapó működés
NL momentschakeling *f*
PO działanie *n* migowe
działanie *n* natychmiastowe
PY быстродействие *n*
SU nopea kaksiasentoto-iminta
SV snabbverkan *c*

3505 snap-action control
DA hurtigtvirkende styring
DE Schnellschluß-regelung *f*
ES control de actuación rápida *m*
F contrôle *m* à déclic
I controllo *m* ad azione rapida
MA gyorsárású szabályozás
NL regelaar *m* met momentschakeling
regeling *f* met momentschakeling
PO sterowanie *n* działaniem natychmiastowym
PY безинерционное регулирование *n*
SU nopea kaksiasentosäätü
SV snabbverkande kontroll *c*

3506 snap-action mechanism
DA hurtigtvirkende mekanisme
DE Schnappvorrichtung *f*
ES mecanismo de acción rápida (electricidad) *m*
F mécanisme *m* à action instantanée
I meccanismo *m* ad azione istantanea
meccanismo *m* a scatto
MA bepattintós szerkezet csapóműküdésű szerkezet
NL momentschakelmechanisme *n*
PO mechanizm *m* działania natychmiastowego
PY быстродействующий механизм *m*
SU nopea kaksiasentomekanismi
SV snabbverkande mekanism *c*

3507 snow melting system
DA snesmeltningssystem
DE Schneeschmelzsystem *n*
ES sistema de fusión de nieve *m*
sistema de licuación de la nieve *m*
F déneigement *m*
I impianto *m* scioglineve
MA hóolvasztó rendszer
NL sneeuwsmeltsysteem *n*
PO instalacja *f* topienia śniegu
PY система *f* таяния снега
SU lumensulatusjärjestelmä
SV snüsmältningssystem *n*

3508 soakaway
DA siveafløb
DE Ablaufkanal *m* Sickergrube *f*
ES drenaje *m* excavación para drenaje de aguas *f*
F puit *m* perdu
I pezzo *m* a perdere
MA elvezetőcsatorna folyóka
NL vuilwatergoot *f* vuilwaterput *m*
PO studnia *f* chłonna
PY поглощающий колодец *m*
SU imeytyskaivo sorapohjalla varustettu kaivo
SV avloppskanal *c*

3509 socket
DA muffe tætning
DE Büchse *f* Fassung *f* Hülse *f*
ES enchufe hembra *m*
F prise *f* de courant
I manicotto *m* zoccolo *m*
MA foglalat lábazat
NL mof *f* voetstuk *n*
PO gniazdo *n* (do wtyczki) tuleja *f*
PY насадка *f*
SU holkki muhvi rasia
SV muff *c* stuts *c*

3510 softening
DA blødgøring
DE Enthärtung *f*
ES ablandamiento *m*
F adoucissement *m*
I addolcimento *m*
MA lágyítás
NL ontharding *f*
PO zmiękczanie *n*
PY умягчение *n*
SU pehmennys
SV mjukgüring *c*

3511 soft soldering
DA tinlodning
DE Weichlütung *f*
ES soldadura blanda *f*
F soudure *f* à l'étain
I brasatura *f* dolce saldatura *f* a stagno
MA lágyforrasztás
NL zachtsolderen *n*
PO lutowanie *n* miękkie
PY мягкая пайка *f*
SU pehmeä juotos
SV mjuklüdning *c*

3512 sol air temperature
DA lufttemperatur i solskin
DE scheinbare Sonnenlufttemperatur *f*
ES temperatura sol-aire *f*
F température *f* fictive extérieure
I temperatura *f* solearia
MA napléghőmérséklet
NL sol air temperatuur *f*
PO temperatura *f* słoneczna powietrza
PY солнечновоздушная температура *f*
SU aurinkoilmalämpütila ekvivalenttilämpütila
SV sol-lufttemperatur *c*

3513 solar collector
DA solfanger
DE Sonnenkollektor *m*
ES colector solar *m*
F capteur *m* solaire
I collettore *m* solare pannello *m* solare
MA napkollektor
NL zonnecollector *m*
PO kolektor *m* słoneczny
PY солнечный коллектор *m*
SU auringonkerääjä aurinkokaha
SV solkollektor *c*

3514 solar constant
DA solkonstant
DE Solarkonstante *f*
ES constante solar *f*
F constante *f* solaire
I costante *f* solare
MA napállandó
NL zonneconstante *f*
PO stała *f* słoneczna
PY солнечная постоянная *f*
SU aurinkovakio
SV solarkonstant *c*

3515 solar energy
DA solenergi
DE Sonnenenergie *f*
ES energía solar *f*
F énergie *f* solaire
I energia *f* solare
MA napenergia
NL zonne-energie *f*
PO energia *f* słoneczna
PY солнечная энергия *f*
SU aurinkoenergia
SV solenergi *c*

3516 solar heat gain
DA varmeoptagelse ved solstråling
DE Wärmegewinn *m* durch Sonnenstrahlung *f*
ES aporte de calor por insolación *m*
F apports *m* de chaleur solaire
I apporto *m* di calore solare
MA szoláris hőnyereség
NL warmtewinst *f* door zoninstraling
PO zysk *m* ciepła od nasłonecznienia
PY приток *m* солнечной теплоты
SU aurinkokuorma
SV solvärmevinning *c*

3517 solar irradiance (*see* solar radiation)

3518 solar radiation
DA solstråling
DE Sonnenstrahlung *f*
ES radiación solar *f*
F rayonnement *m* solaire
I irraggiamento *m* solare
MA napsugárzás
NL zonnestraling *f*
PO promieniowanie *n* słoneczne
PY солнечная радиация *f*
SU auringon säteily
SV solstrålning *c*

3519 solar sensor
DA solføler
DE Sonnenfühler *m*
ES sonda solar *f*
F capteur *m* de rayonnement solaire
I sensore *m* solare
MA napsugárzás érzékelő
NL zonstralings-detector *m* zonvoeler *m*
PO czujnik *m* słoneczny pomiarowy
PY солнечный датчик *m*
SU aurinkotuntoelin
SV solindikator *c*

3520 solder
DA loddemetal
DE Lütmittel *n* Lot *n*
ES soldador *m*
F soudure *f*
I lega *f* per saldatura
MA forrasz
NL soldeer *n*
PO lut *m* stop *m* lutowniczy
PY пайка *f* паяние *n*
SU juotos
SV lüdtenn *n*

3521 solder *vb*
DA lodde
DE lüten
ES soldar
F souder
I brasare saldare
MA forraszt
NL solderen
PO lutować
PY паять
SU juottaa
SV lüda

3522 soldered joint
DA loddested
DE Lütverbindung *f*
ES junta soldada *f*
F joint *m* soudé
I giunzione *m* saldata
MA forrasztott csatlakozás
NL soldeerverbinding *f*
PO złącze *n* lutowane
PY запаянный стык *m*
SU juotosliitos
SV lüdställe *n*

3523 soldering
DA lodning
DE Lütung *f*
ES soldadura (sin metal de aporte) *f*
F soudure *f*
I brasatura *f*
MA forrasztás
NL solderen *n*
PO lutowanie *n*
PY пайка *f* (процесс)
SU juottaminen
SV lüdning *c*

3524 solenoid valve
DA magnetventil
DE Magnetventil *n*
ES válvula de solenoide *f* válvula electromagnética *f*
F électro-vanne *f*
I valvola *f* a solenoide elettromagnetica
MA szolenoid szelep
NL magneetklep *f*
PO zawór *m* elektro-magnetyczny
PY соленоидный вентиль *m*
SU magneettiventtiili
SV solenoidventil *c*

DA	=	Danish
DE	=	German
ES	=	Spanish
F	=	French
I	=	Italian
MA	=	Magyar (Hungarian)
NL	=	Dutch
PO	=	Polish
PY	=	Russian
SU	=	Finnish
SV	=	Swedish

3525 solid phase condensation
DA fastform kondensering
DE Kondensation *f* in fester Phase *f*
ES condensación de fase sólida *f*
F condensation *f* en phase solide
I condensazione *f* in fase solida
MA szilárd fázisú kondenzáció
NL condensatie *f* in vaste fase
PO kondensacja *f* fazy stałej
PY кристаллизация *f*
SU sublimoituminen
SV kondensering *c* till fast fas *c*

3526 solubility
DA opløselighed
DE Lüslichkeit *f*
ES solubilidad *f*
F solubilité *f*
I solubilità *f*
MA oldhatóság
NL oplosbaarheid *f*
PO rozpuszczalność *f*
PY растворимость *f*
SU liukoisuus
SV lüslighet *c*

3527 solution
DA opløsning
DE Lüsung *f*
ES solución *f*
F solution *f*
I soluzione *f*
MA oldat
NL oplossing *f*
PO rozpuszczanie *n* roztwór *m* rozwiązanie *n*
PY раствор *m*
SU ratkaisu
SV lüsning *c*

3528 solvent
DA opløsningsmiddel
DE Lüsungsmittel *n*
ES disolvente *m* solvente *m*
F solvant *m*
I solvente *m*
MA oldószer
NL oplosmiddel *n*
PO rozpuszczalnik *m*
PY растворитель *m*
SU liuotin
SV lüsningsmedel *n*

3529 soot
DA sod
DE Ruß *m*
ES hollín *m*
F suie *f*
I fuliggine *f*
MA korom
NL roet *n*
PO sadza *f*
PY копоть *f* сажа *f*
SU noki
SV sot *n*

3530 soot arrester
DA sodfanger
DE Rußfänger *m*
ES captador de hollín *m* detentor de hollín *m*
F capte-suie *m*
I captatore *m* di fuliggine
MA koromfogó
NL roetvanger *m*
PO chwytacz *m* sadzy
PY уловитель *m* сажи
SU noenerottaja
SV sotavskiljare *c*

3531 soot blower
DA sodblæser
DE Rußbläser *m*
ES soplador de hollín *m*
F souffleur *m* de suies
I soffiatore *m* di fuliggine
MA koromfúvó
NL roetblazer *m*
PO zdmuchiwacz *m* sadzy
PY дымосос *m*
SU nokipuhallin
SV sotningsapparat *c*

3532 soot number
DA sodtal
DE Rußzahl *f*
ES índice de Bacharach *m*
F indice *m* de Bacharach
I indice *m* di Bacharach
MA koromszám
NL roetgetal *n*
PO wskaźnik *m* zawartości sadzy
PY индекс *m* закопченности индекс *m* сажи
SU nokiluku
SV sottal *n*

3533 soot removal
DA sodfjernelse
DE Rußabscheidung *f*
ES deshollinado *m*
F ramonage *m*
I rimozione *f* della fuliggine
MA korom eltávolítás
NL roetverwijdering *f*
PO usuwanie *n* sadzy
PY устранение *n* копоти
SU nuohous
SV sotborttagning *c*

3534 soot test
DA sodprøve
DE Rußtest *m*
ES medición del contenido de hollines *f*
F mesure *f* de la teneur en suies
I prova *f* di fuliggine
MA korompróba
NL roettest *n*
PO pomiar *m* zawartości sadzy
PY испытание *n* сажи
SU nokikuva
SV sottest *n*

3535 sorbent
DA adsorbtionsmiddel
DE Absorptionsmittel *n*
ES sorbente *m*
F absorbant *m*
I assorbente *m*
MA szorbeáló szer
NL absorptiemiddel *n*
PO sorbent *m*
PY поглотитель *m* сорбент *m*
SU sorptioaine
SV sorbator *c*

3536 sound
DA lyd
DE Geräusch *n*
ES sonido *m*
F bruit *m*
son *m*
I suono *m*
MA hang
zaj
NL geluid *n*
klank *m*
PO dzwięk *m*
PY звук *m*
SU ääni
SV ljud *n*

3537 sound absorption
DA lydabsorption
DE Schallabsorption *f*
ES absorción acústica *f*
F absorption *f*
acoustique
I assorbimento *m* del
suono
MA hangelnyelés
NL geluidabsorptie *f*
PO pochłanianie *n*
dźwięku
PY поглощение *n*
звука
SU äänen absorptio
SV ljudabsorption *c*

3538 sound attenuation
DA lyddæmpning
DE Schalldämpfung *f*
ES atenuación
acústica *f*
insonorización *f*
F atténuation *f* du son
I attenuazione *f* del
rumore
MA hangtompítás
NL geluiddemping *f*
PO tłumienie *n* dzwięku
PY затухание *n* звука
SU äänenvaimennus
SV ljuddämpning *c*

3539 sound attenuator
DA lyddæmper
DE Schalldämpfer *m*
ES silenciador *m*
F dispositif *m*
d'insonorisation
silencieux *m*
I dispositivo *m* di
insonorizzazione
silenziatore *m*
MA hangtompító
NL geluiddemper *m*
PO tłumik *m* dzwięku
PY звукопоглотитель
m
SU äänenvaimentaja
SV ljuddämpare *c*

3540 sound control
DA lydovervågning
DE Geräuschüber-
wachung *f*
ES control de sonido *m*
F contrôle *m*
acoustique
I regolazione *f* del
suono
MA zaj szabályozás
NL geluidregeling *f*
PO regulacja *f* dzwięku
PY контроль *m* звука
SU melun torjunta
SV ljudkontroll *c*

3541 sound energy
DA lydenergi
DE Schallenergie *f*
ES energía acústica *f*
F énergie *f* sonore
I energia *f* del suono
MA hangenergia
NL geluidsenergie *f*
PO energia *f* akustyczna
PY энергия *f* звука
SU äänienergia
SV ljudenergi *c*

3542 sound insulation
DA lydisolering
DE Schalldämmung *f*
Schallschutz *m*
ES aislamiento
acústico *m*
F insonorisation *f*
I insonorizzazione *f*
MA hangszigetelés
NL geluidisolatie *f*
PO izolacja *f*
akustyczna
PY звукоизоляция *f*
SU ääneneristys
SV ljudisolering *c*

3543 sound level
DA lydniveau
DE Schallpegel *m*
ES nivel sonoro *m*
F niveau *m* sonore
I livello *m* sonoro
MA zajszint
NL geluidsniveau *n*
PO poziom *m* dzwięku
PY уровень *m* звука
SU äänitaso
SV ljudnivå *c*

3544 sound power level
DA lydstyrkeniveau
DE Schalleistungs-
pegel *m*
ES nivel de potencia
acústica *m*
F niveau *m* de
puissance sonore
I livello *m* di potenza
sonora
MA hangteljesítmény-
szint
NL geluidsvermogen-
niveau *n*
PO poziom *m* natężenia
dzwięku
PY уровень *m*
интенсивности
звука
уровень *m* силы
звука
SU äänen tehotaso
SV ljudeffektnivå *c*

3545 sound pressure
DA lydtryk
DE Schalldruck *m*
ES presión acústica *f*
presión sonora *f*
F pression *f*
acoustique
I pressione *f* sonora
MA hangnyomás
NL geluidsdruk *m*
PO ciśnienie *n*
akustyczne
PY звуковое
давление *n*
SU äänenpaine
SV ljudtryck *n*

3546 sound pressure level
DA lydtryksniveau
DE Schalldruckpegel *m*
ES nivel de presión acústica *m*
nivel de presión sonora *m*
F niveau *f* de pression acoustique
I livello *m* di pressione sonora
MA hangnyomásszint
NL geluidsdrukniveau *n*
PO poziom *m* ciśnienia akustycznego
PY уровень *m* звукового давления
SU äänenpainetaso
SV ljudtrycksnivå *c*

3547 sound proofing
DA lydisolering
DE Schallschutz *m*
ES insonorizado
F insonorisation *f*
I protezione *f* antirumore
MA hanggátlás
NL geluiddichtmaken *n*
PO ochrona *f* przeciwdzwiękowa
PY звукозащита *f*
SU ääneneristys ääneneristäminen
SV ljudisolering *c*

3548 sound propagation
DA lydforplantning
DE Schallausbreitung *f*
ES propagación del sonido *f*
F propagation *f* du son
I propagazione *f* del suono
MA hangterjedés
NL geluidvoort-planting *f*
PO rozchodzenie *n* się dźwięku
rozprzestrzenianie *n* się dźwięku
PY распространение *n* звука
SU äänen eteneminen
SV ljudutbredning *c*

3549 sound reduction
DA lydreduktion
DE Lärmminderung *f*
ES reducción del sonido *f*
F diminution *f* du bruit
I riduzione *f* del rumore
MA hangcsükkentés
NL geluid-vermindering *f*
PO obniżenie *n* poziomu dzwięku
PY ослабление *n* звука
SU melun torjunta
SV ljuddämpning *c*

3550 sound spectrum
DA lydfrekvensområde
DE Geräusch-spektrum *n*
ES espectro acústico *m*
espectro sonoro *m*
F spectre *m* sonore
I spettro *m* sonoro
MA hangspektrum
NL geluidsspectrum *n*
PO widmo *n* dzwięku
PY звуковой спектр *m*
SU äänen spektri
SV ljudspektrum *n*

3551 sound velocity
DA lydhastighed
DE Schallgeschwindig-keit *f*
ES velocidad del sonido *f*
F vitesse *f* du son
I velocità *f* del suono
MA hangsebesség
NL geluidssnelheid *f*
PO prędkość *f* dzwięku
PY скорость *f* звука
SU äänennopeus
SV ljudhastighet *c*

3552 source of sound
DA lydkilde
DE Schallquelle *f*
ES fuente sonora *f*
F source *f* sonore
I fonte *f* sonora
MA hangforrás
NL geluidsbron *f*
PO źródło *n* dzwięku
PY источник *m* звука
SU äänenlähde
SV ljudkälla *c*

3553 space cooling
DA rumkøling
DE Raumkühlung *f*
ES enfriamiento de locales *m*
enfriamiento de recintos (cerrados) *m*
F rafraîchissement *m* de locaux
I raffreddamento *m* ambiente
MA térhűtés
NL ruimtekoeling *f*
PO chłodzenie *n* pomieszczeń
PY охлаждение *n* того или иного объема
SU rakennuksen jäähdytys
SV rumskylning *c*

3554 space heating
DA rumopvarmning
DE Raumheizung *f*
ES calefacción de locales *f*
calefacción de recintos *f*
F chauffage *m* de locaux
I riscaldamento *m* ambiente
MA térfűtés
NL ruimteverwarming *f*
PO ogrzewanie *n* pomieszczeń
PY отопление *n* того или иного объема
SU rakennuksen lämmitys
SV rumsuppvärmning *c*

DA	=	Danish
DE	=	German
ES	=	Spanish
F	=	French
I	=	Italian
MA	=	Magyar (Hungarian)
NL	=	Dutch
PO	=	Polish
PY	=	Russian
SU	=	Finnish
SV	=	Swedish

3555 space requirement
DA pladsbehov
DE Platzbedarf *m*
    Raumbedarf *m*
ES espacio necesario *m*
F encombrement *m*
    place *f* nécessaire
I ingombro *m*
    spazio *m*
MA helyigény
NL ruimtebehoefte *f*
PO przestrzeń *f*
    niezbędna
    przestrzeń *f*
    wymagana
    zapotrzebowanie *n*
    na miejsce
PY требования *n,pl*
    обслуживаемого
    объема
SU tilantarve
SV utrymmesbehov *n*

3556 spacing
DA afstand
DE Abstand *m*
    Entfernung *f*
ES distancia *f*
    paso *m*
    separación *f*
F distance *f*
    écartement *m*
    espacement *m*
    pas *m*
I distanza *f*
MA térküz
NL tussenruimte *f*
PO odstęp *m*
    rozmieszczenie *n*
    rozstawienie *n*
PY расстановка *f*
SU etäisyys
SV avstånd *n*

3557 spanner
DA skruenøgle
DE Schrauben-
    schlüssel *m*
ES llave de tuercas *f*
    llave inglesa *f*
F clé *f* plate
I chiave *f*
    utensile *m*
MA csavarkulcs
NL moersleutel *m*
PO klucz *m* płaski
    (maszynowy)
PY разводной ключ *m*
SU ruuvitaltta
SV skruvnyckel *c*

3558 spare part
DA reservedel
DE Ersatzteil *n*
ES pieza de repuesto *f*
    recambio *m*
F pièce *f* détachée
I parte *f* di ricambio
MA tartalékalkatrész
NL reservedeel *n*
    reserve-onderdeel *n*
PO część *f* zapasowa
PY запасная часть *f*
SU varaosa
SV reservdel *c*

3559 spark
DA gnist
DE Funke *m*
ES chispa *f*
F étincelle *f*
I scintilla *f*
MA szikra
NL vonk *f*
PO iskra *f*
PY искра *f*
SU kipinä
SV gnista *c*

3560 spark igniter
DA gnisttænder
DE Funkenzünder *m*
ES bujía de ignición *f*
F dispositif *m*
    d'allumage
I dispositivo *m* di
    accensione a
    scintilla
MA szikragyújtó
NL ontstekings-
    electrode *f*
PO iskrownik *m*
    zapłonnik *m*
    iskrowy
PY запальник *m*
SU kipinäsytytin
SV gnisttändare *c*

3561 specific *adj*
DA specifik
DE spezifisch
ES específico
F spécifique
I specifico
MA fajlagos
    specifikus
NL specifiek
PO właściwy
PY определенный
    удельный
    характерный
SU ominainen
    ominais-
SV specifik

3562 specification
DA specifikation
DE Angabe *f*
    Spezifizierung *f*
ES especificación *f*
F devis *m* descriptif
I specifica *f*
MA részletes leírás
NL bestek *n*
    specificatie *f*
PO specyfikacja *f*
    wykaz *m*
    wyszczególnienie *n*
PY спецификация *f*
SU erittely
    määrittely
SV specifikation *c*

3563 specific enthalpy
DA specifik entalpi
DE spezifische
    Enthalpie *f*
ES entalpía específica *f*
F enthalpie *f*
    spécifique
I entalpia *f* massica
    entalpia *f* specifica
MA fajlagos entalpia
NL specifieke
    enthalpie *f*
PO entalpia *f* właściwa
PY энтальпия *f*
    (удельная)
SU ominaisentalpia
SV specifik entalpi *c*

3564 specific heat flow
DA specifik varmestrøm
DE Wärmestromdichte *f*
ES densidad del flujo
    térmico *m*
    flujo de calor
    másico *m*
F densité *f* de flux
I flusso *m* di calore
    specifico
MA hőáramsűrűség
NL warmtestroom-
    dichtheid *f*
PO strumień *m* ciepła
    właściwego
PY удельный тепловой
    поток *m*
SU ominaislämpüvirta
SV specifikt
    värmeflüde *n*

3565 speed
DA fart
   hastighed
DE Geschwindigkeit *f*
ES velocidad *f*
F vitesse *f*
I velocità *f*
MA sebesség
NL snelheid *f*
PO prędkość *f*
PY скорость *f*
SU nopeus
   vauhti
SV hastighet *c*

3566 spigot
DA prop
   tap
DE Muffe *f*
   Zapfen *m*
ES articulación de
   encastre *f*
F emboîtement *m*
I parte *f* imboccata
   nel bicchiere
MA karmantyú
   tok
NL spie-eind *n* van pijp
   tapkraan *f* van een
   vat
PO czop *m*
   kielich *m* (rury)
   tuleja *f*
PY втулка *f*
   гладкий конец *m*
   трубы
   пробка *f*
SU tulppa
SV plugg *c*

3567 spigot and socket joint
DA muffesamling
DE Muffenverbindung *f*
ES junta de enchufe y
   cordón *f*
F assemblage *m* à
   emboîtement
I giunto *m* a bicchiere
MA tokos csőkütés
NL mofverbinding *f*
   sokverbinding *f*
PO połączenie *n*
   kielichowe
PY раструбное
   соединение *n*
SU muhviliitos
SV muffürbindning *c*

3568 spinning disc
DA drejeskive
DE Drehscheibe *f*
ES humidificador de
   disco giratorio *m*
F humidificateur *m*
   centrifuge
I disco *m* rotante
MA forgótárcsa
NL sneldraaiende
   schijf *f*
PO tarcza *f* obrotowa
PY вращающийся
   диск *m*
   вращающийся
   увлажнитель *m*
SU pyürivä levy
SV roterande skiva *c*

3569 spiral
DA spiral
DE Spirale *f*
ES en espiral
   espiral *f*
F spirale *f*
I spirale *f*
MA spirál
NL spiraal *f*
PO spirala *f*
PY спираль *f*
SU spiraali
SV spiral *c*

3570 spiral fin tube
DA spiralribberør
DE Lamellenrohr *n*
   Spirallamellenrohr *n*
ES tubo de aletas
   helicoidales *m*
F tuyau *m* à ailettes
   hélicoïdales
I tubo *m* alettato
   spiroidale
MA spirálbordás cső
NL buis *f* met
   spiraalvormige
   ribben
PO rura *f* ożebrowana
   spiralnie
PY труба *f* co
   спиральным
   оребрением
SU ripaputki
SV kamflänsrür *n*

3571 spiral wound duct
DA spiralfalset kanal
DE Spiralkanal *m*
ES conducto
   helicoidal *m*
   espiro-ducto *m*
F conduit *m* hélicoïde
   en tôle
I condotto *m* a spirale
MA spirálvarratos
   csővezeték
NL spiralo-buis *f*
PO przewód *m* spiro
PY спирально-
   навивной
   воздуховод *m*
   спирально-
   навивной
   трубопровод *m*
SU kierresaumakanava
SV spiralfalsad kanal *c*

3572 splash cooling tower packing
DA spredeindsats til
   køletårn
DE Sprühkühlturm-
   einbauten *f*
ES relleno de torre de
   refrigeración de
   salpicadura *m*
F garnissage *m* à
   éclaboussement de
   tour de
   refroidissement
I torre *f* di
   raffreddamento a
   caduta
MA permetezéses
   hűtőtorony tültet
NL sproeikoeltoren-
   vulling *f*
PO agregat *m* wieży
   chłodniczej
   rozpryskowej
PY градирня *f* c
   дроблением
   жидкости на
   пластинах
SU jäähdytystornin
   puinen täyte
SV packning *c* till
   stänkkyltorn *n*

3573 split bush
DA klembøsning
DE Lagerschalenriß *m*
ES cojinete en dos
  piezas *m*
  cojinete partido *m*
F coussinet *m* deux
  pièces
I cuscinetto *m* a
  bussola in due
  pezzi
MA osztott csapágy
NL gedeeld
  kussenblok *n*
PO tuleja *f* dwudzielna
PY распределительный
  вкладыш *m*
SU jaettu laakeriholkki
SV delad bussning *c*

3574 split condenser
DA todelt kondensator
DE getrennter
  Kondensator *m*
ES condensador partido
  m
F condenseur *m*
  multicircuit
I unità *f* condensante
  split (separata)
MA osztott kondenzátor
NL meervoudige
  condensor *m*
PO skraplacz *m* systemu
  split
PY распределительный
  конденсатор *m*
SU moniosainen
  lauhdutin
SV delad kondensor *c*

3575 split system
DA delt system
DE getrenntes System *n*
ES acondicionador con
  evaporador
  remoto *m*
  sistema partido *m*
F conditionneur *m*
  individuel en deux
  parties
I condizionatore *m* a
  due sezioni
  split-system *m*
MA kettéosztott rendszer
  split rendszer
NL gedeeld systeem *n*
  gesplitst systeem *n*
PO system *m* split
PY распределительная
  система *f*
SU jaettu järjestelmä
SV fürdelat system *n*

3576 splitter damper
DA fordelingsspjæld
DE Verteilklappe *f*
ES compuerta
  diversora *f*
F volet *m* de
  répartition
I serranda *f* di
  suddivisione
MA kulisszás
  hangtompító
  osztott csappantyú
NL verdeelklep *f*
PO przepustnica *f*
  dzielona
  przerywacz *m* ciągu
PY жалюзийный
  клапан *m*
SU jakopelti
SV fürdelningsspjäll *n*

3577 spoil draught
DA trækafbryder
DE Zugunterbrecher *m*
ES cortatiros *m*
F coupe-tirage *m*
I taglia-tiraggio *m*
MA huzatmegszakító
NL trekonderbreker *m*
PO przerywacz *m* ciągu
PY прерыватель *m*
  тяги
SU vedon keskeytys
SV dragavbrott *n*

3578 spot cooling
DA punktkøling
DE Punktkühlung *f*
ES enfriamiento
  localizado *m*
F refroidissement *m*
  local
I raffreddamento *m*
  localizzato
MA helyi hűtés
NL plaatselijke koeling *f*
PO chłodzenie *n*
  miejscowe
PY местное
  охлаждение *n*
SU paikallisjäähdytys
SV lokal kylning *c*

3579 spot welding
DA punktsvejsning
DE Punktschweißung *f*
ES soldadura por
  puntos *f*
F soudure *f* par points
I saldatura *f* a punti
MA helyi hegesztés
NL puntlassen *n*
PO spawanie *n*
  punktowe
PY точечная сварка *f*
SU pistehitsaus
SV punktsvetsning *c*

3580 spray *vb*
DA forstøve
  sprøjte
DE sprühen
ES pulverizar
F pulvériser
I polverizzare
  spruzzare
MA permetez
  porlaszt
NL sproeien
  verstuiven
PO natryskiwać
  rozpylać
PY разбрызгивать
SU sumuttaa
SV spruta

3581 spray chamber
DA forstøvningskammer
DE Sprühkammer *f*
ES cámara de
  pulverización *f*
F chambre *f* de
  pulvérisation
I camera *f* di
  polverizzazione
MA nedvesítőkamra
NL sproeikamer *f*
PO komora *f* rozpylania
  komora *f* zraszania
PY камера *f* орошения
  форсуночная
  камера *f*
SU sumutuskammio
SV dyskammare *c*

3582 spray cooling
DA sprøjtekøling
DE Sprühkühlung *f*
ES enfriamiento por
  rociado *m*
F refroidissement *m*
  local
I raffreddamento *m* a
  nebulizzazione
MA porlasztásos hűtés
NL versproeiingskoeling
  f
PO chłodzenie *n*
  natryskowe
PY охлаждение *n*
  путем орошения
  распыливанием
SU suihkutusjäähdytys
SV sprutkylning *c*

3583 spray curtain
DA sprøjtegardin
DE Sprühvorhang *m*
ES cortina de agua
  pulverizada *f*
F rideau *m* de
  pulvérisation
I cortina *f* a getto di
  acqua
MA permetfüggüny
NL regengordijn *n*
PO kurtyna *f* z
  rozpylonej cieczy
PY завеса *f*,
  образуемая
  распыливанием
SU sumuverho
SV sprutridå *c*

3584 sprayed coil
condenser
DA overrislings-
  kondensator
DE Berieselungs-
  verflüssiger *m*
ES condensador de
  lluvia *m*
F condenseur *m* à
  aspersion
I batteria *f* a
  spruzzamento
MA permetezett
  csőkígyós
  kondenzátor
NL besproeide
  condensor *m*
PO skraplacz *m* z
  wężownicą
  zraszaną
PY конденсатор *m* с
  орошаемым
  змеевиком
SU märkälauhdutin
SV sprut-sling-
  kondensor *c*

3585 spray-filled water
cooling tower
DA køletårn uden
  indsats
DE Sprühkühlturm mit
  Einbauten *m*
ES torre de refrigeración
  de rociadores de
  agua *f*
F tour *f* de
  refroidissement à
  pulvérisation
I torre *f* di
  raffreddamento a
  spray
MA permetezett tültetes
  vízhűtőtorony
NL koeltoren *m* zonder
  inbouw
  koeltoren *m* zonder
  vulling
PO wieża *f* chłodnicza
  wodna ze
  zraszanym
  wypełnieniem
PY градирня *f* с
  орошаемой
  насадкой
SU jäähdytystorni ilman
  täytettä
SV kyltorn *n* med
  vatteninsprutning
  c

3586 spray nozzle
DA sprederdyse
  strålespids
DE Einspritzdüse *f*
  Sprühdüse *f*
ES pulverizador *m*
  tobera
  pulverizadora *f*
F buse *f* d'injection
  gicleur *m*
I ugello *m*
  spruzzatore
MA porlasztó fúvóka
NL sproeikop *m*
  verstuiver *m*
PO dysza *f* natryskowa
  dysza *f* rozpylająca
PY распыливающее
  сопло *n*
SU sumutussuutin
SV spridare *c*

3587 spray pipe
DA sprøjterør
DE Einspritzrohr *n*
ES tubo de inyección *m*
F tuyauterie *f*
  d'injection
I tubo *m* spruzzatore
MA porlasztócső
NL sproeipijp *f*
PO rura *f* rozpylająca
PY распыливающая
  труба *f*
SU sumutusputki
SV insprutningsrür *n*

3588 spray point
DA indsprøjtningspunkt
DE Einspritzpunkt *m*
  Sprühpunkt *m*
ES punto de
  aspersión *m*
  punto de
  pulverización *m*
F point *m* de
  pulvérisation
I punto *m* di
  spruzzatura
MA porlasztási pont
NL sproeiplaats *f*
PO punkt *m* rozpylania
PY душевая сетка *f*
SU sumutuspiste
SV insprutningspunkt *c*

DA = Danish
DE = German
ES = Spanish
F = French
I = Italian
MA = Magyar
  (Hungarian)
NL = Dutch
PO = Polish
PY = Russian
SU = Finnish
SV = Swedish

3589 spray pond
DA kølebassin
DE Sprühteich *m*
ES recipiente de
enfriamiento de
agua por
aspersión *m*
F bassin *m* de
refroidisseur d'eau
par pulvérisation
I bacino *m* di
raffreddamento di
acqua a
spruzzamento
MA porlasztó medence
NL sproeivijver *m*
PO basen *m* chłodzący
wodę przez
rozpylanie
PY брызгальный
бассейн *m*
SU suihkuallas
SV sprutbassäng *c*

3590 spring
DA fjeder
kilde
DE Feder *f*
ES muelle *m*
resorte *m*
F ressort *m*
I molla *f*
MA rugó
NL lente *f*
veer *f*
PO resor *m*
sprężyna *f*
PY пружина *f*
peccopa *f*
SU jousi
SV fjäder *c*

3591 spring-loaded *adj*
DA fjederbelastet
DE federbelastet
ES válvula de resorte *f*
F soupape *f* à ressort
I caricato a molla
MA rugóterhelésű
NL veerbelast
PO obciążony sprężyną
sprężynowy
PY пружинный
SU jousikuormitettu
SV fjäderbelastad

3592 stability
DA stabilitet
DE Stabilität *f*
ES estabilidad *f*
F stabilité *f*
I stabilità *f*
MA stabilitás
NL stabiliteit *f*
PO stabilność *f*
stateczność *f*
PY стабильность *f*
SU jäykkyys
stabiliteetti
SV stabilitet *c*

3593 stack
DA skorsten
DE Schornstein *m*
ES chimenea *f*
F souche *f* de
cheminée
I camino *m*
MA kémény
kürtő
NL schoorsteen *m*
PO komin *m* żelazny
stos *m*
PY дымовая труба *f*
SU savupiippu
SV skorsten *c*

3594 stack effect
DA skorstensvirkning
DE Schornstein-
wirkung *f*
ES efecto chimenea *m*
F effet *m* de cheminée
I effetto *m* camino
MA kürtőhatás
NL schoorsteneffect *n*
PO efekt *m* kominowy
PY тяга *f*
эффект *m* дымовой
трубы
SU savupiippuvaikutus
SV skorstenseffekt *c*

3595 stainless steel
DA rustfrit stål
DE rostfreier Stahl *m*
ES acero inoxidable *m*
F acier *m* inoxydable
I acciaio *m*
inossidabile
MA rozsdamentes acél
NL roestvast staal *n*
PO stal *f* nierdzewna
PY нержавеющая
сталь *f*
SU ruostumaton teräs
SV rostfritt stål *n*

3596 stand alone
controller
DA autonom regulator
DE Einzelregler *m*
ES regulador único *m*
F régulateur *m*
autonome
I regolatore *m*
autonomo
MA egyedi szabályozó
NL zelfstandige
regelaar *m*
PO regulator *m*
niezależny
PY отдельно
установленный
регулятор *m*
SU itsenäinen säädin
SV autonom regulator *c*

3597 standard air
DA normalluft
DE Normal-
atmosphäre *f*
Normalluft *f*
ES aire normal *m*
F air *m* normal
I aria *f* normale
aria *f* standard
MA normál levegő
NL standaardlucht *f*
PO powietrze *n*
normalne
powietrze *n*
wzorcowe
PY воздух *m*
нормативного
качества
стандартный
воздух *m*
SU ilma normaalitilassa
SV standardluft *c*

3598 standard colour
DA standardfarve
DE Normfarbe *f*
ES color
normalizado *m*
F couleur *f*
conventionnelle
I colore *f* standard
MA szabványos szín
NL standaardkleur *f*
PO barwa *f* wzorcowa
PY стандартный
цвет *m*
SU standardiväri
SV standardfärg *c*

3599 standard deviation
DA standardafvigelse
DE Standard-
    abweichung *f*
ES desviación normal *f*
F déviation *f* standard
    écart *m* type
I deviazione *f*
    standard
MA normál eltérés
NL standaarddeviatie *f*
PO odchylenie *n*
    standardowe
PY стандартное
    отклонение *n*
SU standardipoikkeama
SV medelavvikelse *c*

3600 stand by
DA reserve
DE Reserve *f*
ES de reserva
    reserva *f*
F réserve *f*
I riserva *f*
MA tartalék
NL reserve *f*
PO rezerwa *f*
    stan *m* pogotowia
PY резерв *m*
SU vara-
SV reserv *c*

**3601 stand-by
    equipment**
DA nødudstyr
    reserveudstyr
DE Reserve-
    ausrüstung *f*
ES equipo de reserva *m*
F équipement *m* de
    réserve
    équipement *m* de
    secours
I apparecchiatura *f* di
    riserva
MA tartalék berendezés
NL reserveapparatuur *f*
    reserveuitrusting *f*
PO wyposażenie *n*
    rezerwowe
PY резервное
    оборудование *n*
SU varalaitteisto
SV reservutrustning *c*

3602 stand-by motor
DA reservemotor
DE Reservemotor *m*
ES motor de reserva *m*
F moteur *m* de réserve
    moteur *m* de secours
I motore *m* di riserva
MA tartalék motor
NL reservemotor *m*
PO silnik *m* rezerwowy
PY резервный
    двигатель *m*
SU varamoottori
SV reservmotor *c*

3603 stand pipe
DA standrør
    stigledning
DE Standrohr *n*
ES tubo de agua de
    alimentación *m*
    tubo vertical *m*
F tube *m* vertical
    d'équilibre
I tubo *m* di livello
MA állványcső
NL standpijp *f*
PO kolumna *f* wodna
    rura *f* ciśnień
    stojak *m*
    hydrantowy
PY стояк *m*
SU nousuputki
SV stigrür *n*

3604 start
DA start
DE Anfang *m*
    Start *m*
ES arranque *m*
F commencement *m*
I avviamento *m*
    partenza *f*
MA indítás
NL aanzetten *n*
    begin *n*
    start *m*
PO początek *m*
    rozruch *m*
    uruchomienie *n*
PY запуск *m*
    пуск *m*
SU alku
    käynnistys
SV start *c*

3605 starter
DA starter
DE Anlasser *m*
ES arrancador *m*
F commutateur *m* de
    démarrage
    démarreur *m*
I avviatore *m*
    starter *m*
MA indító
NL aanloopschakelaar
    m
    starter *m*
    startmotor *m*
PO elektroda *f*
    zapłonowa
    starter *m*
    urządzenie *n*
    rozruchowe
PY пусковое
    устройство *n*
    стартер *m*
SU käynnistin
SV pådrag *n*

3606 starting current
DA startstrøm
DE Anfahrstrom *m*
    Anlaufstrom *m*
ES corriente de
    arranque *f*
F courant *m* de
    démarrage
I corrente *f*
    all'avviamento
    corrente *f* di spunto
MA indítási áram
NL aanloopstroom *m*
PO prąd *m* rozruchu
PY пусковой ток *m*
SU käynnistysvirta
SV startström *c*

3607 starting procedure
DA startprocedure
DE Anfahrvorgang *m*
    Anlaufvorgang *m*
ES operación de puesta
    en marcha *f*
F opération *f* de mise
    en marche
I procedura *f* di
    avviamento
MA indítási eljárás
NL inschakel-
    procedure *f*
PO proces *m* rozruchu
PY пусковая
    процедура *f*
SU käynnistysprosessi
SV startfürfarande *n*

3608 starting switch
DA startknap
DE Anlaufschalter *m*
Ein-Schalter *m*
ES interruptor de
arranque *m*
F interrupteur *m* de
démarrage
I interrutore *m* di
avviamento
MA indítókapcsoló
NL aanloop-
schakelaar *m*
PO wyłącznik *m*
rozruchu
zapłonnik *m*
PY пусковая кнопка *f*
пусковой
рубильник *m*
SU käynnistyskytkin
SV startkontakt *c*

3609 starting time
DA starttid
DE Anlaufzeit *f*
ES tiempo de puesta en
marcha *m*
F temps *m* de mise en
marche
I tempo *m* di messa in
marcia
MA felfutási idő
NL aanlooptijd *m*
PO czas *m* rozruchu
PY время *n* запуска
SU käynnistysaika
SV starttid *c*

3610 starting torque
DA startmoment
DE Anlaufmoment *n*
Anzugsmoment *n*
ES par de arranque *m*
F couple *m* de
démarrage
I coppia *f* di spunto
MA indítónyomaték
NL aanloopkoppel *m*
PO moment *m*
rozruchowy
PY пусковой
момент *m*
SU käynnistysmomentti
SV startmoment *n*

**3611 start under full
load**
DA fuldlaststart
DE Vollast-
einschaltung *f*
ES arranque a plena
carga *m*
F mise *f* en marche à
pleine charge
I avviamento *m* a
pieno carico
MA teljes terheléses
indítás
NL vollastaanloop *m*
vollastinschakeling *f*
PO rozruch *m* przy
pełnym obciążeniu
PY пуск *m* с полной
нагрузкой
SU käynnistys täydellä
kuormalla
SV fullaststart *c*

3612 start-up
DA starte
DE Anlauf *m*
Start *m*
ES arranque *m*
puesta en marcha *f*
F démarrage *m*
mise *f* en service
I avviamento *m*
MA indulás
NL aanloop *m*
inbedrijfstelling *f*
start *m*
PO rozruch *m*
uruchomienie *n*
PY запуск *m*
старт *m*
SU käynnistys
SV start *c*

3613 starved evaporator
DA fordamper med for
lille vandtilgang
DE trockner
Verdampfer *m*
ES evaporador
infraalimentado *m*
F évaporateur *m* sous
alimenté
I evaporatore *m*
sottoalimentato
MA száraz elpárologtató
NL verdamper *m* met
tekort aan
vloeistof
PO parownik *m* o
niewystarczającej
mocy
PY испаритель *m* с
недостаточной
подачей
хладагента
SU vajaasyütteinen
hüyrystin
SV evaporator *c* med
för liten
vattentillförsel *c*

3614 state
DA tilstand
DE Betriebszustand *m*
Zustand *m*
ES estado *m*
F état *m*
régime *m*
I stato *m*
MA állapot
NL staat *m*
toestand *m*
PO stan *m*
PY состояние *n*
SU tila
SV tillstånd *n*

3615 static head
DA statisk loftehøjde
DE statischer Druck *m*
ES presión estática *f*
F hauteur *f* statique
(charge *f* statique)
I carico *m* statico
MA statikus
nyomómagasság
NL statische
drukhoogte *f*
PO wysokość *f*
statyczna
PY статический
напор *m*
SU staattinen paine
SV statiskt tryck *n*

3616 static pressure
DA statisk tryk
DE statischer Druck *m*
ES presión estática *f*
F pression *f* statique
I pressione *m*
MA statikus nyomás
NL statische druk *m*
PO ciśnienie *n* statyczne
PY статическое
   давление *n*
SU staattinen paine
SV statiskt tryck *n*

**3617 static-regain
   method**
DA statisk
   trykgenvindings-
   metode
DE Methode *f* des
   statischen
   Druckrück-
   gewinns *m*
ES método de
   recuperac.de la
   presión estát.(para
   el dimens.de
   cond) *m*
F méthode *f* du regain
   statique
   (dimensionnement
   de conduit)
I metodo *m* a
   recupero di
   pressione statica
MA statikus
   nyomásvisszan-
   yerés módszere
NL statischedrukterug-
   winningsmethode *f*
PO metoda *f*
   odzyskiwania
   ciśnienia
   statycznego
PY метод *m*
   постоянного
   статического
   давления
SU staattisen paineen
   takaisinsaantiin
   perustuva
   menetelmä
SV metod *c* för statisk
   återvinning *c*

3618 station
DA position
   station
DE Standort *m*
   Station *f*
ES central *f*
   estación *f*
F station *f*
I stazione *f*
MA állomás
NL centrale *f*
   station *n*
PO stacja *f*
   stanowisko *n*
PY станция *f*
SU asema
SV station *c*

3619 stator
DA stator
DE Stator *m*
ES estator *m*
F stator *m*
I statore *m*
MA állórész
   sztátor
NL stator *m*
PO stojan *m*
   twornik *m*
PY статор *m*
SU staattori
SV stator *c*

3620 stay
DA ophold
   stag
DE Aufenthalt *m*
ES apoyo *m*
   estancia *f*
   refuerzo *m*
   soporte *m*
F support *m*
I supporto *m*
MA tartózkodás
NL steun *m*
   tui *f*
   verblijf *n*
PO odciąg *m*
   rozpora *f*
   stężenie *n*
   (konstrukcyjne)
PY опора *f*
   стойка *f*
SU harus
   tuki
SV stüd *n*

3621 steady state
DA stationær tilstand
DE Beharrungs-
   zustand *m*
ES régimen estable *m*
   régimen
   permanente *m*
F état *m* de régime
   permanent
   état *m* stationnaire
I regime *m* costante
   regime *m*
   stazionario
MA állandósult állapot
   stacioner állapot
NL stationaire
   toestand *m*
PO stan *m* ustalony
PY стационарное
   состояние *n*
SU jatkuvuustila
SV fosfarighets-
   tillstånd *n*

3622 steam
DA damp
DE Dampf *m*
ES vapor (de agua) *m*
F vapeur *f*
I vapore *m*
MA gőz
NL stoom *m*
PO para *f*
PY пар *m*
SU hüyry
SV ånga *c*

DA	=	Danish
DE	=	German
ES	=	Spanish
F	=	French
I	=	Italian
MA	=	Magyar (Hungarian)
NL	=	Dutch
PO	=	Polish
PY	=	Russian
SU	=	Finnish
SV	=	Swedish

3623 steam atomizing burner
DA dampforstøvnings-
    brænder
DE Dampf-
    Zerstäubungs-
    brenner *m*
ES quemador de
    atomización de
    vapor *m*
F brûleur *m* à
    atomisation par la
    vapeur
I bruciatore *m* con
    atomizzazione a
    vapore
MA gőzporlasztásos égő
NL stoomverstuivings-
    brander *m*
PO palnik *m*
    rozpylający parę
PY горелка *f* с
    распыливанием
    паром
SU hüyryhajoitteinen
    poltin
SV ång-
    finfördelnings-
    brännare *c*

3624 steam boiler
DA dampkedel
DE Dampfkessel *m*
ES caldera de vapor *f*
F chaudière *f* à vapeur
I caldaia *f* a vapore
MA gőzkazán
NL stoomketel *m*
PO kocioł *m* parowy
PY паровой котел *m*
SU hüyrykattila
SV ångpanna *c*

3625 steam chamber
DA damprum
DE Dampfraum *m*
ES cámara de vapor *f*
F chambre *f* à vapeur
I camera *f* a vapore
MA gőzkamra
NL stoomdom *m*
    stoomruimte *f*
PO komora *f* parowa
PY паровая камера *f*
SU hüyrykammio
SV ångrum *n*

3626 steam condenser
DA dampkondensator
DE Dampf-
    kondensator *m*
ES condensador de
    vapor *m*
F condenseur *m* de
    vapeur
I condensatore *m* di
    vapore
MA gőzkondenzátor
NL stoomcondensor *m*
PO skraplacz *m* parowy
PY конденсатор *m*
    пара
SU hüyrylauhdutin
SV ångkondensor *c*

3627 steam connection
DA dampforbindelse
DE Dampfanschluß *m*
ES conexión para el
    vapor *f*
    empalme para el
    vapor *m*
F raccordement *m* de
    vapeur
I attacco *m* di vapore
MA gőzcsatlakozás
NL stoomaansluiting *f*
PO przyłączenie *n* pary
PY паропровод *m*
SU hüyryliitos
SV anslutning *c* för ånga
    c

3628 steam consumption
DA dampforbrug
DE Dampfverbrauch *m*
ES consumo de
    vapor *m*
F consommation *f* de
    vapeur
I consumo *m* di
    vapore
MA gőzfogyasztás
NL stoomverbruik *n*
PO zużycie *n* pary
PY расход *m* пара
SU hüyrynkulutus
SV ångfürbrukning *c*

3629 steam dryer
DA damptørrer
DE Dampftrockner *m*
ES secador de vapor *m*
F sécheur *m* de vapeur
I essiccatore *m* a
    vapore
MA gőzszárító
NL stoomdroger *m*
PO osuszacz *m* pary
    suszarka *f* parowa
PY осушитель *m* пара
SU hüyrynkuivaaja
SV ångtork *c*

3630 steam feed heater
DA dampforvarmer
DE dampfbeheizter
    Erhitzer *m*
ES recalentador de
    vapor *m*
F réchauffeur *m* à
    vapeur
I riscaldatore *m* a
    vapore
MA gőz táphevítő
NL voedingswatervoor-
    verwarmer *m*
PO podgrzewacz *m*
    parowy
PY паровой
    нагреватель *m*
SU hüyrylämmitin
SV ångfürvärmare *c*

3631 steam filter
DA dampfilter
DE Dampffilter *m*
ES filtro de vapor *m*
F filtre *m* à vapeur
I filtro *m* di vapore
MA gőzszűrő
NL stoomfilter *m*
PO filtr *m* pary
PY фильтр *m* для пара
SU hüyrysuodatin
SV ångfilter *n*

3632 steam generator
DA dampgenerator
DE Dampferzeuger *m*
ES generador de
    vapor *m*
F générateur *m* de
    vapeur
I generatore *m* di
    vapore
MA gőzgenerátor
NL stoomgenerator *m*
PO wytwornica *f* pary
PY генератор *m* пара
SU hüyrynkehitin
SV ånggenerator *c*

3633 steam header
DA dampfordeler
    damphætte
DE Dampf-
    Sammelleitung *f*
ES colector de vapor *m*
F collecteur *m* de
    vapeur
I collettore *m* di
    vapore
MA gőzosztó
NL hoofdstoomleiding *f*
    stoomverdeler *m*
PO rozdzielacz *m*
    parowy
PY паропровод *m* для
    отбора пара от
    котла
SU hüyrynjakotukki
SV ångfürdelare *c*

3634 steam heating
DA dampopvarmning
DE Dampfheizung *f*
ES calefacción por
    vapor *f*
F chauffage *m* à
    vapeur
I riscaldamento *m* a
    vapore
MA gőzfütés
NL stoomverwarming *f*
PO ogrzewanie *n*
    parowe
PY паровое отпление
    n
SU hüyrylämmitys
SV ånguppvärmning *c*

3635 steam humidifier
DA dampbefugter
DE Dampfbefeuchter *m*
ES humidificador a
    vapor *m*
F humidificateur *m* à
    vapeur
I umidificatore *m* a
    vapore
MA gőznedvesítő
NL stoombevochtiger *m*
PO nawilżacz *m* parowy
PY паровой
    увлажнитель *m*
SU hüyrykostutin
SV ångfuktare *c*

3636 steam injector
    water-feed pump
DA dampstråle
    fødevandspumpe
DE Dampfstrahl-
    Speisewasser-
    pumpe *f*
ES bomba de
    alimentación de
    agua por eyección
    de vapor *f*
    bomba de eyección
    de vapor *f*
F giffard *m* (injecteur
    à vapeur)
I pompa *f* di
    alimentazione
    acqua a iniezione
    di vapore
MA gőzsugártápszivattyú
NL stoomstraalvoeding-
    waterpomp *f*
PO pompa *f* inżekcyjna
    parowa do
    zasilania wodą
    pompa *f*
    Worthingtona
PY пароструйный
    насос *m* для
    питания котла
SU hüyrykäyttüinen
    syüttüvesipumppu
SV ångdriven
    matarvatten-
    pump *c*

3637 steam jacket
DA dampkappe
DE Dampfmantel *m*
ES camisa de vapor *f*
F enveloppe *f* de
    vapeur
I camicia *f* a vapore
MA gőzküpeny
NL stoommantel *m*
PO płaszcz *m* parowy
PY паровая рубашка *f*
SU hüyryvaippa
SV ångmantel *c*

3638 steam jet air
    ejector
DA dampstråleluft-
    ejektor
DE Dampfstrahl-
    Luftejektor *m*
ES inyector de chorro
    de vapor *m*
F éjecteur *m* d'air à
    vapeur
I eiettore *m* di aria a
    getto di vapore
MA gőzsugár légfúvó
NL stoomstraal-
    vacuumpomp *f*
PO eżektor *m*
    powietrzno-
    parowy
PY пароструйный
    эжектор *m*
    воздуха
SU hüyryejektori
SV ångstrålluftejektor *c*

3639 steam jet
    refrigeration
DA dampstrålekøling
DE Dampfstrahlkälte-
    erzeugung *f*
ES enfriamiento por
    eyección de
    vapor *m*
    sistema frigorífico de
    eyección (de
    vapor) *m*
F réfrigération *f* par
    éjecteur à vapeur
I refrigerazione *f* a
    getto di vapore
MA gőzsugaras hütés
NL stoomstraalkoeling *f*
PO chłodzenie *n*
    strumieniem pary
PY пароструйное
    охлаждение *n*
SU hüyrysuihkujääh-
    dytys
SV ångstrålkylning *c*

3640 steam pipe
DA damprør
DE Dampfrohr *n*
ES tubería de vapor *f*
F conduite *f* de vapeur
    tuyauterie *f* de
    vapeur
I tubazione *f* di
    vapore
MA gőzcső
NL stoomleiding *f*
    stoompijp *f*
PO rura *f* parowa
PY паровая труба *f*
SU hüyryputki
SV ångrür *n*

3641 steam pocket
DA damplomme
DE Dampfblase *f*
ES bolsa de vapor *f*
tapón de vapor (en tuberias) *m*
F bouchon *m* de vapeur
I sacca *f* di vapore
MA gőzzsák
NL stoombel *f*
PO korek *m* parowy
PY паровая пробка *f*
SU hüyrytasku
SV ångficka *c*

3642 steam pressure
DA damptryk
DE Dampfdruck *m*
ES presión de vapor *f*
F pression *f* de vapeur
I pressione *f* di vapore
MA gőznyomás
NL stoomdruk *m*
PO ciśnienie *n* pary
PY давление *n* пара
SU hüyrynpaine
SV ångtryck *n*

3643 steam-pressure reducing valve
DA dampreduktions-ventil
DE Dampfdruckreduzier-ventil *n*
ES válvula reductora de presión de vapor *f*
F détendeur *m* de vapeur
I valvola *f* riduttrice della pressione del vapore
MA gőznyomáscsükk-entő szelep
NL stoomdrukreduceer-ventiel *n*
PO zawór *m* redukcyjny ciśnienia pary
PY паровой редукционный клапан *m*
SU hüyrynpaineenalennus-venttiili
SV ångtrycks-reducerings-ventil *c*

3644 steam tables
DA damptabel
DE Dampftafeln *f,pl*
ES tablas de vapor *f,pl*
F tables *f* de vapeur
I tabelle *f,pl* del vapore
MA gőztáblázatok
NL stoomtabellen *pl*
PO tablice *f* własności pary
PY таблицы *f* свойств пара
SU hüyrytaulukot
SV ångtabeller *c*

3645 steam trap
DA vandudlader
DE Kondenstopf *m*
ES purgador de condensados *m*
F purgeur *m* de vapeur (eau condensée)
I separatore *m* di condensa
MA kondenzedény
NL condenspot *m*
PO garnek *m* kondensacyjny odwadniacz *m*
PY конденсатоот-водчик *m*
SU lauhteen erotin
SV ångfälla *c*

3646 steam turbine
DA dampturbine
DE Dampfturbine *f*
ES turbina de vapor *f*
F turbine *f* à vapeur
I turbina *f* a vapore
MA gőzturbina
NL stoomturbine *f*
PO turbina *f* parowa
PY паровая турбина *f*
SU hüyryturbiini
SV ångturbin *c*

3647 steel
DA stål
DE Stahl *m*
ES acero *m*
F acier *m*
I acciaio *m*
MA acél
NL staal *n*
PO stal *f*
PY сталь *f*
SU teräs
SV stål *n*

3648 steel boiler
DA stålkedel
DE Stahlkessel *m*
ES caldera de acero *f*
F chaudière *f* en acier
I caldaia *f* in acciaio
MA acélkazán
NL plaatstalen ketel *m*
PO kocioł *m* stalowy
PY стальной котел *m*
SU teräslevykattila
SV stålpanna *c*

3649 steel panel radiator
DA stålpladeradiator
DE Stahlplattenheiz-kürper *m*
ES radiador de panel de acero *m*
F radiateur *m* panneau en acier
I radiatore *m* a piastra in acciaio
MA acéllemez radiátor
NL plaatstalen paneelradiator *m*
PO grzejnik *m* płytowy stalowy
PY стальной панельный радиатор *m*
SU teräslevypatteri
SV stålplåtsradiator *c*

3650 steel pipe
DA stålrør
DE Stahlrohr *n*
ES tubo de acero *m*
F tube *m* d'acier
I tubo *m* di acciaio
MA acélcső
NL stalen buis *f*
stalen pijp *f*
PO rura *f* stalowa
PY стальная труба *f*
SU teräsputki
SV stålrür *n*

3651 step
DA trin
DE Schritt *m*
Stufe *f*
ES escalón *m*
etapa *f*
F gradin *m*
marche *f*
I gradino *m*
passo *m*
MA lépcső
NL fase *f*
stap *m*
PO stopień *m*
PY ступень *f*
SU askel
SV steg *n*

3652 step controller
DA trinvis regulator
DE Schrittregler *m*
   Stufenregler *m*
ES regulador de etapas
   m
F régulateur *m* pas à
   pas
I controllore *m* a
   passi
MA lépcsős szabályozó
NL stappenregelaar *m*
PO regulator *m*
   krokowy
   regulator *m*
   schodkowy
   regulator *m*
   stopniowany
PY ступенчатый
   регулятор *m*
SU askelsäädin
SV stegregulator *c*

3653 step grate
DA trapperist
DE Stufenrost *m*
ES parrilla escalonada *f*
F grille *f* à gradins
I griglia *f* a gradini
MA taposórács
NL trappenrooster *m*
PO ruszt *m* schodkowy
PY ярусная
   колосниковая
   решетка *f*
SU porrasarina
SV stegrost *c*

3654 stepped *adj*
DA trinvis
DE schrittweise
   stufenweise
ES de etapas
F échelonné
I a gradini
MA lépcsős
NL trapsgewijs
PO schodkowy
   stopniowany
   stopniowy
PY ступенчатый
SU asteittainen
SV stegvis

3655 stepped control
DA trinvis regulering
DE Stufenregelung *f*
ES regulación paso a
   paso *f*
   regulación por
   etapas *f*
F régulation *f* pas à
   pas
I regolazione *f* a
   gradini
MA lépcsős szabályozás
NL trapsgewijze
   regeling *f*
PO regulacja *f*
   stopniowa
PY ступенчатое
   регулирование *n*
SU askelsäätü
SV stegvis reglering *c*

**3656 stepped piston
   compound
   compressor**
DA differentialestempel
   kompressor
DE Verbundverdichter
   m mit abgestuftem
   Kolben *m*
ES compresor
   compuesto de
   pistón por etapas
   m
F compresseur *m* à
   piston différentiel
I compressore *m* a
   stantuffo
   differenziale
MA tübbfokozatú
   dugattyús
   kompresszor
NL tweetraps-
   compressor *m*
PO sprężarka *f* tłokowa
   wielostopniowa
PY поршневой
   двухступенчатый
   компрессор *m*
SU askelmäntä-
   kompressori
SV

3657 Stirling cycle
DA Stirling cyklus
DE Stirling-
   Kreisprozeß *m*
ES ciclo de Stirling *m*
F cycle *m* de Stirling
I ciclo *m* Stirling
MA Stirling folyamat
NL Stirlingkringloop *m*
PO cykl *m* Stirlinga
PY цикл *m* Стирлинга
SU Stirling-prosessi
SV Stirling-cykel *c*

3658 stirrup bracket
DA bøjlekonsol
DE Steigeisen *n*
ES estribo *m*
   soporte de
   abrazadera *m*
F support *m* à étrier
I sostegno *m* a staffa
MA kengyelbilincs
NL ophangbeugel *m*
PO wspornik *m*
   zaciskowy
PY кронштейн *m* с
   хомутом
SU putken
   heilurikannatin
SV bygelfäste *n*

3659 stoichiometric *adj*
DA støkiometrisk
DE stüchiometrisch
ES estequiométrico
F stoechiométrique
I stechiometrico
MA sztüchiometrikus
NL stöchiometrisch
PO stechiometryczny
PY стехиометрический
SU stükiümetrinen
SV stükiometrisk

3660 stoker
DA stoker
DE Beschicker *m*
   Heizer *m*
ES atizador
   mecánico *m*
   cargador mecánico
   (de parrilla) *m*
   horno mecánico *m*
F chargeur *m* de foyer
   convoyeur *m*
I alimentador *m*
   fuochista *m*
MA tüzelőberendezés
NL stoker *m*
   stookinrichting *f*
PO palacz *m*
   ruszt *m*
   mechaniczny
PY кочегар *m*
SU koneellinen
   polttoaineen
   syüttülaite
SV stoker *c*

3661 stoking
DA påfyldning
DE Beschickung *f*
Schüren *n*
ES carga de horno u
hogar *f*
F chargement *m* de
foyer
I alimentazione *f*
MA adagolás
NL opstoken *n*
stoken *n*
PO zasilanie *n* paleniska
PY загрузка *f* топки
SU polttoaineensyüttü
SV påfyllning *c*

3662 stop
DA stop
DE Halt *m*
Pause *f*
Stillstand *m*
ES interrupción *f*
paro *m*
tope *m*
F arrêt *m*
pause *f*
I arresto *m*
MA megállás
NL begrenzer *m*
stop *m*
PO ogranicznik *m*
zatrzymanie *n*
zderzak *m*
PY остановка *f*
SU pysäytys
SV stopp *n*

3663 stop flange
DA blindflange
stopflange
DE Blindflansch *m*
ES brida ciega *f*
F bride *f* pleine
I flangia *f* cieca
MA záróperem
NL blindflens *m*
PO kołnierz *m* ślepy
PY глухой фланец *m*
SU sokea laippa
umpilaippa
SV blindfläns *c*

3664 stoppage
DA driftforstyrrelse
driftstop
DE Betriebsunter-
brechung *f*
ES interrupción *f*
F interruption *f* de
marche
I fermata *f*
interruzione *f*
MA eltümés
leállás
NL bedrijfs-
onderbreking *f*
stremming *f*
PO przerwa *f*
zatkanie *n*
zatrzymanie *n*
PY засорение *n*
SU käyttühäiriü
pysäytys
SV driftstopp *n*

3665 storage
DA oplagring
DE Lagerung *f*
Speicherung *f*
ES almacenamiento *m*
F accumulation *f*
stockage *m*
I accumulo *m*
immagazzina-
mento *m*
MA tárolás
NL berging *f*
opslag *m*
PO magazynowanie *n*
składowanie *n*
PY аккумуляция *f*
хранение *n*
SU varasto
SV lager *n*

3666 storage heater
DA varmeakkumulator
DE Speicherofen *m*
ES calefactor con
acumulación de
calor *m*
calentador con
almacenamiento
térmico *m*
F appareil *m* de
chauffage à
accumulation
I riscaldatore *m* ad
accumulo
MA tárolós fűtő
NL accumulatiekachel *f*
PO podgrzewacz *m*
akumulacyjny
PY аккумуляционный
нагреватель *m*
SU varaava lämmitin
SV fürrådsvärmare *c*

3667 storage heating
DA opvarmning med
akkumuleret
varme
DE Speicherheizung *f*
ES calefacción por
acumulación de
calor *f*
calentamiento de
acumulación *m*
F chauffage *m* à
accumulation
I riscaldamento *m* ad
accumulo
MA tárolós fűtés
NL accumulatie-
verwarming *f*
PO ogrzewanie *n*
akumulacyjne
PY аккумуляционное
отопление *n*
SU varaava lämmitys
SV ackumulatorvärme *c*

3668 storage room
DA lagerrum
DE Lagerraum *m*
ES almacén *m*
F local *m* de stockage
magasin *m*
I deposito *m*
magazzino *m*
MA raktár
NL opslagruimte *f*
PO magazyn *m*
skład *m*
PY складское
помещение *n*
SU varastohuone
SV lagerrum *n*

3669 storage tank
DA lagertank
DE Lagertank *m*
Vorratsbehälter *m*
ES tanque de
almacenamiento
m
F réservoir *m* de
stockage
I serbatoio *m* di
accumulo
MA tárolótartály
NL opslagtank *m*
PO zasobnik *m*
PY запасной бак *m*
SU varastosäiliü
SV lagringstank *c*

3670 store
DA lager
DE Lager *n*
  Lagerhaus *n*
ES almacén *m*
F magasin *m*
I magazzino *m*
  negozio *m*
MA raktár
NL magazijn *n*
  voorraad *m*
  warenhuis *n*
PO magazyn *m*
  skład *m*
  zapas *m*
PY склад *m*
SU varasto
SV lager *n*

3671 store tank
DA lagertank
DE Lagertank *m*
ES tanque de
  almacena-
  miento *m*
F citerne *f* de stockage
I cisterna *f* di
  stoccaggio
MA tárolótartály
NL voorraadtank *m*
PO zasobnik *m*
PY сборный бак *m*
SU varastosäiliü
SV lagringstank *c*

3672 storey
DA etage
DE Geschoß *n*
  Stockwerk *n*
ES piso *m*
  planta *f*
F étage *m*
I piano *m* di edificio
MA emelet
NL verdieping *f*
PO kondygnacja *f*
  piętro *n*
PY этаж *m*
SU kerros
SV våning *c*

3673 story (*see* storey)

DA = Danish
DE = German
ES = Spanish
F = French
I = Italian
MA = Magyar
  (Hungarian)
NL = Dutch
PO = Polish
PY = Russian
SU = Finnish
SV = Swedish

3674 stove heating
DA kakkelovnsvarme
DE Ofenheizung *f*
ES calefacción por
  estufa *f*
F chauffage *m* par
  poêle
I riscaldamento *m* a
  stufa
MA kályhafűtés
NL kachelverwarming *f*
PO ogrzewanie *n*
  piecowe
PY печное отопление *n*
SU uunilämmitys
SV ugnsuppvärmning *c*

3675 strainer
DA filterdug
DE Filtriertuch *n*
  Schmutzfänger *m*
ES colador *m*
  filtro *m*
F filtre *m* séparateur
  de boue
I filtro *m* a rete
MA szűrő
NL vuilvanger *m*
  zeef *f*
PO filtr *m* (siatkowy)
PY фильтр *m*
SU lian erotin
  suodatin
SV filter *n*

3676 strainer check valve
DA filterkontrolventil
DE Filterprüfventil *n*
ES válvula de retención
  con filtro *f*
F clapet *m* crépine
I valvola *f* di ritegno
  con filtro
MA szűrőellenőrző szelep
NL filtercontrolekraan *f*
  filterterugslagklep *f*
PO zawór *m* zwrotny z
  filtrem
PY клапан *m*
  шламоотде-
  лителя
SU erottimen
  tarkastusventtiili
SV filterkontrollventil *c*

3677 stratification
DA lagdeling
DE Schichtung *f*
ES estratificación *f*
F stratification *f*
I stratificazione *f*
MA rétegződés
NL laagvorming *f*
  stratificatie *f*
PO uwarstwienie *n*
PY расслоение *n*
  стратификация *f*
SU kerrostuminen
SV skiktning *c*

3678 stratified air flow
DA lagdelt luftstrøm
DE geschichtete
  Luftströmung *f*
ES flujo de aire
  estratificado *m*
F écoulement *m* d'air
  stratifié
I flusso *m* di aria
  stratificato
MA réteges légáramlás
NL gelaagde
  luchtstroming *f*
PO uwarstwiony
  przepływ *m*
  powietrza
PY стратифициро-
  ванный поток *m*
  воздуха
SU kerrostunut
  ilmavirtaus
SV skiktat luftflüde *n*

3679 stream
DA strøm
DE Strümung *f*
ES corriente *f*
F courant *m*
I corrente *f*
MA áramlás
NL stroming *f*
  stroom *m*
PO struga *f*
  strumień *m*
PY поток *m*
  струя *f*
SU virta
SV strüm *c*

3680 streamline flow
DA laminar strømning
DE laminare
   Strümung *f*
ES flujo laminar *m*
F écoulement *m*
   laminaire
I flusso *m* laminare
MA áramvonalas áramlás
NL laminaire stroming *f*
PO przepływ *m*
   laminarny
   przepływ *m*
   uwarstwiony
PY ламинарное
   течение *n*
SU laminaarinen virtaus
SV laminär strümning *c*

3681 stress
DA indre spænding
DE Beanspruchung *f*
   Belastung *f*
   Spannung *f*
ES esfuerzo *m*
   fatiga *f*
   tensión *f*
F contrainte *f*
   tension *f*
I sforzo *m*
   tensione *f*
MA feszültség
NL spanning *f*
PO naprężenie *n*
PY нагрузка *f*
   напряжение *n*
SU kuormitus
SV spänning *c*

3682 stroke
DA slag
   slaglængde
DE Hub *m*
ES carrera del émbolo *f*
F course *f* de piston
I corsa *f*
MA ütem
NL slag *m*
   slaglengte *f*
   streep *f*
PO skok *m*
   suw *m*
   udar *m*
PY удар *m*
   ход *m*
SU isku
SV slag *n*

3683 structural member
DA konstruktions-
   element
DE Bauelement *n*
   Bauteil *n*
ES elemento
   estructural *m*
F élément *m* de
   structure
I elemento *m*
   strutturale
MA szerkezeti elem
NL bouwdeel *n*
PO element *m*
   konstrukcyjny
PY конструктивный
   элемент *m*
   структурный
   элемент *m*
SU rakenneosa
SV byggnadselement *n*

3684 stuffing
DA pakning
   tætning
DE Abdichtung *f*
ES estopada *f*
F garniture *f*
   d'étanchéité
I imbottitura *f*
MA tümés
NL afdichting *f*
   opvulling *f*
PO szczeliwo *n*
   uszczelnienie *n*
PY заполнение *n*
SU tiiviste
   täyte
SV packning *c*

3685 stuffing box
DA pakdåse
DE Stopfbüchse *f*
ES caja de estopadas *f*
   prensaestopas *m*
F presse-étoupe *m*
I premistoppa *m*
   pressatreccia *m*
MA tümszelence
NL pakkingbus *f*
   stopbus *f*
PO dławnica *m*
   komora *f*
   dławikowa
PY сальник *m*
SU tiivistysholkki
SV packboc

3686 sub-atmospheric system
DA undertrykssystem
DE Vakuum(heiz)-
   system *n*
ES calefacción bajo
   vacío *f*
F chauffage *m* sous
   vide
I sistema *m*
   sottovuoto
MA vákumrendszer
NL vacuum(verwarmings)-
   systeem *n*
PO system *m*
   podciśnieniowy
PY вакуумметрическая
   система *f*
SU alipainejärjestelmä
SV undertryckssystem *n*

3687 subcontractor
DA underentreprenør
DE Unterlieferant *m*
ES subcontratista *m*
F sous-traitant *m*
I sottocontraente *m*
   subappaltatore *m*
MA alvállakozó
NL onderaannemer *m*
PO podwykonawca *m*
PY субподрядчик *m*
SU aliurakoitsija
SV underentreprenür *c*

3688 subcooled liquid
DA underafkølet væske
DE unterkühlte
   Flüssigkeit *f*
ES líquido subenfriado
   m
F liquide *m* sous-
   refroidi
I liquido *m*
   sottoraffreddato
MA aláhűtütt folyadék
NL onderkoelde
   vloeistof *f*
PO płyn *m* dochładzany
PY переохлажденная
   жидкость *f*
SU alijäähdytetty neste
SV underkyld vätska *c*

3689 sub-cooling

3689 sub-cooling
DA underafkøling
DE Unterkühlung *f*
ES subenfriamiento *m*
F sous-refroidissement *m*
I sottoraffreddamento *m*
MA aláhűtés
NL onderkoeling *f*
PO dochładzanie *n*
PY переохлаждение *n*
SU alijäähdytys
SV underkylning *c*

3690 submaster controller
DA tilbageførings-regulator
DE Regelungsunter-station *f*
ES regulador submaster *m*
F régulateur *m* esclave
I controllore *m* submaster
MA szabályozó alállomás
NL volgregelaar *m*
PO regulator *m* podporządkowany regulator *m* podrzędny
PY дистанционный регулятор *m*
SU kaskadisäädin
SV underkontroll *c*

3691 submerged coil condenser
DA neddykket kondensator
DE überfluteter Kondensator *m*
ES condensador de tubos sumergidos *m*
F condenseur *m* à immersion
I condensatore *m* a serpentino annegato
MA elárasztott csőkígyós kondenzátor
NL dompelcondensor *m*
PO skraplacz *m* z wężownicą zanurzoną
PY затопленный змеевиковый конденсатор *m*
SU uppolauhdutin
SV doppkondensor *c*

3692 subsonic compressor
DA underlyds kompressor
DE Unterschall-verdichter *m*
ES compresor subsonico *m*
F compresseur *m* subsonique
I compressore *m* subsonico
MA csendes kompresszor
NL subsonische compressor *m*
PO sprężarka *f* poddźwiękowa
PY турбокомпрессор *m* с дозвуковой скоростью
SU keskipako-kompressori jossa virtaus on alle äänennopeuden
SV underljuds-kompressor *c*

3693 sub-station
DA understation
DE Unterstation *f*
ES subestación *f*
F sous-station *f*
I sottostazione *f*
MA alállomás
NL onderstation *n*
PO podstacja *f*
PY подстанция *f*
SU alakeskus
SV undercentral *c*

3694 suck *vb*
DA indsuge suge
DE ansaugen
ES aspirar
F aspirer
I aspirare
MA szív
NL aanzuigen zuigen
PO ssać zasysać
PY всасывать
SU imeä
SV suga

3695 suction
DA indsugning
DE Ansaugung *f*
ES aspiración *f* succión *f*
F aspiration *f*
I aspirazione *f*
MA szivás
NL afzuiging *f* zuiging *f*
PO ssanie *n* zasysanie *n*
PY подсос *m*
SU imu
SV insugning *c*

3696 suction capacity
DA sugeevne
DE Saugleistung *f*
ES caudal de aspiración *m* potencia de aspiración *f* presión de aspiración *f*
F débit *m* d'aspiration
I portata *f* di aspirazione
MA szívóteljesítmény
NL aanzuigvermogen *n* zuigvermogen *n*
PO wydatek *m* zasysania
PY эффективность *f* всасывания
SU imuteho
SV sugkapacitet *c*

3697 suction head
DA sugehoved
DE Saughühe *f*
ES altura de aspiración *f*
F hauteur *f* d'aspiration
I pressione *f* di aspirazione
MA szívómagasság szívónyomás
NL zuigdruk *m* zuighoogte *f*
PO wysokość *f* zasysania
PY давление *n* на всасывании (агрегата)
SU imupaine
SV sughüjd *c*

3698 suction hood
DA aftrækshætte
DE Saughaube *f*
ES campana de
aspiración *f*
F hotte *f* d'aspiration
I cappa *f* di
aspirazione
MA elszívóernyő
NL afzuigkap *f*
PO okap *m* odciągowy
PY вытяжной зонт *m*
местный отсос *m*
SU imuhuuva
SV sughuv *c*

3699 suction lift
DA sugehøjde
DE Saugheber *m*
ES altura de succión *f*
F hauteur *f*
d'aspiration
I prevalenza *f* in
aspirazione
MA szívómagasság
NL aanzuighoogte *f*
zuighoogte *f*
PO wysokość *f* ssania
(pompy)
PY высота *f*
всасывания
(насоса)
SU imukorkeus
SV vakuumtransport *c*

3700 suction line filter
DA sugeledningsfilter
DE Ansaugfilter *m*
Saugleitungsfilter *m*
ES filtro de la linea de
succión *m*
F filtre *m* de conduite
d'aspiration
I filtro *m* della linea
di aspirazione
MA szívócső szűrő
NL zuigleidingfilter *m*
PO filtr *m* w przewodzie
ssącym
PY фильтр *m* на линии
всасывания
фильтр *m* на линии
хладагента
SU imupuolen suodatin
SV sugledningsfilter *n*

3701 suction pipe
DA sugerør
DE Saugleitung *f*
Saugrohr *n*
ES tubo de
aspiración *m*
F tuyauterie *f*
d'aspiration
I tubo *m* di
aspirazione
MA szívócső
NL zuigpijp *f*
PO rura *f* ssąca
PY всасывающая
труба *f*
SU imuputki
SV sugledning *c*

3702 suction pressure
DA sugetryk
DE Saugdruck *m*
ES presión de
aspiración *f*
F pression *f*
d'aspiration
I pressione *f* di
aspirazione
MA szívónyomás
NL zuigdruk *m*
PO ciśnienie *n* ssania
PY давление *n*
всасывания
SU imupaine
SV insugningstryck *n*

3703 suction strainer
DA sugefilter
sugekurv
DE Saugfilter *m*
Saugkorb *m*
ES filtro de
aspiración *m*
F crépine *f*
I filtro *m* ad
aspirazione
MA szívószűrő
NL zuigfilter *m*
zuigkorf *m*
PO kosz *m* ssawny
kosz *m* ssący
PY корпус *m* фильтра
SU imusuodatin
SV återsugningsskydd *n*

**3704 sulphur dioxide
(SO₂)**
DA svovldioxid (SO2)
DE Schwefeldioxid *n*
ES anhídrido
sulfuroso *m*
F anhydride *m*
sulfureux
I anidride *f* solforosa
MA kéndioxid
NL zwaveldioxyde *n*
PO dwutlenek *m* siarki
PY двуокись *f* серы
SU rikkidioksidi
SV svaveldioxid *c*

**3705 summer air
conditioning**
DA sommerluft-
konditionering
DE Sommerklimati-
sierung *f*
ES climatización en
verano *f*
F conditionnement *m*
d'air d'été
I condizionamento *m*
estivo
MA nyári
légkondicionálás
NL zomerlucht-
behandeling *f*
PO klimatyzacja *f* w
okresie letnim
PY летнее
кондицион-
ирование *n*
воздуха
SU kesäilmastointi
SV sommarluft-
konditionering *c*

3706 sump
DA samlebrønd
sump
DE Ölwanne *f*
Sumpf *m*
ES colector *m*
sumidero *m*
F pot *m* de
décantation
puisard *m*
I scarico *m*
MA csatornaszem
zsomp
NL vergaarbak *m*
zinkput *m*
PO rząpie *f*
studzienka *f*
zbiornik *m* ściekowy
PY грязевик *m*
отстойник *m*
SU kaivo
SV pumpgrop *c*

3707 super cooling
DA underafkøling
DE Unterkühlung *f*
ES sobrefusión *f*
F surfusion *f*
I sovraraffredda-
mento *m*
MA túlhűtés
NL onderkoeling *f*
PO przechłodzenie *n*
PY переохлаждение *n*
SU alijäähdytys
SV underkylning *c*

3708 superheat
DA overhedning
DE Überhitzung *f*
ES recalentamiento *m*
F surchauffe *f*
I calore *m* di
surriscaldamento
MA túlhevítés
NL oververhittings-
warmte *f*
PO ciepło *n* przegrzania
PY перегрев *m*
SU tulistus
SV überhettning *c*

3709 superheat *vb*
DA overhede
DE überhitzen
ES recalentar
F surchauffer
I surriscaldare
MA túlhevít
NL oververhitten
PO przegrzewać
PY перегревать
SU tulistaa
SV överhetta

3710 superheated *adj*
DA overhedet
DE überhitzt
ES recalentado
F surchauffé
I surriscaldato
MA túlhevített
NL oververhit
PO przegrzany
PY перегретый
SU tulistettu
SV överhettad

3711 superheater
DA overheder
DE Überhitzer *m*
ES recalentador *m*
F surchauffeur *m*
I surriscaldatore *m*
MA túlhevítő
NL oververhitter *m*
PO przegrzewacz *m*
PY перегреватель *m*
SU tulistin
SV överhettare *c*

3712 superheating
DA overhedning
DE Überhitzung *f*
ES recalentamiento *m*
F surchauffe *f*
I surriscaldamento *m*
MA túlfűtés
NL oververhitting *f*
PO przegrzewanie *n*
PY перетоп *m*
SU tulistus
SV överhettning *c*

3713 supersaturate *vb*
DA overmætte
DE übersättigen
ES sobresaturar
F sursaturer
I sovrasaturare
MA túltelít
NL oververzadigen
PO przesycać
PY перенасыщать
SU ylikyllästää
SV övermätta

3714 supersaturated *adj*
DA overmættet
DE übersättigt
ES sobresaturación *f*
F sursaturé
I soprassaturo
MA túltelített
NL oververzadigd
PO przesycony
PY перенасыщенный
SU ylikyllästetty
SV övermättad

3715 supersaturation
DA overmætning
DE Übersättigung *f*
ES sobresaturación *f*
F sursaturation *f*
I sovrasaturazione *f*
MA túltelítés
NL oververzadiging *f*
PO przesycenie *n*
PY перенасыщение *n*
SU ylikyllästys
SV övermättning *c*

3716 supersonic compressor
DA overlydskompressor
DE Überschall-
verdichter *m*
ES compresor
supersónico *m*
F compresseur *m*
supersonique
I compressore *m*
supersonico
MA szuperszónikus
kompresszor
NL supersonische
compressor *m*
PO sprężarka *f*
naddźwiękowa
PY компрессор *m* со
сверхзвуковой
скоростью
движения среды
SU keskipako-
kompressori jossa
virtausnopeus
ylittää äänen
nopeuden
SV überljuds-
kompressor *c*

3717 supervision
DA tilsyn
DE Überwachung *f*
ES supervisión *f*
F contrôle *m*
surveillance *f*
I supervisione *f*
MA felülvizsgálat
NL supervisie *f*
toezicht *n*
PO dozór *m*
nadzór *m*
PY контроль *m*
надзор *m*
SU tarkastus
SV övervakning *c*

DA	=	Danish
DE	=	German
ES	=	Spanish
F	=	French
I	=	Italian
MA	=	Magyar
		(Hungarian)
NL	=	Dutch
PO	=	Polish
PY	=	Russian
SU	=	Finnish
SV	=	Swedish

3718 supplementary heating
DA supplerende
   opvarmning
DE Zusatzheizung *f*
ES calefacción
   suplementaria *f*
   calentamiento
   suplementario *m*
F chauffage *m*
   complémentaire
   chauffage *m*
   d'appoint
I riscaldamento *m*
   supplementare
MA kiegészítő fűtés
NL aanvullende
   verwarming *f*
   bijverwarming *f*
PO ogrzewanie *n*
   dodatkowe
   ogrzewanie *n*
   uzupełniające
PY вспомогательное
   отопление *n*
SU lisälämmitys
SV tillsatsvärme *c*

3719 supply
DA beholdning
   forsyning
DE Versorgung *f*
   Zufuhr *f*
ES alimentación *f*
F approvisionnement
   m
   livraison *f*
I alimentazione *f*
   fornitura *f*
MA ellátás
NL toevoer *m*
   voorraad *m*
PO zaopatrzenie *n*
   zasilanie *n*
PY подвод *m*
   снабжение *n*
SU syüttü-
   tulo-
SV fürråd *n*

3720 supply *vb*
DA forsyne
   tilføre
DE ergänzen
   versorgen
ES suministrar
F alimenter
I alimentare
   fornire
MA ellát
NL aanvoeren
   leveren
   toevoeren
PO dostarczać
   zaopatrywać
   zasilać
PY питать
   подводить
   снабжать
SU jakaa
   syüttää
SV fürsürja

3721 supply air
DA lufttilførsel
DE Zuluft *f*
ES aire suministrado *m*
F air *m* distribué
I aria *f* di immissione
MA légellátás
NL toevoerlucht *f*
PO powietrze *n*
   nawiewne
PY подаваемый
   воздух *m*
SU tuloilma
SV tilluft *c*

3722 supply-air equipment
DA lufttilførselsudstyr
DE Zulufteinrichtung *f*
ES equipo de suministro
   de aire *m*
F équipement *m* de
   distribution d'air
I apparecchiatura *f* di
   immissione di aria
MA légellátó berendezés
NL toevoerinrichting *f*
   toevoerlucht-
   apparatuur *f*
PO urządzenie *n*
   nawiewne
PY устройство *n* для
   подачи воздуха
SU tuloilmalaitteisto
SV tilluftsdon *n*

3723 supply-air opening
DA lufttilførselsåbning
DE Zuluftüffnung *f*
ES abertura para el
   suministro de
   aire *f*
   boca de impulsión
   de aire *f*
F bouche *f* de
   soufflage d'air
I apertura *f* di
   immissione di aria
MA szállított levegő
   nyílás
NL toevoerlucht-
   opening *f*
PO otwór *m* nawiewny
PY отверстие *n* для
   подачи воздуха
SU tuloilma-aukko
SV tilluftsüppning *c*

3724 supply-air outlet
DA lufttilførselsåbning
DE Zuluftauslaß *m*
ES boca de salida
   del aire
   suministrado *f*
   rejilla de salida del
   aire tratado *f*
F bouche *f* de
   soufflage d'air
I uscita *f* dell'aria di
   immissione
MA szállított levegő
   távozónyílás
NL toevoerluchtuit-
   blaasopening *f*
PO nawiewnik *m*
PY выпуск *m*
   приточного
   воздуха
SU tuloilmalaite
SV tilluftsutlopp *n*

3725 supply-air system
DA lufttilførselssystem
DE Zuluftanlage *f*
ES sistema de
distribución del
aire (tratado) *m*
sistema de
suministro de
aire *m*
F système *m* de
soufflage d'air
I impianto *m* di
immissione di aria
MA légellátó rendszer
NL toevoerlucht-
systeem *n*
PO instalacja *f*
nawiewna
PY система *f* воздухос-
набжения
система *f* подачи
воздуха
SU tuloilmajärjestelmä
SV tilluftssystem *n*

3726 supply line
DA fødeledning
DE Vorlaufleitung *f*
Zuleitung *f*
ES conducto de
suministro *m*
línea de
alimentación *f*
F conduite *f*
d'alimentation
I linea *f* di
alimentazione
MA elosztóvezeték
NL aanvoerleiding *f*
voedingsleiding *f*
PO przewód *m*
nawiewny
przewód *m*
zasilający
PY питательная
линия *f*
SU syüttüjohto
SV matarledning *c*

3727 supply manifold
DA fremløbsfordeler
DE Vorlaufsammler *m*
Vorlaufverteiler *m*
ES colector de
alimentación *m*
F collecteur *m* de
départ
collecteur *m* de
distribution
I collettore *m* di
mandata
MA elosztóvezeték osztó
NL aanvoerverdeler *m*
PO rozdzielacz *m*
zasilający
PY питательный
распределитель
m
SU jakokammio
jakotukki
SV fürdelningsledning *c*

3728 supply riser
DA stigrør
DE Vorlaufsteigleitung *f*
ES sobrealimentación *m*
F colonne *f* montante
d'alimentation
I colonna *f* montante
di mandata
MA elosztóvezeték
felszálló
NL aanvoerstrang *m*
voedingsstrang *m*
PO pion *m* zasilający
PY подающий стояк *m*
SU nousulinja
nousuputki

3729 support
DA bære
støtte
DE Auflager *n*
Stütze *f*
ES soporte *m*
F support *m*
I supporto *m*
MA támasz
tartószerkezet
NL ondersteuning *f*
PO podpora *f*
wspornik *m*
PY опора *f*
SU tuki
SV stüd *n*

3730 suppressor
DA dæmper
DE Entstürer *m*
ES dispositivo
antiparasitario *m*
supresor de
perturbaciones *m*
F antiparasite *m*
I antiparassitario *m*
soppressore *m*
MA zavarszűrő
NL storingsonder-
drukker *m*
PO eliminator *m*
tłumik *m*
PY защитная сетка *f*
SU häiriün suojain
vaimennin
SV stürningsskydd *n*

3731 surface
DA overflade
DE Oberfläche *f*
ES superficie *f*
F surface *f* (aire *f*)
I superficie *f*
MA felület
NL oppervlakte *f*
vlak *n*
PO powierzchnia *f*
PY поверхность *f*
SU pinta
SV yta *c*

3732 surface cooling
DA overfladekøling
DE Oberflächenkühlung
f
ES enfriamiento
superficial *m*
F refroidissement *m*
sur surface froide
I raffreddamento *m*
superficiale
MA felületi hűtés
NL oppervlaktekoeling *f*
PO chłodzenie *n*
powierzchniowe
PY поверхность *f*
охлаждения
SU pintajäähdytys
SV ytkylning *c*

3733 surface tension
DA overfladespænding
DE Oberflächen-
spannung *f*
ES tensión superficial *f*
F tension *f*
superficielle
I tensione *f*
superficiale
MA felületi feszültség
NL oppervlakte-
spanning *f*
PO napięcie *n*
powierzchniowe
PY поверхностное
натяжение *n*
SU pintajännitys
SV ytspänning *c*

3734 surroundings
DA omgivelser
DE Umgebung *f*
ES ambiente exterior *m*
medio exterior *m*
F ambiance *f*
environnement *m*
I adiacenze *f,pl*
ambiente *m* esterno
MA kürnyezet
NL omgeving *f*
PO otoczenie *n*
środowisko *n*
PY окружающая
среда *f*
SU ympäristü
SV omgivning *c*

3735 suspended ceiling
DA nedforskallet loft
DE abgehängte Decke *f*
ES techo suspendido *m*
F plafond *m* suspendu
I controsoffitto *m*
MA függesztett
álmennyezet
NL verlaagd plafond *n*
PO sufit *m* podwieszony
PY подвесной
потолок *m*
SU alaslaskettu katto
SV undertak *n*

3736 suspended particle
DA svævende partikel
DE Schwebeteilchen *n*
ES partícula en
suspensión *f*
F particule *f* en
suspension
I particella *f* in
sospensione
MA lebegő részecskék
NL deeltje *n* in
suspensie
PO cząsteczka *f*
zawieszona
PY взвешенная
частица *f*
частица *f*
суспензии
SU leijuva hiukkanen
SV svävande partikel *c*

3737 suspension
DA ophængning
DE Aufhängung *f*
Aufschlämmung *f*
Suspension *f*
ES suspensión *f*
F suspente *f*
I sospensione *f*
MA függesztés
szuszpenzió
NL ophanging *f*
ophangpunt *n*
suspensie *f*
PO podwieszenie *n*
zawieszenie *n*
PY суспензия *f*
SU ripustus
suspensio
SV upphängning *c*

3738 swamp cooler (*see*
evaporative cooler)

**3739 swash plate
compressor** (*see*
wobble plate
compressor)

3740 swell index
DA hævningstal
DE Blähgrad *m*
ES índice de
esponjamiento *m*
F indice *m* de
gonflement
I indice *m* di
rigonfiamento
MA duzzadási hányados
NL zwellingsindex *m*
PO stopień *m*
spęcznienia
stopień *m*
spiętrzenia
PY индекс *m*
вспучивания
SU paisuntaluku
SV svällningstal *n*

3741 switch
DA afbryder
kontakt
DE Schalter *m*
ES interruptor *m*
F interrupteur *m*
I interruttore *m*
MA kapcsoló
NL schakelaar *m*
PO przełącznik *m*
wyłącznik *m*
PY выключатель *m*
SU kytkin
SV strümbrytare *c*

3742 switchboard
DA fordelingstavle
omstillingsbord
DE Schalttafel *f*
ES cuadro de
distribución *m*
cuadro de
mandos *m*
cuadro de
maniobras *m*
F tableau *m* de
commande
I centralino *m*
(pannello di
comando)
quadro *m* di
distribuzione
MA kapcsolótábla
NL bedieningstableau *n*
schakelbord *n*
PO łącznica *f*
telefoniczna
tablica *f* rozdzielcza
PY распределительный
щит *m*
SU kytkentätaulu
SV kopplingstavla *c*

3743 switch box
DA kontaktdåse
DE Schaltkasten *m*
ES caja de
distribución *f*
caja de
interruptores *f*
caja de mandos *f*
F armoire *f* de
commande
boîte *f* de
commande
I cassetta *m* di
distribuzione
quadro *m* di
comando
MA kapcsolószekrény
NL schakelkast *f*
PO łącznica *f*
telefoniczna
szafa *f* rozdzielcza
szafa *f* sterująca
tablica *f* rozdzielcza
PУ щит *m* управления
SU jakorasia
kytkentärasia
SV kopplingsdosa *c*

3744 switch off *vb*
DA afbryde
DE ausschalten
ES desconectar
poner fuera de
circuito
F mettre hors circuit
I disinserire
fermare
interrompere
spegnere
MA kikapcsol
NL uitschakelen
PO wyłączać
PУ выключать
SU kytkeä pois
SV koppla av

3745 switch on *vb*
DA tilslutte
tænde
DE einschalten
ES conectar
poner en circuito
F enclencher
fermer le circuit
mettre en circuit
I accendere
avviare
mettere in moto
MA bekapcsol
NL inschakelen
PO włączać
PУ включать
SU kytkeä päälle
SV koppla på

3746 switch on point
DA indkoblingspunkt
DE Einschaltpunkt *m*
ES punto de
conexión *m*
F point *m*
d'enclenchement
I punto *m* di
inserzione
MA bekapcsolási pont
NL inschakelpunt *f*
PO moment *m*
włączania
punkt *m* włączania
PУ пункт *m*
включения
SU kytkentäpiste
SV tillslagspunkt *c*

3747 symbol
DA symbol
tegn
DE Symbol *n*
ES símbolo *m*
F symbole *m*
I simbolo *m*
MA jelkép
NL symbool *n*
teken *n*
PO symbol *m*
znak *m*
PУ символ *m*
SU symboli
SV symbol *c*

3748 synchronization
DA synkronisering
DE Synchronisation *f*
ES sincronización *f*
F synchronisation *f*
synchronisme *m*
I sincronizzazione *f*
MA szinkronizálás
NL synchronisatie *f*
PO synchronizacja *f*
PУ синхронизация *f*
SU synkronointi
SV synkronisering *c*

3749 synthetic fibre
DA kunstfiber
DE Kunststoffaser *f*
ES fibra sintética *f*
F fibre *f* synthétique
I fibra *f* sintetica
MA műrost
NL kunststofvezel *f*
PO włókno *n*
syntetyczne
PУ пластмассовое
волокно *n*
SU tekokuitu
SV syntetfiber *c*

3750 system
DA system
DE Plan *m*
System *n*
ES instalación *f*
red *f*
sistema *m*
F installation *f*
système *m*
I impianto *m*
sistema *m*
MA rendszer
NL systeem *n*
PO instalacja *f*
system *m*
układ *m*
PУ система *f*
SU järjestelmä
SV system *n*

DA	=	Danish
DE	=	German
ES	=	Spanish
F	=	French
I	=	Italian
MA	=	Magyar (Hungarian)
NL	=	Dutch
PO	=	Polish
PУ	=	Russian
SU	=	Finnish
SV	=	Swedish

T

3751 table
DA bord
    tabel
    tavle
DE Tafel *f*
    Tisch *m*
ES mesa *f*
F table *f*
    tableau *m*
I quadro *m*
    tabella *f*
    tavola *f*
MA asztal
    táblázat
NL tabel *f*
    tafel *f*
PO stół *m*
    tablica *f*
PY таблица *f*
SU taulukko
SV bord *n*
    tabell *c*

3752 take-off piece
DA mellemstykke
DE Zwischenstück *n*
ES tubo de
    derivación *m*
F tubulure *f* de
    dérivation
    tubulure *f* de
    raccordement
I raccordo *m* di
    derivazione
MA küzdarab
NL aftakking *f*
    spruitstuk *n*
PO część *f* łącząca
    łącznik *m*
PY соединительная
    часть *f*
SU haarakappale
    yhdyskappale
SV muff *c*

3753 tandem compressor
DA dobbeltvirkende
    kompressor
DE Tandem-
    verdichter *m*
ES compresor tandem *m*
F compresseur *m* en
    tandem (compres-
    seurs jumelés)
I compressore *m* a
    doppio effetto
    compressore *m*
    duale
MA kettős kompresszor
NL tandem-
    compressor *m*
PO sprężarka *f*
    bliźniacza
PY спаренный
    компрессор *m*
SU kahden
    kompressorin
    yksikkü
SV tandem-
    kompressor *c*

3754 tangential fan
DA tangentiel ventilator
DE Tangential-
    ventilator *m*
ES ventilador
    tangencial *m*
F ventilateur *m*
    tangentiel
I ventilatore *m*
    tangenziale
MA tangenciális
    ventilátor
NL tangentiaal-
    ventilator *m*
PO wentylator *m* o
    przepływie
    poprzecznym
PY тангенциальный
    вентилятор *m*
SU poikittaisvirt-
    auspuhallin
SV tangentialfläkt *c*

3755 tank
DA beholder
    tank
DE Behälter *m*
    Tank *m*
ES depósito *m*
    tanque *m*
F bac *m*
    bâche *f*
    réservoir *m*
I cisterna *f*
    serbatoio *m*
MA tartály
NL reservoir *n*
    tank *m*
PO zbiornik *m*
PY бак *m*
    резервуар *m*
SU astia
    säiliü
SV tank *c*

3756 tanker
DA tankskib
    tankvogn
DE Tankschiff *n*
    Tankwagen *m*
ES buque tanque *m*
    camión cisterna *m*
    vagón cisterna *m*
F citerne *f* de
    transport
I autocisterna *f*
    nave *f* cisterna
    petroliera *f*
    vagone *m* cisterna
MA tartálykocsi
NL tanker *m*
    tankwagen *m*
PO tankowiec *m*
    wagon-cysterna *m*
PY цистерна *f*
SU säiliüalus
SV tankfartyg *n*

3757 tap
DA hane
   tap
DE Zapfhahn *m*
ES defecto *m*
   grifo *m*
   llave de extracción *f*
F robinet *m* à
    boisseau
   robinet *m* de
    puisage
I rubinetto *m* di
    spillamento
MA csap
NL draadtap *m*
   kraan *f*
PO kurek *m* czerpalny
   zawór *m*
PY кран *m*
SU hana
   sekoittaja
SV kran *c*

3758 tap *vb*
DA aftappe
   skære gevind
DE anzapfen
   zapfen
ES derivar (electricidad)
   extraer (mediante
    grifo o llave)
   llave *f*
   roscar
    (interiormente)
F tarauder
I filettare
   maschiare
   spillare
MA csapol
NL draadsnijden
   draadtappen
PO czerpać
   gwintować
   pobierać
PY нарезать
   внутренюю
   резьбу
SU juoksuttaa
SV tappa

3759 tap discharge rate
DA aftapningskapacitet
DE Zapfleistung *f*
ES gasto *m*
F débit *m* de puisage
I portata *f* di
   spillamento
MA csap kifolyási mérték
NL tapcapaciteit *f*
PO wydajność *f* punktu
   poboru
   wydatek *m* zaworu
PY размер *m*
   выпускного
   отверстия
SU juoksutusnopeus
SV tappflüde *n*

3760 tapered thread
DA konisk gevind
DE konisches
   Gewinde *n*
ES rosca cónica *f*
F filetage *m* conique
I filettatura *f* conica
MA kúpos csavarmenet
NL conische
   schroefdraad *m*
PO gwint *m* stożkowy
PY коническая
   резьба *f*
SU kartiomainen kierre
SV konisk gänga *c*

3761 tapped hole
DA indvendigt gevind
DE Gewindebohrung *f*
ES agujero roscado *m*
F trou *m* taraudé
I foro *m* filettato
MA menetes furat
NL gat *n* met
   schroefdraad
PO otwór *m*
   gwintowany
PY отверстие *n* с
   резьбой
SU kierteellinen reikä
SV gängat hål *n*

3762 tapping
DA gevindskæring
DE Gewindebohren *n*
   Gewindeschneiden *n*
   Verschraubung *f*
ES roscado (interior) *m*
   terrajado *m*
F taraudage *m*
I filettatura *f*
   presa *f*
MA menetvágás
NL tappen *n*
PO czerpanie *n*
   gwintowanie *n*
   pobieranie *n*
PY нарезание *n* резьбы
   метчиком
SU kierteitys
SV gängning *c*

3763 tar
DA tjære
DE Teer *m*
ES alquitrán *m*
F goudron *m*
I catrame *m*
MA kátrány
NL teer *m*
PO smoła *f*
PY смола *f*
SU terva
SV tjära *c*

3764 tariff
DA takst
   tarif
DE Tarif *m*
ES arancel *m*
   tarifa *f*
F tarif *m*
I tariffa *f*
MA díjszabás
NL tarief *n*
PO taryfa *f*
PY тариф *m*
SU tariffi
SV tariff *c*

3765 technical *adj*
DA teknisk
DE technisch
ES técnico
F technique
I tecnico
MA műszaki
NL technisch
PO techniczny
PY технический
SU tekninen
SV teknisk

3766 temperature
DA temperatur
DE Temperatur *f*
ES temperatura *f*
F température *f*
I temperatura *f*
MA hőmérséklet
NL temperatuur *f*
PO temperatura *f*
PY температура *f*
SU lämpütila
SV temperatur *c*

3767 temperature controller
DA temperaturregulator
DE Temperaturregler *m*
ES regulador de temperatura *m*
F régulateur *m* de température
I regolatore *m* di temperatura
MA hőmérséklets-zabályozó
NL temperatuur-regelaar *m*
PO regulator *m* temperatury
PY регулятор *m* температуры
SU lämpütilan säätü
SV temperatur-reglering *c*

3768 temperature deviation
DA temperaturafvigelse
DE Temperatur-abweichung *f*
ES desviación de temperatura *f*
F écart *m* de température
I salto *m* di temperatura scarto *m* di temperatura
MA hőmérsékleteltérés
NL temperatuur-afwijking *f*
PO odchylenie *n* temperatury
PY отклонение *n* температуры
SU lämpütilapoikkeama
SV temperatur-avvikelse *c*

3769 temperature difference
DA temperaturforskel
DE Temperatur-differenz *f*
ES diferencia de temperatura *f*
F différence *f* de température
I differenza *f* di temperatura
MA hőmérséklet-külünbség
NL temperatuur-verschil *n*
PO różnica *f* temperatury
PY разность *f* температуры
SU lämpütilaero
SV temperaturskillnad *c*

3770 temperature sensor
DA temperaturføler
DE Temperaturfühler *m*
ES sonda de temperatura *f*
F capteur *m* de température
I sensore *m* di temperatura
MA hőmérsékletérzékelő
NL temperatuur-detector *m* temperatuur-opnemer *m* temperatuur-voeler *m*
PO czujnik *m* temperatury
PY датчик *m* температуры термочувст-вительный элемент *m*
SU lämpütila-anturi
SV temperatur-avkännare *c*

3771 temperature swing
DA temperatursvingning
DE Temperatur-schwankung *f*
ES fluctuación de la temperatura *f* oscilación de la temperatura *f*
F oscillation *f* de température
I oscillazione *f* di temperatura
MA hőmérsékleting-adozás
NL temperatuur-schommeling *f* temperatuur-variatie *f*
PO wahanie *n* temperatury
PY колебание *n* температуры
SU lämpütilavaihtelu
SV temperatur-svängning *c*

3772 tender
DA tilbud
DE Angebot *n* Kostenanschlag *m*
ES oferta *f* presupuesto *m*
F offre *f* proposition *f* soumission *f*
I offerta *f*
MA ajánlat versenytárgyalás
NL aanbieding *f* inschrijving *f* offerte *f*
PO oferta *f*
PY заявка *f* предложение *n*
SU tarjous
SV anbud *n* skütare *c*

3773 tendering
DA afgive tilbud
DE Angebotsabgabe *f*
ES presupuesto *m*
F
I offerta *f*
MA ajánlatkészítés
NL aanbieden *n* inschrijven *n* offreren *n*
PO nadzorowanie *n*
PY работа *f* с заявками и предложениями
SU huoltaminen tarjoaminen
SV

3774 terminal valve
DA afløbsventil
DE Auslaßventil *n*
ES válvula terminal *f*
F robinet *m* terminal
I valvola *f* terminale
MA végszelep
NL eindklep *f*
PO zawór *m* końcowy
PY концевой клапан *m*
SU tyhjennysventtiili
SV utloppsventil *c*

3775 terminal velocity
DA sluthastighed
   udløbshastighed
DE Endgeschwindig-
   keit *f*
ES velocidad final *f*
F vitesse *f* terminale
I velocità *f* terminale
MA végsebesség
NL eindsnelheid *f*
PO prędkość *f* końcowa
PY конечная
   скорость *f*
SU loppunopeus
SV utloppshastighet *c*

3776 test
DA forsøg
   prøve
DE Prüfung *f*
   Versuch *m*
ES ensayo *m*
   prueba *f*
F épreuve *f*
   essai *m*
I collaudo *m*
   prova *f*
MA próba
   teszt
NL keuring *f*
   proef *f*
   test *m*
PO badanie *n*
   próba *f*
   test *m*
PY испытание *n*
SU koe
   testi
SV prov *n*

3777 test bench
DA prøvebænk
DE Versuchswerkbank *f*
ES banco de pruebas *m*
F banc *m* d'essai
I banco *m* di prova
MA próbapad
NL proefbank *f*
   proefstand *m*
   testbank *f*
PO stanowisko *n*
   badawcze
   stół *m* pomiarowy
PY испытательный
   стенд *m*
SU koepenkki
SV provbänk *c*

3778 test booth
DA prøvestand
DE Versuchsraum *m*
   Versuchsstand *m*
ES cámara de ensayos *f*
F chambre *f* d'essai
I camera *f* di prova
MA próbafülke
NL onderzoekruimte *f*
   testruimte *f*
PO kabina *f* badawcza
PY испытательная
   камера *f*
SU koekammio
SV provkammare *c*

3779 test certificate
DA prøveattest
DE Prüfzeugnis *n*
ES certificado de
   pruebas *m*
F certificat *m*
   d'épreuve
   certificat *m* d'essai
I certificato *m* di
   collaudo
MA vizsgálati bizonylat
NL beproevingscertifi-
   caat *n*
   keuringscertificaat *n*
PO świadectwo *n* próby
PY сертификат *m* по
   результатам
   испытаний
SU testaustodistus
SV provningsintyg *n*

3780 test cock
DA prøvehane
DE Prüfhahn *m*
   Prüfventil *n*
ES válvula de prueba *f*
F robinet *m* de
   contrôle
I rubinetto *m* di
   prova
MA próbacsap
NL proefkraan *f*
PO kurek *m* probierczy
PY пробный кран *m*
SU tarkastushana
SV provkran *c*

3781 test dust
DA prøvestøv
DE Prüfstaub *m*
   Teststaub *m*
ES polvo de
   comprobación *m*
   polvo de control *m*
   polvo testigo *m*
F poussière *f* d'essai
I polvere *f* di saggio
MA próbapor
NL standaardstof *f* voor
   beproeven
PO pył *m* testowy
PY калиброванная
   пыль *f*
SU testipüly
SV provstoff *n*

3782 test equipment
DA prøveudstyr
DE Versuchs-
   einrichtung *f*
ES equipo de
   pruebas *m*
F appareillage *m*
   d'essai
I apparecchiatura *f* di
   collaudo
MA vizsgáló berendezés
NL beproevings-
   apparatuur *f*
   testapparatuur *f*
PO aparatura *f*
   badawcza
PY испытательная
   установка *f*
   оборудование *n*
   для испытаний
SU testauslaitteisto
SV provanordning *c*

3783 test method
DA prøvemetode
DE Versuchsmethode *f*
ES método de ensayo *m*
F méthode *f* d'essai
I metodo *m* di
  collaudo
MA vizsgálati módszer
NL beproevings-
  methode *f*
  testmethode *f*
PO metoda *f* badań
PY метод *m*
  испытаний
SU testausmenetelmä
SV provmetod *c*

3784 test pressure
DA prøvetryk
DE Probedruck *m*
  Prüfdruck *m*
ES presión de ensayo *f*
  presión de prueba *f*
F pression *f* d'épreuve
I pressione *m* di
  collaudo
MA próbanyomás
NL beproevingsdruk *m*
  proefdruk *m*
PO ciśnienie *n* próbne
PY испытательное
  давление *n*
SU koepaine
SV provtryck *n*

3785 test result
DA prøveresultat
DE Versuchsergebnis *n*
ES resultado de la
  prueba *m*
F résultat *m* d'essai
I risultato *m* di
  collaudo
MA vizsgálati eredmény
NL beproevings-
  resultaat *n*
  testresultaat *n*
PO wynik *m* badań
PY результат *m*
  испытания
SU koetulos
SV provningsresultat *n*

3786 test rig (*see* test
  equipment)

3787 test room
DA prøverum
DE Prüfraum *m*
  Versuchsraum *m*
ES sala de ensayos *f*
F local *m* d'essai
  salle *f* d'essai
I sala *f* prove
MA vizsgálóhelyiség
NL beproevingsruimte *f*
  proefkamer *f*
PO komora *f* badań
PY помещение *n* для
  испытаний
SU koehuone
SV provrum *n*

3788 test stand
DA prøvestand
DE Prüfstand *m*
ES banco de pruebas *m*
F banc *m* d'essai
I banco *m* di prova
MA próbaállás
NL proefstand *m*
PO stanowisko *n*
  badawcze
PY испытательный
  стенд *m*
SU koepenkki
SV provbänk *c*

3789 thawing
DA optøning
DE Auftauen *n*
  Tauen *n*
ES deshielo *m*
F décongélation *f*
I disgelamento *m*
MA olvadás
NL ontdooiing *f*
PO odtajanie *n*
  rozmrażanie *n*
PY оттаивание *n*
SU sulatus
SV upptining *c*

3790 thermal *adj*
DA termisk
DE thermisch
ES térmico
F thermique *f*
I termico
MA hő
NL thermisch
PO cieplny
  termiczny
PY тепловой
  термический
SU lämpü-
SV termisk

**3791 thermal
  anemometer**
DA varmeanemometer
DE thermisches
  Anemometer *n*
ES anemómetro térmico
  m
F anémomètre *m*
  thermique
I anemometro *m* a
  filo caldo
MA hődrótos
  anemométer
NL hittedraadanemo-
  meter *m*
PO termoanemometr *m*
PY термоанемометр *m*
SU
SV värmeanemometer *c*

3792 thermal balance
DA varmebalance
DE Wärmebilanz *f*
ES balance térmico *m*
F bilan *m* thermique
I bilancio *m* termico
MA hőegyensúly
NL warmtebalans *f*
PO bilans *m* cieplny
PY тепловой баланс *m*
SU lämpötasapaino
SV värmebalans *c*

3793 thermal comfort
DA termisk komfort
DE thermische
  Behaglichkeit *f*
ES confort térmico *m*
F confort *m* thermique
I benessere *m* termico
  confort *m* termico
MA hőkomfort
NL
PO komfort *m* cieplny
PY тепловой
  комфорт *m*
SU lämpüviihtyvyys
SV värmekomfort *c*

DA	=	Danish
DE	=	German
ES	=	Spanish
F	=	French
I	=	Italian
MA	=	Magyar (Hungarian)
NL	=	Dutch
PO	=	Polish
PY	=	Russian
SU	=	Finnish
SV	=	Swedish

3794 thermal conductance
DA varmeledningsevne
DE Wärmeleitung *f*
ES conductancia térmica *f*
F conductance *f* thermique
I conduttanza *f* termica
MA hővezetőképesség
NL warmtegeleiding *f*
PO przewodność *f* cieplna
     współczynnik *m* przewodzenia ciepła
PY тепловая проводимость *f*
SU lämpükonduktanssi
SV värmegenomgång *c*

3795 thermal conductivity
DA varmeledningsevne
DE Wärmeleitfähigkeit *f*
ES coeficiente de conductividad térmica *m*
     conductividad térmica *f*
F conductivité *f* thermique
I conduttività *f* termica
MA fajlagos hővezetőképesség
NL warmtegeleidings-vermogen *n*
PO przewodność *f* cieplna
     współczynnik *m* przewodzenia ciepła
PY теплопроводность *f*
SU lämmünjohtavuus
SV värmelednings-fürmåga *c*

3796 thermal convection
DA termisk konvektion varmeledning
DE thermische Konvektion *f*
ES convección térmica *f*
F convection *f* thermique
I convezione *f* termica
MA hőáramlás
NL natuurlijke convectie *f*
PO konwekcja *f* cieplna
PY тепловая конвекция *f*
SU luonnollinen konvektio
SV värmekonvektion *c*

3797 thermal diffusivity
DA varmediffusion
DE Wärmedurchlässig-keit *f*
ES difusibilidad térmica *f*
F diffusivité *f* thermique
I diffusività *f* termica
MA hődiffundáló képesség
NL temperatuur-vereffening *f*
PO dyfuzyjność *f* cieplna
PY температуропровод-ность *f*
     термодиффузия *f*
SU lämpütilan johtavuus
SV värmediffusivitet *c*

3798 thermal equilibrium
DA termisk ligevægt
DE thermisches Gleichgewicht *n*
ES equilibrio térmico *m*
F équilibre *m* thermique
I equilibrio *m* termico
MA hőegyensúly
NL thermisch evenwicht *n*
PO równowaga *f* cieplna
PY тепловое равновесие *n*
SU lämpütekninen tasapainotila
SV värmejämvikt *c*

3799 thermal insulation
DA varmeisolering
DE Wärmedämmung *f*
ES aislamiento térmico *m*
F isolation *f* thermique
I isolamento *m* termico
MA hőszigetelés
NL warmte-isolatie *f*
PO izolacja *f* cieplna
PY тепловая изоляция *f*
SU lämpüeristys
SV värmeisolering *c*

3800 thermal output
DA varmeydelse
DE Wärmeleistung *f*
ES potencia térmica *f*
F puissance *f* thermique
I potenza *f* termica sviluppata
MA hőteljesítmény
NL thermische produktie *f*
PO wydajność *f* cieplna
PY теплопроизводитель-ьность *f*
SU lämpüteho
SV värmeeffekt *c*

3801 thermal radiation
DA varmestråling
DE Wärmestrahlung *f*
ES radiación térmica *f*
F rayonnement *m* thermique
I irraggiamento *m* radiazione *f* termica
MA hősugárzás
NL warmtestraling *f*
PO promieniowanie *n* cieplne
PY тепловое излучение *n*
SU lämpüsäteily
SV värmestrålning *c*

3802 thermal resistance
DA varmemodstand
DE Wärme-
widerstand *m*
ES resistencia térmica *f*
F résistance *f*
thermique
I resistenza *f* termica
MA hőellenállás
NL warmteweerstand *m*
PO opór *m* cieplny
PY термическое
сопротивление *n*
SU lämmünvastus
SV värmemotstånd *n*

3803 thermal resistivity
DA varmemodstands-
evne
DE Wärmedurchgangswider-
stand *m*
ES resistividad
térmica *f*
F résistivité *f*
thermique
I resistività *f* termica
MA fajlagos hőellenállás
NL warmteweerstands-
vermogen *n*
PO oporność *f* cieplna
właściwa
PY удельное
термическое
сопротивление *n*
SU lämmünvastus
SV värmeresistivitet *c*

3804 thermal shock
DA termochok
DE Wärmeschock *m*
ES choque térmico *m*
F choc *m* thermique
I choc *m* termico
MA hőhatás
NL thermische schok *m*
PO szok *m* cieplny
PY тепловой удар *m*
SU lämpüshokki
SV värmeshock *c*

3805 thermal storage
DA varmelager
DE Wärmespeicherung *f*
ES almacenamiento
térmico *m*
F entreposage *m*
thermique
I accumulo *m* termico
MA hőtárolás
NL warmte-
accumulatie *f*
PO akumulacja *f* ciepła
PY теплое хранение *n*
SU lämpüvarasto
SV värmelagring *c*

**3806 thermal storage
vessel**
DA varmelagertank
DE Wärmespeicher *m*
Wärmespeicher-
behälter *m*
ES recipiente
acumulador de
calor *m*
recipiente de
almacenamiento
térmico *m*
F accumulateur *m* de
chaleur
I accumulatore *m* di
calore
MA hőtároló tartály
NL warmte-
accumulator *m*
warmtebuffer *m*
PO zasobnik *m* ciepła
PY тепловой
аккумулятор *m*
SU lämpüsäiliü
SV värmelagringskärl *n*

3807 thermistor
DA termistor
DE Thermistor *m*
ES termistancia *f*
termistor *m*
F thermistance *f*
I termistore *m*
termometro *m*
elettrico
MA termisztor
NL thermistor *m*
PO termistor *m*
PY термистор *m*
SU termistori
SV termistor *c*

3808 thermocouple
DA termoelement
DE Thermoelement-
paar *n*
ES par termo-
eléctrico *m*
termopar *m*
F thermocouple *m*
I termocoppia *f*
MA hőelem
NL thermoelement *n*
thermokoppel *n*
PO termopara *f*
PY термопара *f*
SU termoelementti
SV termoelement *n*

3809 thermodynamics
DA termodynamik
DE Thermodynamik *f*
ES termodinámica *f*
F thermodynamique *f*
I termodinamica *f*
MA termodinamika
NL thermodynamica *f*
PO termodynamika *f*
PY термодинамика *f*
SU termodynamiikka
SV termodynamik *c*

3810 thermoelectric *adj*
DA termoelektrisk
DE thermoelektrisch
ES termoeléctrico
F thermoélectrique
I termoelettrico
MA termoelektromos
NL thermoelektrisch
PO termoelektryczny
PY термо-
электрический
SU lämpüsähküinen
SV termoelektrisk

3811 thermograph
DA termograf
DE Temperatur-
    schreiber *m*
    Thermograph *m*
ES termógrafo *m*
    termómetro
    registrador *m*
F thermomètre *m*
    enregistreur
I termografo *m*
MA hőmérsékletíró
    termográf
NL schrijvende
    thermometer *m*
    thermograaf *m*
PO termograf *m*
PY термограф *m*
SU piirtävä
    lämpümittari
    termografi
SV termograf *c*

3812 thermo-hygrograph
DA termohygrograf
DE Thermohygro-
    graph *m*
ES termohigrógrafo *m*
F thermohygromètre
    m enregistreur
I termoigrografo *m*
MA hőmérséklet és
    nedvességíró
    termo-hygrográf
NL thermohygrograaf *m*
PO termohigrograf *m*
PY термогигрограф *m*
    терморегулятор *m*
SU termohygrografi
SV termohygrograf *c*

3813 thermometer
DA termometer
DE Thermometer *n*
ES termómetro *m*
F thermomètre *m*
I termometro *m*
MA hőmérő
NL thermometer *m*
PO termometr *m*
PY термометр *m*
SU lämpümittari
SV termometer *c*

3814 thermometer well
DA termometerlomme
DE Thermometer-
    tauchhülse *f*
ES vaina para
    termómetro
    sumergido *f*
F gaine *f* pour
    thermomètre
    plongeant
I pozzetto *m* del
    termometro
MA hőmérőhüvely
NL thermometerhuls *f*
PO osłona *f*
    termometryczna
    tuleja *f* dla
    termometru
    zanurzonego
PY гильза *f*
    термометра
SU lämpümittaritasku
SV termometerficka *c*

3815 thermos flask
DA termoflaske
DE Thermosflasche *f*
ES botella aislante *f*
    termo *m*
F bouteille *f* isolante
I bottiglia *f* termica
MA termoszpalack
NL thermosfles *f*
PO termos *m*
    wkład *m* termosu
PY термос *m*
SU termospullo
SV termosflaska *c*

3816 thermostat
DA termostat
DE Thermostat *m*
ES termostato *m*
F thermostat *m*
I termostato *m*
MA termosztát
NL thermostaat *m*
PO termostat *m*
PY термостат *m*
SU termostaatti
SV termostat *c*

3817 thermostatic *adj*
DA termostatisk
DE thermostatisch
ES termostatico
F thermostatique
I termostatico
MA termosztatikus
NL thermostatisch
PO termostatyczny
PY термостатический
SU termostaattinen
SV termostatisk

3818 thermostatic trap
DA termostatisk
    vandudlader
DE thermostatischer
    Kondenstopf *m*
ES purgador
    termostático *m*
F purgeur *m* à
    dilatation
    purgeur *m*
    thermostatique
I scaricatore *m* termo-
    statico
MA termosztatikus
    kondenzedény
NL thermostatische
    condenspot *m*
PO odwadniacz *m*
    termostatyczny
PY термостатический
    затвор *m*
SU lämpütilaohjattu
    lauhteenerotin
SV termostatisk
    ångfälla *c*

3819 thermostatic valve
DA termostatventil
DE Thermostatventil *n*
ES válvula
    termostática *f*
F vanne *f*
    thermostatique
I valvola *f*
    termostatica
MA termosztatikus szelep
NL thermostatische
    afsluiter *m*
PO zawór *m*
    termostatyczny
PY термостатический
    клапан *m*
SU termostaattinen
    venttiili
SV termostatventil *c*

3820 thermostat night setback
DA nattermostat
DE Nachtabsenkungs-thermostat *m*
ES reducción nocturno del termostato *f*
F ralenti *m* de nuit
I abbassamento *m* notturno del termostato
MA termosztát éjszakai csükkentett üzemre
NL nachtinstelling *f* thermostaat
 nachttemperatuur-verlaging *f*
PO nastawienie *n* na wartość nocną termostatu
PY термостат *m* ночного снижения
SU termostaatin asetusarvon yün aikainen lasku
SV nattinställnings-termostat *c*

3821 thickness
DA godstykkelse tykkelse
DE Dicke *f* Stärke *f*
ES espesor *m*
F épaisseur *f*
I spessore *m*
MA vastagság
NL dikte *f*
PO grubość *f*
PY толщина *f*
SU paksuus
SV tjocklek *c*

3822 thread
DA gevind tråd
DE Gewinde *n*
ES rosca *f*
F filetage *m*
I filetto *m* filetto *m* maschio
MA csavarmenet
NL draad *m* schroefdraad *m*
PO gwint *m*
PY резьба *f*
SU kierre
SV gänga *c*

3823 thread cutter
DA gevindskæremaskine
DE Gewindeklappe *f* Gewindeschneider *m*
ES terraja *f*
F filière *f*
I filiera *f*
MA menetvágó
NL draadsnijder *m* snijijzer *n*
PO frez *m* do gwintu
PY клупп *m*
SU kierteitin
SV gängverktyg *n*

3824 threaded flange
DA gevindflange
DE Gewindeflansch *m*
ES brida roscada *f*
F bride *f* taraudée (filetée)
I flangia *f* filettata
MA menetes karima
NL draadflens *m*
PO kołnierz *m* gwintowany
PY фланец *m* с резьбой
SU kierteellinen laippa
SV gängfläns *c*

3825 threaded sleeve
DA gevindmuffe
DE Gewindemuffe *f*
ES manguito roscado *m*
F manchon *m* fileté
I manicotto *m* filettato
MA menetes karmantyú
NL draadhuls *f*
PO tuleja *f* gwintowana
PY муфта *f* с резьбой
SU kierteellinen muhvi
SV gängad muff *c*

3826 threaded tube
DA gevindrør
DE Gewinderohr *n*
ES tubo roscado *m*
F tube *m* fileté tube *m* taraudé
I tubo *m* filettato
MA menetes cső
NL draadpijp *f*
PO rura *f* gwintowana
PY труба *f* с резьбой (внутренней)
SU kierreputki (sisäkierre)
SV gängat rür *n*

3827 three pass boiler
DA tretrækskedel
DE Dreizugkessel *m*
ES caldera de tres pasos *f*
F chaudière *f* à trois parcours
I caldaia *f* a tre passaggi
MA háromhuzamú kazán
NL drietreksketel *m*
PO kocioł *m* trójkanałowy kocioł *m* trójprzelotowy
PY котел *m* с тремя оборотами дымовых газов
SU kolmivetoinen kattila
SV tredragspanna *c*

3828 three-pipe air conditioning system
DA trekanal-ventilationssystem
DE Dreileiterklima-system *n*
ES sistema de climatización a tres tubos *m*
F conditionnement *m* d'air à trois tuyaux
I impianto *m* di condizionamento a tre tubi
MA háromcsüves légkondicionáló rendszer
NL driepijpsluchtbehandelings systeem *n*
PO system *m* klimatyzacji trzyrurowy
PY трехтрубная система *f* кондицион-ирования воздуха
SU konvektoreiden kolmiputkikyt-kentä
SV trerörs luftkonditionerings-system *n*

3829 three pipe system
DA trerørs system
DE Dreileitersystem *n*
   Dreirohrsystem *n*
ES sistema de tres
   tubos *m*
F distribution *f* à trois
   tuyaux
I impianto *m* a tre
   tubi
MA háromvezetékes
   rendszer
NL driepijpssysteem *n*
PO system *m*
   trzyrurowy
PY трехтрубная
   система *f*
SU kolmiputkijärje-
   stelmä
SV trerörssystem *n*

3830 three port valve (*see*
three-way valve)

3831 three-way cock
DA trevejshane
DE Dreiwegehahn *m*
ES llave de tres vias *f*
F robinet *m* à trois
   voies
I rubinetto *m* a tre vie
MA kétutú csap
NL driewegplugkraan *f*
PO kurek *m*
   trójdrogowy
PY трехходовой
   кран *m*
SU kolmitietulppahana
SV trevägskran *c*

3832 three-way valve
DA trevejsventil
DE Dreiwegeventil *n*
ES válvula de tres vías *f*
F vanne *f* à trois voies
I valvola *f* a tre vie
MA kétutú szelep
NL driewegafsluiter *m*
   driewegklep *f*
PO zawór *m*
   trójdrogowy
PY трехходовой
   вентиль *m*
SU kolmitieventtiili
SV trevägsventil *c*

**3833 threshold limit
value**
DA tærskelværdi
DE zulässiger
   Grenzwert *m*
ES valor de umbral *m*
   valor límite
   admisible *m*
F valeur *f* limite
   admissible
I valore *m* limite
   massimo
MA küszübérték
NL drempelwaarde *f*
PO wartość *f* progowa
PY пороговая
   предельная
   величина *f*
SU suurin sallittu arvo
SV trüskelvärde *n*

**3834 threshold of
hearing**
DA høregrænse
   høretærskel
DE Hürschwelle *f*
ES umbral de
   audibilidad *m*
F seuil *m* d'audibilité
I soglia *m* di udibilità
MA halláküszüb
NL gehoordrempel *m*
PO próg *m* słyszalności
PY порог *m*
   слышимости
SU kuulokynnys
SV hürseltrüskel *c*

3835 throttle
DA spjæld
DE Drossel *f*
   Drosselgerät *n*
ES estrangulador *m*
   mariposa de
   válvula *f*
F étranglement *m*
   étrangleur *m*
I strozzatura *f*
MA fojtószelep
NL smoorklep *f*
PO przepustnica *f*
   zawór *m* dławiący
PY дроссель *m*
SU kuristuspelti
SV spjäll *n*

3836 throttle *vb*
DA drosle
   drøvle
DE drosseln
ES estrangular
F étrangler
I modulare
   strozzare
MA fojt
NL smoren
PO dławić
PY дросселировать
SU kuristaa
SV strypa

3837 throttling
DA drosling
   drøvling
DE Drosselung *f*
ES estrangulación *f*
F étranglement *m*
I modulazione *f*
   strozzamento *m*
   trafilazione *f*
MA fojtás
NL smoren *n*
PO dławienie *n*
PY дросселирование *n*
SU kuristus
SV strypning *c*

3838 throughput
DA gennemstrømning
DE Durchsatz *m*
ES caudal *m*
F débit *m*
I portata *f*
   produzione *f* oraria
MA átmenő teljesítmény
NL debiet *n*
PO przerób *m*
   szybkość *f*
   przesyłania
   danych
   wydajność *f*
   (produkcyjna)
   zdolność *f*
   przepustowa
PY отдача *f*
SU läpivirtaus
SV genomströmning *c*

DA	=	Danish
DE	=	German
ES	=	Spanish
F	=	French
I	=	Italian
MA	=	Magyar (Hungarian)
NL	=	Dutch
PO	=	Polish
PY	=	Russian
SU	=	Finnish
SV	=	Swedish

3839 through-the-wall air conditioner
DA klimaagregat i murgennemføring
DE Wandeinbau-klimagerät *n*
ES acondicionador de aire de muro *m*
F conditionneur *m* d'air de paroi
I condizionatore *m* split
MA fali átvezetésű légkondicionáló
NL gevelkamerkoeler *m*
PO klimatyzator *m* ścienny
PY кондиционер *m*, устанавливаемый в проеме стены
SU seinään asennettu ilmastointikone
SV väggmonterad luftkonditionerings-apparat *c*

3840 through way cock
DA ligeløbsventil
DE Durchgangshahn *m* Durchgangsventil *n*
ES grifo de paso completo *m* grifo de paso directo *m*
F robinet *m* à passage direct
I rubinetto *m* a passaggio diretto
MA áteresztő csap
NL afsluiter *m* met volle doorlaat vrijstroom-afsluiter *m*
PO kurek *m* przelotowy
PY проходной кран *m*
SU täysaukkoinen tulppaventtiili
SV genomgångskran *c*

3841 throw
DA kastelængde slaglængde
DE Kolbenhub *m*
ES recorrido *m* enbolada *f* dardo *m*
F jet *m* lancement *m*
I gittata *f* lancio *m*
MA vetés
NL worp *m*
PO rzut *m*
PY сброс *m* ход *m*
SU ilmasuihkun pituus
SV kast *n* slag *n*

3842 thrust bearing
DA trykleje
DE Drucklager *n*
ES cojinete axial *m*
F palier *m* de butée
I cuscinetto *m* reggispinta
MA talpcsapágy
NL druklager *n*
PO łożysko *n* oporowe łożysko *n* wzdłużne
PY подшипник , воспринимающий осевое усилие
SU aksiaalilaakeri
SV trycklager *n*

3843 tie bar (*see* tie rod)

3844 tie rod
DA stagbolt styrestang
DE Druckstange *f* Zugstange *f*
ES tirante *m*
F barre *f* d'assemblage entretoise *f*
I asta *f* di collegamento tirante *m*
MA üsszekütő rúd
NL trekstaaf *f* trekstang *f*
PO cięgno *n* pręt *m* mocujący ściąg *m*
PY распорка *f* соединительная тяга *f*
SU tuki vetotanko
SV dragstång *c*

3845 tightness
DA tæthed
DE Dichtheit *f* Dichtigkeit *f*
ES estanqueidad *f* hermeticidad *f*
F étanchéité *f*
I tenuta *f*
MA szorosság
NL dichtheid *f*
PO szczelność *f*
PY плотность *n* затяжки
SU tiiveys
SV täthet *c*

3846 time
DA takt tid
DE Dauer *f* Zeit *f*
ES duración *f* tiempo *m*
F durée *f* période *f* temps *m*
I tempo *m*
MA idő
NL maal *f* tijd *m*
PO czas *m*
PY время *n*
SU aika
SV tid *c*

3847 time clock
DA koblingsur
DE Zeitschaltuhr *f*
ES reloj de control *m* reloj programa-dor *m*
F horloge *f* à contacts
I orologio *m* temporizzatore *m*
MA időkapcsolóóra
NL schakelklok *f*
PO zegar *m* czasowy (operacji)
PY часы *n*
SU ohjelmakello
SV kontrollur *n*

3848 time clock operation
DA tidsstyret drift
DE Schaltuhr-
   steuerung *f*
ES accionamiento por
   reloj programa-
   dor *m*
F commande *f* par
   horloge
I comando *m* a
   orologeria
   comando *m* a tempo
MA kapcsolóórás
   műküdés
NL schakelklok-
   bediening *f*
   schakelklokbedrijf *n*
PO działanie *n*
   ogranicznika
   czasowego
PY управление *n* по
   времени
SU aikaohjattu käyttü
SV tidsstyrd drift *c*

3849 time constant
DA tidskonstant
DE Zeitkonstante *f*
ES constante de
   tiempo *f*
F constante *f* de temps
I costante *f* di tempo
MA időállandó
NL tijdconstante *f*
PO stała *f* czasowa
PY постоянная *f*
   времени
SU aikavakio
SV tidskonstant *c*

3850 time defrosting
DA tidstyret afrimning
DE zeitabhängiges
   Abtauen *n*
ES tiempo de
   desescarche *m*
F dégivrage *m* par
   chronorelais
I sbrinamento *m*
   temporizzato
MA időbefagyasztás
NL tijdafhankelijk
   ontdooien *n*
PO rozmrażanie *n*
   czasowe
PY время *n*
   размораживания
SU kellosulatus
SV tidsavfrostning *c*

3851 time delay (*see* time
   lag)

3852 time lag
DA tidsforsinkelse
DE Zeitverzügerung *f*
ES retardo *m*
   tiempo de retardo *m*
   tiempo muerto *m*
F délai *m*
   temps *m* de retard
I sfasamento *m*
   tempo *m* di ritardo
MA időkésleltetés
NL tijdvertraging *f*
PO opóźnienie *n*
   spóźnienie *n*
   zwłoka *f*
PY сдвиг *m* по
   времени
SU viive
SV tidsfürdrüjning *c*

3853 time sharing
DA tidsfordeling
DE Time-Sharing *n*
   Zeitteilung *f*
ES tiempo de reparto *m*
F temps *m* partagé
   utilisation *f*
   collective
   (ordinateur)
I time sharing *m*
MA időosztás
NL time sharing *n*
PO praca *f* (komputera)
   z podziałem czasu
PY временной
   интервал *m*
   отрезок *m* времени
SU
SV tidsdelning *c*

3854 time switch
DA kontaktur
DE Zeitschalter *m*
ES interruptor
   horario *m*
F interrupteur *m* à
   minuterie
   minuterie *f* à
   commande
I interruttore *m*
   orario
MA időkapcsoló
NL schakelklok *f*
   tijdschakelaar *m*
PO wyłącznik *m*
   czasowy
PY реле *n* времени
SU aikakytkin
SV tidstrümställare *c*

3855 tin
DA tin
DE Zinn *n*
ES estaño *m*
F étain *m*
I stagno *m*
MA cin
   ón
NL blik *n*
   tin *n*
PO cyna *f*
PY олово *n*
SU tina
SV tenn *n*

3856 tinning
DA fortinning
DE Verzinnung *f*
ES estañado *m*
F étamage *m*
I stagnatura *f*
MA dobozolás
   ónozás
NL vertinnen *n*
PO cynowanie *n*
PY лужение *n*
SU tinaus
SV fürtenning *c*

3857 tip
DA spids
DE Ende *n*
   Spitze *f*
ES borde (del álabe) *m*
   punta de la pala *f*
F extrémité *f*
I cresta *f*
   inclinazione *f*
   punta *f*
MA csúcs
NL uiteinde *n*
PO końcówka *f*
   zakończenie *n*
PY край *m* лопатки
   вентилятора
SU kärki (puhaltimen
   siiven)
SV topp *c*

3858 tip speed
DA periferihastighed
DE Umfangsgeschwindig-
    keit *f*
ES velocidad de la
    rotación en la
    punta de la pala *f*
    velocidad
    periférica *f*
F vitesse *f*
    périphérique
I velocità *f* periferica
MA kerületi sebesség
NL omtreksnelheid *f*
PO prędkość *f*
    obwodowa
    szybkobieżność *f*
PY окружная скорость
    f рабочего колеса
SU kehänopeus
SV periferihastighet *c*

3859 toggle action
DA vippebevægelse
DE Kipp-
    mechanismus *m*
ES actuación sobre
    interruptor de
    palanca *f*
F action *f* à déclic
I commutazione *f*
MA keresztműküdés
NL doorschakelactie *f*
    omschakelactie *f*
PO działanie *n*
    przegubowe
PY изменение *n*
    положения
    переключателя
SU keinutoiminta
SV knäledsverkan *c*

3860 toggle switch
DA vippeafbryder
DE Kippschalter *m*
ES interruptor de
    palanca *m*
F basculeur *m*
I commutatore *m*
MA tumblerkapcsoló
NL wipschakelaar *m*
PO przełącznik *m*
    dwustabilny
    przełącznik *m*
    migowy
    przechylny
PY положение *n*
    переключателя
SU keinukytkin
SV vippströmbrytare *c*

3861 tongs
DA tang
DE Zange *f*
ES pinza *f*
    tenaza *f*
F pince *f*
I pinza *f*
    tenaglia *f*
MA fogó
NL tang *f*
PO kleszcze *pl*
    szczypce *pl*
PY грейфер *m*
    клещи *n*
SU tongit
SV tång *c*

3862 tongue and groove
    facing flange
DA tap-not-flange
DE Nut-und-Feder-
    Flansch *m*
ES brida de ranura y
    lengüeta encaradas
    m
F bride *f* à
    emboîtement
    double
I flangia *f* con
    giunzione a
    maschio e
    femmina
MA nútféderes perempár
NL tong en groef
    flens *m*
PO kołnierz *m* z
    występem i
    rowkiem
PY фланец *m* с
    ответной
    канавкой и
    выступом
SU ohjausurallinen
    laippa
SV spont-planfläns *c*

3863 tool
DA værktøj
DE Werkzeug *n*
ES herramienta *f*
F outil *m*
I utensile *m*
MA szerszám
NL gereedschap *n*
    werktuig *n*
PO narzędzie *n*
PY инструмент *m*
SU työkalu
SV verktyg *n*

3864 topping up
DA efterfyldning
DE Nachfüllung *f*
ES llenado a tope *m*
F chargement *m*
    complet
I aggiunta *f* di acqua
    distillata
    riempimento *m*
MA feltültés
NL bijvullen *n*
PO dopełnienie *n*
    uzupełnienie *n*
PY пополнение *n*
SU jälkitäyttü
    lisääminen
SV påfyllning *c*

3865 torque
DA vridningsmoment
DE Drehmoment *n*
ES momento de
    torsión *m*
    par de torsión *m*
F couple *m*
I coppia *f*
    momento *m*
    torcente
MA forgatónyomaték
NL draaimoment *n*
    koppel *m*
PO moment *m*
    obrotowy
PY момент *m*
    вращения
SU vääntümomentti
SV vridmoment *n*

3866 total efficiency (*see*
    overall efficiency)

3867 total energy
DA totalenergi
DE Gesamtenergie *f*
ES energía total *f*
F énergie *f* totale
I energia *f* totale
MA teljes energia
NL total-energy *n*
PO energia *f* całkowita
PY полная энергия *f*
SU kokonaisenergia
SV total energi *c*

3868 total energy system
DA totalenergisystem
DE Total-Energie-System *n*
ES sistema de energía total *m*
F système *m* d'énergie totale
I sistema *m* ad energia totale
MA teljes energiarendszer
NL total-energy systeem *n*
PO system *m* wykorzystujący energię całkowitą
PY полная энергия *f* системы
SU kokonaisenergiajärjestelmä
SV totalenergisystem *n*

3869 total head
DA total trykhøjde
DE Gesamtfürderhühe *f*
ES altura de elevación total *f*
altura manométrica total *f*
carga total *f*
F hauteur *f* manométrique totale
I pressione *f* totale
MA üsszes nyomómagasság
NL totale drukhoogte *f* totale opvoerhoogte *f*
PO wysokość *f* całkowita
PY полная высота *f* (водяного столба) полное давление *n* (в метрах водяного столба)
SU kokonaispaine
SV total uppfordringshüjd *c*

3870 total heat output
DA total varmeydelse
DE Gesamtwärmeleistung *f*
ES potencia térmica total *f*
F puissance *f* thermique totale
I potenza *f* termica totale
MA üsszes hőteljesítmény
NL totale warmteproduktie *f*
PO wydajność *f* ciepła całkowita
PY полная теплопроизводительность *f*
SU kokonaislämpüteho
SV total värmeeffekt *c*

3871 total pressure
DA totaltryk
DE Gesamtdruck *m*
ES presión total *f*
F pression *f* totale
I pressione *f* totale
MA üssznyomás
NL totale druk *m*
PO ciśnienie *n* całkowite
PY полное давление *n*
SU kokonaispaine
SV totaltryck *n*

3872 towel rail
DA håndklædeholder
DE Handtuchhalter *m*
ES secador de toallas *m* toallero *m*
F porte-serviette *m*
I portasalviette *m*
MA türülküzőtartó
NL handdoekenrek *n*
PO wieszak *m* do ręczników rurowy
PY полотенцедержатель *m*
SU kuivauspatteri
SV handdukshållare *c*

3873 town gas
DA bygas
DE Stadtgas *n*
ES gas ciudad *m*
F gaz *m* de ville
I gas *m* di città
MA városi gáz
NL stadsgas *n*
PO gaz *m* miejski
PY светильный газ *m*
SU kaupunkikaasu
SV stadsgas *c*

3874 tracer
DA indikator sporstof
DE Indikator *m* Spurengeber *m*
ES indicador *m* trazador *m*
F traceur *m*
I tracciatore *m*
MA előrajzoló nyomjelző
NL indicator *m* tracer *m*
PO traser *m* wskaźnik *m*
PY регистрирующее устройство *n* трассирующий индикатор *m*
SU merkkiaine
SV spårämne *n*

3875 tracer gas
DA sporgas
DE Spurengas *n*
ES gas trazador *m*
F gaz *m* traceur
I gas *m* tracciante
MA nyomjelző gáz
NL tracergas *n*
PO gaz *m* wskaźnikowy
PY трассирующий газ *m*
SU merkkikaasu
SV spårgas *c*

3876 trailing edge thickness
DA bagkanaltykkelse
DE Hinterkantenstärke *f*
ES espesor del borde de salida *m*
F épaisseur *f* du bord aval
I spessore *m* del bordo di uscita
MA kilépő él vastagság
NL achterranddikte *f*
PO grubość *f* krawędzi wypływu
PY толщина *f* кромки ковша
SU jättüreunan paksuus
SV bakkantstjocklek *c*

3877 transducer
DA energioverfører
transducer
DE Meßwertgeber *m*
ES transductor *m*
F transducteur *m*
I trasduttore *m*
MA átalakító
transzduktor
NL signaalomzetter *m*
PO przetwornik *m*
PY преобразователь *m*
SU muunnin
SV transduktor *c*

3878 transfer
DA overføring
DE Übertragung *f*
ES transportar
trasegar
F transfert *m*
I trasferimento *m*
trasporto *m*
MA átadás
küzlés
NL overdracht *f*
PO przekazywanie *n*
przenoszenie *n*
PY перенос *m*
SU siirto
SV övergång *c*

3879 transfer *vb*
DA overføre
DE übertragen
ES transferir
transmitir
F échanger
transférer
transmettre
I trasferire
trasmettere
MA átad
küzül
NL overdragen
PO przekazywać
przenosić
PY передавать
переносить
SU siirtää
SV övergå

DA	= Danish
DE	= German
ES	= Spanish
F	= French
I	= Italian
MA	= Magyar (Hungarian)
NL	= Dutch
PO	= Polish
PY	= Russian
SU	= Finnish
SV	= Swedish

3880 transfer function
DA overføringsfunktion
DE Übertragungs-
funktion *f*
ES función de
transporte/
trasiego *f*
F fonction *f* de
transfert
I funzione *f* di
trasferimento
MA átviteli függvény
NL overdrachtsfunctie *f*
PO funkcja *f*
przeniesienia
przepustowość *f*
PY передаточная
функция *f*
SU siirtofunktio
SV överfürings-
funktion *c*

3881 transfer pump
DA transportpumpe
DE Fürderpumpe *f*
ES bomba de trasiego *f*
F pompe *f* de transfert
I pompa *f* da travaso
MA átemelő szivattyú
NL transportpomp *f*
PO pompa *f*
przetłaczająca
PY перекачивающий
насос *m*
SU üljypumppu
SV oljepump *c*

3882 transformer
DA omformer
transformer
DE Transformator *m*
ES transformador *m*
F transformateur *m*
I trasformatore *m*
MA transzformátor
NL transformator *m*
PO transformator *m*
PY трансформатор *m*
SU muuntaja
SV transformator *c*

3883 transient heat flow
DA variabel varmestrøm
DE veränderlicher
Wärmestrom *m*
ES flujo de calor en
régimen
transitorio *m*
F flux *m* de chaleur
variable
I flusso *m* transitorio
di calore
MA átmeneti hőáram
NL tijdelijke
warmtestroom *m*
PO strumień *m* ciepła
nieustalony
PY нестационарный
тепловой
поток *m*
неустановившийся
тепловой
поток *m*
SU muuttuva lämpüvirta
SV variabelt
värmeflüde *n*

3884 transistor
DA transistor
DE Transistor *m*
ES transistor *m*
F transistor *m*
I transistor *m*
MA tranzisztor
NL transistor *m*
PO tranzystor *m*
PY транзистор *m*
SU transistori
SV transistor *c*

3885 transition flow
DA overgangsstrømning
DE Übergangs-
strümung *f*
ES flujo de transición *m*
F régime *m*
d'écoulement
transitoire
I flusso *m* di
transizione
MA átmenő áramlás
NL overgangsstroming *f*
PO przepływ *m*
przejściowy
ruch *m* przejściowy
PY проходящий
поток *m*
SU ylimenoalueen
virtaus
SV genomstrümning *c*

3886 transition piece
DA overgangsstykke
DE Übergangsstück *n*
ES adaptador *m*
   pieza de reducción *f*
   pieza de transición *f*
F pièce *f* de réduction
I riduzione *f*
MA átmeneti idom
NL overgangsstuk *n*
PO kształtka *f*
   przejściowa
PY переходной
   элемент *m*
SU muunnoskappale
SV üvergångsstycke *n*

3887 transition region
DA overgangsområde
DE Übergangsgebiet *n*
ES area de transición *f*
F domaine *m* de
   régime transitoire
I regione *f* di
   transizione
MA átmeneti tartomány
NL overgangsgebied *n*
PO obszar *m*
   przejściowy
   strefa *f* przejściowa
PY переходная
   область *f*
SU muutosalue
SV üvergångsområde *n*

3888 transmission
DA overføring
   transmission
DE Transmission *f*
   Übertragung *f*
ES transmisión *f*
F transmission *f*
I trasmissione *f*
MA átbocsátás
   átvitel
NL overdracht *f*
   transmissie *f*
PO przekazywanie *n*
   przesyłanie *n*
PY передача *f*
   трансмиссия *f*
SU läpäisy
SV transmission *c*

3889 transmissivity
DA overføringsevne
DE Übertragbarkeit *f*
ES coeficiente de
   transmisión *m*
F transmissivité *f*
I coefficiente *m* di
   trasmissione
MA áteresztőképesség
NL doorgelaten deel *n*
   van opvallende
   straling
   stralings-
   transmissie *f*
PO przepuszczalność *f*
PY проницаемость *f*
SU läpäisevyys
SV transmissions-
   fürmåga *c*

3890 transmitter
DA sendeapparat
DE Sender *m*
   Übermittler *m*
ES transmisor *m*
F émetteur *m*
   transmetteur *m*
I trasmettitore *m*
MA leadó
NL overdrager *m*
   zender *m*
PO nadajnik *m*
PY отправитель *m*
   (груза)
   передаточный
   механизм *m*
SU lähetin
SV givare *c*

3891 transpiration cooling
DA transpirationskøling
DE Verdunstungs-
   kühlung *f*
ES enfriamiento por
   transpiración *f*
F refroidissement *m*
   par transpiration
I raffreddamento *m*
   per traspirazione
MA izzadásos hűtés
NL transpiratiekoeling *f*
PO chłodzenie *n*
   transpiracyjne
PY испарительное
   охлаждение *n*
SU jäähdytys
   hikoilemalla
SV svettkylning *c*

3892 trap
DA fælde
   udlader
DE Geruchverschluß *m*
   Kondenstopf *m*
ES dispositivo de
   retención *m*
   interceptor *m*
   purgador *m*
   separador *m*
F purgeur *m*
   séparateur *m*
   siphon *m*
I scaricatore *m* di
   condensa
   sifone *m*
MA kondenzedény
   vízzár
NL condenspot *m*
   sifon *m*
   stankafsluiter *m*
PO syfon *m*
   kanalizacyjny
PY сифон *m*
   трап *m*
SU erotin
   lauhteen erotin
   vesilukko
SV fälla *c*

3893 trap door
DA falddør
DE Falltür *f*
ES trampilla *f*
F trappe *f*
I botola *f*
MA csapóajtó
NL luik *n*
   valdeur *f*
PO rewizja *f* syfonu
PY сифон-ревизия *m*
SU nosto-ovi
SV fallucka *c*

3894 travelling grate
DA vandrerist
DE Wanderrost *m*
ES parrilla de cadena
   sin fín *f*
F grille *f* mécanique
   ou roulante
I griglia *f* mobile
MA vándorrostély
NL kettingrooster *m*
PO ruszt *m* taśmowy
PY подвижная
   колосниковая
   решетка *f*
SU ketjuarina
SV kedjerost *c*

3895 treated air
DA behandlet luft
DE behandelte Luft *f*
ES aire tratado *m*
F air *m* traité
I aria *f* trattata
MA kezelt levegő
NL behandelde lucht *f*
PO powietrze *n* uzdatnione
PY обработанный воздух *m*
SU käsitelty ilma
SV behandlad luft *c*

3896 treatment
DA behandling
DE Behandlung *f*
ES tratamiento *m*
F traitement *m*
I trattamento *m*
MA kezelés
NL behandeling *f*
PO obróbka *f* uzdatnianie *n*
PY обработка *f*
SU käsittely
SV behandling *c*

3897 trial run
DA prøvedrift prøvekørsel
DE Probelauf *m*
ES funcionamiento de prueba *m* marcha de prueba *f*
F marche *f* d'essai
I corsa *f* di prova
MA próbajáratás
NL proefdraaien *n* proefvaart *f*
PO rozruch *m* próbny ruch *m* próbny
PY пробный ход *m*
SU koekäyttü
SV provkürning *c*

3898 trunnion
DA omdrejningstap tap
DE Zapfen *m*
ES muñón *m* pivote *m* soporte giratorio *m*
F pignon *m*
I perno *m* di articolazione pignone *m*
MA forgócsap
NL taats *f*
PO czop *m* zawieszenia obrotowego
PY шип *m*
SU tappi
SV tapp *c* med invändig gänga *c*

3899 tube
DA rør
DE Rühre *f* Rohr *n*
ES tubo *m*
F tube *m* tuyau *m*
I tubo *m*
MA cső
NL buis *f* pijp *f*
PO rura *f*
PY труба *f*
SU putki
SV rür *n*

3900 tubeaxial fan
DA aksialventilator i rør
DE Axialventilator *m*
ES ventilador tubular-axial *m*
F ventilateur *m* axial à enveloppe
I ventilatore *m* assiale intubato
MA csőventilátor
NL axiaalkanaal-ventilator *m* buisventilator *m*
PO wentylator *m* osiowy kanałowy
PY осевой вентилятор *m* (с двигателем в обечайке)
SU aksiaalipuhallin potkuripuhallin
SV rür-axialfläkt *c*

3901 tube bundle
DA rørbundt
DE Rohrbündel *n*
ES haz de tubo *m* haz tubular *m*
F faisceau *m* tubulaire
I fascio *m* tubiero
MA csőküteg
NL pijpenbundel *m*
PO wiązka *f* rur
PY трубный пучок *m*
SU putkinippu
SV rürknippe *n*

3902 tube friction
DA rørfriktion rørmodstand
DE Rohrreibung *f*
ES rozamiento en el tubo *m*
F perte *f* de charge répartie
I attrito *m* del tubo
MA csősúrlódás
NL leidingweerstand *m* pijpweerstand *m*
PO tarcie *n* w rurze
PY трение *n* в трубах
SU putkikitka
SV rürfriktion *c*

3903 tube plate
DA rørplade
DE Rührenplatte *f*
ES placa tubular *f*
F plaque *f* tubulaire
I piastra *f* tubiera
MA csőfal fenéklap
NL pijpenplaat *f*
PO ściana *f* sitowa
PY трубная доска *f*
SU päätylevy
SV tubplåt *c*

3904 tunnel cooler
DA tunnelkøler
DE Tunnelkühler *m*
ES tunel enfriador *m*
F tunnel *m* de réfrigération
I refrigeratore *m* a tunnel
MA alagúthűtő
NL tunnelkoeler *m*
PO chłodnia *f* tunelowa
PY охладитель *m* с цилиндрической поверхностью
SU tunnelijäähdytin
SV tunnelkyl *c*

3905 tunnel freezer
DA tunnelfryser
DE Tunnelfroster *m*
ES tunel congelador *m*
F tunnel *m* de congélation
I congelatore *m* a tunnel
MA alagútfagyasztó
NL tunnelinvriezer *m*
PO zamrażarnia *f* tunelowa
PY замораживатель *m* с цилиндрической поверхностью
SU tunnelipakastin
SV tunnelfrys *c*

3906 turbine
DA turbine
DE Turbine *f*
ES turbina *f*
F turbine *f*
I turbina *f*
MA turbina
NL turbine *f*
PO turbina *f*
PY турбина *f*
SU turbiini
SV turbin *c*

3907 turbo blower
DA turboblæser
DE Turbogebläse *n*
ES turbosoplante *m* turboventilador *m*
F turbo soufflante *f*
I turbosoffiante *f*
MA turbofúvó
NL turboventilator *m*
PO dmuchawa *f* wirnikowa turbodmuchawa *f*
PY турбовоздуходувка *f*
SU turbopuhallin
SV turbofläkt *c*

3908 turbo compressor
DA turbokompressor
DE Turboverdichter *m*
ES turbo compresor *m*
F turbo compresseur *m*
I turbocompressore *m*
MA turbokompresszor
NL centrifugaal-compressor *m* turbocompressor *m*
PO turbosprężarka *f*
PY турбо-компрессор *m*
SU turbokompressori
SV turbokompressor *c*

3909 turbo generator
DA turbogenerator
DE Turbogenerator *m*
ES turbo generador *m*
F turbo générateur *m*
I turbogeneratore *m*
MA turbogenerátor
NL turbogenerator *m*
PO prądnica *f* turbinowa turbogenerator *m*
PY турбогенератор *m*
SU turbogeneraattori
SV turbogenerator *c*

3910 turbulence
DA hvirvelbevægelse turbulens
DE Turbulenz *f*
ES turbulencia *f*
F turbulence *f*
I turbolenza *f*
MA ürvénylés turbulencia
NL turbulentie *f* werveling *f*
PO burzliwość *f* turbulencja *f*
PY турбулентность *f*
SU turbulenssi
SV turbulens *c*

3911 turbulent flow
DA turbulent strømning
DE turbulente Strümung *f*
ES corriente turbulenta *f* flujo turbulento *m*
F écoulement *m* turbulent
I flusso *m* turbolento
MA turbulens áramlás
NL turbulente stroming *f*
PO przepływ *m* burzliwy przepływ *m* turbulentny
PY турбулентный поток *m*
SU turbulenttinen virtaus
SV turbulent strümning *c*

3912 turning vane
DA drejelig vinge
DE Wendeschaufel *f*
ES álabe direccional *m*
F
I pala *f* rotante
MA terelőlemez
NL
PO łopatka *f* obrotowa
PY поворотная лопатка *f*
SU ohjaussiipi
SV vridbar skovel *c*

3913 twin cylinder compressor
DA to-cylinder kompressor
DE Zwillingszylinder-Verdichter *m*
ES compresor bicilíndrico *m*
F compresseur *m* à deux cylindres
I compressore *m* bicilindrico
MA ikerhengeres kompresszor
NL tweecilinder-compressor *m*
PO sprężarka *f* dwucylindrowa
PY двухцилиндровый компрессор *m*
SU kaksisylinterinen kompressori
SV dubbelcylinder-kompressor *c*

3914 two phase flow
DA tofaset strøm
DE Zweiphasen-strümung *f*
ES flujo de dos fases *m*
F débit *m* biphasique
I flusso *m* bifase
MA kétfázisú áramlás
NL tweefasenstroming *f*
PO przepływ *m* dwufazowy
PY двухфазный поток *m*
SU kaksifaasivirtaus
SV tvåfasflüde *n*

3915 two pipe heating
DA tø-rørs-varmeanlæg
DE Zweirohrheizung *f*
ES calefacción a dos
tubos *f*
F chauffage *m* à deux
tuyaux
I riscaldamento *m* a
due tubi
MA kétcsüves fűtés
NL tweepijpsverwarmings-
systeem *n*
PO ogrzewanie *n*
dwururowe
PY двухтрубное
отопление *n*
SU kaksiputkilämmitys
SV tvårürs
värmesystem *n*

3916 two port valve (*see*
two-way valve)

3917 two-way switch
DA vendekontakt
DE Wechselschalter *m*
ES conmutador *m*
F commutateur *m* de
va et vient
inverseur *m*
I commutatore *m* a
due vie
MA váltókapcsoló
NL omschakelaar *m*
wisselschakelaar *m*
PO wyłącznik *m*
dwupozycyjny
PY двухходовой
выключатель *m*
SU kaksiasentokytkin
SV tvåvägsomkastare *c*

3918 two-way valve
DA tovejsventil
DE Zweiwegeventil *n*
ES válvula de dos vías *f*
F vanne *f* deux voies
I valvola *f* a due vie
MA egyutú szelep
NL tweewegafsluiter *m*
PO zawór *m*
dwudrogowy
PY двухходовой
клапан *m*
SU
SV tvåvägsventil *c*

3919 two wire circuit
DA tolederskreds
DE Zweileiter-System *n*
ES circuito de dos hilos
m
F circuit *m* à deux
conducteurs
I circuito *m* a due fili
MA kétvezetékes
áramkür
NL tweedraadscircuit *n*
PO łącze *n*
dwuprzewodowe
łącze *n* jednofazowe
PY двухпроводная
цепь *f*
SU kaksijohtokytkentä
SV parledning *c*

DA	=	Danish
DE	=	German
ES	=	Spanish
F	=	French
I	=	Italian
MA	=	Magyar (Hungarian)
NL	=	Dutch
PO	=	Polish
PY	=	Russian
SU	=	Finnish
SV	=	Swedish

U

3920 underfeed stoker
DA underfyringsstoker
DE Unterschub-
   feuerung *f*
ES hogar de carga
   inferior *m*
F foyer *m* à
   chargement par
I focolare *m* a
   caricamento dal
   basso
MA alátolásos tüzelés
NL stookinrichting *f*
   met onderaanvoer
PO ruszt *m*
   (mechaniczny)
   podsuwowy
PY подача *f* топлива
   снизу
SU alasyüttüinen
SV stoker *c* med
   undermatning *c*

**3921 under grate air
supply**
DA underblæser
DE Unterwind *m*
ES alimentación de aire
   bajo la parrilla *f*
F alimentation *f* d'air
   sous grille
I alimentazione *m*
   aria sotto
   graticola
MA rostély aláfúvás
NL onderwind *m*
PO podmuch *m* dolny
PY подача *f* воздуха в
   рабочую зону
   подача *f* воздуха
   снизу
SU puhallus arinan alta
SV undre lufttillfürsel *c*

3922 underground tank
DA jordtank
   underjordisk tank
DE unterirdischer
   Lagertank *m*
ES depósito
   enterrado *m*
F réservoir *m* enterré
   (souterrain)
I serbatoio *m*
   interrato
MA füldalatti tartály
NL ondergrondse
   tank *m*
PO zbiornik *m*
   podziemny
PY подземный
   резервуар *m*
SU maanalainen säiliü
SV underjordscistern *c*

**3923 uniflow
compressor**
DA modstrøms-
   kompressor
DE Gleichstrom-
   kompressor *m*
ES compresor
   monoflujo *m*
F compresseur *m* à
   équicourant
I compressore *m*
   monoflusso
MA egyenáramú
   kompresszor
NL gelijkstroom-
   compressor *m*
PO sprężarka *f* tłokowa
   jednostrumien-
   iowa
PY прямоточный
   компрессор *m*
SU yhdensuuntaisvirtaus-
   kompressori
SV likströms-
   kompressor *c*

3924 union
DA fagforening
   forskruning
DE Schraub-
   verbindung *f*
   Verbindung *f*
ES unión *f*
F raccord *m*
   union *f*
I raccordo *m*
MA csőkütés
   egyesítés
NL koppelstuk *n*
   pijpkoppeling *f* met
   wartelmoer
   union *m*
PO połączenie *n*
   złącze *n*
PY соединение *n*
SU putkiliitos
SV anslutning *c*

3925 unit
DA aggregat
   enhed
DE Einheit *f*
ES aparato *m*
   elemento *m*
   grupo *m*
   unidad *f*
F élément *m*
   groupe *m*
   unité *f*
I apparecchio *m*
   unità *f*
MA egység
NL eenheid *f*
   pasklaar apparaat *n*
PO jednostka *f*
   urządzenie *n*
PY агрегат *m*
SU yksikkü
SV enhet *c*

3926 unitary *adj*
DA komplet
DE Einheits-
ES unitario
F unitaire
I unitario
MA egyedi
NL een geheel vormend
PO jednolity
jednostkowy
PY агрегатированный
SU yhtenäinen
SV enhetlig

3927 unitary air conditioner
DA komplet
præfabrikeret
klimaanlæg
DE einbaufertiges
Klimagerät *n*
ES acondicionador de
aire compacto *m*
F conditionneur *m*
d'air indépendant
I condizionatore *m*
autonomo
MA egyedi
légkondicionáló
NL modulair
luchtbehandelings-
apparaat *n*
PO klimatyzator *m*
PY агрегатированный
кондиционер *m*
SU tehtaalla koottu
yhtenäinen
ilmastointikone
SV enhetlig
klimatapparat *c*

3928 unit cooler
DA køleaggregat
DE Kühlgerät *n*
ES enfriador de aire *m*
frigorígeno *m*
F unité *f* de
refroidissement
I unità *f* di
raffreddamento
(autonoma)
MA egységhütő
NL koelapparaat *n*
PO chłodnica *f*
zespolona
PY охладительный
агрегат *m*
SU jäähdytysyksikkü
SV

3929 unit heater
DA luftvarmer
DE Heizgerät *n*
ES aerotermo *m*
F aérotherme *m*
I unità *f* di
riscaldamento
(autonoma)
MA egységfütő
NL luchtverhitter *m*
PO zespół *m* ogrzewczy
PY отопительный
агрегат *m*
SU lämminilmakone
SV luftvärmare *c*

3930 unit of measurement
DA måleenhed
DE Meßeinheit *f*
ES unidad de
medición *f*
F unité *f* de mesure
I unità *f* di misura
MA mértékegység
NL maateenheid *f*
PO jednostka *f* miary
PY единица *f*
измерения
SU mittayksikkü
SV mätenhet *c*

3931 universal burner
DA universalbrænder
DE Allesbrenner *m*
ES quemador
universal *m*
F brûleur *m* tout
combustible
I bruciatore *m*
universale
MA univerzális égő
NL universele
brander *m*
PO palnik *m*
uniwersalny
PY универсальная
горелка *f*
SU yleispoltin
SV universalbrännare *c*

3932 unloaded start (*see* no-load start)

3933 untreated water
DA ubehandlet vand
DE unbehandeltes
Wasser *n*
ES agua no tratada *f*
F eau *f* non traitée
I acqua *f* non trattata
MA kezeletlen víz
NL onbehandeld
water *n*
PO woda *f*
nieuzdatniona
PY сырая вода *f*
SU käsittelemätün vesi
SV obehandlat vatten *n*

3934 upstream
DA modstrøm
opadgående strøm
DE stromaufwärts
ES aguas arriba *f,pl*
corriente arriba *f*
F amont *m*
I a monte
MA áramlással szemben
NL stroomopwaarts
PO przeciwprąd *m*
PY вверх по потоку
прямоток *m*
SU ylävirta
SV uppström *c*

3935 useful capacity
DA nytteeffekt
DE Nutzleistung *f*
ES potencia útil *f*
F puissance *f* utile
I capacità *f* utile
MA hasznos teljesítmény
NL nuttige capaciteit *f*
PO pojemność *f*
użyteczna
pojemność *f*
użytkowa
PY полезная
мощность *f*
SU hyütyteho
SV nyttoeffekt *c*

3936 useful heat
DA nyttiggjort varme
DE Nutzwärme *f*
ES calor útil *m*
F chaleur *f* utile
I calore *m* utile
MA hasznos hő
NL nuttige warmte *f*
PO ciepło *n* użyteczne
PY полезная теплота *f*
SU hyütylämpü
SV nyttigt värme *n*

3937 useful heat output
DA nyttevarme ydelse
DE Nutzwärme-
  leistung f
ES potencia térmica
  útil f
F puissance f
  thermique utile
I potenza f termica
  utile
MA hasznos
  hőteljesítmény
NL nuttige
  warmte-
  productie f
PO wydajność f cieplna
  użyteczna
PY полезная
  теплоотдача f
SU hyütylämpüteho
SV nyttigtvärme-
  effekt c

3938 utilization factor
DA udnyttelsesfaktor
DE Ausnutzungs-
  faktor m
ES coeficiente de
  utilización m
  factor de
  utilización m
F facteur m
  d'utilisation
I fattore m di
  utilizzazione
MA hasznosítási tényező
NL gebruiksfactor m
PO współczynnik m
  użytkowania
PY коэффициент m
  использования
SU käyttüaste
SV utnyttjningsfaktor c

3939 utilization time
DA anvendelsestid
DE Benutzungsdauer f
ES tiempo de
  utilización m
F durée f d'utilisation
I tempo m di
  utilizzazione
MA használati idő
NL gebruikstijd m
PO czas m użytkowania
PY время n
  использования
SU käyttüaika
SV användningstid c

V

3940 vacuum
DA vakuum
DE Unterdruck *m*
   Vakuum *n*
ES gas enrarecido *m*
   vacio *m*
F vide *m*
I vuoto *m*
MA vákum
NL vacuüm *n*
PO próżnia
PY вакуум *m*
SU alipaine
   tyhjü
SV vakuum *n*

3941 vacuum breaker
DA vakuumbryder
DE Vakuumbrecher *m*
ES ruptor del vacio *m*
   válvula reguladora
   del vacio *f*
F soupape *f* anti-
   siphon
   vanne *f*
I valvola *f* antisifone
MA visszaszivást gátló
   légbeszívó szelep
NL vacuümonder-
   breker *m*
PO przerywacz *m*
   próżni
PY устройство *n* для
   прерывания
   вакуума
SU alipaineventtiili
   tyhjüventtiili
SV vakuumbrytare *c*

3942 vacuum chilling (*see*
vacuum cooling)

3943 vacuum cleaner
DA støvsuger
DE Staubsauger *m*
ES aspirador de
   polvo *m*
F aspirateur *m* de
   poussières
I aspirapolvere *m*
MA porszívó
NL stofzuiger *m*
PO odkurzacz *m*
PY вакуумный
   пылесос *m*
SU pülynimuri
SV dammsugare *c*

3944 vacuum cooling
DA vakuumkøling
DE Vakuumkühlung *f*
ES enfriamiento al vacío
   m
F refroidissement *m*
   par le vide
I raffreddamento *m*
   sotto vuoto
MA vákumhűtés
NL vacuümkoeling *f*
PO chłodzenie *n*
   próżniowe
PY вакуумное
   охлаждение *n*
SU alipainejäähdytys
SV vakuumkylning *c*

3945 vacuum gauge
DA vakuummåler
DE Vakuummesser *m*
ES manómetro de vacío
   m
F jauge *f* de vide
I vacuometro *m*
MA vákummérő
NL onderdrukmeter *m*
   vacuümmeter *m*
PO manometr *m*
   próżniowy
   próżniomierz *m*
   wakuometr *m*
PY вакуумметр *m*
SU alipainemittari
   tyhjümittari
SV vakuummät-
   instrument *n*

3946 vacuum pipe
DA vakuumrør
DE Vakuumleitung *f*
ES tubería de aspiración
   por vacio *f*
F conduite *f*
   d'aspiration par le
   vide
I tubo *m* a vuoto
MA vákumcsővezeték
NL vacuümleiding *f*
PO rura *f* próżniowa
PY вакуумпровод *m*
   вытяжная труба *f*
SU alipaineputki
   tyhjüputki
SV vakuumrür *n*

3947 vacuum pump
DA vakuumpumpe
DE Vakuumpumpe *f*
ES bomba de vacio *f*
F pompe *f* à vide
I pompa *f* a vuoto
MA vákumszivattyú
NL vacuümpomp *f*
PO pompa *f* próżniowa
PY вакуумный
   насос *m*
SU tyhjüpumppu
SV vakuumpump *c*

DA	=	Danish
DE	=	German
ES	=	Spanish
F	=	French
I	=	Italian
MA	=	Magyar
		(Hungarian)
NL	=	Dutch
PO	=	Polish
PY	=	Russian
SU	=	Finnish
SV	=	Swedish

3948 vacuum steam heating
DA vakuumdamp-
opvarmning
DE Vakuumdampf-
heizung *f*
ES calefacción por
vapor bajo vacio *f*
F chauffage *m* à
vapeur sous vide
I riscaldamento *m* a
vapore sotto
vuoto
MA vákum gőzfűtés
NL vacuümstoom-
verwarming *f*
PO ogrzewanie *n*
parowe próżniowe
PY вакуум-паровое
отопление *n*
SU alipainehüyry-
lämmitys
SV vakuumång-
uppvärmning *c*

3949 vacuum test
DA vakuumprøve
DE Unterdruck-
versuch *m*
Vakuumversuch *m*
ES ensayo en vacio *m*
F essai *m* sous vide
I collaudo *m*
sottovuoto
MA vákumpróba
NL vacuümproef *f*
vacuümtest *m*
PO badanie *n* w
podciśnieniu
PY испытание *n* на
вакуум
SU alipainekoe
SV vakuumprov *n*

3950 value
DA værdi
DE Wert *m*
ES valor *m*
F valeur *f*
I valore *m*
MA érték
NL waarde *f*
PO wartość *f*
PY величина *f*
значение *n*
SU arvo
SV värde *n*

3951 valve
DA ventil
DE Ventil *n*
ES llave *f*
válvula *f*
F robinet *m* à soupape
soupape *f*
vanne *f*
I valvola *f*
MA szelep
NL afsluiter *m*
PO zawór *m*
PY вентиль *m*
задвижка *f*
клапан *m*
шибер *m*
SU venttiili
SV ventil *c*

3952 valve opening
DA ventilhul
DE Ventilüffnung *f*
ES apertura de
válvula *f*
F ouverture *f* de
soupape
I apertura *f* di valvola
MA szelepnyílás
NL klepopening *f*
PO otwór *m* zaworu
PY отверстие *n*
клапана
SU venttiilin aukaisu
SV ventil üppning *c*

3953 valve seat
DA ventilsæde
DE Ventilsitz *m*
ES asiento de la
válvula *m*
F siège *m* de vanne
I sede *f* di valvola
MA szelepülés
NL klepzitting *f*
PO gniazdo *n* zaworu
PY седло *n* клапана
SU venttiilin istukka
SV ventilsäte *n*

3954 vane
DA skovlhjul
vinge
DE Schaufel *f*
ES álabe *m*
paleta *f*
F aube *f*
pale *f*
I aletta *f*
pala *f*
MA szárnylapát
NL schoep *f*
schoepenblad *n*
waaierblad *n*
PO łopatka *f*
PY лопасть *f*
лопатка *f*
SU siipi
SV ledskena *c*
skovel *c*

3955 vaneaxial fan
DA aksialventilator
DE Axialschaufel-
ventilator *m*
ES ventilador de álabes
axiales *m*
F ventilateur *m*
hélicoïde
I ventilatore *m* assiale
elicoidale
MA csavarszárnyas
ventilátor
NL buisschroefventilator
m met
schoepenkrans
PO wentylator *m*
osiowy z
kierownicą
PY осевой
вентилятор *m*
SU ohjaussiivin
varustettu
aksiaalipuhallin
SV axial fläkt *c*

3956 vaned outlet
DA udløb med
  ledeskinne
DE schaufelfürmiger
  Auslaß *m*
ES salida alabeada *f*
F bouche *f* à ailettes
I diffusore *m* ad alette
MA terelőlemezes
  kifúvócsonk
NL rooster *m* met instel-
  bare schoepen
  schoepenrooster *m*
PO wylot *m* z łopatkami
  kierującymi
PY край *m* лопатки
  (вентилятора,
  насоса)
SU ohjaussiivin
  varustettu
  ulosvirtausaukko
SV ledskensutlopp *n*

3957 vaporization
DA dampforstøvning
DE Verdampfung *f*
ES evaporación *f*
  vaporización *f*
F évaporation *f*
  vaporisation *f*
I vaporizzazione *f*
MA elgőzülügtetés
NL verdamping *f*
PO odparowywanie *n*
  parowanie *n*
PY испарение *n*
SU hüyrystyminen
SV fürångning *c*

3958 vaporizing burner
DA dampforstøvnings-
  brænder
DE Verdampfungs-
  brenner *m*
ES quemador
  nevulizante *m*
  quemador
  vaporizante *m*
F brûleur *m* à
  vaporisation
I bruciatore *m* a
  vaporizzazione
MA elpárologtatásos égő
NL verdampings-
  brander *m*
PO palnik *m*
  odparowujący
PY испарительная
  горелка *f*
SU hüyrystyspoltin
SV fürångnings-
  brännare *c*

3959 vapour
DA damp
DE Dampf *m*
ES vapor *m*
F vapeur *f*
I vapore *m*
MA gőz
  pára
NL damp *m*
PO para *f*
PY пар *m*
SU hüyry
SV ånga *c*

3960 vapour barrier
DA dampspærre
DE Dampfsperre *f*
ES barrera de vapor *f*
  pantalla antivapor *f*
F barrière *f* de vapeur
I barriera *f* al vapore
MA párazáró réteg
NL dampdichte laag *f*
PO izolacja *f*
  paroszczelna
  przegroda *f*
  paroszczelna
PY пароизоляция *f*
SU hüyrysulku
SV ångspärr *c*

**3961 vapour-charged
  power element**
DA dampdrevet maskine
DE dampfgespeiste
  Energie-
  erzeugungseinheit *f*
ES elemento motriz a
  vapor *m*
F
I elemento *m* di
  potenza azionato
  a vapore
MA gőztültetű
  beavatkozó elem
NL dampbekrachtigde
  aandrijving *f*
PO element *m* zasilany
  parą
PY манометрический
  преобразователь
  m
SU kaasupalje
SV bränslecell *c*

3962 vapour detector
DA dampføler
DE Dampffühler *m*
ES detector de vapor *m*
F détecteur *m* de
  vapeur
I rivelatore *m* di
  vapore
MA gőzmutató
NL dampdetector *m*
PO wykrywacz *m* pary
PY паровой
  указатель *m*
SU hüyryn ilmaisin
  kaasun ilmaisin
SV ångdetektor *c*

**3963 vapour-jet
  refrigeration cycle**
DA dampejektor
  køleproces
DE Dampfstrahl-
  Kältekreislauf *m*
ES ciclo de refrigeración
  por chorro de
  agua *m*
F cycle *m* frigorifique
  à éjection
I ciclo *m* frigorifero
  ad eiettore
MA gőzsugaras
  hűtőfolyamat
NL stoomstraalkoude-
  kringloop *m*
PO cykl *m* chłodniczy
  inżekcji pary
PY пароструйный
  холодильный
  цикл *m*
SU ejektorikylmä-
  prosessi
SV ångstrålekylning *c*

3964 vapour lock
DA damplås
DE Dampfsperre *f*
ES llave de cierre de
  vapor *f*
F bouchon *m* de
  vapeur
I bolla *f* di vapore
MA gőzdugó
NL dampbelblokkade *f*
  gasbelblokkade *f*
PO korek *m* parowy
PY паровая пробка *f*
SU kaasutasku
SV ånglås *n*

3965 vapour mixture
DA dampblanding
DE Dampfgemisch *n*
ES mezcla de vapores *f*
F mélange *m* de
vapeurs
I miscela *f* di vapore
MA gőzkeverék
NL dampmengsel *n*
PO mieszanina *f* pary
PY паровая смесь *f*
SU hüyryseos
SV ångblandning *c*

3966 vapour pressure
DA damptryk
DE Dampfdruck *m*
ES presión de vapor *f*
F pression *f* de vapeur
I pressione *f* di
vapore
tensione *f* di vapore
MA gőznyomás
páranyomás
NL dampdruk *m*
PO ciśnienie *n* pary
PY давление *n* пара
SU hüyrynpaine
SV ångtryck *n*

3967 vapour retarder
DA damphærder
DE Dampfsperre *f*
ES retardador de vapor
m
F retardeur *m* de
vapeur
I barriera *f* al vapore
MA gőzfogó
NL dampvertragende
laag *f*
PO spowalniacz *m* pary
PY паровой затвор *m*
SU hüyrysulku
SV ångfälla *c*

3968 variable air volume
DA variabel luftstrøm
DE variables
Luftvolumen *n*
ES volumen de aire
variable *m*
F débit *m* d'air
variable
I portata *f* di aria
variabile
MA változtatható
légmennyiség
NL variabel
luchtvolume *n*
PO zmienny strumień *m*
powietrza
PY переменный расход
m воздуха
SU muuttuva ilmavirta
SV variabel luftvolym *c*

3969 variable air volume
control
DA regulering af
varierende
luftmængde
(VAV)
DE veränderliche
Luftvolumen-
Regelung *f*
ES control de volumen
de aire varible *m*
F régulation *f* de débit
d'air variable
I regolazione *f* varia-
bile di portata
MA változtatható
légmennyiség
szabályozás
NL variabele
debietregeling *f*
variabele
volumeregeling *f*
PO regulacja *f*
zmiennego
stumienia
powietrza
PY регулирование *n*
переменного
расхода воздуха
SU ilmavirtasäätü
muuttuvailmavirtasää-
tüinen
SV variabel luftvolym-
kontroll *c*

3970 variable air volume
system
DA VAV-system
DE veränderliches Luft-
volumensystem *n*
ES sistema (de
acondiciona-
miento) de
caudal variable *m*
F système *m* de
climatisation à
débit d'air
variable
I sistema *m* a portata
di aria variabile
MA változtatható
légmennyiségű
rendszer
NL variabel
debietsysteem *n*
variabel
volumesysteem *n*
PO instalacja *f* ze
zmiennym
strumieniem
powietrza
PY система *f*
кондициониро-
вания с
переменным
расходом
воздуха
SU ilmavirtasäätüinen
järjestelmä
SV variabelt luftvolym-
system *n*

3971 variable flow
DA variabel strømning
DE veränderliche
Strümung *f*
ES flujo variable *m*
F débit *m* variable
I flusso *m* variabile
MA változtatható
áramlás
NL variabele stroming *f*
PO zmienny strumień *m*
PY переменный
поток *m*
SU muuttuva ilmavirta
SV variabelt flüde *n*

DA	=	Danish
DE	=	German
ES	=	Spanish
F	=	French
I	=	Italian
MA	=	Magyar
		(Hungarian)
NL	=	Dutch
PO	=	Polish
PY	=	Russian
SU	=	Finnish
SV	=	Swedish

3972 variable inlet vanes
DA ledeskovlsindløb
DE veränderliche
    Eintrittsleit-
    schaufel *f*
ES álabes de abertura
    regulable *m,pl*
F aubages *m* d'entrée
    à inclinaison
    variable
I palette *f,pl* direttrici
    orientabili
MA változtatható belépő
    terelők
NL verstelbare
    inlaatschoepen *pl*
PO łopatki *f* wlotowe
    nastawne
PY направляющие
    лопатки *f* на входе
SU säädettävät
    johtosiivet
SV variabla
    inloppsskovlar *c*

3973 VAV (*see* variable air
    volume)

3974 VAV control (*see*
    variable air volume
    control)

3975 VAV system (*see*
    variable air volume
    system)

3976 vee belt
DA kilerem
DE Keilriemen *m*
ES correa trapezoidal *f*
F courroie *f*
    trapézoïdale
I cinghia *f*
    trapezoidale
MA ékszíj
NL v-snaar *f*
PO pasek *m* klinowy
PY клиновой
    ремень *m*
SU kiilahihna
SV kilrem *c*

3977 vee belt drive
DA kileremtræk
DE Keilriemenantrieb *m*
ES accionamiento por
    correa
    trapezoidal *m*
    transmisión por
    correa
    trapezoidal *f*
F transmission *f* par
    courroie
    trapézoïdale
I trasmissione *f* a
    cinghia
    trapezoidale
MA ékszíjhajtás
NL v-snaaraandrijving *f*
PO napęd *m* paskiem
    klinowym
PY клиноременная
    передача *f*
SU kiilahihnakäyttü
SV kilremsdrift *c*

3978 velocity
DA hastighed
DE Geschwindigkeit *f*
ES velocidad *f*
F vitesse *f*
I velocità *f*
MA sebesség
NL snelheid *f*
PO prędkość *f*
PY скорость *f*
SU nopeus
SV hastighet *c*

3979 velocity contour
DA hastighedsfordeling
DE Geschwindigkeits-
    linie *f*
ES contorno de
    velocidad *m*
    líneas de velocidades
    iguales *f,pl*
F profil *m* de vitesse
I curva *f* di uguale
    velocità
MA sebességprofil
NL snelheidspatroon *n*
    snelheidsprofiel *n*
PO kontur *m* prędkości
    rozkład *m* prędkości
PY изолиния *f*
    скорости
SU tasanopeusviiva
SV hastighetskurva *c*

3980 velocity head
DA dynamisk tryk
DE Geschwindigkeits-
    hühe *f*
ES altura dinámica *f*
    carga dinámica *f*
    presión dinámica *f*
F pression *f*
    dynamique
I pressione *f* dinamica
MA dinamikus nyomás
    torlónyomás
NL dynamische druk *m*
PO ciśnienie *n*
    dynamiczne
    wysokość *f*
    prędkości
PY динамический
    (скоростной)
    напор *m*
SU dynaaminen paine
SV dynamiskt tryck *n*

3981 velocity of fall
DA faldhastighed
DE Fallgeschwindig-
    keit *f*
ES velocidad de caida *f*
F vitesse *f* de chute
I velocità *f* di caduta
MA esési sebesség
NL valsnelheid *f*
PO prędkość *f* spadania
PY скорость *f* падения
SU putoamisnopeus
SV fallhastighet *c*

3982 velocity profile
DA hastighedsprofil
DE Geschwindigkeits-
    profil *n*
ES perfil de
    velocidades *m*
F profil *m* des vitesses
I profilo *m* della
    velocità
MA sebességprofil
NL snelheidsprofiel *n*
PO profil *m* prędkości
    rozkład *m* prędkości
PY профиль *m*
    скорости
SU nopeusjakauma
SV hastighetsprofil *c*

3983 velocity reduction method
DA hastighedsreduktions-metode
DE Methode f der Geschwindigkeits-abnahme f
ES método de reducción de velocidad (dimensionado de conductos) m
F méthode f de regain statique
I metodo m a riduzione di velocità
MA sebességcsükkentési módszer
NL snelheidsverminderings-methode f
PO metoda f zmniejszania prędkości
PY метод m падения скорости
SU nopeuden alentamiseen perustuva mitoitusmene-telmä
SV hastighetsreducerings-metod c

3984 venetian blind
DA persienne
DE Jalousie f
ES persiana veneciana f
F store m vénitien
I tenda f alla veneziana
MA redőnyzár
NL jalouzie f lamellenzonwering f
PO żaluzja f
PY жалюзи n
SU sälekaihdin
SV persienn

3985 vent
DA aftræk lufthul
DE Ausgang m Öffnung f
ES purgador de aire m respiradero m
F évent m
I ventilazione f
MA szellőztető
NL afvoeropening f ontluchtings-opening f
PO odpowietrznik m
PY вентиляционное отверстие n
SU tuuletusaukko
SV ventil c

3986 vent vb
DA ventilere
DE lüften
ES purgar
F purger
I sfiatare
MA szellőztet
NL ontluchten
PO odpowietrzać
PY вентилировать отводить воздух
SU tuulettaa
SV avlufta

3987 ventilated ceiling
DA ventileret loft
DE Lüftungsdecke f
ES techo ventilado m
F plafond m diffuseur d'air
I soffitto m ventilato
MA szellőztetett mennyezet
NL ventilatieplafond n
PO sufit m wentylacyjny
PY вентилируемый потолок m
SU ilmastoitu (esim. rei'itetty) katto
SV ventilerat tak n

3988 ventilating
DA ventilering
DE Lüftung f
ES ventilación f
F ventilation f
I ventilazione f
MA szellőztetés
NL ventileren
PO wentylowanie n
PY вентилирование n
SU ilmanvaihto
SV ventilation c

3989 ventilating rate (see air change rate)

3990 ventilating unit
DA ventilationsenhed
DE Lüftungseinheit f
ES unidad de ventilación f
F unité f de ventilation
I unità f ventilante
MA szellőztető készülék
NL ventilatieapparaat n
PO urządzenie n wentylacyjne zespół m wentylacyjny
PY вентиляторный агрегат m
SU ilmanvaihtokone
SV ventilations-anläggning c

3991 ventilation
DA ventilation
DE Belüftung f Luftzufuhr f
ES aireación f ventilación f
F aération f ventilation f
I ventilazione f
MA szellőzés
NL luchtverversing f ventilatie f
PO wentylacja f
PY вентиляция f
SU ilmanvaihto
SV ventilation c

3992 ventilation shaft
DA ventilationsskakt
DE Lüftungsschacht m
ES chimenea de ventilación f
F cheminée f d'aération trémie f d'aération
I camino m di aerazione pozzo m di aerazione
MA szellőzőkürtő
NL luchtschacht f ventilatieschacht f
PO szyb m wentylacyjny
PY вентиляционная шахта f
SU ilmanvaihtohormi
SV ventilationsschakt n

3993 ventilator
DA ventilator
DE Ventilator *m*
ES aireador *m*
   ventilador *m*
F aérateur *m*
   bouche *f* d'air
I ventilatore *m*
MA szellőztető nyílás
NL ventilator *m*
PO wentylator *m*
PY вентилятор *m*
SU puhallin
SV fläkt *c*

3994 venting
DA udluftning
DE Entlüftung *f*
ES purga de aire *f*
   salida de gases *f*
F purge *f* d'air
I sfogo *m* di aria
MA szellőztetés
NL ontgassing *f*
   ontluchting *f*
PO odpowietrzanie *n*
PY отвод *m* воздуха
SU tuuletus
SV luftning *c*

3995 venting device
DA ventilationsudstyr
DE Entlüftungsventil *n*
ES dispositivo de
   ventilación *m*
   purgador de aire *m*
   purgador de gases *m*
F purgeur *m* d'air
I valvola *f* di sfogo di
   aria
MA szellőztető szerkezet
NL ontgassings-
   apparaat *n*
   ontluchtings-
   apparaat *n*
PO odpowietrznik *m*
PY воздуховыпускное
   устройство *n*
SU tuuletusventtiili
SV luftningsventil *c*

3996 vent pipe
DA udluftningsrør
DE Entlüftungsrohr *n*
ES tubo de
   ventilación *m*
F tube *m* de sécurité
   tube *m* de sûreté
   tube *m* d'évent
I tubo *m* di
   ventilazione
MA szellőzőcső
NL ontluchtingsleiding *f*
   ontluchtingspijp *f*
PO rura *f*
   odpowietrzjąca
PY выпускная труба *f*
SU tuuletusputki
SV luftningsrür *n*

3997 venturi
DA venturirør
DE Venturidüse *f*
ES tubo de Venturi *m*
F venturi *m*
I tubo *m* di Venturi
MA Venturi cső
NL venturi
PO rurka *f* Venturiego
PY труба *f* Вентури
SU venturiputki
SV venturirür *n*

**3998 vertical
   compressor**
DA vertikal kompresser
DE stehender
   Verdichter *m*
ES compresor vertical *m*
F compresseur *m*
   vertical
I compressore *m*
   verticale
MA függőleges
   kompresszor
NL verticale
   compressor *m*
PO sprężarka *f* pionowa
PY вертикальный
   компрессор *m*
SU pystykompressori
SV vertikal-
   kompressor *c*

**3999 vertical-type
   evaporator**
DA vertikal fordamper
DE stehender
   Verdampfer *m*
ES evaporador tipo
   vertical *m*
F évaporateur *m* à
   tubes verticaux
I evaporatore *m*
   verticale
MA függőleges típusú
   elpárologtató
NL verticale
   verdamper *m*
PO parownik *m* typu
   pionowego
PY испаритель *m*
   вертикального
   типа
SU pystyhüyrystin
SV vertikalevaporator *c*

4000 vessel
DA beholder
   kar
DE Gefäß *n*
ES depósito *m*
   recipiente *m*
F vase *m*
I vaso *m*
MA tartály
NL vat *n*
PO naczynie *n*
PY сосуд *m*
SU astia
SV kärl *n*

4001 vibrating grate
DA rysterist
DE Schüttelrost *m*
ES parrilla vibrante *f*
F grille *f* oscillante
I griglia *f* vibrante
MA rázórostély
NL schudrooster *m*
PO ruszt *m* wibrujący
PY вибрационная
   колосниковая
   решетка *f*
SU täryarina
SV skakrost *c*

DA	=	Danish
DE	=	German
ES	=	Spanish
F	=	French
I	=	Italian
MA	=	Magyar (Hungarian)
NL	=	Dutch
PO	=	Polish
PY	=	Russian
SU	=	Finnish
SV	=	Swedish

4002 vibration
DA svingning
vibration
DE Erschütterung *f*
Schwingung *f*
ES vibración *f*
F vibration *f*
I vibrazione *f*
MA rezgés
NL trilling *f*
PO drganie *n*
wibracja *f*
PY вибрация *f*
SU tärinä
SV vibration *c*

4003 vibration absorber
DA svingningsdæmper
DE Schwingungs-
dämpfer *m*
ES amortiguador de
vibraciones *m*
F amortisseur *m* (de
vibration)
I ammortizzatore *m*
di vibrazioni
smorzatore *m* di
vibrazioni
MA rezgéstompító
NL trillingsdemper *m*
PO tłumik *m* drgań
PY амортизатор *m*
SU tärinänvaimennin
SV vibrationsdämpare *c*

4004 vibration
frequency
DA svingningsfrekvens
DE Schwingungs-
frequenz *f*
ES frecuencia de la
vibración *f*
F fréquence *f* de
vibration
I frequenza *f* di
vibrazione
MA rezgésszám
NL trillingfrequentie *f*
trillingsgetal *n*
PO częstotliwość *f*
drgań
PY частота *f* вибрации
SU värähtelytaajuus
SV vibrationsfrekvens *c*

4005 vice
DA skrue
DE Schraubstock *m*
ES prensa de tornillo *f*
prensa hidráulica
pequeña *f*
tornillo de banco *m*
F étau *m*
I morsa *f*
MA satu
NL bankschroef *f*
PO imadło *n*
PY тиски *n*
SU ruuvipenkki
SV skruvstycke *n*

4006 viscometer
DA viskometer
DE Viskometer *n*
ES viscosímetro *m*
F viscosimètre *m*
I viscosimetro *m*
MA viszkoziméter
NL viscositeitsmeter *m*
PO lepkościomierz *m*
PY вискозиметр *m*
SU viskosimetri
viskositeettimittari
SV viskositetsmätare *c*

4007 viscosity
DA viskocitet
DE Viskosität *f*
Zähigkeit *f*
ES viscosidad *f*
F viscosité *f*
I viscosità *f*
MA viszkozitás
NL viscositeit *f*
PO lepkość *f*
PY вязкость *f*
SU viskositeetti
SV viskositet *c*

4008 viscous filter
DA klæbefilter
tæt filter
DE Viskosefilter *m*
ES filtro viscoso *m*
F filtre *m* à
imprégnation
visqueuse
I filtro *m* viscoso
MA viszkózus szűrő
NL oliegeimpregneerd
filter *m*
PO filtr *m* wiskozowy
PY масляный
фильтр *m*
SU tartunta-aineella
käsitelty suodatin
SV visküst filter *n*

4009 viscous flow
DA tyktflydende strøm
DE zähe Strümung *f*
ES corriente viscosa *f*
flujo viscoso *m*
F écoulement *m*
visqueux
I flusso *m* viscoso
MA viszkózus áramlás
NL laminaire stroming *f*
visceuze stroming *f*
PO przepływ *m* lepki
PY поток *m* вязкой
жидкости
SU viskoosivirtaus
SV visküs strümning *c*

4010 vitiated air
DA forurenet luft
DE schlechte Luft *f*
ES aire viciado *m*
F air *m* vicié
I aria *f* viziata
MA elhasznált levegő
NL bedorven lucht *f*
PO powietrze *n* skażone
PY отработанный
воздух *m*
SU pilaantunut ilma
SV fürorenad luft *c*

4011 volatile *adj*
DA flygtig
DE flüchtig
ES volatil
F volatil
I volatile
MA illó
NL vluchtig
PO lotny
PY летучий
SU haihtuva
SV flyktig

4012 volatile matter
DA flygtig bestanddel
DE flüchtiger
Bestandteil *m*
ES materias
volátiles *f,pl*
F matière *f* volatile
I materia *f* volatile
MA illóanyag
NL vluchtige
bestanddelen *pl*
PO części *f, pl* lotne
substancja *f* lotna
PY летучее вещество *n*
SU haihtuva aine
SV flyktigt ämne *n*

4013 voltage
DA spænding
DE Spannung *f*
ES tensión *f*
F tension *f* (électrique)
I tensione *f*
voltaggio *m*
MA feszültség
NL spanning *f*
voltage *f*
PO napięcie *n*
PY электрическое
напряжение *n*
SU jännite
SV spänning *c*

4014 voltage meter
DA voltmeter
DE Voltmeter *n*
ES voltímetro *m*
F voltmètre *m*
I voltmetro *m*
MA feszültségmérő
NL voltmeter *m*
PO woltomierz *m*
PY вольтметр *m*
SU jännitemittari
SV voltmeter *c*

4015 volume
DA volumen
DE Volumen *n*
ES volumen *m*
F volume *m*
I volume *m*
MA mennyiség
térfogat
NL inhoud *m*
volume *n*
PO objętość *f*
PY объем *m*
SU tilavuus
SV volym *c*

4016 volume control damper
DA mængdestyrings-
spjæld
DE Volumenstell-
klappe *f*
ES compuerta de
regulación de
volumen *f*
F registre *m* de réglage
de débit
I serranda *f* di
regolazione della
portata
MA mennyiségszabá-
lyozó csappantyú
NL luchtdebietregel-
klep *f*
PO przepustnica *f*
regulacji wydatku
PY клапан *m*
регулирования
объема
SU säätüpelti
SV volymkontroll-
spjäll *n*

4017 volumetric efficiency
DA volumetrisk
virkningsgrad
DE volumetrischer
Wirkungsgrad *m*
ES rendimiento
volumétrico *m*
F rendement *m*
volumétrique
I rendimento
volumetrico
MA volumetrikus
hatásfok
NL volumetrisch
rendement *n*
PO skuteczność *f*
objętościowa
współczynnik *m*
napełnienia
PY объемный
коэффициент *m*
полезного
действия
SU tilavuushyütysuhde
SV fyllningsgrad *c*

4018 volumetric pump
DA volumetrisk pumpe
DE volumetrische
Pumpe *f*
ES bomba
volumétrica *f*
F pompe *f*
volumétrique
I pompa *f*
volumetrica
MA volumetrikus
szivattyú
NL doseerpomp *f*
PO pompa *f* dozująca
PY объемный насос *m*
SU syrjäytyspumppu
SV volymetrisk pump *c*

4019 vortex burner
DA hvirvelbrænder
vortex-brænder
DE Wirbelbrenner *m*
ES quemador a
turbulencia *m*
F brûleur *m* à vortex
I bruciatore *m* a
turbolenza
MA örvényégő
NL wervelbrander *m*
PO palnik *m* wirowy
PY вращающаяся
горелка *f*
SU pyürrepoltin
SV vortexbrännare *c*

4020 vortex flow
DA hvirvelstrøm
DE Drallstrümung *f*
ES flujo turbulento *m*
F écoulement *m* en
vortex
I flusso *m* vorticoso
moto *m* turbolento
MA örvényáram
NL wervelstroom *m*
PO przepływ *m* wirowy
PY вращающийся
поток *m*
SU pyürrevirtaus
SV vortexflüde *n*

4021 vortex tube
DA hvirvelrør
DE Wirbelrohr *n*
ES tubo vórtex *m*
F tube *m* spiral
I turbolatore *m*
MA örvénycső
NL wervelbuis *f*
wervelpijp *f*
PO rurka *f* wirowa
wejście *n* do chłodni
PY вихревая труба *f*
SU pyürreputki
SV vortexrür *n*

W

4022 walk-in cooler
DA kølerum, der er til at
gå ind i
DE begehbarer
Kühlraum *m*
ES enfriador de entrada
m
F réfrigérateur *m*
chambre
I raffreddatore *m*
ispezionabile
MA hűtőkamra
NL inloopkoelruimte *f*
PO chłodnia *f*
przechodnia
PY охладитель *m* с
возможностью
входа вовнутрь
SU kylmähuone
SV kylrum *n*

4023 wall
DA mur
væg
DE Wand *f*
ES muro *m*
pared *f*
F cloison *f*
mur *m*
paroi *f*
I muro *m*
parete *f*
MA fal
NL muur *m*
wand *m*
PO ściana *f*
PY стена *f*
SU seinä
SV vägg *c*

DA = Danish
DE = German
ES = Spanish
F = French
I = Italian
MA = Magyar
(Hungarian)
NL = Dutch
PO = Polish
PY = Russian
SU = Finnish
SV = Swedish

**4024 wall below a
window**
DA vinduesbrystning
DE Fensternische *f*
ES antepecho *m*
F allège *f*
I parete *f* sottofinestra
MA ablak alatti fal
mellvéd
parapet
NL borstwering *f*
PO ściana *f* podokienna
PY стена *f* под окном
SU ikkunan alapuolinen
seinä
SV fünsterbrüstning *c*

4025 wall coil
DA varmebatteri i væg
DE wandhängender
Verdampfer *m*
ES serpentín mural *m*
F batterie *f* murale
I serpentino *m* di
raffreddamento a
parete
MA fali csőkígyó
NL muurkoelelement *n*
PO wężownica *m*
PY змеевик *m*,
заделанный в
стену
SU seinälle asennettu
lamellipatteri
SV väggaller *n*

4026 wall fan
DA vægventilator
DE Wandlüfter *m*
Wandventilator *m*
ES ventilador de
pared *m*
F ventilateur *m* mural
I ventilatore *m* a
parete
MA fali ventilátor
NL muurventilator *m*
PO wentylator *m*
ścienny
PY настенный
вентилятор *m*
SU seinäpuhallin
SV väggfläkt *c*

4027 wall grid (*see* wall
coil)

4028 wall heating
DA vægopvarmning
DE Wandheizung *f*
ES calefacción por las
paredes *f*
F chauffage *m* mural
chauffage *m* par les
murs
I riscaldamento *m* a
parete
MA falfűtés
NL wandverwarming *f*
PO ogrzewanie *n*
ścienne
PY панельное (стенное)
отопление *n*
SU seinälämmitys
SV vägguppvärmning *c*

4029 wall heating panel
DA væghængt
varmepanel
DE Wandstrahlungs-
platte *f*
ES panel calefactor de
pared *m*
panel radiante de
pared *m*
F panneau *m*
chauffant mural
I pannello *m* di
riscaldamento a
parete
MA falfűtő panel
NL wandverwarmings-
paneel *n*
PO płyta *f* ogrzewania
ściennego
PY стенная
отопительная
панель *f*
SU (seinä)lämmitys-
paneeli
SV värmepanel *c*

4030 wall-mounted air heater
DA vægmonteret luftopvarmer
DE Wandluftheizer *m*
ES aerotermo de pared *m*
   aerotermo montado en la pared *m*
F aérotherme *m* d'air mural
I riscaldatore *m* di aria a muro aerotermo
MA falra szerelt hőlégfúvó
NL wandlucht-verhitter *m*
PO zespół *m* ogrzewczo-wentylacyjny ścienny
PY настенный воздухонагре-ватель *m*
SU seinälle asennettu ilmanlämmitin
SV väggmonterad luftvärmare *c*

4031 wall outlet
DA udluftningsventil
DE Wandauslaß *m*
ES boca de salida situada en la pared *f*
   rejilla de pared *f*
F bouche *f* de soufflage murale
I bocchetta *f* di immissione di aria a parete
MA fali kifúvónyílás
NL afvoeropening *f* in muur
PO gniazdko *n* wtykowe ścienne
   wylot *m* ścienny
PY выход *m* через стену
SU seinäventtiili
SV frånluftsventil *c* i vägg *c*

4032 wall temperature
DA vægtemperatur
DE Wandtemperatur *f*
ES temperatura de la pared *f*
F température *f* de paroi
I temperatura *f* delle pareti
MA falhőmérséklet
NL wandtemperatuur *f*
PO temperatura *f* ściany
PY температура *f* стены
SU seinän lämpütila
SV väggtemperatur *c*

4033 wall ventilator
DA vægventilator
DE Wandventilator *m*
ES aireador de pared *m*
   ventilador de pared *m*
F ventilateur *m* mural
I ventilatore *m* da parete
MA fali szellőzőnyílás
NL muurventilator *m*
PO wentylator *m* ścienny
PY настенный вентилятор *m*
SU seinätuuletin
SV väggfläkt *c*

4034 warm air
DA varm luft
DE Warmluft *f*
ES aire caliente *m*
F air *m* chaud
I aria *f* calda
MA meleglevegő
NL warme lucht *f*
PO powietrze *n* ciepłe
PY теплый воздух *m*
SU lämmin ilma
SV varmluft *c*

4035 warm air curtain
DA varmlufttæppe
DE Warmluftschleier *m*
ES cortina de aire caliente *f*
F rideau *m* d'air chaud
I cortina *f* di aria calda
MA meleg légfüggüny
NL warmelucht-gordijn *n*
PO kurtyna *f* ciepłego powietrza
PY воздушно-тепловая завеса *f*
SU lämminilmaverho
SV varmluftsridå *c*

4036 warm air duct
DA varmluftkanal
DE Warmluftleitung *f*
ES conducto de aire caliente *m*
F conduit *m* d'air chaud
I condotto *m* per aria calda
MA meleglevegő csatorna
NL warmeluchtkanaal *n*
PO przewód *m* ciepłego powietrza
PY канал *m* теплого воздуха
SU lämminilmakanava
SV varmluftskanal *c*

4037 warm air heater
DA varmluftaggregat
DE Warmluftheizer *m*
ES generador de aire caliente *m*
F générateur *m* d'air chaud
I generatore *m* di aria cadla
MA termoventilátor
NL luchtverhitter *m*
PO podgrzewacz *m* ciepłego powietrza
PY воздухонагреватель *m*
SU lämminilmakone
SV luftvärmare *c*

4038 warm air heating
DA varmluftopvarmning
DE Warmluftheizung *f*
ES calefacción por aire
    caliente *f*
F chauffage *m* à air
    chaud
I riscaldamento *m* ad
    aria calda
MA hőlégfűtés
NL luchtverwarming *f*
PO ogrzewanie *n*
    powietrzne
PY воздушное
    отопление *n*
SU ilmalämmitys
SV varmlufts-
    uppvärmning *c*

4039 warm air heating unit
DA varmluftunit
DE Warmluftheizgerät *n*
ES calefactor de aire *m*
F générateur *m* d'air
    chaud
I apparecchio *m* di
    riscaldamento ad
    aria calda
MA termoventilátor
    egység
NL luchtverwarmings-
    apparatuur *f*
PO zespół *m* ogrzewczo-
    wentylacyjny
PY установка *f*
    воздушного
    отопления
SU ilmalämmityskone
SV varmluftsaggregat *n*

4040 warm air output
DA varmluftafgivelse
DE Warmluftleistung *f*
ES caudal de aire
    caliente *m*
F débit *m* d'air chaud
I portata *f* di aria
    calda
MA meleglevegő
    teljesítmény
NL warmeluchtafgifte *f*
PO wydajność *f* ciepłego
    powietrza
PY выпуск *m* теплого
    воздуха
SU lämminilmateho
SV varmlufts-
    avgivning *c*

4041 warming
DA opvarmning
DE Erwärmung *f*
ES calentamiento *m*
F réchauffage *m*
I riscaldamento *m*
MA melegítés
NL verwarming *f*
PO ogrzewanie *n*
PY нагревание *n*
SU lämpeneminen
SV uppvärmning *c*

4042 warmth
DA varme
DE Hitze *f*
    Wärme *f*
ES calor *m*
F chaleur *f*
I caldo *m*
    calore *m*
    tepore *m*
MA hő
    meleg
NL warmte *f*
PO ciepło *n*
PY теплота *f*
SU lämpimyys
SV värme *c*

4043 warmth index
DA varmekomforttal
DE Wärmeindex *m*
ES índice de confort
    térmico *m*
F indice *m* de confort
    thermique
I indice *m* di tepore
MA hőérzeti jellemző
NL thermische
    behaaglijkheids-
    index *m*
PO wskaźnik *m* ciepła
PY индекс *m*
    теплового
    комфорта
SU lämpimyysindeksi
SV värmeindex *n*

4044 warm water
DA varmt vand
DE Warmwasser *n*
ES agua caliente *f*
F eau *f* chaude
I acqua *f* calda
MA melegvíz
NL warm water *n*
PO woda *f* ciepła
PY горячая вода *f*
SU lämmin vesi
SV varmvatten *n*

4045 warning
DA advarsel
DE Alarm *m*
    Warnung *f*
ES alarma *f*
    aviso *m*
F alerte *f*
    avertissement *m*
I allarme *m*
    avvertenza *f*
    avvertimento *m*
MA figyelmeztető
    jelző
NL waarschuwing *f*
PO ostrzeżenie *n*
PY предупреждение *n*
SU varoitus
SV varning *c*

4046 warning device
DA alarmudstyr
DE Warngerät *n*
ES avisador *m*
    dispositivo de
    alarma *m*
F avertisseur *m*
    dispositif *m* d'alerte
I dispositivo *m* di
    allarme
MA jelzőszerkezet
NL alarminrichting *f*
    waarschuwings-
    apparaat *n*
PO urządzenie *n*
    ostrzegające
PY предупредительное
    устройство *n*
SU varoituslaite
SV larmanordning *c*

4047 warning light
DA signallampe
DE Warnlampe *f*
ES lámpara de aviso *f*
    señal luminosa de
    aviso *f*
F avertisseur *m*
    lumineux
I luce *f* di allarme
MA jelzőfény
NL signaallamp *f*
    signaleringslamp *f*
PO lampa *f*
    ostrzegawcza
    światło *n*
    ostrzegawcze
PY предупредительный
    свет *m*
SU varoitusvalo
SV varningsljus *n*

4048 wash basin
DA vandfad
vaskekumme
DE Waschbecken *n*
ES lavabo *m*
lavadero *m*
F lavabo *m*
I bacino *m* di
lavaggio
MA mosdó
NL wasbak *m*
PO umywalka *f*
PY умывальник *m*
SU pesuallas
SV tvättställ *n*

4049 washer
DA pakning
underlagsskive
DE Dichtungsring *m*
Unterlegscheibe *f*
ES anillo *m*
arandela *f*
F rondelle *f*
I rondella *f*
MA csavaralátét
NL dichtingsring *m*
sluitring *m*
vulring *m*
PO płuczka *f*
podkładka *f*
zespół *m* ogrzewczo-
wentylacyjny
PY прокладка *f*
промывное
устройство *n*
шайба *f*
SU aluslevy
SV bricka *c*
packning *c*

4050 washer (*see* air
washer)

**4051 waste energy
sources**
DA spildenergikilder
DE Abfallenergie-
quellen *f,pl*
ES fuentes de energía
residual *f*
F sources *f* d'énergie
perdue
I fonti *m,pl* di energia
di recupero
MA hulladék
energiaforrások
NL afvalwarmte-
energiebronnen *pl*
PO źródła *n* energii
odpadowej
PY вторичный
источник *m*
теплоты
SU jäte-energian lähteet
SV spillenergikälla *c*

4052 waste heat
DA spildvarme
DE Abwärme *f*
ES calor cedido *m*
calor perdido *m*
F chaleur *f* perdue
I calore *m* disperso
MA hulladékhő
NL afvalwarmte *f*
PO ciepło *n* odpadowe
PY потери *f,pl*
теплоты
SU jätelämpü
SV spillvärme *c*

4053 waste heat boiler
DA spildvarmekedel
DE Abwärmekessel *m*
ES caldera de
recuperación *f*
F chaudière *f* de
récupération
I caldaia *f* a calore di
recupero
MA hulladékhő kazán
NL afvalwarmteketel *m*
PO kocioł *m* na ciepło
odpadowe
kocioł *m*
utylizacyjny
PY котел-
утилизатор *m*
SU jätelämpükattila
SV spillvärmepanna *c*

4054 waste pipe
DA afløbsrør
DE Abflußrohr *n*
ES tubo de desagüe *m*
tubo de escape *m*
F tuyau *m*
d'échappement
tuyau *m*
d'évacuation
I tubo *m* di scarico
MA lefolyócső
NL afvoerpijp *f*
PO rura *f* ściekowa
PY сбросная труба *f*
SU viemäriputki
SV spillrür *n*

4055 waste water
DA asfløbsvand
spildevand
DE Abwasser *n*
ES agua residual *f*
F eau *f* usée
I acqua *f* di rifiuto
MA szennyvíz
NL afvalwater *n*
PO ścieki *m, pl*
woda *f* odpływowa
PY сточная вода *f*
SU jätevesi
SV spillvatten *n*

4056 water
DA vand
DE Wasser *n*
ES agua *f*
F eau *f*
I acqua *f*
MA víz
NL water *n*
PO woda *f*
PY вода *f*
SU vesi
SV vatten *n*

4057 water absorption
DA vandabsorption
DE Wasseraufnahme *f*
ES absorción de agua *f*
F absorption *f* d'eau
I assorbimento *m* di
acqua
MA vízfelvétel
NL waterabsorptie *f*
wateropneming *f*
PO absorpcja *f* wody
PY водопоглощение *n*
SU veden absorptio
SV vattenabsorption *c*

4058 water capacity
DA vandkapacitet
DE Wasserkapazität *f*
ES capacidad de
     absorción de
     agua *f*
   capacidad de carga
     de agua *f*
F capacité *f* d'eau
   volume *m* d'eau
I capacità *f* di acqua
MA vízkapacitás
   vízmennyiség
NL watercapaciteit *f*
PO pojemność *f* wodna
PY водопотребление *n*
   расход *m* воды
SU vesitilavuus
SV vattenkapacitet *c*

4059 water content
DA vandindhold
DE Wassergehalt *m*
ES contenido de
     agua *m*
F contenance *f*
   teneur *f* en eau
I contenuto *m* di
     acqua
MA víztartalom
NL watergehalte *n*
   waterinhoud *m*
PO zawartość *f* wody
PY содержание *n* воды
SU vesisisältü
SV vattenhalt *c*

4060 water-cooled *adj*
DA vandkølet
DE wassergekühlt
ES enfriado por agua
F refroidi par eau
I raffreddato ad acqua
MA vízhűtéses
NL watergekoeld
PO chłodzony wodą
PY водоохлаждаемый
SU vesijäähdytteinen
SV vattenkyld

4061 water cooler
DA vandkøler
DE Wasserkühler *m*
ES enfriador de agua *m*
F refroidisseur *m*
     d'eau
I refrigeratore *m* di
     acqua
MA vízhűtő
NL waterkoeler *m*
PO chłodnica *f* wody
PY водоохладитель *m*
SU vedenjäähdytin
SV vattenkylare *c*

**4062 water cooling
     tower**
DA vandkøletårn
DE Wasserkühlturm *m*
ES torre de refrigeración
     de agua *f*
F tour *f* de
     refroidissement
     d'eau
I torre *f* di
     raffreddamento
MA vízhűtő torony
NL waterkoeltoren *m*
PO chłodnia *f*
     kominowa wodna
   wieża *f* chłodnicza
     wodna
PY градирня *f*
SU vedenjäähdytystorni
SV vattenkyltorn *n*

4063 water defrosting
DA vandafrimning
DE Wasserabtauung *f*
ES desescarche por agua
     m
F dégivrage *m* par eau
I sbrinamento *m* ad
     acqua
MA vízporlasztásos
     leolvasztás
NL ontdooiing *f* met
     water
   waterontdooiing *f*
PO rozmrażanie *n*
     wodne
PY оттаивание *n*
     замерзшей воды
SU sulatus vedellä
SV vattenavfrostning *c*

**4064 water
     demineralization**
DA blødgøring
   demineralisering
DE Wasserentsalzung *f*
ES desmineralización
     del agua *f*
F déminéralisation *f*
     de l'eau
I demineralizzazione *f*
     dell'acqua
MA vízlágyítás
NL water-
     demineralisatie *f*
PO demineralizacja *f*
     roztworów
     wodnych
   odsalanie *n* wody
PY обессоливание *n*
     воды
SU suolan poisto
     vedestä
SV avmineralisering *c* av
     vatten *n*

4065 water discharge
DA vandafløb
DE Entwässerung *f*
ES evacuación del
     agua *f*
F évacuation *f* d'eau
I scarico *m* dell'acqua
MA víztelenítés
NL waterafvoer *m*
PO odwodnienie *n*
PY водоотвод *m*
SU vedenpoisto
SV vattenutlopp *n*

4066 water feed
DA vandtilførsel
DE Wasserzufluß *m*
ES alimentación de
     agua *f*
F alimentation *f* en
     eau
I alimentazione *f* di
     acqua
MA vízbetáplálás
NL watervoeding *f*
PO zasilanie *n* wodą
PY водоснабжение *n*
SU vedensyüttü
SV vattentillfürsel *c*

DA	= Danish
DE	= German
ES	= Spanish
F	= French
I	= Italian
MA	= Magyar (Hungarian)
NL	= Dutch
PO	= Polish
PY	= Russian
SU	= Finnish
SV	= Swedish

4067 water filter strainer
- DA vandfilter
- DE Wasserfilter *m*
- ES filtro de agua *m*
- F filtre *m* à eau
- I filtro *m* di acqua
- MA vízszűrő
- NL waterfilter-zuigkorf *m*
- PO filtr *m* wody siatkowy
- PY сетчатый фильтр *m* для воды
- SU vesisuodatin
- SV vattenfilter *n*

4068 water gauge (*see* Appendix A)

4069 water hammer
- DA vandslag
- DE Wasserschlag *m* Wasserstoß *m*
- ES golpe de ariete (del agua) *m* golpeteo del agua *m*
- F coup *m* de bélier
- I colpo *m* di ariete
- MA vízütés
- NL waterslag *m*
- PO uderzenie *n* wodne
- PY гидравлический удар *m*
- SU paineisku
- SV tryckstüt *c*

4070 water hardness
- DA vands hårhed
- DE Wasserhärte *f*
- ES dureza del agua *f*
- F dureté *f* de l'eau
- I durezza *f* dell'acqua
- MA vízkeménység
- NL waterhardheid *f*
- PO twardość *f* wody
- PY жесткость *f* воды
- SU veden kovuus
- SV vattenhårdhet *c*

4071 water heater
- DA vandopvarmer
- DE Wassererwärmer *m*
- ES calentador de agua *m*
- F chauffe-eau *m* réchauffeur *m* d'eau
- I riscaldatore *m* di acqua
- MA vízmelegítő
- NL warmwater-bereider *m*
- PO podgrzewacz *m* wody
- PY водонагреватель *m*
- SU vedenlämmitin
- SV vattenvärmare *c*

4072 water heating
- DA vandopvarmning
- DE Wasserheizung *f*
- ES calefacción por agua caliente *f*
- F chauffage *m* d'eau
- I riscaldamento *m* ad acqua
- MA vízfűtés
- NL warmwater-bereiding *f* warmwater-verwarming *f*
- PO ogrzewanie *n* wodne
- PY водяное отопление *n*
- SU vedenlämmitys
- SV varmvatten-beredning *c*

4073 water jacket
- DA kølekappe vandkappe
- DE Wassermantel *m*
- ES camisa de agua *f*
- F chemise *f* d'eau
- I camicia *f* di acqua
- MA vízküpeny
- NL watermantel *m*
- PO płaszcz *m* wodny
- PY водяная рубашка *f*
- SU vesivaippa
- SV vattenmantel *c*

4074 water jet pump
- DA vandstrålepumpe
- DE Wasserstrahl-pumpe *f*
- ES ejector hidraulico *m*
- F éjecteur *m* hydraulique
- I pompa *f* per acqua ad eiettore
- MA vízsugárszivattyú
- NL waterstraalpomp *f*
- PO pompa *f* strumieniowa wodna strumienica *f*
- PY водоструйный насос *m*
- SU vesisuihkupumppu
- SV vattenstrålpump *c*

4075 water layer
- DA vandlag
- DE Wasserschicht *f*
- ES lámina de agua *f*
- F lame *f* d'eau
- I lama *f* di acqua
- MA vízréteg
- NL waterlaag *f*
- PO warstwa *f* wody
- PY водный слой *m*
- SU vesikerros
- SV vattenskikt *n*

4076 water leak
- DA utæthed vandlæk
- DE Wasserleck *n*
- ES fuga de agua *f*
- F fuite *f* d'eau
- I perdita *f* di acqua
- MA vízszivárgás
- NL waterlek *n* waterlekkage *f*
- PO przeciek *m* wody
- PY течь *f* воды
- SU vesivuoto
- SV vattenläcka *c*

4077 water level
- DA vandniveau
- DE Wasserstand *m*
- ES nivel de agua *m*
- F niveau *m* d'eau
- I livello *m* dell'acqua
- MA vízszint
- NL waterniveau *n* waterpeil *n*
- PO poziom *m* wody zwierciadło *n* wody
- PY уровень *m* воды
- SU veden pinta
- SV vattennivå *c*

4078 water loss
DA vandtab
DE Wasserverlust *m*
ES pérdida de agua *f*
F perte *f* d'eau
I perdita *f* di acqua
MA vízveszteség
NL waterverlies *n*
PO ubytek *m* wody
PY потеря *f* воды
SU veden häviü
SV vattenfürlust *c*

4079 water outlet
DA vandudløb
DE Wasserablauf *m*
ES evacuador del
agua *m*
salida del agua *f*
toma de agua *f*
F point *m* de puisage
I presa *f* di acqua
MA vízkifolyás
NL tappunt *n*
waterafvoerpunt *n*
PO wypływ *m* wody
PY истечение *n* воды
SU veden ulosjuoksu
SV vattenutlopp *n*

4080 water pocket
DA vandlomme
DE Wassersack *m*
ES bolsa de agua *f*
F poche *f* d'eau
I sacca *f* di acqua
MA vízzsák
NL waterzak *m*
PO kieszeń *f* wodna
PY водяной мешок *m*
(в
трубопроводах)
SU vesitasku
SV vattensäck *c*

4081 waterproof *adj*
DA vandskyende
vandtæt
DE wasserdicht
ES impermeable
F imperméable
I impermeabile
MA vízálló
NL waterdicht
PO wodoodporny
wodoszczelny
PY водонепрони-
цаемый
SU vedenpitävä
SV vattentät

4082 waterproof *vb*
DA gøre vandtæt
imprægnere
DE wasserdicht machen
ES impermeabilizar
F imperméabiliser
I impermeabilizzare
MA vízállóvá tesz
NL waterdicht maken
PO czynić
wodoodpornym
czynić
wodoszczelnym
PY уплотнять против
проникания воды
SU tehdä vedenpitäväksi
SV täta mot vatten

4083 water separator
DA vandudlader
vandudskiller
DE Wasserabscheider *m*
ES separador de
agua *m*
F séparateur *m* d'eau
I separatore *m* di
acqua
MA vízleválasztó
NL waterafscheider *m*
PO oddzielacz *m* wody
odkraplacz *m*
PY водоотделитель *m*
SU vedenerotin
SV vattenavskiljare *c*

4084 water softener
DA vandafsaltnings-
middel
DE Wasserenthärter *m*
ES ablandador de
agua *m*
desendurecedor del
agua *m*
F adoucisseur *m* d'eau
I addolcitore *m* di
acqua
MA vízlágyító
NL waterontharder *m*
PO zmiękczacz *m* wody
PY умягчитель *m*
воды
SU veden pehmennin
SV vattenavhärdare *c*

4085 water softening
DA vandblødgøring
DE Wasserenthärtung *f*
ES ablandamiento del
agua *m*
descalcificación del
agua *f*
desendurecimiento
del agua *m*
F adoucisseur *m* d'eau
I addolcimento *m*
dell'acqua
MA vízlágyítás
NL waterontharding *f*
PO zmiękczanie *n* wody
PY умягчение *n* воды
SU veden pehmennys
SV vattenavhärdning *c*

4086 water-tight *adj*
DA vandtæt
DE wasserdicht
ES estanco al agua
F étanche à l'eau
I a tenuta di acqua
MA víztümür
NL waterdicht
PO wodoszczelny
PY водонепрони-
цаемый
SU vedenpitävä
vesitiivis
SV vattentät

4087 water treatment
DA vandbehandling
DE Wasserbehandlung *f*
ES tratamiento del
agua *m*
F traitement *m* d'eau
I trattamento *m*
dell'acqua
MA vízkezelés
NL waterbehandeling *f*
PO uzdatnianie *n* wody
PY обработка *f* воды
SU vedenkäsittely
SV vattenbehandling *c*

4088 water tube boiler
DA vandrørskedel
DE Wasserrohrkessel *m*
ES caldera tubular *f*
F chaudière *f* à tubes
d'eau
I caldaia *f* a tubi di
acqua
MA vízcsüves kazán
NL waterpijpketel *m*
PO kocioł *m*
wodnorurkowy
PY водотрубный
котел *m*
SU vesiputkikattila
SV tubpanna *c*

4089 water vapour
DA vanddamp
DE Wasserdampf *m*
ES vapor de agua *m*
F vapeur *f* d'eau
I vapore *m* acqueo
MA vízgőz
NL waterdamp *m*
PO para *f* wodna
PY водяной пар *m*
SU vesihüyry
SV vattenånga *c*

4090 water velocity
DA vandhastighed
DE Wasser-
geschwindigkeit *f*
ES velocidad del agua *f*
F vitesse *f* de l'eau
I velocità *f* dell'acqua
MA vízsebesség
NL watersnelheid *f*
PO prędkość *f* wody
PY скорость *f* воды
SU veden nopeus
SV vattenhastighet *c*

4091 Watt (*see* Appendix A)

4092 Watt meter
DA el-måler
wattmeter
DE Wattmesser *m*
ES vatímetro *m*
F Wattmètre *m*
I wattmetro *m*
MA wattmérő
NL Watt-meter *m*
PO watomierz *m*
PY ваттметр *m*
SU wattimittari
SV Wattmätare *c*

4093 wear and tear
DA slitage
DE natürliche
Abnutzung *f*
Verschleiß *m*
ES depreciación por
uso *f*
desgaste *m*
deterioro por el
uso *m*
F usure *f*
I usura *f* e strappo *m*
MA elhasználódás
NL slijtage *f*
PO zużycie *n* w czasie
eksploatacji
PY износ *m*
SU kuluminen
SV fürslitning *c*

4094 weather
DA vejr
DE Wetter *n*
ES tiempo
atmosférico *m*
F temps *m*
I tempo *m*
MA időjárás
NL weer *n*
PO pogoda *f*
PY погода *f*
SU sää
SV väder *n*

4095 weather compensator
DA vejrkompensator
DE Wetter-
kompensator *m*
ES compensador de
tiempo
atmosférico *m*
F
I compensatore *m*
climatico
MA időjárás
kompenzátor
NL weersafhankelijke
regelaar *m*
PO regulator *m*
pogodowy
PY компенсатор *m*
погодных
(ветровых)
условий
SU sääkompensaattori
SV väderkompensator *c*

4096 weather data
DA vejrdata
DE Wetterdaten *f,pl*
ES datos ambientales *m*
F données *f*
météorologiques
I dati *m,pl* climatici
MA időjárási adat
NL meteorologische
gegevens *pl*
PO dane *f* pogodowe
PY данные *n* о погоде
SU säätiedot
SV väderdata *n*

4097 weatherproof *adj*
DA vejrbestandig
DE wetterfest
ES resistente a la
intemperie
F résistant aux
intempéries
I resistente alle
intemperie
MA időjárásálló
NL weerbestendig
PO odporny na wpływy
atmosferyczne
PY непроницаемый
для атмосферных
осадков
SU säänkestävä
SV väderbeständig

4098 weatherproof cover
DA vejrbestandigt
overtræk
DE wasserfester Belag *m*
ES cubierta de
protección contra
la intemperie *f*
F capot *m* de
protection étanche
I coperchio *m* a
tenuta stagna
MA időjárásálló burkolat
NL weerbestendige
afdekking *f*
weerbestendige
behuizing *f*
PO przykrycie *n*
odporne na
wpływy
atmosferyczne
PY защитная завеса *f*
от атмосферных
осадков
SU säänkestävä suoja
SV väderbeständigt
skydd *n*

DA	=	Danish
DE	=	German
ES	=	Spanish
F	=	French
I	=	Italian
MA	=	Magyar (Hungarian)
NL	=	Dutch
PO	=	Polish
PY	=	Russian
SU	=	Finnish
SV	=	Swedish

4099 weatherproof hood
- DA vejrbestandig beskyttelse (hætte)
- DE Wetterschutzhaube f
- ES caperuza de protección contra la intemperie f
- F écran m de protection étanche
- I cuffia f a tenuta stagna
- MA időjárásálló fedél
- NL weerbestendige kap f
- PO okap m odporny na wpływy atmosferyczne
- PY защитный навес m от атмосферных осадков
- SU säänkestävä katos
- SV väderbeständig huv c

4100 weight
- DA vægt
- DE Gewicht n
- ES peso m
- F poids m
- I peso m
- MA súly
- NL gewicht n
- PO ciężar m
- PY вес m
- SU paino
- SV vikt c

4101 weight-loaded valve
- DA vægtbelastet ventil
- DE gewichtsbelastendes Ventil n
- ES válvula de contrapeso f
- F soupape f à contre-poids
- I valvola f a contrappeso
- MA súlyterhelésű szelep
- NL gewichtbelaste klep f
- PO zawór m ciężarkowy
- PY грузовой клапан m
- SU painokuormitettu venttiili
- SV viktbelastad ventil c

4102 weld
- DA svejsesøm svejsning
- DE Schweißnaht f Schweißstelle f
- ES soldadura f
- F cordon m de soudure soudure f
- I cordone m di saldatura giunto m saldato saldatura f
- MA hegesztés
- NL las f lasnaad m
- PO spaw m
- PY шов m
- SU hitsaus
- SV svets c

4103 weld vb
- DA svejse
- DE schweißen
- ES soldar
- F souder
- I saldare
- MA hegeszt
- NL lassen
- PO spawać
- PY сваривать
- SU hitsata
- SV svetsa

4104 welded joint
- DA svejset samling
- DE Schweißverbindung f
- ES unión soldada f
- F assemblage m soudé joint m soudé
- I giunto m saldato
- MA hegesztett kütés
- NL lasverbinding f
- PO połączenie n spawane
- PY соединение n сварное
- SU hitsausliitos
- SV svetsfog c

4105 welder
- DA svejsemaskine svejser
- DE Schweißer m
- ES soldador m
- F soudeur m
- I saldatore m
- MA hegesztő
- NL lasser m
- PO spawacz m spawarka f zgrzewarka f
- PY сварщик m
- SU hitsaaja
- SV svetsare c

4106 welding
- DA svejsning
- DE Schweißung f
- ES soldadura f
- F soudage m
- I saldatura f
- MA hegesztés
- NL lassen n
- PO spawanie n
- PY сварка f
- SU hitsaaminen
- SV svetsning c

4107 welding bend
- DA svejsebøjning
- DE Schweißboden m
- ES codo para conexiones soldadas m codo para soldar m codo soldado m
- F coude m à souder
- I curva f a saldare gomito m a saldare
- MA hegesztett ívcső
- NL lasbocht f
- PO łuk m spawany
- PY сварное колено n
- SU hitsattava käyrä
- SV svetsad büj c

4108 welding elbow
- DA svejsebøjning
- DE Einschweißbogen m
- ES codo para conexiones soldadas m codo para soldar m codo soldado m
- F coude m à souder
- I curva f a saldare gomito m a saldare
- MA hegesztett künyük
- NL laskniebocht f
- PO kolano n spawane
- PY сварной отвод m
- SU hitsattava kulma
- SV svetsad büj c

4109 welding flange
DA svejseflange
DE Schweißflansch *m*
ES brida para soldar *f*
F bride *f* à souder
I flangia *f* a saldare
MA hegesztett karima
NL lasflens *m*
PO kołnierz *m* spawany
PY сварной фланец *m*
SU hitsattava laippa
SV svetsfläns *c*

4110 welding rod
DA svejseelektrode
svejsetråd
DE Schweißdraht *m*
ES varilla de metal de
aporte *f*
varilla de soldar *f*
F baguette *f* de
soudure
I elettrodo *m* di
saldatura
MA hegesztőpálca
NL lasstaaf *f*
PO elektroda *f*
spawalnicza
PY сварочная
проволока *f*
SU hitsauspuikko
SV svetstråd *c*

4111 weld neck flange
DA påsvejst flange
DE Anschweißflansch *m*
ES brida de cuello
soldado *f*
F bride *f* soudée
I flangia *f* a collo
saldato
MA hegesztőtoldatos
karima
NL lasflens *m*
PO kołnierz *m* z
króćcem do
spawania
PY фланец *m* с
канавкой для
приваривания
трубы
SU hitsattava laippa
SV svetsfläns *c*

4112 well
DA brønd
DE Brunnen *m*
ES pocillo *m*
pozo *m*
vaina *f*
F puits *m*
I pozzo *m*
MA kút
NL bron *f* ho
PO studnia *f*
studzienka *f*
ściekowa
PY колодец *m*
SU kaivo
SV brunn *c*

4113 well water
DA brøndvand
DE Brunnenwasser *n*
ES agua de pozo *f*
F eau *f* de puits
I acqua *f* di pozzo
MA kútvíz
NL bronwater *n*
PO woda *f* studzienna
PY колодезная вода *f*
SU kaivovesi
SV brunnsvatten *n*

4114 wet air filter (*see*
capillary air washer)

**4115 wet bulb
depression**
DA psykrometrisk
difference
DE psychrometrische
Differenz *f*
ES diferencia
psicrométrica *f*
F différence *f*
psychrométrique
I differenza *f*
psicrometrica
MA pszichrometrikus
nyomáscsükkenés
NL natteboldepressie *f*
temperatuurverschil
n droge en natte
bol
PO różnica *f*
psychrometryczna
PY психрометрическая
разность *f*
SU kuivan ja märän
lämpütilan erotus
SV våttermometer-
tryckfall *n*

**4116 wet bulb
temperature**
DA vådtemperatur
DE Feuchtkugel-
temperatur *f*
ES temperatura de
bulbo húmedo *f*
temperatura del
termómetro *f*
F température *f*
humide
I temperatura *f* a
bulbo bagnato
(umido)
MA nedves hőmérséklet
NL nattebol-
temperatuur *f*
PO temperatura *f*
termometru
mokrego
PY температура *f*
мокрого
термометра
SU märkälämpütila
SV våttermometer-
temperatur *c*

4117 wet return
DA dykket retur
våd retur
DE nasser Rücklauf *m*
ES tubería de retorno
bajo el nivel del
agua (calderas) *f*
F retour *m* noyé
I ritorno *m* bagnato
MA nedves
kondenzvezeték
NL natte condensaat-
leiding *f*
PO powrót *m* zalany
PY влажный
возврат *m*
SU kostean höyryn
paluujohto
SV fuktreturledning *c*

4118 wet steam
DA våd damp
DE Naßdampf *m*
ES vapor húmedo *m*
F vapeur *f* humide
I vapore *m* umido
MA nedves gőz
NL natte stoom *m*
PO para *f* wilgotna
PY влажный пар *m*
SU kostea höyry
SV våt ånga *c*

4119 wet surface (of boiler)

DA våd overflade (på kedel)
DE nasse Heizfläche *f* des Kessels *m*
ES superficie caliente (de una caldera) *f* superficie mojada *f*
F surface *f* mouillée (de chaudière)
I superficie *f* bagnata
MA nedves felület (kazánnak)
NL nat oppervlak *n* (van ketel)
PO powierzchnia *f* mokra (kotła)
PY смачиваемая поверхность *f* (котла)
SU märkä pinta
SV våt yta *c*

4120 wet-type air cooler

DA vådluftkøler
DE feuchtarbeitender Luftkühler *m*
ES enfriador húmedo de aire *m*
F refroidisseur *m* d'air du type humide
I refrigeratore *m* di aria a umido
MA nedves típusú léghűtő
NL luchtkoeler *m* met direct vloeistofcontact natte luchtkoeler *m*
PO chłodnica *f* powietrza pracująca na mokro
PY контактный воздухоохладитель *m*
SU märkä ilmanjäähdytin
SV våtluftkylare *c*

4121 wheel

DA hjul rat
DE Rad *n*
ES rueda *f* volante *m*
F roue *f*
I ruota *f*
MA kerék
NL wiel *n*
PO koło *n*
PY колесо *n*
SU pyörä
SV hjul *n*

4122 wind

DA vind
DE Wind *m*
ES viento *m*
F vent *m*
I vento *m*
MA szél
NL wind *m*
PO wiatr *m*
PY ветер *m*
SU tuuli
SV vind *c*

4123 window

DA vindue
DE Fenster *n*
ES ventana *f*
F fenêtre *f*
I finestra *f*
MA ablak
NL raam *n* venster *n*
PO okno *n*
PY окно *n*
SU ikkuna
SV fünster *n*

4124 window air conditioner

DA klimaaggregat til vinduesmontering
DE Fensterklimagerät *n*
ES acondicionador de ventana *m*
F conditionneur *m* d'air de fenêtre
I condizionatore *m* da finestra
MA ablakklíma készülék
NL raamkoeler *m*
PO klimatyzator *m* okienny
PY оконный кондиционер *m*
SU ikkunailmastointikone
SV fünster-klimatapparat *c*

4125 window crack

DA vinduesfuge
DE Fensterfuge *f*
ES rendija de una ventana *f*
F joint *m* de fenêtre
I fessura *f* della finestra
MA ablakhézag
NL raamspleet *f*
PO nieszczelność *f* okna
PY неплотность *f* в окне
SU ikkunarako
SV fünsterfog *c*

4126 window-sill heating

DA vinduesbrystnings-opvarmning
DE Fensterbank-heizung *f*
ES calefacción por antepecho *f*
F chauffage *m* en allège
I riscaldamento *m* a soglia sotto finestra
MA parapetfűtés
NL vensterbank-verwarming *f*
PO ogrzewanie *n* podokienne
PY подоконное отопление *n*
SU ikkunalaudan lämmitys
SV fünsterbänks-uppvärmning *c*

4127 wind pressure

DA vindtryk
DE Winddruck *m*
ES empuje del viento *m* presión del viento *f*
F pression *f* du vent
I pressione *f* del vento
MA szélnyomás
NL winddruk *m*
PO ciśnienie *n* wiatru
PY ветровое давление *n*
SU tuulen paine
SV vindtryck *n*

4128 wind tunnel
DA vindtunnel
DE Windkanal *m*
ES tunel
    aerodinámico *m*
F tunnel *m* d'essai
    aéraulique
I galleria *f* del vento
MA szélcsatorna
NL windtunnel *m*
PO tunel *m*
    aerodynamiczny
PY аэродинмическая
    труба *f*
SU tuulitunneli
SV vindtunnel *c*

4129 wind turbine fan
DA vindturbineventilator
DE windgetriebener
    Ventilator *m*
ES turboventilador *m*
F ventilateur *m* à
    turbine éolienne
I turbina *f* eolica
MA szélkerék ventilátor
NL door de wind
    aangedreven
    ventilator *m*
    windventilator *m*
PO silnik *m* wiatrowy
PY ветряной
    вентилятор *m*
SU tuuliroottori
SV vindturbinfläkt *c*

**4130 winter air
conditioning**
DA vinterventilation
DE Winter-
    klimatisierung *f*
ES acondicionamiento
    de aire para
    invierno *m*
F conditionnement *m*
    d'air d'hiver
I condizionamento *m*
    invernale
MA téli légkondicionálás
NL winterlucht-
    behandeling *f*
PO klimatyzacja *f* w
    okresie zimy
PY зимнее
    кондициониро-
    вание *n* воздуха
SU talviajan ilmastointi
SV vinterluftkondition-
    ering *c*

4131 wire
DA tråd
    wire
DE Draht *m*
ES alambre *m*
    hilo *m*
F fil *m*
I filo *m*
MA drót
    huzal
NL draad *n*
PO drut *m*
    przewód *m*
    elektryczny
PY проволока *f*
SU lanka
SV tråd *c*

4132 wiring diagram
DA koblingsdiagram
DE Verdrahtungs-
    schema *n*
ES diagrama de
    circuito *m*
    esquema de
    conexiones *m*
    esquema de montaje
    m
F schéma *m* de
    cablage
I schema *m* elettrico
MA huzalozási vázlat
NL bedradingsschema *n*
PO schemat *m* połączeń
    elektrycznych
PY схема *f*
    электропроводки
SU kytkentäkaavio
SV kopplingsschema *n*

4133 wiring system
DA koblingssystem
DE Verdrahtungs-
    system *n*
ES sistema de
    caldeado *m*
F système *m* de
    cablage
I impianto *m* elettrico
MA huzalozási rendszer
NL bedradingssysteem *n*
PO okablowanie *n*
PY проводная
    система *m*
SU kytkentätapa
SV kopplingssystem *n*

**4134 wobble plate
compressor**
DA slingreplade-
    kompressor
DE Tunnelscheiben-
    Verdichter *m*
ES compresor de disco
    oscilante *m*
F compresseur *m* à
    plateau oscillant
I compressore *m* a
    piatto inclinato
MA bolygótárcsás
    kompresszor
NL tuimelplaat-
    compressor *m*
PO sprężarka *f*
    tarczowa
PY крейцкопфный
    поршневой
    компрессор *m*
SU kalvokompressori
    kiikkulevy-
    kompressori
SV

4135 workable *adj*
DA anvendelig
    bearbejdelig
DE ausführbar
    betriebsfähig
ES capacidad de
    funcionar *f*
    laborable
    practicable
    viable
F capable de
    fonctionner
I lavorabile
    realizzabile
MA megmunkálható
NL werkbaar
PO możliwy do realizacji
    nadający się do
    obróbki
PY работоспособный
SU käytännüllinen
SV användbar

DA	=	Danish
DE	=	German
ES	=	Spanish
F	=	French
I	=	Italian
MA	=	Magyar (Hungarian)
NL	=	Dutch
PO	=	Polish
PY	=	Russian
SU	=	Finnish
SV	=	Swedish

4136 working drawing
DA arbejdstegning
DE Bauplan *m*
 Werkstatt-
  zeichnung *f*
ES plano de
  construcción *m*
F schéma *m* de
  montage
I disegno *m* esecutivo
 schema *m* di
  montaggio
MA kiviteli terv
 szerelési terv
NL werktekening *f*
PO rysunek *m*
  warsztatowy
 rysunek *m*
  wykonawczy
PY рабочий чертеж *m*
SU työpiirustus
SV arbetsritning *c*

4137 working order
DA arbejdsanvisning
 driftsberedskab
DE Arbeitsanweisung *f*
 Betriebs-
  bereitschaft *f*
ES orden de
  ejecución *m*
F en état *m* de marche
 en ordre *m* de
  marche
I in grado di
  funzionare
 ordine *m* di
  lavorazione
MA üzemelési utasítás
NL bedrijfsvaardigheid *f*
 werkinstructie *f*
 werkvolgorde *f*
PO stan *m* gotowości do
  pracy
PY работоспособность
  f
SU työjärjestys
SV användbart skick *n*

4138 working pressure
DA arbejdstryk
DE Betriebsdruck *m*
ES presión de
  funcionamiento *f*
 presión de servicio *f*
F pression *f* de marche
 pression *f* de service
I pressione *f* di
  esercizio
MA üzemnyomás
NL werkdruk *m*
PO ciśnienie *n* robocze
PY рабочее давление *n*
SU työpaine
SV arbetstryck *n*

**4139 working
 temperature**
DA arbejdstemperatur
DE Betriebstemperatur *f*
ES temperatura de
  trabajo *f*
F température *f* de
  travail
I temperatura *f* di
  esercizio
MA üzemi hőmérséklet
NL werktemperatuur *f*
PO temperatura *f*
  robocza
PY рабочая
  температура *f*
SU toimintalämpötila
 työlämpötila
SV arbetstemperatur *c*

4140 works
DA drift
 værk
DE Werk *n*
ES fábrica *f*
F usine *f*
I fabbrica *f*
 stabilimento *m*
MA művek
NL fabriek *f*
PO fabryka *f*
 zakład *m*
  produkcyjny
PY завод *m*
SU tehdas
SV industri *c*

4141 workshop
DA værksted
 workshop
DE Werkstatt *f*
ES taller *m*
F atelier *m*
I officina *f*
 reparto *m* di
  lavorazione
MA műhely
NL werkplaats *f*
PO oddział *m*
  produkcyjny
 warsztat *m*
PY мастерская *f*
SU verstas
SV verkstad *c*

4142 worm
DA snekke
DE Schnecke *f*
ES tornillo sin fín *m*
F vis *f* sans fin
I vite *f* senza fine
MA csiga
NL worm *m*
PO ślimak *m*
PY червяк *m*
SU ruuvi
SV snäckskruv *c*

4143 worm feed
DA snekkefødning
DE Fürderschnecke *f*
 Schnecken-
  fürderer *m*
ES alimentación por
  tornillo sin fín *f*
F vis *f* d'Archimède
I alimentazione *f* a
  coclea
MA adagolócsiga
NL worm-
  transporteur *m*
PO podawanie *n*
  przenośnikiem
  ślimakowym
PY шнековая подача *f*
SU syüttüruuvi
SV kontinuerlig
  matning *c*

4144 wrench
DA skruenøgle
DE Schrauben-
    schlüssel *m*
ES llave para tuercas *f*
F clé *f*
  clé *f*
I chiave *f*
MA csavarkulcs
NL moersleutel *m*
PO klucz *m* maszynowy
PY гаечный ключ *m*
SU jakoavain
SV skruvnyckel *c*

Y

4145 year round
DA året rundt
DE ganzjährig
ES para invierno y
   verano
   para toda la estación
   para todo el año
F toute l'année
I per tutto l'anno
   tutto l'anno
MA egész éves
NL gedurende het gehele
   jaar *n*
PO cały rok *m*
PY круглогодовой *adj*
SU ympärivuotinen
SV året *n* runt

4146 Y-piece (ducting)
DA y-stykke (i kanal)
DE Y-Stück *n*
ES pieza bifurcada *f*
   tubo bifurcado *m*
   tubo en Y *m*
F culotte *f* (conduit
   d'air)
I pezzo *m* a Y
MA ipszilon elágazás
NL broekstuk *n*
PO trójnik *m* w
   kształcie Y
PY тройник *m*
SU kanavahaarakappale
   Y-kappale
SV Y-stycke *n*

DA	=	Danish
DE	=	German
ES	=	Spanish
F	=	French
I	=	Italian
MA	=	Magyar
		(Hungarian)
NL	=	Dutch
PO	=	Polish
PY	=	Russian
SU	=	Finnish
SV	=	Swedish

Z

4147 zero (absolute)
DA nul (absolut)
DE absoluter
    Nullpunkt *m*
ES cero (absoluto) *m*
F zéro *m* absolu
I zero *m* (assoluto)
MA nulla
NL nulpunt *n* (absolute)
PO zero *n* (absolutne)
PY нуль *m*
    (абсолютный)
SU absoluuttinen
    nollapiste
SV nollpunkt *c*
    (absolut)

4148 zero point energy
DA nulpunktsenergi
DE Nullpunktenergie *f*
ES energía del punto
    cero *f*
F énergie *f* au zéro
    absolu
I energia *f* del punto
    di riferimento
MA nullenergia
NL nulpunt-energie *f*
PO energia *f* zerowa
PY абсолютный
    нуль *m*
SU nollapiste-energia
SV nollpunktsenergi *c*

4149 zone
DA zone
DE Zone *f*
ES zona *f*
F zone *f*
I zona *f*
MA zóna
NL gebied *n*
    zone *f*
PO obszar *m*
    strefa *f*
PY зона *f*
SU vyühyke
SV zon *c*

**4150 zone air
    conditioner**
DA zone-luftbehandler
DE Zonenklimagerät *n*
ES climatizador zonal *m*
F conditionneur *m*
    d'air de zone
I condizionatore *m* di
    zona
MA zónás
    légkondicionáló
NL zoneluchtbehandelings-
    apparaat *n*
PO klimatyzator *m*
    strefowy
PY зональный
    кондиционер *m*
SU vyühykeilmasto-
    intikone
SV zonluft-
    konditionering *c*

4151 zoning
DA zoneopdeling
DE Zoneneinteilung *f*
ES zonificación *f*
F zonage *m*
I suddivisione *f* a
    zone
MA szakaszolás
NL zone-indeling *f*
    zoneverdeling *f*
PO podział *m* na strefy
PY зонирование *n*
SU jakaa vyühykkeisiin
SV skiktning *c*

Industrial Property Markets In Western Europe

Edited by **B Wood** and **R H Williams** both of University of Newcastle upon Tyne, UK

> "the book is informative, well presented and succeeded in its aim of introducing the developmental process and industrial markets of Western European nations....This is a good basic text, likely to be of interest to professionals and students alike." - *Journal of Property Valuation & Investment*

Industrial Property Markets in Western Europe sets out in a comparative framework the operation of the industrial property market in seven major West European economies: **Germany, Spain, France, Italy, The Netherlands, Sweden and the United Kingdom**. The book provides a practical introduction to the complexities of operating in the market, in each of the countries, for the practitioner and student of real estate and property development. Each chapter is written by experts from the country concerned and covers planning and property rules and regulations, procedures, and the roles of agents and other participants in the process. Illustrative case studies are used extensively and advice is offered on further reading.

Contents: Introduction - *B Wood and R H Williams*. Key issues and themes - *B Wood and R H Williams*. Germany - *H Dieterich and E Dransfeld*. Spain - *E Calderon and I Espanol*. France - *A Motte*. Italy - *A Fubini, A Landi and R Curti*. Netherlands - *B Needham and B Kruijt*. Sweden - *H Mattsson*. United Kingdom - *B Wood and R H Williams*. Some comparison and contrasts - *B Wood and R H Williams*. Index

June 1992: 234x156: 282pp, b/w photographs, line illustrations
Hardback: 0-419-17050-2

For further information on this and other property related titles, please contact:
The Promotion Dept., E & F N Spon, 2-6 Boundary Row, London SE1 8HN
Tel: 071 865 0066 Fax: 071 522 9623

Part Two

Indexes

Dansk

blandekredsløb, 2695
blandepumpe, 2698
blander, 436
blandet gas, 2692
blandet luft, 2690
blandeventil, 2700
blanding, 436, 2701
blandingsforhold, 2699
blandingsventilator, 2691
blende, 2858
blendplade, 2859
blik, 3419
blindeffekt, 3166
blindflange, 438, 3663
blokere, 439
blokering, 440, 777
blokeringsventil, 443
blokopvarmning, 441
blotlagt, 1762
blus, 1916
bly, 2523
blypakning, 2524
blære, 541
blæse, 444
blæsebælg, 396
blæselampe, 429
blæser, 447, 1790
blæseventilator, 448
blød bøjning, 3501
blødgøring, 3510, 4064
blødstøbte fittings, 2619
boble, 541
boblestrøm, 542
boks, 503
boligopvarmning, 1389, 3250
boligventilation, 3249
bolometer
   (strålevarmemåler), 485
bolt, 486
bord, 3751
boring, 767
bortfjernelse, 3241
Bourdon manometer, 500
bourdonrør, 501
brakvand, 505
brand, 1873
brandalarmsystem, 1874
branddrøj, 1890
branddør, 1882
brandmodstand, 1889
brandsikker, 1886
brandsikre, 1887
brandsikret, 1888
brandspjæld, 1881
brandventil, 1893
bremsekraft, 506
briket, 534
brine, 528
brineanlæg, 532
brinebeholder, 533
brinekredsløb, 529
brinekøler, 530
brineoverrislingskøleanlæg,
   531
britisk standard, 535

brugbarhed, 3397
brugsklar, 3170
brunkul, 537, 2548, 2549
bruttokapacitet, 2195
brystningsmonteret rist, 3446
brændbar, 842
brændbarhed, 841, 1909
brænde, 567
brænde ud, 581
brænder, 568
brænder med flere dyser,
   2745
brænderdyse, 572
brænderflange, 570
brænderhoved, 578
brændermundstykke, 579
brænderplade, 573
brænderregister, 576
brænderrensning, 575
brænder-skueglas, 580
brændertænding, 571
brændsel, 855, 2076
brændselsanalyse, 2078
brændselsart, 2090
brændselsbehandling, 2086
brændselsforbrug, 2080
brændselsforsyning, 2088
brændselshåndtering, 2084
brændselslager, 2087
brændselsolie, 2085
brændselstilførsel, 2081
brændselstransport, 2089
brændstof, 855, 2076
brændstofvalg, 2079
brændtid, 852
brændværdi (øvre), 608
brændværdi (nedre), 609
brænstoffilter, 2082
brønd, 4112
brøndvand, 4113
budget, 1682
bue, 1147
buesvejsning, 261
buffertank, 544
bundblæse, 445
bunden smeltevarme, 2518
bunden varme, 2517
bundfald, 1237, 3377
bundfælde, 1238
bundfældning, 3024
bundt, 565
bunker, 566
bus, el-ledning, 584
butan, 588
butterflyspjæld, 589
butterflyventil, 590
bygas, 3873
byggeelement, 549
byggeleder, 770
byggemateriale, 555, 2637
byggepladsmontage, 3461
byggeprojekt, 556
byggetid, 890
bygherrerepræsentant, 3460
bygning, 545

bygnings varmebelastning,
   558
bygningsafstand, 557
bygningsareal, 259, 546
bygningsautomation, 547
bygningsreguleringssystem,
   548
bygningsstyresystem, 554
bygningsvolumen, 260
by-pass, 594
by-pass-spjæld, 595
byte (EDB), 598
bælg, 396
bælgtætning, 397
bælgventil, 398
bærbar, 3005
bære, 3729
bærefrekvens, 636
bærejern, 504
bøje, 403
bøjelig kanal, 1936
bøjeligt rør, 1939
bøjle, 502, 2214
bøjlekonsol, 3658
bøjning, 1563
børstefilter, 538
bøsning, 585, 587, 3473
båndbredde, 366

CAD, edb-tegnesystem, 923
Carnot proces, 635
celle, 667
cellefilter, 670
celleopbygget isolering af
   gummi, 672
cellestruktur, 673
celletermoisolering, 674
celsiusgrader, 676
cement, 675
centerhastighed, 685
centralbehandlerenhed, 683
centralluftbehandlingsenhed,
   684
centralstyrepanel, 679
centralstyring, 678
centralvarme, 680, 2092
centralvarmeanlæg, 681
centrifugal, 687
centrifugalkompressor, 688
centrifugalkraft, 690
centrifugalpumpe, 691
centrifugalventilator, 689
centrifuge, 693, 1773
certifikat, 695
CFC, 741
chamotte, 1880
chamottesten, 1878
cirkulation, 747
cirkulationspumpe, 749
cirkulationstryk, 748
cirkulationsventilator, 746
cirkulator, 750
cisterne, 752
civilingeniør, 3092

fordampningsmåler, 1702
fordampningstab, 1692
fordampningstemperatur, 1695
fordampningstryk, 1693
fordampningsturbulator, 1706
fordampningsvarme, 2519
fordeler, 1376
fordelerdåse, 1370, 2504
fordeling, 1369
fordelingskanal, 1371
fordelingsledning, 1374
fordelingsledningsnet, 1373
fordelingsspjæld, 3576
fordelingssystem, 1375
fordelingstavle, 3742
forebyggende vedligeholdelse, 3076
foreløbig tegning, 3037
forenelig, kan passe sammen, 884
foretagende, 1644
forfilter, 3027, 3031
forfyring, 3032
forgasning, 2135
forgrening, 2622
forhold, 3164
forindstille, 3039
forindstilling, 3040
foring, 2555, 3473
foringsring, 2906
forkoksning, 814
forkrøppe, 1122
forkrøbning, 403, 1121
forkulning, 632
forkøler, 2022, 3028
forkøling, 3029
forlægning i jord, 2522
formand, 714
formatforhold, 279
formstykke, 2733
foropvarmning, 3035
foropvarmningstid, 3036
forplade, 2069
forrensning, 3027
forrigle, gribe ind i hinanden, 2470
forringelse, 1265
forsegle, 3362
forseglet kondenseringsunit, 3363
forsegling, 3361
forskel, 1288
forskning, 3245
forskrift, 796
forskruning, 3924
forstoppelse, 2797
forstærke, 209, 489, 490, 3230
forstærker, 208, 491
forstærkning, 207, 2107
forstærkningsramme, 3231
forstørret overflade, 1763
forstøve, 299, 3580

forstøvet vand, 2761
forstøvningsbefugter, 300
forstøvningskammer, 3581
forsyne, 3720
forsyning, 3719
forsøg, 1757, 3776
forsøge, 1758
fortinning, 3856
fortrængningsarbejde, 1972
fortrængningskompressor, 3008
fortynder, 1306
fortynding, 302, 1308
fortyndingsmiddel, 1307
fortætning, 949
fortætningstemperatur, 959
forurene, 1005
forurener, 1006, 3000
forurenet, 3001
forurenet luft, 4010
forurening, 2026, 3002
forureningsfaktor, 2027
forvarmer, 3034
forvarmeslange, 3033
forvented varme, 2226
fossilt brændsel, 2025
Foucault strøm, 1544
fragt, 634
frakoble, 1152
frakoblet ledning, 1186
frakoblingsindstilling, 1154
frakoblingspunkt, 1153
frekvens, 2057
frekvensgengivelse, 2058
fremadretter, 238
fremadrettet, 2024
frembringelse, 2161
fremløbsfordeler, 3727
fremløbstemperatur, 1970
freon, 2056
frihedsgrad, 1219
friktion, 2066
friktionsfaktor, 2066
friktionsløs foring, 241
friktionsmodstand, 2065
friktionsring, 2068
friktionstab, 2067
frisk, 2059
frisk damp, 2567
frisk luft, 2060
friskhed, 2062
friskluftandel, 2063
friskluftsupplering, 2061
fritblæsende luftbehandler, 2034
frontareal, 1776
frontelement, 2070
fronthastighed, 1777
frontplade, 2069
frost, 2071
frostbeskyttelse, 2075
frostsikker, 2044
frostsikring, 240
frostskade, 2072
fryse, 2039

fryse tørrer, 2040
fryse tørring, 2041
fryseanlæg, 2051
fryseblanding, 2050
frysehastighed, 2054
frysemedium, 2038
frysemetode, 2053
frysepunkt, 2052
fryser, 2043
fryserum, 2048
frysetid, 2055
fryseudstyr, 2049
frysning, 2047
fuelolie, 2085
fugt, 1165, 2712
fugtbalance, 2713
fugtbeskyttelse, 2716
fugte, 1166
fugte, blive fugtig, 1167
fugtfjerner, 1221
fugtig, 2351, 2711
fugtighed, 1172, 2357
fugtighedsdiagram, 3115
fugtighedsføler, 2360
fugtighedsmåler, 3114
fugtighedsregulering, 2358
fugtindhold, 2714
fugtkatalysator, 2717
fugtmåling, 2715
fugtoverføring, 2719
fugttransmission, 2720
fugtudskiller, 2718
fuld last, 2093
fuld luftbehandling, 2091
fuldlaststart, 3611
fuld-modulerende, 2094
fuldstændig forbrænding, 889, 2932
fundament, 375, 2028
fundamentsplade, 395
funktion, 2852
fyldmasse, 1836
fyldningsgrad, 719
fyr, 2099
fyrboks, 1877, 2100
fyrbro, 1879
fyrdør, 718
fyring med forskellige brændsler, 2744
fyringsolie, 2254
fyringsprogram, 1895
fyringssæson, 2257
fyrrist, 1875
fælde, 3892
fælles, 2989
fælles nulledning, 880
færdigmonteret, 2897
færdigmonteret enhed, 2901
færdigmonteret kedel, 2899
færdigmonteret luftbehandler, 2898
færdigmonteret luftbehandlingsanlæg, 2900
fæstne, 384
fødeledning, 1818, 3726

Deutsch

dämpfen, 1167
Dämpfung *f*, 302, 1171
Dampf *m*, 3622, 3959
Dampfanschluß *m*, 3627
Dampfbefeuchter *m*, 1701,
 3635
dampfbeheizter Erhitzer *m*,
 3630
Dampfblase *f*, 3641
Dampfdruck *m*, 3642, 3966
Dampfdruckreduzierventil *n*,
 3643
Dampferzeuger *m*, 3632
Dampffilter *m*, 3631
Dampffühler *m*, 3962
Dampfgemisch *n*, 3965
dampfgespeiste
 Energieerzeugungseinheit
 f, 3961
Dampfheizung *f*, 3634
Dampfkessel *m*, 3624
Dampfkondensator *m*, 3626
Dampfmantel *m*, 3637
Dampfraum *m*, 3625
Dampfrohr *n*, 3640
Dampf-Sammelleitung *f*,
 3633
Dampfsperre *f*, 3960, 3964,
 3967
Dampfstrahlkälteerzeugung
 f, 3639
Dampfstrahl-Kältekreislauf
 m, 3963
Dampfstrahlkältesystem *n*,
 1560
Dampfstrahl-Luftejektor *m*,
 3638
Dampfstrahl-
 Speisewasserpumpe *f*,
 3636
Dampftafeln *f,pl*, 3644
Dampftrockner *m*, 3629
Dampfturbine *f*, 3646
Dampfverbrauch *m*, 3628
Dampf-Zerstäubungsbrenner
 m, 3623
Daten *f,pl*, 1174
Datenbank *f*, 1175
Datenbasis *f*, 1176
Datensichtmodul *n*, 1178
Datenspeicherung *f*, 1179
Datentafel *f*, 1181
Datenumwandler *m*, 1177
Datenverarbeitungsmaschine
 f, 1180
Dauer *f*, 1509, 3846
Dauer *f* des periodischen
 Betriebs *m*, 1157
Dauerbetrieb *m*, 1009
Dauerhaftigkeit *f*, 1508
DDC-Regelung *f*, 1318
Decke *f*, 658
Deckel *m*, 614, 1113, 2323,
 2537

deckenartige
 Wärmedämmung *f*, 421
Deckenauslaß *m*, 663
Deckenbauart *f*, 664
Deckenheizung *f*, 662, 2234
Deckenhohlraum *m*, 666
Deckenluftauslaß *m*, 660
Deckenrohrschlange *f*, 659
Deckenventilator *m*, 665
Deflektor *m*, 1205
Dehnbarkeit *f*, 1496
Dehnungsausgleicher *m*, 887
Dehnungsbälge *m,pl*, 1742
Dehnungsbogen *m*, 1743,
 1749
Dehnungsfuge *f*, 1748
Deklination *f*, 1196
Dekompressor *m*, 1198
Demodulation *f*, 1234
demontieren, 1336
Desinfektion *f*, 1358
Desinfektionsmittel *n*, 341
Desorption *f*, 1255
Destillation *f*, 1367
destilliertes Wasser *n*, 1368
detaillierter
 Kostenvoranschlag *m*,
 1261
Dezibel *n*, 1194
Dezipol *n*, 1195
Diagramm *n*, 1273, 2177
dicht, 3097
Dichte *f*, 1235
Dichtheit *f*, 3845
Dichtheitsprüfung *f*, 2527
Dichtigkeit *f*, 3845
dichtschließende Tür *f*, 1993
Dichtung *f*, 2136, 2500,
 2903, 3361
Dichtungsmasse *f*, 2501,
 3366
Dichtungsmaterial *n*, 2905
Dichtungsring *m*, 2906, 4049
Dichtungsschutz *m*, 2502
Dicke *f*, 3821
dielektrisch, 1282
dielektrische Konstante *f*,
 1283
dielektrische Kraft *f*, 1284
dielektrische Stärke *f*, 1284
dielektrisches Auftauen *n*,
 1285
Dieselmaschine *f*, 1286
Dieselmotor *m*, 1287
Differential *n*, 1289
Differentialregler *m*, 1290
Differenz *f*, 1288
Differenzdruck *m*, 1292
Differenzdruckmesser *m*,
 1294
Differenzdruckregler *m*, 1293
Differenzdruckschalter *m*,
 3047
Differenztemperatur *f*, 1295
diffundierend, 1296

diffus, 1296
diffuse Reflektion *f*, 1298
Diffusion *f*, 1300
Diffusionsabsorptionssystem
 n, 1303
Diffusionsfläche *f*, 1301
Diffusionszahl *f*, 1302
Diffusor *m*, 1297
Diffusorradius *m* der
 Diffusion *f*, 1299
Digestorium *n*, 2097
digital, 1304
digitale Übertragung *f*, 1305
Dimension *f*, 1309
Direktabsaugungssystem *n*,
 1320
Direktantrieb *m*, 1319
direktbefeuerter Lufterhitzer
 m, 1324
direktbefeuertes Heizgerät *n*,
 1325
direkte Heizfläche *f*, 3080
direkte Kühlung *f*, 1314
direkte
 Warmwasserversorgung *f*,
 1328
direkter Zugriffsspeicher *m*,
 3155
direktes Ausdehnungssystem
 n, 1323
Direktheizungssystem *n*,
 1327
Direktkältesystem *n*, 1333
Direktkontaktwärmetauscher
 m, 1313
Direktkühlungssystem *n*,
 1315
Direktventilationssystem *n*,
 1334
direktverdampfendes
 Kühlsystem *n*, 1322
Direktverdampfer-
 rohrschlange *f*, 1321
Dom *m*, 1387
Doppel *n*, 1507
Doppelboden *m*, 1951
Doppelrohrbündel-
 verflüssiger *m*, 1397
Doppelrohrverflüssiger *m*,
 1401
Doppelrohr-Wärmetauscher
 m, 1402
Doppelscheibenfenster *n*,
 1398
doppelseitiger Ventilator *m*,
 406
Doppelsitzventil *m*, 1404
doppelte Saugleitung *f*, 1407
Doppelthermostat *n*, 1491
doppeltwirkende Regelung *f*,
 1487
doppeltwirkende Verdichtung
 f, 1483
doppeltwirkender Verdichter
 m, 1395, 1486

Doppelunterbrechung *f*,
1396
DOP-Prüfung *f*, 1391
dosieren, 1393
Dosierung *f*, 1392
Dosierwasser *n*, 1394
Draht *m*, 4131
Drallströmung *f*, 4020
drehend, 3306
Drehkolbenverdichter *m*,
3289
Drehmoment *n*, 3865
Drehschalenanemometer *n*,
3277
Drehscheibe *f*, 3568
Drehströmung *f*, 3314
Dreileiterklimasystem *n*,
3828
Dreileitersystem *n*, 3829
Dreirohrsystem *n*, 3829
Dreiwegehahn *m*, 3831
Dreiwegeventil *n*, 3832
Dreizugkessel *m*, 3827
Drossel *f*, 3835
Drosselgerät *n*, 3835
Drosselklappe *f*, 589, 1168
drosseln, 3836
Drosselung *f*, 3837
Drosselventil *n*, 590
Druck *m*, 2217, 3041
Druckabfall *m*, 3048
Druckabsperrventil *n*, 1349
Druckausfall *m*, 3045
Druckausgleich *m*, 1669,
3049
Druckbegrenzer *m*, 3057
Druckbehälter *m*, 908, 2219,
3072
Druckdifferenz *f*, 3046
Druckeinspritzung *f*, 3050
Drucker *m*, 3085
Druckerhöhungspumpe *f*,
493
druckfest, 3071
Druckgas *n*, 896
Druckgefälle *n*, 3054
Druckhaltung *f*, 3073
Druckhöhe *f*, 3055
Druckkessel *m*, 3072
Druckknopfregelung *f*, 3125
Druckkühlung *f*, 3044
Drucklager *n*, 3842
Druckleistung *f*, 3060
Druckleitung *f*, 1344
Druckluft *f*, 895, 2008
Druckmesser *m*, 3053
Druckmessung *f*, 3059
Druckminderer *m*, 3063
Druckprobe *f*, 909, 3069
Druckreduzierpunkt *m*, 3061
Druckreduzierstation *f*, 3062
Druckreduzierung *f*, 3064
Druckreduzierventil *n*, 3063
Druckregler *m*, 3043, 3065
Druckschalter *m*, 3068

Druckschmierung *f*, 2019
Druckstange *f*, 3844
Druckstufe *f*, 3067
Druckverbindung *f*, 904
Druckverlust *m*, 2220, 2587,
3058
Druckzerstäubungsbrenner
m, 3042
Düse *f*, 2791
Düsenauslaß *m*, 2792
Dunst *m*, 2096
Durchbohrung *f*, 2937
Durchbruch *m*, 521
durchdringend, 1296
Durchflußdiagramm *n*, 1959
Durchflußmenge *f*, 1955
Durchflußmesser *m*, 1963
Durchflußrate *f*, 1966
Durchflußregelung *f*, 1957
Durchflußregelventil *n*, 1958,
1968
Durchflußregler *m*, 1967
Durchflußventil *n*, 1326
Durchführung *f* (elektr.),
587
Durchgangshahn *m*, 3840
Durchgangsventil *n*, 3840
Durchlässigkeit *f*, 2942
Durchlauferhitzer *m*, 2166,
2452
durchlöchert, 2933
Durchmesser *m*, 1276
Durchsatz *m*, 3838
durchschnittlich, 314, 2643
Durchschnitts-, 3375
Durchschnittsdruck *m*, 2644,
2659
durchspülen, 1992
dynamisch, 1530
dynamische Grafiken *f*, 1532
dynamische Zähigkeit *f*,
1536
dynamischer Druck *m*, 1535
dynamischer Verlust *f*, 1534
dynamisches Verhalten *n*,
1531

Echtzeit *f*, 3172
Eckbrenner *m*, 1088
Ecke *f*, 1547
Eckventil *n*, 231, 1089
Effekt *m*, 1549
effektiv, 1550
effektive Fläche *f*, 1551
effektive Lebensdauer *f*,
1554
Eichung *f*, 604
Eigenfrequenz *f*, 2759
Einbau-, 560
Einbaueinheit *f*, 2901
einbauen, 2450
einbaufertig, 2897
einbaufertige Einheit *f*, 2901
einbaufertiges Klimagerät *n*,
3927

Einfachfenster *n*, 3452
Einfachtarif *m*, 1932
einfachwirkender Verdichter
m, 3449
Einfassung *f*, 1617
Einflügelklappe *f*, 3453
eingebettet, 1605
eingraben, 583
Einheit *f*, 3925
Einheits-, 3925
Einkanalklimaanlage *f*, 3450
Einkanalsystem *n*, 3451
Einkaufspreis *m*, 1896
Einlaß *m*, 2443
Einlaßventil *n*, 2445
einregulieren, 52
Einrichtung *f*, 1674, 2451,
3406
Einrohrheizung *f*, 2833
Einrohrsystem *n*, 3454
einschalten, 1148, 2992, 3745
Ein-Schalter *m*, 3608
Einschaltpunkt *m*, 1149,
3746
Einschränkung *f*, 3255
Einschweißbogen *m*, 4108
Einspeisung *f*, 1812
Einspritzdüse *f*, 2439, 3586
Einspritzpunkt *m*, 3588
Einspritzrohr *n*, 2440, 3587
einstecken, 2992
einstellbar, 53
Einstellbereich *m*, 55
einstellen, 52, 3404
Einstellgenauigkeit *f*, 3407
Einstellpunkt *m*, 3405
Einstellung *f*, 54, 3406
Einstellwert *m*, 3410
eintauchen, 2392
Einteilung *f*, 758
Eintritt *m*, 1654, 2443
Eintrittsöffnung *f*, 2444
Eintrittsverlust *m*, 1655
Einzelheit *f*, 1258
Einzelheizung *f*, 2414
Einzelregler *m*, 3596
Einzelteilzeichnung *f*, 1260
Einzelwiderstand *m*, 2578
einzufüllende
Wärmedämmung *f*, 1838
Eis *n*, 2372
Eisbehälter *m*, 2375
Eisbildung *f*, 2074, 2380
Eisblock *m*, 2374
Eisen *n*, 2483
Eiserzeugungsanlage *f*, 2382
Eiskristall *m*, 2376
Eisschicht *f*, 2381
Eisspeicher *m*, 2373
Eiswasser *n*, 2378
Eiswasserkühlung *f*, 2379
Eiswürfel *m*, 2377
Ejektor *m*, 1559
Elastizität *f*, 1561
Elastomer *n*, 1562

elektrisch, 1564, 1565
elektrische Strombedarfslast
f, 1580
elektrische
Strombedarfsperiode *f*,
1581
elektrische Stromerzeugung
f, 1583
elektrische
Widerstandsheizung *f*,
3252
elektrischer Leistungsfaktor
m, 1582
elektrischer Stromlastabwurf
m, 1585
elektrischer Stromlastfaktor
m, 1584
elektrischer Widerstand *m*,
1570
elektrisches
Kontaktthermometer *n*,
1571
elektrisches Potential *n*, 1577
elektrisches
Strombedarfsintervall *n*,
1579
Elektrizität *f*, 1575
Elektrode *f*, 1589
Elektrodenkessel *m*, 1590
Elektrofilter *m*, 1601
Elektroheizung *f*, 1574
Elektrokonvektor *m*, 1572
Elektrokühlschrank *m*, 1587
elektrolytisch, 1591
elektrolytisches Paar *n*, 1592
Elektromagnet *m*, 1593
Elektromotor *m*, 1576
elektromotorische Kraft *f*,
1594
Elektronik *f*, 1598
elektronisch, 1595
elektronischer Verstärker *m*,
1596
elektronisch-pneumatisch,
1599
elektro-pneumatische
Regelung *f*, 2997
elektrostatischer Abscheider
m, 1601
elektrostatischer Filter *m*,
1600
Elektrostrahler *m*, 1573
Element *n*, 667, 1602
Emission *f*, 1611
Emissionsvermögen *n*, 1612
Emulsion *f*, 1615
Enddruck *m*, 1866
Ende *n*, 3857
Endgeschwindigkeit *f*, 3775
endothermisch, 1618
energetischer Wirkungsgrad
m, 1626
Energie *f*, 1619, 3012
Energiebilanz *f*, 1620

Energiedurchlaßgrad *m*,
1637
Energieeinsparung *f*, 1622
Energieerhaltungssatz *n*, 987
Energiefaktor *m*, 1627
Energieformen *f*, 2023
Energiekontrollsystem *n*
eines Gebäudes *n*, 550
Energiekosten *f*, 1624
Energiemanagement *n*, 1630
Energiemanagementaufgabe
f, 1632
Energiemanagement-
Kontrollsystem *n*, 1631
Energieniveau *n*, 1628
Energienutzungszahl *f*, 1626
Energiepotential *n*, 1633
Energiequelle *f*, 1635
Energieverbrauch *m*, 1623
Energieverfügbarkeit *f*, 1621
Energieverlust *m*, 1629
Energieverwaltungssystem *n*
eines Gebäudes *n*, 551
Energiewertzahl *n*, 1638
Energiewirtschaft *f*, 1625,
3015
Energiezustand *m*, 1636
Entaschung *f*, 276
Enteisung *f*, 1227
Entfärbung *f*, 1355
Entfernung *f*, 3158, 3556
entfeuchten, 1222
Entfeuchter *m*, 1221, 1436
Entfeuchtung *f*, 1220
Entfeuchtungseffekt *m*, 1223
entflammbar, 1910
entfrosten, 1206
entgasen, 3123
Entgasung *f*, 1213
Enthärtung *f*, 3510
Enthalpie *f*, 1645
Enthalpie-Diagramm *n*, 1647
entladen, 1339
Entladung *f*, 1338, 1353
Entladungskapazität *f*, 1354
entlasteter Anlauf *m*, 2571
Entlastung *f*, 2575
Entlastungsventil *n*, 3236
entleeren, 1417
Entleerung *f*, 1353, 1614
Entleerungsventil *n*, 1424
entlüften, 3123
Entlüfter *m*, 117
Entlüftung *f*, 174, 1737, 3994
Entlüftungsrohr *n*, 3996
Entlüftungsstöpsel *m*, 523
Entlüftungssystem *n*, 1735
Entlüftungsventil *n*, 3995
Entnahmekondensator *m*,
432
Entnahmeleitung *f*, 434
Entnahmestelle *f*, 431
Entnahmeventil *n*, 435
entnehmen, 430, 433
Entropie *f*, 1651

Entropiediagramm *n*, 1653
Entsalzung *f*, 1233
Entschlammen *n*, 3489
entspannter Dampf *m*, 1926
Entspannungskammer *f*,
1922
Entstaubung *f*, 1200, 1521
Entstaubungsanlage *f*, 1523
Entstaubungsausrüstung *f*,
1524
Entstaubungsgrad *m*, 1522
Entstörer *m*, 3730
entwässern, 445, 1224
Entwässerung *f*, 1225, 1418,
1421, 4065
entwerfen, 1246
Entwurf *m*, 1245, 2973
Entwurfsarbeitsdruck *m*,
1254
Entwurfsbedingungen *f*,
1248
Entwurfskriterien *f*, 1249
Entwurfsstudie *f*, 1252
Entzündbarkeit *f*, 1909
Erde *f*, 2196
Erdgas *n*, 2760
Erdgeschoß *n*, 380
Erdleitung *f*, 1538
Erdung *f*, 1538, 1539, 1569
Erdungsfehler *m*, 2197
Erdverlegung *f*, 2522
ergänzen, 3720
erhitzen, 2225
Erhitzer *m*, 2235
Erklärungszeichnung *f*, 757
Erosion *f*, 1679
erproben, 1758
Erregung *f*, 1715
Ersatzteil *n*, 3558
Erschütterung *f*, 4002
Erstbefüllung *f*, 2436
Erwärmung *f*, 4041
Erweiterungswerkzeug *n*,
1920
Erzeugung *f*, 2161
erzwungene Konvektion *f*,
2011
Etagenheizung *f*, 245
eutektisch, 1686
evakuiert, 1687
Evakuierung *f*, 1688
Exergie *f*, 1717
Expansions-Schwimmerventil
n, 1746
Expansionsventil *n*, 1753
Expansionsventilgröße *f*,
1754
Experiment *n*, 1757
Expertensystem *n*, 1759
Explosion *f*, 1760
Exponent *m*, 1761
exzentrisch, 1541

Fabrik *f*, 1781
Fabrikanlage *f*, 2974

Fabrikation *f*, 2629
Fachwerk *n*, 2033
Faktor *m*, 1779
Fallgeschwindigkeit *f*, 3981
Fallhöhe *f*, 1343
Fallstrang *m*, 3266
Fallstudie *f*, 644
Falltür *f*, 3893
Faltenrohrbogen *m*, 1099
Farbe *f*, 831, 1528
Farbnebel *m*, 2908
Farbskala *f*, 832
Farbspritzkabine *f*, 2909
Farbstoff *m*, 1529
Farbtemperatur *f*, 834
Faser *f*, 1828
Faserfilter *m*, 1830
Faseroptik *f*, 1827
Faserstaub *m*, 1829
Fassade *f*, 1775
Fassung *f*, 3509
Feder *f*, 3590
federbelastet, 3591
Fehler *m*, 1680, 1810
Fehler *m* beseitigen, 1190
Fehlerantwort *f*, 1786
Fehlerbreite *f*, 1681
Fehlergröße *f*, 1203
Feinkohle *f*, 3464
Feinmechanik *f*, 3026
Feldgerät *n*, 1831
Fenster *n*, 4123
Fensterbankheizung *f*, 4126
Fensterbauteile *n,pl*, 1826
Fensterfuge *f*, 4125
Fensterklimagerät *n*, 4124
Fensterladen *m*, 3438
Fensternische *f*, 4024
Fensterwerk *n*, 1825
Fernanzeigethermometer *n*, 3239
Fernbedienung *f*, 3238
Fernfühler *m*, 3240
Fernheizsystem *n*, 1382
Fernheizung *f*, 1378
Fernheizzentrale *f*, 1380
Fernkühlung *f*, 1377
Fernthermometer *n*, 1366
Fernwärmehauptleitung *f*, 1379
Fernwärmeversorgung *f*, 1011, 1381
Fertigstellungszeit *f*, 890
Festpunkt *m*, 221, 1900, 3189
feststehendes Gitter *n*, 1899
Fettfilter *m*, 2189
Fettkohle *f*, 2318
feucht, 2351, 2711
feuchtarbeitender Luftkühler *m*, 4120
Feuchte *f*, 1165, 2357, 2712
Feuchteabscheider *m*, 2718
Feuchtebeständigkeit *f*, 1173
Feuchtefühler *m*, 2360, 2368

Feuchtegehalt *m*, 2714
Feuchtegleichgewicht *n*, 2713
Feuchtemessung *f*, 2715
feuchten, 1166
Feuchteregelung *f*, 2358
Feuchteregler *m*, 2359
Feuchteschutz *m*, 2716
Feuchtetransport *m*, 2719
Feuchteübertragung *f*, 2720
Feuchtigkeit *f*, 1172, 2357
Feuchtigkeitssperre *f*, 2717
Feuchtkugeltemperatur *f*, 4116
Feuer *n*, 1873
Feuerbeständigkeit *f*, 1889, 3195
Feuerbrücke *f*, 1879
feuerdämmend, 1890
feuerfest, 1886, 3196
feuerfest machen, 1887
feuerfeste Ausmauerung *f*, 3197
Feuerfestigkeit *f*, 1888
Feuerführung *f*, 1895
Feuerhaken *m*, 2999, 3150
feuerhemmend, 1890
Feuerlöscher *m*, 1883
Feuerrost *m*, 1875
Feuerschutzklappe *f*, 1881
Feuerschutztür *f*, 1882
feuersicher, 1888
Feuerstätte *f*, 1884
Feuertür *f*, 1882
Feuerung *f*, 1894
Feuerwarnsystem *n*, 1874
Feuerzug *m*, 1973
Film *m*, 1839
Filmbildung *f*, 1844
Film-Kühlturm *m*, 1843
Filmsieden *n*, 1840
Filter *m*, 1846
Filter *m* mit trockener Filterschicht *f*, 1472
Filterabscheidegrad *m*, 1861
Filterbecken *n*, 1862
Filterelement *n*, 1857
Filterfläche *f*, 1848
Filterkammer *f*, 1852
Filtermaterial *n*, 1859
filtern, 1847
Filterpapier *n*, 1860
Filterprüfventil *n*, 3676
Filterrahmen *m*, 1858
Filterreinigung *f*, 1853
Filtersack *m*, 1849
Filtersäuberung *f*, 1853
Filtertasche *f*, 1849
Filtertrockner *m*, 1855
Filtertuch *n*, 1854
filtrieren, 1847
Filtriertuch *n*, 3675
Filtrierung *f*, 1864
Filz *m*, 1822
Fitting *n*, 1898

flache Kühlrippe *f*, 2983
Flachheizkörper *m*, 2255
Flachplattenkollektor *m*, 1931
Flachrahmen *m*, 1930
Fläche *f*, 262
Flächenheizung *f*, 2913
Flächenkühlung *f*, 2912
Flamme *f*, 1904
Flammenüberwachung *f*, 1905
Flammkohle *f*, 2661
Flammpunkt *m*, 1885, 1925
Flammrohrkessel *m*, 1892
flammsicher machen, 1887
Flammstoß *m*, 1906
Flankenübertragung *f*, 1914
Flansch *m*, 1911
Flanschverbindung *f*, 1912
Flaschengas *n*, 496
Flexibilität *f*, 1933
flexible Kupplung *f*, 1935
flexible Rohrleitung *f*, 1936
flexible Verbindung *f*, 1938
Fliehkraftabscheider *m*, 692
Flockeneis *n*, 1903
Flockungspunkt *m*, 1946
Flossenventil *n*, 1915
flüchtig, 4011
flüchtiger Bestandteil *m*, 4012
Flügelradanemometer *n*, 1204
Flügelventilator *m*, 125
flüssig, 2559
Flüssigkeit *f*, 1986, 2558
Flüssigkeitsabscheider *m*, 2564
Flüssigkeitseinspritzung *f*, 2560
Flüssigkeitsgemisch *n*, 2563
Flüssigkeitsspiegel *m*, 2561
Flüssigkeitsspiegelanzeiger *m*, 2562
Flugasche *f*, 1997, 2193
Fluid *n*, 1986
Fluide *n,pl*, 1987
fluide Medien *n,pl*, 1987
Fluorchlorkohlenstoff *m*, 741
Fluorkohlenstoff *m*, 1991
Flußmittel *n*, 1996
Förderband *n*, 1043, 1044
Förderpumpe *f*, 3881
Förderschnecke *f*, 4143
Folge *f*, 3391
Formstück *n*, 1898, 2733
Forschung *f*, 3245
Fortluft *f*, 1721, 1738
fossiler Brennstoff *m*, 2025
Fracht *f*, 634
freiausblasender Luftkühler *m*, 2036
freiberuflicher Ingenieur *m*, 3092

gerichteter Luftauslaß *m*, 1330
Geruch *m*, 2803
Geruchsbeseitigungsmittel *n*, 1236
Geruchsfilter *m*, 2804
Geruchverschluß *m*, 3892
Gerüst *n*, 2033, 3347
Gerüstmontage *f*, 3134
gesättigte Luft *f*, 3338
Gesamtbetriebskosten *f,pl*, 2540
Gesamtdruck *m*, 3871
Gesamtenergie *f*, 3867
Gesamtförderhöhe *f*, 3869
Gesamtlänge *f*, 2881
Gesamtwärmeleistung *f*, 3870
geschichtete Luftströmung *f*, 3678
geschlitzter Auslaß *m*, 3486
geschlossener Kreislauf *m*, 778
geschlossener Regelkreis *m*, 779
geschlossener Verdichter *m*, 2301
geschlossenes System *n*, 780, 3364
Geschoß *n*, 1948, 3672
Geschoßfläche *f*, 1949
Geschwindigkeit *f*, 3565, 3978
Geschwindigkeit *f* an der Vorderseite, 1777
Geschwindigkeitshöhe *f*, 3980
Geschwindigkeitslinie *f*, 3979
Geschwindigkeitsprofil *n*, 3982
Geschwindigkeits-verminderung *f*, 1193
Gesetz *n* zur Reinhaltung *f* der Luft *f*, 760
Gesteinsschicht-Regenerierung *f*, 3285
getrennter Kondensator *m*, 3574
getrenntes System *n*, 3575
gewerblich, 873
gewerblicher Kühlraum *m*, 874
gewerbliches System *n*, 875
Gewicht *n*, 2186, 4100
gewichtsbelastendes Ventil *n*, 4101
Gewinde *n*, 3822
Gewindebohren *n*, 3762
Gewindebohrung *f*, 3761
Gewindeflansch *m*, 3824
Gewindeklappe *f*, 3823
Gewindemuffe *f*, 3825
Gewinderohr *n*, 3826
Gewindeschneiden *n*, 3762
Gewindeschneider *m*, 3823

Gewindestopfen *m*, 3358
Gewinn *m*, 2107
Gitter *n*, 2190
(Gitter-)Vorspannung *f*, 405
Gitterwiderstand *m*, 2191
Glas *n*, 2168
Glasfaser *f*, 2169
Glaskolben *m*, 561
Glaswolle *f*, 2170
Glattrohrbogen *m*, 3501
Gleichgewicht *n*, 1672
Gleichrichter *m*, 3184
gleichstellen, 1665
Gleichstrom *m*, 1316
Gleichstromkompressor *m*, 3923
Gleichung *f* 1670
gleichzeitig, 3448
Gleich-zu-Gleich-Kommunikationssystem *n*, 2930
Gleitkompensator *m*, 3479
Gleitschaufelverdichter *m*, 3481
Gleitschieber *m*, 3478
Gleitstütze *f*, 3480
Glied *n*, 3374
Gliederheizkörper *m*, 3147
Globustemperatur *f*, 2171
Globusthermometer *n*, 2172
glühen, 233, 2174
glühend, 2175
Glut *f*, 1873, 2173
Grad *m*, 1214
Gradient *m*, 2176
Gradtag *m*, 1215
Gradtagsverbrauch *m*, 1216
Graphik *f*, 2179
gravimetrischer Wirkungsgrad *m*, 2185
Grenzbedingungen *f,pl*, 498
Grenze *f*, 497
Grenzschicht *f*, 499, 1839
Griff *m*, 2212, 2508
grobkörnige Kohle *f*, 790
Größe *f*, 562, 3462
Größenanordnung *f*, 2856
Grube *f*, 2684, 2969
Grund *m*, 2196
Grundheizung *f*, 327
Grundlaststromerzeugung *f*, 379
Grundplatte *f*, 381, 395
Grundstrahlung *f*, 328
Grundwasser *n*, 2199
Grundwasserspiegel *m*, 2200
Gummidichtung *f*, 3318
Gummischlauch *m*, 3319
Gußeisen *n*, 646
gußeiserner Kessel *m*, 647
gußeisernes Formstück *n*, 648
Gußform *f*, 1281
Gußheizkessel *m*, 647
Gußheizkörper *m*, 649

Haarhygrometer *n*, 2209
Haarriß *m*, 2208
häuslich, 1388
Hahn *m*, 794
Haken *m*, 2324
halbautomatisch, 3382
halbhermetisch, 3384
Halbleiter *m*, 3383
Hallraum *m*, 3269
halogeniert, 2210
Halt *m*, 3662
Halter *m*, 2212
Halterung *f*, 2214
Handbedienung *f*, 2627
Handloch *n*, 2211
Handrad *n*, 2213
Handregelung *f*, 2626
Handtuchhalter *m*, 3872
Hanf *m*, 2296
hartgelötet, 514
hartgelötetes Rohr *n*, 516
hartlöten, 513
Hartlötung *f*, 517, 2216
Hartlötverbindung *f*, 515
Hartschaumstoff *m*, 3278
Haube *f*, 488, 2323
Haupt-, 2608
Hauptabschnitt *m*, 2217
Hauptgasleitung *f*, 2138
Haupthandwerker *m*, 3083
Hauptkanal *m*, 2609
Hauptkontraktor *m*, 3082
Hauptleitung *f*, 879, 2218, 2609
Hauptregler *m*, 2636
Hauptrohr *n*, 2608, 2610
Hauptsteigleitung *f*, 3281
Hausanschluß *m*, 2348, 3398
Hauszuleitung *f*, 3398
Hebel *m*, 2535
heiß, 2332
Heißdampfkühler *m*, 1256
Heißgasleitung *f*, 1344
Heißluft *f*, 2333
Heißluftraum *m*, 2336
Heißwasser *n*, 2309, 2314, 2337
Heißwassererzeuger *m*, 467
Heißwasserheizung *f*, 2310
Heizdraht *m*, 2265
Heizelement *n*, 2248
heizen, 2225
Heizer *m*, 3660
Heizfläche *f*, 2260
Heizflüssigkeit *f*, 2249
Heizgerät *n*, 2242, 3929
Heizgradtag *m*, 2246
Heizkessel-Feuerungsraum *m*, 2100
Heizkörper *m*, 2235, 2242, 3147
Heizkörperventil *n*, 3149

Heizkörperverkleidung *f*, 3148
Heizkosten *f*, 2244
Heizkraftwerk *n*, 840, 2277
Heizkurve *f*, 2245
Heizlast *f*, 2251, 2270
Heizleistung *f*, 2275
Heizmittel *n*, 2249, 2252
Heizöl *n*, 2085, 2254
Heizofen *m*, 2259
Heizperiode *f*, 2257
Heizplatte *f*, 2335
Heizraum *m*, 477
Heizregister *n*, 2236
Heizschlange *f*, 2243
Heizsonne *f*, 1573
Heizung *f*, 2241
Heizungsanlage *f*, 2250, 2256
Heizungsart *f*, 2253
Heizungssystem *n*, 2261
Heizungssystem *n* mit trockenem Rücklauf *m*, 1478
Heizungssystem *n* mit unterer Verteilung *f*, 1410
Heizvorrichtung *f*, 2235
Heizwert *m*, 609, 2264
Heizzeit *f*, 2257
Heizzentrale *f*, 677, 682, 2287
Herd *m*, 2222
Herdfeuer *n*, 2223
hermetisch, 2298
hermetisch abgedichtet, 2300
hermetische Kondensatoreinheit *f*, 3363
hermetischer Verdichter *m*, 3365
herstellen, 2630
Herstellung *f*, 2629
Hilfsbatterie *f*, 333
Hilfskraft-Regelung *f*, 3402
Hilfsmaschine *f*, 312
Hinausschießen *n*, 2890
Hintereinanderschaltung *f*, 1111, 3394
Hinterkantenstärke *f*, 3876
Hitzdrahtanemometer *n*, 2347
Hitze *f*, 2224, 4042
Hochdruck *m*, 2306
Hochdruckdampf *m*, 2312
Hochdruckkessel *m*, 2307
Hochdruckregelung *f*, 2308
Hochdruck-Sicherheitsausschalter *m*, 2311
Hochdruckventilator *m*, 2317
Hochgeschwindigkeits-klimaanlage *f*, 2316
Hochleistungsfilter *m*, 2303
Hochtarifstrom *m*, 2835
Hochvakuum *n*, 1202, 2315

höchstzulässig, 2642
höchstzulässige Konzentration *f*, 2640
Höhe *f*, 2217, 2294
Höhepunkt *m*, 2305
hörbar, 305
Hörbarkeit *f*, 304
Hörschwelle *f*, 3834
Hohlraum *m*, 657
Hohlziegelstein *m*, 78
Hub *m*, 2541, 3682
Hülse *f*, 3473, 3509
hydraulisch, 2362
hydrostatisch, 2364
hygienisch, 2365
Hygrometer *n*, 2366
hygroskopisch, 2367
Hygrostat *m*, 2368
hyperbolisch, 2369
Hypothermie *f*, 2370

Impuls *m*, 2722
in Tätigkeit setzen, 3081
Inbetriebnahme *f*, 876
Inbetriebnahmeplan *m*, 878
Index *m*, 2409
Indikator *m*, 2413, 3874
indirekte Heizfläche *f*, 3373
Induktion *f*, 2418
Induktionsgerät *n*, 2421
Induktionsheizung *f*, 2419
Induktionsverhältnis *n*, 2420
Industrie-Klimatisierung *f*, 2422
Industriewasser *n*, 2423
induziert, 2417
Infektion *f*, 2426
Infiltrierung *f*, 2427
Informationsaustausch *m*, 2431
infrarotdurchlässig, 1280
Infrarotheizung *f*, 2432
Infrarotmessung *f*, 2433
Ingenieur *m*, 1640
Ingenieurwesen *n*, 1641
Inhaber *m*, 2800
Inhalt *m*, 1007
Inhibitor *m*, 2435
Injektionszerstäuber *m*, 2438
Injektor *m*, 2441
Injektorbrenner *m*, 2442
Inkrustation *f*, 2408
innen, 2469
Innendurchmesser *m*, 495
Innengewinde *n*, 1824
Innenluft *f*, 2447
Innenmaß *n*, 2475
Innenrippe *f*, 2446
Innenwand *f*, 2476
innere, 2469
inner(lich), 2474
Installateur *m*, 1897
Installation *f*, 2451
installieren, 2450
Instandhaltung *f*, 2611, 3401

Integralregelung *f*, 2461
Integrator *m*, 2463
integriert, 2462
Intensität *f*, 2464
Interferenz *f*, 2468
intermittierend, 2473
Intervall *n*, 2478
Ion *n*, 2480
Ionenaustausch *m*, 2481
Ionisierung *f*, 2482
Isobare *f*, 2486
isolieren, 2455, 2456
Isoliermantel *m*, 2457
Isoliermaterial *n*, 2454, 2458
Isolierung *f*, 2459
isometrische Zeichnung *f*, 2488
Isotherme *f*, 2489
isothermisch, 2490
Isotop *n*, 2491

Jalousie *f*, 3984
Jalousiegitter *n*, 2591
Jalousieklappe *f*, 2746, 2915
Jalousie-Stellklappe *f*, 3477
Jalousietyp-Stellklappe *f*, 2207
Jalousieverschluß *m*, 2592

Kabel *n*, 599, 971
Kabeleingang *m*, 601
Kabelverbinder *m*, 600
Kälte *f*, 816
Kälteanlage *f*, 1065, 3210
Kältebereich *m*, 1066
Kältebeständigkeit *f*, 822
Kälteerzeugerleistung *f*, 961
Kälteerzeugung *f*, 820, 3207
Kältegradtag *m*, 1056
Kältekompressor *m*, 3204
Kältekreislauf *m*, 3209
Kältelager *n*, 1078
Kältemaschine *f*, 3206
Kältemischung *f*, 2050
Kältemittel *n*, 1064, 3199
Kältemittelkreislauf *m*, 3200
Kältequelle *f*, 824
Kälteraum *m*, 823
Kälteregler *m*, 3208
Kältesatz *m*, 1076
Kälteschutz *m*, 3106
Kältetechnik *f*, 3211
Kältewiderstand *m*, 822
Kaliber *n*, 817
Kalkulation *f*, 603
Kalorie *f*, 606
Kalorimeter *n*, 611
kalt, 817
Kaltlagerung *f*, 825
Kaltluft *f*, 818
Kaltwasser *n*, 728, 827
Kaltwasserversorgung *f*, 828
Kamin *m*, 733, 1884
Kaminwirkung *f*, 737
Kammer *f*, 701

Kanal m, 706, 1492
Kanalabmessung f, 1501
Kanalabschnitt m, 1500
Kanalabstrahlung f, 1499
Kanalbemessung f, 1502
Kanalbogen m, 1495
Kanalgeräusch n, 1493, 1498
Kanalkrümmer m, 1495
Kanalsystem n, 1503, 1505
Kanalübergangsstück n,
   1504
Kanalverteilung f, 1494
Kante f, 1547
Kapazität f, 617
Kapazität f (elektr.), 615
kapillar, 624
Kapillare f, 627
kapillare Aktion f, 625
Kapillarität f, 623
Kapillarluftwäscher m, 626
Kapillarrohr n, 627
Kappe f, 488, 614
Kaskade f, 640
Kaskadenregelung f, 641
Kaskadenregler m, 642
Kaskadenschaltung f, 3395
Kasten m, 503
Kastenkühler m, 725
Katalysator m, 651
Katathermometer n, 2505
Kathode f, 652
kathodischer Schutz m, 653
Kation n, 654
kaufmännisch, 873
Kavitation f, 656
Keilriemen m, 3976
Keilriemenantrieb m, 3977
Keim m, 2163
keimfrei, 2164
keimtötend, 340, 2165
Keller m, 668
Kellergeschoß n, 380
Kennfarbe f, 2383
Kenngröße f, 2919
Kennlinie f, 708
Kennzahl f, 709
Kern m, 1084
Kernfläche f, 1085
Kessel m, 455
Kessel m als betriebsfertige
   Einheit f, 2899
Kessel m mit Koksfeuerung
   f, 811
Kessel m mit Ölfeuerung f,
   2819
Kessel m mit
   Überdruckfeuerung f,
   3051
Kesselanlage f, 473
Kesselaufstellung f, 472
Kessel-Brenner-Einheit f,
   456
Kesselglied n, 478
Kesselhaus n, 470
Kesselheizfläche f, 468

Kesselleistung f, 457, 476
Kesselleistung f (in HP), 469
Kesselmantel m, 458
Kesselplatte f, 474
Kesselreinigung f, 459
Kesselrückwand f, 461
Kesselspeisepumpe f, 463
Kesselspeisewasser n, 464
Kesselspeisewassererhitzer
   m, 465
Kesselspeisung f, 462
Kesselstein m, 2408, 3348
Kesselstein m entfernen,
   1241
Kesselsteinbildung f, 3349
Kesselsteinentfernung f,
   3350
Kesselsteinverhütungsmittel
   n, 242
Kesselwasserabschnitt m,
   480
Kesselwasserlinie f, 481
Kesselwasser-
   temperaturmesser m, 482
Kesselwirkungsgrad m, 460
Kieselgel n, 3445
kinetisch, 2507
Kippmechanismus m, 3859
Kippschalter m, 3860
Kiste f, 643
Kitt m, 3126
Klammer f, 504, 755
Klappe f, 3438
Klappenschalter m, 1169
Klappenstellmotor m, 1170
Klappenventil n, 1915
Klassifizierung f, 758
klebend, 50
Klebstoff m, 49
Kleinrohrsystem n, 2679,
   3491
Klemme f, 1588
Klima n, 771
Klimaanlage f mit
   betriebsfertiger Zentrale f,
   2900
Klimagerät n, 88
Klimakammer f, 772, 1663
Klimaprozeß m, 90
Klimaprüfraum m, 772
Klimasystem n, 91
Klimatisierung f, 89
Klimatisierung f für alle
   Räume m,pl, 2091
Klimaverfahren n, 90
Kniestück n, 2955
Knopf m, 2508
koaxial, 793
kochen, 454
Kode m, 796
Koeffizient m, 797
Körper m, 453
Kohle f, 782
Kohlefilter m, 710
kohlenbeheizt, 786

Kohlenbunker m, 789
Kohlendioxid n, 629
Kohlenfeuerung f, 788
Kohlengruß m, 3464
Kohlenkeller m, 789
Kohlenkessel m, 787
Kohlenlieferung f, 784
Kohlenmonoxid n, 633
Kohlensäure f, 631
Kohlenstaub m, 785, 3118
Kohlenstoff m, 628
Kohlenwasserstoff m, 2363
Koks m, 810
Koksfeuerung f, 812
Kokskohle f, 815
Koksofengas n, 813
Kolben m, 2966
Kolbenhub m, 3841
Kolbenkompressor m, 3175
Kolbenpumpe f, 3176
Kolbenströmung f, 2968
Kolbenverdichter m, 3175
Kolbenverdrängung f, 2967
Kollektor m, 829
kolorimetrisch, 830
Komfort m, 867
Komfortdiagramm n, 869
Komfortkarte f, 869
Komfortklimasystem n, 868
Komfortkühlung f, 870
Kommunikationsnetz n, 881
Kompensator m, 887, 1748
Komponente f, 891
Kompression f, 900
Kompressor m, 912
Kondensat n, 944
Kondensation f, 949
Kondensation f in fester
   Phase f, 3525
Kondensationspunkt m, 950
Kondensationstemperatur f,
   959
Kondensationsturm m, 956
Kondensatleitung f, 946
Kondensatmesser m, 947
Kondensator m, 952
Kondensator m (elektr.), 616
Kondensator-Sammler m,
   3174
Kondensatorschlange f, 953
Kondensatorwärmerück-
   gabeeffekt m, 955
Kondensatpumpe f, 948
Kondensatsammelbehälter
   m, 945
Kondensatsammler m, 1927
kondensieren, 951
Kondenstopf m, 1944, 3645,
   3892
Konfiguration f, 973
konische Schraube f, 1919
konische Verbindung f, 1917
konisches Fitting n, 1918
konisches Gewinde n, 3760
Konsole f, 504

Konsolklimagerät *n*, 988
konstant, 990
Konstantdruckregelventil *n*, 992
Konstante *f*, 989
Konstantniveauventil *n*, 991
Konstantwertregelung *f*, 993
Konstruktion *f*, 994
Kontakt *m*, 999
Kontaktleistung *f*, 1002
Kontaktpunkt *m*, 1001
Kontaktthermometer *n*, 1003
kontinuierliche Regelung *f*, 1008
Kontrollampe *f*, 2949
Kontrolle *f*, 721
kontrollieren, 722
Konvektion *f*, 1034
Konvektionsheizgerät *n*, 1035
konvektiv, 1036
Konvektivstrom *m*, 1037
Konvektor *m*, 1038
Konzentration *f*, 940
Konzentrationsverhältnis *n*, 941
konzentrierender Sonnenkollektor *m*, 939
konzentrische Rohre *n,pl*, 942
koordinierter Entwurf *m*, 1079
Kork *m*, 1086
Korkplattenwärmedämmung *f*, 1087
Korrekturvariable *f*, 1091
korrigierte Effektivtemperatur *f*, 1090
korrodieren, 1093
Korrosion *f*, 1094
Korrosionsschutz *m*, 1096
Korrosionsschutzmittel *n*, 1095
korrosionsverhütend, 239
korrosiv, 1098
Kosten *f*, 1101, 1756
Kostenanschlag *m*, 3772
Kostenvoranschlag *m*, 1682
Kraft *f*, 2007, 3012
Kraftbedarf *m*, 3021
Kraftmesser *m*, 1537
Kraft-Wärme-Kopplung *f*, 804
Kraftwerk *n*, 3019, 3022
Kreide *f*, 700
Kreiselpumpe *f*, 691
Kreiselverdichter *m*, 688
Kreislauf *m*, 744, 1155
Kreuzstrom *m*, 1128, 1129
Kreuzstromkühlturm *m*, 1133
Kreuzstromwärme-austauscher *m*, 1132

Kreuzstromzirkulation *f*, 1130
Kreuzstück *n*, 1134
kritisch, 1126
Krümmer *m*, 1563
kryogen, 1140
Kubikfuß *m* je Minute, 697
Kubikfuß *m* je Sekunde, 698
Kühlaggregat *n*, 1049
Kühleffekt *m*, 1058
Kühlelement *n*, 1060
kühlen, 726, 1045, 3202
Kühler *m*, 729, 1048
Kühlerleitblech *n*, 3213
Kühlfläche *f*, 1068
Kühlfracht *f*, 727
Kühlgerät *n*, 3928
Kühlhaus *n*, 826
Kühlhochhaus *n*, 2313
Kühlkette *f*, 819
Kühlkurve *f*, 1055
Kühllager *n*, 1078
Kühllast *f*, 1062
Kühlleistung *f*, 1053
Kühlluft *f*, 1052
Kühlmantel *m*, 1061
Kühlmittel *n*, 1046, 1051
Kühlraum *m*, 732, 823, 826
Kühlregler *m*, 3208
Kühlschlange *f*, 1054
Kühlschrank *m*, 3212
Kühlsystem *n*, 1069
Kühltunnel *m*, 1075
Kühlturm *m*, 1070
Kühlturmfüllmaterial *n*, 1071
Kühlturmnebel *m*, 1072
Kühlturm-Plattenberieselungseinbau *m*, 2982
Kühlturmschwadenemission *f*, 1074
Kühlung *f*, 730, 1050
Kühlwasser *n*, 1077, 3201
Kühlwirkung *f*, 731, 1058, 3205
Kühlwirkungsgradverhältnis *n*, 1059
künstlicher Zug *m*, 2012
Kugelgelenk *n*, 364
Kugellager *n*, 361
Kugellagerbock *m*, 362
Kunststoff *m*, 2977
Kunststoffaser *f*, 3749
Kunststoffleitung *f*, 2980
Kunststoffrohr *n*, 2980
Kunststoffschaum *m* 2978
Kupfer *n*, 1081
Kupferauflage *f*, 1083
Kupferplattierung *f*, 1083
Kupferrohr *n*, 1082
Kuppel *f*, 1387
Kupplung *f*, 1109
Kurbel *f*, 1121
Kurbelgehäuse *n*, 1123

Kurbelwelle *f*, 1125
Kurbelzapfen *m*, 1124
Kurve *f*, 1147
Kurvenblatt *n*, 2177
Kurzbetrieb *m*, 3433
Kurzschluß *m*, 3431
Kurzschlußbrücke *f*, 2503
Kurzzyklus *m*, 3432

Labor *n*, 2510
Laboratorium *n*, 2510
Labyrinthdichtung *f*, 2511
Lack *m*, 2512
Ladeeinheit *f*, 720
Ladegerät *n*, 720
laden, 712
Laderegler *m*, 713
Ladesatz *m*, 720
Ladestrom *m*, 719
Ladung *f*, 634, 711
Länge *f*, 2531
Längenverhältnis *n*, 279
Längsrippe *f*, 2583
Längsseite *f*, 2584
Lärm *m*, 2772
Lärmminderung *f*, 2777, 3549
Läufer *m*, 3315
Lage *f*, 3006
Lageenergie *f*, 3011
Lager *n*, 391, 3670
Lagerbock *m*, 392
Lagerbüchse *f*, 585, 3474
Lagerhaus *n*, 3670
Lagermetall *n*, 241
Lagerraum *m*, 3668
Lagerschale *f*, 586
Lagerschalenriß *m*, 3573
Lagerstuhl *m*, 3475
Lagertank *m*, 3669, 3671
Lagerung *f*, 3665
Lamelle *f*, 1865, 2981
Lamellenrohr *n*, 1869, 3570
laminar, 2513
laminare Strömung *f*, 3680
Lampe *f*, 2514
Last *f*, 711, 2568
Lastausgleich *m*, 2573
Lastbereich *m*, 2574
lastfreier Anlauf *m*, 2779
lastfreier Zustand *m*, 2778
Lastschwankung *f*, 2576
latent, 2516
Laufrad *n*, 2396
Laufradspitze *f*, 2397
Lautstärke *f*, 2588
Lebensdauer *f*, 2538
Lebenszyklus *m* während der Lebensdauer *f*, 2539
Leck *n*, 2525
lecken, 2526
Leckstelle *f*, 2525
Lecksucher *m*, 2528
Leerlauf *m*, 2385
Legierung *f*, 188

Managementprogramm n, 2620
Mannloch n, 2621
Manometer n, 2623, 3053
manometrisch, 2624
Manschette f, 3473
Mantel m, 2492
manuell, 2625
manuell überlagern, 2628
Masche f, 2670
Maschenweite f, 2671
Maschine f, 1639, 2605
Maschinenbau m, 1641
Maschinenraum m, 1675, 2976
Maß n, 2158, 3462
Maßbild n, 1310
Masse f, 562, 2633
Massenstrom m, 2634
Maßstab m, 3348
Maßzeichnung f, 1310
Materie f, 2638
Matritze f, 1281
Maueraussparungen f,pl, 2322
maximal, 2639
Maximallast f, 2641
mechanisch, 2654
mechanische Dichtung f, 2658
mechanisches Wärmeäquivalent n, 2657
Mehrbrennstoff-Feuerung f, 2744
mehrere Zonen betreffend, 2754
Mehrmantelkondensator m, 2750
Mehrstrahlbrenner m, 2745
mehrstufig, 2751
mehrstufige Verdichtung f, 892
mehrstufiger Verdichter m, 2752
Mehrzugkessel m, 2748
Mehrzweckkühlraum m, 2749
Meißel m, 740
Membran f, 2664
Membranausdehnungsgefäß n, 2665
Membrane f, 1277
Membranventil n, 1279, 2666
Membranverdichter m, 1278
Menge f, 204, 1966, 3131
Meßblende f, 2858
Meßdüse f, 1964
Meßeinheit f, 3930
Meßeinrichtung f, 2650
Messer m, 2158
Meßfühler m, 1262
Meßgerät n, 2158, 2650
Meßglas n, 2159
messingen (aus Messing), 512

Meßinstrument n, 2651
Meßmuffe f, 2652
Meßstab m, 1311
Meßstutzen m, 2652
Meßtechnik f, 2653
Messung f, 2649
Meßwertgeber m, 3877
Metallfilter m, 2673
Metallschlauch m, 2674
Meteorologie f, 2675
Meter m, 2676
Methode f, 2677
Methode f der Geschwindigkeitsabnahme f, 3983
Methode f der gleichen Reibungswiderstände mpl, 1664
Methode f des statischen Druckrückgewinns m, 3617
Methylenblau-Prüfung f, 2678
Mikroklima n, 2680
Mikron n, 2681
Mikroprozessor m, 2682
Mindestluftmenge f, 2687
Mineralöl n, 2685
Mineralwolle f, 2686
Mischbox f, 437, 2693
Mischer m, 436
Mischgas n, 2692
Mischkammer f, 437, 2694
Mischkanal m, 2697
Mischkasten m, 2693
Mischkreislauf m, 2695
Mischluft f, 2690
Mischpumpe f, 2698
Mischregelung f, 2696
Mischströmungsventilator m, 2691
Mischung f, 2701
Mischungsverhältnis n, 2699
Mischventil n, 2700
Mitteldruck m, 2659
Mittellinie f, 320
Mitteltemperatur f, 2660
mit Wasser auffüllen (bei nicht selbstansaugenden Pumpen), 3081
mittler, 2643
mittlere Reparaturdauer f, 2648
mittlere störungsfreie Zeit f, 2647
mittlere Strahlungstemperatur f, 2645
mittlere Temperaturdifferenz f, 2646
Modell n, 2703, 2924
Modellieren n, 2704
Modul m, 2710
modular, 2706
Modulationsgewinn m, 2709

modulierend, 2708
Mollier-Diagramm n, 2721
Moment n, 2722
Montage f, 286, 2734, 3409
Montageanweisung f, 287, 2736
Montageöffnung f, 2735
Monteur m, 1897
montieren, 285, 2450
Motor m, 2726
Motorgeräusch n, 2729
Motorgrundplatte f, 2727
motorisch betrieben, 3016
Motorleistung f, 2730
Motormischventil n, 2728
Motorschrittzeit f, 2731
Motorwicklung f, 2732
Müll m, 1510
Muffe f, 587, 1112, 3473, 3566
Muffenverbindung f, 3567
Muster n, 2703, 2924, 3337

Nabe f, 2350
Nachbilden n, 2704
Nacherhitzer m, 3228
Nacherhitzerschlange f, 3227
Nachfüllung f, 3864
nachgeschalteter Regler m, 3471
nachgeschaltetes Gerät n, 3472
Nachhallzeit f, 3270
nachheizen, 3226
Nachkühler m, 63
Nachprüfung f, 1103
Nachtabsenkung f, 2769
Nachtabsenkungsthermostat m, 3820
Nachwärme f, 3225
Nachwärmer m, 64, 3228
Nadellager n, 2762
Nadelventil n, 2763
Näherung f, 255
Naht f, 3367
nahtlos, 3368
Naßdampf m, 4118
nasse Heizfläche f des Kessels m, 4119
nasser Rücklauf m, 4117
natürliche Abnutzung f, 4093
natürliche Konvektion f, 2755
Naturzugkühlturm m, 2758
Naturzugluftkühler m, 2756
Nebel m, 2003, 2688
Nebenanschluß m, 3434
Nebenkreislauf m, 1386
Nenndruck m, 2781
Nenndurchmesser m, 2780
Nennleistung f, 3160
Nennwärmeleistung f, 3161
Nennwärmestrom m, 3162
Nettoleistung f, 2765

Prüfzeugnis *n*, 3779
Psychrometer *n*, 3114
psychrometrische Differenz
f, 4115
psychrometrisches Diagramm
n, 3115
Puffer *m*, 543
Pufferbehälter *m*, 544
pulsierend, 3116
Pulszählung *f*, 3117
Pumpe *f*, 3119
Pumpengehäuse *n*, 3122
Pumpenkörper *m*, 3122
Punkt *m*, 2998
punktförmige Korrosion *f*,
2972
Punktkühlung *f*, 3578
Punktschweißung *f*, 3579
Pyrometer *n*, 3127

Qualität *f*, 3128
Qualitätskontrolle *f*, 3130
Qualitätssicherung *f*, 3129
Quantität *f*, 3131
Quecksilber *n*, 2667
Quecksilbersäule *f*, 836
Quecksilberschalter *m*, 2668
Quecksilberthermometer *n*,
2669
Querlüftung *f*, 1139
Querschnitt *m*, 1135
Querschnittfläche *f*, 1136
Quersiedelkessel *m*, 1138

Rad *n*, 4121
Radialventilator *m*, 689
Radiator *m*, 3147
Rahmen *m*, 2031
Rahmenkonstruktion *f*, 2032
Rankine-Kreisprozeß *m*,
3159
Rauchfärbung *f*, 833
Rauchfahne *f*, 2993
Rauchfang *m*, 1973
Rauchgas *n*, 1978
Rauchgasabzug *m*, 1981
Rauchgasabzugskanal *m*,
3495
Rauchgasauslaß *m*, 1984
Rauchgasdichte *f*, 3493
Rauchgase *n,pl*, 3497
Rauchgaskasten *m*, 3492
Rauchgasmelder *m*, 3494
Rauchgasprobe *f*, 3499
Rauchhöhe *f*, 2995
Rauchrohr *n*, 1891
Rauchrohrkessel *m*, 1892
Rauchtest *m*, 3499
Rauchverteilung *f*, 2994
Rauhigkeit *f*, 3316
Rauhigkeitskoeffizient *m*,
3317
Raum *m*, 701, 3295
Raumbedarf *m*, 3555
Raumheizer *m*, 3303

Raumheizung *f*, 3554
Raumhöhe *f*, 3304
Raumkenngröße *f*, 3300
Raumklimagerät *n*, 2898,
3297
Raumkonstante *f*, 3301
Raumkühlung *f*, 3553
Raumluft *f*, 2415, 2447, 3296
Raumluftbefeuchter *m*, 3298
Raumluftqualität *f*, 2416
Raumregelung *f*, 3302
Raumsimulator *m*, 1658
Raumtemperatur *f*, 3305
Raumwärmemesser *m*, 3299
Reaktion *f*, 3165
Reaktionskraft *f*, 3166
Reaktor *m*, 3167
Rechenvorschrift *f*, 184
Rechner *m*, 922
reduzierter Druck *m*, 3185
Reduzierventil *n*, 3186
Referenzatmosphäre *f*, 291
Reflektor *m*, 3194
Reflexionsgrad *m*, 3193
Reflexionsvermögen *n*, 3191
Reflexionswärmedämmung
f, 3192
Regelabweichung *f*, 1020
Regelalgorithmus *m*, 1017
Regelbereich *m*, 1032
Regelelement *n*, 1022
Regelfunktion *f*, 1023
Regelgeräteantrieb *m*, 1031
regeln, 1015, 3221
Regelpunkt *m*, 1030
Regelschleife *f*, 1028
Regelstrategie *f*, 1033
Regelung *f*, 1014
Regelung *f* der relativen
Feuchte *f*, 3233
Regelung *f* mit fester
Einstellung *f*, 1901
Regelungsunterstation *f*,
3690
Regelungswirkung *f*, 1016
Regelventil *n* für
Einspeisung *f*, 1816
regenerative Kühlung *f*, 3216
Regenerierung *f*, 3215
Register *n*, 3217
Registerbrenner *m*, 3219
Registrierapparat *m*, 3181
registrieren, 3218
Regler *m*, 1027
regulierbar, 53
regulieren, 3221
Regulier-T-Stück *n*, 3222
Regulierung *f*, 54, 3224
Regulierungsventil *n*, 3223
Reibring *m*, 2068
Reibung *f*, 2064
Reibungskoeffizient *m*, 2066
Reibungsverlust *m*, 2067
Reibungswiderstand *m*, 2065
Reifbildung *f*, 2074

Reihenfolge *f*, 3391
Reihenschaltung *f*, 1111,
3394
reiner Arbeitsplatz *m*, 765
reiner Raum *m*, 764
Reinheitsgrad *m*, 1217
reinigen, 3123
Reinigung *f*, 761, 3124
Reinigungsmittel *n*, 1264
Reinigungstür *f*, 762
Reinraum *m*, 763
Relais *n*, 3234
relativ, 3232
Reserve *f*, 3600
Reserveausrüstung *f*, 3601
Reservemotor *m*, 3602
Resonanz *f*, 3254
resultierend, 3256
Richtung *f*, 1329, 2857
Richtungsschalter *m*, 1332
Riemen *m*, 399
Riementrieb *m*, 400
Ringströmung *f*, 234
Rippe *f*, 1865
Rippenabstand *m*, 1872
Rippenhöhe *f*, 1871
Rippenrohr *n*, 1869
Rippenrohrheizkörper *m*,
1870
Rippenwirkungsgrad *m*,
1867
Riß *m*, 1118
Röhre *f*, 3899
Röhrenplatte *f*, 3903
Rohr *n*, 2950, 3899
Rohrabschnitt *m*, 2960
Rohrabzweigung *f*, 511
Rohraufhängung *f*, 2957,
2962
Rohrbefestigung *f*, 2952
Rohrbogen *m*, 2951
Rohrbündel *n*, 3901
Rohrbündelverdampfer *m*,
3425
Rohrbündelverflüssiger *m*,
3424
Rohrbündelwärme-
austauscher *m*, 3426
Rohrdimensionierung *f*,
2961
Rohrdurchmesser *m*, 2954
Rohrgewinde *n*, 2964
Rohrleitung *f*, 972, 2958
Rohrnetz *n*, 2959
Rohrreibung *f*, 2956, 3902
Rohrschelle *f*, 756
Rohrschlange *f*, 805, 2953
Rohrschlangenkondensator
m, 3422
Rohrschlangenverdampfer
m, 3423
Rohrsuchgerät *n*, 2965
Rohrsystem *n*, 2963
Rohrventil *n*, 3187
Rollbandfilter *m*, 3288

Schmiermittel n, 2604
Schmutz m, 1335, 3502
Schmutzfänger m, 3675
Schmutzfleck m, 3502
Schnappmechanismus m, 3504
Schnappvorrichtung f, 3506
Schnecke f, 4142
Schneckenförderer m, 4143
Schneeschmelzsystem n, 3507
Schnellschlußkupplung f, 3133
Schnellschlußregelung f, 3505
Schnellschlußventil n, 3132
Schnittzeichnung f, 3376
Schornstein m, 733, 3593
Schornsteinaufsatz m, 1116
Schornsteinfuchs m, 525, 738
Schornsteinfundament n, 734
Schornsteinfutter n, 1983
schornsteinlos, 1982
Schornsteinreinigung f, 735
Schornsteinverlust m, 739
Schornsteinwärmeverlust m, 2258
Schornsteinwirkung f, 3594
Schornsteinzug m, 736, 1977
Schotten-Kessel m, 3351
Schrägrohrkessel m, 2403
Schrägrohrmanometer n, 2404
Schranke f, 373
Schraube f, 3353
schrauben, 3354
schraubenförmig, 3354
Schraubenlüfter m, 3100
Schraubenmutter f, 2796
Schraubenschlüssel m, 3557, 4144
Schraubenverdichter m, 3356
Schraubkapsel f, 3355
Schraubstock m, 4005
Schraubverbindung f, 3357, 3924
Schraubzwinge f, 755
Schreiber m, 3181
Schritt m, 3651
Schrittregler m, 3652
schrittweise, 3654
Schrubber m, 3360
Schüreisen n, 2999
Schüren n, 3661
schüren, 3151
Schürhaken m, 3150
Schürung f, 3153
Schüttdichte f, 563
Schüttelrost m, 3286, 4001
Schütz n, 1000
schützen, 3428
schützen gegen
　Sonnenstrahlen m,pl, 3412
schützend, 3108

Schuppeneis n, 1903
Schutz m, 3105, 3108
Schutzgitter n, 3112
Schutzkappe f, 494
Schutzraum m, 3427
Schutzrohr n, 3113
Schutzschicht f, 3111
Schutzüberzug m, 3109
Schutzverkleidung f, 3109
Schutzvorrichtung f, 3110
Schwachlastperiode f, 2806
Schwaden m, 2688
Schwadenabscheider m, 2689
Schwärzung f (nach
　Ringelmann-Skala), 419
schwarzer Körper m, 417
Schwarzkörperäquivalent-
　temperatur f, 418
Schweben n, 1952
Schwebeteilchen n, 3736
Schwebstoffilter m, 7
Schwefeldioxid n, 3704
Schweißboden m, 4107
Schweißdraht m, 4110
schweißen, 4103
Schweißer m, 4105
Schweißflansch m, 4109
Schweißnaht f, 4102
Schweißstelle f, 4102
Schweißung f, 4106
Schweißverbindung f, 4104
Schwerkraftheizung f, 2188
Schwerkraftzirkulation f, 2187
Schwerpunkt m, 686
Schwimmen n, 1952
Schwimmer m, 1940
Schwimmerregelung f, 1942
Schwimmerschalter m, 1943
Schwimmerventil n, 1945
Schwinghebel m, 355
Schwingung f, 2862, 4002
schwingungsdämpfende
　Aufstellung f, 244
Schwingungsdämpfer m, 4003
Schwingungsfrequenz f, 4004
Schwunggrad m, 1998
Scotch-Kessel m, 3351
Sediment n, 3377
Sekundärkondensator m, 3370
Sekundärkühlmittel n, 3371
Sekundärluft f, 1649, 3369
Sekundärtrocknung f, 3372
selbstansaugende Pumpe f, 3381
Selbstkosten f, 2891
selbsttätige Regelung f, 3380
Sender m, 3890
Senke f, 1447
Sensor m, 3387
Sequenz f, 3391
Sequenzregelung f, 3392

serienmäßiger Auslaßstutzen
　m, 3393
serienmäßiger Einlaßstutzen
　m, 3393
Serienschaltung f, 1111
Servomotor m, 3403
Shunt m, 3434
Sicherheitsausdehnungs-
　leitung f, 1750
Sicherheitsdruckhöhe f, 3333
Sicherheitseinrichtung f, 3331
Sicherheitsentlastungsventil
　n, 1609
Sicherheitsfaktor m, 1780, 3332
Sicherheitsleitung f, 3327, 3328
Sicherheitsregelung f, 3329
Sicherheitsscheibe f, 582
Sicherheitsventil n, 3066, 3335
Sicherheitsverbindung f, 3328
Sicherung f, 2101
Sicherungspatrone f, 639
Sichtfeld n, 1834
Sichtglas n, 564
Sick-Building-Syndrom n, 3439
Sickergrube f, 3508
Sieb n, 3352, 3440
sieden, 454
siedend, 483
Siedepunkt m, 484
Signal n, 3442
Signallampe f, 2410
Signalsammelleitung f, 3443
Signalumsetzer m, 2705
Silikagel n, 3445
Simplex-Schaltung f, 3447
simultan, 3448
Sinterung f, 3458
Skala f, 3348
Skizze f, 1433
Sockel m, 375
Solarkonstante f, 3514
Sole f, 528
Solekreislauf m, 529
Solekühler m, 530
Solesprühgefriersystem n, 531
Solesystem n, 532
Soletank m, 533
Sommerklimatisierung f, 3705
Sonde f, 3086
Sonneneinstrahlung f, 2448
Sonnenenergie f, 3515
Sonnenfühler m, 3519
Sonnenkollektor m, 3513
Sonnenschutzvorrichtung f, 3414
Sonnenstrahlung f, 3518
Spalt m, 1118, 3484

System *n*, 3750
System *n* auf Computerbasis
  f, 925
System *n* auf
  Kommunikationsbasis *f*,
  882
System *n* mit werkseitiger
  Montage *f*, 1782

Tafel *f*, 2911, 3751
tafelartige Wärmedämmung
  f, 452
Tagesbehälter *m*, 1163
Tagesgang *m*, 1162
Tagesleistung *f*, 1161
Tagesrate *f*, 1162
Tagesverlängerung *f*, 1182
Tandemverdichter *m*, 3753
Tangentialventilator *m*,
  1131, 3754
Tank *m*, 3755
tanken, 2077
Tankschiff *n*, 3756
Tankwagen *m*, 3756
Tarif *m*, 3764
Tasche *f*, 344
Taschenfilter *m*, 345
Tastatur *f*, 2506
Tau *m*, 1268
tauchbadgalvanisiert, 2334
Tauchsieder *m*, 2393
Tauchtank *m*, 1312
Tauchthermostat *m*, 2394
Tauen *n*, 3789
Taupunkt *m*, 1269
Taupunkterhöhung *f*, 1272
Taupunkterniedrigung *f*,
  1270
Taupunkthygrometer *n*,
  1271
technisch, 3765
Teer *m*, 3763
Teilchen *n*, 2921
Teilchengröße *f*, 2922
Teillast *f*, 2923
Tellerventil *n*, 1356
Temperatur *f*, 3766
Temperatur *f* im
  Einstellpunkt *m*, 353
Temperaturabweichung *f*,
  3768
Temperaturannahme *f*, 1253
Temperaturdifferenz *f*, 3769
Temperaturfühler *m*, 3770
Temperaturregler *m*, 3767
Temperaturschreiber *m*,
  3811
Temperaturschwankung *f*,
  3771
Temperguß *m*, 2618
Tempergußfittings *n,pl*, 2619
Teststaub *m*, 3781
thermisch, 3790
thermische Behaglichkeit *f*,
  3793

thermische Konvektion *f*,
  3796
thermisches Anemometer *n*,
  3791
thermisches Gleichgewicht *n*,
  3798
Thermistor *m*, 3807
Thermodynamik *f*, 3809
thermoelektrisch, 3810
Thermoelementpaar *n*, 3808
Thermograph *m*, 3811
Thermohygrograph *m*, 3812
Thermometer *n*, 3813
Thermometertauchhülse *f*,
  3814
Thermosflasche *f*, 3815
Thermostat *m*, 3816
thermostatisch, 3817
thermostatischer
  Kondenstopf *m*, 3818
Thermostatventil *n*, 3819
Tichelmann-System *n*, 3274
tiefkühlen, 2039
Tiefkühlung *f*, 1201
Time-Sharing *n*, 3853
Tisch *m*, 3751
Tonerde *f*, 192
Tonerde-Trocknungsmittel
  n, 41
Topfbrenner *m*, 3010
Tor *n*, 2157
Torf *m*, 2927
Total-Energie-System *n*,
  3868
Totzeit *f*, 1187
Totzone *f*, 1188
träge, 2424
Trägerfrequenz *f*, 636
Trägheit *f*, 2425
tragbar, 3005
tragbarer Netzanschluß *m*
  (elektr.), 637
Transformator *m*, 3882
Transistor *m*, 3884
Transmission *f*, 3888
Trichter *m*, 1836, 2098, 2325
Trinkwasser *n*, 1440, 3009
Trinkwasserkühler *m*, 1441
trocken, 1457
trockene Luft *f*, 1459
trockene resultierende
  Temperatur *f*, 1476
trockene Verdampfung *f*,
  1464
trockene (ölfreie)
  Verdichtung *f*, 1463
Trockeneis *n*, 1467
trockener Dampf *m*, 1480
trockener Luftkühler *m*,
  1481
trockener Rücklauf *m*, 1477
trockener Verdampfer *m*,
  1482
trockenes Thermometer *n*,
  1462

Trockenfilter *m*, 1466
trockengesättigter Dampf *m*,
  1479
Trockenheit *f*, 1473
Trockenheitsgrad *m*, 1474
Trockenkugeltemperatur *f*,
  1461
Trockenluftkühler *m*, 1460
Trockenschrank *m*, 1469
trocknen, 1458
Trockner *m*, 1226, 1436
trockner Verdampfer *m*,
  3613
Trocknerschlange *f*, 1437
Trocknung *f*, 1225, 1244,
  1468
Trocknungsanlage *f*, 1471
Trocknungsmittel *n*, 1243
Tröpfchen *n*, 1449
Tröpfchenkondensation *f*,
  1450
Trommel *f*, 1455
Trommelkühler *m*, 1456
Tropfen *m*, 1442, 1448
Tropfenabscheider *m*, 1439,
  1451
Tropfengröße *f*, 1452
Tropfschale *f*, 1443
Trübung *f*, 2838
Trübungsfaktor *f*, 2839
Tuchfilter *m*, 1774
Türverschluß *m*, 1390
Tunnelfroster *m*, 3905
Tunnelkühler *m*, 3904
Tunnelscheiben-Verdichter
  m, 4134
Turbine *f*, 3906
Turbogebläse *n*, 3907
Turbogenerator *m*, 3909
Turboverdichter *m*, 3309,
  3908
turbulente Strömung *f*, 3911
Turbulenz *f*, 3910

Überdruck *m*, 1710
Überdruckbrennkammer *f*,
  2014, 3074
Überdruckerzeuger *m*, 3075
Überdruckschalter *m*, 2311
überfließen, 2884
überfluteter Kondensator *m*,
  3691
überfluteter Verdampfer *m*,
  1947
Übergabe *f*, 3242
Übergangsgebiet *n*, 3887
Übergangsströmung *f*, 3885
Übergangsstück *n*, 3886
Übergangszeit *f*, 2683
Übergangszustand *m*, 2789
Überheizen *n*, 2888
überhitzen, 3709
Überhitzer *m*, 3711
überhitzt, 3710
Überhitzung *f*, 3708, 3712

Vielzellenrotationsverdichter
  m, 2753
Vierleitersystem n, 2029
Viskometer n, 4006
Viskosefilter m, 4008
Viskosität f, 4007
Vollast f, 2093
Vollasteinschaltung f, 3611
volle Modulation f, 2094
vollkommen ausgefüllter
  Raum m, 2989
vollkommene Verbrennung
  f, 889, 2932
Voltmeter n, 4014
Volumen n, 4015
Volumenstellklappe f, 4016
volumetrische Pumpe f, 4018
volumetrischer Wirkungsgrad
  m, 4017
Vorabscheider m, 3027
Voranschlag m, 1682
voranschlagen, 1683
Vorarbeiter m, 714
vorausschauende Regelung f,
  3030
Vorbereitung f, 3038
vorbeugende Instandhaltung
  f, 3076
vorbeugende Wartung f,
  1092
Vorderglied n, 2070
voreingestellt, 3039
Voreinstellung f, 3040
Vorentwurfszeichnung f,
  3037
Vorerhitzerschlange f, 3033
Vorfeuerung f, 3032
Vorfilter m, 3031
vorkommend, 2400
Vorkühler m, 2022, 3028
Vorkühlung f, 3029
Vorlauf m, 1953
Vorlaufanschluß m, 1956
Vorlaufleitung f, 3726
Vorlaufsammler m, 3727
Vorlaufsteigleitung f, 3728
Vorlauftemperatur f, 1970
Vorlaufverteiler m, 3727
Vorratsbehälter m, 3669
Vorratswarmwasserbereiter
  m, 2342
Vorrichtung f, 251, 1267
Vorwärmer m, 3034
Vorwärmung f, 3035
Vorwärmzeit f, 3036
vorwärtsgekrümmt, 2024
Vorwegnahme-Regelung f,
  237

waagerecht Feuerzug m,
  2328
Wähler m, 1274
Wählschalter m, 3378
Wählscheibe f, 1274
Wärme f, 2224, 4042

Wärmeabgabe f, 2275
Wärmeanforderung f, 2283
Wärmeaustausch m, 2237
Wärmeaustauscher m, 610,
  2238
Wärmebedarf m, 2283
Wärmebedarfsberechnung f,
  1250, 2272
wärmebeständig, 2284
Wärmebilanz f, 2227, 3792
Wärmebrücke f, 2228
Wärmedämmaterial n, 2267,
  2454
wärmedämmen, 2456
Wärmedämmung f, 2268,
  2459, 3799
Wärmedurchgang m, 2292
Wärmedurchgangswiderstand
  m, 3803
wärmedurchlässig, 1280
Wärmedurchlässigkeit f,
  3797
Wärmeempfindung f, 2226
wärmeerzeugend, 607
Wärmeerzeuger m, 455
Wärmeerzeugung f, 2278
Wärmegewinn m, 2240
Wärmegewinn m durch
  Sonnenstrahlung f, 3516
Wärmeindex m, 4043
Wärmeinhalt m, 2232
Wärmelast f, 2270
Wärmelehre f, 2285
Wärmeleistung f, 2229, 3800
Wärmeleiter m, 2231
Wärmeleitfähigkeit f, 3795
Wärmeleitfähigkeit f der
  Grenzschicht f, 1842
Wärmeleitung f, 969, 2230,
  3794
Wärmeleitungsgewinn m,
  967
Wärmeleitungsverlust m, 968
Wärmeleitwiderstand m der
  Grenzschicht f, 1845
Wärmemesser m, 2274
Wärmepreis m, 2280
Wärmepumpe f, 2279
Wärmepumpenwirkung f des
  Verdichters m, 918
Wärmequelle f, 2286, 2346
Wärmerückgewinnung f,
  2281
Wärmerückgewinnungs-
  system n, 2282
Wärmeschock m, 3804
Wärmeschutz m, 3107
Wärmespeicher m, 3806
Wärmespeicherbehälter m,
  3806
Wärmespeicherung f, 2288,
  3805
Wärmestrahlung f, 3801
Wärmestrom m, 2239
Wärmestromdichte f, 3564

Wärmeträger m, 2252
Wärmeträgeröl n, 2291
Wärmeübergangszahl f bei
  freier Konvektion f, 2035
Wärmeübertragung f, 2289
Wärmeübertragungs-
  koeffizient m der
  Grenzschicht f, 1841
Wärmeübertragungsrohr n,
  2276
Wärmeübertragungszahl f,
  2290
Wärmeverlust m, 2271
Wärmeverlustrate f, 2273
Wärmeverschluß m, 2293
Wärmeverzögerung f, 2269
Wärmewiderstand m, 3802
Wärmewirtschaft f, 2247
Wärmezahl f, 2280
Wärmezufuhr f, 2266
Wand f, 4023
Wandauslaß m, 4031
Wandeinbauklimagerät n,
  3839
Wanderrost m, 699, 3894
wandhängender Verdampfer
  m, 4025
Wandheizung f, 4028
Wandlüfter m, 4026
Wandluftheizer m, 4030
Wandstrahlungsplatte f,
  4029
Wandtemperatur f, 4032
Wandventilator m, 4026,
  4033
Warmluft f, 2333, 4034
Warmluftheizer m, 4037
Warmluftheizgerät n, 4039
Warmluftheizung f, 4038
Warmluftleistung f, 4040
Warmluftleitung f, 4036
Warmluftschleier m, 4035
Warmwasser n, 2337, 3044
Warmwasserbedarf m, 2339
Warmwasserbereiter m, 2344
Warmwasserkessel m, 2338
Warmwasserspeicher m,
  1317, 2341
Warmwassersystem n, 2345
Warmwassertiefdruckheizung
  f, 2601
Warmwasserversorgung f,
  2343
Warmwasserverteilung f,
  2340
Warmwasservorratsbehälter
  m, 1317
Warngerät n, 4046
Warnlampe f, 4047
Warnung f, 4045
Wartung f, 301
Wartungsalarm m, 2612
Waschbecken n, 4048
Wasser n, 4056
Wasserablauf m, 4079

Wasserabscheider *m*, 4083
Wasserabtauung *f*, 4063
Wasseraufnahme *f*, 4057
Wasserbehälter *m*, 752
Wasserbehandlung *f*, 4087
Wasserdampf *m*, 4089
wasserdicht, 4081, 4086
wasserdicht machen, 4082
Wasserenthärter *m*, 4084
Wasserenthärtung *f*, 4085
Wasserentsalzung *f*, 4064
Wassererwärmer *m*, 4071
wasserfester Belag *m*, 4098
Wasserfilter *m*, 4067
Wassergehalt *m*, 4059
wassergekühlt, 4060
Wassergeschwindigkeit *f*, 4090
Wasserhärte *f*, 2215, 4070
Wasserheizung *f*, 4072
Wasserkapazität *f*, 4058
Wasserkühler *m*, 4061
Wasserkühlturm *m*, 4062
Wasserleck *n*, 4076
Wassermangelsicherung *f*, 2594
Wassermantel *m*, 4073
Wasserrohrkessel *m*, 4088
Wassersack *m*, 4080
Wassersäule *f*, 837
Wasserschicht *f*, 4075
Wasserschlag *m*, 4069
Wasserstand *m*, 4077
Wasserstoß *m*, 4069
Wasserstrahlpumpe *f*, 4074
Wassertemperaturregler *m*, 257
Wasserverlust *m*, 4078
Wasserzufluß *m*, 4066
Wattmesser *m*, 4092
Wechselbrandkessel *m*, 1488
Wechselregelung *f*, 704
Wechselschalter *m*, 705, 3917
Wechselstrom *m*, 190
Wegwerffilter *m*, 1363
Weichlötung *f*, 3511
Weißglut *f*, 2399
Wellrohrbogen *m*, 1100
Wendeschaufel *f*, 3912
Werk *n*, 4140
werkseitig eingestellt, 1784
werkseitig montiert, 1783
Werkstatt *f*, 4141
Werkstattzeichnung *f*, 3430, 4136
Werkzeug *n*, 3863
Wert *m*, 3950
Wetter *n*, 4094
Wetterdaten *f,pl*, 4096
wetterfest, 4097
Wetterkompensator *m*, 4095
Wetterschutzhaube *f*, 4099
Widerstand *m*, 1414, 3251
Widerstandsbeiwert *m*, 1415

Widerstandskraft *f*, 3195
Widerstandsschweißung *f*, 3253
wieder einstellen, 3247
Wiedereinschaltzeit *f* nach Brennerausfall *m*, 569
wiedereinzustellender Regler *m*, 3248
wiedererwärmen, 3226
wiederherstellen, 3182
Wiederherstellungskosten *f,pl*, 3244
Wiederumwälzung *f*, 3178
Wind *m*, 4122
Winddruck *m*, 4127
windgetriebener Ventilator *m*, 4129
Windkanal *m*, 4128
Windkessel *m*, 175
Windschatten *m*, 1184
Winkel *m*, 226
Winkeleisen *n*, 228
Winkelverhältnis *n*, 227
Winterklimatisierung *f*, 4130
Wirbelbrenner *m*, 4019
Wirbelrohr *n*, 4021
Wirbelschichtgefrieren *n*, 1989
Wirbelschichtverbrennung *f*, 1988
Wirbelstrom *m*, 1544
Wirbelstromprüfung *f*, 1545
Wirbelstromversuch *m*, 1545
Wirkleistung *f*, 3171
wirklich, 43
Wirkung *f*, 40, 1549
Wirkungsgrad *m*, 1556
Wirkungsgrad *m* des Absetzvermögens *n* atmosphärischen Staubes *m*, 296
Wirtschaft *f*, 1543
Wirtschaftlichkeit *f*, 1543
Wohnungsheizung *f*, 1389, 3250
Wohnungsklimatisierung *f*, 3249
Würfel *m*, 1281
Würfeleis *n*, 1141
Wurfweite *f*, 2498

Y-Stück *n*, 4146

zähe Strömung *f*, 4009
Zähigkeit *f*, 4007
Zähler *m*, 2676
Zahl *f*, 2794
Zahnradpumpe *f*, 2160
Zange *f*, 3861
Zapfen *m*, 2990, 3566, 3898
zapfen, 3758
Zapfhahn *m*, 3757
Zapfleistung *f*, 3759
Zapfstelle *f*, 1434
Zapfventil *n*, 1435

zeichnen, 2178
Zeichnung *f*, 1433
Zeigerthermometer *n*, 1275
Zeilendrucker *m*, 2554
Zeit *f*, 3846
zeitabhängiges Abtauen *n*, 3850
Zeitkonstante *f*, 3849
Zeitschalter *m*, 3854
Zeitschaltuhr *f*, 3847
Zeitteilung *f*, 3853
Zeitverzögerung *f*, 3852
Zelle *f*, 667
Zellenbauweise *f*, 673
Zellenfilter *m*, 670
Zellgummi-Wärmedämmung *f*, 672
zellulare Wärmedämmung *f*, 674
Zement *m*, 675
Zentigrad *m*, 676
Zentrale *f*, 1675
zentrale Leittechnik *f*, 548
zentrales Luftbehandlungsgerät *n*, 684
Zentralheizung *f*, 680, 2092
Zentralheizungsanlage *f*, 681
Zentralprozessor *m*, 683
Zentralregelung *f*, 678
Zentralschaltwarte *f*, 679
zentrifugal, 687
Zentrifugalkraft *f*, 690
Zentrifuge *f*, 693
Zerfall *m*, 1265
Zerfallsrate *f*, 1192
Zerlegung *f*, 1191
zerreibend, 5
Zersetzung *f*, 1197
zerstäuben, 299
Zerstäubungsbefeuchter *m*, 300
Zertifikat *n*, 695
Zeugnis *n*, 695
Ziegelstein *m*, 526
Zinn *n*, 3855
Zirkulation *f*, 747
Zirkulationsventilator *m*, 746
Zirkulator *m*, 750
Zisterne *f*, 752
Zone *f*, 262, 3374, 4149
Zoneneinteilung *f*, 4151
Zonenklimagerät *n*, 4150
Zubehör *n*, 29, 256
Zubehörteil *n*, 29
Zündelektrode *f*, 2387
Zündfähigkeit *f*, 1909
Zündpunkt *m*, 2388
Zündsicherheitskontrolle *f*, 1907
Zündsicherung *f*, 2389
Zündtemperatur *f*, 2390
Zündung *f*, 2386, 3084

zufälliger Wärmegewinn *m*, 650, 2401
Zufallsfehler *m*, 3156
Zufluß *m*, 2430
zufrieren, 2045
zuführen, 1813
Zufuhr *f*, 2430, 3719
Zug *m*, 1425
Zugbedarf *m*, 28
Zugangskontrollsystem *n*, 26
Zugangstür *f*, 27
Zuganzeiger *m*, 1428
Zugausgleich *m*, 349
Zugbedarf *m*, 1431
Zugbegrenzer *m*, 1430
zugelassen, 2536
Zugerscheinung *f*, 1409
Zugerzeuger *m*, 1429
Zugmesser *n*, 1426
Zugregler *m*, 1432
Zugstange *f*, 3844
Zugunterbrecher *m*, 3577
zulässige Beanspruchung *f*, 3336
zulässige Stromstärke *f*, 205
zulässiger Grenzwert *m*, 3833
Zuleitung *f*, 3726

Zuluft *f*, 2460, 3721
Zuluftanlage *f*, 3725
Zuluftauslaß *m*, 3724
Zulufteinrichtung *f*, 3722
Zuluftöffnung *f*, 3723
Zusammenbau *m*, 286
zusammenbauen, 285
Zusammensetzung *f*, 2701
Zusatz *m*, 47, 491
Zusatzelektrosystem *n*, 311
Zusatzheizung *f*, 3718
Zusatzluft *f*, 310, 2613
Zusatzverdichter *m*, 492
Zusatzwärmequelle *f*, 313
Zusatzwasser *n*, 2615
Zustand *m*, 3614
Zustandsgleichung *f*, 1671
Zustandslinie *f*, 963
zustopfen, 2991
Zuverlässigkeit *f*, 3235
Zwangsumlauf *m*, 2009
Zweidruckregelung *f*, 1490
zweifach ansaugender Verdichter *m*, 1405
Zweifachschalter *m*, 1408
zweigliedrig, 411
Zweiglinie *f*, 510

Zweikanalklimasystem *n*, 1484
Zweikanalsystem *n*, 1485
Zweikontaktschalter *m*, 1403
Zweileiter-System *n*, 3919
Zweileiter-Unterbrechung *f*, 1400
Zweiphasenströmung *f*, 3914
Zweipunktregelung *f*, 2304
Zweirohrheizung *f*, 3915
zweiseitig saugender Ventilator *m*, 1406
Zweistoffbrenner *m*, 1489
Zweistoff-Dampf-Kreislauf *m*, 412
Zweiwegeventil *n*, 3918
Zwillingszylinder-Verdichter *m*, 3913
Zwischenboden *m*, 1789
Zwischenkühler *m*, 2467
Zwischenraum *m*, 766
Zwischenschicht *f*, 2471
Zwischenstück *n*, 3752
Zwischenüberhitzer *m*, 2472
Zyklon *m*, 1158
Zyklus *m*, 1155
Zylinder *m*, 1159
Zylinderkopf *m*, 1160

The Multilingual Dictionary of Real Estate

FINANCIAL AND PROFESSIONAL SUPPORT PROVIDED BY GOOCH AND WAGSTAFF AND NORWICH UNION

A guide for the property professional in the Single European Market

L van Breugel, Gooch and Wagstaff, UK, **B Wood**, and **R H Williams**, both of University of Newcastle upon Tyne, UK

* Dutch *

* English *

* French *

* German *

* Italian *

* Spanish *

The Multilingual Dictionary of Real Estate is *more than* merely a word-for-word dictionary of specialist and technical terminology; it offers explanations of terms which cannot be translated exactly or are liable to cause confusion. A supplementary section of the dictionary provides detailed notes as well as translations of particularly tricky terms. There are also invaluable sections outlining the real estate and planning hierarchies and real estate associations throughout Europe.

March 1993: 234x170: 414pp
Hardback: 0-419-18020-6

Here's what is included in *The Multilingual Dictionary of Real Estate* in all 6 languages * how to use this dictionary * objectives * word index * translations * explanations * key questions * government and planning hierarchies

For further information on this and other property related titles, please contact: **The Promotion Dept., E & F N Spon,** 2-6 Boundary Row, London SE1 8HN Tel 071 865 0066 Fax 071 522 9623

Illustrated Encyclopedia of Building Services

Eur Ing David Kut, Consulting Engineer, UK

"it contains much which the practising engineer needs when creating a specification or a purchase order...easy-to-read book...For anyone associated with the building and construction industry this is a very practical book." - *Building Services*

"A clear and comprehensive dictionary of building services terms, with some illustrations" - *the architects' journal*

An explanatory, highly-illustrated encyclopedia of terms used in the following engineering services in buildings:

* air conditioning

* heating and ventilating

* hot and cold water supply

* fire-fighting and protection

* drainage and sanitation

* electrical services.

This book explains over **3,000 terms (over 100,000 words)** and contains over 200 line illustrations drawn by professional technicians. This practical handbook is intended for day to day use as a reference or as a source of enlightenment for anyone associated with the building and construction industry. It also provides comprehensive practical explanations of the many terms listed, providing guidance, examples of use and, in certain cases, cautionary remarks concerning aspects of the applications.

December 1992: 246x189: 368pp
Hardback: 0-419-17680-2

For further information and to order please contact: **The Promotion Dept., E & F N Spon**, 2-6 Boundary Row, London SE1 8HN Tel 071 865 0066 Fax 071 522 9623

Español

a prueba de presión, 3071
ábaco *m*, 2177
abertura *f*, 2841
abertura (de válvula) *f*, 3004
abertura de extracción *f*, 1733
abertura de salida *f*, 2871
abertura en la envolvente del edificio *f*, 553
abertura para el montaje *f*, 2735
abertura para el suministro de aire *f*, 3723
aberturas en serie *f*, 3393
ablandador de agua *m*, 4084
ablandamiento *m*, 3510
ablandamiento del agua *m*, 4085
abrasión *f*, 3
abrasivo, 5
abrasivo *m*, 4
abrazadera *f*, 504, 755
abrazadera (tubos) *f*, 2952
abrigar, 3428
absoluto, 6
absorbedor *m*, 12
absorbente *m*, 11, 56
absorber, 9
absorción *f*, 15, 1303
absorción acústica *f*, 3537
absorción de agua *f*, 4057
accesorio *m*, 29, 256
accesorio de caldera *m*, 472
accesorio de desviación *m*, 1383
accesorios de fundición maleable *m,pl*, 2619
accesorios de hierro colado *m,pl*, 648
accesorios de hierro fundido *m,pl*, 648
accesorios del ventilador *m*, 1792
accidente causado por congelación *m*, 2072
acción *f*, 40
acción bacteriana *f*, 337
acción de control *f*, 1016
acción flotante *f*, 1941
accionador *m*, 44
accionamiento *m*, 1444
accionamiento a mano *m*, 2627

accionamiento por acoplamiento directo *m*, 1319
accionamiento por correa *m*, 400
accionamiento por correa trapezoidal *m*, 3977
accionamiento por reloj programador *m*, 3848
aceite *m*, 2809
aceite combustible *m*, 2085
aceite de calefacción *m*, 2254
aceite mineral *m*, 2685
aceite transmisor de calor *m*, 2291
aceleración *f*, 21
acelerador *m*, 750
acelerador (bomba) *m*, 491
acero *m*, 3647
acero inoxidable *m*, 3595
acetileno *m*, 33
acidez *f*, 38
ácido *m*, 35
ácido carbónico *m*, 631
aclimatación *f*, 30
acometida *f*, 997, 2348, 2808, 3398
acometida de corriente *f*, 3018
acondicionador con evaporador remoto *m*, 3575
acondicionador de aire *m*, 88
acondicionador de aire autónomo *m*, 2898
acondicionador de aire compacto *m*, 3927
acondicionador de aire de muro *m*, 3839
acondicionador de aire de tejado *m*, 3293
acondicionador de aire doméstico *m*, 3297
acondicionador de ventana *m*, 4124
acondicionador evaporativo *m*, 1696
acondicionamiento de aire *m*, 89
acondicionamiento de aire a alta velocidad *m*, 2316
acondicionamiento de aire industrial *m*, 3088

acondicionamiento de aire para invierno *m*, 4130
acondicionamiento del aire para buques *m*, 2632
acondicionar, 87
acoplamiento *m*, 1109
acoplamiento abocardado *m*, 1918
acoplamiento de desembrague rápido *m*, 3133
acoplamiento directo *m*, 1319
acoplamiento elástico *m*, 885
acoplamiento en paralelo *m*, 1110
acoplamiento en serie *m*, 1111, 3394
acoplamiento flexible *m*, 1935
acoplamiento macho *m*, 2616
acoplamiento por bridas *m*, 1912
actuación rápida *f*, 3504
actuación sobre interruptor de palanca *f*, 3859
actuador *m*, 44
actuador de una compuerta *m*, 1169
acumulación de agua caliente *f*, 2341
acumulación de calor *f*, 2288
acumulador *m*, 31, 3246
acumulador de agua caliente *m*, 2342
acústica *f*, 39
adaptador *m*, 3886
adhesivo, 50
adhesivo *m*, 49
adiabático, 51
aditivo *m*, 47
admisión *f*, 2430
admisión por la base de la chimenea *f*, 738
adsorción *f*, 57
aerodinámico, 60
aerosol *m*, 62
aerotermo *m*, 3929
aerotermo de combustión directa *m*, 1324
aerotermo de gas *m*, 2127
aerotermo de pared *m*, 4030
aerotermo montado en la pared *m*, 4030
afluencia *f*, 2430

aflujo *m*, 2430
agente a pie de obra *m*, 3460
agitador *m*, 70
agotamiento *m*, 1614
agregado *m*, 66
agresividad *f*, 68
agresivo, 67
agua *f*, 4056
agua caliente *f*, 2337, 4044
agua caliente a alta presión *f*, 2309
agua de alimentación *f*, 1819
agua de alimentación de la caldera *f*, 464
agua de aportación *f*, 1394
agua de aporte *f*, 2615
agua de enfriamiento *f*, 1077, 3201
agua de pozo *f*, 4113
agua de red *f*, 753
agua de refrigeración *f*, 1077
agua de relleno *f*, 1394
agua de reposición *f*, 1394, 2615
agua de retorno *f*, 3268
agua destilada *f*, 1368
agua enfriada *f*, 728
agua fría *f*, 827
agua helada *f*, 2378
agua no tratada *f*, 3933
agua para usos industriales *f*, 2423
agua potable *f*, 1440, 3009
agua refrigerada *f*, 728
agua refrigerante *f*, 3201
agua residual *f*, 4055
agua salobre *f*, 505
agua sobrecalentada *f*, 2309, 2314
agua suplementaria *f*, 1394, 2615
aguas abajo *f,pl*, 1411
aguas arriba *f,pl*, 3934
aguas freáticas *f,pl*, 2199
aguas residuales *f*, 1557
aguas subterráneas *f,pl*, 2199
agujero *m*, 2321
agujero de lavado *m*, 2211
agujero roscado *m*, 3761
ahorro de energía *m*, 3015
aire *m*, 71
aire acondicionado *m*, 962
aire acondicionado industrial *m*, 2422
aire adicional *m*, 2613
aire ambiente *m*, 195
aire ambiente interior *m*, 2415, 3296
aire auxiliar *m*, 310
aire caliente *m*, 2333, 4034
aire comburente *m*, 866
aire comprimido *m*, 895
aire de admisión *m*, 2460
aire de aporte *m*, 2613
aire de combustión *m*, 845

aire de enfriamiento *m*, 1052
aire de entrada *m*, 2460
aire de mezcla *m*, 2690
aire de renovación *m*, 2864
aire de retorno *m*, 3258
aire de salida *m*, 1738, 2868
aire enfriado *m*, 1047
aire evacuado *m*, 1721
aire exterior *m*, 2060, 2864, 2875
aire extraído *m*, 1721
aire fresco *m*, 2060, 2864
aire frío *m*, 818
aire impulsado *m*, 2008
aire interior *m*, 2447
aire limpio *m*, 759
aire mixto *m*, 2690
aire normal *m*, 3597
aire primario *m*, 3077
aire recirculado *m*, 3177
aire refrigerado *m*, 1047, 1052
aire saturado *m*, 3338
aire seco *m*, 1459
aire secundario *m*, 3369
aire sobrante *m*, 1709
aire suministrado *m*, 3721
aire suplementario *m*, 2613
aire tratado *m*, 3895
aire viciado *m*, 1184, 4010
aireación *f*, 59, 3991
aireador *m*, 3993
aireador de pared *m*, 4033
airear, 58
aislamiento *m*, 2459
aislamiento acústico *m*, 3542
aislamiento de la chimenea *m*, 1983
aislamiento térmico *m*, 2268, 3799
aislamiento térmico celular *m*, 674
aislamiento térmico con coquillas de corcho *m*, 1087
aislamiento térmico de bloques *m*, 442
aislamiento térmico de goma celular *m*, 672
aislamiento térmico de mantas *m*, 421
aislamiento térmico de paneles *m*, 452
aislamiento térmico de relleno *m*, 1838
aislamiento térmico expandido *m*, 2001
aislamiento térmico reflectivo *m*, 3192
aislante *m*, 2454
aislante térmico *m*, 2267
aislar (acústicamente), 2455
aislar (eléctricamente), 2455
aislar (térmicamente), 2456
ajustable, 53

ajustar, 52, 3404
ajuste *m*, 54, 3406
al descubierto, 1762
álabe *m*, 420, 3954
álabe del ventilador *m*, 1794
álabe direccional *m*, 3912
álabe director *m*, 2203
álabe director (de un ventilador) *m*, 1793
álabes de abertura regulable *m,pl*, 3972
álabes direccionales *m*, 161
alambre *m*, 4131
alarma *f*, 178, 4045
alarma de mantenimiento *f*, 2612
albañilería *f*, 527
alcalinidad *f*, 186
alcalino, 185
alcance del chorro *m*, 2498
alcance del chorro de aire *m*, 168
alcantarilla *f*, 3411
alcantarillado *m*, 3411
alcohol *m*, 180
aleación *f*, 188
aleación bimetálica *f*, 408
aleta *f*, 1865
aleta interior *f*, 2446
aleta longitudinal *f*, 2583
aleta plana *f*, 2983
aleteado, 1868
alfanumérico, 189
algas *f,pl*, 181
algicida *m*, 182
algoritmo *m*, 184
algoritmo de control *m*, 1017
alimentación *f*, 1812, 3719
alimentación (elec) *f*, 1818
alimentación a presión *f*, 3050
alimentación de agua *f*, 4066
alimentación de agua caliente *f*, 2343
alimentación de aire bajo la parrilla *f*, 3921
alimentación de aire para la combustión *f*, 864
alimentación de la caldera *f*, 462
alimentación por tornillo sin fín *f*, 4143
alimentar, 1813
almacén *m*, 3668, 3670
almacén frigorífico *m*, 826
almacenamiento *m*, 3665
almacenamiento al aire libre *m*, 2878
almacenamiento de calor *m*, 2288
almacenamiento de frío *m*, 1078
almacenamiento de gas *m*, 2148

almacenamiento del ordenador *m*, 937
almacenamiento frigorífico *m*, 825
almacenamiento térmico *m*, 3805
alquitrán *m*, 3763
alta presión *f*, 2306
altar del horno *m*, 1879
alto vacío *m*, 2315
altura *f*, 2294
altura de aspiración *f*, 3697
altura de descarga *f*, 1343
altura de elevación total *f*, 3869
altura de impulsión *f*, 1343
altura de penachos de humos *f*, 2995
altura de succión *f*, 3699
altura del local *f*, 3304
altura dinámica *f*, 3980
altura estática *f*, 2217
altura manométrica *f*, 3055
altura manométrica total *f*, 3869
altura piezométrica *f*, 3055
alúmina *f*, 192
aluminio *m*, 193
ambiente, 194
ambiente *m*, 1657
ambiente exterior *m*, 3734
amianto *m*, 265
amoniaco *m*, 200
amortiguación *f*, 302, 1171
amortiguador *m*, 12, 543, 3429
amortiguador de aire *m*, 102
amortiguador de vibraciones *m*, 4003
amortiguar, 1167
amortización *f*, 202, 1240
amortizar, 203
amperaje *m*, 206
amperímetro *m*, 199
amplificación *f*, 207
amplificador *m*, 208
amplificador electrónico *m*, 1596
amplificar, 209
amplitud *f*, 210
amplitud de la banda *f*, 366
análisis *m*, 220
análisis de gases *m*, 2112
análisis de los gases de combustión *m*, 861
análisis del aire *m*, 73
análisis del caso *m*, 644
análisis del combustible *m*, 2078
analizador *m*, 219
análogo, 212
anclaje *m*, 221
anclar, 222
ancho de batería *m*, 809
ancho de malla *m*, 2671

andamiaje *m*, 3347
anemómetro *m*, 223
anemómetro de copa *m*, 1142
anemómetro de corriente lateral *m*, 1204
anemómetro de hilo caliente *m*, 2347
anemómetro de vena giratoria *m*, 3277
anemómetro térmico *m*, 3791
angular de acero *m*, 228
ángulo *m*, 226
ángulo de admisión *m*, 24
ángulo de descarga *m*, 229
ángulo de salida *m*, 230
anhídrido carbónico *m*, 629
anhídrido carbónico, sólido *m*, 1467
anhídrido sulfuroso *m*, 3704
anillo *m*, 4049
anillo de estanqueidad *m*, 2906
anillo de fricción *m*, 2068
anión *m*, 232
anodo *m*, 235
ánodo protector fungible *m*, 3326
antepecho *m*, 4024
anticipación térmica *f*, 2226
anticipador *m*, 238
anticongelante *m*, 240
anticorrosivo, 239
anticorrosivo (resistente a la corrosión), 1097
antiincrustante *m*, 242
antracita *f*, 236
apagafuegos *m*, 1883
apagar, 1770
aparato *m*, 247, 251, 3925
aparato de acondicionamiento de aire *m*, 88
aparato de calefacción *m*, 2242
aparato de campo *m*, 1831
aparato de medición *m*, 2158
aparato de Orsat (humos) *m*, 2861
aparato de regulación *m*, 1027
aparato instalado en obra *m*, 1833
aparato para el exterior *m*, 2865
aparato registrador *m*, 3181
aparato respiratorio *m*, 524
aparato terminal de un circuito de aire *m*, 166
aparatos de medición *m,pl*, 2650
apertura de válvula *f*, 3952
aplicación *f*, 252
aportación modulante *f*, 2709
aporte *m*, 2107
aporte de calor *m*, 2240

aporte de calor por insolación *m*, 3516
aporte fortuito del calor *m*, 2401
aporte (de calor) por conducción *m*, 967
apoyo *m*, 3620
aprendiz *m*, 254
aproximación *f*, 255
arancel *m*, 3764
arandela *f*, 4049
árbol *m*, 1446, 3415
arcilla refractaria *f*, 1880
arco *m*, 502
arcón congelador *m*, 725
área *f*, 262
área de sección transversal *f*, 1136
area de transición *f*, 3887
area efectiva *f*, 1551
árido (hormigón) *m*, 66
aritmético, 264
armar, 285
armario secador *m*, 1469
armazón *f*, 2033
arquitecto *m*, 258
arquitecto proyectista *m*, 3096
arrabio *m*, 646
arrancador *m*, 3605
arranque *m*, 3604, 3612
arranque a plena carga *m*, 3611
arranque en vacío *m*, 2571
arranque sin carga *m*, 2779
arrastrar, 1648
arrastre *m*, 1414, 1649
articulación de encastre *f*, 3566
asbesto *m*, 265
ascensor *m*, 2541
aséptico, 2164
asfalto *m*, 280
asiento de la válvula *m*, 3953
aspiración *f*, 282, 3695
aspirador *m*, 284, 1727, 1773
aspirador de polvo *m*, 3943
aspirar, 3694
atadura *f*, 1808
ataque por un ácido *m*, 36
atenuación *f*, 302
atenuación acústica *f*, 3538
ático *m*, 303
atizador *m*, 2999
atizador mecánico *m*, 3660
atmósfera *f*, 290
atmósfera controlada *f*, 1024
atmósfera de referencia *f*, 291
atmosférico, 292
atomizador por inyección *m*, 2438
atomizar, 299
atornillar, 3354
audibilidad *f*, 304
audible, 305

auditoria energética *f*, 1620
aumentar (la presión), 490
automación *f*, 309
automático, 308
autónomo de tipo monobloque *m*, 2897
autoridad *f*, 306
autorizado, 2536
aventador *m*, 447
avería *f*, 518
avisador *m*, 4046
aviso *m*, 4045
axial, 315
azeotrópico, 323

bacterial, 336
bacterias *f,pl*, 335
bactericida, 340
bactericida *m*, 341
baja presión *f*, 2595
bajante de desagüe *m*, 1421
balance de humedad *m*, 2713
balance térmico *m*, 2227,
3792
balasto *m*, 360
bancada *f*, 381, 1951, 2028
bancada del motor *f*, 2727
banco de datos *m*, 1175
banco de pruebas *m*, 3777,
3788
banda *f*, 3158
banda proporcional *f*, 3102
bandeja de desagüe *f*, 1422
bandeja de goteo *f*, 1443
baño *m*, 382
baño de aceite *m*, 2810
baño de inmersión *m*, 1312
barniz *m*, 2512
barómetro *m*, 368
barómetro aneroide *m*, 224
barra *f*, 367
barra de hielo *f*, 2374
barra de suspensión *f*, 2214
barrear térmica *f*, 289
barrera *f*, 373
barrera de vapor *f*, 792, 3960
barrido de gases *m*, 1981
barrido de la cámara de
combustión *m*, 575
barrote de parrilla *m*, 1876,
2182
base *f*, 375
base de datos *f*, 1176
base de la chimenea *f*, 734
bastidor *m*, 2031, 2033
bastidor de hierro plano *m*,
1930
bastidor de refuerzo *m*, 3231
bastidor del filtro *m*, 1858
batería *f*, 385, 805
batería de calefacción *f*, 2236
batería de calefacción para
aire caliente *f*, 128
batería de
desrecalentamiento *f*, 1257

bateria de expansión *f*, 1745
batería de expansión directa
f, 1321
batería de precalentamiento
f, 3033
batería de recalentamiento *f*,
3227
batería de refrigeración *f*,
1049
batería de reserva *f*, 333
batería de secado *f*, 1437
batería frigorífica *f*, 98
batería horizontal de techo *f*,
659
batería refrigerante *f*, 98
biela *f*, 981
bienestar *m*, 867
bifurcación *f*, 407
bilámina *f*, 409
bimetal *m*, 408
binario, 411
bióxido de carbono *m*, 629
bisagra *f*, 2319
bit *m*, 413
bloque de hielo *m*, 2374
bloquear, 439
bloqueo *m*, 440
bobinado del motor *m*, 2732
boca de admisión del
ventilador *f*, 1798
boca de descarga *f*, 1345,
2867
boca de extracción de aire *f*,
1725
boca de hombre *f*, 2621
boca de impulsión de aire *f*,
3723
boca de salida *f*, 2867
boca de salida de aire
orientable *f*, 1330
boca de salida del aire
suministrado *f*, 3724
boca de salida situada en la
pared *f*, 4031
boca del quemador *f*, 578
bodega *f*, 668
bolómetro (medidor de
radiación) *m*, 485
bolsa *f*, 344
bolsa de agua *f*, 4080
bolsa de aire *f*, 79, 144
bolsa de vapor *f*, 3641
bolsa filtrante *f*, 1849
bomba *f*, 3119
bomba aceleradora *f*, 750
bomba aspiranteimpelente *f*,
3176
bomba autoaspirante *f*, 3381
bomba centrífuga *f*, 691
bomba de aire *f*, 149
bomba de alimentación de
agua por eyección de
vapor *f*, 3636
bomba de alimentación de la
caldera *f*, 463

bomba de aspiración del
condensado *f*, 948
bomba de calor *f*, 2279
bomba de circulación *f*, 749
bomba de émbolo *f*, 3176
bomba de engranajes *f*, 2160
bomba de eyección de vapor
f, 3636
bomba de retorno del
condensador *f*, 948
bomba de trasiego *f*, 3881
bomba de vacio *f*, 3947
bomba elevadora de presión
f, 493
bomba mezcladora *f*, 2698
bomba para pruebas de
presión *f*, 3070
bomba rotativa *f*, 3312
bomba volumétrica *f*, 4018
boquilla de inyección *f*, 2439
boquilla de quemador *f*,
2806, 2813
boquilla de salida *f*, 2870
boquilla del quemador *f*, 572
borde *m*, 1547
borde (del álabe) *m*, 3857
borde de los álabes (del
rodete) *m*, 2397
borna *f*, 1588
borna de conexión *f*, 982
borne *m*, 1588
botella aislante *f*, 3815
bouillant, 483
brida *f*, 1911
brida ciega *f*, 422, 438, 3663
brida de ajuste *f*, 756
brida de apriete *f*, 756
brida de conexión *f*, 978
brida de cuello soldado *f*,
4111
brida de ranura y lengüeta
encaradas *m*, 3862
brida del quemador *f*, 570
brida deslizante *f*, 3483
brida móvil *f*, 3483
brida para soldar *f*, 4109
brida roscada *f*, 3824
brillar, 2174
brillo *m*, 2173
briqueta *f*, 534
bromuro de litio *m*, 2565
bucle *m*, 2585
bucle de compensación *m*,
886
bujía *f*, 2990
bujía de ignición *f*, 3560
bulbo *m*, 561
bulón *m*, 486, 1124
buque tanque *m*, 3756
burbuja *f*, 541
burbuja de aire *f*, 79
buscador de tubos *m*, 2965
buscatubos *m*, 2965
butano *m*, 588
byte *m*, 598

caballo vapor *m*, 2329
cabezal *m*, 2217, 2218
cabezal de seguridad *m*, 3333
cabina de control *f*, 1018
cabina de mandos *f*, 1018
cabina para pintar a pistola *f*, 2909
cable *m*, 599
cadena de ancla *f*, 599
cadena del frío *f*, 819
caída *f*, 1447, 1448
caída de presión *f*, 3048
caida del chorro de aire *f*, 112
caja *f*, 503, 643
caja (de un aparato) *f*, 1617
caja de conexiones *f*, 2504
caja de distribución *f*, 1370, 3743
caja de empaquetadura *f*, 2902
caja de estopadas *f*, 3685
caja de humos *f*, 3492
caja de interruptores *f*, 3743
caja de mandos *f*, 3743
caja de mezcla *f*, 2693
caja del ventilador *f*, 1795
caja mezcladora *f*, 2693
calafateado *m*, 655
cálculo *m*, 603, 1682
cálculo de las pérdidas de calor *f*, 2272
cálculo detallado *m*, 1261
caldeo *m*, 1894
caldeo con aceite combustible *m*, 2820
caldeo con alimentación por tolva *m*, 2327
caldeo con petroleo *m*, 2820
caldeo por combustión de gas *m*, 2131
caldeo por coque *m*, 812
caldera *f*, 455
caldera a sobrepresión *f*, 3051
caldera caldeada con coque *f*, 811
caldera caldeada por aceite combustible *f*, 2819
caldera caldeada por petróleo *f*, 2819
caldera calentada por coque *f*, 811
caldera con carga por tolva *f*, 2326
caldera convertible *f*, 1042
caldera de acero *f*, 3648
caldera de agua caliente *f*, 2338
caldera de alta presión *f*, 2307
caldera de baja presión *f*, 2597
caldera de carbón *f*, 787

caldera de electrodos *f*, 1590
caldera de fundición *f*, 647
caldera de gas *f*, 2113
caldera de hierro fundido *f*, 647
caldera de pasos múltiples *f*, 2748
caldera de recuperación *f*, 4053
caldera de tres pasos *f*, 3827
caldera de tubos de agua *f*, 1892
caldera de tubos inclinados *f*, 2403
caldera de vapor *f*, 3624
caldera monoblor *f*, 2899
caldera para dos combustibles *f*, 1488
caldera policombustibles *f*, 2743
caldera tubular *f*, 4088
caldera vertical con el hogar atravesado por tubos de gas *f*, 1138
calefacción *f*, 2241
calefacción a dos tubos *f*, 3915
calefacción bajo el pavimento *f*, 1950
calefacción bajo vacío *f*, 3686
calefacción central *f*, 680, 2092
calefacción central a gas *f*, 2129
calefacción de base *f*, 327
calefacción de confort por radiación *f*, 3136
calefacción de gas *f*, 2134
calefacción de locales *f*, 3554
calefacción de recintos *f*, 3554
calefacción del aire *f*, 129
calefacción doméstica *f*, 1389
calefacción eléctrica *f*, 1574
calefacción individual *f*, 245, 2414
calefacción monotubo *f*, 2833
calefacción para un grupo de viviendas *f*, 441
calefacción por aceite *f*, 2822
calefacción por acumulación de calor *f*, 3667
calefacción por acumulación de calor en horas valle *f*, 2807
calefacción por agua caliente *f*, 4072
calefacción por agua caliente a baja presión *f*, 2601
calefacción por agua caliente a gas *f*, 2130
calefacción por agua sobrecalentada *f*, 2310
calefacción por aire caliente *f*, 2188, 4038

calefacción por aire caliente a gas *f*, 2128
calefacción por antepecho *f*, 4126
calefacción por calentamiento del piso *f*, 1950
calefacción por contrato (servicio contratado) *f*, 1011
calefacción por el techo mediante paneles suspendidos *f*, 662
calefacción por estufa *f*, 3674
calefacción por inducción *f*, 2419
calefacción por infrarrojos *f*, 2432
calefacción por las paredes *f*, 4028
calefacción por paneles *f*, 2913
calefacción por resistencia eléctrica *f*, 3252
calefacción por termosifón *f*, 2188
calefacción por vapor *f*, 3634
calefacción por vapor a baja presión *f*, 2600
calefacción por vapor bajo vacio *f*, 3948
calefacción radiante *f*, 3142
calefacción residencial *f*, 3250
calefacción suplementaria *f*, 3718
calefacción urbana por agua caliente *f*, 1378
calefactor con acumulación de calor *m*, 3666
calefactor de aire *m*, 4039
calefactor de zócalo *m*, 377
calentador *m*, 2235
calentador con almacenamiento térmico *m*, 3666
calentador de agua *m*, 2344, 4071
calentador de agua instantáneo *m*, 2452
calentador de aire *m*, 127
calentador de aire de caldeo directo *m*, 1324
calentador de inmersión *m*, 2393
calentador de llama directa *m*, 1325
calentador por convección *m*, 1035
calentador por infrarrojos a gas *m*, 2146
calentador soplante *m*, 429
calentamiento *m*, 4041
calentamiento anormal, sobrecalentamiento *m*, 2888

calentamiento de
  acumulación *m*, 3667
calentamiento del aire *m*, 129
calentamiento dieléctrico *m*,
  1285
calentamiento por gas *m*,
  2131
calentamiento suplementario
  m, 3718
calentar, 2225
calibración *f*, 604
calibre *m*, 495, 2158
calibre de descarga *m*, 1342
calidad *f*, 3128
calidad de gas *f*, 2145
calidad del aire *f*, 151
calidad del aire interior *f*,
  2416
caliente, 2332
calor *m*, 2224, 4042
calor absorbido *m*, 2266
calor cedido *m*, 4052
calor de combustión *m*, 854
calor de fusión *m*, 2104
calor latente *m*, 2517
calor latente de fusión *m*,
  2518
calor latente de vaporización
  m, 2519
calor metabólico *m*, 2672
calor perdido *m*, 4052
calor radiante *m*, 3140
calor útil *m*, 3936
caloría *f*, 606
calorífero *m*, 2242
calorífico, 607
calorímetro *m*, 611, 2274
calorímetro local *m*, 3299
cama *f*, 394
cámara *f*, 701, 3295
cámara ambiental *f*, 1658
cámara de aire *f*, 159
cámara de carga *f*, 2572
cámara de combustión *f*, 847,
  859, 1877, 2099, 2100
cámara de combustión a
  sobrepresión *f*, 3074
cámara de combustión de
  tiro forzado *f*, 2014
cámara de distribución de
  aire *f*, 2989
cámara de ensayos *f*, 3778
cámara de ensayos climáticos
  f, 1663
cámara de ensayos
  climatológicos *f*, 772
cámara de filtrado *f*, 1852
cámara de mezcla *f*, 437,
  2694
cámara de mezcla de aire *f*,
  2989
cámara de pruebas de
  climatización *f*, 772
cámara de pulverización *f*,
  3581

cámara de refrigeración *f*,
  1048
cámara de reverberación *f*,
  3269
cámara de separación (detrás
  de la válvula de expansión)
  f, 1922, 1927
cámara de sobrepresión *f*,
  2989
cámara de vapor *f*, 3625
cámara frigorífica *f*, 732, 823
cámara frigorífica comercial
  f, 1359
cámara frigorífica polivalente
  f, 2749
cambiador (de calor) *m*, 1714
cambiador de calor
  (calorífero) *m*, 610
cambiador de calor a
  contracorriente *m*, 1107
cambiador de calor
  intermedio *m*, 2472
cambiador de calor rotativo
  m, 3311
cambiador multitubular de
  envolvente *m*, 3426
cambiar, 1713
cambio *m*, 1712
cambio de color *m*, 1355
cambio de estado *m*, 702
cambio de fase *m*, 702
camión cisterna *m*, 3756
camión cisterna (para el
  transporte de petróleo) *m*,
  2830
camisa *f*, 2492
camisa aislante *f*, 2457
camisa de agua *f*, 4073
camisa de caldera *f*, 458
camisa de vapor *f*, 3637
camisa enfriadora *f*, 1061
campana de aspiración *f*,
  3698
campana de extracción *f*,
  1732
campana de humos *f*, 2097
campo de aplicación *m*, 253
campo de control *m*, 1032
campo de enfriamiento *m*,
  1066
campo de regulación *m*, 55
campo de visión *m*, 1834
canal *m*, 706
canal de desagüe *m*, 3411
canaleta *f*, 742
canalización *f*, 972, 1492
canalización de gas *f*, 2138
canalización principal de
  calefacción urbana *f*, 1379
cantidad *f*, 204, 3131
cantidad de movimiento *f*,
  2722
cantidad mínima de aire *f*,
  2687
canto *m*, 1547

cáñamo *m*, 2296
cañería *f*, 2950
capa *f*, 2520
capa de hielo *f*, 2381
capa de pintura *f*, 791
capa intermedia *f*, 2471
capa límite *f*, 499
capa protectora *f*, 3111
capacidad *f*, 617, 2195
capacidad calorífica *f*, 2232
capacidad de absorción *f*, 16
capacidad de absorción de
  agua *f*, 4058
capacidad de acumulación
  térmica *f*, 716
capacidad de carga de agua *f*,
  4058
capacidad de contacto *f*,
  1002
capacidad de descarga *f*,
  1354
capacidad de flujo *f*, 1955
capacidad de funcionar *f*,
  4135
capacidad de la unidad
  condensadora *f*, 961
capacidad de la válvula de
  expansión *f*, 1754
capacidad de retención de
  polvo *f*, 1517
capacidad del compresor *f*,
  913
capacidad neta *f*, 2765
capacidad térmica *f*, 2229
capacitancia *f*, 615
capataz *m*, 714, 3083
caperuza *f*, 488
caperuza de escape de gases
  f, 1732
caperuza de evacuación *f*,
  1732
caperuza de protección
  contra la intemperie *f*,
  4099
capilar, 624
capilar *m*, 627
capilaridad *f*, 623
cápsula aneroide *f*, 225
cápsula para medición *f*,
  2652
captador de hollín *m*, 2194,
  3530
caracola del ventilador *f*,
  1795
característica *f*, 707
características del local *f,pl*,
  3300
características *f,pl*, 1242
carbón *m*, 782
carbón activado *m*, 42
carbón antracitoso *m*, 783
carbón bituminoso *m*, 415,
  537
carbón con proporción
  media de volátiles *m*, 2661

carbón de alto contenido de
  volátiles *m*, 2318
carbón de gas *m*, 2117
carbón de grano basto *m*,
  790
carbón de grano grueso *m*,
  790
carbón de llama larga *m*,
  2582
carbón magro *m*, 2782
carbón no bituminoso *m*,
  2782
carbón no coquizable *m*,
  2784
carbón para coque *m*, 815
carbón pulverizado *m*, 3118
carbonera *f*, 566, 789
carbonización *f*, 632
carbono *m*, 628
carcaja de bomba *f*, 3122
carcasa *f*, 645, 1617
carga *f*, 634, 711, 712, 2568
carga (cantidad de) *f*, 719
carga (de un horno) *f*, 715
carga admisible *f*, 2944
carga calorífica *f*, 2251, 2270
carga conectada *f*, 976
carga de combustible *f*, 2081
carga de demanda eléctrica *f*,
  1580
carga de horno u hogar *f*,
  3661
carga de la parrilla *f*, 2183
carga de proyecto *f*, 1251
carga de régimen *f*, 2848
carga de servicio *f*, 2848
carga de trabajo admisible *f*,
  3336
carga dinámica *f*, 3980
carga en porcentaje fijo *f*,
  1932
carga fraccionada *f*, 2923
carga frigorífica *f*, 1062
carga incompleta *f*, 2923
carga inicial *f*, 2436
carga límite *f*, 2574, 2641
carga máxima *f*, 2641
carga normal de trabajo *f*,
  2848
carga por calefacción *f*, 2251
carga punta *f*, 2926
carga reducida *f*, 2923
carga refrigerada *f*, 727, 3203
carga térmica del edificio *f*,
  558
carga total *f*, 3869
carga uniforme *f*, 1932
carga útil *f*, 2874
cargador automático de
  parrilla articulada
  (calderas) *m*, 699
cargador mecánico (de parrilla) *m*, 3660
carrera *f*, 2541
carrera del émbolo *f*, 3682

carril de guía *m*, 2204
cárter *m*, 1123
cascada *f*, 640
casquillo *m*, 3473
casquillo de cojinete *m*, 585
casquillo lámparas *m*, 614
catalizador *m*, 651
catatermómetro *m*, 2505
catión *m*, 654
cátodo *m*, 652
caudal *m*, 1966, 3838
caudal de aire *m*, 123, 167
caudal de aire caliente *m*,
  4040
caudal de aspiración *m*, 3696
caudal de descarga *m*, 1347
caudal másico *m*, 2634
cavidad *f*, 657
cavidad de aire *f*, 159
cavitación *f*, 656
cazuela del quemador *f*, 574
cebado *m*, 3084
cebar, 3081
celdilla *f*, 667
célula *f*, 667
cemento *m*, 675
cenicero *m*, 271
ceniza *f*, 270
cenizas *f,pl*, 743
cenizas volantes *f,pl*, 1997,
  2193
centígrado, 676
central *f*, 3618
central combinada para
  calefacción y energía *f*, 840
central de calefacción *f*, 677,
  681, 682, 2287
central de calefacción urbana
  f, 1380
central de reducción de
  presión *f*, 3062
central eléctrica *f*, 3019, 3022
central térmica *f*, 2287
central termoeléctrica *f*, 2277
centrífugo, 687, 693
centro de gravedad *m*, 686
cero (absoluto) *m*, 4147
cerradura *f*, 2579
cerrar herméticamente, 3362
cerrojo *m*, 1390
certificado *m*, 695
certificado de pruebas *m*,
  3779
ciclaje *m*, 1156
ciclo *m*, 1155
ciclo cerrado *m*, 778
ciclo de Carnot *m*, 635
ciclo de compresión *m*, 901
ciclo de desescarche *m*, 1209
ciclo de Rankine *m*, 3159
ciclo de referencia *m*, 3188
ciclo de refrigeración por
  chorro de agua *m*, 3963
ciclo de Stirling *m*, 3657
ciclo de trabajo *m*, 1527

ciclo de vapor binario *m*, 412
ciclo frigorífico *m*, 3209
ciclo interrumpido *m*, 3432
ciclo inverso *m*, 3271
ciclo Otto *m*, 2863
ciclón *m*, 1158
cielo raso *m*, 658
cielo raso perforado *m*, 2934
cierre *m*, 2579, 3438
cierre de cigüeñal *m*, 3361
cigüeñal *m*, 1125
cilindro *m*, 1159
cimentación *f*, 2028
cimiento *m*, 375
cimientos *m,pl*, 2028
cincel *m*, 740
cinético, 2507
cinta calefactora *f*, 3144
cinta transportadora *f*, 1044
circuito *m*, 744
circuito de control *m*, 1028
circuito de desviación *m*,
  1386
circuito de dos hilos *m*, 3919
circuito de mezcla *m*, 2695
circuito de salmuera *m*, 529
circuito de seguridad *m*, 3327
circuito del fluido frigorífico
  m, 3200
circuito del refrigerante *m*,
  3200
circuito frigorífico *m*, 3200
circuito primario *m*, 3078
circuito simple *m*, 3447
circulación *f*, 747
circulación de burbujas *f*, 542
circulación del aire *f*, 83
circulación en
  contracorriente *f*, 1106
circulación en corrientes
  cruzadas *f*, 1130
circulación forzada *f*, 2009
circulación por gravedad *f*,
  2187
circunferencia *f*, 751
cisterna *f*, 752
cisterna de inodoro *f*, 1994
clasificación *f*, 758
clave del ordenador *f*, 926
clima *m*, 771
climatización *f*, 89
climatización de buques *f*,
  2632
climatización en verano *f*,
  3705
climatización residencial *f*,
  3249
climatización total *f*, 2091
climatizador *m*, 88, 126
climatizador central *m*, 684
climatizador de condal
  variable *m*, 2034
climatizador tipo cónsola *m*,
  988
climatizador zonal *m*, 4150

climatizar, 87
clorofluorcarbono *m*, 741
coaxial, 793
cobre *m*, 1081
código *m*, 796
código de colores *m*, 832
codo *m*, 502, 1563, 2955
codo corrugado *m*, 1100
codo de 180° *m*, 3260
codo de dilatación *m*, 1743
codo de tubo *m*, 2951
codo de un conducto *m*, 1495
codo en U *m*, 3260
codo para conexiones soldadas *m*, 4107, 4108
codo para soldar *m*, 4107,
   4108
codo soldado *m*, 4107, 4108
codo suave *m*, 3501
coeficiente *m*, 797
coeficiente de absorción *m*,
   17
coeficiente de amortiguación
   m, 1199
coeficiente de arrastre *m*,
   1650
coeficiente de
   compresibilidad *m*, 899
coeficiente de conductividad
   m, 799
coeficiente de conductividad
   térmica *m*, 3795
coeficiente de demora *m*,
   1415
coeficiente de descarga *m*,
   800, 1340
coeficiente de descarga de
   aire *m*, 108
coeficiente de difusión *m*,
   1302
coeficiente de dilatación *m*,
   801
coeficiente de eficacia
   energética *m*, 2939
coeficiente de eficiencia
   energética *m*, 1626
coeficiente de ensuciamiento
   m, 2027
coeficiente de expansión *m*,
   1744
coeficiente de opacidad *m*,
   2839
coeficiente de película *m*, 803,
   1841
coeficiente de reflexión *m*,
   3193
coeficiente de rendimiento *m*,
   2939
coeficiente de rozamiento *m*,
   2066
coeficiente de rugosidad *m*,
   3317
coeficiente de seguridad *m*,
   3332

coeficiente de sombra *m*,
   3413
coeficiente de transmisión *m*,
   3889
coeficiente de transmisión de
   calor *m*, 2290
coeficiente de transmisión
   energética *m*, 1637
coeficiente de transmisión
   superficial *m*, 803
coeficiente de transmisión
   superficial (de calor) *m*,
   1841
coeficiente de utilización *m*,
   3938
cogeneración *f*, 804
cojinete *m*, 391
cojinete axial *m*, 3842
cojinete de agujas *m*, 2762
cojinete de bolas *m*, 361
cojinete de fricción *m*, 3474
cojinete en dos piezas *m*,
   3573
cojinete de bolas *m*, 3287
cojinete liso *m*, 3474
cojinete partido *m*, 3573
colador *m*, 3675
colchón de aire *m*, 102
colector *m*, 829, 2218, 2622,
   3173, 3706
colector (elec) *m*, 584
colector de alimentación *m*,
   3727
colector de placa plana *m*,
   1931
colector de retorno *m*, 3264
colector de señales *m*, 3443
colector de vapor *m*, 3633
colector múltiple (tubos) *m*,
   2622
colector solar *m*, 3513
colector solar concentrador
   m, 939
colocación bajo tierra *f*, 2522
color *m*, 831, 1528
color de identificación *m*,
   2383
color normalizado *m*, 3598
colorante *m*, 1529
colorimétrico, 830
columna *f*, 835
columna de agua *f*, 837
columna de mercurio *f*, 836
columna de retorno *f*, 3266
columna vertical *f*, 3280
collar de chimenea *m*, 1975
collar de fijación (tubo) *m*,
   2952
combustibilidad *f*, 841
combustible, 842
combustible de baja calidad
   m, 2593
combustible de bajo poder
   calorífico *m*, 2593
combustible fósil *m*, 2025

combustible líquido *m*, 2076
combustible pobre *m*, 2593
combustión *f*, 844
combustión a carbón, 786
combustión a carbón *f*, 788
combustión completa *f*, 889
combustión descendente *f*,
   1412
combustión en lecho
   fluidificado *f*, 1988
combustión en sentido
   descendente *f*, 1412
combustión incompleta *f*,
   2406
combustión pelicular *f*, 1840
combustión perfecta *f*, 2932
comercial, 873
compartimiento congelador
   m, 2048
compatible, 884
compensado ambientalmente,
   196
compensador *m*, 357, 887
compensador (articulado) *m*,
   888
compensador axial *m*, 316
compensador de dilatación
   deslizante *m*, 3479
compensador de tiempo
   atmosférico *m*, 4095
componente *m*, 891
componentes de ventanaje *m*,
   1826
comportamiento dinámico *m*,
   1531
compresibilidad *f*, 898
compresión *f*, 900
compresión compuesta *f*, 892
compresión doble *f*, 1483
compresión escalonada *f*, 892
compresión seca *f*, 1463
compresor *m*, 912
compresor abierto *m*, 2842
compresor alternativo *m*,
   3175
compresor axial *m*, 318
compresor bicilíndrico *m*,
   3913
compresor centrífugo *m*, 688
compresor compuesto de
   pistón por etapas *m*, 3656
compresor de acción simple
   m, 3449
compresor de álabe
   deslizante *m*, 3481
compresor de desplazamiento
   positivo *m*, 3008
compresor de diafragma *m*,
   1278
compresor de disco oscilante
   m, 4134
compresor de doble acción
   m, 1395
compresor de doble
   aspiración *m*, 1405

corriente turbulenta *f*, 3911
corriente viscosa *f*, 4009
corroer, 1093
corrosión *f*, 1094, 3325
corrosión alveolar superficial *f*, 2972
corrosión puntual *f*, 2972
corrosivo, 1098
cortatiros *m*, 3577
corte *m*, 1151
corte de reproducción *m*, 3244
corte de sustitución *m*, 3243
cortina de agua pulverizada *f*, 3583
cortina de aire *f*, 101
cortina de aire caliente *f*, 4035
cortocircuito *m*, 745, 3431
coste(s) *m*, 1101
coste de calefacción *m*, 2244
coste de energía *m*, 1624
coste de explotación *m*, 1102, 2844, 3322
coste de instalación *m*, 1896
coste de la ayuda *m*, 1778
coste de la instalación *m*, 2975
coste del periodo de duración *m*, 2540
costo(s) *m*, 1101
costo de adquisición *m*, 2891
costura *f*, 3367
criba *f*, 3440
criogénico, 1140
cristal (de ventana) *m*, 2910
cristal de hielo *m*, 2376
criterios del proyecto *m,pl*, 1249
crítico, 1126
croquis *m*, 1433
cruceta *f*, 1134
cuadro central de control *m*, 679
cuadro de distribución *m*, 3742
cuadro de mandos *m*, 1029, 3742
cuadro de maniobras *m*, 3742
cuarto de baño *m*, 383
cuarto estufa *m*, 2336
cubierta *f*, 1113, 1114
cubierta de protección contra la intemperie *f*, 4098
cubito de hielo *m*, 1141, 2377
cubo (rueda) *m*, 2350
cubrerradiador *m*, 3148
cuelgatubos *m*, 2957
cuelgue *m*, 2214
cuerpo *m*, 453
cuerpo de bomba de acción directa *m*, 1317
cuerpo negro *m*, 417
culata *f*, 1160
cuota *f*, 1811

curva *f*, 403, 1147, 3501
curva característica *f*, 708
curva característica del ventilador *f*, 1797, 1801
curva corrugada *f*, 1099
curva de calefacción *f*, 2245
curva de enfriamiento *f*, 1055
curva de potencia *f*, 3014
curva de saturación *f*, 3341
curva de tubo *f*, 2951
curvado hacia atrás, 334
curvado hacia delante, 2024
curvas (NC) *f,pl*, 2774
curvatura *f*, 502

chapa *f*, 2981, 3419
chapa de recubrimiento *f*, 1115
chapa para calderas *f*, 474
chapa perforada *f*, 2936
chapa repujada *f*, 1607
chapado de cobre *m*, 1083
chimenea *f*, 700, 733, 3495, 3593
chimenea abierta *f*, 2223
chimenea de evacuación de gases quemados *f*, 2132
chimenea de ventilación *f*, 177, 3992
chimenea francesa *f*, 1884, 2223
chispa *f*, 3559
chisporroteo en la caldera *m*, 475
choque térmico *m*, 3804
chorro *m*, 2494
chorro de aire *m*, 135

daño *m*, 1164
dardo *m*, 3841
datos *m*, 1174
datos ambientales *m*, 4096
datos analógicos *m*, 213
datos de explotación *m,pl*, 2845
de descarga, 1353
de etapas, 3654
de reserva, 332, 3600
de un álabe, 3456
de un etapa, 3455
de varios escalones, 2751
de varios saltos, escalonado múltiple, 2751
deceleración *f*, 1193
decibelio *m*, 1194
decipol *m*, 1195
declinación *f*, 1196
decoloración *f*, 1355
defecto *m*, 3757
deflector *m*, 342, 1205, 2203
degradación bacteriana *f*, 338
demanda *f*, 1232
demanda de agua caliente *f*, 2339
demanda eléctrica *f*, 1566

densidad *f*, 1235
densidad aparente *f*, 563
densidad de aire en ventilador *f*, 1791
densidad de empaquetadura *f*, 2904
densidad de relleno *f*, 2904
densidad del flujo térmico *f*, 3564
densidad del humo *f*, 3493
depositar, 1238
depósito *m*, 1237, 3024, 3755, 4000
depósito (de sedimento) *m*, 1239
depósito a presión *m*, 908
depósito cilíndrico *m*, 1455
depósito de agua caliente *m*, 2346
depósito de aire (comprimido) *m*, 175
depósito de combustible *m*, 2087
depósito de equilibrado *m*, 356
depósito de escarcha *m*, 2073
depósito de filtración *m*, 1862
depósito de polvo *m*, 1512
depósito enterrado *m*, 3922
depósito equilibrador *m*, 1667
depósito pulmón *m*, 544
depreciación *f*, 1240
depreciación por uso *f*, 4093
depresión *f*, 2764
deprimómetro *m*, 1426
depuración *f*, 3124
depuración de gases *f*, 2116
depuración del aire *f*, 86
derivación *f*, 507, 509, 593, 2808, 3434
derivación ciega *f*, 1186
derivar (electricidad), 3758
derrame *m*, 2883
desaireación *f*, 1189
desarrollo de bacterias *m*, 339
descalcificación del agua *f*, 4085
descarbonizador *m*, 3360
descarga *f*, 445, 1338
descarga de aire *f*, 116
descarga del compresor *f*, 914
descargador de compresor *m*, 921
descargar, 1339
descenso del punto de rocío *m*, 1270
descomposición *f*, 1191, 1197
descompresor *m*, 1198
desconectar, 1152, 3744
desconector de presión diferencial *m*, 3047
desconector de seguridad de alta presión *m*, 2311

dispositivo de retención *m*, 3892

dispositivo de seguridad *m*, 3331

dispositivo de seguridad contra la falta de gas *m*, 2124

dispositivo de seguridad en la ignición *m*, 2389

dispositivo de ventilación *m*, 3995

dispositivo (de seguridad) para el caso de apagarse la llama *m*, 1905

dispositivo terminal de un circuito de aire *m*, 166

dispositivos de descarga de aire *m*, 165

distancia *f*, 1365, 3556

distribución *f*, 1369

distribución de abajo arriba *f*, 1372

distribución de agua caliente *f*, 2340

distribución de aire *f*, 109

distribución de gas *f*, 2121, 2150

distribución desde abajo *f*, 1372

distribución por conductos *f*, 1494

distribución por la parte superior *f*, 2887

distribución urbana de frío *f*, 1377

distribuidor *m*, 1376, 2218, 2622

disyuntor *m*, 519

doble *m*, 1507

doble pared *f*, 2492

doméstico, 1388

domo *m*, 488, 1387

dosificación *f*, 1392

dosificar, 1393

drenaje *m*, 1416, 3508

drenar, 1417

ductibilidad *f*, 1496

durabilidad *f*, 1508

duración *f*, 1509, 2538, 3846

duración de la combustión *f*, 852

duración de la etapa de un motor *f*, 2731

duración efectiva *f*, 1554

duración real *f*, 1554

dureza del agua *f*, 2215, 4070

ebullición *f*, 1540

economía *f*, 1543

economía energética *f*, 1625, 3015

economía térmica *f*, 2247

economización de energía *f*, 1625

economizador *m*, 1542

economizador de alimentación de agua *m*, 1820

economizador del compresor *m*, 917

ecuación *f*, 1670

ecuación de estado *f*, 1671

edificio *m*, 545

edificio de calderas *m*, 470

efectividad de la aleta *f*, 1867

efectivo, 43, 1550

efecto *m*, 1549

efecto barométrico *m*, 371

efecto calefactor del compresor (bomba de calor) *m*, 918

efecto capilar *m*, 625

efecto chimenea *m*, 3594

efecto de chimenea *m*, 737

efecto de disipación de calor del condensador *m*, 955

efecto de refrigeración *m*, 731

efecto deshumidificador *m*, 1223

efecto Peltier *m*, 2931

eficiencia de eliminación de polvo *f*, 1522

eficiencia de filtrado (método opacimétrico) *f*, 296

eficiencia de filtrado *m*, 1861

eje *m*, 384, 3415

eje (en general:geometría, geología, matemáticas, etc.) *m*, 320

eje (vehículos, máquinas) *m*, 321

eje motor *m*, 1446

ejector hidraulico *m*, 4074

ejecución *f*, 1716

elasticidad *f*, 1561

elastómetro *m*, 1562

electricidad *f*, 1575

eléctrico, 1564, 1565

electrodo *m*, 1589

electrodo de cebado *m*, 2387

electrodo de encendido *m*, 2387

electroimán *m*, 1593

electrolítico, 1591

electro-neumático, 1599

electrónica *f*, 1598

electrónico, 1595

elemento *m*, 891, 1602, 3374, 3925

elemento bimetálico *m*, 409

elemento calefactor *m*, 2248

elemento de construcción *m*, 549, 995

elemento de control *m*, 1022

elemento de la caldera *m*, 478

elemento detector *m*, 1262

elemento enfriador *m*, 1060

elemento estructural *m*, 3683

elemento filtrante *m*, 1857

elemento motriz a vapor *m*, 3961

elevación *f*, 2541

elevación del punto de rocío *f*, 1272

eliminación *f*, 3241

eliminación de gases *f*, 1213

eliminación de gases de la combustión *f*, 1980

eliminación de humos *f*, 1980

eliminación del aire *f*, 1189

eliminación del polvo *f*, 1521

eliminador de aire *m*, 117

eliminador de gotas *m*, 1604

eliminador de polvo *m*, 1513

eliminador de sedimentos (torre de refrigeración) *m*, 1439

embarque *m*, 1649

embolada de compresión *f*, 907

embolada de descarga *f*, 1350

émbolo *m*, 2966

embudo *m*, 2098

emisión *f*, 1611

emisión térmica direccional *f*, 1331

emisividad (total) *f*, 1612

emitancia *f*, 1613

empalmar, 975

empalme *m*, 986, 2500

empalme para el vapor *m*, 3627

empañamiento *m*, 522

empaquetadora de membrana *f*, 397

empaquetadura *f*, 2136, 2903

empaquetadura de caucho *f*, 3318

empaquetadura de eje *f*, 3416

emplazamiento *m*, 3459

empotrado, 1605

empresa *f*, 1644

empuje *m*, 489

empuje del viento *m*, 4127

empuñadura *f*, 2212, 2508

emulsión *f*, 1615

en espiral, 3569

enbolada *f*, 3841

encargado *m*, 714

encargado de la obra *m*, 770

encargado de las obras *m*, 3460

encendido *m*, 1894, 2386

encendido del quemador *m*, 571

enchufar (electricidad), 2992

enchufe hembra *m*, 3509

enchufe macho *m*, 2990

endotérmico, 1618

energía *f*, 1619, 3012

energía acústica *f*, 3541

energía consumida *f*, 3013

energía del punto cero *f*, 4148

energía eléctrica absorbida *f*, 1578
energía potencial *f*, 1633, 3011
energía solar *f*, 3515
energía total *f*, 3867
enfriado por agua, 4060
enfriado por aire, 94
enfriador *m*, 729, 1048
enfriador de aceite *m*, 2815
enfriador de agua *m*, 4061
enfriador de agua potable *m*, 1441
enfriador de aire *m*, 97, 3928
enfriador de aire de capacidad variable *m*, 2036
enfriador de aire de convección natural *m*, 2756
enfriador de aire forzado *m*, 2010
enfriador de aire seco *m*, 1460
enfriador de aire tipo seco *m*, 1481
enfriador de entrada *m*, 4022
enfriador de gases *m*, 2120
enfriador de placas *m*, 2985
enfriador de salmuera *m*, 530
enfriador de tambor *m*, 1456
enfriador evaporativo *m*, 1698
enfriador húmedo de aire *m*, 4120
enfriador instantáneo *m*, 1924
enfriador intermedio *m*, 2467
enfriador pelicular *m*, 1787
enfriador por aire forzado *m*, 424
enfriamiento *m*, 1050, 1057
enfriamiento (artificial) *m*, 3207
enfriamiento a chorro *m*, 2495
enfriamiento a presión *m*, 3044
enfriamiento al vacío *m*, 3944
enfriamiento de aceite *m*, 2816
enfriamiento de locales *m*, 3553
enfriamiento de recintos (cerrados) *m*, 3553
enfriamiento de tipo forzado *m*, 2016
enfriamiento del aire *m*, 99
enfriamiento directo *m*, 1314
enfriamiento evaporativo *m*, 1699
enfriamiento localizado *m*, 3578
enfriamiento para bienestar *m*, 870
enfriamiento para confort *m*, 870

enfriamiento por agua helada *m*, 2379
enfriamiento por aire *m*, 103
enfriamiento por aire forzado *m*, 75
enfriamiento por evaporación de agua *m*, 1699
enfriamiento por eyección de vapor *m*, 3639
enfriamiento por paneles *m*, 2912
enfriamiento por radiación *m*, 3137
enfriamiento por rociado *m*, 3582
enfriamiento por rociado de techo *m*, 3292
enfriamiento por transpiración *m*, 3891
enfriamiento regenerativo *m*, 3216
enfriamiento superficial *m*, 3732
enfriamiento urbano *m*, 1377
enfriar, 726, 1045
enfriar (artificialmente), 3202
enlace *m*, 983, 986
enlace rápido *m*, 3133
enlucido (edif) *m*, 3242
enriquecimiento *m*, 1643
ensanchamiento cónico *m*, 1916
ensayo *m*, 1757, 3776
ensayo de humos *m*, 3499
ensayo en vacio *m*, 3949
ensuciamiento *m*, 2026
entalpía *f*, 1645
entalpía específica *f*, 3563
enterrar, 583
entrada *f*, 1654, 2107, 2443
entrada a ordenador *f*, 929
entrada analógica *f*, 216
entrada de aire *f*, 133, 164, 2434
entrada de cable *f*, 601
entradas de calor *f,pl*, 2240
entradas fortuitas del calor *f,pl*, 2401
entrega *f*, 1228
entrepaño *m*, 2910
entretenimiento *m*, 2611, 3401
entretiempo *m*, 2683
entronque *m*, 511
entropía *f*, 1651
envejecimiento *m*, 65
envolvente *f*, 1615, 1656, 3427
envolvente aislante *f*, 2457
envolvente de caldera *f*, 458
envolvente del edificio *f*, 552
envolvente del ventilador *f*, 1795, 1805

envuelta *f*, 645, 1617, 1656, 2492
época entre dos estaciones *f*, 2683
época entre dos temporadas *f*, 2683
equilibrado, 347
equilibrado de presiones *m*, 3049
equilibrador *m*, 357, 1666
equilibrador de flujo *m*, 1961
equilibrar, 346, 1665
equilibrio *m*, 1672
equilibrio de evaporación *m*, 1700
equilibrio de humedad *m*, 2713
equilibrio térmico *m*, 3798
equipo(s) *m*, 1674
equipo congelador *m*, 2049
equipo de congelación *m*, 2049
equipo de evacuación de aire *m*, 1722
equipo de extracción de aire *m*, 1722
equipo de humidificación *m*, 2355
equipo de medida *m*, 2650
equipo de pruebas *m*, 3782
equipo de reserva *m*, 3601
equipo(s) de separación de polvo *m*, 1524
equipo de suministro de aire *m*, 3722
equipo frigorífico *m*, 1076
equipo separador de polvo *m*, 1524
equivalente, 1676
equivalente mecánico del calor *m*, 2657
equivocación *f*, 1680
erosión *f*, 1679
erosionar, 2
error *m*, 1680
error aleatorio *m*, 3156
escala de ajuste *f*, 55
escala de Beaufort *f*, 393
escala de indicación *f*, 2412
escalón *m*, 3651
escalón de presión *m*, 3067
escape *m*, 1719
escape de aire *m*, 137
escape de gas *m*, 1642
escarcha *f*, 2071
escarchado *m*, 2380
escobillón *m*, 1974
escoplo *m*, 740
escorias *f,pl*, 743
escorias (de hierro o de carbón) *f,pl*, 773
escorificación *f*, 3467
esfera (de un medidor) *f*, 1274
esfuerzo *m*, 3681

espacio cerrado *m*, 1617
espacio edificado *m*, 557
espacio hueco sobre el falso
 techo *m*, 666
espacio muerto *m*, 768
espacio necesario *m*, 3555
espacio ocupado *m*, 2801
especificación *f*, 3562
específico, 3561
espectro acústico *m*, 3550
espectro sonoro *m*, 3550
espesor *m*, 3821
espesor de chapa *m*, 3421
espesor de la capa *m*, 2521
espesor del borde de salida
 m, 3876
espiral *f*, 3569
espiro-ducto *m*, 3571
espita *f*, 1810
espuma *f*, 1999
espuma de vidrio *f*, 2002
espuma plástica *f*, 2978
espuma plástica rígida *f*,
 3278
espumaje en la caldera *m*,
 466
espumar, 2000
esquema *m*, 2924
esquema de conexiones *m*,
 4132
esquema de montaje *m*, 4132
esquema eléctrico *m*, 1567
estabilidad *f*, 3592
estabilizador *m*, 355
estabilizador de tiro *m*, 1432
estable al calor, 2284
estación *f*, 3618
estación de equilibrado *f*, 358
estado *m*, 3614
estado del aire *m*, 964
estancia *f*, 3620
estanco a los gases, 2151
estanco al agua, 4086
estanco al aire, 169
estanco al polvo, 1520
estanco (a la presión), 3071
estanqueidad *f*, 3845
estanquidad al aire *f*, 170
estañado *m*, 3856
estaño *m*, 3855
estator *m*, 3619
estequiométrico, 3659
estéril, 2164
esterilizado, 2164
estimación *f*, 1682, 1684
estimar, 1683
estopada *f*, 3684
estrangulación *f*, 3837
estrangulador *m*, 3835
estrangular, 3836
estrategia de control *f*, 1033
estratificación *f*, 3677
estratificación del aire *f*, 163
estribo *m*, 3658
estructura *f*, 2032, 2033

estructura celular *f*, 673
estructura de refuerzo *f*, 3231
estructura de techo *f*, 664
estufa *f*, 2126, 2259, 3303
estufa de gas *f*, 2149
estufa de petróleo *f*, 2828
estufa eléctrica *f*, 1573
estufa escocesa *f*, 3351
etapa *f*, 3651
etapa de compresión *f*, 906
eupateoscopio *m*, 1685
eupateóscopo *m*, 1685
eutéctico, 1686
evacuación *f*, 1558, 1688,
 3241
evacuación de escorias *f*,
 3466
evacuación del agua *f*, 4065
evacuación del aire *f*, 1189
evacuado, 1687
evacuador del agua *m*, 4079
evacuar, 1339, 1720
evaporación *f*, 1690, 3957
evaporador *m*, 1703
evaporador de expansión
 seca *m*, 1465
evaporador de placas *m*,
 2987
evaporador infraalimentado
 m, 3613
evaporador inundado *m*,
 1947
evaporador multitubular *m*,
 3423
evaporador multitubular de
 envolvente *m*, 3425
evaporador tipo seco *m*, 1482
evaporador tipo vertical *m*,
 3999
evaporar, 1689
evaporímetro *m*, 1702
exactitud *f*, 32
excavación para drenaje de
 aguas *f*, 3508
excedente *m*, 1708
excéntrico, 1541
exceso *m*, 1708, 2890
exceso de temperatura *m*,
 1711
excitación *f*, 1715
exento de aceite, 2821
exergía *f*, 1717
exfiltración *f*, 1718
exfiltración de aire *f*, 118
exhibir, 1362
exigencia de tiro *f*, 1431
expansión *f*, 1741
expansión seca (directa) *f*,
 1464
experimentar, 1758
experimento *m*, 1757
explosión *f*, 1760
explosión de gas *f*, 2123
explotación *f*, 2852

explotación de la instalación
 f, 3323
exponente *m*, 1761
expuesto, 1762
extensión diurna *f*, 1182
exterior, 1764
externo, 1765
extinguir, 1770
extintor *m*, 1771
extintor de incendios *m*, 1883
extracción de aire *f*, 119,
 1737
extracción de escorias *f*, 763,
 775
extracción de fangos *f*, 3489
extracción de polvo *f*, 1515
extracción del polvo *f*, 1521
extractor *m*, 1727, 1773
extraer, 1720, 1772
extraer (mediante grifo o
 llave), 3758
extremo de los álabes (del
 rodete) *m*, 2397
eyector *m*, 1559

fábrica *f*, 1781, 4140
fábrica de ladrillo *f*, 527
fabricación *f*, 2629
fabricación robotizada *f*, 924
fabricar, 2630
factor *m*, 1779
factor angular *m*, 227
factor de amortiguación *m*,
 1199
factor de carga *m*, 2570
factor de carga (elec) *m*, 1584
factor de compresión *m*, 903
factor de flujo *m*, 1962
factor de forma *m*, 227, 279,
 974
factor de potencia *m*, 620
factor de potencia (elec) *m*,
 1582
factor de rendimiento *m*, 802
factor de seguridad *m*, 1780
factor de utilización *m*, 1384,
 3938
factor energético *m*, 1627
factoría *f*, 1781
fachada *f*, 1775
fachada longitudinal *f*, 2584
fallo de potencia *m*, 3017
fallo eléctrico *m*, 1568
fancoil *m*, 1796
fango *m*, 3487
fase *f*, 2946
fatiga *f*, 3681
fiabilidad *f*, 3235
fibra *f*, 1828
fibra de vidrio *f*, 2169
fibra óptica *f*, 1827
fibra sintética *f*, 3749
fieltro *m*, 1822
fijación *f*, 1808
fijar, 3404

filtración *f*, 1864
filtración del aire *f*, 86
filtrado *m*, 1864
filtrado para la eliminación del polvo *m*, 1515
filtrar, 1847
filtro *m*, 1846, 3025, 3675
filtro absoluto *m*, 7
filtro de agua *m*, 4067
filtro de aire *m*, 120
filtro de aire de extracción *m*, 1723
filtro de aire en un baño de aceite *m*, 2811
filtro de alta eficacia *m*, 2303
filtro de alto rendimiento *m*, 2303
filtro de aspiración *m*, 3703
filtro de bolsa *m*, 345
filtro de capa seca *m*, 1472
filtro de carbón vegetal *m*, 710
filtro de cartucho *m*, 670
filtro de cepillos *m*, 538
filtro de combustible *m*, 2082
filtro de fangos *m*, 3488
filtro de fibras *m*, 1830
filtro de grasas *m*, 2189
filtro de impacto *m*, 2395
filtro de la línea de succión *m*, 3700
filtro de manga *m*, 3476
filtro de polvo *m*, 1516
filtro de repuesto *m*, 1363
filtro de tela *m*, 1774
filtro de tubo flexible *m*, 3476
filtro de vapor *m*, 3631
filtro deshidratador *m*, 1855
filtro desodorizante *m*, 2804
filtro electrostático *m*, 1600, 1601
filtro enrrolable *m*, 3288
filtro metálico *m*, 2673
filtro para gases *m*, 2125
filtro rotativo *m*, 3310
filtro seco *m*, 1466
filtro tubular flexible *m*, 3476
filtro viscoso *m*, 4008
fisura *f*, 1118
fisura capilar *f*, 2208
fisuración *f*, 1119
flexibilidad *f*, 1933
flexible, 1934
flotación *f*, 1952
flotador *m*, 1940
fluctuación *f*, 2361
fluctuación de la temperatura *f*, 3771
fluido *m*, 1986
fluido calefactor *m*, 2249
fluido frigorífico *m*, 1051, 3199
fluidos *m*, 1987
flujo *m*, 1953, 2968

flujo (de calor, magnético, etc.) *m*, 1996
flujo anular *m*, 234
flujo cruzado *m*, 1129
flujo de aire *m*, 121
flujo de aire de proyecto *m*, 1247
flujo de aire estratificado *m*, 3678
flujo de calor *m*, 2239, 3162
flujo de calor en régimen transitorio *m*, 3883
flujo de calor másico *m*, 3564
flujo de dos fases *m*, 3914
flujo de transición *m*, 3885
flujo disperso *m*, 1360
flujo equicorriente *m*, 2917
flujo equilibrado *m*, 350
flujo laminar *m*, 3490, 3680
flujo rotativo *m*, 3314
flujo térmico *m*, 3162
flujo turbulento *m*, 3911, 4020
flujo variable *m*, 3971
flujo viscoso *m*, 4009
fluorcarbono *m*, 1991
fluxómetro *m*, 1963
foco de calor *m*, 2286
foco frío *m*, 824
fondo bombeado *m*, 1357
forjado *m*, 1948
formación de algas *f*, 183
formación de escarcha *f*, 2074
formación de escorias *f*, 3467
formación de hielo *f*, 2380
formación de película *f*, 1844
formas de energía *f,pl*, 2023
formón *m*, 740
forro *m*, 2555, 3473
fortuito, 2400
foso *m*, 2969
foso de cenizas *m*, 274
fracción máxima de humedad *f*, 2714
frecuencia *f*, 2057
frecuencia de la vibración *f*, 4004
frecuencia portante *f*, 636
frecuencia propia *f*, 2759
fregadero *m*, 3457
freno de doble línea *m*, 1400
freno doble *m*, 1396
freón (inadecuada) *m*, 2056
fresco, 2059
frescura *m*, 2062
frigorífico comercial *m*, 874
frigorífico de absorción *m*, 18
frigorífico eléctrico *m*, 1587
frigorígeno *m*, 3928
frío, 817
frío *m*, 816
frío artificial *m*, 3207
fuego *m*, 1873
fuelle *m*, 396, 447

fuelle de dilatación *m*, 1742
fuelle de expansión *m*, 1742
fuente de calor *f*, 2286
fuente de energía *f*, 1635
fuente de energía alternativa *f*, 191
fuente de frío *f*, 824
fuente sonora *f*, 3552
fuente térmica auxiliar *f*, 313
fuentes de energía residual *f*, 4051
fuerza *f*, 2007, 3012
fuerza ascensional *f*, 268
fuerza centrífuga *f*, 690
fuerza electromotriz *f*, 1594
fuerza motriz *f*, 2725
fuerza motriz del control *f*, 1031
fuga *f*, 2525
fuga a tierra *f*, 2197
fuga de agua *f*, 4076
fuga de aire *f*, 136
fugar, 2526
función *f*, 646
función de control *f*, 1023
función de gestión energética *f*, 1632
función de transporte/trasiego *f*, 3880
funcionamiento *m*, 2852
funcionamiento cíclico *m*, 1156
funcionamiento continuo *m*, 1009
funcionamiento de corta duración *m*, 3433
funcionamiento de prueba *m*, 3897
funcionamiento del ordenador *m*, 934
funcionamiento en serie *m*, 3395
funda *f*, 1114
fundación *f*, 375
fundente *m*, 1996
fundición maleable *f*, 2618
fundir, 2662
fusible *m*, 2101, 2102
fusible de cartucho *m*, 639
fusión *f*, 2103

galga *f*, 2158
galvanización *f*, 2110
galvanizado, 2109
galvanizado en caliente, 2334
galvanizar, 2108
gama *m*, 3158
ganancia *f*, 2107
ganancia de calor *f*, 2240
ganancia de calor imprevista *f*, 650
ganancia fortuita del calor *f*, 2401
ganancia (de calor) por conducción *f*, 967

gancho *m*, 2324
garantía *f*, 2202
gas *m*, 2111
gas ciudad *m*, 3873
gas combustible *m*, 2083
gas comprimido *m*, 896
gas de combustión *m*, 1978
gas de escape *m*, 1728
gas de gasógeno *m*, 2529
gas de hornos de coque *m*, 813
gas disuelto *m*, 1364
gas embotellado *m*, 496
gas enrarecido *m*, 3940
gas instantáneo (desprendido por el refrig.al expansionarse) *m*, 1923
gas mixto *m*, 2692
gas natural *m*, 2760
gas pobre *m*, 2529, 3090
gas trazador *m*, 3875
gaseoso, 2122
gasificación *f*, 2135
gasómetro *m*, 2140
gasto(s) *m*, 1101
gasto *m*, 1756, 3759
gastos de explotación *m,pl*, 2844
gel de sílice *f*, 3445
gel de sílice *m*, 61
gelser *m*, 2166
generación de base *f*, 379
generación de electricidad *f*, 1583
generación in situ *f*, 2837
generador *m*, 2162
generador de aire caliente *m*, 4037
generador de aire caliente por gas *m*, 2133
generador de vapor *m*, 467, 3632
germen *m*, 2163
germicida, 2165
gestión energética *f*, 1630
girar, 1122
golpe de ariete (del agua) *m*, 4069
golpe de expansión *m*, 1751
golpeteo del agua *m*, 4069
gota *f*, 1448
gota (fina) *f*, 1449
gotear, 1453
goteo *m*, 1442
gotita *f*, 1449
gradiente *m*, 2176
gradiente de presión *m*, 3054
grado *m*, 1214
grado de deterioro *m*, 1192
grado de evaporación *m*, 1694
grado de presión *m*, 3060
grado de pureza *m*, 1217
grado de recalentamiento *m*, 1218

grado de saturación *m*, 3343
grado de sequedad *m*, 1474
grado-día *m*, 1215
grado-día de enfriamiento *m*, 1056
grados de libertad *m*, 1219
grados-día de calefacción *m*, 2246
gráfico *m*, 2177
gráficos *m*, 2179
gráficos dinámicos *m*, 1532
gravedad *f*, 2186
grieta capilar *f*, 2208
grieta fina *f*, 2208
grifo *m*, 794, 795, 1810, 3757
grifo de paso completo *m*, 3840
grifo de paso directo *m*, 3840
grifo de purga *m*, 1420
grifo de purga de condensador *m*, 432
grifo de vaciado *m*, 1420
grupo *m*, 3925
grupo autónomo *m*, 2901
grupo compresor-condensador *m*, 960
guarnición *f*, 2903

habitación *f*, 3295
habitación caliente *f*, 2336
halogenado, 2210
hardware *m*, 928
haz *m*, 565
haz de tubo *m*, 3901
haz tubular *m*, 3901
haz tubular principal de la caldera *m*, 478
hélice *f*, 3099
helicoidal, 2295
herméticamente sellado, 2300
hermeticidad *f*, 3845
hermético, 2298
hermético a los gases, 2151
hermético al aire, 169
hermético al polvo, 1520
herramienta *f*, 3863
herrumbre *f*, 3325
hervir, 454
hidráulico, 2362
hidrocarburo *f*, 2363
hidroestático, 2364
hidrómetro (presión medida en columna de agua) *m*, 4068
hidrostato *m*, 2368
hielo *m*, 2071, 2372
hielo en escamas *m*, 1903
hielo fragmentado *m*, 2030
hielo raso *m*, 1454
hielo seco *m*, 1467
hierro *m*, 2483
hierro angular *m*, 228
hierro colado *m*, 646
hierro fundido *m*, 646
higiénico, 2365

higrómetro *m*, 2366
higrómetro de aspiración *m*, 281
higrómetro de cabello *m*, 2209
higrómetro de honda *m*, 3482
higrómetro de punto de rocío *m*, 1271
higroscópico, 2367
hilada saliente (arquitectura) *f*, 399
hilo *m*, 4131
hilo calefactor *m*, 2265
hilo térmico *m*, 2265
hiperbólico, 2369
hipotermia *f*, 2370
hogar *m*, 1884, 2099, 2222
hogar con alimentación superior *m*, 2882
hogar de carga inferior *m*, 3920
hogar multicombustibles *m*, 2744
hoja *f*, 3418
holgura *f*, 768
hollín *m*, 3529
honorarios *m*, 1811
hormigón *m*, 943
hornacina *f*, 2768
horno *m*, 2099
horno de condensación *m*, 956
horno de convección *m*, 2021
horno de flujo equicorriente *m*, 2918
horno de gas *m*, 2133
horno de petróleo *m*, 2828
horno mecánico *m*, 3660
hostilación periódica anormal *f*, 2361
hulla *f*, 782
hulla de llama larga *f*, 2582
hulla para coque *f*, 815
hulla seca *f*, 2782
humectación del aire *f*, 130
humectar, 1167
humedad *f*, 1165, 2357, 2712
humedad específica *f*, 2714
humedad relativa *f*, 3233
humedecer, 1167
húmedo, 1166, 2351, 2711
humidificación *f*, 2352
humidificación del aire *f*, 130
humidificador *m*, 2353
humidificador a vapor *m*, 3635
humidificador de aire *m*, 131, 3298
humidificador de disco giratorio *m*, 3568
humidificador por atomización *m*, 300
humidificador por capilaridad *m*, 626

inversión f, 703
investigación f, 3245
inyección de líquido f, 2560
inyector m, 2441
inyector de chorro de vapor m, 3638
ión m, 2480
ionización f, 2482
irradiación f, 2484
isóbara f, 2486
isoentálpico, 2485
isotérmico, 2490
isotermo, 2489
isótopo m, 2491

jefe de equipo m, 714
junta f, 1898, 2136, 2500
junta abocardada f, 1917
junta de caucho f, 3318
junta de compresión f, 904
junta de dilatación f, 1748
junta de dilatación deslizante f, 3479
junta de enchufe y cordón f, 3567
junta de estanqueidad f, 3361
junta de plomo f, 2524
junta de rótula f, 364
junta elástica f, 885
junta estanca al aire f, 157
junta flexible f, 1938
junta hermética f, 3361
junta roscada f, 3357
junta soldada f, 515, 3522
junta tórica f, 2860

laborable, 4135
laboratorio m, 2510
laca f, 2512
ladrillo m, 526
ladrillo (hueco) m, 78
ladrillo (perforado) m, 78
ladrillo holandés m, 773
ladrillo recocido m, 773
ladrillo refractario m, 1878, 3197
lámina de agua f, 4075
laminar, 2513
lámpara f, 2514
lámpara de aviso f, 4047
lámpara indicadora f, 2410
lámpara piloto f, 2949
lana de escorias f, 2686, 3470
lana de vidrio f, 2170
lana mineral f, 2686
latente, 2516
latón m, 512
lavabo m, 4048
lavadero m, 4048
lavador (de aire, de gases) m, 3360
lavador de aire m, 176
lectura indicación f, 3168
lecho m, 394

lecho de escorias m, 774, 3468
leva f, 612
levantamiento (de un edificio) m, 1677
ley de conservación de la energía f, 987
leyes contra la contaminación atmosférica f,pl, 760
leyes físicas de los ventiladores f,pl, 1800
liberado, 2536
licuación f, 2556
licuador m, 2557
licuefacción f, 2556
lignito m, 537, 2548
limar, 3152
limitación f, 3255
limitador m, 2551
limitador de presión m, 3057
limitador de tiro m, 1430
límite m, 497
límite de carga m, 2574
límite de potencia del ventilador m, 1799
límite máximo de carga m, 2641
límites m,pl, 3158
limpiar a presión (mediante un líquido o un gas), 1992
limpiar con un fluido a presión, 1992
limpieza f, 761
limpieza de caldera f, 459
limpieza de chimeneas f, 735
limpieza de filtros f, 1853
línea de alimentación f, 3726
línea de condensados f, 946
línea de descarga f, 1344
línea de saturación f, 3342
líneas de velocidades iguales f,pl, 3979
liofilización f, 2041
liofilización atmosférica f, 297
líquido, 2559
líquido m, 2558
líquido comprimido m, 897
líquido subenfriado m, 3688
lira f, 2585
lira de dilatación f, 1749
listo para funcionar, 3170
listo para su utilización, 3170
litro m, 2566
local m, 3295
longitud f, 2531
longitud de batería f, 808
longitud de fisura f, 1120
longitud total f, 2881
lubricación forzada f, 2019
lubricante m, 2604
lugar de elevada humedad m, 1172
luminosidad f, 2546

luz f, 2542
luz indicadora f, 2410

llama f, 1904
llamarada f, 1916
llave f, 794, 3758, 3951
llave angular f, 1089
llave de cierre de vapor f, 3964
llave de extracción f, 3757
llave de tres vias f, 3831
llave de tuercas f, 3557
llave inglesa f, 3557
llave para la tuerca de la tobera f, 579
llave para tuercas f, 4144
llenado a tope m, 3864

malla f, 2670
mampostería f, 527
mando a distancia m, 3238
mando por pulsador m, 3125
manga filtradora f, 1849
manga para filtrar f, 345, 1849
mango m, 2508
manguera f, 2330
manguera de caucho f, 3319
manguera metálica f, 2674
manguito m, 587, 3473
manguito de acoplamiento m, 1112
manguito roscado m, 2770, 3825
maniobra manual f, 2627
maniobra por pulsadores f, 3125
manipulación del combustible f, 2084
manivela f, 1121
manométrico, 2624
manómetro m, 2158, 2623, 3053
manómetro absoluto m, 894
manómetro de Bourdon m, 500
manómetro de tubo inclinado m, 2404
manómetro de vacío m, 3945
manómetro diferencial m, 1294
mantenimiento m, 301, 2611
mantenimiento correctivo m, 1092
mantenimiento preventivo m, 3076
manual, 2625
manual de servicio m, 3400
manubrio m, 1121
máquina f, 1639, 2605
máquina auxiliar f, 312
máquina frigorífica f, 3206
marcha continua f, 1009
marcha de prueba f, 3897
marcha lenta f, 2385

número de revoluciones *m*, 3276

objetivo de utilidad energética *m*, 1638
obstrucción *f*, 777, 2797
obturador *m*, 438, 2990
obturador de persiana *m*, 2592
ocupante *m*, 2800
oferta *f*, 3772
oído *m*, 1798
oído del ventilador *m*, 1807
ojo de buey *m*, 564
olf *m*, 2832
olor *m*, 2803
opacidad *f*, 2838
opacidad de la mancha de polvo *f*, 1526
opaco, 2840
operación de carga *f*, 715
operación de puesta en marcha *f*, 3607
orden de ejecución *m*, 4137
orden de magnitud *m*, 2856
ordenador *m*, 922
órgano de seguridad *m*, 3331
orientación *f*, 2857
orificio *m*, 2858
orificio calibrado *m*, 2859
orificio de admisión *m*, 2444
orificio de boquilla *m*, 2792
orificio de descarga del aceite *m*, 2817
orificio de entrada *m*, 2444
orificio de escape de gases *m*, 1642
orificio de ventilación *m*, 174
oscilación *f*, 2862
oscilación de la temperatura *f*, 3771
oxidación *f*, 2892
oxígeno *m*, 2894
ozono *m*, 2896

pala *f*, 420
palanca *f*, 2535
paleta *f*, 3954
paleta del ventilador *f*, 1794
panel *m*, 2911
panel calefactor de pared *m*, 4029
panel embebido *m*, 1606
panel empotrado *m*, 1606
panel encastrado *m*, 1606
panel radiante *m*, 2255, 2914, 3143
panel radiante de pared *m*, 4029
pantalla *f*, 1361
pantalla analógica *f*, 214
pantalla antivapor *f*, 3960
pantalla de datos *f*, 1178
pantalla de la boquilla del quemador *f*, 579

pantalla protectora *f*, 3112
papel bituminoso *m*, 414
papel de filtrar *m*, 1860
papel filtro *m*, 1860
par de arranque *m*, 3610
par de torsión *m*, 3865
par electrolítico *m*, 1592
par termoeléctrico *m*, 3808
para invierno y verano, 4145
para toda la estación, 4145
para todo el año, 4145
parámetro *m*, 2919
parasol *m*, 3414
parcial, 3375
pared *f*, 4023
pared exterior *f*, 1769
pared interior *f*, 2476
paro *m*, 3436, 3662
parrilla *f*, 2180, 2190
parrilla de cadena sin fín *f*, 3894
parrilla de registro *f*, 3217
parrilla escalonada *f*, 3653
parrilla hogar *f*, 1875
parrilla oscilante *f*, 3286, 3417
parrilla vibrante *f*, 4001
partícula *f*, 2921
partícula en suspensión *f*, 3736
partículas aerodifusas *f*, 76
partículas arrastradas por los humos de combustión *f,pl*, 2193
partículas de hollín *f,pl*, 3503
partículas de polvo *f,pl*, 1519
pasador *m*, 486
pasador de fijación *m*, 1902
pasamuros *m*, 2322
paso *m*, 2970, 3556
paso de aire *m*, 143
paso de calor *m*, 2239
paso de las aletas *m*, 1871
patinillo de ventilación *m*, 177
patrón *m*, 2924
pegamento *m*, 675
película *f*, 1839
pelo (piezas fundidas) *m*, 2208
penacho de chimenea *m*, 2993
penacho de humos *m*, 2993
penacho de la torre de refrigeración *m*, 1074
percha para tubos *f*, 2957
pérdida *f*, 1756, 2586
pérdida de agua *f*, 4078
pérdida de aire *f*, 137
pérdida de calor *f*, 2271
pérdida(s) de calor calculadas *f,pl*, 1250
pérdida de calor en chimenea *f*, 2258

pérdida(s) de calor previstas (en el proyecto) *f,pl*, 1250
pérdida de carga *f*, 2587
pérdida de energía *f*, 1629
pérdida de frío *f*, 1063
pérdida de presión *f*, 2220, 2587, 3058
pérdida de presión a través de la rejilla *f*, 2191
pérdida de presión dinámica *f*, 1534
pérdida dinámica *f*, 1534
pérdida en la entrada *f*, 1655
pérdida(s) por conducción *f,pl*, 968
pérdida(s) por gases de la combustión *f,pl*, 1979
pérdida por rozamiento *f*, 2067
pérdidas de chimenea *f,pl*, 739
perfil de la distribución del aire *m*, 110
perfil de velocidades *m*, 3982
perforación *f*, 521, 2937
perforado, 2933
periférico, 2941
periódico, 2940
periodo *m*, 1155
periodo de demanda de electricidad *m*, 1581
periodo de duración *m*, 2539
período punta *m*, 2836
permeabilidad *f*, 2942
permiso de funcionamiento *m*, 877
perno *m*, 486
perno de sujeción *m*, 1902
persiana veneciana *f*, 3984
peso *m*, 4100
petroleo para calefacción *m*, 2254
picadura *f*, 2972
pie cúbico por minuto *m*, 697
pie cúbico por segundo *m*, 698
pieza bifurcada *f*, 4146
pieza de conexión *f*, 979
pieza de reducción *f*, 3886
pieza de repuesto *f*, 3558
pieza de transición *f*, 3886
pieza de unión *f*, 979
pieza transversal *f*, 1134
pilar (edit) *m*, 2948
pintura bituminosa *f*, 416
pinza *f*, 3861
pirómetro, 3127
piso *m*, 3672
pistón *m*, 2966
pivote *m*, 3898
placa (metálica) *f*, 2981
placa anterior (de hogar) *f*, 2069
placa calefactora *f*, 2335
placa caliente *f*, 2335

placa de apoyo (hornos) *f*,
  1121
placa de asiento *f*, 2222
placa de base *f*, 381
placa de cierre *f*, 1115
placa de fondo *f*, 395, 461
placa de fundación, 395
placa de guía *f*, 2203
placa de la caldera (para el
  quemador) *f*, 573
placa deflectora *f*, 343
placa directriz *f*, 2203
placa frontal (de hogar) *f*,
  2069
placa perforada *f*, 2936
placa terminal abombada *f*,
  1357
placa tubular *f*, 3903
placa tubular (calderas) *f*,
  461
plan *m*, 2973
plan de puesta en marcha *m*,
  878
plancha *f*, 3419
plancha de caldera *f*, 474
plancha de fondo *f*, 375
plancha de montaje *f*, 394
plano *m*, 2973
plano acotado *m*, 1310
plano axonométrico *m*, 322
plano de alzado *m*, 1603
plano de construcción *m*,
  4136
plano de detalle *m*, 1260
plano de perspectiva *m*, 2945
plano de sección *m*, 3376
plano de taller *m*, 3430
plano del expediente *m*, 3180
plano isométrico *m*, 2488
plano preliminar *m*, 3037
plano real definitivo *m*, 267
planta *f*, 2974, 3672
plástico *m*, 2977
plástico celular *m*, 2978
plástico celular rígido
  (aislamientos) *m*, 3278
plásticos expandidos *m,pl*,
  1740
plazo de ejecución *m*, 890
plazo de entrega *m*, 1230
plazo de realización *m*, 890
plena carga *f*, 2093
plénum *f*, 2989
plomo *m*, 2523
pocillo *m*, 4112
poder absorbente *m*, 16
poder calorífico *m*, 608, 609
poder refractario *m*, 3195
politrópico, 3003
polvo *m*, 1510
polvo atmosférico *m*, 295
polvo de carbón *m*, 785
polvo de comprobación *m*,
  3781
polvo de control *m*, 3781

polvo de fibras *m*, 1829
polvo testigo *m*, 3781
pomo *m*, 2508
poner en circuito, 3745
poner fuera de circuito, 3744
porcentaje *m*, 204
porcentaje de cenizas *m*, 272
porcentaje de vapor seco *m*,
  1474
portátil, 3005
portillo de servicio *m*, 27
posición *f*, 3006
post-enfriamiento *m*, 3179
potencia *f*, 617, 3012
potencia al freno *f*, 506
potencia aparente *f*, 249
potencia calorífica *f*, 2275
potencia calorífica nominal *f*,
  3161
potencia de aspiración *f*,
  3696
potencia de caldera *f*, 469
potencia de calderas *f*, 476
potencia de la caldera *f*, 457
potencia de régimen *f*, 3163
potencia de servicio *f*, 3163
potencia del régimen del
  motor *f*, 2730
potencia del ventilador *f*,
  1802
potencia eléctrica abosorbida
  f, 1578
potencia frigorífica *f*, 1053,
  3205
potencia frigorífica
  (específica) *f*, 1058
potencia frigorífica nominal
  f, 821
potencia nominal *f*, 3160
potencia nominal del motor
  f, 2730
potencia reactiva *f*, 3166
potencia real *f*, 3171
potencia requerida *f*, 3021
potencia sonora del
  ventilador *f*, 1806
potencia térmica *f*, 3800
potencia térmica total *f*, 3870
potencia térmica útil *f*, 3937
potencia útil *f*, 3935
potenciado, 3016
potencial eléctrico *m*, 1577
pozo *m*, 2969, 4112
pozo de ventilación *m*, 177
practicable, 4135
preajuste *m*, 3040
precalentador *m*, 3034
precalentador de aire *m*, 147
precalentamiento *m*, 3035
precipitación *f*, 1239, 3024
precipitación de polvo *f*, 1512
precipitador *m*, 3025
precisión *f*, 32
precisión de ajuste *f*, 3407
predepuración *m*, 3027

preencendido *m*, 3032
preenfriador *m*, 1256
pre-enfriador *m*, 2022
pre-enfriamiento *m*, 3029
prefiltro *m*, 3028
pre-filtro *m*, 3031
prendimiento en la
  producción de gas *m*, 2155
prensa de tornillo *f*, 4005
prensa hidráulica pequeña *f*,
  4005
prensaestopas *m*, 2167, 3685
prensaestopas de dilatación
  m, 1747
preparación *f*, 3038
presión *f*, 2217, 3041
presión acústica *f*, 3545
presión ambiental *f*, 197
presión atmosférica *f*, 298
presión barométrica *f*, 372
presión de aspiración *f*, 3696,
  3702
presión de baja *f*, 2595
presión de circulación *f*, 748
presión de condensación *f*,
  957
presión de descarga *f*, 1343,
  1346
presión de ensayo *f*, 3784
presión de equilibrio *f*, 354,
  1669
presión de evaporación *f*,
  1693
presión de funcionamiento *f*,
  2849, 4138
presión de impulsión *f*, 1343,
  1346
presión de prueba *f*, 3784
presión de régimen *f*, 2781
presión de saturación *f*, 3345
presión de servicio *f*, 2849,
  4138
presión de trabajo de
  proyecto *f*, 1254
presión de vapor *f*, 3642,
  3966
presión de vaporización *f*,
  1693
presión del aire *f*, 148
presión del viento *f*, 4127
presión diferencial *f*, 1292
presión dinámica *f*, 1535,
  3980
presión estática *f*, 3615, 3616
presión final *f*, 1866
presión intermedia *f*, 2659
presión límite de seguridad *f*,
  3333
presión media *f*, 2644, 2659
presión nominal *f*, 2781
presión parcial *f*, 2920
presión reducida *f*, 3185
presión sonora *f*, 3545
presión total *f*, 3871
presostato *m*, 3045, 3068

quemador de cazoleta
(cubeta) *m*, 1143
quemador de crisol *m*, 3010
quemador de chorro a
presión *m*, 3056
quemador de flujo
equicorriente *m*, 2918
quemador de gas *m*, 2114
quemador de gas universal
m, 2115
quemador de horno *m*, 3010
quemador de inyección *m*,
2442
quemador de petroleo *m*,
2812
quemador de pulverización
mecánica *m*, 2655
quemador de tiro forzado *m*,
2013
quemador de varias toberas
m, 2745
quemador inversor *m*, 1040
quemador nevulizante *m*,
3958
quemador policombustible
m, 1489
quemador rotativo *m*, 3308
quemador universal *m*, 3931
quemador vaporizante *m*,
3958
quemadura *f*, 581
quemar, 567

R-12, 2056
rácor *m*, 1898, 2770
rácor hembra *m*, 1823
rácor rápido *m*, 3133
radiación *f*, 3146
radiación de fondo *f*, 328
radiación directa *f*, 389
radiación en conductos *f*,
1499
radiación solar *f*, 3518
radiación térmica *f*, 3801
radiador *m*, 3141, 3147
radiador de columnas *m*, 838
radiador de fundición *m*, 649
radiador de gas *m*, 2147,
3139
radiador de panel de acero
m, 3649
radiador de tubos aleteados
m, 1870
radiador de zócalo *m*, 378
radiante, 3135
radio de difusión diseminada
m, 1299
ralantí *m*, 2385
ramal *m*, 507, 510
ramificación *f*, 509
rampa *f*, 742
ranura *f*, 3484
ranura de salida *f*, 3486
rastrillado *m*, 3153
rastrillar, 3151

rastrillo *m*, 3150
rayo de luz *m*, 388
razón *f*, 3164
reacción *f*, 3165
reactor *m*, 3167
real, 43
realimentación *f*, 1814
realización del vacío *f*, 1688
rearmar, 3247
rebosadero *m*, 2883
rebosar, 2884
recalentado, 3710
recalentador *m*, 64, 3228,
3711
recalentador de intermedio
m, 2472
recalentador de vapor *m*,
3630
recalentador del agua de
alimentación de la caldera
m, 465
recalentamiento *m*, 3225,
3708, 3712
recalentar, 3226, 3709
recambio *m*, 3558
recepción *f*, 23
recinto cerrado *m*, 1617
recipiente *m*, 1004, 3173,
4000
recipiente a presión *m*, 908,
3072
recipiente acumulador de
calor *m*, 3806
recipiente colector del
condensado *m*, 945
recipiente de almacenamiento
térmico *m*, 3806
recipiente de enfriamiento de
agua por aspersión *m*,
3589
recipiente de expansión *m*,
1752
recipiente de líquido *m*, 3173,
3174
recirculación *f*, 3178
recirculación de los gases de
escape *f*, 1731
recocer, 233
recorrido *m*, 3841
rectificador *m*, 3184
recubrimiento *m*, 791, 1114
recubrimiento con chapa
metálica *m*, 754
recubrimiento protector *m*,
3109
recuperable, 3182
recuperación *f*, 3183, 3214
recuperación de calor *f*, 2281
recuperación energética *f*,
1634
recuperador por baterías *m*,
3321
red *f*, 2766, 3750
red de conductos *f*, 1505
red de distribución *f*, 1373

red de tuberías *f*, 2959
red local *f*, 2577
reducción *f*, 494, 3255
reducción de carga *f*, 2575
reducción de presión *f*, 3064
reducción de sonido *f*, 2777
reducción de un conducto *f*,
1504
reducción del sonido *f*, 3549
reducción nocturna *f*, 2769
reducción nocturno del
termostato *f*, 3820
reductor de capacidad *m*, 621
reflectancia *f*, 3191, 3193
reflector *m*, 3194
reflexión difusa *f*, 1298
reforzar, 3230
refractaria, 2284
refractario, 3196
refractario *m*, 2269
refrigeración (enfriamiento) *f*,
730
refrigeración del aire *f*, 99
refrigeración por absorción *f*,
18
refrigeración por inversión
del ciclo *f*, 3273
refrigerador *m*, 97, 3212
refrigerador por hielo *m*,
2375
refrigerante *m*, 1046, 1051
refrigerante-12 *m*, 2056
refrigerante *m*, 3199
refrigerante primario *m*, 3079
refrigerante secundario *m*,
3371
refrigerar, 726, 1045, 3202
refuerzo *m*, 3620
regeneración *f*, 3183, 3215
regeneración de fondos *f*,
3285
régimen de funcionamiento
m, 2843
régimen estable *m*, 3621
régimen permanente *m*, 3621
régimen variable *m*, 2789
región *f*, 262
registrador *m*, 3181
registro *m*, 1168, 2211, 3217,
3438
registro de antirretorno *m*,
2787
registro de derivación *m*, 595
registro de evacuación de
gases *m*, 1730
registro de gases de escape *m*,
1730
registro de gases evacuados
m, 1730
registro de guillotina *m*, 3477
registro de lamas opuestas *m*,
2855
registro de mariposa *m*, 589
registro de quemador *m*,
3219

registro de retención *m*, 2787
registro equilibrador *m*, 1668
reglamentación contra la
contaminación atmosférica
f, 760
regulable, 53
regulación *f*, 54, 1014, 2834,
3224
regulación de alimentación
de aceite *f*, 2818
regulación de alimentación
de fueloil *f*, 2818
regulación de alimentación
de petroleo *f*, 2818
regulación de la alimentación
f, 1817
regulación de la humedad *f*,
2358
regulación de la mezcla *f*,
2696
regulación en cascada *f*, 641
regulación integral *f*, 2461
regulación manual *f*, 2626
regulación paso a paso *f*,
3655
regulación por etapas *f*, 3655
regulación proporcional *f*,
3104
regulador *m*, 1027
regulador de capacidad *m*,
619, 622
regulador de carga *m*, 713
regulador de caudal *m*, 1967
regulador de combustión *m*,
576, 849, 857
regulador de etapas *m*, 3652
regulador de nivel *m*, 2533
regulador de presión *m*, 3043,
3065
regulador de presión de
evaporación *m*, 1705
regulador de refrigeración *m*,
3208
regulador de temperatura *m*,
3767
regulador de temperatura de
la caldera *m*, 482
regulador de tiro *m*, 370
regulador en cascada *m*, 642
regulador esclavo *m*, 3471
regulador master *m*, 2636
regulador submaster *m*, 3690
regulador único *m*, 3596
regular, 52, 1015, 3221
rejilla *f*, 2190
rejilla (de paso de aire) *f*,
2590
rejilla apersianada *f*, 2591
rejilla con deflectores fijos *f*,
1899
rejilla de aire *f*, 2190
rejilla de antepecho *f*, 3446
rejilla de pared *f*, 4031
rejilla de persiana *f*, 2591

rejilla de retorno del aire *f*,
3259
rejilla de salida *f*, 2869
rejilla de salida del aire
tratado *f*, 3724
rejilla de techo *f*, 663
rejilla diferencial *f*, 1291
rejilla fija *f*, 3176
rejilla perforada *f*, 2935
relación *f*, 3164
relación de compresión *f*, 905
relación de concentración *f*,
941
relación de inducción *f*, 2420
relación de mezcla *f*, 2699
relación de sequedad *f*, 1474
relación de volumen de
compresión *f*, 911
relativo, 3232
relé *m*, 3234
relé posicionador *m*, 3007
relevador *m*, 3234
reloj de control *m*, 3847
reloj programador *m*, 3847
relleno *m*, 1836, 2903
relleno de la torre de
refrigeración *m*, 1071
relleno de placas de torre de
refrigeración *m*, 2982
relleno de torre de
refrigeración de
salpicadura *m*, 3572
remachar, 3284
remache *m*, 3283
remanente *m*, 3237
rendija de una ventana *f*,
4125
rendimiento *m*, 1556, 2938
rendimiento de caldera *m*,
460
rendimiento de compresión
m, 902
rendimiento de enfriamiento
m, 1059
rendimiento de la aleta *m*,
1867
rendimiento de la
combustión *m*, 853
rendimiento de un filtro *m*,
3389
rendimiento diario *m*, 1161
rendimiento efectivo *m*, 2880
rendimiento gravimétrico *m*,
2185
rendimiento total *m*, 2880
rendimiento volumétrico *m*,
4017
renovación de aire *f*, 80
renovación de aire fresco *f*,
2061
renovación del aire *f*, 155
renovaciones de aire por
hora *f*, 82
representar gráficamente,
2178

reserva *f*, 3600
residuo *m*, 3237
residuos de combustión *m,pl*,
858
resistencia *f*, 1414, 3251
resistencia al flujo de aire *f*,
124
resistencia al frío *f*, 822
resistencia al fuego *f*, 1889,
3195
resistencia al rozamiento *f*,
2065
resistencia de flujo *f*, 1969
resistencia eléctrica *f*, 1570
resistencia local *f*, 2578
resistencia peculiar *f*, 1845
resistencia térmica *f*, 3802
resistente a la intemperie,
4097
resistente al fuego, 1886
resistividad térmica *f*, 3803
resonancia *f*, 3254
resorte *m*, 3590
respiración *f*, 522
respiradero *m*, 3985
respuesta a fallo *f*, 1786
respuesta de frecuencia *f*,
2058
resto *m*, 3237
restricción *f*, 3255
restricción de la carga *f*, 2575
resultado de la prueba *m*,
3785
resultante, 3256
retardador de humedad *m*,
2717
retardador de incendios *m*,
1890
retardador de vapor *m*, 3967
retardador del fuego *m*, 1890
retardo *m*, 3852
retén *m*, 3361
retroceso *m*, 3257
reversible, 3275
revestido, 2493
revestimiento *m*, 1114, 2555
revestimiento (calderas) *m*,
645
revestimiento (tubular) de la
chimenea *m*, 1983
revestimiento antifricción *m*,
241
revestimiento de caldera *m*,
458
revestimiento de la cámara
de combustión *m*, 3198
revestimiento interior de la
cámara de combustión *m*,
848
revestimiento metálico *m*,
754
revestimiento protector *m*,
791
revestimiento refractario *m*,
3198

riostra *f*, 221
rocío *m*, 1268
rodamiento *m*, 361, 391
rodete *m*, 2396
rodete del ventilador *m*, 1804
rollo de alambre *m*, 565
rosca *f*, 3822
rosca cónica *f*, 3760
rosca hembra *f*, 1824
rosca interior *f*, 1824
rosca macho *f*, 2617
rosca para tubos *f*, 2964
roscado (interior) *m*, 3762
roscar, 3354
roscar (interiormente), 3758
rotación *f*, 3313
rotativo, 3306
rotor *m*, 3315
rótula *f*, 364, 367
rotura *f*, 518
roya *f*, 3325
rozamiento *m*, 2064
rozamiento dentro del tubo *m*, 2956
rozamiento en el tubo *m*, 3902
rueda *f*, 4121
rugosidad *f*, 3316
rugosidad absoluta *f*, 8
ruido *m*, 2772
ruido de aire en un conducto *m*, 1493
ruido de desconexión *m*, 520
ruido del motor *m*, 2729
ruido en conductos *m*, 1498
ruido llevado por el aire *m*, 77
ruptor del vacio *m*, 3941

saco *m*, 344
sala blanca *f*, 764
sala de aparatos *f*, 1675
sala de calderas *f*, 470, 477
sala de ensayos *f*, 3787
sala de herrametal *f*, 2976
sala de maquinaria *f*, 2976
sala de máquinas *f*, 1675
sala de material *f*, 2976
salida *f*, 2867
salida alabeada *f*, 3956
salida analógica *f*, 217
salida de aire *f*, 142
salida de gases *f*, 1984, 3994
salida de humos *f*, 3495
salida del agua *f*, 4079
salida del ordenador *f*, 931
salmuera *f*, 528
sangrador *m*, 431
sangrar, 430
sangría de la escoria (hornos) *f*, 763, 775
saturación *f*, 3340
secado *m*, 1468
secado al aire *m*, 111, 114
secado por aire *m*, 111

secado secundario *m*, 3372
secador *m*, 1436
secador congelador *m*, 2040
secador de aire *m*, 113
secador de toallas *m*, 3872
secador de vapor *m*, 3629
secar, 1458
sección *f*, 3374
sección de agua de la caldera *f*, 480
sección de tubos *f*, 2960
sección del conducto *f*, 1500
sección frontal *f*, 2070
sección transversal *f*, 1135
seccional, 3375
seco, 1457
secuencia *f*, 3391
sedimentación *f*, 1239
sedimento *m*, 1237, 3377, 3487
sedimentos *m*, 1438
seguridad de fallo *f*, 1785
seguridad de funcionamiento *f*, 3334
seguro de calidad *m*, 3129
selección del combustible *f*, 2079
sellar, 3362
sello *m*, 3361
sello laberíntico *m*, 2511
sello mecánico *m*, 2658
semiautomático, 3382
semicojinete *m*, 586
semiconductor *m*, 3383
semihermético, 3384
sensible, 3385
sensor *m*, 3386
sentido *m*, 1329
sentido de flujo *m*, 1960
señal *f*, 3442
señal luminosa de aviso *f*, 4047
separación *f*, 2970, 3388, 3556
separación de cenizas *f*, 276
separación de las aletas *f*, 1872
separación del aire *f*, 158
separación entre aletas *f*, 1871
separador *m*, 3025, 3390, 3892
separador centrífugo *m*, 692
separador de aceite *m*, 2824, 2826
separador de agua *m*, 4083
separador de cenizas *m*, 278
separador de frigorífico *m*, 3213
separador de gotas *m*, 1451, 2718
separador de líquido *m*, 2564
separador de niebla *m*, 2689
separador de polvo *m*, 1525
sequedad *f*, 1473

serpentín *m*, 805, 2953
serpentín condensador *m*, 953
serpentín de aire forzado *m*, 423
serpentín de calefacción *m*, 2243
serpentín de enfriamiento *m*, 1054
serpentín evaporador *m*, 1691
serpentín mural *m*, 4025
servicialidad *f*, 3397
servicio *m*, 301, 2852, 3396
servicio permanente *m*, 1009
servocontrol *m*, 3402
servomotor *m*, 3403
sesgo *m*, 405
sicrómetro *m*, 3114
sicrómetro de aspiración *m*, 283
silenciador *m*, 2740, 3444, 3539
silencio *m*, 2775
silo *m*, 2325
silo frigorífico *m*, 2313
símbolo *m*, 3747
simulación por ordenador *f*, 935
simultáneo, 3448
sin empaquetadura *f*, 2907
sin soldadura, 3368
sin tubos de gases *m*, 1982
sincronización *f*, 3748
sincronizador *m*, 2470
síndrome de edificio enfermo *m*, 3439
sinterización *f*, 3458
sistema *m*, 3750
sistema a cuatro tubos *m*, 2029
sistema (frigorífico) autónomo *m*, 2901
sistema comercial *m*, 875
sistema computerizado *m*, 925
sistema de acondicionamiento de aire a baja presión *m*, 2596
sistema de agitación del aire *m*, 72
sistema de agua caliente *m*, 2345
sistema de alarma contra incendio *m*, 1874
sistema de automatización de un edificio *m*, 547
sistema de caldeado *m*, 4133
sistema de calefacción *m*, 2261
sistema de calefacción de retorno seco *m*, 1478
sistema de calefacción descendente *m*, 1410
sistema de calefacción urbana *m*, 1382

sistema de calentamiento
directo *m*, 1327
sistema (de
acondicionamiento) de
caudal variable *m*, 3970
sistema de ciclo cerrado *m*,
780
sistema de circuito cerrado
m, 780
sistema de climatización *m*,
91
sistema de climatización a
tres tubos *m*, 3828
sistema de climatización con
unidades autónomas *m*,
2900, 3379
sistema de climatización de
doble conducto *m*, 1484
sistema de climatización
modular *m*, 2707
sistema de climatización para
confort *m*, 868
sistema de climatización
todo-aire *m*, 187
sistema de comunicación *m*,
881
sistema de comunicación de
igual a igual *m*, 2930
sistema de conductos *m*, 1503
sistema de control de la
gestión energética *m*, 1631
sistema de control de un
edificio *m*, 548
sistema de control y gestión
energética de un edificio *m*,
550
sistema de descarga directa
m, 1320
sistema de desescarche *m*,
1211
sistema de difusión *m*, 1303
sistema de distribución *m*,
1375
sistema de distribución del
aire (tratado) *m*, 3725
sistema (de
acondicionamiento) de
doble conducto *m*, 1485
sistema de drenaje *m*, 1418
sistema de energía total *m*,
3868
sistema de enfriamiento *m*,
1069
sistema de enfriamiento
directo *m*, 1315
sistema de enfriamiento por
radiación *m*, 3138
sistema de evacuación *m*,
1735
sistema de expansión directa
m, 1323
sistema de extracción *m*, 1735
sistema de extracción de aire
m, 1724

sistema de fusión de nieve *m*,
3507
sistema de gestión
centralizada *m*, 882
sistema de gestión de un
edificio *m*, 554
sistema de gestión energética
de un edificio *m*, 551
sistema de licuación de la
nieve *m*, 3507
sistema de realimentación *m*,
1815
sistema de recuperación de
calor *m*, 2282
sistema de refrigeración de
ciclo de eyección *m*, 1560
sistema de refrigeración
directa *m*, 1333
sistema de refrigeración por
compresión *m*, 910
sistema de refrigeración por
expansión directa *m*, 1322
sistema de refrigeración por
pulverización de salmuera
m, 531
sistema de refrigerante
expandible *m*, 1755
sistema de salmuera *m*, 532
sistema de suministro de aire
m, 3725
sistema de tres tubos *m*, 3829
sistema de tubería única *m*,
3454
sistema de tuberías *m*, 2963
sistema de tubos de diámetro
muy fino *m*, 2679
sistema de tubos de pequeño
diámetro *m*, 3491
sistema de un solo conducto
m, 3451
sistema de ventilado directo
m, 1334
sistema eléctrico auxiliar *m*,
311
sistema estanco *m*, 3364
sistema experimentado *m*,
1759
sistema frigorífico *m*, 1069
sistema frigorífico de
eyección (de vapor) *m*,
3639
sistema hermético *m*, 3364
sistema monotubular *m*, 3454
sistema montado en fábrica
(autónomo) *m*, 1782
sistema partido *m*, 3575
sistema Tichelmann (de
retorno invertido) *m*, 3274
situación *f*, 3006
situación de carga nula *f*,
2778
situación energética *f*, 1636
sobre el suelo, 1
sobrealimentación *m*, 3728
sobrealimentar, 490

sobrecalentamiento *m*, 1711
sobrecarga *f*, 2889
sobrecarga del ordenador *f*,
932
sobrefusión *f*, 3707
sobrepresión *f*, 1679, 1710
sobrerrecorrido de maniobra
manual *m*, 2628
sobresaturación *f*, 3714, 3715
sobresaturar, 3713
sobretensión del compresor
m, 919
software *m*, 936
soldado, 514
soldador *m*, 3520, 4105
soldadura *f*, 3367, 4102, 4106
soldadura (sin metal de
aporte) *f*, 3523
soldadura a tope *f*, 592
soldadura blanda *f*, 3511
soldadura con metal de
aportación *f*, 517
soldadura eléctrica por
resistencias *f*, 3253
soldadura fuerte *f*, 517, 2216
soldadura oxiacetilénica *f*,
2893
soldadura por arco *f*, 261
soldadura por llama de gas *f*,
2154
soldadura por puntos *f*, 3579
soldar, 3521, 4103
soldar (con latón), 513
solera de horno *f*, 2222
solicitud de ofertas *f*, 605
solubilidad *f*, 3526
solución *f*, 3527
solución amoniacal *f*, 201
solución incongelable *f*, 240
solvente *m*, 3528
sombrear, 3412
sombrerete *m*, 2323
sombrerete de chimenea *m*,
1116, 1427
sonda *f*, 3086, 3386, 3387
sonda de calidad del aire *f*,
153
sonda de combustión *f*, 850
sonda de humedad *f*, 2360
sonda de temperatura *f*, 3770
sonda exterior *f*, 1767, 2877
sonda lumínica *f*, 2547
sonda remota *f*, 3240
sonda solar *f*, 3519
sonido *m*, 3536
sonido transmitido por el
aire *m*, 77
sonoridad *f*, 2588
soplado *m*, 449
soplador *m*, 1790
soplador de hollín *m*, 3531
soplar, 430, 444
soportatubos *m*, 2962
soporte *m*, 504, 3620, 3729

soporte de abrazadera *m*, 3658
soporte de cojinete *m*, 392, 2928
soporte de cojinete liso *m*, 3475
soporte de rodamiento *m*, 362
soporte del cojinete de bolas *m*, 362
soporte deslizante *m*, 3480
soporte giratorio *m*, 3898
soporte para tubos *m*, 2962
sorbente *m*, 3535
sótano *m*, 380, 668
subcontratista *m*, 3687
subenfriador *m*, 63
subenfriamiento *m*, 3689
subestación *f*, 3693
subir, 3279
succión *f*, 3695
suciedad *f*, 1335, 2026, 3348
suciedad adherente *f*, 3502
suelo *m*, 2196
suelo falso *m*, 1789
sujeción *f*, 1808
sumergir, 583, 2392
sumidero *m*, 3457, 3706
suministrar, 3720
suministro *m*, 1228
suministro de aceite *m*, 2829
suministro de agua fría *m*, 828
suministro de aire *m*, 164
suministro de calefacción urbana *m*, 1381
suministro de carbón *m*, 784
suministro de combustible *m*, 2081, 2088
suministro de gas *m*, 2150
suministro de petróleo *m*, 2829
suministro directo de agua caliente *m*, 1328
suministro instantáneo de agua caliente *m*, 1328
superficial, 1
superficie *f*, 262, 1949, 3731
superficie adicional *f*, 1763
superficie calefactora *f*, 2260
superficie caliente (de una caldera) *f*, 4119
superficie construida *f*, 259, 546
superficie de absorción *f*, 13
superficie de apertura *f*, 246
superficie de calefacción de la caldera *f*, 468
superficie de calefacción directa *f*, 3080
superficie de calefacción indirecta *f*, 3373
superficie de difusión *f*, 1301
superficie de enfriamiento *f*, 1068

superficie de flujo *f*, 1954
superficie de parrilla *f*, 2181
superficie del núcleo *f*, 1085
superficie filtrante *f*, 1848
superficie frontal *f*, 1776
superficie frontal de batería *f*, 807
superficie mojada *f*, 4119
supervisión *f*, 3717
supresor de parásitos (elec) *m*, 637
supresor de perturbaciones *m*, 3730
suspensión *f*, 3737
sustancia absorbida *f*, 10

T de regulación *f*, 3222
tabique deflector *m*, 343
tabla *f*, 384
tabla de datos *f*, 1181
tablas de vapor *f,pl*, 3644
tablero de base (montajes) *m*, 376
taller *m*, 4141
taller (fábrica) de calderas *m*, 473
tamaño *m*, 562, 3462
tamaño de la gota *m*, 1452
tamaño de las partículas *m*, 2922
tamaño del conducto *m*, 1501
tambor *m*, 1455
tamiz *m*, 3352, 3440
tanque *m*, 3755
tanque colector de condensados *m*, 945
tanque de alimentación por gravedad *m*, 2219
tanque de almacenamiento *m*, 3669, 3671
tanque de almacenamiento de aceite *m*, 2827
tanque de almacenamiento de petróleo *m*, 2827
tanque de expansión *m*, 1752
tanque de salmuera *m*, 533
tanque de servicio (diario) *m*, 1163
tanque del agua de alimentación *m*, 1821
tapa *f*, 1113, 2537
tapa abisagrada *f*, 2320
tapajuntas *m*, 2502
tapón de desaire *m*, 523
tapón de vapor (en tuberias) *m*, 3641
tapón hembra *m*, 614
tapón macho *m*, 2990
tapón roscado *m*, 781, 3355, 3358
taponar, 2991
tarado *m*, 604
tarifa *f*, 3764
tasa de calefacción *f*, 2264

tasa de circulación del aire *f*, 84
tasa de contaminación *f*, 146
tasa de deficiencia *f*, 1203
tasa de renovación del aire *f*, 81
tasa diaria *f*, 1162
teclado *m*, 2506
técnica de medición *f*, 2653
técnica del frío *f*, 3211
técnica frigorífica *f*, 3211
técnico, 3765
techo *m*, 658, 3291
techo calefactado *m*, 2234
techo calefactor *m*, 2234
techo perforado *m*, 2934
techo suspendido *m*, 3735
techo ventilado *m*, 3987
tejado *m*, 3291
tejuelo *m*, 2006
tela filtrante *f*, 1854
telemando *m*, 3238
teletermómetro *m*, 1366, 3239
temperatura *f*, 3766
temperatura ambiental *f*, 198
temperatura ambiente *f*, 1662
temperatura aparente *f*, 250
temperatura de alimentación *f*, 1229
temperatura de autoignición *f*, 307
temperatura de bulbo húmedo *f*, 4116
temperatura de combustión *f*, 860
temperatura de condensación *f*, 959
temperatura de descarga *f*, 1351
temperatura de equilibrio *f*, 353, 1673
temperatura de evaporación *f*, 1695
temperatura de evaporización *f*, 1695
temperatura de humos *f*, 1985
temperatura de ignición *f*, 2390
temperatura de la pared *f*, 4032
temperatura de operación *f*, 2853
temperatura de proyecto *f*, 1253
temperatura de régimen *f*, 2850
temperatura de retorno *f*, 3267
temperatura de rocío del aparato *f*, 248
temperatura de salida (flujo) *f*, 1970
temperatura de servicio *f*, 2850

temperatura de termómetro
de globo *f*, 2171
temperatura de trabajo *f*,
4139
temperatura del bulbo seco *f*,
1461
temperatura del color *f*, 834
temperatura del local *f*, 3305
temperatura del medio *f*,
2660
temperatura del termómetro
f, 4116
temperatura del termómetro
seco *f*, 1461
temperatura diferencial *f*,
1295
temperatura efectiva
corregida *f*, 1090
temperatura equivalente del
cuerpo negro *f*, 418
temperatura exterior *f*, 1768,
2866
temperatura inicial *f*, 2437
temperatura inicial (flujo) *f*,
1970
temperatura radiante *f*, 3145
temperatura radiante media
f, 2645
temperatura seca *f*, 1461
temperatura seca resultante *f*,
1476
temperatura sol-aire *f*, 3512
temporada de calefacción *f*,
2257
tenaza *f*, 3861
tenazas para escorias *f,pl*,
3469
tendido subterráneo *m*, 2522
tensión *f*, 3681, 4013
tensión superficial *f*, 3733
térmico, 3790
terminal *m*, 1588
terminal de aire orientable *m*,
1330
terminal de conexión *m*, 982
término medio *m*, 314
termistancia *f*, 3807
termistor *m*, 3807
termo *m*, 3815
termodinámica *f*, 3809
termoeconomía *f*, 2247
termoeléctrico, 3810
termofluido (aceite) *m*, 2291
termógrafo *m*, 3811
termohigrógrafo *m*, 3812
termómetro *m*, 3813
termómetro a distancia *m*,
1366, 3239
termómetro bimetálico *m*,
410
termómetro de bulbo seco *m*,
1462
termómetro de contacto *m*,
1003

termómetro de contactos
eléctricos *m*, 1571
termómetro de cuadrante *m*,
1275
termómetro de globo *m*, 2172
termómetro de lectura a
distancia *m*, 1366, 3239
termómetro de mercurio *m*,
2669
termómetro registrador *m*,
3811
termopar *m*, 3808
termorresistente, 2284
termostatico, 3817
termostato *m*, 3816
termostato de inmersión *m*,
257, 2394
termostato doble *m*, 1491
termotecnia *f*, 2285
terraja *f*, 3823
terrajado *m*, 3762
tiempo *m*, 3846
tiempo atmosférico *m*, 4094
tiempo de congelación *m*,
2055
tiempo de desescarche *m*,
1212, 3850
tiempo de funcionamiento *m*,
2851
tiempo de funcionamiento de
un accionador *m*, 45
tiempo de paro *m*, 2384
tiempo de precalentamiento
m, 3036
tiempo de puesta en marcha
m, 3324, 3609
tiempo de puesta en régimen
de la calefacción *m*, 2263
tiempo de reparto *m*, 3853
tiempo de respuesta al fallo
de llama *m*, 569
tiempo de retardo *m*, 3852
tiempo de reverberación *m*,
3270
tiempo de servicio *m*, 2851
tiempo de utilización *m*, 3939
tiempo medio de reparación
m, 2648
tiempo medio entre fallos *m*,
2647
tiempo muerto *m*, 1187, 3852
tiempo real *m*, 3172
tierra *f*, 2196
timbre de alarma *m*, 179
tinte *m*, 1528
tipo de combustible *m*, 2090
tirante *m*, 3844
tiro *m*, 1425
tiro de chimenea *m*, 736
tiro de la chimenea *m*, 1977
tiro de sentido descendente
m, 1409
tiro equilibrado *m*, 349
tiro forzado por impulsión
m, 2012

título del vapor *m*, 1474
tiznadura de hollín o de
carbón *f*, 3502
toallero *m*, 3872
tobera *f*, 1964, 2791
tobera de inyección *f*, 2439
tobera pulverizadora *f*, 3586
tobogán *m*, 742
tolerancia *f*, 766
tolerancia de un cilindro *f*,
767
tolva *f*, 566, 2098, 2325
toma *f*, 2808
toma de agua *f*, 4079
toma de aire *f*, 134
toma de fuerza *f*, 3020
toma de tierra *f*, 1539, 1569
toma en la base de la
chimenea *f*, 738
tope *m*, 3662
tope cónico *m*, 1920
tornillo *m*, 3353
tornillo de banco *m*, 4005
tornillo sin fín *m*, 4142
torre a contracorriente *f*,
1133
torre de enfriamiento *f*, 1070
torre de refrigeración *f*, 1070
torre de refrigeración de agua
f, 4062
torre de refrigeración de agua
de tiro forzado *f*, 2017
torre de refrigeración de agua
de tiro mecánico *f*, 2656
torre de refrigeración de agua
de tiro natural *f*, 2758
torre de refrigeración de
rociadores de agua *f*, 3585
torre de refrigeración
pelicular *f*, 1843
trabajo a corto plazo *m*, 3433
trabajo de precisión *m*, 3026
tramo *m*, 3374
tramo de tubería *m*, 2960
trampa de calor *f*, 2293
trampilla *f*, 3893
tranco *m*, 1879
transcripción de datos *f*, 1179
transductor *m*, 3877
transferir, 3879
transformador *m*, 1041, 3882
transistor *m*, 3884
transmisión *f*, 1444, 1446,
3888
transmisión analógica *f*, 218
transmisión de calor *f*, 2289,
2292
transmisión de la humedad *f*,
2720
transmisión de masa *f*, 2635
transmisión digital *f*, 1305
transmisión lateral *f*, 1914
transmisión por correa
trapezoidal *f*, 3977
transmisión térmica *f*, 2289

transmisor *m*, 3890
transmitir, 3879
transportador *m*, 1043
transportar, 3878
transporte de combustible *m*, 2089
transporte de humedad *m*, 2719
transporte de masa *m*, 2635
trasegar, 3878
tratamiento *m*, 3896
tratamiento del agua *m*, 4087
tratamiento del aire *m*, 171
tratamiento previo del combustible *m*, 2086
travesaño *m*, 1134
trazador *m*, 3874
trazador (para buscar tubos) *m*, 2965
triturar, 2192
troquel *m*, 1281
tubería *f*, 972, 2950
tubería ascendente *f*, 3280
tuberia de aspiración por vacio *f*, 3946
tubería de caucho *f*, 3319
tubería de conducción (para líquidos y gases) *f*, 2958
tubería de descarga *f*, 2872
tubería de distribución *f*, 1374
tubería de expansión *f*, 1750
tubería de gas *f*, 2142
tubería de plástico *f*, 2980
tubería de retorno *f*, 3265
tubería de retorno bajo el nivel del agua (calderas) *f*, 4117
tubería de vapor *f*, 3640
tubería del retorno sobre el nivel del agua (caliente) *f*, 1477
tubería principal *f*, 2608, 2610
tubo *m*, 2950, 3899
tubo acodado *m*, 404
tubo ascendente *m*, 3282
tubo bifurcado *m*, 4146
tubo capilar *m*, 627
tubo compensador *m*, 352, 1750
tubo curvado *m*, 404
tubo de acero *m*, 3650
tubo de agua de alimentación *m*, 3603
tubo de aletas *m*, 1869
tubo de aletas helicoidales *m*, 3570
tubo de aspiración *m*, 3701
tubo de Bourdon *m*, 501
tubo de calor *m*, 2276
tubo de cobre *m*, 1082
tubo de comunicación *m*, 980
tubo de conexión *m*, 980

tubo de derivación *m*, 596, 3752
tubo de desagüe *m*, 4054
tubo de escape *m*, 4054
tubo de escape de gases *m*, 1729
tubo de extracción *m*, 434
tubo de humos *m*, 1891, 1973
tubo de inyección *m*, 2440, 3587
tubo de Pitot *m*, 2971
tubo de protección *m*, 3113
tubo de purga *m*, 434
tubo de salida *m*, 2872
tubo de vaciado *m*, 1423
tubo de ventilación *m*, 3996
tubo de Venturi *m*, 3997
tubo en Y *m*, 4146
tubo equilibrador *m*, 352
tubo flexible *m*, 1939
tubo fluorescente *m*, 2545
tubo metálico flexible *m*, 2674
tubo roscado *m*, 3826
tubo soldado *m*, 516
tubo vertical *m*, 3603
tubo vórtex *m*, 4021
tubos concéntricos *m,pl*, 942
tuerca *f*, 2796
tuerca cónica *f*, 1919
tunel aerodinámico *m*, 4128
tunel congelador *m*, 3905
tunel de refrigeración *m*, 1075
tunel enfriador *m*, 3904
turba *f*, 2927
turbina *f*, 2396, 3906
turbina de combustión *f*, 862
turbina de combustión interna *f*, 2152
turbina de gases *f*, 2152
turbina de vapor *f*, 3646
turbo compresor *m*, 3908
turbo generador *m*, 3909
turbosoplante *m*, 3907
turboventilador *m*, 3907, 4129
turbulador de evaporación *m*, 1706
turbulencia *f*, 3910

ultracongelación *f*, 1201
ultrafiltro *m*, 7
ultravacío *m*, 1202
umbral de audibilidad *m*, 3834
unidad *f*, 3925
unidad compresora, 920
unidad condensadora *f*, 960
unidad condensadora enfriada por aire *f*, 96
unidad condensadora hermética *m*, 3363
unidad de medición *f*, 3930

unidad de mezcla de aire *f*, 139
unidad de renovación de aire *f*, 2614
unidad de ventilación *f*, 3990
unidad esclava *f*, 3472
unidad evaporadora *f*, 1707
unión *f*, 983, 2500, 3924
unión soldada *f*, 4104
unión soldada a tope *f*, 591
unitario, 3926

vaciado *m*, 1416, 1614
vaciar, 1417
vacio *m*, 3940
vagón cisterna *m*, 3756
vaho *m*, 522
vaina *f*, 4112
vaina para termómetro sumergido *f*, 3814
valor *m*, 3950
valor de ajuste *m*, 3410
valor de umbral *m*, 3833
valor límite admisible *m*, 3833
valoración *f*, 1684
válvula *f*, 3951
válvula antirretorno *f*, 2788
válvula antisifón *f*, 243
válvula contrafuegos *f*, 1893
válvula de acción rápida *f*, 3132
válvula de admisión *f*, 2445
válvula de aguja *f*, 2763
válvula de aislamiento *f*, 2487
válvula de alimentación de cierre automático *f*, 1816
válvula de asiento doble *f*, 1404
válvula de aspiración *f*, 331, 2445
válvula de cierre *f*, 1150, 3437
válvula de cierre de descarga *f*, 1349
válvula de compuerta *f*, 2156
válvula de contrapeso *f*, 4101
válvula de corte *f*, 1150
válvula de derivación *f*, 597, 3435
válvula de descarga *f*, 446, 1352
válvula de descarga automática *f*, 1995
válvula de diafragma *f*, 1279
válvula de disco *f*, 1356, 1915
válvula de dos vías *f*, 3918
válvula de equilibrado *f*, 359
válvula de escape *f*, 1736, 3236
válvula de escape de emergencia *f*, 1609
válvula de escuadra *f*, 231
válvula de expansión *f*, 1753

volante de mano *m*, 2213
volatil, 4011
voltear, 1122
voltímetro *m*, 4014
volumen *m*, 562, 4015
volumen construido *m*,
   260

volumen de aire variable *m*,
   3968
volumen de combustión *m*,
   865
volumen muerto *m*, 769

zona *f*, 262, 3158, 4149

zona de bienestar *f*, 872
zona de confort *f*, 872
zona de trabajo blanca *f*, 765
zona muerta *f*, 1184, 1188
zona neutra *f*, 2767
zona ocupada *f*, 2802
zonificación *f*, 4151

The Right Environment

There is only one condition for obtaining the ideal air in technological rooms and environments with electronic equipment: choosing a **Hiross Close Control Air Conditioning** unit. Because for over 25 years Hiross has been producing the widest range of machines for conditioning in data processing centres and technological areas in general, guaranteeing air treatment systems with total reliability and maximum efficiency. All the units are equipped with **Hiromatic**, the latest generation of microprocessor controls specially designed by Hiross Research & Development for its conditioning units. So clean air, a constant temperature and the right relative humidity are maintained 24 hours a day, 365 days a year.

From the perfect air conditioning of technological environments to the maximum comfort and flexibility of the **Flexible Space System**: Hiross innovative capacities are really unconditioned. The **Flexible Space System** allows all the plant engineering and the conditioned air circulating system to be placed in the plenum under the floor. In this way, besides the extreme versatility of space, excellent air quality is obtained. So this is the ideal choice for creating office spaces that are in the lead, functional, healthy and versatile. All main **Flexible Space System** components are manufactured at Hiross production centres. **Flexible Space System**: one more reason for letting yourself be conditioned by Hiross.

HIROSS SpA - Zona Industriale Tognana - Via Leonardo da Vinci, 8 - 35028 Piove di Sacco PD
Phone 049/9719111 - Telex 430093 Icepi I - Telefax 049/5841257/5841421
Italian Branches: MILANO, PADOVA, TORINO, ROMA, NAPOLI
Hiross in the World: AUSTRIA, CANADA, CZECHOSLOVAKIA, FRANCE, GERMANY, ITALY, POLAND, PORTUGAL,
SINGAPORE, SPAIN, SWEDEN, SWITZERLAND, UNITED KINGDOM, UNITED STATES
Production Centres: AUSTRIA, CANADA, GERMANY, ITALY, PORTUGAL, UNITED STATES

Français

brouillard *m* de tour de refroidissement, 1072
bruit *m*, 2772, 3536
bruit *m* aérien, 77
bruit *m* de moteur, 2729
bruit *m* émis, 520
bruit *m* émis par les conduits d'air, 1493
brûler, 567
brûler complètement, 581
brûleur *m*, 568
brûleur *m* à air soufflé, 2013
brûleur *m* à atomisation, 74
brûleur *m* à atomisation mécanique, 2655
brûleur *m* à atomisation par la vapeur, 3623
brûleur *m* à coupelle, 3010
brûleur *m* à coupelle rotative, 3308
brûleur *m* à courants parallèles, 2918
brûleur *m* à deux combustibles, 1489
brûleur *m* à fuel, 2812
brûleur *m* à gaz, 2114
brûleur *m* à injection, 2442, 3056
brûleur *m* à plusieurs becs, 2745
brûleur *m* à pot, 3010
brûleur *m* à vaporisation, 3958
brûleur *m* à vortex, 4019
brûleur *m* atmosphérique, 293
brûleur *m* d'angle, 1088
brûleur *m* polycombustible, 1040
brûleur *m* rotatif à atomisation, 3307
brûleur *m* sous pression à atomisation, 3042
brûleur *m* tous gaz, 2115
brûleur *m* tout combustible, 3931
bulbe *m*, 561
bulle *f*, 541
bulle *f* d'air, 79
burin *m*, 740
bus *m* de données, 3443
bus *m* (électrique), 584
buse *f* de soufflage, 2870
buse *f* d'injection, 3586
butane *m*, 588
by-pass *m*, 593

cabine *f* de commande, 1018
cabine *m* de peinture, 2909
câble *m*, 599
cache-radiateur *m*, 645, 3148
cadran *m*, 1274
cadre *m*, 2031
cadre *m* de filtre, 1858
cadre *m* de renfort, 3231

cadre *m* en fers plats, 1930
caillebotis *m*, 1951
caisse *f*, 643
calcaire *m*, 700
calcul *m*, 603
calcul *m* des déperditions calorifiques, 2272
calibrage *m*, 604
calibre *m* (de tuyau), 495
caloduc *m*, 2276
calorie *f*, 606
calorifique, 607
calorifuge *m*, 2459
calorifugeage *m* réflectif, 3192
calorifuger, 2456
calorimètre *m*, 611
came *f*, 612
camion-citerne *m* de fuel, 2830
canal *m*, 706, 1492
canal *m* de décharge, 2157
canal *m* de fuite, 2157
canalisation *f*, 972, 1505
canalisation *f* de soutirage, 431
canalisation *f* principale, 2608
caniveau *m*, 1492
capable de fonctionner, 4135
capacitance *f*, 615
capacité *f*, 617
capacité *f* calorifique, 2232
capacité *f* calorifique (stockage thermique), 716
capacité *f* (condensateur), 616
capacité *f* d'absorption, 16
capacité *f* d'eau, 4058
capacité *f* de rétention de poussières, 1517
capacité *f* de stockage (stockage thermique), 1354
capacité *f* thermique, 2229
capillaire, 624
capillarité *f*, 623, 625
capot *m*, 1113
capot *m* de protection étanche, 4098
capsule *f* anéroïde, 225
capte-suie *m*, 2194, 3530
capteur *m*, 3387
capteur *m* à distance (décentralisé), 3240
capteur *m* d'humidité, 2360
capteur *m* de la qualité de l'air, 153
capteur *m* de rayonnement solaire, 3519
capteur *m* de température, 3770
capteur *m* extérieur, 1767, 2877
capteur *m* solaire, 3513

capteur *m* solaire à concentration, 939
caractéristique *f*, 707
caractéristique *f* du local, 3300
caractéristiques *f* du projet, 1249
carbone *m*, 628
carbonisation *f*, 632
cargo *m*, 634
cargo *m* froid, 727
cargo *m* réfrigéré, 3203
carneau *m*, 738, 1973, 1976, 2328
carneau *m* de fumée, 3495
carneaux *m* de fumée, 525
carter *m*, 1123
cascade *f*, 640
casing *m*, 458
catalyseur *m*, 651
catathermomètre *m*, 2505
cathode *f*, 652
cation *m*, 654
cave *f*, 380, 668
cavitation *f*, 656
cavité *f*, 657
cellule *f*, 667
cendre *f*, 270
cendres *f* volantes, 1997, 2193
cendrier *m*, 271
centigrade *m*, 676
centrale *f*, 2974
centrale *f* combiné chaleur force, 840
centrale *f* de chaleur/force, 2277
centrale *f* de chauffage, 2287
centrale *f* de chauffage à distance centrale *f*, 1380
centrale *f* de production d'énergie, 3019
centrale *f* de traitement d'air, 684
centrale *f* électrique, 3022
centrale *f* modulaire de climatisation, 2707
centre *m* de gravité, 686
centrifuge, 687, 693
certificat *m*, 695
certificat *m* d'épreuve, 3779
certificat *m* d'essai, 3779
chaîne *f* du froid, 819
chaleur *f*, 2224, 4042
chaleur *f* de combustion, 854
chaleur *f* de fusion, 2104
chaleur *f* latente, 2517
chaleur *f* latente de fusion, 2518
chaleur *f* latente d'évaporation, 2519
chaleur *f* perdue, 4052
chaleur *f* présumée, 2226
chaleur *f* rayonnée, 3140
chaleur *f* utile, 3936

chambre *f*, 701, 3295
chambre *f* à vapeur, 3625
chambre *f* calorimétrique, 3299
chambre *f* climatique, 1658
chambre *f* de chargement, 2572
chambre *f* de combustion, 847, 859, 1877, 2099
chambre *f* de combustion d'un générateur de chaleur, 2100
chambre *f* de détente, 1922
chambre *f* de filtration, 1852
chambre *f* de mélange, 2694
chambre *f* de presse-étoupe, 2902
chambre *f* de pulvérisation, 3581
chambre *f* d'essai, 3778
chambre *f* d'essais climatiques, 772, 1663
chambre *f* frigorifique de grandeur hauteur, 2313
chambre *f* froide, 732, 823
chambre *f* froide pour distribution locale, 1359
chambre *f* propre, 763
chambre *f* réverbérante, 3269
chamotte *f*, 1880
champ *m* visuel, 1834
changement *m* de point de consigne, 3248
changement *m* de température été-hiver, 703
changement *m* d'état, 702
changement *m* été-hiver, 703
chanvre *m*, 2296
chapeau *m*, 614
chapeau *m* de cheminée, 1116
chapeau *m* de presse-étoupe, 2167
charbon *m*, 782
charbon à coke, 815
charbon *m* à gaz, 2117
charbon *m* à haute teneur en matières volatiles, 2318
charbon *m* actif, 42
charbon *m* bitumeux, 415
charbon *m* cokéfiant, 815
charbon *m* de gros calibre, 790
charbon *m* flambant, 2661
charbon *m* maigre, 2782
charbon *m* non cokéfiant, 2784
charbon *m* pulvérisé, 3118
charge *f*, 711, 2568
charge *f* admissible, 2944, 3336
charge *f* calorifique, 2251
charge *f* connectée, 976

charge *f* de la demande de puissance électrique, 1580
charge *f* de la grille, 2183
charge *f* de pointe, 2926
charge *f* frigorifique, 1062
charge *f* initiale, 2436
charge *f* maximale, 2641
charge *f* partielle, 2923
charge *f* prévisionnelle, 1251
charge *f* thermique, 2270
chargement *m*, 715, 720
chargement *m* complet, 3864
chargement *m* de foyer, 3661
charger, 712
chargeur *m* de foyer, 3660
charnière *f*, 2319
charpente *f*, 2033
chaud, 2332
chaudière *f*, 455, 467
chaudière *f* à bouilleurs croisés, 1138
chaudière *f* à coke, 811
chaudière *f* à condensation, 956
chaudière *f* à deux combustibles, 1488
chaudière *f* à eau chaude, 2338
chaudière *f* à électrodes, 1590
chaudière *f* à foyer surpressé, 3051
chaudière *f* à fuel, 2819
chaudière *f* à gaz, 2113
chaudière *f* à plusieurs parcours, 2748
chaudière *f* à trémie de combustible, 2326
chaudière *f* à trois parcours, 3827
chaudière *f* à tube de fumée, 3351
chaudière *f* à tubes de fumée, 1892
chaudière *f* à tubes d'eau, 4088
chaudière *f* à tubes inclinés, 2403
chaudière *f* à vapeur, 3624
chaudière *f* au charbon, 787
chaudière *f* basse pression, 2597
chaudière *f* convertible, 1042
chaudière *f* de préparation d'eau chaude sanitaire, 2344
chaudière *f* de récupération, 4053
chaudière *f* en acier, 3648
chaudière *f* en fonte, 647
chaudière *f* équipée, 2899
chaudière *f* haute pression, 2307
chaudière *f* multi-combustible, 2743

chauffage *m*, 2241
chauffage *m* à accumulation, 3667
chauffage *m* à air chaud, 4038
chauffage *m* à air chaud au gaz, 2128
chauffage *m* à deux tuyaux, 3915
chauffage *m* à distance par eau chaude sous pression, 1378
chauffage *m* à eau chaude au gaz, 2130
chauffage *m* à eau chaude basse pression, 2601
chauffage *m* à eau chaude sous pression, 2310
chauffage *m* à un tuyau, 2833
chauffage *m* à vapeur, 3634
chauffage *m* à vapeur basse pression, 2600
chauffage *m* à vapeur sous vide, 3948
chauffage *m* au fuel, 2822
chauffage *m* au gaz, 2134
chauffage *m* central, 680
chauffage *m* central au gaz, 2129
chauffage *m* central général, 2092
chauffage *m* complémentaire, 3718
chauffage *m* d'air, 129
chauffage *m* d'appartement, 245
chauffage *m* d'appoint, 3718
chauffage *m* de base, 327
chauffage *m* de confort par rayonnement, 3136
chauffage *m* d'eau, 4072
chauffage *m* d'îlot, 441
chauffage *m* de locaux, 3554
chauffage *m* domestique, 1389
chauffage *m* électrique, 1574
chauffage *m* électrique par résistance, 3252
chauffage *m* en allège, 4126
chauffage *m* individuel, 2414
chauffage *m* infra-rouge, 2432
chauffage *m* monotube, 2833
chauffage *m* mural, 4028
chauffage *m* par accumulation d'énergie hors pointe, 2807
chauffage *m* par éjecto-convecteur, 2419
chauffage *m* par le plafond, 662
chauffage *m* par le sol, 1950
chauffage *m* par les murs, 4028

combustion *f* inversée (de haut en bas), 1412
combustion *f* par lit fluidisé, 1988
combustion *f* parfaite, 2932
commande *f*, 1014, 1444
commande *f* à distance, 3238
commande *f* centrale, 678
commande *f* d'alimentation, 1817
commande *f* d'alimentation en huile, 2818
commande *f* de changement de température été-hiver, 704
commande *f* directe, 1319
commande *f* en cascade des chaudières, 1895
commande *f* manuelle, 2627
commande *f* par bouton poussoir, 3125
commande *f* par courroie, 400
commande *f* par horloge, 3848
commander, 1015
commencement *m*, 3604
commercial, 873
commission *f* de réception, 877
communication *f* sur systèmes de même nature, 2930
commutateur *m*, 1332, 3378
commutateur *m* de démarrage, 3605
commutateur *m* de va et vient, 3917
compartiment *m* de congélation, 2048
compatible, 884
compensateur *m*, 887
compensateur *m* à joints glissants, 3479
compensateur *m* articulé, 888
compensateur *m* axial, 316
compensateur *m* de dilatation, 1742
comportement *m* dynamique, 1531
composants *m* de surface vitrée, 1826
compresseur *m*, 912, 920
compresseur *m* à contre-courant (à flux inversé), 3262
compresseur *m* à deux cylindres, 3913
compresseur *m* à double aspiration, 1405, 1486
compresseur *m* à double effet, 1395
compresseur *m* à équicourant, 3923

compresseur *m* à flux axial, 318
compresseur *m* à membrane, 1278
compresseur *m* à palettes, 3481
compresseur *m* à piston, 3175
compresseur *m* à piston différentiel, 3656
compresseur *m* à piston rotatif, 3289
compresseur *m* à piston sec, 1475
compresseur *m* à plateau oscillant, 4134
compresseur *m* à simple effet, 3449
compresseur *m* bi-étagé (compound), 893
compresseur *m* centrifuge, 688
compresseur *m* de type ouvert, 2842
compresseur *m* en tandem (compresseurs jumelés), 3753
compresseur *m* frigorifique, 3204
compresseur *m* hélicoïdal, 3356
compresseur *m* hermétique, 2301, 3365
compresseur *m* multi-étage, 2752
compresseur *m* rotatif, 3309
compresseur *m* rotatif multicellulaire, 2753
compresseur *m* rotatif volumétrique, 3309
compresseur *m* subsonique, 3692
compresseur *m* supersonique, 3716
compresseur *m* surpresseur, 492
compresseur *m* vertical, 3998
compresseur *m* volumétrique, 3008
compressibilité *f*, 898
compression *f*, 900
compression *f* à double aspiration, 1483
compression *f* à plusieurs étages, 892
compression *f* en régime de surchauffe, 1463
comptage *m* d'impulsions, 3117
compteur *m*, 2676
compteur *m* à gaz, 2140
compteur *m* de calories, 2274
compteur *m* de chaleur, 2274

compteur *m* de chaleur à évaporation, 1702
compteur *m* d'eau condensée, 947
concentrateur *m* de données, 1179
concentration *f*, 940
concentration *f* (d'un produit), 941
concentration *f* de gaz, 2118
concentration *f* maximum admise, 2640
conception *f*, 1245
conception *f* assistée par ordinateur, 923
conception *f* environnementale, 1660
condensat *m*, 944
condensateur *m* à double faisceau, 1397
condensateur *m* à plaques, 2986
condensation *f*, 949
condensation *f* de gouttelettes, 1450
condensation *f* en phase solide, 3525
condenser, 951
condenseur *m*, 952
condenseur *m* à air, 95
condenseur *m* à air à convection forcée, 2015
condenseur *m* à aspersion, 3584
condenseur *m* à calandre et serpentin, 3422
condenseur *m* à calendres multiples, 2750
condenseur *m* à contact direct, 369
condenseur *m* à faisceau tubulaire, 3424
condenseur *m* à immersion, 3691
condenseur *m* à soutirage, 432
condenseur *m* à tubes concentriques, 1401
condenseur *m* atmosphérique, 294
condenseur *m* de vapeur, 3626
condenseur *m* évaporatif, 1697
condenseur *m* multicircuit, 3574
condenseur *m* réservoir (condenseur bouteille), 3174
condenseur *m* secondaire, 3370
conditionnement *m* d'air, 89
conditionnement *m* d'air à haute vitesse, 2316

contrôle *m* par pressostat de
  sécurité d'huile, 2825
contrôle *m* par thermostat
  de sécurité d'huile, 2831
contrôle *m* prédictif, 3030
contrôle *m* qualité, 3130
contrôler, 722
contrôleur *m* de chantier,
  3460
controleur *m* différentiel,
  1290
convecteur *m*, 1038
convecteur *m* électrique,
  1572
convection *f*, 1034
convection *f* forcée, 2011
convection *f* naturelle, 2755
convection *f* thermique, 3796
convective, 1036
conversion *f*, 1039
convertisseur *m*, 1041
convertisseur *m* de données,
  1177
convoyeur *m*, 3660
cordon *m* chauffant, 3144
cordon *m* d'amiante, 266
cordon *m* de soudure, 4102
cornière *f*, 228
corps *m*, 453
corps *m* de chauffe, 2242,
  3147
corps *m* de pompe, 3122
corps *m* noir, 417
corroder, 1093
corrosif, 1098
corrosion *f*, 1094
corrosion *f* ponctuelle, 2972
cote *f* extérieure, 2876
couche *f*, 2520
couche *f* de glace, 2381
couche *f* de mâchefer, 774
couche *f* de protection, 3111
couche *f* intermédiaire, 2471
couche *f* limite, 499
coude *m*, 403, 502, 1495,
  1563, 2955
coude *m* à souder, 4107,
  4108
coude *m* de dilatation, 1743
coude *m* de tuyauterie, 2951
coude *m* en tube lisse, 3501
coude *m* en tube plissé, 1099,
  1100
coude *m* en U, 3260
couleur *f*, 831
couleur *f* conventionnelle,
  3598
coup *m* de bélier, 4069
coupelle *f*, 1143
couper, 1152
coupe-tirage *m*, 3577
couplage *m*, 1109
couplage *m* en parallèle,
  1110, 2916

couplage *m* en série, 1111,
  3394
couple *m*, 3865
couple *m* de démarrage,
  3610
couple *m* électrolytique,
  1592
coupure *f*, 1151
coupure *f* bipolaire, 1400
coupure *f* thermique, 519
courant *m*, 1144, 3679
courant *m* alternatif, 190
courant *m* continu, 1316
courant *m* d'air, 1425
courant *m* d'air froid
  descendant contre-tirage,
  1409
courant *m* de convection,
  1037
courant *m* de démarrage,
  3606
courant *m* de pointe, 2835
courant *m* parallèle, 2917
courants *m* croisés, 1128,
  1129
courbe *f*, 1147
courbe *f* caractéristique, 708
courbe *f* caractéristique de
  ventilateur, 1797, 1801
courbe *f* de chauffage, 2245
courbe *f* de puissance, 3014
courbe *f* de refroidissement,
  1055
courbe *f* de saturation, 3341
courbes *f* de crières
  acoustiques, 2774
courette *f*, 177
courroie *f*, 399
courroie *f* trapézoïdale, 3976
course *f* de compression, 907
course *f* de détente, 1751
course *f* de piston, 3682
course *f* de refoulement,
  1350
course *f* de refoulement d'un
  compresseur, 915
course *f* d'un compresseur,
  916
court-circuit *m*, 3431
coussinet *m*, 585, 3474
coussinet *m* deux pièces,
  3573
coût *m*, 1101, 1756
coût *m* de centrale, 2975
coût *m* de remplacement,
  3243
coût *m* de reproduction,
  3244
coût *m* d'exploitation, 1102,
  3322
coût *m* du cycle de vie, 2540
couvercle *m*, 614, 1113,
  2320, 2537
couverture *f*, 1114
couvre-joint *m*, 2502

craie *f*, 700
crapaudine *f*, 2006
crépine *f*, 3703
cristal *m* de glace, 2376
critiqué, 1126
crochet *m*, 2324
croix *f*, 1134
cryodessiccation *f*
  atmosphérique, 297
cryogénique, 1140
cube *m* de glace, 1141, 2377
cuivre *m*, 1081
culotte *f* (conduit d'air),
  4146
cycle *m*, 1155
cycle *m* court, 3432
cycle *m* de Carnot, 635
cycle *m* de compression, 901
cycle *m* de dégivrage, 1209
cycle *m* de Rankine, 3159
cycle *m* de référence, 3188
cycle *m* de réfrigération
  inversé, 3273
cycle *m* de Stirling, 3657
cycle *m* de vapeur binaire,
  412
cycle *m* de vie, 2539
cycle *m* d'Otto, 2863
cycle *m* fermé, 778
cycle *m* frigorifique, 3209
cycle *m* frigorifique à
  éjection, 3963
cycle *m* inversé, 3271
cyclone *m*, 1158
cylindre *m*, 1159
cylindrée *f*, 2967

débit *m*, 1953, 2634, 3838
débit *m* biphasique, 3914
débit *m* calorifique, 2275
débit *m* d'air, 123, 167
débit *m* d'air chaud, 4040
débit *m* d'air minéral, 2687
débit *m* d'air variable, 3968
débit *m* d'aspiration, 3696
débit *m* d'écoulement, 1966
débit *m* d'évacuation, 1347
débit *m* de puisage, 3759
débit *m* équilibré, 350
débit *m* frigorifique d'un
  détenteur, 1754
débit *m* journalier, 1161
débit *m* nominal, 1247
débit *m* nominal (d'un filtre),
  1955
débit *m* variable, 3971
débitmètre *m*, 1963
débordement *m* (surcharge)
  d'ordinateur, 932
déborder, 2884
débouché *m* de conduit de
  fumée, 1984
décélération *f*, 1193
décharge *f*, 1338, 1416

décharge *f* (stockage
    thermique), 1353
décibel *m*, 1194
décipol *m*, 1195
déclenchement *m*, 1148
déclinaison *f*, 1196
décollement *m* de flamme,
    1908
décoloration *f*, 1355
décomposition *f*, 1191, 1197
décompresseur *m*, 1198
décongélation *f*, 3789
décongélation *f* diélectrique,
    1285
décrassage *m*, 775, 3153
décrasser, 3151
décret *m* sur la pollution de
    l'air, 760
défaut *m*, 1164, 1810
défaut *m* dans l'enveloppe
    du bâtiment, 553
défaut *m* de mise à la terre,
    2197
défaut *m* électrique, 1568
défectuosité *f*, 1810
déflagration, 1916
déflagration *f* de gaz, 2123
déflecteur *m*, 342, 2203, 2590
déflecteur *m* (aube de
    guidage), 1205
dégager, 3152
dégât *m* dû au gel, 2072
dégazage *m*, 1213
dégivrage *m*, 1208
dégivrage *m* naturel
    cyclique, 2805
dégivrage *m* par
    chronorelais, 3850
dégivrage *m* par eau, 4063
dégivrage *m* par inversion de
    cycle, 3272
dégivrer, 1206
degré *m*, 1214
degré *m* de pureté, 1217
degré *m* de saturation, 3343
degré *m* de surchauffe, 1218
degré-jour *m*, 1215
degré-jour *m* de chauffage,
    2246
degrés *m* de liberté, 1219
degrés jours de
    refroidissement, 1056
délai *m*, 3852
délai *m* d'exécution, 890
délai *m* de livraison, 1230
délétère, 2790
demande *f* de chaleur, 2283
démarrage *m*, 3612
démarrage *m* à vide, 2571,
    2779
démarreur *m*, 3605
demi-coussinet *m*, 586
déminéralisant, 1233
déminéralisation *f* de l'eau,
    4064

demi-saison *f*, 2683
démodulation *f*, 1234
démonter, 1336
déneigement *m*, 3507
densité *f*, 1235
densité *f* d'air du ventilateur
    à l'entrée, 1791
densité *f* de flux, 3564
densité *f* de fumée, 3493
densité *f* de garniture, 2904
densité *f* en vrac, 563
déodorant *m*, 1236
dépense *f*, 1101, 1756
dépenses *f* d'investissement,
    1896
déperdition *f*, 2586
déperdition *f* calorifique,
    2271
déperditions *f* calorifiques
    du bâtiment, 558
déperditions *f* thermiques
    calculées, 1250
déposer, 1238
dépôt *m*, 1237, 3024, 3348,
    3377
dépôt *m* de givre, 2073
dépôt *m* de poussières, 1512
dépôt *m* de suie ou de
    poussières, 3502
dépoussiérage *m*, 1200, 1515,
    1521
dépoussiéreur *m*, 85, 1513,
    1525
dépoussiéreur *m*
    électrostatique, 1601
dépréciation *f*, 1240
dépression *f*, 2764
déprimomètre *m*, 1426
dérivation *f*, 507, 593
désaération *f*, 1189
description *f*, 1242
déshuileur *m*, 2826
déshumidificateur *m*, 1221
déshumidificateur *m* d'air,
    106
déshumidification *f*, 1220
déshumidification *f* d'air,
    105
déshumidifier, 1222
déshydratant, 1243
déshydratant *m* à l'alumine
    activée, 41
déshydratation *f*, 1225
déshydrater, 1224
déshydrateur *m* (sécheur),
    1226
desincrustation *f*, 3350
désincruster, 1241
désinfection *f*, 1358
désodorisant *m*, 1236
désorption *f*, 1255
désséchant *m*, 1243
dessication *f*, 1244
dessication *f* secondaire,
    3372

dessin *m*, 1245
dessin *m* axonométrique, 322
dessin *m* côté, 1310
dessin *m* d'atelier, 3430
dessin *m* de détail, 1260
dessin *m* de projet, 3037
dessin *m* de recollement, 267
dessin *m* en coupe, 3376
dessin *m* en élévation, 1603
dessin *m* en perspective,
    2945
dessin *m* isométrique, 2488
désurchauffeur *m*, 1256
détail *m*, 1258
détailler, 1259
détartrage *m*, 3350
détecteur *m*, 1263
détecteur *m* de combustion,
    850
détecteur *m* de fuite, 2528
détecteur *m* de fumée, 3494
détecteur *m* de gaz, 2137
détecteur *m* de gaz
    combustible, 843
détecteur *m* de lumière, 2547
détecteur *m* de présence,
    2798
détecteur *m* de vapeur, 3962
détendeur *m*, 3186
détendeur *m* de vapeur, 3643
détente *f*, 3064
détente *f* directe, 1464
détergent *m*, 1264
détérioration *f*, 1265
détremper (le métal), 233
développement *m* des
    bactéries, 339
déverminer, 1190
déverrouillage *m* manuel,
    2628
déviation *f*, 405, 1266
déviation *f* standard, 3599
devis *m*, 1682
devis *m* descriptif, 3562
devis *m* détaillé, 1261
diagramme *m*, 1273, 2177
diagramme *m* de
    combustion, 851
diagramme *m* de confort,
    869
diagramme *m* de Mollier,
    2721
diagramme *m* électrique,
    1567
diagramme *m* enthalpique,
    1647
diagramme *m* entropique,
    1653
diagramme *m*
    psychrométrique, 3115
diamètre *m*, 1276
diamètre *m* de tube, 2954
diamètre *m* (de tuyau), 495
diamètre *m* extérieur, 1766
diamètre *m* nominal, 2780

diaphragme *m*, 2858, 2859
diathermane, 1280
diélectrique, 1282
différence *f*, 1288
différence *f* de pression, 3046
différence *f* de température, 3769
différence *f* psychrométrique, 4115
différentiel *m*, 1289
différentiel *m* de fonctionnement, 2846
différentiel *m* de température, 1295
diffus, 1296
diffuseur *m*, 1297
diffuseur *m* à fente, 3485
diffuseur *m* de plafond, 660
diffuseur *m* linéaire, 2552
diffuseur *m* plafonnier, 660
diffusion *f*, 1300
diffusion *f* d'air, 107
diffusivité *f* thermique, 3797
digital, 1304
dilatation *f*, 1741
dilatation *f* linéaire, 2553
diluant *m*, 1306
dilution *f*, 1308
dimension *f*, 1309, 3462
dimension *f* de conduit, 1498, 1501
dimension *f* de gouttelettes, 1452
dimension *f* de maille, 2671
dimension *f* intérieure, 2475
dimensionnement *m* de conduits, 1502
dimensionnement *m* de tuyauterie, 2961
diminution *f* du bruit, 3549
diode *f* lumineuse, 2543
direction *f*, 1329
direction *f* de l'écoulement, 1960
dispersion *f* des fumées, 2994
disponibilité *f* en énergie, 1621
disponibilité *f* technique, 3397
dispositif *m*, 1267
dispositif *m* antivibratoire, 244
dispositif *m* d'alerte, 4046
dispositif *m* d'allumage, 3560
dispositif *m* de combustion, 846
dispositif *m* de décharge, 1341
dispositif *m* de délestage, 745
dispositif *m* de délestage de compresseur, 921
dispositif *m* d'insonorisation, 3539

dispositif *m* de protection, 3110
dispositif *m* de rejet d'air, 1722
dispositif *m* de sécurité, 3331
dispositif *m* de sécurité d'allumage, 2389
dispositif *m* de sécurité de manque d'eau, 2594
dispositif *m* de sécurité en cas de manque de gaz, 2124
dispositif *m* de sécurité par coupure, 3329
dispositif *m* de variation de puissance, 622
dispositif *m* pare-soleil, 3414
disque *m*, 1337
disque *m* de sécurité, 582
distance *f*, 1365, 3556
distance *f* d'espacement, 2970
distillation *f*, 1367
distributeur *m*, 1376, 2218
distribution *f*, 1369
distribution *f* à trois tuyaux, 3829
distribution *f* d'air, 109
distribution *f* d'air par réseau aéraulique, 1494
distribution *f* de chaleur à distance, 1381
distribution *f* d'eau chaude, 2340
distribution *f* de froid à distance refroidie, 1377
distribution *f* de gaz, 2150, 2121
distribution *f* en parapluie, 1410, 2887
distribution *f* inférieure, 1372
distribution *f* par le haut, 2887
distribution *f* urbaine de froid, 1377
domaine *m*, 3158
domaine *m* de régime transitoire, 3887
dôme *m*, 1387
domestique, 1388
dommage *m*, 1164
donnée *f*, 1174
données *f* analogiques, 213
données *f* du projet, 1248
données *f* météorologiques, 4096
dosage *m*, 1392
doser, 1393
double coupure *f*, 1396
double fenêtre *f*, 1398
douille *f*, 3473
drain *m*, 1423
drainage *m*, 1418
drainer, 1417

droite *f* de soufflage, 963
ductilité *f*, 1496
dudgeonnière *f*, 1920
duplex *m*, 1507
durabilité *f*, 1508
durée *f*, 1509, 3846, 2538
durée *f* de combustion, 852
durée *f* de dégivrage, 1212
durée *f* de la demande de puissance électrique, 1581
durée *f* de marche, 2851
durée *f* de mise en température, 2263, 3036
durée *f* d'utilisation, 3939
durée *f* de vie, 1157, 2851
durée *f* de vie réelle, 1554
durée *f* moyenne de réparation, 2648
dureté *f* de l'eau, 2215, 4070
dynamique, 1530
dynamomètre *m*, 1537

eau *f*, 4056
eau *f* chaude, 2337, 4044
eau *f* chaude haute pression, 2309
eau *f* condensée, 944
eau *f* d'alimentation, 1819
eau *f* d'alimentation de chaudière, 464
eau *f* d'appoint, 1394, 2615
eau *f* de puits, 4113
eau *f* de réfrigération, 3201
eau *f* de refroidissement, 1077
eau *f* de retour, 3268
eau *f* de ville, 753
eau *f* distillée, 1368
eau *f* entraînée, 1438
eau *f* froide, 827
eau *f* glacée, 728, 2378
eau *f* haute température, 2314
eau *f* industrielle, 2423
eau *f* non traitée, 3933
eau *f* potable, 1440, 3009
eau *f* saumâtre, 505
eau *f* souterraine, 2199
eau *f* surchauffée, 2309
eau *f* usée, 1557, 4055
ébullition *f*, 1540
ébullition *m* par film, 1840
écart *m* de réglage, 1020
écart *m* de température, 3768
écart *m* du point de rosée, 1270
écart *m* moyen de température, 2646
écart *m* moyen logarithmique de température, 2581
écart *m* type, 3599
écartement *m*, 1365, 3556
échafaudage *m*, 2033, 3347
échange *m*, 1712

échange *m* de chaleur, 2237
échange *m* d'information,
2431
échange *m* d'ions, 2481
échanger, 1713, 3879
échangeur *m*, 1714
échangeur *m* à courants
croisés, 1132
échangeur *m* à faisceau
tubulaire, 3426
échangeur *m* à tubes
concentriques, 1402
échangeur *m* de chaleur, 610,
2238
échangeur *m* de chaleur à
contre-courant, 1107
échangeur *m* rotatif, 3311
échangeur *m* thermique à
contact direct, 1313
échantillon *m*, 3337
échauffer, 2225
échelle *f* de Beaufort, 393
échelle *f* (dessin), 3348
échelonné, 3654
éclairagisme *m*, 2544
économie *f*, 1543
économie *f* d'énergie, 1625,
3015
économie *f* du chauffage,
2247
économiseur *m*, 1542
économiseur *m* d'eau
alimentaire, 1820
écoulement *m*, 1416, 1953
écoulement *m* à bulles, 542
écoulement *m* annulaire, 234
écoulement *m* d'air, 100, 121
écoulement *m* d'air stratifié,
3678
écoulement *m* diphasique
ondulé et à bouchons,
3490
écoulement *m* dispersé, 1360
écoulement *m* en spirales,
3314
écoulement *m* en vortex,
4020
écoulement *m* laminaire,
3680
écoulement *m* turbulent,
1544, 3911
écoulement *m* vers
l'extérieur, 2879
écoulement *m* visqueux,
4009
écouvillon *m*, 1974
écran *m*, 3352
écran *m* anti-rayonnement,
289
écran *m* de protection, 3112
écran *m* de protection
étanche, 4099
écrou *m*, 2796
écrou *m* borgne, 3355
écume *f*, 1999

écumer, 2000
édition *f* d'ordinateur, 933
effect *m* de
déshumidification, 1223
effectif, 43, 1550
effet *m*, 1549
effet *m* barométrique, 371
effet *m* de cheminée, 737,
3594
effet *m* de refroidissement,
731, 1058, 3205
effet *m* Peltier, 2931
efficace, 1550
égalisateur *m*, 1961
égalisation *f*, 1670
égaliser, 1665
égaliseur *m*, 1666
égout *m*, 3411
égoutter, 1453
égouttoir *m* (bac de
condensation), 1443
éjecteur *m*, 1559
éjecteur *m* d'air à vapeur,
3638
éjecteur *m* hydraulique, 4074
éjecto-convecteur *m*, 2421
élasticité *f*, 1561
élastomère *m*, 1562
électricité *f*, 1575
électrique, 1564, 1565
électro-aimant *m*, 1593
électrode *f*, 1589
électrode *m* d'allumage, 2387
électrolytique, 1591
électronique, 1595
électronique *f*, 1598
électropneumatique, 1599
électro-vanne *f*, 3524
élément *m*, 891, 1602, 3374,
3925
élément *m* chauffant, 2248
élément *m* de chaudière, 478
élément *m* de construction,
549, 995
élément *m* de façade, 2070
élément *m* de refroidissement, 1060
élément *m* de régulation de
puissance, 1031
élément *m* de structure, 3683
élément *m* détecteur, 1262
élément *m* filtrant, 1857
élément *m* sensible, 3386
élément *m* terminal de
commande, 1022
élévation *f* du point de rosée,
1272
élimination *f*, 3241
embase *f* de cheminée, 734
emboîtement *m*, 3566
embranchement *m*, 509
émetteur *m*, 3890
émission *f*, 1611
émissivité *f*, 1612
émittance *f*, 1613

émittance *f* thermique
directionnelle, 1331
émulsion *f*, 1615
en état *m* de marche, 3170,
4137
en ordre *m* de marche, 4137
encastré, 1605
encastrer, 387
enceinte *f*, 1617
enclencher, 3745
encombrement *m*, 3555
encrassement *m*, 2026, 777
endothermique, 1618
enduit *m*, 791
enduit *m* (bâtiment), 3242
enduit *m* bitumeux, 416
enduit *m* de protection, 3109
énergie *f*, 1619
énergie *f* au zéro absolu,
4148
énergie *f* consommée, 3013
énergie *f* d'écoulement, 1972
énergie *f* électrique requise
(ou puissance etc...), 1566
énergie *f* potentielle, 1633,
3011
énergie *f* récupérée, 1634
énergie *f* rejetée au
condensateur, 955
énergie *f* solaire, 3515
énergie *f* sonore, 3541
énergie *f* totale, 3867
enfoncé, 1605
enfouir, 583
enlever, 3152
enregistreur *m*, 3181
enrichissement *m*, 1643
enrobage *m*, 791
ensemble *m* brûleur-
chaudière, 456
ensemble *m* d'aubes
directrices, 2206
ensoleillement *m*, 2448
entartrage *m*, 3349
enterrer, 583
enthalpie *f*, 1645
enthalpie *f* spécifique, 3563
entonnoir *m*, 2098
entraînement *m*, 1444
entraînement *m* d'air
(induction), 1649
entraînement *m* direct, 1319
entraîner, 1445, 1648
entrée *f*, 2443, 1654
entrée *f* analogique, 216
entrée *f* d'air, 133
entrée *f* de câble, 601
entrée *f* d'ordinateur, 929
entreposage *m* thermique,
3805
entrepôt *m* frigorifique, 826
entrepôt *m* frigorifique
polyvalent, 2749
entrepreneur *m*, 1012
entreprise *f*, 1644

entreprise *m* principale, 3082
entretien *m*, 2611, 3401
entretoise *f*, 3844
entropie *f*, 1651
enveloppe *f*, 645, 754, 1617, 1656, 2492
enveloppe *f* de vapeur, 3637
enveloppe *f* de ventilateur, 1795
enveloppe *f* du bâtiment, 552
environnement *m*, 1657, 3734
épaisseur *f*, 3821
épaisseur *f* de couche, 2521
épaisseur *f* de tôle, 3421
épaisseur *f* du bord aval, 3876
épreuve *f*, 3776
épreuve *f* de pression, 3069
épreuve *f* d'étanchéité, 2527
épuration *f*, 761, 3124
épuration *f* de gaz, 2116
épuration *f* de l'air, 86
équation *f*, 1670
équation *f* d'état, 1671
équilibrage *m*, 357
équilibrage *m* de pression, 1669, 3049
équilibré, 347
équilibre *m*, 1672
équilibre *m* d'humidité, 2713
équilibre *m* thermique, 3798
équilibrer, 346
équipé, 2897
équipement *m*, 1674
équipement *m* de commande, 1021
équipement *m* de distribution d'air, 3722
équipement *m* de mesure, 2650
équipement *m* de réserve, 3601
équipement *m* de secours, 3601
équipement *m* extérieur, 2865
équipement *m* installé sur site, 1833
équipements *m* de diffusion d'air, 165
équivalent, 1676
équivalent *m* mécanique de la chaleur, 2657
érosion *f*, 1679
erreur *f*, 1680
erreur *f* aléatoire, 3156
espace *m*, 3295
espace *m* d'air, 159
espace *m* de bâtiment, 557
espace *m* entre plafond et faux plafond, 666
espace *m* nuisible, 769
espace *m* nuisible (du cylindre), 767

espacement *m*, 3556
essai *m*, 1757, 3776
essai *m* au bleu de méthylène, 2678
essai *m* de compression, 909
essai *m* de pression, 909, 3069
essai *m* de réception, 25, 876
essai *m* de turbulence, 1545
essai *m* d'étanchéité, 2527
essai *m* DOP, 1391
essai *m* d'opacité, 419
essai *m* d'opacité de fumée, 3499
essai *m* sous vide, 3949
essais *m* de combustion (gaz), 861
essais *m* de ventilateur, 1803
essieu *m*, 321
estimation *f*, 1684
estimer, 1683
étage *m*, 3672
étage *m* de compression, 906
étage *m* de pression, 3067
étain *m*, 3855
étamage *m*, 3856
étanche, 3071, 3097
étanche à la poussière, 1520
étanche à l'air, 169
étanche à l'eau, 4086
étanche au gaz, 2151
étanchéité *f*, 3845
étanchéité *f* à l'air, 170
état *m*, 3614
état *m* d'air, 964
état *m* de régime permanent, 3621
état *m* d'énergie, 1628, 1636
état *m* satisfaisant (du point de vue du fonctionnement), 3397
état *m* stationnaire, 3621
étau *m*, 4005
éteignoir *m* (appareil), 1771
éteindre, 1770
étincelle *f*, 3559
étouffer, 1166
étranglement *m*, 3835, 3837
étrangler, 3836
étrangleur *m*, 3835
étude *f*, 1245
étude *f* de cas, 644
étude *f* du projet, 1252
étude *f* par ordinateur, 927
étuve *f* de séchage, 1469
eupathéoscope *m*, 1685
eutectique *m*, 1686
évacuation *f*, 1338, 1558, 1688
évacuation *f* de scories, 3466
évacuation *f* de vapeur, 1734
évacuation *f* d'eau, 4065
évacuation *f* des boues, 3489
évacuation *f* des cendres, 276

évacuation *f* des fumées, 1980
évacuation *f* des gaz, 1980
évacué, 1687
évacuer, 1339, 1720
évaluation *f*, 1684
évaporateur *m*, 1703
évaporateur *m* à calandre et serpentin, 3423
évaporateur *m* à détente directe, 1465
évaporateur *m* à faisceau tubulaire, 3425
évaporateur *m* à plaques, 2987
évaporateur *m* à tubes verticaux, 3999
évaporateur *m* noyé, 1947
évaporateur *m* sec, 1482
évaporateur *m* sous alimenté, 3613
évaporation *f*, 1690, 3957
évaporation *f* à détente directe, 1321
évaporation *f* causée par chute brusque de pression, 1923
évaporer, 1689
évent *m*, 523, 3985
évier *m*, 3457
examen *m* infra-rouge, 2433
excentrique, 1541
excès *m*, 1708
excès *m* d'air, 1709
excès *m* de température, 1711
excitation *f*, 1715
exécution *f*, 1716
exécution *f* d'un programme informatique, 934
exempt d'huile, 2821
exergie *f*, 1717
exfiltration *f*, 1718
exfiltration *f* d'air, 118
expansion *f*, 1741
expérience *f*, 1757
expérimenter, 1758
explicateur *m*, 1761
exploitation *f* (contractuelle) de chauffage, 1011
explosion *f*, 1760
explusé, 1687
exposé, 1762
exsudat *m*, 1442
extérieur *m*, 1764
externe, 1765
extincteur *m*, 1883, 1771
extracteur *m*, 1773
extraction *f*, 1719
extraire, 1772
extrémité *f*, 3857
extrémité *f* d'aube, 2397

fabrication *f*, 2629

fuite f d'eau, 4076
fumée f, 1978, 2096
fumées f, 3497
fumées f colorées, 833
fusible, 2102
fusible m, 2101
fusible m à cartouche, 639
fusion f, 2103
fusion f de glace, 1227

gain m, 2107
gain m de chaleur, 2240
gain m de chaleur
    occasionnel, 650
gain m modulateur, 2709
gain m par conduction, 967
gaine f, 1492
gaine f de distribution, 1371
gaine f d'évacuation, 1348
gaine f pour thermomètre
    plongeant, 3814
gains m de chaleur variable,
    2401
galvanisation f, 2110
galvanisé, 2109
galvanisé au bain, 2334
galvaniser, 2108
gamme f, 3158
garantie f, 2202
garnissage m, 2555
garnissage m à
    éclaboussement de tour de
    refroidissement, 3572
garnissage m de chaudière,
    3198
garnissage m de tour de
    ruissellement, 1071
garnissage m réfractaire,
    848, 3198
garniture f, 2903
garniture f à soufflet, 397
garniture f de joint, 3366
garniture f d'étanchéité,
    3684
garniture f en caoutchouc,
    3318
gaz m, 2111
gaz m carbonique, 629
gaz m combustible, 2083
gaz m comprimé, 896
gaz m d'échappement, 1642,
    1728
gaz m d'épreuve, 2529
gaz m de four à coke, 813
gaz m de gazogène, 3090
gaz m de ville, 3873
gaz m dissous, 1364
gaz m en bouteille, 496
gaz m extrait, 1728
gaz m naturel, 2760
gaz m rejeté, 1728
gaz m sous pression, 896
gaz m traceur, 3875
gazéification f, 2135
gazeux, 2122

gazière f, 2143
gazogène m (appareil), 2143
gel m, 2071
gelé, 2038
générateur m, 2162
générateur m d'air chaud,
    4037, 4039
générateur m d'air chaud à
    air pulsé, 2021
générateur m de vapeur,
    3632
génie m climatique, 1661
germe m, 2163
germicide, 2165
gestion f de l'énergie, 1630,
    1632
gestion f technique de
    bâtiment, 554
gicleur m, 2438, 2439, 2791,
    3586
gicleur m de brûleur à fuel,
    2813
giffard m (injecteur à
    vapeur), 3636
givrage m, 2074
glace f, 2372
glace f en éclats, 1903
glace f en paillettes, 2030
gond m, 2319
goudron m, 3763
goulotte f, 742
goutte f, 1448
gouttelette f, 1449
gradient m, 2176
gradient m de pression, 3054
gradin m, 3651
grandeur f, 3462
graphique f, 2179
gravité f, 2186
grenier m, 303
grillage m, 2190
grille f, 2180, 2190
grille f à gradins, 3653
grille f de cheminée, 1875
grille f de soufflage, 2869,
    2871
grille f en tablette de fenêtre,
    3446
grille f mécanique ou
    roulante, 3894
grille f oscillante, 3286, 3417,
    4001
grille f perforée, 2935
groupe m, 3925
groupe m compresseur-
    condenseur hermétique,
    3363
groupe m condenseur, 960
groupe m de positions
    binaires, de bits
    consécutifs, 598

halogène, 2210
haute pression f, 2306
hauteur f, 2294

hauteur f d'aspiration, 3697,
    3699
hauteur f de charge, 2217,
    3055
hauteur f de local, 3304
hauteur f de refoulement,
    1343
hauteur f du panache de
    fumée, 2995
hauteur f manométrique,
    3055
hauteur f manométrique
    totale, 3869
hauteur f statique (charge f
    statique), 3615
hélice f, 3099
hélicoïde, 2295
hermétique, 2298
honoraire m, 1811
horloge f à contacts, 3847
hotte f, 2323
hotte f d'aspiration, 3698
hotte f d'évacuation, 1732
hotte f d'extraction, 1732
hotte f de laboratoire, 2097
houille f, 782
hublot m, 564
huile f, 2809
huile f minérale, 2685
humide, 2351, 2711
humidificateur m, 2353, 2355
humidificateur m à
    évaporation, 1701
humidificateur m à vapeur,
    3635
humidificateur m centrifuge,
    3568
humidificateur m d'air, 131
humidificateur m de pièce,
    3298
humidification f, 2352
humidification f d'air, 130
humidifier, 2354
humidité f, 1165, 1172, 2357,
    2712
humidité f relative, 3233
hydraulique, 2362
hydrocarbure m, 2363
hydrostatique, 2364
hygiénique, 2365
hygromètre m, 2366
hygromètre m à aspiration,
    281
hygromètre m à cheveux,
    2209
hygromètre m à point de
    rosée, 1271
hygroscopique, 2367
hygrostat m, 2368
hyperbolique, 2369
hypothermie f, 2370

ignifugeage m, 1888
ignifuger, 1887
imbrûlés m, 858

joint *m* mandriné, 1917
joint *m* mécanique, 2658
joint *m* soudé, 3522, 4104
joint *m* torique, 2860

laboratoire *m*, 2510
laine *f* de roche, 2686
laine *f* de scories, 3470
laine *m* de verre, 2170
laine *f* minérale, 2686
laiton *m*, 512
lame *f*, 2520, 3418
lame *f* d'eau, 4075
lame *f* d'eau de chaudière,
    480
lamelle *f*, 1865
laminaire, 2513
lampe *f*, 2514
lampe *f* de signalisation,
    2410
lampe *f* témoin, 2410
lancement *m*, 3841
laque *f*, 2512
largeur *f* de bande, 366
largeur *f* de batterie, 809
latent, 2516
latte *f*, 384
lattes *f* d'arrimage, 1506
lavabo *m*, 4048
laveur *m*, 3360
laveur *m* d'air, 176
laveur *m* d'air par
    capillarité, 626
lecture *f*, 3168
levier *m*, 2532, 2535
licencié, 2536
liège *m*, 1086
lieu *m* de mise en place, 3459
ligne *f* de saturation, 3342
lignite *m*, 537, 2548, 2549
limite *f*, 497
limite *f* de charge, 2574
limite *f* de puissance, 2574
limiteur *m*, 2551
limiteur *m* de pression, 3057
limiteur *m* de tirage, 1430
liquéfacteur *m*, 2557
liquéfaction *f*, 2556
liquide, 2559
liquide *m*, 2558
liquide *m* comprimé, 897
liquide *m* sous-refroidi, 3688
lit *m*, 394
lit *m* de mâchefer, 774
lit *m* de scories, 3468
litre *m*, 2566
livraison *f*, 1228, 3719
local *m*, 3295
local *m* de centrale, 2976
local *m* de stockage, 3668
local *m* d'essai, 3787
local *m* occupé, 2801
local *m* technique, 1675
logiciel *m*, 936

loi *f* de la conservation de
    l'énergie, 987
lois *f* fondamentales des
    ventilateurs, 1800
longueur *f*, 2531
longueur *f* de batterie, 808
longueur *f* de fentes, 1120
longueur *f* totale, 2881
lubrifiant *m*, 2604
lubrification *f* forcée, 2019
lumière *f*, 2542
lumière *f* de contrôle, 2949
luminance *f*, 2391
luminosité *f*, 2546
lyophilisateur *m*, 2040
lyophilisation *f*, 2041
lyre *f* de dilatation, 1743,
    1749

mâchefer *m*, 773
machine *f*, 1639, 2605
machine *f* auxiliaire, 312
machine *f* frigorifique, 3206
maçonnerie *f*, 527
maçonnerie *f* réfractaire,
    3197
magasin *m*, 3668, 3670
maille *f*, 2670
maintenance *f* corrective,
    1092
maintenance *f* préventive,
    3076
mamelon *m*, 781
manche *m*, 2212
manchette *f*, 3473
manchon *m*, 3473
manchon *m* de
    raccordement, 1112
manchon *m* fileté, 3825
maneton *m*, 1124
manette *f*, 2212
manivelle *f*, 1121
manomètre *m*, 2623, 3053
manomètre *m* à tube incliné,
    2404
manomètre *m* de
    refoulement, 1342
manomètre *m* différentiel,
    1294
manométrique, 2624
manostat *m*, 3068
manovacuomètre *m*, 894
manuel, 2625
manuel *m* de service, 3400
manutention *f* de
    combustible, 2084
marche *f*, 2724, 2852, 3651
marche *f* à vide, 2385
marche *f* continue, 1009
marche *f* d'essai, 3897
masse *f*, 562, 2633, 2634
mastic *m*, 3126
mastic *m* pour joint, 2501
matage *m*, 655
matelas *m* d'air, 102

matériau *m* isolant, 2458
matériel *m* assemblé en
    usine, 1782
matériel *m* de construction,
    555, 2637
matériel *m* de garniture,
    2905
matériel *m* informatique, 928
matière *f*, 2638
matière *f* absorbante, 19
matière *f* adhésive, 49
matière *f* colorante, 1529
matière *f* filtrante, 1859
matière *f* volatile, 4012
matrice *f*, 1281
maximum, 2639
maximun *m* admissible, 2642
mécanique, 2654
mécanique *f* de précision,
    3026
mécanique *f* des fluides, 1990
mécanisme *m* à action
    instantanée, 3506
mélange *m*, 2701
mélange *m* d'air, 2690
mélange *m* de gaz, 2141,
    2692
mélange *m* de liquides, 2563
mélange *m* de vapeurs, 3965
mélange *m* frigorifique, 2050
mélanger, 2662
mélangeur *m*, 436
mélangeur *m* d'air, 139
membrane *f*, 1277, 2664
mémoire *f* d'ordinateur, 930
mémoire *f* morte, 3169
mémoire *f* vive, 3155
mercure *m*, 2667
mesure *f*, 2649, 3168, 3462
mesure *f* de la teneur en
    poussières, 1518
mesure *f* de la teneur en
    suies, 3534
mesure *f* de pression, 3059
mesure *f* d'humidité, 2715
mesures *f* automatiques,
    2723
métabolisme *m*, 2672
métal *m* d'apport pour
    soudure, 1996
météorologie *f*, 2675
méthode *f*, 2677
méthode *f* de calcul à pertes
    de charges réparties
    constantes, 1664
méthode *f* d'essai, 3783
méthode *f* de regain statique,
    3983
méthode *f* du regain statique
    (dimensionnement de
    conduit), 3617
mettre en circuit, 3745
mettre en marche à la
    manivelle, 1122
mettre en place, 3247

mettre hors circuit, 3744
mettre sous forme graphique,
  2178
meuler, 2192
microclimat *m*, 2680
micron *m*, 2681
microprocesseur *m*, 2682
milieu *m* contrôlé, 1025
mine *f*, 2684
minuterie *f* à commande,
  3854
mise *f* à la terre, 1538, 1539
mise *f* au point, 3406
mise *f* en marche à pleine
  charge, 3611
mise *f* en place, 1677, 3406
mise *f* en service, 3612
mise *f* en température, 2262
mitigeur *m*, 436
mnémonique *f*, 2702
mode *m* de chauffage, 2253
mode *m* d'emploi, 2453
modèle *m*, 2703, 2924
modélisation *f*, 2704
Modem, 2705
modulaire, 2706
modulateur-démodulateur,
  2705
modulation *f* complète, 2094
modulé, 2708
module *m*, 2710
module *m* d'affichage de
  données, 1178
moite, 2711
mono-étagé, 3455
mono-pale, 3456
montage *m*, 1677, 2734, 3409
montage *m* en rack, 3134
montage *m* sur chantier,
  3461
monté en usine, 1783, 1784
monter, 3279
monteur *m*, 1897
morceau *m*, 413
moteur *m*, 2726
moteur *m* à air, 140
moteur *m* de registre, 1169,
  1170
moteur *m* de réserve, 3602
moteur *m* de secours, 3602
moteur *m* diesel, 1286, 1287
moteur *m* électrique, 1576
moteur *m* pas à pas, 2731
mousse *f*, 1999
mousser, 2000
mouvement *m*, 2724, 2737
mouvement *m* d'air, 141
moyen, 2643
moyenne *f*, 314
moyeu *m*, 2350
multiplet *m*, 598
multi-zones, 2754
mur *m*, 4023
mur *m* extérieur, 1769

nappe *f* aquifère, 2199
nappe *f* phréatique, 2199
nébulisation *f*, 2004
nébuliser, 2761
neige *m* carbonique, 1467
nettoyage *m*, 761
nettoyage *m* de chaudière,
  459
nettoyage *m* de filtre, 1853
neutre *m* commun, 880
nez *m* de brûleur, 578
niche *f*, 2768
niveau *m* d'accès, 28
niveau *m* d'eau, 4077
niveau *m* d'eau de chaudière,
  481
niveau *m* d'huile, 2823
niveau *m* de liquide, 2561
niveau *f* de pression
  acoustique, 3546
niveau *m* de puissance
  sonore, 3544
niveau *m* phréatique, 2200
niveau *m* sonore, 2589, 2776,
  3543
nivelage *m* de charge, 2573
nombre *m*, 2794
nombre *m* caractéristique,
  709
nombre *m* d'ailettes par
  unité de longueur, 1872
non enterré, 1
non équipé, 2907
non permutable, 2783
norme *f* anglaise, 535
notice *f* d'assemblage, 287
notice *f* de montage, 287
noyau *m*, 1084
nucléaire, 2793
numérique, 2795, 1304
numéro *m*, 2794

obstruction *f*, 2797, 777
obturateur *m*, 3438
obturation *f* par congélation,
  2045
occupant *m*, 2800
odeur *f*, 2803
offre *f*, 3772
olf *m*, 2832
opacité *f*, 2838
opacité *f* à la tache, 1526
opacité *f* de fumée, 3493
opaque, 2840
opération *f*, 2852
opération *f* cyclique, 1156
opération *f* de mise en
  marche, 3607
opération *f* périodique, 1156
opérations *f* en cascade,
  3395
ordinateur *m*, 922
ordinogramme *m*, 1959
ordre *m* de grandeur, 2856
organe *m* d'équilibrage, 355

organigramme *m*, 1959
orientation *f*, 2857
orifice *m*, 2858
orifice *m* d'entrée, 2444
orifice *m* de soufflage, 2871
oscillation *f*, 2862
oscillation *f* de température,
  3771
ossature *f*, 2032
ouïe *f* d'aspiration de
  ventilateur, 1798
outil *m*, 3863
ouverture *f*, 2841, 2858
ouverture *f* de soupape, 3952
ouverture *f* d'entrée, 2444
ouverture *f* pour montage,
  2735
oxydation *f*, 2892
oxyde de carbone *m*, 633
oxygène *m*, 2894
ozone *m*, 2896

pale *f*, 3954
palier *m*, 391, 392
palier *m* à billes, 362
palier *m* de butée, 3842
palier *m* lisse, 3475
palier *m* support auxiliaire,
  2928
panache *m* de fumée, 2993
panache *m* de tour de
  refroidissement, 1074
panne *f*, 518
panne *f* de courant, 3017
panneau *m*, 2911
panneau *m* chauffant, 2255
panneau *m* chauffant mural,
  4029
panneau *m* d'accès, 27
panneau *m* de régulation,
  1029
panneau *m* incorporé, 1606
panneau *m* rayonnant, 3143
panneaux *m* radiants au gaz,
  2146
papier *m* bitumé, 414
papier *m* filtrant, 1860
paramètre *m*, 2919
parefeu, 1890
parevapeur *f*, 792
paroi *f*, 4023
paroi *m* d'échange à grande
  surface de contact, 1763
paroi *f* extérieure, 1769
paroi *f* intérieure, 2476
particule *f*, 2921
particule *f* en suspension,
  3736
particules *f* aériennes, 76
particules *f* de poussières,
  1519
pas *m*, 2970, 3556
pas *m* d'ailettes, 1871
passage *m* d'air, 143
pastille *f*, 582

pause *f*, 3662
peinture *f* bitumeuse, 416
pellicule *f* de film, 1839
percements *m*, 2322
perforation *f*, 2937
perforé, 2933
performance *f*, 2938
performance *f* énergétique
  prévue, 1638
performance *f* pratique, 253
période *f*, 1155, 3846
période *f* de pointe, 2836
période *f* hors pointe, 2806
périodique, 2940
périphérique, 2941
perméabilité *f*, 2942
perte *f*, 2586
perte *f* à la cheminée, 739
perte *f* de chaleur, 2271
perte *f* de chaleur par unité
  de temps, 2273
perte *f* de charge, 2587
perte *f* de charge à l'entrée,
  1655
perte *f* de charge dans un
  tube, 2956
perte *f* de charge de bouche
  d'air, 2191
perte *f* de charge répartie,
  3902
perte *f* d'eau, 4078
perte *f* d'énergie, 1629
perte *f* de froid, 1063
perte *f* de pression, 2587,
  3058
perte *f* de pression
  dynamique, 1534
perte *f* par conduction, 968
perte *f* par évaporation (tour
  de refroidissement), 1692
perte *f* par frottement, 2067
pertes *f* à la cheminée, 1979
pertes *f* d'énergie aux fumées,
  2258
pertes *f* par les fumées, 1979
phase *f*, 2946
pièce *f*, 3295, 413
pièce *f* chaude, 2336
pièce *f* de raccordement
  aéraulique, 1504
pièce *f* de réduction, 587,
  3886
pièce *f* de serrage, 755
pièce *f* de transformation,
  494
pièce *f* détachée, 3558
pièce *f* moulée, 2733
pieds *m* cubes par minute,
  697
pieds *m* cubes par seconde,
  698
pignon *m*, 3898
pilier *m*, 2948
pince *f*, 3861
pince *f* à machefer, 3469

pipe-line *f*, 2958
piquage *m*, 511
piston *m*, 2966
pitting *m*, 2972
place *f* nécessaire, 3555
plafond *m*, 658
plafond *m* chauffant, 2234
plafond *m* diffuseur d'air,
  3987
plafond *m* perforé, 2934
plafond *m* suspendu, 3735
plage *f*, 2412, 3158
plage *f* de réglage, 55
plage *f* de régulation, 1032
plage *f* d'erreur, 1681
plan *m*, 1245, 2973
plan *m* côté, 1310
plan *m* d'atelier, 3430
plan *m* de mise en route, 878
plan *m* d'ensemble, 1079
plancher *m* (bas), 1948
plancher *m* (haut), 664
plaque *f* ailette, 2983
plaque *f* arrière de
  chaudière, 461
plaque *f* avant de foyer,
  2069
plaque *f* chauffante, 2335
plaque *f* d'assise, 381, 395,
  2222
plaque *f* de base, 2222
plaque *f* de brûleur, 570
plaque *f* de chaudière, 474
plaque *f* de chaudière (de
  brûleur), 573
plaque *f* de fond, 1357
plaque *f* de fond de
  chaudière, 461
plaque *f* d'obturation, 2859
plaque *f* (métal), 2981
plaque *f* perforée, 2936
plaque *f* tubulaire, 3903
plaques *f* de remplissage de
  tour de refroidissement,
  2982
plastique *m*, 2977
plastique *m* expansé
  (isolation), 2978
plastique *m* expansé rigide,
  3278
plastiques *m* expansés,
  1740
pleine charge *f*, 2093
plenum *m*, 2989
plénum *m* de mélange, 2694
plinthe *f* chauffante, 376,
  378, 377
plomb *m*, 2523
plonger, 2392
pneumatique, 2996
poche *f* d'air, 1184
poche *f* d'eau, 4080
poêle *m*, 3303
poêle *m* à gaz, 2147
poêle *m* à mazout, 2828

poêle *m* de chauffage, 2259
poids *m*, 4100
poignée *f*, 2212, 2508
point *m*, 2998
point *m* de condensation,
  950
point *m* de congélation, 2052
point *m* de consigne, 3405
point *m* de consigne fixe,
  1900
point *m* de contact, 1001
point *m* de contrôle, 1030
point *m* de déclenchement,
  1153
point *m* de détente, 3061
point *m* d'ébullition, 484
point *m* d'éclair, 1925
point *m* d'enclenchement,
  1149, 3746
point *m* de floc, 1946
point *m* de fusion, 2663
point *m* d'inflammabilité,
  1885
point *m* d'inflammation,
  2388
point *m* de puisage, 1434,
  4079
point *m* de pulvérisation,
  3588
point *m* de référence, 3189
point *m* de rosée, 1269
point *m* de rosée acide, 37
point *m* de saturation, 3344
point *m* fixe, 221
point *m* haut, 2305
pointe *f*, 2925
polluant, 3001
polluant *m*, 1005, 3000
polluant *m* de l'air, 93
polluer, 1006
pollution *f*, 3002
pollution *f* de l'air, 145
polytropique, 3003
pompage *m* en régulation,
  2361
pompe *f*, 3119
pompe *f* à air, 149
pompe *f* à amorçage
  automatique, 3381
pompe *f* à chaleur, 2279
pompe *f* à engrenage, 2160
pompe *f* à piston, 3176
pompe *f* à vide, 3947
pompe *f* auxiliaire, 493
pompe *f* centrifuge, 691
pompe *f* d'alimentation de
  chaudière, 463
pompe *f* de circulation, 749
pompe *f* d'épreuve, 3070
pompe *f* de mélange, 2698
pompe *f* de retour d'eau
  condensée, 948
pompe *f* de surpression, 493
pompe *f* de transfert, 3881
pompe *f* rotative, 3312

sans huile, 2821
sans soudure f, 3368
saturation f, 3340
saumure f, 528
saut m, 2503
sauvegardé, 332
scellé hermétiquement, 2300
scellement m, 3361
sceller, 3362
schéma m, 1433
schéma m de cablage, 4132
schéma m d'écoulement, 1965
schéma m d'écoulement d'air, 122
schéma m de montage, 4136
schéma m explicatif, 757
scorie f, 773
sec, 1457
séchage m, 1468
séchage m à l'air, 114
séchage m d'un bâtiment, 1470
séché à l'air, 111
sécher, 1458
sécheresse f, 1473
sécheur m, 1243, 1436
sécheur m d'air, 113
sécheur m de vapeur, 3629
section f, 3374
section f de chaudière, 478
section f de passage, 1954
section f transversale, 1135, 1136
sectionnel, 3375
sécurité f, 3235
sécurité f d'exploitation, 3334
sédiment m, 3377
semi-automatique, 3382
semiconducteur m, 3383
semi-hermétique, 3384
sensible, 3385
séparateur m, 1604, 3025, 3390, 3892
séparateur m centrifuge, 692
séparateur m de cendres, 278
séparateur m d'eau, 4083
séparateur m de goutellettes, 1451, 2689, 2718
séparateur m de goutelettes (tour de refroidissement), 1439
séparateur m d'huile, 2826
séparateur m de liquide, 2564
séparateur m de poussières, 1525
séparation f, 3388
séparation f d'air, 158
séparer par congélation, 2042
séquence f, 3391
serpentin m, 805, 2953
serpentin m à détente directe, 1321

serpentin m de refroidissement, 1054
serrure f, 2579
service m, 301, 3396
service m de courte durée, 3433
service m d'eau chaude, 2345
servocommande f, 3402
servo-moteur m, 3403
seuil m d'audibilité, 3834
shunt m, 3434
siccité f, 1473
siège m de vanne, 3953
signal m, 3442
signalisation f, 2411
silencieux m, 2740, 2741, 2775, 3444, 3539
silicagel m (gel de silice), 61, 3445
silo m, 2325
simulation f numérique, 935
simultané, 3448
siphon m, 3892
site m, 3459
situation f, 3006
socle m, 375, 381, 395
socle m de chaudière, 472
socle m de moteur, 2727
sol m, 2196
solubilité f, 3526
solution f, 3527
solvant m, 3528
son m, 3536
sonde f, 3086, 3386
sonnette f d'alarme, 179
sortie f, 2867
sortie f analogique, 217
sortie f d'air, 142
sortie f d'ordinateur, 931
souche f de cheminée, 3593
soudage m, 4106
soudage m au chalumeau, 2154
souder, 3521, 4103
soudeur m, 4105
soudure f, 3367, 3520, 3523, 4102
soudure f à l'arc, 261
soudure f à l'étain, 3511
soudure f bout à bout, 592
soudure f électrique par résistance, 3253
soudure f oxy-acétylénique, 2893
soudure f par points, 3579
soufflage m, 449
souffler, 444
soufflet m, 396
soufflet m de dilatation, 1742
souffleur m de suies, 3531
soulever par derrière, 490
soumission f, 3772
soupage f de décharge, 1352
soupape f, 3951

soupape f à action rapide, 3132
soupape f à contre-poids, 4101
soupape f à languettes, 1915
soupape f à membrane, 2666
soupape f à ressort, 3591
soupape f anti-siphon, 3941
soupape f anti-siphon (casse-vide), 243
soupape f d'admission, 2445
soupape f de décharge, 3236
soupape f d'échappement, 1736
soupape f d'évacuation, 1736
soupape f de retenue, 2788
soupape f de sécurité, 1753, 3066
soupape f de sûreté, 3335
source f alternative d'energie, 191
source f de chaleur, 2286
source f de chaleur souterraine, 2346
source f d'énergie, 1635
source f froide, 824
source f sonore, 3552
source f thermique auxiliaire, 313
sources f d'énergie perdue, 4051
sous-refroidissement m, 3689
sous-sol m, 380, 668
sous-station f, 3693
sous-traitant m, 3687
soute f, 566
soute f à charbon, 789
soute f à combustible, 2087
soutirage m, 433
soutirer, 430
spécifique, 3561
spectre m sonore, 3550
spirale f, 3569
stabilité f, 3592
station f, 3618
station f de détente, 3062
stator m, 3619
stérile, 2164
stockage m, 3665
stockage m de gaz, 2148
stockage m de glace, 2373
stockage m d'eau chaude, 2341
stockage m extérieur, 2878
stockage m froid, 1078
stoechiométrique, 3659
store m vénitien, 3984
stratégie f de contrôle, 1033
stratification f, 3677
stratification f de l'air, 163
structure f cellulaire, 673
substance f absorbée, 10
substance f adhésive, 49
substance f diluante, 1307
suie f, 3529

ventilateur *m* basse pression, 2599
ventilateur *m* bifurqué, 406
ventilateur *m* brasseur d'air, 746
ventilateur *m* centrifuge, 689
ventilateur *m* d'extraction d'air, 1727
ventilateur *m* de soufflage, 448
ventilateur *m* de tirage, 1429
ventilateur *m* de toiture, 3294
ventilateur *m* hélico-centrifuge, 2691
ventilateur *m* hélicoïde, 1804, 3100, 3955
ventilateur *m* mural, 4026, 4033
ventilateur *m* tangentiel, 1131, 3754
ventilation *f*, 3988, 3991
ventilation *f* à aubes profilées, 125
ventilation *f* par entraînement, 2499
ventilation *f* par extraction, 1737
ventilation *f* par infiltration, 2428
ventilation *f* transversale, 1139
ventilo-convecteur *m*, 1796
venturi *m*, 3997
vernis *m*, 2512
verre *m*, 2168

verrouillage *m* de porte, 1390
vibration *f*, 4002
vidange *f*, 1416, 1614
vidange *f* de retour, 1419
vidange *m* d'huile, 2817
vide *m*, 3940
vide *m* poussé, 1202, 2315
vie *f*, 2538
vieillissement *m*, 65
vilebrequin *m*, 1125
vis *f*, 3353
vis *f* d'Archimède, 4143
vis *f* sans fin, 4142
viscosimètre *m*, 4006
viscosité *f*, 4007
viscosité *f* dynamique, 1536
viseur *m*, 3441
visser, 3354
vitesse *f*, 3565, 3978
vitesse *f* axiale, 319, 685
vitesse *f* d'air, 160
vitesse *f* de chute, 3981
vitesse *f* de combustion, 863
vitesse *f* de congélation, 2054
vitesse *f* d'échappement de sortie, 2873
vitesse *f* d'écoulement, 1971
vitesse *f* de l'air, 160
vitesse *f* de l'eau, 4090
vitesse *f* de refroidissement, 1067
vitesse *f* de sortie, 1739
vitesse *f* du son, 3551
vitesse *f* frontale, 1777
vitesse *f* périphérique, 3858
vitesse *f* terminale, 3775

vitre *f*, 2910
volant *m* à main, 2213
volant *m* mécanique, 1998
volatil, 4011
volet *m*, 3438
volet *m* anti-retour, 2787
volet *m* de dérivation, 595
volet *m* d'équilibrage, 1668
volet *m* de réglage, 1168
volet *m* de réglage d'air, 154
volet *m* de répartition, 3576
voltmètre *m*, 4014
volume *m*, 617, 4015
volume *m* architectural, 260
volume *m* brut de chambre froide, 2195
volume *m* de combustion, 865
volume *m* d'eau, 4058
volume *m* utile d'une chambre froide, 2765
voyant *m* de brûleur, 580

Watt *m*, 4091
Wattmètre *m*, 4092

zéro *m* absolu, 4147
zonage *m*, 4151
zone *f*, 262, 4149
zone *f* centrale, 1085
zone *f* de confort, 872
zone *f* d'occupation, 2802
zone *f* de travail blanche, 765
zone *f* morte, 1188
zone *f* neutre, 2767

The *AutoFlow* ® cartridge

provides dependable, dynamic control of liquid flow for applications ranging from radiators to industrial processes.

AutoFlow maintains the design flow of fluid without regard to pipe size, orifice or pump speed.

The AutoFlow piston responds instantly to variations in the hydronic system compensating for pressure changes which occur as a result of pump speed or other flow modulations.

The AutoFlow cartridge is removable after installation for inspection and cleaning.

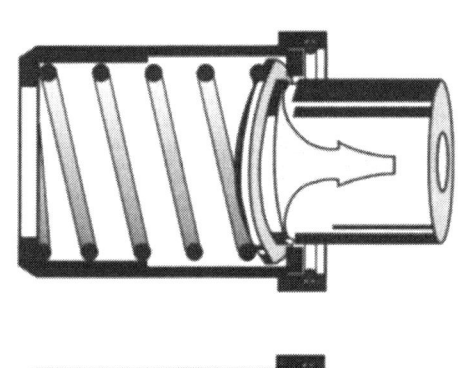

☆ BELOW THE CONTROL RANGE

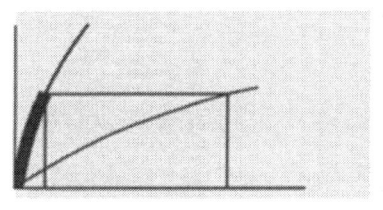

☆ WITHIN CONTROL RANGE

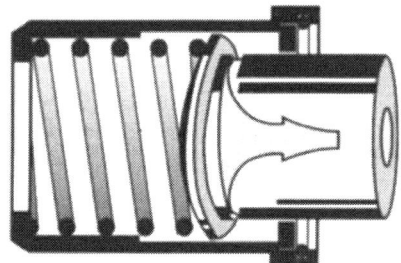

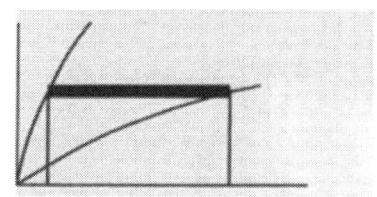

☆ ABOVE CONTROL RANGE

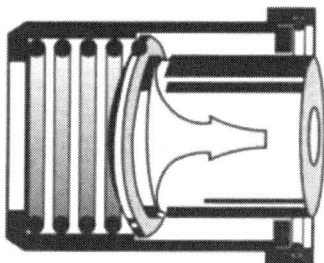

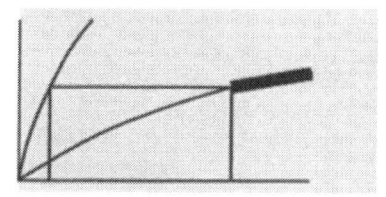

☆ Reference ASHRAE HVAC Systems Manual-page 43.9

CALEFFI
componenti idrotermici

FLOW DESIGN INC.

CALEFFI S.P.A.   28010 FONTANETO D'AGOGNA (NO)   S.S. 229 - ITALY -
TEL. +39 322 8491 R.A.   TELEX 200346 CALEFF I FAX +39 322 - 863305

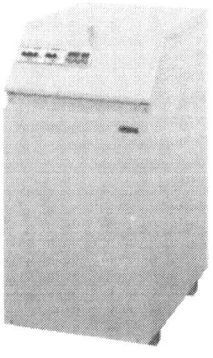
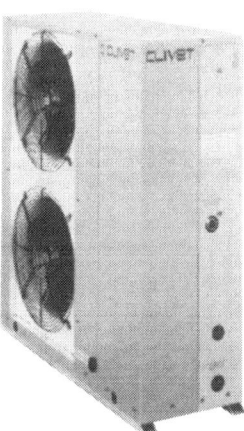

Italiano

abbassamento *m* del punto di rugiada, 1270
abbassamento *m* notturno del set-point del termostato, 2769
abbassamento *m* notturno del termostato, 3820
abilitato, 2536
abrasione *f*, 3
abrasivo *m*, 4
abrasivo, 5
absorbire, 9
accelerazione *f*, 21
accendere, 3745
accensione *f*, 1894, 2386
accensione *f* del bruciatore, 571
accessori *m,pl* del ventilatore, 1792
accessorio *m*, 29, 256
accettazione *f*, 23
acciaio *m*, 3647
acciaio *m* inossidabile, 3595
acclimatazione *f*, 30
accoppiamento *m*, 1109
accoppiamento *m* elastico, 1935
accoppiamento *m* flessibile, 1935
accoppiamento *m* in parallelo, 1110
accoppiamento *m* in serie, 1111, 3394
accoppiamento *m* svasato, 1918
accumulatore *m*, 31
accumulatore *m* di calore, 3806
accumulo *m*, 3665
accumulo *m* di acqua calda, 2341
accumulo *m* di calore, 2288
accumulo *m* di freddo, 1078
accumulo *m* di gas, 2148
accumulo *m* di ghiaccio, 2373
accumulo *m* di polvere, 1512
accumulo *m* termico, 3805
acetilene *m*, 33
acidità *f*, 38
acido *m*, 35
acido *m* carbonico, 631
acqua *f*, 4056

acqua *f* ad alta temperatura, 2314
acqua *f* calda, 2337, 4044
acqua *f* calda a bassa pressione, 2601
acqua *f* di acquedotto, 753
acqua *f* di alimentazione, 1819
acqua *f* di alimentazione della caldaia, 464
acqua *f* di pozzo, 4113
acqua *f* di rabbocco, 2615
acqua *f* di raffreddamento, 1077, 1394
acqua *f* di refrigerazione, 3201
acqua *f* di ricambio, 2615
acqua *f* di ricircolo, 3268
acqua *f* di rifiuto, 4055
acqua *f* di sottosuolo, 2199
acqua *f* distillata, 1368
acqua *f* effluente, 1557
acqua *f* fredda, 827
acqua *f* gelata, 2378
acqua *f* industriale, 2423
acqua *f* non trattata, 3933
acqua *f* potabile, 1440, 3009
acqua *f* refrigerata, 728
acqua *f* salmastra, 505
acqua *f* surriscaldata, 2309
acquisizione *f* dati, 1179
acustica *f*, 39
additivo *m*, 47
addolcimento *m*, 3510
addolcimento *m* dell'acqua, 4085
addolcitore *m* di acqua, 4084
adescamento *m* (di una pompa), 3084
adescare (una pompa), 3081
adesivo *m*, 49
adesivo, 50
adiabatico, 51
adiacenze *f,pl*, 3734
adsorbente *m*, 56
adsorbimento *m*, 57
aerare, 58
aerazione *f*, 59
aerodinamico, 60
aerosol *m*, 62
aerotermo *m* a combustione diretta, 1324
aerotermo *m* a gas, 2127

affidabilità *f*, 3235
aggiunta *f* di acqua distillata, 3864
aggregato *m*, 66
aggressività *f*, 68
aggressivo, 67
agitatore *m*, 70
a gradini, 3654
albero *m*, 3415
albero *m* a gomiti, 1125
albero *m* motore, 1446
alcalinità *f*, 186
alcalino, 185
alcool *m*, 180
aletta *f*, 1865, 3954
aletta *f* direttrice, 161
aletta *f* interna, 2446
aletta *f* longitudinale, 2583
alettato, 1868
alettatura *f* a piastra, 2983
alette *f,pl* di regolazione dell'aria di combustione, 576
alette *f,pl* direzionali, 2206
alfanumerico, 189
alga *f*, 181
algicida *m*, 182
algoritmo *m*, 184
algoritmo *m* di controllo, 1017
alimentare, 1813, 3720
alimentato, 3016
alimentato a carbone, 786
alimentatore *m*, 3660
alimentatore *m* del fuoco (caricatore superiore), 2882
alimentatore *m* (elettrico), 1818
alimentazione *f*, 1812, 3661, 3719
alimentazione *f* a carbone, 788
alimentazione *f* a coclea, 4143
alimentazione *m* aria sotto graticola, 3921
alimentazione *f* del combustibile, 2081
alimentazione *f* della caldaia, 462
alimentazione *f* di acqua, 4066

alimentazione *f* di acqua
  calda, 2343
alimentazione *f* di acqua
  fredda, 828
alimentazione *f* di aria
  comburente, 864
alimentazione *f* di nafta,
  2829
alimentazione *f* di potenza
  del ventilatore, 1799
alimentazione *f* di pressione,
  3050
alimentazione *f* di
  riscaldamento
  centralizzato, 1381
allarme *m*, 178, 4045
allarme *m* di manutenzione,
  2612
allumina *f*, 192
alluminio *m*, 193
alogenato, 2210
alta pressione *f*, 2306
altare *m* del focolare, 1879
altezza *f*, 2294
altezza *f* del locale, 3304
altezza *f* di mandata, 1343
altezza *f* manometrica, 3055
ambiente, 194
ambiente *m*, 1657
ambiente *m* edificato, 557
ambiente *m* esterno, 1764,
  3734
ambiente *m* interno, 2469
amianto *m*, 265
ammoniaca *f*, 200
ammortamento *m*, 202, 1240
ammortare, 203
ammortizzamento *m*, 202
ammortizzare, 203, 1166,
  1167
ammortizzatore *m*, 3429
ammortizzatore *m* di
  vibrazioni, 4003
a monte, 3934
amperaggio *m*, 206
amperometro *m*, 199
ampiezza *f*, 210
ampiezza *f* di banda, 366
amplificare, 209
amplificatore *m*, 208
amplificatore *m* elettronico,
  1596
amplificazione *f*, 207
analisi *f*, 220
analisi *f* dei gas di
  combustione, 861
analisi *f* del caso, 644
analisi *f* del combustibile,
  2078
analisi *f* del gas, 2112
analisi *f* dell'aria, 73
analizzatore *m*, 219
analogico, 212
ancoraggio *m*, 221
ancorare, 222

andamento *m* dei filetti
  fluidi, 1965
andamento *m* del flusso di
  aria, 122
anello *m*, 2585
anello *m* di guarnizione,
  2906
anello *m* di tenuta, 2068,
  2860
anemometro *m*, 223
anemometro *m* a coppa,
  1142
anemometro *m* a filo caldo,
  2347, 3791
anemometro *m* a pale
  orientabili, 1204
anemometro *m* a ventola,
  3277
angolare *m* di ferro, 228
angolo *m*, 226
angolo *m* di accettazione, 24
angolo *m* di efflusso, 229
angolo *m* di uscita, 230
anidride *f* carbonica, 629
anidride *f* solforosa, 3704
anima *f*, 1084
anione *m*, 232
annegato, 1605
anodo *m*, 235
anodo *m* solubile, 3326
anticipo *m* del calore, 2226
anticongelante, 2044
anticorrosivo, 239
antigelo *m*, 240
antincrostante *m*, 242
antiparassitario *m*, 3730
antracite *f*, 236, 783
a pale in avanti, 2024
a pale rovesce, 334
apertura *f*, 2841
apertura *f* di aerazione, 523
apertura *f* di espulsione,
  1733
apertura *f* di immissione di
  aria, 3723
apertura *f* di ingresso, 2444
apertura *f* di uscita, 2871
apertura *f* di valvola, 3952
apertura *f* nell'involucro
  edilizio, 553
appaltatore *m*, 1012
appannatura *f*, 522
apparato *m*, 247
apparecchiatura *f*, 251, 1674
apparecchiatura *f* di campo,
  1831, 1833
apparecchiatura *f* di
  collaudo, 3782
apparecchiatura *f* di
  immissione di aria, 3722
apparecchiatura *f* di Orsat,
  2861
apparecchiatura *f* di riserva,
  3601

apparecchiatura *f* di
  umidificazione, 2355
apparecchio *m*, 247, 3925
apparecchio *m* ad induzione,
  2421
apparecchio *m* a scarico
  libero, 1982
apparecchio *m* di misura,
  2650
apparecchio *m* di
  riscaldamento a
  convezione, 1035
apparecchio *m* di
  riscaldamento ad aria
  calda, 4039
apparecchio *m* monoblocco,
  2901
apparecchio *m* per
  installazione esterna, 2865
apparecchio *m* per la
  respirazione, 524
apparecchio *m* per la
  separazione delle polveri,
  1524
applicazione *f*, 252
apporto *m*, 2107
apporto *m* di calore, 2240
apporto *m* di calore
  accidentale, 2401
apporto *m* di calore
  occasionale, 650
apporto *m* di calore per
  conduzione, 967
apporto *m* di calore solare,
  3516
apprendista *m*, 254
a prova di guasto, 1785
architetto *m*, 258
architetto *m* capoprogetto,
  3096
archivio *m*, 1176
area *f*, 262
area *f* dell'assorbitore, 13
area *f* dell'edificio, 546
area *f* del pavimento, 1949
area *f* di apertura, 246
area *f* di diffusione, 1301
area *f* di flusso, 1954
area *f* frontale, 1776
area *f* frontale di un
  serpentino, 807
area *f* interna (di un edificio
  climatizzato), 1085
area *f* trasversale, 1136
argilla *f* refrattaria, 1880
aria *f*, 71
aria *f* ambiente, 195, 3296
aria *f* ausiliaria, 310
aria *f* calda, 2333, 4034
aria *f* comburente, 866
aria *f* compressa, 895
aria *f* condizionata, 962
aria *f* di combustione, 845
aria *f* di estrazione, 1721

aria *f* di immissione, 2868, 3721
aria *f* di miscela, 2690
aria *f* di raffreddamento, 1052
aria *f* di ricambio, 2613
aria *f* di ricircolo, 3258
aria *f* di rinnovo, 2060
aria *f* di ritorno, 3258
aria *f* espulsa, 1738
aria *f* esterna, 2060, 2864, 2875
aria *f* forzata, 2008
aria *f* fredda, 818
aria *f* interna, 2415
aria *f* interna (ambiente), 2447
aria *f* introdotta, 2460
aria *f* normale, 3597
aria *f* primaria, 3077
aria *f* pulita, 759
aria *f* raffreddata, 1047
aria *f* ricircolata, 3177
aria *f* satura, 3338
aria *f* secca, 1459
aria *f* secondaria, 3369
aria *f* standard, 3597
aria *f* trattata, 3895
aria *f* viziata, 4010
aritmetico, 264
armadio *m* di essiccazione, 1469
arresto *m*, 1151, 3436, 3662
arresto *m* per basso livello, 2594
arresto *m* per pressione differenziale, 3047
arricchimento *m*, 1643
ascensore *m*, 2541
ASCII, 269
asciutto, 1457
asfalto *m*, 280
ASHRAE, 275
a singola aletta, 3456
aspirapolvere *m*, 3943
aspirare, 3694
aspiratore *m*, 284
aspiratore *m* statico, 1116
aspirazione *f*, 282, 3695
asportazione *f* delle scorie, 775
assale *m*, 321
asse *m*, 320, 384
assemblaggio *m*, 286, 1677
assemblare, 285
assemblato in fabbrica, 1783
assiale, 315
assicurazione *f* di qualità, 3129
assoluto, 6
assonometria *f*, 322
assonometria *f* isometrica, 2488
assorbente *m*, 11, 3535
assorbimento *m*, 15

assorbimento *m* del suono, 3537
assorbimento *m* di acqua, 4057
assorbire, 9
assorbitore *m*, 12
asta *f* di collegamento, 3844
ASTM, 288
a tenuta di acqua, 4086
a tenuta di aria, 169
a tenuta di gas, 2151
a tenuta di polvere, 1520
a tenuta stagna, 3071
atmosfera *f*, 290
atmosfera *f* controllata, 1024
atmosfera *f* di riferimento, 291
atmosferico, 292
attacco *m* acido, 36
attacco *m* a flangia, 1912
attacco *m* di mandata, 1956
attacco *m* di servizio, 3398
attacco *m* di vapore, 3627
attacco *m* femmina, 1823
attacco *m* maschio, 2616
attenuazione *f*, 302
attenuazione *f* del rumore, 3538
attenuazione *f* sonora, 2777
attestato *m*, 695
attico *m*, 303
attizzatoio *m*, 2999
attrezzatura *f*, 1674
attrito *m*, 2064
attrito *m* aerodinamico, 1414
attrito *m* del tubo, 2956, 3902
attuale, 43
attuatore *m*, 44
attuatore *m* della serranda, 1169
attuatore *m* di controllo, 1022
aumentare, 3279
autocisterna *f*, 2830, 3756
automatico, 308
automazione *f*, 309
autorità *f*, 306
a valle, 1411
avvertenza *f*, 4045
avvertimento *m*, 4045
avviamento *m*, 3604, 3612
avviamento *m* a pieno carico, 3611
avviamento *m* a vuoto, 2571, 2779
avviare, 1122, 3745
avviatore *m*, 3605
avvitare, 3354
avvolgimenti *m,pl* del motore, 2732
azeotropico, 323
azione *f*, 40
azione *f* batterica, 337
azione *f* di regolazione, 1016

azione *f* flottante, 1941
azione *f* rapida, 3504
azoto *m*, 2771
azzerare, 3247

bacino *m* di lavaggio, 4048
bacino *m* di raffreddamento di acqua a spruzzamento, 3589
bagnare, 1166, 1167
bagno *m*, 382
bagno *m* di olio, 2810
banca *f* dati, 1175
banchina *f*, 2948
banco *m* di prova, 3777, 3788
banda *f* di tolleranza, 1681
banda *f* morta, 1188
banda *f* proporzionale, 3102
barometro *m*, 368
barometro *m* aneroide, 224
barra *f*, 367
barriera *f*, 373
barriera *f* all'umidità, 2717
barriera *f* al vapore, 792, 3960, 3967
barriera *f* isolante, 289, 519
barrotto *m* di griglia, 1876, 2182
basamento *m*, 381, 2727
basamento *m* della caldaia, 472
base *f*, 375
base *f* del camino, 734
bassa *f* pressione, 2595
batteria *f*, 385, 805
batteria *f* a espansione diretta, 1321
batteria *f* a spruzzamento, 3584
batteria *f* di post-riscaldamento, 3227
batteria *f* di raffreddamento, 1049, 1054
batteria *f* di raffreddamento aria, 98
batteria *f* di riscaldamento, 2243
batteria *f* di riscaldamento di aria, 128
batteria *f* di riserva, 333
batteria *f* riscaldante, 2236
battericida, 340
battericida *m*, 341
batterico, 336
batterie *f,pl* gemelle, 3321
batterio *m*, 335
benessere *m*, 867
benessere *m* termico, 3793
bias *m*, 405
biella *f*, 981
biforcazione *f*, 407
bilancia *f*, 3348
bilanciere *m*, 355
bilancio *m* di umidità, 2713

bilancio *m* termico, 2227,
  3792
bimetallo *m*, 408
binario, 411
binario *m* di guida, 2204
bit *m*, 413
bloccaggio *m*, 440
bloccare, 439
blocco *m* di ghiaccio, 2374
blocco *m* frigorifero, 1076
bocca *f* a soffitto, 665
bocca *f* di aspirazione conica
  del ventilatore, 1807
bocca *f* di scarico, 1345
bocca *f* di uscita, 2867
bocchetta *f*, 3217
bocchetta *f* di estrazione di
  aria, 1725
bocchetta *f* di immissione di
  aria a parete, 4031
bocchetta *f* di mandata
  orientabile, 1330
boccola *f*, 585
bolla *f*, 541
bolla *f* di aria, 79
bolla *f* di vapore, 3964
bollente, 2332
bollire, 454
bolometro *m*, 485
bordo *m*, 1547
botola *f*, 3893
bottiglia *f* termica, 3815
bottone *m*, 2508
braccio *m* morto, 1186
brasare, 513, 3521
brasato, 514
brasatura *f*, 2216, 3523
brasatura *f* dolce, 3511
brasatura *f* forte, 517
brillare, 2174
brina *f*, 2071, 2688
British Standard, 535
bromuro *m* di litio, 2565
bronzina *f*, 587
bruciare, 567, 581
bruciatore *m*, 568
bruciatore *m* a coppa, 1143
bruciatore *m* a correnti
  parallele, 2918
bruciatore *m* ad angolo,
  1088
bruciatore *m* ad aria soffiata,
  2013
bruciatore *m* ad
  atomizzazione meccanica,
  2655
bruciatore *m* ad iniezione,
  2442
bruciatore *m* a gas, 2114
bruciatore *m* a gas
  universale, 2115
bruciatore *m* a getti multipli,
  2745
bruciatore *m* a iniezione
  pneumatica, 3042

bruciatore *m* a pentola, 3010
bruciatore *m* a pressione,
  3056
bruciatore *m* a registro, 3219
bruciatore *m* a tazza, 1143
bruciatore *m* atmosferico,
  293
bruciatore *m* a turbolenza,
  4019
bruciatore *m* a
  vaporizzazione, 3958
bruciatore *m* con
  atomizzazione ad aria, 74
bruciatore *m* con
  atomizzazione a vapore,
  3623
bruciatore *m* con
  atomizzazione rotativa,
  3307
bruciatore *m* di nafta, 2812
bruciatore *m* dual-fuel, 1489
bruciatore *m* per
  combustibile diverso da
  quello originario, 1040
bruciatore *m* pressurizzato,
  2013
bruciatore *m* rotativo, 3308
bruciatore *m* universale,
  3931
BTU, 540
buco *m*, 2321
buffer *m*, 543
bulbo *m*, 561
bullone *m*, 486
bullone *m* di fissaggio, 1902
bus *m*, 584
bus *m* di segnale, 3443
busta *f*, 1656
butano *m*, 588
by-pass *m*, 593
byte *m*, 598

cabina *f* di comando, 1018
cabina *f* di controllo, 1018
cabina *f* di verniciatura,
  2909
caduta *f*, 1448
caduta *f* del getto di aria,
  112
caduta *f* di pressione, 3048
calcare *m*, 700
calcestruzzo *m*, 943
calcolatore *m*, 922
calcolo *m*, 603
calcolo *m* con elaboratore,
  934
calcolo *m* delle dispersioni,
  2272
caldaia *f*, 455, 2099
caldaia *f* a bassa pressione,
  2597
caldaia *f* a calore di
  recupero, 4053
caldaia *f* a carbone, 787
caldaia *f* a coke, 811

caldaia *f* a condensazione,
  956
caldaia *f* ad alta pressione,
  2307
caldaia *f* ad elettrodi, 1590
caldaia *f* a doppia
  combustione, 1488
caldaia *f* a gas, 2113
caldaia *f* alimentata da
  tramoggia, 2326
caldaia *f* a nafta, 2819
caldaia *f* a passaggi multipli,
  2748
caldaia *f* a tre passaggi, 3827
caldaia *f* a tubi di acqua,
  4088
caldaia *f* a tubi di fumo,
  1892
caldaia *f* a tubi inclinati,
  2403
caldaia *f* a tubi sovrapposti,
  1138
caldaia *f* a vapore, 3624
caldaia *f* convertibile, 1042
caldaia *f* in acciaio, 3648
caldaia *f* in ghisa, 647
caldaia *f* monoblocco, 2899
caldaia *f* per acqua calda,
  2344
caldaia *f* policombustibile,
  2743
caldaia *f* pressurizzata, 3051
caldaia *f* scozzese, 3351
caldo, 2332
caldo *m*, 4042
calibratura *f*, 604
calibro *m*, 2158
calibro *m* di profondità,
  1311
calore *m*, 2224, 4042
calore *m* di combustione,
  854
calore *m* di fusione, 2104
calore *m* disperso, 4052
calore *m* di
  surriscaldamento, 3708
calore *m* latente, 2517
calore *m* latente di
  evaporazione, 2519
calore *m* latente di fusione,
  2518
calore *m* metabolico, 2672
calore *m* radiante, 3140
calore *m* utile, 3936
caloria *f*, 606
calorifico, 607
calorimetro *m*, 611
calorimetro *m* ambiente,
  3299
cambiamento *m* di stato, 702
camera *f*, 503, 701, 3295
camera *f* a vapore, 3625
camera *f* bianca, 763, 764,
  765
camera *f* calda, 2336

ciclo *m* di servizio, 1527
ciclo *m* di vita, 2539
ciclo *m* frigorifero, 3209
ciclo *m* frigorifero a
  compressione, 901
ciclo *m* frigorifero ad
  eiettore, 3963
ciclo *m* inverso, 3271
ciclone *m*, 1158
ciclo *m* Otto, 2863
ciclo *m* Rankine, 3159
ciclo *m* Stirling, 3657
cilindrata *f*, 2967
cilindrata *f* del compressore,
  916
cilindro *m*, 1159
cilindro *m* diretto, 1317
ciminiera *f*, 2132
cinetico, 2507
cinghia *f*, 399
cinghia *f* trapezoidale, 3976
circolatore *m*, 750
circolazione *f*, 747
circolazione *f* a flusso
  incrociato, 1130
circolazione *f* a gravità, 2187
circolazione *f*
  controcorrente, 1106
circolazione *f* dell'aria, 83
circolazione *f* forzata, 2009
circonferenza *f*, 751
circuito *m*, 744, 2585
circuito *m* a due fili, 3919
circuito *m* a singolo
  conduttore, 3447
circuito *m* del refrigerante,
  3200
circuito *m* di distribuzione,
  1374
circuito *m* di miscela, 2695
circuito *m* di salamoia, 529
circuito *m* di sicurezza, 3327
circuito *m* divergente, 1386
circuito *m* equilibrato, 351
circuito *m* primario, 3078
cisterna *f*, 752, 3755
cisterna *f* di stoccaggio, 3671
classificazione *f*, 758
clima *m*, 771
climatizzazione *f*, 89
clinker *m*, 773
clo *m*, 776
clorofluorocarburo *m*, 741
coassiale, 793
coclea *f* del ventilatore, 1805
codice *m*, 796
codice *m* di calcolo, 926
codice *m* di colore, 832
coefficiente *m*, 797
coefficiente *m* di
  assorbimento, 17
coefficiente *m* di attrito, 2066
coefficiente *m* di attrito
  aerodinamico, 1415

coefficiente *m* di
  conduttanza superficiale,
  803
coefficiente *m* di diffusione,
  1302
coefficiente *m* di dilatazione,
  801
coefficiente *m* di efflusso,
  800, 1340
coefficiente *m* di efflusso
  dell'aria, 108
coefficiente *m* di
  ombreggiamento, 3413
coefficiente *m* di prestazione,
  802
coefficiente *m* di resa, 2939
coefficiente *m* di
  trascinamento, 1415
coefficiente *m* di
  trasmissione, 3889
coefficiente *m* di trasmissione
  del calore, 2290
coefficiente *m* di trasmissione
  del calore liminare, 1841
coefficiente *m* di trasmissione
  della radiazione, 1637
coefficiente *m* di
  utilizzazione, 620
cogenerazione *f*, 804
coibentare, 2455
coibentazione *f*, 2459
coibentazione *f* termica,
  2268
coibente *m*, 2454
coke *m*, 810
coke *m* frantumato, 536
cokificazione *f*, 814
collare *m* di sbocco dei fumi,
  1975
collare *m* di tubazione, 2952
collaudo *m*, 876, 3776
collaudo *m* sottovuoto, 3949
collegamento *m*, 983
collegamento *m* di edificio,
  2348
collegamento *m* di ritorno,
  3261
collegamento *m* in parallelo,
  984
collegamento *m* in serie, 985,
  3394
collegamento *m* parallelo,
  2916
collegamento *m* volante,
  2503
collegare, 975
collettore *m*, 829, 1376,
  2218, 2622
collettore *m* a piastra piana,
  1931
collettore *m* dei fumi, 525
collettore *m* di mandata,
  3727
collettore *m* di ritorno, 3264
collettore *m* di vapore, 3633

collettore *m* solare, 3513
collettore *m* solare a
  concentrazione, 939
colonna *f*, 835, 2608
colonna *f* di acqua, 837,
  4068
colonna *f* di mercurio, 836
colonna *f* di ritorno, 3266
colonna *f* montante, 3280,
  3281
colonna *f* montante di
  mandata, 3728
colorante *m*, 1528
colore *m*, 831
colore *m* di identificazione,
  2383
colore *f* standard, 3598
colorimetrico, 830
colpo *m* di ariete, 4069
comando *m*, 1444
comando *m* a distanza, 3238
comando *m* a orologeria,
  3848
comando *m* a tempo, 3848
comando *m* diretto, 1319
comando *m* manuale, 2627
comburente *m*, 855
combustibile, 842
combustibile *m*, 2076
combustibile *m* fossile, 2025
combustibile *m* povero, 2593
combustibilità *f*, 841
combustione *f*, 844, 1894
combustione *f* a carbone,
  788
combustione *f* a coke, 812
combustione *f* a gas, 2131
combustione *f* a letto fluido,
  1988
combustione *f* a nafta, 2820
combustione *f* a tramoggia,
  2327
combustione *f* completa,
  889, 2932
combustione *f* dall'alto in
  basso, 1412
combustione *f* discendente,
  1412
combustione *f* incompleta,
  2406
commerciale, 873
commissione *f* di collaudo,
  877
commutatore *m*, 705, 1332,
  3860
commutatore *m* a due vie,
  3917
commutazione *f*, 3859
compartimentato, 3375
compatibile *m*, 884
compensato in base alle
  condizioni ambiente, 196
compensatore *m*, 887
compensatore *m* articolato,
  888

condotto *m*, 972, 1492
condotto *m* ascendente, 1976
condotto *m* a spirale, 3571
condotto *m* dei fumi, 1976
condotto *m* di miscela, 2697
condotto *m* orizzontale, 2328
condotto *m* per aria, 115
condotto *m* per aria calda, 4036
conducibilità *f* termica, 799
condurre, 1445
conduttanza *f*, 965
conduttanza *f* liminare, 1842
conduttanza *f* termica, 3794
conduttività *f*, 970
conduttività *f* termica, 799, 3795
conduttore *m*, 971
conduzione *f*, 301, 966
conduzione *f* del calore, 969, 2230
conduzione *f* dell'impianto, 3323
configurazione *f*, 973
confort *m*, 867
confort *m* termico, 3793
congelamento *m*, 2047
congelamento *m* a bassa temperatura, 1201
congelamento *m* in superficie, 2045
congelamento *m* per espansione, 427
congelante *m*, 2038
congelare, 2039, 2042
congelatore *m*, 2043
congelatore *m* a letto fluido, 1989
congelatore *m* a piatto, 2984
congelatore *m* a tunnel, 3905
congelatore *m* per espansione, 426
connettere, 975
connettore *m*, 986
connettore *m* del cavo, 600
consegna *f*, 1228
conservazione *f* dell'energia, 1622
consumo *m*, 998
consumo *m* di aria, 92
consumo *m* di combustibile, 2080
consumo *m* di energia, 1623, 3013
consumo *m* di gas, 2119
consumo *m* di nafta, 2814
consumo *m* di vapore, 3628
consumo *m* per grado giorno, 1216
contaminante *m*, 1005, 3000
contaminante, 3001
contaminante *m* dell'aria, 93
contaminare, 1006
contatore *m*, 2676

contatore *m* di calore, 2274
contatore *m* di condensa, 947
contatore *m* di gas, 2140
contatto *m*, 999
contatto *m* di fiamma, 1906
contattore, 44
contattore *m*, 1000
conteggio *m* di impulsi, 3117
contenitore *m*, 1004
contenuto *m*, 1007
contenuto *m* di acqua, 4059
contenuto *m* di umidità, 2714
contenuto *m* in cenere, 272
contenuto *m* in polvere, 1511
contenuto *m* termico, 2232
contratto *m*, 1010
controflangia *f*, 1104
controllare, 722, 1015
controllo *m*, 721
controllo *m* a configurazione fissa, 1901
controllo *m* adattativo, 46
controllo *m* ad azione rapida, 3505
controllo *m* a logica programmabile, 3094
controllo *m* autoattuato, 3380
controllo *m* con retroazione, 779
controllo *m* continuo, 1008
controllo *m* dei fumi, 3499
controllo *m* del congelamento in superficie, 2046
controllo *m* della qualità dell'aria, 152
controllo *m* dello sbrinamento, 1207
controllo *m* dell'umidità, 2358
controllo *m* di accesso, 26
controllo *m* di combinazione, 839
controllo *m* di flusso, 1957
controllo *m* di inversione, 704
controllo *m* di limite, 2550
controllo *m* di minima e di massima pressione, 1490
controllo *m* di oscillazione, 2696
controllo *m* di potenza, 618
controllo *m* di processo, 3089
controllo *m* di qualità, 3130
controllo *m* di sicurezza di fiamma, 1907
controllo *m* flottante, 1942
controllo *m* predittivo, 3030
controllo *m* previsionale, 237
controllore *m* a passi, 3652

controllore *m* submaster, 3690
controllo *m* sequenziale, 3392
contropendenza *f*, 1108
contropressione *f*, 329
controprova *f*, 1103
controrotazione *f*, 1013
controsoffitto *m*, 3735
conversione *f*, 1039
convertitore *m*, 1041
convertitore *m* di dati, 1177
convettivo, 1036
convettore *m*, 1038
convettore *m* elettrico, 1572
convezione *f*, 1034
convezione *f* forzata, 2011
convezione *f* naturale, 2755
convezione *f* termica, 3796
coperchio *m*, 488, 614, 1113, 2323, 2537
coperchio *m* a tenuta stagna, 4098
copertura *f*, 1114
coppia *f*, 3865
coppia *f* di spunto, 3610
coppia *f* elettrolitica, 1592
copri-calorifero *m*, 3148
coprigiunto *m*, 2502
cordone *m* di amianto, 266
cordone *m* (di saldatura), 3367
cordone *m* di saldatura, 4102
cornice *f*, 2031
corpo *m*, 453
corpo *m* di pompa, 3122
corpo *m* nero, 417
corpo *m* scaldante, 2242, 3147
corrente *m*, 1144
corrente *f*, 1953, 3679
corrente *f* all'avviamento, 3606
corrente *f* alternata, 190
corrente *f* a stantuffo, 3490
corrente *f* a tappi, 3490
corrente *f* (bifase) a bolle, 542
corrente *f* continua, 1316
corrente *f* di aria, 100
corrente *f* di convezione, 1037
corrente *f* di punta, 2835
corrente *f* discendente di aria, 1409
corrente *f* di spunto, 3606
corrente *f* incrociata, 1128
corrente *f* turbolenta, 1544
corrodere, 1093
corrosione *f*, 1094
corrosione *f* puntiforme, 2972
corrosivo, 1098
corsa *f*, 3682
corsa *f* di prova, 3897

differenziale *m*, 1289
differenziale *m* di lavoro,
2846
differenziale *m* di pressione
al turbolatore, 2191
diffusione *f*, 1300
diffusione *f* dell'aria, 107
diffusività *f* termica, 3797
diffuso, 1296
diffusore *m*, 1297
diffusore *m* ad alette, 3956
diffusore *m* a feritoia, 3486
diffusore *m* a soffitto, 660
diffusore *m* lineare, 2552,
3485
digitale, 1304
dilatazione *f*, 1741
dilatazione *f* lineare, 2553
diluente *m*, 1306, 1307
diluizione *f*, 1308
dimensionamento *m* del
canale, 1502
dimensionamento *m* del
tubo, 2961
dimensione *f*, 1309, 3462
dimensione *f* del canale,
1501
dimensione *f* della goccia,
1452
dimensione *f* esterna, 2876
dimensione *f* interna, 2475
dinamico, 1530
dinamometro *m*, 1537
diodo *m* luminoso, 2543
diramazione *f*, 509, 2808
direzione *f*, 1329
direzione *f* del flusso, 1960
di riserva, 332
dischetto *m* di sicurezza, 582
disco *m*, 1337
disco *m* rotante, 3568
disegno *m*, 1433, 2973
disegno *m* assonometrico,
322
disegno *m* dell'eseguito, 267
disegno *m* di archivio, 3180
disegno *m* dimensionale,
1310
disegno *m* di officina, 3430
disegno *m* esecutivo, 4136
disegno *m* esplicativo, 757
disegno *m* in prospettiva,
2945
disegno *m* particolareggiato,
1260
disegno *m* preliminare, 3037
disgelamento *m*, 3789
disidratare, 1224
disidratazione *f*, 1225
disincrostare, 1241, 3151,
3152
disincrostazione *f*, 3153,
3350
disinfezione *f*, 1358
disinserimento *m*, 1151

disinserire, 1152, 3744
disossidante *m*, 1996
disperdimento *m* di calore
per conduzione, 968
disperdimento *m* secondo
progetto, 1250
dispersione *m* dei fumi, 2994
dispersione *f* del getto di
aria, 162
dispersione *f* di calore, 2271
display *m*, 1361
display *m* analogico, 214
disponibilità *f* energetica,
1621
dispositivi *m,pl* di
immissione dell'aria, 165
dispositivo *m*, 1267
dispositivo *m* di accensione a
scintilla, 3560
dispositivo *m* di allarme,
4046
dispositivo *m* di blocco, 2470
dispositivo *m* di
caricamento, 720
dispositivo *m* di
combustione, 846
dispositivo *m* di controllo
dello spegnimento della
fiamma, 1905
dispositivo *m* di estrazione
di aria, 1722
dispositivo *m* di
insonorizzazione, 3539
dispositivo *m* di protezione,
3110
dispositivo *m* di scarico,
1341
dispositivo *m* di sicurezza,
3331
dispositivo *m* di sicurezza
per l'accensione, 2389
dispositivo *m* di sicurezza
per mancata erogazione
del gas, 2124
dispositivo *m* di tenuta, 2903
dispositivo *m* limitatore,
2551
dispositivo *m* per aumentare
le prestazioni, 491
dispositivo *m* per
l'elaborazione dei dati,
1180
dissipazione *f* termica al
condensatore, 955
distacco *m*, 1151
distanza *f*, 1365, 3556
distanza *f* del getto, 2498
distanziatura *f* delle alette,
1872
distillazione *m*, 1367
distributore *m*, 1376, 2218
distribuzione *f*, 1369
distribuzione *f* a pioggia,
2887

distribuzione *f* dal basso,
1372
distribuzione *f* del gas, 2121
distribuzione *f* dell'aria, 109
distribuzione *f* di acqua
calda, 2340
distribuzione *f* in canali,
1494
domanda *f*, 1232
domestico, 1388
doppia interruzione *f*, 1396
doppia interruzione *f* di
linea, 1400
dosaggio *m*, 1392
dosare, 1393
drenaggio *m*, 1418
duplex *m*, 1507
durabilità *f*, 1508, 3397
durata *f*, 1509, 1554, 2538
durata *f* del ciclo, 1157
durata *f* del dì, 1182
durata *f* della combustione,
852
durata *f* di servizio, 2851
durezza *f* dell'acqua, 2215,
4070
duttilità *f*, 1496

ebollizione *f*, 1540
ebollizione *m* a film, 1840
eccentrico, 1541
eccesso *m*, 1708
eccesso *m* di aria, 1709
eccitazione *f*, 1715
economia *f*, 1543
economia *f* di energia, 3015
economia *f* energetica, 1625
economizzatore *m*, 1542
economizzatore *m* del
compressore, 917
economizzatore *m* dell'acqua
di alimentazione, 1820
edificio *m*, 545
effettivo, 1550
effetto *m*, 1549
effetto *m* barometrico, 371
effetto *m* camino, 737, 3594
effetto *m* deumidificante,
1223
effetto *m* di raffreddamento,
1058
effetto *m* frigorifero, 1059,
3205
effetto *m* frigorigeno
specifico, 1626
effetto *m* Peltier, 2931
effetto *m* refrigerante, 731
effetto *m* utile di una pompa
di calore, 918
efficienza *f*, 1556
efficienza *f* della
depolverazione, 1522
efficienza *f* di filtrazione,
1861

efficienza *f* di filtrazione
(metodo opacimetrico),
296
efficienza *f* di separazione,
3389
efflusso *m*, 1558
eiettore *m*, 1559
eiettore *m* di aria a getto di
vapore, 3638
elaboratore *m*, 922
elasticità *f*, 1561
elastomero *m*, 1562
elemento *m*, 1602, 3374
elemento *m* bimetallico, 409
elemento *m* da costruzione,
549
elemento *m* della caldaia,
478
elemento *m* di controllo
della potenza, 1031
elemento *m* di costruzione,
995
elemento *m* di potenza
azionato a vapore, 3961
elemento *m* filtrante, 1857
elemento *m* raffreddante,
1060
elemento *m* riscaldante, 2248
elemento *m* rivelatore, 1262
elemento *m* sensibile, 3386
elemento *m* strutturale, 3683
elettricità *f*, 1575
elettrico, 1564, 1565
elettrodo *m*, 1589
elettrodo *m* di accensione,
2387
elettrodo *m* di saldatura,
4110
elettrolitico, 1591
elettromagnete *m*, 1593
elettronica *f*, 1598
elettronico, 1595
elettro-pneumatico, 1599
elica *f*, 3099
elicoidale, 2295
eliminare gli errori di un
codice di calcolo, 1190
emettenza *f*, 1613
emettenza *f* termica
direzionale, 1331
emissione *f*, 1611
emissività *f*, 1612
emulsione *f*, 1615
endotermico, 1618
energia *f*, 1619, 3012
energia *f* del punto di
riferimento, 4148
energia *f* del suono, 3541
energia *m* elettrica assorbita,
1578
energia *f* potenziale, 3011
energia *f* recuperata, 1634
energia *f* solare, 3515
energia *f* totale, 3867
energy *f* audit, 1620

entalpia *f*, 1645
entalpia *f* massica, 3563
entalpia *f* specifica, 3563
entrata *f*, 2157, 2443
entropia *f*, 1651
equalizzare, 1665
equalizzatore *m*, 1666
equalizzatore *m* di flusso,
1961
equazione *f*, 1670
equazione *f* di stato, 1671
equilibrare, 346, 1665
equilibrato, 347
equilibratura *f*, 357
equilibratura *f* della
pressione, 3049
equilibrio *m*, 1672
equilibrio *m* termico, 3798
equivalente, 1676
equivalente *m* meccanico del
calore, 2657
ermetico, 2298
erodere, 2
erogazione *f* di aria, 164
erosione *f*, 1679
errore *m*, 1680
errore *m* casuale, 3156
escursione *f* termica
giornaliera, 1162
esecuzione *f*, 1716
esercitare azione abrasiva, 2
esercizio *m*, 2852
esfiltrazione *f* dell'aria, 118
espansione *f*, 1741
espansione *f* secca (diretta),
1464
esperimento *m*, 1757
esplosione *f*, 1760
esplosione *f* di gas, 2123
esponente *m*, 1761
esposto, 1762
espulsione *f*, 1719
espulsione *f* dell'aria, 119
essere incandescente, 2174
essiccamento *m*, 1470
essiccamento *m* secondario,
3372
essiccante *m*, 1243
essiccante *m* a biossido di
silicio, 61
essiccante *m* ad allumina
attivata, 41
essiccare, 1458
essiccato in aria, 111
essiccatore *m*, 1226, 1436
essiccatore *m* a vapore, 3629
essiccatore *m* del filtro, 1855
essiccatore *m* di aria, 113
essiccazione *f*, 1244, 1468
essiccazione *f* dell'aria, 114
essiccazione *f* per
congelamento, 2041
esterno, 1765
estinguere, 1770
estintore *m*, 1771, 1883

estrarre, 1772
estrattore *m*, 1773
estrazione *f*, 1719
estrazione *f* delle polveri,
1515
estremità *f* della girante,
2397
eupateoscopio *m*, 1685
eutettico, 1686
evacuare, 1339
evacuazione *f*, 1338, 1353,
1688, 3241
evaporare, 1689
evaporatore *m*, 1703
evaporatore *m* ad espansione
secca, 1465
evaporatore *m* a fascio
tubiero, 3423
evaporatore *m* allagato, 1947
evaporatore *m* a piastra,
2987
evaporatore *m* a secco, 1482
evaporatore *m* a tubi e
mantello, 3425
evaporatore *m* (batteria),
1691
evaporatore *m*
sottoalimentato, 3613
evaporatore *m* verticale,
3999
evaporazione *f*, 1690
exergia *f*, 1717
exfiltrazione *f*, 1718

fabbisogno *m*, 1232
fabbisogno *m* di acqua
calda, 2339
fabbisogno *m* di aria, 156
fabbisogno *m* di calore, 2283
fabbisogno *m* di energia,
3021
fabbrica *f*, 1781, 4140
fabbricare, 2630
fabbricazione *f*, 2629
facciata *f*, 1775
falda *f* acquifera, 2199
fango *m*, 3487
fardaggio *m*, 1506
far girare, 1122
fascio *m*, 565
fascio *m* tubiero, 3901
fase *f*, 2946
fase *f* di compressione (di un
ciclo), 907
fase *m* di espansione, 1751
fase *f* di scarico, 1350
fase *f* di scarico del
compressore, 915
fattore *m*, 1779
fattore *m* di ammortamento,
1199
fattore *m* di carico, 1584,
2570
fattore *m* di compressione,
903

fattore *m* di comprimibilità, 899
fattore *m* di energia, 1627
fattore *m* di flusso, 1962
fattore *m* di forma, 227, 974
fattore *m* di incrostazione, 2027
fattore *m* di opacità, 2839
fattore *m* di potenza, 1582
fattore *m* di rugosità, 3317
fattore *m* di sicurezza, 1780, 3332
fattore *m* di simultaneità, 1384
fattore *m* di utilizzazione, 3938
feedback *m*, 1814
feltro *m*, 1822
fermare, 3744
fermata *f*, 3436, 3664
ferro *m*, 2483
fessura *f*, 1118, 3484
fessura *f* capillare, 2208
fessura *f* della finestra, 4125
fessurazione *f*, 1119
fiamma *f*, 1904
fibra *f*, 1828
fibra *f* di vetro, 2169
fibra *f* ottica, 1827
fibra *f* sintetica, 3749
figura *f* di merito, 1835
filettare, 3758
filettatura *f*, 3762
filettatura *f* conica, 3760
filettatura *f* femmina, 1824
filetto *m*, 3822
filetto *m* del tubo, 2964
filetto *m* maschio, 2617, 3822
filiera *f*, 1281, 3823
filo *m*, 4131
filtrare, 1847
filtrazione *f*, 1864
filtro *m*, 1846
filtro *m* a carbone attivo, 710
filtro *m* a cartuccia, 670
filtro *m* ad alta efficienza, 2303
filtro *m* ad aspirazione, 3703
filtro *m* a manica, 1774, 3476
filtro *m* antigrasso, 2189
filtro *m* a perdere, 1363
filtro *m* a rete, 3675
filtro *m* a rullo, 3288
filtro *m* a sacco, 345
filtro *m* a spazzola, 538
filtro *m* assoluto, 7
filtro *m* a strato secco, 1472
filtro *m* da polvere, 1516
filtro *m* dei fanghi, 3488
filtro *m* del combustibile, 2082
filtro *m* della linea di aspirazione, 3700
filtro *m* deodorante, 2804

filtro *m* di acqua, 4067
filtro *m* di aria a bagno di olio, 2811
filtro *m* di aria a secco, 1466
filtro *m* di aria di estrazione, 1723
filtro *m* di vapore, 3631
filtro *m* elettrostatico, 1600
filtro *m* inerziale, 2395
filtro *m* in materiale fibroso, 1830
filtro *m* metallico, 2673
filtro *m* per aria, 120
filtro *m* per gas, 2125
filtro *m* rotativo, 3310
filtro *m* viscoso, 4008
finestra *f*, 4123
finestra *f* a doppi vetri, 1398
finestra *f* a vetro semplice, 3452
finestratura *f*, 1825
fini *m,pl* di carbone, 3464
fiocchi *m,pl* di fuliggine, 3503
fissaggio *m*, 1808
fissare, 222, 3404
flangia *f*, 1911
flangia *f* a collo saldato, 4111
flangia *f* a saldare, 4109
flangia *f* cieca, 438, 3663
flangia *f* con giunzione a maschio e femmina, 3862
flangia *f* del bruciatore, 570
flangia *f* di boccaporto, 883
flangia *f* di collegamento, 978
flangia *f* filettata, 3824
flangia *f* folle, 3483
flangia *f* non forata, 422
flessibile, 1934
flessibilità *f*, 1933
fluidica *f*, 1987
fluido *m*, 1986
fluido *m* refrigerante secondario, 3371
fluido *m* scaldante, 2249
fluorocarburo *m*, 1991
flusso *m*, 1953, 1996
flusso *m* anulare, 234
flusso *m* a pistone, 2968
flusso *m* bifase, 3914
flusso *m* bilanciato, 350
flusso *m* contrapposto, 1105
flusso *m* di aria, 121
flusso *m* di aria stratificato, 3678
flusso *m* di calore, 2239, 3162
flusso *m* di calore specifico, 3564
flusso *m* disperso, 1360
flusso *m* di transizione, 3885
flusso *m* incrociato, 1129
flusso *m* laminare, 3680

flusso *m* parallelo, 2917
flusso *m* rotazionale, 3314
flusso *m* transitorio di calore, 3883
flusso *m* turbolento, 3911
flusso *m* variabile, 3971
flusso *m* verso l'esterno, 2879
flusso *m* viscoso, 4009
flusso *m* vorticoso, 4020
focolare *m*, 1877, 2100, 2133, 2222, 2223
focolare *m* a caricamento dal basso, 3920
focolare *m* a combustibili multipli, 2744
focolare *m* a griglia meccanica, 699
foglio *m*, 3418
fogna *f*, 3411
fondamenta *f,pl*, 2028
fondazione *f*, 2028
fondere, 581, 2662
fonte *f* di energia alternativa, 191
fonte *f* fredda, 824
fonte *f* sonora, 3552
fonti *m,pl* di energia di recupero, 4051
fori *m,pl* passanti, 2322
formazione *f* di alghe, 183
formazione *f* di brina, 2074
formazione *f* di film, 1844
formazione *f* di ghiaccio, 2074, 2380
formazione *f* di incrostazione, 3349
formazione *f* di nebbia, 2004
formazione *f* di schiuma in un generatore di vapore, 466
formazione *f* di scorie, 3467
forme *f,pl* di energia, 2023
fornire, 3720
fornitura *f*, 1228, 3719
fornitura *m* di calore, 2266
fornitura *f* di carbone, 784
fornitura *f* di gas, 2150
forno *m*, 2099
forno *m* a gas, 2133, 2149
foro *m*, 2321, 2621
foro *m* di ispezione, 2449, 2929
foro *m* di montaggio, 2735
foro *m* di passaggio, 2211
foro *m* filettato, 3761
forza *f*, 2007
forza *f* ascensionale, 268
forza *f* centrifuga, 690
forza *f* del getto, 2496
forza *f* elettromotrice, 1594
forza *f* motrice, 2725
foschia *f*, 2688
fossa *f*, 2969
fossa *f* del ceneraio, 274

frazione *f* dello spazio
morto, 768
freddo *m*, 816
freddo, 817
freon *m*, 2056
frequenza *f*, 2057
frequenza *f* di vibrazione,
4004
frequenza *f* naturale, 2759
frequenza *f* portante, 636
freschezza *f*, 2062
fresco, 2059
fuga *f* di aria, 137
fuliggine *f*, 2193, 3529
fumi *m,pl*, 3497
fumo *m*, 2096
fumo *m* colorato, 833
funzionamento *m*, 2852
funzionamento *m* continuo,
1009
funzionamento *m* di breve
durata, 3433
funzione *f* di controllo, 1023
funzione *f* di gestione
dell'energia, 1632
funzione *f* di trasferimento,
3880
fuochista *m*, 3660
fuoco *m*, 1873
fuori terra, 1
fusibile *m*, 2101
fusibile, 2102
fusibile *m* a cartuccia, 639
fusione *f*, 2103, 2733

galleggiamento *m*, 1952
galleggiante *m*, 1940
galleria *f* del vento, 4128
gamma *f*, 3158
gancio *m*, 2324
gara *f* di appalto, 605
garanzia *f*, 2202
gas *m*, 2111
gas *m* combustibile, 2083
gas *m* compresso, 896
gas *m* compresso (in
bombole), 496
gas *m* di accensione, 1923
gas *m* di città, 3090, 3873
gas *m* di cokeria, 813
gas *m* di combustione, 1978
gas *m* di scarico, 1728
gas *m* di scarico del motore,
1642
gas *m* disciolto, 1364
gasificazione *m*, 2135
gas *m* naturale, 2760
gas *m* povero, 2529
gassoso, 2122
gas *m* tracciante, 3875
gel *m* di silice, 3445
gelo *m*, 2071
generatore *m*, 2162
generatore *m* ad acqua
calda, 2338

generatore *m* di aria cadla,
4037
generatore *m* di aria calda a
tiraggio forzato, 2021
generatore *m* di gas, 2143
generatore *m* di vapore, 467,
3632
generazione *f* di potenza
elettrica, 1583
generazione *f* di vapore di
flash, 1928
generazione *f* elettrica di
base, 379
generazione *f* locale, 2837
germe *m*, 2163
germicida, 2165
gesso *m*, 700
gestione *f* dell'energia, 1630
getto *m*, 2494, 2733
getto *m* di aria, 135
ghiacciaia *f*, 2375
ghiaccio *m*, 2372
ghiaccio *m* frammentato,
2030
ghiaccio *m* in cubetti, 1141
ghiaccio *m* in scaglie, 1903
ghiaccio *m* secco, 1467
ghisa *f*, 646
ghisa *f* malleabile, 2618
gioco *m*, 766
girante *f* del ventilatore,
1804
giro *m*, 3276
gittata *f*, 3841
gittata *f* dell'aria, 168
giunto *m*, 1109, 2500, 3361
giunto *m* a bicchiere, 3567
giunto *m* a compressione,
904
giunto *m* a sfera, 364
giunto *m* brasato, 515
giunto *m* di collegamento,
979
giunto *m* di dilatazione, 885,
1748
giunto *m* di testa, 591
giunto *m* saldato, 3367,
4102, 4104
giunto *m* svasato, 1917
giunzione *f* di testa, 592
giunzione *f* saldata, 3522
globotermometro *m*, 2172
goccia *f*, 1442, 1448
gocciolare, 1453
gocciolatoio *m*, 1443
gocciolina *f*, 1449
gola *f* del bruciatore, 578
gomito *m*, 403, 502, 1563
gomito *m* a saldare, 4107,
4108
gomito *m* corrugato, 1099,
1100
gomito *m* di canale, 1495
gradi *m,pl* di libertà, 1219
gradiente *m*, 2176

gradiente *m* di pressione,
3054
gradino *m*, 3651
grado *m*, 1214
grado *m* di contatto, 1002
grado *m* di purezza, 1217
grado *m* di saturazione, 3343
grado *m* di
surriscaldamento, 1218
grado-giorno *m*, 1215
grado-giorno *m* di
raffreddamento, 1056
grado-giorno *m* di
riscaldamento, 2246
grafica *f*, 2179
grafica *f* animata, 1532
grafica *f* dinamica, 1532
grafico *m*, 2177
grandezza *f*, 3462
grandezza *f* della particella,
2922
gravità *f*, 2186
grembiale *m*, 742
griglia *f*, 2180, 2190
griglia *f* a barrotti, 3446
griglia *f* a barrotti del
focolare, 1875
griglia *f* ad alette fisse, 1899
griglia *f* a gradini, 3653
griglia *f* a lamiera forata,
2935
griglia *f* a scuotimento, 3417
griglia *f* a sezione variabile,
1291
griglia *f* di aerazione, 2591
griglia *f* di focolare, 2069
griglia *f* di uscita, 2869
griglia *f* mobile, 3894
griglia *f* montante, 3280
griglia *f* oscillante, 3286
griglia *f* vibrante, 4001
guadagno *m*, 2107
guadagno *m* modulante,
2709
guarnizione *f*, 2136, 2500,
2903, 3361
guarnizione *f* dell'albero,
3416
guarnizione *f* di gomma,
3318
guarnizione *f* di piombo,
2524
guarnizione *f* di tenuta, 3366
guasto *m*, 518
guasto *m* elettrico, 1568
guidare, 1445

hardware *m* di un
elaboratore, 928

idraulico, 2362
idrocarburo *m*, 2363
idrostatico, 2364
igienico, 2365
ignifugare, 1887

integrato, 2462
integratore *m*, 2463
intelaiatura *f*, 2033
intensità *f*, 2464
intensità *f* luminosa, 2465,
  2546
intensità *f* sonora, 2588
intercambiabilità *f*, 2466
interfaccia *f* operatore, 2854
interferenza *f*, 2468
intermittente, 2473
interno, 2474
interrare, 583
interrompere, 3744
interrupt *m*, 2477
interrutore *m* di avviamento,
  3608
interruttore *m*, 3741
interruttore *m* a doppia
  posizione, 1408
interruttore *m* a galleggiante,
  1943
interruttore *m* a mercurio,
  2668
interruttore *m* bipolare, 1403
interruttore *m* del computer,
  938
interruttore *m* orario, 3854
interruttore *m* selettore, 3378
interruzione *f*, 1151, 3664
interruzione *f*
  dell'erogazione di energia
  elettrica, 3017
interruzione *f* di pressione,
  3045
interruzione *f* di un circuito,
  745
intervallo *m*, 2478
intervallo *m* di indicazione,
  2412
intervallo *m* di richiesta di
  potenza elettrica, 1579
intervento *m*, 1148
inumidire, 1166, 1167
invecchiamento *m*, 65
inversione *f*, 703
inviluppo *m* della
  distribuzione dell'aria, 110
involucro *m*, 1617, 1656
involucro *m* dell'edificio, 552
ione *m*, 2480
ionizzazione *f*, 2482
iperbolico, 2369
ipotermia *f*, 2370
irradianza *f* di fondo, 328
irradianza *f* diretta, 389
irraggiamento *m*, 2484, 3146,
  3801
irraggiamento *m* solare, 3518
isentalpico, 2485
isobara *f*, 2486
isolamento *m*, 2459
isolamento *m* termico, 2268,
  3799

isolamento *m* termico a
  cappotto, 421
isolamento *m* termico
  cellulare, 674
isolamento *m* termico di
  schiuma formata in loco,
  2001
isolamento *m* termico in
  blocchi, 442
isolamento *m* termico in
  gomma cellulare, 672
isolamento *m* termico in
  pannelli rigidi, 452
isolamento *m* termico
  riflettente, 3192
isolante *m*, 2454
isolante *m* termico in
  pannelli di sughero, 1087
isolare, 2455, 2456
isoterma *f*, 2489
isotermico, 2490
isotermo, 2490
isotopo *m*, 2491
ispettore *m* dei lavori, 3460
istruzioni *f,pl* di servizio,
  3399
istruzioni *f,pl* per il
  montaggio, 287, 2736
istruzioni *f,pl* per l'uso,
  2453, 2847

katatermometro *m*, 2505

laboratorio *m*, 2510
lama *f* di acqua, 4075
lama *f* di aria, 101
lamiera *f* di metallo, 3419
lamiera *f* forata, 2936
laminare, 2513
lampada *f*, 2514
lampada *f* di segnalazione,
  2410
lana *f* di roccia, 2686
lana *f* di scorie, 3470
lana *f* di vetro, 2170
lancio *m*, 3841
lancio *m* dell'aria, 168
larghezza *f* di un serpentino,
  809
lastra *f* di metallo, 3419
latente, 2516
lato *m* longitudinale, 2584
lavare, 1992
lavatore *m*, 3360
lavatore *m* di aria, 176
lavatore *m* (umidificatore) di
  aria a pacco evaporante,
  626
lavorabile, 4135
lavoro *m* di precisione, 3026
lavoro *m* di spostamento,
  1972
LED *m*, 2543
lega *f*, 188
lega *f* per saldatura, 3520

legge *f* antinquinamento, 760
legge *f* di conservazione
  dell'energia, 987
leggi *f,pl* dei ventilatori, 1800
letto *m*, 394
letto *m* di scoria, 774
lettura *f*, 3168
leva *f*, 2535
lignite *f*, 537, 2548, 2549
limitatore *m* del tiraggio,
  1430
limitatore *m* di pressione,
  3057
limitazione *f* della pressione
  dell'olio, 2825
limitazione *f* della
  temperatura dell'olio, 2831
limitazione *f* di pressione,
  3045
limitazione *f* di sicurezza
  della pressione, 2311
limitazione *f* di sicurezza
  della pressione minima,
  2602
limitazione *f* di sicurezza
  della temperatura di
  scarico, 2302
limite *m*, 497
limite *m* di carico, 2574
limiti *m,pl* di impiego, 253
linea *f* di alimentazione,
  3726
linea *f* di alimentazione di
  potenza, 3018
linea *f* di saturazione, 3342
linea *f* di scarico, 1344
liofilizzazione *f*, 2041
liofilizzazione *f* a pressione
  ambiente, 297
liquefattore *m*, 2557
liquefazione *f*, 2556
liquido *m*, 2558
liquido, 2559
liquido *m* in pressione, 897
liquido *m* sottoraffreddato,
  3688
litro *m*, 2566
livella *f*, 2532
livellamento *m* del carico,
  2573
livellamento *m* del carico
  elettrico, 1585
livello *m*, 2532
livello *m* dell'acqua, 4077
livello *m* dell'acqua in una
  caldaia, 481
livello *m* del liquido, 2561
livello *m* dell'olio, 2823
livello *m* di accesso, 28
livello *m* di falda sotterranea
  (freatica), 2200
livello *m* di intensità sonora,
  2589
livello *m* di potenza sonora,
  3544

livello *m* di pressione
  sonora, 3546
livello *m* energetico, 1628
livello *m* sonoro, 2776, 3543
locale *m* caldaia, 470, 477
lubrificante *m*, 2604
lubrificazione *f* forzata, 2019
luce *f*, 2542
luce *f* di allarme, 4047
luce *f* netta, 766
lunghezza *f*, 2531
lunghezza *f* della maglia,
  2671
lunghezza *f* di incrinatura,
  1120
lunghezza *f* di un serpentino,
  808
lunghezza *f* totale, 2881
luogo *m*, 3459

macchina *f*, 1639, 2605
macchina *f* di riserva, 312
macchina *f* frigorifera, 3206
macchinario *m*, 1674
macchinario *m* per il
  congelamento, 2049
macinare, 2192
magazzino *m*, 3668, 3670
magazzino *m* frigorifero,
  825, 2313
magazzino *m* frigorifero con
  diverse temperature di
  stoccaggio, 2749
magazzino *m* frigorifero di
  distribuzione, 1359
maglia *f*, 2670
mandata *f* dell'aria, 164
manichetta *f*, 2330
manicotto *m*, 1112, 3473,
  3509
manicotto *m* filettato, 781,
  3825
maniglia *f*, 2212
manometrico, 2624
manometro *m*, 1426, 2623,
  3053
manometro *m* a tubi
  inclinati, 2404
manometro *m* Bourdon, 500
manometro *m* differenziale,
  1294
manometro *m* di scarico,
  1342
manovella *f*, 1121
mantello *m*, 645, 2492
mantello *m* della caldaia,
  458
mantello *m* isolante, 2457
manuale, 2625
manuale *m* di servizio, 3400
manufatturare, 2630
manutenzione *f*, 2611, 3401
manutenzione *f* correttiva,
  1092

manutenzione *f* preventiva,
  3076
marcia *f*, 2724
marcia *f* a vuoto, 2385
marmitta *f*, 2741
maschiare, 3758
massa *f*, 562, 2633
massimo, 2639
massimo *m* permissibile,
  2642
mastice *m*, 3126
mastice *m* per giunzione,
  2501
materia *f*, 2638
materiale *m* coibente, 2267
materiale *m* da costruzione,
  555, 2637
materiale *m* di guarnizione,
  2905
materiale *m* di riempimento,
  1836
materiale *m* filtrante, 1859
materiale *m* isolante, 2458
materiale *m* isolante sciolto,
  1838
materia *f* volatile, 4012
materie *f,pl* plastiche
  espanse, 1740
matrice *f*, 1281
mattone *m*, 526
mattone *m* di ventilazione,
  78
mattonella *f* di carbone, 534
mattone *m* refrattario, 1878
meccanica *f* dei fluidi, 1990
meccanico, 2654
meccanismo *m* ad azione
  istantanea, 3506
meccanismo *m* a scatto, 3506
medio, 314, 2643
membrana *f*, 1277, 2664
memoria *f* ad accesso
  casuale, 3155
memoria *f* di sola lettura,
  3169
memoria *f* di un elaboratore,
  930, 937
memoria *f* ram, 3155
mensola *f*, 504
mensola *f* per tubo, 2957
mercantile *m*, 634
mercantile *m* refrigerato,
  727, 3203
mercurio *m*, 2667
messa *f* a punto, 3406
messa *f* a regime, 2262
messa *f* a terra, 1539, 1569,
  2197
messa *f* in funzione, 876
metallo *m* di apporto per
  saldatura, 1996
meteorologia *f*, 2675
metodo *m*, 2677
metodo *m* ad attrito
  costante, 1664

metodo *m* a recupero di
  pressione statica, 3617
metodo *m* a riduzione di
  velocità, 3983
metodo *m* di collaudo, 3783
metro *m*, 2676
mettere in moto, 3745
mezza stagione *f*, 2683
mezzo *m* assorbente, 19
mezzo *m* controllato, 1025
mezzo *m* di raffreddamento,
  1051, 1064
mezzo *m* scaldante, 2252
microclima *m*, 2680
micron *m*, 2681
microprocessore *m*, 2682
miniera *f*, 2684
miscela *f*, 2701
miscela *f* congelante, 2050
miscela *f* di gas, 2692
miscela *f* di vapore, 3965
miscela *f* gassosa, 2141
miscela *f* liquida, 2563
miscelatore *m*, 436
miscelatore *m* di aria, 139
misura *f*, 2158, 2649, 3462
misura *f* dell'umidità, 2715
misuratore *m* del tiraggio,
  1426
misuratore *m* di energia
  raggiante, 485
misuratore *m* di
  evaporazione, 1702
misuratore *m* di portata,
  1963
misuratore *m* di pressione
  assoluta, 894
misurazione *f*, 2649
misurazione *f* del contenuto
  in polvere, 1518
misurazione *f* della
  pressione, 3059
mnemonica *f*, 2702
modellizzazione *f*, 2704
modello *m*, 2703, 2924
modem *m*, 2705
modulante, 2708
modulare, 2706, 3836
modulazione *f*, 3837
modulazione *f* completa,
  2094
modulo *m*, 2710
modulo *m* di visualizzazione
  dati, 1178
molla *f*, 3590
momento *m* torcente, 3865
monitoraggio *m*, 2723
monoblocco, 2897
monossido *m* di carbonio,
  633
monostadio, 3455
montaggio *m*, 286, 1677,
  2734, 3409
montaggio *m* a cremagliera,
  3134

montaggio *m* in cantiere,
  3461
montaggio *m* in scaffale,
  3134
montante *m* a doppia
  aspirazione, 1407
montare, 285
montato in loco, 560
montatore *m*, 1897
morsa *f*, 4005
morsettiera *f*, 982
morsetto *m*, 986
morsetto *m* elettrico, 1588
moto *m* dell'aria, 141
motore *m*, 1639, 2726
motore *m* ad aria, 140
motore *m* Diesel, 1286, 1287
motore *m* di riserva, 3602
motore *m* elettrico, 1576
moto *m* turbolento, 4020
movimento *m*, 2724, 2737
mozzo *m*, 2350
multizone, 2754
muratura *f*, 527
muratura *f* refrattaria, 3197
muro *m*, 4023
muro *m* esterno, 1769

nafta *f*, 2809
nafta *f* nebulizzata, 2824
nastro *m* trasportatore, 1044
nave *f* cisterna, 3756
nebbia *f*, 2003, 2688
nebbia *f* di torre di
  raffreddamento, 1072
nebulizzare, 299, 2761
negozio *m*, 3670
neutro *m* comune, 880
nicchia *f*, 2768
nocivo, 2790
non accettabile, 2479
non inversione *f*, 2783
non valido, 2479
nucleare, 2793
nucleo *m*, 1084
numerico, 2795
numero *m*, 2794
numero *m* caratteristico, 709
numero *m* di Bacharach, 324
numero *m* di Grashof, 2035
numero *m* di incombusti
  nella cenere, 277
numero *m* di ranghi di un
  serpentino, 806

occhio *m* di bue, 564
occupante *m*, 2800
odore *m*, 2803
offerta *f*, 3772, 3773
offerta *f* dettagliata, 1261
officina *f*, 4141
olf *m*, 2832
olio *m*, 2809
olio *m* combustibile, 2085
olio *m* diatermico, 2291

olio *m* di riscaldamento,
  2254
olio *m* minerale, 2685
opacità *f*, 2838
opacità *f* del fumo, 3493
opacità *f* (metodo Ashrae
  dust spot), 1526
opaco, 2840
operazione *f* in sequenza,
  3395
ordine *m* di grandezza, 2856
ordine *m* di lavorazione,
  4137
orientamento *m*, 2857
orifizio *m*, 2858
O-ring *m*, 2860
orologio *m*, 3847
oscillazione *f*, 1156, 2862
oscillazione *f* di temperatura,
  3771
oscillazione *f* (di un
  regolatore), 2361
ossatura *f*, 2032
ossidazione *f*, 2892
ossigeno *m*, 2894
ostruzione *f*, 2797
otturare, 2991
otturatore *m*, 2990
overflow *m* di un
  elaboratore, 932
overshoot *m*, 2890
ozono *m*, 2896

pala *f*, 420, 3954
pala *f* del ventilatore, 1794
pala *f* rotante, 3912
palette *f,pl* direttrici
  orientabili, 3972
pannello *m*, 2911
pannello *m* annegato, 1606
pannello *m* di riscaldamento,
  2255
pannello *m* di riscaldamento
  a parete, 4029
pannello *m* radiante, 3143
pannello *m* radiante a
  soffitto, 659
pannello *m* solare, 3513
parametro *m*, 2919
parete *f*, 4023
parete *f* esterna, 1769
parete *f* interna, 2476
parete *f* sottofinestra, 4024
parte *f*, 891
parte *f* di ricambio, 3558
parte *f* imboccata nel
  bicchiere, 3566
partenza *f*, 3604
particella *f*, 2921
particella *f* in sospensione,
  3736
particelle *f,pl* di polvere,
  1519
particolato *m* aerodisperso,
  76

passaggio *m*, 2157
passaggio *m* di aria, 143
passo *m*, 2970, 3651
passo *m* dell'alettatura, 1871
passo *m* d'uomo, 2621
pavimento *m*, 1948
pavimento *m* sopraelevato,
  1789
pellicola *f*, 1839
pennacchio *m* della torre di
  raffreddamento, 1074
pennacchio *m* di fumo, 2993
perdere, 2526
perdita *f*, 2525, 2586
perdita *f* al camino, 739,
  2258
perdita *f* all'ingresso, 1655
perdita *f* di acqua, 4076,
  4078
perdita *f* di aria, 136
perdita *f* di carico, 2220,
  2587
perdita *f* di energia, 1629
perdita *f* di freddo, 1063
perdita *f* di fumi al camino,
  1979
perdita *f* di pressione, 3058
perdita *f* di pressione
  dinamica, 1534
perdita *f* per attrito, 2067
perdita *f* per evaporazione,
  1692
perforato, 2933
perforazione *f*, 2937
periferico, 2941
periodico, 2940
periodo *m* di picco, 2836
periodo *m* di richiesta di
  potenza elettrica, 1581
periodo *m* di sbrinamento,
  1212
periodo *m* fuori picco, 2806
permeabilità *f*, 2942
perno *m* di articolazione,
  3898
perno *m* di biella, 1124
persiana *f*, 2590
per tutto l'anno, 4145
peso *m*, 4100
petroliera *f*, 3756
pezzo *m* a perdere, 3508
pezzo *m* a Y, 4146
piano *m*, 2973
piano *m* di edificio, 3672
pianta *f*, 2973
piastra *f*, 2981
piastra *f* a fondazione, 395
piastra *f* del bruciatore, 573
piastra *f* del deflettore, 343
piastra *f* di caldaia, 474
piastra *f* di copertura, 1115
piastra *f* di otturazione, 2859
piastra *f* goffrata, 1607
piastra *f* scaldante, 2335

piastra *f* terminale concava,
1357
piastra *f* terminale della
caldaia, 461
piastra *f* tubiera, 3903
picco *m*, 2925
piedi *m,pl* cubici per minuto,
697
piedi *m,pl* cubici per
secondo, 698
pieno carico *m*, 2093
pignone *m*, 3898
pilastro *m*, 2948
pilone *m*, 2948
pinza *f*, 3861
pinza *f* per scorie, 3469
piombo *m*, 2523
pirometro *m*, 3127
pistone *m*, 2966
plastica *f*, 2977
plenum *m*, 2989
pneumatico, 2996
politropico, 3003
polvere *f*, 1510
polvere *f* di carbone, 785,
3118
polvere *f* di saggio, 3781
polvere *f* fibrosa, 1829
polverizzare, 299, 3580
pomo *m*, 2508
pompa *f*, 3119
pompa *f* a ingranaggio, 2160
pompa *f* a pistone, 3176
pompa *f* ausiliaria, 493
pompa *f* autoadescante,
3381
pompa *f* a vuoto, 3947
pompa *f* centrifuga, 691
pompa *f* da travaso, 3881
pompa *f* di alimentazione,
493
pompa *f* di alimentazione
acqua a iniezione di
vapore, 3636
pompa *f* di alimentazione
della caldaia, 463
pompa *f* di calore, 2279
pompa *f* di circolazione, 749
pompa *f* di miscela, 2698
pompa *f* di prova di
pressione, 3070
pompa *f* di ricircolazione
condensa, 948
pompa *f* per acqua ad
eiettore, 4074
pompa *f* per aria, 149
pompa *f* rotativa, 3312
pompa *f* volumetrica, 4018
ponte *m* termico, 2228
porta *f*, 3004
porta *f* a scomparsa, 1993
portasalviette *m*, 3872
porta *f* seriale, 3393
portata *f*, 1966, 3838

portata *f* del flusso di aria,
123
portata *f* del ventilatore,
1802
portata *f* di aria, 167
portata *f* di aria calda, 4040
portata *f* di aria di progetto,
1247
portata *f* di aria variabile,
3968
portata *f* di aspirazione,
3696
portata *f* di massa, 2634
portata *f* di scarico, 1347
portata *f* di spillamento,
3759
portata *f* di un conduttore,
205
portata *f* giornaliera, 1161
porta *f* tagliafuoco, 1882
portata *f* minima dell'aria,
2687
portatile, 3005
portello *m* del ceneraio, 273
portello *m* di accesso, 27
portello *m* di caricamento,
718
portello *m* di pulitura, 762
portone *m*, 2157
posa *f* sottoterra, 2522
posizione *m*, 3006
posizione *f* abbassata, 1447
posto *m*, 3459
postraffreddamento *m*, 3179
postraffreddatore *m*, 63
postriscaldamento *m*, 3225
postriscaldare, 3226
postriscaldatore *m*, 64, 3228
potenza *f*, 617, 3012, 3163
potenza *f* all'albero, 506
potenza *f* apparente, 249
potenza *f* attiva, 3171
potenza *f* calorifica
nominale, 3161
potenza *f* del motore, 2730
potenza *f* frigorifera, 821
potenza *f* nominale, 3160
potenza *f* reattiva, 3166
potenza *f* sonora del
ventilatore, 1806
potenza *f* termica, 2280
potenza *f* termica dispersa,
2273
potenza *f* termica sviluppata,
3800
potenza *f* termica totale,
3870
potenza *f* termica utile, 3937
potenziale *m* di energia, 1633
potenziale *m* elettrico, 1577
potenzialità *f*, 617, 2275
potenzialità *f* del
compressore, 913
potenzialità *f* del
condensatore, 961

potenzialità *f* della caldaia,
457, 476
potenzialità *f* della valvola di
espansione, 1754
potenzialità *f* di
raffreddamento, 1053
potere *m* calorifico, 2264
potere *m* calorifico inferiore,
609
potere *m* calorifico
superiore, 608
pozzetto *m* del termometro,
3814
pozzo *m*, 4112
pozzo *m* caldo, 2346
pozzo *m* di aerazione, 3992
pozzo *m* di ventilazione, 177
pre-accensione *f*, 3032
precipitatore *m*, 3025
precipitatore *m*
elettrostatico, 1601
precipitazione *f*, 3024
precisione *f*, 32
precisione *f* di taratura, 3407
prefiltro *m*, 3027, 3031
prefissato, 3039
prelevare, 433
prelievo *m* da campioni,
3157
premistoppa *m*, 3685
preparazione *f*, 3038, 3409
preparazione *f* del
combustibile, 2086
preraffreddamento *m*, 3029
preraffreddatore *m*, 3028
prerefrigeratore *m*, 2022
pre-riscaldamento *m*, 3035
preriscaldatore *m*, 3034
preriscaldatore *m* di aria,
147
presa *f*, 3762
presa *f* di acqua, 4079
presa *f* di aria, 134
presa *f* di aria secondaria,
580
presa *f* di corrente a soffitto,
663
presa *f* di misura, 2652
presa *f* di ritorno, 3259
pressatreccia *m*, 3685
pressione *f*, 2217, 3041, 3055
pressione *m*, 3616
pressione *f* ambiente, 197
pressione *f* atmosferica, 298
pressione *f* barometrica, 372
pressione *f* dell'aria, 148
pressione *f* del vento, 4127
pressione *f* di arresto, 3045
pressione *f* di aspirazione,
3697, 3702
pressione *f* di circolazione,
748
pressione *m* di collaudo,
3784

pressione *f* di condensazione, 957
pressione *f* di equilibrio, 354, 1669
pressione *f* di esercizio, 2849, 4138
pressione *f* di evaporazione, 1693
pressione *f* differenziale, 1292
pressione *f* di lavoro di progetto, 1254
pressione *f* dinamica, 1535, 3980
pressione *f* di picco del compressore, 919
pressione *f* di saturazione, 3345
pressione *f* di scarico, 1346
pressione *f* di targa, 3060
pressione *f* di vapore, 3642, 3966
pressione *f* finale, 1866
pressione *f* media, 2644, 2659
pressione *f* nominale, 2781
pressione *f* parziale, 2920
pressione *f* ridotta, 3185
pressione *f* sonora, 3545
pressione *f* totale, 3869, 3871
pressostato *m*, 3068
pressotreccia *f*, 2167
pressurizzatore *m*, 3075
pressurizzazione *f*, 3073
prestazione *f*, 2938
pre-taratura *f*, 3040
prevalenza *f*, 1343
prevalenza *f* in aspirazione, 3699
preventilazione *f* della camera di combustione, 575
preventivo *m*, 1682
prezzo *m* di costo, 2891
procedura *f* di avviamento, 3607
processo *m*, 3087
processo *m* di condizionamento, 90
processo *m* di congelamento, 2053
processo *m* di sbrinamento, 1210
processore *m* centrale, 683
prodotti *m,pl* chimici, 724
prodotto *m* della combustione, 856
produzione *f*, 2161
produzione *f* assistita dal calcolatore, 924
produzione *f* di calore, 2278
produzione *f* di freddo, 820
produzione *f* di gas, 2144
produzione *f* oraria, 3838

professionista *m* capocommessa, 3083
profilo *m* della velocità, 3982
profondità *f* di un serpentino, 806
progettare, 1246
progettazione *f* ambientale, 1660
progettazione *f* assistita dal calcolatore, 923
progettazione *f* del calcolatore, 927
progetto *m*, 1245
progetto *m* coordinato, 1079
progetto *m* di un edificio, 556
programmabile, 3093
programma *m* di accensione caldaia, 1895
programma *m* di collaudo, 878
programma *m* di fabbricazione, 3091
programma *m* di gestione, 2620
programma *m* per elaboratore, 936
pronto all'uso, 3170
propagazione *f* del suono, 3548
propano *m*, 3098
proporzionale, 3101
prospetto *m*, 1603
proteggere, 3428
proteggere antifiamma, 1887
protettivo, 3108
protezione *f*, 3105
protezione *f* anticorrosione, 1096
protezione *f* antifiamma, 1888
protezione *f* antigelo, 2075
protezione *f* antirumore, 3547
protezione *f* catodica, 653
protezione *f* contro l'umidità, 2716
protezione *f* dal caldo, 3107
protezione *f* dal freddo, 3106
protezione *f* dall'umidità, 1173
prova *f*, 3776
prova *f* al blu di metilene, 2678
prova *f* a pressione, 909
prova *f* di accettazione, 25
prova *f* di compressione, 909
prova *f* di flusso turbolento, 1545
prova *f* di fuliggine, 3534
prova *f* di pressione, 3069
prova *f* di Ringelmann, 419
prova *f* di tenuta, 2527
prova *f* DOP, 1391
psicrometro *m*, 3114

psicrometro *m* ad aspirazione, 283
psicrometro *m* a fionda, 3482
pulire, 1992
pulitura *f* della caldaia, 459
pulizia *f*, 761
pulizia *f* del camino, 735
pulizia *f* del filtro, 1853
pulsante, 3116
pulviscolo *m* atmosferico, 295
punta *f*, 2925, 3857
punto *m*, 2998
punto *m* alto, 2305
punto *m* di accensione, 1925, 2388
punto *m* di arresto, 1153, 1154
punto *m* di condensazione, 950
punto *m* di congelamento, 2052
punto *m* di contatto, 1001
punto *m* di controllo, 1030
punto *m* di ebollizione, 484
punto *m* di fusione, 2663
punto *m* di inserzione, 3746
punto *m* di intervento, 1149
punto *m* di intorbidamento, 1946
punto *m* di prelievo, 1434
punto *m* di presa energia, 3020
punto *m* di riduzione di pressione, 3061
punto *m* di rugiada, 1269
punto *m* di rugiada acida, 37
punto *m* di saturazione, 3344
punto *m* di spruzzatura, 3588
punto *m* di taratura, 3405
punto *m* di uscita, 2872
punto *m* fisso, 221, 1900
punto *m* fisso di riferimento, 3189
purezza *f* dell'aria, 150
purgare, 3123

quadrante *m*, 1274
quadro *m*, 3751
quadro *m* centrale di controllo, 679
quadro *m* di comando, 1019, 1029, 3743
quadro *m* di distribuzione, 3742
qualità *f*, 3128
qualità *f* del gas, 2145
qualità *f* dell'aria, 151
qualità *f* dell'aria interna, 2416
quantità *f*, 204, 3131
quantità *f* di moto, 2722

registrare, 3218
registratore *m*, 3181
registro *m*, 3217
regolabile, 53
regolamento *m*, 796
regolare, 52, 1015, 3221, 3404
regolato in fabbrica, 1784
regolatore *m*, 1027
regolatore *m* asservito, 3471
regolatore *m* autonomo, 3596
regolatore *m* del tiraggio, 1432
regolatore *m* di anticipo, 238
regolatore *m* di carica, 713
regolatore *m* di combustione, 849, 857
regolatore *m* di contropressione (pressione dell'evaporatore), 330
regolatore *m* differenziale, 1290
regolatore *m* di livello, 2533
regolatore *m* di portata, 1967
regolatore *m* di potenza, 622
regolatore *m* di potenzialità, 619
regolatore *m* di pressione, 3043, 3065
regolatore *m* di pressione dell'evaporatore, 1705
regolatore *m* di programma, 3095
regolatore *m* di refrigerazione, 3208
regolatore *m* di temperatura, 3767
regolatore *m* di umidità, 2359
regolatore *m* in cascata, 642
regolatore *m* principale, 2636
regolazione *f*, 54, 1014, 3224
regolazione *f* a banda proporzionale, 3103
regolazione *f* a bassa pressione, 2598
regolazione *m* a doppio effetto, 1487
regolazione *f* a gradini, 3655
regolazione *f* alto-basso, 2304
regolazione *f* ambiente, 3302
regolazione *f* a pulsante, 3125
regolazione *f* a punto fisso, 993
regolazione *f* a valore costante, 993
regolazione *f* centrale, 678
regolazione *f* della portata di nafta, 2818

regolazione *f* della temperatura del bollitore, 482
regolazione *f* del suono, 3540
regolazione *f* di alta pressione, 2308
regolazione *f* digitale diretta, 1318
regolazione *f* di portata, 1817
regolazione *f* di pressione differenziale, 1293
regolazione *f* di sicurezza, 3329
regolazione *f* elettro-pneumatica, 2997
regolazione *f* in cascata, 641
regolazione *f* integrale, 2461
regolazione *f* manuale, 2626
regolazione *f* proporzionale, 3104
regolazione *f* proporzionale-integrale, 2947
regolazione *f* tutto-niente, 2834
regolazione *f* variabile di portata, 3969
regole *f,pl* dell'arte, 796
REHVA, 3229
rel *m*, 3234
relativo, 3232
rel *m* di posizionamento, 3007
rendimento *m*, 1556
rendimento *m* della caldaia, 460
rendimento *m* della combustione, 853
rendimento *m* dell'aletta, 1867
rendimento *m* di compressione, 902
rendimento *m* di separazione, 3389
rendimento *m* globale, 2880
rendimento *m* ponderale, 2185
rendimento *m* volumetrico, 4017
reparto *m* di lavorazione, 4141
resa *f*, 1556
resa *f* del gas, 2155
resa *f* di separazione, 3389
resa *f* termica, 2275
residuo *m*, 3237
residuo *m* della combustione, 858
resina *f* plastica schiumosa, 2978
resistente al fuoco, 1890
resistente al gelo, 2044
resistente alla corrosione, 1097

resistente alle intemperie, 4097
resistenza *f*, 1570, 3251
resistenza *f* ad immersione, 2393
resistenza *f* al flusso, 1969
resistenza *f* al freddo, 822
resistenza *f* al fuoco, 1889, 3195
resistenza *f* allo scorrimento, 1414
resistenza *f* al moto dell'aria, 124
resistenza *f* dielettrica, 1284
resistenza *f* localizzata, 2578
resistenza *f* pellicolare, 1845
resistenza *f* per attrito, 2065
resistenza *f* termica, 3802
resistività *f* termica, 3803
respiro *m*, 522
restituzione *f* grafica, 3242
resto *m*, 3237
restrizione *f*, 3255
rete *f*, 2766, 2958
rete *f* di canali, 1503
rete *f* di distribuzione, 1373, 1374
rete *f* di telecomunicazione, 881
rete *f* di tubazioni, 2959, 2963
rete *f* locale, 2577
retroazione *f*, 1814
retta *f* di carico, 963
reversibile, 3275
ricambi *m,pl* di aria orari, 82
ricambio *m* di aria, 80, 155, 2061
ricerca *f*, 3245
ricevitore *m*, 3173
ricircolazione *f*, 3178
ricircolazione *f* del gas di scarico, 1731
ricircolo *m*, 3178
ricovero *m* antiaereo o antiatomico, 566
ricuocere, 233
riduttore *m* di capacità, 621
riduttore *m* di potenza, 621
riduttore *m* di pressione, 3186
riduzione *f*, 3886
riduzione *f* del rumore, 3549
riduzione *f* di carico, 2575
riduzione *f* di pressione, 3064
riduzione *f* di temperatura, 1057
riempimento *m*, 3864
riempimento *m* della torre evaporativa, 1071
riflettanza *f*, 3191
riflettanza *f* diffusa, 1298
riflettività *f*, 3193
riflettore *m*, 3194

rifornimento *m* di
combustibile, 2088
rifornire (di combustibile),
2077
rifugio *m*, 3427
rigenerazione *f*, 3215
rigenerazione *f* a letto di
ghiaia, 3285
rilevamento *m* all'infrarosso,
2433
rimozione *f* dei fanghi, 3489
rimozione *f* del ghiaccio,
1227
rimozione *f* della cenere, 276
rimozione *f* della fuliggine,
3533
rinforzare, 3230
rinnovabile, 2785
riparare, 3428
riparare dal sole, 3412
riparo *m*, 3427
riposizionamento *m*, 3248
riscaldamento *m*, 2241, 4041
riscaldamento *m* a
circolazione naturale, 2188
riscaldamento *m* ad
accumulo, 3667
riscaldamento *m* ad
accumulo al di fuori delle
ore di picco, 2807
riscaldamento *m* ad acqua,
4072
riscaldamento *m* ad acqua
calda a gas, 2130
riscaldamento *m* ad acqua
surriscaldata, 2310
riscaldamento *m* ad aria
calda, 4038
riscaldamento *m* ad
infrarossi, 2432
riscaldamento *m* ad
irraggiamento, 3142
riscaldamento *m* a due tubi,
3915
riscaldamento *m* a gas, 2134
riscaldamento *m* a
induzione, 2419
riscaldamento *m* ambiente,
3554
riscaldamento *m* a nafta,
2822
riscaldamento *m* a pannelli,
2913
riscaldamento *m* a pannelli a
pavimento, 1950
riscaldamento *m* a pannelli
radianti, 3142
riscaldamento *m* a parete,
4028
riscaldamento *m* a
resistenza, 3252
riscaldamento *m* a soffitto,
662
riscaldamento *m* a soglia
sotto finestra, 4126

riscaldamento *m* a stufa,
3674
riscaldamento *m* a vapore,
3634
riscaldamento *m* a vapore a
bassa pressione, 2600
riscaldamento *m* a vapore
sotto vuoto, 3948
riscaldamento *m* centrale,
680
riscaldamento *m* centrale a
gas, 2129
riscaldamento *m*
centralizzato, 2092
riscaldamento *m*
centralizzato (di quartiere),
1378
riscaldamento *m*
contrattuale, 1011
riscaldamento *m* di
appartamento, 245
riscaldamento *m* di aria, 129
riscaldamento *m* di aria a
gas, 2128
riscaldamento *m* di base, 327
riscaldamento *m* domestico,
1389
riscaldamento *m* elettrico,
1574
riscaldamento *m* individuale,
2414
riscaldamento *m* isolato, 441
riscaldamento *m* per
irraggiamento, 3136
riscaldamento *m*
residenziale, 3250
riscaldamento *m*
supplementare, 3718
riscaldare, 2225
riscaldatore *m*, 2235
riscaldatore *m* a
combustione diretta, 1325
riscaldatore *m* ad accumulo,
3666
riscaldatore *m* a energia
radiante, 3141
riscaldatore *m* ambiente,
3303
riscaldatore *m* a vapore,
3630
riscaldatore *m* dell'acqua di
alimentazione della
caldaia, 465
riscaldatore *m* di acqua,
4071
riscaldatore *m* di aria, 127
riscaldatore *m* di aria a
muro aerotermo, 4030
riscaldatore *m* di aria a tubo
alettato, 429
riscaldatore *m* perimetrale,
377
riscaldatore *m* radiante a
gas, 3139

riscaldatore *m* radiante a
strisce, 3144
riserva *f*, 3600
risonanza *f*, 3254
risorsa *f* energetica, 1635
risparmio *m* energetico, 1625
risposta *f* a guasto, 1786
risposta *f* in frequenza, 2058
risultante, 3256
risultato *m* di collaudo, 3785
ritarare, 3247
ritaratura *f*, 3248
ritardo *m* di un fenomeno
termico, 2269
ritorno *m*, 3257
ritorno *m* a secco, 1477
ritorno *m* bagnato, 4117
ritorno *m* invertito, 3274
rivelatore *m* di fumo, 3494
rivelatore *m* di gas
combustibile, 843
rivelatore *m* di perdite, 2528
rivelatore *m* di vapore, 3962
rivestimento *m*, 754, 791,
2555
rivestimento *m* antifrizione,
241
rivestimento *m* della camera
di combustione, 848
rivestimento *m* della canna
fumaria, 1983
rivestimento *m* di
protezione, 3109
rivestimento *m* refrattario,
3198
rivestito con materiale
isolante, 2493
rivettare, 3284
rivetto *m*, 3283
rondella *f*, 4049
rotativo, 3306
rotazione *f*, 3313
rotolo *m*, 805
rotore *m*, 3315
rubinetto *m*, 794
rubinetto *m* a collo, 1435
rubinetto *m* a disco, 1356
rubinetto *m* a passaggio
diretto, 3840
rubinetto *m* a tre vie, 3831
rubinetto *m* di arresto, 2487,
3437
rubinetto *m* di prova, 3780
rubinetto *m* di scarico, 1349
rubinetto *m* di spillamento,
3757
rubinetto *m* di spurgo, 1420
ruggine *f* (corrosione), 3325
rugiada *f*, 1268
rugosità *f*, 3316
rugosità *f* assoluta, 8
rumore *m*, 2772
rumore *m* di motore, 2729
rumore *m* di sbocco del
canale, 1493

rumore *m* nei canali, 1498
ruota *f*, 4121

sacca *f* di acqua, 4080
sacca *f* di aria, 144
sacca *f* di vapore, 3641
sacco *m*, 344
sacco *m* filtrante, 1849
sala *f* impianti, 2976
sala *f* macchine, 1675
salamoia *f*, 528
sala *f* prove, 3787
saldare, 3521, 4103
saldatore *m*, 4105
saldatura *f*, 4102, 4106
saldatura *f* a cannello, 2154
saldatura *f* ad arco, 261
saldatura *f* a punti, 3579
saldatura *f* a resistenza, 3253
saldatura *f* a stagno, 3511
saldatura *f* autogena, 2893
salire, 3279
salto *m* di temperatura, 3768
saracinesca *f*, 3437, 3438
saturazione *f*, 3340
sbaffo *m*, 3502
sbarra *f*, 585
sbocco *m* della canna
  fumaria, 1984
sbrinamento *m*, 1208
sbrinamento *m* ad acqua,
  4063
sbrinamento *m* a inversione
  di ciclo, 3272
sbrinamento *m* con
  interruzione del ciclo
  frigorifero, 2805
sbrinamento *m*
  temporizzato, 3850
sbrinare, 1206
scala *f* di Beaufort, 393
scala *f* (disegno), 3348
scaldabagno *m* istantaneo,
  2166
scaldacqua *m* istantaneo,
  1328, 2452
scalpello *m*, 740
scambiare, 1713
scambiatore *m*, 1714
scambiatore *m* a tubi e
  mantello, 3426
scambiatore *m* di calore,
  610, 2238
scambiatore *m* di calore a
  contatto diretto, 1313
scambiatore *m* di calore ad
  aria a convezione forzata,
  423
scambiatore *m* di calore a
  flussi incrociati, 1132
scambiatore *m* di calore a
  tubi concentrici, 1402
scambiatore *m* di calore in
  controcorrente, 1107

scambiatore *m* di calore
  rotativo, 3311
scambio *m*, 1712
scambio *m* di informazioni,
  2431
scambio *m* di ioni, 2481
scambio *m* termico, 2237
scantinato *m*, 380, 668
scaricare, 1339, 1720, 3123
scaricatore *m* di aria, 117
scaricatore *m* di condensa,
  3892
scaricatore *m* di condensa a
  galleggiante, 1944
scaricatore *m* termostatico,
  3818
scarico *m*, 1338, 1353, 1688,
  1719, 3457, 3706
scarico *m* dei gas di
  combustione (fumi), 1980
scarico *m* dei prodotti di
  combustione, 1981
scarico *m* del compressore,
  914
scarico *m* dell'acqua, 4065
scarico *m* dell'aria, 116
scarico *m* dell'ugello, 2792
scarico *m* di nafta, 2817
scarico *m* di olio, 2817
scarico *m* di scorie, 3466
scarto *m* di regolazione,
  1020
scarto *m* di temperatura,
  3768
scarto *m* medio di
  temperatura, 2646
scatola *f*, 503
scatola *f* di derivazione, 2504
scatola *f* di distribuzione,
  1370
scelta *f* del combustibile,
  2079
schema *m* di montaggio,
  4136
schema *m* elettrico, 1567,
  4132
schermo *m*, 3352, 3414
schermo *m* di protezione,
  3112
schiuma *f*, 1999
schiuma *f* plastica rigida,
  3278
schiumare, 2000
scintilla *f*, 3559
scoloramento *m*, 1355
scongelamento *m* dielettrico,
  1285
scoperta *f* innovativa, 521
scoperto, 1762
scoppio *m*, 520
scoppio *m* di gas, 2123
scoria *f*, 773
scovolo *m*, 3150
scovolo *m* per canna
  fumaria, 1974

secchezza *f*, 1473
secco, 1457
secondo corrente, 1411
sede *f* di valvola, 3953
sedimento *m*, 3377
segnale *m*, 2411, 3168, 3442
semiautomatico, 3382
semi-bussola *f*, 586
semiconduttore *m*, 3383
semiermetico, 3384
sensibile, 3385
senso *m*, 1329
sensore *m*, 3387
sensore *m* della qualità
  dell'aria, 153
sensore *m* di fiamma, 850
sensore *m* di luce, 2547
sensore *m* di occupazione,
  2798
sensore *m* di temperatura,
  3770
sensore *m* di umidità, 2360
sensore *m* esterno, 1767,
  2877
sensore *m* remoto, 3240
sensore *m* solare, 3519
senza lubrificazione, 2821
senza riempimento, 2907
senza saldatura, 3368
separatore *m*, 1604, 3390
separatore *m* centrifugo, 692
separatore *m* di acqua, 4083
separatore *m* di cenere, 278
separatore *m* di condensa,
  3645
separatore *m* di gocce, 637,
  1439, 1451, 2689, 2718
separatore *m* di liquidi, 2564
separatore *m* di olio, 2826
separatore *m* di polvere,
  1513, 1525
separatore *m* di prodotti
  nebulizzati, 2689
separazione *f*, 3388
separazione *f* dell'aria, 158
separazione *f* di fiamma,
  1908
sequenza *f*, 3391
serbatoio *m*, 1455, 3246,
  3755
serbatoio *m* a evaporazione
  rapida, 1927
serbatoio *m* del filtro, 1862
serbatoio *m* dell'acqua
  calda, 2342
serbatoio *m* di accumulo,
  3669
serbatoio *m* di aria, 175
serbatoio *m* di
  bilanciamento, 356
serbatoio *m* di carico, 2219
serbatoio *m* di compressione,
  908
serbatoio *m* di nafta, 2827

serbatoio *m* di raccolta
condensa, 945
serbatoio *m* di salamoia, 533
serbatoio *m* equalizzatore,
1667
serbatoio *m* giornaliero,
1163
serbatoio *m* interrato, 3922
serbatoio *m* polmone, 544
serpentina *f* di
riscaldamento, 2243
serpentino *m*, 805, 2953
serpentino *m* del
condensatore, 953
serpentino *m* dell'essiccatore,
1437
serpentino *m* di
desurriscaldamento, 1257
serpentino *m* di espansione,
1745
serpentino *m* di
raffreddamento a parete,
4025
serpentino *m* preriscaldatore,
3033
serrafilo *m*, 986
serranda *f*, 590, 1168, 3438
serranda *f* a coulisse
(scorrevole), 3478
serranda *f* ad alette multiple,
2746
serranda *f* a farfalla, 589
serranda *f* a ghigliottina,
3477
serranda *f* a lama singola,
3453
serranda *f* a lame
contrapposte, 2855
serranda *f* a lame parallele,
2915
serranda *f* di by-pass, 595
serranda *f* di non ritorno,
326
serranda *f* di regolazione
aria, 428
serranda *f* di regolazione
della portata, 4016
serranda *f* di regolazione
dell'aria, 154
serranda *f* di ritegno, 2787
serranda *f* di suddivisione,
3576
serranda *f* di taratura, 1668
serranda *f* motorizzata, 1170
serranda *f* per aria, 104
serranda *f* per il gas di
scarico, 1730
serranda *f* tagliafuoco, 1881
serratura *f*, 2579
serratura *f* del portello, 1390
servizio *m*, 2852, 3396
servocontrollo *m*, 3402
servo-motore *m*, 3403
setaccio *m*, 3352, 3440
set point *m*, 3405

settaggio *m*, 3410
sezionatore *m*, 519
sezione *f*, 3374, 3376
sezione *f* di miscela, 2693
sezione *f* di transizione del
canale, 1504
sezione *f* di tubo, 2960
sezione *f* frontale, 2070
sezione *f* trasversale, 1135
sfasamento *m*, 3852
sfasamento *m* di un
fenomeno termico, 2269
sfiatare, 3986
sfiato *m* dell'aria, 174
sfogo *m* di aria, 174, 3994
sforzo *m*, 3681
shunt *m*, 3434
sicurezza *f*, 3235
sicurezza *f* di impiego, 3334
sifone *m*, 3892
sigillare, 2991, 3362
sigillato ermeticamente, 2300
sigillatura *f*, 655
silenziatore *m*, 2740, 3444,
3539
silenzioso, 2775
silo *m*, 2325
simbolo *m*, 3747
simulazione *f* all'elaboratore,
935
simultaneo, 3448
sincronizzazione *f*, 3748
sindrome *f* dell'edificio
malato, 3439
sinterizzazione *f*, 3458
sistema *m*, 3750
sistema *m* ad energia totale,
3868
sistema *m* ad espansione
diretta, 1323
sistema *m* a diffusione-
assorbimento, 1303
sistema *m* a portata di aria
variabile, 3970
sistema *m* a quattro tubi,
2029
sistema *m* a raffreddamento
diretto, 1315, 1333
sistema *m* a retroazione,
1815
sistema *m* a riscaldamento
diretto, 1327
sistema *m* a salamoia, 532
sistema *m* a sbocco diretto,
1334
sistema *m* a scarico diretto,
1320
sistema *m* a sfogo diretto,
1334
sistema *m* a tenuta, 3364
sistema *m* basato su
elaboratore, 925
sistema *m* chiuso, 780, 3364
sistema *m* di agitazione
dell'aria, 72

sistema *m* di allarme
antincendio, 1874
sistema *m* di automazione
dell'edificio, 547
sistema *m* di comunicazione
da pari a pari, 2930
sistema *m* di controllo della
gestione energetica
dell'edificio, 550
sistema *m* di controllo
dell'edificio, 548
sistema *m* di distribuzione,
1375
sistema *m* di drenaggio del
collettore solare (drain
back), 1419
sistema *m* di drenaggio del
collettore solare (drain
down), 1421
sistema *m* di gestione
dell'edificio, 554
sistema *m* di
raffreddamento, 1069
sistema *m* di recupero del
calore, 2282
sistema *m* di refrigerazione a
compressione, 910
sistema *m* di refrigerazione
ad eiettore, 1560
sistema *m* di refrigerazione
ad espansione diretta, 1322
sistema *m* di refrigerazione a
perdita totale, 1755
sistema *m* di riscaldamento,
2253, 2256, 2261
sistema *m* di riscaldamento a
distribuzione dal basso,
1410
sistema *m* di riscaldamento a
ritorno secco, 1478
sistema *m* di riscaldamento a
tubo di piccolo diametro,
2679
sistema *m* di riscaldamento
minitubo, 3491
sistema *m* di sbrinamento,
1211
sistema *m* di scarico, 1735
sistema *m* di supervisione,
1631
sistema *m* di supervisione
dell'edificio, 551
sistema *m* esperto, 1759
sistema *m* monocondotto,
3451
sistema *m* monotubo, 3454
sistema *m* sottovuoto, 3686
sistema *m* telematico, 882
smaltimento *m*, 1719
smantellare, 1336
smontare, 1336
smorzamento *m*, 1171
smorzare, 1166, 1167
smorzatore *m* di vibrazioni,
4003

soffiaggio *m*, 449
soffiante *f*, 447, 448
soffiare, 444
soffiatore *m* di fuliggine, 3531
soffietto *m*, 396
soffietto *m* di dilatazione, 1742
soffitto *m*, 658
soffitto *m* forato, 2934
soffitto *m* ribassato, 1454
soffitto *m* riscaldato, 2234
soffitto *m* ventilato, 3987
soglia *m* di udibilità, 3834
solaio *m*, 303, 664
solubilità *f*, 3526
soluzione *f*, 3527
soluzione *f* ammoniacale, 201
solvente *m*, 3528
sonda *f*, 1263, 3086, 3386
soppressore *m*, 3730
sopraelevazione *m* del pennacchio, 2995
soprassaturo, 3714
sorgente *f* di calore, 2286
sorgente *f* fredda, 824
sorgente *f* termica ausiliaria, 313
sorveglianza *f*, 301
sospensione *f*, 3737
sostanza *f* assorbita, 10
sostanza *f* colorante, 1529
sostegno *m*, 2214, 2734
sostegno *m* a staffa, 3658
sottocontraente *m*, 3687
sottoraffreddamento *m*, 3689
sottostazione *f*, 3693
sotto vuoto, 1687
sovraccarico *m*, 2889
sovralimentare, 490
sovralimentazione *f*, 489
sovrapressione *f*, 1710
sovraraffreddamento *m*, 3707
sovrasaturare, 3713
sovrasaturazione *f*, 3715
sovratemperatura *f*, 1711
sovrintendente *m*, 770
spandere, 2526
spazio *m*, 3555
spazio *m* morto, 767
spazio *m* occupato, 2801
spazio *m* tra controsoffitto e intradosso del solaio, 666
specifica *f*, 3562
specifico, 3561
spegnere, 1770, 3744
sperimentare, 1758
spesa *f*, 1101, 1756
spessore *m*, 3821
spessore *m* del bordo di uscita, 3876
spessore *m* della lamiera, 3421

spessore *m* dello strato, 2521
spettro *m* sonoro, 3550
spezzone *m* di tubo, 2960
spia *f* luminosa, 2949
spigolo *m*, 1547
spillare, 3758
spina *f*, 2990
spina *f* a vite, 3355
spingere, 490
spinta *f*, 489, 1438
spirale *f*, 3569
split-system *m*, 3575
sporcamento *m*, 2026
sporcizia *f*, 1335
sportello *m* a cerniera, 2320
spruzzare, 3580
spurgare, 430, 1417
spurgo *m*, 445, 1416
spurgo *m* di aria, 1189
stabile *m*, 545
stabilimento *m*, 1781, 4140
stabilità *f*, 3592
stadi multipli, 2751
stadio *m* di compressione, 906
stadio *m* di pressione, 3067
staffa *f*, 504, 755, 2214
staffa *f* del collare, 756
staffa *f* di collegamento, 977
staffa *f* di serraggio, 756
stagione *f* di riscaldamento, 2257
stagnatura *f*, 3856
stagno, 3097
stagno *m*, 3855
stampante *f*, 2554, 3085
stampante *f* di resoconto, 2580
stampo *m*, 1281
stanza *f*, 3295
stanza *f* da bagno, 383
starter *m*, 3605
starter *m* di lampade a scarica, 360
stato *m*, 3614
stato *m* dell'aria, 964
stato *m* energetico, 1636
statore *m*, 3619
stato *m* transitorio, 2789
stazione *f*, 3618
stazione *f* di bilanciamento, 358
stazione *f* di riduzione di pressione, 3062
stechiometrico, 3659
sterile, 2164
stima *f*, 1682, 1684
stimare, 1683
strategia *f* di controllo, 1033
stratificazione *f*, 3677
stratificazione *f* dell'aria, 163
strato *m*, 2520
strato *m* di ghiaccio, 2381
strato *m* di scorie, 3468
strato *m* intermedio, 2471

strato *m* limite, 499
strato *m* protettivo, 3111
strozzamento *m*, 3837
strozzare, 3836
strozzatura *f*, 3255, 3835
strumento *m* di misura, 2651
strumento *m* di regolazione, 1021
struttura *f* cellulare, 673
struttura *f* del soffitto, 664
struttura *f* di rinforzo, 3231
studio *m* di progetto, 1252
studio *m* economico del riscaldamento, 2247
stufa *f*, 2259
stufa *f* a gas, 2147, 2149
stufa *f* a gasolio, 2828
subappaltatore *m*, 3687
suddivisione *f* a zone, 4151
sughero *m*, 1086
suola *f*, 2222
suono *m*, 3536
suono *m* trasmesso attraverso l'aria, 77
superficie *f*, 262, 3731
superficie *f* bagnata, 4119
superficie *f* costruita, 259
superficie *f* di griglia, 2181
superficie *f* di raffreddamento, 1068
superficie *f* effettiva, 1551
superficie *f* estesa, 1763
superficie *f* filtrante, 1848
superficie *f* riscaldante della caldaia, 468
superficie *f* riscaldante primaria, 3080
superficie *f* scaldante, 2260
superficie *f* scaldante secondaria, 3373
supervisione *f*, 3717
supporto *m*, 504, 2214, 3620, 3729
supporto *m* antivibrante, 244
supporto *m* per cuscinetto, 392
supporto *m* per cuscinetto a bronzine, 3475
supporto *m* per cuscinetto a sfere, 362
supporto *m* per tubo, 2957, 2962
supporto *m* scorrevole, 3480
surriscaldamento *m*, 2888, 3712
surriscaldare, 3709
surriscaldato, 3710
surriscaldatore *m*, 3711
surriscaldatore *m* intermedio, 2472
svasatura *f*, 1916
sviluppo *m* batterico, 339
svuotamento *m*, 1614
switch *m* del computer, 938

torre *f* di raffreddamento,
1070, 4062
torre *f* di raffreddamento a
caduta, 3572
torre *f* di raffreddamento ad
aria forzata, 2017
torre *f* di raffreddamento a
flussi incrociati, 1133
torre *f* di raffreddamento a
pacco evaporante, 1843
torre *f* di raffreddamento a
piatti, 2982
torre *f* di raffreddamento a
spray, 3585
torre *f* di raffreddamento a
ventilazione meccanica,
2656
torre *f* di raffreddamento a
ventilazione naturale, 2758
torrino *m*, 3294
traboccare, 2884
traccia *f* di sporco, 3502
tracciante *m* per tubi, 2965
tracciare, 2178
tracciatore *m*, 3874
trafilazione *f*, 3837
traliccio *m*, 2033
tramoggia *f*, 2098, 2325
transistor *m*, 3884
trappola *f* di calore, 2293
trascinamento *m*, 1414, 1649
trascinare, 1648
trasduttore *m*, 3877
trasferimento *m*, 3878
trasferimento *m* di calore,
2292
trasferire, 3879
trasformatore *m*, 3882
trasmettere, 3879
trasmettitore *m*, 3890
trasmissione *f*, 3888
trasmissione *f* a cinghia, 400
trasmissione *f* a cinghia
trapezoidale, 3977
trasmissione *f* analogica, 218
trasmissione *f* del calore,
2289
trasmissione *f* digitale, 1305
trasmissione *f* di massa, 2635
trasmissione *f* di umidità,
2720
trasmissione *f* laterale, 1914
trasportatore *m*, 1043
trasporto *m*, 3878
trasporto *m* di combustibile,
2084, 2089
trasporto *m* di umidità, 2719
trasporto *m* eccessivo di
gocce di acqua in un
generatore, 475
trattamento *m*, 3896
trattamento *m* dell'acqua,
4087
trattamento *m* dell'aria, 171
trave *f*, 387

trazione *f* diretta, 1319
tronco *m* di canale, 1500
troppo-pieno *m*, 2883
tubatura *f*, 972
tubazione *f*, 2950, 2958
tubazione *f* della condensa,
946
tubazione *f* di by-pass, 596
tubazione *f* di gas, 2142
tubazione *f* di spillamento,
431
tubazione *f* di vapore, 3640
tubazione *f* principale, 2608,
2610
tubi *m,pl* concentrici, 942
tubo *m*, 2950, 3899
tubo *m* a gomito, 404
tubo *m* alettato, 1869
tubo *m* alettato spiroidale,
3570
tubo *m* a vuoto, 3946
tubo *m* Bourdon, 501
tubo *m* brasato, 516
tubo *m* capillare, 627
tubo *m* di acciaio, 3650
tubo *m* di aspirazione, 3701
tubo *m* di bilanciamento,
352
tubo *m* di calore, 2276
tubo *m* di collegamento, 980
tubo *m* di equalizzazione,
352
tubo *m* di fumo, 1891, 3495
tubo *m* di iniezione, 2440
tubo *m* di livello, 3603
tubo *m* di plastica, 2980
tubo *m* di protezione, 3113
tubo *m* di rame, 1082
tubo *m* di ritorno, 3265
tubo *m* di scarico, 1423,
4054
tubo *m* di sicurezza, 1750
tubo *m* di spurgo, 434
tubo *m* di ventilazione, 3996
tubo *m* di Venturi, 3997
tubo *m* filettato, 3826
tubo *m* flessibile, 1936, 1939,
2330
tubo *m* flessibile metallico,
2674
tubo *m* fluorescente, 2545
tubo *m* in gomma, 3319
tubo *m* montante, 3282
tubo *m* Pitot, 2971
tubo *m* spruzzatore, 3587
tunnel *m* frigorifero, 1075
turbina *f*, 3906
turbina *f* a combustione, 862
turbina *f* a gas, 2152
turbina *f* a vapore, 3646
turbina *f* eolica, 4129
turbocompressore *m*, 3908
turbogeneratore *m*, 3909
turbolatore *m*, 4021

turbolatore *m*
dell'evaporatore, 1706
turbolenza *f*, 3910
turbosoffiante *f*, 3907
tutto l'anno, 4145

udibile, 305
udibilità *f*, 304
ugello *m*, 2438, 2791
ugello *m* del bruciatore, 572,
579
ugello *m* del bruciatore di
nafta, 2813
ugello *m* di flusso, 1964
ugello *m* di iniezione, 2439
ugello *m* di uscita, 2870
ugello *m* spruzzatore, 3586
umidificare, 2354
umidificatore *m*, 2353
umidificatore *m* ad
evaporazione, 1701
umidificatore *m* a
nebulizzazione, 300
umidificatore *m* a vapore,
3635
umidificatore *m* di aria, 131
umidificatore *m* di aria
ambiente, 3298
umidificazione *f*, 2352
umidificazione *f* di aria, 130
umidità *f*, 1165, 1172, 2357,
2712
umidità *f* relativa, 3233
umido, 2351, 2711
uncino *m*, 2324
unità *f*, 3925
unità *f* asservita, 3472
unità *f* condensante, 960
unità *f* condensante
raffreddata ad aria, 96
unità *f* condensante split
(separata), 3574
unità *f* di misura, 3930
unità *f* di misura pari a
9809.5 W, 469
unità *f* di raffreddamento
(autonoma), 3928
unità *f* di reintegro aria,
2614
unità *f* di riscaldamento
(autonoma), 3929
unità *f* di trattamento
dell'aria, 126
unità *f* evaporante, 1707
unità *f* monoblocco caldaia -
bruciatore, 456
unitario, 3926
unità *f* terminale (di un
circuito di aria), 166
unità *f* ventilante, 3990
uscita *f*, 2867
uscita *f* analogica, 217
uscita *f* a stampa (tabulato)
di un elaboratore, 933
uscita *f* dell'aria, 142

uscita *f* dell'aria di
   immissione, 3724
usura *f* e strappo *m*, 4093
utensile *m*, 3557, 3863
utensile *m* per svasare, 1920

vacuometro *m*, 3945
vagone *m* cisterna, 3756
valore *m*, 3950
valore *m* di default, 1203
valore *m* di taratura, 3410
valore *m* limite massimo,
   3833
valore *m* obiettivo di
   prestazione energetica,
   1638
valutare, 1683
valutazione *m*, 1684
valvola *f*, 3951
valvola *f* a clapet, 1915
valvola *f* a contrappeso,
   4101
valvola *f* ad azione rapida,
   3132
valvola *f* a disco, 1356
valvola *f* a doppia sede, 1404
valvola *f* a due vie, 3918
valvola *f* a farfalla, 590
valvola *f* a flusso diretto,
   1326
valvola *f* a fuoco, 1893
valvola *f* a galleggiante, 1945
valvola *f* a ghigliottina, 2207
valvola *f* a lamella, 3187
valvola *f* a membrana, 1279,
   2666
valvola *f* antisifone, 243,
   3941
valvola *f* a pressione
   costante, 992
valvola *f* a saracinesca, 2156
valvola *f* a sfera per liquido,
   795
valvola *f* a shunt, 3435
valvola *f* a soffietto, 398
valvola *f* a solenoide, 2607
valvola *f* a solenoide
   elettromagnetica, 3524
valvola *f* a spillo, 2763
valvola *f* a squadra, 231
valvola *f* a tre vie, 3832
valvola *f* a tre vie di
   regolazione, 3222
valvola *f* barometrica, 370
valvola *f* di alimentazione e
   ritegno, 1816
valvola *f* di ammissione,
   2445
valvola *f* di angolo, 1089
valvola *f* di arresto, 1150
valvola *f* di blocco, 443
valvola *f* di by-pass, 597
valvola *f* di controllo di
   flusso, 1958

valvola *f* di contropressione,
   331
valvola *f* di deviazione, 1385
valvola *f* di espansione, 1753
valvola *f* di espansione a
   galleggiante, 1746
valvola *f* di livello costante,
   991
valvola *f* di non ritorno,
   2788
valvola *f* di regolazione,
   3223
valvola *f* di regolazione della
   pressione di
   condensazione, 958
valvola *f* di regolazione di
   portata, 1968
valvola *f* di riduzione di
   pressione, 3063
valvola *f* di ritegno, 723
valvola *f* di ritegno con
   filtro, 3676
valvola *f* di scarico, 446,
   1352, 1736, 1995, 3236
valvola *f* di scarico del
   compressore, 921
valvola *f* di scarico di
   emergenza, 1609
valvola *f* di sfiato, 435, 450
valvola *f* di sfogo di aria,
   3995
valvola *f* di sicurezza, 3066,
   3335
valvola *f* di spurgo, 1424
valvola *f* di taratura, 359
valvola *f* di troppo pieno,
   2885
valvola *f* miscelatrice, 2700
valvola *f* motorizzata, 2728
valvola *f* per radiatore, 3149
valvola *f* riduttrice della
   pressione del vapore, 3643
valvola *f* terminale, 3774
valvola *f* termostatica, 3819
vano *m* tecnico, 3415
vapore *m*, 3622, 3959
vapore *m* a bassa pressione,
   2603
vapore *m* acqueo, 4089
vapore *m* ad alta pressione,
   2312
vapore *m* di scarico, 1734
vapore *m* nascente, 1926
vapore *m* saturo, 3339
vapore *m* saturo secco, 1479
vapore *m* secco, 1480
vapore *m* umido, 4118
vapore *m* vivo, 2567
vaporizzazione *f*, 3957
variabile *f* controllata, 1026
variabile *f* di correzione,
   1091
variazione *f* di carico, 2576
vasca *f* di disgelo, 1312
vaschetta *f* di raccolta, 1422

vaso *m*, 4000
vaso *m* di alimentazione,
   1821
vaso *m* di espansione, 1752
vaso *m* di espansione a
   membrana, 2665
velocità *f*, 3565, 3978
velocità *f* assiale, 319, 685
velocità *f* dell'acqua, 4090
velocità *f* dell'aria, 160
velocità *f* del suono, 3551
velocità *f* di accumulo, 719
velocità *f* di caduta, 3981
velocità *f* di combustione,
   863
velocità *f* di congelamento,
   2054
velocità *f* di decadimento,
   1192
velocità *f* di flusso, 1971
velocità *f* di raffreddamento,
   1067
velocità *f* di uscita, 1739,
   2873
velocità *f* frontale, 1777
velocità *f* periferica, 3858
velocità *f* terminale, 3775
ventilatore *m*, 1790, 3993
ventilatore *m* a bassa
   pressione, 2599
ventilatore *m* a biforcazione,
   406
ventilatore *m* a doppia
   aspirazione, 1399, 1406
ventilatore *m* a flusso misto,
   2691
ventilatore *m* a parete, 4026
ventilatore *m* assiale, 317
ventilatore *m* assiale
   elicoidale, 3955
ventilatore *m* assiale
   intubato, 3900
ventilatore *m* a tetto, 3294
ventilatore *m* centrifugo, 689
ventilatore *m* con pale a
   profilo alare, 125
ventilatore *m* da parete, 4033
ventilatore *m* di circolazione,
   746
ventilatore *m* di estrazione di
   aria, 1727
ventilatore *m* di tiraggio ad
   induzione, 1429
ventilatore *m* elicoidale,
   3100
ventilatore *m* per impianto
   ad alta velocità, 2317
ventilatore *m* tangenziale,
   1131, 3754
ventilazione *f*, 3985, 3988,
   3991
ventilazione *f* a getto, 2499
ventilazione *m* per
   depressione, 1737

ventilazione *f* trasversale, 1139
ventilconvettore *m*, 1796
vento *m*, 4122
ventola *f*, 2396
vernice *f*, 2512
vernice *f* bituminosa, 416
vernice *f* nebulizzata, 2908
vetrata *f*, 2910
vetro *m*, 2168
vetro *m* cellulare, 2002
vetro *m* spia, 3441
vibrazione *f*, 4002
viscosimetro *m*, 4006
viscosità *f*, 4007
viscosità *f* dinamica, 1536
visore *m*, 1361
visualizzare, 1362

vita *f* effettiva, 1554
vite *f*, 3353
vite *f* senza fine, 4142
vitone *m*, 488
volano *m*, 1998
volantino *m*, 2212, 2213
volatile, 4011
voltaggio *m*, 4013
voltmetro *m*, 4014
volume *m*, 4015
volume *m* costruito, 260
volume *m* dello spazio morto, 769
volume *m* di acqua di una caldaia, 480
volume *f* di combustione, 865
voluta *f* del ventilatore, 1805

vuoto *m*, 3940
vuoto *m* spinto, 1202, 2315

Watt, 4091
wattmetro *m*, 4092

zero *m* (assoluto), 4147
zincare, 2108
zincato, 2109
zincato a caldo, 2334
zincatura *f*, 2110
zoccolo *m*, 375, 376, 3509
zona *f*, 4149
zona *f* di benessere, 872
zona *f* neutra, 2767
zona *f* occupata, 2802
zone multiple, 2754

FOR *a Better* *Future*

New TECHNOLOGY

RC GROUP

CFC FREE

Attilio Bascioli

Magyar

A szavak felsorolásánál a rövid és hosszú magánhangzók között nem tesz különbséget a szerkesztés, igy előfordulhat, hogy a szavak nem szoros ABC ben követik egymást./Például: ágy, agy, ágyazat, ajánlat, stb/

szabadkifúvású léghűtő, 2036
szabadkifúvású
 légkondicionáló, 2034
szabadságfok, 1219
szabályoz, 52, 1015, 3221
szabályozás, 54, 1014, 3224
szabályozás működtető elem,
 1031
szabályozási eltérés, 1020
szabályozási feladat, 1023
szabályozási pont, 1030
szabályozási stratégia, 1033
szabályozási tartomány, 1032
szabályozható, 53
szabályozó, 1027
szabályozó alállomás, 3690
szabályozó algoritmus, 1017
szabályozó elem, 1022
szabályozó hurok, 1028
szabályozó készülék, 1021
szabályozó T idom, 3222
szabályozószelep, 3223
szabályozott közeg, 1025
szabályozott légállapot, 1024
szabályozott változó, 1026
szabványos szín, 3598
szag, 2803
szagszűrő, 2804
szagtalanító, 1236
szakasz, 3374
szakaszolás, 4151
szakaszos, 2473, 3375
szakaszos üzemelés, 1156
szakértői rendszer, 1759
szakmérnök, 3092
szál, 1828
szálanyagos szűrő, 1830
szálas por, 1829
szállítás, 1228
szállítási határidő, 1230
szállítási hőmérséklet, 1229
szállítmány, 634
szállító szalag, 1044
szállítóberendezés, 1043
szállított levegő nyílás, 3723
szállított levegő távozónyílás,
 3724
szállópernye, 1997
száloptika, 1827
szám, 2794
számítás, 603
számítógép, 922
számítógép bemeneti adat,
 929
számítógép futás, 934
számítógép kapcsoló, 938
számítógép kimeneti adat,
 931
számítógép kód, 926
számítógép memória, 930,
 937
számítógép túlterhelés, 932
számítógépes hardver, 928
számítógépes központi
 vezérlő rendszer, 882

számítógépes nyomtatás, 933
számítógépes rendszer, 925
számítógépes szimuláció, 935
számítógépes szoftver, 936
számítógépes tervezés, 927
számítógéppel segített
 gyártás, 924
számítógéppel segített
 tervezés, 923
számjegyes, 1304
számlap, 1274
számlapos hőmérő, 1275
számtani, 264
száraz, 1457
száraz elpárologtató, 3613
száraz eredő hőmérséklet,
 1476
száraz expanzió, 1464
száraz expanziós
 elpárologtató, 1465
száraz gőz, 1480
száraz hőmérő, 1462
száraz hűtő, 1466
száraz kompresszió, 1463
száraz kondenzvezeték, 1477
száraz kondenzvezetékes
 fűtési rendszer, 1478
száraz levegő, 1459
száraz réteges hűtő, 1472
száraz telített gőz, 1479
száraz típusú elpárologtató,
 1482
száraz típusú léghűtés, 1481
szárazdugattyús
 kompresszor, 1475
szárazgőztartalom, 1474
szárazhőmérséklet, 1461
szárazjég, 1467
szárazlevegős hűtő, 1460
szárazság, 1473
szárít, 1224, 1458
szárítás, 105, 1244, 1468
szárító, 106, 1436
szárító csökígyó, 1437
szárítóberendezés, 1471
szárítóközeg, 41, 1243
szárítószekrény, 1469
szárny, 420
szárnykerék, 3099
szárnykerekes anemométer,
 1204
szárnylapát, 3954
szegecs, 3283
szegecsel, 3284
szegély, 376, 1547
szegélyfűtő, 377
szegélyfűtőtest, 378
szegélyléc, 384
szekrény, 503
szekunder fűtőfelület, 3373
szekunder hűtőfolyadék,
 3371
szekunder kondenzátor, 3370
szekunder levegő, 3369
szekunder szárítás, 3372

szél, 4122
szélcsatorna, 4128
szelep, 3951
szelepnyílás, 3952
szelepülés, 3953
szélkerék ventilátor, 4129
szellőzés, 3991
szellőzőcső, 3996
szellőzőkürtő, 3992
szellőztet, 58, 3986
szellőztetés, 3988, 3994
szellőztetett mennyezet, 3987
szellőztetett padló, 1951
szellőztető, 3985
szellőztető készülék, 3990
szellőztető nyílás, 3993
szellőztető szerkezet, 3995
szélnyomás, 4127
szemcse, 2193
szemcseleválasztó, 2194
szemétégető, 2402
szén, 628, 782
széndioxid, 629
szenesedés, 632
szénhidrogén, 2363
szénmonoxid, 633
szennyez, 1006
szennyezés, 3002
szennyező, 3001
szennyezőanyag, 1005, 3000
szennyeződés, 1335
szennyvíz, 1557, 4055
szennyvízcsatorna, 3411
szénpor, 785
szénsav, 631
szénszállítás, 784
szénszűrő, 710
széntároló, 789
széntüzelés, 788
széntüzelésű, 786
széntüzelésű kazán, 787
szerel, 285
szerelés, 286, 1677
szerelési terv, 4136
szerelési utasítás, 287
szerelési utasítások, 2736
szerelő, 1897
szerelőnyílás, 2735
szerelvény, 1898
szerkezet, 994, 1267
szerkezeti anyag, 2637
szerkezeti elem, 995, 3683
szerszám, 3863
szerviz, 3396
szervizelés, 3401
szervizelhetőség, 3397
szervizkapcsolás, 3398
szervomotor, 3403
szervoszabályozás, 3402
szerződés, 1010
szétszerel, 1336
szétszóródás, 1300
szétszóródási terület, 1301
szétszórt, 1296
szétterjedő füstcsóva, 2994

Nederlands

controle *f*, 721, 1014
controlekamer *f*, 1018
controleren, 722, 1015
controleur *m*, 1027
convectie *f*, 1034
convectief, 1036
convectiestroom *m*, 1037
convectieverwarmingstoestel
   n, 1035
convector *m*, 1038
conversie *f*, 1039
correctief onderhoud *m*,
   1092
corrigerende variabele *f*,
   1091
corrigeren van fouten, 1190
corroderen, 1093
corrosie *f*, 1094
corrosiebescherming *f*, 1096
corrosiebestendig, 239
corrosiebestrijder *m*, 1095
corrosief, 1098
corrosie-inhibitor *m*, 1095
corrosievast, 1097
corrosiewerend, 239
cos phi *n*, 1582
cryogeen, 1140
cupbrander *m*, 3307, 3308
curve *f*, 1147
curvimeter *m*, 2965
cycloon *m*, 1158
cyclus *m*, 1155

dagproduktie *f*, 1161
dagtank *m*, 1163
dagtemperatuur-bereik *n*,
   1162
dagverlenging *f*, 1182
dak *n*, 3291
dakberegeningskoeling *f*,
   3292
dakluchtbehandelingscentrale
   f, 3293
dakventilator *m*, 3294
daling *f* van de luchtstroom,
   112
damp *m*, 2096, 3959
dampbekrachtigde
   aandrijving *f*, 3961
dampbelblokkade *f*, 3964
dampbelvorming *f*, 1540
dampdetector *m*, 3962
dampdichte bekleding *f*, 792
dampdichte laag *f*, 3960
dampdruk *m*, 3966
dampen *pl*, 3497
dampkring *m*, 290
dampmengsel *n*, 3965
dampremmende laag *f*, 792
dampvertragende laag *f*,
   3967
dauw *m*, 1268
dauwpunt *n*, 1269
dauwpunthygrometer *m*,
   1271

dauwpuntsverhoging *f*, 1272
dauwpuntsverlaging *f*, 1270
dauwpunt *n* van uitgaande
   koelerlucht *m*, 248
debiet *n*, 1955, 1966, 3838
debietmeter *m*, 1963
debietregelaar *m*, 1967
debietregelafsluiter *m*, 1958,
   1968
debietregeling *f*, 1957
decibel *m*, 1194
decipol, 1195
declinatie *f*, 1196
decompressieklep *f*, 921
decompressietoestel *n*, 1198
deellast *m*, 2923
deeltje *n*, 2921
deeltje *n* in suspensie, 3736
deeltjesgrootte *f*, 2922
defect *n*, 518, 1810
defect *n* door verbranding,
   581
deflector *m*, 2203
dehydratie *f*, 1225
dekenisolatie *f*, 421
deklaag *f*, 791
dekplaat *f*, 1115
deksel *n*, 488, 1113, 2537
demineralisatie, 1233
demodulatie *f*, 1234
demonteren, 1336
dempen, 1166, 1167
demping (trilling) *f*, 1171
desorptie *f*, 1255
detail *n*, 1258
detailleren, 1259
detailtekening *f*, 1260
detector *m*, 1263, 3386, 3387
deurvergrendeling *f*, 1390
diafragma *n*, 1277
diagram *n*, 1273
diameter *m*, 1276
dichtheid *f*, 1235, 3845
dichtheidsonderzoek *n*, 2527
dichtheid
   ventilatorinlaatlucht *f*,
   1791
dichtingsring *m*, 4049
dichtstoppen, 2991
dichtvriezen, 2045
diëlectrisch, 1282
diëlektrische constante *f*,
   1283
diëlektrische ontdooiing *f*,
   1285
dienst *m*, 3396
dienstaansluiting *f*, 3398
dienstverlening *f*, 3396
diepvacuüm *m*, 2315
diepvriesfabriek *f*, 2051
diepvriezen, 1201
dieselmotor *m*, 1286, 1287
diffuse reflectie *f*, 1298
diffusie *f*, 1300

diffusie-absorptiesysteem *n*,
   1303
diffusiecoëfficiënt *m*, 1302
diffuus, 1296
digitaal, 1304
digitale overdracht *f*, 1305
digitale regeling en sturing *f*
   (DDC), 1318
dikte *f*, 3821
dilatatievoeg *f*, 1748
directe aandrijving *f*, 1319
directe expansie *f*, 1464
directe expansie batterij *f*,
   1321
directe expansie spiraal *f*,
   1321
directe koeling, 1314
directe rookgasafvoer *m*,
   1320
direct-expansie koeling *f*,
   1315
direct expansie systeem *n*,
   1323
directgestookte luchtverhitter
   m met ventilator, 2021
directgestookte
   luchtverwarmer *m*, 1324
directuitblazende luchtkoeler
   m, 2036
directuitblazend
   luchtbehandelingsapparaat
   n, 2034
direct-verwarmd
   warmwatervoorraadvat *n*,
   1317
direct verwarmingsoppervlak
   n, 3080
direct warmte-uitwisselend
   systeem *n*, 1327
dispersiestroming *f*, 1360
distillatie *f*, 1367
distributiekoelhuis *n*, 1359
dode leiding *f*, 1186
dode ruimte *f*, 767
dode tijd *m*, 1187
dode zone *f*, 1188
doekfilter *m*, 1774
dompelcondensor *m*, 3691
dompelen, 2392
dompeltank *m*, 1312
dompelthermostaat *m*, 2394
door de wind aangedreven
   ventilator *m*, 4129
doorgelaten deel *n* van
   opvallende straling, 3889
doorlaat *m*, 3004
doorlaatbaarheid *f*, 2942
doorschakelactie *f*, 3859
doorslagwaarde *f*, 1284
doorsnede *f*, 1135, 3374
doorsnede-oppervlak *n*, 1136
doorsnedetekening *f*, 3376
doorspoelen, 1992, 3123
doorstralend geluid *n*, 520

doorstromende hoeveelheid
 f, 1966
doorstroomafsluiter *m*, 1326
doorstroom-
 warmwaterbereider *m*,
 2452
doorstroomwarmwater-
 voorziening *f*, 1328
doorverbinding *f*, 2503
doos *f*, 503, 643
DOP test *m*, 1391
doseerpomp *f*, 4018
doseren, 1393
dosering *f*, 1392
doven, 1770
draad *m*, 3822
draad *n*, 4131
draadflens *m*, 3824
draadhuls *f*, 3825
draadpijp *f*, 3826
draadsnijden, 3758
draadsnijder *m*, 3823
draadtap *m*, 3757
draadtappen, 3758
draagbaar, 3005
draaggolffrequentie *f*, 636
draagvlak *n*, 391
draaiend, 3306
draaiing *f*, 3313
draaimoment *n*, 3865
drempelwaarde *f*, 3833
driepijpsluchtbe-
 handelingssysteem *n*, 3828
driepijpssysteem *n*, 3829
drietreksketel *m*, 3827
driewegafsluiter *m*, 3832
driewegklep *f*, 3832
driewegplugkraan *f*, 3831
drijvende kracht *f*, 2725
drinkwater *n*, 753, 1440,
 3009
drinkwaterkoeler *m*, 1441
drogebol temperatuur *f*,
 1461
drogebol thermometer *m*,
 1462
droge condensaatleiding *f*,
 1477
droge filter *m*, 1466, 1472
droge lucht *f*, 1459
droge luchtkoeler *m*, 1460
droge luchtkoeler *m* (niet
 bevochtigd), 1481
drogen, 1458
droger *m*, 1226, 1436
droge resulterende
 temperatuur *f*, 1476
droge stoom *m*, 1480
droge verdamper *m*, 1465,
 1482
droge verzadigde stoom *m*,
 1479
droging *f*, 1244, 1468
droging, 2041
drogingsfilter *m*, 1855

droog, 1457
droogheidsgraad *m* (stoom),
 1474
droog ijs *n* (vast koolzuur),
 1467
drooginstallatie *f*, 1471
droogkast *f*, 1469
droogmiddel *n*, 1243
droogte *f*, 1473
druipend water *n*, 1442
druk *m*, 3041
drukbegrenzingsapparaat *n*,
 3057
drukbestendig, 3071
drukbeveiligingsschakelaar
 m, 3045
drukegalisatie *f*, 3049
drukgradiënt *m*, 3054
drukhoogte *f*, 3055
drukker *m*, 3085
drukketel *m*, 3072
drukklasse *f*, 3060
drukknopbediening *f*, 3125
druklager *n*, 3842
druklucht *f*, 895
drukluchtverstuivingsbrander
 m, 74
drukmeter *m*, 3053
drukmeting *f*, 3059
drukproef *f*, 909, 3069
drukreduceertoestel *n*, 3062
drukregelaar *m*, 3043, 3065
drukregelventiel *n*, 992
drukschakelaar *m*, 3068
druksmering *f*, 2019
druktrap *m*, 3067
drukval *m*, 3048
drukvat *n*, 908, 3072
drukveiligheidsklep *f*, 3066
drukvereffening *f*, 3049
drukverhoger *m*, 3075
drukverhogingspomp *f*, 493
drukverlaging *f*, 3064
drukverlies *n*, 2220, 2587,
 3058
drukverminderingspunt *n*,
 3061
drukverschil *n*, 3046
drukverschilbe-
 veiligingsschakelaar *m*,
 3047
drukverschil *n* over rooster,
 2191
drukverstuivingsbrander *m*,
 2655, 3042, 3056
drukverval *n*, 3054
druppel *m*, 1448
druppelcondensatie *f*, 1450
druppelgrootte *f*, 1452
druppeltje *n*, 1449
druppelvanger *m*, 637, 1439,
 1451
dubbel, 1507
dubbelaanzuigende ventilator
 m, 1406

dubbele aanzuigstijgleiding *f*,
 1407
dubbele bocht *f*, 3260
dubbele vork *m*, 407
dubbelglas raam *n*, 1398
dubbelthermostaat *m*, 1491
dubbelwerkende compressor
 m, 1395
dubbelzittingafsluiter *m*,
 1404
duur *m*, 1509, 2531
duurzaam, 2785
duurzaamheid *f*, 1508
dwarspijpketel *m*, 1138
dwarsstroom *m*, 1128
dwarsventilatie *f*, 1139
dynamisch, 1530
dynamisch drukverlies *n*,
 1534
dynamische druk *m*, 1535,
 3980
dynamische grafische
 afbeeldingen *pl*, 1532
dynamische viscositeit *f*,
 1536
dynamisch gedrag *n*, 1531
dynamo *m*, 2162
dynamometer *m*, 1537

economie *f*, 1543
economiser, 1542
een geheel vormend, 3926
eenheid *f*, 3925

éénkanaalslucht-
 behandelingsinstallatie *f*,
 3450
éénkanaalssysteem *n*, 3451
éénpijpssysteem *n*, 3454

eenpijpsverwarming *f*, 2833

ééntraps, 3455

eerste vulling *f*, 2436
effect *n*, 1549
effectief, 1550
effectief oppervlak *n*, 1551
eigendomskosten *pl*, 2891
eigen frequentie *f*, 2759
eigen trillingsgetal *n*, 2759
einddruk *m*, 1866
eindklep *f*, 3774
eindsnelheid *f*, 3775
ejecteur *m*, 1559
elasticiteit *f*, 1561
elastomeer *f*, 1562
elekrische koelkast *f*, 1587
elektriciteit *f*, 1575
elektriciteitsbehoefte *f*, 1566
elektrisch, 1564, 1565
elektrische centrale *f*, 3019,
 3022

elektrische
  contactthermometer *m*,
  1571
elektrische convector *m*,
  1572
elektrische kachel *f*, 1573
elektrische storing *f*, 1568
elektrische verwarming *f*,
  1574
elektrische weerstand *m*,
  1570
elektrische
  weerstandsverwarming *f*,
  3252
elektrisch hulpsysteem *n*, 311
elektrisch lassen *n*, 261
elektrisch potentiaal *n*, 1577
elektrisch schema *n*, 1567
elektrode *f*, 1589
elektrode ketel *m*, 1590
elektrolytisch, 1591
elektrolytisch koppel *n*, 1592
elektromagneet *m*, 1593
elektromotor *m*, 1576
elektromotorische kracht
  (emk) *f*, 1594
elektronica *f*, 1598
elektronisch, 1595
elektronische versterker *m*,
  1596
elektropneumatisch, 1599
elektrostatisch filter *n*, 1600,
  1601
element *n*, 1602
emissie *f*, 1611
emulsie *f*, 1615
endothermisch, 1618
energie *f*, 1619
energiebeheer *n*, 1630
energie beheer- en
  regelsysteem *n*, 1631
energiebeheerfunctie *f*, 1632
energiebeschikbaarheid *f*,
  1621
energiebesparing *f*, 1622
energiebron *f*, 1635
energie-economie *f*, 1625
energie-factor *m*, 1627
energiegebruik *n*, 1623
energiegebruiksonderzoek *n*,
  1620
energiekosten *pl*, 1624
energie-niveau *n*, 1628, 1636
energie-opwekking *f* op
  eigen terrein, 2837
energieoverdracht-
  sverhouding *f*, 1637
energiepotentieel *n*, 1633
energie-rendementsfactor *m*,
  1626
energieverlies *n*, 1629
energievormen *pl*, 2023
enkelbeglaasd raam *n*, 3452
enkelbladig, 3456
enkelglas raam *n*, 3452

enkelschots, 3456
enkeltraps, 3455
enkelvoudig circuit *n*, 3447
enkelvoudigwerkende
  compressor *m*, 3449
enthalpie *f*, 1645
enthalpie diagram *n*, 1647
entropie *f*, 1651
entropie diagram *n*, 1653
erosie *f*, 1679
etageverwarming *f*, 245
eupatheoscoop *m*, 1685
eutectisch, 1686
evenwicht *n*, 1672
evenwichtig, 347
evenwichtsdruk *m*, 354
evenwichtsleiding *f*, 352
evenwichtstemperatuur *f*,
  353, 1673
excentrisch, 1541
exergie *f*, 1717
expansie *f*, 1741
expansiebocht *f*, 886, 1743
expansiebocht *m*, 1749
expansiekoelsysteem *n*
  (gesloten), 1333
expansiekoelsysteem *n*
  (open), 1322
expansieleiding *f*, 1750
expansiespiraal *f*, 1745
expansiestuk *n*, 885
expansietank *m*, 1752
expansievat *n* (open), 2219
expansieventiel *n*, 1753
expansieventielcapaciteit *f*,
  1754
expansievlotterventiel *n*,
  1746
experiment *n*, 1757
experimenteren, 1758
expert systeem *n*, 1759
explosie *f*, 1760
exponent *m*, 1761
extern, 1765

fabriceren, 2630
fabriek *f*, 1781, 4140
fabrieksgemonteerd, 1783
fabrieksingesteld, 1784
fabrikaat *n*, 2629
factor *m*, 1779
fakkel *f*, 1916
fase *f*, 2946, 3651
fijnkool *f*, 3464
film *m*, 1839
filmvorming *f*, 1844
filter *n*, 1846
filtercontrolekraan *f*, 3676
filterdoek *m*, 1854
filterelement *n*, 1857
filteren, 1847
filterframe *n*, 1858
filterkamer *f*, 1852
filtermateriaal *n*, 1859
filtermedium *n*, 1859

filteroppervlak *n*, 1848
filterpapier *m*, 1860
filterraam *n*, 1858
filterreiniging *f*, 1853
filterrendement *n*, 1861
filterterugslagklep *f*, 3676
filterzak *m*, 1849
filtratie *f*, 1864
filtreertank *m*, 1862
filtreren, 1847
fitting *m*, 1898
flankerende
  geluidsoverdracht *f*, 1914
flareblok *n*, 1920
flarefitting *f*, 1918
flare-ijzer *n*, 1920
flaremoer *f*, 1919
flareverbinding *f*, 904, 1917
flashgas *n*, 1923
flashstoom *m*, 1926
flatverwarming *f*, 245
flens *m*, 1911
flensverbinding *f*, 1912
flessengas *n*, 496
flexibel, 1934
flexibele pijp *f*, 1939
flexibele verbinding *f*, 1935,
  1938
flexibel kanaal *n*, 1936
flexibiliteit *f*, 1933
flotatie *f*, 1952
fluidica *pl*, 1987
fluorkoolstof *f*, 1991
fossiele brandstof *f*, 2025
fout *f*, 1680, 1810
frame *n*, 2031
frame uit staalstrip *m*, 1930
freon *n*, 2056
frequentie *f*, 2057
frequentie responsie *f*, 2058
frictiering *m*, 2068
fris, 2059
frisheid *f*, 2062
frisheidsindex *m*, 2063
frontale snelheid *f*, 1777
frontoppervlak *n*, 1776
frontplaat *f*, 2069
fundament *n*, 2028
fundatie *f*, 375
fundatieplaat *f*, 381, 395

galvaniseren, 2108
galvanisering, 2110
garantie *f*, 2202
gas *n*, 2111
gasachtig, 2122
gasanalyse *f*, 2112
gasbelblokkade *f*, 3964
gasbrander *m*, 2114
gasconcentratie *f*, 2118
gasdetector *m*, 843
gasdicht, 2151
gasdistributie *f*, 2121
gasexplosie *f*, 2123
gasfilter *m*, 2125

kanaal *n*, 706
kanaalverloopstuk *n*, 494
kanaalwerk *n*, 1505
kant *m*, 226, 1547
kap *f*, 488, 614, 2323
karakteristiek *f*, 707
karakteristieke kromme *f*,
    708
katalysator *m*, 651
katathermometer *m*, 2505
kathode *f*, 652
kathodische bescherming *f*,
    653
kation *n*, 654
keerplaat *f*, 343
keerschot *n*, 1205
kelder *m*, 380, 668
kengetal *n*, 709
kenkleur *f*, 2383
kennissysteem *n*, 1759
kern *f*, 1084
kerngebied *n*, 1085
kernoppervlak *n*, 1085
kernsnelheid *f*, 685
ketel *m*, 455, 2899
ketel-brander eenheid *f*, 456
ketelcapaciteit *f*, 457, 476
ketelfrontplaat *f*, 461
ketelfundatie *f*, 472
ketelhuis *n*, 470, 473, 477
ketelinstallatie *f*, 473, 2256
ketellid *n*, 478
ketelmantel *m*, 458
ketel *m* met
    brandstofvultrechter, 2326
ketel *m* met schuinliggende
    waterpijpen, 2403
ketelmontage *f*, 472
ketelplaat *f*, 474
ketelreiniging *f*, 459
ketelrendement *n*, 460
ketelsteen *n*, 3348
ketelsteen verwijderen, 1241
ketelsteenverwijdering *f*,
    3350
ketelsteenvorming *f*, 2408,
    3349
ketelvermogen *n*, 469
ketelvoeding *f*, 462
ketelvoedingspomp *f*, 463
ketelvoedingswater *n*, 464
ketelvoedingswaterverwarmer
    m, 465
ketel *m* voor diverse
    brandstoffen, 1042, 2743
ketel *m* voor twee
    brandstoffen, 1488
ketel-watermantel *m*, 480
ketelwaterniveau *n*, 481
ketelwatertemper-
    atuurregeling *f*, 482
kettingrooster *m*, 3894
kettingroosterstookinrichting
    f, 699
keuring *f*, 23, 3776

keuringscertificaat *n*, 3779
keuringsproef *f*, 25
keuzeschakelaar *m*, 3378
kiem *f*, 2163
kiemdodend, 340, 2165
kiemvrij, 2164
kijkgat *n*, 2449, 2929
kijkglas *n*, 564, 3441
kinetisch, 2507
kist *f*, 503
klamp *m*, 755
klank *m*, 3536
klankleer *f*, 39
kleefmiddel *n*, 49
klembeugel *m*, 756
klemkoppeling *f*, 904
klepopening *f*, 3952
klepzitting *f*, 3953
kleur *f*, 831
kleurcode *m*, 832
kleurstof *f*, 1528
kleurtemperatuur *f*, 834
klevend, 50
klimaat *n*, 771
klimaatkamer *f*, 772, 1658,
    1663
klinken, 3284
klinknagel *m*, 3283
knaldemper *m*, 2740
knalpot *m*, 2740, 2741
kniestuk *n*, 1563
kniestuk *n* (van pijp), 2955
knop *m*, 2508
koelapparaat *n*, 1076, 3928
koelbatterij *f*, 1049, 1054
koelbereik *n*, 1066
koelcapaciteit *f*, 821, 1053
koelcircuit *n*, 3200
koelcompressor *m*, 3204
koelcurve *f*, 1055
koelcyclus *m*, 3209
koeleffect *n*, 731, 1058
koeleffect *m*, 3205
koelelement *n*, 1060, 3212
koelen, 726, 1045, 3202
koelend oppervlak *n*, 1068
koeler *m*, 729, 1048
koelgraaddag *m*, 1056
koelhuis *n*, 826
koelhuis *n* voor diverse
    producten, 2749
koeling *f*, 1050, 3207
koeling *f* door middel van
    panelen, 2912
koeling *f* met gedwongen
    luchtcirculatie, 2016
koeling *f* met lucht onder
    druk, 3044
koeling *f* met
    warmteterugwinning, 3216
koelinrichting *f*, 1065
koelinstallatie *f*, 3210
koelkast *f*, 3212
koellast *m*, 1062
koellucht *f*, 1052

koelmachine *f*, 3206
koelmachineruimte *f* 1065
koelmantel *m*, 1061
koelmedium *n*, 1064
koelmiddel *n*, 1046, 1051,
    3199
koelrendement *n*, 1059
koelruimte *f*, 732, 823, 3212
koelsnelheid *f*, 1067
koelspiraal *f*, 1054
koelsysteem *n*, 1069
koelsysteem *n* met verbruik
    van koelmedium, 1755
koeltechniek *f*, 3211
koeltoren *m*, 1070
koeltoren met
    filmverdamping *f*, 1843
koeltoren *m* met gedwongen
    luchtcirculatie, 2017
koeltoren *m* met natuurlijke
    luchtcirculatie, 2758
koeltorenmist *m*, 1074
koeltorenvulling *f*, 1071
koeltoren *m* zonder inbouw,
    3585
koeltoren *m* zonder vulling,
    3585
koeltunnel *m*, 1075
koelunit *n*, 1076
koelverlies *n*, 1063
koelvermogen *n*, 1067
koelvloeistof *f*, 1051
koelwater *n*, 1077, 3201
koepel *m*, 1387
kogelgewricht *n*, 364
kogellager *n*, 361
kogellagerstoel *m*, 362
koken, 454
kokend, 483
koken *n* in grenslaag, 1840
koken *n* van ketelleden, 655
koker *m*, 2098
kolengestookt, 786
kolengestookte ketel *m*, 787
kolengruis *n*, 3464
kolenleverantie *f*, 784
kolenopslag *m*, 789
kolenopslagplaats *f*, 789
kolenstof *n*, 785
kolenstoken *n*, 788
kolom *f*, 835
kolomradiator *m*, 838
kookplaat *f*, 2335
kookpunt *n*, 484
kool *f*, 782
koolfilter *m*, 710
koolmonoxyde *n*, 633
koolstof *f*, 628
koolstofdioxyde *n*, 629
koolwaterstof *f*, 2363
koolzuur *n*, 631
koperbekleding *f*, 1083
koppel *m*, 3865
koppelen, 975
koppeling *f*, 983, 1109

maateenheid *f*, 3930
maatschets *f*, 1310
maattekening *f*, 1310
machine *f*, 1639, 2605
machinekamer *f*, 2976
machinist *m*, 1640
MAC-waarde *f*, 2640
magazijn *n*, 3670
magere kool *f*, 2782
magneetklep *f*, 2607, 3524
managementprogramma *n*,
    2620
manchet *f*, 585, 3473
mangat *n*, 2621
manometer *m*, 2623, 3053
manometrisch, 2624
mano-vacuümmeter *m*, 894
mantel *m*, 643, 645, 2492
mantelbuis *f*, 3113
massa *f*, 562, 2633
massastroom *m*, 2634
materie *f*, 2638
mathematische
    processimulatie *f*, 2704
matrijs *f*, 1281
maximaal, 2639
maximaal toelaatbaar, 2642
maximaal toelaatbare
    concentratie *f*, 2640
maximale belasting *f*, 2641
mechanisch, 2654
mechanische asafdichting *f*,
    2658
mechanische circulatie *f*,
    2009
mechanische trek *m*, 2012
mechanische ventilatie *f*,
    2008
mechanisch warmte-
    equivalent *n*, 2657
meegevoerde waterdruppels
    pl, 1438
meenemen, 1649
meerfunctie regelaar *m*, 839
meertraps, 2751
meertrapscompressie *f*, 892
meertrapscompressor *m*,
    893, 2752
meertreksketel *m*, 2748
meervoudige condensor *m*,
    2750, 3574
meeslepen, 1649
meesleuren, 1648
meesleuren van water *n* met
    stoom, 475
meetaansluiting *f*, 2652
meetapparatuur *f*, 2650
meetinrichting *f*, 2650
meetinstrument *n*, 2158,
    2651, 2676
meetperiode *f* opgenomen
    electrisch vermogen, 1579
meetperiode *f* opgenomen
    elektrisch vermogen, 1581
meetplug *f*, 2652

meetschijf *f*, 2859
meettechniek *f*, 2653
meevoeren, 1648
membraan *n*, 1277, 2664
membraanafsluiter *m*, 1279,
    2666
membraanbarometer *m*, 224
membraancompressor *m*,
    1278
membraandoos *f*, 225
membraan-expansievat *n*,
    2665
mengbox *m*, 437
mengcircuit *n*, 2695
mengkamer *f*, 2694
mengkanaal *n*, 2697
mengkast *f*, 437, 2693
mengklep *f*, 2700
menglucht *f*, 2690
mengpomp *f*, 2698
mengregeling *f*, 2696
mengsel *n*, 2701
mengtoestel *n*, 436
mengverhouding *f*, 2699
messing, 512
metabolische warmte *f*, 2672
metalen filter *n*, 2673
metalen slang *f*, 2674
meteorologie *f*, 2675
meteorologische gegevens *pl*,
    4096
meter *m*, 2158, 2676
met gelijke enthalpie *f*, 2485
methode *f*, 2677
meting *f*, 2649
met meer zones, 2754
metselsteen *m*, 526
metselwerk *n*, 527
microklimaat *n*, 2680
micrometer *m*, 2681
middeldruk *m*, 2659
middellijn *f*, 1276
middelpuntvliedend, 687
middelpuntvliedende kracht
    f, 690
middeltemperatuur *f*, 2660
mijn *f*, 2684, 2969
milieu *n*, 1657
milieubewust ontwerp *n*,
    1660
milieutechniek *f*, 1661
minerale olie *f*, 2685
minerale wol *f*, 2686
minimale luchthoeveelheid *f*,
    2687
mist *m*, 2003, 2688
mistvorming *f* rond
    koeltoren, 1072
miswijzing *f*, 1680
model *n*, 2703, 2924
modelleren *n*, 2704
modem *n*, 2705
modulair, 2706

modulair
    luchtbehandelingsapparaat
    n, 3927
modulair opgebouwd
    luchtbehandelingssysteem
    n, 2707
modulatie versterking *f*,
    2709
modulerend, 2708
moduul *m*, 2710
moer *f*, 2796
moersleutel *m*, 3557, 4144
mof *f*, 585, 3473, 3509
mofverbinding *f*, 3567
Mollierdiagram *n*, 2721,
    3115
momentane elektrische
    belasting *f*, 1580
momenteel, 43
momentschakeling *f*, 3504
momentschakelmechanisme
    n, 3506
monster *n*, 3337
montage *f*, 1677, 2734, 3409
montage *f* op de
    bouwplaats, 3461
montage-opening *f*, 2735
montagevoorschrift *n*, 287
montagevoorschriften *pl*,
    2736
monteren, 285
monteur *m*, 1897
motor *m*, 1639, 2726
motorblok *n*, 453
motorfundatieplaat *f*, 2727
motorgeluid *n*, 2729
motoruitlaat *m*, 1642
motorvermogen *n*, 2730
motorwikkeling *f*, 2732
muur *m*, 4023
muurkoelelement *n*, 4025
muurventilator *m*, 4026,
    4033
mythyleen-blauwproef *f*,
    2678

naad *m*, 3367
naadloos, 3368
naadpijp *f*, 516
naaf *f*, 2350
naaldlager *n*, 2762
naaldventiel *n*, 2763
nachtinstelling *f*
    thermostaat, 3820
nachtstroom *m*
    bufferverwarming, 2807
nachttemperatuurverlaging *f*,
    3820
nachtverlaging *f*, 2769
nagalmkamer *f*, 3269
nagalmtijd *m*, 3270
nakoeler *m*, 63
nakoeling *f*, 3179
nameten, 722

uitleesvenster *n*, 1361
uitrusting *f*, 1674
uitschakelen, 1152, 3744
uitschakeling *f*, 1151
uitschakelpunt *n*, 1153
uitschakelwaarde *f*, 1153
uitslag *m*, 210
uitslijping *f*, 1679
uitstallen, 1362
uitstoot *m*, 1611
uitstromende stof *n*, 1558
uitstroming *f*, 1338, 1558, 1718
uitstroomcoëfficiënt *m*, 800, 1340
uitstroommondstuk *n*, 1964
uitstroomtuit *f*, 1964, 2791
uittrededruk *m*, 1343
uittredehoek *m*, 230
uittredende lucht *f*, 1738
uittredetemperatuur *f*, 1351
uittreesnelheid *f*, 1739
uitvlokkingspunt *n*, 1946
uitvoering *f*, 1716
uitvoering *f*
    computerprogramma, 934
uitvoeringsperiode *f*, 890
uitvriezen, 2042
uitwendig, 1765
uitwendige diameter *m*, 1766
uitwisselbaarheid *f* ho , 2466
uitwisselen, 1713
uitwisseling *f*, 1712
uitzending *f*, 1741
uitzettingscoëfficiënt *m*, 801, 1744
uniform tarief *n*, 1932
union *m*, 3924
universeel gasbrander *m*, 2115
universele brander *m*, 3931

vaan-anemometer *m*, 1204
vacumeren *n*, 1688
vacuum(verwarm-
    ings)systeem *n*, 3686
vacuüm *n*, 3940
vacuümkoeling *f*, 3944
vacuümleiding *f*, 3946
vacuümmeter *m*, 3945
vacuümonderbreker *m*, 3941
vacuümpomp *f*, 3947
vacuümproef *f*, 3949
vacuümstoomverwarming *f*, 3948
vacuümtest *m*, 3949
valdeur *f*, 3893
valsnelheid *f*, 3981
van lamellen voorzien, 1868
variabel debietsysteem *n*, 3970
variabele debietregeling *f*, 3969
variabele stroming *f*, 3971

variabele volumeregeling *f*, 3969
variabel luchtvolume *n*, 3968
variabel volumesysteem *n*, 3970
vast ingesteld, 3039
vast punt *m*, 1900
vastschroeven, 3354
vat *n*, 1455, 4000
veer *f*, 3590
veerbelast, 3591
veermanometer *m*, 500
veilig bij storing *f*, 1785
veilige overdruk *m*, 3333
veiligheidscircuit *n*, 3327
veiligheidsfactor *m*, 1780, 3332
veiligheidsklep *f*, 1609, 3066, 3335
veiligheidskoppeling *f*, 3328
veiligheidsschakelaar *m*, 519
veiligheidsschakelaar *f*, 3329
vel *n*, 3418
venster *n*, 4123
vensterbankrooster *m*, 3446
vensterbankverwarming *f*, 4126
vensterindeling *f*, 1825
ventilatie *f*, 3991
ventilatieapparaat *n*, 3990
ventilatie *f* door middel van injectie, 2499
ventilatieplafond *n*, 3987
ventilatierooster *m*, 3217
ventilatieschacht *f*, 3992
ventilatiesteen *m*, 78
ventilator *m*, 447, 448, 1790, 3993
ventilatoraanzuigopening *f*, 1798
ventilatorconvector *m*, 1796
ventilatorgeleideschot *n*, 1793
ventilatorgeluidsvermogen *n*, 1806
ventilatorhuis *n*, 1795, 1805
ventilatorkarakteristiek *f*, 1797
ventilator *m* met
    geprofileerde schoepen, 125
ventilator *m* met tangentiale en radiale stroming, 2691
ventilatorprestatiecurve *f*, 1801
ventilatorschoep *f*, 1794
ventilatortheorie *f*, 1800
ventilatortoebehoren *n*, 1792
ventilatorvermogen *n*, 1802
ventilatorwaaier *m*, 1804
ventilatorwaaierkonus *m*, 1807
ventilatorwetten *pl*, 1800
ventileren, 3988
venturi, 3997

verankerd, 1605
verankeren, 222
verbinden, 975
verbinding *f*, 983, 2500
verbinding met buitendraad *m*, 2616
verbindingsflens *m*, 978
verbindingsmanchet *n*, 1112
verbindingsmof *f*, 1112
verbindingspijp *f*, 980
verbindingsstaaf *f*, 981
verbindingsstang *f*, 981
verbindingsstuk *n*, 979, 986
verbintenis *f*, 1010
verblijf *n*, 3620
verblijfsgebied *n*, 2802
verblijfszone *f*, 2802
verbranden, 567
verbranding *f*, 844
verbranding *f* met
    theoretische
    luchthoeveelheid, 2932
verbrandingsdiagram *n*, 851
verbrandingsduur *m*, 852
verbrandingsinrichting *f*, 846
verbrandingskamer *f*, 847, 2100
verbrandingskamer *f* met
    mechanische trek, 2014
verbrandingslucht *f*, 866
verbrandingslucht *f*ho , 845
verbrandingsluchttoevoer *m*, 864
verbrandingsmiddel *n*, 855
verbrandingsoven *m*, 2402
verbrandingsprodukt *n*, 856
verbrandingsregelaar *m*, 849, 857
verbrandingsrendement *n*, 853
verbrandingsrest *f*, 858
verbrandingsruimte *f*, 859
verbrandingssnelheid *f*, 863
verbrandingstemperatuur *f*, 860
verbrandingsvolume *n*, 865
verbrandingswaarde *f*, 2264
verbrandingswarmte *f*, 854
verbruik *n*, 998
verbruikersaansluiting *f*, 997
vercokesing, 544
verdampen, 1689
verdamper *m*, 1703
verdamper-
    compressoreenheid *f*, 1707
verdamperdrukregelaar *m*, 1705
verdamper *m* met
    pijpenbundel, 3425
verdamper *m* met spiraal, 3423
verdamper *m* met tekort aan vloeistof, 3613
verdamperturbulator *m*, 1706

Spon's Asian and Pacific Construction Costs Handbook

Edited by **Davis Langdon & Seah**, Chartered Quantity Surveyors, Singapore

The economies of the Pacific rim are among the fastest growing of the world. Construction is therefore booming. This unique reference covers construction costs and procedures in the main countries of the area. It will be required reading for construction professionals in the Far East, for those contemplating work or investment there and for multinational companies, development agencies and others needing a world overview.

Contents: Introduction. The South East Asian construction industries. Country sections: Australia, Brunei, Canada, China, Hong Kong, Indonesia, Japan, Malaysia, New Zealand, Philippines, Singapore, South Korea, Thailand, UK, USA. Comparative data. Index.

September 1993: 234x156: c.400pp
Hardback: 0-419-17570-9

For further information and to order please contact: **The Promotion Dept., E & F N Spon**, 2-6 Boundary Row, London SE1 8HN Tel 071 865 0066 Fax 071 522 9623

Polski

gaz *m* miejski, 3090, 3873
gaz *m* niskokaloryczny, 2529
gaz *m* opałowy, 2083
gaz *m* płynny, 496
gaz *m* rozpuszczony, 1364
gaz *m* spalinowy, 1978
gaz *m* sprężony, 896
gaz *m* ubogi, 2529
gaz *m* wskaźnikowy, 3875
gaz *m* ziemny, 2760
gaz *m* zmieszany, 2692
gazomierz *m*, 2140
gazoszczelny, 2151
gazowy, 2122
gazy *m*, *pl* spalinowe, 1728
gazy *m*, *pl* wydechowe, 1728
gazyfikacja *f*, 2135
gejzer *m*, 2166
generator *m*, 2162
gęstość *f*, 1235
gęstość *f* dymu, 3493
gęstość *f* nasypowa, 563
gęstość *f* powietrza w
    wentylatorze, 1791
gęstość *f* zapisu, 2904
giętki, 1934
giętkość *f*, 1933
glina *f* ogniotrwała, 1880
glon *m*, 181
głęboka próżnia *f*, 1202,
    2315
głębokościomierz *m* zwilżany
    prętowy, 1311
głębokość *f* wężownicy, 806
głośność *f*, 2588
głowica *f*, 2217
głowica *f* cylindra, 1160
główny regulator *m*, 2636
główny układ *m* sterowniczy,
    2636
główny wykonawca *m*, 3082
gniazdko *n* wtykowe ścienne,
    4031
gniazdo *n* (do wtyczki), 3509
gniazdo *n* pomiarowe, 2652
gniazdo *n* zaworu, 3953
gnicie *n*, 1191, 1197
gorące powietrze *n*, 2333
gorący, 2332
gospodarka *f*, 1543
gospodarka *f* cieplna, 2247
gospodarka *f* energią, 1625
gospodarowanie *n* energią,
    1630
gotować, 454
gotowy do pracy, 3170
grabie *pl*, 3150
gradient *m*, 2176
gradient *m* ciśnienia, 3054
grafika *f*, 2179
grafika *f* dynamiczna, 1532
granica *f*, 497
granica *f* obciążenia, 2574
grubość *f*, 3821
grubość *f* blachy, 3421

grubość *f* krawędzi
    wypływu, 3876
grubość *f* warstwy, 2521
grunt *m*, 2196
grys *m*, 2193
grzałka *f* nurnikowa, 2393
grzejnik *m*, 2235, 3147
grzejnik *m* bezpośrednio
    ogniowy, 1325
grzejnik *m* gazowy
    pokojowy, 2147
grzejnik *m* gazowy
    promiennikowy, 2146
grzejnik *m* konwekcyjny,
    1035
grzejnik *m* listwowy, 376,
    377, 378
grzejnik *m* nurnikowy, 2393
grzejnik *m* płaszczyznowy,
    2914
grzejnik *m* płytowy, 2255,
    2914
grzejnik *m* płytowy stalowy,
    3649
grzejnik *m* powietrzny
    bezpośrednio ogniowy,
    1324
grzejnik *m* promieniujący,
    3141
grzejnik *m* słupowy, 838
grzejnik *m* z rur
    ożebrowanych, 1870
grzejnik *m* żeliwny, 649
gwarancja *f*, 2202
gwarancja *f* jakości, 3129
gwint *m*, 3822
gwint *m* rurowy, 2964
gwint *m* stożkowy, 3760
gwint *m* wewnętrzny, 1824
gwint *m* zewnętrzny, 2617
gwintować, 3758
gwintowanie *n*, 3762

hak *m*, 2324
halogenowy, 2210
hałas *m*, 2772
hałas *m* od wibracji
    przewodów, 1493
hałas *m* pochodzący z
    kanałów w czasie
    przepływu powietrza, 1498
hałas *m* silnika, 2729
hałas *m* spowodowany
    wybuchem, 520
handlowy, 873
harmonijka *f*, 396
hermetycznie szczelny, 2300
hermetyczny, 2298
higieniczny, 2365
higrometr *m*, 2366
higrometr *m* punktu rosy,
    1271
higrometr *m* respiracyjny,
    281
higrometr *m* włosowy, 2209

higroskopijny, 2367
higrostat *m*, 2368
hiperboliczny, 2369
hipotermia *f*, 2370
humidostat *m*, 2368
hydrauliczny, 2362
hydrostatyczny, 2364

ilość *f*, 204, 3131
iluminator *m*, 564
imadło *n*, 4005
indeks *m*, 2409
indukcja *f*, 2418
indukowany, 2417
infekcja *f*, 2426
infiltracja *f*, 2427
infiltracja *f* powietrza, 132
infiltracja *f* przez
    nieszczelności, 2428
inhibitor *m*, 2435
inhibitor *m* korozji, 1095
inspektor *m* nadzoru na
    budowie, 3460
instalacja *f*, 2451, 2974, 3750
instalacja *f* alarmu
    pożarowego, 1874
instalacja *f* centralnego
    ogrzewania, 681
instalacja *f* chłodnicza, 3210
instalacja *f* do usuwania
    pyłu, 1523
instalacja *f* do wytwarzania
    lodu, 2382
instalacja *f* klimatyzacji
    jednoprzewodowej, 3450
instalacja *f* mrożeniowa,
    2051
instalacja *f* nawiewna, 3725
instalacja *f* odpylająca, 1523
instalacja *f* ogrzewania, 2250
instalacja *f* ogrzewcza, 2256
instalacja *f* prefabrykowana,
    1782
instalacja *f* topienia śniegu,
    3507
instalacja *f* wody gorącej,
    2345
instalacja *f* wywiewna, 1724
instalacja *f* z rurociągami o
    małej średnicy, 2679
instalacja *f* ze zmiennym
    strumieniem powietrza,
    3970
instalować, 2450
instalowanie *n*, 2734
instrukcja *f* montażowa, 287
instrukcja *f* montażu, 2736
instrukcja *f* obsługi, 2453,
    3399, 3400
instrukcje *f* eksploatacyjne,
    2847
instrukcje *f* obsługi, 2847
integrator *m*, 2463
intensywność *f*, 2464
interferencja *f* (fal), 2468

inżektor *m*, 2441
inżynier *m*, 1640
inżynier *m* budowy, 1832
inżynier *m* dyplomowany, 3092
inżynier *m* projektant, 996
inżynieria *f*, 1641
inżynieria *f* środowiska, 1661
iskra *f*, 3559
iskrownik *m*, 3560
izentalpowy, 2485
izobara *f*, 2486
izolacja *f*, 2459
izolacja *f* akustyczna, 3542
izolacja *f* cieplna, 2268, 3799, 3106
izolacja *f* cieplna komórkowa, 674
izolacja *f* cieplna obudowująca, 442
izolacja *f* cieplna odblaskowa, 3192
izolacja *f* cieplna piankowa natryskiwana na budowie, 2001
izolacja *f* cieplna płaszczowa, 421
izolacja *f* cieplna płytowa, 452
izolacja *f* cieplna wypełniająca, 1838
izolacja *f* cieplna z gumy piankowej, 672
izolacja *f* cieplna z płyt korkowych, 1087
izolacja *f* paroszczelna, 3960
izolować (elektrycznie-akustycznie), 2455
izolować (cieplnie), 2456
izolowany, 2493
izoterma *f*, 2489
izotermiczny, 2490
izotop *m*, 2491

jakość *f*, 3128
jakość *f* gazu, 2145
jakość *f* powietrza, 151
jakość *f* powietrza wewnętrznego, 2416
jarzenie *n*, 2173
jawny, 3385
jednoczesny, 3448
jednolity, 3926
jednołopatkowy, 3456
jednostka *f*, 3925
jednostka *f* miary, 3930
jednostka *f* podporządkowana, 3472
jednostkowe zużycie *n* ciepła, 2280
jednostkowy, 3926
jednostopniowy, 3455
jon *m*, 2480
jon *m* dodatni, 654

jon *m* ujemny, 232
jonizacja *f*, 2482

kabel *m*, 599
kabina *f* badawcza, 3778
kabina *f* sterownicza, 1018
kadłub *m*, 453
kalibrowanie *n*, 604
kaloria *f*, 606
kalorymetr *m*, 611
kalorymetr *m* pokojowy, 3299
kamień *m* kotłowy, 2408, 3348
kanał *m*, 706, 972, 1492
kanał *m* blaszany, 3420
kanał *m* dymowy, 1973, 3495
kanał *m* elastyczny, 1936
kanał *m* główny, 2609
kanał *m* mieszający, 2697
kanał *m* odgałęźny, 508
kanał *m* odpływowy, 1348
kanał *m* powietrzny, 115
kanał *m* rozdzielczy, 1371
kanał *m* spalinowy, 2132
kanał *m* ściekowy, 3411
kanał *m* wywiewny, 1726
kapacytancja *f*, 615
kapanie *n*, 1442
kapilara *f*, 627
kapilarność *f*, 623
kapilarny, 624
karter *m*, 1123
kaseta *f*, 643
kaskada *f*, 640
kaskadowy, 2751
katalizator *m*, 651
katatermometr *m*, 2505
kation *m*, 654
katoda *f*, 652
kawitacja *f*, 656
kąpiel *f*, 382
kąpiel *f* olejowa, 2810
kąt *m*, 226
kąt *m* dopuszczalny, 24
kąt *m* natarcia, 3150
kąt *m* wylotu, 230
kąt *m* wypływu, 229
kątownik *m* stalowy, 228
kielich *m* (rury), 3566
kielich *m* rury, 2350
kierownica *f* wentylatora, 1793
kierownik *m* robót, 770
kierownik *m* zespołu, 714
kierunek *m*, 1329
kierunek *m* przepływu, 1960
kierunki *m*, *pl* regulacji, 1033
kieszeń *f* wodna, 4080
kinetyczny, 2507
kit *m*, 3126
klamka *f*, 1993
klasyfikacja *f*, 758
klawiatura *f*, 2506

klej *m*, 49
klejący, 50
kleszcze *pl*, 3861
klimat *m*, 771
klimatyzacja *f*, 89
klimatyzacja *f* dla okrętów, 2632
klimatyzacja *f* komfortu, 868
klimatyzacja *f* pełna, 2091
klimatyzacja *f* pomieszczeń mieszkalnych, 3249
klimatyzacja *f* przemysłowa, 2422
klimatyzacja *f* w okresie letnim, 3705
klimatyzacja *f* w okresie zimy, 4130
klimatyzacja *f* wysokoprędkościowa, 2316
klimatyzator *m*, 88, 3927
klimatyzator *m* dachowy, 3293
klimatyzator *m* indywidualny, 988
klimatyzator *m* okienny, 4124
klimatyzator *m* pokojowy, 3297
klimatyzator *m* strefowy, 4150
klimatyzator *m* ścienny, 3839
klimatyzator *m* z czynnikiem odparowującym, 1696
klimatyzator *m* ze swobodnym wylotem powietrza, 2034
klimatyzator *m* zwarty, 2898
klinkier *m*, 773
klucz *m* maszynowy, 4144
klucz *m* płaski (maszynowy), 3557
kocioł *m*, 455
kocioł *m* elektrodowy, 1590
kocioł *m* gazowy, 2113
kocioł *m* na ciepło odpadowe, 4053
kocioł *m* na dwa rodzaje paliwa, 1488
kocioł *m* na różne paliwa, 1042
kocioł *m* niskiego ciśnienia, 2597
kocioł *m* opalany koksem, 811
kocioł *m* opalany olejem, 2819
kocioł *m* opalany pod ciśnieniem, 3051
kocioł *m* opalany węglem, 787
kocioł *m* parowy, 3624
kocioł *m* płomienicowo-płomieniówkowy, 3351

korek *m* odpowietrznika,
523
korek *m* parowy, 3641, 3964
korek *m* powietrzny, 144
korkować, 2991
korodować, 1093
korozja *f*, 1094
korozja *f* punktowa, 2972
korozja *f* wżerowa, 2972
korozyjny, 1098
korpus *m*, 453
korpus *m* pompy, 3122
kostka *f* lodu, 2377
kostka *f* prasowana, 534
kosz *m* ssawny, 3703
kosz *m* ssący, 3703
koszt *m*, 1756
koszt *m* (koszty), 1101
koszt *m* eksploatacji, 1102,
2844, 3322
koszt *m* energii, 1624
koszt *m* instalacji, 2975
koszt *m* inwestycyjny, 1896
koszt *m* odtworzenia, 3244
koszt *m* okresu
użytkowania, 2540
koszt *m* urządzenia, 1778
koszt *m* urządzeń, 2975
koszt *m* użytkowania, 1102
koszt *m* własny, 2891
koszt *m* wymiany, 3243
kosztorys *m* szczegółowy,
1261
koszty *m* ogrzewania, 2244
kotew *m*, 221
kotłownia *f*, 470, 473, 477
kotłownia *f* centralna, 677
kotłownia *f* centralnego
ogrzewania, 682
kotrola *f* zamrażania, 2046
kotwica *f*, 221
kowalność *f*, 1496
kółko *n* ręczne, 2213
krata *f*, 2180
krata *f* z żaluzjami, 2591
kratka *f*, 2190
kratka *f* perforowana, 2935
kratka *f* podokienna, 3446
kratka *f* różnicowa, 1291
kratka *f* wywiewna, 2869
kratka *f* z kierownicami
nieruchomymi, 1899
kratownica *f*, 2033
krawędź *f*, 1547
krążek *m*, 1337
krążenie *n*, 747
kreda *f*, 700
kreślić, 2178
kropelka *f*, 1449
kropla *f*, 1448
krotność *f* wymiany
powietrza, 81
krotność *f* wymiany
powietrza na godzinę, 82

króciec *m* przyłączny
powrotu, 3261
króciec *m* wodny kotła, 480
króciec *m* wylotowy, 2872
krótkie spięcie *n*, 3431
kryształ *m* lodu, 2376
kryterium *n* jakości, 1835
krytyczny, 1126
kryza *f*, 1277, 2858
krzywa *f*, 1147
krzywa *f* charakterystyczna,
708
krzywa *f* charakterystyki,
708
krzywa *f* charakterystyki
wentylatora, 1797
krzywa *f* chłodzenia, 1055
krzywa *f* mocy, 3014
krzywa *f* nasycenia, 3341
krzywa *f* ogrzewania, 2245
krzywa *f* regulacyjna
ogrzewania, 2245
krzywa *f* sprawności
wentylatora, 1801
krzywe *f* charakterystyczne
hałasu, 2774
krzywka *f*, 612
krzyżak *m*, 1134
kształt *m*, 973
kształtka *f* przejściowa, 3886
kubatura *f*, 159
kubatura *f* budynku, 557
kurek *m*, 794, 795
kurek *m* czerpalny, 3757
kurek *m* probierczy, 3780
kurek *m* przelotowy, 3840
kurek *m* spustowy, 1420
kurek *m* trójdrogowy, 3831
kurtyna *f* ciepłego
powietrza, 4035
kurtyna *f* powietrzna, 101
kurtyna *f* z rozpylonej
cieczy, 3583
kurz *m*, 1510
kwas *m*, 35
kwas *m* węglowy, 631
kwasowość *f*, 38
kwaśność *f*, 38
kwota *f*, 204

laboratorium *n*, 2510
lakier *m*, 2512
lamelka *f*, 2983
laminarny, 2513
lampa *f*, 2514
lampa *f* ostrzegawcza, 4047
lampa *f* sygnalizacyjna, 2410
lej *m*, 2098
lej *m* samowyładowczy, 2325
lej *m* zasypowy, 2325
lepki, 50
lepkościomierz *m*, 4006
lepkość *f*, 4007
lepkość *f* dynamiczna, 1536
libella *f*, 2532

licencjonowany, 2536
liczba *f*, 2794
liczba *f* Bacharacha, 324
liczba *f* charakterystyczna,
709
liczba *f* swobodnej
konwekcji, 2035
liczbowy, 2795
liczenie *n* impulsów, 3117
licznik *m* ciepła, 2274
licznik *m* przepływu, 1963
lignit *m*, 537, 2548
lina *f* stalowa, 599
linia *f* boczna, 510
linia *f* elektroenergetyczna,
3018
linia *f* nasycenia, 3342
linia *f* wodna kotła, 481
linia *f* zasilająca, 1818
lista *f* rozkazów komputera,
926
listwa *f*, 384
listwa *f* przypodłogowa, 376
litr *m*, 2566
lodówka *f*, 2375
lodówka *f* elektryczna, 1587
lotny, 4011
lód *m*, 2372
lód *m* rozdrobniony, 2030
lód *m* suchy, 1467
lód *m* w kostkach, 1141
lód *m* w postaci płatków,
1903
lut *m*, 3520
lutować, 3521
lutować lutem twardym, 513
lutowanie *n*, 3523
lutowanie *n* miękkie, 3511
lutowanie *n* twarde, 517,
2216
lutowany lutem twardym,
514
luz *m*, 766
luz *m* ujemny, 2468

ładować, 712
ładowanie *n* (paleniska lub
pieca), 715
ładunek *m*, 634, 2568, 711
ładunek *m* chłodzony, 3203
łata *f*, 384
łatwotopliwy, 2102
łazienka *f*, 383
łaźnia *f*, 382
łącze *n* dwuprzewodowe,
3919
łącze *n* jednofazowe, 3919
łącznica *f* telefoniczna, 3743,
3742
łącznik *m*, 979, 981, 983,
986, 1898, 3752, 2503
łącznik *m* kablowy, 600
łączyć, 975
łączyć śrubami, 3354
łeb *m*, 2217

stabilizator *m* przepływu, 1961
stabilność *f*, 3592
stacja *f*, 3618
stacja *f* redukcyjna ciśnienia, 3062
stacja *f* wyrównawcza, 358
stal *f*, 3647
stal *f* nierdzewna, 3595
stała *f*, 989
stała *f* czasowa, 3849
stała *f* dielektryczna, 1283
stała *f* pomieszczenia, 3301
stała *f* regulacja wartości, 993
stała *f* słoneczna, 3514
stały, 990
stan *m*, 3614
stan *m* energetyczny, 1636
stan *m* gotowości do pracy, 4137
stan *m* jałowy, 2778
stan *m* niestały, 2789
stan *m* nieustalony, 2789
stan *m* pogotowia, 3600
stan *m* powietrza, 87, 964
stan *m* przejściowy, 2789
stan *m* równowagi przy odparowaniu, 1700
stan *m* równowagi wilgoci, 2713
stan *m* suchy, 1473
stan *m* ustalony, 3621
stanowisko *n*, 3618
stanowisko *n* badawcze, 3777, 3788
stanowisko *n* pracy czyste, 765
starter *m*, 3605
starzenie, 65
stateczność *f*, 3592
stechiometryczny, 3659
sterować, 1015
sterowanie *n*, 1014
sterowanie *n* astatyczne, 1942
sterowanie *n* całkowe, 2461
sterowanie *n* centralne, 678
sterowanie *n* ciągłe, 1008
sterowanie *n* cyfrowe bezpośrednie, 1318
sterowanie *n* działaniem natychmiastowym, 3505
sterowanie *n* elektro-pneumatyczne, 2997
sterowanie *n* graniczne, 2550
sterowanie *n* kaskadowe, 641
sterowanie *n* niskociśnieniowe, 2598
sterowanie *n* procesem, 3089
sterowanie *n* procesem rozmrażania, 1207
sterowanie *n* programowe, 3094

sterowanie *n* przy użyciu siłownika, 3402
sterowanie *n* przyciskowe, 3125
sterowanie *n* ręczne, 2626
sterowanie *n* ręczne kasujące nastawienie urządzenia, 2628
sterowanie *n* w obwodzie zamkniętym, 779
sterowanie *n* wyprzedzające, 237
sterowanie *n* zdalne, 3238
sterowanie *n* ze sprzężeniem zwrotnym, 779
stężenie *n*, 940
stężenie *n* (konstrukcyjne), 3620
stężenie *n* dopuszczalne maksymalne, 2640
stężenie *n* gazu, 2118
stężenie *n* graniczne, 3344
stężenie *n* zanieczyszczenia powietrza, 146
stojak *m*, 835
stojak *m* hydrantowy, 3603
stojan *m*, 3619
stop *m* (metali), 188
stop *m* lutowniczy, 3520
stopienie *n*, 2103
stopień *m*, 1214, 3651
stopień *m* Celsjusza, 676
stopień *m* chłodzenia, 1067
stopień *m* ciśnienia, 3067
stopień *m* czystości, 1217
stopień *m* indukcji, 2420
stopień *m* koncentracji, 941
stopień *m* nasycenia, 3343
stopień *m* odparowania, 1694
stopień *m* porywania, 1650
stopień *m* przegrzania, 1218
stopień *m* spęcznienia, 3740
stopień *m* spiętrzenia, 3740
stopień *m* sprężania, 905
stopień *m* zmieszania, 2699
stopnie *m* swobody, 1219
stopniodzień *m*, 1215
stopniodzień *m* ogrzewania, 2246
stopniowany, 3654
stopniowy, 3654
stos *m*, 3593
stosunek *m*, 3164
stosunek *m* indukcji, 2420
stosunek *m* mieszania, 2699
stosunek *m* objętości sprężonej, 911
stosunek *m* wydajności chłodzenia, 1059
stożkowa przekładnia *f* wentylatora, 1807
stół *m*, 3751
stół *m* pomiarowy, 3777
strata *f*, 2586

strata *f* chłodzenia, 1063
strata *f* ciepła, 2271
strata *f* ciepła kominowa, 2258
strata *f* ciepła w warunkach obliczeniowych, 1250
strata *f* ciśnienia, 2220, 3058
strata *f* ciśnienia dynamicznego, 1534
strata *f* ciśnienia hydrostatycznego, 2587
strata *f* energii, 1629
strata *f* kominowa, 739
strata *f* na tarcie, 2067
strata *f* na wlocie, 1655
strata *f* przez przewodzenie, 968
strata *f* przy przepływie spalin, 1979
strategia *f* sterowania, 1033
straty *f*, *pl* wskutek odparowania, 1692
strefa *f*, 4149
strefa *f* komfortu, 872
strefa *f* martwa (powietrza), 1184
strefa *f* neutralna, 2767
strefa *f* nieczułości, 1188, 2767
strefa *f* przejściowa, 3887
strefa *f* zajęta, 2802
strona *f* zewnętrzna, 1764
strop *m*, 664
strop *m* perforowany, 2934, 1454
struga *f*, 3679
strumienica *f*, 4074
strumienica *f* ssąca, 1559
strumień *m*, 1953, 1996, 2494, 3679
strumień *m* ciepła, 2239
strumień *m* ciepła nieustalony, 3883
strumień *m* ciepła właściwego, 3564
strumień *m* natężenia napromieniowania, 389
strumień *m* pierścieniowy, 234
strumień *m* powietrza, 121, 135, 168
strumień *m* rozproszony, 1360
strych *m*, 303
studium *n* projektowe, 1252
studnia *f*, 4112
studnia *f* chłonna, 3508
studzienka *f*, 3706
studzienka *f* rewizyjna, 2621
studzienka *f* ściekowa, 4112
stycznik *m*, 1000
styk *m*, 999
stykowa wartość *f* znamionowa, 1002
substancja *f*, 2638

wełna *f* mineralna, 2686
wełna *f* szklana, 2170
wełna *f* żużlowa, 3470
wentylacja *f*, 3991
wentylacja *f* poprzeczna,
 1139
wentylacja *f* strumieniowa,
 2499
wentylacja *f* wyciągowa,
 1737
wentylacja *f* wywiewna, 1737
wentylator *m*, 447, 448,
 1790, 3993
wentylator *m* dachowy, 3294
wentylator *m* dwustronnie
 ssący, 1399, 1406
wentylator *m* łopatkowy,
 125
wentylator *m*
 niskociśnieniowy, 2599
wentylator *m* o dużej
 prędkości, 2317
wentylator *m* o przepływie
 mieszanym, 2691
wentylator *m* o przepływie
 poprzecznym, 1131, 3754
wentylator *m* obiegowy, 746
wentylator *m* odśrodkowy,
 689
wentylator *m* osiowy, 317,
 3100
wentylator *m* osiowy
 kanałowy, 3900
wentylator *m* osiowy z
 kierownicą, 3955
wentylator *m* równolegle
 pracujący (bliźniaczy), 406
wentylator *m* sufitowy, 665
wentylator *m* ścienny, 4026,
 4033
wentylator *m* śmigłowy,
 3100
wentylator *m* wyciągowy,
 1727
wentylator *m* wywiewny,
 1727
wentylowanie *n*, 3988
wetknąć, 2992
wewnętrzny, 2474
węgiel *m*, 628, 782
węgiel *m* aktywowany, 42
węgiel *m* bitumiczny, 415
węgiel *m* brunatny, 537
węgiel *m* chudy, 2782
węgiel *m* długopłomienny,
 2582
węgiel *m* drobny, 3464
węgiel *m* gazowany, 2117
węgiel *m* gruboziarnisty, 790
węgiel *m* kamienny, 783
węgiel *m* koksowniczy, 815
węgiel *m* lignitowy, 2549
węgiel *m* niebitumiczny,
 2782
węgiel *m* niekoksujący, 2784

węgiel *m* o dużej zawartości
 części lotnych, 2318
węgiel *m* średniolotny, 2661
węglowodór *m*, 2363
węzeł *m* ciepłowniczy, 1380
wężownica *f*, 805, 2953
wężownica *m*, 4025
wężownica *f* bezpośredniego
 odparowania, 1321
wężownica *f* chłodząca, 1054
wężownica *f* grzejna, 2243
wężownica *f* osuszacza, 1437
wężownica *f* parownika,
 1691
wężownica *f* schładzacza,
 1257
wężownica *f* skraplacza, 953
wężownica *f* wyparki, 1691
wężownica *f* z wymuszonym
 przepływem, 423
wężownica *f* zatopiona w
 suficie, 659
wgłębienie *n*, 657
wiatr *m*, 4122
wiązka *f*, 565
wiązka *f* rur, 3901
wibracja *f*, 4002
wibroizolacja *f*, 244
widmo *n* dźwięku, 3550
wieko *n*, 2537
wielkość *f*, 3462
wielkość *f* błędu, 1203
wielkość *f* ciśnienia
 bezpiecznego, 3333
wielkość *f* cyrkulacji
 powietrza, 84
wielkość *f* kropelki, 1452
wielkość *f* ładowania, 719
wielkość *f* maksymalna
 dopuszczalna, 2642
wielkość *f* nastawy, 3410
wielkość *f* odpływu, 1347
wielkość *f* przepływu, 1966
wielkość *f* straty ciepła, 2273
wielkość *f* strumienia ciepła,
 3162
wielkość *f* strumienia
 powietrza, 123
wielkość *f* wypływu, 1347
wielkość *f* zysków ciepła od
 urządzeń elektrycznych,
 1584
wielostopniowy, 2751
wielostrefowy, 2754
wieszak *m*, 2214
wieszak *m* do ręczników
 rurowy, 3872
wieszak *m* rury, 2957
wieża *f* chłodnicza, 1070
wieża *f* chłodnicza o ciągu
 naturalnym, 2758
wieża *f* chłodnicza płytowa,
 2982
wieża *f* chłodnicza wodna,
 4062

wieża *f* chłodnicza wodna ze
 sztucznym ciągiem, 2656
wieża *f* chłodnicza wodna ze
 zraszanym wypełnieniem,
 3585
wieża *f* chłodnicza z
 wymuszonym ciągiem,
 2017
wieża *f* o przepływie
 krzyżowym, 1133
wilgoć *f*, 1165, 1172, 2712
wilgotność *f*, 1165, 1172,
 2357, 2712
wilgotność *f* względna, 3233
wilgotny, 2351, 2711
winda *f*, 2541
wirnik *m*, 2396, 3315
wirnik *m* wentylatora, 1804
wirówka *f*, 693
wkład *m* filtracyjny, 1857
wkład *m* termosu, 3815
wkładka *f* topikowa
 zamknięta, 639
wkręcać śrubę, 3354
wkręt *m*, 3353
wkrętka *f*, 3358
wlot *m*, 2443, 1654
wlot *m* do komina przy
 podstawie, 738
wlot *m* do wentylatora, 1798
wlot *m* kablowy, 601
wlot *m* powietrza, 133, 2434
właściwy, 3561
właz *m*, 2621
włączać, 3745
włókno *n* szklane, 2169
włoskowatość *f*, 623
włoskowaty, 624
włókno *n*, 1828
włókno *n* syntetyczne, 3749
wnęka *f*, 657, 2768
wnętrze *n*, 2469
woda *f*, 4056
woda *f* chłodnicza, 3201
woda *f* chłodząca, 728, 1077
woda *f* ciepła, 4044
woda *f* dawkowana, 1394
woda *f* destylowana, 1368
woda *f* gorąca, 2337
woda *f* gorąca wysokiego
 ciśnienia, 2309
woda *f* gruntowa, 2199
woda *f* lodowa, 2378
woda *f* miejska, 753
woda *f* nieuzdatniona, 3933
woda *f* o wysokiej
 temperaturze, 2314
woda *f* odpływowa, 1557,
 4055
woda *f* pitna, 1440, 3009
woda *f* powrotna, 3268
woda *f* przemysłowa, 2423
woda *f* słonawa, 505
woda *f* studzienna, 4113
woda *f* uzdatniona, 2615

woda *f* zasilająca, 1819
woda *f* zasilająca kocioł, 464
woda *f* zimna, 827
wodoodporny, 4081
wodoszczelny, 4081, 4086
wojłok *m*, 1822
wolnolodowy zasobnik *m*
  zimna, 2373
wolny od drobnoustrojów,
  2164
woltomierz *m*, 4014
worek *m*, 344
worek *m* filtracyjny, 1849
wrzący, 483
wrzeć, 454
wrzenie *n*, 1540
wrzenie *n* błonkowe, 1840
wskazanie *n*, 2411
wskazanie *n* analogowe, 215
wskazanie *n* danych, 3168
wskazywanie *n*, 2411
wskaźnik *m*, 2158, 2409,
  2413, 3874
wskaźnik *m* ciągu, 1428
wskaźnik *m* ciepła, 4043
wskaźnik *m* ciśnienia, 3053
wskaźnik *m* komfortu, 871
wskaźnik *m* odniesienia,
  3190
wskaźnik *m* poziomu, 2534
wskaźnik *m* poziomu cieczy,
  2562
wskaźnik *m* świeżości, 2063
wskaźnik *m* zapełnienia,
  2798
wskaźnik *m* zawartości
  sadzy, 3532
wspomagający, 332
wspornik *m*, 504, 3729
wspornik *m* rury, 2952
wspornik *m* zaciskowy, 756,
  3658
współczynnik *m*, 797, 1779
współczynnik *m* absorpcji,
  17
współczynnik *m*
  bezpieczeństwa, 1780, 3332
współczynnik *m*
  chropowatości, 3317
współczynnik *m* dyfuzji,
  1302
współczynnik *m* jakości,
  1835
współczynnik *m* kątowy, 227
współczynnik *m* kształtu,
  974
współczynnik *m* mocy
  elektrycznej, 1582
współczynnik *m* napełnienia,
  4017
współczynnik *m*
  niejednoczesności
  (obciążenia), 1384
współczynnik *m*
  nieprzezroczystości, 2839

współczynnik *m*
  nierównomierności
  (rozbioru), 1384
współczynnik *m* obciążenia,
  2570
współczynnik *m* ocienienia,
  3413
współczynnik *m* odbicia,
  3191, 3193
współczynnik *m* oporu, 1415
współczynnik *m*
  prawidłowości działania,
  2939
współczynnik *m* przenikania
  ciepła, 2290
współczynnik *m* przepływu,
  1962
współczynnik *m*
  przewodności, 799
współczynnik *m*
  przewodzenia ciepła, 3795,
  3794
współczynnik *m*
  przewodzenia
  powierzchniowego, 803
współczynnik *m*
  rozszerzalności, 801, 1744
współczynnik *m* sprawności,
  802
współczynnik *m* sprężania,
  903
współczynnik *m* szorstkości,
  3317
współczynnik *m* ściśliwości,
  899
współczynnik *m* tarcia, 2066
współczynnik *m* ubytku,
  1199
współczynnik *m*
  użytkowania, 3938
współczynnik *m* wydajności,
  620
współczynnik *m* wydajności
  chłodzenia, 1059
współczynnik *m* wydatku,
  800, 1340
współczynnik *m* wydatku
  powietrza, 108
współczynnik *m* wymiany
  ciepła przez błonkę, 1841
współczynnik *m* wypływu,
  1340
współczynnik *m*
  zanieczyszczenia, 2027
współczynnik *m* zdolności
  sprężania, 899
współczynnik *m*
  zmniejszenia, 1199
współosiowy, 793
współwytwarzanie *n*, 804
wstępnie nastawiony, 3039
wstępnie regulowany, 3039
wsysanie *n*, 282
wtrysk *m* cieczy, 2560
wtryskiwacz *m*, 2441

wtyczka *f* (elektr.), 2990
wtykać, 2992
wybierak *m*, 3378
wybór *m* paliwa, 2079
wybuch *m*, 1760, 1916
wybuch *m* gazu, 2123
wyciąg *m*, 1719, 1773
wyciąg *m* laboratoryjny,
  2097
wyciąg *m* powietrza, 119
wyciągać, 1772
wyciek *m*, 1558
wycięcie *n*, 1151
wycinać, 1152
wycior *m* czopucha, 1974
wyczystka *f*, 762, 2211
wydajność *f*, 617, 1338, 1556
wydajność *f* (produkcyjna),
  3838
wydajność *f* chłodnicza, 821
wydajność *f* chłodzenia,
  1053
wydajność *f* cieplna, 2275,
  3800
wydajność *f* cieplna
  użyteczna, 3937
wydajność *f* ciepła
  całkowita, 3870
wydajność *f* ciepła
  nominalna, 3161
wydajność *f* ciepłego
  powietrza, 4040
wydajność *f* dobowa, 1161
wydajność *f* gazu, 2155
wydajność *f* kotła, 457
wydajność *f* pracy, 2938
wydajność *f* punktu poboru,
  3759
wydajność *f* sprężarki, 913,
  914
wydajność *f* znamionowa,
  3160
wydajność *f* znamionowa
  kotła, 476
wydatek *m*, 1756
wydatek *m* zasysania, 3696
wydatek *m* zaworu, 3759
wydłużalnik *m* rurowy
  pętlicowy zamknięty, 1749
wydłużka *f*, 887
wydłużka *f* lirowa, 886
wydłużka *f* mieszkowa, 1742
wydłużka *f* osiowa, 316
wydłużka *f* przesuwna, 3479
wydłużka *f* segmentowa, 888
wydłużka *f* ślizgowa, 3479
wydmuch *m*, 445
wydmuchiwać, 444
wydmuchiwanie *n*, 449
wydruk *m* komputera, 933
wygarniać popiół, 3152
wygaszać, 1770
wygięcie *n*, 502
wygięty do przodu, 2024
wyjście *n* analogowe, 217

zasuwa *f*
  jednopłaszczyznowa, 2207
zasuwa *f* kominowa, 3217
zasuwa *f* palnika, 576
zasuwa *f* ślizgowa, 3478
zasysać, 3694
zasysanie *n*, 282, 3695
zaślepka *f*, 2990
zaślepka *f* kołnierzowa, 422,
  438
zatkanie *n*, 3664
zatrzymanie *n*, 3436, 3662,
  3664
zatrzymywać, 722
zatykać, 2991
zatykanie *n* się przewodów,
  777
zawartość *f*, 1007
zawartość *f* części
  niepalnych w popiele, 277
zawartość *f* popiołu, 272
zawartość *f* pyłu, 1511
zawartość *f* tlenu, 2895
zawartość *f* wilgoci, 2714
zawartość *f* wody, 4059
zawiasa *f*, 2319
zawieszenie *n*, 3737, 3134
zawór *m*, 3757, 3951, 1151
zawór *m* bezpieczeństwa,
  3066, 3335
zawór *m* bocznikowy, 3435
zawór *m* ciężarkowy, 4101
zawór *m* ciśnieniowy
  kondensatu, 958
zawór *m* czerpalny, 1435
zawór *m* dławiący, 3835
zawór *m* dwudrogowy, 3918
zawór *m* dwugniazdowy,
  1404
zawór *m* dźwigniowy, 3187
zawór *m*
  elektromagnetyczny, 3524
zawór *m* grzejnikowy, 3149
zawór *m* iglicowy, 2763
zawór *m* jednokierunkowy,
  723
zawór *m* kątowy, 231, 1089
zawór *m* klapowy, 1915
zawór *m* końcowy, 3774
zawór *m* kurkowy, 795
zawór *m* magnetyczny, 2607
zawór *m* membranowy,
  1279, 2666
zawór *m* mieszający, 2700
zawór *m* motylkowy, 590
zawór *m* nadmiarowy, 3236
zawór *m* nadmiarowy
  ciśnieniowy, 3066
zawór *m* nadmiarowy
  szybkiego działania, 1609
zawór *m* obejściowy, 597
zawór *m* odciążający, 3236
zawór *m* odcinający, 443,
  1150, 2487, 3437

zawór *m* odcinający na
  odpływie, 1349
zawór *m* odpływowy, 1352
zawór *m* pływakowy, 1945
zawór *m* pożarowy, 1893
zawór *m* przeciwsyfonowy,
  243
zawór *m* przelewowy, 2885
zawór *m* przepływowy
  prosty, 1326
zawór *m* redukcyjny, 3063,
  3186
zawór *m* redukcyjny
  ciśnienia pary, 3643
zawór *m* regulacyjny, 3223
zawór *m* regulacyjny
  przepływu, 1958
zawór *m* regulujący
  przepływ, 1968
zawór *m* rozdzielczy, 1385
zawór *m* rozprężny, 1753
zawór *m* rozprężny
  pływakowy, 1746
zawór *m* spłukujący, 1995
zawór *m* spustowy, 435,
  1424
zawór *m* spustowy
  (kotłowy), 450, 446
zawór *m* stałego ciśnienia,
  992
zawór *m* stałego poziomu,
  991
zawór *m* szybkiego
  działania, 3132
zawór *m* talerzowy, 1356
zawór *m* tarczowy, 1356
zawór *m* termostatyczny,
  3819
zawór *m* trójdrogowy, 3832
zawór *m* upustowy, 1352
zawór *m* wlotowy, 2445
zawór *m* wydmuchowy, 446,
  450
zawór *m* wyrównawczy, 359
zawór *m* wywiewny, 1736
zawór *m* z dławicą
  mieszkową, 398
zawór *m* z siłownikiem, 2728
zawór *m* zwrotny, 331, 723,
  2788
zawór *m* zwrotny na
  zasileniu, 1816
zawór *m* zwrotny z filtrem,
  3676
zbiornik *m*, 752, 2325, 3173,
  3246, 3755
zbiornik *m* bezpośredni,
  1317
zbiornik *m* ciśnieniowy, 908
zbiornik *m* filtracyjny, 1862
zbiornik *m* kondensatu, 945,
  2346
zbiornik *m* o pojemności
  dobowej, 1163
zbiornik *m* oleju, 2827

zbiornik *m* opadowy, 2219
zbiornik *m* podziemny, 3922
zbiornik *m* powietrza, 175
zbiornik *m* skroplin, 2346
zbiornik *m* solanki, 533
zbiornik *m* spłukujący
  (klozetowy), 1994
zbiornik *m* szybkiego
  odparowania, 1927
zbiornik *m* ściekowy, 3706
zbiornik *m* termometru, 561
zbiornik *m* wody zasilającej,
  1821
zbiornik *m* wstępny, 544
zbiornik *m* wyrównawczy,
  356, 544, 1667, 1752, 2219
zbiornik *m* zamokowy, 1312
zbiornik *m* zanurzeniowy,
  1312
zbliżanie *n*, 255
zblokowany, 2897
zderzak *m*, 543, 3662
zdmuchiwacz *m* sadzy, 3531
zdolność *f*, 617
zdolność *f* absorpcyjna, 16
zdolność *f* chłodzenia, 1053
zdolność *f* emisyjna, 1612
zdolność *f* ładowania
  (akumulacja cieplna), 716
zdolność *f* odbijania, 3193
zdolność *f* przepustowa,
  1955, 3838
zdolność *f* przepustowa
  powietrza, 167
zdolność *f* przesyłania
  energii, 1637
zdolność *f* rozładowania,
  1354
zdolność *f* sprężania, 898
zdolność *f* zatrzymywania
  pyłu, 1517
zegar *m* czasowy (operacji),
  3847
zero *n* (absolutne), 4147
zerowanie *n*, 880
zespolony przyrząd *m*
  pomiarowy, 894
zespół *m*, 66, 286
zespół *m* centralnego
  uzdatniania powietrza, 684
zespół *m* chłodzący, 1049,
  1076
zespół *m* do uzdatniania
  powietrza, 126
zespół *m* ładowania, 720
zespół *m* mieszający
  powietrza, 139
zespół *m* ogrzewczo-
  wentylacyjny, 4049, 4039
zespół *m* ogrzewczo-
  wentylacyjny ścienny, 4030
zespół *m* ogrzewczy, 3929
zespół *m* palnikowy kotła,
  456
zespół *m* silnikowy, 3019

Русский

абразив *m*, 4
абразивный, 5
абсолютная шероховатость
 f, 8
абсолютно черное тело *n*,
 417
абсолютный, 6
абсолютный нуль *m*, 4148
абсолютный фильтр *m*, 7
абсорбент *m*, 11
абсорбер *m*, 12
абсорбировать, 9
абсорбируемая среда *f*, 10
абсорбирующая среда *f*, 19
абсорбционное охлаждение
 n, 18
абсорбция *f*, 15
аварийный звуковой сигнал
 m, 520
аварийный
 предохранительный
 клапан *m*, 1609
аварийный прибор *m*, 519
авария *f*, 518
автоматизация *f*, 309
автоматический, 308
автономная система *f*
 кондиционирования
 воздуха, 3379
автономное управление *n*,
 3380
автономный кондиционер
 m, 3379
агрегат *m*, 66, 3925
агрегат *m* с приводом,
 3472
агрегатированная
 установка *f*
 кондиционирования
 воздуха, 2900
агрегатированный, 3926
агрегатированный
 кондиционер *m*, 3927
агрессивность *f*, 68
агрессивность *f* кислоты,
 36
агрессивный, 67
адиабатный, 51
адсорбент *m*, 56
адсорбция *f*, 57
азеотропный, 323
азот *m*, 2771
акклиматизация *f*, 30

аккумулятор *m*, 31
аккумулятор *m* горячей
 воды, 2342
аккумулятор *m* холода,
 1078
аккумуляционная емкость
 f, 716
аккумуляционная
 установка *f*, 720
аккумуляционное
 отопление *n*, 3667
аккумуляционное
 отопление *n* за счет
 использования провала
 нагрузки, 2807
аккумуляционный
 нагреватель *m*, 3666
аккумуляция *f*, 3665
аккумуляция *f* горячей
 воды, 2341
аккумуляция *f* теплоты,
 2288
аксонометрический чертеж
 m, 322
активированный уголь *m*,
 42
активная мощность *f*, 3171
акустика *f*, 39
алгоритм *m*, 184
алгоритм *m* управления,
 1017
алкоголь *m*, 180
альтернативный источник
 m энергии, 191
альфаметрический, 189
алюминий *m*, 193
Американский кодекс *m*
 стандартных
 обозначений, 269
Американское общество *n*
 инженеров по
 отоплению, хоиодиьной
 технике и
 кондиционированию
 воздуха, 275
Американское общество *n*
 испытаний и материалов,
 288
аммиак *m*, 200
аммиачный раствор *m*, 201
амортизатор *m*, 3429, 4003
амортизационное
 устройство *n*, 244

амортизация *f*, 202, 1171
амортизировать, 203
амперметр *m*, 199
амплитуда *f*, 210, 3158
анализ *m*, 220
анализ *m* воздуха, 73
анализ *m* газа, 2112
анализ *m* топлива, 2078
анализатор *m*, 219
аналоговая индикация *f*,
 215
аналоговая передача *f*, 218
аналоговое представление
 n, 214
аналоговые данные *n*,pl,
 213
аналоговый, 212
анемометр *m*, 223
анемометр *m* с нитью
 накаливания, 2347
анероидная капсула *f*, 225
анероидный барометр *m*,
 224
анион *m*, 232
анод *m*, 235
антикоррозийная защита *f*,
 1096
антифриз *m*, 240
антрацит *m*, 236
аппарат *m*, 247
аппарат *m* с орошаемой
 насадкой, 626
арифметический, 264
арматура *f*, 29
архитектор *m*, 258
архитектор-проектировщик
 m, 3096
архитектурная площадь *f*
 здания (общая полезная
 площадь), 259
архитектурный объем *m*,
 260
асбест *m*, 265
асбестовый шнур *m*, 266
аспиратор *m*, 284
аспирационный гигрометр
 m, 281
аспирационный
 психрометр *m*, 283
аспирация *f*, 282
астатическое действие *n*,
 1941

взвешенная частица *f*, 3736
взрыв *m*, 1760
взрыв *m* газа, 2123
вибрационная
  колосниковая решетка *f*,
  4001
вибрация *f*, 4002
вид *m* снаружи, 1764
вид *m* топлива, 2090
винт *m*, 3099, 3353
винтовое соединение *n*,
  3357
винтовой компрессор *m*,
  3356
вискозиметр *m*, 4006
вихревая труба *f*, 4021
вихревое течение *n*, 1544
включать, 3745
влагопередача *f*, 2720
влагопоглотитель *m*, 2717
влагосодержание *n*, 2714
влажностный баланс *m*,
  2713
влажность *f*, 1165, 1172,
  2357, 2712
влажный, 2351, 2711
влажный возврат *m*, 4117
влажный пар *m*, 4118
влияние *n* внешних
  условий, 196
внепиковый период *m*,
  2806
внешнее давление *n*, 197
внешний, 1765
внутренне ребро *n*, 2446
внутреннее сопротивление
  n, 1823
внутренний, 1388, 2474
внутренний воздух *m*, 2415,
  2447
внутренний размер *m*, 2475
внутренняя резьба *f*, 1824
внутренняя стена *f*, 2476
внутридомовой, 1388
вода *f*, 4056
вода *f* в качестве
  холодильного агента,
  3201
вода *f* с содержанием соли
  больше, cem v pitçevoij,
  505
водный слой *m*, 4075
водо-водяной
  пластинчатый
  охладитель *m*, 2985
водомер *m*, 4068
водонагреватель *m*, 4071
водонепроницаемый, 4081,
  4086
водоотвод *m*, 4065
водоотделитель *m*, 4083
водоохладитель *m*, 729,
  4061
водоохлаждаемый, 4060
водопад *m*, 640

водопоглощение *n*, 4057
водопотребление *n*, 4058
водоросли *f,pl*, 181
водоснабжение *n*, 4066
водоструйный насос *m*,
  4074
водотрубный котел *m*,
  4088
водяная рубашка *f*, 4073
водяное отопление *n*, 4072
водяное отопление *n*
  высокого давления, 2310
водяное отопление *n*
  низкого давления, 2601
водяное отопление *n* с
  использованием газа в
  виде топлива, 2130
водяной мешок *m* (в
  трубопроводах), 4080
водяной пар *m*, 4089
водяной столб *m*, 837
возбуждение *n*, 1715
возврат *m*, 3214, 3257
возврат *m* части, 1814
воздух *m*, 71
воздух *m* для горения, 845,
  866
воздух *m* нормативного
  качества, 3597
воздух *m* помещения, 3296
воздуховод *m*, 115
воздуховод *m* из листового
  металла, 3420
воздуховыбросное
  устройство *n*, 1722
воздуховыпускное
  устройство *n*, 3995
воздуходувка *f*, 447, 1790
воздухонагреватель *m*, 127,
  4037
воздухонагреватель *m*
  предварительного
  подогрева, 147
воздухонаревательная
  установка *f*, 128
воздухообмен *m*, 80
воздухообмен *m* в час, 82
воздухоосушитель *m*, 106,
  113
воздухоохладитель *m*, 97
воздухоохладитель *m*,
  работающий в условиях
  естественной конвекции,
  2756
воздухоохладитель *m*
  рециркуляционного типа,
  2036
воздухоохладитель *m* с
  принудительной
  циркуляцией, 2010
воздухоохладительная
  установка *f*, 98
воздухоочиститель *m*, 85
воздухопроизво-
  дительность *f*, 167

воздухопроинцаемость *f*,
  170
воздухопромыватель *m*,
  176
воздухопроницаемый, 169
воздухораспределительное
  устройство *n*, 166
воздухосмесительная
  система *f*, 72
воздушная диффузия *f*, 107
воздушная завеса *f*, 101
воздушная заслонка *f*, 101
воздушная направляющая
  лопатка *f*, 161
воздушная неплотность *f*,
  137
воздушная подушка *f*, 102
воздушная прослойка *f*,
  159
воздушная смесь *f*, 2690
воздушная струя *f*, 135
воздушная тень *f*, 1184
воздушное отопление *n*,
  129, 4038
воздушное охлаждение *n*,
  99
воздушное пространство *n*,
  159
воздушное уплотнение *n*,
  157
воздушно-сухой, 111
воздушно-тепловая завеса
  f, 4035
воздушный (дроссельный)
  клапан *m*, 104
воздушный затвор *m*, 144
воздушный насос *m*, 149
воздушный поток *m*, 121
воздушный пузырек *m*, 79
воздушный регулирующий
  клапан *m*, 154
воздушный сосуд *m*, 175
воздушный фильтр *m*, 120
воздушный фильтр *m* с
  поддоном для масла,
  2811
воздушный холодильный
  цикл *m*, 103
воздушный шум *m*, 77
возобновляемый, 2785
войлок *m*, 1822
волокнистая пыль *f*, 1829
волокнистый фильтр *m*,
  1830
волокно *n*, 1828
волоконная оптика *f*, 1827
волосяная трещина *f*, 2208
волосяной гигрометр *m*,
  2209
вольтметр *m*, 4014
воронка *f*, 2098, 2325
вороток *m*, 2508
воспламеняемость *f*, 841
впадина *f*, 657

вытяжной шкаф *m*, 2097
вытяжной штуцер *m*, 2872
выхлоп *m*, 1338
выхлоп *m* двигателя, 1642
выхлоп *m* компрессора, 914
выход *m* через стену, 4031
выходная величина *f*, 217
выходное отверстие *n*, 2867
выходное отверстие *n* дымовой трубы, 1984
выходное щелевое отверстие *n*, 3486
вычислительная машина *f*, 922
вязкость *f*, 1496, 4007

гаечный ключ *m*, 4144
газ *m*, 2111
газ *m* в емкости, 496
газификация *f*, 2135
газовая горелка *f*, 2114
газовая печь *f*, 2133, 2149
газовая сварка *f*, 2154
газовая смесь *f*, 2692
газовая труба *f*, 2142
газовая турбина *f*, 862, 2152
газовое отопление *n*, 2134
газовое хранилище *n*, 2148
газовоздушное отопление *n*, 2128
газовый, 2122
газовый воздухонагреватель *m*, 2127
газовый излучатель *m*, 2146
газовый излучающий нагреватель *m*, 3139
газовый комнатный нагреватель *m*, 2147
газовый котел *m*, 2113
газовый нагреватель *m*, 2126
газовый счетчик *m*, 2140
газовый фильтр *m*, 2125
газовый холодильник *m*, 2120
газообразный, 2122
газоплотный, 2151
газоснабжение *n*, 2150
газоход *m*, 2132
гайка *f*, 2796
галогенный, 2210
гарантия *f*, 2202
гарантия *f* качества, 3129
генеральный подрядчик *m*, 3082
генеральный эксплуатационник *m*, 3083
генератор *m*, 2162

генератор *m* абсорбционной холодильной машины, 219, 467
генератор *m* льда, 2382
генератор *m* пара, 3632
генераторный газ *m*, 3090
герметизация *f*, 3073
герметизированная камера *f*, 2902
герметически заваренный, 2300
герметичная система *f*, 3364
герметичный, 2298, 3071
герметичный компрессор *m*, 2301, 3365
герметичный компрессорно-конденсаторный агрегат *m*, 3363
гибкая вставка *f*, 1936
гибкая труба *f*, 1939
гибкий, 1934
гибкий стык *m*, 1938
гибкий участок *m* трубопровода, 1936
гибкое соединение *n*, 1935, 2331
гибкость *f*, 1933
гигиенический, 2365
гигрометр *m*, 2366
гигрометрическая разность *f*, 1270
гигроскопический, 2367
гигростат *m*, 2368
гидравлический, 2362
гидравлический затвор *m* котла, 475
гидравлический затвор *m* (на котле), 1956
гидравлический удар *m*, 4069
гидроаэродинамика *f*, 1990
гидромеханика *f*, 1990
гидростатический, 2364
гильза *f*, 585
гильза *f* для прохода коммуникаций, 1504
гильза *f* термометра, 3814
гильотинный клапан *m*, 2207
гиперболический, 2369
гипотермия *f*, 2370
главный стояк *m*, 3281
главный трубопровод *m*, 879, 2608
гладкий конец *m* трубы, 3566
глубина *f* змеевика, 806
глубокий вакуум *m*, 1202, 2315
глубокое замораживание *n*, 1201
глухое ответвление *n*, 1186

глухой фланец *m*, 422, 3663
глушитель *m*, 2740, 3444
глушитель *m* вентилятора, 1793
головка *f* болта, 3355
головка *f* цилиндра, 1160
горелка *f*, 568
горелка *f* с механическим распыливанием, 2655
горелка *f* с принудительным дутьем, 2013
горелка *f* с распыливанием паром, 3623
горение *n*, 844
горение *n* в псевдосжиженном слое, 1988
гореть, 567
горизонтальный газоход *m*, 2328
городская вода *f*, 753
горючая среда *f*, 855
горючее вещество *n*, 855
горючесть *f*, 841, 1909
горючий, 842
горячая вода *f*, 2337, 4044
горячая вода *f* под высоким давлением, 2309
горячее водоснабжение *n*, 2343
горячий, 2332
горячий воздух *m*, 2333
горячий цех *m*, 2336
гофрированное колено *n*, 1099
гофрированный отвод *m*, 1100
гоючий газ *m*, 2083
гравитационная циркуляция *f*, 2187
гравитационное отопление *n*, 2188
гравитационный пленочный охладитель *m*, 1787
гравитация *f*, 2186
градиент *m*, 2176
градиент *m* давления, 3054
градирня *f*, 4062
градирня *f* с дроблением жидкости на пластинах, 3572
градирня *f* с естественным движением воздуха, 2758
градирня *f* с механическим побуждением (движения воздуха), 2656
градирня *f* с орошаемой насадкой, 3585
градирня *f* с охлаждением вынужденным потоком воздуха, 2017
градирня *f* с перекрестным током, 1133

изолирующий клапан *m*,
2487
изолятор *m*, 373, 2454
изоляционная оболочка *f*,
2457
изоляционный материал
m, 2458
изоляция *f*, 2459
изометрический чертеж *m*,
2488
изотерма *f*, 2489
изотермный, 2490
изотоп *m*, 2491
изоэнтальпийный, 2485
изучение *n* и
использование *n* свойств
жидких и газовых-
потоков, 1987
имеющий разрешение, 2536
импульс *m*, 2722
импульс *m* струи, 2496
ингибитор *m*, 2435
индекс *m*, 2409
индекс *m* вспучивания,
3740
индекс *m* закопченности,
3532
индекс *m* комфорта, 871
индекс *m* сажи, 3532
индекс *m* свежести, 2063
индекс *m* теплового
комфорта, 4043
индивидуальное отопление
n, 2414
индикатор *m*, 2413
индикатор *m* горения, 850
индикатор *m* давления,
3053
индикатор *m* утечки, 2528
индикаторная лампа *f*,
2410
индикаторная лампа *f*
электрической нагрузки,
2543
индикация *f*, 2411
индукционное устройство
n, 2421
индукционный нагрев *m*,
2419
индукция *f*, 2418
индуцированный, 2417
инертный, 2424
инерционный
пылеотделитель *m*, 2395
инерция *f*, 2425
инжектор *m*, 2441
инжекторная труба *f*, 2440
инжекторное сопло *n*, 2439
инжекторный распылитель
m, 2438
инжекционная горелка *f*,
2442
инжекционная форсунка *f*,
3056
инженер *m*, 1640

инженер-инспектор *m*, 1832
инженерная система *f*
общественного здания,
875
инсоляция *f*, 2448
инспектор *m* рабочего
места (охраны труда),
3460
Институт *m*
кондиционирования
воздуха и холодильной
техники, 263
инструкция *f* по монтажу,
287, 2736
инструкция *f* по
обслуживанию, 2453,
2847, 3399
инструмент *m*, 3863
интегральное управление
n, 2461
интегральный, 2462
интегратор *m*, 2463
интенсивность *f*, 2464
интенсивность *f* испарения,
1694
интенсивность *f*
освещения, 2546
интервал *m*, 2478
интервал *m* проведения
обслуживания
электроустановки, 1579
интерьер *m*, 2469
инфекция *f*, 2426
инфильтрация *f*, 2427
инфильтрация *f* воздуха,
132
инфильтрация *f* через
неплотности, 2428
информационный обмен
m, 2431
инфракрасная съемка *f*,
2433
инфракрасное отопление *n*,
2432
ион *m*, 2480
ионизация *f*, 2482
ионный обмен *m*, 2481
ископаемое топливо *n*,
2025
искра *f*, 3559
испарение *n*, 1690, 3957
испаритель *m*, 1703
испаритель *m*
вертикального типа, 3999
испаритель *m*
затопленного типа, 1947
испаритель *m* полного
испарения, 1465
испаритель *m* с
недостаточной подачей
хладагента, 3613
испарительная горелка *f*,
3958
испарительное охлаждение
n, 1699, 3891

испарительное равновесие
n, 1700
испарительный агрегат *m*,
1707
испарительный
конденсатор *m*, 1697
испарительный
кондиционер *m*, 1696
испарительный охладитель
m, 1698
испарительный
увлажнитель *m*, 1701
испарять, 1689
исполнение *n*, 1716, 2938
исполнительный механизм
m (в системе
регулирования), 1031
исполнительный чертеж *m*,
267
испытание *n*, 3776
испытание *n* здания
(помещения) на
герметичность, 1803
испытание *n* индуктивным
методом, 1545
испытание *n* на вакуум,
3949
испытание *n* на горение,
861
испытание *n* на плотность,
2527
испытание *n* на
прозрачность (мутность),
419
испытание *n* на сжатие, 909
испытание *n* под
давлением, 3069
испытание *n* сажи, 3534
испытательная камера *f*,
3778
испытательная установка
f, 3782
испытательное давление *n*,
3784
испытательный стенд *m*,
3777, 3788
исследование *n*, 3245
исследование *n* корпуса
(оболочки), 644
истечение *n*, 1558, 2879
истечение *n* воды, 4079
источник *m* воздуха, 177
источник *m* звука, 3552
источник *m* теплоты, 2286,
2346
источник *m* холода, 824
источник *m* энергии, 1635
исходная (фиксированная)
точка *f*, 3189
исходные данные *n,pl*
компьютера, 929
исходный указатель *m*,
3190
исходный цикл *m*, 3188

кабель *m*, 599
кабельная жила *f*, 1084
кабельная муфта *f*, 600
кабельный ввод *m*, 601
кабина *f* управления, 1018
кавитация *f*, 656
кажущаяся мощность *f*, 249
калач *m*, 3260
калиброванная пыль *f*, 3781
калибровка *f*, 604
калориметр *m*, 611
калорифер *m* (теплообменник), 610
калория *f*, 606
каменный уголь *m*, 783
камера *f*, 701
камера *f* орошения, 3581
камера *f* подачи воздуха на горение, 580
камера *f* сгорания, 847, 859
камин *m*, 1884, 2223
канал *m*, 706, 1492, 2098
канал *m* для выхода дымовых газов в атмосферу, 1981
канал *m* теплого воздуха, 4036
капелька *f*, 1449
капельная конденсация *f*, 1450
капиллярное действие *n*, 625
капиллярный, 624
капиллярный эффект *m*, 623
каплеотделитель *m*, 117, 1451
капля *f*, 1442, 1448
карбонизация *f*, 632
карданный вал *m*, 1446
каркас *m* конструкции, 2033
каркас *m* фильтра, 1858
карта *f* комфортных условий, 869
картер *m*, 1123
каскад *m*, 640
каскадное регулирование *n*, 641
каскадное сжатие *n*, 906
каскадный регулятор *m*, 642
катализатор *m*, 651
кататермометр *m*, 2505
катион *m*, 654
катод *m*, 652
катодная защита *f*, 653
качающаяся колосниковая решетка *f*, 3286
качество *n*, 3128
качество *n* внутреннего воздуха, 2416
качество *n* газа, 2145

квартирное отопление *n*, 245, 441
кинетический, 2507
кипеть, 454
кипящий, 483
кирпич *m*, 526
кирпичная кладка *f*, 527
кислород *m*, 2894
кислота *f*, 35
кислотность *f*, 38
клапан *m*, 3951
клапан *m*, регулирующий течение, 1968
клапан *m* высокого давления в конденсаторе, 958
клапан *m* для перекрытия выброса, 1349
клапан *m* на вытяжке газа, 1730
клапан *m* постоянного давления, 992
клапан *m* постоянного уровня, 991
клапан *m* регулирования объема, 4016
клапан *m* регулирования течения, 1958
клапан *m* с двигателем, 2728
клапан *m* с двумя седлами, 1404
клапан *m* с параллельными лопатками, 2915
клапан *m* шламоотделителя, 3676
классификация *f*, 758
клей *m*, 49
клейкий, 50
клемма *f*, 982
клещи *n*, 3861
климат *m*, 771
климатическая испытательная камера *f*, 772
климатическая исследовательская камера *f*, 1663
климатическая камера *f*, 1658
клиновой ремень *m*, 3976
клиноременная передача *f*, 3977
кло (характеристика одежды), 776
клупп *m*, 3823
кнопочное управление *n*, 3125
коагулянт *m*, 3025
коаксиальный, 793
ковкий чугун *m*, 2618
код *m*, 796
кожух *m*, 2492
кожух *m* вентилятора, 1795

кожухотрубный змеевиковый испаритель *m*, 3423
кожухотрубный змеевиковый конденсатор *m*, 3422
кожухотрубный испаритель *m*, 1745, 3425
кожухотрубный конденсатор *m*, 3424
кожухотрубный теплообменник *m*, 3426
кокс *m*, 810
коксовая топка *f*, 812
коксовый, 814
коксовый газ *m*, 813
коксовый щебень *m*, 536
коксующийся уголь *m*, 815, 2117
колба *f*, 561
колебание *n*, 2361, 2862
колебание *n* температуры, 3771
колено *n*, 1563
колено *n* трубопровода, 1495
колено *n* трубы, 2951, 2955
коленчатый вал *m*, 1125
колесо *n*, 4121
количество *n*, 204, 3131
количество *n* воздуха, 151
количество *n* движения, 2722
количество *n* движения в струе, 2496
коллектор *m*, 829, 1004, 2218, 2622
коллектор *m* канализации, 3411
колодезная вода *f*, 4113
колодец *m*, 4112
колонный радиатор *m*, 838
колосник *m*, 2180
колотый лед *m*, 2030
колпак *m*, 488, 614, 2323
кольцевой поток *m*, 234
комбинированная горелка *f*, 1040
комбинированное регулирование *n*, 839
коммерческая система *f*, 875
коммерческий, 873
коммутатор *m*, 2506
комната *f*, 3295
комнатный калориметр *m*, 3299
комнатный кондиционер *m*, 3297
комнатный нагреватель *m*, 3303
комнатный увлажнитель *m* воздуха, 3298
компактный, 2897

окружная скорость *f*
　рабочего колеса, 3858
олово *n*, 3855
олф *m* (антропогенное
　загрязнение воздуха
　одним человеком), 2832
омываемая поверхность *f*
　змеевика, 423
оперативное
　запоминающее
　устройство *n*, 3155
описание *n*, 1242
оплата *f*, 1811
опора *f*, 391, 3620, 3729
опора *f* для трубы, 2962
опора *f* для шарикового
　подшипника, 362
опора *f* муфтового
　подшипника, 3475
опора *f* подшипника, 392
опорный подшипник *m*,
　2006, 2928
опорожнение *n*, 1614, 1688
оправка *f*, 1920
определение *n* давления,
　razvivaemogo nasosom,
　3070
определенный, 3561
оребренная труба *f*, 1869
оребренный, 1868
ориентация *f*, 2857
оросительный конденсатор
　m, 432
орошаемая насадка *f*
　градирни, 1071
осадка *f*, 3377
осадок *m*, 1237, 3377
осадок *m* пыли, 1512
осаждать, 1238
осаждение *n*, 1239, 3024
освещение *n*, 2542
освещенность *f*, 2465
осевая скорость *f*, 319, 685
осевой, 315
осевой вентилятор *m*, 317,
　3100, 3955
осевой вентилятор *m* (с
　двигателем в обечайке),
　3900
осевой вентилятор *m* с
　двигателем вне потока,
　406
осевой компрессор *m*, 316
ослабление *n*, 302
ослабление *n* звука, 3549
осмечивать, 1683
основание *n*, 375, 394, 2028,
　2222
основание *n* дымовой
　трубы, 734
основное отопление *n*, 327
остановка *f*, 3662
остаток *m*, 3237
остаток *m* горения, 858

остаточная стоимость *f*,
　2540
остаточное давление *n*,
　1866
остекление *n*, 1826
острый пар *m*, 2567
остывание *n*, 1057
осушать, 1222
осушение *n*, 1220, 1244,
　1470
осушение *n* воздуха, 105,
　114
осушитель *m*, 1221, 1226,
　1243, 1436
осушитель *m* пара, 3629
осушитель-
　вымораживатель *m*, 2040
осушительный аппарат *m*,
　1469
осушительный змеевик *m*,
　1437
осушительный эффект *m*,
　1223
осциллирование *n*, 2862
ось *f*, 320, 321
отапливать, 2077
отбирать, 430, 433
отбойный слой *m* (в
　градирнях), 1439
отбойный щиток *m*, 343
отверстие *n*, 2321, 2841
отверстие *n* (в трубе), 495
отверстие *n* для подачи
　воздуха, 3723
отверстие *n* клапана, 3952
отверстие *n* с резьбой, 3761
ответвительная линия *f*,
　510
ответвление *n*, 507, 2808,
　3434
ответвление *n*
　воздуховода, 508
ответвление *n* провода, 511
ответный патрубок *m*, 3261
ответный фланец *m*, 883
отвод *m*, 404, 502, 509,
　1495, 2951
отвод *m* воздуха, 3994
отводить воздух, 3986
отводной клапан *m*, 435
отводной контур *m*, 1386
отдача *f*, 3838
отделение *n*, 3388
отделение *n* воздуха, 158
отделение *n* для
　замораживания, 2048
отделение *n* пламени, 1908
отделение *n* шлака, 3466
отделитель *m*, 1604
отдельно установленный
　регулятор *m*, 3596
отказ *m*
　терморегулирующего
　вентиля, 2045
откаченный, 1687

откачивание *n*, 1688
откидной клапан *m*, 1915
отклонение *n*, 405, 1266
отклонение *n* от
　установленного
　значения, 2890
отклонение *n*
　температуры, 3768
отключатель *m* давления,
　3045
отключатель *m* по
　разности давления, 3047
отключающий клапан *m*,
　1150
открытый, 1762
отлаживать, 1190
отличительная окраска *f*,
　2383
отложение *n*, 1239
отложение *n* водорослей,
　183
отложение *n* инея, 2073
относительная влажность
　f, 3233
относительный, 3232
отношение *n*, 3164
отношение *n*
　концентраций, 941
отогрев *m*, 3225
отогреватель *m*, 3228
отогревать, 3226
отопительная нагрузка *f*,
　2251
отопительная панель *f*,
　2255
отопительная печь *f*, 2259
отопительная установка *f*,
　2250, 2256
отопительное масло *n*,
　2254
отопительный агрегат *m*,
　2256, 3929
отопительный градусодень
　m, 2246
отопительный график *m*,
　2245
отопительный змеевик *m*,
　2243
отопительный прибор *m*,
　2242
отопительный сезон *m*,
　2257
отопительный эффект *m*
　компрессора (в режиме
　теплового насоса), 918
отопление *n*, 2241
отопление *n* жилых
　зданий, 3250
отопление *n* низкого
　давления, 2600
отопление *n* с естественной
　циркуляцией, 2188
отопление *n* того или
　иного объема, 3554
отпотевание *n*, 1268

приводить в движение, 1445
приводить в движение рукояткой, 1122
приглашение *n* подавать заявки, 605
пригодность *f*, 3397
пригодный для утилизации, 3182
приготовление *n* свежего воздуха, 2061
приготовлять, 3081
придавать огнеупорность, 1887
прием *m*, 23
приемка *f*, 23
приемное испытание *n*, 25
приложение *n*, 252
применение *n*, 252
примесь *f* к воздуху, 93
принадлежности *f* вентилятора, 1792
принадлежность *f*, 29
принтер *m*, 3085
принудительная циркуляция *f*, 2009
принятая схема *f*, 3180
природный газ *m*, 2760
прирост *m* бактерий, 339
присоединение *n* абонента, 997
присоединение *n* для вытяжки газа, 1729
присоединительная коробка *f*, 2504
приспособление *n* для чистки дымовой трубы, 1974
приток *m*, 2430
приток *m* солнечной теплоты, 3516
приточный воздух *m*, 3077
проба *f*, 3337
проба *f* дыма, 3499
проба *f* с помощью метиловй синьки, 2678
пробка *f*, 777, 1086, 2990, 3566
пробковая теплоизоляция *f*, 1087
пробковый кран *m*, 795
пробный кран *m*, 3780
пробный ход *m*, 3897
проветривать, 58
проводимость *f*, 965, 966
проводная система *m*, 4133
проводник *m*, 971
проводник *m* теплоты, 2231
проволока *f*, 4131
прогнозирование *n* окружающих условий, 1660
программа *f* производства, 3091

программированное логическое регулирование *n*, 3094
программированный, 3093
программное обеспечение *n* компьютера, 936
программное регулирование *n*, 3030
программный регулятор *m*, 3095
продолжительность *f*, 1509
продолжительность *f* горения, 852
продолжительность *f* дня, 1182
продольное ребро *n*, 2583
продольный фасад *m*, 2584
продувать, 444, 3123
продувка *f*, 445
продувочный клапан *m*, 446
продукт *m* сгорания, 856
преобразование *n*, 804
проект *m*, 1245
проект *m* здания, 556
проект *m* пуско-наладочных работ, 878
проектирование *n* окружающих условий, 1660
проектирование *n* с использованием компьютеров, 923
проектировать, 1246
проектная документация *f*, 3430
проектные требования *n*, 1249
проектные условия *n*, 1248
проем *m*, 3004
прозрачность *f* пылевого пятна, 1526
производитель *m* газа, 2143
производительность *f*, 1556
производительность *f* вентилятора, 1802
производительность *f* компрессора, 913
производительность *f* конденсационной установки, 961
производительность *f* котла, 460
производительность *f* по холоду, 821
производительность *f* терморегулирующего вентиля, 1754
производить, 2630
производственное кондиционирование *n* воздуха, 2422
производство *n*, 2629

производство *n* газа, 2144
производство *n* холода, 820
производство *n* электроэнергии, 1583
прокаливать, 233
прокладка *f*, 2136, 2903, 4049
прокладка *f* в гильзе, 587
прокладка *f* трубопровода, 1502
промежуток *m*, 3436
промежуточный охладитель *m*, 2467
промежуточный перегреватель *m*, 2472
промежуточный слой *m*, 2471
промывать, 1992
промывное устройство *n*, 4049
промывной клапан *m*, 1995
промышленная вода *f*, 2423
проницаемость *f*, 2942, 3889
пропан *m*, 3098
пропеллерный, 2295
пропорциональное регулирование *n*, 3104
пропорциональный, 2708, 3101
прорезь *f*, 3484
просвет *m*, 766
пространство *n* для нагревания воды в котле, 480
пространство *n* за потолком, 666
противодавление *n*, 329
противооборот *m*, 1013
противоток *m*, 1105
противоточное течение *n*, 1103, 1106
противоточный теплообменник *m*, 1107
профессиональный (дипломированный) инженер *m*, 3092
профилактическое обслуживание *n*, 3076
профиль *m*, 1603, 2733
профиль *m* скорости, 3982
проход *m*, 2157, 3004
проход *m* воздуха, 143
проходной клапан *m*, 2156
проходной клапан *m* (без изменения направления течения), 1326
проходной кран *m*, 3840
проходящий поток *m*, 3885
процент *m* насыщения, 3343
процесс *m*, 3087

скорость *f* воздушного
　потока, 123
скорость *f* горения, 863
скорость *f* замораживания,
　2054
скорость *f* звука, 3551
скорость *f* истечения, 1347
скорость *f* на выходе, 1739,
　2873
скорость *f* падения, 3981
скорость *f* течения, 1971
скорректированная
　эффективная
　температура *f*, 1090
скребок *m*, 3150
скруббер *m*, 3360
скрытая теплота *f*, 2517
скрытая теплота *f*
　испарения, 2519
скрытая теплота *f*
　плавления, 2518
скрытый, 2516
следящая автоматическая
　система *f* управления,
　3402
слесарь-монтажник *m*,
　1897
слив *m*, 3457
слой *m*, 2520
слой *m* льда, 2381
слой *m* шлака, 774, 3468
случайная погрешность *f*,
　3156
случайное поступление *n*
　теплоты, 650
случайное
　теплопоступление *n*,
　2401
случайный, 2400
слышимость *f*, 304
слышимый, 305
смазка *f*, 2604
смазка *f* под давлением,
　2019
смачиваемая поверхность *f*
　(котла), 4119
смачивать, 1166, 1167
смесительная камера *f*,
　437, 2694
смесительное устройство
　n, 2693
смесительный клапан *m*,
　2700
смесительный контур *m*,
　2695
смесительный насос *m*,
　436, 2698
смесительный трубопровод
　m (канал), 2697
смесь *f*, 2701
смесь *f* газов, 2141
смесь *f* жидкости, 2563
смета *f*, 1682
смешивающий вентилятор
　m, 2691

смола *f*, 3763
смотровое отверстие *n*,
　2211
смотровое стекло *n*, 3441
смотровой колодец *m*,
　2621
смотровой люк *m*, 2929
смыкание *n*, 2470
снабжать, 3720
снабжение *n*, 3719
снижение *n* давления, 3064
сниженное давление *n*, 3185
собранный на заводе, 1783
собственная частота *f*, 2759
совместимый, 884
согласование, 884
содержание *n*, 1007
содержание *n* в золе
　частиц неполного
　сгорания, 277
содержание *n* воды, 4059
содержание *n* золы, 272
содержание *n* кислорода,
　2895
содержание *n* пыли, 1511
соединение *n*, 983, 1109,
　3924
соединение *n* с дымовой
　трубой, 1976
соединение *n* сваркой, 515
соединение *n* сварное, 4104
соединение *n* типа 'труба в
　трубе', 2616
соединитель *m*, 986
соединительная деталь *f*,
　979
соединительная муфта *f*,
　1112
соединительная труба *f*,
　980
соединительная тяга *f*,
　3844
соединительная часть *f*,
　3752
соединительный зажим *m*,
　982
соединительный участок
　m, 979
соединительный фланец *m*,
　978
соединительный шток *m*,
　981
соединять, 975
соленоидный вентиль *m*,
　3524
солнечная постоянная *f*,
　3514
солнечная радиация *f*, 3518
солнечная энергия *f*, 3515
солнечно-воздушная
　температура *f*, 3512
солнечный датчик *m*, 3519
солнечный коллектор *m*,
　3513
солончаковая вода *f*, 505

соотношение *n* асбеста, 279
сопло *n*, 2791
сопло *n* горелки, 572, 579,
　1143
сопло *n* мазутной
　форсунки, 2813
сопловое отверстие *n*, 2792
сопротивление *n*, 1414,
　3251
сопротивление *n* течению,
　1969
сопротивление *n* трения,
　2065
сорбент *m*, 3535
сорбент *m* из
　активированной окиси
　алюминия, 41
составляющая *f*
　воздушной струи, 168
составная часть *f*, 891
состояние *n*, 3614
состояние *n* воздуха, 87,
　964
сосуд *m*, 4000
сосуд *m* под давлением,
　3072
сотовая структура *f*, 673
спаивать, 513
спаренный компрессор *m*,
　3753
спаянный, 514
спекание *n*, 3458
спецификация *f*, 3562
спираль *f*, 3569
спирально-навивной
　воздуховод *m*, 3571
спирально-навивной
　трубопровод *m*, 3571
спирт *m*, 180
сплав *m*, 188
способ *m*, 2677
способ *m* отопления, 2253
спуск *m*, 1416
спуск *m* масла, 2817
спускной кран *m*, 1420
среднее время *n* между
　отказами, 2647
среднее давление *n*, 2644,
　2659
среднее *n* за сезон, 2683
среднерадиационная
　температура *f*, 2645
средний, 314, 2643
средняя
　продолжительность *f*
　ремонта, 2648
средняя разность *f*
　температуры, 2646
средняя температура *f*,
　2660
средство *n*
　антикоррозийной
　защиты, 1095
средство *n* против накипи,
　242

топливный фильтр *m*, 2082
топливо *n*, 2076
топочная дверь *f*, 1882
топочная заслонка *f*, 1881
топочная камера *f*, 1877
топочный порог *m*, 1879
торговый холодильник *m*, 874
торф *m*, 2927
точечная коррозия *f*, 2972
точечная сварка *f*, 3579
точка *f*, 2998
точка *f* воспламенения (вспышки), 1925
точка *f* зажигания, 2388
точка *f* замерзания, 2052
точка *f* кипения, 484
точка *f* конденсации, 950
точка *f* контакта, 1001
точка *f* насыщения, 3344
точка *f* отбора, 1434
точка *f* отбора мощности, 3020
точка *f* отключения, 1153
точка *f* плавления, 2663
точка *f* росы, 1269
точка *f* росы кислоты, 37
точка *f* снижения давления, 3061
точная (прецизионная) работа *f*, 3026
точность *f*, 32
точность *f* настройки, 3407
точность *f* термостатирования, 2226
точность *f* установки, 3407
тощий уголь *m*, 2782
транзистор *m*, 3884
трансмиссия *f*, 3888
транспортер *m*, 1043
транспортировка *f* и уход *m* за топливом, 2084
транспортировка *f* топлива, 2089
трансформатор *m*, 3882
трап *m*, 3892
трассирующий газ *m*, 3875
трассирующий индикатор *m*, 3874
требования *n*,pl обслуживаемого объема, 3555
требуемая величина *f* энергии, 1638
требуемая тяга *f*, 1431
трение *n*, 2064
трение *n* в трубах, 3902
трение *n* в трубе, 2956
трехтрубная система *f*, 3829
трехтрубная система *f* кондиционирования воздуха, 3828
трехходовой вентиль *m*, 3832

трехходовой кран *m*, 3831
трещина *f*, 1118
тройник *m*, 2622, 4146
тройник *m* (на трубопроводе), 1383
труба *f*, 2950, 3899
труба *f* Вентури, 3997
труба *f* с резьбой (внутренней), 3826
труба *f* со спиральным оребрением, 3570
трубка *f* Бурдона, 501
трубка *f* Пито, 2971
трубка *f* уровня, 564
трубная доска *f*, 3903
трубная доска *f* (в теплообменнике), 1357
трубная резьба *f*, 2964
трубный пучок *m*, 3901
трубный разметчик *m*, 2965
трубопровод *m*, 972, 1492, 2958
трубопровод *m* для отбора, 434
трубопровод *m* для отвода конденсата, 431
трубопровод *m* из листового металла, 3420
тугоплавкий, 3196
тугоплавкость *f*, 3195
туман *m*, 2003, 2688
турбина *f*, 3906
турбовоздуходувка *f*, 3907
турбогенератор *m*, 3909
турбокомпрессор *m*, 318, 688, 3908
турбокомпрессор *m* с дозвуковой скоростью, 3692
турбулентность *f*, 3910
турбулентный поток *m*, 3911
турбулизатор *m* испарителя, 1706
тушить, 1770
тяга *f*, 1425, 3594
тяга *f* дымовой трубы, 736, 1977
тягомер *m*, 1426, 1428

увеличение *n* влагосодержания, 1272
увеличивать, 209
увлажнение *n*, 2352
увлажнение *n* воздуха, 130
увлажнитель *m*, 2353
увлажнитель *m* воздуха, 131
увлажнять, 1166, 1167, 2354
увлекать, 1648
угарный газ *m*, 633
углеводород *m*, 2363
углекислый газ *m*, 629
углерод *m*, 628

угловая горелка *f*, 1088
угловой вентиль *m*, 231
угловой вентиль *m* (клапан), 1089
угловой коэффициент *m*, 227
угол *m*, 226
угол *m* выпуска, 229
угол *m* выхода, 230
уголковое железо *n*, 228
уголь *m*, 782
уголь *m* с большим содержанием летучих, 2318
уголь *m* с большим содержанием смолы, 415
уголь *m* со средним содержанием летучих, 2661
угольная кислота *f*, 631
угольная мелочь *f*, 3464
угольная пыль *f*, 785
угольная топка *f*, 788
угольный бункер *m*, 789
угольный фильтр *m*, 710
удаление *n*, 3241
удаление *n* воздуха, 174
удаление *n* дымовых газов, 1980
удаление *n* котельной накипи, 3350
удаление *n* пыли, 1515, 1521
удаление *n* шлака, 775
удаление *n* шлама, 3489
удалять (воздух), 1720
удалять отложения, 1241
удар *m*, 3682
удельная проводимость *f*, 970
удельное термическое сопротивление *n*, 3803
удельный, 3561
удельный поток *m* массы, 2634
удельный тепловой поток *m*, 3564
указатель *m*, 2409
уклон *m*, 405, 3150
укомплектованный на заводе, 1784
укрывать, 3428
укрытие *n*, 3427
улитка *f* компрессора, 3359
уловитель *m* сажи, 3530
уменьшение *n* шума, 2777
умывальник *m*, 4048
умягчение *n*, 3510
умягчение *n* воды, 4085
умягчитель *m* воды, 4084
универсальная газовая горелка *f*, 2115
универсальная горелка *f*, 3931
унос *m*, 1649

Suomi

749

askelmäntäkompressori, 3656
askelsäädin, 3652
askelsäätö, 3655
aste, 1214
asteittainen, 3654
astepäivä, 1215
astia, 3755, 4000
asuinrakentamiseen liittyvä,
  1388
asukas, 2800
asumiseen liittyvä, 1388
asuntoilmastointi, 3249
asuntolämmitys, 1389, 3250
atmosfäärinen, 292
aukko, 2841
aukon ala, 246
auktoriteetti, 306
auringonkerääjä, 3513
auringonkerääjän tehollinen
  ala, 246
auringon suoja, 3414
auringon säteily, 2448, 3518
aurinkoenergia, 3515
aurinkoilmalämpötila, 3512
aurinkokaha, 3513
aurinkokuorma, 3516
aurinkotuntoelin, 3519
aurinkovakio, 3514
automaatio, 309
automaattinen, 308
automaattinen lauhteen
  erotin, 1944
avautumispaine, 3333
avoin jäähdytysjärjestelmä,
  1755
avoin kompressori, 2842
azeotrooppinen, 323

Bacharach-luku, 324
bakteereita tappava, 340,
  2165
bakteeri, 335, 2163
bakteerikasvu, 339
bakteeripitoinen, 336
bakteeritoiminta, 337
bakteeriton, 2164
Beaufort asteikko, 393
bensiini, 2111
betoni, 943
bi-metalli, 408
bi-metallielementti, 409
bimetallilämpömittari, 410
binääri, 411
binääriseosprosessi, 412
bitumimaali, 416
bitumipaperi, 414
bolometri, 485
bourdonkaaripainemittari,
  500
Bourdon-putki, 501
briketti, 534
bruttoala, 259
bruttoteho, 2195
bruttotilavuus, 260
butaani, 588

Carnot-prosessi, 635
celsiusaste (°C), 676
comfort
  ilmastointijärjestelmä, 868

datamuunnin, 1177
deklinaatio, 1196
deodorantti, 1236
desibeli, 1194
desinfiointi, 1358
desinfioiva aine, 341
desipol, 1195
desorptio, 1255
detaljipiirustus, 757, 1260
diagrammi, 1273, 2177
dielektrinen, 1282
dielektrinen sulatus, 1285
dielektrisyysvakio, 1283
dieselmoottori, 1286, 1287
differentiaali-, 1289
diffuusi, 1296
diffuusiheijastus, 1298
diffuusio, 1300
diffuusiokerroin, 1302
diffuusiokoneisto, 1303
diffuusori, 1297
digitaalinen, 1304
digitaalinen tiedonsiirto,
  1305
DOP-testi
  (mikrosuodattimelle), 1391
dynaaminen, 1530
dynaaminen grafiikka, 1532
dynaaminen käyttäytyminen,
  1531
dynaaminen paine, 1535,
  3980
dynaaminen suodatin, 2395
dynaaminen viskositeetti,
  1536
dynaamisen paineen häviö,
  1534
dynamometri, 1537

edullisuus, 1835
efektiivinen, 1550
ehdotus, 1252
ehkäisevä huolto, 1092
ehtymätön, 2785
ejektori, 1559
ejektorijäähdytys, 1560
ejektorikylmäprosessi, 3963
ekvivalenttilämpötila, 3512
elastinen materiaali, 1562
elastisuus, 1561
elektrodi, 1589
elektrodikattila, 1590
elektrolyyttinen, 1591
elektroniikka, 1598
elektroninen, 1595
elektroninen vahvistin, 1596
elementti, 1602
elinikä, 2538
elinikäinen kustannus, 2540

elinkaari, 2539
elohopea, 2667
elohopeakytkin, 2668
elohopealämpömittari, 2669
elohopeapatsas, 836
emissiivisyys, 1612, 1613
emissio, 1611
emulsio, 1615
emäksinen, 185
emäksisyys, 186
endoterminen, 1618
energia, 1619
energiahäviö, 1629
energiakatselmus, 1620
energiakustannus, 1624
energialähde, 1635
energiamuodot, 2023
energian hallinta, 1630
energian hallintajärjestelmä,
  1631
energian häviämättömyyden
  laki, 987
energian kulutus, 1623
energian käyttötavoite, 1638
energian läpäisysuhde, 1637
energian saatavuus, 1621
energian säästö, 1622
energian tehokas käyttö,
  1632
energiapotentiaali, 1633
energiatalous, 1625
energiataso, 1628
energiatehokkuus, 1626, 1627
energiatila, 1636
englantilainen standardi, 535
enimmäispitoisuus, 2640
ennakointivastus, 238
ennakointivastus
  termostaatissa, 2226
ennakoiva säätö, 237, 3030
ensitäyttö, 2436
entalpia, 1645
entalpiapiirros, 1647
entropia, 1651
entropiapiirros, 1653
epäjatkuvuustila, 2789
epäkeskeinen, 1541
epäpuhtaus, 1005
epäsuora lämmönsiirrin,
  3321
epätäydellinen palaminen,
  2406
eriste, 2454
eristysaine, 2458
eristäminen, 2459
eristävä peite, 2457
eristää, 2455, 2456
erittely, 3562
ero, 1288
ero-, 1289
eroalue, 2846
eroosio, 1679
erosäätö, 1290
erotin, 1604, 3025, 3390,
  3892

ilmatyyny, 102
ilmaus, 431, 1189
ilmaverho, 101
ilmavirran suuruus, 123
ilmavirta, 100, 121
ilmavirtasäätö, 3969
ilmavirtasäätöinen
　järjestelmä, 3970
ilmavirtaus, 100
ilmavuoto, 136, 137
ilmaylimäärä, 1709
ilmaääni, 77
imeytyskaivo, 3508
imeä, 3694
imu, 282, 3695
imuhuuva, 3698
imukorkeus, 3699
imupaine, 3697, 3702
imupuolen suodatin, 3700
imuputki, 3701
imusuodatin, 3703
imuteho, 3696
indeksi, 2409
induktio, 1649, 2418
induktiolämmitin, 2419
induktiolämmitys, 1285
induktiosuhde, 1650, 2420
induktioyksikkö, 2421
indusoitua, 1648
indusoitunut, 2417
inertia, 2425
inertti, 2424
infektio, 2426
infrapunakuvaus, 2433
infrapunalämmitys, 2432
inhibiittori, 2435
injektori, 2441
injektoripoltin, 2442
insinööri, 1640
insinööritiede, 1641
insinöörityö, 1641
integraattori, 2463
integroitu, 2462
integroitu säätö, 2461
intensiteetti, 2464
interferenssi, 2468
ioni, 2480
ionisaatio, 2482
ionivaihtaja, 2481
irtolaippa, 3483
irtonainen lämpöeriste, 1838
isku, 3682
iskunvaimentaja, 3429
iskupora, 2503
iskutilavuus, 2967
isobaari, 2486
isometrinen piirustus, 2488
isotermi, 2489
isoterminen suihku, 2490
isotooppi, 2491
itseimevä pumppu, 3381
itsenäinen säädin, 3596
itsestään tyhjentyvä (esim.
　auringonkeräysjär-
　jestelmästä), 1419

jaettu järjestelmä, 3575
jaettu laakeriholkki, 3573
jakaa, 3720
jakaa vyöhykkeisiin, 4151
jakaja, 1376
jakelujärjestelmä, 1375
jako, 1369, 2970
jakoastia, 2219
jakoavain, 4144
jakokammio, 3727
jakokanava, 1371
jakolaatikko, 1370
jakopelti, 3576
jakopiiri, 1386
jakoputkisto, 1374
jakorasia, 3743
jako(säätö)venttiili, 1385
jakotukki, 2218, 3727
jakoverkko, 1373
jakso, 1155
jaksottainen, 1156, 2940
jaksottainen lämmitys, 2473
jatkuva käyttö, 1009
jatkuva säätö, 1008
jatkuvuustila, 3621
johdin, 971
johtaa keskipakopumppuun
　vettä sen käyttöönotossa,
　3081
johtavuus, 970
johto, 972
johtosiipisäätölaitteisto, 2206
johtuminen, 966
johtumislämpöhäviö, 968
johtumislämpökuorma, 967
jousi, 3590
jousikuormitettu, 3591
joustava, 1934
joustava kanava, 1936
joustava liitos, 1935, 1938
joustava putki, 1939
joustavuus, 1933
joutoaika, 2384
julkisivu, 1775
juoksusiiven kärki, 2397
juoksuttaa, 3758
juoksutusnopeus, 3759
juoksutusputki, 434
juomaveden jäähdytin, 1441
juomavesi, 1440
juotava vesi, 3009
juotos, 3520
juotosliitos, 3522
juottaa, 3521
juottaminen, 3523
jälkijäähdytin, 63, 3371
jälkijäähdytys, 3179
jälkikaiunta-aika, 3270
jälkikuivaus, 3372
jälkilauhdutin, 3370
jälkilämmitin, 64, 3228
jälkilämmittää, 3226
jälkilämmitys, 3225
jälkilämmityspatteri, 3227

jälkitarkastus, 1103
jälkitäyttö, 3864
jännite, 4013
jännitekestoisuus, 1284
jännitemittari, 4014
järjestelmä, 3750
järjestys, 3391
jäte-energian lähteet, 4051
jätelämpö, 4052
jätelämpökattila, 4053
jätevesi, 1557, 4055
jättöreunan paksuus, 3876
jäykkyys, 3592
jäykkä vaahtomuovieriste,
　3278
jää, 2372
jääastia, 2375
jäähdyke, 1046, 2038
jäähdytetty ilma, 1047
jäähdytetty lasti, 727, 3203
jäähdytetty vesi, 728
jäähdyttimen vaippa, 3148
jäähdyttäjä, 1048
jäähdyttää, 726, 1045, 3202
jäähdytyksen astepäiväluku,
　1056
jäähdytyksen säätö, 3208
jäähdytys, 730, 870, 1050,
　1057, 3207
jäähdytysaine, 1051, 1064
jäähdytyselementti, 1060
jäähdytys hikoilemalla, 3891
jäähdytyshäviö, 1063
jäähdytysilma, 1052
jäähdytysjärjestelmä, 1069
jäähdytys jäävedellä, 2379
jäähdytyskuivaus
　normaalipaineessa, 297
jäähdytyskuorma, 1062
jäähdytyskäyrä, 1055
jäähdytyslaitos, 1065
jäähdytyslamelli, 3213
jäähdytysnesteen
　lämpötilaero, 1066
jäähdytys paineellisella
　ilmalla, 3044
jäähdytyspaneeli, 3213
jäähdytyspatteri, 98, 1049,
　1054
jäähdytyspinta, 1068
jäähdytysteho, 1053, 1058
jäähdytystorni, 1070, 1843
jäähdytystorni ilman täytettä,
　3585
jäähdytystornin
　poistovirtaus, 1074
jäähdytystornin puinen täyte,
　3572
jäähdytystornin
　sumunmuodostus, 1072
jäähdytystornin täyte, 1071
jäähdytystornin täytelevyt,
　2982
jäähdytystunneli, 1075
jäähdytysvaikutus, 731

rei'itetty, 2933
rei'itetty hajottaja, 2935
rei'itetty katto, 2934
rei'itetty levy, 2936
rei'itys, 2937
reikä, 2321
reikätiili, 78
rekisteröidä, 3218
rele, 3234
rengasvirtaus, 234
resonanssi, 3254
resultoiva, 3256
resultoiva lämpötila, 1476
reuna, 1547
reunaehdot, 498
rikastus, 1643
rikkidioksidi, 3704
rinnakkainen virtaus, 2917
rinnakkaisvirtauspoltin, 2918
rinnankytkentä, 984, 1110, 2916
ripa, 1763, 1865
ripa-, 1868
ripahyötysuhde, 1867
ripajako, 1871
ripalevy, 2983
ripaputki, 1869, 3570
ripaputkipatteri, 1870
ripaväli, 1872
ripustus, 3737
ristikappale, 1134
ristiveto, 1139
ristivirta, 1128, 1129
ristivirtajäähdytystorni, 1133
ristivirtakierto, 1130
ristivirtalämmönsiirrin, 1132
ritilä, 2180
rivallinen, 1868
rivikirjoitin, 2554
rivoitettu pinta, 423
roihu, 1916
ROM-muisti, 3169
roottori, 3315
rotaatiopumppu, 3312
routa, 2071
routavaurio, 2072
ruiskumaalauskoppi, 2909
rullalaakeri, 3287
rullasuodatin, 3288, 3310
rumpu, 1455
rumpujäähdytin, 1456
runko, 453, 2033
ruoste, 3325
ruostumaton teräs, 3595
ruskohiili, 537, 2548
ruuvata, 3354
ruuvi, 3353, 4142
ruuvikompressori, 3356
ruuviliitos, 3357
ruuvipenkki, 4005
ruuvitaltta, 3557
räjähdys, 1760

saaste, 3000
saasteinen, 3001

saastuminen, 3002
saastuttaa, 1006
sade, 3024
sademäärä, 3024
sairasrakennusongelma, 3439
sakka, 1237, 3377
sakkautua, 1238
sallittu kuormitus, 2944, 3336
samakeskeiset putket, 942
samanaikainen, 3448
samanaikaisuuskerroin, 1384
samanarvoinen, 1676
samepiste, 1946
sammutin, 1771
sammuttaa, 1770
sarana, 2319
saranoitu kansi, 2320
sarjaankytkentä, 985, 1111, 3394
sarjakäyttö, 3395
sarjaportti, 3393
sarjasäätö, 3392
satunnainen lämpökuorma, 2401
satunnaisotanta, 3157
satunnaisvirhe, 3156
sauma, 3367
saumaton, 3368
savu, 3497
savuhormi, 1981
savuhormin liitoskaulus, 1975
savuhälytin, 3494
savukaasu, 1973, 1978
savukaasuanalyysi, 861, 3499
savukaasuhäviö, 739, 1979, 2258
savukaasukammio, 3492
savukaasukanava, 3495
savukaasun lämpötila, 1985
savukaasun poisto, 1980
savukaasun poistokohta, 1984
savukaasun poistoputki, 2132
savun tiheys, 3493
savupiippu, 733, 3593
savupiippuvaikutus, 737, 3594
savupiipun hattu, 1116
savupiipun perusta, 734
savupiipun sisäänmeno, 738
savupiipun verhous, 1983
savupiipun veto, 736, 1977
savupiiputon lämmitin, 1982
savuvana (konvektio), 2993
savuvanan leviäminen, 2994
savuvanan nousukorkeus, 2995
seinä, 4023
seinälle asennettu ilmanlämmitin, 4030
seinälle asennettu lamellipatteri, 4025
seinälämmitys, 4028
(seinä)lämmityspaneeli, 4029

seinän lämpötila, 4032
seinän läpäisevä käyttövalmis ilmastointikone, 2900
seinäpinnan tasoon sovitettu ovi, 1993
seinäpuhallin, 4026
seinätuuletin, 4033
seinäventtiili, 4031
seinään asennettu ilmastointikone, 3839
seisokkiajan sulatus, 2805
sekavirtauspuhallin, 2691
sekoitettu ilma, 2690
sekoitin, 436
sekoittaja, 3757
sekoitus, 2701
sekoitusilmakanava, 2697
sekoituskammio, 2694
sekoituskone, 70
sekoituslaatikko, 437, 2693
sekoituspiiri, 2695
sekoituspumppu, 2698
sekoitussuhde, 2699
sekoitussäätö, 2696
sekoitusventtiili, 2700
sekundääri-ilman syöttökotelo, 580
sekundäärinen lämpöpinta, 3373
sementti, 675
sentrifugi, 692, 693
servomoottori, 3403
servosäätö, 3402
seula, 3440
signaali, 3442
siilo, 742, 2325
siipi, 420, 3954
siipianemometri, 1204
siipipyörä, 2396
siipipyöräanemometri, 3277
siirrettävä, 3005
siirto, 3878
siirtofunktio, 3880
siirtymisenestolaite, 637
siirtäjä, 1714
siirtää, 1713, 3879
silmukka, 2585
sintraus, 3458
sisustus, 2469
sisä-, 2469, 2474
sisähalkaisija, 495
sisäilma, 2415, 2447
sisäilman laatu, 2416
sisäinen, 2474
sisäinen lukitus, 2470
sisäkierre, 1824
sisällys, 1007
sisällysluettelo, 1007
sisältö, 1007
sisämitta, 2475
sisäpuolinen ripa, 2446
sisäseinä, 2476
sisäänrakennettu, 560
sisääntulo, 3004
sisäänvirtaus, 2430

yliampuminen, 2890
yli- ja alipainemittari, 894
ylijäämä, 1708
ylikorjaus, 2890
ylikuorma, 2889
ylikuumenemissuoja, 2302
ylikyllästetty, 3714
ylikyllästys, 3715
ylikyllästää, 3713
ylilämmitys, 2888
ylilämpötila, 1711
ylimenoalueen virtaus, 3885
ylimenokausi, 2683
ylimenokytkin, 705
ylimäärä, 1708
ylipaine, 1710
ylipainekattila, 3051
ylipainepoltin, 2013
ylipainesuoja, 2311, 3045
ylipainetulipesä, 2014
ylipaineventtiili, 3066, 3236
ylivirtaus, 2883
ylivirtauspelti, 1668
ylivirtausventtiili, 450, 2885
ylivuoto, 932
yläjako, 2887
yläjakoinen
    lämmitysjärjestelmä,
    1410
yläpaineen säätö, 2308
yläsyöttöinen, 2882
ylävirta, 3934
ympäristö, 1657, 3734

ympäristön lämpötila, 198,
    1662
ympäristön paine, 197
ympäristön vaikutuksilta
    kompensoitu, 196
ympäristöolot, 1659
ympäristösuunnittelu, 1660
ympäristötekniikka, 1661
ympäriverhottu, 2493
ympärivuotinen, 4145
ympäröivä, 194
ympäröivä ilma, 195
yritys, 1644
yöaikainen lämpötilan lasku,
    2769

äkillisen paineen alenemisen
    vuoksi höyrystynyt kaasu,
    1923
äänen absorptio, 3537
ääneneristys, 3542, 3547
ääneneristäminen, 3547
äänen eteneminen, 3548
äänenlähde, 3552
äänennopeus, 3551
äänenpaine, 3545
äänenpainetaso, 3546
äänen spektri, 3550
äänentaso, 2589
äänen tehotaso, 3544
äänenvaimennin, 2740, 2741
äänenvaimennus, 3538
äänenvaimentaja, 3444, 3539

äänenvoimakkuus, 2588
ääni, 3536
äänienergia, 3541
äänitaso, 3543

öljy, 2809
öljykamiina, 2828
öljykattila, 2819
öljykylpy, 2810
öljylämmitys, 2822
öljynerotin, 2826
öljynjäähdytin, 2815
öljynjäähdytys, 2816
öljynkulutus, 2814
öljyn lämpötilan säätö, 2831
öljynpaineen säätö, 2825
öljynpinnan korkeus, 2823
öljynpoltto, 2820
öljynsyöttö, 2829
öljynsyötön säätö, 2818
öljyn tyhjennys, 2817
öljypoltin, 2812
öljypolttimen suutin, 2813
öljypumppu, 3881
öljysumu, 2824
öljysäiliö, 2827
öljytetty ilmasuodatin, 2811
öljytön, 2821
öljytön kompressori, 1475
öljyvapaa, 2821

o-potentiaali, 1569
o-vaihe, 880

Spon's European Construction Costs Handbook

Edited by **Davis Langdon & Everest**, Chartered Quantity
Surveyors, UK

A unique source of information on the world's largest construction
market.
27 countries arranged in alphabetical order, each have their own
chapter containing the following information.

* key data on the main economic and construction industries

* an outline of the national construction industry, covering structure,
  tendering and contract procedures, regulations and standards

* labour and material costs data

* measured rates (in local currency) for up to 63 construction operations

* costs per unit area for a range of building types from housing to offices
  and factories

* regional variations percentages, tax details, cost and price indices,
  exchange rates with £ sterling and $US

* addresses of authorities, professional institutions, trade associations etc.

* multilingual glossary with fully detailed specifications, in 5 languages for
  the operations priced as measured rates

Countries covered in detail: Austria * Belgium * Cyprus * Denmark * Finland
* France * Germany * Greece * Hungary * Ireland * Italy * Japan *
Luxembourg * Malta * Netherlands * Norway * Poland * Portugal * Spain *
Sweden * Switzerland * Turkey * UK * USA * USSR(CIS) * Yugoslavia *
Japan and the USA are included for the purpose of comparison

March 1992: 234x156: 544pp
Hardback: 0-419-17480-X

For further information and to order please contact: **The Promotion Dept.**, **E & F N
Spon**, 2-6 Boundary Row, London SE1 8HN Tel 071 865 0066 Fax 071 522 9623

Svenska

avfrostningstid *c*, 1212
avfrostning *c* under
stillestånd *n*, 2805
avfukta, 1222, 1224
avfuktare *c*, 1221, 1226,
1436
avfuktning *c*, 1220, 1225
avfuktningseffekt *c*, 1223
avgas *c*, 1728
avgasanslutning *c*, 1729
avgasning *c*, 1213
avgasrecirkulation *c*, 1731
avgasspjäll *n*, 1730
avgasventil *c*, 1736
avgasöppning *c*, 1733
avgivning *c*, 1613
avgrening *c*, 511, 509
avisning *c*, 1227
avkännare *c*, 3387
avlasta, 1339
avlastad start *c*, 2571
avlastning *c*, 1338, 1353,
2575
avlastningsventil *c*, 3236
avlastningsvinkel *c*, 229
avledningsarmatur *c*, 1383
avledningsförmåga *c*, 1354
avlopp *n*, 1416, 3457
avloppskanal *c*, 3508
avloppsledning *c*, 3411
avloppsvatten *n*, 1557
avlufta, 3986
avluftare *c*, 117
avluftning *c*, 174
avlägsnande *n* av pannsten *c*,
3350
avlägsnande *n* av slagg *c*,
3466
avläsning *c*, 3168
avläst temperatur *c*, 250
avmineralisering *c* av vatten
n, 4064
avrinning *c* bakåt, 1419
avrinning *c* nedåt, 1421
avsaltning *c*, 1233
avskiljare *c*, 1604, 3025
avskiljning *c*, 3388
avskiljningseffektivitet *c*,
3389
avskrivning *c*, 1240
avstånd *n*, 1365, 3556
avstängningsventil *c*, 1150,
2487, 3437
avsätta, 1238
avsättning *c*, 1237, 1239
avtappa, 433
avtappningsanordning *c*,
1341
avtappningskondensor *c*,
432
avtappningskran *c*, 1420
avtappningskärl *n*, 1422
avtappningsventil *c*, 435,
1424
avvikelse *c*, 1020

avvikelse *c* från börvärde,
405
axel *c*, 320, 321
axeltätning *c*, 3416
axialfläkt *c*, 317
axial fläkt *c*, 3955
axialflödeskompressor *c*, 318
axial hastighet *c*, 319
axialhastighet *c*, 685
axialkompensator *c*, 316
axialt, 315
azeotropisk, 323

bacharachtal *n*, 324
backspjäll *n*, 2787
backventil *c*, 2788
bad *n*, 382
badrum *n*, 383
bakdrag *n*, 1409
bakdragsspjäll *n*, 326
bakgrundsstrålning *c*, 328
bakkantstjocklek *c*, 3876
baktericid *c*, 341
bakterie *c*, 335, 2163
bakterieaktivitet *c*, 337
bakteriedödande, 340, 2165
bakteriell, 336
bakteriesönderfall *n*, 338
bakterietillväxt *c*, 339
bakåtböjd, 334
balansera, 346
balanserad, 347
balanserad ventilation *c*,
349, 351
balanserat flöde *n*, 350
balanstemperatur *c*, 353
ballast *c*, 360
bandbredd *c*, 366
barometer *c*, 368
barometereffekt *c*, 371
barometertryck *n*, 372
bas *c*, 375
baslastalstring *c*, 379
batteri *n*, 385
batteriarea *c*, 807
batteridjup *n*, 806
Beaufort-skala *c*, 393
befukta, 2354
befuktare *c*, 2353
befuktning *c*, 1171, 2352
befuktningsutrustning *c*,
2355
begränsningsdon *n*, 2551
behandlad luft *c*, 962, 3895
behandling *c*, 3896
behov *n*, 1232
behållare *c*, 1004
beklädnad *c*, 1114, 2555
belasta, 712
belastning *c*, 2568
belastning *c* under drift *c*,
2848
belastningsfaktor *c*, 2570
belastningsgräns *c*, 2574
belastningsnivellering *c*, 2573

belastningsregulator *c*, 713
belastningsvariation *c*, 2576
belysning *c*, 2391
belysningsstyrka *c*, 2465
beläggning *c*, 791
beläggningsspärr *c*, 792
beredare *c*, 1328
beredning *c* med direkt
varmvatten *n*, 3038
beräkna, 1683
beräknat luftflöde *n*, 1247
beräkning *c*, 603, 1682, 1684
beskickningskapacitet *c*, 716
beskrivning *c*, 1242
beståndsdel *c*, 891
betong *c*, 943
bifurkation *c*, 407
bildskärm *c*, 1361
bimetall *c*, 408
bimetallelement *n*, 409
bimetalltermometer *c*, 410
binär, 411
binär ångcykel *c*, 412
bit *c*, 413
bitumenfärg *c*, 416
bitumenpapper *n*, 414
bituminöst kol *n*, 415
bjälke *c*, 387
blandad luft *c*, 2690
blandare *c*, 70, 436
blandarpump *c*, 2698
blandgas *c*, 2692
blandning *c*, 2701
blandningsbox *c*, 437
blandningsfläkt *c*, 2691
blandningsförhållande *n*,
2699
blandningskammare *c*, 2694
blandningskanal *c*, 2697
blandningslåda *c*, 2693
blandningsregulator *c*, 2696
blandningsventil *c*, 2700
blindfläns *c*, 422, 438, 3663
blockera, 439
blockering *c*, 440
blockventil *c*, 443
bly *n*, 2523
blydiktning *c*, 2524
blåsa, 444
bolometer *c*, 485
bom *c*, 367
bord *n*, 3751
borstfilter *n*, 538
bortförande *n* av rökgas *c*,
1980
bortförd, 1687
borttagning *c*, 3241
bostadsuppvärmning *c*, 1389
bottenblåsningsventil *c*, 446
bottenplatta *c*, 381, 395
bottenplåt *c*, 1357
bourdonrör *n*, 501
brand *c*, 1873
brandalarmsystem *n*, 1874
branddörr *c*, 1882

brandmotstånd *n*, 1889
brandskydd *c*, 1888
brandsläckare *c*, 1883
brandspjäll *n*, 1881
brandsäker, 1886
brandventil *c*, 1893
bricka *c*, 4049
brikett *c*, 534
brinnpunkt *c*, 1885
brinntid *c*, 852
brittisk standard *c*, 535
bromskraft *c*, 506
bruksanvisningar *c*, 2453
brunkol *n*, 537, 2548, 2549
brunn *c*, 4112
brunnsvatten *n*, 4113
bruttokapacitet *c*, 2195
brygga *c*, 2948
bryta, 1152
brytare *c*, 519
brytläge *n*, 1154
brytpunkt *c*, 1153
bräckt vatten *n*, 505
bränna, 567
brännare *c*, 568
brännare *c* med flera
    munstycken *n*, 2745
brännare *c* med fläkt *c*, 2013
brännarfläns *c*, 570
brännargenomspolning *c*,
    575
brännarhals *c*, 578
brännarmunstycke *n*, 572,
    579
brännarplåt *c*, 573
brännarregister *n*, 576
brännartändning *c*, 571
brännjbar, 842
brännbarhet *c*, 841
brännkammare *c*, 1877, 2100
bränsle *n*, 2076
bränsleanalys *c*, 2078
bränslecell *c*, 3961
bränslefilter *n*, 2082
bränsleförbrukning *c*, 2080
bränslehantering *c*, 2084
bränslelager *n*, 2087
bränslepreparering *c*, 2086
bränsletillförsel *c*, 2081, 2088
bränsletransport *c*, 2089
bränsletyp *c*, 2090
bränsleval *n*, 2079
B th u, 540
bubbelflöde *n*, 542
bubbla *c*, 541
buffert *c*, 543
bufferttank *c*, 544
buller *n*, 520, 2772
bullerkriteriekurvor *c*, 2774
bullerminskning *c*, 2777
bullernivå *c*, 2776
bult *c*, 486
bunker *c*, 566
bunt *c*, 565
buss *c*, 584

bussning *c*, 2167
butan *c*, 588
bygelfäste *n*, 3658
byggelement *n*, 549
byggherre *c*, 3460
byggledare *c*, 770
byggledningssystem *n*, 554
byggmaterial *n*, 555
byggnad *c*, 545
byggnadsautomationssystem
    n, 547
byggnadselement *n*, 995,
    3683
byggnadsenergistyrsystem *n*,
    550
byggnadsmaterial *n*, 2637
byggnadsregleringssystem *n*,
    548
(byggnads *c*) ställning *c*, 3347
byggnadsvolym *c*, 260
byggnads *c* värmelast *c*, 558
byggnadsyta *c*, 259, 546
byggprojekt *n*, 556
byggtid *c*, 890
byggyta *c*, 557
byte *c*, 598
bågsvetsning *c*, 261
bädd *c*, 394
bälg *c*, 396
bärfrekvens *c*, 636
böj *c*, 403, 502, 1563

carnotprocess *c*, 635
cell *c*, 667
cellbyggnad *c*, 673
cellgummi värmeisolering *c*,
    672
cellvärmeisolering *c*, 674
celsiusgrad *c*, 676
cement *c*, 675
centralenhet *c*, 683
central kontroll *c*, 678
centralkontrollpanel *c*, 679
central
    luftbehandlingsanläggning
    c, 684
centraluppvärmning *c*, 441
centralvärme *c*, 680, 2092
centralvärme *c* med gas, 2129
centralvärmeanläggning *c*,
    681
centrifug *c*, 693
centrifugal, 687
centrifugalfläkt *c*, 689
centrifugalkompressor *c*, 688
centrifugalkraft *c*, 690
centrifugalpump *c*, 691
certifikat *n*, 695
cirkulation *c*, 747
cirkulationsfläkt *c*, 746
cirkulationspump *c*, 749, 750
cirkulationstryck *n*, 748
cistern *c*, 752
clo *c*, 776
colorimetrisk, 830

computer-aided design *c*,
    923
computer-aided
    manufacturing *c*, 924
cykel *c*, 1155
cyklon *c*, 1158
cyklonavskiljare *c*, 692
cylinder *c*, 1159
cylinderhuvud *n*, 1160

dagg *c*, 1268
daggpunkt *c*, 1269
daggpunktshygrometer *c*,
    1271
daggpunktsstegring *c*, 1272
daggpunktssänkning, 1270
daglig mängd *c*, 1162
dagtank *c*, 1163
dammsugare *c*, 3943
dammtät, 1520
data *c*, 1174
databank *c*, 1175
databas *c*, 1176
datadisplaymodul *c*, 1178
dataomvandlare *c*, 1177
dataregistrering *c*, 1179
datatabell *c*, 1181
datautskrift *c*, 933
dator *c*, 922, 1180
datorbaserat system *n*, 925
datorhårdvara *c*, 928
datorinternminne *n*, 937
datorkod *c*, 926
datorkonstruktion *c*, 927
datorkontakt *c*, 938
datorkörning *c*, 934
datorminne *n*, 930
datormjukvara *c*, 936
datorsimulation *c*, 935
dators *c* ingångsdata, 929
dators *c* utgångsdata *c*, 931
datoröverflöde *n*, 932
decibel *c*, 1194
decipol *c*, 1195
deflektor *c*, 1205
deklination *c*, 1196
dekompressionsanordning *c*,
    1198
dekrementfaktor *c*, 1199
delad bussning *c*, 3573
delad kondensor *c*, 3574
dellast *c*, 2923
delning *c*, 2970
demodulering *c*, 1234
demontera, 1336
deodorant *c*, 1236
desickant aktiverad
    aluminiumoxid *c*, 41
desinfektion *c*, 1358
desorption *c*, 1255
destillation *c*, 1367
destillerat vatten *n*, 1368
detalj *c*, 1258
detaljera, 1259
detaljerad kalkyl *c*, 1261

infiltration c, 2427, 2428
informationsutbyte n, 2431
infraröd bild c, 2433
infravärmning c, 2432
infrysning c, 2045
infrysningskontroll c, 2046
ingenjör c, 1640
ingenjörskonst c, 1641
ingående eleffekt c, 1578
ingång c, 1654
ingångsdata c, 929
ingångsförlust c, 1655
inhibitor c, 2435
initialfyllning c, 2436
initialtemperatur c, 2437
injektor c, 2441
injektorbrännare c, 2442
inkapsling c, 1617
inklädnad c, 754
inkoppling c, 1148
inkopplingspunkt c, 1149
inlopp n, 2443
inloppsventil c, 2445
inloppsöppning c, 2444
innehåll n, 1007
inneluft c, 2415, 2447
inneluftkvalitet c, 2416
innergänga c, 1824
innermått n, 2475
innervägg c, 2476
inre, 2474
inre n, 2469
inre fläns c, 2446
inspektionsglas n, 3441
inspektionshål n, 2211
inspektionskontroll c, 26
inspektionslucka c, 2449
inspektionsnivå c, 28
insprutningsmunstycke n,
    2439
insprutningspunkt c, 3588
insprutningsrör n, 2440,
    3587
installation c, 2451
installera, 2450
inströmning c, 2430
inställning c, 3406
inställningsnoggrannhet c,
    3407
inställningspunkt c, 3405
inställningsrelä n, 3007
inställningsvisare c, 3190
inställningsvärde n, 3410
insugning c, 3695
insugningstryck n, 3702
integralstyrning c, 2461
integrator c, 2463
intensitet c, 2464
intermittent, 2473
intervall n, 2478
invalidiserad, 2479
is c, 2372
isbehållare c, 2375
isbildning c, 2380
isblock n, 2374

isentalpisk, 2485
iskristall c, 2376
iskub c, 2377
islagring c, 2373
isobar c, 2486
isolera, 2455, 2456
isolering c, 2459
isoleringskonstant c, 1283
isolermaterial n, 2454, 2457,
    2458
isometrisk projektion c, 322
isometrisk ritning c, 2488
isoterm c, 2489
isotermisk, 2490
isotop c, 2491
isskikt n, 2381
istillverkningsanläggning c,
    2382
isvatten n, 2378
isvattenkylning c, 2379

jalusispjäll n, 1291, 2592,
    2915
jetventilation c, 2499
jon c, 2480
jonbyte n, 2481
jonisering c, 2482
jord (el) c, 1538
jordning c, 1539, 1569
jämn böj c, 3501
jämvikt c, 357, 1672
jämviktstemperatur c, 1673
jämviktstryck n, 354
järn n, 2483

kabel c, 599
kabelförbindning c, 600
kabelinföring c, 601
kalibrering c, 604
kall, 817
kalluft c, 818
kallvatten n, 827
kallvattenförsörjning c, 828
kalori c, 606
kalorimeter c, 611
kam c, 612
kamflänsrör n, 1869, 3570
kamflänsrörradiator c, 1870
kanal c, 706, 1492
kanalbuller n, 1498
kanalböj c, 1495
kanaldel c, 1500
kanaldimension c, 1501
kanaldimensionering c, 1502
kanaldistribution c, 1494
kanalljud n, 1493
kanals c övergångsdel c, 1504
kanalstrålning c, 1499
kanalsystem n, 1503, 1505
kant, 1547
kapacitans c, 615
kapacitansfaktor c, 620
kapacitet c, 617
kapacitetskontroll c, 618
kapillaritet c, 623, 625

kapillarlufttvättare c, 626
kapillär, 624
kapillärrör n, 627
karaktäristik c, 707
karaktäristisk kurva c, 708
karaktäristiskt tal n, 709
kaskad c, 640
kassettfilter n, 670
kast n, 3841
kastlängd c, 2498
katalysator c, 651
katatermometer c, 2505
katjon c, 654
katod c, 652
katodiskt skydd n, 653
kavitation c, 656
kavitet c, 657
kedjerost c, 699, 3894
kemikalier c, 724
kilrem c, 3976
kilremsdrift c, 3977
kinetisk, 2507
kitt n, 3126
klaffventil c, 1915
klammer c, 755
klassifikation c, 758
klenrörssystem n, 2679, 3491
klibbig, 50
klimat n, 771
klimatapparat c, 88
klimatkammare c, 772
klimatskärm c, 552
klimatskärmens c termiskt
    motstånd n, 553
klinker c, 773
klorfluorkarbon n, 741
klämma c, 755
knopp c, 2508
knäledsverkan c, 3859
knärör n, 404
koaxial, 793
kod c, 796
koefficient c, 797
koka, 454
kokande, 483
kokning c, 1540
kokpunkt c, 484
koks c, 810
kokseldad panna c, 811
kokseldning c, 812
kokskol n, 815
koksugnsgas c, 813
kol n, 628, 782
koldioxid c, 629
koleldad, 786
koleldad panna c, 787
koleldning c, 788
kolfilter n, 710
kollager n, 789
kolleverans c, 784
koloxid c, 633
kolpulver n, 785, 3118
kolstybb c, 3464
kolsyra c, 631
kolsyre is c, 1467

kolv *c*, 561, 2966
kolvkompressor *c*, 3175
kolvpump *c*, 3176
kolväte *n*, 2363
kombinationskontroll *c*, 839
komfort *c*, 867
komfortdiagram *n*, 869
komfortindex *n*, 871
komfortklimatanläggning *c*, 868
komfortkylning *c*, 870
komfortmätare *c*, 1685
komfortzon *c*, 872
kommersiell, 873
kommersiell kylare *c*, 874
kommersiellt system *n*, 875
kommunalt vatten *n*, 753
kommunikationsbaserat system *n*, 882
kommunikationsnät *n*, 881
kompatibilitet *c*, 884
kompensator *c*, 398, 887
kompensatorkoppling *c*, 885
kompensatortätning *c*, 397
kompressibilitet *c*, 898
kompressibilitetsfaktor *c*, 899
kompression *c*, 900
kompressionscykel *c*, 901
kompressionsfaktor *c*, 903
kompressionsförhållande *n*, 905
kompressionskoppling *c*, 904
kompressionskylsystem *n*, 910
kompressionsprov *n*, 909
kompressionsrum *n*, 769
kompressionssteg *n*, 906
kompressionstakt *c*, 907
kompressionstank *c*, 908
kompressionsverkningsgrad *c*, 902
kompressions-volymförhållande *n*, 911
kompressionsvärmeeffekt *c*, 918
kompressor *c*, 912
kompressoravlastare *c*, 921
kompressoravlastning *c*, 914
kompressoravlastningsslag *n*, 915
kompressordeplacement *n*, 916
kompressorenhet *c*, 920
kompressorförvärmning *c*, 917
kompressorkapacitet *c*, 913
kompressoröverbelastning *c*, 919
komprimerad gas *c*, 896
komprimerad vätska *c*, 897
koncentration *c*, 940
koncentrationsförhållande *n*, 941

koncentrationssolkollektor *c*, 939
koncentriska rör *n*, 942
kondensat *n*, 944
kondensation *c*, 949
kondensationspunkt *c*, 950
kondensatledning *c*, 946
kondensatmätare *c*, 947
kondensator *c*, 616
kondensatpump *c*, 948
kondensatsamlare *c*, 945
kondensera, 951
kondensering *c*, 2556
kondenseringspanna *c*, 956
kondensering *c* till fast fas *c*, 3525
kondensor *c*, 952, 2557
kondensorenhet *c*, 960
kondensorkapacitet *c*, 961
kondensor *c* med två rörknippen *n*, 1397
kondensorslinga *c*, 953
kondensortryck *n*, 957
kondensortryckventil *c*, 958
kondenstemperatur *c*, 959
konduktans *c*, 965
konduktivitet *c*, 970
konfiguration *c*, 973
konisk gänga *c*, 3760
konsollager *n*, 2928
konstant, 990
konstant *c*, 989
konstantnivåventil *c*, 991
konstantvärdeskontroll *c*, 993
konstgjord snö *c*, 1903
konstruera, 1246
konstruktion *c*, 994, 1245
konsulterande ingenjör *c*, 996
kontakt *c*, 999
kontaktor *c*, 1000
kontaktpunkt *c*, 1001
kontakttermometer *c*, 1003
kontinuerlig drift *c*, 1009
kontinuerlig kontroll *c*, 1008
kontinuerlig matning *c*, 4143
kontrakt *n*, 1010
kontroll *c*, 721, 1014
kontrollalgoritm *c*, 1017
kontrollera, 722, 1015
kontrollerad atmosfär *c*, 1024
kontrollfunktion *c*, 1023
kontrollmetod *c*, 1033
kontrollomfång *n*, 1032
kontrollpunkt *c*, 1030
kontrollslinga *c*, 1028
kontrollsystem *n* för energidrift *c*, 1631
kontrollur *n*, 3847
kontrollvariabel *c*, 1026, 1091
kontrollventil *c*, 723
kontrollåtgärd *c*, 1016

konvektion *c*, 1034
konvektionsström *c*, 1037
konvektionsvärmare *c*, 1035
konvektiv, 1036
konvektor *c*, 1038
konverteringsbrännare *c*, 1040
koppar *c*, 1081
kopparplätering *c*, 1083
kopparrör *n*, 1082
koppla av, 3744
koppla in, 2992
koppla på, 3745
koppling *c*, 1109, 2500
koppling *c* (handel), 2616
kopplingsdosa *c*, 2504, 3743
kopplingsfläns *c*, 883
kopplingshylsa *c*, 1112
kopplingsschema *n*, 4132
kopplingssystem *n*, 4133
kopplingstavla *c*, 3742
kopplingstätning *c*, 2501
kork *c*, 1086
korrigera, 1190
korrigerad effektiv temperatur *c*, 1090
korrodera, 1093
korrosion *c*, 1094
korrosionshindrande, 239
korrosionshärdig, 1097
korrosionsinhibitor *c*, 1095
korrosionsskydd *n*, 1096
korrosiv, 1098
korrugerad böj *c*, 1099, 1100
korsström *c*, 1128
korsströmsvärmeväxlare *c*, 1132
kort cykel *c*, 3432
kortslutning *c*, 3431
korttidsdrift *c*, 3433
kostnad *c*, 1756
kostnad(er) *c*, 1101
kraft *c*, 2007, 3012
kraftbrist *c*, 3017
kraftelement *n*, 1031
kraftförbrukning *c*, 3013
kraftsituation *c*, 1636
kraftstation *c*, 3019, 3022
kraftstyrning *c*, 1002
kraftuttag *n*, 3020
kraftvärmeverk *n*, 840
kran *c*, 794, 3757
kranventil *c*, 795
kretsavbrott *n*, 745
kretslopp *n*, 778, 1156
kristallis *c*, 2030
krita *c*, 700
kritisk, 1126
krok *c*, 2324
kropp *c*, 453
krossad koks *c*, 536
kryogenisk, 1140
kryssförbindelse *c*, 2503
kubikfot *c* per minut *c*, 697
kubikfot *c* per sekund *c*, 698

Appendix A   Units

BTU   British thermal unit: $1/\mathrm{BTU} = 1{\cdot}06 \times 10^3$ Joule
cfm   cubic feet per minute: 1 cfm $= 0{\cdot}4719 \times 10^3 \mathrm{\ m^3/s}$
cfs   cubic feet per second: 1 cfs $= 28{\cdot}32 \times 10^3 \mathrm{\ m^3/s}$
clo   non-SI unit of clothing insulation:
                1 clo = the thermal insulation necessary to keep a
                        sitting person comfortable in a normally
                        ventilated room at 21 degrees Centigrade and
                        50% relative humidity
                1 clo $= 0{\cdot}155 \mathrm{\ m^2\ K/W}$
Watt   Energy flow at the rate of one Joule per second
Water gauge   (Column of water): $1''$ water $= 249$ Pa;
                                    1 cm water $= 98$ Pa

Appendix B Abbreviations used for administrative bodies

ARI   Airconditioning and Refrigeration Institute
ASCII   American Standard Code for Information Interchange
ASHRAE   American Society of Heating, Refrigeration and Air-conditioning Engineers
ASTM   American Society for Testing and Materials
REHVA   Federation of European Heating and Airconditioning Associations

Printed and bound by CPI Group (UK) Ltd, Croydon, CR0 4YY

01/11/2024

01782640-0002